exploring
EARTH
SCIENCE

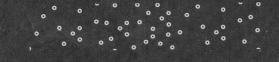

About the Cover

The cover photograph by well-known photographer Michael Collier features the Grand Teton, its top shrouded in clouds. The peaks stand 13,775 feet above sea level, and are within one of the youngest mountain ranges of the extensive Rocky Mountain system. Although the mountain range is very young in geologic terms, it is cored with ancient metamorphic and igneous rocks, some of which formed as much as 2.6 billion years ago, at a time when the North American continent was first forming. The entire Teton Range, looming more than a mile above Jackson Hole, Wyoming, is along the rising edge of a fault-bounded crustal block that has been tilted — up on the east side and down on the west. This tilting and uplift started only five million years ago, and continues today. The Tetons have been thrust up into a windy world of intense weather, where clouds often gather and storms sometimes rage. During the Ice Ages, glaciers left gouges and imparted onto the peaks their overall shapes, and now precipitation pummels the peaks, helping carve the rock into spires and cliffs.

Michael Collier received his BS in geology at Northern Arizona University, MS in structural geology at Stanford, and MD from the University of Arizona. He rowed boats commercially in the Grand Canyon in the 1970s and '80s, then practiced family medicine in northern Arizona. Collier published books about the geology of Grand Canyon, Death Valley, Denali, and Capitol Reef national parks. He has done books on the Colorado River basin, glaciers of Alaska, climate change in Alaska, and a three-book series on American mountains, rivers, and coastlines. As a special projects writer with the USGS, he produced books about the San Andreas fault, the downstream effects of dams, and climate change. Collier's photography has been recognized with awards from the USGS, National Park Service, American Geosciences Institute, and National Science Teachers Association.

SECOND EDITION

exploring
EARTH
SCIENCE

Stephen J. Reynolds
Arizona State University

Julia K. Johnson
Arizona State University

CONTRIBUTING AUTHORS

Robert V. Rohli
Louisiana State University

Peter R. Waylen
University of Florida

Mark A. Francek
Central Michigan University

Cynthia C. Shaw
Lead Illustrator, Art Director

Mc
Graw
Hill
Education

EXPLORING EARTH SCIENCE, SECOND EDITION

Published by McGraw-Hill Education, 2 Penn Plaza, New York, NY 10121. Copyright © 2019 by McGraw-Hill Education. All rights reserved. Printed in the United States of America. Previous edition © 2016. No part of this publication may be reproduced or distributed in any form or by any means, or stored in a database or retrieval system, without the prior written consent of McGraw-Hill Education, including, but not limited to, in any network or other electronic storage or transmission, or broadcast for distance learning.

Some ancillaries, including electronic and print components, may not be available to customers outside the United States.

This book is printed on acid-free paper.

4 5 6 7 8 9 LWI 21 20

ISBN 978-1-259-63861-9
MHID 1-259-63861-8

Portfolio Manager: *Michael Ivanov, Ph.D.*
Product Developers: *Jodi Rhomberg*
Marketing Manager: *Kelly Brown*
Content Project Managers: *Laura Bies, Tammy Juran, and Sandy Schnee*
Buyer: *Sandy Ludovissy*
Design: *Matt Backhaus*
Content Licensing Specialists: *Lori Hancock*
Cover Image: *©Michael Collier*
Layout: *Stephen J. Reynolds, Julia K. Johnson, Cynthia C. Shaw, and SPi Global*
Compositor: *SPi Global*

All credits appearing on page or at the end of the book are considered to be an extension of the copyright page.

Library of Congress Cataloging-in-Publication Data

Names: Reynolds, Stephen J., author. | Johnson, Julia K., author. | Rohli, Robert V., contributing author. | Waylen, Peter R., contributing author. | Francek, Mark A., contributing author.
Title: Exploring earth science / Stephen J. Reynolds, Arizona State University, Julia K. Johnson, Arizona State University; contributing authors, Robert V. Rohli, Louisiana State University, Peter R. Waylen, University of Florida, Mark A. Francek, Central Michigan University, Cynthia C. Shaw, lead illustrator, art director.
Description: 2e. | New York, NY: McGraw-Hill Education, [2019] | Includes index.
Identifiers: LCCN 2017046084 | ISBN 9781259638619 (alk. paper)
Subjects: LCSH: Earth sciences.
Classification: LCC QE28.3 .R49 2019 | DDC 550—dc23
LC record available at https://lccn.loc.gov/2017046084

The Internet addresses listed in the text were accurate at the time of publication. The inclusion of a website does not indicate an endorsement by the authors or McGraw-Hill Education, and McGraw-Hill Education does not guarantee the accuracy of the information presented at these sites.

mheducation.com/highered

 **connect**®

McGraw-Hill Connect® is a highly reliable, easy-to-use homework and learning management solution that utilizes learning science and award-winning adaptive tools to improve student results.

Homework and Adaptive Learning

- Connect's assignments help students contextualize what they've learned through application, so they can better understand the material and think critically.
- Connect will create a personalized study path customized to individual student needs through SmartBook®.
- SmartBook helps students study more efficiently by delivering an interactive reading experience through adaptive highlighting and review.

Over **7 billion questions** have been answered, making McGraw-Hill Education products more intelligent, reliable, and precise.

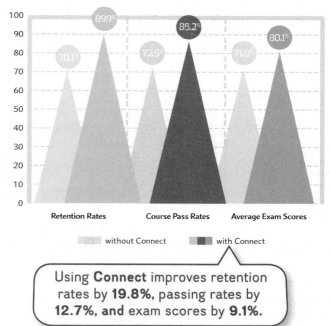

Connect's Impact on Retention Rates, Pass Rates, and Average Exam Scores

	without Connect	with Connect
Retention Rates	70.1%	89.9%
Course Pass Rates	72.5%	85.2%
Average Exam Scores	71.0%	80.1%

Using **Connect** improves retention rates by **19.8%**, passing rates by **12.7%, and** exam scores by **9.1%**.

73% of instructors who use **Connect** require it; instructor satisfaction **increases** by 28% when **Connect** is required.

Quality Content and Learning Resources

- Connect content is authored by the world's best subject matter experts, and is available to your class through a simple and intuitive interface.
- The Connect eBook makes it easy for students to access their reading material on smartphones and tablets. They can study on the go and don't need internet access to use the eBook as a reference, with full functionality.
- Multimedia content such as videos, simulations, and games drive student engagement and critical thinking skills.

©McGraw-Hill Education

Robust Analytics and Reporting

- Connect Insight® generates easy-to-read reports on individual students, the class as a whole, and on specific assignments.

- The Connect Insight dashboard delivers data on performance, study behavior, and effort. Instructors can quickly identify students who struggle and focus on material that the class has yet to master.

- Connect automatically grades assignments and quizzes, providing easy-to-read reports on individual and class performance.

©Hero Images/Getty Images

Impact on Final Course Grade Distribution

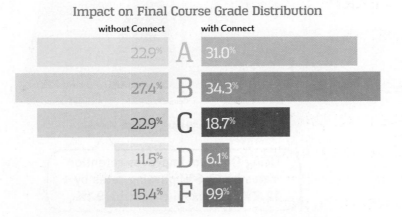

without Connect		with Connect
22.9%	A	31.0%
27.4%	B	34.3%
22.9%	C	18.7%
11.5%	D	6.1%
15.4%	F	9.9%

> More students earn **As** and **Bs** when they use **Connect**.

Trusted Service and Support

- Connect integrates with your LMS to provide single sign-on and automatic syncing of grades. Integration with Blackboard®, D2L®, and Canvas also provides automatic syncing of the course calendar and assignment-level linking.

- Connect offers comprehensive service, support, and training throughout every phase of your implementation.

- If you're looking for some guidance on how to use Connect, or want to learn tips and tricks from super users, you can find tutorials as you work. Our Digital Faculty Consultants and Student Ambassadors offer insight into how to achieve the results you want with Connect.

BRIEF CONTENTS

CONTENTS

CHAPTER 4: EARTH HISTORY 98

CHAPTER 5: PLATE TECTONICS 126

CHAPTER 6: VOLCANISM AND OTHER IGNEOUS PROCESSES 160

CHAPTER 7: DEFORMATION AND EARTHQUAKES   192

CHAPTER 8: MOUNTAINS, BASINS, AND CONTINENTAL MARGINS   224

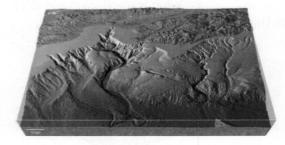

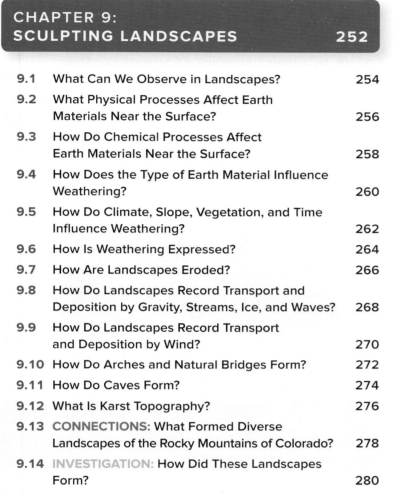

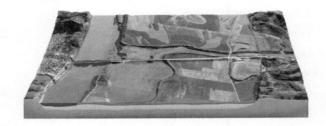

CHAPTER 12:
STREAMS, LAKES, AND GROUNDWATER 352

CHAPTER 13:
ENERGY AND MATTER IN THE ATMOSPHERE 390

CHAPTER 14:
ATMOSPHERIC MOTION 424

CHAPTER 15:
ATMOSPHERIC MOISTURE 448

CHAPTER 16:
WEATHER AND STORMS 476

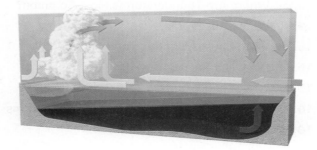

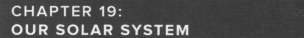

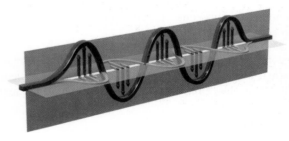

TELLING THE STORY . . .

WE WROTE *EXPLORING EARTH SCIENCE* so that students could learn from the book on their own, freeing up instructors to teach the class in any way they want. I (Steve Reynolds) first identified the need for this type of book while I was a National Association of Geoscience Teachers' (NAGT) distinguished speaker. As part of my NAGT activities, I traveled around the country conducting workshops on how to infuse active learning and scientific inquiry into introductory college science courses, including those with upwards of 200 students. In the first part of the workshop, I asked the faculty participants to list the main goals of an introductory science course, especially for nonmajors. At every school I visited, the main goals were similar to those listed below:

- to engage students in the process of scientific inquiry so that they learn what science is and how it is conducted,
- to teach students how to observe and interpret landscapes and other aspects of their physical environment,
- to enable students to learn and apply important concepts of science,
- to help students understand the relevance of science to their lives, and
- to enable students to use their new knowledge, skills, and ways of thinking to become more informed citizens.

I then asked faculty members to rank these goals and estimate how much time they spent on each goal in class. At this point, many instructors recognized that their activities in class were not consistent with their own goals. Most instructors were spending nearly all of class time teaching content. Although this was one of their main goals, it commonly was not their top goal.

Next, I asked instructors to think about why their activities were not consistent with their goals. Inevitably, the answer was that most instructors spend nearly all of class time covering content because (1) textbooks include so much material that students have difficulty distinguishing what is important from what is not, (2) instructors needed to lecture so that students would know what is important, and (3) many students have difficulty learning independently from the textbook.

In most cases, textbooks drive the curriculum, so my coauthor (Julia Johnson) and I decided that we should write a textbook that (1) contains only important material, (2) indicates clearly to the student what is important and what they need to know, and (3) is designed and written in such a way that students can learn from the book on their own. This type of book would give instructors freedom to teach in a way that is more consistent with their goals, including using local examples to illustrate concepts and their relevance. Instructors would also be able to spend more class time teaching students to observe and interpret landscapes, tectonics, and atmospheric or astronomic phenomena, and to participate in the process of scientific inquiry, which represents the top goal for many instructors.

COGNITIVE AND SCIENCE-EDUCATION RESEARCH

To design a book that supports instructor goals, we delved into cognitive and science-education research, especially research on how our brains process different types of information, what obstacles limit student learning from textbooks, and how students use visuals versus text while studying. We also conducted our own research on how students interact with textbooks, what students see when they observe photographs showing landscape features, and how they interpret different types of scientific illustrations, including maps, cross sections, and block diagrams that illustrate evolution of environments. *Exploring Earth Science* is the result of our literature search and of our own science-education and cognitive research. As you examine *Exploring Earth Science*, you will notice that it is stylistically different from most other textbooks, which will likely elicit a few questions.

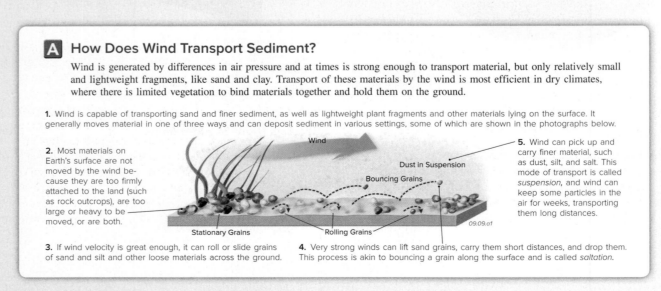

A **How Does Wind Transport Sediment?**

Wind is generated by differences in air pressure and at times is strong enough to transport material, but only relatively small and lightweight fragments, like sand and clay. Transport of these materials by the wind is most efficient in dry climates, where there is limited vegetation to bind materials together and hold them on the ground.

1. Wind is capable of transporting sand and finer sediment, as well as lightweight plant fragments and other materials lying on the surface. It generally moves material in one of three ways and can deposit sediment in various settings, some of which are shown in the photographs below.

2. Most materials on Earth's surface are not moved by the wind because they are too firmly attached to the land (such as rock outcrops), are too large or heavy to be moved, or are both.

Wind

Dust in Suspension

Bouncing Grains

5. Wind can pick up and carry finer material, such as dust, silt, and salt. This mode of transport is called *suspension,* and wind can keep some particles in the air for weeks, transporting them long distances.

Stationary Grains

Rolling Grains

09.09.a1

3. If wind velocity is great enough, it can roll or slide grains of sand and silt and other loose materials across the ground.

4. Very strong winds can lift sand grains, carry them short distances, and drop them. This process is akin to bouncing a grain along the surface and is called *saltation.*

Exploring Earth Science promotes inquiry and science as an active process. It encourages student curiosity and aims to activate existing student knowledge by posing the title of every two-page spread and every subsection as a question. In addition, questions are dispersed throughout the book. Integrated into the book are opportunities for students to observe patterns, features, and examples before the underlying concepts are explained. That is, we employ a *learning-cycle approach* where student exploration precedes the introduction of new terms and the application of knowledge to a new situation. For example, chapter 10 on slope stability begins with a three-dimensional image of northern Venezuela, pictured above, and asks readers to observe where people are living in this area and what natural processes might have formed these sites.

Wherever possible, we introduce terms after students have an opportunity to observe the feature or concept that is being named. This approach is consistent with several educational philosophies, including a learning cycle and just-in-time teaching. Research on learning cycles shows that students are more likely to retain a term if they already have a mental image of the thing being named (Lawson, 2003). For example, this book presents students with maps showing the spatial distribution of earthquakes, volcanoes, and mountain ranges and asks them to observe the patterns and think about what might be causing the patterns. Only then does the textbook introduce the concept of tectonic plates.

Also, the figure-based approach in this book allows terms to be introduced in their context rather than as a definition that is detached from a visual representation of the term. We introduce new terms in italics rather than in boldface, because boldfaced terms on a textbook page cause students to immediately focus mostly on the terms, rather than build an understanding of the concepts. The italics, however, let a student know when they have encountered an important term during their reading. The book includes a glossary for those students who wish to look up the definition of a term to refresh their memory. To expand comprehension of the definition, each entry in the glossary references the pages where the term is defined in the context of a figure.

WHY ARE THE PAGES DOMINATED BY ILLUSTRATIONS?

Earth science is a visual science. Earth science textbooks contain a variety of photographs, maps, cross sections, block diagrams, and other types of illustrations. These diagrams help portray the spatial distribution and geometry of features in the landscape, atmosphere, oceans, and universe in ways words cannot. In earth sciences, a picture really is worth a thousand words.

Exploring Earth Science contains a wealth of figures to take advantage of the visual nature of earth science and the efficiency of figures in conveying earth science concepts. This book contains few large blocks of text—most text is in smaller blocks that are specifically linked to illustrations. Examples of our integrated figure-text approach are shown throughout the book. In this approach, each short block of text is one or more complete sentences that succinctly describe a feature, process, or both of these. Most of these text blocks are connected to their illustrations with leader lines so that readers know exactly which feature or part of the diagram is being referenced in the text block. A reader does not have to search for the part of the figure that corresponds to a text passage, as occurs when a student reads a traditional textbook with large blocks of text referencing a figure that may appear on a different page. Most short blocks are numbered to guide students to read the blocks in a specific order.

This approach is especially well suited to covering earth science topics because it allows the text to have a precise linkage to the features and geographic location of the aspect being described. A text block discussing the Intertropical Convergence Zone can have a leader that specifically points to the location of this feature. A cross section of atmospheric circulation, such as those related to El Niño conditions, can be accompanied by short text blocks that describe each part of the system and that are linked by leaders directly to specific locations on the figure. This allows the reader to concentrate on the concepts being presented, not deciding what part of the figure is being discussed.

The approach in *Exploring Earth Science* is consistent with the findings of cognitive scientists, who conclude that our minds have two different processing systems, one for processing pictorial information (images) and one for processing verbal information (speech and written words), as illustrated below. Images enter our consciousness through our eyes, and text can enter either through our eyes, such as when we read, or through our ears, as occurs during a lecture. Research into learning and cognition shows that having text enter via our ears, while our eyes examine an image, is among the best ways to learn. Cognitive scientists also speak about two types of memory: *working memory* holds information and actively processes it, whereas *long-term memory* stores information until we need it (Baddeley, 2007). Both the verbal and pictorial processing systems have a limited amount of working memory, and our minds have to use much of our mental processing space to reconcile the pictorial and verbal types of information in working memory. For information that has both pictorial and verbal components, as most earth-science information does, the amount of knowledge we retain depends on reconciling these two types of information, on transferring information from working memory to long-term memory, and on linking the new information with our existing mental framework. For this reason, this book integrates text and figures, as in the example shown here.

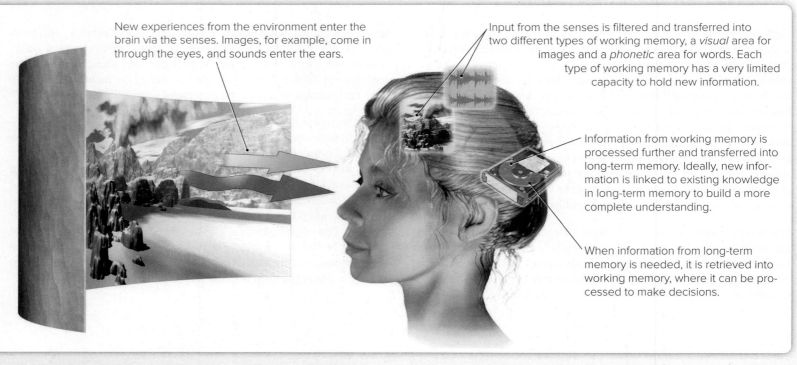

New experiences from the environment enter the brain via the senses. Images, for example, come in through the eyes, and sounds enter the ears.

Input from the senses is filtered and transferred into two different types of working memory, a *visual* area for images and a *phonetic* area for words. Each type of working memory has a very limited capacity to hold new information.

Information from working memory is processed further and transferred into long-term memory. Ideally, new information is linked to existing knowledge in long-term memory to build a more complete understanding.

When information from long-term memory is needed, it is retrieved into working memory, where it can be processed to make decisions.

WHY ARE THERE SO MANY FIGURES?

This textbook contains more than 2,500 figures, which is two to three times the number in most earth science textbooks. One reason for this is that the book is designed to provide a concrete example of each process, environment, or feature being illustrated. Research shows that many college students require concrete examples before they can begin to build abstract concepts (Lawson, 1980). Also, many students have limited travel experience, so photographs and other figures allow them to observe places, environments, and processes they have not been able to observe firsthand. The numerous photographs, from geographically diverse places, help bring the sense of place into the student's reading. The inclusion of an illustration for each text block reinforces the notion that the point being discussed is important. In many cases, as in the example below, conceptualized figures are integrated with photographs and text so that students can build a more coherent view of the environment or process.

Exploring Earth Science focuses on the most important earth science concepts and makes a deliberate attempt to eliminate text that is not essential for student learning of these concepts. Inclusion of information that is not essential tends to distract and confuse students rather than illuminate the concept; thus, you will see fewer words. Cognitive and science-education research has identified a redundancy effect, where information that restates and expands upon a more succinct description actually results in a decrease in student learning (Mayer, 2001). Specifically, students learn less if a long figure caption restates information contained elsewhere on the page, such as in a long block of text that is detached from the figure. We avoid the redundancy effect by including only text that is integrated with the figure.

The style of illustrations in *Exploring Earth Science* was designed to be more inviting to today's visually oriented students who are used to photo-realistic, computer-rendered images in movies, videos, and computer games. For this reason, many of the figures were created by world-class scientific illustrators and artists who have worked on award-winning textbooks, on Hollywood movies, on television shows, for *National Geographic,* and in the computer-graphics and gaming industry. In most cases, the figures incorporate real data, such as satellite images, aerial photographs, weather and climatological data, and locations of earthquakes and volcanoes. Our own research shows that many students do not understand cross sections and other subsurface diagrams, so nearly every cross section in this book has a three-dimensional aspect, and many maps are presented in a perspective view that incorporates topography. Research findings by us and other researchers (Roth and Bowen, 1999) indicate that including people and human-related items in photographs and figures attracts undue attention, thereby distracting students from the features being illustrated. As a result, our photographs have nondistracting indicators of scale, like dull coins and plain marking pens. Figures and photographs do not include people or human-related items unless we are trying to (1) illustrate how geoscientists study earth science processes and features (2) reinforce the relevance of the processes on humans, or (3) help students appreciate that geoscience can be done by diverse types of people, potentially including them, as depicted in our photographs.

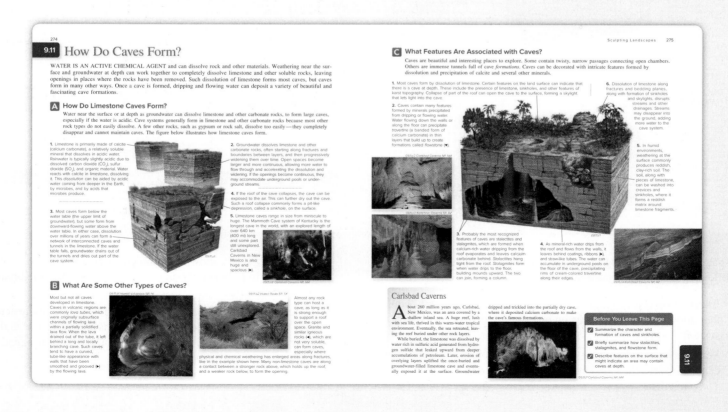

WHY DOES THE BOOK CONSIST OF TWO-PAGE SPREADS?

This book consists of two-page spreads, most of which are further subdivided into sections. Research has shown that because of our limited amount of working memory, much new information is lost if it is not incorporated into long-term memory. Many students keep reading and highlighting their way through a textbook without stopping to integrate the new information into their mental framework. New information simply displaces existing information in working memory before it is learned and retained. This concept of cognitive load (Sweller, 1994) has profound implications for student learning during lectures and while reading textbooks. Two-page spreads and sections help prevent cognitive overload by providing natural breaks that allow students to stop and consolidate the new information before moving on.

Each spread has a unique number, such as 17.9 for the ninth topical two-page spread in chapter 17. These numbers help instructors and students keep track of where they are and what is being covered. Each two-page spread, except for those that begin and end a chapter, contains a *Before You Leave This Page* checklist that indicates what is important and what is expected of students before they move on. This list contains learning objectives for the spread and provides a clear way for the instructor to indicate to the student what is important. The items on these lists are compiled into a master *What-to-Know List* provided to the instructor, who then deletes or adds entries to suit the instructor's learning goals and distributes the list to students before the students begin reading the book. In this way, the *What-to-Know List* guides the students' studying.

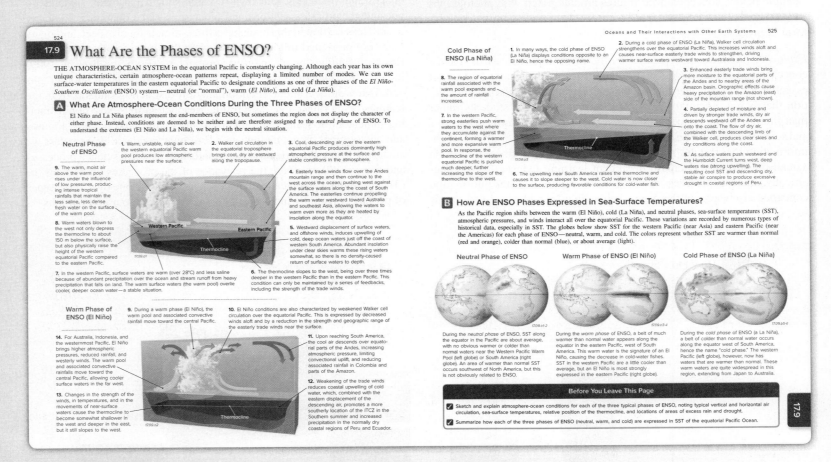

Two-page spreads and integrated *Before You Leave This Page* lists offer the following advantages to the student:

- Information is presented in relatively small and coherent chunks that allow a student to focus on one important aspect or earth system at a time.
- Students know when they are done with this particular topic and can self-assess their understanding with the *Before You Leave This Page* list.

- Two-page spreads allow busy students to read or study a complete topic in a short interval of study time, such as the breaks between classes.
- All test questions and assessment materials are tightly articulated with the *Before You Leave This Page* lists so that exams and quizzes cover precisely the same material that was assigned to students via the *What-to-Know* list.

The two-page spread approach also has huge advantages for the instructor. Before writing this book, the authors wrote most of the items for the *Before You Leave This Page* lists. We then used this list to decide what figures were needed, what topics would be discussed, and in what order. In other words, *the textbook was written from the learning objectives*. The *Before You Leave This Page* lists provide a straightforward way for an instructor to tell students what information is important. Because we provide the instructor with a master *What-to-Know* list, an instructor can selectively assign or eliminate content by providing students with an edited *What-to-Know* list. Alternatively, an instructor can give students a list of assigned two-page spreads or sections within two-page spreads. In this way, the instructor can identify content for which students are responsible, even if the material is not covered in class. Two-page spreads provide the instructor with unparalleled flexibility in deciding what to assign and what not to cover. It allows this book to be easily used for one-semester and two-semester courses.

This textbook, which is purposely designed to allow students to learn on their own, in combination with the components of McGraw-Hill's online learning system (*Connect, SmartBook, LearnSmart*), allows an instructor to offload much student learning to out-of-class times. An instructor can rely on the book to teach students content that the instructor does not wish to cover in class, opening up class time for teaching the most important aspects or those with local relevance, for modelling scientific reasoning, or for having students solve earth-science problems. The book and online materials easily provide the necessary components to allow an instructor to use a flipped-class approach, to teach a hybrid class, or for a fully online course. We do all of these.

CONCEPT SKETCHES

Most items on the *Before You Leave This Page* list are by design suitable for student construction of concept sketches. Concept sketches are sketches that are annotated with complete sentences that identify features, describe how the features form, characterize the main processes, and summarize histories (Johnson and Reynolds, 2005). An example of a concept sketch is shown to the right.

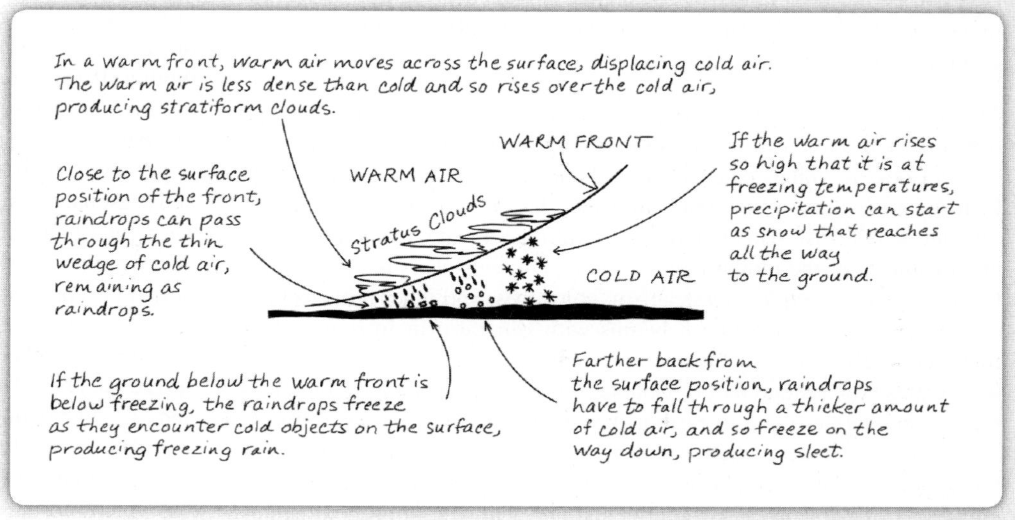

Concept sketches are an excellent way to actively engage students in class and to assess their understanding of earth science features, processes, and history. Concept sketches are well suited to the visual nature of earth science, especially cross sections, maps, and block diagrams. Earth scientists are natural sketchers using field notebooks,

In a warm front, warm air moves across the surface, displacing cold air. The warm air is less dense than cold and so rises over the cold air, producing stratiform clouds.

Close to the surface position of the front, raindrops can pass through the thin wedge of cold air, remaining as raindrops.

WARM FRONT

WARM AIR

Stratus Clouds

COLD AIR

If the warm air rises so high that it is at freezing temperatures, precipitation can start as snow that reaches all the way to the ground.

If the ground below the warm front is below freezing, the raindrops freeze as they encounter cold objects on the surface, producing freezing rain.

Farther back from the surface position, raindrops have to fall through a thicker amount of cold air, and so freeze on the way down, producing sleet.

blackboards, publications, and even napkins, because sketches are an important way to record observations and thoughts, organize knowledge, and try to visualize tectonic processes, the subsurface geometry of rock units, the evolution of landscapes, circulation in the atmosphere and oceans, and motions of astronomical objects. Our research data show that a student who can draw, label, and explain a concept sketch generally has a good understanding of that concept. In our classes, exams are two concept sketches out of a list of 10 to 12 possible questions provided to students ahead of time.

REFERENCES CITED

Baddeley, A. D. 2007. *Working memory, thought, and action*. Oxford: Oxford University Press, 400 p.

Johnson, J. K., and Reynolds, S. J. 2005. Concept sketches—Using student- and instructor-generated annotated sketches for learning, teaching, and assessment in geology courses. *Journal of Geoscience Education*, v. 53, pp. 85–95.

Lawson, A. E. 1980. Relationships among level of intellectual development, cognitive styles, and grades in a college biology course. *Science Education*, v. 64, pp. 95–102.

Lawson, A. 2003. *The neurological basis of learning, development & discovery: Implications for science & mathematics instruction*. Dordrecht, The Netherlands: Kluwer Academic Publishers, 283 p.

Mayer, R. E. 2001. *Multimedia learning*. Cambridge: Cambridge University Press, 210 p.

Roth, W. M., and Bowen, G. M. 1999. Complexities of graphical representations during lectures: A phenomenological approach. *Learning and Instruction*, v. 9, pp. 235–255.

Sweller, J. 1994. Cognitive Load Theory, learning difficulty, and instructional design. *Learning and Instruction*, v. 4, pp. 295–312.

HOW IS THIS BOOK ORGANIZED?

Two-page spreads are organized into 20 chapters that are arranged into five major parts: (1) introduction to earth systems, earth materials, and geologic time; (2) tectonic processes and features; (3) landscapes; (4) the atmosphere and oceans; and (5) the solar system and universe. The first chapter provides an overview of earth science, how we represent location and geologic features, the scientific approach, and an introduction to *earth systems*—a unifying theme interwoven throughout the rest of the book. Chapter 2 introduces minerals and mineral resources, providing an example of our approach in this book of presenting information about mineral, energy, and water resources in the chapter that is most pertinent to each topic. Chapter 3 follows with an introduction to earth materials and to the processes that form the main families of rocks. Part one of the book ends with Chapter 4, which presents the important concepts about determining sequences of events, ages of rocks, and other aspects of geologic time.

The second part of the book covers various aspects of tectonics. Chapter 5 begins with having students observe large-scale features on land and the seafloor, as well as patterns of earthquakes and volcanoes, as a lead-in to tectonic plates. Integrated into the chapter are two-page spreads on continental drift, paleomagnetism, continental and oceanic hot spots, and evolution of the modern oceans and continents. This is followed by Chapter 6, which explores volcanism, volcanoes, and other igneous processes and features. Chapter 7 begins with general principles of deformation and geologic structures, emphasizing how these are expressed in landscapes. The second half of Chapter 7 takes these principles of deformation and applies them to earthquakes, including their causes, settings, and resulting damage. Chapter 8, the final chapter in the second part of the book, explores explanations for mountains and other regions of high elevations, the formation of continents, and features along continental margins. It also explores the origin of local mountains and basins, a topic unique to this textbook, and provides an introduction to oil and natural gas, including shale gas and oil.

The third part of the book focuses on the broad field of geomorphology—the form and evolution of landscapes. It begins with Chapter 9, a visually oriented introduction to processes that sculpt landscapes and redistribute earth materials. This chapter presents a brief introduction to weathering, erosion, and transport. Wind erosion, transport, deposition, and resulting landforms are integrated into Chapter 9, rather than being a separate, sparse-content chapter that forcibly brings in non-wind topics, as is done in other textbooks. This chapter also illustrates the formation of arches, natural bridges, karst topography, and caves, all of which are topics of interest to many students.

The remaining chapters in the third part of the book cover different aspects of geomorphology. Chapter 10 treats the formation, description, and classification of soils, followed by a figure-based presentation of mass wasting and slope stability. Chapter 11 integrates information about glaciers, coasts, and changing sea level, to highlight the interactions among different earth systems. It introduces glacial movement, landforms, and deposits, along with the causes of glaciation. This chapter then moves to coastal processes, landforms, and hazards, and it ends with the consequences of changing sea level on landforms and humans, emphasizing the role of glaciers in raising and lowering sea level. Chapter 12, the final chapter in this third part, is about various topics involving water, including the hydrologic cycle, water use, streams, stream processes, different types of streams, and flooding. The second half of the chapter explores the relationship between surface water and groundwater, including the important topics of contamination and overpumping of groundwater.

The fourth part of the book is about the atmosphere and oceans. It begins with Chapter 13, an introduction to energy, matter, and the atmosphere, providing a solid background for later chapters. Chapter 14 follows this up with coverage about the processes and manifestations of atmospheric motion. It features separate two-page spreads on circulation in the tropics, high latitudes, and mid-latitudes, allowing students to concentrate on one part of the system at a time, leading to a synthesis of lower-level and upper-level winds. Chapter 14 also covers air pressure, the Coriolis effect, and seasonal and regional winds. These topics lead naturally into Chapter 15, which is an introduction to atmospheric moisture, including clouds and various forms of precipitation. Within the chapter are globes and other maps presenting global, regional, and seasonal patterns of humidity and precipitation. Chapter 16 follows this with a visual, map-oriented discussion of weather and storms, including cyclones, tornadoes, and other severe weather. The next chapter (Chapter 17) is devoted to the oceans and their interactions with the atmosphere and cryosphere. It features sections on ocean currents, sea-surface temperatures, ocean salinity, and a thorough treatment of ENSO. The chapters in this part build into Chapter 18, which presents various aspects of climate, including the controls on climate and climate classification. Chapter 18 features a two-page spread on each of the main climate groups, illustrated with a rich blend of globes, process-oriented figures, climographs, and photographs. These spreads are built around the globes, each of which portray a few related climate types, enabling students to concentrate on the distribution and controls of each climate type. The climate chapter also has a data-oriented presentation on the important topic of climate change, especially the data for climate change, the controlling factors, and predicted consequences. It ends with a two-page spread on alternative (non–fossil fuel) energy sources.

The fifth and final group of chapters focuses on the solar system and the rest of the universe. Chapter 19 presents a highly visual introduction to various objects in the solar system and how we study and investigate them. It is followed by Chapter 20, the final chapter in the book, which explores the rest of the universe. It begins with a treatment of how we observe the universe and our framework for referencing these observations. It introduces forces, motions, and light, presenting the laws of motion of Newton and Kepler. The chapter successively explores stars, stellar evolution, stellar remnants, and galaxies, ending with a discussion of cosmology and the early history of the universe.

TWO-PAGE SPREADS

Most of the book consists of *two-page spreads*, each of which is about one or more closely related topics. Each chapter has four main types of two-page spreads: opening, topical, connections, and investigation.

Opening Two-Page Spread

Opening spreads introduce the chapter, engaging the student by highlighting some interesting and relevant aspects and posing questions to activate prior knowledge and curiosity.

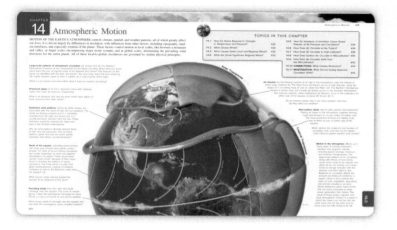

Topical Two-Page Spread

Topical spreads comprise most of the book. They convey the content, help organize knowledge, describe and illustrate processes, and provide a spatial context. The first topical spread in a chapter usually includes some aspects that are familiar to most students, as a bridge or scaffold into the rest of the chapter. Most chapters have at least one two-page spread illustrating how earth science processes impact society and commonly another two-page spread that specifically describes how earth scientists study typical problems.

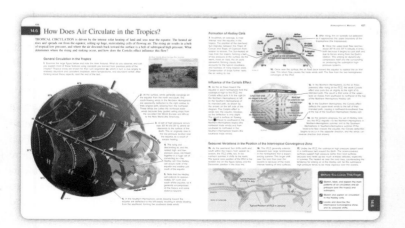

Connections Two-Page Spread

The next-to-last two-page spread in each chapter is a *Connections spread* designed to help students connect and integrate the various concepts from the chapter and to show how these concepts can be applied to an actual location. *Connections* are about real places that illustrate the concepts and features covered in the chapter, often explicitly illustrating how we investigate an earth science problem and how these problems have relevance to society.

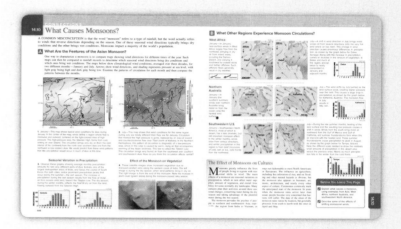

Investigation Two-Page Spread

Each chapter ends with an *Investigation* spread that is an exercise in which students apply the knowledge, skills, and approaches learned in the chapter. These exercises mostly involve virtual places that students explore and investigate to make observations and interpretations and to answer a series of relevant questions. Some involve actual global data. Investigations are modeled after the types of problems geoscientists investigate, and they use the same kinds of data and illustrations encountered in the chapter. The Investigation includes a list of goals for the exercises and step-by-step instructions, including calculations and methods for constructing maps, graphs, and other figures. These investigations can be completed by students in class, as part of a laboratory exercise, as worksheet-based homework, or as online activities.

NEW IN THE SECOND EDITION

The second edition of *Exploring Geology* represents a significant revision, with every chapter receiving additions and improvements, including 188 new photographs. Some changes will be obvious, while others are more subtle but nevertheless substantial. The style, approach, and sequence of chapters is unchanged, but every chapter received new photographs, many revised figures, major to minor editing of text blocks, and, in some cases, reorganization. We revised many text blocks to improve clarity and conciseness, or to present recent discoveries and events. Most chapters contain the same number and order of two-page spreads, but one chapter gained two two-page spreads and another had two spreads completely revised. Many sections of two-page spreads were revised in content and layout, such as by the addition of a photograph not in the previous edition. Nearly all changes were made in response to comments by reviewers and students. The most important revisions are listed below:

- This edition features completely different fonts from the previous edition. The new fonts were chosen partly to improve the readability on portable electronic devices, while retaining fidelity to a quality printed book. This font replacement resulted in countless small changes in the layout of individual text blocks on every two-page spread. In addition to replacing all of the fonts within the text, all figure labels were replaced with the new font, a process that required opening, editing, and commonly resizing every illustration that had text, as in the axes of graphs. In addition, all labels were incorporated into the actual artwork, rather than overlaying them on the artwork using the page-layout program, as was done for many figures in the previous edition. This involved adding labels to hundreds of illustrations, but it has the benefit of having every label as an integral part of its associated art file, a useful feature for constructing PowerPoint files.

- This edition contains 188 new photographs, with a deliberate intention to represent a wider geographic diversity and to provide more detail and clarity about various processes and features, whether on land, in the atmosphere, and in the water. Most of these new photographs represent upgrades of existing photographs from the previous edition, but a number are new photographs in the layout. Also, for this edition, we individually reprocessed nearly all photographs that were derived from scanned slides, using technology and techniques that were not available when the original scanned versions were generated and processed. This reprocessing involved opening up the original high-resolution scans or digital photograph and using modern image-processing software to correct brightness, contrast, and color balance, and to remove visual noise. The resulting improvements will be noticeable for many images in the printed book, but they are more conspicuous in the digital e-book and especially in the high-resolution images we provide instructors for use in classrooms.

- This edition contains many new and replaced figures and even more that were lightly revised, such as replacing fonts. Figures from the first edition were replaced with new versions to update information so that it is more recent, to improve student understanding of certain complex topics, and for improved appearance. All fonts were replaced in every figure that has text.

- This edition contains two new two-page spreads on sedimentary environments and a new section on impact craters. We also thoroughly revised the coverage of climate change, more prominently featuring recent climate change at the start of the discussion. This is followed by a new section that discusses the types of climate proxies, using a more geologic, photograph-based approach in place of the previous collection of small graphs of proxies. In the next spread, which covers factors that could cause climate change, the role of CO_2 was moved up front to again start the discussion focussed on factors involved in recent climate change, followed by those that affect climate on geologic time scales.

- Many two-page spreads have been extensively revised with improved layout, illustrations, and text. In addition to the new or revised illustrations, we updated text to reflect new ideas or data. For example, we updated information on Pluto, comets, satellite temperatures, sea-level rise, and many other relatively minor data points.

- Throughout the book, we added numbers to most text boxes to guide students to read the text boxes in a specific order. We also renumbered many figure numbers so that they are in the same order as the newly numbered text boxes. For all chapter-ending Investigations, we replaced numbers with letters in the Procedures lists to avoid confusion with newly numbered text boxes.

- Every box with the learning objectives was changed from "Before You Leave This Page Be Able To" to simply "Before You Leave This Page." This is more concise, and opened up room on nearly every two-page spread.

CHAPTER 1 received a moderate revision, mostly involving nine new photographs (five replacing existing ones) and the reprocessing of most other photographs. The investigation received four additional photographs to depict important features students need to consider in their deliberations. The chapter also has one revised illustration that now incorporates an actual photograph of Pluto.

CHAPTER 2 received a light revision, with eight new photographs, mostly of rocks and mineral resources. Some other photographs were processed from the original scans. Fonts were replaced throughout, resulting in many small changes in wording and layout, as occurred in every chapter.

CHAPTER 3 was heavily revised, featuring 36 new photographs and two new two-page spreads that present an early, visual overview of sedimentary environments on land, near shorelines, and in the ocean. Two new page-spanning illustrations and 14 additional photos accompany this new material. Two of the new photographs are from the Franciscan of California, and are accompanied by a new, brief introduction to melange. Other new photographs are mostly from Florida, Texas, and New Mexico.

CHAPTER 4 contains nine new photographs of rocks, fossils, and environments. It has a new section on impact craters accompanied by three new illustrations. It also now incorporates a new photograph and modified discussion of concretions. Several sections received significant edits.

CHAPTER 5 on plate tectonics is mostly unchanged, but every illustration with text was edited to replace the fonts. Several maps were revised, including the one showing North American transform faults.

CHAPTER 6 has 14 new or replaced photographs representing more diverse locations, including Joshua Tree National Park. It has a new photograph of the Valles Caldera and a number of reprocessed ones. Two photographs of Augustine pyroclastic eruptions were reprocessed and recropped to better convey the vertical extent of the eruptions. In addition to font changes, the chapter has two rebuilt illustrations.

CHAPTER 7 includes eight new photographs of structures and landscape features, including ones in a heavily revised section showing features related to erosion of tilted layers. There is also a new short section on erosion of fault scarps, accompanied by a new photograph. Several illustrations were moderately revised, mostly ones showing seismic waves.

CHAPTER 8 contains eight new photographs illustrating the landscape appearance of different types of rocks. Several sections were reordered and heavily edited around the new photographs. A map of sedimentary basins was revised to better display the geographic features in the area covered by each basin.

CHAPTER 9 on sculpting landscapes was heavily revised, with 24 new photographs of weathered limestone, caves, karst topography, and problem soils. The new photographs are mostly from Florida, Texas, and Carlsbad Caverns National Park. The new photographs of Carlsbad illustrate the size of the cavern better than most textbook images we have seen. The section on trading location for time was vastly improved through changes in layout and a new photograph from Monument Valley that perfectly illustrates the concept.

CHAPTER 10 contains 10 new photographs of soils and slopes, including new photographs specifically retaken of the Slumgullion Landslide and a new computer-generated 3D-perspective showing the 2017 Big Sur landslide.

CHAPTER 11 has 25 new photographs, mostly of coastal regions of Florida and glacial features of the western U.S. and Alaska. The chapter was renamed using coasts instead of shorelines, and text and headings were changed throughout to reflect this change.

CHAPTER 12 on streams, flooding, and groundwater now features 15 new photographs of streams and stream-related features. There are newly inserted photographs of cutbanks, point bars, and entrenched meanders, accompanied by changes in layout and text editing to accomodate the new images. One photograph of a spring was deleted. All the graphs and maps were revised for new fonts and other improvements, such as arrow colors depicting groundwater flow.

CHAPTER 13 is the first of the atmosphere chapters, and received only minor revisions. It contains five new photographs. Each illustration, including each of the many graphs, was edited for the change in fonts and incorporation of labels in the art files.

CHAPTER 14 was for the most part lightly revised, except for font changes in all the illustrations. There are three new versions of figures showing upper-level polar circulation and two new versions of global wind patterns and the Coriolis effect. There is one new photograph, and the investigation has a new layout.

CHAPTER 15 has a number of new versions of illustrations and one new photograph. Globes showing humidity were rebuilt and rerendered, with a clearer legend. There are new versions of figures showing Rossby waves, clouds, freezing rain, and sleet. Global amounts of precipitation are now shown with three globes rather than one flat map. As for all the globes in the book, the authors provide a media file for every globe shown.

CHAPTER 16 had minor revisions on most two-page spreads, but major revisions on some. Major revisions included sections about lightning and upper-level lightning phenomena, where in both cases text was separated

from the figures to improve readability and for use in presentation software. The Investigation was heavily revised to include data and discussion of upper-level airflow patterns. The chapter has three new photographs.

CHAPTER 17 features, in additon to all the font changes, several newly redone illustrations, such as on global wind directions and the Southern Oscillation. We revised the labels, layout, and order of globes showing ocean currents. There are two new photographs.

CHAPTER 18 displays major revisions to figures and some two-page spreads. All the climate globes, of which there are many, were rebuilt using new 3D files and rerendered. Likewise, all figures with part of a globe were redrawn using the new renders. We updated a number of figures to show the most current data, such as on sea-level rise, extent of Arctic sea ice, global temperatures, and CO_2. We redid the first two spreads on climate change, aiming to consolidate the discussion of recent climate change at the start of the discussion, followed by climate change over geologic timescales. Graphs of proxies were replaced with photographs to provide a more geologic approach and to better convey the diversity of types of climate proxy data. The role of CO_2 was moved up in the discussion of the possible causes of climate change, with the intent of again leading with recent climate change and then moving to long-term climate changes. A graph of sunspot data was replaced with a photograph.

CHAPTER 19 had moderate revisions, with the addition of four new images depicting more recent images of Pluto, nebulae, and a comet. We added or refined the discussions of Pluto, Ceres, comets, the age of the solar system, and the number of moons of Jupiter, each reflecting current information.

CHAPTER 20 on the Universe had very minor revisions, but fonts on the many text-rich illustrations were replaced, which often involved repositioning and changing layout of the text. Some figures received minor additional revisions.

FRONT AND BACK MATTER, including the *Preface*, *Glossary*, and *Index*, were revised and updated to reflect the revised table of contents and changes in page numbers due to reorganizations.

ACKNOWLEDGMENTS

Writing a totally new type of introductory earth science textbook would not be possible without the suggestions and encouragement we received from instructors who reviewed various incarnations of text and artwork in this book. We are especially grateful to people who contributed entire days either reviewing our books or attending symposia to openly discuss the vision, challenges, and refinements of this kind of new approach. We also appreciate the support of hundreds of instructors who have reported great success with using our books in their classrooms, validating our unusual approach and encouraging us to extend our original vision into various fields of earth science.

This book is a decidedly collaborative effort, incorporating material from our two other textbooks. Our colleagues Chuck Carter, Mike Kelly, and Paul Morin contributed materials to our *Exploring Geology* textbook, and some of this content is included here. Likewise, we have greatly benefitted from our collaboration with geographers Bob Rohli, Peter Waylen, and Mark Francek on our *Exploring Physical Geography* textbook, which provided the starting materials for chapters on the atmosphere and oceans. We gratefully acknowledge the words, figures, organization suggestions, and friendship provided by these colleagues.

This book contains over 2,500 figures, several times more than a typical introductory earth science textbook. This massive art program required great effort and artistic abilities from the illustrators and artists who turned our vision and sketches into what truly are pieces of art. We are especially appreciative of Cindy Shaw, who was lead illustrator, art director, and a steady hand that helped guide the project. For many figures, she extracted data from NOAA and NASA websites and then converted the data into exquisite maps and other illustrations. Cindy also fine-tuned or extensively reworked the authors' layouts, standardized various aspects of the illustrations, and prepared the final figures for printing. This second edition was especially onerous for Cindy, because the two main fonts were replaced throughout the book and in every figure label. Cindy also moved labels from the page-layout program into the art files, partly so that the labels are present in PowerPoint images. Chuck Carter produced many spectacular pieces of art, including virtual places featured in the chapter-ending Investigations. Susie Gillatt contributed many of her wonderful photographs from around the world, photographs that helped us tell the story in a visual way. She also color corrected and retouched most of the photographs in the book, including every image in this edition that was derived from a scanned color slide, and there were many. We also used visually unique artwork by Daniel Miller, David Fierstein, and Susie Gillatt. Suzanne Rohli performed magic with GIS files and helped in many other ways. We were ably assisted in data compilation and other tasks by students and former students Cheryl Replogle, Jenna Donatelli, Emma Harrison, Abeer Hamden, Peng Jia, Javier Vázquez, and Courtney Merjil. Terra Chroma, Inc., of Tucson, Arizona, supported many aspects in the development of this book, including funding parts of the extensive art program and maintaining the *ExploringEarthScience.com* website.

Many people went out of their way to provide us with photographs, illustrations, and advice. These helpful people included Susie Gillatt, Cindy Shaw, Vladimir Romanovsky, Paul McDaniel, Lawrence McGhee, Charles Love, Ramón Arrowsmith, Dan Trimble, Bixler McClure, Michael Forster, Vince Matthews, Ron Blakey, Doug Bartlett, Phil Christensen, Scott Johnson, Peg Owens, Skye Rodgers, Steve Semken, Jonathan Warrick, and Michael Vanden Berg. We are extremely grateful to Thomas Arny and Stephen Schneider for the use of many illustrations from their excellent astronomy textbooks. Our astronomy figures and text benefitted from thorough reviews by Steven Desch, William Karl Pitts, and David A. Williams.

We used a number of data sources to create many illustrations. We extensively used the Blue Marble and Blue Marble Next Generation global satellite composites. We are very appreciative of the NOAA Reanalysis Site, which we used extensively, and for other sites of the USDA, NASA, USGS, and NPS.

We have treasured our interactions with the wonderful Iowans at McGraw-Hill Education, who enthusiastically supported our vision, needs, and progress. We especially thank our current and previous publishers Michael Ivanov, Michelle Vogler, Ryan Blankenship, and Marge Kemp for their continued encouragement and excellent support. Jodi Rhomberg, Laura Bies, and others skillfully and cheerfully guided the development of the book during the publication process, making it all happen. Lori Hancock helped immensely with our ever-changing photographic needs, and Janet Robbins and Erica Gordon guided us through the obstacle-laden arena of photo permissions. We also appreciate the support, cooperation, guidance, and enthusiasm of Thomas Timp, Marty Lange, Kelly Brown, Matthew Backhaus, Tammy Ben, and many others at McGraw-Hill who worked hard to make this book a reality. Kevin Campbell provided thorough copy editing and also compiled the index and glossary. Our wonderful colleague Gina Szablewski expertly directed the development of LearnSmart materials, suggested helpful changes to the text, and provided general encouragement. She also helped update, refine, and improve the test bank.

Finally, a project like this is truly life consuming, especially when the authors do the writing, illustrating, photography, near-final page layout, media development, and development of assessments, teaching ancillaries, and the instructor's website. We are extremely appreciative of the support, patience, and friendship we received from family members, friends, colleagues, and students who shared our sacrifices and successes during the creation of this new vision of a textbook. Steve Reynolds thanks the ever-cheerful, supportive, and talented Susie Gillatt; John and Kay Reynolds; and our mostly helpful book-writing companions, Widget, Jasper, and Ziggy. Julia Johnson thanks Annabelle Louise and Hazel Johnson, and the rest of her family for enthusiastic support and encouragement. Steve and Julia appreciate the support of their wonderful colleagues at ASU and elsewhere.

The authors are grateful for the many students who have worked with us on projects, infused our classrooms with energy and enthusiasm, and provided excellent constructive feedback about what works and what doesn't work. We wrote this book to help instructors, including us, make students' time in our classes even more interesting, exciting, and informative. Thank you all!

REVIEWERS

Special thanks and appreciation go out to all reviewers of *Exploring Earth Science*, *Exploring Geology*, and *Exploring Physical Geography*. This book was improved by many beneficial suggestions, new ideas, and invaluable advice provided by the reviewers of these three books. We appreciate all the time they devoted to reviewing manuscript chapters, attending focus groups, surveying students, and promoting this text to their colleagues.

We would like to thank the following individuals who wrote and/or reviewed learning goal-oriented content for *LearnSmart*.

Northern Arizona University, Sylvester Allred
Youngstown State University, Ray Beiersdorfer
Georgia Southern University, Michelle Cawthorn
Western Kentucky University, Margaret E. Crowder
Roane State Community College, Arthur C. Lee
University of North Carolina–Chapel Hill, Trent McDowell
Florida Atlantic University, Jessica Miles
State University of New York at Cortland, Noelle J. Relles
University of Wisconsin–Milwaukee, Gina Seegers Szablewski
Elise Uphoff

REVIEWERS

Antelope Valley College, Michael W. Pesses
Appalachian State University, Cynthia M. Liutkus
Appalachian State University, Ellen A. Cowan
Appalachian State University, Johnny Waters
Appalachian State University, Sarah K. Carmichael
Arizona State University, Bohumil Svoma
Arizona State University, David A. Williams
Arizona State University, Ed Garnero
Arizona State University, Ronald Greeley
Arizona State University, Steve Desch
Auburn University, David T. King, Jr.
Auburn University, Lorraine W. Wolf
Auburn University, Ming-Kuo Lee
Auburn University, Willis Hames
Austin Peay State University, Robert A. Sirk
Ball State University, David A. Call
Bethel University, Bryan Anderson
Blinn College, Amanda Palmer Julson
Boston College, Yvette Kuiper
Bowling Green State University, Paula J. Steinker
Brigham Young University, Dan Moore
Brigham Young University, William W. Little
Brown University, Jan Tullis
Buffalo State College, Elisa Bergslein
Buffalo State College, Kyle C. Fredrick
California State University–Long Beach, Stan Finney
California State University–Los Angeles, Steve LaDochy
California State University–Fresno, Stephen D. Lewis

California State University–Fullerton, Diane Clemens-Knott
California State University–Fullerton, Jeffrey Knott
California State University–Fullerton, Stephen Wareham
California State University–Northridge, George E. Davis
California State University–Northridge, Karen L. Savage
California State University–Sacramento, Tomas Krabacker
Cape Fear Community College, James Criswell
Clemson University, Richard Warner
Cleveland State University, Heather Gallacher
College of Charleston, Cassandra J. Runyon
College of Charleston, James L. Carew
College of Charleston, Michael Katuna
College of Charleston, Steven C. Jaume
College of Charleston, T. J. Callahan
College of DuPage, Mark J. Sutherland
College of Southern Idaho, Shawn Willsey
College of Southern Nevada, Barry Perlmutter
Columbus State Community College, Jeffery G. Richardson
Duke University, Bruce H. Corliss
East Carolina University, Richard L. Mauger
East Tennessee State University, Yongli Gao
Eastern Kentucky University, Stewart S. Farrar
Eastern Washington University, Jennifer Thomson
Eastern Washington University, Richard Orndorff
Edinboro University of Pennsylvania, Eric Straffin
El Paso Community College, Kathleen Devaney
Florida International University, Dean Whitman
Florida International University, Grenville Draper
Florida International University, Neptune Srimal
Florida State University, Holly M. Widen
Florida State University, Neil Lundberg
Florida State University, Victor Mesev
Florida State University, William C. Parker
Frostburg State University, Phillip P. Allen
Frostburg State University, Tracy L. Edwards
Fullerton College, Richard Lozinsky
Fullerton College, Sean Chamberlin
George Mason University, Patricia Boudinot
George Mason University, Stacey Verardo
Georgia Institute of Technology, Meg Grantham
Georgia Perimeter College, Deniz Z. Ballero
Georgia Perimeter College, Gerald D. Pollack
Georgia Perimeter College, John R. Anderson, II
Georgia Perimeter College, Pamela J. W. Gore
Georgia State University, Pamela C. Burnley
Georgia State University, Seth Rose
Glendale Community College, J. Robert Thompson
Glendale Community College, Pamela Nelson
Glendale Community College, Steven D. Kadel
Grand Valley State University, John Weber
Grand Valley State University, Kevin Cole
Grand Valley State University, Stephen Mattox
Guilford Tech Community College, Steve Adams
Gustavus Adolphus College, Alan D. Gishlick
Hofstra University, J. Bret Bennington
Hunter College, Randye L. Rutberg
Indiana State University, James H. Speer
Indiana State University, Jennifer Latimer
Indiana University Northwest, Zoran Kilibarda

Iowa State University, Cinzia Cervato
Iowa State University, Kenneth Windom
Jacksonville State University, David A. Steffy
Kansas State University, Mary Hubbard
Keene State College, Peter A. Nielsen
Las Positas College, Thomas Orf
Lehman College, CUNY, Stefan Becker
Long Beach City College, Douglas Britton
Long Island University, Margaret F. Boorstein
Louisiana State University, Jeffrey A. Nunn
Louisiana Tech University, Maureen McCurdy Hillard
Mansfield University of Pennsylvania, Nicole Wilson
Marshall University, Dewey D. Sanderson
Mesa Community College, Clemenc Ligocki
Mesa Community College, Robert A. Leighty
Mesa Community College, Steve Bass
Metro State, Jon Van de Grift
Metro State, Kenneth Engelbrecht
Metropolitan State College of Denver, Jason Janke
Miami University of Ohio, Elizabeth Widom
Michigan Technical University, Theodore Bornhorst
Middle Tennessee State University, Michael Westphal Hiett
Minnesota State University, Forrest D. Wilkerson
Mississippi State University, John E. Mylroie
Mississippi State University, Renee M. Clary
Missouri State University, Kevin Ray Evans
Missouri State University, Melida Gutierrez
Monroe Community College, Amanda Colosimo
Monroe Community College, Jessica Barone
Monroe Community College, SUNY, Jonathon Little
Montgomery College, Nathalie Nicole Brandes
Moorpark College, Michael T. Walegur
Morehead State University, Eric Jerde
New Mexico State University, Jeffrey M. Amato
New Mexico State University, Marilyn C. Huff
Normandale Community College, Dave Berner
Northern Arizona University, Abe Springer
Northern Arizona University, David M. Best
Northern Arizona University, Nancy Riggs
Northern Essex Community College, Mark E. Reinhold
Northern Illinois University, David Goldblum
Northern Illinois University, Mark Fischer
Northern Illinois University, Mark Frank
Northern Virginia Community College, Kenneth Rasmussen
Ohio University, Alycia L. Stigall
Ohio University, Daniel I. Hembree
Oklahoma State University, Jianjun Ge
Oregon State University, Roy Haggerty
Pasadena City College, James R. Powers
Pennsylvania State University, Timothy Bralower
Portland Community College, Frank D. Granshaw
Purdue University–West Lafayette, Yuch-Ning Shieh
Rhodes College, David Shankman
Rutgers University, Mark D. Feigenson
Rutgers University, Roy Schlische
Saint Louis University, John Encarnacion
Samford University, Jennifer Rahn
San Francisco State University, Barbara A. Holzman
San Francisco State University, Bridget Wyatt
San Jose State University, Richard Sedlock
San Juan College, John H. Burris
South Dakota State University, Trisha Jackson
South Dakota State University, Jim Peterson

Southern Illinois University–Edwardsville, Michael
 Grossman
Southern Utah University, Paul R. Larson
Southwestern Illinois College, Stanley C. Hatfield
St. Cloud State University, Kate Pound
St. Norbert College, Nelson R. Ham
State University of New York, College at Potsdam,
 Michael Rygel
State University of New York at New Paltz, Ronald G.
 Knapp
Sul Ross State University, G. David Mattison
Syracuse University, Henry T. Mullins
Syracuse University, Jeffrey A. Karson
Syracuse University, Suzanne L. Baldwin
Tarleton State University, Bethany D. Rinard
Tarleton State University, Carol Thompson
Tarrant County College, Clair Russell Ossian
Tennessee Tech University, Michael J. Harrison
Texas A&M University, Bruce Herbert
Texas A&M University–Corpus Christi, Thomas Naehr
Texas State University–San Marcos, David R. Butler
Texas Tech University, Aaron Yohsinobu
The College of New Jersey, Margaret H. Benoit
The University of Akron, Wayne College, Adil M. Wadia
The University of Memphis, Hsiang-te Kung
Towson University, Kent Barnes
Tulane University, Mead A. Allison
Tulane University, Sadredin C. Moosavi
United States Military Academy, Peter Siska
University of Alabama, Andrew M. Goodliffe
University of Alabama, Nathan L. Green
University of Alabama–Tuscaloosa, Delores Robinson
University of Alaska–Anchorage, Derek Sjostrom
University of Alaska–Fairbanks, Paul McCarthy
University of Arkansas, Steve Boss
University of Arkansas–Little Rock, Wendi J. W. Williams
University of Calgary, Lawrence Nkemdirim
University of Cincinnati, Craig Dietsch
University of Cincinnati, Teri Jacobs
University of Colorado–Boulder, Eric Small
University of Colorado–Boulder, Jake Haugland
University of Colorado–Boulder, Kevin H. Mahan
University of Colorado–Colorado Springs, Steve Jennings
University of Florida, Joseph G. Meert
University of Georgia, Andrew Grundstein
University of Hawaii at Manoa, Scott Rowland
University of Houston, Michael A. Murphy
University of Houston, Peter Copeland
University of Houston, William Dupre
University of Idaho, Dennis Geist
University of Illinois–Chicago, Roy E. Plotnick
University of Louisiana–Lafayette, Brian E. Lock
University of Louisiana–Lafayette, Carl Richter
University of Maryland –College Park, Christine
 A. M. France
University of Memphis, Lensyl Urbano

University of Memphis, Randel Tom Cox
University of Memphis, Steven R. Newkirk
University of Michigan–Dearborn, Jacob A. Napieralski
University of Minnesota, Kent C. Kirkby
University of Mississippi, Cathy A. Grace
University of Missouri, C. Mark Cowell
University of Missouri–Columbia, Francisco Gomez
University of Missouri–Columbia, Martin Appold
University of Nevada–Reno, Franco Biondi
University of New Orleans, Mark A. Kulp
University of North Carolina–Chapel Hill, Trent
 McDowell
University of North Carolina–Charlotte, William Garcia
University of North Carolina–Wilmington, John R.
 Huntsman
University of North Dakota, Dexter Perkins
University of North Dakota, Paul Todhunter
University of North Texas, Harry Williams
University of North Texas, Paul F. Hudak
University of Northern Colorado, Jared R. Morrow
University of Northern Colorado, Michael Taber
University of Oklahoma, Barry Weaver
University of Oklahoma, David E. Fastovsky
University of Oklahoma, G. Randy Keller
University of Oklahoma, Scott Greene
University of Saskatchewan, Dirk de Boer
University of South Carolina, Robert Thunell
University of South Carolina, Scott White
University of South Carolina–Aiken, Allen Dennis
University of South Dakota, Timothy Heaton
University of South Florida, Chuck Connor
University of South Florida, Judy McIlrath
University of Southern Mississippi, David Harms Holt
University of Tennessee, Derek J. Martin
University of Tennessee–Chattanooga, Ann Holmes
University of Tennessee–Knoxville, Julie Y. McKnight
University of Tennessee–Knoxville, Kula C. Misra
University of Texas–Brownsville, Ravi Nandigam
University of Texas–Brownsville, Elizabeth Heise
University of the Pacific, Lydia K. Fox
University of Toledo, James Martin-Hayden
University of West Georgia, Curtis L. Hollabaugh
University of West Georgia, James R. Mayer
University of West Georgia, Julie K. Bartley
University of West Georgia, Philip M. Novack-Gottshall
University of West Georgia, Rebecca L. Dodge
University of Wisconsin–Eau Claire, Christina M. Hupy
University of Wisconsin–Eau Claire, Donald J. Sidman
University of Wisconsin–Eau Claire, Garry Leonard
 Running
University of Wisconsin–Eau Claire, Joseph P. Hupy
University of Wisconsin–Madison, Steven Ralser
University of Wisconsin–Milwaukee, Gina Seegers
 Szablewski
University of Wisconsin–Platteville, Mari Vice
Wake Technical Community College, Adrianne

A. Leinbach
Wake Technical Community College,
 Gretchen Miller
Wayne State University, Mark Baskaran
Weber State University, Eric C. Ewert
West Virginia University, John Renton
Western Illinois University, Kyle Mayborn
Western Illinois University, Steven W. Bennett
Western Kentucky University, Aaron J. Celestian
Western Kentucky University, Margaret E. Crowder
Western Michigan University, Duane Hampton
Western Michigan University, G. Michael Grammer
Western Washington University, Bernard A. Housen
Western Washington University, David Hirsch
Western Washington University, Scott R. Linneman
Western Washington University, Thor A. Hansen
Wharton County Junior College, Danny Glenn
Wright State University, Michael G. Bourne, Jr.
Wright State University, Stacey A. Hundley
Youngstown State University, Shane V. Smith

FOCUS GROUP AND SYMPOSIUM PARTICIPANTS

Ball State University, Petra Zimmermann
Blinn College, Rhonda Reagan
California State University–Los Angeles, Steve LaDochy
Florida State University, Neptune Srimal
Georgia State University, Leslie Edwards
Indiana Purdue University–Indianapolis (IUPUI),
 Andrew Baker
Kansas State University, Doug Goodin
Mesa Community College, Steven Bass
Miami University of Ohio, Elizabeth Widom
Minnesota State University, Ginger L. Schmid
Northern Illinois University, Lesley Rigg
Northern Illinois University, Mike Konen
South Dakota State University, Bruce V. Millett
Tarleton State University, Carol Thompson
Texas A&M University, Steven Quiring
University of Alabama, Amanda Epsy-Brown
University of Colorado–Boulder, Peter Blanken
University of North Carolina–Greensboro,
 Michael Lewis
University of North Dakota, Dexter Perkins
University of Oklahoma, Scott Greene
University of Texas-Brownsville, Elizabeth Heise
University of Wisconsin-Milwaukee, Gina Seegers
 Szablewski
University of Wisconsin–Oshkosh, Stefan Becker
Youngstown State University, Shane V. Smith

ABOUT THE AUTHORS

STEPHEN J. REYNOLDS

Stephen J. Reynolds received an undergraduate geology degree from the University of Texas at El Paso, and M.S. and Ph.D. degrees in structure/tectonics and regional geology from the University of Arizona. He then spent ten years directing the geologic framework and mapping program of the Arizona Geological Survey, where he completed the 1988 *Geologic Map of Arizona*. Steve is a professor in the School of Earth and Space Exploration at Arizona State University, where he teaches Physical Geology, Structural Geology, Advanced Field Geology, Orogenic Systems, Cordilleran Regional Geology, Teaching Methods in the Geosciences, and others. He helped establish the ASU Center for Research on Education in Science, Mathematics, Engineering, and Technology (CRESMET), and was President of the Arizona Geological Society.

Steve has authored or edited over 200 geologic maps, articles, and reports, including the 866-page *Geologic Evolution of Arizona*. He also coauthored *Structural Geology of Rocks and Regions*, a structural geology textbook, and *Observing and Interpreting Geology*, a laboratory manual for physical geology. Working with a team of geographers, he authored *Exploring Physical Geography*, which follows the style and approach of his award-winning *Exploring Geology* textbook.

His current geologic research focuses on structure, tectonics, stratigraphic correlations, landscape evolution, and mineral deposits of the Southwest. For several decades, he has conducted science-education research on student learning in college geoscience courses, especially the role of visualization. He was the first geologist with his own eye-tracking laboratory, where he and his students demonstrated that students learn more when using the unique design, layout, and approach of this textbook, compared to how much (or little) they learn from a traditional textbook.

Steve is known for innovative, award-winning teaching methods, and has a widely used website. He was a National Association of Geoscience Teachers (NAGT) distinguished speaker, and he often travels across the country presenting talks and workshops on visualization and on how to infuse active learning and inquiry into large introductory geoscience classes. He is commonly an invited speaker to national workshops and symposia on active learning, visualization, and teaching methods in college geoscience courses. He has been a long-time industry consultant in mineral, energy, and water resources, and has received outstanding alumni awards from UTEP and the University of Arizona.

JULIA K. JOHNSON

Julia K. Johnson is a full-time faculty member in the School of Earth and Space Exploration at Arizona State University. She received an undergraduate degree from Ricks College in eastern Idaho and undergraduate and graduate degrees from Arizona State University. Her graduate research involved structural geology and geoscience education research.

She teaches introductory geoscience to more than 2,000 students per year, both online and in person. She also supervises the associated in-person and online labs, and is the lead author of both the online and in-person versions of *Observing and Interpreting Geology, a Laboratory Manual for Physical Geology*, an innovative laboratory manual in which all learning is built around a virtual world (Painted Canyon). She helps coordinate the introductory geoscience teaching efforts of the School of Earth and Space Exploration (SESE), guiding other instructors as they incorporate active learning, inquiry, and online materials into large lecture classes. She is responsible for the design and deployment of online resources for SESE's introductory physical geology courses, including the lecture and lab components.

Julia coordinated an innovative project focused on redesigning introductory geology classes so that they incorporated more online content and asynchronous learning. This project was very successful in improving student performance, mostly due to the widespread implementation of concept sketches and partly due to Julia's approach of decoupling multiple-choice questions and concept-sketch questions during exams and other assessments. As a result of these successful and innovative efforts, Julia was asked to be a Redesign Scholar for the National Center for Academic Transformation.

Julia is recognized as one of the best science teachers at ASU, and has received student-nominated teaching awards and very high teaching evaluations in spite of her challenging classes. Her efforts have helped dramatically increase enrollments in introductory geoscience courses and the number of majors. She coauthored the widely used *Exploring Geology* and *Exploring Physical Geography* textbooks. Julia has authored publications on geology and science-education research, including a widely cited article in the *Journal of Geoscience Education* on concept sketches. She also developed a number of websites used by students around the world, including the *Visualizing Topography* and *Biosphere 3D* websites. She leads many geologic field trips for schools and professional groups in the Phoenix region.

CONTRIBUTING AUTHORS AND ILLUSTRATORS

ROBERT V. ROHLI

Robert Rohli, coauthor of *Exploring Physical Geography* and *Louisiana Weather and Climate*, received a B.A. in geography from the University of New Orleans, an M.S. degree in atmospheric sciences from The Ohio State University, and a Ph.D. in geography from Louisiana State University (LSU). He is a professor of geography at LSU and also faculty director of the LSU Residential Colleges Program. His teaching and research interests are in synoptic and applied meteorology/climatology, atmospheric circulation variability, and hydroclimatology. Major themes in his teaching include the systems approach to earth systems, collaboration among students from different disciplines, and development of applied problem-solving skills. He has been an active supporter of undergraduate education initiatives, including the Louisiana Geographic Education Alliance. He has authored more than 60 research articles, encyclopedia articles, and proceedings papers, mostly on topics related to synoptic or applied climatology.

CYNTHIA SHAW

Cindy Shaw, lead illustrator and art director for this book, holds a B.A. in zoology from the University of Hawaii–Manoa as well as a master's in education from Washington State University, where she researched the use of guided illustration as a teaching and learning tool in the science classroom. Now focusing on earth science, mapping, and coral reef ecology, she writes and illustrates for textbooks and museums, and develops ancillary educational materials through her business, Aurelia Press. Her kids' novel, *Grouper Moon*, is used in many U.S. and Caribbean science classrooms, and is making an impact on shaping childrens' attitudes toward fisheries conservation. Currently landlocked in Richland, Washington, Cindy escapes whenever possible to travel, hike, and dive the reefs to field-sketch and do reference photography for her projects. She is also lead illustrator and art director for recent editions of *Exploring Geology* and *Exploring Physical Geography*. In this role, Cindy generates new illustrations, often researching the necessary data for those illustrations, standardizes illustrations contributed by other artists, finalizes page layout, and prepares files for electronic delivery of all materials to the compositors.

PETER R. WAYLEN

Peter Waylen, coauthor of *Exploring Physcial Geography*, is professor of geography at the University of Florida, where he served as chair for seven years. He holds a B.Sc. in geography from the London School of Economics, England, and a Ph.D. from McMaster University, Canada. He was also assistant professor at the University of Saskatchewan. Peter's teaching and research interests are in the fields of hydrology and climatology, particularly the temporal and spatial variability of risks of such hazards as floods, droughts, freezes, and heat waves, and the way in which these vary in the long run, driven by global-scale phenomena like ENSO. He teaches Introductory Physical Geography, Principles of Geographic Hydrology, and Models in Hydrology, and was selected University of Florida Teacher of the Year in 2002. His research is principally interdisciplinary and collaborative with colleagues and students. His research results appear in over 100 refereed articles and book chapters on geography, hydrology, and climatology.

PAUL J. MORIN

Paul Morin, coauthor of *Exploring Geology*, is director of the Polar Geospatial Center at the University of Minnesota, which supports National Science Foundation scientific and research operations through remote sensing and other geospatial data. Paul co-founded the GeoWall Consortium, which promoted the use of visualization and stereo projection in the classroom. Over the past five years, Paul has been instrumental in bringing earth science visualization to science museums around the world.

SUSANNE GILLATT

Susie Gillatt received a bachelor of arts degree from the University of Arizona and is president of Terra Chroma, Inc., a multimedia studio. She focuses on scientific illustration and photo preparation for academic books and journals. Many photographs in this book were contributed by Susie from her travels to experience different landscapes, ecosystems, and cultures around the world. Her award-winning art, which combines photographs of nature with digital painting, has been displayed in galleries in Arizona, Colorado, and Texas.

MARK A. FRANCEK

Mark Francek, coauthor of *Exploring Physical Geography*, is a geography professor at Central Michigan University (CMU). He earned his doctorate in geography from the University of Wisconsin-Milwaukee, his master's from the University of South Carolina, and his bachelor's degree from the State University College at Geneseo, New York. He has teaching and research interests in earth science education, physical geography, and soil science. He has authored and coauthored more than 30 scholarly papers, and has presented his research at numerous conferences. At CMU, Mark served as director of the Environmental Studies Program and director of the Science and Technology Residential College. He has received numerous teaching awards, including the Carnegie Foundation for the Advancement of Teaching Michigan Professor of the Year and the Presidents Council of State Universities of Michigan Distinguished Professor of the Year. His "Earth Science Sites of the Week" Listserv highlights the best earth science websites.

CHUCK CARTER

Chuck Carter has worked in the science and entertainment industries for three decades. He developed the innovative video game, *Myst*, and more than two dozen other video games in a variety of art, animation, and management roles, including computer graphics supervisor and art director. His illustrations and animations have appeared in *National Geographic*, *Scientific American*, *Wired*, the BBC, NASA, and Disney's *Mission to Mars*. Chuck is president of Eagre Games, designing fully immersive adventures, including *ZED*.

DANIEL MILLER

Daniel Miller is a self-taught artist, beginning his career as a silversmith, then goldsmith, painter and sculptor, designer, and art director. Attaining his goal of working in the film industry, he created notable sculptural elements for many major films, including *Stargate* and *Chronicles of Riddick*. He was a concept artist and matte painter for films and video games. He completed large-scale sculptural installments, including *Fountains of the Gods* at Caesars Palace in Las Vegas, where he lives and pursues his passion for oil painting.

exploring

Earth
Science

The Nature of Earth Science

EARTH SCIENCE FOCUSES ON THE FOUR COMPONENTS of the earth system—land, water, air, and life—and their interactions. Processes within the solid earth cause volcanoes and earthquakes, form mountains, and rearrange continents. Most of Earth's surface is covered by oceans, but water also forms ecologically important lakes, rivers, and wetlands. Above the surface, the atmosphere contains gases essential to life, as well as clouds, precipitation, wind, and storms. Living things depend upon and interact with the land, water, atmosphere, and energy from space. Together, the various components of our planet control the climate and overall suitability for life, the distribution of natural resources, and the susceptibility for floods, landslides, and other natural disasters. Earth science is the study of the solid earth, oceans, atmosphere, life, and our setting in space.

North America and Central America have a wealth of interesting features. The large image below (▼) is computer-generated and combines different types of data to show features on the land, in the oceans, and in the atmosphere. The shading and colors on land are from space-based satellite images, and whitish colors in the atmosphere are clouds of various types and heights. Can you find the region where you live or visit? What types of landscape features, water bodies, and clouds are there?

◄ **The dramatic scenery of Banff, Alberta**, in the Canadian Rockies, features spectacularly tilted and folded rock layers expressed in various shades of gray. Exposed to sunlight, moisture, plants, animals, and the downward pull of gravity, the rock layers begin to disintegrate, some forming precipitous cliffs and others wearing away into slopes covered by loose pieces. The mountains interact with the atmosphere, causing clouds, rain, and snow to be concentrated over the mountains, influencing the growth of trees and other plants, and affecting the lives of the various creatures, including mammals and birds.

What processes form rock layers, sculpt the land surface, and produce such beautiful scenery?

01.00.a2 Banff, Alberta, Canada

The 1980 eruption of Mount St. Helens in southwestern Washington (▼) ejected huge amounts of volcanic ash into the air, toppled millions of trees, unleashed large floods and mudflows down nearby valleys, and killed 57 people. Earth scientists study volcanic phenomena to determine how and when volcanoes erupt and what hazards volcanoes pose to humans and other creatures.

How do studies of the Earth's surface and subsurface help us determine where it is safe to live?

01.00.a3 Mount St. Helens, WA

01.00.a1

Banff, Alberta

Mount St. Helens

Tornado Alley

TOPICS IN THIS CHAPTER

Earth's atmosphere is dynamic, with constantly moving air masses and clouds that produce precipitation (rain, snow, and hail) and locally cause severe storms, like the ominous line of clouds shown below (▼). Heating from the Sun provides the energy for winds, as well as atmospheric moisture for clouds and storms, so Earth's setting in space influences important aspects of our planet, including climate and seasons. Severe weather, including tornadoes, is especially common in the center of the U.S. along a north-south region often called "Tornado Alley."

How do clouds and storms form, and how does the Sun influence winds?

01.00.a4 Tornado Alley, Central U.S.

The oceans, like the atmosphere, are dynamic, with waves and ocean currents that move water and energy from one region to another, and various manifestations of life, such as coral reefs (▼). The major influences on the oceans are the Sun, the rotation of Earth around its axis, and the configuration of the seafloor and continents.

How do the oceans interact with and influence the land, seafloor, atmosphere, and life?

Cayman Islands

01.00.a5 Cayman Islands

A View of North America

North America is a diverse continent, ranging from the low, tropical rain forests of Central America to the high Rocky Mountains of western Canada. In the large image of North America on the left, the colors on land are from satellite images that show the distribution of rock, soil, plants, and lakes. Green colors represent dense vegetation, including forests shown in darker green, and fields and grassy plains shown in lighter green. Brown colors represent deserts and other regions that have less vegetation, including regions where rock and sand are present. Lakes are shown with a solid blue color. Clouds for a single day are overlain on Earth's surface, but should be viewed as one snapshot of a continuously playing movie—the clouds will have moved by the next day. The image is computer generated from several data sets and is not an actual photograph.

The colors of the ocean reflect the depth to the underlying seafloor, but the actual shape of the seafloor is not shown. Light blue colors represent shallow areas, such as those flanking the continent, whereas dark blue represents places where the seafloor is deep. Observe the larger features on land, at sea, and in the atmosphere. Ask yourself the following questions: What is this feature? Why is it located here? How did it form? In short, what is its story?

Notice that the two sides of North America are very different from each other and from the middle of the continent. The western part of North America has many rugged mountains and deep valleys. The mountains in the eastern United States are more subdued, and the East Coast is surrounded by a broad shelf (shown in a light blue-gray) that continues out beneath the Atlantic Ocean. The center of the continent has no mountains but has broad plains, hills, river valleys, and large lakes. These variations in landscapes greatly affect weather and regional climates.

All of the features on this image of Earth are part of earth science. Earth science explains why the mountains on the two sides of the continent are so different and when and how the mountains formed. It explains processes that operate within the waters of the oceans. Earth science addresses climate, weather, water resources, and landscapes, and the impact of these aspects on life. The land, oceans, atmosphere, and life are greatly affected by the Sun, Moon, and certain other features in the universe, so earth science also involves many aspects of *astronomy*. Earth science especially deals with the interactions of these various components, focusing on the Earth as a series of systems, an approach often called *earth-system science*. As shown throughout this book, earth's systems affect many aspects of our society.

1.0

1.1 How Do Earth's Features and Processes Influence Where and How We Live?

EARTH PROCESSES INFLUENCE OUR LIVES IN MANY WAYS. A major influence is the shape and character of the land on which we live. This aspect of earth science is part of the discipline of *geology*, the study of the earth. Geologic features and processes constrain where people can live because they determine whether a site is safe from landslides, floods, or other natural hazards. Some areas are suitable building sites, but others are underlain by unstable earth materials that could cause damage to any structure built there. Geologic factors also control the distribution of energy and mineral resources and croplands. The land is imprinted by interactions with the Sun, weather, water, and life, and the entire system is investigated by many types of scientists, including geographers, climatologists, oceanographers, and ecologists. Collectively, we call such scientists *earth scientists* or *geoscientists*.

A Where Is It Safe to Live?

The landscape around us contains many clues about whether a place is relatively safe or whether it is a natural disaster waiting to happen. What important clues should guide our choice of a safe place to live?

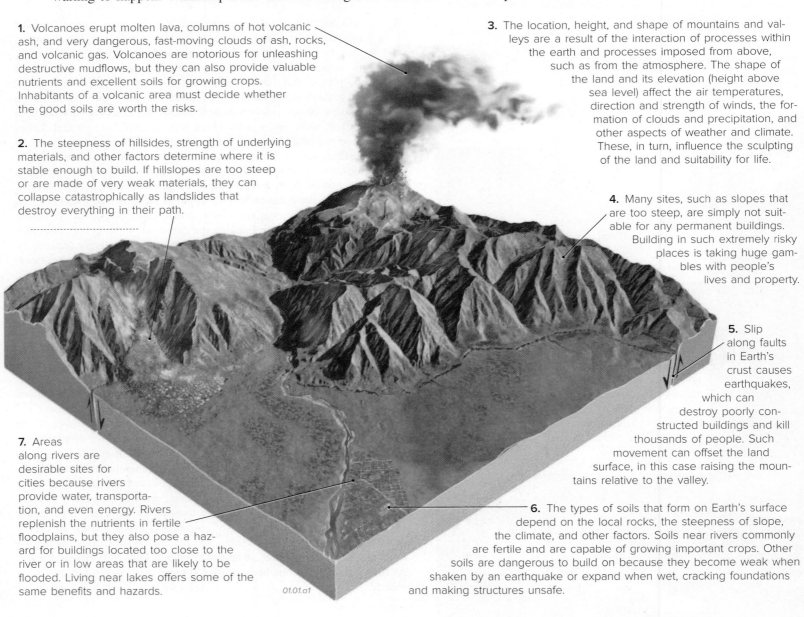

1. Volcanoes erupt molten lava, columns of hot volcanic ash, and very dangerous, fast-moving clouds of ash, rocks, and volcanic gas. Volcanoes are notorious for unleashing destructive mudflows, but they can also provide valuable nutrients and excellent soils for growing crops. Inhabitants of a volcanic area must decide whether the good soils are worth the risks.

2. The steepness of hillsides, strength of underlying materials, and other factors determine where it is stable enough to build. If hillslopes are too steep or are made of very weak materials, they can collapse catastrophically as landslides that destroy everything in their path.

3. The location, height, and shape of mountains and valleys are a result of the interaction of processes within the earth and processes imposed from above, such as from the atmosphere. The shape of the land and its elevation (height above sea level) affect the air temperatures, direction and strength of winds, the formation of clouds and precipitation, and other aspects of weather and climate. These, in turn, influence the sculpting of the land and suitability for life.

4. Many sites, such as slopes that are too steep, are simply not suitable for any permanent buildings. Building in such extremely risky places is taking huge gambles with people's lives and property.

5. Slip along faults in Earth's crust causes earthquakes, which can destroy poorly constructed buildings and kill thousands of people. Such movement can offset the land surface, in this case raising the mountains relative to the valley.

6. The types of soils that form on Earth's surface depend on the local rocks, the steepness of slope, the climate, and other factors. Soils near rivers commonly are fertile and are capable of growing important crops. Other soils are dangerous to build on because they become weak when shaken by an earthquake or expand when wet, cracking foundations and making structures unsafe.

7. Areas along rivers are desirable sites for cities because rivers provide water, transportation, and even energy. Rivers replenish the nutrients in fertile floodplains, but they also pose a hazard for buildings located too close to the river or in low areas that are likely to be flooded. Living near lakes offers some of the same benefits and hazards.

01.01.a1

B How Do Earth Processes Influence Our Lives?

To explore how earth features and processes affect our lives, observe this photograph, which shows a number of different features, including clouds, snowy mountains, slopes, and a grassy field with horses and cows (the small, dark spots). For each feature you recognize, think about what is there and what processes might be occurring. Then, think about how these features and processes influence the life of the animals and how they would influence your life if this was your home. Think broadly, considering aspects of the land, atmosphere, and any expressions of water or life. Think about this before reading on.

In the distance are snow-covered mountains partially covered with clouds. Snow and clouds both indicate the presence of water, an essential ingredient for life. The mountains have a major influence on water in this scene. As the snow melts, water flows downhill toward the lowlands, to the horses and cows.

The horses and cows roam on a flat, grassy pasture, avoiding slopes that are steep or barren of vegetation. The steepness of slopes reflects the strength of the rocks and soils, and the flat pasture resulted from loose sand and other materials that were laid down during flooding along a desert stream. Where is the likely source of the water needed to grow grass in the pasture?

01.01.b1 Henry Mtns., UT

C What Factors Control Where We Can Grow Food?

The suitability of an area for growing crops and raising livestock depends on many factors, especially the overall climate, which incorporates the temperatures, amount and timing of precipitation, and various seasonal effects, such as number of days without freezing temperatures. The climate, steepness of slopes, type of earth materials, types of plants and animals, and other factors in turn control the type and thickness of soil.

The two globes below show important controls on agriculture and ranching. The left globe represents the average amount of precipitation, with darker purple indicating higher amounts of precipitation. The right globe depicts the length of the growing season and how warm it is, using a calculated number called the growing degree days, with darker orange representing conditions with longer, warmer days (generally more favorable for growing plants). What are the conditions where you live, and can you explain some of the patterns?

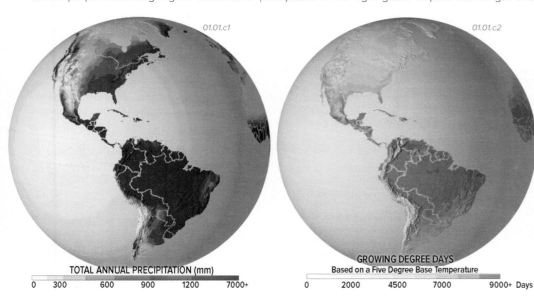

01.01.c1

01.01.c2

TOTAL ANNUAL PRECIPITATION (mm)
0   300   600   900   1200   7000+

GROWING DEGREE DAYS
Based on a Five Degree Base Temperature
0   2000   4500   7000   9000+ Days

Before You Leave This Page

✓ Sketch or list some ways that earth processes control where it is safe to live and the landscape around us.

✓ Explain some factors that influence where we can grow food.

1.1

1.2 How Does Earth Science Explain Our World?

THE WORLD HAS INTERESTING FEATURES at all scales. Views from space show oceans, continents, mountain belts, and clouds. Traveling through the countryside, we notice smaller things—a beautiful rock formation or soft, green hills. Upon closer inspection, the rocks may include fossils that provide evidence of ancient life and past climates. Here, we give examples of how earth science explains big and small features of our world.

A How Do Continents Differ from Ocean Basins?

Examine the figure below, which is a computer-generated view of the continent of Australia and the surrounding ocean basins. Colors on land show vegetation, rocks, soil, and sand, whereas colors in the oceans indicate depth, with darker blue being the deepest seafloor. Note the main features, especially those on the seafloor.

AUSTRALIA

Deep Ocean

01.02.a1

1. This map illustrates one of the most important distinctions on Earth — our planet is divided into *continents* and *oceans*.

2. The boundary between the blue colors of the oceans and the greens and browns of the land is the coastline, which outlines the familiar shape of Australia as seen on world maps.

3. Surrounding the land is a fringe of seafloor that is not very deep, represented on this map by light blue colors. This fringe of shallow seafloor, called the *continental shelf*, is wider on the north side of the continent than on the other three sides. Earth scientists consider the continent to continue past the coastline and to the outer edge of the continental shelf.

4. The seafloor beneath deep parts of the ocean is locally complex, containing chains of submarine mountains east of Australia and long features that look like large scratch marks south of the continent. The deep parts of the seafloor are much rougher than the smooth-appearing continental shelf. Seafloor depths greatly influence ocean temperatures and currents.

5. The distinction between continents and oceans is a reflection of differences in the types and thicknesses of the rocks and in how each formed. The land varies in elevation and character, such as higher, vegetation-covered mountains in eastern Australia than in the rest of the continent. Within the oceans are major variations in the depth and character of the seafloor from place to place. There is a large variation in the temperature of the ocean in this region, with very warm seawater temperatures to the north and very cold seawater to the south. The landscape, oceans, and other factors affect the weather, climate, abundance and movement of water, and the overall conditions for life.

B What Stories Do Landscapes Tell?

Observe this photograph of a canyon wall. After you have done this, think of at least two questions about what you notice, before you read the text. Go ahead, try it!

1. The landscape has cliffs and slopes composed of rock units that are shades of tan, brown, and yellow. There are not many plants.

2. In the bottom half of the image, some large, angular blocks of brownish rock are perched near the edge of a lower cliff.

3. Several questions about the landscape come to mind. What types of rocks are exposed here? How did the large, brownish blocks get to their present position? How long will it take for the blocks to fall or slide off the lower cliff? Why are there few plants?

4. The answer to each question helps explain part of the scene. The first and last questions are about the *present*, the second is about the *past*, and the third is about the *future*. The easiest questions to answer are usually about the present, and the hardest ones are about the past or the future.

5. All of the rocks in this view are volcanic rocks, typical of those formed during a very explosive type of volcanic eruption. There are few plants because this region is relatively dry and hot (a desert).

6. The large blocks are composed of the same material as the upper brown cliff and were part of that cliff before falling or sliding downhill.

7. It is difficult to predict when the blocks will fall off the lower cliff. Some blocks near the edge could fall in the next rainstorm, but others will probably be there for hundreds of thousands of years.

01.02.b1 Superstition Mtns., AZ

C How Has the Global Climate Changed Since the Ice Age?

These computer-generated images show where glaciers and large ice sheets were during the last ice age and where they are today. Note how the extent of these features changed in this relatively short period of time. What caused this change, and what might happen in the future because of global warming or cooling?

28,000 Years Ago

Twenty-eight thousand years ago, Earth's climate was slightly cooler than it is today. Cool climates permitted continental ice sheets to extend across most of Canada and into the upper Midwest of the United States. Ice sheets also covered parts of northern Asia and Europe.

01.02.c1

01.02.c2

Today

Since 20,000 years ago, Earth's climate warmed enough to melt back the ice sheets to where they are today. Our knowledge of the past extent of ice sheets comes from earth scientists who examine the landscape for appropriate clues, including glacial features and deposits that remained after the glaciers retreated.

D What Is the Evidence That Life in the Past Was Different from Life Today?

Museums and action movies contain scenes, like the one below, of dinosaurs lumbering or scampering across a land covered by exotic plants. Where does the evidence for these strange creatures come from?

▶ This mural, painted by artist Karen Carr, is two stories tall and shows what types of life are interpreted to have been on Earth during the Jurassic Period, approximately 160 million years ago. Dinosaurs roamed the landscape, while the ancestors of birds began to take flight. Flowering plants were not yet abundant and grasses had not yet appeared, so non-flowering trees, bushes, and ground cover dominated the landscape.

01.02.d2 Dinosaur NP, UT

01.02.d1

◀ Fossil bones of Jurassic dinosaurs are common in Dinosaur National Park, Utah. From such bones and other information, we can infer how long ago these creatures roamed the planet, what the creatures looked like, how big they were, how they lived, and why they died. Studying the rock layers that enclose the bones provides clues to the local and global environments at the time of the dinosaurs. Rocks and fossils are the record of past geologic events, environments, and prehistoric creatures.

Before You Leave This Page

✓ Explain the difference in appearance between continents and oceans.

✓ Describe some things we can learn about Earth's past by observing its landscapes, rocks, and fossils.

1.2

1.3 What Forces and Processes Affect Our Planet?

EARTH IS SUBJECT TO VARIOUS FORCES. Some forces arise within Earth, and others come from the Sun and Moon. Especially important is gravity, the mutual attraction that any two objects exert on one another. The interactions between these forces and Earth's land, water, air, and inhabitants control most natural processes and influence our lives in many ways.

A How Do Forces and Processes Affect Earth?

1. Earth's *gravity* causes air in the atmosphere to press down on Earth's surface and on its inhabitants. The weight of this air causes *atmospheric pressure*, which generally is greater at sea level than at high elevations—there is less air on top of high elevations than at sea level.

2. *Water*, in either liquid or frozen forms, moves downhill in streams and glaciers, transporting rocks and other debris and carving downward into the landscape. The downward movement of ice and water is driven by the pull of Earth's gravity.

3. The Sun and Moon exert a *gravitational pull* on Earth. Although the Sun is much larger, it exerts less force on the Earth because it is so far away compared to the Moon.

4. *Cosmic rays*, high-energy radiation mostly originating deep into the universe, strike the Earth's atmosphere and surface. Cosmic rays pose danger to electronics and people outside the atmosphere and also influence some processes in the atmosphere.

5. *Electromagnetic energy,* including visible light and ultraviolet energy, radiates from the Sun to Earth. The Sun provides in excess of 99% of Earth's surface-energy budget and so drives surface temperatures, wind, and other processes.

6. Uneven solar heating results in variations in water and air temperatures across the surface of Earth, causing *wind* and *ocean currents*. Blowing wind picks up and moves sand and dust across Earth's surface and makes waves on the surface of oceans and lakes. Ocean currents can carry huge amounts of warm or cold water across the ocean. Rotation of Earth around its axis helps guide the direction of wind and ocean currents as they distribute thermal energy from one part of Earth to another.

7. The mass of Earth causes a downward pull of gravity, which attracts objects toward the center of Earth. Earth's gravity is the force that makes water, ice, and rocks move downhill.

8. Earth's gravity causes the weight of rocks to exert a downward force on underlying rocks. These rocks in turn push against adjacent rocks, causing squeezing of rocks from all directions. This force increases deeper into the interior because more rocks lie above. In many parts of Earth, forces compress the rocks equally from all directions, but additional forces arise by other processes deep within Earth, such as from the subsurface movement of rocks and magma. Forces generated in one area can be transferred to an adjacent area, causing sideways pushing, pulling, or shearing on the rocks.

9. *Radioactive decay* of naturally occurring uranium, potassium, and certain other elements produces heat, especially in the crust where these radioactive elements are concentrated.

10. Temperature increases downward into Earth's interior. *Heat* from deeper in Earth rises upward toward the cooler surface. Some heating is by direct contact between a hotter rock and a cooler rock, whereas other transfer of heat occurs via a moving material, especially rising molten rock (magma).

01.03a1

B How Do Earth's Surface and Atmosphere Interact with Solar Energy?

Critical interactions occur between radiative energy from the Sun (insolation) and Earth's atmosphere, oceans, and land. These interactions express themselves in wind, clouds, rain, snow, and the climate of an area. Our atmosphere shields Earth from cosmic radiation, transfers water from one place to another, and permits life to exist. Like the oceans, the atmosphere is constantly moving, producing winds, clouds, and storms that impact Earth's surface.

1. The atmosphere is mostly gaseous nitrogen and oxygen, but it includes a low, but important, percentage of water vapor, most of which *evaporated* from Earth's oceans. Under certain conditions, the water vapor condenses to produce clouds, which are made of tiny water droplets or ice crystals. Rain, snow, and hail may fall from clouds back to the surface as *precipitation*.

2. The Sun produces vast amounts of energy, including *ultraviolet radiation* and visible light. In the upper levels of the atmosphere, oxygen absorbs most of the Sun's harmful ultraviolet radiation and prevents it from reaching Earth's surface, where it would have a detrimental effect on many forms of life. Most of the Sun's energy, including light and other forms of radiation, passes through the atmosphere, eventually reaching Earth, warming the planet and providing light for plants and animals.

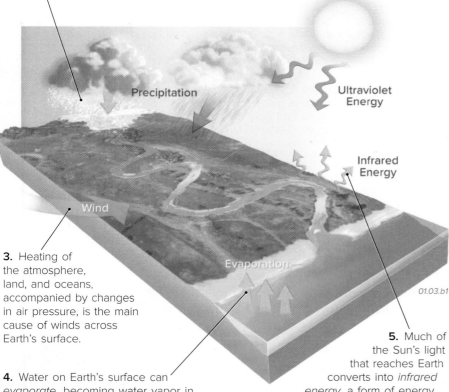

Precipitation

Ultraviolet Energy

Infrared Energy

Wind

Evaporation

01.03.b1

3. Heating of the atmosphere, land, and oceans, accompanied by changes in air pressure, is the main cause of winds across Earth's surface.

4. Water on Earth's surface can *evaporate*, becoming water vapor in the atmosphere. Most *water vapor* comes from evaporation in the oceans, but some also comes from evaporation of lakes, rivers, irrigated fields, and other sites of surface water. Some comes from evaporation of water drops in clouds. Plants take moisture from soils, surface waters, or air, and release water vapor into the atmosphere through the process of *evapotranspiration*.

5. Much of the Sun's light that reaches Earth converts into *infrared energy*, a form of energy related to heat. Some of this energy radiates upward and is trapped by the atmosphere, which warms in a process called the *greenhouse effect*. This process regulates global temperatures, which are moderate enough to allow water to exist as liquid water, gaseous water vapor, and solid ice. Water is a key requirement for life.

Energy and Forces

Earth's energy supply originates from internal and external sources. *Internal energy* comes from within Earth and includes heat energy trapped when the planet formed and heat that is produced by radioactive decay. This heat drives many internally generated processes, including mountain building and the melting of rocks at depth to produce magma and volcanoes.

The most significant source of *external energy* is the Sun, which bathes Earth in light, thermal energy, and other electromagnetic energy. Thermal energy and light from the Sun are more intense in equatorial areas of Earth than in polar areas, causing temperature differences in the atmosphere, oceans, and on land. The resulting temperature differences help drive wind and ocean currents. Sunlight is also the primary energy source for plants, through the process of *photosynthesis*.

Early in Earth's history, meteoroids and other objects left over from the formation of the solar system bombarded the planet. During the impacts, *kinetic energy* (energy due to movement of an object) changed into thermal energy, adding a tremendous amount of heat, some of which remains stored in Earth's hot interior.

Internal forces also affect Earth. All objects that have mass exert a gravitational attraction on other masses. If a mass is large and close, the pull of gravity is relatively strong. Earth's gravity acts to pull objects toward the center of the planet. Gravity is probably the most important agent on Earth for moving material from one place to another. It causes loose rocks, flowing glaciers, and running water to move downhill from higher elevations to lower ones, and it drives ocean currents and wind. Moving water, ice, air, and rocks can etch down into Earth's surface, shaping landscapes.

Objects on Earth also feel an *external* pull of gravity from the Sun and Moon. Gravity between the Sun and Earth maintains our planet's orbit around the Sun. The Moon's pull of gravity on Earth is stronger than that of the Sun and causes more observable effects, especially the rise and fall of ocean tides. Many earth scientists and astronomers conclude that long-term changes in Earth's orbit around the Sun and in the tilt of Earth's rotation axis result in large climatic changes, helping start and stop episodes of glaciation.

Before You Leave This Page

☑ Describe the different kinds of energy that impact Earth from the outside, and what effects they have on our planet.

☑ List the different kinds of energy that arise within Earth's interior and explain their origins.

☑ Sketch and explain how Earth's surface and atmosphere interact with solar energy.

1.3

1.4 How Do Natural Systems Operate?

EARTH HAS A NUMBER OF SYSTEMS in which matter and energy are moved or transformed. These involve processes of the solid Earth, water in all its forms, the structure and motion of the atmosphere, and how these three domains (Earth, water, and air) influence life. Such systems are *dynamic*, responding to any changes in conditions, whether those changes arise internally *within* the system or are imposed externally, from *outside* the system.

A What Are the Four Spheres of Earth?

Earth consists of four overlapping spheres—the atmosphere, biosphere, hydrosphere, and lithosphere—each of which interacts with the other three spheres. The atmosphere is mostly gas, but includes liquids (e.g., water drops) and solids (e.g., ice and dust). The hydrosphere represents Earth's water, and the lithosphere is the solid Earth. The biosphere includes all the places where there is life—in the atmosphere, on and beneath the land, and on and within the oceans.

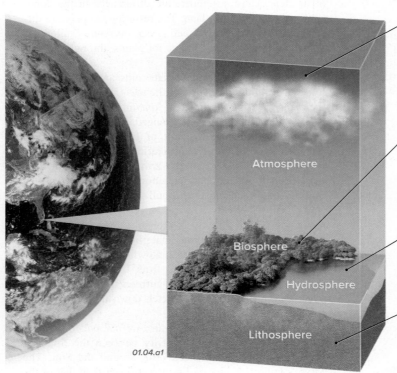

01.04.a1

1. The *atmosphere* is a mix of mostly nitrogen and oxygen gas that surrounds Earth's surface, gradually diminishing in concentration out to a distance of approximately 100 km, the approximate edge of outer space. In addition to gas, the atmosphere includes clouds, precipitation, and particles such as dust and volcanic ash. The atmosphere is approximately 78% nitrogen, 21% oxygen, less than 1% argon, and smaller amounts of carbon dioxide and other gases. It has a variable amount of water vapor, averaging about 1%, mostly in the lower atmosphere.

2. The *biosphere* includes all types of life (including humans) and all of the places it can exist on, above, and below Earth's surface. In addition to the abundant life on Earth's surface, the biosphere extends about 10 km up into the atmosphere, to the bottom of the deepest oceans, and downward into the cracks and tiny spaces in the subsurface. In addition to visible plants and animals, Earth has a large population of diverse microorganisms.

3. The *hydrosphere* is water in oceans, glaciers, lakes, streams, wetlands, groundwater, moisture in soil, and clouds. Over 96% of water on Earth is salt water in the oceans, and most fresh water is in ice caps, glaciers, and groundwater, not in lakes and rivers.

4. The *lithosphere* refers generally to the solid upper part of the earth, including Earth's crust and uppermost mantle. Water, air, and life extend down into the lithosphere, so the boundary between the solid earth and other spheres is not distinct, and the four spheres overlap.

B What Are Open and Closed Systems?

Many aspects of Earth can be thought of as a system—a collection of matter, energy, and processes that are somehow related and interconnected. For example, an air-conditioning system consists of some mechanical apparatus to cool the air, ducts to carry the cool air from one place to another, a fan to move the air, and a power source. There are two main types of systems: *open systems* and *closed systems*.

01.04.b1 Noxubee NWR, MS

An *open system* allows matter and energy to move into and out of the system. A tree (◄), is an open system, taking in water and nutrients from the soil, extracting carbon dioxide from the air to make the carbon-rich wood and leaves, sometimes shedding those leaves during the winter as shown here, and expelling oxygen as a by-product of photosynthesis, fueled by externally derived energy from the Sun.

A *closed system* does not exchange matter, or perhaps even energy, with its surroundings. The Earth as a whole (►) is fundamentally a closed system with regard to matter, except for the escape of some light gases into space, the arrival of occasional meteorites, and the exit and return of spacecraft and astronauts. Earth is an open system with regard to energy, which continuously enters and exits the planet.

01.04.b2

C How Do Earth Systems Operate?

Systems consist of matter and energy, and they respond to internally or externally caused changes in matter and energy, as a tree responds to a decrease in rain (matter) or colder temperatures during the winter (energy). Systems can respond to such changes in various ways, either reinforcing the change or counteracting the change.

System Inputs and Responses

1. One of Earth's critical systems involves the interactions between ice, surface water, and atmospheric water. This complex system, greatly simplified here (▶), remains one of the main challenges for computer models attempting to analyze the causes and possible consequences of climate change.

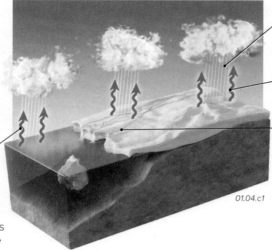

01.04.c1

2. Liquid water on the surface evaporates (represented by the upward-directed blue arrows), becoming water vapor in the atmosphere. If there is enough water vapor, small airborne droplets of water accumulate, forming these low-level clouds.

3. Under the right conditions, the water freezes, becoming snowflakes or hail, which can fall to the ground. Over the centuries, if snow accumulates faster than it melts, the snow becomes thick and compressed into ice, as in glaciers.

4. The water molecules in snow and ice can return directly to the atmosphere via several processes.

5. If temperatures are warm enough, snow and ice can melt, releasing liquid water that can accumulate in streams and flow into the ocean or other bodies of surface water. Alternatively, the meltwater can evaporate back into the atmosphere. Melting also occurs when icebergs break off from the glacier.

6. The movement of matter and energy carried in the various forms of water is an example of a *dynamic system*—a system in which matter, energy, or both are constantly changing their position, amounts, or form.

Feedbacks

7. The system can respond to changes in various ways, which either can reinforce the effect, causing the overall changes to be more severe, or can partially or completely counteract the effect, causing changes to be less severe. Such reinforcements or inhibitors are called *feedbacks*.

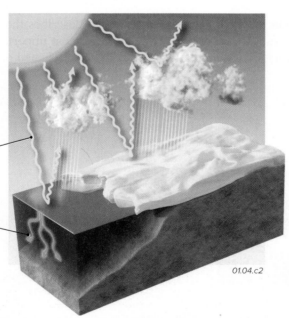

01.04.c2

8. In our example, sunlight shines on the ice and water. The ice is relatively smooth and light-colored, reflecting much of the Sun's energy upward, into the atmosphere or into space. In contrast, the water is darker and absorbs more of the Sun's energy, which warms the water.

9. If the amount of solar energy reaching the surface or trapped near the surface increases for whatever reason, this may cause more melting of the ice. As the front of the ice melts back, it exposes more dark water, which absorbs more heat and causes even more warming of the region. In this way, an initial change (warming) triggers a response that causes even more of that change (more warming). Such a reinforcing result is called a *positive feedback*.

10. The warming of the water results in more evaporation, moving water from the surface to the atmosphere. The increase in water vapor may result in more clouds. Low-level clouds are highly reflective, so an increase in cloud cover intercepts more sunlight, leading to less warming. This type of response does not reinforce the change but instead dampens it and diminishes its overall effect. This dampening and resultant counteraction is called a *negative feedback*.

11. As this overly simplified example illustrates, a change in a system can be reinforced by positive feedbacks or stifled by negative ones. Both types of feedbacks are likely and often occur at the same time, each nudging the system toward opposite behaviors (e.g., overall warming or overall cooling). Feedbacks can leave the system largely unchanged, or the combined impact of positive and negative feedbacks can lead to a stable but gradually changing state, a condition called *dynamic equilibrium*.

Before You Leave This Page

✓ Describe Earth's four spheres.

✓ Explain what is meant by open and closed systems.

✓ Sketch and explain examples of positive and negative feedbacks.

1.4

1.5 What Are Some Important Earth Cycles?

MATTER AND ENERGY MOVE within and between each of the four spheres. A fundamental principle of all natural sciences is that energy and matter can be neither created nor destroyed, but only transferred from one form to another. Also, energy and matter tend to become dispersed into a more uniform spatial distribution. As a result, matter and energy are stored, moved, dispersed, and concentrated as part of dynamic earth systems, where material and energy move back and forth among various sites within the four spheres.

A How Are Energy and Matter Moved in the Atmosphere?

Atmospheric processes involve the redistribution of *energy* and *matter* from one part of the atmosphere to another. The processes by which this occurs cause various types of weather and other phenomena described elsewhere in this book.

Energy

01.05.a1

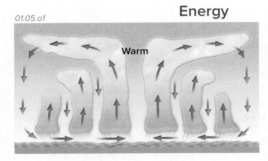

Warm

Storage and transfer of energy are the drivers of Earth's climate and weather. Energy can be moved from one part of the atmosphere to another, such as by air currents associated with storms (◄). Also, energy is released or extracted from the local environment when water changes from one state of matter to another, such as from a liquid to a gas during evaporation.

Matter

01.05.a2

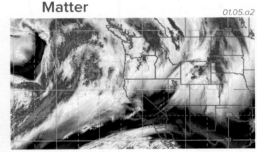

Water in all of its forms, along with other matter, moves globally and tends to disperse, but other factors prevent an even spatial distribution. As a result, some regions, like here in the western U.S., are more humid and cloudy than others, as shown in this satellite image of water vapor (blue is more, brown is less). Water can also cycle between vapor, liquid, and solid states.

B How Are Matter and Energy Moved in the Hydrosphere?

Many processes in the Earth occur as part of a cycle, a term that describes the movement of matter and energy between different sites in Earth's surface, subsurface, and atmosphere. The most important of these is the *hydrologic cycle*, which involves local- to global-scale storage and circulation of water and associated energy near Earth's surface.

1. Water vapor in the atmosphere can condense to form drops, which can remain suspended in the air as clouds or can fall to the ground as *precipitation*. Once on the surface, water can coat rocks and soils, and it causes the solid earth materials to dissolve, decompose, and erode.

2. When winter snows don't melt completely, as is common at higher elevations and at polar latitudes, ice accumulates in glaciers, which are huge, flowing fields of ice. Glaciers transport sediment and carve the underlying landscape.

3. Moving water on the surface can flow downhill in *streams*, encountering obstacles like solid rock and loose debris. The flowing water and the material it carries breaks these obstacles apart and picks up and transports the pieces. Flowing water, whether in large rivers or in thin sheets spread out across the surface, is the most important agent for sculpting Earth.

6. Water evaporates from the oceans, surface waters on land, and from soils and plants, returning to the atmosphere and completing the hydrologic cycle. Phase changes between a solid, liquid, and gas also involve the exchange of energy.

5. Water can sink into the ground and travel through cracks and other empty spaces in rocks and soils, becoming *groundwater*. Groundwater can react chemically with some rocks through which it flows, dissolving or depositing material. It typically flows toward lower areas, where it may emerge back on Earth's surface as springs.

01.05.b1

4. The uppermost part of oceans is constantly in motion, partly due to friction between winds in the atmosphere and the surface of the oceans. Winds in the oceans cause waves that erode and shape shorelines, especially during powerful storms.

C How Does the Rock Cycle Affect Materials of the Lithosphere?

Rocks and other solid materials are moved on Earth's surface and subsurface. The *rock cycle*, summarized below and discussed in more detail later, describes the movement of earth materials on and below Earth's surface at timescales from seconds to billions of years, involving such processes as erosion, burial, melting, and uplift of mountains.

1. *Weathering:* Rock is broken apart into pieces or can be altered by chemical reactions when exposed to sunlight, rain, wind, plants, and animals at or near the surface. This weathering creates loose pieces of rock called *sediment*.

8. *Uplift:* Rock may be uplifted back to the surface where it is again exposed to weathering.

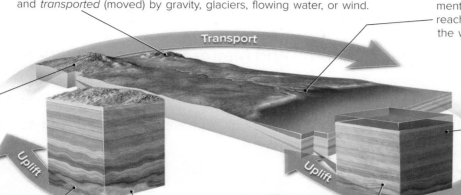

2. *Erosion and Transport:* Sediment is then stripped away by *erosion,* and *transported* (moved) by gravity, glaciers, flowing water, or wind.

3. *Deposition:* After transport, the sediment is laid down, or *deposited* when it reaches the sea or at any point along the way, such as beside the stream or in a sand dune.

4. *Burial and Formation of Rock:* Sediment is eventually buried, compacted by the weight of overlying sediment, and perhaps cemented together by chemicals in the water to form a hard *sedimentary rock.*

7. *Solidification:* As magma cools, either at depth or after being erupted onto the surface, it begins to crystallize and solidify into rock. A rock formed by solidification of magma is an *igneous rock.*

6. *Melting:* A rock exposed to high temperatures may melt and become molten, forming a mass of magma, which can remain in place or rise toward the surface.

5. *Deformation and Related Processes:* Rock can be subjected to strong forces that squeeze, bend, and break the rock, causing it to *deform*. The rock might also be heated and deformed so much that it is changed (*metamorphosed*) into a new kind of rock, called a *metamorphic rock.*

D What Cycles and Processes Are Important in the Biosphere?

The biosphere includes life and all of the places it exists. It overlaps the atmosphere, hydrosphere, and lithosphere, extending from more than 10 km above the surface to more than 10 km beneath sea level, both on the seafloor and within Earth's subsurface. Life interacts with the other three spheres, forming a number of important cycles.

1. *Atmosphere*: Plants exchange gases with the atmosphere. They extract carbon dioxide (CO_2) from the atmosphere and use the carbon for their leaves, stems, roots, spines, and other leafy or woody parts. Plants also release oxygen, a key ingredient in life. Life on Earth is currently the main source of the oxygen and CO_2 in the atmosphere.

2. *Hydrosphere*: Life interacts with the hydrologic cycle. Plants take in water from the rocks and soil, which may have arrived from lakes and other bodies of surface water, or directly from the atmosphere, as during precipitation. Plants then release some water back into the environment. The Sun is the ultimate source of energy for photosynthesis, as well as for movement of matter and energy in the atmosphere and for most movement of material on Earth's surface.

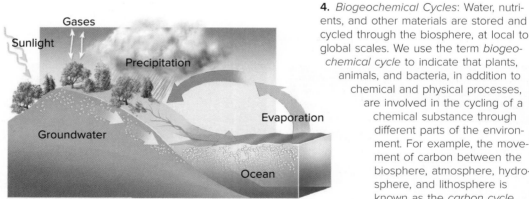

4. *Biogeochemical Cycles*: Water, nutrients, and other materials are stored and cycled through the biosphere, at local to global scales. We use the term *biogeochemical cycle* to indicate that plants, animals, and bacteria, in addition to chemical and physical processes, are involved in the cycling of a chemical substance through different parts of the environment. For example, the movement of carbon between the biosphere, atmosphere, hydrosphere, and lithosphere is known as the *carbon cycle.*

3. *Lithosphere*: Life also interacts with aspects of the rock cycle. Plants help break down materials on Earth's surface, such as when a plant's roots push open fractures in rocks and soil. Plants help stabilize soils, inhibiting erosion by slowing down the flow of water. The slowed water remains in contact with rocks and soil longer, increasing the rate at which weathering breaks down materials. Organisms can also add material to the lithosphere, such as when corals, sponges, and other organisms build a coral reef.

Before You Leave This Page

☑ Describe some examples of transfer of energy and matter in the atmosphere.

☑ Sketch and describe the hydrologic cycle, rock cycle, and how the biosphere exchanges matter with the other spheres.

1.6 How Do Earth's Four Spheres Interact?

EARTH'S DYNAMIC SYSTEMS move energy and matter between the land, water, atmosphere, and biosphere—between the four spheres. There are various expressions of these interactions, many of which we can observe in our daily lives. In addition to natural interactions, human activities, such as the clearing of rain forests, can affect interactions between the spheres. Changes in one component of one sphere can cause impacts that affect components of other spheres.

A What Are Some Examples of Energy and Matter Exchanges Between Two Spheres?

The four spheres interact in complex and sometimes unanticipated ways. As you read each example below, think of other interactions—observable in your typical outdoor activities—that occur between each pair of spheres.

Atmosphere-Hydrosphere

01.06.a1 Indonesia

The Sun's energy evaporates water from the ocean and other parts of the hydrosphere, moving the water molecules into the atmosphere. The water vapor can remain in the atmosphere or can condense into tiny drops that form most clouds. Under the right conditions, the water returns to the surface as precipitation, usually rain or snow.

Atmosphere-Lithosphere

01.06.a2 Mount Veniaminof, AK

Active volcanoes emit gases into the atmosphere, and major eruptions release huge quantities of steam, sulfur dioxide, carbon dioxide, and volcanic ash. In contrast, weathering of rocks removes gas and moisture from the atmosphere. Precipitation accumulates on the land, where it can form standing water, groundwater, or erosion-causing runoff.

Atmosphere-Biosphere

01.06.a3 Indonesia

Plants and animals exchange carbon with the atmosphere in the form of CO_2, and use precipitation from the atmosphere. Some plants can extract moisture directly out of the air without precipitation. Large-scale circulation patterns in the atmosphere influence an area's climate, which controls the types of plants and animals that inhabit a region.

Hydrosphere-Lithosphere

01.06.a4 Glacier Parkway, Alberta, Canada

Channels within a stream generally bend back and forth as the water flows downhill. The water is faster and more energetic in some parts of the stream than in others, and erodes into the streambed and riverbank. In less energetic sections, sediment will be deposited on the bed, like the gravel in this photograph. The flowing water can dissolve soluble materials in bedrock along the channel or in loose stream sediment, and carry away the liberated chemical constituents.

Hydrosphere-Biosphere

01.06.a5 Indonesia

Oceans contain a diversity of life, from whales to algae, and everything in between. Coral reefs, shown above, represent an especially life-rich environment formed when living organisms extract materials dissolved in or carried by seawater to produce the hard parts of corals, shells, and sponges. At greater ocean depths, where waters are colder, shells and similar biologic materials dissolve, transferring material back to the seawater.

Lithosphere-Biosphere

01.06.a6 Near Badlands National Park, SD

The clearest interaction between the lithosphere and biosphere is the relationship between plants and soils. The type of soil helps determine the type of plants that can grow, and in turn depends on the type of starting materials (rocks and sediment), the geographic setting of the site (e.g., slope versus flat land), climate, and other factors. Plants remove nutrients from the soil but return material back to the soil through roots, annual leaf fall, or plant death and decay.

B To What Extent Do Humans Influence Interactions Between the Spheres?

Anyone who has flown in an airplane or viewed the landscape from a hill or building appreciates the amazing amount of human influence on the landscape. The intent of development is almost always to improve the human condition, but the complex chain reaction of impacts that cascade through the system can cause unintended and often harmful impacts elsewhere in one or more of the four spheres, as illustrated in the examples below. Some consequences of human impacts are not felt immediately but only appear much later, after the activity has continued for many years.

01.06.b1

01.06.b2 Grand Coulee Dam, WA

01.06.b3 Phoenix, AZ

Humans clear forests, a critical part of the biosphere, to provide lumber and grow food. In addition to the loss of habitat for plants and animals, deforestation reduces the amount of CO_2 that can be extracted out of the atmosphere and stored in the carbon-rich trunks, branches, and leaves of plants. Removing plant cover also causes increased runoff, which enhances soil erosion and leads to the additional loss of plant cover—an unintended consequence and a positive feedback.

Over 80,000 dams exist in the U.S., providing water supplies, generating electricity, protecting towns from flooding, and providing recreational opportunities. Dams also alter the local water balance by interrupting the normal seasonal variations in flows of water and by capturing silt, sand, wood, and other materials that would normally go downstream. Construction and filling of the reservoir disrupts ecosystems, displaces people, and threatens or destroys plant and animal communities.

Local warming of the atmosphere occurs near cities because of normal urban activities (lighting, heating, etc.) and because many urban materials, like dark asphalt, capture and store more heat than natural open space. Heat is also released from car exhausts and industrial smokestacks. Non-natural drainage systems cause rapid accumulation and channeling of water. Development infringes on natural plant and animal communities, disturbs or covers soil, and alters erosion rates.

Learning with Concept Sketches

Earth science is a very visual science, relying on photographs, graphs, and many types of illustrations to show features and processes, and to represent how various earth systems operate. Suppose you wanted to summarize some of the interactions between Earth's spheres, as in the hydrologic cycle (introduced earlier in the book), which describes movement of water on our planet. It can be difficult to keep track of all the processes that are occurring, such as precipitation, runoff, and evaporation. We have a great suggestion to help you learn, retain, and apply any new information—construct a concept sketch!

A concept sketch, like the one shown below, is a simplified sketch annotated with labels and complete sentences that describe the features, processes, and relationships between different aspects. By constructing a concept sketch, you are putting the information into your own words and drawings. Research and our experience with teaching thousands of students show that this process greatly aids learning and improves student performance in the classroom, in labs, and on field trips.

To construct a concept sketch, begin by listing the important features, processes, and relationships you want to describe. Then, decide what features you need to show on your sketch in order to represent these features, processes, and relationships. Draw your sketch and write complete sentences describing the sketch, including labels and leader lines as needed. Try it—you will learn more and get better grades!

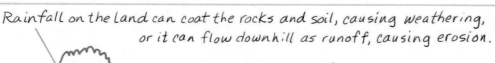

Rainfall on the land can coat the rocks and soil, causing weathering, or it can flow downhill as runoff, causing erosion.

Water in oceans, lakes, and rivers can evaporate into water vapor in the atmosphere as part of the hydrologic cycle.

Water can sink into the ground, forming groundwater, which flows toward lower areas and can emerge on the surface as springs.

01.06.t1

Before You Leave This Page

☑ Provide an example of an interaction between each pair of spheres.

☑ Describe examples of how humans can affect the natural system in each of the four spheres.

☑ Produce a concept sketch for some earth system in this chapter.

1.6

1.7 How Do We Depict Earth's Surface?

THE SURFACE OF EARTH displays various features, including mountains, hillslopes, and river valleys. We commonly represent such features on the land surface with *topographic maps* and *shaded-relief maps*. To depict the types of materials on Earth's surface, we use *satellite images* and *geologic maps*. A geologic map is the most important piece of geologic information for an area, because it shows the ages and types of rocks and sediment, as well as geologic features, some of which could pose a natural hazard.

A How Do Maps and Satellite Images Help Us Study Earth's Surface?

Satellite images and various types of maps are the primary ways we portray the land surface. Some maps depict the shape and elevation of the land surface, whereas others represent the materials on that surface. Views and maps of *SP Crater* in northern Arizona provide a particularly clear example of the relationship between geologic features, the land surface, and different types of maps.

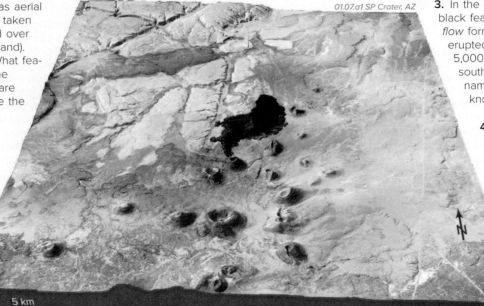

01.07.a1 SP Crater, AZ

5 km

1. This perspective view has aerial photography (photographs taken from the air) superimposed over topography (shape of the land). Take a minute and look! What features do you observe in the topography? Which areas are high in elevation? What are the most distinctive features?

2. The area has distinct, cone-shaped hills surrounded by broad, less steep areas. The hills are small volcanoes called *scoria cones*, which form when fragments of molten rock are ejected into the air and settle around a volcanic vent.

3. In the center of the area is a nearly black feature, which is a solidified *lava flow* formed when fluid magma erupted onto the surface in the last 5,000 years. The scoria cone at the southern end of the lava flow is named *SP Crater*, and is well known to most geologists.

4. Examine other features in the scene. Note the light-gray areas in the upper left parts of the image, and the linear features that cut across the gray rocks. This entire area is dry, with few trees to obscure the geology. There is a clear correspondence between the topographic and geologic features.

01.07.a2 SP Crater, AZ

01.07.a3 SP Crater, AZ

5. This photograph (▲), taken from the large crater south of SP Crater, shows the crater (on the left) and several other scoria cones. The view is toward the north.

6. This photograph (◄), taken from the air, shows SP Crater and the dark lava flow that erupted from the base of the volcano.

Before You Leave These Pages

✓ Describe how each of the four types of maps and images depicts Earth's surface.

✓ Describe what contours on a topographic map represent and how contour spacing indicates the steepness of a slope.

✓ Briefly describe what a geologic map shows, using the area around SP Crater as an example.

7. A *shaded-relief map* (▼) emphasizes the shape of the land by simulating light and dark shading on the hills and valleys. The individual hills on this map are *scoria cones*. The area is cut by straight and curving stream valleys that appear as gouges in the landscape. Simulated light comes from the upper left corner of the image.

8. A *topographic map* (▼) shows the elevation of the land surface with a series of lines called *contours*. Each contour line follows a specific elevation on the surface. Standard shaded-relief maps and topographic maps depict the shape of the land surface but give no specific information about the types of earth materials on the surface.

01.07.a4

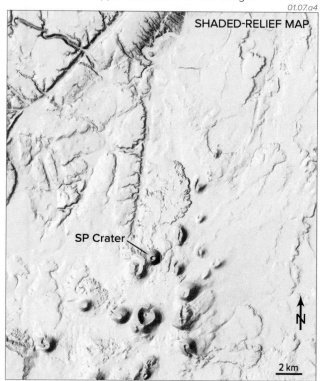

SHADED-RELIEF MAP

SP Crater

N

2 km

01.07.a5

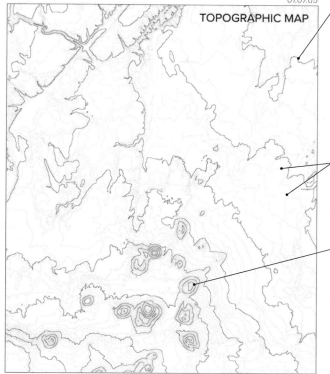

TOPOGRAPHIC MAP

9. Most topographic maps show every fifth contour with a darker line, to help emphasize the broader patterns and to allow easier following of lines across the map. These dark lines are called *index contours*.

10. Adjacent contour lines are widely spaced where the land surface is fairly flat (has a gentle slope).

11. Contour lines are more closely spaced where the land surface is steep, such as on the slopes of the scoria cones. Note how the shapes of the contours reflect the shapes of the different scoria cones.

12. A *satellite image* (▼) commonly uses measurements of different wavelengths of light reflecting from a land surface. The computer-processed image below shows the distribution of different types of plants, rocks, and other features. The dark area in the center of the image is the black, solidified lava flow that erupted from the base of SP Crater, and reddish areas are scoria cones.

13. A *geologic map* (▼) represents the distribution of rock units and geologic features exposed on the surface. This one shows SP Crater and its associated lava flow and older rock units. Compare the colored areas on this geologic map with the different areas visible on the satellite image to the left. Each color on this geologic map represents areas that have a certain type of rock or feature.

01.07.a6

SATELLITE IMAGE

01.07.a7

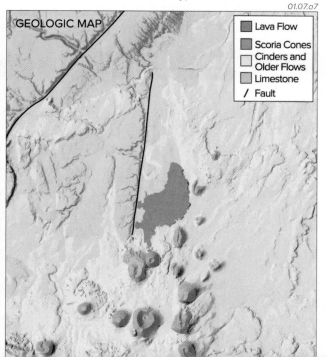

GEOLOGIC MAP

| Lava Flow |
| Scoria Cones |
| Cinders and Older Flows |
| Limestone |
| / Fault |

14. The gray area in the center of the map marks the SP lava flow, and light pinkish-brown areas are scoria cones. Light pink represents volcanic cinders and older lava flows. Lavender indicates areas with light-colored rock (limestone) at the surface.

15. Compare the four maps to match specific features of the area. Which map or image gives you the best information about the shape of the landscape, and which gives you the best information about geology?

1.7

1.8 How Do We Depict Earth's Heights, Slopes, and Subsurface Aspects?

DIAGRAMS OF THE LAND SURFACE AND THE SUBSURFACE are essential tools for visualizing and understanding Earth. We use two-dimensional and three-dimensional diagrams to depict the steepness of slopes, the thickness and subsurface geometry of rock units, and how these units interact with the surface. Some diagrams show interpretations of how present-day landscapes formed via a sequence of geologic events.

A How Do We Refer to Differences in Topography?

Earth's surface is not flat and featureless but instead has high and low parts. The variation in the height and steepness of the land—that is, the shape of the land—is called *topography*. Topography is steep in some areas but nearly flat in others. We use common terms to refer to the height of the land and the steepness of slopes.

1. The height of a feature above sea level is its *elevation*. Scientists describe elevation in *meters* or *kilometers* above sea level, but some maps and most signs list elevation in feet.

2. Beneath water, we talk about *depth,* generally expressing it as depth below sea level. We use *meters* for shallow depths and *kilometers* for deep ones.

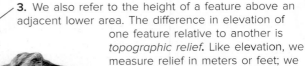

01.08.a1

3. We also refer to the height of a feature above an adjacent lower area. The difference in elevation of one feature relative to another is *topographic relief.* Like elevation, we measure relief in meters or feet; we refer to rugged areas as having *high relief* and to topographically subdued areas as having *low relief*.

4. Cliffs and slopes that drop sharply in elevation are *steep* slopes, whereas topography that is less steep is referred to as being *gentle*, as in a gentle slope.

B How Do We Represent Topographic Slopes?

We can depict steepness of the land surface with an imaginary slice through a terrain, like one through SP Crater and its surroundings (▾). This type of portrayal of ups and downs of the land surface is a *topographic profile*.

1. The front of this figure shows the change in elevation across the land surface. It is a type of topographic profile.

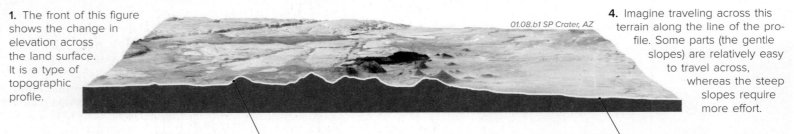

01.08.b1 SP Crater, AZ

4. Imagine traveling across this terrain along the line of the profile. Some parts (the gentle slopes) are relatively easy to travel across, whereas the steep slopes require more effort.

2. Steeper parts of the profile represent steep slopes on the sides of the small volcanoes. There is moderate relief between the peaks and surrounding plains.

3. Other parts of the profile are less steep, including lower elevation plains surrounding the volcanoes. There is only low relief from one part of the plains to another.

5. Some topographic profiles are simple plots of height of the topography versus distance across the land, like the black line that traces a profile across SP Crater (▶). The profile runs from west (on the left) to east (on the right), so it is an east-west profile. Most topographic profiles have such directions labeled directly on the plot, along with scales for elevation and for horizontal distances.

01.08.b2 SP Crater, AZ

W   E

26°

6. We describe steepness of a slope in *degrees* from horizontal. The eastern slope of SP Crater has a 26-degree slope (26° slope). We also talk about *gradient*—a 26° slope drops 480 meters over a distance of 1,000 meters (one kilometer), which is typically expressed as 0.48, or 480 m/1,000 m.

C How Do We Represent Features in the Subsurface?

Most of our planet's geology is beneath Earth's surface, hidden from view. We are most aware of rock units and other earth materials if they are exposed in a natural exposure (an *outcrop*) on a mountainside or in a deep canyon, or perhaps in a roadside cut created during road construction (a *roadcut*). However, such units are also present beneath areas of relatively gentle topography that lack any natural or human-created exposures. To portray the surface and subsurface, we use diagrams, many specialized to earth science, to envision and understand the thicknesses, orientations, and subsurface distributions of materials. Such diagrams are important ways earth scientists document and communicate their understanding of an area.

Block Diagram

1. A *block diagram* portrays in three dimensions the shape of the land surface and the subsurface distributions of rock units. It also shows the location and orientation of faults, folds, and other geologic features (if present).

Cross Section

2. A *cross section* shows surficial and subsurface features as a two-dimensional slice through the land. This example is equivalent to the front-left side of the block diagram.

Stratigraphic Section

3. A *stratigraphic section* shows the rock units stacked on top of one another (with appropriate relative thicknesses).

4. Commonly, the patterns within each rock unit visually represent the character of the unit, such as the blocky fractures in the gray unit or the rounded pebbles in this orange-colored sedimentary unit.

5. One edge of the diagram (here the left edge) typically conveys the relative resistance of the different rock units to weathering and erosion. A more easily eroded unit is recessed, like the orange unit with the pebbles, whereas more resistant units protrude farther out, like the two gray units.

01.08.c1

Evolutionary Diagrams

6. *Evolutionary diagrams* (▶) are block diagrams, cross sections, or maps that show the history of an area as a series of steps, proceeding from the earliest stages to the most recent one. Here, the upper tan rock layer depicted in the figures above is deposited on the gray layer and later eroded.

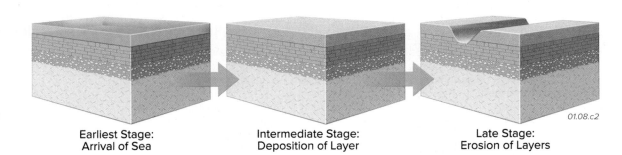

Earliest Stage:
Arrival of Sea

Intermediate Stage:
Deposition of Layer

Late Stage:
Erosion of Layers

01.08.c2

Sketching the Earth

A challenge of earth science is trying to visualize how features exposed at the surface continue at depth. Sketches drawn in the field while studying landscapes and other natural features capture one's thoughts while they are still fresh and while ideas can be tested by making additional field observations. The sketch to the right is a simplified cross section drawn to summarize field relationships for some offset rock layers. A sketch is an excellent way to conceptualize and think about earth science—whether in the field, on a boat, looking through a telescope, or reading a textbook—because it highlights the most important features.

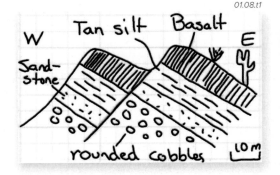

01.08.t1

Before You Leave This Page

☑ Sketch and describe what we mean by elevation, depth, relief, and slope.

☑ Sketch or describe the types of diagrams earth scientists use to represent subsurface geology and the sequence of rock units.

☑ Sketch or describe what is shown by a series of evolutionary diagrams.

1.8

1.9 How Do We Describe Locations on Earth?

TO DESCRIBE LOCATION ON A SPHERE, we need a frame of reference—some specific features from which to reference any location. For Earth, that frame of reference involves the equator, the center of the Earth, and an imaginary north-south line through England. Using this frame of reference, we can devise systems of imaginary gridlines that encircle the Earth. The most commonly used imaginary gridlines are *latitude* and *longitude*, which are displayed on many maps and are provided by the location capabilities of many cellular phones.

A How Do We Represent Locations on a Globe?

If you were trying to convey the location of a specific place on a sphere, or the location of a city on our nearly spherical planet, a good place to begin visualizing the problem is to establish a framework of imaginary gridlines. Another important aspect is to consider how lines and planes interact with a sphere.

Parallels

1. We could draw lines that circle the globe, each staying the same distance from the North Pole or South Pole. The lines are parallel to one another and remain the same distance apart, and so are called *parallels*. In addition, these lines are parallel to imaginary cuts through the Earth, perpendicular to Earth's spin axis (which goes through the North and South poles). The parallel that is halfway between the North and South poles is the *equator*.

01.09.a1

2. If we traveled along one of these lines (i.e., along a parallel), we would stay at the same distance from the pole as we encircled the planet. In other words, our position in a north-south framework would not change.

Meridians

01.09.a2

3. Lines that encircle the globe from North Pole to South Pole are called *meridians*. Meridians do not stay the same distance apart and are not parallel. Instead, meridians are widest at the equator and converge toward each pole. A meridian would be the path you would travel if you took the most direct route from the North Pole to the South Pole, or from south to north.

4. The term *meridian* comes from a Latin term for midday because, at any place, the Sun is along a meridian (i.e., is due south or north) at approximately noon. The terms A.M. (for before noon) and P.M. (for after noon) are also derived from this Latin term (e.g., post meridiem).

Latitude

5. The *latitude* of a location indicates its position north or south of the equator. Lines of latitude are *parallels* that encircle the globe east-west.

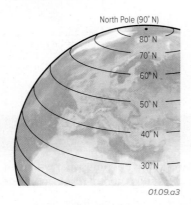
01.09.a3

6. The zero line of latitude is the equator, with the values increasing to 90 at the north and south poles. Locations near the equator are said to have a *low latitude*, and those near the pole are a *high latitude*; areas in between, such as central Europe, are in the *mid-latitudes*.

Longitude

7. The *longitude* of a location indicates its east-west position. Lines of longitude are *meridians* that encircle the globe north-south.

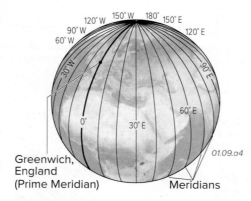
01.09.a4

Greenwich, England (Prime Meridian)

Meridians

8. For meridians of longitude, a starting, zero-degree longitude is defined as going through Greenwich, U.K. Meridians of longitude are widest at the equator and converge toward the poles. Starting at the zero longitude through Greenwich, values increase to 180° E and 180° W at the *International Date Line*, an imaginary line that runs through the middle of the Pacific Ocean (not shown; on the opposite side of the globe).

Important Lines of Latitude

9. In addition to the equator, there are a few lines of latitude that are especially important. These include the *Tropic of Cancer* and *Tropic of Capricorn*, which are 23.5° north and south of the equator, respectively. Areas near the equator are the *tropics* and those between the tropics and mid-latitudes are the *subtropics*.

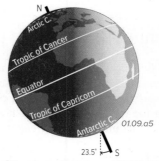
01.09.a5

10. Other important lines of latitude are the *Arctic Circle* and *Antarctic Circle*, which are 66.5° north and south of the equator (23.5° away from the corresponding pole). Areas near the poles are described as *polar*. As discussed later, the 23.5° angle is how much Earth's axis is tilted with respect to the Sun.

B How Do We Use Latitude and Longitude?

If you were a pilot flying from New York City to Moscow, Russia, how would you know which way to go? We could describe locations using latitude and longitude, which are expressed in degrees. Fractions of a degree are expressed as decimal degrees (e.g., 9.73°) or as minutes and seconds, where there are 60 minutes (indicated by ') in a degree and 60 seconds (") in a minute (e.g., 9° 43' 48").

This map (▶) illustrates the nature of the problem. If we want to navigate from New York to Moscow, we can tell from this map that we need to go a long way to the east and some amount to the north. These directions, although accurate, would not be good enough to guide us to

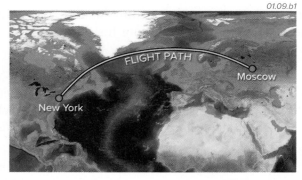
01.09.b1

Moscow. Fortunately, we can find on the Internet that the location of New York City, as given by latitude and longitude, is 40.7142° N, 74.0064° W. The location of Moscow is 55.7517° N, 37.6178° E. From these more precise ways of expressing location, we could fly from one city to another.

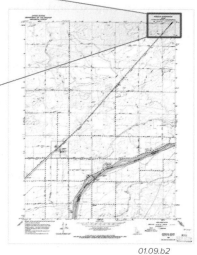

A road map, a world atlas, or any other type of map usually includes latitude and longitude. Topographic maps (▶) from the U.S. Geological Survey (USGS), used by hikers and others, are defined by boundaries that are latitude and longitude, as shown in the enlargement above. For example, a 7.5° quadrangle is 7.5° on a side.

01.09.b2

C What Is GPS?

Most people think of GPS as a navigation system in our cars or embedded in our cellular telephones, or a handheld device used for location and guidance while hiking across the countryside. GPS, short for *global positioning system*, provides the accurate position on Earth's surface including latitude, longitude, elevation, and even how fast we are traveling. This information comes from a series of satellites orbiting Earth that send radio signals to ground-based receivers, like the ones on our dashboards, in our phones, or in our hand-held GPS when we go hiking.

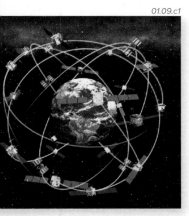
01.09.c1

1. The U.S. government launches, controls, and monitors a constellation of dozens of satellites orbiting around Earth (◀). Two generations of satellites currently operate in the GPS constellation (▶). A third type will soon be deployed to improve the accuracy and reliability of satellite signals.

2. The time required for a radio signal from a satellite to reach a receiver on Earth is related to its distance to the receiver. A GPS receiver "knows" where each satellite is located in space at the instant the GPS unit receives the signal. Calculating the distances from four or more satellites allows the GPS unit to calculate its own position, commonly with a precision and accuracy of several meters (for a handheld GPS unit). Higher precision can be achieved by occupying a single site for a long interval of time and then averaging the measurements.

01.09.c2

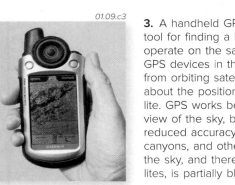
01.09.c3

3. A handheld GPS device (◀) is a navigation tool for finding a location. These instruments operate on the same principles as all other GPS devices in that they receive radio signals from orbiting satellites that contain information about the position and distance of the satellite. GPS works best outside and with a clear view of the sky, but it can operate with reduced accuracy in forested areas, deep canyons, and other settings where parts of the sky, and therefore the view of the satellites, is partially blocked.

Before You Leave This Page

☑ Sketch and explain a parallel, meridian, latitude, and longitude, indicating where the zero value is for each measurement and identifying important lines of latitude.

☑ Briefly explain GPS.

1.9

1.10 How Do We Describe Time and Rates?

WE REFER TO TIME all the time in our daily lives. But how do we define time, and what do we mean when we say someone lives in a different time zone? The Earth rotates about its axis, causing locations on the surface to pass from day to night and back again. Not everyone witnesses sunrise at the same time, because Earth's rotation causes the Sun to rise at different times in different locations. The concept of longitude helps us understand these differences and describe time so that society can operate in a more orderly manner. Most of us think of time as the hours, minutes, and seconds on a clock, but much longer units of time are used when considering Earth's long history.

A How Do We Define Time Globally?

Some units of time, like a year or length of day, arise from natural progressions of the Earth as it orbits around the Sun in a year and completes a full daily rotation in 24 hours. Other expressions of time, such as "noon," depend on location, so in the 1800s the world agreed on an international system for defining time, based on the *Prime Meridian* and the *International Date Line*.

1. *Prime Meridian* is defined as the 0° longitude measurement on the Earth, passing through the British Royal Observatory in Greenwich, U.K. This location was also chosen as the reference point for world time, a time called *Greenwich Mean Time* (GMT). Time anywhere in the world is referenced relative to time at Greenwich. You may have seen the initials GMT when setting time on your computer, tablet, or cellular phone.

2. The globe shown here has meridians spaced equally apart so that there are 24 zones centered on the lines, one for each of the 24 hours in a day. If you could instantaneously travel from one meridian to the next, there would be a one-hour time difference. If political and other considerations did not intervene, the distribution of time zones could precisely follow lines of longitude, each 15° apart.

3. The *International Date Line* (IDL) is defined as the 180° measurement of longitude in the Pacific Ocean—the meridian on the exact opposite side of the Earth from Greenwich. Segments of the IDL have been shifted east and west to accommodate the needs of some nations in the Pacific so that travel and trade among those islands is less problematic.

4. If you cross the International Date Line, you cross into a different calendar day. Traveling westward across the IDL puts you one day later in the calendar, relative to immediately east of the IDL; this is described as "losing a day." Moving eastward across the line, you move to the previous day on the calendar, so we say you "gain a day."

01.10.a1

B How Are Time Zones Defined?

The world is divided into 24 time zones, based loosely on longitude. This map color-codes these 24 time zones, most of which have irregular boundaries because they follow natural or political boundaries or try to keep some population center in a single zone. The boundaries between the four time zones covering the contiguous U.S. are mostly drawn along state or county boundaries or natural features.

As Earth rotates around its spin axis and the Sun remains stationary relative to Earth, the Sun illuminates different longitudes with the passage of time. Earth spins at a rate of 15° of longitude per hour, so we divide the planet into 24 time zones, each about 15° of longitude wide. Areas within a time zone adopt the same time, and there is a one-hour jump from one time zone to the next. If you are in one time zone, the time zone to the west is one hour earlier, and the time zone to the east is one hour later.

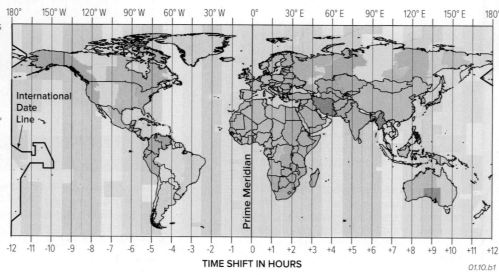

TIME SHIFT IN HOURS

01.10.b1

In most of the U.S. and Canada, clocks during the summer are set to one hour later for much of the year. This *Daylight Savings Time* (DST) provides daylight for an extra hour during the evening and one less hour of daylight in the morning. Some areas, like Saskatchewan and most of Arizona, do not observe DST, remaining instead on "standard time."

How Do We Refer to Rates of Events and Processes?

Many aspects of earth science involve the *rates* of processes, such as how fast a hurricane is moving toward a coastline or how fast water in a stream is flowing. We calculate the rates of such processes in a similar way to how we calculate the speed of a car or a runner. Some rates, like wind velocities inside a tornado, are very rapid, whereas other rates, like the uplift of a mountain range, are very slow. For scientific work, units are metric, so we talk about millimeters per year, kilometers per hour, or similar units of distance and time.

Calculating Rates

1. A runner (▶) provides a good reminder of how to calculate rates. A rate is how much something changed divided by the time required for the change to occur. Generally, we are referring to how much distance someone or something travels in a given amount of time, but rate can also describe other processes, such as how fast air cools with increasing altitude.

01.09.c1

2. If this runner sprinted 40 meters in 5 seconds, the runner's average speed is calculated as follows:

distance/time = 40 m/5 s = 8 m/s

3. What is the runner's average speed if she runs 400 meters in 80 seconds? Go ahead, try it. We don't need to provide you the answer to this one.

Relatively Rapid Earth Processes

4. Some earth processes are relatively rapid, occurring within seconds, minutes, or days. Relatively rapid natural processes include the velocity of upper-level winds (100s km/hr), speeds of winds inside a severe storm (100s km/hr), motion of the ground during earthquakes (5 km/s), movement of an earthquake-generated wave (tsunami) across the open ocean (100s km/hr), and the catastrophic advance of an explosive volcanic eruption (100s km/hr).

5. Hurricane Sandy, a huge and incredibly destructive storm that occurred in October 2012, is an example of a rapid natural process. Sandy originated as a tropical storm but migrated up the East Coast of the U.S. until it turned inland and struck New Jersey. While in the tropics, Sandy had winds estimated at 185 km/hr (115 mi/hr), but it had weakened considerably by the time the storm came ashore. The storm killed nearly 300 people along its path and caused damages of over $70 billion, mostly in New Jersey and New York.

01.10.c2-3 Seaside Heights, NJ

May 21, 2009

November 5, 2012   USGS

6. The two photographs above show Seaside Heights, N.J., before and after the storm. The yellow and red arrows point to the same houses in both photographs. Although the photographs were taken several years apart, nearly all the damage occurred within a 24-hour period. In this short time, the shape of the coastline was extensively rearranged, houses were destroyed, and the entire neighborhood was covered in a layer of beach sand washed in by the waves. Such storm-related erosion and deposition of sediment (and houses) are rapid earth processes.

Relatively Slow Earth Processes

7. Other earth processes are very slow, occurring over decades, centuries, or millions of years. Natural processes that are fairly slow include movement of groundwater (m/day), motion of continents (cm/yr), and uplift and erosion of the land surface (as fast as mm/yr, but typically much slower). Although these processes are relatively slow, the Earth's history is long (4.55 billion years), so there is abundant time for slow processes to have big results, such as uplift of a high mountain range.

01.10.c4 Arches NP, UT

8. Observe this photograph of Delicate Arch, taken in Arches National Park, Utah. The arch is a remnant of a once-continuous rock layer that has been mostly eroded away. How long do you think it took natural erosion to erode away most of a hard rock layer, leaving this beautiful landform behind?

9. Several questions about the rates of processes come to mind. How old are the rock layers, and how long did it take to form all of the layers in the photograph? These layers are approximately 160 million years old, and accumulated slowly, one layer at a time, over millions of years. The landscape currently is being eroded, and this process has been occurring in this area for millions of years. Not much rock is eroded away each year, but even at very slow rates, significant erosion can occur in a million years. A final question might be about how long the arch will remain standing. This arch is relatively sturdy, but other arches in the region have collapsed in the last 100 years.

Before You Leave This Page

✓ Explain what GMT is and where the starting point is.

✓ Describe why we have time zones, and what influences time-zone boundaries and width.

✓ Explain how we calculate rates, giving some examples of relatively fast and slow natural processes.

1.11 What Is Earth's Place in the Solar System?

EARTH IS NOT ALONE IN SPACE. It is part of a system of planets and moons associated with the Sun, which together comprise a *solar system*. Our solar system is just one of countless solar systems in the many other galaxies in the universe. The Sun is the most important object for Earth and our solar system because it provides light and heat, without which life would be difficult if not impossible. Earth has a number of neighbors, including the Moon.

A What Are Earth's Nearest Neighbors?

Earth has five nearby neighbors—three other planets, one moon, and the Sun. The Sun and the Moon have direct effects on Earth. Both exert gravitational pull on our planet, and the Sun is our primary source of energy. The three planets (Mercury, Mars, and Venus), while not affecting us directly, provide glimpses of how Earth might have turned out. Earth and these three other planets are rocky and are called *terrestrial planets*, or simply the *inner planets*.

1. The *Sun* is the center of our solar system. It is by far the largest object in our solar system, but it is only a medium-sized star compared with other suns in our galaxy. The Sun's gravity is strong enough to keep all the planets orbiting around it. On Earth, a year is defined as the time it takes Earth to complete one orbit around the Sun.

2. The Sun creates light, heat, and other types of energy by fusing together hydrogen atoms in the process of *nuclear fusion*. This process is different than the process of nuclear fission, which causes atoms to break apart and is how Earth generates much of its internal heat. The Sun is the only object in our solar system that generates its own light—all the planets and moons, including our own, are bright because they reflect the Sun's light.

5. *Mars* is farther from the Sun and is smaller than Earth. Recent exploration of Mars reveals that water once flowed on the planet's surface.

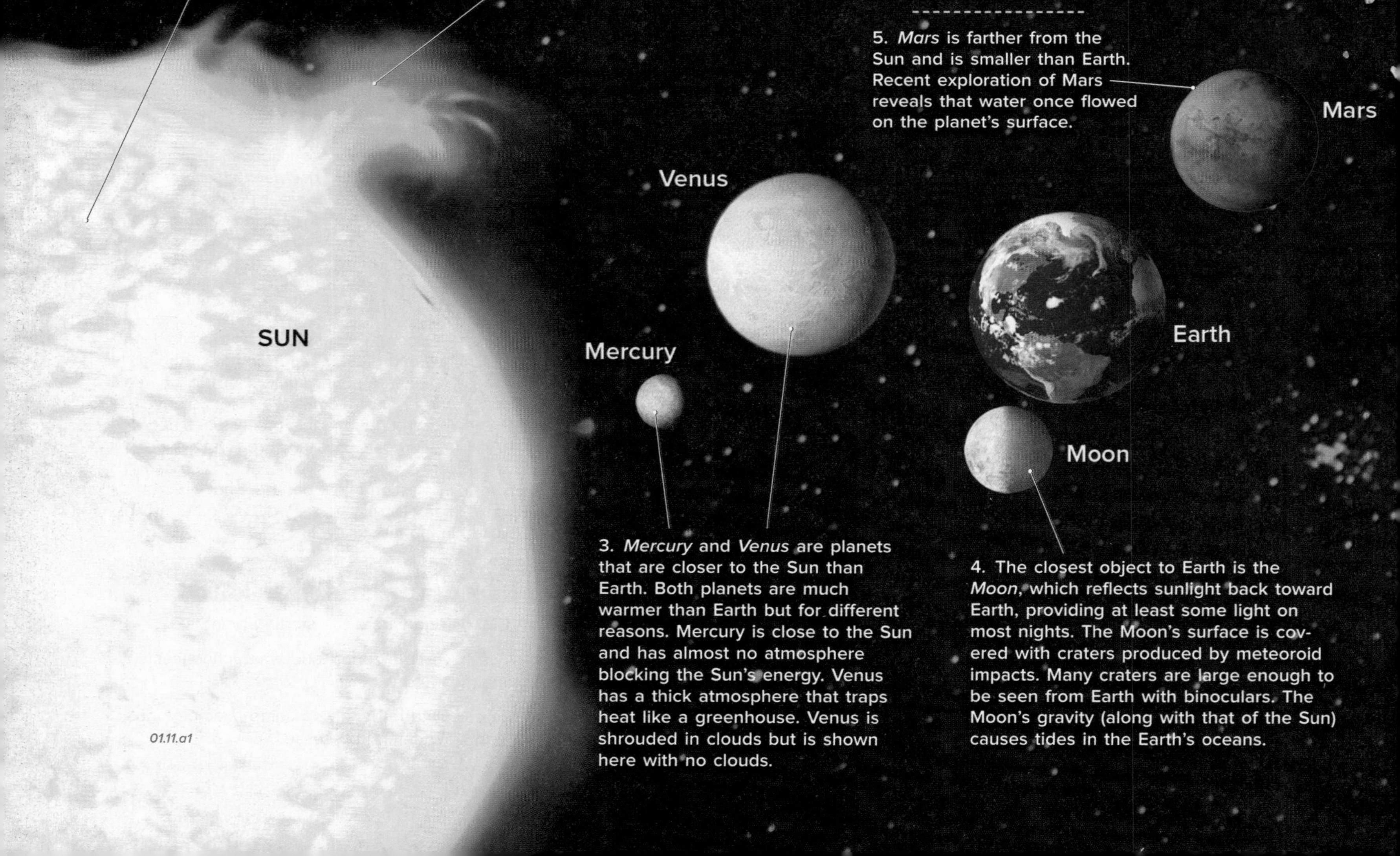

Mars

Venus

Earth

SUN

Mercury

Moon

01.11.a1

3. *Mercury* and *Venus* are planets that are closer to the Sun than Earth. Both planets are much warmer than Earth but for different reasons. Mercury is close to the Sun and has almost no atmosphere blocking the Sun's energy. Venus has a thick atmosphere that traps heat like a greenhouse. Venus is shrouded in clouds but is shown here with no clouds.

4. The closest object to Earth is the *Moon*, which reflects sunlight back toward Earth, providing at least some light on most nights. The Moon's surface is covered with craters produced by meteoroid impacts. Many craters are large enough to be seen from Earth with binoculars. The Moon's gravity (along with that of the Sun) causes tides in the Earth's oceans.

B What Are Some Characteristics of the Outer Planets?

The four outer planets, Pluto, and many asteroids are farther from the Sun than Mars. The outer planets are called *gas giants* because of their large size and gas-rich character. Pluto is a small, distant object that is no longer considered to be a planet by many astronomers.

1. Hundreds of times larger than Earth, *Jupiter* is the largest planet in the solar system. Like the Sun, Jupiter is composed mostly of hydrogen and helium. It has a distinctly banded, swirly atmosphere with what appears to be a huge red storm. Jupiter and the other gas giants are much larger than the inner planets.

2. *Saturn* is a gas giant similar to Jupiter in composition and atmosphere. Saturn has huge, beautiful rings, composed mostly of small chunks of ice and dust (as observed by the Cassini spacecraft).

3. *Uranus* and *Neptune* are smaller gas-giant planets, but these planets are still much larger than Earth. Their atmospheres contain significant methane, which causes their bluish color.

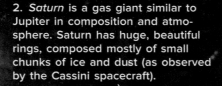

Jupiter

Uranus

Neptune

Pluto

Saturn

4. Far from the Sun, *Pluto* is a small, icy object. Most astronomers no longer consider Pluto to be a planet, but instead think it is related to similar icy objects farther out in the solar system. This designation leaves our solar system with eight true planets instead of the nine we have traditionally considered. Pluto's size is greatly exaggerated here compared with the Sun and eight planets.

5. *Asteroids* are rocky fragments left over from the formation of the solar system. Most orbit between Mars and Jupiter and have a composition that is similar to certain meteorites.

C What Is the Shape and Spacing of the Orbits of the Planets?

1. If we step back from the solar system so we can observe the shape of each planet's orbit, the view below is what the orbits of the inner planets and Jupiter look like. In other words, if you traveled straight up from the Arctic, perpendicular to the orbit of the planets, this is the view you would have.

2. Observe that all of the planets' orbits, including Earth's orbit, are almost circular. In other words, Earth is at about the same distance from the Sun during all times of the year.

3. Earth's orbit is essentially *circular*, so Earth receives nearly the same amount of light and heat at all times of the year. Earth's seasons (summer and winter), therefore, are not caused by changes in the distance between Earth and the Sun. The seasons have another explanation, which involves the tilt of Earth's spin axis relative to its orbit, a topic we explore later in this book.

----

4. Note how far from the Sun Jupiter is compared with the inner planets. The distance from Jupiter to Saturn is greater than the distance from the Sun to Mars, and the distances to Uranus and Neptune (not shown) are even larger. Jupiter is much larger, relative to the other planets, than shown here, but is much smaller than the Sun.

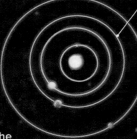

5. The sizes of the planets are greatly exaggerated here relative to the size of the Sun.

Before You Leave This Page

☑ Sketch a view of the solar system, from the Sun outward to Jupiter.

☑ Explain why the Sun and the Moon are the most important objects to Earth.

☑ Summarize how the outer planets are different from the inner planets.

1.12 How Do We Approach Earth Science Problems?

EARTH SCIENTISTS APPROACH PROBLEMS in many ways, asking questions about Earth's features and processes and then collecting data that help document what is there and what is happening. Some questions can be addressed by making general observations with our human senses, whereas others require *quantitative* data, which are numeric and are typically visualized and analyzed using data tables, calculations, equations, and graphs.

A What Is the Difference Between Qualitative and Quantitative Data?

01.12.a1 Augustine Island, AK

01.12.a2 Augustine Island, AK

01.12.a3 Augustine Island, AK

When Augustine volcano in Alaska erupts, earth scientists make various types of observations and measurements. Some observations are *qualitative*, like simple descriptions, and others are measurements that are *quantitative*. Both types of data are essential for documenting natural phenomena.

Qualitative data include descriptive words, labels, sketches, or other images. We can describe this picture of Augustine volcano with phrases like "contains large, angular fragments," "releases steam," or "the rocks are mostly gray." Such phrases can convey very important information about the site.

Quantitative data involve numbers that represent measurements. Most come from scientific instruments, such as this thermal camera that records temperatures on Augustine volcano, or from simple measuring devices like a thermometer. We can collect quantitative data in many settings.

B What Quantitative Properties Do We Directly Measure While in the Environment?

Earth scientists often describe features qualitatively, but they also collect quantitative data, which consist of numeric values measured with scientific tools or instruments. Some measurements are collected out in the environment.

01.12.b1 Lake Pleasant, AZ

◄ ORIENTATION: Geologists observe and measure the orientation of geologic features, such as layers, fractures, and folds. In this view, a geologist is using a level on a hand-held compass to measure how much the sedimentary layers have been tilted.

▶ SURFACE FEATURES: Most earth scientists use topographic and other maps to mark locations of data, but some, especially those earth scientists who study landscapes, use precise surveying instruments to study landforms. Such measurements can document the movement of the land surface before or after an earthquake or volcanic eruption.

01.12.b2 Kyrgyzstan

01.12.b3 NASA Global Hawk Aircraft

◄ GAS COMPOSITION: Understanding of atmospheric processes is strengthened by having precise estimates of the amount of water vapor and other gases in the atmosphere, as is being measured by this airplane.

▶ WATER FLOW AND CHEMISTRY: We can measure the velocity and volume of flowing water in streams and groundwater. Chemical analyses, including some performed in the field, document what the water contains.

01.12.b4 Yellowstone NP, WY

C   What Quantitative Properties Do We Measure Using Laboratories and Sensors?

Some data collection is done in laboratory environments, by collecting data with sensors on satellites, or by observing through telescopes. These data often include quantitative measurements done with sophisticated scientific instruments that can measure physical properties, chemical composition, light spectra, and ages of rocks, soils, and other materials.

01.12.c1 State College, PA

01.12.c2 Solar Dynamics Observatory, NASA

01.12.c3 Syracuse, NY

PHYSICAL PROPERTIES: Density, strength, and other physical properties of earth materials, as measured in the laboratory, form the basis for evaluating how rocks behave when subjected to forces, such as during earthquakes.

COMPOSITION: We conduct chemical analyses on samples of earth materials, water, and air, and also use telescopes and satellites to analyze remote objects. This image shows our Sun, as viewed in ultraviolet frequencies.

AGE: Certain materials can be dated using precise analytical instruments that measure the ratios between different types of radioactive elements. These can be on materials from the Earth, Moon, or meteorites.

D   How Do We Calculate Density, and How Does It Differ from Weight?

Density is a very important quantitative property for understanding the Earth and other objects in the solar system and universe. It controls regional elevations and causes forces that result in earthquakes. It is a key aspect in the rising and sinking of air in the atmosphere and water in oceans and lakes. It influences the gravitational attraction between objects in space. We determine or estimate density by directly measuring a material in the laboratory, by using instruments to measure the pull of gravity, or by numerically analyzing how fast seismic waves pass through materials.

Density

1. *Density* refers to how much mass (substance) is present in a given volume. Below, a wooden block, a "cube" of water, and a stone block all have the same shape and volume but different amounts of mass. The wood is less dense than water and floats, but the stone is more dense and sinks. The cube of water has the same density as the surrounding water and so does not sink to the bottom or float on the surface.

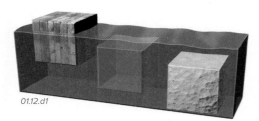
01.12.d1

2. We calculate density using the following formula:

$$density = mass/volume$$

We measure mass in grams or kilograms, and volume in cubic centimeters, cubic meters, or liters. Density is therefore in units of g/cm^3 or g/L. Water has a density of 1 g/cm^3 at room temperature and pressure, whereas ice, which floats on water, is slightly less dense at 0.92 g/cm^3. Granite has a higher average density of 2.65 g/cm^3.

Weight

3. The *weight* of an object is how much downward force it exerts under the pull of gravity. Weight depends on how much mass the object contains and the strength of the gravity field.

Moon

Earth

01.12.d3

01.12.d2

4. A person has the same mass, whether standing on Earth or on the Moon. If the person weighs 180 pounds on Earth, the person will weigh only 30 pounds on the Moon (because of the Moon's lower gravity). When addressing scientific issues, scientists rarely talk about weight, instead referring to mass, generally using metric-system units like g/cm^3.

Before You Leave This Page

- ✓ Explain how qualitative data differ from quantitative data.
- ✓ Describe some types of quantitative data that earth scientists use.
- ✓ Describe what density is, how it is calculated, and how it differs from weight.

1.13 How Do We Develop Scientific Explanations?

EARTH SCIENCE IS A FIELD OF SCIENCE that aims to explain Earth's features and processes. Every land region, ocean basin, part of the atmosphere, and planetary object has a wealth of interesting questions with answers of importance to society. To answer the questions, earth scientists use their senses and scientific instruments to observe features and processes on Earth and elsewhere in the universe. They use the resulting observations to answer questions and then, through a series of logical steps, build from observations to explanations.

A What Are Observations?

We learn about our world by making *observations*. We look, listen, smell, and feel so we can record and analyze what is around us. Scientific instruments provide additional information about aspects of the world that we cannot sense, and they allow us to discriminate fine details. For example, we might sense that the temperature outside is near freezing, but if we use a thermometer we can measure a precise value. Every day we make judgments about whether our observations are worth remembering and reliable enough to plan a course of action.

1. Scientists take special care to make valid *observations*, such as when examining these layers of volcanic ash. An observation that is judged to be valid becomes a piece of *data* that can be used to develop possible explanations.

01.13.a1 Gray Mtn., AZ

3. Evaluating the validity of observations is critical, so scientists commonly repeat measurements to compare values. They may invite other scientists to check and discuss their observations, measurements, and ideas (▾).

2. Scientific instruments provide quantitative information, provided they are checked and calibrated to ensure that measurements represent valid and trustworthy data. This typically involves reanalysis of standard materials for which the physical properties have already been well determined. Once the instruments are calibrated, scientists can measure and record data in a field notebook, tablet, or computer and collect samples to permit later reexamination and analysis.

01.13.a2 Gray Mtn., AZ

B How Are Interpretations Different from Data?

Data by themselves are not very useful until we analyze them in the context of existing ideas. Perhaps the data will confirm old ideas, or perhaps they will point out a need for a new interpretation. The recent history of volcanic eruptions near Yellowstone National Park illustrates the difference between data and interpretations.

DATA: This map shows a belt of relatively smooth, lower elevation terrain that trends northeast across the mountains of southern Idaho and northern Nevada. It contains mostly volcanic rocks.

INTERPRETATION: Some process related to volcanism formed a belt of low topography and volcanic rocks in the belt outlined in red.

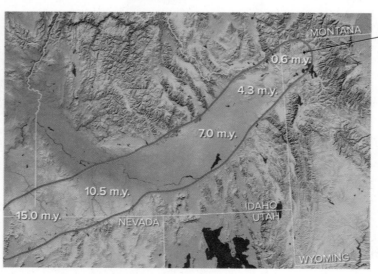
01.13.b1

DATA: The belt of smooth topography ends near Yellowstone, a recently active volcanic area in the northwestern corner of Wyoming.

INTERPRETATION: Recent volcanism at Yellowstone may be related to the process that smoothed the topography of the belt.

DATA: Samples of volcanic rock analyzed in the laboratory provide ages for when the rock formed. The ages, shown in white as millions of years (m.y.), get younger toward the northeast, from 15 million years in Nevada to less than one million years near Yellowstone.

INTERPRETATION: The very recent volcanism at Yellowstone occurred for the same reasons as the older volcanism to the southwest.

C What Is an Explanation?

When scientists examine a collection of related data, several interpretations may fit together to make a coherent story or *explanation.* The table below summarizes data-interpretation pairs from part B. The bottom row in the table is a new piece of data obtained from other studies. These data and interpretations combine to form a possible explanation, or *hypothesis,* for how the belt of smooth topography formed.

Data	Interpretation	A Possible Explanation
A belt of smoothed topography, mostly in volcanic rocks, extends in a northeast direction and cuts across the region.	The belt of smoothed topography is related to some process that also produced volcanic eruptions.	For 15 million years, North America and its lithosphere have been moving southwest over a deep thermal disturbance called a *hot spot.* The hot spot involves melting of rocks at depth, resulting in volcanism on the surface. As North America moves southwestward over the hot spot, new volcanoes erupt and then become inactive once that area moves past the hot spot. If North America continues to move southwestward, the hot spot may cause new volcanoes northeast of Yellowstone.
The belt ends at Yellowstone National Park.	Volcanism at Yellowstone is related to the smoothed topography.	
Volcanic rocks along the belt get younger to the northeast.	The smoothed belt did not form all at once but rather sequentially, from southwest to northeast.	
The North American continent is moving slowly to the southwest based on satellite observations, and the age progression of volcanism is related to this motion.	There is a source of magma beneath Earth's crust. The continent has moved over the magma source, causing volcanic activity to occur in a narrow belt. Volcanism initially occurred in the southwest but migrated to the northeast over time as North America moved to the southwest.	

D What Is Happening at Yellowstone National Park?

Geoscientists have long recognized that several huge volcanic eruptions occurred in Yellowstone during the past two million years. Geoscientist Bob Smith studied Yellowstone for decades and in 1973 noticed that lake levels along the southern side of Yellowstone Lake had risen, drowning trees. How did he investigate his observations?

1. To check his observation, Smith and colleagues conducted a new, detailed survey of the area's topography using high-precision surveying equipment.

01.13.d1 Yellowstone NP, WY

2. This view shows Yellowstone Lake, with north to the left. When Smith compared the new survey with the last survey done in 1920, he discovered that the elevation of the area shaded orange had increased (the area had risen) in a remarkably short period of time. What was causing this area along the north side of the lake to rise in elevation?

3. Smith concluded that the rising area north of the lake caused the lake to spill over its southern shoreline, drowning trees in the areas shown in purple.

←*N*

01.13.d2

Observations, Interpretations, and Hypotheses

Discovery of drowned trees along Yellowstone Lake and the follow-up studies illustrate how we develop and investigate questions. An observation (the drowned trees) led to the question, *What is going on here?* The question led to a possible interpretation that parts of the land around the lake may be actively rising or sinking. An explanation that is developed to explain observations and that allows testing is a *hypothesis.* To test his predictions, Smith used precise surveying equipment to collect new observations of the land surface. He scrutinized and validated the new data and proposed a new hypothesis that rising land beneath the northern part of the lake had

displaced water that drowned trees along the southern shore.

This example illustrates the strategy of considering different types of data and different scales of observation. Smith interpreted local uplift north of Yellowstone Lake to be the result of a large magma chamber beneath the surface. One interpretation is that the magma originated by deep melting related to a hot spot in the mantle. As the North American continent and lithosphere moved over the hot spot, volcanism and faulting formed the belt of smoothed topography along the *Snake River Plain.* Many geoscientists accept this explanation but still consider and test other explanations.

Before You Leave This Page

✓ Explain what observations are and how they become valid.

✓ Describe how data differ from an interpretation, and provide one example of each.

✓ Summarize how data and interpretations lead to new explanations.

✓ Describe how a series of observations led to an explanation for regional and local processes at Yellowstone.

How Do Scientific Ideas Get Established?

HOW DOES A SCIENTIFIC EXPLANATION move through the steps from an initial idea to a testable hypothesis and finally to a widely accepted *theory* supported by a rigorous body of knowledge? Scientists begin with observations, propose possible explanations (hypotheses), make predictions based on each hypothesis, and conduct investigations to test each prediction. Science is a way to evaluate which hypotheses are most likely to be correct and which are not. It is a body of knowledge based on supported hypotheses and on accepted theories that have been examined and tested many times.

A How Do We Test Alternative Explanations?

Science proceeds as scientists explore the unknown—making observations and then systematically investigating questions that arise from observations that are puzzling or unexpected. Often, we try to develop several possible explanations and then devise ways to test each one. The normal steps in this process are illustrated below, using an investigation of groundwater contaminated by gasoline.

Steps in the Investigation

Observations

1. Someone makes the *observation* that groundwater from a local well (on the left side of the figure) contains gasoline near an old buried gasoline tank. The first step in any investigation is to make observations, recognize a problem, and state the problem clearly and succinctly. Stating the problem as simply as possible simplifies it into a more manageable form and helps focus our thinking on its most important aspects.

Questions Derived from Observations

2. The observation leads to a *question*—Did the gasoline in the groundwater come from a leak in the buried tank? Questions may be about what is happening currently, what happened in the past, or, in this case, who or what caused a problem.

Proposed Explanations and Predictions from Each Explanation

3. Scientists often propose several explanations, referred to as *hypotheses*, to explain what they observe. A hypothesis is a causal explanation that can be tested, either by conducting additional investigations or by examining data that already exist. Drawing a sketch or other type of figure often helps us better conceptualize the alternatives.

4. One explanation is that the buried tank is the source of contamination.

5. Another explanation is that the buried tank is not the source of the contamination. Instead, the source is somewhere else, and contamination flowed into the area.

6. We develop *predictions* for each explanation. For the first option, the tank should have some kind of leak and should be surrounded by gasoline. Also, the type of gasoline in the tank should be the same as in the groundwater. Next, we plan some way to *test* the predictions, such as by inspecting the tank or analyzing the gasoline in the tank and groundwater.

Results of Investigation

7. The investigation discovered no holes in the tank or any gasoline in the soil around the tank. Records show that the tank held unleaded gasoline, but gasoline in the groundwater is leaded. We compare the results of any investigation with the predictions to determine which possible explanation is most consistent with the new data.

Conclusions

8. Data collected during the investigation support the conclusion that the buried tank is not the source of contamination. Any explanation that is inconsistent with data is probably incorrect, so we pursue other explanations. In this example, a nearby abandoned gas station may be the source of the gasoline. We can devise ways to evaluate this new hypothesis by investigating the site near the older gas station. We also can revisit the previously rejected hypothesis if we discover a new way in which it might explain the data.

01.14.a1

B How Does a Hypothesis Become an Established Theory?

A hypothesis that survives scientific scrutiny can be elevated to the higher standard of acceptability of a *theory*. Like a hypothesis, a theory explains existing data and helps predict data not yet collected, but a theory encompasses a more extensive body of knowledge. The scientific process rejects many hypotheses, and few hypotheses survive the intense investigation, experimentation, and testing of predictions to become theories accepted by a majority of scientists. The testing and rejecting of ideas distinguishes science from ways of knowing based on faith.

1. Scientists found fossils of the same land animals in South America and Africa, even though these continents are separated by an ocean. To explain these observations, scientists proposed a hypothesis that long ridges of land, called *land bridges,* once linked the two continents but were now under water. The hypothetical land bridges would have allowed land animals to walk from one continent to the other. According to the hypothesis, the bridges later collapsed or were submerged beneath the oceans.

AFRICA

Hypothetical
Land
Bridge

SOUTH
AMERICA

01.14.b1

2. If the land bridges once existed, then the South Atlantic Ocean should contain submerged ridges, or remnants of ridges, that once connected the two continents. When surveys of the ocean floor failed to find land bridges, the hypothesis had to be abandoned. So scientists had to look for another way to explain the similarity of fossils in South America and Africa.

3. A land-bridge hypothesis was also proposed as a way to explain the migration of animals and humans from Asia to North America during the Ice Ages. This hypothesis, unlike the hypothesis about the Atlantic Ocean, is supported by a lot of data and by a credible explanation of why a land bridge existed. A submerged ridge does link Alaska and Asia, and it would have been dry land when the growth of glaciers lowered sea level. So the hypothesis that a land bridge existed off Alaska evolved into a theory, while the hypothesis that one existed in the South Atlantic was rejected.

How and Why Scientific Understandings Change Over Time

Science is a way of investigating the world around us. It is an evolving framework of knowledge and methods, not a static collection of facts. Explanations and theories accepted by the scientific community can change over time as new data, new scientific instruments, and new ideas become available.

Although many scientific explanations are considered to be "correct" and are supported by many lines of evidence, the history of science warns us not to trust any explanation as "final truth." There is so much evidence supporting some theories that they probably will never be shown to be wrong. On the other hand, scientific scrutiny has caused many proposed hypotheses or theories to be rejected or greatly modified based on new data. Some accepted scientific explanations needed only to be revised slightly to account for new data or other scientific advances. In other cases, the science of the time was not sophisticated enough to produce explanations that could hold up under scrutiny. Scientists operate under the principle that no explanation in science is ever *proven*, but some are eliminated. There are no final answers, but science proceeds through a series of logical steps toward a well-tested explanation that

best explains all the available observations and other data.

In the 1700s, for example, the most influential scientists of the time could not accept that stones (*meteorites*), such as the one shown here, fell out of the sky. For a time, scientists and others believed that meteors and meteorites resulted from lightning that fused dust with other particles in

01.14.t1

the air. This explanation was rejected when chemists noted that some meteorites consisted of iron-nickel alloys that were not found in any Earth rocks. Also, some meteorites fell in plain view when there were no lightning storms. Stones really were falling from the sky!

We have gained much understanding using the methods of science, but we still do not know many things about the uni-

verse. There are countless interesting questions left to investigate, and many important theories left to imagine. We not only lack reasonable explanations for many scientific phenomena, in many cases we do not yet know the right questions to ask or what data to collect. Some hypotheses we currently accept will be proven wrong by future studies. There is still much to learn.

Before You Leave This Page

☑ Explain the logical steps taken to evaluate an explanation.

☑ Describe how a hypothesis becomes an established theory.

☑ Describe what causes changes in scientific understandings, and discuss why scientific explanations are never proven to be "true."

1.14

How Are Earth-System Processes Expressed in the Black Hills and in Rapid City?

THE BLACK HILLS OF SOUTH DAKOTA AND WYOMING are a natural wonder. The area is home to three national parks and one national monument. It is famous for its gold and for the presidents' faces carved into granite cliffs at Mount Rushmore. Rapid City, at the foot of the mountains, was devastated by a flash flood caused by unstable atmospheric conditions. In this region, the impacts of nature are dramatic and provide an opportunity for us to examine how concepts presented in this chapter connect together and how they apply to a real place.

A   What Is the Setting of the Black Hills?

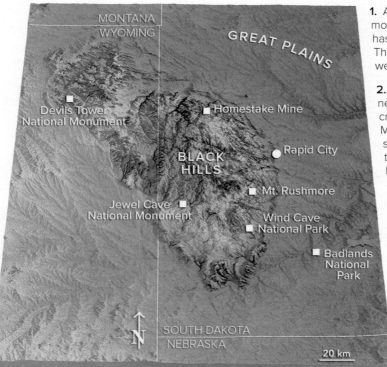

01.15.a1

1. As seen in this shaded relief perspective (◀), the *Black Hills* are an isolated mountainous area that rises above the surrounding *Great Plains*. The region has a moderately high elevation, more than 1,000 m (3,000 ft) above sea level. The highest parts of the Black Hills consist of erosionally resistant rocks that were uplifted to the surface.

2. The famous gold deposits of the *Homestake Mine* formed on the seafloor nearly 1.8 billion years ago. The rocks were then buried deep within the crust, where they were heated, strongly deformed, and metamorphosed. Much later, uplift of the Black Hills brought the rocks and gold closer to the surface. The Homestake Mine produced 39 million ounces of gold, more than any other mine in the Western Hemisphere. The deep mine has also been used as a site to measure high-energy particles from space.

3. Rapid City is on the eastern flank of the Black Hills. To the south, *Badlands National Park*, known for its intricately eroded landscapes, is carved into soft sedimentary rocks. The Black Hills is home to many caves, including at *Wind Caves National Park* and *Jewel Caves National Monutment*.

01.15.a2 Mount Rushmore, SD

4. The presidents' faces at *Mount Rushmore* (▶) were chiseled into a granite that solidified in an underground magma chamber 1.7 billion years ago. The granite and surrounding metamorphic rocks were cooled, uplifted, and overlain by a sequence of rock layers. More recently, the rocks were uplifted to the surface when the Black Hills formed 60 million years ago.

01.15.a3 Devils Tower, WY

5. *Devils Tower* (◀) is a well-known landmark that rises out of the Black Hills. The rock formed by solidification of a magma chamber at depth, followed by uplift and erosion to bring the rocks to the surface. The distinctive columns are the result of fracturing as the hot rock cooled.

6. This figure (▼) shows the geometry of rock units beneath the Black Hills. The Black Hills rose when horizontal forces squeezed the area and warped its rock layers. As the mountains were uplifted, erosion stripped off upper layers of sedimentary rock (shown in purples, blues, and greens), exposing an underlying core of ancient igneous and metamorphic rocks (shown in brown). Rapid City is near the boundary between the hard, ancient bedrock in the center of the mountains and younger sedimentary rocks of the plains.

01.15.a4

Sedimentary Rocks

Igneous and Metamorphic Rocks

Sedimentary Rocks

B   What Geologic Processes Affect the Rapid City Area?

This view of Rapid City is an aerial photograph superimposed over topography. Where do you think erosion is occurring? Where is sediment being deposited? Which places are most susceptible to landslides? Examine this scene and think about the earth processes that might be occurring in each part of the area.

1. Rapid City is located along the mountain front, partly in the foothills and partly on the plains. Some parts of the city are on low areas next to Rapid Creek, which begins in the Black Hills and flows eastward through a gap in a ridge and then through the center of the city.

2. Upturned rock layers form a ridge that divides the city into two halves. Some of the homes are right along the creek, whereas others are on the steep hillslopes.

3. The plains contain sedimentary rocks, some of which were deposited in a great inland sea and then buried by other rocks. With uplift of the mountains and erosion, the rocks came back to the surface where they are weathered and eroded today.

01.15.b1

4. This part of the Black Hills consists of hard igneous and metamorphic rocks that form steep mountains and canyons. Farther northwest (not in this view) is the world-famous *Homestake Mine*, a former producer of gold. The underground mine is no longer operating, but it reached depths of more than 2.5 km (8,000 ft)!

5. *Rapid Creek* drains a large area of the Black Hills and flows through the middle of Rapid City. A small dam forms Canyon Lake (the blue-green area) just above the city.

6. A flash flood in 1972 destroyed buildings, bridges, and roads along Rapid Creek, leaving the creek littered with shattered houses and other debris, an example of the hazards of living too near flowing water. After the flood, the city decided to restrict building in areas most likely to be flooded, instead turning the flood-prone areas into parks and trails as part of a greenbelt along the creek, a wise and less risky use of such space.

The Rapid City Flash Flood of 1972

In June 1972, winds pushed moist air westward up the flanks of the Black Hills, forming severe thunderstorms. The huge thunderstorms remained over the mountains, where they dumped as much as 15 inches of rain in one afternoon and evening. This downpour unleashed a *flash flood* down Rapid Creek that was ten times larger than any previously recorded flood on the creek. The swirling floodwaters breached the dam at Canyon Lake, which increased the volume of the flood downstream through Rapid City. The floodwaters raced toward the center of the city. They killed 238 people, destroyed more than 1,300 homes, and

caused $160 million in damage. Most of the damage occurred along the creek channel, where many homes had been built too close to the creek, and in areas low enough to be flooded by this large volume of water. Since the flood, the city has removed buildings on many flood-prone sites and developed a wide greenbelt in the areas most likely to flood. Some buildings were allowed to remain in flood-prone areas, but many were raised above likely levels of future floods. As for all streams, the occurrence of an exceptionally large flood causes us to reassess what size floods are possible and how often such floods are likely to reoccur.

Before You Leave This Page

✓ Briefly sketch the landscape around Rapid City and explain how earth processes affect this landscape.

✓ Identify and explain ways that earth features and processes affect the people of Rapid City.

✓ Describe the events that led to the Rapid City flood and explain why there was so much damage.

How Are Earth Processes Affecting This Place?

EARTH SCIENCE HAS A MAJOR ROLE, from global to local scales, in the well-being of our society. The image below shows an aerial photograph superimposed on topography for an area near St. George, Utah. In this investigation, you will observe and identify some landscape features, consider possible earth processes operating in this region, and think about how the landscape, weather, climate, and various processes affect the people who live here.

Goals of This Exercise:

- Determine where important earth processes are occurring.
- Interpret how earth processes are affecting the people who live here.
- Identify a relatively safe place to live that is away from geologic hazards, such as volcanism, earthquakes, landslides, and flooding.

Begin by reading the procedures list on the next page. Then examine the figure and read the descriptions flanking the figure.

1. The main feature is the Virgin River, which receives water from precipitation in mountains around Zion National Park. It enters the valley through a narrow gorge. Hot springs provide recreation at the end of the gorge, where the river flows through the cliffs.

2. Most of this region receives only a small amount of rain and is fairly dry. The low areas are part of a desert that has little vegetation and is hot during the summer. The dry climate, coupled with erosion, provides dramatic exposures of the various rocks. People living here rely on water from wells, reservoirs, and the rivers that flow into the area from distant mountains that receive more rain and snow than this low, dry area.

3. A high, pine-covered mountain range and steep, rocky cliffs flank the valley. The cliffs and mountains receive abundant winter snow and torrential summer rains, which cause flash flooding down canyons that lead into the valley. This photograph (▼) shows the valley, mountains, and cliffs, viewed toward the northwest. The high mountains in the photograph are outside of the area shown in the main figure. They commonly have some puffy clouds over their peaks, while the sky over the valley has few or no clouds. As a result of its lower elevation and fewer clouds, the valley is much hotter and drier than the mountains.

01.16.a2 St. George, UT

Reservoir

N

0.5 km

4. This figure exaggerates the height of the land surface to better show the features. It shows the mountains twice as high and twice as steep as they really are. Exaggerating the topography in this way is called *vertical exaggeration*.

Procedures

Use the figure and descriptions to complete the following steps. Record your answers in the worksheet, which will be provided by your instructor in paper form, as a printable file, or as an activity you complete online.

A. Using the image below, explore this area. Make observations about the landscapes and earth processes that are likely on land and in the atmosphere. Next, mark on the provided worksheet at least one location where the following processes would likely occur: locally intense rainfall, weathering, erosion, transport of sediment, deposition of sediment, volcanic activity, landsliding, and flooding.

B. Using your observations and interpretations, indicate on the worksheet all the ways that earth processes might influence the lives of the people who live here. Think about each landscape feature and earth process, and then decide whether it has an important influence on the people. Where would you look for water? Is there a higher potential for a certain type of natural hazard (flooding, earthquakes, etc.) in a particular part of the area? How might the mountains influence the amount of clouds and rainfall?

C. Using all your information, select a location away from natural hazards that would be a relatively safe place to live compared to more hazardous sites in the area. Mark this location on your worksheet with the word *Here*.

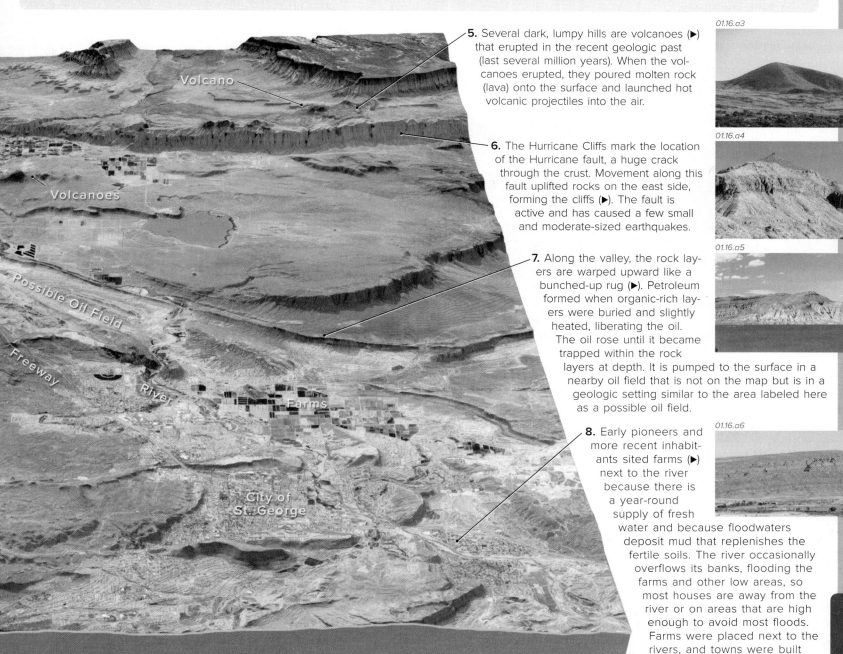

01.16.a3

5. Several dark, lumpy hills are volcanoes (▶) that erupted in the recent geologic past (last several million years). When the volcanoes erupted, they poured molten rock (lava) onto the surface and launched hot volcanic projectiles into the air.

01.16.a4

6. The Hurricane Cliffs mark the location of the Hurricane fault, a huge crack through the crust. Movement along this fault uplifted rocks on the east side, forming the cliffs (▶). The fault is active and has caused a few small and moderate-sized earthquakes.

01.16.a5

7. Along the valley, the rock layers are warped upward like a bunched-up rug (▶). Petroleum formed when organic-rich layers were buried and slightly heated, liberating the oil. The oil rose until it became trapped within the rock layers at depth. It is pumped to the surface in a nearby oil field that is not on the map but is in a geologic setting similar to the area labeled here as a possible oil field.

01.16.a6

8. Early pioneers and more recent inhabitants sited farms (▶) next to the river because there is a year-round supply of fresh water and because floodwaters deposit mud that replenishes the fertile soils. The river occasionally overflows its banks, flooding the farms and other low areas, so most houses are away from the river or on areas that are high enough to avoid most floods. Farms were placed next to the rivers, and towns were built near the farms.

01.16.a1

1.16

2 Minerals and Mineral Resources

EARTH'S SURFACE IS COMPOSED of many kinds of materials: black lava flows, white sandy beaches, red cliffs, and gray granite hills. Such landscapes are composed of rocks, sediment, and soil, which are all largely composed of *minerals*. Some regions of Earth provide a treasury of gemstones and other mineral resources, many of which are essential to modern society. What kinds of materials are common on Earth, and how did the less common ones, such as gemstones, form? Here we explore minerals, from landscapes to atoms.

This perspective view (▶) shows satellite data superimposed over topography for southernmost California and adjacent Baja California, Mexico. The Peninsular Ranges, a forested mountainous area east of San Diego, are in greens and browns in the center of the image. The white line across the image, added for reference, marks the border between the United States and Mexico.

What are the rocks that make up the hills and mountains of the Peninsular Ranges (▼)?

02.00.a1

02.00.a2 Jacumba, CA

Peninsular Ranges

San Diego

UNITED STATES

MEXICO

10 km

The Peninsular Ranges contain many outcrops of grayish-colored rocks, most of which are igneous rocks like granite. When viewed up close (▼), the granite displays four different kinds of crystals: whitish, light pink, transparent gray, and black.

What are rocks made of, and what controls the color and other properties of a rock?

02.00.a3 Anza-Borrego Desert SP, CA

02.00.a4

San Diego County is a famous source of beautiful minerals, including tourmaline crystals (◀), that can be pink, purple, green, or all three colors.

What are crystals, how do they form, and where do we find them?

TOPICS IN THIS CHAPTER

Salton Sea

Salton Trough

Peninsular Ranges

East of the Peninsular Ranges, the land drops down into the lowlands of the Salton Trough, which is characterized by sandy deserts, farmlands of the Imperial Valley, and several large, salty lakes, including the blue Salton Sea. The sand and pebbles in the Salton Trough were eroded from the adjacent mountains and carried to the area by streams or strong winds.

What are most sand grains composed of?

The Peninsular Ranges

The area called the Peninsular Ranges is a broad, upland region that stretches 1,500 km across southernmost California and southward into the Baja Peninsula of Mexico. In this image, the mountains appear green because they are mostly covered by forests and other types of vegetation. The lowlands of the Salton Trough east (right) of the mountains receive much less rain and have a lighter color in this image because vegetation is sparse, and sand and rocks cover the surface.

The contrast between the Peninsular Ranges and the Salton Trough illustrates some important aspects to consider when observing and thinking about landscapes. First, landscapes develop from the materials that are available. The mountains contain granite because that is the type of rock that was there, long before the mountains were uplifted. The mountains would have a different appearance if they instead consisted of a material much weaker than granite and, in this case, might not even be mountains today. In addition to the type of materials, the geometry of the rock units and other materials can also impart a distinct style to the landscape. An appropriate place to begin thinking about landscapes is, therefore, to examine the repertoire of earth materials, starting with minerals.

Once the materials are at the surface, they are acted on by the atmosphere, moisture in the soil, and other components of the environment that tend to break down rocks into loose materials. The kind of climate greatly influences this loosening process. Once loosened, these materials can be stripped off the land surface and transported away by running water, wind, and ice, eventually being deposited as sediment someplace else, such as sedimentary rocks in the Salton Trough, as in the photograph below. This sequence of events is what shaped the mountains of the Peninsular Ranges and also deposited the resulting loose materials in the Salton Trough. These processes are still occurring, so the landscape continues to evolve.

02.00.a5 Rancho Vallecito, CA

2.0

2.1 What Is the Difference Between a Rock and a Mineral?

WHAT MATERIALS MAKE UP THE WORLD around us? What do we see if we look closely at a rock outcrop? How does the rock look when viewed with a magnifying glass or microscope? We investigate these questions using the beautiful scenery of Yosemite National Park in California.

A What Materials Make Up a Landscape?

1. Observe this photograph of Yosemite Valley, the heart of Yosemite National Park in the Sierra Nevada range of California. What do you notice about the landscape?

2. This landscape is dominated by dramatic cliffs and steep slopes of massive gray rock perched above a green, forested valley. The valley is famous for waterfalls and huge rock faces. The appropriately named Half Dome is in the right side of the photograph. What would we see if we got closer to this landscape?

02.01.a1 Yosemite NP, CA

02.01.a2 Yosemite NP, CA

3. From several meters away, the rock forming Yosemite's cliffs looks fairly homogeneous. It all seems to be the same kind of gray rock, a kind of igneous rock called *granite*. The granite is cut by fractures.

02.01.a3 Sierra Nevada, CA

4. Closer examination reveals several different-colored grains in the rock: whitish, pinkish, clear gray, and black. To better observe a rock at this scale, a geoscientist may collect a hand-sized piece, called a *hand specimen*.

02.01.a4

5. When examined with a magnifying glass or *hand lens*, the rock contains different minerals with distinct appearances. The clear gray crystals all have similar chemical composition and physical properties, and so represent one kind of *mineral*. The whitish crystals are a different kind of mineral, as are the pink and black crystals.

6. To examine the rock in even more detail, a geoscientist will cut a very thin slice from a hand specimen and glue it to a glass slide to make a *thin section*. The slice is so thin that light can pass through it. Geoscientists then examine the thin section using a microscope that has polarizing filters.

02.01.a5

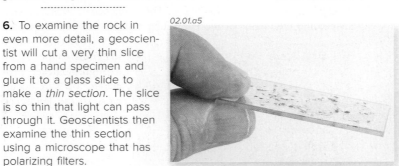

02.01.a6

7. When polarized light shines through a thin section and optical filters, the internal structure of crystals interacts with the light in ways that allow us to observe diagnostic characteristics, to identify minerals, and to estimate percentages of each mineral.

B What Is and What Is Not a Mineral?

What characteristics define a mineral? To be considered a mineral, a substance must fulfill all of the criteria listed below. A mineral is a naturally occurring, inorganic, crystalline solid with a relatively consistent composition.

Natural

02.01.b1 Apophyllite/Stilbite   02.01.b2

1. A mineral must be *natural*. Crystals on the left grew naturally from warm water flowing through a rock, but synthetic crystals on the right grew in a laboratory. Natural diamonds are minerals, but synthetic diamonds grown in the lab are not.

Inorganic

02.01.b3 Calcite

2. The crystal on the left is *inorganic* and a mineral. The shells on the right have the same composition as the crystal, but they were made by creatures; the shells are not considered to be a mineral by most geoscientists because the shells are not inorganic.

Solid

02.01.b5   02.01.b6

3. All minerals are *solid*, not liquid or gaseous. Ice, a solid, is a mineral, but liquid water is not, even though it has the same composition. Liquid mercury, although natural and found in rocks, is not considered a mineral.

Ordered Internal Structure

02.01.b7 Calcite   02.01.b8 Obsidian

4. A mineral has an *ordered internal structure,* which means that atoms are arranged in a regular, repeating way. Such substances are considered to be *crystalline*, and they can form well-defined geometric crystals. The mineral on the left is crystalline, and the shape of the crystals reflects the internal arrangement of its atoms. The volcanic glass (obsidian) on the right is not crystalline. Its atoms are arranged in a random way, so volcanic glass is not a mineral.

Specific Chemical Composition

02.01.b9 Halite and Table Salt   02.01.b10 Conglomerate

5. Minerals are homogeneous and so have specific chemical compositions that do not depend on the size of the sample that is analyzed. Table salt, which is the mineral *halite*, contains atoms of the chemical elements *sodium* (Na) and *chlorine* (Cl) in equal proportions, no matter how big or small the specimen. The rock on the right is not a mineral because different parts of the rock have very different compositions. Most minerals have a specific chemical formula, like NaCl for halite.

Rocks, Minerals, and "Minerals"

When we hear the word *mineral* used in the context of *vitamins and minerals*, are these minerals the same as the minerals described above? The answer is *no*. In a kitchen or pharmacy, the term *mineral* refers to a chemical *element*, such as potassium (K). This type of (nutritional) mineral is different from the crystalline mineral of geoscientists.

In earth science, most minerals consist of at least two different chemical elements, i.e., naturally occurring chemical compounds, such as the *sodium* (Na) and *chlorine* (Cl) that make up the mineral *halite* (salt). Many minerals have three, four, or even more chemical elements. A few minerals, however, include only one chemical element, and these are called *native elements*. The mineral *diamond*, for example, consists entirely of the element *carbon*.

Some rocks contain only a single mineral. Limestone may be 100% of the mineral calcite. Sandstone may be 100% quartz. Most rocks, like the granites from Yosemite National Park, include several different minerals, and each mineral is made of one or more elements. So, rocks are made of minerals, and minerals are made of elements. Although the vitamin pill you take with breakfast may not contain any geologic minerals, most of the nutritional elements in the pill were extracted from geologic minerals.

2.1

2.2 How Are Minerals Put Together in Rocks?

THERE ARE MANY KINDS OF ROCKS, varying greatly in texture, color, and the minerals they contain. Geoscientists use the term *texture* to refer not only to the roughness or smoothness of a rock, but also to the way its grains and minerals are arranged. What controls the texture of a rock? How are minerals in a rock connected to one another? What can we determine about a rock, including how it formed, from its texture and the types of minerals it contains?

A How Are Minerals Put Together in Rocks?

Minerals compose igneous, sedimentary, metamorphic, and hydrothermal rocks. The beautiful and geologically interesting Engineer Mountain in the San Juan Mountains of southwestern Colorado contains two different types of rocks, providing examples of the two main ways that minerals occur in rocks.

02.02.a1 Engineer Mtn., CO

1. The main mountain has an upper gray part and a lower reddish-brown part (◄). Loose pieces of the upper gray part tumble down the hillside, forming gray slopes that cover some of the red rocks.

02.02.a2 Engineer Mtn., CO

2. A closer view of the mountainside (▲) reveals differences between the gray and reddish parts. Both parts contain vertical fractures, but the reddish-brown part also has well-defined, nearly horizontal layers, whereas the gray part does not.

Two Types of Rocks

02.02.a3 Engineer Mtn., CO

3. *Crystalline*—The gray rock displays light-colored crystals surrounded by a gray, fine-grained material (called a *matrix*) with crystals too small to see in this photograph. A rock composed of interlocking minerals that grew together is a *crystalline rock*. Crystalline rocks typically form in high-temperature environments by crystallization of magma, by metamorphism, or by precipitation from hot water. Some crystalline rocks consist of crystals formed from the precipitation of minerals in cooler waters, such as when a lake evaporates.

02.02.a4 Engineer Mtn., CO

4. *Clastic*—A close look at the reddish-brown layer reveals that it includes distinct pieces of rock derived from older weathered and eroded rocks. These pieces, called *clasts*, range from small sand grains to larger pebbles. A rock consisting of pieces derived from other rocks is a *clastic rock*. Most clastic rocks form on Earth's surface in low-temperature environments, such as sand dunes, streams, glaciers, and beaches—any place sediment is deposited. Clasts also compose some volcanic rocks, including those formed by explosive volcanic eruptions.

B What Different Attributes Do Rocks Display?

Crystalline rocks and clastic rocks can have various attributes, depending on the sizes and shapes of crystals and clasts in the rocks and on how the crystals and clasts are arranged. The photographs below display some common rock textures. Some are of natural exposures of rocks and some are cut and polished rock slabs, like you would find on a kitchen countertop. The parts of the slabs shown are 5 to 30 cm (2 to 12 in.) across.

Crystalline Rocks

Clastic Rocks

Types of Minerals

02.02.b1 Brazil

Rocks can consist of one mineral or many minerals, but most contain more than one mineral. This crystalline rock (◄) is an igneous rock with several types of minerals, each having a distinctly different color. This clastic sedimentary rock (►) also includes different types of minerals, including small, partially rounded pebbles of quartz in various shades of gray.

02.02.b2 Moab, UT

Sizes of Crystals or Clasts

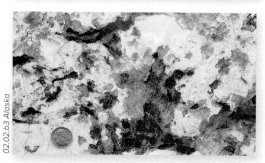

02.02.b3 Alaska

Rocks can contain various sizes of crystals and clasts. This granitic crystalline rock (◄) has coarse crystals, including some whitish ones that are more than 5 to 10 cm (2 to 4 in.) in diameter. This clastic sedimentary rock (►) includes various sizes of larger clasts in a matrix of smaller pebbles and sand.

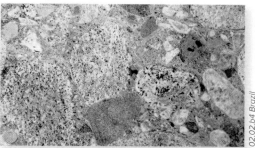

02.02.b4 Brazil

Shapes of Crystals or Clasts

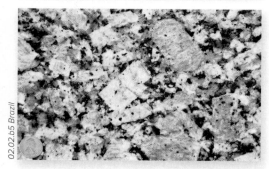

02.02.b5 Brazil

Crystals and clasts in rocks can have various shapes. This crystalline igneous rock (◄) includes some large, nearly rectangular crystals surrounded by smaller, more irregularly shaped crystals. This clastic sedimentary rock (►) includes mostly rounded pebbles (clasts) in a matrix of sand and smaller clasts. Some clastic rocks contain clasts that are sharp and angular.

02.02.b6 Smoky Mtns., TN

Layers or No Layers

02.02.b7 Stone Mtn. SP, GA

Crystalline rocks and clastic rocks may or may not have distinct layers. Most granites (◄) do not have obvious layers or compositional variations. In contrast, this clastic rock (►), which is mostly composed of sand grains, displays a sequence of layers. An important first observation about any rock is whether it has layers and a variable composition or lacks layers or variation.

02.02.b8 Gaviota Beach, CA

Before You Leave This Page

☑ Explain the difference between a clastic rock and a crystalline rock and the differences between the general environments in which clastic and crystalline rocks form.

☑ Describe or sketch four general characteristics to observe in crystalline and clastic rocks.

2.2

2.3 How Do We Distinguish One Mineral from Another?

MINERALS HAVE MANY PROPERTIES that allow us to distinguish them from each other. Some properties are reflected by the shape of the mineral or the way the mineral breaks. Others are physical properties, like hardness and magnetism, that we can evaluate with simple tests.

A What Clues Does the Appearance of a Mineral Provide?

The first thing that we notice about a mineral is usually its outward appearance. We may note its size, shape, color, or how light reflects from its surface. These properties provide clues about the identity of a mineral. The figures below illustrate some physical properties that are relevant for identifying a mineral.

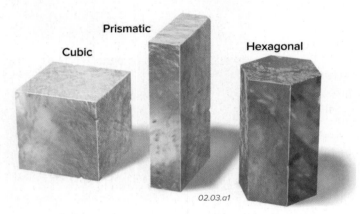

02.03.a1

1. *Crystal Shape*—A mineral that grows unobstructed by its surroundings can have a distinctive geometric shape. The shape of the crystal reflects the arrangement of atoms within the mineral and therefore provides a clue about the mineral's identity. Common crystal shapes include, but are not limited to, cubes, rectangular prisms, and hexagons (six-sided shapes).

2. *Cleavage*—Some minerals break in specific ways because of their internal arrangement of atoms. If a mineral breaks preferentially along a specific set of planes, the mineral has *cleavage*. Some minerals, like the ones shown here, break along one set of cleavage planes and cleave into thin sheets, but other minerals break along several sets of cleavage planes having different orientations.

02.03.a2 Mica

3. *No Cleavage*—Some minerals have an internal atomic arrangement that does not contain planes along which the mineral breaks. These minerals do not have cleavage but instead break along *fractures*. Fractures in minerals tend to have rough, irregular surfaces, like ones shown here cutting a quartz crystal. In contrast, cleavage planes tend to be more planar and regular.

02.03.a3 Quartz

02.03.a4 Quartz

4. *Color*—The color of a mineral is a useful, but not always reliable, property for mineral identification. Bright or unique colors are easily noticed, but a mineral can occur in several color varieties, such as the different-colored versions of quartz shown here. Other minerals always have the same color. It is the color and the crystal shape that make some minerals so beautiful and so highly valued as gemstones or mineral specimens.

02.03.a5 Talc/Galena/Hematite

5. *Luster*—The way that light bounces off a mineral is a property called *luster*. A mineral can be highly reflective, dull, or somewhere in between. It can be partially transparent or opaque. It can look like metal, a pearly shell, a silky material, or a simple piece of earth. The names of different types of luster reflect the quality and intensity of the reflection. Luster terms include metallic, nonmetallic, glassy, pearly, silky, resinous, and earthy.

02.03.a6 Olivine/Pyroxene

6. *Microscopic Observations*— To identify minerals in rocks, especially fine-grained rocks, we often examine a *thin section* (◄) using a microscope. When a thin section is viewed between two polarizing filters, light shining through the thin section causes the different minerals to exhibit distinctive and diagnostic colors.

B What Tests Can We Perform to Help Us Identify a Mineral?

We determine some mineral properties by conducting tests. We may touch a mineral with a magnet to check its magnetism, or we may try to scratch it to determine its hardness.

1. *Hardness*—Some minerals are very hard, and some are quite soft. A mineral can be scratched by a material that is harder than the mineral but not by one that is softer. To estimate mineral hardness, we often conduct scratch tests using common objects, such as a fingernail, penny, or knife blade. We may also use other minerals of known hardness for comparison.

02.03.b1 Gypsum

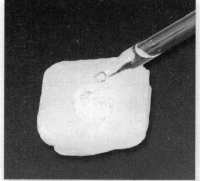

02.03.b3 Calcite

3. *Effervescence*—If a drop of dilute hydrochloric acid (HCl) is placed on a mineral, a reaction may cause vigorous bubbling, or *effervescing*. The mineral *calcite*, which is the main mineral in limestone, effervesces strongly with HCl, but no other common minerals do. Any sample that effervesces when tested with hydrochloric acid is most likely to be calcite.

2. *Streak*—If a mineral is rubbed against an unglazed porcelain plate (called a streak plate), it may leave a trail of powdered material called a *streak*. Some minerals have a diagnostic streak color. The iron-oxide mineral *hematite,* for example, has a reddish streak.

02.03.b2 Hematite

02.03.b4 Magnetite

4. *Magnetism*—A few iron-bearing minerals are naturally magnetic. The mineral *magnetite* is the strongest natural magnet. It is attracted to other magnets, and its magnetism can be strong enough to deflect a compass needle. Other magnetic minerals are less magnetic than magnetite, but magnetism may still help identify them.

5. *Density*—Some minerals are more dense than others. This property can often be detected by simply holding a mineral and noting how heavy it feels. We call this approach a *heft test*. In the lab, scientists precisely determine the ratio of the density of a substance to the density of freshwater, a property called *specific gravity*.

6. In this example (▶), crushed crystals are placed into a glass beaker on one side of a balance scale and weigh as much as two beakers of water. This sample of dry, crushed crystals is therefore twice as dense as water and has a specific gravity of 2. It would be more dense if it were a solid crystal without air between the crushed pieces. A typical specific gravity of rock is 2.7 (i.e., 2.7 times more dense than freshwater).

02.03.b5

Mohs Hardness Scale

Mohs Hardness Scale (▶) consists of 10 familiar minerals ranked in order of hardness, from 1 to 10. The softest mineral (talc) is 1, and the hardest mineral (diamond) is 10. These numbers describe the *relative* hardnesses of the minerals, but the numbers do not provide a real comparison of their actual hardnesses. Quartz (hardness of 7) is twice as hard as apatite (hardness of 5), and diamond (hardness of 10) is about five times as hard as corundum (hardness of 9).

Hardness and Mineral	Common Objects
1 Talc	
2 Gypsum	
3 Calcite	Fingernail (2.5)
4 Fluorite	Copper wire (3.5)
5 Apatite	
6 K-feldspar	Window glass or knife blade (5.5)
7 Quartz	
8 Topaz	
9 Corundum	
10 Diamond	

Before You Leave This Page

✓ Explain the properties of a mineral that can be observed without using a test.

✓ Describe how to test for hardness, streak, effervescence, and magnetism.

✓ Explain the meaning of a mineral's specific gravity.

✓ Explain the Mohs Hardness Scale.

2.3

2.4 What Controls a Crystal's Shape?

CRYSTALS CAN HAVE BEAUTIFUL SHAPES. The outward shape of a crystal reflects a combination of factors, including the arrangement of atoms in the crystal and how the crystal's growth was affected by the material around it. What, at an atomic scale, controls the shape of a crystal?

A How Is the Shape of a Mineral Related to Its Internal Structure?

If the growth of a mineral is unconstrained by surrounding materials, the outward shape of the crystal mimics the mineral's internal structure of atoms. The relatively simple outward shape and internal structure of halite nicely illustrate this relationship between the interior and the exterior of a mineral.

1. The photograph below shows natural crystals of table salt, which is the mineral *halite* (NaCl). These crystals grew together and look like a number of cubes connected together.

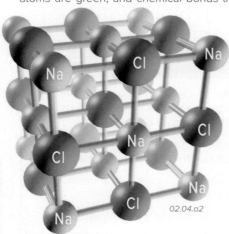

02.04.a1 Halite

2. Mineralogists (geologists and other scientists who study minerals) have documented that halite consists of equal proportions of sodium (Na) and chlorine (Cl) atoms. It has the chemical formula NaCl. Mineralogists have investigated the atomic arrangement of atoms within halite and find that sodium and chlorine atoms have a geometric arrangement that is like a cube. In the figure below, sodium atoms are yellow, chlorine atoms are green, and chemical bonds that link adjacent atoms are represented by stick-like connectors. Note that the green chlorine atom in the center of the structure is surrounded by and bonded with six sodium atoms. Other minerals have more complicated shapes or chemical formulas, but we use halite here because it is so simple.

02.04.a2

3. In a crystal, one part of the atomic arrangement repeats indefinitely to make the entire crystal. In halite, the smallest part is one pair of sodium (Na) and chlorine (Cl) atoms. Sodium and chlorine atoms alternate in three perpendicular directions. Note that in this figure, whether you go up-down, left-to-right, or front-to-back, Na and Cl alternate in the crystalline structure.

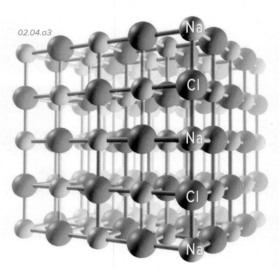

02.04.a3

02.04.a4

4. A different way to represent crystals is to show atoms as spheres that fit together and touch (◄). This type of model more accurately represents the relationship between adjacent atoms and their electrons, but it is more difficult to see the internal structure. Note that for halite, the relative sizes of sodium (Na) and chlorine (Cl) atoms allow them to pack together tightly in a cube-shaped arrangement. The atoms in halite are so tiny that a one-inch cube of halite contains more than 100,000,000,000,000,000,000,000 [10^{23}] pairs of Na and Cl atoms.

5. In addition to growing as cubic crystals, halite will also cleave into cube-shaped or shoebox-shaped fragments. If you examine table salt with a magnifying glass, you will observe that most salt grains are tiny cubes or slightly elongated boxes, like the halite crystal shown here. Note the cube-shaped to rectangular "steps" where pieces have broken off the front corner of the crystal, a result of the cleavage characteristic of halite.

02.04.a5 Halite

B How Are Atoms Arranged in a Mineral?

Atoms fit together in a limited number of ways. How closely atoms can be packed together depends on their electrical charge (positive versus negative) and the relative sizes of different kinds of atoms (e.g., smaller Na atoms fit between larger Cl atoms). A single atom typically bonds to 3, 4, 6, 8, or even more atoms. Atoms of similar charge repel each other, whereas atoms of opposite charge attract, and so atoms are generally arranged in geometric patterns. Three common arrangements of atoms are shown below, but other, more complicated arrangements are common.

Atoms can be arranged in the shape of a cube. This type of structure is referred to as *cubic*.

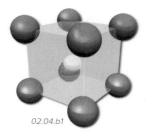

02.04.b1

One atom can be surrounded by four other atoms, arranged as a pyramid with three sides and a base. This arrangement and four-sided shape is called a *tetrahedron* (tetra = four).

02.04.b2

Atoms can be arranged in a shape that is like two oppositely pointing, four-sided pyramids joined at their bases. This shape is an *octahedron* (octa = eight).

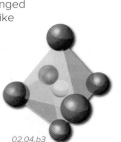

02.04.b3

C How Is the Shape of a Crystal Affected by the Environment in Which It Grows?

For a crystal to attain a perfect shape, it must grow unimpeded by surrounding material. Most nicely shaped crystals grew in an open space, in water or in magma; their growth was not constrained by other, preexisting crystals. When crystals grow within solid rock or around preexisting crystals, they generally do not have such well-formed shapes.

These crystals have well-defined shapes, flat crystal faces, and sharp ends called *terminations*. Most such crystals grew into a space filled with hot or cold water rather than solid rock.

02.04.c1 Calcite

02.04.c2 Polished Slab

In this rock, partially transparent gray quartz is in irregularly shaped masses that fill the spaces between and around the white and red minerals. The quartz grew after the other minerals were already there, so it had to conform to their shapes.

States of Matter: Solids, Liquids, and Gases

The most fundamental attribute of a material is its *state of matter:* whether it is a solid, liquid, or gas. Materials that are *solid* have a relatively fixed shape and volume because their atoms are packed closely together and connected, or bonded, to one another, like the crystals shown below. Rock, minerals, and volcanic glasses are solids and retain their shape and volume unless they are being actively deformed, dissolved, or perturbed in some other way.

In contrast, a *liquid* easily changes shape, conforming to its surroundings, as when water fills a glass. Atoms in a liquid are held together, but weakly enough that the material is mobile and can change shape. A liquid maintains a relatively constant volume, unless it is subjected to changes in temperature (heated or cooled) or in pressure, as occurs to waters below the earth's surface.

In a gas, atoms and molecules are even less connected and more mobile. A gas does not have a constant shape or constant volume; it will conform to the shape of its container and expand or contract according to how much space is available.

02.04.t1 Apophyllite/Stilbite

Before You Leave This Page

✓ Explain what it means to say that crystals have an ordered atomic arrangement, using the crystal form of halite as an example.

✓ Sketch and describe three common ways in which atoms are arranged in a mineral.

✓ Explain how the shape of a crystal is affected by the environment in which the crystal grows.

✓ Summarize the three states of matter.

2.4

2.5 What Causes Cleavage in Minerals?

CLEAVAGE IS THE TENDENCY OF MINERALS TO BREAK along parallel planes. Some minerals cleave into cubes, and others cleave into thin sheets. Still other minerals break along irregular fractures instead of cleavage planes. Cleavage is controlled by the arrangement of atoms in a mineral and the strengths of the bonds between atoms.

A What Happens at an Atomic Scale When a Mineral Cleaves?

The same orderly arrangement of atoms that causes crystals to form with specific shapes can also affect the way crystals break. Breaking a mineral requires applying enough force to break the links—bonds—between adjacent atoms. In many minerals, different bonds have different strengths, so the mineral breaks preferentially (cleaves) along the easiest directions and through the weakest bonds.

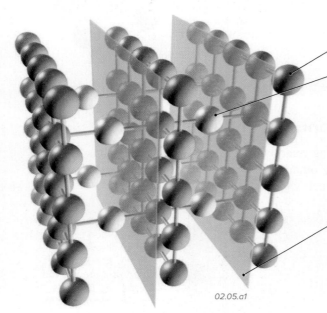

02.05.a1

1. This mineral consists of three kinds of atoms shown here in brown, blue, and gray.

2. The brown atoms are linked with (bonded to) the blue atoms, forming flat sheets.

3. Adjacent sheets are joined together by long bonds between the gray atoms and the brown atoms in the sheets on either side.

4. Bonds between the brown and blue atoms (within the sheets) are stronger than bonds between the brown and gray atoms (linking the sheets). If the mineral is subjected to sufficient force, the force will break the weakest bonds (those between the brown and gray atoms). The breaks will occur along the cleavage planes shown in yellow.

5. With this type of arrangement of atoms and bonds, the mineral will cleave along one set of planes, splitting into thin sheets, like the cleaved pieces of mineral shown here (▶).

02.05.a2 Biotite

B What Happens if All of the Bonds Have the Same Strength?

In the example above, one set of bonds is relatively weak and so forms a natural place for breaking across the mineral. How does the mineral break if all the bonds have similar strengths or if the arrangement of atoms and bonds does not allow the crystal to break along any planes?

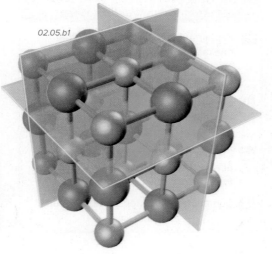

02.05.b1

The bonds in this mineral (◀) all have a similar strength but are arranged in such a way that the mineral can break along three sets of planes without passing through an atom. In this example, the three planes are mutually perpendicular (at 90° to each other).

02.05.b2 Quartz

The bonds in this mineral (◀) are not arranged into a configuration that allows any cleavage planes to form. Instead, the crystal has broken like glass, along irregular curved *fractures* instead of along cleavage planes. A crystal that fractures in this irregular way can still contain well-defined planes, called *crystal faces*, that formed during the growth of the crystal. In such minerals, the way in which the crystal grows can be different than the way in which it breaks.

C What Are Some Common Types of Cleavage?

If a mineral has cleavage, it can cleave along one or more sets of parallel planes. Two sets of planes might be perpendicular (90°) to one another or might intersect at some other angle. In the diagrams below, colored planes show the orientation of possible cleavage planes. The specific mineral groups mentioned below are shown and described on later pages.

One Direction of Cleavage

02.05.c1

If a mineral has a single direction of cleavage, it cleaves along one set of parallel planes, forming thin sheets. Examples of a single direction of cleavage are members of the mica group of minerals.

Two Perpendicular Directions of Cleavage

02.05.c2

Many minerals cleave along two sets of planes that are perpendicular to one another. This type of cleavage results in right-angle (90°) steps along broken crystal faces. The pyroxene mineral group has right-angle cleavage.

Two Non-Perpendicular Directions of Cleavage

02.05.c3

Two planes of cleavage can intersect at angles other than 90°. Minerals with this type of cleavage can break into pieces having corners that do not form right angles. The amphibole group of minerals has this type of cleavage.

Three Perpendicular Directions of Cleavage

02.05.c4

If a mineral cleaves along three perpendicular sets of planes, broken faces have a stair-step geometry and the mineral commonly breaks into cubes, as is typical of halite.

Three Non-Perpendicular Directions of Cleavage

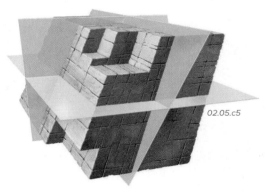

02.05.c5

Minerals that cleave along three directions of planes that are not mutually perpendicular break into pieces that are shaped like a *rhomb*, or a sheared box. Calcite is the most common mineral that cleaves into rhombs.

Before You Leave This Page

✓ Explain or sketch the relationship between cleavage and the arrangement and strengths of bonds.

✓ Explain what happens if a mineral lacks planes along which it may cleave.

✓ Sketch and describe five types of cleavage.

2.5

2.6 How Are Minerals Classified?

WITH NEARLY 100 NATURALLY OCCURRING ELEMENTS, it should not be a surprise that there are thousands of different minerals. Some minerals are so rare that they occur only in unusual environments, but others are so common they are almost everywhere on Earth's surface. Here, we concentrate on minerals that are very common and are critical to our understanding of Earth's landscapes and processes.

A How Are Similar Chemical Elements Grouped in the Periodic Table?

Chemical elements are the fundamental building blocks of minerals, so we classify minerals into several *mineral groups* based on the main chemical components within those minerals. Before discussing these mineral groups, we take a tour of the chemical elements via the *Periodic Table*, a useful way to organize the elements.

1. Each element in the Periodic Table has an *atomic symbol*, one or two letters representing the name of the element (commonly the name in *Latin*) and an *atomic number* (shown to the upper left of the symbol). Elements that share a background color on the table share some similar chemical properties.

2. The table begins with hydrogen (H), the lightest element, and advances to higher atomic numbers and heavier elements from left to right and from top to bottom.

3. Elements shaded orange are the *alkali* and *alkali earth metals* and include sodium (Na), potassium (K), calcium (Ca), and magnesium (Mg) on the left side of the table and aluminum (Al) and some other elements in the right half of the table.

4. Elements colored yellow are called *transition metals*. They include many familiar metals, such as chromium (Cr), iron (Fe), nickel (Ni), copper (Cu), zinc (Zn), silver (Ag), and gold (Au).

5. The elements colored green are *nonmetals* and include carbon (C), silicon (Si), and oxygen (O). The nonmetals typically bond with both types of metallic elements to form minerals.

6. The last column includes elements called *noble gases* because they are gases that do not readily combine with other elements.

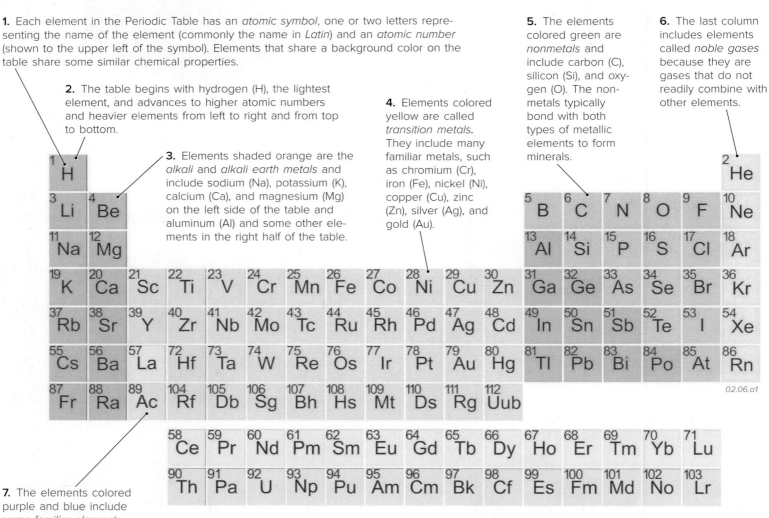

02.06.a1

7. The elements colored purple and blue include some familiar elements, such as uranium (U), and many that are less familiar. Elements with atomic numbers higher than 92 are not known in natural settings (these are produced only in the laboratory), except for plutonium (Pu), which is produced naturally only in unusual circumstances by natural nuclear reactions.

8. The lightest and simplest elements, hydrogen (H) and helium (He), are the most abundant elements in the universe. The elements oxygen (O) and silicon (Si) make up 74% of Earth's crust, with the rest being mostly aluminum (Al), iron (Fe), calcium (Ca), sodium (Na), potassium (K), and magnesium (Mg). Consequently, the most common minerals that we see are made of oxygen and silicon, with lesser amounts of the other common elements.

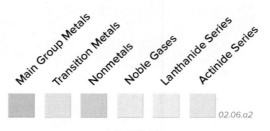

02.06.a2

LEGEND

B What Are the Major Classes of Rock-Forming Minerals?

The most important rock-forming minerals can be divided into several classes based on their chemistry. The Periodic Table provides a useful framework.

02.06.b2 Quartz

1. *Silicates*, including the mineral *quartz* (◄), are the most important mineral group on Earth. They contain silicon and oxygen, the two most abundant elements in the crust, and so are very common. In silicates, each silicon atom is bonded only to oxygen. In most silicate minerals, the silicon-oxygen units are linked by bonds to metals, such as Fe, Mg, Na, K, Ca, and Al.

2. *Carbonates* contain carbon and oxygen bonded together in a triangular arrangement. The triangles are linked together by other elements, most commonly the metal calcium (Ca). An example is the mineral *calcite* ($CaCO_3$). ►

02.06.b3 Calcite

3. *Oxides* consist of oxygen bonded with a metal, such as iron in the mineral *hematite* (Fe_2O_3). ►

02.06.b4 Hematite

02.06.b1

7. *Native minerals* are minerals that contain only a single element. The metals copper (Cu), silver (Ag), and gold (Au) can occur as native minerals or in combinations, called *alloys*, with other metals. Non-metallic elements that occur as native minerals include sulfur (S) as *native sulfur* and carbon (C) as *graphite* and *diamond*.

4. *Halides* contain chlorine (Cl) or fluorine (F), both of which are nonmetals that typically bond with a metal from the left side of the table. *Halite* (NaCl) is a halide mineral (►).

02.06.b5 Halite

6. *Sulfides* contain sulfur (S) bonded with a metal, such as iron (Fe) or copper (Cu). The mineral *pyrite* (FeS_2), also called "fool's gold," is a common sulfide (►).

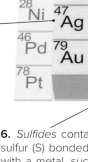
02.06.b7 Pyrite

5. *Sulfates* contain sulfur (S) that is only bonded to oxygen. The sulfur-oxygen units are bonded to a metal, such as calcium (Ca) or iron (Fe). *Gypsum* ($CaSO_4 \cdot 2H_2O$) is a common sulfate (►).

02.06.b6 Gypsum

Asbestos: The Importance of Mineral Classification

Asbestos has long had the reputation of being a dangerous material. Asbestos was used on pipes, ceilings, and ships because of its excellent fire-retarding and insulating properties. Some health studies have shown that asbestos dust, when inhaled, can cause cancer and other serious health problems. But what is asbestos, and is the health story simple?

The term *asbestos* does not refer to a single mineral, but instead refers to a number of silicate minerals whose common characteristic is that they tend to form fibers. Five types of asbestos belong to a group of silicate minerals called *amphiboles*, including a variety known as "blue" asbestos, which forms individual, straight fibers. Medical studies have shown conclusively that "blue" asbestos poses a severe health risk.

In contrast, the type of asbestos most commonly used in the United States is *chrysotile*, a fibrous form of the silicate mineral *serpentine*. Chrysotile has a totally different mineral structure than the other types of asbestos and occurs in fibers that are curved and form interlocking bundles. Studies of this "white" chrysotile asbestos indicate that it is much less hazardous than blue asbestos, mostly causing problems if breathed in large amounts for a long time, as occurs with chrysotile miners. The medical community and Environmental Protection Agency (EPA) are at odds with many scientists because the medical studies and the EPA traditionally have lumped all asbestos into the same category rather than considering the health risks of each kind separately. As a result, hundreds of billions of dollars may be spent, perhaps needlessly, removing

chrysotile asbestos from schools, businesses, and other facilities.

Before You Leave This Page

☑ Describe the Periodic Table, including the locations of the main groups of chemical elements (metals, transition metals, nonmetals, and noble gases).

☑ List the major classes of minerals and discuss the main chemical characteristic of each class.

☑ Describe two types of asbestos and the health controversy over asbestos.

2.6

What Is the Crystalline Structure of Silicate Minerals?

SILICATE MINERALS ARE THE MOST IMPORTANT rock-forming minerals because they comprise most of Earth's crust and mantle. There are different groups of silicate minerals, and the different groups have distinctive cleavage and other mineral characteristics. Here, we explore the types of silicate minerals, from atoms to crystals.

A What Do Silicate Minerals Contain?

In most silicates, one silicon atom is bonded with four oxygen atoms to form the negatively charged SiO_4^{4-} complex. This SiO_4^{4-} complex has a very important shape, called a *tetrahedron*, that controls many aspects of silicate minerals. The silicon-oxygen tetrahedron forms a building block for the vast majority of the minerals on Earth.

Silicon-Oxygen Tetrahedron

1. The four oxygen atoms and one silicon atom combine in an SiO_4^{4-} complex, which can be represented by a four-pointed pyramid called a *tetrahedron*.

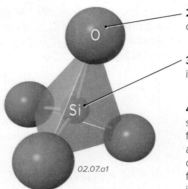

2. An oxygen atom is at each corner of the pyramid.

3. A much smaller silicon atom is in the center of the pyramid.

4. The SiO_4^{4-} complex takes the shape of a tetrahedron because the four oxygen atoms have similar atomic charges and so repel each other. The oxygen atoms move as far as possible from each other, taking positions that define the shape of a tetrahedron.

02.07.a1

Linked Silicon-Oxygen Tetrahedra

5. A silicon-oxygen tetrahedron has a negative electric charge that allows it to bond with other tetrahedra. Each oxygen atom in the tetrahedron is a naturally protruding site ready to bond to other elements and chemical complexes.

6. An oxygen atom can be shared by two adjacent tetrahedra. In this manner, silicon-oxygen tetrahedra can link together to form different types of silicate minerals.

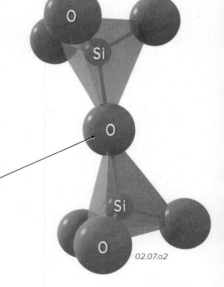

02.07.a2

Silicon-Oxygen Tetrahedra and Metallic Elements

7. In addition to bonding to one another, silicon-oxygen tetrahedra bond with other elements, such as the green atoms shown in this figure. The bonds to the green atoms are not shown.

8. Silicon-oxygen tetrahedra have a negative electrical charge and so attract positively charged atoms, called *cations* (the green atoms).

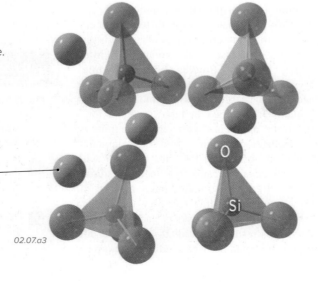

02.07.a3

9. A huge variety of minerals results from the ability of silicon-oxygen tetrahedra to bond with other silicon-oxygen tetrahedra, with various cations, and with other chemical substances. There are thousands of known minerals, but most are uncommon to rare. Several dozen minerals, many of which are silicate minerals, compose most rocks we encounter at the surface. A typical rock contains one to five main minerals, with a small number of less abundant minerals, many of which are visible only under a microscope. So learning only a few minerals helps you to identify most rocks.

B What Are the Different Types of Silicate Minerals?

Silicon-oxygen tetrahedra can be connected in five main ways, each producing a major group of silicate minerals that share common characteristics. Bonds that link one tetrahedron to another are strong, but bonds to other elements between tetrahedra provide planes of weaker bonds, allowing most silicate minerals to cleave. A silicate mineral's cleavage, or lack of cleavage, reflects how the tetrahedra are arranged. The silicon-oxygen tetrahedra are shown below with the oxygen atoms on each corner.

Independent Tetrahedra

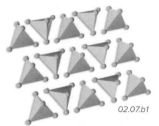

02.07.b1

1. Some minerals contain silicon-oxygen tetrahedra that are bonded to other elements (not shown), but not to other tetrahedra. Minerals in this group, including *olivine*, do not break along clearly defined planes because bonds are more or less equally strong in all directions.

Single Chains

02.07.b2

2. Tetrahedra may form *single chains* by sharing two oxygen atoms. The chains are strongly bonded and difficult to break, so cleavage cuts parallel to, rather than across, the chains. This results in two planes of cleavage that are nearly perpendicular (90° angle) to each other. Minerals that belong to this group are called *pyroxene minerals*, or simply *pyroxenes*.

Double Chains

02.07.b3

3. Tetrahedra can also form *double chains* if half the tetrahedra share two oxygen atoms and half share three, as shown here. Such minerals cleave parallel to the double chains and along two planes of cleavage separated by angles of 60° and 120°. Minerals of this group are called *amphibole minerals*, or simply *amphiboles*.

Sheets

02.07.b4

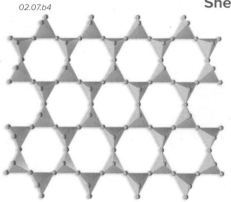

4. In *sheet silicates*, tetrahedra share three oxygen atoms to form continuous sheets. Other elements and water molecules can fit between the sheets, forming minerals with layered structures. Bonds between sheets are weak, so these minerals have one main direction of cleavage parallel to the sheet structure. The most common sheet-silicate minerals are *micas* and *clay minerals*.

Frameworks

02.07.b5

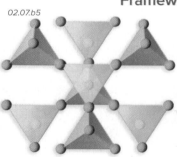

5. Tetrahedra in *framework silicates* share all four oxygen atoms, forming a structure bonded well in three dimensions. *Quartz*, and a few rarer framework silicates, contain only silicon-oxygen tetrahedra bonded to each other. Quartz is hard and has no cleavage, so it fractures instead. Some framework silicates have other elements in the structure between the silicon-oxygen tetrahedra, providing several planes of cleavage. Minerals belonging to the *feldspar group* are good examples of framwork silicate minerals with cleavage.

Silicon, Silica, and Silicone

These three similar words can be confusing, so let's explore what each one means.

Silicon is the fourteenth element of the Periodic Table, having atomic symbol Si. The name silicon also is used for a *synthetic* material — a material produced by humans that does not occur naturally. Synthetic silicon is a semiconductor used to make computer chips.

Silica refers to a compound containing only silicon and oxygen in a ratio of 1:2, so it has the formula SiO_2. Quartz is 100% silica.

Although each silicon atom in quartz is bonded to four oxygen, each oxygen is shared between two silicon, so the ratio Si:O is 1:2. Geoscientists speak about silica more than about silicon because silicon is nearly always bonded with oxygen in rocks and minerals.

Silicone is a *synthetic* material in which carbon is bonded to silicon atoms to keep the material in long chains. These chains make silicone a material that can be used as a type of grease or as caulk for sealing around windows and doors.

Before You Leave This Page

☑ Sketch or explain a silicon-oxygen tetrahedron and how one can join with another tetrahedron or a cation.

☑ Explain or sketch how silicon-oxygen tetrahedra link in five different geometries to produce five silicate mineral groups.

☑ Explain the differences between silicon, silica, and silicone.

2.7

2.8 What Are Some Common Silicate Minerals?

SILICATE MINERALS ACCOUNT FOR OVER 90% of the minerals in Earth's crust. Most silicate minerals also contain other elements, commonly aluminum (Al), calcium (Ca), sodium (Na), potassium (K), iron (Fe), and magnesium (Mg). The presence and amounts of these elements influence the crystalline structure, which in turn determines mineral properties, such as color and cleavage.

A What Are Some Light-Colored Silicate Minerals?

The most common silicate minerals in the upper part of the continental crust have light colors and typically are white, light gray, and light pink. Some of these minerals are almost transparent, and some have a reflective, silvery color. Light-colored silicate minerals predominate in the upper continental crust and are present in smaller amounts in rocks of the oceanic crust and the mantle. Such light-colored silicate minerals are called *felsic*, a term combining the words "feldspar" (a family of minerals described below) and "silica."

Quartz—This very common mineral, with a formula of SiO_2, is generally transparent to nearly white, but it can be pink, brown, or purple. Its silicon (Si) and oxygen (O) atoms are strongly bonded in a tight, three-dimensional *framework*, so quartz is hard (Mohs hardness of 7) and does not cleave. Instead, it breaks along fractures that have irregular or smoothly curving surfaces that are described as being *conchoidal*, as on the broken, front face of the right crystal. The front crystal has well-developed crystal faces that formed during growth of the crystal, but the bottom is a rough fracture.

02.08.a1

02.08.a2

Potassium Feldspar— Potassium feldspar, often just called *K-feldspar*, contains potassium (K), aluminum (Al), silicon, and oxygen, with lesser amounts of sodium (Na) and calcium (Ca). It generally is a pink-to cream-colored mineral, but in volcanic rocks, it can be nearly transparent. Many K-feldspar crystals display two directions of cleavage, and some show wavy, light-colored lines on crystal surfaces, as shown here. K-feldspar is abundant in all granites, and it is common in many other igneous, sedimentary, and metamorphic rocks.

Plagioclase—Plagioclase is one of the two most common feldspar minerals. Feldspars are a group of framework silicates that contain varying amounts of potassium (K), sodium (Na), calcium (Ca), and aluminum (Al), in addition to silicon and oxygen. In plagioclase, the potassium content is close to zero percent. Plagioclase exhibits a complete gradation from Na-rich varieties, which are nearly white to cream colored, to Ca-rich varieties, which are dark gray or brown. However, most plagioclase has a whitish to light-gray color. Some crystal faces display straight lines called *striations*, as shown on the left front crystal here.

02.08.a3

02.08.a4

Muscovite—This sheet-silicate mineral is part of the *mica* family, whose members all have one direction of cleavage and so break into flakes and sheets. It typically is partially transparent, a clear to silvery-gray color, and somewhat shiny because the flat surfaces of the sheets reflect light. Muscovite contains potassium (K) and aluminum (Al), in addition to silicon and oxygen. Its atomic structure contains a component of water, expressed in its chemical formula as $(OH)^-$. The bonds holding the sheets together are stronger than the bonds between sheets, so sheets can be peeled apart with your fingers.

Before You Leave These Pages

☑ Describe the main light- and dark-colored silicate minerals, including their general characteristics, such as cleavage and main elements.

☑ Discuss the characteristics of clay minerals and how they form.

B What Are Some Dark-Colored Silicate Minerals?

Dark silicate minerals predominate in dark igneous and metamorphic rocks and also are many of the dark crystals scattered within otherwise light-colored rocks. They form most of the oceanic crust and the mantle and are present in variable amounts in continental crust, especially the lower crust. Dark-colored silicate minerals are also called *mafic minerals* to acknowledge their high magnesium (Mg) and iron (Fe) content.

Amphibole—The term *amphibole* refers to a group of related silicate minerals. Amphibole minerals can contain magnesium, iron, calcium, sodium, and aluminum, in addition to silicon and oxygen. They can be black, dark green, pale green, or nearly white. They commonly form crystals that

02.08.b1

are long compared to their width, like the long green crystals that are present in the back specimen. Amphiboles are double-chain silicates and so cleave along planes that meet to form angles of 60° and 120°.

02.08.b2

Pyroxene—The term *pyroxene* refers to a group of single-chain silicate minerals that share a similar crystal structure. Pyroxene minerals can include various amounts of calcium, sodium, aluminum, iron, and magnesium in addition to silicon and oxygen. Their color can be black, dark brown, green, or nearly white. Most pyroxenes tend to form crystals that are roughly *equant*, meaning that all dimensions are about the same. Pyroxenes have two nearly perpendicular directions of cleavage (90° angles), which helps distinguish them from amphiboles.

02.08.b3

Olivine—Olivine is the most common mineral in the upper mantle and usually has a distinctive olive-green color. It has independent tetrahedra linked by iron or magnesium, and no cleavage. Its composition varies between iron-rich and magnesium-rich end members, but samples from the mantle are magnesium rich.

02.08.b4

Garnet—Garnets are silicates that can be just about any color, but a deep red color is very common. The crystals are distinctive, having 12 diamond-shaped faces when perfectly formed. The reason color is so variable is because chemistry is variable. Garnets contain silica with variable amounts of calcium, iron, magnesium, manganese, and aluminum.

02.08.b5

Biotite—Biotite is a dark-colored mica (sheet silicate) that is typically black or brown. All biotite contains potassium, aluminum, silicon, and oxygen, with variable amounts of iron and magnesium. Brown biotite, commonly having a tint of bronze, is rich in magnesium and contains little iron. Like all micas, biotite has one dominant direction of cleavage.

Clay Minerals

The term *clay* is used in two ways in earth science. It refers either to a family of minerals or to any very fine sedimentary particles that are less than 0.002 millimeters in diameter. Clay minerals have a sheet-silicate structure similar to that of mica, but the bonds holding the atoms together are much weaker. The sheets in clays are weakly held together, so they easily slip past one another, giving clays their slippery feel.

When some clay minerals get wet, water pushes apart the weakly bonded sheets, causing some clay minerals to expand.

Most clay minerals have light colors but may appear dark if mixed with other material, especially dark minerals or organic debris. Most clay minerals form by weathering of rocks at Earth's surface or from chemical reactions that occur when hot water interacts with rocks containing

feldspar, volcanic ash, and other reactive materials. Fine grain size and low density mean that clay particles are easily transported. Fine particles of clay can be picked up by wind and water and then transported long distances. Clay can be deposited on land by streams, wind, and other agents of transport, but some clay makes it to the open ocean, where it finally settles to the ocean floor, forming extensive deposits of submarine mud.

02.08.t1

2.8

2.9 What Are Some Common Nonsilicate Minerals?

MANY MINERALS DO NOT INCLUDE SILICON and so are classified as *nonsilicates*. Some of the most common nonsilicate minerals are *carbonates* and *halides,* which typically form by precipitation from water. *Oxides* and *sulfides* form when metal atoms bond with oxygen or sulfur, respectively. Nonsilicate minerals are an important resource for our society and are used widely in industry, highways, and homes.

Carbonates

Carbonate minerals contain a metallic element, such as calcium (Ca) or magnesium (Mg), linked with a carbon-oxygen combination called *carbonate* $(CO_3)^{2-}$. The most common carbonate minerals are *calcite* and *dolomite*. Others include *malachite* and *azurite,* striking green and blue copper carbonates. *Trona,* a sodium carbonate, is an important mineral used to manufacture many products. Carbonates typically precipitate from water or have an organic origin (e.g., corals).

02.09.a1

02.09.a2

02.09.a3 Hunt Valley, MD

Calcite—This mineral is the most common calcium-carbonate mineral $(CaCO_3)$ and occurs in a variety of water-related environments. It may be almost clear but commonly has a cream to light gray color. It is the only common mineral that effervesces with dilute hydrochloric acid (HCl) because HCl breaks bonds in calcite and releases carbon dioxide (CO_2) gas.

Dolomite—This mineral is similar to calcite, but magnesium (Mg) substitutes in the structure for some calcium (Ca). It has the formula $CaMg(CO_3)_2$. The mineral is cream-colored, light gray, tan, or brown and may not effervesce with HCl unless pulverized into a fine powder. A rock composed mostly of the mineral dolomite is a *dolostone*. Rocks that contain dolomite also commonly contain calcite.

Most limestones are nearly 100% calcite, and some carbonate rocks contain a mix of calcite and dolomite. Carbonate minerals also occur in coral and shells, including the mineral *aragonite,* which has the same composition as calcite but a different atomic arrangement. When limestone is heated and metamorphosed, calcite grows into larger crystals and the limestone becomes *marble* (▲).

Oxides

Oxide minerals consist of oxygen bonded with iron (Fe), titanium (Ti), aluminum (Al), or other metals. Iron-oxide minerals are the most common oxides, except for ice, which is a hydrogen-oxide mineral (the solid phase of H_2O).

02.09.a4

02.09.a5

02.09.a6 Sherman Mine, Ontario, Canada

Hematite—This iron oxide (Fe_2O_3) can be black, brown, silvery gray, or earthy red, but it consistently has a red streak. Hematite is the red color in rust, provides color in some paints, and is responsible for many red-rock landscapes. It commonly forms when other iron-bearing minerals oxidize.

Magnetite—This iron oxide (Fe_3O_4) is typically black and is strongly magnetic, here attracting a circular magnet. It is present as small black grains in many kinds of igneous, sedimentary, and metamorphic rocks, as well as in beach sands and other sediments.

Magnetite and hematite occur together in beautifully layered sedimentary rocks called *banded iron formations*. Some Precambrian iron formations are mined for iron in the Great Lakes region of the United States and Canada and elsewhere.

Sulfides

Sulfide minerals contain sulfide ions $(S)^{2-}$ bonded with iron (Fe), lead (Pb), zinc (Zn), or copper (Cu). Sulfide minerals, including the copper-iron sulfide mineral *chalcopyrite*, are the principal metal ores in many large mines. Most sulfide minerals have a metallic luster and can occur as well-formed crystals or irregular masses.

Pyrite — Pyrite is a common iron-sulfide mineral (FeS_2). It has a pale bronze to brass-yellow color for which it earns the name "fool's gold." It commonly forms cube-shaped crystals with faces showing straight lines (*striations*).

02.09.a7

02.09.a9

Galena — This mineral is a lead sulfide (PbS). It forms distinctive metallic-gray cubes with a cubic cleavage. It has a high density (specific gravity), which can be felt easily by picking up a sample (i.e, a heft test). In the United States, many galena crystals are from lead mines near the Mississippi Valley.

Small crystals of brass-colored pyrite, as shown here, are commonly deposited by hot (hydrothermal) water. Weathering of pyrite can cause adjacent rocks to become coated with yellow and orange, sulfur-rich material, like the stained quartz on the left side of this photograph.

02.09.a8, Nova Scotia, Canada

02.09.a10

There are many other important sulfide minerals, including iron sulfides, copper sulfides, lead sulfides, and zinc sulfides. We mine sulfides because of their high metal contents. Most sulfide-rich mineral deposits formed when hydrothermal fluids passed through rock.

Salt and Related Minerals (Halides and Sulfates)

Halide minerals (salts) consist of a metallic element, such as sodium (Na) or potassium (K), and a halide element, usually chlorine (Cl). *Sulfate minerals*, especially *gypsum*, commonly occur with salt. They consist of an element such as calcium (Ca) and a sulfur-oxygen complex ion called sulfate $(SO_4)^{2-}$. Many halides and sulfates form when water evaporates in a lake or from precipitation in a shallow sea with limited connection to the ocean.

02.09.a11

02.09.a12

Gypsum — This hydrated calcium-sulfate mineral $(CaSO_4 \cdot 2H_2O)$ is typically gray, white, or clear and can be scratched with a fingernail. Most gypsum forms in environments similar to those in which halite forms (evaporation of salty water), and the two minerals commonly occur together. Gypsum also precipitates from hot or warm water that circulated underground through fractures in rocks. Like salt, it is an important mineral to society, being used in wall board to sheath framed walls, in plaster, and as a component of certain types of cement.

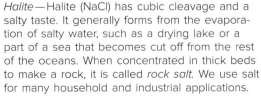

Halite — Halite (NaCl) has cubic cleavage and a salty taste. It generally forms from the evaporation of salty water, such as a drying lake or a part of a sea that becomes cut off from the rest of the oceans. When concentrated in thick beds to make a rock, it is called *rock salt*. We use salt for many household and industrial applications.

Before You Leave This Page

☑ Discuss the key chemical constituents for each of the five nonsilicate mineral groups.

☑ Describe the major nonsilicate minerals, including their general characteristics such as color, cleavage, and any diagnostic attributes.

2.9

2.10 What Are the Building Blocks of Minerals?

MINERALS ARE COMPOSED OF CHEMICALLY BONDED ELEMENTS. An element is a type of atom that has a specific number of protons (e.g., all hydrogen atoms have one proton, whereas all oxygen atoms have eight protons). The mineral halite can be broken into smaller pieces of halite, but if separated into its chlorine and sodium atoms, it is no longer halite.

 ## How Are Minerals Related to Elements and Atoms?

An atom is the smallest unit of an element that retains the characteristics of the element. Atoms are made of even smaller particles (including electrons, protons, and neutrons), but if, for example, a single atom of gold could be broken apart, its pieces would no longer be gold.

02.10.a1 Halite

◄ The mineral halite consists of atoms of two chemical elements—chlorine and sodium. If halite is dissolved in water, it dissolves (►) to produce salt water containing individual atoms of chlorine and sodium.

02.10.a2

Chlorine (Cl) and sodium (Na) atoms (►) each have a small central nucleus surrounded by electrons at various distances from the nucleus.

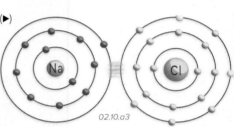

02.10.a3

 ## What Is a Model for the Structure of an Atom?

Atoms are too small to observe, so we use conceptual models to visualize them. The simple two-dimensional model of atoms shown in the previous figure does not fully represent atoms, which are three-dimensional spheres.

1. Atoms have a tiny central core called the *nucleus*. The nucleus is much smaller than the entire atom but is shown enlarged here.

2. The nucleus has two kinds of fundamental particles—*protons* and *neutrons*. Protons, shown in blue, have a positive (+) electrical charge, and neutrons, shown in reddish purple, do not have a charge.

3. The number of protons is called the *atomic number* of an element. The number of neutrons and protons is the atom's *atomic mass*. For any element, the number of protons is consistent, but the number of neutrons can vary.

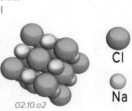

02.10.b1

4. Negatively charged (−) electrons, shown in red, surround and can be thought of as orbiting the nucleus. To be electrically neutral, an atom must have the same number of electrons (−) and protons (+). The proton's positive charges attract the atom's electrons, keeping them associated with the nucleus. The area where the electrons travel is called the *electron cloud*, but it really is not a cloud. It is simply a way of showing the area in which the electrons can reside. The outer edge of the cloud defines the size of an atom, but nearly all of the atom is empty space.

Electron Shells

5. Groups of electrons orbit the nucleus at different distances, called *electron shells*. Each shell has a different level of energy, increasing away from the nucleus. The atom below has three shells, numbered 1, 2, and 3.

6. The inner shell (1), closest to the nucleus, can hold two electrons. Moving outward, successive shells can hold 8, 18, and 32 electrons. Electrons fill inner shells before they fill outer shells, so the inner shells are full but the outermost shell may only be partially full.

7. Atoms are most stable when their outermost shell is full, so atoms with only a few electrons in an unfilled outer shell may donate electrons to another atom in order to become more stable. Alternatively, atoms with a nearly full outer shell may borrow electrons from another atom to get a full shell and become stable.

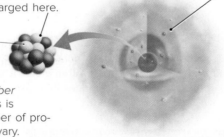

02.10.b2

8. This chlorine atom has seven electrons in its outer shell and so can accommodate one more. It would try to gain an electron to fill its outer shell (3).

9. This atom (sodium) has only one electron in its outer shell and so has a tendency to lose or loan this electron, perhaps donating it to an atom such as the chlorine atom on the left.

10. If an atom *gains* an electron, it acquires an overall *negative* charge. If it *loses* an electron, it acquires a *positive* charge. Charged atoms are called *ions*.

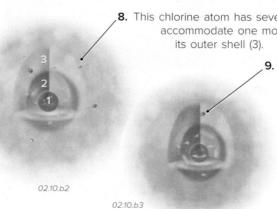

02.10.b3

C How Does the Periodic Table Organize the Characteristics of Elements?

Chemists use the Periodic Table to organize the elements according to the elements' *atomic number* and *electron shells*. The table begins with the lightest element (hydrogen) and advances to the heaviest elements. Below we consider the two left-hand columns and the six right-hand columns because these are the most straightforward.

1. The columns are numbered from I to VII with Roman numerals to indicate the *number of electrons* in the outermost shell.

2. The rows correspond to the *number of electron shells*. Elements in the top row have one shell, those in the second row have two shells, and so forth.

3. Elements in the first column have only one electron in their outer shell. Hydrogen (H) only has one shell (it is in row one), whereas sodium (Na) is in the third row and so has three shells. Recall that the number of outer electrons influences whether an atom loses or gains electrons.

4. The first shell can hold only two electrons, which is why a large gap exists between the right and left sides of the first row.

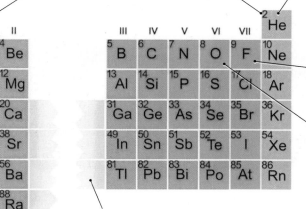

02.10.c1

5. The last column is reserved for noble gases, which do not easily gain or lose an electron because they have complete outer shells. This column could be numbered with both II and VIII—helium (He) only has two electrons, enough to fill its outer shell, whereas other noble gases in the column have eight or more electrons filling their outer shell.

6. Fluorine (F), in the second row, has two shells. It has seven electrons in its outer shell and so is in column VII. If it could borrow another electron, its outer shell would be full.

7. Oxygen (O) has two shells and six electrons in its outer shell; it needs two more electrons to fill this shell. Two oxygen atoms can fill their outer shells by bonding with a silicon atom, which has four electrons in its outer shell.

8. Transition metals, such as iron, occupy columns in the central part of the table (not shown). They lose and gain electrons from several shells, not just the outermost shell.

Some Practice with the Periodic Table

We can use the Periodic Table to predict how many shells each element has and how many electrons are in its outer shell. Try this for lithium, magnesium, nitrogen, and potassium. After you are done, check your answers in the table to the right. We try this out with chlorine, to show how it works.

Chlorine (Cl): Chlorine is in the third row, so it has three shells. It is in the seventh column (VII), so has seven electrons in its outer shell. It seeks one more electron to complete its outer shell, and this can be accomplished by borrowing an electron from sodium (Na), as in halite (NaCl).

Lithium (Li) — 2 shells, 1 electron in outer shell

Magnesium (Mg) — 3 shells, 2 electrons in outer shell

Nitrogen (N) — 2 shells, 5 electrons in outer shell

Potassium (K) — 4 shells, 1 electron in outer shell

Portraying the Atom

Atoms are tiny, but they can be detected by high-powered electron microscopes. We cannot *look* down an optical microscope and see an atom with its nucleus and electrons. Our view of an atom is a *model*—a human-generated representation, or approximation, of what we think is there. Many tests have confirmed that the basic model is valid, but there are limitations. In particular, drawing an atom presents unavoidable problems.

The first problem is one of scale. The nucleus is so tiny compared to the size of the atom that you cannot show an accurately scaled nucleus and still fit the atom on a page. A hydrogen atom, for example, is nearly 150,000 times larger than its nucleus. The electrons, too, are extremely small when compared to an entire atom and so cannot be plotted to scale.

A second problem is how to show the electrons. They are in motion but do not travel around the nucleus in a regular manner. It is tempting to draw electron orbitals the same way we draw planets orbiting our Sun. In reality, an electron can be nearly anywhere within the cloud of electrons, although at any instant it is most likely to be somewhere near the center of its shell. Electron shells represent different *energy levels* more than they represent specific distances away from the nucleus.

Finally, atoms are not hard spheres with well-defined edges. Atoms are more empty space than matter, and their edges are defined by how far out the outermost electrons travel away from the nucleus. We often show atoms as hard-edged spheres because it is much easier to see relationships between solid objects than between fuzzy, partially overlapping clouds. Such depictions, however, are incorrect in detail.

Before You Leave This Page

☑ Describe the relationship between a mineral and the elements of which it is composed.

☑ Explain or sketch the structure of an atom, including its main particles.

☑ Sketch the general shape of the Periodic Table and explain the significance of its rows and columns.

2.10

2.11 How Do Atoms Bond Together?

ELEMENTS COMBINE TO FORM MINERALS. The kind of bond that develops to hold two atoms together depends on the way in which the two atoms borrow, donate, or share electrons. The Periodic Table helps explain whether an element will gain, lose, or share electrons, and therefore how the element will bond with other atoms to form a mineral.

A How Do Atoms Bond Together?

Two atoms bond together by sharing, donating, or borrowing electrons from their outermost orbital shells. This process, called *chemical bonding*, can be illustrated by the ways people in a room might share or transfer money.

02.11.a1–4

Two people in the same room could both hold, or *share*, a single bill. The people have to stay close together to hold the bill, so sharing a bill forms a strong bond.

One person could loan one (or more) bills to another person. A bond formed by *loaning* money is not as strong as one formed by sharing money, but it is still a bond.

If we had many bills, we could keep passing (transferring) them around so that each person had a few. People and bills could move around freely while staying in the room.

If we stack the bills they may cling together a little, but we can easily pull the bills apart. This type of bond, where the bills are neither shared nor loaned, is very weak.

Covalent Bond

When two atoms *share* an electron, the bond is a *covalent bond*. The figure below shows a covalent bond between hydrogen and oxygen. Together the two hydrogen and one oxygen atom make the compound H_2O, or water.

Ionic Bond

An *ionic bond* forms because of the attraction of two oppositely charged ions, such as when one atom *loans* one or more electrons to another atom.

Metallic Bond

Electrons in a *metallic bond* are *shared widely* by many atoms. This holds the material together in a nonrigid way, which is why many metals are pliable (bendable).

Intermolecular Force

Several types of weak bonds can attract a molecule (a combination of atoms) to another molecule. Such bonds are relatively weak, such as those that connect sheets in micas and clays.

02.11.a5

Electron
Nucleus
Orbital Shells
Shared Electrons

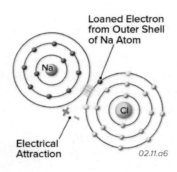

Loaned Electron from Outer Shell of Na Atom
Na
Cl
Electrical Attraction
02.11.a6

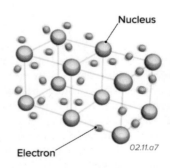

Nucleus
Electron
02.11.a7

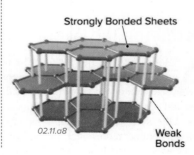

Strongly Bonded Sheets
02.11.a8
Weak Bonds

Bond Type	Strength	Mechanism	Mineral Examples
Covalent Bond	Strongest	*Sharing* of electrons between atoms	Diamond, bonds within sheets of graphite
Ionic Bond	Moderate	*Transfer (loaning)* of electrons from one atom to another, resulting in attraction between opposite charges	Halite, fluorite
Metallic Bond	Low	Widespread *sharing* of electrons among many atoms	Gold, copper
Intermolecular Force	Lowest	Attraction due to polarity of molecules, which are bonded combinations of atoms	Water ice and bonds between sheets of mica, clay, and graphite

B How Does the Periodic Table Reflect These Bond Types?

The Periodic Table provides general guidance about which type of bond two elements are likely to form. If one element is much better at attracting and holding onto electrons than the other element, the bond will be *ionic*. When two elements have nearly equal ability to attract and hold electrons, they *share* the electrons and the bond is *covalent*. A measure of this ability to attract electrons is called *electronegativity*. Electronegativity changes predictably across the Periodic Table, as shown by the relative height of each element in the figure below. A greater height indicates a greater ability to pull electrons (higher electronegativity).

1. Elements on the left side have only one or two electrons in their outer shell. So, except for hydrogen, they easily give up those electrons and do not have a strong ability to attract other electrons. When they combine with elements on the right side (group VII), which do exert a strong attraction, an *ionic* bond forms. Halite (NaCl) forms when sodium (Na) transfers an electron to chlorine (Cl), resulting in two oppositely charged ions that attract each other. Halite is one of the minerals that is nearly 100% ionically bonded.

02.11.b2

2. A water molecule (H_2O) forms when two hydrogen (H) atoms share electrons with one oxygen (O) atom. The electronegativities of hydrogen and oxygen are about the same, so the bond is covalent. When frozen, water molecules combine in an orderly structure to form the mineral *ice*.

02.11.b3

3. Elements on the right side, except for the very last column, have a strong ability to attract electrons because their outer shells have one or two vacancies available for electrons. The mineral fluorite forms when a calcium atom transfers electrons to two fluorine atoms, and the bonds are ionic. Elements in the last column to the right have completely filled shells and so do not attract electrons.

02.11.b4

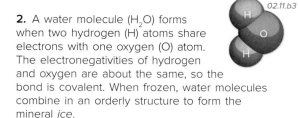
02.11.b1

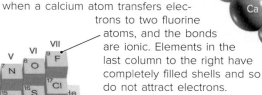
02.11.b5

4. Some elements have similar abilities to attract electrons (similar heights in this table) and can join via covalent bonds. A silicon atom can share electrons with four oxygen atoms, forming strong, covalent bonds in an SiO_4 tetrahedron.

6. Metals in the center of the table, like gold (Au) and copper (Cu), share electrons freely among their atoms to form metallic bonds. This makes many of these elements, especially copper, good conductors of electricity.

02.11.b7

5. In calcite ($CaCO_3$), bonds between the carbon atom and three surrounding oxygens are covalent, but bonds between the calcium atom and the carbonate group (CO_3) are ionic because the calcium atom transfers two electrons to the carbonate group.

02.11.b6

C How Do Bonds Explain the Difference Between Diamond and Graphite?

The type and arrangement of bonds that hold a mineral together control many of the mineral's physical properties, like hardness. The minerals *diamond* and *graphite* both consist solely of carbon. Diamond is a very hard crystal, whereas graphite is soft and feels greasy. Why are these two minerals so different?

Diamond contains strong covalent carbon-to-carbon bonds that form a strong interconnected framework. Diamond is the hardest naturally occurring mineral, with a Mohs hardness of 10, and is suitable for cutting and polishing into sparkling gemstones.

Graphite has several kinds of bonds. It contains sheets that are covalently bonded and therefore strong, but weak intermolecular bonds hold adjacent sheets together. These weak bonds allow sheets to slide apart easily, making graphite feel slippery and soft enough to use as the "lead" in pencils and as an industrial lubricant.

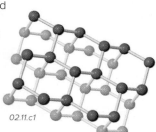
02.11.c1

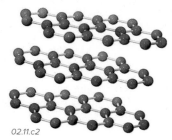
02.11.c2

Before You Leave This Page

- ☑ Explain the different types of bonds and how electrons cause each type.

- ☑ Explain how the Periodic Table helps predict which kind of bond will form, and provide a mineral example for each kind of bond.

- ☑ Explain how differences in bonds cause diamond and graphite to have very different properties.

2.11

2.12 How Do Chemical Reactions Help Minerals Grow or Dissolve?

ELEMENTS COMBINE TO FORM MINERALS under various temperatures and pressures, but minerals also can be destroyed under a wide variety of conditions. The types of chemical bonds in a mineral determine whether the mineral grows, dissolves, or is unaffected by its physical and chemical environment. Water is often a key factor, so here we discuss how minerals crystallize or dissolve in water.

A What Are the Properties of Water, and How Does It Dissolve Some Solid Materials?

H_2O can be a gas, a liquid, or a solid. It has some unusual properties, setting it apart from many other chemical compounds. A water molecule consists of two hydrogen atoms covalently bonded to one oxygen atom.

The Water Molecule and Its Polarity

1. In a water molecule, the two hydrogen atoms are on one side of the oxygen. The water molecule has no overall charge (because the 10 protons are balanced by 10 electrons), but the electrons are not evenly distributed.

2. Oxygen more strongly attracts electrons, so the shared electrons spend more time around the oxygen.

3. The molecule therefore has a polarity, with a negative side and two positive ends on the other side (▼). This polarity causes water molecules to be attracted to charged atoms (*ions*).

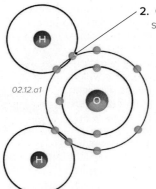
02.12.a1

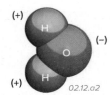
02.12.a2

Hydrogen Bonding

4. Water molecules are attracted to each other as well as to ions. In water, a weak bond called a *hydrogen bond* forms between one molecule's hydrogen atom and another molecule's oxygen atom. The hydrogen bond is responsible for some of water's unique properties (e.g., viscosity, surface tension, etc.).

5. This bond forms as hydrogen is attracted in two different directions. The covalent bond *inside* the water molecule pulls the hydrogen inward, keeping the molecule together.

6. Hydrogen bonds from another water molecule pull the atoms outward, causing a weak bond to form between the molecules.

02.12.a3

The Ability of Water to Dissolve Minerals

7. We all know that water can dissolve some solid materials, like salt, but how does water do this? When we add salt crystals, like the one in the center of this block, to a pot of water, each crystal is surrounded by water molecules. In halite, sodium atoms have loaned an electron and so have a positive charge (Na^+). Such positively charged ions are called *cations*. Chlorine has gained an electron and so has a negative charge (Cl^-). Such negatively charged ions are called *anions*.

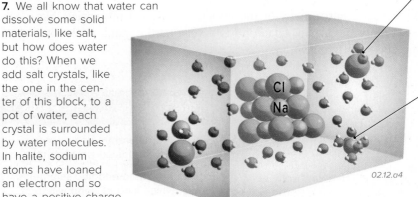
02.12.a4

8. The negatively charged chlorine anion is attracted to the positively charged (H) end of the water molecule. If this attraction is strong enough, it can pull the chlorine away from the halite crystal and into the water.

9. In a similar manner, sodium in halite is a positively charged cation (Na^+) and so is attracted to the negative side of any adjacent water molecule. This attraction can pull the sodium ion away from the halite crystal and into the water.

10. Once dissolved in water, the positively charged sodium cation (Na^+) will be surrounded by the negative sides of water molecules. The encircling water molecule may prevent sodium from rejoining the halite crystal.

11. The negatively charged chlorine anion (Cl^-) is likewise surrounded by the positive side of the encircling water molecules. It is water's polarity that enables it to dissolve some solid materials and makes it a good solvent and cleaning agent.

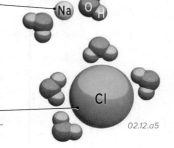
02.12.a5

B How Do Minerals Precipitate from Water?

To understand how a mineral can grow, we need to think about the environment where it is growing. Minerals on Earth are always surrounded by other materials, perhaps other minerals, magma, water, or air. Minerals contain specific atoms, and a crystal needs a nearby source of these atoms in order to grow. A growing crystal of halite (NaCl) needs additional sodium and chlorine ions. So the environment in which crystallization occurs is important, because it places physical constraints and chemical constraints on crystal growth.

1. What happens if salty water evaporates? Over time, salt crystals precipitate (grow) on the bottom and sides of the container. How did the salt crystals form, and where did the material come from?

2. If there is lots of water, the few sodium (Na⁺) ions and chloride (Cl⁻) ions are kept apart by the water molecules and the constant movement of the ions. As a result, the ions rarely come into direct contact with each other and so remain dissolved in the water.

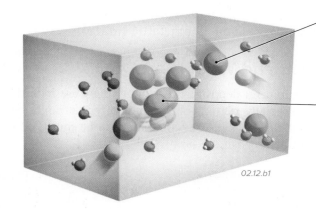

02.12.b1

3. As the water evaporates into the air, sodium and chlorine are left behind and so become more concentrated in the remaining water. As a result, the two ions begin to find each other and begin to bond.

4. The resulting NaCl pairs combine and, to keep local charges balanced, begin to organize into an ordered structure with alternating cations and anions, forming a salt crystal.

Writing a Chemical Reaction for Salt in Water

5. Dissolving halite in water, or precipitating halite crystals from water, like any chemical reaction, can be expressed with an equation that describes what is happening. For this reaction, we put halite (NaCl) on one side of the equation and the sodium (Na⁺) and chlorine (Cl⁻) ions on the other side.

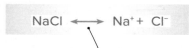

$$NaCl \longleftrightarrow Na^+ + Cl^-$$

6. Because this reaction can go either way (halite may dissolve or precipitate), we link the two sides of the equation with a two-headed arrow.

7. Whether halite precipitates or dissolves depends on the proportion of halite (NaCl) versus Na⁺ and Cl⁻ ions in the water. If we add more NaCl, the reaction indicates that halite crystals will dissolve into Na⁺ and Cl⁻ ions, unless the water solution is already saturated with the two ions. If there is too much dissolved Na⁺ and Cl⁻ in the solution, halite will not dissolve and can instead crystallize. Temperature is also important. Most minerals dissolve more easily in hot water than in cold water.

The Properties of Ice

Depending on pressure and temperature, water molecules can exist as a gas, liquid, or solid. The solid form of water (H₂O) is ice, which is a mineral with an orderly crystalline structure. Ice is clear when pure but generally is cloudy or blue because of trapped air bubbles. It is a soft mineral, but it can erode into landscapes if it carries harder rocks, as it does in glaciers.

When water freezes, the weak hydrogen bonds form as the molecules rearrange to form a crystal. This rearrangement of bonds results in water molecules that are farther away from each other in ice than they are in liquid water. As a result, ice is less dense than water and so floats in water, whether it is ice cubes in our glass or icebergs adrift in the sea (▶). Water is one of the few compounds that has a solid form that is less dense than the liquid form. Most other materials, including most rocks, become less dense as they melt, so the liquid form is less dense than the solid form.

The lower density of ice has many important implications for landscapes and life. Water that freezes and expands can pry apart rocks and soil, loosening pieces that can be transported away. If ice was more dense than water, ice that formed on the surface of a lake would sink to the bottom, allowing new ice to form on the surface. This process would repeat until the lake was frozen solid, from top to bottom. Few creatures could survive freezing of all the water in a lake. Our world would be very different if ice was more dense than water and sank rather than floated.

02.12.t1 Antarctica

Before You Leave This Page

☑ Sketch a water molecule and illustrate why it has polarity.

☑ Describe the properties of water that are attributable to polarity and those that are attributable to hydrogen bonding.

☑ Describe how halite dissolves and crystallizes in water.

☑ Describe why ice is less dense than water and why this is important.

2.12

2.13 What Are Mineral Deposits and How Do They Form?

PHYSICAL AND CHEMICAL PROCESSES concentrate and disseminate elements and minerals, causing rocks to have a higher content of some elements and minerals, and a lower content of others. If a volume of rock is enriched enough in an element or mineral to be potentially valuable, we call it a *mineral deposit*. Materials extracted from mineral deposits provide the very foundation for our modern world. Some mineral deposits are explored for their content of metallic elements, like copper, and some for their nonmetallic minerals.

A What Is a Mineral Deposit and What Is an Ore?

Most rocks are not considered to be mineral deposits, even though they are indeed composed of minerals. Instead, the term *mineral deposit* means the rock is especially rich in some commodity that might be valuable, and such rocks are said to be *mineralized*. If a mineral deposit contains enough of a commodity to be mined at a profit, it is an *ore deposit*, and the valuable rocks or other materials in that deposit are *ore*.

An outcrop of plain white quartz is not a mineral deposit unless it contains flecks of gold. If rich enough in gold, like the fist-sized, gold-rich sample in the inset photograph, the piece of quartz may also be ore.

02.13.a1 Murchison, South Africa

02.13.a2 Malartic, Quebec, Canada

02.13.a3 Hemlo Mine, Ontario, Canada

Ore can be conspicuous, like this rock that contains shiny, brass-colored sulfide minerals. Some ore is much more subtle, being enriched in some element but otherwise looking like a typical igneous, metamorphic, or sedimentary rock.

B What Determines Whether a Mineral or Rock Is an Ore?

Many factors, some of them nongeologic, determine whether a rock or other material is considered ore. These include concentration of the commodity (valuable material) in the rock, how easily the commodity is extracted from the rock, the proximity to markets, and the economics that controls prices, especially supply and demand.

02.13.b1 Kidd Creek, Ontario, Canada

02.13.b2 Morenci, AZ

Grade of Ore—The percentage or concentration of the valuable commodity in a rock is called the *grade*. A rock that is very rich in the commodity, like this sulfide-rich copper ore (◄), is *high grade;* the opposite is *low grade.*

Type of Ore—A commodity can occur in different types of minerals, such as copper in these blue-green copper-oxide minerals (►). In the case of copper, it is cheaper to extract copper from oxide minerals than from the copper sulfide minerals shown in the left photograph.

Size and Depth of Mineral Deposit—The size of a deposit determines if it is worth mining, because of the large cost of setting up a mining operation. If a deposit is small, the investment in equipment may not be worthwhile unless the ore is very rich. A shallow mineral deposit is cheaper to mine than a deeper one. A large, open-pit mine (◄) is more economical to operate than a small mine or a deep, underground one.

Location of Deposit—A deposit that is close to markets and to infrastructure, such as railroads and electrical lines, will be more economical than one that is far from civilization or in an environmentally sensitive place. Economic factors, such as the price for which the commodity can be sold, and political factors, especially whether the area has a stable government, can determine whether a deposit can be mined (►).

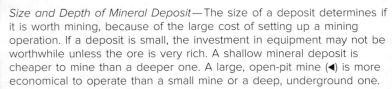
02.13.b3 Morenci Mine, AZ

02.13.b4 Bingham Canyon Mine, UT

What Are the Main Types of Mineral Deposits?

There are a number of ways in which to classify mineral deposits, such as by the kinds of minerals and elements they contain or by the environment in which the deposits form. Many geologic processes can concentrate minerals or chemical elements enough to make a mineral deposit, so classifications based on the environment of formation can be very detailed. From a general perspective, we subdivide mineral deposits based on the content of minerals and chemical elements. In this system, there are four general types of mineral deposits—*precious metals*, *base metals*, *industrial rocks and minerals*, and *gemstones*.

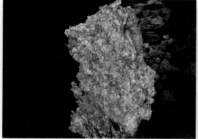

02.13.c1 Large Gold Nugget, Murphys, CA

Precious Metals—As the name implies, precious metal deposits contain a high-value metal, such as gold, silver, and platinum. These metals are widely used for industrial and monetary purposes in addition to jewelry, but they occur only in relatively minor concentrations in Earth's crust and so are high-cost materials.

02.13.c3 Gypsum, Gypsum, CO

Industrial Rocks and Minerals—Some minerals and rocks are referred to as being "industrial" because we use them in many industrial processes, such as making cement, concrete, and wallboard. Such rocks and minerals include limestone (cement), gypsum (wallboard), salt, and large volumes of sand and gravel (concrete).

02.13.c2 Iron Formation, Jasper Knob, Ishpeming, MI

Base Metals—Base metals are much more common than precious metals and, unlike gold, most tarnish in air. Base metals include iron, nickel, copper, lead, zinc, and aluminum. They are fundamental to our daily lives, being the chemical elements in steel, automobiles, cans, wiring, plumbing fixtures, and coins.

02.13.c4 Peridot, Peridot Mesa, AZ

Gemstones—We collect some minerals, such as gemstones, for their aesthetic properties. Precious gemstones include diamond, ruby, sapphire, and emerald. Some minerals and mineral-like materials are more common and less expensive, and are therefore called semiprecious gemstones; these include amethyst, garnet, peridot, and topaz.

Natural Resources and Our Modern Society

We use many different types of natural resources, including energy resources, mineral resources, soil resources, water resources, biologic resources, and others. Energy resources include oil and gas, as well as coal, nuclear fuels, and energy derived from dams, wind, and the Sun. These resources are not equally distributed in every part of the world. Some regions are rich in energy resources, whereas others are rich in mineral resources. Some areas have neither. Variations in earth history from region to region lead to differences in the abundance and kinds of energy and mineral resources found in different places.

Our society, and we as individuals, use large quantities of energy, mineral, and water resources. Some resources, such as solar power, are *renewable*, meaning that they are somehow replenished by natural processes. Other resources, like oil, are *nonrenewable*, meaning there is a finite supply, which we diminish as we consume that resource. Most people are unaware of the amounts of energy and mineral resources, mostly nonrenewable, consumed in the United States per person. These amounts include materials used to construct roads, gypsum in wallboard, and copper in wiring. As summarized in the table included here, the National Mining Association estimates that average consumption in the United States in 2013 was 38,500 pounds of minerals per person! Nearly 40% of this amount was sand, gravel, and stone. We also consume large amounts of coal, natural gas, and petroleum used for fuels and to make plastics and many other items.

Finding mineral resources is essential to support our modern society, so geoscientists study many aspects of resources, from their general characteristics to the processes by which they form. Certain processes form oil, and others form copper deposits. For many important resources, we are at the mercy of events that happened millions of years ago. The political and economic systems of the world must function around this reality.

Material	Per Capita Consumption (pounds)
Sand, Gravel, Stone	15,093
Natural Gas	8,643
Petroleum Products	6,542
Coal	5,842
Cement	573
Salt	349
Iron Ore	314
Phosphate Rock	220
Clays	154

Before You Leave This Page

✓ Explain the meaning of mineral deposit, mineralization, and ore.

✓ Summarize geologic and non-geologic factors that determine whether a mineralized body can be mined.

✓ Summarize the four types of mineral deposits and how we use them.

How Are Minerals Used in Society?

PEOPLE HAVE ALWAYS USED MINERALS, and minerals have become essential to our modern society. We need minerals to build our houses, cars, roadways, and buildings. Sometimes we use minerals in their natural form. Other times, we extract key elements from the minerals and use them for manufacturing. On average, each American uses, either directly or indirectly, 22 metric tons of minerals and rock per year.

A How Are Minerals Used for Their Chemical Components?

One of the major uses of minerals is as a source of elements and compounds that we then use to manufacture other products. We mine minerals and process them to extract the required elements or compounds. The resulting materials are used in the manufacture of materials, such as glass, metals, and computers, or are then combined with other elements to create new useful compounds.

1. *Iron*—This element is mostly mined from iron formations (▶) containing *hematite* and *magnetite*—both minerals of which are iron oxides. Iron is the main ingredient in steel, which is used in many products, from kitchen utensils and appliances to automobiles, construction equipment, skyscrapers, railroads, and ships.

02.14.a1 Jasper Knob, Ishpeming, MI

02.14.a4 Kidd Creek Mine, Ontario, Canada

4. *Copper*—Copper conducts electricity and so is used for electrical wires in telephones, computers, automobiles, and nearly everything that is electric. It is also used in brass and bronze. Most copper comes from copper-sulfide minerals (◀) and from various blue-green copper minerals.

2. *Sodium and Halite*—Sodium has some uses as a pure element, and it is extracted from *halite* (NaCl) and from the mineral *trona*, a sodium carbonate. Halite is used in human and animal diets, as a highway deicer, and in water softeners. It and trona are used to make soaps, metals, and many household items. Trona is important for manufacturing glass.

02.14.a2 Puna de Atacama, Argentina

02.14.a5 Florida

5. *Phosphorus*—We use phosphorus in fertilizers, soft drinks, and consumer devices, including some televisions. Many large phosphate mineral deposits form by accumulations of marine sediments. The main phosphorus ore mineral is *apatite*, a calcium phosphate mineral similar in composition to human teeth.

3. *Silicon*—The element silicon, used to create computer chips (▶) and solar panels, is mostly derived from *quartz*, a very common silicon-oxide mineral. Although quartz is present in granite and other rocks, it is most concentrated in certain sandstones, loose sands, and quartz veins, so that is where most silicon-rich materials are mined.

02.14.a3

02.14.a6

6. *Calcium*—*Calcite*, the most common carbonate mineral, is the chief source of calcium and is used to help construct many parts of our infrastructure. Calcite is processed into the main ingredient in cement, which is used in roadways, sidewalks, bridges, airports, large buildings, the foundations of homes, and backyard patios (◀).

Minerals in Your Medicine Cabinet

Minerals are used to make many items you find in a house, and you may be surprised at some of the unexpected places minerals show up, such as in your medicine cabinet. Most toothpastes contain calcite as the "scouring agent," and they also contain fluoride, derived from the mineral *fluorite*, and various sodium compounds derived from *trona*. The major ingredient in many antacids is calcium carbonate, in some cases derived from ground-up calcite. The abrasive material on nonmetallic nail files, or emery boards, is finely ground *garnet* or the mineral *corundum,* which also occurs as rubies and sapphires. Most makeup consists of *clay minerals* but may also include small flakes of mica as a glitter. Foot and body powders, also called talcum powder, may contain *kaolinite* (a clay mineral) and *talc*, the softest mineral in the Mohs Hardness Scale. Finally, the medicine cabinet is composed of steel, derived from *hematite* and *magnetite*, or aluminum, derived from fine-grained, clay-like materials. The mirror of the cabinet consists mostly of glass derived from *quartz* and is coated with a silver compound or some other reflective substance.

B How Are Minerals Used for Their Physical Properties?

In addition to being sources of chemicals, we use many minerals intact because of some special property they possess. Minerals are used because of their color, density, resistance to heat or abrasion, shininess, or the ease with which they can be shaped. We use minerals for ceramics and as fillers to thicken and extend the volume of materials like paint, and we use huge volumes of crushed stone from rocks containing quartz, feldspar, or calcite.

Quartz—Large quantities of quartz are melted and mixed with other materials to make glass windows and glass block (▶). Quartz is used as filler materials in paint, paper, and in some food and vitamin products (listed as silicon dioxide). Synthetically grown quartz crystals are used in halogen bulbs and timing devices.

02.14.b1

02.14.b3

Clay Minerals— We use many clay minerals, all of which are sheet silicates. Clay is used to produce brick, cement, and ceramics (such as bathtubs), as well as tile for roofs (◀), floors, and walls. Large quantities are also used for cat litter and as fillers in paper, paint, and food products.

Feldspars—These very common silicate minerals are used in ceramics, including tile (▶) and china, and in glass-fiber insulation. They are also used in glass production to improve hardness and durability. In the United States, most feldspar is mined from granite and other igneous rocks.

02.14.b2

02.14.b4

Gypsum—Gypsum is a sulfate mineral that mostly forms from the evaporation of water in salty lakes or some inland sea. It is mostly used in construction (◀) for wallboard (sheetrock) and plaster products. Additionally, it is used for cement production, agricultural applications, glass making, and other industrial processes.

C In What Geologic Environments Do Gem-Quality Minerals Form?

Most gems are minerals—very beautiful ones! Some gems are not minerals because they are not natural (e.g., cubic zirconium) or they do not have an ordered crystalline structure. Some organic materials, including pearls and amber, are sometimes considered gems. What environments enable beautiful gem minerals to grow?

Diamonds form deep in the mantle under conditions of high temperature and extremely high pressures. They are brought to the surface through volcanic conduits called *kimberlite pipes*. Diamond is mined in pipes or in sediment eroded from diamond pipes.

Opal shimmers as it moves in light, showing various shades of blue, green, red, and other colors. Opal is not a mineral because it does not have an orderly crystalline structure. It consists of microscopic spheres of silica that include trapped water, giving opals their distinctive spectrum of color. Opal commonly forms in volcanic rocks and some sedimentary rocks that have fractures and other natural openings (called *voids*). Silicon-rich water fills these fractures and voids, and deposits the opal.

Ruby and *sapphire* are both varieties of the mineral *corundum* (aluminum oxide). *Emerald* and *aquamarine* are varieties of the mineral *beryl* (an aluminum silicate). All four gemstones mostly form in *pegmatites*, which are coarse-grained igneous rocks that crystallized from magma containing relatively high amounts of water. Extra water promotes the growth of very large crystals. Ruby, sapphire, emerald, and aquamarine may also form during metamorphism of some sedimentary rocks.

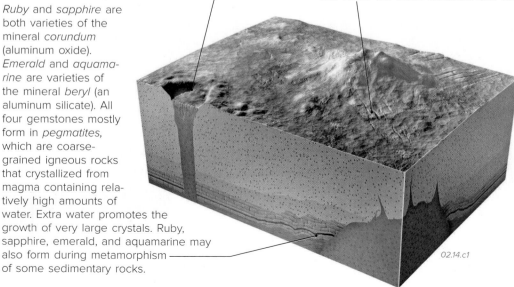
02.14.c1

Before You Leave This Page

✓ Distinguish the two main ways that minerals are used in society.

✓ Describe some chemical uses of common minerals.

✓ Describe how minerals are used in some of the products that are in your medicine cabinet.

✓ Describe some ways we use the physical properties of minerals.

✓ Describe the geologic environments in which some gemstones form.

2.14

2.15 What Minerals Would You Use to Build a House?

MINERALS AND ROCKS ARE USED to make many objects around us. Minerals, rocks, and products derived from them compose our homes, cars, streets, buildings, electrical grid, and water-supply system. If something is not grown, it comes from rocks, minerals, or petroleum. In this exercise, you will decide what minerals are used in materials to build the important parts of a house.

Goals of This Exercise:

- Make some observations about minerals based on their appearance in a photograph or from samples provided by your instructor.
- Identify minerals based on their appearance and diagnostic properties.
- Determine, based on each mineral's characteristics and how it is commonly used, which mineral(s) to use for each part of a house.

A Describe and Identify These Minerals

Examine each mineral in the photographs below or from samples provided by your instructor. For each mineral, make observations, such as crystal form, luster, color, and cleavage. Write these observations on the accompanying worksheet or a sheet of paper. Then, read the accompanying text blocks that provide additional information about each mineral. If you have access to mineral samples, perform tests, such as determining hardness, on each mineral. For each mineral, the worksheet contains additional important information that will help in identification.

02.15.a1

1. These six-sided crystals have a hardness of 7 and a conchoidal fracture instead of cleavage. The mineral does not effervesce.

02.15.a2

2. This mineral is partially transparent, has a hardness of 3, cleaves into rhombs, and effervesces with dilute HCl.

02.15.a3

3. This mineral is very soft, feels sticky when wet, and does not effervesce. It contains very fine material. It is not talc or graphite.

4. Each of these spherical masses consists of a number of intergrown crystals of a cream-colored to partially transparent mineral. The mineral can be scratched with a fingernail and does not effervesce (▶).

02.15.a4

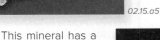

02.15.a5

5. This mineral (◀) has one direction of cleavage and flakes into thin sheets. It is nonmagnetic and does not effervesce. When held up to the light, thin sheets are partially transparent and have a silvery-gray color.

02.15.a6

Copper Carbonate and Other Copper Minerals

Native Copper

6. These blue-green and copper-colored minerals (◀) contain copper. They include copper-carbonate minerals, such as malachite (green) and azurite (blue). The metallic material is native copper. These minerals were not discussed in detail.

7. This mineral has a metallic luster and a distinctive red streak. It is nonmagnetic and in some samples has a reddish tint (▶).

02.15.a7

B   Devise Ways to Build a House Using Minerals and Mineral Products

This illustration shows parts of a house for which you need to find a mineral or mineral-derived product. Using the minerals that you identified in part A, along with information about the uses of minerals in the chapter and in the worksheet, consider options for which mineral or mineral product you will use to construct different parts of the house. Identify the mineral by name and list the properties this mineral had that were useful for the house.

Roof—A roof is a barrier to rain and snow. Some type of mineral product is used to cover the plywood sheets on the roof.

Mineral Name and Useful Mineral Properties:

_____

_____

Insulation—To keep the house at a comfortable temperature, a material that conducts heat slowly is placed outside, inside, or within the exterior walls. Commonly, this material is fiberglass, which is produced by melting a common and inexpensive silicate rock and turning the melt into glass fibers.

Mineral Name and Useful Mineral Properties:

_____

_____

Exterior Walls—The outside walls act as a barrier to rain and snow, and support the roof and the rest of the structure.

Mineral Name and Useful Mineral Properties:

_____

_____

Windows—These let in visible light and other solar energy, and provide visibility to the outside.

Mineral Name and Useful Mineral Properties:

_____

_____

Electrical Wiring—A material that conducts electricity is used for electrical wiring. Most wire is made from a metal because metals are conductive and ductile (can be shaped easily into wire).

Mineral Name and Useful Mineral Properties:

_____

_____

02.15.b1

Cement Slab—Cement is used to make a fairly smooth, stable base for floor tile, wood, or carpet. It is also used as a foundation to support the walls.

Mineral Name and Useful Mineral Properties:

_____

_____

Inside of Walls—Interior walls separate the house into rooms but commonly do not support the structure. They typically have vertical beams (called studs) of a strong material that supports sheets of wallboard that form the actual wall. The covering sheets should be soft enough so that holes can be cut for electrical outlets and switches.

Mineral Name and Useful Mineral Properties:

_____

_____

Plumbing—Metal pipes are commonly used to carry freshwater into the house and from one part of the house to another.

Mineral Name and Useful Mineral Properties:

_____

_____

EARTH'S NATURAL LANDSCAPES HAVE FOUR MAIN COMPONENTS—solid bedrock, loose pieces of rock (sediment), soil, and various types of vegetation. Landscapes vary greatly in the relative amounts of these four components, and each component likewise displays great diversity. Some bedrock is gray, cliff-forming, and massive, whereas other bedrock is reddish, thinly layered, and slope-forming. What are these different materials, and how did each form? Here, we explore various types of earth materials, the processes that form them, and the characteristic landscapes each type of material produces.

This image shows satellite data and topography for the Coast Range and the Fraser River Valley of southwestern British Columbia. The river enters the Strait of Georgia, an inlet of the Pacific Ocean. The large lavender area along the river includes the cities of Vancouver, British Columbia, Bellingham, Washington, and neighboring communities.

Valleys in these mountains were originally carved by glaciers but now contain lakes and steep mountain streams. The streams are eroding into the mountains and transporting sand, pebbles, and other materials toward the Fraser River.

Where is the material carried by streams and glaciers derived, and how is it transported?

Mountains north of Vancouver contain large masses of granite, which formed from the subsurface cooling and solidification of molten rock (magma). Other rocks represent materials that were buried deep in Earth's crust, where they were heated, squeezed, and changed into a new rock.

What types of rocks form at depth, and what types of landscapes do such rocks form?

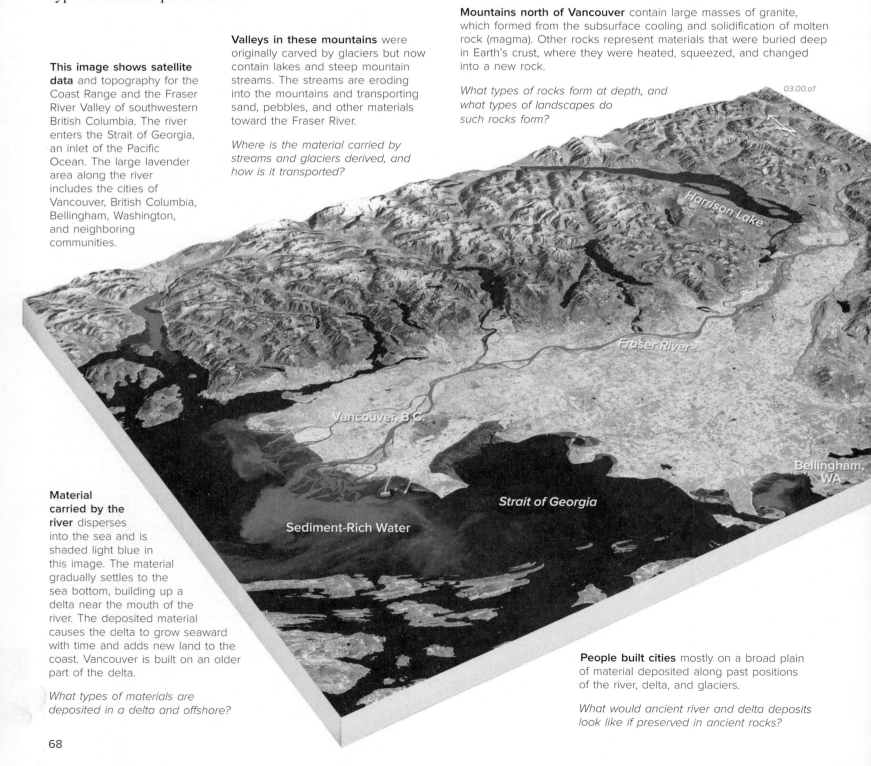

03.00.a1

Harrison Lake

Fraser River

Vancouver, B.C.

Strait of Georgia

Sediment-Rich Water

Bellingham, WA

Material carried by the river disperses into the sea and is shaded light blue in this image. The material gradually settles to the sea bottom, building up a delta near the mouth of the river. The deposited material causes the delta to grow seaward with time and adds new land to the coast. Vancouver is built on an older part of the delta.

What types of materials are deposited in a delta and offshore?

People built cities mostly on a broad plain of material deposited along past positions of the river, delta, and glaciers.

What would ancient river and delta deposits look like if preserved in ancient rocks?

TOPICS IN THIS CHAPTER

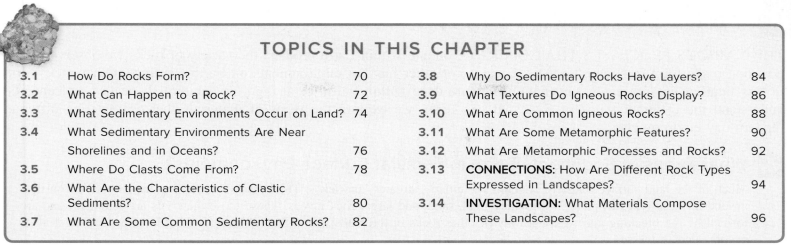

03.00.a2 Seton Lake,, British Columbia, Canada

Mt. Baker

20 km

Mountains encircling Vancouver rise above the surrounding landscapes and shed loose material along their steep flanks (▲).

What types of material are deposited near steep mountain fronts, and how would we recognize ancient deposits formed in similar settings?

Mount Baker (▶), located in northernmost Washington, is an active volcano that last erupted in the 1800s. The volcano emits hot gases and poses a threat to people living nearby and along stream valleys originating from the volcano.

What types of rocks and other materials are formed by volcanoes, and how do we recognize a volcanic landscape from a distance?

03.00.a3 Mount Baker, WA

Sediment, Soil, and Rock

The diversity of earth materials around Vancouver nicely illustrates the different types of materials that exist. We can think of natural landscapes as having four main components. One component is *bedrock*, the rocky materials that are solidly attached to the Earth and its subsurface. The second component consists of loose pieces that formed or accumulated on or near the Earth's surface. Most loose pieces, and the deposits they form, are called *sediment*. Sediment can come in various sizes, from huge boulders to the finest mud. Sediment includes material carried and deposited in streams, glaciers, sand dunes, deltas, beaches, lakes, and oceans.

Soil, the near-surface material in which we grow crops and on which we build most buildings and other human constructions, is the third component and forms from the physical and chemical disintegration of materials more or less in place. Most soil is loose and easily removed, but some can be somewhat consolidated. Sediment is generally transported, but undisturbed soil is not, although it can incorporate sediment. Ash and other materials explosively erupted from volcanoes can be loose, providing material for dangerous volcano-related landslides. Soil generally forms the foundation for *vegetation*, the fourth component of natural landscapes.

Earth materials that are consolidated are called *rock*. Some rocks represent sediment that originally was loose, but later was buried and hardened into rock. Rocks also form when magma cools and solidifies, either on the surface or at depth. Other rocks form deep within the Earth, by high pressures and by temperatures that are high, but not hot enough to melt the rock. Finally, some rocks form by precipitation from hot waters. Each type of sediment, soil, and rock informs us about earth processes and is the starting material for the natural landscapes around us.

3.0

3.1 How Do Rocks Form?

THE VARIOUS PROCESSES THAT OPERATE on and within Earth produce the variety of rocks we observe. Many common rocks form in stream bottoms, beaches, or other familiar environments on Earth's surface. Other rocks form in less familiar environments, under high pressure deep within Earth, or at high temperatures beneath a volcano. To understand the different kinds of rock that can form, we explore the types of materials that characterize different modern-day environments.

A What Types of Sediments Form in Familiar Surface Environments?

Much of the land surface of Earth contains mountains, streams, and lakes. Think back to what you have observed on the ground in these types of places—probably mud, sand, and larger rock pieces. These loose materials are *sediment,* and are formed by the breaking and wearing away of other rocks in the landscape. Although more hidden from us, sediment also occurs beneath the sea.

03.01.a2 Switzerland

03.01.a3 Granite, CO

1. *Glaciers* can incorporate rock debris (◄) into their flowing, icy masses. They carry a wide variety of sediment, from large, angular boulders to fine rock powder. They ultimately deposit the sediment along the edges of the melting ice.

2. *Steep mountain fronts* exhibit large, angular rocks that broke away from bedrock and moved downhill under the influence of gravity. Steep mountains may produce landslides and unstable, rocky slopes covered with angular blocks (►).

3. *Sand dunes* (►) are mostly sand, which has been shifted along the ground by the wind. They contain sand because wind cannot pick up larger particles, but it blows away any smaller particles.

03.01.a4 Namibia

03.01.a1

4. *Beaches* (▼) typically have waves, sand, broken shells, and rounded, well-worn stones. Some beaches are mostly sand, and others are mostly stones.

6. *Stream channels* contain sand, pebbles, and cobbles, whereas low areas beside the channel accumulate silt and clay. Some streams flow into *lakes,* which have a muddy bottom with sand around the lake shore.

5. In deeper water, the *seafloor* consists of mud and the remains of swimming and floating creatures that eventually settle to the bottom. Seafloor closer to the land receives a greater contribution of sand and other sediment derived from the land. Streams and wind are especially effective in delivering this sediment from the land to the sea.

03.01.a5 Naxos, Greece

B What Types of Rocks Form in Hot or Deep Environments?

Some rocks form in environments that are foreign to us and hidden from view, deep within Earth. Others form at very high temperatures associated with volcanic eruptions. Distinct families of rocks result from these rock-forming processes, which include solidification of magma, precipitation of minerals from hot water, or the action of high temperatures and pressures that transform one type of rock into another type of rock.

03.01.b2 Philippines

03.01.b3 Hawaii

3. In many volcanoes, magma flows onto the surface, creating *lava* that flows downhill or piles up around a vent (▶).

1. Explosive volcanoes erupt *volcanic ash* (▲), which can fall back to Earth and blanket the terrain or can rapidly and dangerously surge down the flanks of a volcano.

4. Distinctive rocks form when *hot waters* cool and precipitate minerals. This may occur beneath the surface or on the surface in hot springs (▼).

2. Magma that does not erupt may cool and solidify in a *magma chamber,* forming granite or other rocks at depth. Heat from the magma chamber may bake adjacent rocks, changing them into different kinds of rocks.

Magma

Heating of Rocks

Force

Force

03.01.b1

5. Deep within Earth where temperatures and pressures are high, *forces* can squeeze and deform rocks into new arrangements and into new types of rocks. Under such force, solid rocks slowly flow, shear, and bend. Changing a rock by heat, pressure, or deformation is the process of *metamorphism.*

03.01.b4 Yellowstone NP, WY

Families of Rocks

Diverse environments shown on these pages produce many different types of rocks that, depending on the classification scheme, are grouped into three or four families. To interpret how rocks form, we observe modern environments and note the dominant types of sediment, lava, or other material. We infer that these same types of materials would have been produced in older, prehistoric versions of that environment. By doing this, we use modern examples to interpret ancient rocks and to understand how they formed. In this way, *the present is the key to the past.*

Sedimentary rocks form on Earth's surface, mostly from loose sediment that is deposited by moving water, air, or ice. If loose sediment is buried, it can become consolidated into hard rock over time. Other types of sedimentary rocks form by precipitation of minerals from water or by coral and other organisms that extract material directly from water.

Rocks formed from cooled and solidified magma are *igneous rocks.* These form when volcanoes erupt ash and lava, or when molten rock crystallizes in magma chambers at depth.

Rocks changed by temperatures, pressures, or deformation are *metamorphic rocks.* Metamorphism can change sedimentary or igneous rocks, or even preexisting metamorphic rocks. Finally, rocks that precipitate directly from hot water are *hydrothermal rocks.* Some geoscientists classify these rocks with metamorphic rocks.

Before You Leave This Page

☑ Distinguish the four families of rocks by describing how each type forms.

☑ For each family of rocks, describe two settings where such rocks form and the processes that take place in each setting.

☑ Describe what we mean by "the present is the key to the past" and how it is used to interpret the origin of rocks and sediment.

3.1

3.2 What Can Happen to a Rock?

MANY THINGS CAN HAPPEN TO A ROCK after it forms. It can break apart into sediment or be buried deeply and metamorphosed. If temperatures are high enough, a rock can melt and then solidify to form an igneous rock. Uplift can bring metamorphic and igneous rocks to the surface, where they break down into sediment. Examine the large figure below and think of all the things that can happen to a rock.

1. Weathering

A rock on the surface interacts with sunlight, rain, wind, plants, and animals. As a result, it may be mechanically broken apart or altered by chemical reactions via the process of *weathering*. Weathering creates sediment, which ranges from very fine clay to the large boulders shown here (▼).

2. Erosion and Transport

Rock pieces loosened or dissolved by weathering can be stripped away by *erosion* and moved away from their source. Glaciers, flowing water, wind, and the force of gravity on hillslopes can *transport* eroded material away from where it originated.

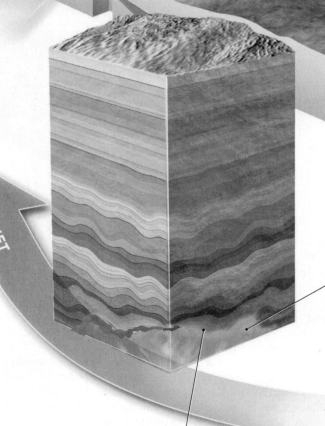

03.02.a2 Central CO

03.02.a1

TRANSPORT

8. Uplift

At any point during its history, a rock may be *uplifted* back to the surface where it is again exposed to weathering. Uplift commonly occurs in mountains, but it can also occur over broad regions that lack mountains.

UPLIFT

6. Melting

A rock exposed to high temperatures may *melt* to produce a magma. Melting usually occurs at great depth, in the lower crust or the mantle.

03.02.a5 Acadia NP, ME

7. Solidification

As magma cools, either at depth or after being erupted onto the surface, it will solidify and harden, a process called *solidification*. If crystals form during solidification, the process is called *crystallization*. Crystallization that occurs by slow cooling of magma at depth can form large, well-formed crystals, such as those displayed by a typical granite (◄). This granite crystallized at depth and was uplifted to the surface much later.

The Life and Times of a Rock — The Rock Cycle

This process, in which a rock may be moved from one place to another or even converted into a new type of rock, is the *rock cycle*. Scottish physician James Hutton first conceived of the rock cycle in the late 1700s as a way to explain the recycling of older rocks into new sediment. Most rocks do not go through the entire cycle, but instead move through only part of the cycle. Importantly, the different steps in the rock cycle can happen in almost any order. Steps are numbered on this page only to guide your reading and follow *possible* sequences of events for a single rock.

Suppose that uplift brings up a rock and exposes it at Earth's surface. Weathering dissolves and breaks up the rock into smaller pieces that can be eroded and transported at least a short distance before being deposited. Under the right conditions, the rock fragments will be buried beneath other sediment or perhaps beneath volcanic rocks that are erupted onto the surface. Many times, however, sediment is not buried, but only weathered, eroded, transported, and deposited again. As an example of this circumstance, imagine a rounded rock in a stream. When the stream currents are strong enough, they pick up and carry the rock downstream, perhaps depositing it within or near the channel, where it may remain for years, centuries, or even millions of years. Later, a flood that is larger than the last one may pick up the rock and transport it farther downstream.

If the rock is buried, it has two possible paths. It can be buried to some depth and then uplifted back to the surface to be weathered, eroded, and transported again. Alternatively, it may be buried so deeply that it is metamorphosed under high temperatures and pressures. Later uplift can bring the metamorphic rock to the surface.

If the rock remains at depth and is heated to even higher temperatures, it can melt. The magma that forms may remain at depth or be erupted onto the surface. In either case, the magma eventually will cool and solidify into an igneous rock. Igneous rocks formed at depth may later be uplifted to the surface or remain at depth, where the rocks can be metamorphosed or even remelted.

A key point to remember is that the rock cycle illustrates the possible things that can happen to a rock. Most rocks do not complete the cycle because of the many paths, interruptions, backtracking, and shortcuts a rock can take. The path a rock takes through the cycle depends on the specific geologic events that happen and the order in which they occur. There are many possible variations in the path a rock can take.

3. Deposition

When transportation energy decreases sufficiently, water, wind, and ice *deposit* their sediment. Sediment carried by streams can reside within or next to the channel, or collect near the stream's mouth. The stream gravels in the photograph below are at rest for now but could be picked up and moved by a large flood. Some sediment rocks are deposited from the *precipitation of ions* from water or by the actions of organisms.

03.02.a3 Tibet

4. Burial and Lithification

Once deposited, sediment can be buried and compacted by the weight of overlying material. Chemicals in groundwater can coat sedimentary grains with minerals and deposit natural cements that bind adjacent grains. The process of sediment turning into rock is *lithification*.

UPLIFT

BURIAL

5. Deformation and Metamorphism

After a rock forms, strong forces can squeeze the rock and fold its layers, a process called *deformation*. If buried deeply enough, a rock can be heated and deformed to produce a *metamorphic* rock. The rock in the photograph to the left began as some other type of rock but was strongly deformed, metamorphosed, and converted into a metamorphic rock.

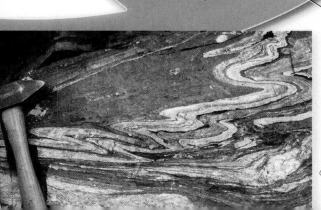

03.02.a4 Kettle Falls, WA

Before You Leave This Page

☑ Sketch a simple version of the rock cycle, labeling and explaining, in your own words, the key processes.

☑ Describe why a rock might not experience the entire rock cycle.

3.2

3.3 What Sedimentary Environments Occur on Land?

EARTH'S LAND INCLUDES DIFFERENT SEDIMENTARY ENVIRONMENTS, each characterized by distinctive kinds of sediment. The environments differ because of variations in topography, local geology, and the amount of available water. Examining these environments, both on land and at sea, helps us interpret modern landscapes, ancient sedimentary rocks, and energy, mineral, and water resources. These two pages identify the most common sedimentary environments on land, which are discussed in detail on following pages. Observe the large figure below and envision what types of materials (sediment) would be present in each setting.

03.03.a2 South Fork, CO

03.03.a3 Denali NP, AK

1. *Mountain* environments are characterized by steep slopes developed on bedrock. Many mountains, but not all, also have high elevation. Erosion is vigorous on such steep slopes and provides abundant sediment, such as the large, angular blocks in this photograph (◄). Once it is produced, the sediment can be transported out of the mountains and into other settings.

03.03.a1

Mountains

Braided River

Sand Dunes

03.03.a4 Death Valley, CA

2. *Streams* in mountains typically have steep gradients and are confined by bedrock canyons. As streams leave the mountains, they can develop a *braided* appearance defined by channels that split apart and rejoin (▲). We use the adjective *fluvial* to refer to the processes and sediment of streams, as in fluvial processes.

3. In dry climates, wind picks up and moves sand grains and finer particles. The moving grains form fields of *sand dunes* (►), composed almost entirely of sand, with almost no other sizes of particles.

4. In high mountains or at high latitudes (close to the North Pole or South Pole), snow can accumulate faster than it is removed by melting or other processes. Over time, the snow becomes compacted into ice, which may flow downhill as a *glacier* (◀). As glaciers move, they erode underlying materials and carry sediment away. The sediment and water are released upon melting of the ice, mostly at the end of the glacier.

03.03.a5 Chamonix, French Alps

5. Streams that flow over gentle terrain commonly meander gracefully (▶). Most streams are flanked by relatively flat land (a floodplain) that may be covered when the stream floods. *Floodplains* of meandering streams are built, layer upon layer, by mud and sand carried by floodwaters.

03.03.a6

6. Where a stream enters a standing body of water, such as an ocean or lake, its current slows, which causes most of its sediment to spread out and be deposited. The sediment piles up and forms a *delta* that builds out into the ocean or lake.

03.03.a7 Lake Superior, Marquette, MI

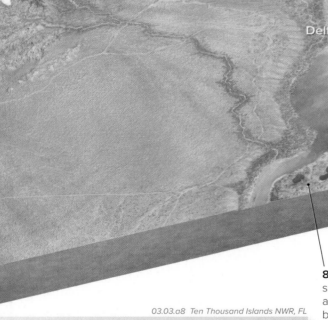

Meandering River

Glaciers

Delta

Lake

Wetlands

03.03.a8 Ten Thousand Islands NWR, FL

8. In very wet environments, such as those adjacent to lakes and in delta areas, the soil may become saturated with water, allowing lakes, swamps, bogs, and ponds to form (◀). Such wetlands typically have abundant water-dependent vegetation, which may become an important component of the sediment. The water may protect the underlying sediment from the atmosphere, limiting the amount of oxidation.

7. Lakes contain a range of environments (▲), from quiet, deep water in the center, to more active water with wind-driven waves along the shoreline. Beaches along the shoreline can be sandy or rocky. Some lakes are always filled with water, but others dry completely when the water evaporates or when it seeps into underlying materials.

Before You Leave This Page

 Sketch or describe the main sedimentary environments on land, and describe some characteristics of each.

3.3

3.4 What Sedimentary Environments Are Near Shorelines and in Oceans?

OCEANS AND THEIR SHORELINES are dynamic environments with wind, waves, and ocean currents transporting sediment eroded from the coastline or brought in from elsewhere. The characteristics of each environment, especially the types of sediment, depend mostly on the proximity to shore, the availability of sediment, and the depth, temperature, and clarity of the water. Examine the large figure below and envision what you would expect in each setting, including the type of sediment that would occur there.

03.04.a2 Carmel, CA

1. *Beaches* are stretches of coastline along which sediment has accumulated (◄). Most beaches consist of sand, pieces of shells, and rounded gravel, cobbles, or boulders. The specifics of the topographic and geologic setting determines which of these components is most abundant. Some shorelines have bedrock all the way to the ocean and so have little or no beach.

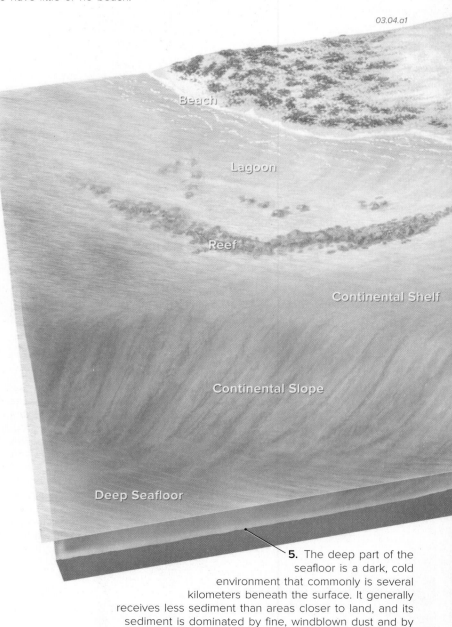
03.04.a1

03.04.a3 Akumal, Mexico

2. The water near the shoreline may be sheltered by offshore reefs or islands. The sheltered water, called a *lagoon* (◄), is commonly shallow, calm, and perhaps warm. The near-shore parts of lagoons contain sand, mud, and stones derived from land, whereas the outer parts may have sand and pieces of coral eroded from a reef.

03.04.a4 Raja Ampat, Indonesia

3. Where ocean water is shallow, warm, and clear, coral and other marine creatures construct *reefs* (◄), which can parallel the coast, encircle islands, or form irregular mounds and platforms. Reefs typically protect the shoreline from the energetic, big waves of the deeper ocean.

4. Away from the shoreline, many landmasses are flanked by continental shelves and slopes consisting of layers of mud, sand, and carbonate minerals. Material from these sites can move down the slope in landslides or in turbulent, flowing masses of sand, mud, and water called *turbidity currents*. The slopes of some continents are incised by branching *submarine canyons* (not shown here) that funnel sediment toward deeper waters.

5. The deep part of the seafloor is a dark, cold environment that commonly is several kilometers beneath the surface. It generally receives less sediment than areas closer to land, and its sediment is dominated by fine, windblown dust and by remains of mostly single-celled marine organisms.

03.04.a5 Namibia

6. Sandy dunes that are inland from beaches are called *coastal dunes* (◄). These dunes commonly form where sand and finer sediment from the beach are blown or washed inland and reshaped by the wind. When strong winds blow onto land, sand can move from the beach to the dunes, and sand can move back toward the beach when winds blow toward the sea or lake.

03.04.a6 Olympic Peninsula, WA

7. Some shorelines include low areas, called *tidal flats*, that are flooded by the seas during high tide but exposed to the air during low tide (►). Most tidal flats are covered by mud and sand or are rocky. Some low parts of the land adjacent to tidal flats can accumulate salt and other *evaporite minerals* as seawater and terrestrial (on-land) waters evaporate under hot, dry conditions.

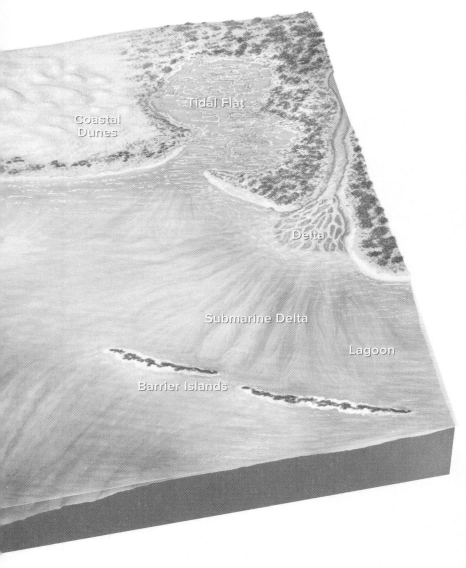

Coastal Dunes

Tidal Flat

Delta

Submarine Delta

Lagoon

Barrier Islands

03.04.a7 Mississippi Delta, LA

8. In addition to the parts of deltas overlapping the shore, *submarine deltas* (▲) extend in some places for tens of kilometers offshore. The muddy or sandy front of the delta may be unstable, and material can slide or tumble down the slope, sending sediment into deeper water.

9. Other accumulations of sand rise above the shallow coastal waters as long, narrow islands, called *barrier islands*. Most barrier islands, such as the one below (▼), are only hundreds of meters wide. The areas between barrier islands and the shoreline are commonly shallow lagoons or saltwater marshes.

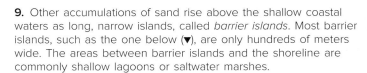

Before You Leave This Page

☑ Sketch and describe the main sedimentary environments in oceanic and near-shore environments.

03.04.a8 Santa Rosa Island, FL

3.5 Where Do Clasts Come From?

SEDIMENTARY ROCKS CONSIST OF MATERIALS that came mostly from other locations. Most sediment consists of pieces of other rocks, or *clasts*, formed by weathering and transport. Other sediment is extracted from water when dissolved material is *precipitated* by chemical reactions or is used by coral and other aquatic creatures for their habitats.

A How Do Physical and Chemical Weathering Produce Sediment?

Most sediment forms by *weathering*, which involves physical and chemical processes that act on rocks at or near Earth's surface, loosening pieces and dissolving some material. Different sizes, shapes, and types of sediment form, depending on the material that is weathered and the environmental conditions during weathering. The processes of physical and chemical weathering are summarized below and discussed in detail later.

Physical Weathering

Physical weathering is the physical breaking apart of rocks that are exposed to the environment. There are four major causes of physical weathering.

1. *Near-Surface Fracturing*—Many processes on or near the surface break rock into smaller pieces. For example, rocks are fractured when they break away from a steep cliff. Fractures also result when rocks expand as they are uplifted toward the surface and are progressively exposed to less pressure.

2. *Frost and Mineral Wedging*—Rocks can be broken as water freezes and expands in fractures. When the ice melts, the fractured pieces may become dislodged from the bedrock. Crystals of salt and other minerals that grow in thin fractures can also cause rocks to break apart.

3. *Thermal Expansion*—Rocks are heated by wildfires and by the sun during the day. As rocks heat up, they expand, often irregularly, and may crack. This process probably plays a relatively minor role in weathering, and geoscientists currently debate its importance.

4. *Biological Activity*—Roots can grow downward into fractures and pry rocks apart as the root diameter increases. Burrowing animals can transport rock and soil from depth and move it to the surface where it is exposed to the elements, weathered, and eroded.

03.05.a1

Chemical Weathering

Chemical weathering includes several types of chemical reactions that affect a rock by breaking down minerals, which causes new minerals to form, or by removing soluble material from the rock. Chemical weathering attacks both solid rock and loose rock fragments, producing loose grains and other pieces, a covering of clay-rich soil, and ions in solution.

1. *Dissolution*—Some minerals are soluble in water, especially the weakly acidic waters that are common in nature. These minerals, along with the rocks, sediment, and soil that contain them, can dissolve. The dissolved material may be carried away in streams or groundwater, or used locally by plants.

2. *Oxidation*—Some minerals, especially those containing iron, are unstable when exposed to Earth's atmosphere. These minerals can combine with oxygen to form oxide minerals, such as iron oxides, which compose the reddish and yellowish material that forms when metal rusts.

3. *Hydrolysis*—When silicate minerals are exposed to water, especially water that is somewhat acidic, the water reacts chemically with the minerals. This process commonly converts the original materials to clay minerals and produces leftover dissolved material that is carried away by the water. Hydrolysis is responsible for the formation of many clay-rich soils.

4. *Biological Reactions*—Decaying plants produce acids that can attack rocks, and some bacteria consume certain parts of rocks. These biological processes cause minerals to break down into their constituent elements.

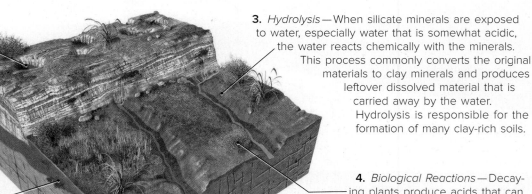

03.05.a2

B   Why Are There So Many Different Products of Weathering?

Physical and chemical weathering affect various starting materials with different compositions, grain size, and solubility. The two kinds of weathering work with each other, but one or the other may dominate, depending on the climate and other conditions. The interplay results in various sizes and compositions of sediment.

| Type of Material | Different Parts of Material | Importance of Fracturing |

03.05.b1 San Juan River, UT

03.05.b2 South Park, CO

03.05.b3 Echo Cliffs, AZ

Rocks, sediment, and soils exposed on Earth's surface have various compositions, which affects how they respond to physical and chemical processes. Materials that are soluble in water and weak acids commonly weather to a pitted or grooved appearance, as displayed by the weathered limestone above. The water dissolves the soluble material and carries it away as dissolved ions, which can later be precipitated in minerals. Other rocks, especially quartz-rich ones, are much less soluble during weathering.

Many rocks, sediments, and soils contain more than one mineral, and each mineral reacts differently to weathering. Pink and cream-colored feldspar crystals in the weathered granite shown here chemically weather to clay minerals or physically weather to sand grains. In general, quartz crystals physically weather to sand grains, whereas dark, mafic minerals chemically weather into clay and iron oxides. As shown here, the mineralogy, grain size, texture, and other aspects of a rock can affect how it weathers.

Rocks that are fractured are weaker and more easily weathered than rocks that are intact. Fracturing increases the amount of surface area that is exposed to the environment. Fractures permit water, air, and organisms to invade the rock, which causes more chemical and physical weathering. Parts of a rock unit that are more fractured will weather faster than less fractured parts. As a result, highly fractured areas tend to form low parts of the topography, or they may form linear notches across ridges and hills.

C   How Do Transportation and Erosion Affect Sediment?

Once weathering has loosened pieces of rock, the pieces can be transported by streams, glaciers, waves, wind, and other forces. During transportation, larger clasts are broken and abraded (scraped off) to produce smaller ones.

03.05.c1 San Juan River, UT

03.05.c2 Westwater Canyon, UT

Silt, sand, and larger clasts carried in water can cause abrasion of other clasts in a channel or of bedrock along the channel (◀). This process is akin to sandpapering, and so it smooths rough edges, scours pits and other recesses into bedrock, and removes small pieces of bedrock that become additional sediment. Note how the lower rocks in this photograph have been smoothed and sculpted more than the upper rocks.

As large clasts, such as boulders and cobbles (▲), are transported, they can break when they collide with, or grind against, other large clasts. Through this process, boulders can become rounded cobbles, and cobbles can break down into smaller pebbles. Some clasts end up as sand or even smaller grains. All of those sizes are present in the stream gravels shown above.

Before You Leave This Page

☑ Describe the main processes of physical and chemical weathering.

☑ Describe how the type of material and degree of fracturing influence the type of sediment that results.

☑ Describe how rocks can be broken during transport.

3.5

What Are the Characteristics of Clastic Sediments?

SEDIMENT CONSISTS OF LOOSE FRAGMENTS of rocks and minerals, or *clasts*. When *clastic* sediment becomes sedimentary rock, the name assigned to the rock depends on the size and shape of the clasts. Other characteristics of the clasts, such as sorting of clasts, can be used to further describe the resulting rock.

A How Are Clastic Sediments Classified?

We primarily use the sizes of clasts to classify loose particles of sediment and the resulting sedimentary rocks into which they are lithified. In addition to clast size, we use the shapes of clasts to further characterize sediment.

Sediment Name	Size Range (millimeters)	Particle Name
Gravel	larger than 256 mm	Boulder
	64 to 256 mm	Cobble
	4 to 64 mm	Pebble
	2 to 4 mm	Granule
Sand	1/16 to 2 mm	Sand
Mud	1/256 to 1/16 mm	Silt
	less than 1/256 mm	Clay

Sizes of Clasts

1. The table to the left (◀) shows the familiar names applied to different sizes of sediment.

2. The largest clasts are *boulders*, which are more than 0.25 m long. Note the rock hammer near the center (▶).

3. *Cobbles* are smaller than boulders, being about the size of a softball. The next smaller sediment size is a *pebble*, which has a size that can be held comfortably in one hand. A pebble can be a little longer than 6 cm. This pen rests on a cobble, and the smaller nearby stones are pebbles (▶).

4. *Sand* is smaller than 2 mm, is larger than 1/16 mm and can have a coarse-, medium-, or fine-grain size. *Medium sand* has grains between 1/4 and 1/2 mm in diameter, and *fine sand* has diameters smaller than 1/10 mm. In these photographic enlargements, the left image is coarse sand, the middle one is medium-grained sand, and the right one is fine-grained sand (▶).

03.06.a1 Baja California, Mexico

03.06.a2 Salt River, AZ

03.06.a3–5

03.06.a6 Salt River, AZ

5. *Clay* and *silt* are particles finer than sand; together they compose *mud* (◀). The term *clay* refers both to grain size and to a group of minerals. Besides being very small, clay-sized grains are usually made of clay minerals. Clay minerals commonly feel slippery between fingertips. In contrast, silt, which is generally made of quartz and is coarser than clay, feels gritty.

03.06.a7 Wolf Creek Pass, CO

Shapes of Clasts

6. *Angular clasts* (◀) have sharp corners and edges. They can be blocky, triangular, or shaped like chips or plates with angular edges. Most clasts, when first loosened from bedrock, start with angular shapes.

7. Many clasts have an angular shape with edges and corners that have been partially rounded (▶).

03.06.a8 Fairplay, CO

03.06.a9 San Juan Islands, WA

8. *Rounded clasts* (◀) have smooth, curved surfaces and shapes like eggs, flattened balloons, or objects that are nearly spheres.

Before You Leave These Pages

☑ Sketch and describe how sediments are classified according to size, sorting, and shape of clasts.

☑ Describe how clast transport affects the size, shape, and sorting of clasts.

☑ Explain four factors that influence the type of sediment transported.

Amount of Sorting

03.06.a10 La Sal Mtns., UT

03.06.a11 Ios, Greece

03.06.a12 Namibia

9. *Sorting* describes the size range of clasts in sediment. *Poorly sorted* sediment, like this example, contains a wide range of clast sizes.

10. Many sediments have moderate sorting of clasts, perhaps containing sand and small pebbles, or perhaps silt and clay.

11. *Well-sorted* sediment consists of clasts that all have the same size. Sand on dunes (shown here) is typically well-sorted sand.

B   What Controls the Size, Shape, and Sorting of Clasts?

How do clasts of such different sizes and shapes form? Once they are eroded from a bedrock source, clasts are transported by water, wind, or ice. During transport, the tumbling, collisions, and abrasion may reduce clast size, increase roundness, and sort the clasts by size. The distance that a clast has been transported is therefore a key factor. This loosened and transported clastic material is sometimes called *detritus*, and the resulting deposits can be described with the adjective *detrital*, as in detrital sediments. To most geologists, detrital means the same thing as clastic.

1. Bedrock exposed in mountains or along cliffs breaks off to form blocks the size of boulders and cobbles. Clasts near their source are usually large, angular, and poorly sorted.

2. As boulders and cobbles are transported by streams, their sharp corners break off because they are the most exposed and weakest parts of a clast. The clasts become more rounded and smaller as pieces break off.

3. Far from their bedrock source, the clasts are worn into well-rounded pebbles and sand grains. Stream currents, beach waves, and wind separate clasts by size, eventually producing better-sorted sediments. Coarse materials are slowly left behind, and only the smaller clasts are carried far from the sediment source.

03.06.b1

03.06.b2 San Rafael Swell, UT

03.06.b3 Cascade Creek, CO

4. *Steepness of Slope*—Steep slopes, such as the one shown here (◄), commonly have clasts that are larger, more angular, and more poorly sorted than the clasts observed on gentle (less steep) slopes.

5. *Strength of Current*—Strong, turbulent stream currents and ocean waves can move large clasts, such as these meter-wide boulders, but slow, less turbulent currents move only fine-grained sediment (◄).

6. *Sediment Supply*—A river, beach, or other agent of transport can only move the sediment that is available. Bedrock on this beach provides large boulders that can be worn down into smaller clasts (►).

7. *Agents of Transport*—Wind can carry only sand and finer particles, but streams, glaciers, mudflows, and other agents of transport can pick up and carry large clasts. These dunes consist of nothing but well-sorted sand because wind cannot bring larger clasts to the area, and smaller material is blown away (►).

03.06.b4 Naxos, Greece

03.06.b5 Morocco

3.6

3.7 What Are Some Common Sedimentary Rocks?

SEDIMENTARY ROCKS ARE THE MOST COMMON rock in many regions, forming many of the landscapes we encounter. They form when loose sediment is converted into a rock through the process of *lithification*, which involves compaction by overlying materials and cementation by calcium carbonate or other materials.

A What Are Some Common Clastic Sedimentary Rocks?

Some sedimentary rocks are composed of rock and mineral pieces (clasts) and are called *clastic rocks*. We describe and classify clastic sedimentary rocks based primarily on the sizes of clasts, along with other aspects such as clast roundness. Common clastic sedimentary rocks are shown below and described using these criteria of classification.

Gravel-Sized Clasts

◄ *Conglomerate* has large, rounded clasts (pebbles, cobbles, or boulders) with sand and other fine sediment between the large clasts. This conglomerate has well-rounded pebbles in a matrix of mostly quartz sand.

► *Breccia* is similar to conglomerate except that the clasts are angular. Breccia usually has a jumbled appearance because most has a range of clast sizes, such as boulders in a mud-rich matrix, as shown here.

03.07.a1 San Rafael Swell, UT

03.07.a2 Wickenburg, AZ

Sand-Sized Clasts

◄ *Sandstone* consists of sand-sized grains. It can contain some larger and smaller clasts, but is mostly composed of sand. It generally has better-defined layers than conglomerate or breccia.

Mud-Sized Clasts

► *Shale* consists mostly of fine-grained clasts (i.e., mud), especially very fine-grained clay minerals. The minerals are aligned, so the rock breaks in sheetlike pieces or chips, as displayed here.

03.07.a3 Durango, CO

03.07.a4 Boulder, CO

B How Do Clastic Sediments Become Clastic Sedimentary Rock?

Compaction

As sediment is buried, increasing pressure pushes clasts together, a process called *compaction*. Compaction forces out excess water and causes sediments to lose as much as 40% of their volume. Originally loose sediment becomes more dense and compact.

Sediments near Earth's surface, such as these sand grains, are a loose collection of clasts. The grains rest on one another but do not fit together tightly, so spaces, called *pore spaces*, exist between the grains. Pore spaces are generally filled with air and water.

As sand grains are buried, the weight of overlying sediment and other materials forces the grains closer together. The amount of pore space decreases as air or water is expelled, so the layer loses thickness. In this manner, sediment becomes compacted into sedimentary rock.

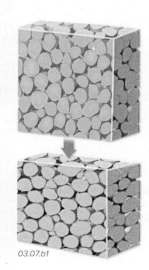

03.07.b1

Cementation

Even after sediment is compacted, adjacent clasts do not fit together perfectly. The remaining pore spaces are commonly filled with water that contains dissolved materials. The dissolved materials can *precipitate* (come out of solution) to form minerals that act as a natural *cement* that holds the sediment together.

When sand grains and other sediment are deposited, abundant pore spaces exist between the grains, even after compaction. If these spaces are interconnected, as they are with sand grains, water can flow slowly through the sediments carrying chemical components into or out of the sediment.

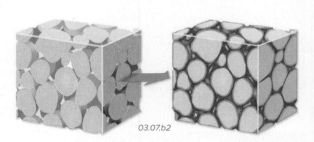

03.07.b2

As the sediment is buried, the water moving through the pore spaces can precipitate minerals in the pore spaces. Materials that accumulate in pore spaces are called *natural cement*, and they help turn the sediment into hard sedimentary rock.

C   What Are Some Common Nonclastic Sedimentary Rocks?

Some sedimentary rocks are not composed of clasts and are therefore *nonclastic rocks*. Processes capable of producing nonclastic sedimentary rocks may be chemical or biological, or have both chemical and biological aspects. The photographs below show common nonclastic sedimentary rocks.

1. *Rock salt* refers to halite (NaCl) or to a rock mostly composed of halite. Halite commonly precipitates as the water that dissolved it evaporates. Most table salt forms in this way.

03.07.c1 Paradox, CO

03.07.c3 San Juan, River, UT

3. *Limestone,* made mostly of calcium carbonate, forms when shells and coral skeletons are cemented together. Many limestones, like the one here, contain fossils, a record of past life.

2. *Gypsum* refers to a specific mineral or a type of rock. Like halite, it mostly forms when water evaporates in tidal flats, narrow seas, and lakes. It is the material in wallboard (sheetrock) in our houses.

03.07.c2 Eagle, CO

03.07.c4 Colorado Springs, CO

4. *Chalk* is a soft, very fine-grained limestone that forms from the accumulation of the calcium carbonate remains of microscopic organisms that float in the sea. Chalk forms the famous White Cliffs of Dover, England.

03.07.c5 Marin Headlands, CA

03.07.c6 Jasper Knob, Ishpeming, MI

03.07.c7 Witbank, South Africa

5. *Chert* is a silica-rich rock that forms in several ways. One way chert forms is in layers from the accumulation and compaction of tiny, silica-rich plankton shells that fall to the ocean bottom. It can also occur as layers and irregular masses within limestone that form from the mixing of water with different chemistries.

6. *Iron formation* is a rock composed of centimeter-thick layers of iron oxide, iron carbonate, and iron silicate minerals, commonly intermixed with very fine-grained quartz. Most iron formations precipitated from seawater early in Earth's history. It is the main source of iron used in steel.

7. *Coal* is formed from wood and other plant parts that have been buried, compacted, and heated enough to drive off most of its water and oxygen. Depending on the amount of heat and pressure, coal can be soft and dull or hard and shiny. There are different kinds of coal, each with a different quality.

Types of Natural Cement

There are four main types of natural cement that hold grains together: calcite, silica, clay minerals, and iron oxides. Other materials, such as gypsum, can function as cement but are less common.

Calcite ($CaCO_3$) is a common cement in sandstone and other sedimentary rocks. It holds grains together moderately well, but it is easily redissolved, so a calcite-cemented sandstone may become friable (crumbly).

Silica (SiO_2) acts as a cement in some sandstone and other sedimentary rocks. It forms a strong cement that can tightly bind grains, forming a tough, resistant rock.

Clay minerals can cement together larger grains, including sand. They may have been deposited with the sediment or formed from the alteration of feldspar or volcanic ash.

Hematite and other iron oxide minerals precipitate from water as a natural cement between the grains. Iron oxide minerals commonly give sediment deposited on land a reddish color, as displayed in the spectacular red-rock landscapes of the Desert Southwest.

3.7

3.8 Why Do Sedimentary Rocks Have Layers?

MOST SEDIMENT IS DEPOSITED IN LAYERS. A sedimentary unit, such as a limestone or a sandstone, may be a single thick layer or may include many smaller layers called *beds*. What defines such layers, how do the layers form, and what do they tell us about the conditions that existed during deposition?

A What Types of Layers Do Sedimentary Rocks Contain?

Almost all sedimentary rocks contain layers. The layers vary greatly in thickness, lateral continuity, and characteristics that define the layers. Boundaries between layers also vary; some are quite sharp and others are more gradational.

Thickness of Layers

1. Sedimentary deposits and rocks commonly have layers that may be millimeters to many meters thick. These layers are referred to as *beds* or *bedding*. A rock unit that is distinct and laterally traceable across the landscape, like many of the layers shown in the photograph above, can be called a *formation*.

Definition of Layers

2. Adjacent layers may be distinct because they have different grain sizes, such as a conglomerate bed next to one of sandstone. Layers can also have different compositions, such as a quartz-rich sandstone layer next to a limestone layer. The distinct beds shown above vary in grain size and composition.

Boundaries of Layers

3. The boundary between two layers can be very sharp or gradational. A sharp contact generally records an abrupt event of change in environment, whereas a gradational contact records a more gradual change. In the photograph above, a gray shale grades progressively upward into a tan sandstone.

Graded Beds

03.08.a4

4. Fill a jar with water and sediment of mixed size, then shake the jar. The coarser material will settle first, followed by successively finer sediment, as shown here. Such variation, called a *graded bed*, forms if a strong current loses velocity and drops progressively finer sediment. The beds below grade from tan, coarse sand upward to gray, finer sand and mud. Graded beds, such as these, indicate that the current slowed over time and deposited finer sediment over coarser sediment.

Cross Beds

03.08.a6

5. When sand and silt move over a dune or underwater ripple, grains accumulate in thin beds on the down-current side of the dune or ripple, as shown above. Such beds are at an angle to other beds in the same rock and so are called *cross beds*. Cross beds that are centimeters to more than a meter high form within streams, deltas, and shorelines. Larger cross beds (▼) typically form in large sand dunes. Cross beds preserve the curved profile of the dune and the direction the wind was blowing when the sediment was deposited.

Parallel Beds

03.08.a8 Labyrinth Canyon, UT

6. Most beds are parallel. Parallel beds form under a variety of conditions, and in many cases simply reflect the piling of one layer on another.

03.08.a5 Point Lobos, CA

03.08.a7 Sedona, AZ

Before You Leave These Pages

✓ Sketch or describe the types of layers that sedimentary rocks contain, including what defines the layers, and whether their boundaries are sharp or gradational.

✓ Describe how layers, including graded beds and cross beds, form.

B How Do Layers in Sedimentary Rocks Form?

03.08.b1 Elkins, WV

03.08.b2 Punchbowl, CA

03.08.b3 Southwestern NM

1. *Discrete Event*—Some individual layers, or boundaries between layers, mark a discrete event, like a flood. Each 10-centimeter-thick, light-colored layer above represents a single, rapid influx of sand onto a muddy seafloor.

2. *Change in Current*—The change from layer to layer may reflect a change in the strength or direction of the current that deposited the sediment. Coarse-grained sediment is deposited by strong, turbulent currents.

3. *Sediment Supply*—Some layers record a change in the type or amount of sediment being supplied. The tan, quartz-rich sandstone in this photograph is between layers of conglomerate that contains limestone clasts.

4. *Sea-Level Change*—A global rise or fall in sea level can cause sedimentary environments to change. Here, sea level rose and fell many times, causing ledge-forming marine limestone to alternate with slope-forming layers of marine and nonmarine shale and siltstone.

03.08.b4 Goosenecks SP, UT

03.08.b5 Vermilion Cliffs, AZ

5. *Climate Change*—Some boundaries between layers reflect regional or global changes of climate. The lower gray layers of rock here formed from clay deposited during a wet period, and the top layers formed from sands deposited in a desert.

C Why Do Layers Vary from Place to Place and with Time?

Layers in sedimentary rocks vary greatly in thickness, composition, whether they are sharp or gradational, and whether they are parallel, graded, or cross bedded. The reason for this is the diversity in types of sedimentary environments and in how those environments may shift and otherwise change over time. Examine the figure below and think about how the types of layers might vary between the different environments, and what would happen if the environments shifted laterally, for example, if a rise in sea level flooded most of this area.

1. In the deep ocean, most sediment comes from clays blown as dust over the water or as small particles wash in from the land. Such fine sediment forms shale layers. Other sediment is organic in origin, such as from the remains of tiny organisms that float in the water and settle to the bottom when they die. Conditions in the deep ocean away from land remain relatively constant over time, so the same kind of sediment is deposited again and again, with few changes to form different types of layers.

6. Sand dunes are nearly all sand, and so deposit nothing but sand for millions of years. As a result, they can deposit a very thick formation of sand over a large area. Within the thick layer of sand, however, the shape of the dunes can be recorded by countless cross beds, inclined, on average, in the direction in which the wind was blowing.

5. Steep mountain fronts are also sites subject to rapidly changing conditions, such as the sudden collapse of the mountain front in a landslide or a brief but intense rainfall that causes a flash flood.

Ocean   Delta   Sand Dunes   River Deposits

03.08.c1

2. Closer to the shoreline, the ocean receives more input of sediment, so the sediment generally has layers of sand. Sand is also the dominant sediment along most beaches. These environments along the shoreline are subject to abrupt changes, such as those caused by a hurricane in a few days, or gradual changes, such as sea level rising or falling a few millimeters a year.

3. On land, sediment is deposited in a delta, which forms when a stream empties into the sea or a lake. During flooding, the flowing water deposits large amounts of sand, silt, and other sediment across a broad area. Between the times of flooding, not much sediment accumulates on the land except for the remains of plants. Since flooding may occur often, many layers are deposited.

4. Likewise, streams farther inland can remain within their channels during normal conditions or can flood adjacent land with water and sediment. These fluctuations in the strength of current and amount of water produce many layers.

3.8

3.9 What Textures Do Igneous Rocks Display?

IGNEOUS ROCKS FORM BY SOLIDIFICATION OF MAGMA. Most igneous rocks have millimeter- to centimeter-sized crystals, but some have meter-long crystals, and others are noncrystalline glass. Igneous rocks vary from nearly white to nearly black, or they can have mixed colors. They may contain holes, fragments, or ash that has been compacted. What do the different textures tell us about how the magma solidified?

A What Textures Are Common in Igneous Rocks?

The *texture* of a rock refers to the sizes, shapes, and arrangement of different components. The texture of an igneous rock depends mostly on overall crystal size, the variation in crystal size within that rock, and the presence of other features, such as holes and rock fragments.

1. The most obvious textural distinction among igneous rocks is whether or not a rock has crystals that are visible to the unaided eye (i.e., without using a hand lens or microscope). The crystals in the rock shown here are large and easily observed without a hand lens or microscope. Igneous rocks with crystals that are visible to the unaided eye are called *phaneritic*.

03.09.a1 Polished Slab

03.09.a2 Wickenburg, AZ

2. Some igneous rocks, like the one shown here, do not contain crystals that are visible to the unaided eye. Instead, such rocks consist of microscopic crystals, fine-grained volcanic ash, volcanic glass without any crystals, or a combination of these. These rocks are *aphanitic* and result from magma that solidifies too rapidly to grow crystals visible in outcrop.

03.09.a3 Black Hills, SD

3. Some igneous rocks contain very large crystals, which may be centimeters to meters long. We call very coarse igneous rocks, like the one shown above, *pegmatite*.

03.09.a4 Harquahala Mtns., AZ

4. This rock is *coarsely crystalline* (also described as being *coarse grained*). Most crystals are larger than several millimeters, and many are several centimeters across.

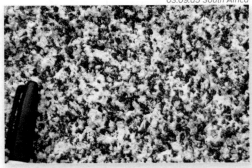
03.09.a5 South Africa

5. *Medium-grained* rocks have crystals that are easily visible to the unaided eye. Crystals in such rocks are typically millimeters across, but not centimeters across.

03.09.a6 White Tank Mtns., AZ

6. Crystals in *fine-grained* igneous rocks can be too small to see without a hand lens. In some fine-grained rocks, the crystals are visible only with a microscope.

03.09.a7 Greece

7. Some igneous rocks consist of glass rather than crystals of minerals. A rock may be 100% *volcanic glass* or may be mostly *glassy* with some crystals or rock fragments.

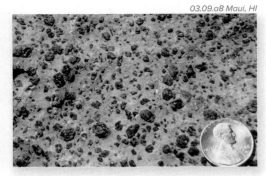
03.09.a8 Maui, HI

8. Igneous rocks that include larger crystals in a finer grained matrix are *porphyritic*. The crystals in a porphyritic rock are termed *phenocrysts*.

03.09.a9 Northern NM

03.09.a10 Superior, AZ

03.09.a11 Mule Creek, AZ

9. Many volcanic rocks contain small holes known as *vesicles*, and we describe such rocks with the adjective *vesicular*.

10. Volcanic ash and pumice, when still hot, can be compacted by overlying materials, becoming a hard rock with a *welded* texture, marked by flattened, lens-shaped objects.

11. Some volcanic rocks contain angular fragments in a finer matrix and are called a *volcanic breccia*.

B In What Settings Do the Different Igneous Textures Form?

The different textures of igneous rocks reflect the environment in which the magma solidified. Magma can solidify at depth, erupt onto the surface as molten *lava*, or be explosively erupted as volcanic ash. Examine the figure below and think about where each texture in the photographs on these two pages might form. Then read on.

1. *Vesicles* form when gases dissolved in magma accumulate as bubbles. They can form only under low pressures on the surface or very near the surface. Many lavas are vesicular, and much of the material in volcanic ash forms when the thin walls between vesicles burst, shattering partially solidified magma into sharp particles. Most volcanic ash is broken vesicles.

2. *Volcanic breccia* can form in many ways, including from explosive eruptions of ash and rock fragments, from a lava flow that breaks apart as it partially solidifies while flowing, or from volcano-triggered mudflows and landslides on the steep and unstable slopes of the volcano.

3. *Volcanic glass* forms when magma erupts on the surface and cools so quickly that crystals do not have time to form. This can happen in a lava flow or in volcanic ash.

4. *Fine-grained* igneous rocks form if the magma only has enough time to grow small crystals. This commonly occurs when magma solidifies on the surface in a thick lava flow or at shallow depths beneath the surface, because cooling in these settings is fairly rapid. Medium-grained rocks form deeper, where cooling occurs more slowly.

5. *Coarse-grained* igneous rocks form at greater depths, where magma cools at a rate that is slow enough to allow large crystals to grow.

6. Some *volcanic ash* erupts vertically in a column and settles back to Earth. This ash cools significantly before accumulating on the surface. Because it is relatively cool and strong, the ash may not become welded; thus it is said to be *nonwelded*.

7. Other volcanic ash erupts in thick clouds of hot gas, ash, and rock fragments, called *pyroclastic flows,* that flow rapidly downhill under the influence of gravity. The ash deposited by pyroclastic flows is very hot, and so most parts are *welded* to some extent.

8. For a *porphyritic* texture to form, magma needs sufficient time in a subsurface magma chamber to grow visible crystals. Later, the magma rises to just below or on the surface, where the remaining magma solidifies rapidly into the fine-grained matrix around the larger crystals (phenocrysts).

9. *Pegmatite* may form if magma is relatively water rich. The dissolved water allows atoms to migrate farther and faster and so helps large crystals to grow. This generally occurs near the sides and top of a magma chamber and in local pockets within the magma. Most pegmatite forms at moderate to deep levels within Earth's crust.

03.09.b1

Before You Leave This Page

✓ Sketch or describe the various textures displayed by igneous rocks.

✓ Sketch an igneous system and show where the main igneous textures form.

3.9

3.10 What Are Common Igneous Rocks?

IGNEOUS ROCKS VARY IN chemical composition and therefore in mineral content. Some are composed entirely of dark minerals, whereas others contain only light-colored minerals. We classify igneous rocks so that we can use a single name to identify rocks that form in a similar way and have a similar composition.

A How Does the Composition of Igneous Rocks Vary?

Geoscientists organize igneous rocks according to the *size of crystals* and the *kinds of minerals* in a rock. Below, images in the left column feature rocks with coarse crystals. Each rock in the right column has a composition similar to the rock on the left but has a smaller grain size. From top to bottom, the rocks contain lower percentages of light-colored minerals. Rocks with a light color and abundant quartz and feldspar are *felsic* rocks, whereas rocks that are dark and contain minerals rich in magnesium and iron are *mafic* or *ultramafic* rocks. *Intermediate* rocks are in between felsic and mafic in composition.

Coarsely Crystalline

Finely Crystalline or Glassy

Felsic

Granite is a coarsely crystalline, light-colored igneous rock. The light color is due to an abundance of the light-colored, felsic minerals feldspar and quartz. Most granites also contain some biotite (black mica), and some contain light-colored mica and garnet.

03.10.a1 Enchanted Rock SP, TX

03.10.a2 Superstition Mtns. AZ

Rhyolite is the fine-grained equivalent of granite. It is mostly a finely crystalline rock, but it can contain glass, volcanic ash, pieces of pumice, and variable amounts of visible crystals (phenocrysts) of quartz, K-feldspar, or biotite.

Intermediate

Diorite contains more mafic minerals than does granite. It is *intermediate* between felsic and mafic compositions. It generally contains plagioclase feldspar and amphibole, and it can contain biotite or pyroxene. A rock between a granite and diorite is a *granodiorite*.

03.10.a3 Smarthville, CA

03.10.a4 Flagstaff, AZ

Andesite is the fine-grained equivalent of diorite. It is commonly gray or greenish, but it can also have a slight maroon or purplish tint. Andesite commonly has phenocrysts of cream-colored feldspar or dark amphibole.

Mafic

Gabbro is a coarsely crystalline, mafic rock. It typically is dark and consists of pyroxene and other mafic minerals, along with light-gray, calcium-rich plagioclase feldspar. Feldspar-rich varieties are lighter colored, and some gabbro has olivine.

03.10.a5 Selway, ID

03.10.a6 Grants, NM
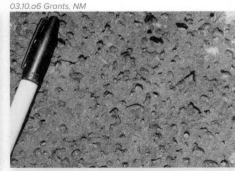

Basalt is a dark mafic lava rock. Most basalt is dark gray to nearly black, and many outcrops have vesicles, as shown here. Basalts can contain some phenocrysts of dark pyroxene, green olivine, or cream-colored plagioclase feldspar.

Ultramafic

Peridotite is the main coarsely crystalline *ultramafic* rock. Compared to mafic rocks, it contains more magnesium-rich and iron-rich minerals, such as olivine.

03.10.a7 San Carlos, AZ

03.10.a8 South Africa

Ultramafic lavas erupted early in Earth's history. The magma was so hot that it grew olivine or pyroxene crystals that are unusually long for a lava flow.

B What Are Some Other Common Igneous Rocks?

Some common igneous rocks fit into the classification system presented on the previous two pages, appropriately being called granite or basalt, but they possess some attribute that caused geologists to assign the rock a special name to convey the texture and, by inference, the specific way in which the rock formed.

Obsidian is a shiny volcanic glass that is normally a gray to black color. Most obsidian has a composition equivalent to that of rhyolite. It forms when a felsic lava flow cools too rapidly to form crystals. Obsidian can contain phenocrysts, fragments, or some swirly bands.

03.10.b1 Newberry Volcano, OR

Tuff is a volcanic rock composed of a mix of volcanic ash, pumice, crystals, and rock fragments. If the particles of ash and pumice cool before being buried by overlying materials, the rock remains only weakly consolidated and is *nonwelded tuff*, as photographed here.

03.10.b5 Northern NM

Volcanic glass is unstable, eventually changing from noncrystalline glass into a fine-grained rock consisting of very small crystals, generally too small to see. The conversion to small crystals can produce a mottled coloring, with subtle patches or layers.

03.10.b2 Wickenburg, AZ

If tuff gets buried while still hot, as within a thick pyroclastic flow, the weight of overlying materials compacts ash and pumice into lenses, forming *welded tuff*. Tuff commonly contains angular fragments of older rocks, which do not compact, as shown here.

03.10.b6 Hieroglyphic Mtns., AZ

Pumice is a volcanic rock containing many vesicles (holes). The holes are so numerous that most pumice floats on water. The solid material in pumice begins as volcanic glass, but over time it can convert into microscopic crystals.

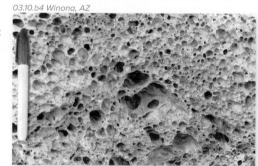

03.10.b3 Katmai, AK

Volcanic rocks with fragments form in other ways, such as the breaking apart of lava that solidifies during flow. Fragmental rocks also form from mixtures of volcanic rock, ash, and mud. In either case, the resulting fragmental rock is a *volcanic breccia*.

03.10.b7 Tushar Mtns., UT

Scoria is a dark gray, black, or reddish volcanic rock that contains many vesicles. It usually has the composition of basalt or andesite. In outcrops, scoria consists of a jumbled mass of rock fragments as large as several meters across.

03.10.b4 Winona, AZ

When magma crystallizes deep within the crust, a crystallizing magma may contain enough dissolved water that it grows exceptionally large crystals. If the crystals are larger than several centimeters (they can be meters across), the rock is *pegmatite*.

03.10.b8 Polished Slab

Before You Leave This Page

☑ Describe how igneous rocks are classified, summarizing the main differences between felsic, intermediate, mafic, and ultramafic rocks.

☑ List some common igneous rocks and a few characteristics of each.

3.11 What Are Some Metamorphic Features?

DEFORMATION CAN BE ACCOMPANIED BY METAMORPHISM, during which temperature and pressure can cause the rearrangement of existing materials and the formation of new minerals and new structural features. What characteristics of metamorphic structures can we observe in exposures of metamorphic rock?

A What Is Rock Cleavage?

When rocks are deformed under conditions of low to moderate temperature (less than about 300°C), they may develop a planar fabric along which they break, or *cleave*. This fabric, called *rock cleavage* or simply *cleavage*, is not a fracture but a type of structural discontinuity in the rock. It also is not related to the way minerals cleave (*mineral cleavage*).

03.11.a3 Great Smoky Mtns., TN

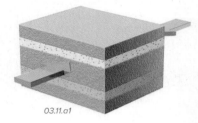
03.11.a1

1. When stress is applied to a rock, the rock can become strained and change shape. The rock typically begins to squash or shorten in the direction of the maximum applied stress (◄).

3. Some cleavage is expressed as closely spaced planes that cause the rock to cleave into thin sheets or slivers. In this photograph (▶), the marking pen is aligned parallel to cleavage. The folded layers are bedding.

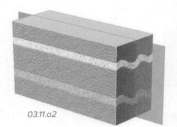
03.11.a2

2. As the rock shortens, it can develop new planar features (marked by the purple plane) that are oriented perpendicular to the direction in which the rock is shortened the most. One type of a planar feature formed in this way is rock *cleavage*, which is a weakness along which the rock breaks (◄).

4. Cleavage typically cuts across bedding, as in this photograph and the one above. Here (▶), cleavage is parallel to the red knife and cuts across the gray and brown layers, forcing a sliver of hard, brown rock out of the way.

03.11.a4 Inyo Mtns., CA

B What Is Foliation and How Is It Expressed in Rocks?

Foliation is used as a general term for any planar metamorphic fabric, including cleavage, but it commonly is reserved for more strongly metamorphosed rocks. Foliation forms because of differential stress and is expressed in a variety of ways, depending on factors including temperature, the amount of shearing, and the starting rock types.

One type of foliation occurs in metamorphic rocks that are rich in mica minerals. This type of foliation, called *schistosity,* is defined by a parallel orientation of mica and other platy minerals. Schistosity makes most rocks shiny, and the rock is *schist.*

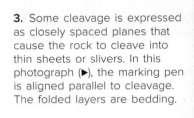
03.11.b1 Hunt Valley, MD

03.11.b3 Swiss Alps

Where metamorphic rocks have alternating lighter and darker-colored *bands,* with different proportions of light and dark minerals, the foliation is a *gneissic* (pronounced "nice-ick") foliation. A metamorphic rock with this type of foliation is a *gneiss* or a *banded gneiss.*

Foliation can be defined by the flattened shapes of deformed objects, like these light-colored pebbles in a metamorphosed conglomerate. Foliation is parallel to the flattened shapes of the pebbles.

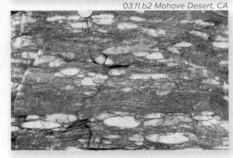
03.11.b2 Mohave Desert, CA

03.11.b4 San Gabriel Mtns., CA

Shearing can form a foliation by flattening and smearing out mineral grains. This rock has a foliation largely defined by lens-shaped, deformed crystals and smeared-out light and dark (gneissic) bands.

C What Is Lineation and How Is It Expressed in Metamorphic Rocks?

In addition to planar fabrics (foliation), metamorphism and deformation can form *linear features*, such as aligned minerals or long, deformed pebbles. A linear feature in metamorphic rocks is called *lineation*.

03.11.c1 Naxos, Greece

03.11.c2 Harquahala Mtns., AZ

03.11.c3 Chemehuevi Mtns., CA

Lineation in this metamorphic rock is defined by elongated, blue crystals that grew in a preferred, linear orientation during metamorphism. The mineral is kyanite.

This lineation is defined by the long axes of light-colored feldspar crystals that were sheared and stretched out in a horizontal direction. The rock is a metamorphosed granite.

These linear streaks formed as minerals were smeared out during metamorphism and ductile shearing. This type of lineation is parallel to the direction of shearing.

D What Are Some Other Features in Metamorphic Rocks?

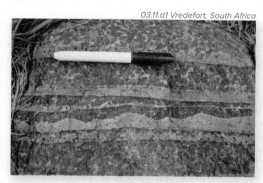
03.11.d1 Vredefort, South Africa

03.11.d2 Rincon Mtns., AZ

03.11.d3 Harcuvar Mtns., AZ

1. If they are not too strongly deformed, metamorphic rocks can preserve features that existed in the rock before it was metamorphosed. This metamorphic rock has tan layers that were sandstone beds with curved tops that represent sedimentary ripples.

2. This rock began as conglomerate with pebbles. The tan and gray lenses are stretched pebbles, but the pebbles became flattened and folded during deformation and metamorphism, and are not easily recognized as stretched pebbles.

3. Metamorphic rocks can contain zones that show evidence of intense shearing. Such features are *shear zones*, and are partly a structural feature and partly a metamorphic one. In this photograph, a thin shear zone cuts across metamorphic layers.

03.11.d4 Old Woman Mtns., CA

4. Most metamorphic rocks have folds, some small and some large and spectacular. This spectacular fold (◀) is in banded, gneissic rocks. For scale, note the geologist examining the bottom of the outcrop.

03.11.d5 Santa Catalina Mtns., AZ

5. Metamorphic rocks, especially those with gneissic foliation, commonly are intimately associated with layers, lenses, pods, and dikes of igneous material. Here, light-colored granite occurs between dark metamorphic layers (▲).

Before You Leave This Page

✓ Summarize how cleavage and foliation are expressed in common metamorphic rocks.

✓ Summarize the types of features that define lineation and how each type of lineation forms.

✓ Describe some other features that may be present in metamorphic rocks.

3.12 What Are Metamorphic Processes and Rocks?

METAMORPHISM INVOLVES CHANGING A ROCK that has become unstable. A rock may be unstable because the minerals it contains are unstable and as a result it develops new (metamorphic) minerals. It may be unstable because of the way the grains or layers are arranged, so it develops a new (metamorphic) texture. Depending on the conditions, a rock may develop new metamorphic minerals and a new metamorphic texture.

A What Causes Metamorphism?

For a rock to be metamorphosed, it must be subjected to conditions of temperature, pressure, and fluid chemistry that make it unstable. This generally occurs when temperature (T) and pressure (P) increase, possibly accompanied by forces that cause the rock to *deform* or by an influx of deep fluids. There are two main kinds of metamorphism: *contact metamorphism* is caused by local heating by magma, typically without deformation, whereas *regional metamorphism* involves deformation along with heating over a broader region.

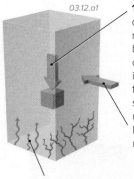
03.12.a1

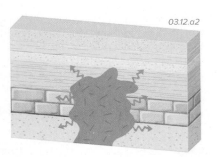
03.12.a2

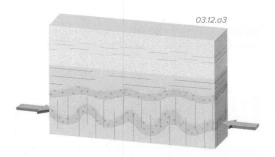
03.12.a3

1. *Pressure* increases with depth in Earth because rocks are more deeply buried. Higher pressures compress the rocks and, in combination with high temperatures, may cause some minerals to be unstable. Forces deep in Earth can cause rocks to move and deform.

2. *Temperature* increases with depth and near magma. An increase in temperature usually causes new minerals to grow or existing minerals to grow larger. It can also weaken the rocks, allowing them to deform. The crust contains abundant water and other *fluids*, shown here as blue in water-filled fractures. Such fluids interact with minerals and carry material into, through, and out of rocks.

3. *Contact Metamorphism*—Rising magma efficiently brings thermal energy higher into the crust, heating the wall rocks. This is called contact metamorphism because it occurs near contacts (boundaries) of magma. Heating causes new minerals to grow or existing minerals to increase in size. Heating may not be accompanied by deformation, so contact metamorphism commonly forms metamorphic rocks with little evidence of deformation.

4. *Regional Metamorphism*—In regional metamorphism, heating is accompanied by enough force to cause deformation. The imposed forces can result from tectonics or burial. Regional metamorphism causes new minerals to grow and existing minerals to increase in size, while deformation during metamorphism generally results in the rock developing planar or linear metamorphic fabrics during metamorphism.

B What Processes Occur During Metamorphism and Deformation?

Many processes that operate during metamorphism change the minerals in a rock or change the arrangement of those minerals. Some metamorphic processes are related to heating or are chemical in character. Such processes cause grains to grow, recrystallize, redistribute themselves, or even dissolve in response to temperature, pressure, and any imposed forces. Some processes are physical and may deform or rotate individual crystals, grains, and layers, producing planar or linear metamorphic fabrics.

Chemical Processes

During metamorphism, movement of chemical elements can cause existing minerals to grow larger or can form new minerals. Minerals can grow in fairly random orientations, as occurs during contact metamorphism, which is due to an increase in temperature but not in forces. Or, minerals can grow with a preferred orientation, becoming aligned as in the example shown here. If minerals are arranged in lines, the rock has a *lineation*. If they are arranged in planes, the rock has a *foliation*. Many metamorphic rocks have both lineation and foliation.

03.12.b1

Physical Processes

Metamorphism is generally accompanied by deformation of any original constituents of the rocks, like crystals or pebbles. Deformation can flatten grains and clasts that were initially somewhat spherical into shapes like pancakes or the thin, long top of a skateboard. If a rock is flattened during metamorphism, the deformed objects will define a foliation. If a rock is stretched in one direction, it acquires a lineation, commonly along with a foliation. Metamorphism usually involves both physical and chemical processes.

03.12.b2

C What Rocks Form When Sedimentary Rocks Are Metamorphosed?

There are many types of metamorphic rocks, and these rocks indicate diverse starting rock types and various conditions under which rocks can be deformed and metamorphosed. Some metamorphic rocks form when sedimentary rocks are subjected to a change in temperature, pressure, or both. Photographs below show what happens to three common sedimentary rocks—shale, sandstone, and limestone. Igneous rocks can also be metamorphosed with similar results, and metamorphic rocks can be metamorphosed several times under varying conditions of temperature and pressure.

Shale

Increasing Temperatures

Slate

03.12.c1 Larder Lake, Ontario, Canada

When a shale is metamorphosed at low to moderate temperature, it can develop cleavage (a type of foliation) and become *slate* (◄). Slates are dull (not shiny) and commonly dark.

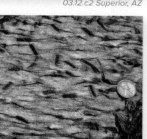

Phyllite

03.12.c2 Superior, AZ

At slightly higher temperatures, microscopic mica crystals give the rock a shiny aspect or *sheen* (◄). Such a rock is *phyllite*.

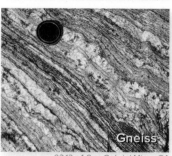

Schist

03.12.c3 Pioneer Mtns., ID

At higher temperatures, crystals of mica and other minerals become large enough to see. The resulting rock is a shiny rock called a *schist* (◄).

Gneiss

03.12.c4 San Gabriel Mtns., CA

At even higher temperatures, chemical constituents are mobilized and light- and dark-colored minerals separate, forming a foliation and banded rock called *gneiss* (◄).

Sandstone

Quartzite

03.12.c5 Baraboo, WI

Most sandstones are predominantly quartz (SiO_2), a mineral that is stable over a wide range of temperature and pressure conditions. During metamorphism, quartz grains grow together and become so tightly bonded that fractures break across the grains rather than around them. This type of rock is a *quartzite* (▲). Quartzite is made of quartz, just like the original sandstone, and can preserve beds and other original sedimentary features.

Coarse Quartzite

03.12.c6 Joshua Tree, CA

With higher temperatures of metamorphism, the quartz in the rock begins to merge into larger crystals and can become a coarser grained quartzite (▲), in some cases with no individual grains left. Quartz is soluble and mobile in metamorphic fluids and so at high temperatures can be redistributed into quartz veins, which are common in metamorphic rocks.

Limestone

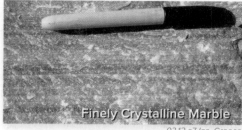

Finely Crystalline Marble

03.12.c7 Ios, Greece

Limestone consists mostly of calcite, a chemically reactive mineral. Low-temperature (<250°C) metamorphism of limestone causes calcite to recrystallize slightly, forming a finely crystalline *marble* (▲).

03.12.c8 Franklin Mtns., El Paso, TX

Impure Marble

At moderate temperatures (~400°C), marble becomes medium grained (▲). When metamorphosed, impurities in the limestone, like clay and chert, may produce various new metamorphic minerals, like these dark garnets.

03.12.c9 Naxos, Greece

Coarse Marble

At higher temperatures, calcite crystals grow coarser to produce a coarsely crystalline marble (▲). The one shown here is nearly 100% calcite, but coarse marbles commonly also contain other minerals.

Before You Leave This Page

☑ Summarize causes of metamorphism, and describe or sketch a chemical and physical process that can accompany metamorphism.

☑ Describe the changes different sedimentary rocks undergo as they metamorphose, and the metamorphic rocks they become.

How Are Different Rock Types Expressed in Landscapes?

SEDIMENTARY, IGNEOUS, AND METAMORPHIC ROCKS can each have a distinctive appearance in the landscape, often allowing us to recognize these rocks from a distance. With some practice, we can drive down a highway, observe the characteristics of a hill or mountain, and make an educated guess about what types of rocks are exposed. Here, we provide a brief introduction into interpreting the rock types of landscapes from a distance.

A How Are Sedimentary Rocks Expressed in the Landscape?

There are diverse types of sedimentary rocks, displaying a wide variety in colors, thickness of layers, and resistance to erosion. The unifying feature of most sedimentary rocks is the presence of visible layers. Each main type of sedimentary rock has certain characteristics that allow us to identify it from a distance.

Sandstone

03.13.a1 McElmo Canyon, UT

03.13.a2 Ruby Canyon, CO

1. Some sandstone layers appear massive because they have little variation of grain size, as in the sandstone in the lower part of this cliff. Thick layers of massive sandstone generally were deposited by wind as sand dunes. The top of the cliff has layered sandstone deposited by streams.

2. Most sandstone has layers that differ from one another in color, grain size, or composition of the grains. Such layers can be parallel beds, as shown here, or cross beds centimeters to tens of meters high. Numerous layers mean many changes in conditions during deposition.

Fine-Grained Clastic Rocks

3. Shale and other fine-grained clastic sedimentary rocks are relatively easily eroded rocks. Where exposed, these rocks typically form soft slopes covered by small, loose chips derived from weathering of the thinly bedded rocks.

03.13.a3 Grand Junction, CO

4. Shale and associated fine-grained rocks form another distinctive type of landform—*badlands*. Badlands have a soft, rounded appearance that exhibits the softness of the rocks. Badlands also have an intricate network of small drainages and eroded ridges carved into the soft rocks.

03.13.a4 Petrified Forest, AZ

Limestone

03.13.a5 Guadalupe Mtns., TX

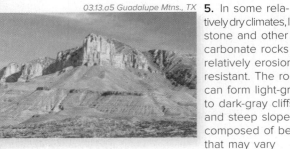

5. In some relatively dry climates, limestone and other carbonate rocks are relatively erosion-resistant. The rocks can form light-gray to dark-gray cliffs and steep slopes composed of beds that may vary slightly in thickness and color. Most limestone has visible layers, on either a small or a large scale.

03.13.a6 Austrian Alps

6. Limestone is very soluble, and so in very wet climates it dissolves and weathers quickly, especially along fractures and layers, forming caves, sinkholes, pits, and depressions.

03.13.a7 Guilin, China

7. Weathered and dissolved limestone forms distinctive landforms with caves, sinkholes, and pillars. Such landscapes formed by dissolution of limestone are called *karst terrain*.

Before You Leave These Pages

☑ Describe the characteristics of some common sedimentary rocks, including their expression in landscapes.

☑ Describe the appearance of some common igneous rocks in landscapes.

☑ Describe some characteristics displayed by metamorphic rocks, as exposed in landscapes.

B How Are Igneous Rocks Expressed in the Landscape?

Igneous rocks form from magma, either from magma that solidifies below the surface, forming *intrusive* (plutonic) rocks, or magma that erupts onto the surface, forming *extrusive* (volcanic) rocks.

Granite makes up the bulk of the continental crust, and most granite is fractured. These fractures help speed up the weathering process, which wears away the edges and corners of fractures, resulting in rounded shapes in many granite exposures, a process called *spheroidal weathering*.

03.13.b1 Joshua Tree NP, CA

Basaltic lava flows are the most common type of volcanic rock in many landscapes. Basalt flows are dark and spread out easily, forming distinct layers. Most basaltic sequences have reddish zones from pyroclastic rocks that accompanied the eruption of lava, as shown in this photograph.

03.13.b3 Grand Coulee, WA

Volcanic rocks commonly have some sort of layers, and they also commonly have fractures perpendicular to the layers (▶). Depending on the rocks' compositions and the type of eruption that formed them, volcanic rocks can be various shades of gray, green, brown, tan, and cream colored.

03.13.b2 Copper Creek, AZ

03.13.b4 Grand Canyon, AZ

Many volcanic units, especially basaltic lava flows, have distinctive columnar fractures called *columnar joints*. Such joints form during the cooling and contraction of solidified igneous rocks. The size and orientation of the columns indicate how the rock cooled, but most columns are steep.

C How Are Metamorphic Rocks Expressed in Landscapes?

Rocks that have been changed (or metamorphosed) by increased temperature and pressure are metamorphic rocks. Different types of metamorphic rocks result from different starting rock types and different conditions under which rocks can be metamorphosed and deformed. These differences, in turn, are expressed by a somewhat variable appearance of metamorphic rocks in the landscape.

03.13.c1 Patagonia, Argentina

3. Especially intense shearing and deformation can fold and slice apart originally intact layers into discontinuous lenses within a matrix of highly sheared rock (▼). We call such folded, sheared, and sliced rocks that have intermixed a variety of preexisting rock types a *melange*, which is common in coastal California and adjacent areas.

03.13.c2 Aurland Trail, Norway

03.13.c3 Marin Headlands, CA

1. Metamorphic rocks can be shiny, even from a distance, if their mica minerals share a similar orientation and reflect light. This schist (▲) is shiny, even from a distance, and has large folds, each tens of meters high.

2. Many metamorphic rocks have layers that form platy, jagged outcrops and tabular slabs of rock (▶). Such rocks generally have folds, even if they are not obvious.

3.14

What Materials Compose These Landscapes?

Imagine you are riding across the country, observing landscapes as you travel from one area to another. What do you observe, and what do your observations tell you about each landscape? To answer such questions, it helps to have a broad view of the landscape and a close-up view of the earth materials that compose that landscape. In this exercise, you will observe a series of landscapes and close-ups of the associated rocks. From your observations, you will interpret what type of rock is present and the general way in which that rock formed.

Goals of This Exercise:

• Using photographs, make observations of landscapes and their associated rocks, identify the type of rock that is represented, and interpret how that rock formed.

• Propose what earth processes are occurring in each scene and what processes occurred in the past.

Procedures

For each set of photographs below, follow these steps, writing your observations and interpretations on the worksheet or entering them online. Your instructor may provide samples of similar rocks.

A. Observe the general features of the landscape. Some questions to ask include, but are not limited to, the following: What color are the rocks? Are there layers, and if so, how thick and how continuous are the layers? Do the rocks form steep, hard exposures, or do they form soft-appearing slopes? Are there loose materials on the hillsides? Record your observations on the worksheet.

B. Examine the smaller, close-up photograph of a representative rock from the landscape. Observe the characteristics of this rock and write your observations on the worksheet. Compare your observations of the rock with photographs and descriptions of different rock types earlier in the chapter, make an interpretation of what type of rock (sedimentary, igneous, or metamorphic) is shown by the close-up, and assign a name (such as granite) to the rock.

C. Consider the implications of everything you observe, including the type of rock, and what processes were involved in forming the landscape. Did the rocks form from a volcanic eruption or some other igneous process? If the rock is sedimentary, what are some options for the environment in which this type of rock formed? Was the rock formed at the surface or at depth? If it formed at depth, what must have happened to have the rock be on the surface today? What processes appear to be occurring today?

Landscape 1

1. This landscape is in a dry desert, so the earth materials are only partly obscured by plants and soil. Many of the rocks are loose, and some have rolled or slid downhill a short distance. Many of the rocks have not moved downhill, such as rocks that form the large rock faces in the lower left of the photograph.

2. The photograph below shows a close-up view of a fresh surface of the rock, revealed by breaking open one of the smaller rocks with a

rock hammer. The dark crystals are biotite (a dark mica), whereas the light-colored minerals are feldspar and quartz.

03.14.a1

03.14.a2

Landscape 2

03.14.a3

3. This landscape is exposed in canyon walls next to a river. The rocks of most interest are those that form a series of ledges. The dark patches of rock below the ledges are loose and are composed of the same type of rock as in the ledges. The small trees and bushes are in poorly developed soil.

4. The photograph below shows a close-up view of a fresh piece of the rock, taken from one of the ledges.

03.14.a4

Landscape 3

03.14.a5

5. This landscape shows one wall of a deep, rugged canyon. The main rock type is gray and is cut by some lighter-colored sheets of granite.

6. The photograph below shows a close-up view of an exposure of the main gray rock. The gray layers are composed of quartz, feldspar, and mica minerals, whereas the dark layers contain darker colored minerals, such as a dark amphibole and biotite mica.

03.14.a6

Landscape 4

03.14.a8

7. This landscape has two main parts: an upper tan cliff, and a lower gray slope. There is very little soil and vegetation. Shown are close-up photographs of rocks from the upper part of the cliff, the lower part of the cliff, and the slopes. The rock in the lower cliff has abundant sand grains, and the rock in the slope has clay-size particles.

03.14.a7

From Slope

03.14.a9

From Lower Cliff

03.14.a10

3.14

4 Earth History

EARTH HAS A LONG HISTORY. Rocks around the world contain evidence that seas advanced and retreated across the land many times, that vast mountains were uplifted and eroded away, and that various types of creatures arose, left their remains preserved in the rock record, and became extinct. This chapter explains the story of geologic time and how we use geologic principles to reconstruct Earth history.

Siccar Point, east of Edinburgh, Scotland, is one of the most important geologic sites in the world. Scottish geologist James Hutton realized that rock layers exposed at Siccar Point (▼) require Earth to have a long and protracted history. Observe the photograph below and make observations about the rocks, geometry of layers, and other features.

04.00.a1 Siccar Point, Scotland

The geologic feature for which Siccar Point is famous is a boundary, or contact, that separates two chapters in Earth's history. This contact goes through the center of this photograph, sloping from right to left. Examine the nature of this contact and the features in the rock types on either side of the contact. Below the contact are gray sandstone and shale, whose beds are nearly vertical. Above the contact are beds of reddish sandstone, breccia, and conglomerate that dip gently to the left in this view.

How does a contact like the one exposed at Siccar Point form, and what does it tell us about the geologic history of the area?

TOPICS IN THIS CHAPTER

04.00.a2 Siccar Point, Scotland

In this close-up view at Siccar Point, the contact crosses the photograph from lower left to upper right (◀). Below the contact, beds of gray sandstone and shale are nearly vertical. Where the gray beds end in the middle and upper part of the photograph, they are overlain by reddish-brown sandstone and breccia that contain angular pieces of the underlying gray sandstone. The contact is irregular in shape, with the reddish rocks filling in the spaces between underlying resistant beds of gray sandstone. As described below, the features along this contact record a special sequence of geologic events that happened here in the distant geologic past. The contact between these two rock sequences with very different geologic histories is what inspired Hutton's profound insight.

What would you infer about this contact? What does a contact like the one exposed at Siccar Point imply about the length of Earth history?

Ruins of an Earlier World

As James Hutton explored the rocky coasts of Scotland in the late 1700s, he encountered the remarkable geologic exposures at Siccar Point. The insight he gained on that day in 1788 changed the world. James Hutton's profound realizations provided a new way to think about Earth.

At Siccar Point, James Hutton's attention was drawn to the enigmatic contact, which even from a distance is striking, with vertical gray beds below and gently inclined red beds above. Hutton pondered what had happened to produce such an arrangement of rock types. He wondered if the ancient contact represented the same processes currently occurring on the beach next to the outcrop—modern beach sand was being deposited in horizontal layers over the vertical beds of gray sandstone. In other words, Hutton's insight was that you might be able to use modern processes to interpret events that had occurred in Earth's past. This principle, today called *uniformitarianism,* was the key step in the development of geology as a science. Uniformitarianism is an important tenet of the modern science of geology, being based on the logical idea that processes operating today are the same or are similar to processes that operated in the past. Uniformitarianism is often stated as "*the present is the key to the past.*"

Following this new logic, Hutton realized that to explain the relationships at Siccar Point, the gray sandstones below the contact must have been deposited, tilted, and eroded before the red sandstone was deposited across the upturned layers. In essence, Hutton realized that this contact represented an *ancient erosion surface,* which we now call an *unconformity.* Hutton concluded that the gray rocks below the unconformity represented a mountainous landscape that had been eroded away, and he called these rocks "the ruins of an earlier world."

Hutton noted that erosion and many other geologic processes could be observed to occur relatively slowly compared to the life span of a human, so he realized that the contact at Siccar Point required Earth to have a very long history, much longer than was perceived at the time. Hutton concluded that the history of Earth was very long and partially shrouded, with "no vestige of a beginning, no prospect of an end." The ideas of Hutton were elaborated in books and other writings by Scottish Professor John Playfair, a contemporary of James Hutton, and later by Sir Charles Lyell, a Scottish geologist, who published very influential books in the 1830s and 1840s. Hutton and Lyell are among those people cited as the founders of geology.

4.0

4.1 How Do We Infer the Relative Ages of Events?

TO DECIPHER THE GEOLOGIC HISTORY OF AN AREA, geoscientists use several strategies to determine the ages of geologic units, features, and events. The first strategy is to determine the age of one rock relative to another, using a series of commonsense approaches collectively called *relative dating*. Geoscientists then use analytical laboratory methods, or *isotopic dating*, to assign actual numbers, in thousands to billions of years, to this relative chronology. We also use *fossils* to compare ages of different rock layers and to correlate one rock unit to another. We start here with five main principles of relative dating. We think going to a new place and reconstructing the sequence of geologic events, using these principles, is one of the most fun and interesting aspects of geology.

Principle 1: Most Sediments Are Deposited in Horizontal Layers

Most sediments and many volcanic units are deposited in layers that originally are more or less horizontal, a principle called *original horizontality*. If layers are no longer horizontal, some event affected the layers after they formed. The few exceptions to the principle are small in scale and in special environments, such as the face of a sand dune or the undersea slopes of a delta.

1. These canyon walls expose horizontal gray and reddish layers. These layers were deposited horizontally in Paleozoic time and have remained nearly so for 300 million years.

04.01.a1 Goosenecks of the San Juan, UT

04.01.a2 Mexican Hat, UT

2. Just to the east, the same gray and reddish layers are folded. They are no longer horizontal, so something must have happened, like tilting associated with folding or faulting.

Principle 2: A Younger Sedimentary or Volcanic Unit Is Deposited on Top of Older Units

When a layer of sediment is deposited, any rock unit on which it rests must be older, a concept called the *principle of superposition*. This principle is illustrated below.

04.01.a3

1. A layer of tan sediment is deposited over older rocks.

04.01.a4

2. A series of horizontal red layers are then deposited over the first tan layer.

04.01.a5

3. A third series of layers is deposited last, and is on top. In this sequence, the oldest layer is on the bottom and the youngest layer is on the top.

4. Observe all the different layers in this rock sequence. The sediments were deposited and then lithified to form sedimentary rock long before the river eroded the canyon. Which exposed layer is oldest, and where would you predict is the youngest rock layer (▶)?

5. Where is the oldest unit in these tilted rock layers (▶)? It is most likely on the left, in the lowest part of the section. However, tectonic forces in some exceptional places have actually overturned layering, placing the oldest rock on top instead of on the bottom, but not here.

04.01.a6 Dead Horse Point, UT

04.01.a7 San Juan River, UT

Principle 3: A Younger Sediment or Rock Can Contain Pieces of an Older Rock

When a rock or sedimentary deposit forms, it may incorporate pieces, or clasts, of older rock. A cobble eroded from bedrock and carried by a stream cannot exist unless the bedrock already was there. The presence of clasts of an older rock in a younger rock clearly indicates the relative ages, even if you cannot see the two rock units in contact with one another.

1. The dark, lower basalt contributed clasts into an over-lying layer of tan conglomerate. The conglomerate contains *clasts* of—and is therefore younger than—the basalt. The con-glomerate also filled fractures in the basalt (▶).

04.01.a8 Lake Pleasant, AZ

2. A light-colored granite contains dark pieces, called *inclusions*, of older metamorphic rocks that became incor-porated into the magma. The meta-morphic rocks, and their metamorphic layering, are con-tained within, and are older than, the granite (▶).

04.01.a9 Harcuvar Mtns., AZ

Principle 4: A Younger Rock or Feature Can Cut Across Any Older Rock or Feature

Many rocks are crosscut by fractures, so the rocks were there before the fractures formed, a *cross-cutting relation*. There are two main types of fractures: *joints* formed by slight pulling apart of the rock and *faults* along which rocks on opposite sides have moved up and down, side to side, or some combination. Either type of fracture can cut a preexisting rock. A fracture can be occupied by magma, forming a sheetlike igneous feature called a *dike* or a *sill*. A dike cuts across layers, but a sill is injected parallel to layers; both exhibit cross-cutting relations with their host rock.

1. Several fractures cut across the lime-stone layers, so they formed after the rock already existed. The frac-tures crosscut the limestone or are said to be *cross cutting* (▶).

04.01.a10 Little Colorado River, AZ

2. Light-colored dikes of granite crosscut through darker igneous rocks (▶). The cross-cutting gran-ite is younger than the darker gray igneous host rocks.

04.01.a11 Santa Catalina Mtns., AZ

Principle 5: Younger Rocks and Features Can Cause Changes Along Their Contacts with Older Rocks

Magma comes into contact with preexisting rocks when it erupts onto the surface or solidifies at depth. In either setting, the magma may locally bake adjacent rock, or fluids from the magma may chemically alter nearby rocks. These changes, called *contact effects*, indicate that the magma is younger than the rocks that were altered.

1. A dike of basalt intrudes across a grayish sedimentary rock. Heat and fluids from the magma affected the older sedimentary rock, causing a reddish baked zone next to the dike (▼). If a magma is injected between exist-ing layers, it bakes rocks above and below the magma body.

04.01.a12 Bloody Basin, AZ

2. A lava flow or hot pyroclastic flow can bake and redden older underlying rocks, as shown here (▼). Sediments deposited on top of the volcanic unit after the eruption will not be baked. This con-trasts with a sill injected between preexisting lay-ers, which bakes rocks above and below.

04.01.a13 Lewiston, WA

Before You Leave This Page

☑ Sketch and explain each of the five principles of relative dating, providing an example of each principle.

☑ Apply the principles of relative dating to a photograph or sketch showing geologic relations among several rock units, or among rock units and structures.

4.1

4.2 What Is the Significance of an Unconformity?

EROSION SURFACES CAN BE BURIED AND PRESERVED beneath later deposits. These buried erosion surfaces, called *unconformities*, can represent large intervals of time missing from a rock sequence. They provide a glimpse of the shape and longevity of ancient landscapes, as was recognized by James Hutton at Siccar Point. There are three types of unconformities: *angular unconformities, nonconformities,* and *disconformities*. What are the characteristics of these features, how does an unconformity form, and what does an unconformity tell us about past geologic events?

A What Does an Angular Unconformity Represent?

Erosion surfaces, formed in the past, can be buried and preserved within a sequence of rocks. If underlying rocks are tilted before formation of the erosion surface, the unconformity is an *angular unconformity.*

1. A gray limestone is deposited under the sea in nearly horizontal layers. The blue in the figure below represents water in which the limestone is being deposited.

2. Later, the sea withdraws and the limestone beds are folded. As the folded beds are uplifted, they are beveled by erosion, in this case by two streams. This results in tilted beds being exposed on the surface.

3. A conglomerate is deposited over the eroded beds, forming an unconformity. If the underlying layers have been tilted, as in this example, it is an angular unconformity. Tilting can be associated with anticlines, synclines, and faults.

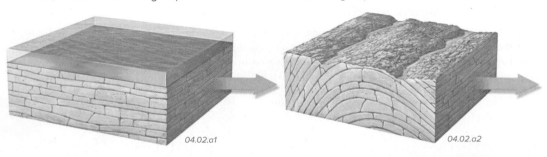

04.02.a1

04.02.a2

04.02.a3

04.02.a4 Grand Canyon, AZ

Horizontal Layers Above Unconformity

Unconformity

Tilted Layers Below Unconformity

4. Examine this photograph of the eastern Grand Canyon (▲). There is an angular unconformity between tilted layers below and nearly flat-lying layers (beds) above. The rocks below the unconformity are approximately 1,100 million years (1.1 billion years) old, whereas those above are younger than 541 million years old. There is a long time span represented by the unconformity, for which there is no record, except that there was tilting followed by erosion, before deposition of the upper layers. Other events could have—and did—occur, but we have no record of them at this site.

04.02.a5 Ouray, CO

Rocks Above Unconformity

Unconformity

Rocks Below Unconformity

5. The same unconformity (separating rocks that are more than 1 billion years old from those that are less than 541 million years old) is exposed across many parts of the United States and is called the *Great Unconformity.* In this view (▲), gently tilted sedimentary layers unconformably overlie vertical layers within underlying metamorphic rocks. The time represented by the unconformity is 1.3 billion years (from 1.7 billion years for the metamorphic rocks below to 400 million years for the sedimentary rocks above).

B How Does a Nonconformity Form?

Some erosion surfaces form on top of rocks that are not layered, especially igneous rocks like granite. This type of unconformity is called a *nonconformity*.

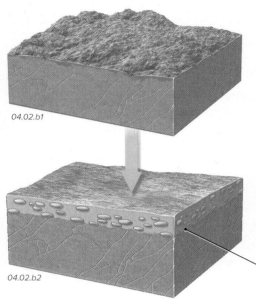

04.02.b1

04.02.b2

1. The formation of a nonconformity begins when a granite, or other nonlayered rock, is formed at depth and uplifted (◄). Material that was once on top is eroded away, exposing the granite to weathering and erosion at the surface. Weathering can form soils and other weathering products, like the reddish zone of sand, clay, and iron oxides shown on top of the granite in this figure.

2. Subsequently, conditions change, and the erosion surface is buried by sand and cobbles (◄), perhaps derived in part from weathering of the granite. Ultimately, the sediment lithifies into sandstone and conglomerate. The contact between the granite and overlying sedimentary rock is a nonconformity.

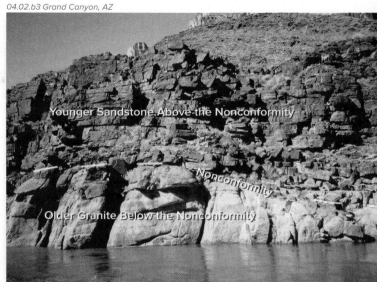

04.02.b3 Grand Canyon, AZ

Younger Sandstone Above the Nonconformity

Nonconformity

Older Granite Below the Nonconformity

3. This nonconformity (▲) has dark sandstone over tan granite. It is the same Great Unconformity as shown on the previous page, but here it overlies granite instead of layered and tilted metamorphic or sedimentary rocks.

C What Is a Disconformity, and What Does It Indicate About an Area's History?

If rock layers are not tilted before they are overlapped by younger layers, but the boundary still represents millions of years of time, the contact is a *disconformity*. A disconformity can involve erosion or just a long time period with little or no deposition. Disconformities may be overlooked because they are parallel to rock layers.

04.02.c1

04.02.c2

04.02.c3

1. The first step in the development of a disconformity is deposition of horizontal layers producing sedimentary rock. In the figure above, the layers are limestone but could be any type of sedimentary rock.

2. Next, the rock is exposed at the surface because the region is uplifted or because sea level drops. Sedimentation stops, and weathering and erosion affect the now-exposed land surface.

3. After some time, sedimentation resumes, and the surface of older rock is eventually buried by a younger layer of sediment, forming a disconformity. This new layer can be deposited by water or can be deposited on land, perhaps as sand dunes in a desert.

04.02.c4 Buckfarm Canyon, Grand Canyon, AZ

Younger Limestone

Lens of Sedimentary Rocks

Older Limestone

4. This photograph, colored lightly for emphasis, shows three series of rock layers separated by two disconformities. Can you find them? One disconformity is below the reddish lens of sedimentary rocks, and a younger disconformity is above the reddish lens. The lower disconformity is more obvious than the upper one.

Before You Leave This Page

 Sketch an angular unconformity, a nonconformity, and a disconformity, and describe what sequence of events is implied by each.

4.2

4.3 How Are Ages Assigned to Rocks and Events?

DETERMINING THE RELATIVE AGES OF ROCKS, and the order of events, is only one part of deciphering the geologic history of a location. We also want to know how long ago the rocks formed and when the events occurred, and to assign *ages* in hundreds, thousands, millions, or billions of years. This is done by using *isotopic dating methods*, most of which involve chemically analyzing a rock for the products of natural radioactive decay. Determining the ages of rocks using analytical measurements is called *isotopic dating*.

A How Does Radioactive Decay Occur?

All atoms of any given element must have the same number of protons, but some differ in the number of neutrons they contain. Thus, different varieties of the same element may have different *atomic weights;* these varieties of the same element are called *isotopes*. Some isotopes are unstable through time, changing into a new element or isotope by the process of *radioactive decay*.

04.03.a1–3

1. This schematic figure shows atoms before any radioactive decay. These starting atoms are called the *parent atoms* or *parent isotopes*. Over time, some of the parent isotope will decay into a different element called the *daughter product or daughter atom*.

2. At a later time, half of the parent atoms (green) will have decayed into the daughter product (purple). The amount of time it takes for this to occur is called the *half-life*. After one half-life, there are an equal number of parent and daughter atoms.

3. After a time equal to another half-life has passed, half of the remaining parent atoms have decayed into daughter atoms. That is, after two half-lives, 3/4 of the parent atoms have decayed and 1/4 remain.

4. This table summarizes the radioactive decay shown in the figures above. If the number of parent atoms was initially 100, half of the parent atoms (50) will have decayed to atoms of the daughter product after one half-life. After two half-lives, only 25 parent atoms remain, alongside 75 daughter atoms.

	Before Any Decay	After One Half-Life	After Two Half-Lives
Atoms of Parent	100	50	25
Atoms of Daughter	0	50	75

5. Decay rates are different for different radioactive elements, but for any given isotope, the decay rate is always the same, predictable, and measurable in the laboratory. Geoscientists, therefore, can calculate the age of a rock by measuring the ratio of parent atoms to daughter atoms in the rock. Dating rocks using radioactive decay is called *isotopic dating*.

Measuring and Calculating Isotopic Ages

Geologists, working alongside chemists and physicists, use an instrument called a *mass spectrometer,* shown below, to measure the ratio of parent isotopes to daughter product in the rock or the mineral to be dated.

When some minerals form, they incorporate atoms of the parent isotope, especially if this ele-

ment fits in the mineral's crystalline structure. The mineral typically does not contain daughter atoms, which have a different atomic size and atomic charge than the parent atoms. Over time, radioactive decay converts parent atoms into daughter atoms, producing a specific and predictable proportion of parent and daughter atoms.

Geoscientists prepare a rock or mineral sample and place it in a mass spectrometer, where the sample is ionized and propelled down a tube toward a very strong electromagnet. The magnet pulls lighter atoms with lower atomic weights more than atoms that are heavier. The strength of the magnet can be altered by adjusting the amount of electric current passing through it. With the proper settings, only atoms of the desired atomic weight reach a detector at the end of the tube, which counts the number of arriving atoms. Mass spectrometers measure ratios of isotopes more easily than absolute amounts

of isotopes, so most results and calculations use ratios between isotopes. The results are calculated using equations and computer programs, and are commonly plotted on a graph.

Isotopes that decay quickly are used to date young rocks and archaeological artifacts. They are not appropriate for older materials because all of the parent element disappears quickly. Isotopes that decay very slowly are used to date ancient rocks. They cannot be used to date young materials because only minute amounts of daughter products will have formed in that short period of time. The dating process involves many potential complications and assumptions, so geoscientists select the correct isotope, and they consider and evaluate each assumption before applying the determined age to a rock. A key aspect to report, along with the age, is how precise the measurement and lab are—the analytical uncertainty.

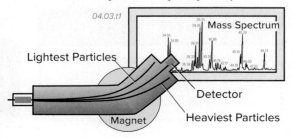

04.03.t1

Lightest Particles

Mass Spectrum

Detector

Magnet

Heaviest Particles

B What Isotopic Systems Do We Use to Determine Ages of Rocks and Events?

A number of elements have radioactively unstable isotopes, and geoscientists use these different isotopic systems to measure different types of ages on rocks, minerals, sediment, and other materials. Which isotopic system is used depends on what datable materials are present, the likely age of the materials, and the geologic history of the unit. Sedimentary rocks are the most difficult to date because most are composed of pieces of older rocks.

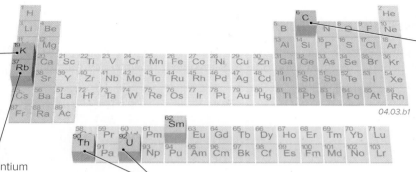

1. One isotope of potassium (K) decays to the noble gas argon (Ar, which is on the far right side of the periodic table), and to an isotope of calcium. We use K-Ar dating and the related Ar-Ar (pronounced argon-argon) method to date volcanic rocks and the cooling of deep rocks brought to the surface. These methods are most useful for dating Cenozoic and Mesozoic rocks, but they are also used for older rocks.

2. An isotope of rubidium (Rb) decays to strontium (Sr), providing the basis for Rb-Sr dating. This dating method provided some of the first ages for old granites and metamorphic rocks, and was key in demonstrating that some rocks on Earth and some meteorites were billions of years old.

3. A number of thorium (Th) and uranium (U) isotopes decay through a series of steps to different isotopes of lead (Pb), element number 82. We use these isotopes to date many kinds and ages of rocks, including sand-sized grains of certain minerals in sediments. Samarium (Sm) decays to Neodymium (Nd; element 60), and we use Sm-Nd dating mostly to date old rocks and to investigate sources of magma.

4. One isotope of carbon (C), carbon-14, is used to date wood, charcoal, bones, shells, and carbon-rich rocks and water. Carbon-14 has a relatively short half-life that makes it useful only for dating materials that are hundreds to thousands of years old.

04.03.b1

C What Can Isotopic Ages Tell Us?

We use different isotopic systems for isotopic dating, and the different types of isotopic ages do not all provide the same kind of information. Some record when the rock formed, whereas others record later cooling. Also, some methods are more precise for more recent rocks, and other methods are better for older rocks.

1. We date volcanic units using a variety of isotopic systems, like K-Ar and Ar-Ar. Volcanic rocks form on the surface and cool rapidly, so an age of the rock is typically the age of *eruption*.

2. Solidified magma bodies lose certain isotopes until they cool to a certain temperature, so we determine the age of such bodies using only those minerals that retain isotopes and provide the age of *crystallization*. Today, we mostly use U-Pb dating of the mineral zircon.

3. Some minerals, such as biotite mica in a granite body, tell us when a rock *cooled* through a specific temperature, as when it was being uplifted to the surface.

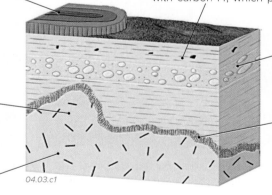

04.03.c1

4. Black pieces of charcoal incorporated into recent sediment can be dated with carbon-14, which provides an age for *deposition* of the sediment.

5. Dates from individual boulders, cobbles, or even sand-sized grains in a sedimentary rock help us to infer the age of the *source rocks* from which the sediment was eroded. The oldest ages ever measured, more than 4.4 billion years old, are for individual grains in sedimentary rocks from Australia.

6. We investigate the age of a *metamorphic event,* like baking next to a magma, using minerals that formed during metamorphism or minerals that record certain metamorphic temperatures. We can do this to date contact or regional metamorphism.

7. In many cases, geologists use different methods on a single rock to obtain information about different parts of the rock's geologic history (▶). For granite, a U-Pb age on zircon, a uranium-bearing mineral, can provide the age when the magma solidified, and an Ar-Ar age on biotite provides the time when the rock cooled below 300°C. By using different dating methods on the same rock, geologists can show when the rock formed and how fast it cooled through time, as plotted here.

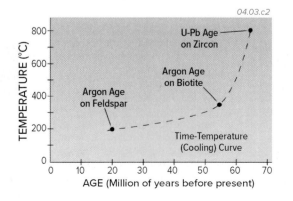

04.03.c2

Before You Leave This Page

✓ Explain how to determine how many half-lives have passed based on the ratio of parent to daughter atoms.

✓ Describe the different ways that isotopic dating is used for dating geologic events.

✓ Describe how a mass spectrometer is used to determine isotopic ages.

4.3

4.4 What Are Fossils?

ROCKS CONTAIN FOSSILS—EVIDENCE OF ANCIENT LIFE. Rocks of appropriate age and type preserve shells, coral, bone, petrified wood, leaf impressions, dinosaur tracks, and features created by burrowing animals. What kinds of creatures left these traces of past life, and how are they preserved in the rock record?

A What Are Fossils and How Are They Preserved?

Fossils are any remains, traces, or imprints of a plant or animal that are preserved in a rock or sediment. Fossils can be of different types depending on what type of life is involved, in what environment the plant or animal lived, and how the remains were buried and preserved.

04.04.a1 Caesar Creek SP, OH

 04.04.a2 Hot Springs, SD

 04.04.a3 Florissant Fossil Beds NM, CO

1. Most fossils found in the field are preserved *hard parts*, or parts that have been replaced by hard minerals of marine organisms, including shellfish and coral. The photograph above shows several types of fossilized shells of marine creatures.

2. Vertebrate animals have hard parts, most significantly *bones* and teeth, that can be preserved. Most bones are found as fragments instead of complete skeletons because of the destruction and dispersal caused by scavengers, weathering, erosion, and transport.

3. Some fossils are preserved because the original organic material is replaced by silica, pyrite, or some other material. One example is wood from trees that is replaced by fine-grained silica, forming *petrified wood,* in the case above, preserving a 34 m.y.-old stump.

 04.04.a4 El Paso, TX

 04.04.a5

 04.04.a6

4. Another type of fossil forms when an animal is buried and decays. This leaves a cavity in the rock that mimics the animal's shape. The cavity is a *mold* if unfilled (▲) and is a *cast* if it is later filled by minerals.

5. After burial, some carbon-rich plants and animals become *thin films* of carbon or other materials that preserve the original shape of the plant or animal. This fossil fern is almost 300 million years old.

6. Fish and other soft creatures can be preserved as *impressions,* especially when the remains come to rest in quiet waters of a lake or deep sea. Such fossils can preserve amazing details, including fins and scales.

 04.04.a7

7. Animals and plants can become fossils in other ways, such as becoming trapped in tree sap, which through time hardens into golden-brown *amber* (◄). Such preservation can preserve fragile features of the animal, like wings, legs, and antennae on these insects.

8. Some fossils do not preserve the actual organism, but instead represent something that the organism constructed. The mound-like features shown here (►), called *stromatolites,* were built by ancient microscopic algae hundreds of millions of years ago. To interpret how these ancient mounds formed, we compare them to mounds constructed by modern-day algae.

 04.04.a8 Carlsbad, NM

B What Traces Do Creatures Leave in the Rock Record?

In addition to preserved remains of organisms, rocks contain other features made by animals that moved across the surface or burrowed into soft sediment. Geoscientists call these features *trace fossils*.

Creatures that walk on land, such as reptiles, or on the sea bottom, like crabs, can leave *footprints* behind, such as this one from a Jurassic dinosaur. Most footprints are indentations in sediment that are filled by later sediment. A trail of related footprints is called a *trackway,* from which geoscientists can infer how the creature moved, how much it weighed, and whether it traveled alone or in a group.

04.04.b1 Moenave, AZ

04.04.b2 Grand Canyon, AZ

Worms and other creatures wriggle, dig, or tunnel into mud, forming cavities that can be filled by other sediment, producing a trace fossil. The type of trace fossil shown here is a *burrow*. The creatures were too soft to be preserved but still left behind a record.

C What Determines Whether a Fossil Is Preserved?

Most creatures are never preserved as fossils because fossil preservation requires certain favorable circumstances. The most important factors include the existence of hard body parts and rapid burial after death.

Hard Parts—Preservation as a fossil is much more likely if a creature, like this crinoid, has a shell, bones, teeth, or some other hard part. Only 30% of modern animals have hard parts, but such animals are overrepresented in the fossil record compared to animals like insects or jellyfish that lack hard parts. Soft parts of creatures can be eaten by scavengers, crushed during sediment compaction, dissolved by chemical reactions, or otherwise destroyed. Some ancient creatures with only soft parts were likely never preserved as fossils.

Rapid Burial—A fossil cannot be preserved unless it is buried. If a creature's remains are left on the surface, whether on land or in the sea, they can decompose due to exposure to the atmosphere and water, or can be scavenged by other creatures. Rapid burial means less opportunity for destruction. Preservation is easier beneath the sea than on land because burial is generally more rapid, and because a lower content of oxygen in the deep sea slows decay.

04.04.c1

Features That Look Like Fossils but Are Not

Some natural geologic features look like fossils but are not fossils. These features form through *inorganic* processes and do not represent the remains or traces of any organism. The most common features mistaken for fossils are the dark, branching mineral growths (▼). These growths, called *dendrites,* typically consist of very dark gray manganese-oxide minerals that grow in branching patterns along fractures and on beds within sedimentary rocks.

Spherical features, called *concretions,* which grow in sediment during cementation, are also commonly mistaken for fossils. These weather out of sediment as small spheres (▼), lenses, or oddly shaped objects that can look organic. Formation of concretions can involve some biologic processes, but concretions are not fossils.

04.04.t1

04.04.t1

Before You Leave This Page

☑ Describe the different ways in which a plant or animal can be preserved as a fossil.

☑ Describe two types of commonly encountered trace fossils.

☑ Describe the two main factors that influence whether a creature is preserved as a fossil.

☑ Describe a feature that can be mistaken for a fossil.

4.4

4.5 How and Why Did Living Things Change Through Geologic Time?

DIFFERENT FOSSILS OCCUR IN DIFFERENT ROCK UNITS. Some of these differences reflect variations in sedimentary facies (different depositional environment), for example, between reefs and streams, but most reflect the systematic way that living things and their fossils varied over geologic time. Why did these changes in fossils occur, and what do they tell us about how life on our planet has changed through time?

A How Do Fossils Vary with Age?

Early geoscientists recognized that fossils change upward from older layers of sedimentary rock to younger layers. This systematic change of fossils with age, called *faunal succession,* helped geoscientists identify time periods defined by major changes in life on Earth. Using the principles of relative dating and faunal succession, geoscientists subdivided geologic time into four major chapters, each with subdivisions. Later, results from isotopic dating provided numeric ages, in millions of years before present, for when each chapter started and ended.

04.05.a1

| Cenozoic |
| 66 Ma |
| Mesozoic |
| 252 Ma |
| Paleozoic |
| 541 Ma |
| Precambrian |
| (Started about 4500 Ma) |

1. The *Cenozoic Era,* meaning *recent life,* spans the last 66 million years. It is called the *age of mammals* because mammals, such as these (fossilized) mammoths (▼), became a dominant type of life on Earth.

04.05.a2 Hot Springs, SD

2. The *Mesozoic Era (middle life)* is known as the *age of dinosaurs* because dinosaurs (▼) rose to dominance during this era. The end of the Mesozoic Era, at 66 Ma, is marked by the extinction of dinosaurs.

04.05.a3 Dinosaur National Monument, UT

3. *The Paleozoic Era (ancient life)* was dominated by several major groups of marine animals, including coral, creatures like clams that had hard shells (▼), and various types of fish. Plants, insects, and amphibians also colonized the land during this era. The end of the Paleozoic Era is marked by a major time of extinction called the *Great Dying.* This extinction killed off many species of animals in the seas and to a lesser extent on land.

4. The *Precambrian (before the Cambrian Period)* comprises nearly 90% of geologic time. For most of this time, only simple life forms existed, such as bacteria and algae that formed stromatolites like those shown here (▼).

04.05.a5 Grand Canyon NP, AZ

04.05.a4 Caesar Creek SP, OH

5. The colors on this timescale, corresponding to the different eras, are fairly standard in geologic publications. They are commonly used on geologic maps and other figures to convey the ages of sedimentary and volcanic rock units. Granites and related rocks are usually shown in red, irrespective of age.

B What Factors Determine Whether a Species Survives or Becomes Extinct?

Boundaries between the major chapters of the geologic timescale are defined either by the emergence of new life-forms, by massive extinctions, or both. Almost all species that ever existed are now extinct. What factors influence whether a species survives or becomes extinct?

04.05.b1 Namibia

04.05.b2 Amboseli NP, Kenya

04.05.b3 Vulture Mtns., AZ

Environmental Setting—Animals, plants, and other organisms have certain ways they live, and certain survival needs. Some organisms thrive because they developed along with other plants or animals that provide them with essential food, habitat, or other needs.

Climate—Changes in climate resulting in loss of water and food sources, along with other changes in critical habitat, can threaten a regional population of animals. Environmental stresses, including disease, can eliminate entire classes of animals.

Reproductive Strategy—Different plants and animals reproduce in different ways. These plants flower and produce seeds, whereas the cacti grow new versions from small parts of the original plant. Some reproductive strategies will be more successful than others.

04.05.b4 Galápagos Islands

04.05.b5 Namibia

04.05.b6 Namibia

Adaptability—The more adaptable a species is, the more likely it will survive changes in the environment, such as increases of temperature or salinity of water. These marine iguanas started as land animals but developed the ability to also forage in the sea.

Competition—If two or more species are competing for the same sparse resources, there are likely to be winners and losers. Competition between members of a single species can also be a problem if it means that needed resources are in short supply for all.

Predators and Prey—Being a food source for some other creature is never a good survival strategy. The opposite is also true—if an animal relies on only one kind of food, survival becomes problematic if that food source becomes scarce or even disappears.

Evolution: Observed Changes and Possible Explanations

The term *evolution* is used in two ways. First, it refers to *observed changes* in the fossil record or documented changes in more recently living animals. This is commonly called the "fact of evolution." Second, evolution refers to the *theories* that help explain the observed changes.

Observed changes in the fossil record over time are well documented and can be verified by anyone who studies fossil-bearing rocks from different geologic times. For more than a hundred years, geologists and paleontologists (geoscientists who study fossils) have used fossils to compare life-forms from rocks of different ages around the world. These comparisons are supported by many isotopic ages.

The *theory of natural selection* originated with Charles Darwin to explain the birds and other animals of the Galápagos Islands.

Using this and other evolutionary theories, paleontologists try to explain how a Paleozoic fish developed front fins strong enough to support its weight on land, a mutation many paleontologists accept as having eventually led to amphibians.

One evolutionary hypothesis, called *punctuated equilibrium,* explains how new organisms, or new characteristics of an existing organism, appear rather suddenly in geologic terms, instead of evolving more gradually. Proponents argue that new and favorable mutations are more likely to succeed in small, isolated populations than in large populations. After a favorable change develops fully in a small group, the group may rejoin the larger population and outcompete the other individuals, causing an observed evolutionary change. Evolution can proceed slowly or more rapidly, in geologic terms, depending on the environmental setting and ecological pressures.

Before You Leave This Page

✓ Describe the four chapters of Earth history and how the boundaries are defined.

✓ Describe some factors that affect survival and extinction.

✓ Describe the difference between observed fossil changes (evolution) and evolutionary theory.

4.5

4.6 How Was the Geologic Timescale Developed?

GEOSCIENTISTS DEVELOPED THE GEOLOGIC TIMESCALE to help them correlate rock units across regions and continents and to have a standard vocabulary for describing geologic time. The timescale was devised by using fossils or by noting the absence of fossils. Geologists and paleontologists commonly establish boundaries between units at places in the rock section where fossils record major changes in the types of life. Later, geoscientists and chemists assigned numeric ages to the timescale by using carefully calculated isotopic ages at key localities.

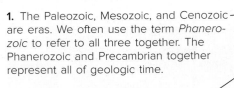

A What Are the Main Subdivisions of the Geologic Timescale?

After it was established that fossil assemblages change upward through sections of rock, geoscientists concluded that two different sites that have matching fossil assemblages were the same age. They recognized sequences of related layers across Europe and in North America and named different geologic time periods after places where rocks of that age are well exposed. The largest time intervals shown below are *eras,* and include the Paleozoic, Mesozoic, and Cenozoic eras, from oldest to youngest. Boundaries between eras are marked by major changes in the fossils, specifically the disappearance (extinction) of many species and families of creatures. Such major extinctions are referred to as *mass extinctions.* Geoscientists subdivided each era into several *periods,* shown below (and at the very back of the book) with the derivation of the name of each period shown to the right of the column.

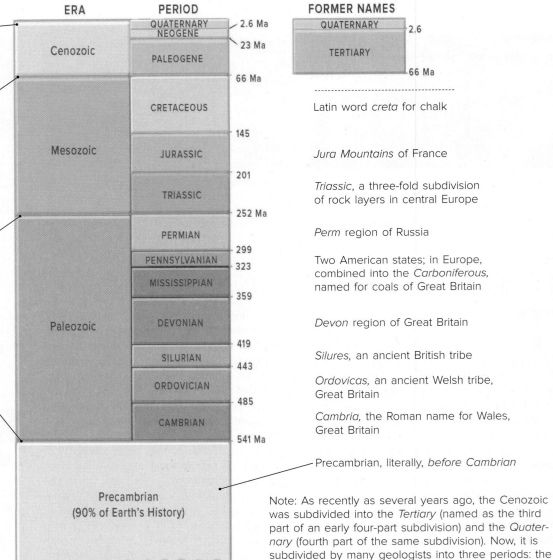

1. The Paleozoic, Mesozoic, and Cenozoic are eras. We often use the term *Phanerozoic* to refer to all three together. The Phanerozoic and Precambrian together represent all of geologic time.

2. This boundary was based on the disappearance of many life-forms (mass extinction) at the end of the Mesozoic Era. This boundary also is called the **K-P** boundary because it separates the Cretaceous (whose abbreviation is K) from the Paleogene. This boundary is well dated at 66 Ma (million years before present).

3. This boundary was based on a mass extinction called the Great Dying, which took place at the end of the Paleozoic Era, currently dated at 252 Ma.

4. This boundary, between the Precambrian below and Paleozoic above, was based on the widespread appearance of hard-shelled organisms at the beginning of the Cambrian Period, an event called the *Cambrian explosion.* The boundary is currently dated at 541 Ma.

5. Although not shown here, the Precambrian is also subdivided into three parts (eons). The earliest part of the Precambrian is called the *Hadean* and refers to Earth's earliest history, from 4.6 to 4 billion years ago (Ga). The middle part is the *Archean,* which spans from 4 billion to 2.5 Ga. The most recent part of the Precambrian is the *Proterozoic,* which represents the time from 2.5 Ga until the start of the Cambrian at 541 Ma.

Latin word *creta* for chalk

Jura Mountains of France

Triassic, a three-fold subdivision of rock layers in central Europe

Perm region of Russia

Two American states; in Europe, combined into the *Carboniferous,* named for coals of Great Britain

Devon region of Great Britain

Silures, an ancient British tribe

Ordovicas, an ancient Welsh tribe, Great Britain

Cambria, the Roman name for Wales, Great Britain

Precambrian, literally, *before Cambrian*

Note: As recently as several years ago, the Cenozoic was subdivided into the *Tertiary* (named as the third part of an early four-part subdivision) and the *Quaternary* (fourth part of the same subdivision). Now, it is subdivided by many geologists into three periods: the Quaternary, Neogene (*Neo* for *new*), and Paleogene (*Paleo* for *ancient*), as shown in the main column.

04.06.a1

B How Are Numeric Ages Assigned to the Geologic Timescale?

Geoscientists have assigned numeric ages to geologic periods and their subdivisions by studying localities where isotopically dated igneous rocks have a clear relationship to fossil-bearing layers.

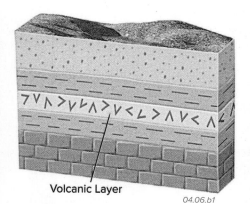

The isotopic age of a volcanic layer interbedded with fossil-bearing sedimentary rocks (◄) can be used to assign a numeric age to the geologic period during which these types of fossils formed. Also, the age of a fossil-bearing bed can be bracketed by dating volcanic units above and below it; the bed is younger than a volcanic unit beneath it and older than a volcanic unit above.

Volcanic Layer

04.06.b1

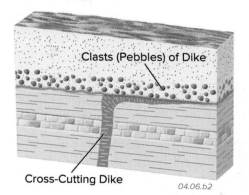

Clasts (Pebbles) of Dike

Cross-Cutting Dike

04.06.b2

Fossils must be older than the ages of dikes that cross cut the fossil-bearing layer (◄), like the lower green shale. In contrast, the upper tan-colored, fossil-bearing layer is younger than clasts of the dike included in the layer.

C How Is the Timescale Used to Assign Numeric Ages to Rocks and Events?

Once the ages of the periods and shorter units of the geologic timescale were constrained, these ages could be used to estimate numeric ages of fossil-bearing units that lack datable igneous rocks.

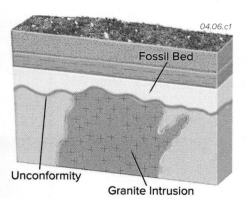

04.06.c1

Fossil Bed

Unconformity

Granite Intrusion

The age of this unconformity, shown by the squiggly line (◄), can be bracketed by using both isotopic ages and fossil ages. The unconformity is younger than the isotopic age of the granite. It is older than the age assigned to the fossils in the overlying bed based on their position in the geologic timescale.

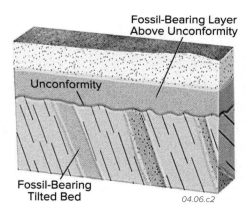

Fossil-Bearing Layer Above Unconformity

Unconformity

Fossil-Bearing Tilted Bed

04.06.c2

We can bracket the age of an unconformity (◄) by using fossils and the timescale to assign ages to rocks above and below the unconformity. The unconformity is younger than tilted sedimentary beds below, but older than rocks above.

How to Remember the Geologic Timescale

Students have developed many techniques to help them remember the names and numbers of the geologic timescale. What do we recommend? Many students use a *mnemonic* device in which the first letters of the mnemonic words match those of the names of the subdivisions of the timescale. One mnemonic for the periods of the Paleozoic, Mesozoic, and Cenozoic eras is the following:

Cuddly Old Sheep Dogs Make Perfect Pets; They Just Crowd People Nearby Quietly.

This mnemonic stands for Cambrian, Ordovician, Silurian, Devonian, Mississippian, Pennsylvanian, Permian, Triassic, Jurassic, Cretaceous, Paleogene, Neogene, and Quaternary. Envision the sheepdog and each part of the sentence. You can also use the mnemonic to help you draw a visual representation of the geologic timescale. Practice filling in the associated numeric ages until it becomes easier to draw and thus remember the names of the time periods and the numeric ages that mark the boundaries between the Precambrian, Paleozoic, Mesozoic, and Cenozoic.

Before You Leave This Page

✓ Briefly summarize how the geologic timescale was developed.

✓ From oldest to youngest, list the four main geologic chapters and periods.

✓ Explain or sketch how numeric ages are assigned to the timescale and how the timescale is used to assign numeric ages to fossil-bearing rocks.

4.6

4.7 What Is the Evidence for the Age of Earth?

EARTH IS 4.55 BILLION YEARS OLD. Early geoscientists suspected that Earth had a long history and devised several approaches to estimate the age of Earth. The advent of isotopic dating techniques finally provided the tool needed to demonstrate that Earth is indeed very old. What evidence indicates that Earth is billions of years old and not several thousand years old?

A What Were the Early Attempts to Estimate the Age of Earth?

For centuries, scientists tried to figure out ways to date the age of Earth. Many of these ideas were reasonable for their time, but key information was missing and so early estimates of Earth's age were too young. Radioactive decay was not discovered until the late 1800s, but its use revolutionized thought about the age of Earth.

04.07.a1 Brazos River, Bryan, TX

An early method to determine the age of Earth calculated how fast salt had accumulated in the oceans. By measuring the salinity and volume of water flowing in rivers (◄) and estimating evaporation rates, scientists could calculate how much time it would take for ocean water to attain its present salt content. The calculation yielded an age of 90 million years, which was far too young, because the estimate did not consider salt lost from the oceans to salt beds.

04.07.a2

Lord Kelvin, a late 1800s scientist, estimated Earth's age by calculating how long a molten Earth should take to cool to its present temperature. Using thermal properties of rocks and estimates of Earth's internal temperature, he calculated an age of 100 million years. This estimate was done before the discovery that radioactive decay adds internal heat, partly explaining why Lord Kelvin's estimate is too young.

B What Is the Evidence That Earth's History Is Not Short?

Events that are seasonal and leave a physical record, such as the growth rings in trees, are easily observed. Using such features to estimate age requires only that we assume that processes that are occurring today also occurred in the past. Satellite data provide another way to check rates of geologic processes.

04.07.b1

1. *Tree rings* (◄), which record annual growth cycles, have textures that vary between seasons and widths that vary from dry years to wet ones. Thickness patterns in bristlecone pine rings from the American West can be correlated from living trees to dead ones to form a continuous record back to 9,000 years ago. This approach is independent of, but strongly supported by, many carbon-14 ages.

2. *Ice cores* (▶), cylinders of ice drilled from ices sheets in Greenland and Antarctica, contain thin layers produced by yearly cycles of the seasons. Some ice-core records are thousands of meters long, representing tens of thousands of years to well over 100,000 years. Determining the age is done by counting the rings, along with determining isotopic ages for layers of volcanic ash within the ice.

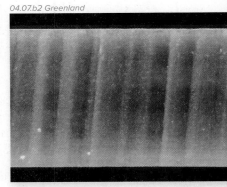

04.07.b2 Greenland

04.07.b3 Connecticut Valley, NH

3. *Varves* are alternating light and dark sediment layers (◄) that form in lakes because of seasonal variations of sedimentation and biologic activity. Lighter layers represent increased accumulations of sand and silt during the summer, whereas darker layers record the slower deposition of mud during the winter. There are more than 4,800 varves in glacial lakes that existed in New England around 10,000 to 14,000 years ago, and the Green River Formation in Wyoming has millions of varves that represent at least several million years.

4. Current rates of plate motion can be measured precisely using high-precision GPS data (▶) to be on the order of centimeters per year. At these rates, it would have taken the Atlantic Ocean more than 100 million years to form as Africa and the Americas moved apart. Similar rates of motion explain the ages of the Hawaiian Islands and displacement on the San Andreas fault.

04.07.b4

C   What Are the Oldest Dated Rocks on Earth?

Geoscientists have dated rocks from several places, including Australia, Canada, and Greenland, at more than 3.5 billion years old. The oldest dates (4.4 b.y.) are on individual grains in metamorphic rocks.

04.07.c1 Front Range, CO

The oldest rocks in any region are generally metamorphic rocks (▲) that display evidence of a protracted and complex geologic history. One of the oldest dated rocks on Earth is a metamorphic rock, named the Acasta Gneiss, in northwestern Canada. It is interpreted to have started as a granite that was later metamorphosed. Its age of nearly 4.0 billion years, when the granite solidified, was determined using uranium dating on zircon. Other examples of Earth's most ancient rocks are in western Australia and in northern Quebec, Canada. Some of these rocks contain even older zircon crystals derived from even older rocks.

Many parts of North America have Precambrian metamorphic and granitic rocks beneath a younger sedimentary cover. The crystalline rocks, including those shown here (▶), have been dated at thousands of sites and provide a systematic regional pattern of ages across North America. The oldest rocks, some more than 3 billion years old, are in crystalline rocks exposed across a broad expanse of Canada and the Great Lakes region, an area known as the *Canadian Shield*. Similarly old rocks are exposed in the uplifted mountains of Wyoming and adjacent areas. From these old centers of the continent, the ages systematically decrease to the southeast, being about 1 billion years in Texas and the southeastern United States. The pattern of ages indicates that the southern part of North America was added to an older, northern part. The consistency of the pattern is strong verification of the methodology and supporting knowledge that are the basis of isotopic dating.

04.07.c2 Grand Canyon, AZ

Where We Get the 4.55 Billion-Year Age of the Earth, and the Concept of Deep Time

Earth and the solar system are interpreted to be the same age, as measured from the radioactive decay of isotopes in Moon rocks and in meteorites. The Moon is interpreted to have formed early in Earth's history, and meteorites are presumed to represent shattered rocky fragments formed at the same time as Earth.

Geologists and planetary scientists collect meteorites around the world, especially in Antarctica, where the dark rocks stand out on the light-colored snow and ice. They have dated various types of meteorites. The oldest ages are interpreted to represent the time just before the planets cooled. The meteorite analyses support formation of the solar system and Earth between 4.53 and 4.58 billion years ago. The meteorite shown here is the Allende meteorite, dated at 4.56 billion years old.

Nine missions to the Moon returned a limited number of samples of Moon rocks for isotopic dating. The oldest Moon rocks are 4.4 to 4.5 billion years old.

The Milky Way Galaxy currently is estimated to be approximately 13 billion years old, nearly as old as the Universe (14 billion years). This age is based on several methods, including the stage of evolution of certain features in our galaxy compared to similar features in other galaxies.

Geoscientists and chemists have obtained isotopic ages for many different meteorites, several Moon rocks, and countless Earth rocks. The oldest dates converge on 4.5 billion years before present, even though the rocks are from very different places and more than one dating method was used. The slightly younger age, 4.0 billion years, for the "oldest dated rock" on Earth is expected because erosion, deposition, and tectonic activity remove rocks, bury them, deform and metamorphose them, and even melt them. Some geologists are surprised that such an old rock was able to survive at all. A 4.4 billion-year-old zircon grain from Australia is even closer to the age of the meteorites.

There is remarkable consistency between ages from meteorites, the Moon, and Earth. This consistency strongly supports the 4.55 billion-year age for Earth, the Moon, and the solar system.

04.07.t1

We can understand that the age of Earth is well determined at 4.55 billion years old, but it is still very difficult, even for geologists, to fully grasp the incredible duration of geologic time. For this reason, geologists often talk about geologic history in terms of the concept of *deep time*. This is an acknowledgement that geologic time is long and nearly impossible for humans to fully appreciate, given our direct experience with events that only last seconds to years. The uplift and eroding away of entire mountain ranges seems like an impossible task, but it is possible with the very long time available over geologic time—deep time.

Before You Leave This Page

☑ Describe early methods for determining the age of Earth and why they proved to be inaccurate.

☑ Describe evidence that suggests Earth has a long history, including isotopic ages on basement rocks in North America.

☑ Describe how meteorites and Moon rocks help constrain the age of Earth.

4.7

4.8 What Events Occurred Early in Earth's History and How Did Earth Change Over Time?

THE EARLIEST CHAPTERS OF EARTH'S HISTORY featured many processes and events similar to those observed today, but at vastly different rates and intensities. Some early processes and events are not similar to those observed today because of changes through time in the amount of our planet's internal heat, in the formation and evolution of the atmosphere and oceans, and in the changing role of life. What do you know about Earth's earliest history?

A What Are Impacts and How Are They Expressed on the Surface?

Our Solar System has a huge number of asteroids, comets, and other extraterrestrial objects (i.e., coming from outside planet Earth), which have the potential of colliding at very high velocities with other objects, including planets and moons. During such collisions, called *impacts*, the energy of the collision causes the formation of an *impact crater* on the larger object, if the objects do not simply break apart. There were vast numbers of such objects early in the Solar System's history, so Earth's earliest chapters were dominated by impact cratering. The formation of impact craters is illustrated below.

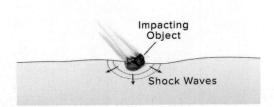

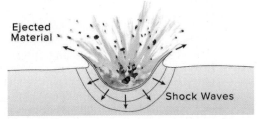

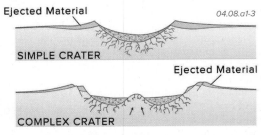

04.08.a1-3

During an impact, much kinetic energy of the fast-moving object is transferred onto the rocks near the impact, sending shock waves through the subsurface. This process fractures and shatters the adjacent rocks and can even cause some of the rocks to melt (an *impact melt*). The impacting object breaks apart, scattering pieces across the surface, or the object can be totally vaporized.

The energy of the impact excavates the subsurface, throwing (ejecting) pieces of the impacting object and of the underlying material in all directions from the impact site. The ejected material generally forms a ringlike mound or collar of rock fragments around the crater. Steep walls of the crater can collapse downward in landslides and talus slopes, as is observed on Mars, the Moon, and elsewhere.

Two main types of craters are formed, depending on the size of the impact. If the impact event is not too large, it forms a relatively simple crater surrounded by a fringe of ejected material. If the collision is larger, it forms a complex crater that has a central peak formed by shock waves converging back toward the center of the crater. We can observe both types of craters on our Moon.

B What Types of Rocks Formed Early in Earth's History?

After the surface of Earth had solidified, its surface was bombarded by impacts, covering it with craters, much like on our Moon. Early in its history, Earth's interior is inferred to have been much hotter than today because of heat left over from its formation, heating from impacts, and from a higher concentration of radioactive isotopes. As a result of these factors, different types or abundances of rocks formed early in Earth's history (i.e., during the Archean) than form today.

During the earliest part of Earth's history, much loose space debris was left over from the solar nebula. Some of this material collided with Earth, forming large impact craters. An impact causes shattering of rocks near the crater and the huge imposed stresses can cause

04.08.b1 Vredefort, South Africa

04.08.b2 Noranda, Quebec, Canada

rocks to melt. The photograph here (▲) shows fragments of shattered rocks surrounded by a dark matrix that was originally melt that rapidly solidified to glass. The frequency of these collisions was very high early on, but decreased over time, as less debris remained in space.

Another rock type common in Archean exposures is *greenstone*, representing slightly metamorphosed lava flows. We interpret volcanism to have been very active in Earth's early, hot times, erupting ultramafic lavas that were hotter than those erupting today. Some copper deposits were formed at this time from hot water associated with submarine volcanism. Plate tectonics may not have operated during these hot early times, but the planet progressively lost more heat and plate tectonics dominated Earth's later chapters.

C How Did Earth's Atmosphere and Oceans Form and Change Over Time?

Earth's earliest atmosphere is interpreted to have contained hydrogen, methane, ammonia, and water vapor, but the lightest of these gases were lost relatively easily from a planet of Earth's size. Other gases, such as nitrogen gas and carbon dioxide, accumulated early in Earth's history, but the early atmosphere lacked significant oxygen. The amount of oxygen increased in two steps or transitions, called *oxygenation events*. These had dramatic effects on the planet.

04.08.c1 Mt. Bromo, Java, Indonesia

04.08.c2 Johannesburg, South Africa

04.08.c3 Sherman Mine, Ontario, Canada

1. The earliest atmosphere of Earth had light gases that were part of the accretion from the solar nebula. As they were lost, other gases, including carbon dioxide, water vapor, and sulfur-bearing gases, were produced from within the Earth and released into the atmosphere, such as during volcanism.

2. Early life on Earth consisted of cyanobacteria, often called blue-green algae, which formed mound-like stromatolites (▲). Cyanobacteria conduct photosynthesis, producing oxygen gas as a by-product. By this process, these tiny organisms oxygenated (added oxygen to) Earth's early atmosphere and oceans.

3. Earth originally lacked an ocean, such as during its mostly molten phase, but it gradually accumulated enough water to fill low spots to form early oceans. Halfway through Earth's history, enough oxygen was added to the atmosphere and oceans to allow iron to become oxidized, forming *iron formation* (▲), red- and gray-banded rocks composed mostly of iron oxide minerals and quartz.

Major Oxygenation Events

4. By studying the character and isotopic content of sedimentary rocks through time, we can estimate how the composition of the atmosphere has changed over the last 3 to 4 billion years. The graph below shows how the oxygen content of Earth's atmosphere is interpreted to have changed over time, based on various types of data. For example, some materials, like the sulfide mineral pyrite, are not stable on the surface if the atmosphere contains much oxygen, such as the conditions today. These materials, however, are present in Archean sedimentary rocks. This observation and other data indicate that Earth's atmosphere contained almost no oxygen during most or all of the Archean.

5. The lack of oxygen in Earth's early history has many implications. In addition to having no oxygen, the atmosphere contained methane, ammonia, and other gases that would have been toxic to most modern organisms, including humans. Methane is a powerful greenhouse gas, probably helping to keep Earth warm at that time. With no oxygen but abundant carbon dioxide in the atmosphere, weathering processes would have been very different. For example, there was not enough oxygen to form the distinctive red-colored iron oxide minerals that we commonly associate with soils and oxidation on and near the surface.

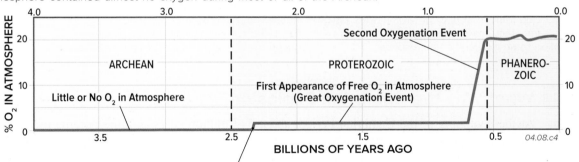
04.08.c4

6. Various types of evidence show that free oxygen (oxygen as a gas or dissolved in water, not oxygen tied up in minerals) first appeared in the atmosphere and oceans at about 2.3 to 2.4 billion years ago, in what is called the *Great Oxygenation Event* (GOE). The first appearance of free oxygen had many ramifications. Prior to this time, large quantities of iron were able to remain dissolved in the oceans, but the increase in oxygen during the GOE caused the iron to be precipitated in the oceans, forming large volumes of iron formation between 2.4 and 1.8 billion years ago. The oldest reddish sedimentary rocks (red beds) are this same age, as the increase in oxygen in the atmosphere allowed iron minerals to begin oxidizing and turning red. The only plausible source of oxygen is from photosynthesis by cyanobacteria.

7. Free oxygen first appeared during the GOE, but the levels were still very low, about 1% of the total atmospheric gas content. The levels stayed low for the next 1.7 billion years, during most of Proterozoic time. A second oxygenation event, bringing oxygen levels to near-modern levels of about 20%, occurred late in the Proterozoic, well before the start of the Cambrian (541 Ma), which was marked by the appearance of animals with shells and other hard parts. Geoscientists are actively researching the cause and timing of this second oxygenation event and the role that this event may have played in the appearance and rapid diversification of complex life after that time. Unlike cyanobacteria, which produce oxygen, most modern life-forms require oxygen to live. Such life-forms, including humans, were not possible until the oxygen levels rose.

Before You Leave This Page

☑ Sketch and describe how an impact crater forms and when in Earth's history they were most abundant.

☑ Summarize changes in composition of the atmosphere and oceans, what caused them, and how they were expressed in the rock record.

4.8

4.9 What Were Some Milestones in the Early History of Life on Earth?

LIFE ON EARTH BEGAN before there was much oxygen in the atmosphere. Very simple, but successful, Precambrian organisms evolved, started to photosynthesize, and eventually produced an oxygen-rich atmosphere. During the Paleozoic Era, from 541 to 252 Ma, more complex organisms appeared, eventually including fishes, plants, amphibians, reptiles, insects, and various types of marine organisms.

A What Were the Earliest Forms of Life on Earth?

Some of Earth's first inhabitants were early forms of algae, represented by fossils called *stromatolites* found in Australian rocks that are 3.5 billion years old, and rocks elsewhere that are only slightly younger. These organisms lived when Earth had little or no oxygen in its atmosphere. Some still live today, in certain special environments.

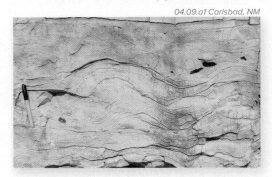
04.09.a1 Carlsbad, NM

04.09.a2 Australia

04.09.a3 Rockwood, CO

This rock contains *stromatolites,* the earliest type of non-microscopic fossils identified on Earth. The cyanobacteria that form modern stromatolites use photosynthesis to make food, and if those in ancient stromatolites did the same, they may have begun the transformation to an oxygen-rich atmosphere.

Ancient stromatolites probably had a similar appearance and structure to modern stromatolites in Australia (▲). Today's stromatolites live in an oxygen- and nitrogen-rich atmosphere, whereas the atmosphere 3.5 billion years ago had more carbon dioxide, which was produced mostly by outgassing from volcanoes.

About 2 billion years ago, cyanobacteria had produced enough oxygen through photosynthesis to increase the amount of oxygen in the atmosphere and form a protective ozone layer. Sometime after 1.5 billion years ago, organisms began to reproduce sexually, which led eventually to complex, multicellular organisms.

B What Was the Cambrian Explosion?

At the beginning of the Cambrian Period, about 542 million years ago, life became more diverse and included organisms with hard protective coverings. Creatures with shells had an obvious advantage over earlier soft-bodied organisms. Over a period of about 20 to 30 million years, many new shelled organisms appeared on Earth. This period of rapid evolutionary change is called the *Cambrian explosion.*

04.09.b1 House Range, UT

04.09.b2

04.09.b3

Trilobites were one of the dominant organisms of the Cambrian Period. They had external skeletons and diverse appearances, and lived in a wide range of environments. Many are index fossils for Paleozoic rocks.

The Cambrian seas produced simple forms of marine animals related to clams, starfish, sponges, and crabs, including *brachiopods* like those shown above.

Some of the best examples of Cambrian fossils come from a shale in the Canadian Rockies. The shale preserves more than 150 species, including impressions of the soft parts of some rather odd creatures, like these (▲).

C What Life Existed During the Paleozoic Era?

During the Paleozoic Era, an extraordinary diversity of life evolved—both in the seas and on land. Artistic reconstructions of three times of the Paleozoic were produced by Karen Carr, whose work is featured in museums.

Early Paleozoic
04.09.c1

In the early Paleozoic, corals, crinoids (which look like platy underwater lilies), and mollusks were anchored to the seafloor, trilobites and snails moved across the seafloor, and shelled creatures with tentacles propelled themselves through the water.

Middle Paleozoic
04.09.c2

In the middle of the Paleozoic, corals built large reefs, and pieces of crinoid stems littered the seafloor. Fish became diverse and abundant. On land, many forms of insects appeared, and plants included ferns and seedless trees.

Late Paleozoic
04.09.c3

Amphibians and early reptiles evolved during this time, with a dramatic rise of reptile groups and a continued diversity of marine life. Land plants, insects, and marine life continued to diversify until a major extinction (the Great Dying) at the end of the Paleozoic.

Possible Causes of the Great Dying

The end of the Paleozoic Era marks Earth's greatest extinction, called the *Great Dying*. On land, about 70% of all species, including many invertebrates, amphibians, and reptiles, went extinct. The event took a huge toll in the oceans, extinguishing almost 90% of marine species, including trilobites. Geologists are still actively investigating a number of possible causes.

A great outpouring of lava occurred at the end of the Paleozoic. Large volumes of basalt erupted in northern Asia, in a region called the *Siberian Traps* (*trap* is an old word used to describe basalt). Such eruptions expel volcanic ash and gases, including water vapor, carbon dioxide, and sulfur dioxide. The ash and gases have the potential to warm or cool the planet and possibly cause other catastrophic effects, such as changing circulation patterns in the oceans. Results of recent isotopic dating demonstrate that these eruptions and the extinctions occurred at the same time. There is some debate about whether extinction on land occurred at the same time as in the oceans.

There is some evidence, currently being debated, for a large meteorite impact at the end of the Paleozoic. Geologists have proposed that an impact can explain unusual carbon molecules found in rocks of this age. Geologists have found several suitably large impact craters, but none directly tied to the extinction. A huge impact could have triggered the massive eruptions of the Siberian Traps, but this connection remains conjectural.

Throughout most of the Paleozoic, continents were separated by warm, shallow seas. By Permian time, all the continents were connected as part of the supercontinent *Pangaea*, whose formation closed seas that had once nourished Paleozoic life. The supercontinent became more arid, and vast evaporite deposits formed that could have changed the salt concentrations in seawater. These and other effects of the formation of Pangaea may have helped kill off specialized organisms and set the stage for a more dramatic event.

An alternative explanation is that conditions in the atmosphere and oceans led to a massive overturn of ocean water, causing deep, oxygen-poor water to be brought to the surface. This could have caused a dramatic change in shallow ocean temperatures and in the amount of CO_2 in the atmosphere, leading to sudden and catastrophic climate changes. Such changes could affect the entire planet, resulting in a mass extinction on the land and in the oceans. This theory, like the others, is unproven, and the Great Dying remains a geologic mystery with several possible explanations for its cause.

Before You Leave This Page

☑ Describe the environments of early life and some important evolutionary events that took place during Earth's early history.

☑ Briefly describe what happened during the Cambrian explosion.

☑ Explain four possible causes for the Great Dying, the largest extinction event in Earth history.

4.9

4.10 What Were Some Milestones in the Later History of Life on Earth?

MASS EXTINCTION AT THE END OF THE PALEOZOIC ERA provided evolutionary opportunities for new life-forms. The organisms that repopulated the early Mesozoic seas and lands were very different from Paleozoic organisms. Diverse life existed during the Mesozoic Era (252 to 66 Ma), including dinosaurs. The end of the Mesozoic Era is defined by another major extinction event, which gave rise to yet another evolutionary chapter, the ascent of mammals during the Cenozoic Era. The artwork of Karen Carr provides us with one interpretation of the scenes represented by the bones, shells, leaves, and other fossils.

A What Life Was Abundant During the Mesozoic Era?

Diverse life existed during the Mesozoic Era, but it is known as the *age of dinosaurs*, the best known creatures of this time. The Mesozoic has three periods: Triassic, Jurassic, and Cretaceous, from oldest to youngest.

Early Mesozoic: Triassic   04.10.a1

Middle Mesozoic: Jurassic   04.10.a2

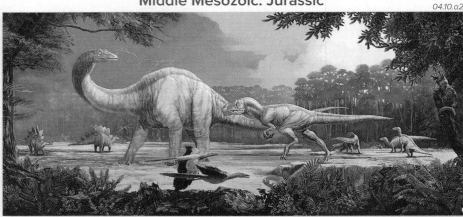

During the Triassic Period, small and nimble dinosaur-like creatures and mammals appear beneath the seed-bearing conifer forests. In the seas, shallow-sea niches left open by the Permian extinction were occupied by coiled ammonites and other marine animals.

Dinosaurs diversified and many new species appeared during the Jurassic Period, including *Stegosaurus* with plates on its back and the huge plant-eating *Apatosaurus*. Carnivorous predators, like *Allosaurus,* stalked the landscape. The Jurassic Period also featured *Archaeopteryx,* an early bird. The seas flourished with many diverse creatures, including ammonites, starfish, and large marine reptiles.

Late Mesozoic: Cretaceous

04.10.a3   04.10.a4

During the Cretaceous Period, dinosaurs remained diverse, and included various plant-eating dinosaurs that walked on four or two legs, as well as predators like the raptors lurking in the bushes. Flying reptiles and birds graced the skies. Not shown is the fearsome *Tyrannosaurus rex*. For the first time, flowering plants, called *angiosperms,* became abundant on land. Insects remained a vibrant and diverse group, and most mammals continued a rather low-key existence.

During the Cretaceous Period, animals similar to those of the Jurassic thrived in the seas, including fish of many kinds, straight and coiled nautiloids, large marine reptiles, and turtles. Not shown because of their tiny size are countless floating and drifting organisms called *plankton*.

B   What Were Dinosaurs and What Caused Their Demise?

Dinosaurs evolved from Permian ancestors and existed on Earth for 165 million years, throughout most of the Mesozoic. By the middle of the Mesozoic, they dominated the land, but they and many other animals went extinct at the end of the Mesozoic Era, at what has traditionally been called the *K-T extinction*. With changes in names on the geologic timescale, it has been called the *K-P extinction* by some scientists, because it separates the Cretaceous Period (K) from the Paleogene (P) Period. Geoscientists and other scientists have proposed numerous hypotheses to explain the extinction.

1. There were two types of dinosaurs, differing in their hip structure. One group of dinosaurs had a hip structure similar to lizards and included a diverse group of carnivores, such as *Tyrannosaurus rex,* and herbivores, such as *Apatosaurus.* Some walked slowly on

04.10.b1

four legs; others walked and ran on two legs. Another group of dinosaurs had a birdlike hip structure (but were not related to birds) and were herbivores. Some, like *Stegosaurus* and *Triceratops,* walked and grazed on four legs. Others, like duck-billed dinosaurs, could move on two legs.

2. A well-known hypothesis for the K-T extinction involves a huge comet or asteroid striking Earth, sending massive amounts of dust and gas into the atmosphere and blocking sunlight. Earth's surface would have been cold for decades. Many geoscientists conclude that the impact site, 66 million years ago, was the Chicxulub crater (▶) on the Yucatán Peninsula in Mexico (shown by the red circle).

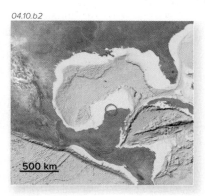
04.10.b2

500 km

3. Another possible cause of the extinction was massive outpourings of basalt in the *Deccan Traps* in India (not shown). Huge eruptions could have put enough sulfur dioxide gas into the atmosphere to cause a winter that lasted decades.

C   What Life Appeared During the Cenozoic Era?

The Cenozoic Era is also called the *age of mammals*. After dinosaurs went extinct, mammals were able to diversify rapidly and fill many niches left behind by the K-T extinction.

04.10.c2

1. By early Cenozoic time, the ancestors of modern mammals, including bats, rodents, primates, sloths, whales, hoofed animals, and carnivores, were abundant and lived in a variety of habitats. Marsupial mammals, represented by modern kangaroos, thrived on the isolated southern continents of South America and Australia.

2. Although they lived 20 million years ago, many of the mammals shown here (▼) may be familiar to you because they are fairly similar to their modern descendants. Each type of mammal, however, underwent many changes between then and now. Horses, for example, changed dramatically in size. These changes are well recorded by bones and teeth of different species of horses found at thousands of sites around the world.

3. Late in the Cenozoic, during the Ice Age, a number of large mammals roamed the continents (▶). Many of these animals, like the mammoth, saber-toothed cats, and giant beaver, went extinct as the Ice Age ended and humans spread across the globe. The first humans (*Homo sapiens*) appeared before 300,000 years ago, based on fossil evidence. Human-migration data are still controversial, but by at least 50,000 years ago *Homo sapiens* populated several parts of the planet, having left their sites of origin in Africa. The details of human history are continuously refined by discoveries of new archeological sites and even older ancestors.

04.10.c1

Before You Leave This Page

✓ Contrast the kinds of organisms that lived during the Mesozoic Era with those that lived during the Cenozoic Era.

✓ Describe some of the variety observed in dinosaurs, and summarize two theories for why dinosaurs became extinct.

4.10

4.11 How Do We Study Ages of Landscapes?

KNOWING THE RELATIVE AGES OF ROCKS AND STRUCTURES provides only one piece of the geologic story. We also need to understand when and how landscape features, such as mountains and valleys, formed.

A How Does a Typical Landscape Form?

Most landscapes have a similar history—rocks form and then are eroded. The histories of many regions typically include the deposition of a sequence of sedimentary layers, lithification into rocks, and later erosion of the rocks.

1. The sequence begins (▼) with deposition of a new sedimentary unit on top of preexisting metamorphic and igneous rocks. Most sediments, such as the layer of tan-colored sand shown, are deposited as nearly horizontal layers.

04.11.a1

04.11.a2

2. Through time, the depositional environment changes and a series of different sedimentary layers accumulate (◄), with each younger layer being deposited on top.

7. The canyon exposes five or six main sedimentary units and a number of smaller layers. In the middle is a dominant light-colored cliff of sandstone. Layers below the sandstone are the oldest in this area, and red cliffs in the distance expose higher and younger layers. The far mountains are igneous intrusions that baked, and are younger than, the youngest layers in the cliffs. All the layers were deposited and lithified, and the intrusions were emplaced, before erosion began carving the canyon.

3. Over time, the layers are lithified. At some point, deposition stops, and all the layers that will be deposited are there (►). Weathering and erosion can begin.

04.11.a3

04.11.a6 Grand View Point, Canyonlands NP, UT

4. If the region is uplifted or the seas withdraw, the area can begin to be eroded by streams, glaciers, the wind, and gravity. Erosion can more or less uniformly strip the entire land surface, removing the top layers. More likely, erosion will be faster in some areas, like along a river cutting downward in a small canyon (►).

04.11.a4

6. Observe this photograph and think about the likely sequences of events that formed the landscape before reading the text above the image. How many main rock layers do you observe? Use relative dating principles, like superposition, to infer the relative ages of the different rock layers you noted. Then, guided by the sequence of figures to the right, visualize how the area probably evolved over time.

5. Erosion by a river cuts downward, carving a deeper canyon. The canyon widens as small stream drainages erode outward from the main river and as the steep canyon walls move downhill in landslides and slower movements. The combination of downcutting, widening, and development of subsidiary drainages, called *tributaries*, sculpts a deeper, wider, and more intricate canyon (►).

04.11.a5

B | How Do We Infer the Age of a Landscape Surface?

To investigate when a landscape surface formed, we commonly try to find a rock unit or other geologic feature that was there before the surface formed or one that came after the surface already existed.

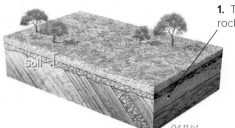

04.11.b1

1. The age of a landscape surface must be younger than any rocks on which it is carved. In this example (◄), erosion beveled across an older series of tilted layers, which were then covered by a thin veneer of sediment and soil. The surface is younger than the tilted layers.

2. A landscape surface is older than any rock that is deposited on top of the surface. A lava flow (▶) is ideal for dating a surface because it formed during a short time and its age can usually be determined by isotopic dating methods. This lava flow is on top of, and younger than, the underlying surface.

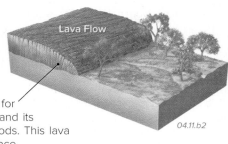

Lava Flow

04.11.b2

3. Sometimes, the age of a landscape surface cannot be dated directly, but we can infer its age relative to other features. Many streams are flanked by raised, gentle surfaces called *terraces*. A terrace was formed sometime in the past, before the stream eroded down to its present level (◄). The terrace shown here is older than the modern channel.

Terrace

04.11.b3

Well-developed Soil

No Soil

04.11.b4

4. A landscape surface progressively develops more soil if it remains undisturbed by erosion and deposition. A surface with well-developed soil, such as the uplands shown with thick red clay and white carbonate accumulations (▶), must be thousands of years old. Recent sediment along the stream has no soil. The high soil-covered surface is older than the low surface with no soil.

C | What Are Some Approaches to Investigating Earth History?

We study landscapes using relative dating, isotopic dating, fossils, and the principles summarized on this page. The figure below shows a landscape with various features, numbered in the order in which they formed, from oldest (1) to youngest (13). Examine this figure and think about why the features are interpreted to have formed in this order.

1. Cliffs expose horizontal rock layers, capped by a gray limestone (7). At the base of the cliff, the land surface has been offset by movement on a fault (a fracture along which there has been movement), forming a step in the land—a fault scarp (11). The valley is covered with fairly recent sediment (9).

2. The front of the block depicts a sequence of layers (4–7), some of which are also exposed in the cliff. The lowest layer in the series is the oldest (4), and the highest layer, the recent sediments (9), is the youngest. The contact between the limestone (7) and the sediments is a disconformity (8).

3. At the base of the block are metamorphic rocks (1), the oldest rocks in the area. The contact between the metamorphic rocks and layer 4 (a conglomerate) is depositional, with pieces of metamorphic rocks in the conglomerate. The contact between the steep metamorphic rocks and conglomerate is an angular unconformity (3).

5. A river valley cuts through the cliffs, forming a canyon (12). In the valley, the river contains a thin veneer of river gravels (13). A fault has offset the surface, forming a fault scarp (11), but it does not offset the gravel, so the fault is older.

6. A recent-looking volcano (10) has erupted lava (also numbered 10) that flowed downhill. The lava flow poured over the cliff, but is offset by the latest movement along the fault scarp (11). Some fault movement probably predated the eruption to produce the cliffs. The cone and lava flow are somewhat weathered and eroded.

7. The subsurface conduit for the volcano is a now-solidified, sheetlike igneous body called a dike (10). The dike cuts across all rock units it encounters on the side of the block.

8. In the subsurface, a granite (2) cuts across the metamorphic rocks (1), but is truncated by the lower angular unconformity (3). The dark dike (10) cuts the granite.

7
6
12–Canyon
10–Volcano
11–Fault Scarp
13–River Gravel
10–Lava Flow
10–Dike
8–Disconformity
9
11–Fault
2
3–Unconformity
7
6
5
4
1
04.11.c1

4. As shown on the side of the block, the fault cuts (displaces), and so is younger than, the metamorphic rocks, layers 4–7, and the uppermost sediments (9).

Before You Leave This Page

✓ Describe the sequence of events in a typical landscape of flat-lying sedimentary rocks, and describe how to constrain the ages of surfaces.

✓ Describe or sketch how you could reconstruct the history of landscapes.

4.11

What Is the History of the Grand Canyon?

GEOLOGICALLY, THE GRAND CANYON HAS IT ALL. It contains some of the best exposed and studied, as well as the most beautiful, rock sequences in the world. It is discussed in almost every geoscience class because it so clearly expresses a history of geologic events over the last 1.7 to 1.8 billion years.

1. This computer-generated perspective of the Grand Canyon region is viewed toward the north. The Colorado River, which formed the canyon, flows from right to left, exits the canyon through high cliffs, and enters Lake Mead. The dark east-west line shows the location of a geologic cross section from A to B.

2. The Grand Canyon cuts through the Colorado Plateau, a region of broad plateaus, mesas, and deep canyons, which expose a mostly flat-lying sequence of Mesozoic and Paleozoic sedimentary rocks.

3. The river flows southwest across the area, cutting across nearly horizontal to locally tilted layers. The deepest part of the canyon is where the Colorado River erodes through the uplifted Kaibab Plateau.

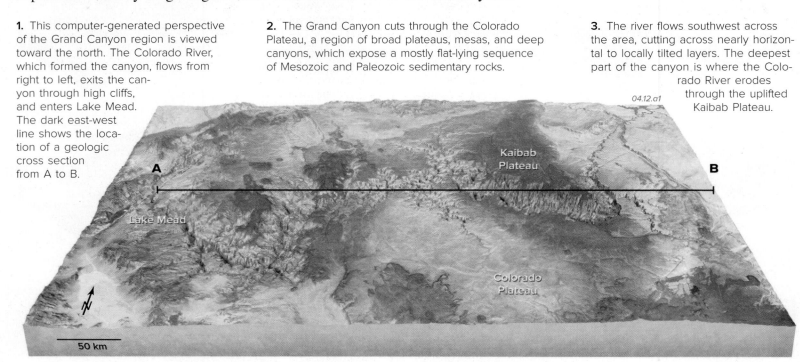

04.12.a1

4. Basalt flows cap some plateaus and predate formation of the main canyon. They are dated by K-Ar methods to be 8 million years old.

5. Large faults, like the Hurricane fault, cut across the region, downdropping rocks to the west. These faults cut basalt flows that are less than 1 to 2 million years old.

6. Some older basalt flows flowed down into the already-carved canyon, demonstrating that much of the canyon is older than 4 to 5 million years.

7. Paleozoic sedimentary layers cap most plateaus and are warped over a few broad, one-sided folds called monoclines, which downfold rocks on the east side of the folds.

8. Mesozoic sedimentary rocks are preserved on the down-folded sides of monoclines and contain famous dinosaur tracks and petrified wood in the Painted Desert.

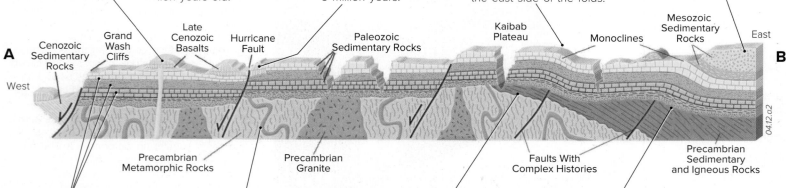

04.12.a2

9. The colorful walls of the canyon expose a flat-lying sequence of, from top to bottom, late, middle, and early Paleozoic rocks. There are disconformities within the Paleozoic section, each representing tens of millions of years of missing time.

10. The oldest rocks are metamorphic rocks and granites that are 1.7 billion years old, with the granites being just slightly younger than the metamorphic rocks. Both these rocks are exposed in the bottom of the canyon.

11. The near-vertical metamorphic rocks are overlain by tilted late Precambrian sedimentary and volcanic rocks, shown in purple. The contact is an angular unconformity and is called the *lower unconformity*. Where this unconformity overlies granite plutons, it is a nonconformity.

12. A separate, angular unconformity marks where gently dipping Paleozoic layers, shown in blue and red, overlie the moderately tilted late Precambrian layers, shown in purple. This is called the *upper unconformity*, and to the west it cuts across the lower unconformity.

13. The Grand Canyon exposes all three types of unconformities: angular unconformity, nonconformity, and disconformity. Photographs of each type from the Grand Canyon are in section 4.2.

Sequence of Rocks

04.12.a3 Grand Canyon, AZ

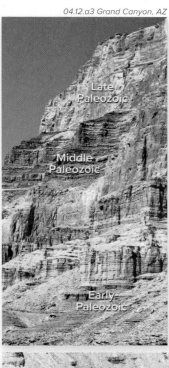

Late Paleozoic

Middle Paleozoic

Early Paleozoic

Early Paleozoic

Upper Unconformity

Late Precambrian

Lower Unconformity

Precambrian Basement

04.12.a4 Grand Canyon, AZ

04.12.a5

Geologic History and Key Age Constraints

Geologists reconstructed the sequence of events in the Grand Canyon using relative dating, fossils, and various isotopic dating methods. The geologic history resulting from these studies is summarized below. Read this history from bottom to top, from oldest to youngest.

7. *Deformation, Uplift, and Erosion*—The Paleozoic strata largely have escaped deformation and remain nearly flat, except near a few faults and folds, including some monoclines visible in cross section A–B. The monoclines are bracketed, using relative-dating methods, to between 80 and 40 million years ago. The region was uplifted by some amount at this same time, but the modern canyon was not carved until much later, mostly within the last 5 million years. Some faults, like the Hurricane fault, remain active.

6. *Deposition of Late Paleozoic Layers*—Overlying sedimentary layers (shown in red, pink, tan, and blue-green) record a wide range of environments, including shallow marine, shorelines, rivers, and a dune-covered desert. These rocks are dated with marine and nonmarine fossils as late Paleozoic (Pennsylvanian and Permian). Disconformities (mostly not shown) separate some of the formations and represent time when the region was above sea level.

5. *Deposition of Early and Middle Paleozoic Units*—After erosion carved the upper unconformity, seas covered the land and deposited sandstone, shale, and limestone (shown in brown and blue). These deposits are dated by trilobites and other fossils as early and middle Paleozoic (Cambrian, Devonian, and Mississippian). Later, the seas left and, in several instances, formed disconformities within the limestones.

4. *Tilting and Upper Unconformity*—Layers in the Late Precambrian rocks were gently to moderately tilted and then beveled by erosion. This produced the *upper unconformity*. As this unconformity is followed west, it truncates the *lower unconformity* beneath the Kaibab Plateau (see the cross section A–B). To the west, these combined unconformities represent even more missing time (from 1.7 billion years to 540 million years, or more than 1.1 billion years); it is appropriately called the *Great Unconformity* and can be followed eastward to the Great Lakes region.

3. *Late Precambrian Rocks and Lower Unconformity*—In the Late Precambrian, sedimentary and volcanic rocks were deposited in horizontal layers across the upturned basement layers. This formed the *lower unconformity* (above the metamorphic rocks). The lower parts of these late Precambrian rocks are dated by several isotopic methods at 1.1 billion years. Since the underlying basement rocks are 1.7 billion years old, the lower unconformity represents 600 million years of time not recorded by any rocks!

2. *Uplift and Erosion of the Basement*—After the metamorphism, the basement rocks cooled as they were uplifted and eroded over a period that lasted for hundreds of millions of years. Erosion beveled across the steep metamorphic layers.

1. *Basement Rocks*—Metamorphic and granitic rocks in the bottom of the canyon represent the oldest events. They were formed, metamorphosed, and deformed to near-vertical orientations, all between 1.76 and 1.70 billion years ago.

The Percentage of Geologic Time That the Canyon Records

Although the canyon is a classic geologic locality with a thick sequence of formations, it represents a relatively small amount of geologic time. The oldest rocks are "only" about 1.7 billion years old, so the area contains no record for 2.8 billion years of Earth history (4.5–1.7 billion years). Next, the two unconformities together cut out another 700 to 800 million years of history, or one-sixth of Earth's history.

Even the Paleozoic sequence is missing more time than it records! The formations only represent five out of the seven geologic periods (rocks of the Ordovician and Silurian periods are not present), none of the formations span an entire period, and there are major disconformities. Mesozoic and Cenozoic rocks are largely absent in the canyon, so yet more time is not represented by rocks in the canyon walls.

Before You Leave This Page

☑ Describe examples of how different methods of dating events and rocks were used to reconstruct the geologic history of the Grand Canyon.

☑ Describe why the canyon does not represent all of geologic time.

What Is the Geologic History of This Place?

This terrain exposes various geologic relationships that have been documented in the field and recorded as descriptions. Samples collected from the area were analyzed either for their isotopes or their characteristic fossils. You will use this information to reconstruct the sequence and ages of events that produced features exposed in the landscape today. A block diagram of the area is on the next page, and key observations are listed below.

Goals of This Exercise:

- Observe the distribution of different rock types exposed in the terrain to characterize the sequence of rocks and the geologic features that are present.
- Use descriptions of units and of key contact relationships, along with fossils, to infer the relative sequence of events.
- Calculate isotopic ages for key igneous rocks to help constrain when important events occurred.

Procedures

Use your observations to complete the following steps. Your instructor may provide you with rock or fossil specimens.

A. Observe the terrain to understand the overall pattern of rocks. Based on this pattern, use the associated descriptive text to determine in what order the units formed and where in that sequence different geologic features, such as a fault and dike, developed.

B. Examine the six fossils in the table below, and the geologic period to which each is assigned; complete the stratigraphic section on the worksheet, listing the units in the order in which the units formed, from bottom to top in the section.

C. Use the table of isotopic measurements below to calculate the age of a sample of granite and a sample of the dike.

D. Summarize the geologic history by arranging the different events in their proper order on the worksheet or online.

Field Notes

The units and features are described below. Each unit or feature has a letter assigned to it, but these do not reflect the order in which the features formed. Some letters were skipped so that some features would have letters that were easy to remember, such as V for the volcano.

Unit A—Tan sandstone with land fossils, including plants of Permian age.

Unit B—Greenish shale with marine fossils, including Ordovician trilobites. The top of the unit was weathered and eroded prior to deposition of unit A, but the layers in the two units are parallel to each other and to their mutual contact.

Unit C—Coarse sandstone and beach conglomerate that contains Cambrian trilobites. The base contains clasts derived from the underlying granite (G).

Unit D—Finely crystalline dike that has baked units A, B, C, and G.

Feature F—Fault that cuts units B, C, and G. Some units are not near the fault.

Unit G—Coarse granite that is weathered near the contact with unit C.

Unit K—Gray limestone with marine fossils of Cretaceous age.

Units L and V—Unweathered lava flow (L) associated with a volcano (V).

Feature N—Narrow canyon.

Unit R—Partly consolidated river gravels with a thick, well-developed soil. Contains land mammals of middle Cenozoic age.

Unit S—Reddish and pinkish sandstone that was deposited by rivers and in lakes. It contains Jurassic dinosaur bones.

Identification of Fossils

Rock Unit	Fossil	Period
R	Mammals	Cenozoic
K	Fish	Cretaceous
S	Dinosaurs	Jurassic
A	Plants	Permian
B	Trilobites	Ordovician
C	Trilobites	Cambrian

Table of Isotopic Measurements

Rock Unit	Half-Life of Isotope	Number of Parent Atoms	Number of Daughter Atoms
G	500 Million Years	125	875
D	40 Million Years	500	500

This view shows a landscape with various rocks and features. There is a central plateau (high flat area) flanked by several mountains, an obvious volcano, a canyon, and a number of lines and curved features that cross the landscape. The geology in the subsurface is shown on the sides of the block. Any type of unconformity is shown with a squiggly line, reflecting some topographic relief along the erosion surface represented by the unconformity. Normal depositional contacts are shown by thin lines, and a fault is marked by a thicker line.

1. A section of layers forms a series of cliffs and slopes on three corners of the block. These were encountered first and so are lettered A, B, and C, not in the order in which the units were formed. Unit A is a brown sandstone that was deposited on land and contains Permian plant fossils. Unit B is greenish marine shale and contains Ordovician trilobites. Unit C is a coarse sandstone and beach conglomerate that contains Cambrian trilobites.

2. A dark dike (D) forms a linear wall across the landscape. It mostly is uninterrupted by other geologic features, except for one obvious gap near a belt of some tan-colored soils (associated with unit R). The dike consists of dark basalt and was dated by isotopic methods.

3. A series of old river channels (R) cross the plateau and form low troughs in the topography. One channel goes all the way to the edge of the canyon, where it stops abruptly, evidently having been cut off. Along their lengths, the channels are partially filled by river gravels and are characterized by well-developed, tan soils. They contain bones of small horses and other fossils from the middle Cenozoic.

4. The top of one mountain in the area (right corner of this figure) exposes higher layers than are preserved elsewhere. There is a red sandstone (S) that contains bones of Jurassic dinosaurs. The sandstone is overlain by a gray limestone (K) that has fish and other marine fossils from the last part of the Mesozoic (Cretaceous).

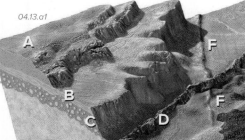

04.13.a1

5. There is a cone-shaped volcano (V) surrounded by a black lava flow (L). Neither the volcanic deposits (scoria) on the volcano nor the lava flow has developed any soil.

6. A fault (F) forms an obvious line across parts of the area, but is not continuous. It is also shown in cross section on the side of the block. It has not formed a fault scarp, but is expressed in the topography because it is the boundary between rock types that erode in slightly different ways. In a nearby area, the fault cuts the main sequence of layers, including layers C, B, A, S, and K.

7. The lowest unit in the area is a gray granite (G). Geoscientists determined an isotopic age on a sample of the granite, and these results are in the table on the previous page.

8. A narrow canyon (N) cuts through the area. The canyon is especially narrow in one segment where dark lava flows (L) have poured from the plateau and into the already formed canyon.

9. Reconstruct the history using superposition, cross-cutting relationships, and the relationship of different features to the landscape. Be systematic, focusing your attention on any pair of objects that are in contact. For example, does the dike crosscut the fault or vice versa? Is unit A above or below unit B? Some objects may not be in direct contact with each other, but their relative age can be determined by comparing their ages relative to some other feature.

4.13

5 Plate Tectonics

THE SURFACE OF EARTH IS NOTABLE for its dramatic mountains, beautiful valleys, and intricate coastlines. Beneath the sea are unexpected features, such as undersea mountain ranges, deep ocean trenches, and thousands of submarine mountains. In this chapter, we examine the distribution of these features, along with the locations of earthquakes and volcanoes, to explore the *theory of plate tectonics*.

These images of the world show large topographic features on the land, colored using satellite data that show areas of vegetation, rocks, and sand. Colors on the seafloor indicate depths below sea level, ranging from light blue for seafloor that is at relatively shallow depths to dark blue for seafloor that is deep. Observe the main features of the continents and seafloor, as shown on the large figure, then examine the more detailed views of some important regions.

The seafloor west of North America displays a long, fairly straight fracture that trends east-west and ends abruptly at the coastline. North of this fracture, a ridge called the Juan de Fuca Ridge zigzags across the seafloor.

What are these features on the seafloor, and how did they form?

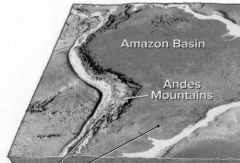

South America has two very different sides. The mountainous Andes parallel the western coast, but a wide expanse of lowlands, including the Amazon Basin, makes up the rest of the continent. The western edge of the continent drops steeply into the Pacific Ocean and is flanked by a deep trench. The eastern edge of the continent continues well beyond the shoreline and forms a broad bench covered by shallow waters (shown in light blue).

Why are the two sides of the continent so different?

A huge mountain range, longer than any on land, is hidden beneath the waters of the Atlantic Ocean. The part of the range shown here is halfway between South America and Africa. The ridge zigzags across the seafloor, mimicking the shape of the two continents.

What is this underwater mountain range, and why is it almost exactly in the middle of the ocean?

TOPICS IN THIS CHAPTER

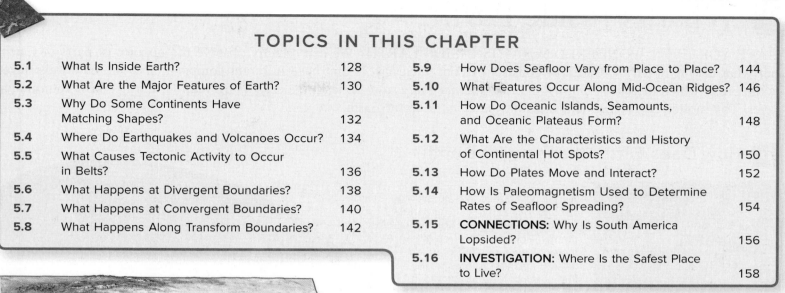

The Tibetan Plateau of southern Asia rises many kilometers above the lowlands of India and Bangladesh to the south. The Himalaya mountain range with Mount Everest, the highest mountain on Earth, is perched on the southern edge of this plateau.

Why does this region have such a high elevation?

Japan lies along the intersection of large, curving ridges that are mostly submerged beneath the ocean. Each ridge is flanked to the east by a deep trench in the seafloor. This area is well known for its destructive earthquakes and for Japan's picturesque volcano, Mount Fuji.

Do submarine ridges and trenches play a role in earthquake and volcanic activity?

The Arabian Peninsula and nearby areas provide much of the world's oil. East of the peninsula, the Persian Gulf has a shallow and smooth seafloor, and is flanked by the world's largest oil fields. West of the peninsula, the Red Sea has a well-defined trough or fissure down its center.

How did the Red Sea form, and what processes are causing its seafloor to be disrupted?

5.0

5.1 What Is Inside Earth?

HAVE YOU EVER WONDERED WHAT IS INSIDE EARTH? We can directly observe the uppermost parts of Earth, but what else is down there? Earth consists of concentric layers that have different compositions. The outermost layer is the *crust*, which includes *continental crust* and *oceanic crust*. Beneath the crust is the *mantle*, Earth's most voluminous layer. The molten *outer core* and the solid *inner core* are at Earth's center.

A How Does Earth Change with Depth?

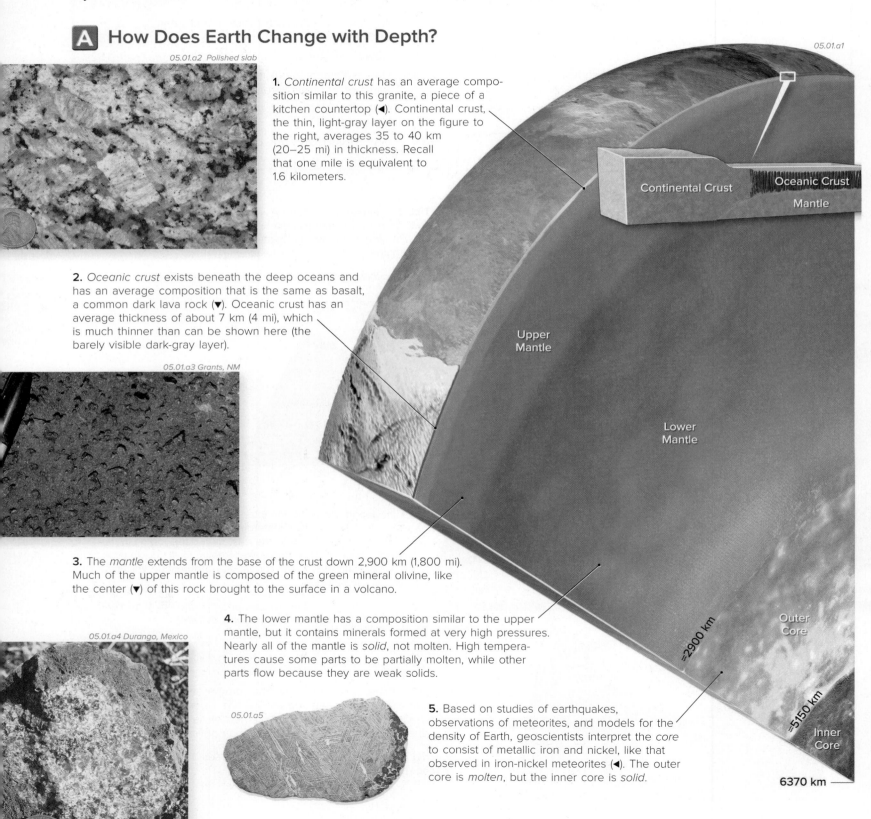

05.01.a2 Polished slab

1. *Continental crust* has an average composition similar to this granite, a piece of a kitchen countertop (◄). Continental crust, the thin, light-gray layer on the figure to the right, averages 35 to 40 km (20–25 mi) in thickness. Recall that one mile is equivalent to 1.6 kilometers.

2. *Oceanic crust* exists beneath the deep oceans and has an average composition that is the same as basalt, a common dark lava rock (▼). Oceanic crust has an average thickness of about 7 km (4 mi), which is much thinner than can be shown here (the barely visible dark-gray layer).

05.01.a3 Grants, NM

3. The *mantle* extends from the base of the crust down 2,900 km (1,800 mi). Much of the upper mantle is composed of the green mineral olivine, like the center (▼) of this rock brought to the surface in a volcano.

4. The lower mantle has a composition similar to the upper mantle, but it contains minerals formed at very high pressures. Nearly all of the mantle is *solid*, not molten. High temperatures cause some parts to be partially molten, while other parts flow because they are weak solids.

05.01.a4 Durango, Mexico

05.01.a5

5. Based on studies of earthquakes, observations of meteorites, and models for the density of Earth, geoscientists interpret the *core* to consist of metallic iron and nickel, like that observed in iron-nickel meteorites (◄). The outer core is *molten*, but the inner core is *solid*.

05.01.a1

Continental Crust

Oceanic Crust

Mantle

Upper Mantle

Lower Mantle

Outer Core

Inner Core

≈2900 km

≈5150 km

6370 km

B Are Some Layers Stronger Than Others?

In addition to layers with different compositions, Earth has layers that are defined by strength and by how easily the material in the layers fractures or flows when subjected to forces.

The uppermost part of the mantle is relatively strong and solidly attached to the overlying crust. The crust and uppermost mantle together form an upper, rigid layer called the *lithosphere* (*lithos* means *"stone"* in Greek). The part of the uppermost mantle that is in the lithosphere is the *lithospheric mantle*.

The mantle directly beneath the lithosphere is mostly solid, but it is hotter than the rock above and can flow under pressure. This part of the mantle, called the *asthenosphere*, functions as a soft, weak zone over which the lithosphere may move. The word *asthenosphere* is from a Greek term for *"not strong."* The asthenosphere is approximately 80 to 150 km thick, so its base can be as deep as about 250 km.

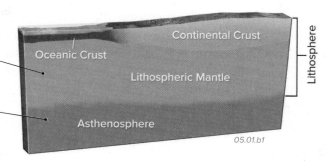

C Why Do Some Regions Have High Elevations?

Why is the Gulf Coast of Texas near sea level, while the Colorado mountains are 3 to 5 km (2 to 3 mi) above sea level? Why are the continents mostly above sea level, but the ocean floor is below sea level? The primary factor controlling the elevation of a region is the thickness of the underlying crust.

1. The granitic crust is less dense than the underlying mantle, and so rests, or floats, on top of the mantle. The underlying lithospheric mantle is mostly solid, not liquid.

2. The thickness of continental crust ranges from less than 25 km (16 mi) to more than 60 km (37 mi). Regions that have high elevation generally have thick crust. The crust beneath the Rocky Mountains of Colorado is commonly more than 45 km (28 mi) thick.

3. The crust beneath low-elevation regions like Texas is thinner. If the crust is thinner than 30 to 35 km (18 to 20 mi), the area will probably be below sea level, but it can still be part of the continent, like on a continental shelf.

4. Most islands are volcanic mountains built on oceanic crust, but some are small pieces of continental crust.

5. Oceanic crust is thinner than continental crust and consists of denser rock than continental crust. As a result, regions underlain only by oceanic crust are well below sea level.

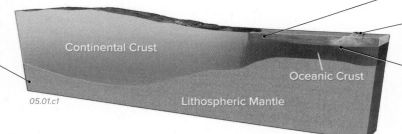

Density and Isostasy

The relationship between regional elevation and crustal thickness is similar to that of wooden blocks of different thicknesses floating in water (▾). Wood floats on water because it is less dense than water. Ice floats on water because it is less dense than water, although ice and water have the same composition. Thicker blocks of wood, like thicker parts of the crust, rise to higher elevations than do thinner blocks of wood.

For Earth, we envision the crust being supported by mantle that is solid, unlike the liquid water used in the wooden-block example. This concept of different thicknesses of crust riding on the mantle is called *isostasy*. Isostasy explains most of the variations in elevation from one region to another, and it is

commonly paraphrased by saying *mountain belts have thick crustal roots*. Observe the figure above to envision this. As in the case of the floating wooden blocks, most of the change in crustal thickness occurs at depth and less occurs near the surface. Smaller, individual mountains do not necessarily have thick crustal roots. They can be supported by the strength of the crust, like a small lump of clay riding on one of the wooden blocks.

The *density* of the rocks also influences regional elevations. The fourth block shown here has the same thickness as the third block, but it consists of a denser type of wood. It therefore floats lower in the water. Likewise, a region of Earth underlain by especially dense crust or mantle is lower in elevation than a

region with less dense crust or mantle, even if the two regions have similar thicknesses of crust. Temperature controls the thickness of the lithosphere, which also affects a region's elevation. If the lithosphere in some region is heated, it expands, becoming less dense, and so the region rises in elevation. Thinner lithosphere also yields higher elevations.

Before You Leave This Page

✓ Sketch the major layers of Earth.

✓ Sketch and describe differences in thickness and composition between continental crust and oceanic crust, and contrast lithosphere and asthenosphere.

✓ Sketch and discuss how the principle of isostasy can explain differences in regional elevation.

5.1

5.2 What Are the Major Features of Earth?

OCEANS COVER 71% OF EARTH'S SURFACE. Seven major continents make up most of the rest of the surface, and islands account for less than 2%. We are all familiar with the continents and their remarkable diversity of landforms, from broad coastal plains to steep, snow-capped mountains. Features of the ocean floor, not generally seen by people, are just as diverse and include deep trenches and submarine mountain ranges. Islands exhibit great diversity, too. Some are large and isolated, but other islands define arcs, ragged lines, or irregular clusters. What are the characteristics of each type of feature, and how did these features form?

05.02.a1

1. This map shows large features on land and on the seafloor. The colors on land are from images taken by satellites orbiting Earth and show vegetated areas (green), rocky areas (brown), and sandy areas (tan). Greenland and Antarctica are white and light gray because they are mostly covered with ice and snow. Ocean colors show the depth of the seafloor and range from light blue where the seafloor is shallow to darker blue where it is deep.

2. Parts of the seafloor have mountains, the largest of which form islands like Hawaii. Most mountains on the seafloor do not reach sea level and are termed *seamounts*. Some islands and seamounts, like Hawaii, are in long belts, which we refer to as *island and seamount chains*. Other islands and seamounts are isolated or form irregular clusters.

3. Some large islands, such as New Zealand, look like a small version of a continent. Some geoscientists think that New Zealand and some adjacent relatively shallow parts of the seafloor represent a mostly submerged continent.

4. Much of the ocean floor is moderately deep—3 to 5 km (9,800–16,000 ft)—and has a fairly smooth surface. These deep, smooth regions are *abyssal plains*.

5. *Mid-ocean ridges* are broad, symmetrical ridges that cross the ocean basins. They are 2 to 3 km (6,600–9,800 ft) higher than the average depth of the seafloor. One long ridge, named the East Pacific Rise, crosses the eastern Pacific and heads toward North America. Another occupies the middle of the Atlantic Ocean.

6. Cracks and steps cross the seafloor mostly at right angles to the mid-ocean ridges. These features are *oceanic fracture zones*.

7. Some continents continue outward from the shoreline under shallow seawater (light blue in this image) for hundreds of kilometers, forming submerged benches known as *continental shelves*. Observe which coastlines have broad continental shelves, like those surrounding Great Britain.

8. All continents contain large interior regions with gentle topography. Some continents have flat coastal plains, while others have mountains along their edges. Some mountains, like the Ural Mountains, are in the middle of continents.

9. Most continental areas have elevations of less than 1 to 2 km (3,300 to 6,600 ft). Broad, high regions, called *plateaus*, reach higher elevations, such as the Tibetan Plateau of southern Asia. Continents also contain mountain chains and individual mountains. Mount Everest, the highest point in the world, is almost 9 km (30,000 ft) in elevation.

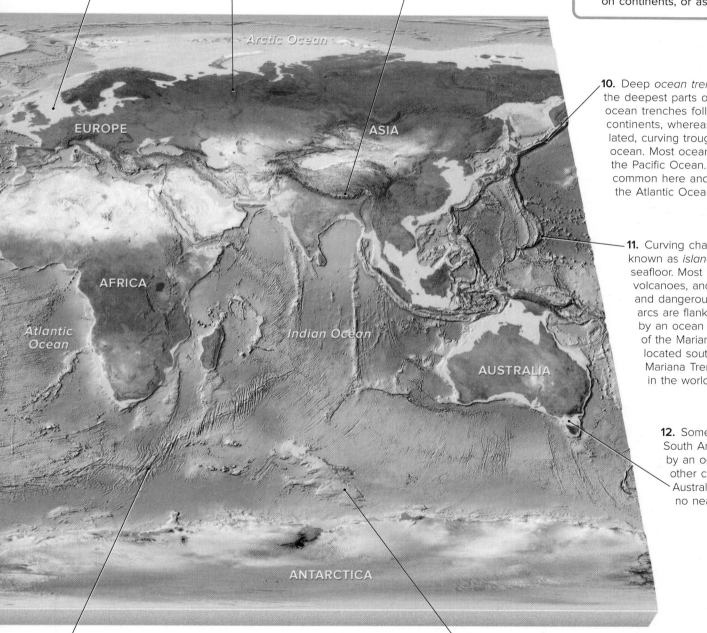

10. Deep *ocean trenches* make up the deepest parts of the ocean. Some ocean trenches follow the edges of continents, whereas others form isolated, curving troughs out in the ocean. Most ocean trenches are in the Pacific Ocean. Why are they so common here and less common in the Atlantic Ocean?

11. Curving chains of islands, known as *island arcs*, cross the seafloor. Most of the islands are volcanoes, and many are active and dangerous. Most island arcs are flanked on one side by an ocean trench. Offshore of the Mariana island arc, located south of Japan, is the Mariana Trench, the deepest in the world.

12. Some continents (such as South America) are flanked by an ocean trench, but other continents, such as Australia and Africa, have no nearby trenches.

14. Mid-ocean ridges and their associated fracture zones encircle much of the globe. In the Atlantic and Southern oceans, they occupy a position halfway between the adjacent continents.

13. The oceans contain several broad, elevated regions called *oceanic plateaus*. The Kerguelen Plateau near Antarctica is one example, and another oceanic plateau lies northeast of Australia.

5.2

5.3 Why Do Some Continents Have Matching Shapes?

SOME CONTINENTS HAVE MATCHING SHAPES that appear to fit together like the pieces of a giant jigsaw puzzle. Alfred Wegener (1880–1930) observed the fit of these continents and tried to explain this and other data with a new hypothesis called *continental drift*. Wegener argued that the continents were once joined together but later drifted apart. The hypothesis of continental drift was an important historical step that led to current theories that explain the distribution and shapes of the continents.

A Were the Continents Once Joined Together?

Fairly accurate world maps became available during the 1800s, and scientists, including Alfred Wegener, noted that some continents, especially the southern continents, appeared to fit together. After considering many types of data, Wegener arrived at a creative explanation for this pattern.

05.03.a1

1. This figure shows how the southern continents are interpreted to have fit together 150 million years ago. In this figure, we included the *continental shelves* because they are parts of continents that are currently underwater. In this arrangement, the bulge on the eastern side of South America fits nicely into the embayment on the western coast of Africa.

2. The fit of the continents and other supporting evidence preserved in rocks and fossils inspired Wegener and others to suggest that South America, Africa, Antarctica, Australia, and most of India were once joined but later drifted apart. Even Madagascar can fit into the puzzle.

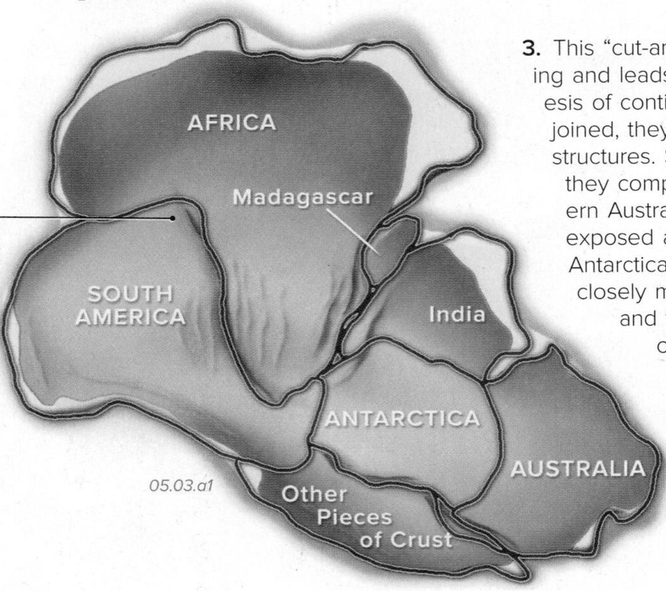

3. This "cut-and-paste" fit of the continents is intriguing and leads to predictions for testing the hypothesis of continental drift. If continents were once joined, they should have similar rocks and geologic structures. Scientists find such similarities when they compare the rocks and structures in southern Australia with the rocks and structures exposed around the edges of ice sheets on Antarctica. Similarly, the geology of western Africa closely matches that of eastern South America, and these two areas are adjacent to each other in Wegener's reconstruction.

4. Geoscientists gave the name *Gondwana* to this hypothetical combination of the southern continents into a single large supercontinent.

B Is the Distribution of Fossils Consistent with Continental Drift?

Another piece of evidence supporting continental drift is the correspondence of the fossils of plants and land animals on continents now several thousand kilometers apart and separated by wide oceans.

05.03.b1

1. This figure illustrates that fossils of some land animals exist on several continents that are now separated by wide oceans. The animals, including a reptile named *Mesosaurus*, lived more than 150 million years ago and are now extinct. These land animals could not swim across the wide oceans that currently separate the continents. Another key fossil linking the land areas of Gondwana are fossilized leaves of *Glossopteris*, a seed-bearing plant that was widespread during late Paleozoic time (before 252 Ma).

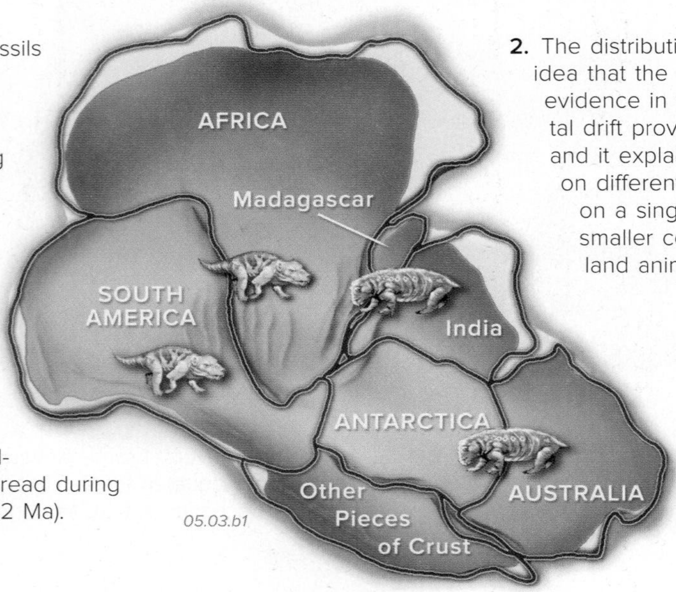

2. The distribution of plant and animal fossils is consistent with the idea that the continents were once joined. It was a key piece of evidence in favor of continental drift. The hypothesis of continental drift provided an alternative to the hypothesis of *land bridges*, and it explained why identical plant and animal fossils are found on different continents. The plants and animals were originally on a single huge supercontinent that later split into separate smaller continents. Two continents could share plants and land animals before they split, but not after.

3. Other fossil data suggest that Antarctica was once farther north, away from the South Pole. Such data include coal beds interpreted to have formed from plants that grew in warm-weather swamps. One explanation is that Antarctica moved to its present polar location after the coal formed more than 150 million years ago.

C How Did Continental Drift Explain Glacial Deposits in Unusual Places?

Geoscientists studying continents in the Southern Hemisphere were puzzled by evidence that ancient glaciers had once covered places that today are close to the equator, and much too warm to have major glaciers.

05.03.c1 Kimberly, South Africa

05.03.c2 Kimberly, South Africa

1. This rounded outcrop in South Africa has a polished and scratched surface that is identical to those observed at the bases of modern glaciers. This observation is surprising because South Africa is currently a fairly warm and dry region without any glaciers.

2. Sedimentary rocks above the polished surface contain an unsorted collection of rocks of various sizes. Some of the rocks have scratch marks, like those seen near modern-day glaciers.

3. The scratch marks on the polished surface tell geoscientists the direction that glaciers moved across the land as they gouged the bedrock. Geoscientists interpret the scratch marks and other observations as evidence that glaciers moved across the area about 280 million years ago.

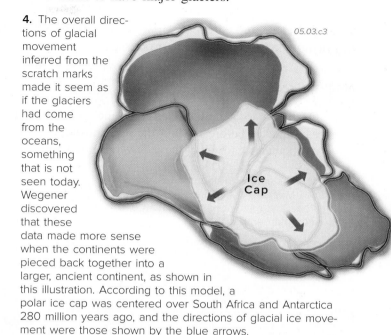

05.03.c3

Ice Cap

4. The overall directions of glacial movement inferred from the scratch marks made it seem as if the glaciers had come from the oceans, something that is not seen today. Wegener discovered that these data made more sense when the continents were pieced back together into a larger, ancient continent, as shown in this illustration. According to this model, a polar ice cap was centered over South Africa and Antarctica 280 million years ago, and the directions of glacial ice movement were those shown by the blue arrows.

Old and New Ideas About Continental Drift

The hypothesis of *continental drift* received mixed reviews at first from geologists and other scientists. Geologists working in the Southern Hemisphere were intrigued by the idea because it explained the observed similarities in rocks, fossils, and geologic structures on opposite sides of the Atlantic and Indian Oceans. Geologists working in the Northern Hemisphere were more skeptical, in part because many had not seen the Southern Hemisphere data for themselves.

We now know that Wegener, with the evidence he considered, was on the right track. A crucial weakness of his hypothesis was that he could not explain how or why the continents moved. Wegener imagined that continents plowed through or over oceanic crust in the same way that a ship plows through the ocean. Scientists of his day, however, could demonstrate that this mechanism was not feasible. Continental crust is not strong enough to survive the forces needed to move a large mass across such a great distance while pushing aside oceanic crust. Because scientists of Wegener's time could show with experiments and calculations that this mechanism was unlikely, they practically

abandoned the hypothesis, in spite of its other appeals. The hypothesis probably would have been more widely accepted if Wegener or another scientist of that time had proposed a viable mechanism that explained how continents could move.

In the late 1950s, the idea of drifting continents again surfaced with the availability of new information about the topography (▼), age, and magnetism of the seafloor. The magnetic data had largely been acquired in the search for enemy submarines during World War II. These data showed, for the

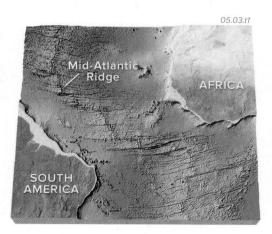

05.03.t1

Mid-Atlantic Ridge

AFRICA

SOUTH AMERICA

first time, that the ocean floor had long submarine mountain belts, such as the Mid-Atlantic Ridge in the middle of the Atlantic Ocean. Harry Hess and Robert Dietz, two geologists familiar with Wegener's work, examined the new data on ocean depths, and also new data on magnetism of the seafloor. Hess and Dietz both proposed that oceanic crust was spreading apart at underwater mountain belts, carrying the continents apart. This process of *seafloor spreading* rekindled interest in Alfred Wegener's idea of continental drift. Wegener's hypothesis morphed into the *theory of plate tectonics*, which is described later in this chapter.

Before You Leave This Page

☑ Describe observations Wegener used to support continental drift.

☑ Discuss why the hypothesis was not widely accepted.

☑ List some discoveries about the seafloor that brought a renewed interest in the idea of continental drift.

5.4 Where Do Earthquakes and Volcanoes Occur?

EARTHQUAKES AND VOLCANOES are spectacular manifestations of geology. Many of these are in distant places, but some are close to where we live. The distributions of earthquakes and volcanoes are not random, but instead define clear patterns and show a close association with mountain belts and other regional features. These patterns reflect important, large-scale Earth processes.

A Where Do Most Earthquakes Occur?

On this map, yellow circles show the locations of recent moderate to strong earthquakes. Observe the distribution of earthquakes before reading the text surrounding the map below. What patterns do you notice? Which regions have many earthquakes and which have few? Are earthquakes associated with certain types of features?

1. Earthquakes are not distributed uniformly across the planet. Most are concentrated in discrete belts, such as one that runs along the western coasts of North and South America.

2. Most earthquakes in the oceans occur along the winding crests of mid-ocean ridges. Where the ridges curve or zigzag, so do the patterns of earthquakes.

3. Earthquakes are sparse in some continental interiors but are abundant in others, like the Middle East, China, and Tibet.

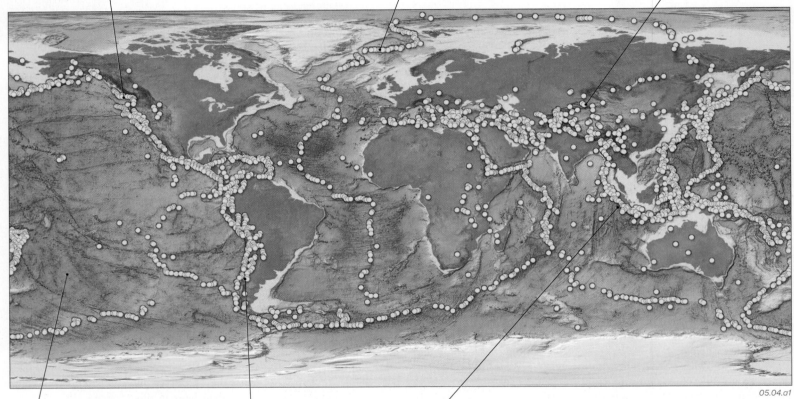

05.04.a1

4. Large areas of the seafloor, especially the abyssal plains, have few earthquakes. Volcanically active islands in the middle of the Pacific Ocean, like Hawaii, do have earthquakes.

5. Some continental edges experience many earthquakes, but other edges have few. Earthquakes are common along the western coasts of South America and North America, and these edges also have narrow continental shelves. There are few earthquakes along the eastern coasts of the Americas, where the continental shelves are wide.

6. Ocean trenches and associated island arcs have numerous earthquakes. In fact, many of the world's largest and most deadly earthquakes occur near ocean trenches. Recent examples were the large earthquakes that produced deadly ocean waves (tsunamis) in the Indian Ocean in 2004 and in Japan in 2011.

Before You Leave These Pages

☑ Show on a world relief map the major belts of earthquakes and volcanoes.

☑ Describe how the distribution of volcanoes corresponds to that of earthquakes.

☑ Compare the distributions of earthquakes, volcanoes, and high elevations.

B Which Areas Have Volcanoes?

On the map below, orange triangles show the locations of volcanoes that have been active in the last several million years. Observe the distribution of volcanoes and note which areas have volcanoes and which have none. How does this distribution compare with the distribution of earthquakes?

1. Volcanoes, like earthquakes, are widespread, but commonly occur in belts. One belt extends along the western coasts of North and South America.

2. Some volcanoes occur in the centers of oceans, such as the volcanoes near Iceland. Iceland is a large volcanic island along the mid-ocean ridge in the center of the North Atlantic Ocean.

3. Volcanoes occur along the western edge of the Pacific Ocean, extending from north of Australia through the Philippines and Japan. Many are part of island arcs, associated with ocean trenches and earthquakes.

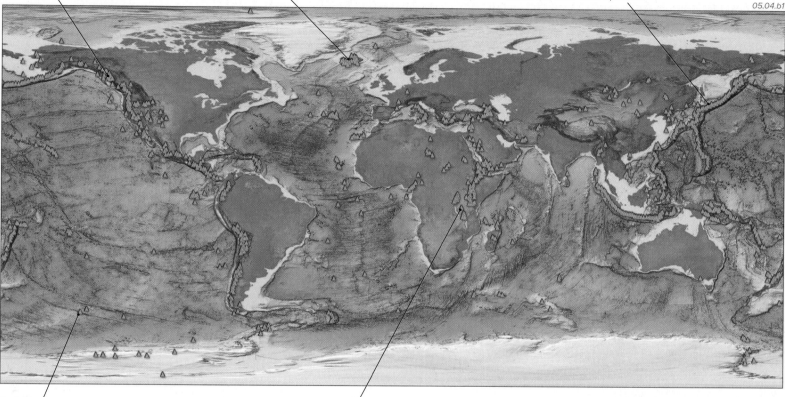

05.04.b1

4. Volcanic eruptions occur beneath the oceans, but this map shows only the largest submarine volcanic mountains. Volcanism is widespread along mid-ocean ridges, but it generally does not form mountains.

5. Some volcanoes form in the middle of continents, such as in the eastern part of Africa and China.

6. This map (▼) shows the topography of Earth's surface and seafloor, with high elevations in brown, low land elevations in green, shallow seafloor in light blue, and deep seafloor in dark blue. Using the three maps shown here, compare the distributions of earthquakes, volcanoes, and high elevations. Identify areas where there are (1) mountains but no earthquakes, (2) mountains but no volcanoes, and (3) earthquakes but no volcanoes. Make a list of these areas, or mark the areas on a map.

05.04.b2

5.5 What Causes Tectonic Activity to Occur in Belts?

WHY DO EARTHQUAKES AND VOLCANOES occur in belts around Earth's surface? Why are there vast regions that have comparatively little of this activity? What underlying processes cause these observed patterns? These and other questions helped lead to the *theory of plate tectonics*.

A What Do Earthquake and Volcanic Activity Tell Us About Earth's Lithosphere?

1. Examine the map below, which shows the locations of recent earthquakes (yellow circles) and volcanoes (orange triangles). After noting the patterns, compare this map with the lower map and then read the associated text.

2. On the upper map, there are large regions that have few earthquakes and volcanoes. These regions are relatively stable and intact pieces of Earth's outer layers. There are a dozen or so of these regions, each having edges defined by belts of earthquakes and, to a lesser extent, volcanoes.

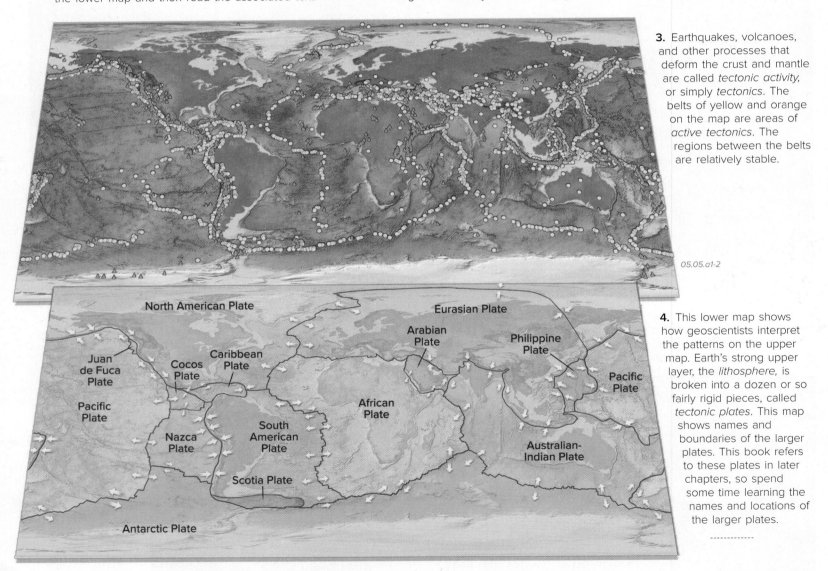

3. Earthquakes, volcanoes, and other processes that deform the crust and mantle are called *tectonic activity*, or simply *tectonics*. The belts of yellow and orange on the map are areas of *active tectonics*. The regions between the belts are relatively stable.

05.05.a1-2

4. This lower map shows how geoscientists interpret the patterns on the upper map. Earth's strong upper layer, the *lithosphere*, is broken into a dozen or so fairly rigid pieces, called *tectonic plates*. This map shows names and boundaries of the larger plates. This book refers to these plates in later chapters, so spend some time learning the names and locations of the larger plates.

5. Compare the two maps and note how the distribution of tectonic activity, especially earthquakes, outlines the shapes of the plates. Earthquakes are a better guide to plate boundaries than are volcanoes. Most, but not all, volcanoes are near plate boundaries, but many plate boundaries have no volcanoes.

6. Some earthquakes occur in the middle of plates, so the situation is more complicated than a simple plate-tectonic model, in part because some parts of a plate are weaker than others. Forces can be transmitted through the strong parts, causing weaker parts to break and slip, generating an earthquake within the plate. Generally, though, most tectonic activity occurs near plate boundaries.

B How Do Plates Move Relative to One Another?

Plate boundaries have tectonic activity because plates are moving *relative to one another*. For this reason, we talk about the *relative motion* of plates across a plate boundary. Two plates can move away, toward, or sideways relative to one another, resulting in three types of plate boundaries: *divergent*, *convergent*, and *transform*.

Divergent Boundary

05.05.b1

Convergent Boundary

05.05.b2

Transform Boundary

05.05.b3

At a *divergent boundary*, two plates move apart relative to one another. In most cases, magma fills the space between the plates.

At a *convergent boundary*, two plates move toward one another. A typical result is that one plate slides under the other.

At a *transform boundary*, two plates move horizontally past one another, as shown by the white arrows on the top surface.

C Where Are the Three Types of Plate Boundaries?

This map shows plate boundaries according to type. Compare this map with the maps in part A and with those shown earlier in the chapter. For each major plate, note the types of boundaries between this plate and other plates it contacts. Then use the various maps to determine whether each type of plate boundary has the following features:

- Earthquakes
- Volcanoes
- Mountain belts
- Mid-ocean ridges
- Ocean trenches

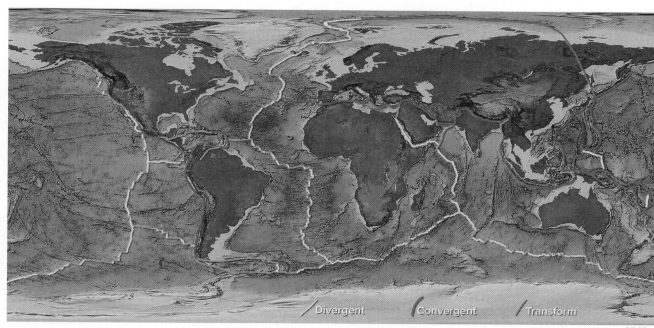

Divergent   Convergent   Transform

05.05.c1

Before You Leave This Page

✓ Describe plate tectonics and how it explains the distribution of tectonic activity.

✓ Sketch and explain the three types of plate boundaries.

✓ Compare the three types of plate boundaries with the distributions of earthquakes, volcanoes, mountain belts, mid-ocean ridges, and ocean trenches.

5.6 What Happens at Divergent Boundaries?

AT MID-OCEAN RIDGES, Earth's tectonic plates diverge (move apart). Ridges are the sites of many small to moderate-sized earthquakes and much submarine volcanism. On the continents, divergent motion can split a continent into two pieces, commonly forming a new ocean basin as the pieces move apart.

A What Happens at Mid-Ocean Ridges?

Mid-ocean ridges are divergent plate boundaries where new oceanic lithosphere forms as two oceanic plates move apart. These boundaries are also called *spreading centers* because of the way the plates spread apart.

1. A narrow trough, or *rift*, runs along the axis of most mid-ocean ridges. The rift forms because large blocks of crust slip down as spreading occurs. The divergence and movement of fault blocks causes faulting, resulting in frequent small to moderate-sized earthquakes.

2. As the plates move apart, solid mantle in the asthenosphere rises toward the surface. It partially melts in response to a decrease in pressure. The molten rock (magma) rises along narrow conduits, accumulates in magma chambers beneath the rift, and eventually becomes part of the oceanic lithosphere.

3. Much of the magma solidifies at depth, but some erupts onto the seafloor, forming submarine lava flows. These eruptions create new ocean crust that is incorporated into the oceanic plates as they move apart.

4. Mid-ocean ridges are elevated above the surrounding seafloor because they consist of hotter, less dense materials, including magma. They also are higher because the underlying lithosphere is thinner beneath ridges than beneath typical seafloor. Lower density materials and thin lithosphere mean that the plate "floats" higher above the underlying asthenosphere. The elevation of the seafloor decreases away from the ridge because the rock cools and contracts, and because the less dense asthenosphere cools enough to become part of the more dense lithosphere.

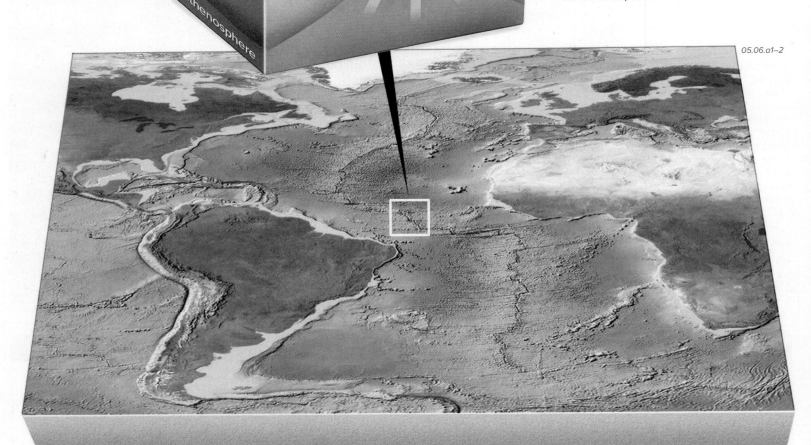

05.06.a1–2

B What Happens When Divergence Splits a Continent Apart?

Most divergent plate boundaries are beneath oceans, but a divergent boundary may also form within a continent. This process, called *continental rifting,* creates a *continental rift,* such as the Great Rift Valley in East Africa. Rifting can lead to seafloor spreading and formation of a new ocean basin, following the progression shown here.

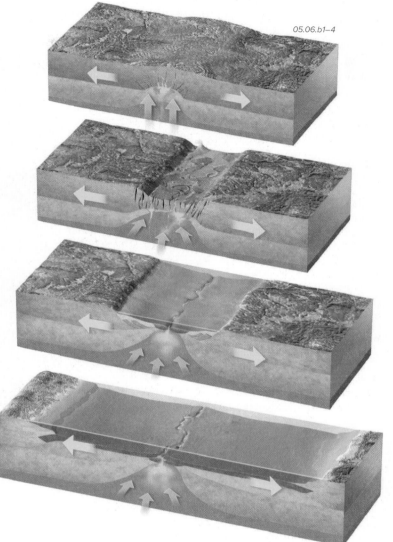

05.06.b1–4

1. The initial stage of continental rifting commonly includes broad uplift of the land surface as mantle-derived magma ascends into and pushes up the crust. The magma heats and can melt parts of the continental crust, producing additional magma. Heating of the crust causes it to expand, which results in further uplift.

2. Stretching of the crust causes large crustal blocks to drop down along faults, forming a *continental rift,* like in the Great Rift Valley. The downdropped blocks may form relatively low areas, called *basins,* that can trap sediment and water, resulting in lakes. Deep rifting causes solid mantle material in the asthenosphere to flow upward and partially melt. The resulting magma may solidify beneath the surface or may erupt from volcanoes and long fissures on the surface. The entire crust thins as it is pulled apart, so the central rift becomes lower in elevation over time.

3. If rifting continues, the continent splits into two pieces, and a narrow ocean basin forms as seafloor spreading takes place. A modern example of this is the narrow Red Sea, which runs between Africa and the Arabian Peninsula. As the edges of the continents move away from the heat associated with active spreading, the thinned crust cools and drops in elevation, eventually dropping below sea level. The continental margin ceases to be an active plate boundary. A continental edge that lacks tectonic activity (i.e., is not a plate boundary) is called a *passive margin.*

4. With continuing seafloor spreading, the ocean basin becomes progressively wider, eventually becoming a broad ocean like the modern-day Atlantic Ocean. The Atlantic Ocean basin formed when North and South America rifted away from Europe and Africa, following the sequence shown here. Continental edges on both sides of the Atlantic Ocean are currently passive margins. Seafloor spreading continues today along the ridge in the middle of the Atlantic Ocean, so the Americas continue moving away from Europe and Africa, and the Atlantic Ocean widens.

Continental Rifting in East Africa

5. East Africa and adjacent seas illustrate the different stages of continental rifting. Here, a piece of continent has been rifted away from Africa, and another piece is in the early stages of possibly doing the same (▶).

6. Early stages of rifting occur along the East African Rift, a long continental rift that begins near the Red Sea and extends into central Africa. The rift is within an elevated (uplifted) region and has several different segments, each featuring a downdropped rift. Some parts of the rift contain large lakes.

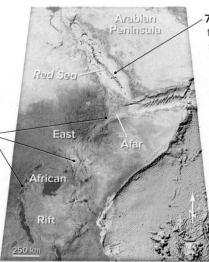

Arabian Peninsula

Red Sea

East

Afar

African

Rift

250 km

05.06.b5

7. The Red Sea represents the early stages of seafloor spreading. It began forming about 50 million years ago when the Arabian Peninsula rifted away from Africa. The Red Sea continues to spread and slowly grow wider.

5.6

5.7 What Happens at Convergent Boundaries?

CONVERGENT BOUNDARIES FORM when two plates move toward each other. Convergence can involve two oceanic plates, an oceanic plate and a plate that carries a continent, or two continent-carrying plates. Oceanic trenches, island arcs, and Earth's largest mountain belts form at convergent boundaries. Many of Earth's most dangerous volcanoes and largest earthquakes also occur along these boundaries.

A What Happens When Two Oceanic Plates Converge?

1. Convergence of two oceanic plates forms an *ocean-ocean convergent boundary*. One plate bends and slides beneath the other plate along an inclined zone. The process of one plate sliding beneath another plate is *subduction*, and the zone around the downward-moving plate is a *subduction zone*. Many large earthquakes occur in association with subduction zones.

2. An *oceanic trench* forms as the subducting plate bends down. Sediment and slices of oceanic crust collect in the trench, forming a wedge called an *accretionary prism*. This name signifies that material is being added *(accreted)* over time to the wedge- or prism-shaped region.

3. As the plate subducts, its temperature increases, releasing water from minerals in the oceanic crust. This water causes melting in the overlying asthenosphere, and the resulting magma is buoyant and rises into the overlying plate.

4. Some magma erupts, initially under the ocean and later as dangerous, explosive volcanoes that rise above the sea. With continued activity, the erupted lava and exploded volcanic fragments construct a curving belt of islands—an *island arc*. An example is the arc-shaped belt of the Aleutian Islands of Alaska. The area between the island arc and the ocean trench accumulates sediment, most of which comes from volcanic eruptions and from the erosion of volcanic materials in the arc.

5. Magma that solidifies at depth adds to the volume of the crust. Over time, the crust gets thicker and becomes transitional in character between oceanic and continental crust. Volcanic islands join to form more continuous strips of land, as occurred to form the island of Java in Indonesia.

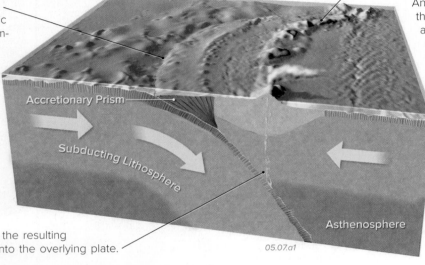

05.07.a1

B What Happens When an Oceanic Plate Subducts Beneath a Continent?

1. The convergence of an oceanic plate and one that contains a continent forms an *ocean-continent convergent boundary*. Along this boundary, the denser oceanic plate subducts beneath the more buoyant continental material.

2. An oceanic trench marks the plate boundary and receives sediment from the adjacent continent. This sediment and material scraped off the oceanic plate form an accretionary prism.

3. Volcanoes form on the surface of the overriding continent in the same way the volcanoes form in an ocean-ocean convergent boundary. These volcanoes erupt, often violently, producing large amounts of volcanic ash, lava, and mudflows, which pose a hazard for people who live nearby. Examples include large volcanoes of the Andes of South America and the Cascade Range of Washington, Oregon, northern California, and southern British Columbia.

4. Compression associated with the convergent boundary squeezes the crust for hundreds of kilometers into the continent. The crust deforms and thickens, resulting in uplift of the region. Uplift, volcanism, and magmatic additions at depth may produce a high mountain range, such as the Andes.

5. Magma forms by melting of the asthenosphere above the subduction zone. It can solidify at depth, rise into the overlying continental crust before solidifying, or reach the surface and cause a volcanic eruption.

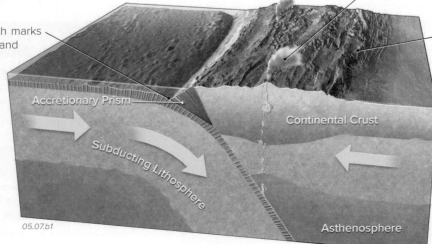

05.07.b1

C What Causes the Pacific Ring of Fire?

Volcanoes surround the Pacific Ocean, forming the *Pacific Ring of Fire*, as shown in the map below. The volcanoes extend from the southwestern Pacific, through the Philippine Islands, Japan, and Alaska, and then down the western coasts of the Americas. The Ring of Fire results from subduction on both sides of the Pacific Ocean.

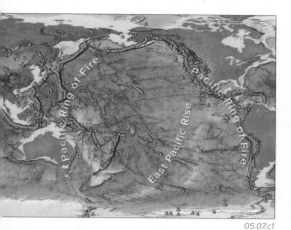

05.07.c1

1. In the Pacific, new oceanic lithosphere forms along a mid-ocean ridge, the East Pacific Rise. Once formed, new lithosphere moves away from the ridge as seafloor spreading continues.

2. Oceanic lithosphere subducts beneath the Americas, forming oceanic trenches on the seafloor and volcanoes on the overriding, mostly continental, plates.

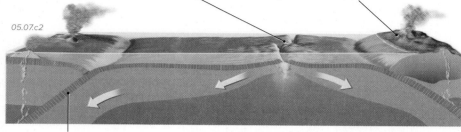

05.07.c2

3. Subduction of oceanic lithosphere also occurs to the west, beneath Japan and island arcs of the western Pacific.

4. More oceanic plate is subducted than is produced along the East Pacific Rise, so the width of the Pacific Ocean is shrinking with time.

D What Happens When Two Continents Collide?

Two continental masses may converge along a *continent-continent convergent boundary*. This type of boundary is commonly called a *continental collision,* and it produces huge mountain ranges.

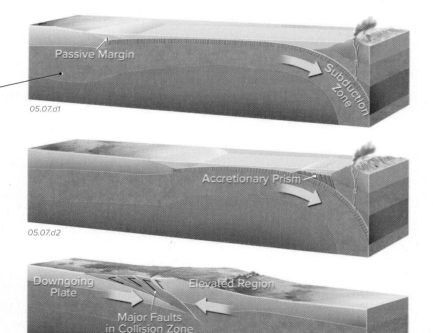

05.07.d1

05.07.d2

05.07.d3

1. The large plate in the figure to the right is partly oceanic and partly continental, and the oceanic part is being subducted to the right, under another continent at a convergent boundary.

2. As the oceanic part of the plate continues to subduct, the two continents become closer to each other. Magmatic activity occurs in the overriding plate above the subduction zone. The edge of the approaching continent has no such activity because it is not yet a plate boundary.

3. When the converging continent arrives at the subduction zone, it may partially slide under the other continent or simply clog the subduction zone as the two continents collide. The two continents are thick and have the same density, so neither can be easily subducted beneath the other and into the asthenosphere. Along the boundary, faults slice up the continental crust, stacking one slice on top of another. These slices are distinct from the accretionary prism that formed along the convergent boundary prior to the actual continental collision. Continental collisions form enormous mountain belts and high plateaus, such as the Himalaya and Tibetan Plateau of southern Asia. The Himalaya and Tibetan Plateau are still forming today, as continental crust of India collides with the southern edge of Asia.

Before You Leave This Page

☑ Sketch, label, and explain the features and processes associated with ocean-ocean and ocean-continent convergent boundaries.

☑ Sketch, label, and explain the steps leading to a continental collision (continent-continent convergent boundary).

5.7

5.8 What Happens Along Transform Boundaries?

AT TRANSFORM BOUNDARIES, PLATES SLIP HORIZONTALLY past each other along *transform faults*. In the oceans, transform faults are associated with mid-ocean ridges. Transform faults combine with spreading centers to form a zigzag pattern on the seafloor. A transform fault can link different types of plate boundaries, such as a mid-ocean ridge and an ocean trench. Some transform boundaries occur beside or within a continent, sliding one large crustal block past another, as occurs along the San Andreas fault in California.

A Why Do Mid-Ocean Ridges Have a Zigzag Pattern?

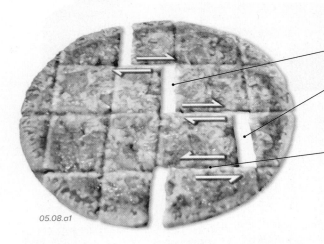

05.08.a1

1. To understand the zigzag character of mid-ocean ridges, examine how the two parts of this pizza have pulled apart, just like two *diverging* plates.

2. The break in the pizza did not follow a straight line. It took jogs to the left and the right, following cuts where the pizza was the weakest.

3. Openings created where the pizza pulled apart represent the segments of a mid-ocean ridge that are spreading apart. However, unlike a pizza, no open gaps exist at a mid-ocean ridge because new material, much of it molten, derived from the underlying mantle fills the space as fast as the plates spread apart, forming new oceanic crust.

4. The openings are linked by breaks, or *faults*, where the two parts of the pizza simply slide by one another. There are no gaps along these breaks, only *horizontal* movement of one plate sliding past the other. Arrows show the direction of relative motion. A fault that accommodates the horizontal movement of one tectonic plate past another is a *transform fault*. The spreading direction must be parallel to the transform faults and perpendicular to the spreading segments, so a zigzag pattern is required to allow a plate boundary to be curved.

Transform Faults Along the Mid-Ocean Ridge

5. Mid-ocean ridges, such as this one in the southern Atlantic Ocean, have a zigzag pattern similar to the broken pizza.

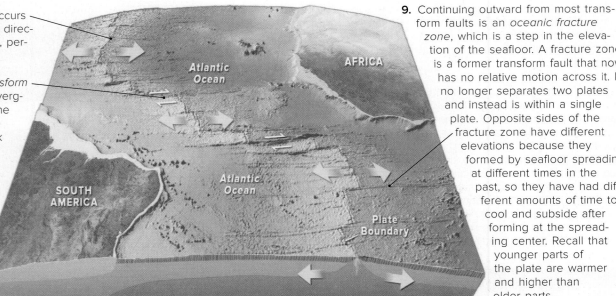

05.08.a2

6. In this region, spreading occurs along north-south ridges. The direction of spreading is east-west, perpendicular to the ridges.

7. East-west offsets are *transform faults* along which the two diverging plates simply slide past one another, like the breaks in the pizza. The transform faults link the spreading segments and have the relative motion shown by the small, white arrows.

8. Transform faults along mid-ocean ridges are generally perpendicular to the axis of the ridge. As in the pizza example, transform faults are parallel to the direction in which the two plates are spreading apart.

9. Continuing outward from most transform faults is an *oceanic fracture zone*, which is a step in the elevation of the seafloor. A fracture zone is a former transform fault that now has no relative motion across it. It no longer separates two plates and instead is within a single plate. Opposite sides of the fracture zone have different elevations because they formed by seafloor spreading at different times in the past, so they have had different amounts of time to cool and subside after forming at the spreading center. Recall that younger parts of the plate are warmer and higher than older parts.

10. The zigzag pattern of mid-ocean ridges reflects the alternation of spreading segments with transform faults. In this example, the overall shape of the ridge mimics the edges of Africa and South America and so was largely inherited from the shape of the original continental rift that split the two continents apart.

B   What Are Some Other Types of Transform Boundaries?

The Pacific seafloor and western North America contain several different transform boundaries. The boundary between the Pacific plate and the North American plate is mostly a transform boundary, with the Pacific plate moving northwest relative to the main part of North America.

1. The Queen Charlotte transform fault, shown as a long green line, lies along the edge of the continent, from north of Vancouver Island to southeastern Alaska.

2. The zigzag boundary between the Pacific plate and the small Juan de Fuca plate has three transform faults, shown here as green lines. These transform faults link three ridge segments that are spreading (shown here as yellow lines).

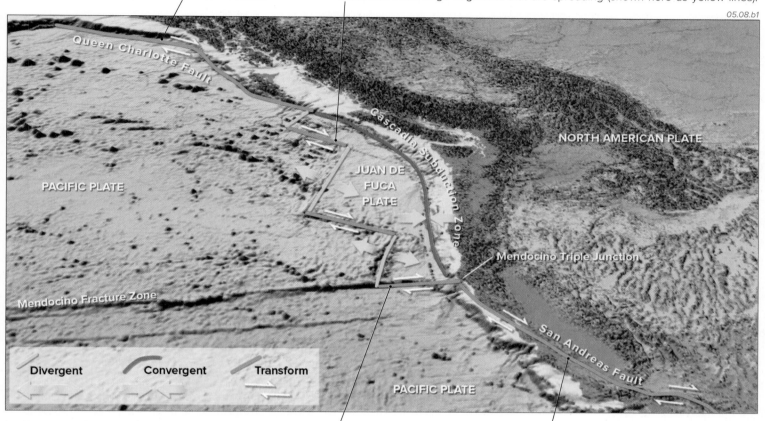

05.08.b1

3. The Mendocino fracture zone originated as a transform fault, but it is now entirely within the Pacific plate and is no longer active. Oceanic crust to the north is higher because it is younger and warmer than oceanic crust to the south.

4. A transform fault links a spreading center (between the Pacific plate and the Juan de Fuca plate) with the Cascadia subduction zone and the San Andreas fault. The place where the three plate boundaries meet is a *triple junction*. The Mendocino triple junction is the meeting place of two different transform faults and a subduction zone.

5. The San Andreas transform fault extends from north of San Francisco to southeast of Los Angeles. The part of California west of the fault is on the Pacific plate and is moving approximately 5 cm/yr to the northwest relative to the rest of North America. South of this map area, the transform boundary continues across southern California and into the Gulf of California.

6. Californians have a transform fault in their backyard. In central California, the San Andreas fault forms linear valleys, abrupt mountain fronts, and lines of lakes. In the Carrizo Plain (▶), the fault is a linear gash in the topography. Some streams follow the fault and others jog to the right as they cross the fault, recording relative movement of the two sides. In this view, the North American plate is to the left, and the Pacific plate is to the right and is being displaced toward the viewer at several centimeters per year.

05.08.b2 Carrizo Plain, CA

Before You Leave This Page

✓ Sketch, label, and explain an oceanic transform boundary related to seafloor spreading at a mid-ocean ridge.

✓ Sketch, label, and explain the motion of transform faults along the west coast of North America.

How Does Seafloor Vary from Place to Place?

THE SEAFLOOR VARIES IN DEPTH, in age, and in the thickness of its sediment cover, and these three factors are all related. How do depth, age, and sediment thickness vary from one part of the seafloor to another, and how are these variations explained by plate tectonics?

Depth and Age of the Seafloor

1. This map shows depths of the seafloor. Darker gray areas are continents, and the light to dark blue regions are oceans and seas. Darker blue shows the deepest areas of the seafloor, and light bluish gray shows the shallowest. Letters identify some of the ridges (R), trenches (T) along convergent boundaries, and passive margins (P) that are not plate boundaries.

2. The deepest part of the oceans is at trenches (T) along active margins, where one plate subducts beneath another plate. The shallowest parts of the seafloor are on continental shelves, many of the widest of which are along passive margins (P). Mid-ocean ridges (R) are intermediate in depth.

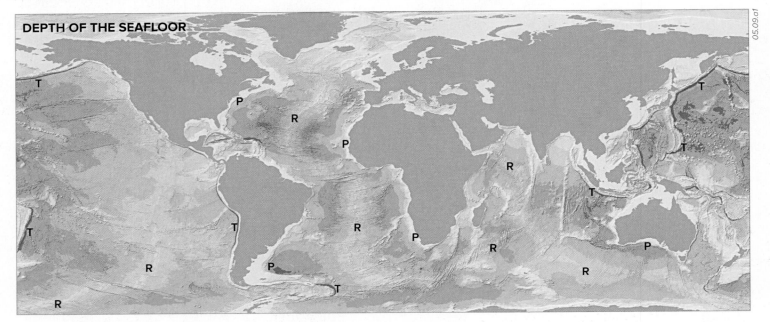

3. The map below shows the age of the seafloor. Purple represents the oldest areas (about 180 million years), and the darkest orange represents very young oceanic crust. Compare this map with the one above.

4. The youngest oceanic crust is near mid-ocean-ridge spreading centers (R). These areas are also higher than most of the ocean floor.

5. The oldest oceanic crust in any ocean is the most distant from mid-ocean ridges. None is older than about 180 million years, because all older oceanic crust has been subducted back into the mantle. The oldest seafloor is much younger than the oldest continental rocks.

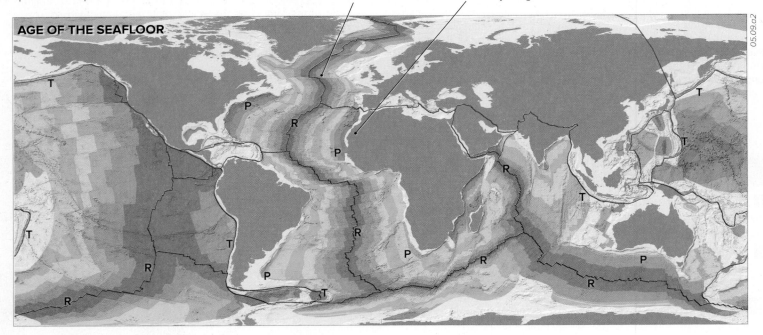

Thickness of Sediment on the Seafloor

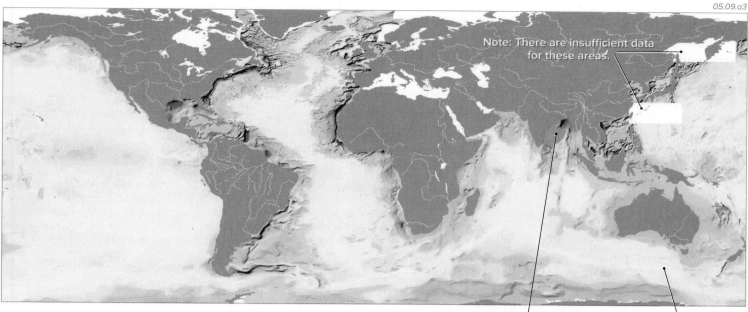

05.09.a3

Note: There are insufficient data for these areas.

6. This map shows sediment thickness on the seafloor. It ranges from light blue, where thickness is less than 200 m, to orange and red, where sediment is over 5 to 10 km (3–6 mi) thick. White colors indicate locations where there are insufficient data to show on this map. The white lines on the continents are rivers. What patterns do you observe on this map? Where is sediment on the seafloor thickest, where is it thinnest, and how does thickness relate to major rivers?

7. The thickest sediment is along continental margins, especially those that were formed by rifting. Seafloor sediment is also thickest near the mouths of rivers or where the oceanic crust is relatively old.

8. There is virtually no sediment cover over the youngest crust at the mid-ocean ridges, which here are along the belts of light blue.

A Plate-Tectonic Explanation for the Variations in Thickness of Sediment on the Seafloor

According to plate tectonics, oceanic crust forms from upwelling magma and spreading at a mid-ocean ridge and then moves away from the ridge with further spreading. If so, the crust should be youngest near the ridge, where it was just formed, and should be progressively older away from the ridge. Also, oceanic crust near the ridge will not have had time to accumulate much sediment, but the sediment cover should thicken outward from the ridge.

1. Since 1968, ocean-drilling ships have drilled thousands of holes into the seafloor. Geoscientists use drill cores and other drilling results to measure the thickness of sediment and examine the underlying volcanic rocks (basalt). They analyze samples of sediment and rock, study fossils, and measure magnetic characteristics to determine the age, character, and origin of the materials. From this, they interpret the geologic history.

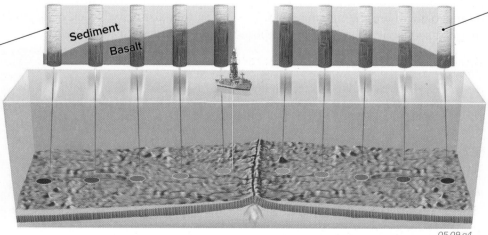

Sediment

Basalt

05.09.a4

2. Drill core samples reveal that sediment is thin or absent on a mid-ocean ridge but becomes thicker away from the ridge. Age determinations from fossils in the sediment and from underlying volcanic rocks show that oceanic crust gets systematically older away from mid-ocean ridges. Drilling results from many parts of the oceans therefore strongly support the theory of plate tectonics.

Before You Leave This Page

☑ Describe and explain how the age of the seafloor relates to mid-ocean ridges, depths of seafloor, and sediment thicknesses.

☑ Sketch and explain why sediment thickness on the seafloor increases away from a mid-ocean ridge.

5.10 What Features Occur Along Mid-Ocean Ridges?

MID-OCEAN RIDGES FORM where two oceanic plates diverge. Magma ascending from the mantle erupts onto the seafloor or solidifies at depth, making new oceanic crust. Heat associated with the hot rocks and magma produces undersea vents of hot water that nourish unique life-forms on the seafloor.

A What Happens When Plates Spread Apart?

As two oceanic plates move apart, solid rock and magma rise from the mantle to occupy the space between the plates. The cooling and solidifying magma forms new oceanic crust, which then gets transported away from the mid-ocean ridge as a new plate is formed along the spreading center. Slices of this rock sequence can be scraped off and preserved on land, allowing geoscientists to study examples of oceanic crust without diving to the bottom of the ocean.

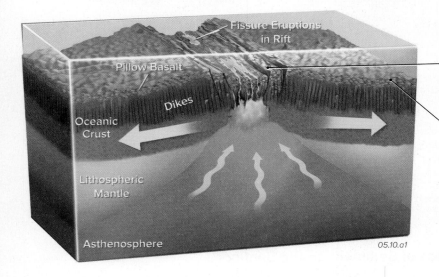

05.10.a1

1. As oceanic crust stretches apart, basaltic lava erupts within the rift. As the magma encounters cold seawater, it forms pillow-shaped features, called *pillow basalts*, on the seafloor. Some magma solidifies within large chambers and in magma-filled fissures parallel to the mid-ocean ridge (perpendicular to plate movement). The magma-filled fissures solidify into sheetlike *dikes*.

2. At many mid-ocean ridges, faults allow blocks of crust to be displaced downward, forming a fault-bounded *rift*.

3. As the cooled crust moves away from the ridge, it is progressively covered with deep-sea sediment. Over time, the sediment tends to smooth over the rough topography formed in the rift. As a result, older oceanic crust tends to have relatively smooth topography.

4. Below a depth of about 4,500 to 5,000 m, sediment is dominated by clay and silica-rich materials. At these depths and greater, carbonate minerals dissolve into seawater as fast as they accumulate, because calcite is more soluble at lower temperatures and at higher pressures. The depth at which calcite disappears in the oceans is the *carbonate compensation depth* (CCD). The actual depth of the CCD varies somewhat from place to place within the ocean, mostly due to temperature variations of seawater.

B What Types of Igneous Rocks Form Along Mid-Ocean Ridges?

New oceanic crust formed at mid-ocean ridges consists of several different kinds of igneous rocks. The rocks are all mafic (rich in magnesium and iron), but they have different textures and features depending on how and where the magma solidified. When this distinctive stack of igneous rocks and overlying seafloor sediment is uplifted and exposed on land, we call it an *ophiolite*.

05.10.b1

1. The upper part of oceanic crust consists of basaltic lava flows. When these lavas erupted into water, they formed a series of overlapping mounds called *pillows*. These distinctive rocks, called *pillow basalts* (▶), have in some cases become uplifted above sea level, where we can now observe them. The pillows in this photograph are about 0.5 to 1 meter across. On the seafloor, the pillows are covered by deep-sea sediment.

2. Countless thin, vertical intrusions of finely crystalline basalt cut across the pillow basalts from below. These thin intrusions, called *dikes*, are so closely spaced that they are called *sheeted dikes*. Each dike represents a thin, tabular conduit through which magma passed. Most dikes are oriented parallel to the oceanic rift and perpendicular to the direction of spreading.

05.10.b2 San Juan Islands, WA

3. Sheeted dikes merge downward into *gabbro*, the coarsely crystalline equivalent of basalt. The gabbro represents magma chambers beneath the rift and locally displays layers (▶) formed by accumulation of light-colored and dark-colored crystals on the floor of the magma chamber.

4. The base of the gabbro is the base of the oceanic crust, below which are *ultramafic* rocks of the mantle. The mantle rocks show evidence of having been partially melted to form all of the overlying mafic rocks in the crust (pillow basalt, sheeted dikes, and gabbro).

05.10.b3 Smartville, CA

C What Are Submarine Black Smokers and How Do They Form?

05.10.c1

1. Mid-ocean ridges and other sites of submarine volcanism locally contain features called *black smokers,* shown here in a photograph taken from a submersible (◀). Black smokers are *hydrothermal vents,* where hot water from within the rock jets out into the cold seawater. As the hot water cools, the metals, sulfur, and other elements dissolved in the hot water form small crystals that make the water black and cloudy.

05.10.c3

2. Sulfur-bearing minerals precipitate around the vent, forming a hollow, circular column called a *chimney.* Some chimneys are more than 5 m (16 ft) high and a meter across, and can grow tens of centimeters per day. Black smokers and sulfide-rich chimneys are interpreted to have formed on mid-ocean ridges and other submarine volcanoes throughout much of Earth's history, forming mineral deposits rich in copper, zinc, and other valuable elements, like these metal-rich sulfide layers (▶).

05.10.c2 Kidd Creek, Ontario, Canada

3. Black smokers form when water in rocks above a magma chamber is heated and rises toward the surface (▲). As the water rises, seawater from nearby areas flows in to take its place. This seawater heats up and leaches metals and other chemical elements from rocks through which it passes, becoming rich in dissolved chemicals. The heated seawater rises toward the surface along faults and other pathways, eventually venting in a black smoker. The water is very hot, commonly over 350°C, but it does not boil because of the pressure exerted by the deep water.

Life at Hydrothermal Vents

Deep-sea hydrothermal vents associated with black smokers support a unique community of unusual creatures. Scientists are actively exploring the ecosystems of these vents, in part to study the unique life-forms that congregate around the vents.

05.10.t1

The photograph included here was taken by scientists using a submersible to investigate these vents and their unusual marine inhabitants. The underwater exploration of deep-sea hydrothermal vents continues to result in major discoveries.

Sunlight is the energy source for green plants, which provide the bulk of food for animals living on Earth's surface. No sunlight reaches the deep seafloor. Instead, life around the hydrothermal vents uses a completely different energy source. Here, life is dependent on somewhat unusual bacteria that are able to break down hydrogen sulfide (H_2S), one of the chemical compounds common within black smokers. These bacteria produce sugars, which feed giant (meter-long) red tube worms. The worms can tolerate the hot water and live close to the vents, where the bacteria are abundant. In fact, many bacteria live within the worms' tissues. The worms in turn form the main food for an assembly of scavenging animals, including white crabs, one of which is shown in this photograph. Large clams also live around hydrothermal vents and draw nutrients by extracting small bits of material from the water and from the bacteria. Fossils of tube worms in ancient hydrothermal vent

deposits show that such communities have existed for millions of years. Some geoscientists think that deep hydrothermal vents could have been the site where early primitive life originated on Earth. We only discovered these unusual ecosystems in the 1980s.

Before You Leave This Page

☑ Describe or sketch the processes that accompany the formation of new oceanic crust at mid-ocean ridges.

☑ Describe black smokers, how they form, and where the hot water originates and how it gets heated.

☑ Describe the type of life that exists around hydrothermal vents and where the different creatures derive their food.

5.10

5.11 How Do Oceanic Islands, Seamounts, and Oceanic Plateaus Form?

SUBMARINE MOUNTAINS, called *seamounts*, rise above the seafloor. In some places, they reach the surface and make islands. These islands include Tahiti, the Galápagos, and other exotic places. The seafloor also has relatively high and broad areas that are *oceanic plateaus*. How are seamounts and oceanic plateaus formed?

A How Does Plate Tectonics Help Explain Island and Seamount Chains?

Fairly straight lines of oceanic islands and submarine mountains (*seamounts*) cross some parts of the ocean floor. These *island and seamount chains* are different in character and origin from curved *island arcs*, which are related to subduction. How do linear chains of islands and seamounts form?

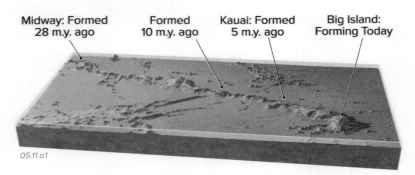

Midway: Formed 28 m.y. ago
Formed 10 m.y. ago
Kauai: Formed 5 m.y. ago
Big Island: Forming Today

05.11.a1

1. Most island and seamount chains are in the Pacific Ocean. One begins on Hawaii's Big Island and continues northwest more than 2,000 km, passing through Midway Island, the site of a pivotal air and sea battle during World War II.

2. Volcanoes are active on the Big Island today, but not on most of the other Hawaiian islands, including Kauai, the northwestern-most island. Ages of volcanic rocks in the island and seamount chain increase systematically to the northwest. When we plot the ages of these rocks as a function of distance from Kilauea (the active volcano on the Big Island), there is a clear relationship between age and distance. How can we explain this pattern?

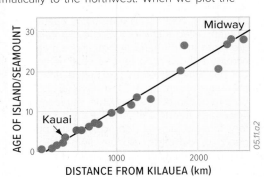

05.11.a2

A Model for the Formation of Island and Seamount Chains

Island and seamount chains and most clusters of islands in the oceans have two key things in common: they were formed by volcanism and they are near sites that geoscientists interpret to be above unusually high-temperature regions in the deep crust and upper mantle. Geoscientists refer to these anomalously hot regions as *hot spots*.

1. This figure shows how linear island and seamount chains can be related to a plate moving over a hot spot. At a hot spot, hot mantle rises and melts, forming magma that ascends into the overlying plate. If the plate above the hot spot is moving relative to the hot spot, volcanism constructs a chain of volcanoes.

2. Magma generated by a hot spot may solidify at depth or form a volcanic mountain on the ocean floor. If the submarine volcano grows high enough above the seafloor, it becomes a volcanic island. Each of the Hawaiian islands consists of volcanoes. Geoscientists consider the hot spot to be currently below or near the eastern side of the Big Island, near Kilauea volcano.

Hot Spot

05.11.a3

4. If a plate is not moving or is moving very slowly, the hot spot forms a cluster of volcanic islands and seamounts instead of a linear chain. The Galápagos, a cluster of volcanic islands in the eastern Pacific, are interpreted to be above a hot spot.

3. As an area on the plate moves beyond the hot spot, it cools, subsides, and erodes, so volcanoes that start out as islands may sink beneath the sea to become seamounts. In this way, a hot spot makes a chain of volcanic islands and seamounts, each created when it was over the hot spot. According to this model, volcanoes above the hot spot may be erupting today, those close to but not above the hot spot are relatively young, and those farthest from the hot spot are older. The present volcanic activity and pattern of ages on the graph presented earlier are consistent with the hot-spot model and with the calculated motion of the Pacific plate on which Hawaii rides. Lines of hot-spot-generated seamounts allow us to determine past directions and rates of plate motions.

B How Do Some Flat-Topped Seamounts Form?

Once an oceanic island is constructed over a hot spot, cooling and subsidence of the region can produce a distinctive, flat-topped seamount, called a *guyot*. The figures below illustrate the sequence of events that form a guyot.

05.11.b1

05.11.b2

05.11.b3

Magmatism caused by an underlying hot spot begins building a submarine volcano by eruption of lava flows onto the seafloor. Magmatism related to hot spots is usually basaltic in composition.

Continued eruptions build up the volcano until it may eventually rise above the sea as an island. Once magmatism ceases, perhaps when an island moves off a hot spot, the oceanic plate cools and subsides.

The top of the mountain is beveled off by wave erosion as the mountain continues subsiding, becoming a submarine, flat-topped seamount. Over time, it is covered by layers of marine sediments.

C How Do Oceanic Plateaus Form?

Some large regions of the seafloor rise a kilometer or more above their surroundings, forming *oceanic plateaus*. These plateaus are largely composed of basalts and, like the seafloor in general, are mostly late Mesozoic and Cenozoic in age (mostly 130 million years ago to the present).

1000 km

05.11.c1

05.11.c2

05.11.c3

05.11.c4

1. This perspective (▲) shows the Kerguelen oceanic plateau, which rises above the surrounding seafloor in the southern Indian Ocean. The plateau is several thousand kilometers long, but it only reaches sea level in a few small islands. The small sliver of land showing in the lower right corner is part of Antarctica.

2. Geoscientists interpret oceanic plateaus as forming at hot spots, above rising mantle plumes. The plumes travel through the mantle as solid masses, not liquids.

3. When the top of a plume encounters the base of the lithosphere, it causes widespread melting. Submarine basalts pour out onto the seafloor through fissures and central vents.

4. Immense volumes of basalt (as much as 50 million cubic kilometers) erupt onto the seafloor over millions of years. This volcanism creates a broad, high oceanic plateau.

05.11.c5

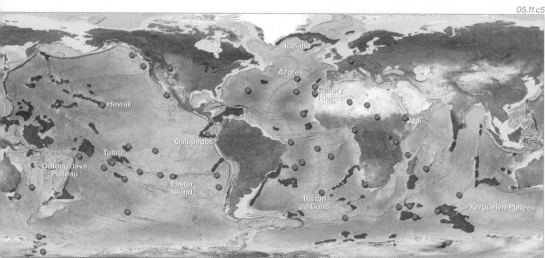

Iceland

Azores

Canary Islands

Hawaii

Afar

Galapagos

Tahiti

Ontong Java Plateau

Easter Island

Tristan da Cuna

Kerguelen Plateau

5. On this map, red dots show the locations of likely hot spots (◀), many of which are located at the volcanically active ends of linear island chains. Dark gray areas mark especially thick accumulations of hot spot-related basalt.

Before You Leave This Page

✓ Describe an island and seamount chain and how it is interpreted to be related to a hot spot and plume.

✓ Explain how a mantle plume is interpreted to form an oceanic plateau.

5.11

5.12 What Are the Characteristics and History of Continental Hot Spots?

A HOT SPOT WITHIN A CONTINENT is marked by high elevations, abundant volcanism, and continental rifting. Hot spots can facilitate complete rifting and separation of a continent into two pieces and can help determine where the split occurs. Several continental hot spots are active today.

A What Features Are Typical of Continental Hot Spots?

Hot spots are volcanic areas interpreted to be above rising mantle plumes. Continental hot spots are associated with certain characteristics, including high elevations, volcanism, and the presence of rifts. Two examples are the Afar region of East Africa and the Yellowstone region of the western United States.

Afar Region, East Africa

1. Continental hot spots have high elevations largely because of heating and thinning of the lithosphere by a rising plume of hot mantle. Many geologists interpret the Afar region of eastern Africa to be located above a hot spot that is currently active.

2. The East African Rift is within the African plate. It may or may not evolve into a full rift that fragments the continent into two parts and that leads to seafloor spreading.

3. Near the hot spot, the Arabian Peninsula has pulled away from Africa along the Red Sea and the Gulf of Aden. Beneath these seas, seafloor spreading generates new oceanic crust.

4. The Red Sea, Gulf of Aden, and East African Rift come together in the Afar region, branching off like three spokes on a wheel. The Afar region is among the most volcanically active areas on Earth and has experienced recent volcanic eruptions. Volcanism has been so prolific here that it has created a triangular area of new land in the corner of Africa from which the Arabian Peninsula pulled away. Remove the volcanically constructed Afar region, and the Arabian Peninsula fits back in Africa.

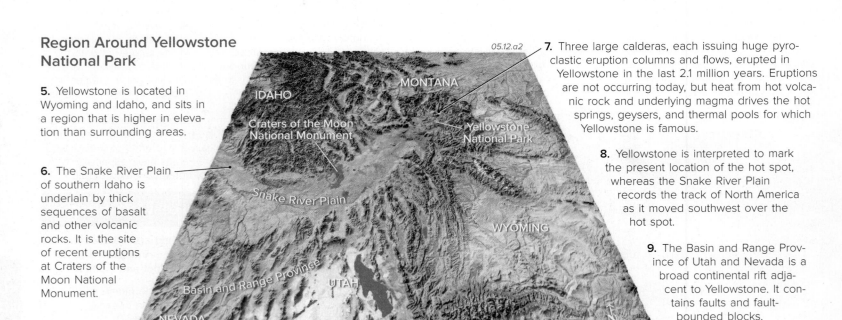

Region Around Yellowstone National Park

5. Yellowstone is located in Wyoming and Idaho, and sits in a region that is higher in elevation than surrounding areas.

6. The Snake River Plain of southern Idaho is underlain by thick sequences of basalt and other volcanic rocks. It is the site of recent eruptions at Craters of the Moon National Monument.

7. Three large calderas, each issuing huge pyroclastic eruption columns and flows, erupted in Yellowstone in the last 2.1 million years. Eruptions are not occurring today, but heat from hot volcanic rock and underlying magma drives the hot springs, geysers, and thermal pools for which Yellowstone is famous.

8. Yellowstone is interpreted to mark the present location of the hot spot, whereas the Snake River Plain records the track of North America as it moved southwest over the hot spot.

9. The Basin and Range Province of Utah and Nevada is a broad continental rift adjacent to Yellowstone. It contains faults and fault-bounded blocks.

B How Do Continental Hot Spots Evolve?

Many continental hot spots underwent a similar sequence of events. They started with doming and ended with the formation of a new continental margin and a new ocean formed by seafloor spreading.

1. Hot spots mark where a mostly solid, hot mass rises, probably from the lower mantle, and encounters the base of the lithosphere. The rising material melts as it rises and also causes melting of nearby lithosphere.

----

2. As the upper mantle and crust heat up, a broad, domal uplift forms on the surface. Doming is accompanied by stretching of the crust, which commonly begins to break apart along three rifts that radiate out from the hot spot.

3. All three parts, or arms, of the rift are bordered by faults, which downdrop long fault blocks. The downdropped blocks form basins that contain lakes and are partially filled by sediment and rift-related volcanic rocks.

----

4. Some mantle-derived magma escapes to the surface and erupts commonly as voluminous basalts. More felsic magma (more granitic in composition) forms where mantle-derived magma causes melting of the crust.

5. Complete rifting of the continent occurs along two arms of the rift. This results in a new continental margin and seafloor spreading in the new ocean basin. At the onset of spreading, the edge of the continent is uplifted because the lithosphere is heated and thinned due to the rifting.

----

6. The third arm of the rift begins to become less active and fails to break up the continent into more pieces. This *failed rift* is lower than the surrounding continent and commonly becomes the site of major rivers.

7. Sediment transported by rivers down the failed rift will form a delta at the bend in the continent. This is currently occurring along the western coast of Equatorial Africa at the large inward bend in the coast (see the figure and text below).

8. As seafloor spreading continues, the generation of new oceanic lithosphere causes the mid-ocean ridge to move farther out to sea. The continental margin cools and subsides, and is covered by marine sediment on the newly formed continental shelf. This continental margin is no longer a plate boundary and is now a passive margin.

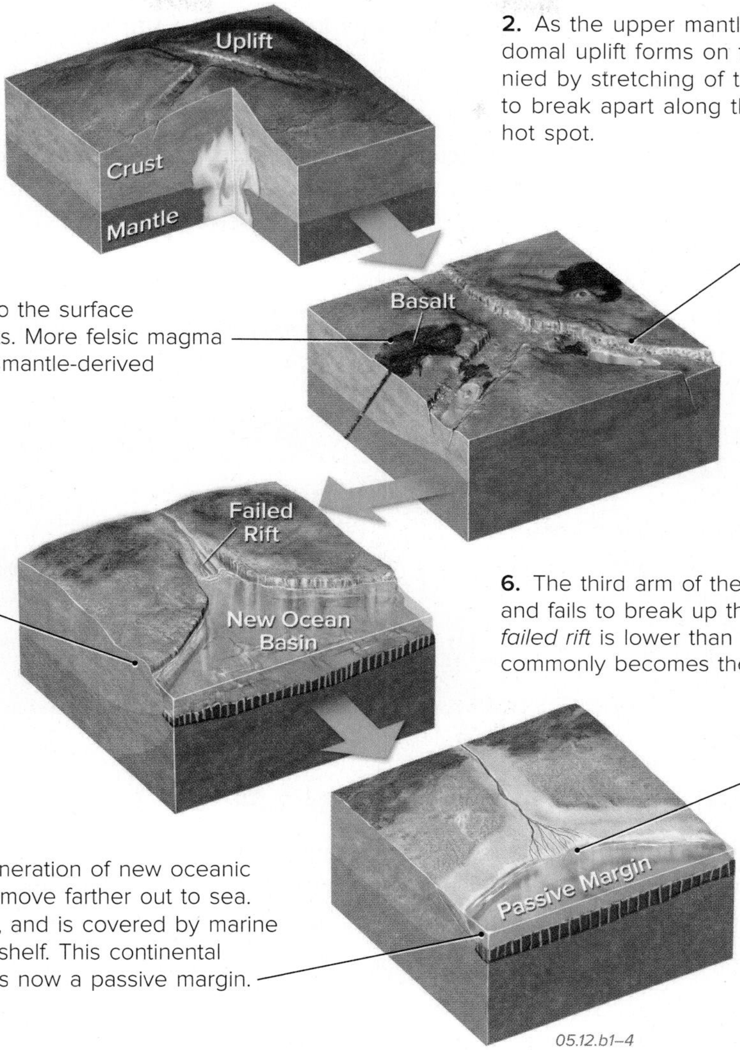

05.12.b1–4

Hot Spots and Continental Outlines

Geologists conclude that hot spots have helped define the outlines of the continents by shaping the boundary along which continents separate from one another. The best example of this is the inward bend of the western coast of Africa. This bend occurs at the intersection of three arms of a rift, two of which led to the opening of the South Atlantic Ocean. The third *failed arm* cuts northeastward into Africa and is the site of several major rivers (the rivers are not shown in this figure). Large eruptions of basalt occurred along the rifts, and active volcanism near the failed rift may mark the location of a hot spot. This figure shows what the area may have looked like 110 million years ago, after the continents first started to rift apart.

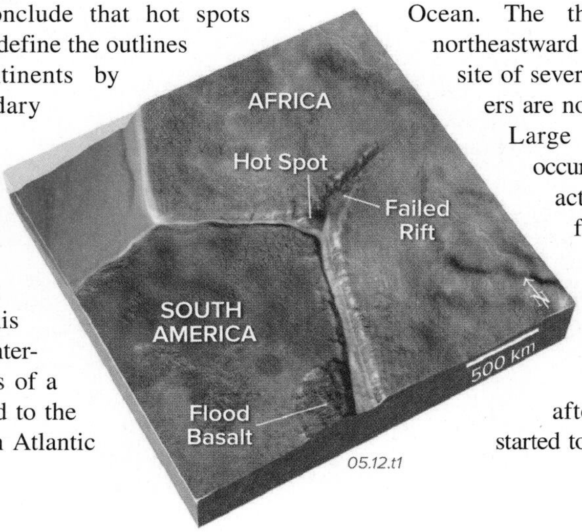

05.12.t1

Before You Leave This Page

☑ Summarize the features that are typical of continental hot spots, providing an example of each type of feature.

☑ Summarize or sketch how continental hot spots evolve over time.

☑ Describe or sketch how hot spots influence continental outlines, providing an example.

5.12

5.13 How Do Plates Move and Interact?

THE PROCESS OF PLATE TECTONICS circulates material back and forth between the asthenosphere and the lithosphere. Some asthenosphere becomes lithosphere at mid-ocean spreading centers and then takes a slow trip across the ocean floor before going back down into the asthenosphere at a subduction zone. Besides creating and destroying lithosphere, this process is the major way that Earth transports heat to the surface.

A What Moves the Plates?

How exactly do plates move? To move, an object must be subjected to a *driving force* (a force that drives the motion). The driving force must exceed the *resisting forces* (those forces that resist the movement), such as friction and any resistance from other material that is in the way. What forces drive tectonic plates?

1. *Slab Pull*—Subducting oceanic lithosphere is more dense than the adjacent asthenosphere, so gravity pulls the plate downward into the asthenosphere. This pulling, called *slab pull*, is a significant force, and a plate being subducted generally moves faster than a plate not being subducted. Subduction sets up other forces in the mantle that can work with or against slab pull.

2. *Ridge Push*—The mid-ocean ridge is higher than the ocean floor away from the ridge because lithosphere near the ridge is thinner and hotter. Gravity causes the plate to slide away from the topographically high ridge and push the plate outward—*ridge push*.

3. *Mantle Convection*—The asthenosphere, although a solid, is capable of flow. It experiences *convection*, where hot material rises due to its lower density, while cold material sinks because it is more dense. Hot material rises at mid-ocean ridges, cools, and eventually sinks back into the asthenosphere at a subduction zone. *Convection* also occurs at centers of upwelling mantle material called *hot spots*, and it can help or hinder the motion of a plate. Another important source of forces is the motion of a plate with respect to the underlying mantle.

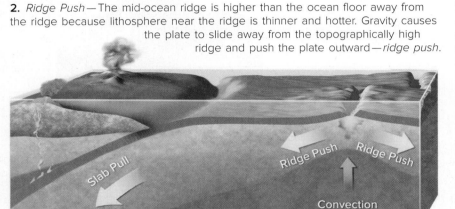

05.13.a1

B How Fast and in What Directions Do Plates Move Relative to One Another?

Plates move at 1 to 15 cm/yr, about as fast as your fingernails grow. This map shows velocities and relative motions along major plate boundaries, based on long-term rates. Arrows indicate whether the plate boundary has divergent (outward pointing), convergent (inward pointing), or transform (side by side) motion.

05.13.b1

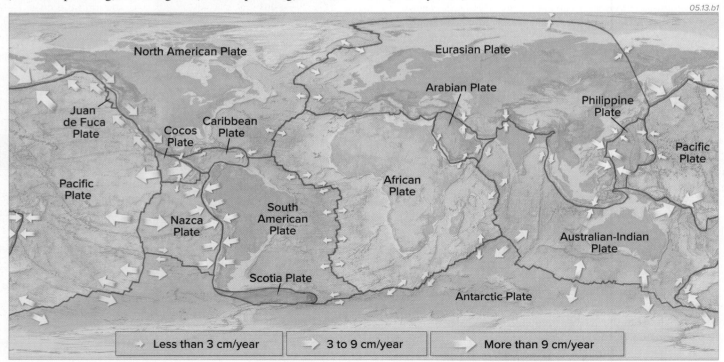

| → Less than 3 cm/year | ⇒ 3 to 9 cm/year | ⟹ More than 9 cm/year |

C Is There a Way to Directly Measure Plate Motions?

Modern technology allows direct measurement of plate motions using satellites, lasers, and other tools. The measured directions and rates of plate motions are consistent with our current concept of lithospheric plates and with the theory of plate tectonics.

The global positioning system (GPS) is an accurate location technique that uses small radio receivers to record signals from several dozen Earth-orbiting satellites. By attaching GPS receivers to sites on land and then monitoring changes in position over time, scientists produce maps showing motions for each plate. Arrows point in the direction of motion, and longer arrows indicate faster motion.

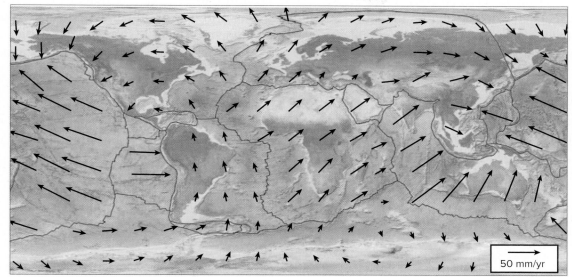

50 mm/yr

05.13.c1

Note the motions of different plates. Africa is moving to the northeast, away from South America. North America is moving westward and rotating counterclockwise in this view. These motions match predictions from the theory of plate tectonics.

D What Happens Where Plate Boundaries Change Their Orientation?

The boundary between two plates may be of a different type in different places. It can change from divergent to transform or from convergent to transform, for example, as its orientation varies compared to the direction of relative plate movement. Nearly all plate boundaries contain curves or abrupt bends, so most boundaries change type as they cross Earth's surface.

1. As these two interlocking blocks pull apart (▼), two gaps (equivalent to spreading centers) form, linked by a transform boundary where the blocks slip horizontally by one another. The boundary changes its type as it changes its orientation.

3. The boundary between the North American and Pacific plates illustrates how a plate boundary changes its character as it changes orientation. In most of Alaska, the two plates converge, and the Pacific plate subducts beneath North America.

4. To the southeast, the plate boundary bends and becomes parallel to the edge of the continent. As it bends, the plate boundary changes from being convergent to largely having a transform motion. The transform fault, named the Queen Charlotte fault, allows the Pacific plate to slide northwestward past North America at a rate of 5 cm/yr.

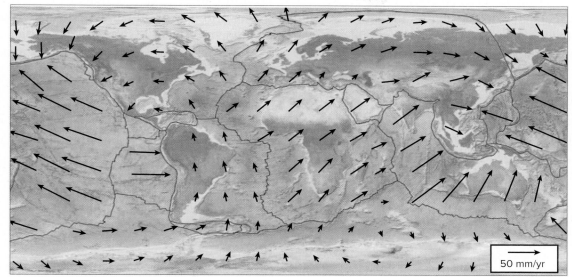

05.13.d1

200 km

05.13.d3

2. A small-scale example of this type of change in motion occurred along a fault (▶) in Alaska, where lateral motion on the fault during an earthquake caused local pulling apart of the rock and ice as the fault curved around several bends.

Blocks are pulled apart along the bend in the fault.

Before You Leave This Page

✓ Sketch and explain the driving forces of plate tectonics.

✓ Describe the typical rates of relative motion between plates.

✓ Describe one way to directly measure plate motion.

✓ Sketch, label, and explain how a plate boundary can change its type as its orientation changes.

5.13

5.14 How Is Paleomagnetism Used to Determine Rates of Seafloor Spreading?

PALEOMAGNETISM IS THE RECORD OF PAST CHANGES in Earth's magnetic field, as recorded in rocks and other materials. The magnetic field is strong enough to orient magnetism in certain minerals, especially the iron-rich mineral *magnetite,* in the direction of the prevailing magnetic field. Magnetic directions preserved in volcanic rocks, intrusive rocks, and some sedimentary rocks provide an important way to determine the rates of seafloor spreading.

A What Causes Earth's Magnetic Field?

Earth has a metallic iron core, which is composed of a solid inner core surrounded by a liquid outer core. The liquid core flows and behaves like a *dynamo* (an electrical generator), creating a magnetic field around Earth.

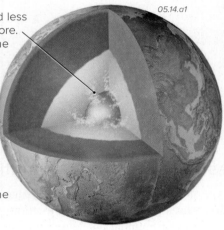

05.14.a1

1. The inner core transfers heat and less dense material to the liquid outer core. This transfer causes liquid in the outer core to rise, forming *convection currents.* These convection currents are limited to the molten outer core and are not the same as those in the upper mantle.

2. Movement of the molten iron is affected by forces associated with Earth's rotation. The resulting movement of liquid iron and electrical currents generates the magnetic field.

3. Earth's magnetic field currently flows from south to north (▶), causing the magnetic ends of a compass needle to point toward the north. This orientation is called a *normal polarity.*

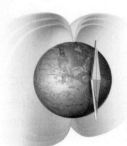

05.14.a2

4. Many times in the past, the magnetic field has had a *reversed polarity* (◀), so that a compass needle would point south. The switch between normal polarity and reversed polarity is a *magnetic reversal.*

05.14.a3

B How Do Magnetic Reversals Help Us Infer the Age of Rocks?

The north and south magnetic poles have switched many times, typically remaining either normal or reversed anywhere from 100,000 years to a few million years. Geoscientists have constructed a *magnetic* timescale by isotopically dating sequences of rocks that contain magnetic reversals. This *geomagnetic polarity timescale* then serves as a reference to compare against other sequences of rocks.

1. Geoscientists measure the direction and strength of the magnetism preserved in rocks with an instrument called a *magnetometer* (▼). With this device, we can tell whether the magnetic field had a normal polarity or a reversed polarity when an igneous rock solidified and cooled or when a sedimentary layer accumulated.

2. This figure (▶) shows the series of magnetic reversals during the last 10 million years, the most recent part of the Cenozoic Era. This time period is within the *Neogene* and *Quaternary periods.* The timescale shows periods of normal magnetization (N) in black and those of reversed magnetization (R) in white. Examine this figure and note how many magnetic reversals occurred just within the last 10 million years. Dividing the number of reversals by 10 m.y. provides a measure of how often Earth's magnetic field reverses—4 or 5 times per m.y.

3. Variability in the spacing and duration of magnetic reversals produced a unique pattern through time. Geologists can measure the pattern of reversals in a rock sequence and compare this pattern to the magnetic timescale to see where the patterns match. This allows an estimate of the age of the rock or sediment. Geoscientists use other age constraints, including isotopic ages or fossils, to further refine the age of the magnetized rocks. The magnetic timescale is best documented for the last 180 million years because seafloor of this age is widely preserved and can have its magnetic polarity measured.

05.14.b1

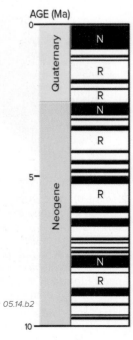

05.14.b2

How Are Magnetic Reversals Expressed at Mid-Ocean Ridges, and How Do Magnetic Patterns on the Seafloor Help Us Study Plate Tectonics?

In the 1950s, geocientists discovered that the ocean floor displayed magnetic variations in the form of matching stripes on either side of the mid-ocean ridge. Geoscientists Frederick Vine, Drummond Mathews, and Lawrence Morley interpreted the patterns to represent a magnetic field that had reversed its polarity. This Vine-Mathews-Morley hypothesis led to the theory of plate tectonics. Magnetic patterns allow us to estimate the ages of large areas of seafloor and to calculate the rates at which two diverging oceanic plates spread apart.

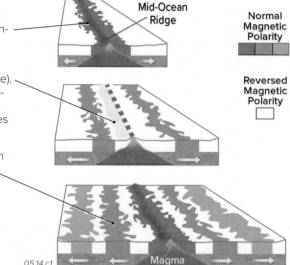

1. As the oceanic plates spread apart at a mid-ocean ridge, basaltic lava erupts onto the surface or solidifies at depth. As the rocks cool, the orientation of Earth's magnetic field is recorded by the iron-rich mineral magnetite contained within the rocks. In this example, the magnetite records normal polarity (shown with a reddish color) at the time the rock forms.

2. If the magnetic field reverses, new rocks that form will acquire a reversed polarity (shown in white). Rocks forming all along the axis of the mid-ocean ridge will have the same magnetic direction, forming a stripe of similarly magnetized rocks parallel to the ridge. Once the rocks have cooled, they retain their original magnetic direction, unless they are heated significantly or altered by certain types of fluids. In most cases, the magnetic polarity is preserved by the seafloor.

3. The magnetic polarity switches many times, and continued seafloor spreading produces a pattern of alternating magnetic stripes on the ocean floor. This pattern is strong enough to be detected by magnetic instruments, called *magnetometers,* towed behind a ship or a plane.

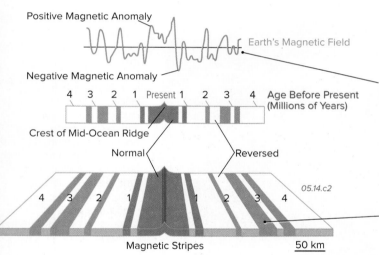

4. As magnetic instruments are towed behind a ship, the strength of the magnetic field is measured and plotted. Stronger measurements plot high on the graph and are called *positive magnetic anomalies.* The stronger magnetic signal occurs because the normal-polarity magnetization of the rocks is adding to the strength of the Earth's modern magnetic field in that area. The magnetic signal is weaker over crust that was formed under a reversed magnetic field, because the magnetic direction in such rocks is opposite to—and works to counteract—the modern magnetic field. The reverse magnetization of the rocks slightly weakens the measured magnetic signal and will plot low on the graph, forming a *negative magnetic anomaly.*

5. The patterns of positive and negative magnetic anomalies on the seafloor are compared with the patterns on the geomagnetic polarity timescale to assign ages to each reversal. Geologists simplify and visualize these data as reversely and normally magnetized stripes on the seafloor, as shown in this illustration.

6. We can calculate rates of seafloor spreading by measuring the width of a specific magnetic stripe in map or cross-section view and then dividing that distance by the length of time the stripe represents:

rate of spreading for stripe = width of stripe / time duration

7. If a magnetic stripe is 60 km wide and formed over 2 million years, then the average rate at which spreading formed the stripe was 30 km/m.y. This rate is equivalent to 3 cm/yr. Spreading added an equal width of oceanic crust to a plate on the other side of the mid-ocean ridge, so the total rate of spreading across the ridge was 60 km/m.y. (6 cm/yr), a typical rate of seafloor spreading.

8. The magnetic patterns on the seafloor, in addition to magnetic measurements on sequences of rocks and sediment on the seafloor and on land, demonstrate that Earth's magnetic field has reversed many times. Scientists are currently debating the possible causes of the reversals, with most explanations attributing reversals to chaotic flow in the molten outer core, which add to or subtract from the patterns caused by the dynamo, disrupting the prevailing magnetic field and causing a magnetic reversal.

Before You Leave This Page

✓ Describe how Earth's magnetic field is generated.

✓ Describe how magnetic reversals help with determining the age of rocks.

✓ Describe or sketch how magnetic patterns develop on the seafloor.

✓ Calculate the rate of seafloor spreading if given the width and duration of a magnetic stripe.

5.14

Why Is South America Lopsided?

THE TWO SIDES OF SOUTH AMERICA are very different. The western margin is mountainous while the eastern side and center of the continent have much less relief. The differences are a reflection of the present plate boundaries and of the continent's history during the last 200 million years. South America nicely illustrates many aspects of plate tectonics, including the connections between tectonics of the land and seafloor. It also is an excellent example of how to analyze the major features of a region.

A What Is the Present Setting of South America?

The perspective view below shows South America and the surrounding oceans. Observe the topography of the continent, its margins, and the adjacent seafloor. See if you can find mid-ocean ridges, transform faults, ocean trenches, oceanic fracture zones, and other plate-tectonic features. From these features, infer the locations of plate boundaries in the oceans, and predict what type of motion (divergent, convergent, or transform) is likely along each boundary. Make your observations and predictions before reading the accompanying text.

1. The Galápagos Islands are located in the Pacific Ocean, west of South America. They consist of a cluster of about 20 volcanic islands, flanked by seamounts. Some of the islands are volcanically active and are interpreted to be over a hot spot.

9. The center of the South American continent has low, subdued topography because it is away from any plate boundaries. It is a relatively stable region that has no large volcanoes and few significant earthquakes. It is not tectonically active.

8. In the South Atlantic, the Mid-Atlantic Ridge is a divergent boundary between the South American and African plates. Seafloor spreading creates new oceanic lithosphere and moves the continents farther apart at a rate of 3 cm/yr.

2. The Andes mountain range follows the west coast of the continent and is the site of many dangerous earthquakes and volcanoes. A deep ocean trench along the edge of the continent marks where an oceanic plate subducts eastward beneath South America.

7. The eastern side of South America has a continental shelf that slopes gently toward the adjacent seafloor. There is no trench, no significant tectonic activity, or other evidence for a plate boundary. Instead, the continent and adjacent seafloor to the east are part of the same plate, and this edge of the continent is a *passive margin.*

3. The Pacific seafloor contains mid-ocean ridges with the characteristic zigzag pattern of a divergent boundary with offsets along transform faults. The Nazca plate lies north of this ridge, and the Antarctic plate is to the south.

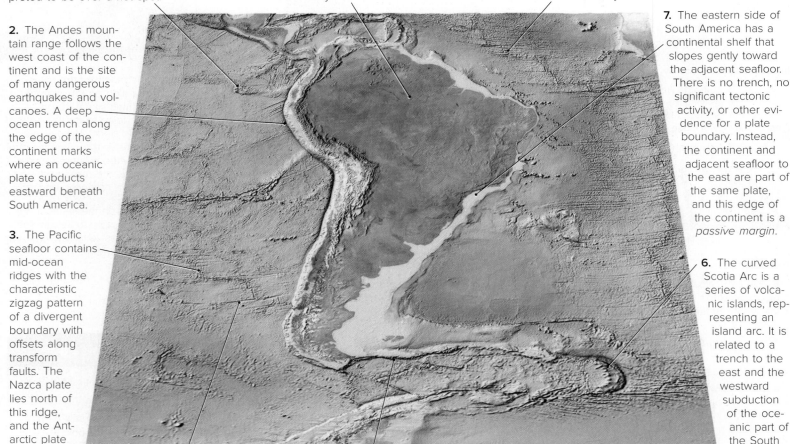

05.15.a1

6. The curved Scotia Arc is a series of volcanic islands, representing an island arc. It is related to a trench to the east and the westward subduction of the oceanic part of the South American plate.

4. Many oceanic fracture zones cross the seafloor and were formed along transform faults, but they are no longer plate boundaries.

5. The southern edge of the continent is very abrupt and has a curving "tail" extending to the east. This edge of the South American plate is a transform boundary where South America is moving westward relative to oceanic plates to the south.

B What Is the Geometry of the South American Plate and Its Neighbors?

This cross section shows how geoscientists interpret the configuration of plates beneath South America and the adjacent oceans. Compare this cross section with the plate boundaries you inferred in part A.

05.15.b1

1. At the Mid-Atlantic Ridge, new oceanic lithosphere is added to the African and South American plates as they move apart. As this occurs, the oceanic parts of the South American and African plates get wider. The Atlantic Ocean is therefore also getting wider with time.

2. Along the eastern edge of South America, continental and oceanic parts of the plate are simply joined together along a passive margin. There is no subduction, no seafloor spreading, or any type of plate boundary. As a result, the eastern continental margin of South America lacks volcanoes, earthquakes, and mountains.

3. A subduction zone dips under western South America, carrying oceanic lithosphere beneath the continent. Subduction causes large earthquakes and produces magma that feeds dangerous volcanoes in the Andes.

C How Did South America Develop Its Present Plate-Tectonic Situation?

If South America is on a moving plate, where was it in the past? When did it become a separate continent, and when did its current plate boundaries develop? Here is one commonly agreed-upon interpretation.

1. Around 140 million years ago, Africa and South America were part of a single large *supercontinent* called *Gondwana*. At about this time, a continental rift developed, starting to split South America away from the rest of Gondwana and causing it to become a separate continent. Rifting was instigated by one or more hot spots.

2. By 100 million years ago, Africa and South America were completely separated by the south Atlantic Ocean. Spreading along the Mid-Atlantic Ridge moved the two continents farther apart with time. While the Atlantic Ocean was opening, oceanic plates in the Pacific were subducting beneath western South America. This subduction thickened the crust by compressing it horizontally and by adding magma, resulting in the formation and rise of the Andes mountain range.

3. Today, Africa and South America are still moving apart at a rate of several centimeters per year. As spreading along the mid-ocean ridge continues, the Atlantic Ocean gets wider. Earth, however, is not growing through time, and the expanding Atlantic Ocean is balanced by shrinking of the Pacific Ocean, whose oceanic lithosphere disappears into subduction zones along the Pacific Ring of Fire, including along the western side of South America.

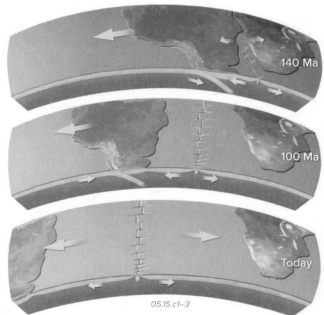

05.15.c1–3

4. These photographs contrast the rugged Patagonian Andes of western South America with landscapes farther east that have more gentle relief and are not tectonically active.

05.15.c4 Cuernos del Paine, Chile

05.15.c5 Central Argentina

Before You Leave This Page

✓ Sketch and describe the present plate-tectonic setting of South America, and explain the main features on the continent and adjacent seafloor.

✓ Discuss the plate-tectonic evolution of South America over the last 140 million years.

5.16

Where Is the Safest Place to Live?

AN UNDERSTANDING OF PLATE TECTONICS allows us to predict which places are at most risk from earthquakes and volcanoes. The most important things to know in this context are the locations and types of plate boundaries. In this exercise, you will examine an unknown ocean between two continents, make observations of the land and seafloor, and identify plate boundaries and other features. Using this information, you will predict the risk for earthquakes and volcanoes, and determine the safest places to live.

Goals of This Exercise:

- Use the features of an ocean and two continental margins to identify possible plate boundaries and their types.
- Use the types of plate boundaries to predict the likelihood of earthquakes and volcanoes.
- Determine the safest sites for two cities, considering the potential for earthquakes and volcanic eruptions.
- Draw a cross section that shows the geometry of the plates at depth.

Procedures for the Map

This perspective view shows two continents, labeled Continent A and Continent B, and an intervening ocean. Use the topography on the land and seafloor to identify possible plate boundaries and then complete the following steps. Mark your answers on the map on the worksheet, which will be provided to you by your instructor in either paper or electronic form. Alternatively, your instructor may have you complete the investigation online.

A. Use the topographic features on land and the depths of the seafloor to identify possible plate boundaries. Draw lines showing the location of each plate boundary on the map in the worksheet. Label the boundaries as either divergent, convergent, or transform. Use colored pencils or different types of lines to better distinguish the different types of boundaries. Provide a legend that explains your colors and lines.

B. Draw circles [O] or use color shading to show places, on land or in the ocean, where you think earthquakes are likely.

C. Draw triangles [▲] at places, on land or in the ocean, where you think volcanoes are likely. Remember that not all volcanoes form *directly on* the plate boundary; some form off to one side. For different plate-tectonic settings, consider where volcanoes form relative to that type of plate boundary.

D. Determine a relatively safe place to build one city on each continent. Show each location with a large plus sign [+] on the map. On the worksheet, explain your reasons for choosing these as the safest sites.

Oceanic Trench

CONTINENT A

Procedures for the Cross Section

The worksheet contains a modified version of this figure for you to use as a starting point for making a cross section. Add lines and colors to the front of the diagram to show the configuration of the plates in the subsurface. Use other figures in this chapter as guides to the thicknesses of the lithosphere and to the subsurface geometries typical for each type of plate boundary. Your cross section should only show features on the front of the block diagram, not features that do not reach the front edge. Your cross section should clearly do the following:

A. Identify the crust, mantle, lithosphere, and asthenosphere, and show an accurate representation of their relative thicknesses.

B. Show the locations and relationships between lithospheric plates at any spreading center or subduction zone.

C. Include arrows to indicate which way the plates are moving relative to each other.

D. Show where melting is occurring at depth to form volcanoes on the surface.

05.16.a1

Volcanism and Other Igneous Processes

A VOLCANIC ERUPTION is one of nature's most spectacular events. Volcanoes blast scalding volcanic ash into the air as hot streams of molten rock pour down the volcano's flank. Volcanoes represent an obvious geologic hazard, and eruptions can claim the lives of tens of thousands of people at a time. In this chapter, we explore volcanoes and other igneous processes, and associated landforms and hazards.

Mount St. Helens in southwestern Washington was once one of the most beautiful and symmetrical high peaks (▼) in the Cascade Range of the Pacific Northwest. Its shape changed forever in May 1980 when the sleeping volcano erupted violently. The eruption blew apart the volcano's north flank and excavated a huge crater where the mountain peak used to be. Within the newly formed crater, continuing eruptions built the steaming lava dome shown in the larger photograph below.

What is a volcano, and how do we recognize one?

06.00.a3

Mount St. Helens

06.00.a2

Pre-1980 View from West

06.00.a1

2005 View from North

TOPICS IN THIS CHAPTER

06.00.a4

The May 1980 eruption started with a northward-directed blast that knocked over millions of trees and unleashed a *pyroclastic flow,* a swirling, hot cloud of dangerous gases, volcanic ash, and angular rock fragments. The pyroclastic flow swept downhill and across the landscape, burying and killing almost all living things in its path. This was followed immediately by a huge column of volcanic ash that rose 25 km (15 mi) into the atmosphere (◄). The ash was carried eastward by the wind and blocked sunlight as it settled back to Earth across a large area of Washington, Idaho, and Montana.

What are the different ways that volcanoes erupt, and what hazards are associated with each type of eruption?

06.00.a5

Since the main eruption, magma rising through the throat of the volcano has collected on top of the vent, forming a series of *lava domes* (▲). Periodic collapse of part of the unstable domes unleashes explosions or avalanches of hot volcanic ash and rocky fragments.

What factors determine whether magma erupts as an explosion of hot ash or a slow outpouring of lava?

The May 1980 Eruption

With eruptions continuing into 2008, Mount St. Helens is the most active of the 15 large volcanoes that crown the Cascade Range of the Pacific Northwest. The mountain is the youngest volcano in the range, being entirely constructed during the last 40,000 years. Before 1980, a team of geologists from the U.S. Geological Survey studied the geology of the mountain and recognized that past eruptions had unleashed vast amounts of volcanic ash, lava, and volcanic mudflows. Prior to 1980, the volcano last erupted in the mid-1800s.

After more than 100 years of quiescence, the volcano reawakened in March 1980 when it vented steam, shook the area with many earthquakes, and pushed out an ominous bulge of rock on its north flank. At 8:32 a.m. on May 18, 1980, an earthquake caused the oversteepened north flank to collapse downhill in a huge landslide that carried rock pieces as large as buildings. This catastrophic removal of rock released pressure on the magma inside the volcano, which exploded northward in a cloud of scalding and suffocating volcanic ash. The pyroclastic flow raced across the landscape at speeds of up to 500 km (300 mi) per hour. The eruption blasted away most of the north flank of the mountain and forever changed the peak's appearance. It turned the surrounding countryside into a barren wasteland smothered by a thick blanket of volcanic ash. Early evacuations helped limit the loss of life, but 57 people perished, and damage estimates for the eruption exceeded $1 billion, making it the most expensive and deadly volcanic eruption in U.S. history.

Although the level of activity at Mount St. Helens greatly diminished in 2007 and 2008, geoscientists continue to monitor the volcano by keeping track of any ongoing volcanic and seismic activity, and carefully measuring changes in the mountain. Remotely operated cameras keep watch over the crater and the domes. Geoscientists also use instruments to monitor temperatures, gas emissions, and tilting of the land surface. Swarms of small earthquakes in 2016 were interpreted as a new batch of magma rising below the mountain. The volcanic history of this very active volcano is clearly not over yet.

6.0

6.1 How Does Magma Form?

VOLCANIC ERUPTIONS INVOLVE MAGMA (MOLTEN ROCK), but where and how do magmas form? Since magma is molten rock, how do rocks melt, under what conditions do they melt, and where in Earth does melting occur? How do the melting processes influence the type of magma, and therefore the type of volcano and eruptions that result?

A What Happens When a Substance Changes from a Solid to a Liquid?

What is the difference between a solid and a liquid, and what happens at a molecular level when a solid melts?

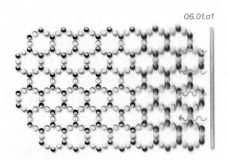
06.01.a1

06.01.a2

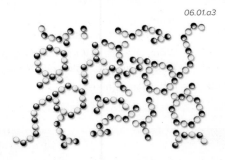
06.01.a3

In solids, atoms and bonds are always vibrating. A temperature increase causes the vibrations to increase, eventually to the point where bonds break and melting begins. An increase in pressure has the opposite effect, compressing the solid and making it more resistant to melting.

A mineral or rock will remain solid if the pressure and bond strength are sufficient to overcome vibrations due to temperature. Beneath Earth's surface, pressure arises mostly from the weight of overlying rocks, gradually increasing with depth of burial.

When bond strength and pressure are inadequate to hold a solid material together, melting will begin. Different bonds break at different temperatures, so magma generally contains some partially bonded, or weakly bonded, molecules and material within the melt.

B Under What Conditions Is a Material Solid or Liquid?

If pressure tends to keep a rock solid while increasing temperature causes it to melt, which one prevails? The graph below shows temperatures and pressures under which a material can exist either as a solid or as a liquid.

1. Temperature is plotted on the horizontal axis, and pressure is plotted on the vertical axis. The conditions for any place within Earth can be shown as a point, such as point A, that represents a specific temperature and pressure.

2. Pressure increases downward within the earth, so pressure is plotted on this graph as increasing from top to bottom. In this manner, the graph mimics the earth, with pressure increasing with depth, but it is a graph, not a cross section.

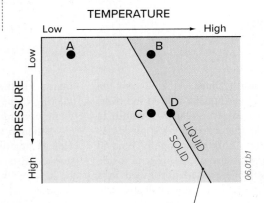
06.01.b1

4. A rock at the low temperature and low pressure represented by point A is solid (it is in the "solid" field).

5. A rock at point B is at the same low pressure as the rock at point A, but it has a higher temperature. It plots in the liquid field and so under these conditions is liquid (magma).

6. A rock at point C is at the same high temperature as the magma at point B, but it is solid because the higher pressure helps hold the atoms together and prevents melting.

7. If pressure-temperature conditions for a rock plot directly on the melting curve, like at point D, the rock is in the process of melting or solidifying. Under these conditions, some solid rock and some magma are present.

3. A line, called the *melting curve* or *solidus*, divides the graph into two areas, called *fields*. If a rock is at a pressure and temperature that plot to the left of the line, the rock remains solid. If the pressure-temperature conditions plot to the right of the line, the rock will be completely melted (magma). The melting curve slopes down to the right because higher temperatures are needed to melt a rock under higher pressure. The position of the melting curve depends on the composition of the rock, shifting to the left (lower melting temperature) for more felsic compositions. Also, different minerals have different melting curves, so not all minerals melt at once.

Before You Leave These Pages

✓ Describe how thermal vibrations and pressure affect a crystalline structure.

✓ Draw graphs showing how increasing temperature, decreasing pressure (decompression), or adding water to hot rocks causes melting.

✓ Describe partial versus complete melting and how the source area influences the type of magma.

What Causes a Rock to Melt?

When we think of melting, we normally think of heating something, an ice cube for example, until it turns into a liquid. Heating does cause rocks to melt, but there are other factors. Rock melting is influenced by three main factors: temperature, pressure, and water content.

Melting by Heating

1. When a rock is heated, some or all of its minerals can melt. On this graph, melting would occur if a rock were heated so that its temperature increased from point *A* to point *B*. Therefore, a temperature increase caused by heating can melt a rock. Most rocks contain different minerals with different melting temperatures, so an increase in temperature causes only partial melting, unless temperature becomes very high.

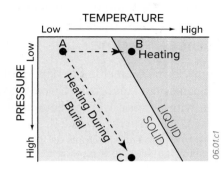

2. If an increase in temperature is accompanied by an increase in pressure, as from point *A* to point *C*, the higher pressure may be enough to keep the rock from melting. The path from point *A* to point *C* is similar to the change in conditions that occurs when a rock is simply buried—temperature increases, but the rock does not melt because pressure also increases.

Melting by Decompression

3. Pressure decreases if a rock moves up from depth, getting closer to the surface. So, a rock that is uplifted will experience a decrease in pressure, as from point *C* to point *B*. If the rock is already hot (point *C*), it may melt as the pressure decreases (to point *B*), a process called *decompression melting*.

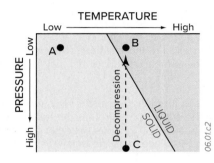

4. For decompression melting to occur, the rock has to be fairly hot and must be uplifted fast enough so that it cannot cool significantly during uplift. If a rock is uplifted slowly, it can cool enough to stay solid. A hot, deeply buried rock following a path from point *C* to point *A* (cooling during uplift) would stay within the solid field on the diagram. In other words, the rock would not melt.

Melting by Adding Water

5. Adding water can significantly lower, by as much as 500°C, the temperature at which a rock will melt. On this graph, the dashed line shows the position of the melting curve if the rock contains water. Adding water moves the melting curve to lower temperatures. So, adding water to a dry rock at point *E* puts it on the liquid side of the melting curve, and the rock will melt.

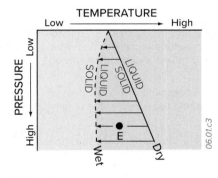

6. A hot rock can melt, therefore, if water moves into the system, even with no change in pressure or temperature. A rock at point *E* will be solid under normal, dry conditions (it is to the left of the non-dashed melting curve). If a small amount of water is added to the small spaces within and between crystals, the boundary between solid and liquid shifts position to the dashed curve, so the rock at point *E* now begins to melt.

Partial and Nearly Complete Melting

7. If a magma was generated by *complete melting* of the source region, it would have a composition identical to that of the source. For a number of reasons, complete melting is not common.

8. Most rocks melt by partial melting as some minerals melt before others. Felsic minerals melt at lower temperatures than mafic minerals, so partial melting produces a magma that is *more felsic* than the source. For example, partial melting of a mafic source can yield an intermediate magma.

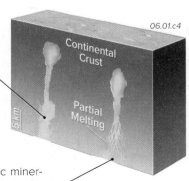

Type of Source Area

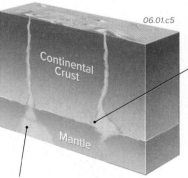

9. The place where a magma is generated is called the *source area*. If a more felsic source area, such as continental crust, is melted, the magma will be felsic. If an intermediate source is almost completely melted, the magma will have an intermediate composition, but partial melting of such a source more commonly produces a felsic magma.

10. The overall composition of the mantle is ultramafic but, due to partial melting, magmas generated in the mantle are mostly mafic. Most mafic magma is derived by partial melting of the mantle.

6.2 How Does Magma Move?

ONCE FORMED, MAGMA CAN RISE TOWARD THE SURFACE or remain close to where it formed. Magma rises through the crust and upper mantle in several ways, and how easily it moves is controlled by its temperature, composition, and other factors. The cooling history of the magma influences the resulting type of igneous rock.

A What Is the History of a Typical Magma?

The igneous process begins when magma forms by melting at depth, followed by movement of the magma toward the surface and then solidification of the magma into solid rock. Given this sequence of processes, igneous systems are best described from the bottom up, so begin with number 1 at the bottom of this section.

6. Magma that reaches the surface erupts as lava (molten rock that flows on the surface) or as volcanic ash. Volcanic ash forms when dissolved gases in the magma expand and blow the magma apart into small fragments of volcanic glass. Any igneous rock that forms on the surface is called an *extrusive* rock because it forms from magma *extruded* onto the surface. More commonly, we simply call it a *volcanic* rock.

5. Many magma chambers are only several kilometers below the surface, as beneath a volcano. Magma may be added a little at a time to the chamber, and some magma may solidify before the next batch arrives. Some of the magma may crystallize in the chamber, while some rises to the surface. In this case, the rising magma may carry some of the early-formed crystals all the way to the surface, forming a porphyritic volcanic rock.

4. As magma rises through the crust, it may stop in, or pass through, a series of magma chambers. A body of molten rock in the subsurface is referred to as an *intrusion* because of the way the magma intrudes into (invades) the surrounding rocks. Any igneous rock that solidifies below the surface is called an *intrusive* rock. Although there is a subtle difference, most geoscientists use the terms *plutonic* and *intrusive* synonymously.

3. Magma can accumulate to form a *magma chamber*. Some magma chambers represent a large batch of magma emplaced at approximately the same time, but most grow in discrete increments, from the injection of smaller batches of magma in sheetlike bodies. The magma may solidify in this chamber and never reach the surface, or it may reside in the chamber temporarily before continuing its journey upward. An igneous rock that solidified at a considerable depth (more than several kilometers) is referred to as a *plutonic* rock, and the body of rock that forms is called a *pluton*. Granite and granodiorite are very common plutonic rock and form plutons.

2. Once magma begins to form, separate pockets of magma may accumulate to make a larger volume of magma. The magma rises because it is less dense than rocks around it.

1. The first stage in the formation of an igneous rock is melting, typically 40–150 km beneath the surface, in the deeper parts of the crust or in the mantle. The place where melting occurs is called the *source area*. Complete melting is rare, and most magmas result from *partial melting*, leaving most of the source area unmelted.

5 km

06.02.a1

Cooling History of a Magma

7. This graph shows conditions of temperature versus depth for a rising magma. Pressure is proportional to depth, increasing with greater depths. We can track the history of a magma by plotting temperature-depth points that show the path it follows as it cools and reaches the surface. Follow the numbered changes by starting at the bottom.

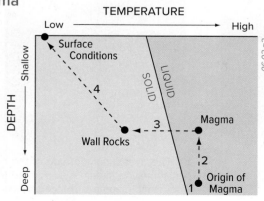

TEMPERATURE

Low ─────────────→ High

DEPTH

Shallow

Deep

Surface Conditions

4

SOLID

LIQUID

Wall Rocks

3

Magma

2

1 Origin of Magma

06.02.a2

11. At some later time, the now-solidified magma and its wall rocks are uplifted along path 4 to the surface, where they cool to low temperature and we can inspect them for clues about their history.

10. The magma cools by losing thermal energy to the surrounding wall rocks. It crosses the melting curve along path 3 on the graph, and so it crystallizes.

9. Some of the magma rises some distance in the mantle or crust, along path 2 on the graph, and so it is in a place where it is surrounded by cooler rocks.

8. Magma forms at depth, at point 1, where temperature is high enough to overcome pressure and cause melting.

B How Does Magma Rise Through the Crust?

Magma can travel through and fill a fracture, forming a *fissure*. If magma solidifies within the fissure, it forms a sheetlike feature called a *dike*, like the light-colored ones shown here. A fissure generally forms when the pressure of the magma is enough to push apart the rocks, forming a fracture into

06.02.b1 Franklin Mtns., El Paso, TX

which the magma can then flow. Such fractures can also be caused, or aided, by tectonic forces, like the pulling forces that accompany stretching of the crust. Most dikes are steeply oriented, but many are not.

06.02.b2 Joshua Tree NP, CA

Magma can move through the crust by removing overlying rocks piece by piece. Solid rocks above a magma chamber can break off and drop down into the magma, providing space and letting the magma move up.

The dark fragments shown here broke off the walls of a magma chamber and were incorporated into the magma that solidified into the light-colored granite.

C What Controls How Easily Magma Moves?

Viscosity is a measure of a material's resistance to flow. A *viscous* magma does not flow easily, whereas a *less viscous* (more fluid) magma flows more easily. Magmas are considerably more viscous than most other hot liquids with which you are familiar. A magma's viscosity is controlled by its temperature, composition, and crystal content.

1. *Viscous magma* strongly resists flowing. When viscous magma erupts on the surface, it does not spread out but piles up, forming mounds or domes of lava.

2. *Less viscous* (more fluid) *magma* flows more easily and may spread out in thin layers on the surface. This magma can travel longer distances from its source and cover large areas with lava.

Temperature

3. *Low Temperature*— The temperature of a magma is the most important control of viscosity. Magma at a relatively low temperature, such as one barely hot enough to be molten, flows only with great difficulty—it is very viscous.

4. *High Temperature*— Magma that is very hot has low viscosity and so flows very easily. Mafic magma is hotter and less viscous than felsic magma, even if the two magmas are at the same temperature.

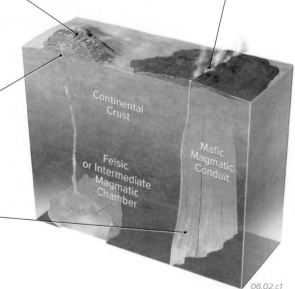

06.02.c1

Composition

5. *Abundant Silicate Chains*—Silicon and oxygen tetrahedra in magma can link into long silicate chains that do not bend or move easily out of the way of one another. Felsic and most intermediate magmas have a high silicon and oxygen content, and the resulting long silicate chains make the magmas very viscous.

6. *Few Silicate Chains*—Mafic magma contains less silicon and oxygen than intermediate or felsic magma. Consequently, silicon-oxygen tetrahedra are less connected or are in short chains. This allows the mafic magma to flow more easily—it is *less* viscous.

7. *Volatiles*—Silicon content and the abundance of silicate chains may be the most significant compositional variable controlling viscosity, but water dissolved in magma disrupts long chains, decreasing the viscosity. Water and other volatiles decrease viscosity in other ways not discussed here. A *volatile* is an element or chemical compound that tends to convert to a gas.

Percentage of Crystals

8. *Abundant Crystals*—As a magma cools, crystals begin to form. The crystals in the flowing magma get in each other's way and cause the magma to flow more slowly. A magma with abundant crystals is more viscous (has more resistance to flow) than a magma with fewer crystals.

9. *Few Crystals*—A magma that has few crystals has few internal obstructions and flows more easily (is *less* viscous). Such magma flows more smoothly and thus can flow faster and farther. The amount of crystals is related to the cooling history of the magma, including any pauses in the magma's rise.

Abundant Crystals
06.02.c2

Few Crystals
06.02.c3

Before You Leave This Page

✔ Sketch and describe the history of a typical magma, including its cooling.

✔ Describe ways in which magmas rise through the crust.

✔ Explain the factors that control the viscosity of a magma.

6.2

6.3 What Is and Is Not a Volcano?

AN ERUPTING VOLCANO IS UNMISTAKABLE—glowing orange lava cascading down a hillside, molten fragments blasting into the air, or an ominous, billowing plume of gray volcanic ash rising into the atmosphere. But what if a volcano is not erupting? How do we tell whether a mountain is, or was, a volcano?

A What Are the Characteristics of a Volcano?

How would you describe a volcano to someone who had never seen one? Examine the photographs of two volcanoes below and look for common characteristics, then read the accompanying text.

06.03.a1 Pu'u 'Ō 'ō volcano

1. A volcano is a *vent* where magma and other volcanic products erupt onto the surface. The volcano on the left (the Pu'u 'Ō 'ō volcano in Hawaii Volcanoes National Park) produces large volumes of molten lava, whereas the one on the right (Kanaga volcano, Alaska) is erupting volcanic ash. Many geoscientists reserve the term *volcano* for hills or mountains that have been *constructed* by volcanic eruptions. Some eruptions do not produce hills or mountains, and we consider them to be volcanoes, too.

06.03.a2 Mount Kanaga , AK

2. Most volcanoes have a *crater*, a roughly circular depression usually located near the top of the volcano. Other volcanoes have no obvious crater or are nothing but a crater.

3. Volcanoes consist of *volcanic rocks,* which form from lava, pumice, volcanic ash, and other products of volcanism.

4. Besides erupting from volcanoes that have the classic shape of a cone, magma erupts from fairly linear cracks called *fissures* and from huge circular depressions called *calderas.*

5. Many volcanoes display evidence of having been active during the last several hundred to several million years, or even during the last several days. Such evidence can include a layer of volcanic ash on hillslopes (left side of volcano shown above) or lava and ash flows that are relatively unweathered and that lack a well-developed soil. Over time, erosion degrades and disguises volcanoes and volcanic craters, making them less obvious.

B Is Every Hill Composed of Volcanic Rocks a Volcano?

If a landscape feature lacks most of the diagnostic features described above, it is probably not a volcano. Many mountains and hills are not volcanoes, and some volcanoes do not make mountains or hills.

1. The flat-topped hill below (▼), called a *mesa*, has a cap of volcanic rocks, but it is not a volcano. It did not form over a volcanic vent. Instead, it is an eroded remnant of a once extensive lava flow that covered the region, as shown in the figures to the right.

2. Lava erupts from a central volcanic vent or from a linear vent called a *fissure*. Once erupted, the lava spreads outward and cools into a solid rock.

3. Erosion removes the edges of the lava flow and works inward toward a central remnant.

06.03.b1 Hopi Buttes, AZ

4. The past location of the fissure is marked by a dike that cuts through the rocks and across the landscape.

5. The lava flow is more resistant to erosion than underlying rocks and so forms a steep-sided, flat-topped mesa. It is a hill composed of volcanic rocks, but it is not over the vent and is not a volcano.

06.03.b2–4

C What Are Some Different Types of Volcanoes?

Volcanoes have different sizes and shapes, and contain different types of rocks. These variations reflect differences in the composition of the magmas and the style of the eruptions. There are four common types of volcanoes that are shaped like hills and mountains: *scoria cones, shield volcanoes, composite volcanoes,* and *volcanic domes.* Later in this chapter, we describe other types of volcanoes that are not hills or mountains.

Scoria Cone

06.03.c1

◀ *Scoria cones* are cone-shaped hills several hundred meters high, or higher, usually with a small crater at their summit. They also are called *cinder cones* because they contain loose black or red, pebble-sized volcanic *cinder* (scoria), along with larger volcanic *bombs*. Most scoria is basaltic or, less commonly, andesitic in composition. Some scoria cones form next to or on the flanks of composite and shield volcanoes (described below).

Shield Volcano

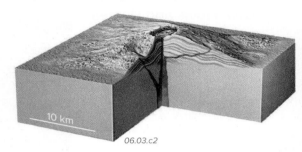

06.03.c2

▶ *Shield volcanoes* have broad, gently curved slopes and can be relatively small (less than a kilometer across) or can form huge mountains tens of kilometers wide and thousands of meters high. They commonly contain a crater or line of craters and have *fissures* along their summit. Shield volcanoes consist mostly of basaltic lava flows with smaller amounts of scoria and volcanic ash.

Composite Volcano

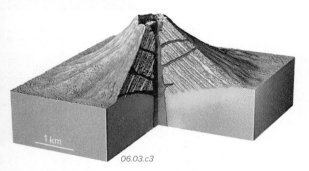

06.03.c3

◀ *Composite volcanoes* are typically fairly symmetrical mountains thousands of meters high, with moderately steep slopes and commonly a crater at the top. They may be large but are, on average, much smaller than shield volcanoes. Their name derives from the interlayering of lava flows, pyroclastic deposits, and volcanic mudflows. They consist mostly of intermediate-composition rocks, such as andesite, but can also contain felsic and mafic rocks. Mount St. Helens is a composite volcano.

Volcanic Dome

06.03.c4

▶ *Volcanic domes* are dome-shaped features that may be hundreds of meters high. They consist of solidified lava, which can be highly fractured or mostly intact. Domes include some volcanic ash intermixed with rock fragments derived from solidified lava in the dome. They form where felsic or intermediate magma erupts and is so viscous that it piles up around a vent. Many domes are within craters of composite volcanoes.

The Relative Sizes of Different Types of Volcanoes

Volcanoes vary from small hills less than a hundred meters high to broad mountains tens of kilometers across. Although sizes vary quite a bit, we can make some generalizations about the relative sizes of the different volcano types.

The figure below illustrates that some types of volcanoes are larger than others. The volcanoes on this figure cannot be drawn to their true scale relative to one another because the largest shield volcanoes are so large that we cannot show them on the same drawing with small scoria cones. The figure does accurately show which volcanoes are the largest and which ones are the smallest.

Scoria cones and domes, which typically form during a single eruptive episode, are the smallest volcanoes. Shield volcanoes and composite volcanoes are much larger because they are constructed, layer by layer, by multiple eruptions. Shield volcanoes have more gentle slopes than scoria cones, domes, or composite volcanoes.

Before You Leave This Page

☑ Sketch or describe the diagnostic characteristics of a volcano.

☑ Describe or sketch why every hill composed of volcanic rocks is not a volcano.

☑ Sketch and describe the four main types of volcanoes that construct hills and mountains.

☑ Sketch or describe the relative sizes of different types of volcanoes.

6.3

Scoria
Cone
Dome   Small
Shield   Composite Volcano   Large Shield

06.03.t1

6.4 What Controls the Style of Eruption?

THE DIFFERENT SHAPES OF VOLCANOES reflect differences in the style of eruption. Some eruptions are explosive, whereas others are comparatively calm. What causes these differences? The answer involves magma chemistry and gas content, both of which control how magma behaves near the surface.

A What Are Ways That Magma Erupts?

Magma may behave in several different ways once it reaches Earth's surface. Explosive *pyroclastic eruptions* throw bits of lava, volcanic ash, and other particles into the atmosphere. During nonexplosive eruptions, lava issues from a vent and flows onto the surface. Both types of eruptions can occur from the same volcano.

Lava Flows and Domes

1. When magma erupts onto the surface and flows away from a vent, it creates a *lava flow* (▶). Erupted lava can be fairly fluid, flowing downhill like a fast river of molten rock. Some lava flows are not so fluid and travel only a short distance before solidifying.

06.04.a1 Kilauea, HI

2. A *lava dome* (▶) forms from the eruption of highly viscous lava. The high viscosity of the lava is generally due to a high silica content and causes the lava to pile up around the vent instead of flowing away. Domes are often accompanied by several types of explosive eruptions.

06.04.a2 Mount St. Helens, WA

Pyroclastic Eruptions

3. Some explosive eruptions send molten lava into the air. A *lava fountain* (▶), such as shown here, can accompany basaltic volcanism and results from a high initial gas content in a less viscous lava. The gas propels the lava and separates it into discrete pieces.

06.04.a3 Kilauea, HI

4. Other explosive eruptions eject a mixture of volcanic ash, pumice, and rock fragments into the air (▶). Such airborne material is called *tephra*, and tephra particles that are sand-sized or smaller are *volcanic ash*. Ash mostly forms when bubbles blow apart bits of magma. Tephra is derived from pumice, fragmented volcanic glass, and shattered preexisting rocks.

06.04.a4 Redoubt volcano, AK

Two Different Eruptions of Tephra from the Same Volcano

The Augustine volcano in Alaska produces tephra in two eruptive styles—an *eruption column* and a *pyroclastic flow*.

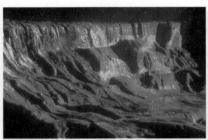
06.04.a5

06.04.a6

1. *Eruption Column*—Tephra, which forms when magma is blown apart by volcanic gases, can erupt high into the atmosphere, forming an *eruption column* (▲). The tephra falls back to Earth as solidified and cooled pieces of rock. Finer particles of ash drift many kilometers away from the volcano and slowly settle down to the ground.

2. *Pyroclastic Flow*—Some ash does not jet straight up but collapses down the side of the volcano as a dense, hot cloud of ash particles and gas (◀). This eruption style is a *pyroclastic flow* or simply an *ash flow*. A pyroclastic flow can be devastating because of its high speeds (more than 100 km/hr) and very high temperatures (exceeding 500°C).

3. The two kinds of eruptions differ primarily because of the gas content of the magma. An *eruption column* forms when large volumes of volcanic gas come out of the magma and overcome gravity to carry the cloud of tephra up into the atmosphere.

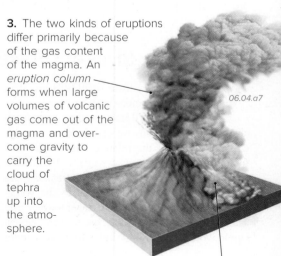
06.04.a7

4. A *pyroclastic flow* forms when the amount of gas is less and cannot support the eruption column, so the column rapidly collapses and flows downhill under the force of gravity.

B How Do Gases Affect Magma?

1. To envision dissolved gas in magma, think what happens when you open a bottle or can of soda. The liquid may have no bubbles until it is opened, at which time bubbles appear in the liquid, rise to the top, and perhaps cause the soda to spill out. The dissolved gas was always in the liquid, but it only became visible when you opened the top and released the pressure that held the gas in solution.

06.04.b1

2. Magma, like the soda, contains some dissolved gases, including H_2O (water vapor), CO_2 (carbon dioxide), and SO_2 (sulfur dioxide). These gases have a critical effect on eruption style and help the magma rise toward the surface.

3. As shown in this enlargement of the magma, confining pressure at depth keeps most of the gases in solution and keeps bubbles from forming.

06.04.b2

4. As the magma approaches the surface, the pressure decreases and the gases cannot remain in solution. Bubbles of gas form in the magma. If enough bubbles form quickly, the expanding bubbles cause the magma to be more buoyant and help it rise toward the surface and erupt out of the volcano.

C How Does Viscosity Affect Gases in Magma?

Viscosity, the resistance to flow, dictates how fast a magma can flow and how fast crystals and gas can move through the magma. When gas in a magma comes out of solution, movement of the resulting bubbles is resisted by the magma's viscosity. If the bubbles cannot escape, the magma can build up and potentially be more explosive.

06.04.c1 Mount St. Helens, WA

More Viscous
◄ Felsic magmas contain a lot of silica, and so they are relatively viscous. The high viscosity prevents gas from escaping easily. Gas builds up in the magma and, when it expands, greatly increases the pressure on the surrounding rock. This can cause explosive eruptions.

Less Viscous
► Less viscous magma, such as one with a basaltic composition, allows gas bubbles to escape relatively easily. This can lead to a fairly nonexplosive eruption, such as this basaltic lava flow that flows smoothly downhill from the vent.

06.04.c2 Kilauea, HI

Composition, Viscosity, and Eruptive Style

Composition of magma is the main control on a volcano's eruptive style, shape, associated rock types, and potential hazards. This is because composition, especially the amount and length of silicate chains in the melt, controls viscosity and whether gas builds up in the magma.

Mafic (basaltic) magma has fewer and shorter silicate chains than felsic or intermediate magma, and so is relatively less viscous. The lower viscosity allows mafic magma to flow from the volcano in a relatively fluid lava flow. The fluidity of mafic lavas accounts for the relatively gentle slopes of shield volcanoes, which largely consist of basaltic (mafic) lava flows. Explosive gases can build up in mafic magma, as demonstrated by lava fountains, but the resulting explosive eruptions are relatively small and localized, scattering basaltic scoria and ash close to the scoria cone.

Felsic and intermediate magma have more silicate chains, and the chains are longer, restricting the flow of the magma and making the magma more viscous. The high viscosity of felsic and intermediate lavas produces steep volcanic domes and steep composite volcanoes. Magma in domes, composite volcanoes, and large volcanic calderas can trap gas and erupt explosively, producing gas-propelled pyroclastic eruptions of volcanic ash, tephra, and rock fragments. As a result, these volcanoes produce a mix of pyroclastic rocks and lava flows, mostly of felsic and intermediate composition. Composition controls viscosity, eruptive style, the shape of the volcano, and the rock types that compose that volcano.

Before You Leave This Page

✓ Describe four ways that magma erupts.

✓ Describe the difference between an eruption column and a pyroclastic flow, and the role that gas plays in eruptive style.

✓ Explain how gas behaves at different depths in a magma and how it influences eruptive style.

✓ Describe how viscosity influences how explosive an eruption is and the type of volcano that results.

6.4

6.5 What Hazards Are Associated with Volcanoes?

VOLCANIC ERUPTIONS can affect areas very close to the vent or can spread damaging volcanic ash over huge regions. The relative hazard depends on the type of eruption. Eruptions of pyroclastic flows have killed more than 30,000 people at a time, all within minutes. Surprisingly, volcanic eruptions can also cause huge floods.

A What Is Meant by a Hazard and a Risk?

The terms *hazard* and *risk* may seem more appropriate for a lesson about insurance, but geoscientists frequently apply these terms when discussing the effects that natural (and human-caused) events can have on humans and society. What is the difference between a hazard and a risk?

06.05.a1 Kilauea, HI

06.05.a2 Kilauea, HI

A *hazard* is the existence of a potentially dangerous situation or event, such as a potential landslide of a steep slope or a lava flow erupting from a volcano. The hazard in this photograph was a basaltic lava flow (◀).

Risk is an assessment of whether the hazard might have some *societal impact*, such as loss of life, damage to property, loss of employment, destruction of fields and forests, or implications for local or global climates. Remnants of destroyed houses, cars, and roads (◀) demonstrate that this area had a high risk for volcanic hazards.

The risk was extreme for people living in the Royal Gardens subdivision (▲) on the flanks of Kilauea, one of the most active volcanoes on the planet. Eruptions of basaltic lava progressively overran and destroyed more and more of the subdivision, until the last house was abandoned.

B What Hazards Are Associated with Eruptions of Cinder, Ash, and Gas?

Basaltic eruptions can be deadly and destructive, but mostly to nearby areas. They can hurl lava and solid rock into the air and spew out dangerous gases. Fine ash ejected high into the air can cause damage that is more widespread.

Falling Objects

During the formation of lava fountains (◀), most scoria falls back to Earth near the vent and piles up in a scoria cone. Hazards include being struck and burned by scoria and being struck by blobs of lava and other projectiles. Larger ejected pieces, called *volcanic bombs* (▼), pose a severe hazard for anyone close to the erupting cone.

06.05.b1 Hawaii

06.05.b2 Flagstaff, AZ

Volcanic Ash from Scoria Cones

Sand-sized cinders and finer particles of ash can bury nearby structures, and they may cause breathing problems for people and livestock. In March and April of 2010, a shield volcano in Iceland, called Eyjafjallajökull, erupted large amounts of volcanic ash that drifted over Europe, shutting down most air travel and stranding hundreds of thousands of passengers. Eruption columns from scoria cones typically reach lesser heights of several kilometers, and only affect areas nearby.

06.05.b3 Iceland

Gases

Volcanic gases are a significant hazard associated with many types of volcanoes. Gases such as carbon dioxide (CO_2) cause asphyxiation if concentrated. Other gases, including hydrogen sulfide (H_2S), cause death by paralysis. Gaseous sulfur dioxide (SO_2), hydrochloric acid (HCl), sulfuric acid (H_2SO_4), and fluorine compounds expelled during eruptions can destroy crops, kill livestock, and poison drinking water for people and animals. Gas eruptions can kill hundreds of people.

06.05.b4 Krafla, Iceland

C What Are the Hazards Associated with Pyroclastic Eruptions?

Lava flows usually move slowly enough that people can get out of the way, but such flows can completely destroy any structures in their path. In contrast, pyroclastic flows race downhill and across the landscape at hundreds of kilometers per hour, incinerating and smothering any creature in their path. They can cause massive destruction. Eruption columns spread their damage over wider areas and over longer time periods.

06.05.c1 Mount St. Helens, WA

One type of pyroclastic eruption forms an eruption column, a nearly vertical, billowing mass of volcanic ash and other debris that can rise to great heights in the atmosphere. The example shown here (◀), from the eruption of Mount St. Helens in 1980, rose 25 km (15 mi) into the atmosphere. The ash was carried eastward by the wind and blocked sunlight as it cooled and settled back to Earth across a large area of Washington, Idaho, and Montana. Eruption columns, because of their great height, can affect large regions, causing collapse of roofs overloaded with volcanic ash, respiratory problems to people and animals, and damage to jet engines, automobiles, and anything with an air intake.

After the main eruption, pyroclastic flows (▶) and other eruptions continued on Mount St. Helens until 1986. After a lull of several years, the volcano built, destroyed, and rebuilt several new volcanic domes. The domes would grow at some times and collapse at others, unleashing smaller pyroclastic flows, like the one shown here. Activity subsided in 2008, but the volcano is still being closely monitored by instruments designed to detect changes in temperature, gas emissions, or steepness of slopes.

06.05.c2 Mount St. Helens, WA

D What Hazards Accompany Volcanic Collapse?

Various other hazards can accompany the catastrophic collapse of a volcano. One of the largest historic volcanic eruptions struck the Indonesian Island of Krakatau in 1883. A large eruption and collapse of a volcano unleashed pyroclastic flows and huge waves that spread out across the sea, resulting in the deaths of almost 40,000 people.

06.05.d1

06.05.d2

06.05.d3

1. Before 1883, the area contained three islands, the largest of which, *Krakatau*, was made of three volcanoes. The region was densely populated, and many people lived along the coast of these and neighboring islands. Both factors contributed to the heavy death toll from the 1883 eruption.

2. When the eruption began, massive amounts of magma erupted from a magma chamber beneath the islands, forming a high eruption column and pyroclastic flows. The eruption was accompanied by huge explosions, landslides, caldera collapse, and destruction of two of the volcanoes and nearly half of another. Large waves struck ships and adjacent coasts.

3. After the eruption, only part of Krakatau remained. In 1927, a small volcano began to grow within the caldera, forming a new island called *Anak Krakatau* (child of Krakatau). Today, Indonesia is densely populated, with more than 150 volcanoes in a curved line across Sumatra, Java, and smaller islands. It faces a high risk of future deadly eruptions.

Floods of Water, Ice, and Debris

Iceland is a land of ice and fire due to its location near the Arctic Circle and its position on top of a mid-ocean ridge and an oceanic hot spot. It has many glaciers (the Icelandic term for glacier is *jökull*), including a large ice sheet that covers 25% of the country. Beneath the ice are a half dozen shield volcanoes, several of which have erupted in the last several decades.

In 1996, a volcanic eruption beneath an ice sheet melted the ice, releasing a catastrophic flood of meltwater (called a *jökulhlaup* in Icelandic). The huge flood carried blocks of ice, rock, and other debris, causing widespread damage by destroying bridges and houses and by covering vast areas with sediment, including the dark, newly deposited sediment shown below.

06.05.t1 Southern Iceland

Before You Leave This Page

☑ Explain how risk differs from hazard, and provide an example of each.

☑ Describe and compare the hazards associated with lava flows, lava fountains, volcanic gases, and various types of pyroclastic eruptions.

☑ Describe the eruptions that occurred at Mount St. Helens and Krakatau, and what caused most deaths.

☑ Explain eruption-caused floods.

6.5

6.6 What Volcanic Features Consist of Basalt?

BASALT IS THE MOST ABUNDANT VOLCANIC ROCK on Earth's surface. Dark-colored basalt covers large areas on every continent and forms the upper part of oceanic crust. When magma with a composition of basalt erupts, it can produce a variety of landforms. It forms scoria cones and associated dark lava flows, and it also produces shield volcanoes and lava flows that cover huge areas, called flood basalts. Which type of volcanic feature forms is largely controlled by the gas in the magma, the total volume of magma, and how fast the magma erupts.

A How Are Scoria Cones and Basalt Flows Expressed in Landscapes?

Basaltic magma has a relatively low viscosity compared to other magmas, and it erupts in characteristic ways. A basaltic eruption can form a lava flow or throw pieces of molten rock into the air. The lava can flow smoothly, like a hot, glowing stream, or it can partially solidify and break apart. Lava erupted into water forms distinctive features.

06.06.a1 Hawaii

1. *Basaltic Eruptions*—At the beginning of many basaltic eruptions, gases carry bits of lava into the air, forming a lava fountain. The airborne bits of lava cool and then fall around the vent as loose pieces of cinder. The lava fountain may be followed by or accompanied by eruption of a basaltic lava flow.

2. *Scoria Cones*—Pieces of cinder from the lava fountain gradually create a cone-shaped hill called a scoria cone. Ejected fragments can be as small as sand grains or as large as huge boulders. Scoria cones typically form in a short amount of time, from a few months to a few years, and generally are no more than 300 m (about 1,000 ft) high.

3. *Basaltic Lava Flows*—Fluid basaltic lava pours from the vent and flows downhill. Sometimes, as shown here, the lava fills up and overtops the crater in the scoria cone. At other times, a lava flow issues from cracks near the base of the scoria cone after most of the cone has been constructed.

Scoria Cones

106.06.a2 Sunset Crater, AZ

4. Most scoria cones begin with a conical shape (▲) and a central crater at the top of the cone. Young scoria cones have little soil or vegetation on them and commonly are associated with dark, fresh-looking lava flows.

06.06.a3 Springerville, AZ

5. Over time, erosion wears away the summit of a scoria cone, breaching the crater (▲) and later making the cone into a rounded hill without a crater. Erosion cuts into the slopes, which gradually build up a veneer of soil.

Types of Lava Flows

06.06.a4 Hawaii

6. *Aa* lava (pronounced "ah-ah") is a type of rough-surfaced lava flow, formed when the lava breaks apart into a mass of jumbled, angular blocks of hardened lava that tumble down the front of the flow as it moves (▲). Aa forms a very rough, jagged surface.

7. *Pahoehoe* (pronounced "pa-hoy-hoy") is a type of lava flow that has an upper surface with small, billowing folds that form a "ropy" texture. A pahoehoe lava flow is usually fed by a lava tube and grows as a series of lobes (▼). As the front of the flow solidifies, the lava breaks out and forms a new lobe, as shown in this photograph. Pahoehoe lava moves relatively smoothly and easily compared to aa flows.

06.06.a5 Hawaii

Tubes and Pillows

106.06.a6 Hawaii

8. *Lava tubes* form when the surface of a lava flow solidifies to form an insulating roof over the hot, still-moving interior of the flow (▲). Lava flows insulated by lava tubes flow farther than lava flows on the surface because the lava stays hotter longer. If the lava tube drains, it becomes a curving, tube-shaped cave.

06.06.a7 San Juan Islands, WA

9. When fluid lava erupts into water, the lava grows forward as small, individual lumps that form rounded shapes called *pillows* (▶). Pillows are reliable evidence that lava erupted into water. Basaltic lava flows on the seafloor contain countless pillows.

B What Is a Shield Volcano?

Shield volcanoes have a broad, *shield-shaped* form and fairly gentle slopes when compared to other volcanoes, because their eruptions are dominated by relatively nonexplosive outpourings of low-viscosity lava from fissures and vents. Shield volcanoes form in various settings, but the largest ones, including those in Hawaii, formed at oceanic hot spots.

1. This image shows satellite data superimposed on topography of the Big Island of Hawaii. The island consists mainly of three large volcanoes. Green areas are heavily vegetated, and recent lava flows are brown or dark gray.

2. Mauna Loa, the central mountain, is the world's largest volcano. It rises 9,000 m (29,500 ft) above the seafloor and is 4,170 m (13,680 ft) above sea level. From seafloor to peak, Mauna Loa is Earth's tallest mountain. Nearby Mauna Kea is an inactive shield volcano and the site of astronomical observatories.

3. Kilauea volcano, probably the most active volcano in the world, is on the southeastern side of the island. Recent lava flows (shown in dark grayish brown) flowed eastward, destroying roads and housing subdivisions.

06.06.b2 Mauna Loa, HI

4. In shield volcanoes, magma rises through a fracture and erupts onto the surface from a long fissure (▲). Large volumes of lava can flow out of the fissure, and escaping gas throws smaller amounts of molten rock into the air as a fiery curtain.

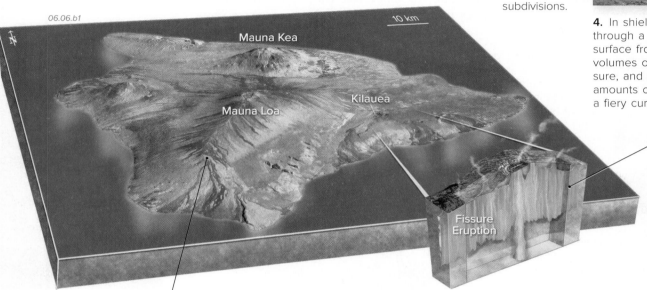

06.06.b1

Mauna Kea

10 km

Mauna Loa

Kilauea

Fissure Eruption

5. Eruptions also occur in more centralized vents, such as those on Kilauea (▼). These vents are interpreted to overlie fissures.

06.06.b3 Kilauea, HI

6. The spine of Mauna Loa is a fissure from which lava flows erupted as recently as 1984. The fissure is the surface expression of one or more magma-filled fissures at depth.

C What Are Flood Basalts and How Are They Expressed in Landscapes?

Flood basalts are basaltic lava flows covering vast areas and are commonly several kilometers thick. They generally involve multiple eruption events, but individual lava flows can cover thousands of square kilometers and contain more than 1,000 cubic kilometers of magma! Flood basalts are fed by a series of long fissures.

06.06.c1

A fissure forms when pressure from a magma pushes outward against the wall rocks, holding them apart while the magma passes through. A wide fissure allows faster eruption rates, which result in a large volume of erupted magma that can remain hot and travel long distances. Narrower fissures restrict the eruption rates, leading to lower volume lava flows that cool and solidify before they can travel as far.

06.06.c2 Palouse Canyon, WA

Some of the most famous flood basalts are in the Columbia Plateau of Washington, Oregon, and western Idaho. Cut into the plateau are canyons that expose multiple basalt flows, each forming a ledge or step in the canyon walls. Each flow represents a single eruption, separated by thousands of years in which not much happened. One basalt flow on the Columbia Plateau covers more than 130,000 km^2 (50,000 mi^2) and probably erupted very quickly, perhaps in only several decades.

Before You Leave This Page

- ✓ Summarize the features of scoria cones and basalt flows.

- ✓ Describe the general characteristics of a shield volcano and how the eruptions occur.

- ✓ Describe flood basalts and explain why they can cover huge areas.

6.6

6.7 What Are Composite Volcanoes and Volcanic Domes?

COMPOSITE VOLCANOES FORM STEEP CONICAL MOUNTAINS that are hard to mistake for anything other than a volcano. They are common above subduction zones, especially along the Pacific Ring of Fire. Composite volcanoes are extremely dangerous. Volcanic domes consist of smaller, dome-shaped masses of highly viscous lava that accumulated over and close to the associated volcanic vent—they are also very dangerous.

A What Are Some Characteristics of a Composite Volcano?

Composite volcanoes are constructed of interlayered volcanic material formed by lava flows, pyroclastic flows, falling ash, and volcano-related mudflows and landslides. Composite volcanoes, also called stratovolcanoes, erupt over long time periods, which explains their large size and complex internal structure.

06.07.a2 Mount St. Helens, WA

1. *Eruption Column*—Composite volcanoes produce a distinctive column of pumice, ash, and gas that rises upward many tens of kilometers into the atmosphere (◄). Coarser pieces settle around the volcano, but finer particles (volcanic ash) can drift hundreds of kilometers in the prevailing winds.

2. *Pyroclastic Flows*—These are the most violent eruptions from the volcano. They form when the eruption column collapses downward as a dense, swirling cloud (►) of hot gases, volcanic ash, and angular rocks. Pyroclastic flows are one of the main mechanisms by which these volcanoes are constructed.

06.07.a3 Mount Mayon, Philippines

6. *Shape*—Composite volcanoes display the classic volcano shape because most material erupts out of a central vent and then settles nearby. They have steep slopes because they form from small eruptions of viscous lava flows that pile up on the flanks of the volcano and help protect pyroclastic material from erosion. The shape represents one snapshot in a series of stacked volcanic mountains that have been built over time.

3. *Landslides and Mudflows*—Composite volcanoes can be large mountains that collect rain or snow. Rain and snowmelt mix with loose ash and rocks on the volcano's flanks, causing a volcano-derived mudflow called a *lahar*. There is a high hazard for landslides and lahars (▼) because of the steep slopes, loose rocks, and abundant slippery clay minerals produced when hot water interacts with the volcanic rocks.

5. *Lava Domes and Flows*— Lava domes and flows can erupt from any level of a composite volcano. Lava may erupt from the summit crater or escape through vents on the volcano's sides or base. Lavas associated with composite volcanoes are moderately to highly viscous, and so they move slowly and with difficulty. The lava may break into blocks that fall, slide, or roll downhill, forming a tongue or apron (▼) of jumbled pieces from the lava.

06.07.a1

4. *Rocks*—Composite volcanoes consist of alternating layers of pyroclastic rocks, lava flows, and deposits from landslides and mudflows. Most rocks are intermediate to felsic in composition. The volcanoes we see today, formed during eruptions from long-lived vents, are built on and around earlier versions of the volcanoes. A large composite volcano can be constructed in tens of thousands of years or a few hundred thousand years.

06.07.a5 Augustine volcano, AK

06.07.a4 Mount St. Helens, WA

B What Are Some Characteristics of a Volcanic Dome?

Volcanic domes form when viscous lava mounds around a vent. Some volcanic domes have a nearly symmetrical dome shape, but most have an irregular shape because some parts of the dome have grown more than other parts or because one side of the dome has collapsed. Domes may be hundreds of meters high and one or several kilometers across, but they can be much smaller, too.

06.07.b1 Novarupta, AK

Dome

1. This rubble-covered dome (▲) formed near the end of the 1912 eruption in the Valley of Ten Thousand Smokes in Alaska. Volcanic domes commonly have this type of rubbly appearance because their outer surface consists of angular, broken pieces of the dome.

Growth of a Dome

2. Domes mostly grow from the inside as magma injects into the interior of the dome. This new material causes the dome to expand upward and outward, fracturing the partially solidified outer crust of the dome. This process creates the blocks of rubbly, solidified lava that coat the outside of the dome. The blocks and other pieces, which are as small as several centimeters across to as large as houses, commonly tumble or slide down the steep slopes on the side of a dome.

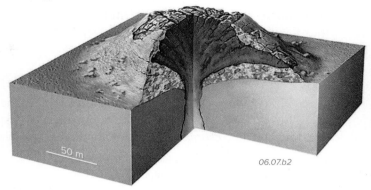
50 m
06.07.b2

3. Domes also grow as magma breaks through to the surface and flows outward as thick, slow-moving lava. As the magma advances, the front of the flow cools and solidifies, and can collapse into angular blocks and ash. Most domes display both modes of growth.

Collapse or Destruction of a Dome

4. Domes can be partially destroyed when steep flanks of the dome collapse and break into a jumble of blocks and ash that flow downhill as small-scale pyroclastic flows (▶).

06.07.b3

06.07.b4

5. Domes can also be destroyed by explosions originating within the dome (◀). These typically occur when magma solidifies in the conduit and traps gases that build up until the pressure can no longer be held.

Deadly Collapse of a Dome at Mount Unzen, Japan

Mount Unzen towers above a small city in southern Japan. The top of the mountain contains a steep volcanic dome that formed and collapsed repeatedly between 1990 and 1995. The collapsing domes unleashed more than 10,000 small pyroclastic flows (top photograph) toward the city below. In 1991, the opportunity to observe and film these small pyroclastic flows attracted volcanologists and other onlookers to the mountain. Unfortunately, partial collapse of the dome caused a pyroclastic flow larger than had occurred previously. This larger flow killed 43 journalists and volcanologists, and left a path of destruction through the valley (lower photograph). Note that damage was concentrated along valleys that drain the mountain.

06.07.t1 Mount Unzen, Japan

06.07.t2 Mount Unzen, Japan

Before You Leave This Page

✓ Describe or sketch the characteristics of a composite volcano and how such a volcano forms.

✓ Sketch and explain the characteristics of a volcanic dome and the two ways by which a volcanic dome can grow.

✓ Explain or sketch how a volcanic dome can collapse or be destroyed by an explosion.

6.7

6.8 What Disasters Were Caused by Composite Volcanoes and Volcanic Domes?

COMPOSITE VOLCANOES AND DOMES ARE DANGEROUS because they can be very explosive and can unleash pyroclastic flows, toxic gases, and other deadly materials. They are responsible for horrific human disasters, including the destruction of Pompeii in Italy, St. Pierre in Martinique, and the area around Mount St. Helens.

A How Did Vesuvius Destroy Pompeii?

Vesuvius is an active composite volcano near the city of Naples in southwestern Italy. In A.D. 79, a series of pyroclastic flows moved down the flank of the volcano, destroyed the coastal towns of Pompeii and Herculaneum, and killed the cities' inhabitants, estimated at 20,000 and 5,000 people, respectively.

1. This image is an artist's conception of an explosive Vesuvius eruption striking the city of Naples, which covers most of the region shown. Naples and the surrounding area currently are home to over 3 million people.

2. The dashed red line marks the outward limit of pyroclastic flows from Vesuvius, but tephra from the eruption column covered a wider area. Note how much of the present city of Naples is within the area devastated by the eruption of 79 A.D.

Herculaneum

Limit of Pyroclastic Flows

Pompeii

10 km

06.08.a1

3. Archeologic and geologic evidence from Pompeii indicates that the catastrophe began with earthquakes and the formation of an eruption column that deposited a layer of loose tephra over Pompeii, killing some inhabitants.

4. The tephra fall was immediately followed by six pyroclastic flows that raced down the mountainside. Three of these flows hit Pompeii. The first probably burned most of the remaining survivors, and the last was strong enough to complete the destruction of standing buildings. People smothered, suffocated, died from thermal shock, or were crushed by collapsing buildings. The bodies of victims in the ash decomposed, leaving mostly hollow molds, which archeologists filled with plaster to make models of the victims' last moments (▶).

06.08.a2

B What Happened at St. Pierre, Martinique?

The Caribbean island of Martinique consists of composite volcanoes and a series of volcanic domes. On May 8, 1902, Mount Pelée, one of the active volcanoes, erupted and sent a pyroclastic flow into the town of St. Pierre.

1. This view shows the island of Martinique, which consists of several distinct volcanoes, including Mount Pelée, the northernmost peak. Mount Pelée is a composite volcano.

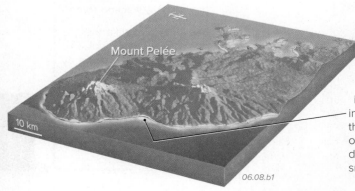

Mount Pelée

10 km

06.08.b1

2. The coastal town of St. Pierre is in a bay, at the foot of Mount Pelée. Before the main eruption, the volcano gave obvious warning signs, including noisy explosions, earthquakes, sulfurous gases, and small eruptions that dusted nearby areas with ash. People from the surrounding countryside sought shelter in the town of St. Pierre, where they witnessed minor eruptions of ash, the formation of a lava dome in the crater, and some small pyroclastic flows.

3. During the main eruption, a massive pyroclastic flow, estimated to have traveled at 500 km/hr, entered the town. Every building was mostly or completely destroyed (▼). Almost all of the 30,000 residents died within minutes. Most deaths were probably caused by asphyxiation as people breathed hot gas and ash. After the main eruption, additional eruptions formed an eruption column and more pyroclastic flows. The lessons learned from Mount Pelée and other eruptions saved lives in 1995, when volcanic eruptions started on Montserrat, a similar volcanic island to the south.

06.08.b2

C What Events Accompanied the Mount St. Helens Eruption?

The Cascade Range of the Pacific Northwest has produced some large and notable eruptions. Geoscientists consider these volcanoes to be dangerous, and native people remember, through oral traditions, other cataclysmic eruptions at places like Crater Lake in the Cascades. The eruption of Mount St. Helens in Washington was the first major composite volcano eruption to occur in the age of television, and the world watched the event.

1. Geologists studied Mount St. Helens before the eruption. They mapped the volcano and constructed a geologic cross section through the mountain. This cross section shows that, before the eruption, the volcano consisted of interlayered lava flows, pyroclastic rocks, and mudflows, and had a dome-like central conduit (▶). In other words, it was a typical composite volcano.

Mount St. Helens

WSW   ENE

1 km

Dome and Conduit
Lava Flows and Dikes
Pyroclastic Rocks and Mudflows

06.08.c1

06.08.c4-6

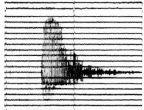

06.08.c2

2. In March 1980, Mount St. Helens began to shake from earthquakes that geoscientists interpreted to be caused by magma moving beneath the mountain. These moderate quakes, including the one recorded on this seismogram (◀), were the signal that something was going to happen at Mount St. Helens.

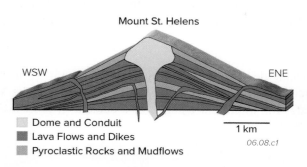

06.08.c3

3. In April 1980, a bulge formed on the north side of the mountain and then continued to grow. Geologists inferred that the bulge was caused by upward-moving magma, and they recognized that the bulge was dangerous and unstable. They monitored its growth carefully. David Johnston, a 30-year-old USGS volcanologist, was monitoring Mount St. Helens from a nearby ridge at the moment of the eruption. His observation post was considered to have low risk, provided that the volcano erupted out the top as expected.

4. On May 18, 1980, at 8:32 a.m., an earthquake triggered a massive avalanche as the bulge slid off the north side of the mountain (▲). As this sequence of images above shows, the lowering of pressure on the magma caused a lateral blast and an upward growth of an eruption column that spread ash over several states. Pyroclastic flows moved horizontally and ravaged the landscape near the volcano. As David Johnston watched the events unfold, he spoke the words "Vancouver! Vancouver! This is it!" His last recorded scientific observation, spoken at the catastrophic start of the eruption, was undeniably correct. This was it! In addition to David Johnston, 56 other people died in the eruption, mainly from asphyxiation by hot gas and ash. The eruption also unleashed floods and mudflows that devastated areas far from the volcano.

Crater Lake and the Eruption of Mount Mazama

Crater Lake National Park contains a spectacular caldera, now filled by a beautiful deep-blue lake. The caldera formed about 7,700 years ago when a huge eruption and associated collapse destroyed the top of a large composite volcano called *Mount Mazama*. The eruption was more than 50 times larger than the 1980 eruption of Mount St. Helens. As the magma erupted in a huge eruption column, the roof of the magma chamber subsided, forming the main crater (caldera). Ash from this eruption can be traced all the way to southern Canada, 1,200 km (750 mi) from its source. A small scoria cone, named Wizard Island, grew on one side of the floor of the caldera after the main explosive eruption. The beauty and serenity of Crater Lake today seem incompatible with its fiery, cataclysmic origin 7,700 years ago.

06.08.t1

Before You Leave This Page

☑ Describe the type of deadly eruption that occurred at Vesuvius.

☑ Discuss the eruption at Mount Pelée and events that preceded it.

☑ Discuss the eruption of Mount St. Helens and the data that warned of its dangers.

6.8

6.9 What Are Calderas?

CALDERA ERUPTIONS ARE AMONG NATURE'S most violent phenomena. They can spread volcanic ash over huge areas, and the largest erupt more than 1,000 cubic kilometers of magma. As the magma vacates the magma chamber during an eruption, the roof of the chamber collapses to form a depression tens of kilometers across. The depression may fill with ash, lava flows, and sediment. On these two pages, we focus on large calderas, but smaller calderas can form on composite volcanoes or when magma withdraws beneath a shield volcano. Calderas are also called "supervolcanoes."

A What Is a Caldera?

A *caldera* is a large, basin-shaped volcanic depression, which typically has a low central part surrounded by a topographic escarpment, referred to as the *wall* of the caldera (▶). The Valles Caldera of New Mexico nicely illustrates the important features of a caldera. It is relatively young and uneroded because it formed only two million years ago.

06.09.a2 Valles Caldera, NM

1. This image shows satellite data superimposed on topography. The circular Valles Caldera contains a central depression, about 22 km (14 mi) across, surrounded by steep walls. The caldera formed when a huge volume of magma erupted from a shallow magma chamber, producing a large eruption column and pyroclastic flows.

2. Internally, the caldera contains a series of faulted blocks that have been downdropped relative to rocks outside the caldera. Faulting and ground subsidence occurred at the same time as the main eruption, so relatively thick amounts of ash and other materials accumulated inside the caldera.

Valles Caldera

3. Small, rounded volcanic domes formed within the caldera after the main pyroclastic eruption. The domes were fed by fissures that tapped leftover magma. Some of this magma has solidified at depth, forming granite, but some may still be molten. The area has been explored for geothermal energy associated with the hot rocks at depth.

06.09.a1

B How Does a Caldera Form?

The formation of a caldera and the associated eruption occur simultaneously—the caldera subsides in response to rapid removal of magma from the underlying chamber. The largest caldera eruptions produced volcanic ash layers more than 1,000 m thick.

5 km

06.09.b1

06.09.b2

1. The first stage in the formation of a caldera is the generation of felsic magma. The magma rises and accumulates in magma chambers that can be kilometers thick and tens of kilometers across. The chambers may be within several kilometers of the surface.

2. Next, some magma reaches the surface (at which point it is called lava) and eruptions begin. As the magma chamber loses material, the roof of the chamber subsides to occupy the space that is being vacated. Curved fractures allow crustal blocks to drop, outlining the edges of the caldera and providing many conduits for magma to reach the surface.

3. The eruption forms columns and flows of pyroclastic material, much of which falls back into the caldera, creating a thick pile of volcanic ash. Landslides from steep caldera walls produce large blocks and smaller clasts that become part of the caldera deposit. Some ash escapes the caldera and covers surrounding areas.

4. As the eruption lessens, magma rises through fissures along the edges or in the caldera, erupting onto the surface as volcanic domes.

06.09.b3

06.09.b4

C Did the Eruption of a Caldera Destroy a Civilization Near Greece?

Santorini, east of the Greek mainland, is a group of volcanic islands with geology that records a major caldera collapse only 3,500 years ago, about the time of the collapse of the Minoan civilization on the island of Crete to the south.

06.09.c1 Santorini, Greece

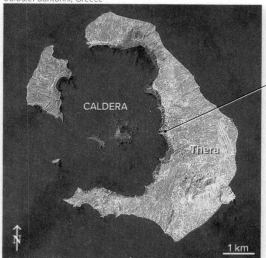

1. This satellite view (◄) shows the islands of Santorini, including *Thera,* the largest island. Thera and the other islands encircle a submerged caldera that formed when the center of a larger volcanic island collapsed, leaving the modern islands as remnants. The steep cliffs around the caldera are eroded segments of the original wall of the caldera. The curving cliffs (►) expose volcanic layers, products of explosive volcanism that began one to two million years ago. Islands in the middle of the caldera were constructed by more recent eruptions.

06.09.c2 Santorini, Greece

2. The main eruption produced an ash column perhaps 40 km (25 mi) high, followed by pyroclastic flows. The erupted ash buried towns, now excavated (►), with up to 50 m of pumice and ash. Caldera collapse occurred as the eruption of ash emptied a large magma chamber.

3. The collapse of the caldera evidently unleashed a large destructive wave that traveled southward across the sea, probably helping lead to the downfall of the Minoan civilization on the island of Crete. This destruction of the civilization on Santorini and collapse of the volcanic island into the sea may have started legends about the sinking of a landmass and city (Atlantis) into the sea.

06.09.c3 Akrotiri, Santorini, Greece

D Why Is Yellowstone Caldera Considered to Be Dangerous?

Yellowstone is one of the world's largest active volcanic areas and site of a "supervolcano." Abundant geysers, hot springs, and other hydrothermal activity are leftovers from its recent volcanic history. During the last two million years, the Yellowstone region experienced three huge, caldera-forming eruptions. What is the possibility that Yellowstone could erupt again and rain destructive ash over the Rocky Mountains and onto the Great Plains?

06.09.d1

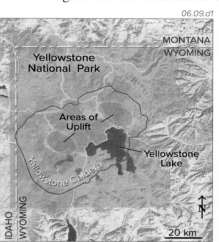

◄ This image shows the outline of the youngest Yellowstone caldera, which formed 640,000 years ago. The boundaries of the caldera have been partially obscured by erosion, deposition of sediment, and lava flows that erupted after the caldera formed. Several areas within the caldera have been experiencing recent uplift. Magma and hot rocks at relatively shallow depth heat groundwater to form the spectacular geysers and hot springs for which Yellowstone is famous.

Ash from the three Yellowstone eruptions was carried by the wind and deposited over a huge area (►) that extends from northern Mexico to southern Canada and as far east as the Mississippi River. A repeat of such an eruption could devastate the region around Yellowstone and cause extensive crop loss in the farmlands of the Great Plains and Midwest. It would truly be catastrophic.

06.09.d2

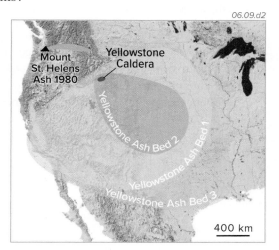

Before You Leave This Page

☑ Sketch and explain the characteristics of a caldera and the four stages by which one forms.

☑ Explain how a volcanic eruption destroyed Santorini.

☑ Describe what type of volcanic feature Yellowstone is, how it forms, and how a future eruption could impact the Great Plains.

6.9

6.10 What Types of Volcanism and Other Igneous Processes Occur Along Plate Boundaries?

MOST VOLCANOES AND ERUPTIONS OF MAGMA occur at or near boundaries of tectonic plates. They occur along mid-ocean ridges, during continental rifting, above subduction zones, and within continent-continent collisions. Each setting produces a distinctive suite of magmas and types of volcanic eruptions.

A What Causes Melting Along Mid-Ocean Ridges?

Nearly 60% of magma is generated along mid-ocean ridges, mostly on the seafloor and out of our view. In such settings, two plates move away from one another (diverge) along mid-ocean ridges. To understand how melting occurs here, examine the magmatic system, beginning with processes at the bottom, within the mantle.

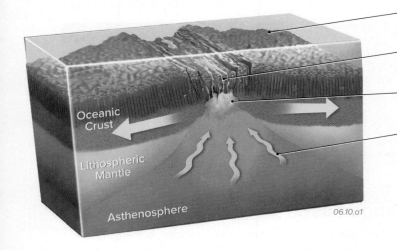

06.10.a1

5. Older oceanic crust moves away from the ridge in a conveyor-belt manner as new oceanic crust forms along the axis of the ridge.

4. Magma rises upward through magma-filled fractures that form as the plates *pull apart*. Some magma erupts as lava within the rift.

3. The buoyant, mafic magma rises away from the unmelted residue in the mantle and accumulates in magma chambers in the crust and upper mantle.

2. As the plates separate, solid asthenosphere rises to fill the area between the plates. As the asthenosphere rises, pressure decreases and the rock partially melts (decompression melting).

1. Mantle rocks, including those in the asthenosphere, are mostly solid and crystalline, not molten. The mantle's high pressures and temperatures allow these rocks to flow as a weak solid while maintaining a crystalline structure. Parts of the asthenosphere are close to their melting temperature.

B How Are Magmas Generated in Continental Rifts?

Continental rifts form where tectonic forces attempt, perhaps successfully, to split a continent apart. Such rifts have a central trough where faults drop down huge crustal blocks. In some cases, a rift becomes a divergent plate boundary, but others simply disrupt part of a continent, without causing breakup of the continent and a true divergent boundary. Continental rifts are characterized by a diverse suite of igneous rocks because melting takes place in both the mantle and the crust. The sequence of events begins in the mantle, so begin your reading at the bottom.

4. Some felsic and intermediate magmas solidify underground as granite and related igneous rocks, while others erupt on the surface in potentially explosive volcanoes.

3. Heat from the hot mafic magma melts the adjacent continental crust, producing felsic magma. Intermediate magma forms from mixing of felsic and mafic magmas or from the melting of continental crust by a mafic magma.

2. The mantle-derived mafic magma rises into the upper mantle and lower continental crust, and accumulates in large magma chambers. Some of the mafic magma reaches the surface and erupts as mafic (basaltic) lava flows.

1. Solid asthenosphere rises beneath the rift and undergoes *decompression melting*. Partial melting of the ultramafic mantle source rock yields mafic magma.

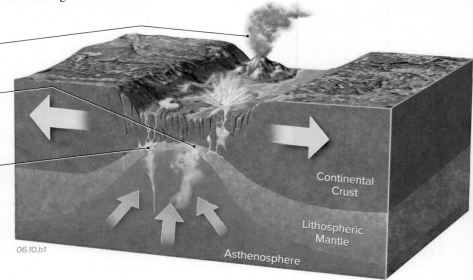

06.10.b1

C How Is Magma Generated Along Subduction Zones?

About one-fifth of Earth's magma forms where an oceanic plate subducts into the mantle, at an ocean-ocean or ocean-continent convergent boundary. This is also the setting where most composite volcanoes form.

1. When an oceanic plate composed of oceanic crust and lithospheric mantle converges with another oceanic plate or with a continental plate, subduction occurs. As the subducted plate descends, both pressure and temperature gradually increase.

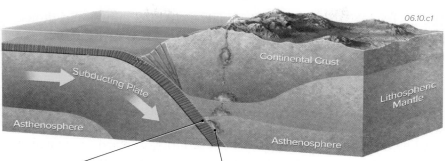

06.10.c1

4. Most subduction-related magma probably solidifies at depth, but some erupts, forming clusters or belts of volcanoes. If the overlying crust is continental, the volcanoes are usually part of a mountain belt. If the overlying crust is oceanic, subduction-generated magma creates individual volcanoes along an *island arc*. In both settings, the subduction-generated magma mostly has an intermediate composition (andesite). Island arcs can erupt mafic magma, and continent-derived magma can be felsic. In both cases, magmas added at depth and on the land surface thicken the crust.

2. In response to the changes in pressure and temperature, existing minerals in the subducting plate convert into new ones through the process of *metamorphism*. Water-bearing minerals, such as mica, break down, which forces water out of the crystalline structures. The water liberated from minerals then rises into the overlying asthenosphere.

3. The water released from the subducting slab reduces the melting temperature and causes melting of the overlying mantle. Most magma generated in this way begins with a *mafic* composition because it forms by partial melting of the mantle. Less commonly, partial melting may generate magma of *intermediate* composition. If the overriding plate is a continental plate, the rising magma encounters thick continental crust that slows its upward journey. The magma heats the surrounding rocks, commonly causing localized *partial melting* that produces *felsic* or *intermediate* magma.

D What Magmatism Accompanies Continental Collisions?

When two continents, such as Asia and India, converge, the encounter is best described as a *collision* because continental crust is buoyant and difficult to subduct. One plate may partially slide beneath the other but, because the descending plate is continental rather than oceanic, continental collisions result in different types of magmas than those in a typical subduction zone. Begin reading from the bottom, where one plate is subducted beneath the other.

3. Magmas produced by continental collisions typically do not reach the surface, partly because they have to pass through thick continental crust. Also, some magmas that are produced have a relatively high water content compared to mantle-derived magmas, and this causes them to solidify as they rise. So, continental collisions, unlike other convergent boundaries, do not have many volcanoes.

2. Water may be released by metamorphism of water-bearing minerals and, if the descending continental crust gets hot enough, it undergoes partial melting, producing *felsic* magmas.

1. During a continental collision, one continental plate may slide beneath another continental plate. The descending continental crust gets hotter and experiences increased pressure.

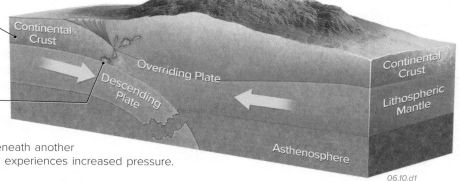

06.10.d1

How Water Gets into Subduction Zones

During subduction and collision, water is released by sediments and minerals in the descending crust. Where does the water come from? Any oceanic plate being subducted formed originally along a mid-ocean ridge, where seawater flows into the hot crust and forms water-bearing metamorphic minerals. These water-bearing minerals, shown with blue spots in the oceanic crust, travel with the plate as it moves away from the ridge. Once formed, oceanic crust is slowly covered by sediment derived from continents, islands, and creatures living in the sea. This sediment contains trapped seawater and minerals, including clay, that have water in their mineral structure.

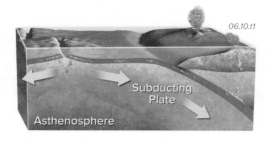

06.10.t1

Before You Leave This Page

☑ Sketch or describe why melting occurs along mid-ocean ridges and where those magmas go.

☑ Describe melting in continental rifts and the resulting magmas.

☑ Sketch and describe magma generation in subduction zones.

☑ Describe how magma forms in a continent-continent collision.

6.11 How Do Large Magma Chambers Form and How Are They Expressed in Landscapes?

MAGMA OFTEN ACCUMULATES UNDERGROUND IN CHAMBERS containing thousands of cubic kilometers of molten rock. How do these chambers form, what are their shapes, and what processes occur within them? What do they look like after they have solidified and are uplifted to the surface?

A In What Settings Do Large Magma Chambers Form?

A *magma chamber* is an underground body of molten rock. Think of it as an always-full reservoir or holding tank that allows magma to enter from below and perhaps exit out the top. Magma chambers are very dynamic, with magmas evolving, crystallizing, and replenished by additions of new magma. A large influx of magma is required to form a large magma chamber. This, in turn, requires melting on a large scale that is possible only in certain tectonic settings.

In oceanic lithosphere, large magma chambers form above hot spots and within mid-ocean ridges. In both cases, the mantle-derived magma is mafic.

Large magma chambers of intermediate and felsic composition form above subduction zones, either within magmatic arcs on continents or within oceanic island arcs.

Hot spots and rifts within continents produce large amounts of mantle-derived magma that can erupt as basalt flows or melt continental crust to form large felsic magmas and plutons.

Continental collisions cause crustal thickening, which can lead to melting of continental crust. Large amounts of felsic magma may be trapped at depth.

06.11.a1

B What Type of Magmatic Activity Occurs at Hot Spots?

A hot spot is a site of intense magmatic activity that cannot be explained by its plate-tectonic setting (e.g., not near a subduction zone or mid-ocean ridge). Many hot spots, like Hawaii, do not coincide with a plate boundary, but some are close to a plate boundary. Hot spots have different igneous manifestations, depending on whether the hot spot is within a continental or oceanic plate. Some igneous activity is not related to a plate boundary or a hot spot (not shown here).

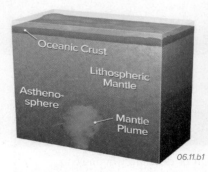

06.11.b1

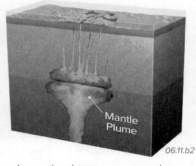

06.11.b2

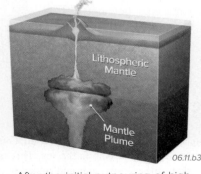

06.11.b3

06.11.b4

Most hot spots are considered to be a crustal expression of a rising *plume* of hot, but solid, mantle material. When material in the rising plume melts due to decompression, the generated magma encounters the lithosphere, where it spreads out along the boundary and causes melting of the lower lithosphere.

A mantle plume can generate an incredibly large volume of mafic magma that can erupt onto the surface, producing flood basalt, either on the seafloor or on land. If erupted on the seafloor, the lava flows pile on top of one another, building a large, broad feature on the seafloor, an *oceanic plateau*.

After the initial outpouring of high volumes of magma, the hot spot produces mafic magma at a slower rate, constructing large volcanoes on the seafloor, as in Hawaii. The eruptions produce a cluster of islands and seamounts if the plate overlying the plume is not moving or a line of islands and seamounts if it is moving.

When a rising mantle plume encounters continental lithosphere, its high temperatures cause melting of the lithosphere. Melting in the mantle part of the lithosphere produces mostly mafic magma, but melting of continental crust can generate felsic magma and explosive calderas, such as is occurring at Yellowstone.

C How Are Large, Solidified Magma Chambers Exposed at the Surface?

A solidified magma chamber is called a *pluton*. A pluton can be cylindrical, sheetlike, or very irregular in shape. Several generations of magma may intrude into the same region, forming a complex mass of plutons with various compositions, textures, and shapes. Plutons are classified according to their size and geometry.

Irregular Plutons

Many plutons have irregular shapes, somewhat like vertical cylinders. A pluton with an exposed area of less than 100 km² is a *stock*.

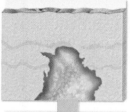

Most stocks are irregularly shaped. Many have a shape like a steeply oriented cylinder or downward-widening, bumpy mass.

On the surface, most stocks have steep boundaries and may resist erosion more than surrounding rocks.

1 km
06.11.c1–2

06.11.c3 Joshua Tree NP, CA

The top of a pluton of light-colored granite protruded into overlying dark-colored metamorphic rocks, and both rocks were later uplifted and exposed by erosion.

Sheetlike Plutons

Some plutons have a tabular shape, like a thin or thick sheet. The sheet can be vertical, horizontal, or at some other angle.

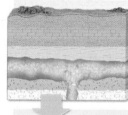
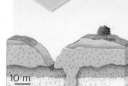

Plutonic sheets can be horizontal, vertical, or inclined, and may be parallel to or cutting across layers in the wall rocks.

When horizontal sheets are exposed at the surface, their tops and bottoms may be visible.

10 m
06.11.c4–5

06.11.c6 Cuernos del Paine, Chile

The gray granitic rocks were a horizontal sheet of magma that squeezed between dark metamorphic rocks that are above and below.

Batholiths

A *batholith* is one or more contiguous plutons that cover more than 100 km². Most batholiths include a number of rock types.

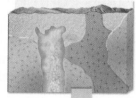
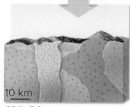

Most batholiths form from multiple magmas emplaced into the same part of the crust over a long time.

Exposed batholiths are characterized by plutonic rocks that cover a huge region.

10 km
06.11.c7-8

06.11.c9 Sierra Nevada, CA

A huge expanse of gray granite characterizes the Sierra Nevada batholith of California, seen here from the east.

The White Mountain Batholith of New England

The White Mountain batholith is centered in the middle of New Hampshire. Granitic rocks of the batholith form high peaks of the White Mountains and many of the area's scenic landmarks.

The batholith consists of several dozen individual plutons (shown in red and yellow) that were emplaced between 200 and 155 million years ago. The plutons represent separate injections of magma, some emplaced at somewhat different times. Some plutons are cylindrical; others are like curved dikes.

20 km
06.11.t1

Geoscientists interpret the White Mountain batholith as being related to a hot spot that melted its way into continental crust. The age of the batholith coincides with rifting as North America pulled away from Africa when the central Atlantic Ocean formed. A line of submerged volcanic mountains in the Atlantic Ocean, called the *New England Seamount Chain*, is interpreted to mark the path of the North American plate over the hot spot.

Before You Leave This Page

- ✓ Sketch or summarize the tectonic settings in which large magma chambers form.
- ✓ Sketch and explain magmatism at oceanic and continental hot spots.
- ✓ Sketch the different geometries of large magma chambers and summarize how these are expressed in the landscape.
- ✓ Describe the character of the White Mountain batholith and how it is interpreted to have formed.

6.11

6.12 How Are Small Intrusions Formed and Expressed in Landscapes?

MANY INTRUSIONS ARE RELATIVELY SMALL OR THIN FEATURES, small enough to be exposed on a single small hill or in a roadcut. Small intrusions can have a sheetlike, pipelike, or even lumpy geometry. Where exposed at the surface, small intrusions can form distinctive landscape features, like a natural wall or a volcanic neck.

A What Features Form When Magma Is Injected as Sheets?

Many small intrusions have the shape of thin or thick sheets, typically ranging in thickness from several centimeters to several tens of meters. These form when underground forces allow magma to generate new fractures or to open up and inject into existing fractures. In some cases, magma squeezes between preexisting layers in the wall rocks, commonly between the horizontal layers of sedimentary rocks.

Dike

06.12.a1

A *dike* is a sheetlike intrusion that cuts across any layers present in the host rocks. Most dikes are steep (◀) because the magma pushes apart the rocks in a horizontal direction as it rises vertically and fills the resulting crack to form a dike. In some dikes, magma flows into the dike horizontally, and the dike grows sideways with time. Dikes are also common within larger plutons.

In front of this mountain is a dike (▶) that formed under the surface but was later uncovered by erosion, forming a resistant natural wall.

06.12.a2 Spanish Peaks, CO

Sill

06.12.a3

An intrusion that is parallel to layers (◀) in the host rocks is called a *sill*. Most sills are subhorizontal and form by pushing adjacent rocks upward rather than sideways.

These dark-colored mafic sills (▶) intruded parallel to layers of light-colored, sedimentary wall rocks. Like most sills, these contain steep fractures formed by cooling of the sills after they solidified.

06.12.a4 Salt River Canyon, AZ

Laccolith

In some areas, ascending magma encounters gently inclined layers and begins squeezing parallel to them, forming a *sill*. The magma then begins inflating a lump- or bulge-shaped magma body called a *laccolith*. As the magma chamber grows, the layers over the laccolith tilt outward and eventually define a dome-shaped feature (▼).

06.12.a5

The Four Corners region of the American Southwest contains some of the world's most famous stocks and laccoliths, including these in the Henry Mountains of southern Utah (▶). The laccoliths formed 25 million years ago at a depth of several kilometers and were later uncovered by erosion. Igneous rocks of the laccolith are medium grained and porphyritic, and have an intermediate composition.

06.12.a6 Henry Mtns, UT

B What Kinds of Magma Chambers Form Within and Beneath Volcanoes?

Magma that erupts from volcanoes is fed through conduits that may be circular, dike shaped, or both. After the volcano erodes away, the solidified conduit can form a steep topographic feature called a *volcanic neck*.

06.12.b1

06.12.b2

1. A small volcano has been partially eroded, revealing a cross section through the volcano (▶). A resistant and jointed volcanic conduit marks the center of the volcano.

2. Many volcanic necks, like the one to the right, form as erosion wears down a volcano (◀), exposing the harder, more resistant rocks that solidified inside the magmatic conduit of the volcano.

06.12.b3 Mount Taylor, NM

3. Shiprock is a famous volcanic neck that rises above the landscape of New Mexico (▼). It consists of fragmented mafic rocks (breccia), and it connects to dikes (not shown) that radiate out from the conduit.

06.12.b4

06.12.b5

4. Some volcanic necks, including Shiprock, were not originally *inside* a volcano but instead were magmatic conduits that formed well *beneath* the volcano. The volcano above Shiprock was not a mountain, but a crater (pit) excavated by a violent explosion (◀). The explosion occurred when magma ascending up a conduit encountered groundwater and generated huge amounts of steam that expanded violently, causing an explosion. After the volcanic eruption, erosion removed the crater and hundreds of meters of rock that once overlay the area around the conduit.

06.12.b6 Shiprock, NM

Columnar Joints

Many igneous rock bodies display distinctive fracture-bounded columns of rocks, like the ones shown here (▶). These fractures, known as *columnar joints,* form when a hot but solid igneous rock contracts as it cools. The fractures carve out columns that commonly have five or six sides. Columnar joints are common in basaltic lava flows, felsic ash flows, sills, dikes, and some laccoliths. In a tabular unit, like a flow, sill, or dike, columnar joints tend to be perpendicular to the tabular unit—they are vertical in a horizontal lava flow, ash flow, or sill, but horizontal in a vertical dike.

06.12.t1 Columbia River Gorge, OR

Before You Leave This Page

- ✓ Sketch the difference between a dike and a sill, and how each relates to layers in the host rocks.

- ✓ Sketch or discuss the geometry of a laccolith.

- ✓ Sketch and explain two ways that a volcanic neck can form.

- ✓ Describe how columnar joints form.

6.12

6.13 What Areas Have the Highest Potential for Volcanic Hazards?

IN SOME PLACES, THE RISK POSED by volcanic hazards is great. In others, it is inconsequential. Volcanic eruptions are more likely in Indonesia than in Nebraska. Additionally, different types of volcanoes have different eruptive styles, so some volcanoes are more dangerous than others. What factors should we consider when determining which areas are the most dangerous and which are the safest?

A How Do We Assess the Danger Posed by a Volcano?

Potential hazards of a volcano depend on the type of volcano, which we can infer from its shape and rock types, and on its history. Examine the volcano below for clues about what type of volcano is present and how it might erupt. If your only piece of data was this photograph, how much could you infer about the volcano's behavior and hazards?

06.13.a1 Augustine volcano, AK

1. *Shape*—The shape of a volcano provides important clues about how dangerous the volcano might be. Volcanoes with steep slopes, such as composite volcanoes, are more dangerous because they form from potentially explosive, viscous magma. Also, steep volcanoes are prone to landslides. Volcanoes that have relatively gentle slopes, like most shield volcanoes, result from less explosive basaltic eruptions and are less likely to have landslides.

2. *Rock Type*—The types of rocks on a volcano reflect the magma composition and style of eruption. If a volcano contains welded tuffs, it has erupted felsic pyroclastic flows. If it consists of rhyolite or andesite, it is more dangerous than a volcano composed of basalt. We can use chemical analyses to help classify the rocks and thereby assess a volcano's potential danger. The volcano shown above is a steep, composite volcano composed of andesitic lava and pyroclastic rocks.

3. *Age and History*—The age of a volcano is essential information. If the volcano has not erupted for a long time, maybe it is dormant. The shape of a volcano, especially whether it still has a fresh-looking volcano shape or has been eroded, is one indicator of a volcano's age. Important clues are also provided by a volcano's history if recorded in historical records, including oral histories from nearby people. Isotopic measurements on volcanic units can provide an accurate indication of a volcano's age. Geologic studies of the sequence of volcanic layers, combined with isotopic ages, provide insight into how often eruptions recur and how often certain types of eruptions occur. The volcano above clearly has recent activity, as expressed by the recent dark deposits and the steam and other gases escaping from the summit dome.

B What Areas Around a Volcano Have the Highest Risk?

Once we have determined the type of volcano that is present, we consider other factors that help identify which areas near the volcano have the highest potential risk.

06.13.b1

1. *Proximity*—The biggest factor determining potential risk is proximity, closeness to the volcano. The most hazardous place is inside an active crater. The potential risk decreases with increasing distance away from the volcano.

2. *Valleys*—Lava flows, small pyroclastic flows, and mudflows are channeled into valleys carved into the volcano and surrounding areas. Such valleys are more dangerous than nearby ridges.

3. *Wind Direction*—Volcanic ash and pumice that are thrown from the volcano are carried farthest in the direction that the wind is blowing at the time of the eruption. Most regions have a prevailing wind direction, so a greater hazard of falling material exists in this direction from a volcano.

4. *Particulars*—Each volcano has its own peculiarities, and these influence which part of the volcano is most dangerous. Steeper parts of a volcano pose special risks, and one side of a volcano may contain a dome that could collapse and form pyroclastic flows. This image shows three small villages around a volcano. Is one village at greater risk than the others? Which one is in the least hazardous place, and what ideas led you to this conclusion?

C What Regions Have the Highest Risk for Volcanic Eruptions?

We can think on a broader regional scale about which regions are most dangerous. In North America, volcanoes are relatively common along the west coast and virtually absent east of the Rocky Mountains. *Tectonic setting*, especially proximity to certain types of plate boundaries, is the major factor making some places more prone to volcanic hazards than others. The map below shows locations of recently active volcanoes (red triangles).

1. The largest concentration of composite volcanoes is along the Pacific Ring of Fire. The volcanoes form above subduction zones, either in island arcs or in mountain ranges along active continental margins. Some subduction-zone volcanoes erupt so vigorously that they form calderas.

2. Much fluid basaltic lava erupts on the seafloor at mid-ocean ridges. Such eruptions pose little risk to humans because almost all of these occur at the bottom of the ocean. The island of Iceland, where a mid-ocean ridge coincides with a hot spot, is an exception.

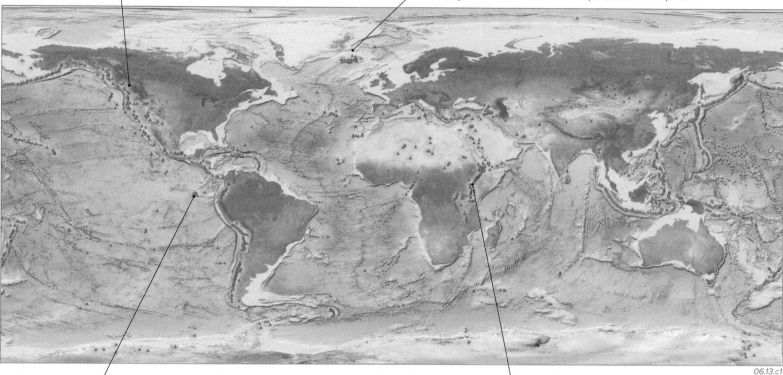

06.13.c1

3. Many shield volcanoes occur along lines of islands and submarine mountains in the Pacific and other oceans. Most of these linear island chains, and a few other clusters of islands, formed above hot spots. Hawaii and the Galápagos Islands are good examples. Shield volcanoes also occur in other settings, including on continents.

4. Some volcanic features, including basalt flows, scoria cones, and composite volcanoes, are in the middle of continents. Most of these form over hot spots or in continental rifts, like the East African Rift.

Forecasts, Policy, and Publicity

Predicting volcanic eruptions is currently an imprecise science, but it has greatly improved over the last several decades. There have been some fabulous successes and some disappointing failures. Volcanologists have successfully predicted some eruptions by studying clusters of small earthquakes generated as magma rises through the crust, by measuring changes in the amount of gas released by volcanoes, and through other types of investigations. Some predictions (e.g., Mount St. Helens and Mount Pinatubo) have saved lives because government officials and citizens acted on the scientific evidence. Some predictions have been unsuccessful because an eruption that was con-

sidered possible or even likely did not occur. In other cases, predictions, policy, and publicity interacted in a bad way, with deadly results. In 1985, geologists working on Nevado del Ruiz, a composite volcano in the Colombian Andes, warned of an impending eruption. The city of Armero, with an estimated 29,000 inhabitants, lay in a valley that drained the steep volcano. Local government officials downplayed the risk and assured the citizens that there was no danger. A pyroclastic eruption occurred at night, melting snow and ice on the volcano and unleashing a mudflow that moved at hundreds of miles per hour, engulfing most of Armero and killing more than 20,000 people.

Before You Leave This Page

☑ Summarize ways to assess the potential danger of a volcano based on its characteristics.

☑ Describe ways to identify which areas around a volcano have the highest potential hazard.

☑ Describe how the plate-tectonic setting of a region influences its potential for volcanic hazards.

6.13

What Volcanic Hazards Are Posed by Mount Rainier?

MOUNT RAINIER IS PART OF A CHAIN OF VOLCANOES above the Cascadia subduction zone of the Pacific Northwest. What kind of volcano is Mount Rainier, how did it form, when and how did it last erupt, and what risks does it pose to the many people living in the valleys below? Mount Rainier provides an opportunity to examine some important aspects of volcanoes, connecting eruption styles, tectonic setting, and potential hazards.

A What Kind of Volcano Is Mount Rainier?

Mount Rainier rises ominously above the city of Tacoma (▶). The steep, symmetrical shape of the mountain identifies it as a dangerous composite volcano. A composite volcano plus a city equals high risk.

06.14.a2

06.14.a1

This image (▲) shows the position of Mount Rainier and the suburbs of Tacoma. The top of the volcano is covered by glacial ice and snow. River valleys, two of which are labeled, begin on the flanks of the volcano and continue into the suburbs. These provide a pathway for mudflows and pyroclastic flows from the volcano to the people.

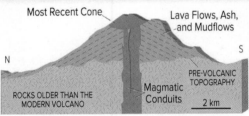
06.14.a3

A geologic cross section of Mount Rainier (▲) shows that the andesitic composite volcano was built on an eroded surface of granitic rocks and was fed by a pipelike magmatic conduit. The top of the mountain, largely covered by ice, is a younger volcanic cone that was constructed within an older crater.

B What Is the Plate-Tectonic Setting of Mount Rainier?

Mount Rainier is one of the volcanoes that cap the Cascade Range. The Cascade volcanoes exist because of melting at depth associated with the Cascadia subduction zone, which is an ocean-continent convergent boundary.

1. The large composite volcanoes of the Cascades are related to a plate boundary between the continental North American plate to the east and the small oceanic Juan de Fuca plate to the west. The Juan de Fuca plate is moving eastward with respect to North America and subducting into the mantle, beneath the edge of the continent.

06.14.b1

2. The Cascade Range is a north-south belt of mountains and is capped by snow-covered composite volcanoes, including the three labeled here. Additional Cascade volcanoes are in Canada to the north and in Oregon and California to the south. Mount Rainier and the other Cascade volcanoes have historically erupted mostly viscous, intermediate (andesitic) magmas that form thick, slow-moving lava flows, domes, and explosive pyroclastic materials.

3. Magma forms by partial melting of mantle above the subduction zone and rises to interact with overlying continental crust. The result is intermediate-composition magma and dangerous composite volcanoes.

C What Hazards Does Mount Rainier Pose to the Surrounding Area?

Mount Rainier is considered to be a very dangerous volcano. It has had at least eleven significant pyroclastic eruptions in the last 10,000 years. The most recent occurred in 1820. According to the U.S. Geological Survey, mudflows from Mount Rainier constitute the greatest volcanic hazard in the Cascade Range.

1. This large figure shows hazard zones for lava flows, pyroclastic flows, and mudflows. The green zone around Mount Rainier has the highest hazard for lavas and pyroclastic flows. The yellow and orange colors show the potential for mudflows of different sizes and different recurrence intervals (how often, on average, they occur). Yellow is used for least frequent but large mudflows, whereas reddish orange is used for more frequent but small mudflows, especially near the volcano.

2. The hazards from lava flows are mostly near the volcano. Small explosive eruptions also have the most impact close to the summit, but pyroclastic flows can travel tens of kilometers away from the summit, in part following valleys. During a major eruption, a large eruption column could spread ash and pumice across the region. Prevailing winds would probably spread the ash to the east but could blow in any direction depending on the weather conditions at the time of the eruption.

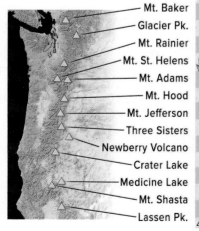

Tacoma

Suburbs of Tacoma

MT RAINIER

06.14.c2

10 km

06.14.c1

▮ Small mudflow event, not necessarily associated with volcanism

▮ Moderate-size mudflow event

▮ Large mudflow event

▮ Lava flows and pyroclastic flows

4. Mudflows could flow northwest all the way into Tacoma and its suburbs. Houses have been built directly in potential mudflow paths and even on top of a huge mudflow that occurred only 600 years ago. Besides eruptions, some of these mudflows were caused by avalanches of rock and ice unrelated to volcanic activity. The risk posed by mudflows is very great! Would you buy one of these houses? How would you know if a house was at risk? Hazard maps like the one shown here are a good place to start.

3. *Mudflows* have formed where eruptions melted the ice cap or where the steep slopes of the volcano produced large landslides. Mudflows form on the volcano and then flow down valleys as thick slurries of mud, pyroclastic material, rocks and almost anything else that gets in the way. The hazards are greatest close to the volcano and in valleys that drain the volcano.

Recent Eruptions of Volcanoes in the Cascade Range

Mount Rainier and Mount St. Helens are only two of the dangerous volcanoes in the Cascades. Ten other volcanoes in the U.S. part of the Cascades erupted during the last 4,000 years. Cascade volcanoes also continue a short distance into Canada.

This figure (▶) shows the locations of the large Cascade volcanoes and when they erupted during the last 4,000 years. Although Mount St. Helens is the most active of the Cascade volcanoes, Glacier Peak, Medicine Lake, and Mount Shasta have each erupted six or more times during the last 4,000 years. Seven of the volcanoes, including Mount Rainier, have erupted during the last 200 years. Nearly all of

these volcanoes are dangerous composite volcanoes, so living on the flanks or in the valleys below one of these volcanoes carries the risk of mudflows, ash falls, and even pyroclastic flows.

All of the volcanoes and the associated hazards exist because the Juan de Fuca plate is subducting beneath the continental crust of North America. Farther south, toward central California, and farther north, just a bit into Canada, the convergent Cascade boundary becomes a transform boundary and the string of Cascade volcanoes ends.

06.14.t1

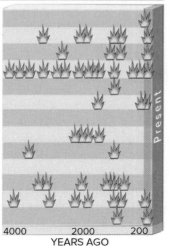

CASCADE ERUPTIONS DURING THE PAST 4,000 YEARS

— Mt. Baker
— Glacier Pk.
— Mt. Rainier
— Mt. St. Helens
— Mt. Adams
— Mt. Hood
— Mt. Jefferson
— Three Sisters
— Newberry Volcano
— Crater Lake
— Medicine Lake
— Mt. Shasta
— Lassen Pk.

Present

4000   2000   200
YEARS AGO

Before You Leave This Page

☑ Describe the type of volcano that Mount Rainier represents.

☑ Explain the plate-tectonic setting of Mount Rainier.

☑ Discuss the volcanic hazards near Mount Rainier.

☑ Briefly summarize how active the Cascade volcanoes have been during the last 4,000 years.

6.14

How Would You Assess Hazards on This Volcano?

DECIDING WHERE TO LIVE requires careful consideration. An overriding factor is whether a place is safe. In this exercise, you will investigate a volcanic island to determine what types of eruptions have occurred in the past, to assess the volcanic hazards, and to find the least dangerous place on the island to live.

Goals of This Exercise:

- Observe the physical characteristics of the volcano and the rock types it contains.
- Use your observations to determine what type of volcano is present and how it would likely erupt.
- Assess the potential for volcanic hazards in different parts of the island and determine the least dangerous place to live.

A Observing the Characteristics of the Volcano

The study of a volcano begins by observing its physical characteristics, such as its size, shape, steepness of slopes, locations of ridges and valleys, and any unusual topographic features. The next step is to observe the types of rocks that are present, determine the aerial distribution of each rock type, and interpret how and in what order each rock formed. Follow the steps below, and record your answers from each step in the accompanying worksheet or online.

1. Observe the image to the right. Record any important characteristics of the volcano on the worksheet or in your notes.

2. Observe the photographs of rock samples and describe each rock's key attributes, such as whether it contains fragments. Use these attributes to identify the rock types (basalt, rhyolite and tuff, for example) by comparing the photographs and descriptions to those in Chapter 3. Alternatively, your instructor may provide hand specimens of the rocks for you to observe and identify.

3. Use your rock identifications to infer the *style of eruption* by which each rock formed.

B Assessing the Volcanic Hazards of the Island

Assess the general volcanic hazards of the volcano, and then assess the relative hazards of each part of the island compared to the others. Using your hazard assessments, determine the most dangerous places and the relatively least dangerous place to live.

1. Consider your rock identifications in the context of the topography and geologic features in the areas where the samples were collected. From these combined data, interpret what types of volcanic features, such as craters and domes, are present in different parts of the island. A newspaper account of previous eruptions (bottom of the next page) provides some useful clues.

2. Assess how each volcanic feature contributes to the hazard potential in different parts of the island. On the map in the worksheet, draw boundaries around and label those areas that have high, medium, or low hazard potential compared to the rest of the island. The differences between the three hazard zones will be fairly subjective. Use your best judgment and be consistent.

3. From your investigations, identify the areas you interpret to be the most dangerous and the least dangerous places to live. When choosing between two sites that are equally safe, you may consider other factors, such as the scenery, whether the sites are level enough to build on safely, and whether they are subject to storms, landslides, floods, and other natural hazards.

06.15.a2

Rock 1: Dark gray, glassy igneous rock (▲) with some bands produced by flow when the rock was molten. A chemical analysis indicates that this rock has a felsic composition.

06.15.a3

Rock 2: Hard, igneous rock (▲) that contains flattened pieces of pumice (light gray) and small crystals of quartz and feldspar. It does not seem to be a recently formed rock.

06.15.a4

Rock 3: Unit consisting of angular pieces of a light- to dark-gray igneous rock in a matrix of powdery volcanic ash and smaller rock pieces (▲). Many of the dark pieces are glassy and banded, and contain scattered vesicles. The volcanic deposit has baked (heated up) the underlying soil, reddening it.

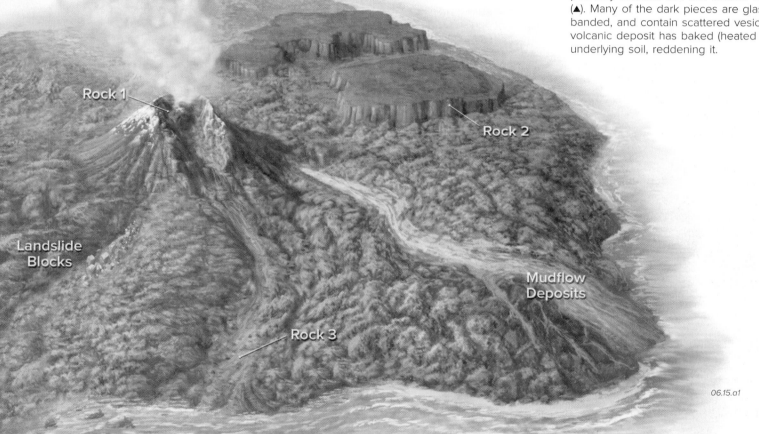

Rock 1

Landslide Blocks

Rock 2

Mudflow Deposits

Rock 3

06.15.a1

The following is a newspaper account:

Volcano Erupts!

The *Juanannita volcano* began erupting in early September of 1952, and dozens of small eruptions have occurred since that time. For 10 years before the 1952 eruption, residents of the area observed plumes of white steam rising from the summit of the crater. In the summer of 1952, local inhabitants reported an increase in the output of steam, an increased smell of sulfur, and a series of small earthquakes.

The first eruption was a single explosive burst that lasted about three hours and that was accompanied by clouds of ash that rose kilometers into the air. Heavy ash fell around the volcano, and a light dusting of ash was reported on adjacent islands up to 20 kilometers away. The eruption melted snow and ice high on the crater, forming a mudflow that moved along stream channels and inundated many areas in valleys downstream from the volcano. After the main eruption, a lava dome started growing in the crater.

All subsequent eruptions have been smaller and of a different style. They have been similar to one another. In each eruption, a cloud of ash and rocks moves rapidly downhill and is mostly restricted to stream channels. After each eruption, geologists noted that one side of the dome in the crater had collapsed into a pile of ash and rocks.

Deformation and Earthquakes

TECTONIC FORCES DEFORM THE LITHOSPHERE, causing movement of materials in the subsurface and often displacing the ground surface, such as during an earthquake. Earthquakes are among Earth's deadliest natural phenomena, having the ability to topple buildings, liquefy normally solid ground, cause landslides, and unleash massive ocean waves that wipe out coastal cities. One earthquake can kill thousands—or hundreds of thousands—of people. What causes earthquakes, where are they most common, and how do we determine where an earthquake occurred? In this chapter, we explore various manifestations of deformation, including earthquakes.

The world's strongest earthquake in 40 years struck Indonesia on December 26, 2004. The magnitude 9.1 earthquake occurred west of Sumatra and was caused by movement on a fault, shown by the red line on this map. The red line shows the length of the fault that ruptured during the earthquake. Movement on a fault is one expression of deformation, but others include the formation of folds, mountain ranges, and many valleys. Yellow dots adjacent to the red line on the map show the locations of smaller, related earthquakes, outlining smaller faults that slipped before, during, or after the main earthquake.

What causes earthquakes, faulting, and other types of deformation?

The earthquake occurred beneath the ocean, where the Indian-Australian plate is being subducted beneath the Eurasian plate. Sudden faulting abruptly uplifted the Eurasian plate, pushing up a large region of seafloor and displacing overlying seawater. Movement on the fault propagated laterally for several minutes. Uplift of the seafloor caused a massive wave, called a *tsunami,* that spread across the Indian Ocean as a low wave, traveling at speeds approaching 800 km/hr (500 mi/hr)! The curved lines show a model of the wave's position by hour (numbers in small white circles).

What happens when an earthquake occurs under the sea, rather than on land?

The tsunami increased in height as it approached the coasts of Indonesia, Thailand, Sri Lanka, India, east Africa, and various islands. Low coastal areas were inundated by as much as 20 to 30 m of water (65 to 100 ft) in Indonesia and 12 m (40 ft) in Sri Lanka. Cities and villages were completely demolished along hundreds of kilometers of coastline, leaving more than 220,000 people dead or missing. The numbers below show casualties by location.

How does a tsunami form, how does it move through the sea, and what determines how destructive it is?

07.00.a1

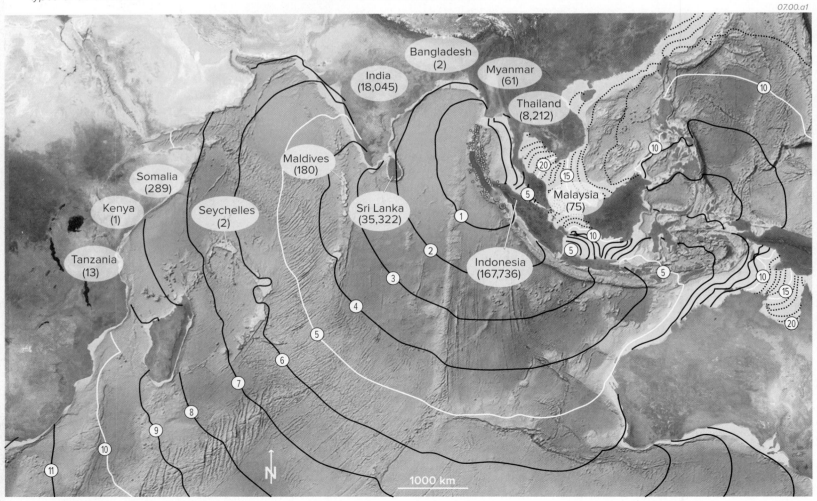

TOPICS IN THIS CHAPTER

07.00.a2

The tsunami caused damage to low-lying coastlines around the Indian Ocean, reaching as far away as the eastern coast of Africa. The destructive power of the tsunami is clear from this photograph of Banda Aceh, the regional capitol of Sumatra's northernmost province. This city of 320,000 people was reduced to rubble, and nearly a third of its inhabitants were killed or reported missing.

What parts of the world are most likely to have a tsunami, and are certain types of earthquakes more likely to cause a tsunami than other types of earthquakes?

The aerial photograph below, taken from a low-flying plane, shows an area near Banda Aceh after the tsunami. Nearly all buildings and vegetation were stripped bare by the water's rush onto the land and the subsequent retreat back to the sea. The hills in the background of the photograph escaped damage and retained their forest, because they were either high enough above sea level or far enough from the shoreline to escape the rising waters. As this photograph indicates, low coastal areas are more susceptible to tsunamis than areas that are higher or farther inland.

What factors affect a coastal area's risk for a tsunami?

07.00.a3

2004 Sumatran Earthquake and Indian Ocean Tsunami

The 2004 Sumatran earthquake struck on the morning of December 26, violently shaking the region and triggering the massive Indian Ocean tsunami. It ranks as one of the largest earthquakes ever recorded. The magnitude of the earthquake is variably estimated at 9.0 to 9.3, depending on how the calculations are done. Large aftershocks followed the main quake, including one with a surprisingly large magnitude of 8.7. The earthquake resulted from slip along a fault—a fracture along which rocks and other materials slip past one another. From the seismic records of the main quake and aftershocks, it is estimated that a fault surface 1,220 km (760 mi) in length slipped by as much as 10 m during the earthquake. The earthquake lasted over 8 minutes, an unusually long duration.

The earthquake started at a depth of 30 km (19 mi) and ruptured all the way up the fault to the seafloor. It lifted a large section of seafloor several meters in height, displacing tens of cubic kilometers of seawater. The displaced water spread out in all directions, forming a wave called a *tsunami*. The tsunami rose to heights of more than 30 m (100 ft) when it came ashore, and in many places it washed inland for more than a kilometer.

As a result of the earthquake, parts of the Andaman Islands northwest of Sumatra were changed forever. Coral reefs, which had been undersea, were lifted above sea level. A lighthouse that was originally on land is now surrounded by seawater one meter deep. The changes to the land seem insignificant compared to the massive loss of life in this event, one of the deadliest natural disasters in world history.

7.0

What Is Deformation and How Is It Expressed in Landscapes?

ROCKS SEEM LIKE STRONG MATERIALS, but they can deform in many settings and under many conditions. Rock can be subjected to forces resulting from burial, tectonic activity, heating or cooling, and other processes. If the forces result in enough stress, they cause a rock to deform—move, rotate, change shape, or some combination of these. What kinds of stress affect rocks, what types of deformation can result from this stress, and how is deformation expressed in landscapes?

A What Are Force and Stress?

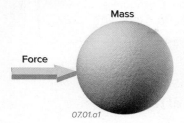

Mass

Force

07.01.a1

Force is a push or a pull that causes, or tends to cause, change in the motion of a body (◄). It is commonly expressed as the amount of acceleration experienced by a mass.

07.01.a2

The amount of force divided by the area upon which the force is applied is called the *stress*. The force from a metal weight (◄) is distributed evenly across the top of a broad, wooden pillar.

07.01.a3

If the same amount of weight is on a much thinner pillar (◄), the stress (force per unit area) on the pillar is greater. It might cause the pillar to splinter or break.

B How Do Materials Respond to Differential Stress?

Rocks, sediment, and other materials may be subject to three types of differential stress: *compression*, *tension*, or *shear*. The way in which rocks respond to these stresses varies as a function of depth, because rock strength changes with depth as pressure and temperature increase downward in the crust. The examples below show how rocks respond to different types of stress under shallow levels of the crust, where materials are brittle—that is, they fracture.

Type of Stress

Compression

07.01.b1

1. When stress pushes in on rock, the stress is called *compression*, shown by the inward-directed arrows above.

Tension

07.01.b3

3. When stress is directed outward, pulling the rock, the stress is called *tension*. Tension is shown with stress arrows pointing away from the rock.

Shear

07.01.b5

5. A third type of stress acts to *shear* the rock as if stresses on the edges of a block were applied in opposite directions.

Type of Structure Formed

07.01.b2

2. Compression in shallow levels of the crust can cause rocks to fracture and slip past one another, like pushing one block over another. In this figure, the rocks are responding to compression by trying to get out of the way. A fracture along which two rock masses slip past each other is a *fault*.

07.01.b4

4. Tension can form fractures that help the rock stretch as it is pulled apart. Fluids, if present, can deposit minerals in the fracture, forming a mineral-filled *vein*. Tension may also cause slip along fractures, pulling one block past another (like in the figure to the left for compression, but with the blocks moving in the opposite direction).

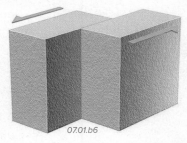

07.01.b6

6. Shearing in shallow parts of the crust usually forms a fault. Depending on the direction of shearing, one block can move up, down, or sideways relative to the block on the other side of the fault. In the case shown here, the two blocks are moving horizontally (sideways) relative to one another.

C How Do Materials Respond to Force and Stress?

Materials within Earth are subjected to forces from the weight of overlying rocks, from tectonic forces pushing or pulling on the rocks, from cooling and heating, and from pressurized fluids, such as water and magma. Just like the wooden pillar on the previous page, if a force is concentrated (i.e., high stress), a rock can break or otherwise deform. As a result, we normally talk about stress instead of force. These figures show stress with a blue arrow.

07.01.c1

1. A block of material may remain unchanged if subjected to only a small amount of stress (►). If the imposed stresses are greater, three things can happen. The material may be *displaced* from one place to another, it may be *rotated*, or it may have its shape *modified*, or *strained*. All three responses may occur at the same time.

Displacement

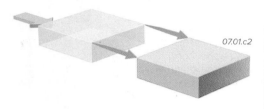
07.01.c2

2. In response to stress, a material may be moved, or displaced, from one place to another. During *displacement*, a material can behave as a rigid object or change shape as it moves. In the photograph below, a thin sheet of light-colored granite has been displaced by movement along fractures (▼).

07.01.c3 Tortolita Mtns., AZ

Rotation

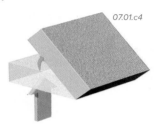
07.01.c4

3. A material may be rotated in response to stresses. Rotation can be expressed by tilting, folding, or a partial horizontal spin of the material. The sedimentary layers in the photograph below (▼) were deposited as horizontal layers, but the layers have since been tilted (rotated) so that they are inclined to the right in this view.

07.01.c5 El Dorado Canyon, CO

Strain

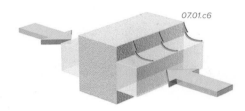
07.01.c6

4. A material can respond to stress by deforming internally—changing size or shape by ductile or brittle deformation. A change of size or shape is called *strain*. Stress is the cause, and strain is the effect. Below, originally rounded pebbles in a conglomerate were strained in response to stress squeezing the rock under moderately high temperatures (▼).

07.01.c7 Granite Wash Mtns., AZ

Strength of Continental Crust at Depth

5. The strength of continental crust varies as a function of depth because temperature and pressure both increase downward (▼). At shallow levels of the crust, rocks deform by fracturing and other types of *brittle deformation*. Rocks in the upper crust become stronger with depth because increasing confining pressure acts to hold rocks together and makes slip along any fractures more difficult.

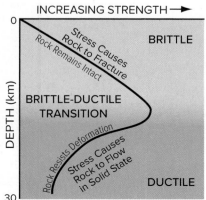
07.01.c8

6. Deeper, where pressure and temperature are greater, rocks may deform by flowing as a solid but soft material—*ductile deformation*. There is a gradational transition between the *upper brittle* and *lower ductile* parts of the crust. This typically occurs at a depth of approximately 15 km and temperatures of more than 300°C. At greater depths, the effects of temperature dominate over the effects of pressure, and rocks become progressively weaker, beginning to flow as a weak solid material. The strength of the crust decreases rapidly downward.

Before You Leave This Page

✓ Describe or illustrate the concept of stress.

✓ Sketch and explain the three types of stress, providing examples of the types of structures each forms.

✓ Sketch or describe the three ways that a material can respond to stress.

✓ Sketch and explain how crustal strength varies with depth.

7.1

7.2 How Are Fractures Expressed in Landscapes?

FRACTURES ARE THE MOST COMMON geologic structures. They range from countless small cracks visible in an outcrop (an exposure of rock) to huge faults hundreds or thousands of kilometers long. Movement along fractures can uplift an entire mountain range or downdrop an area, forming a valley. What are the different types of fractures, how do different types of fractures form, and how are fractures expressed in landscapes?

A In What Different Ways Do Rocks Fracture?

There are two main types of fractures: *joints* and *faults*. Joints and faults both result from stress but have different kinds and amounts of movement across the fracture.

Joints

07.02.a1

1. Most fractures form as simple cracks representing places where the rock has pulled apart by a small amount (◄). These cracks are called *joints* and are the most common type of fracture. They form under tension, where the rock is pulled apart.

2. This exposure of massive granite (►) is cut by a series of near-vertical joints. The granite is not offset up and down or side to side by the joints, but is simply pulled apart by a very small amount.

07.02.a2 Joshua Tree NP, CA

Faults

3. A *fault* is a fracture where rocks have slipped past one another (►). Rocks across a fault can slip up and down, as shown here, or they can slip sideways or at some other angle. A fault displaces the rocks on one side relative to the other side.

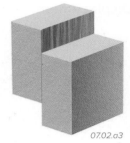
07.02.a3

4. The long fault in the center of this photograph (◄) cuts across and offsets the rock layers that are gently inclined toward the right. The layers across the fault have been displaced relative to each other.

07.02.a4 Moab, UT

B What Stresses Form Joints?

07.02.b1

The stresses that form joints arise from many sources, but they are mostly due to burial and tectonic forces. Tectonic forces may push, pull, or shear the rock. These volcanic rocks (▼) are cut by vertical joints formed by tectonic stresses.

07.02.b3

Stresses build up as rocks get warmer or cooler. As some igneous rocks cool, they contract into polygon-shaped columns bounded by joints that commonly meet at 120° angles. The photograph below (▼) shows an example of such *columnar joints*.

07.02.b5

Stresses also arise during uplift of buried rocks, causing rocks to fracture due to reduced pressure. These joints, called *unloading joints*, form parallel to the surface and slice off thin sheets of rock (▼).

07.02.b2 South Fork, CO

07.02.b4 Iceland

07.02.b6 Estes Park, CO

C   What Are the Main Types of Faults?

We classify faults based on the motion of one block relative to the other. There are three main types of faults: *normal faults,* *reverse faults,* and *strike-slip faults*. Black arrows show relative movement, and blue arrows show stress. When a fault is inclined (not vertical), we call the block above the fault the *hanging wall* and the block below the fault the *footwall*.

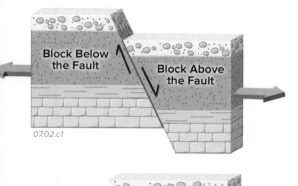

07.02.c1

Normal Fault—In a fault, if the block above the fault (the hanging wall) moves down relative to the block below the fault (the footwall), the fault is a *normal fault* (◄). A normal fault forms when the rock units are pulled apart and lengthened, as for example by tension. The direction in which a fault (or layer) is inclined is called its *dip*. This normal fault dips to the right.

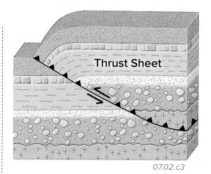

07.02.c3

Thrust Fault—A reverse fault that has a gentle dip (◄) is a *thrust fault*. The rock above the fault is called a *thrust sheet* and is pushed up and over the block below the fault. On maps and other figures, teeth indicate that a fault is a thrust.

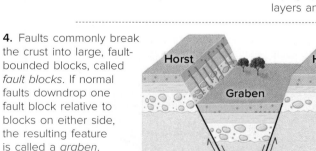

07.02.c2

Reverse Fault—If the block above the fault moves *up* relative to the block below the fault (◄), the fault is a *reverse fault*. A reverse fault forms as a result of *horizontal compression* and shortens the rock units in a horizontal direction. This reverse fault dips to the right.

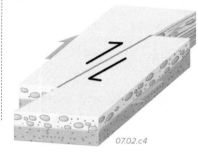

07.02.c4

Strike-Slip Fault—When rocks along a fault move with a side-to-side motion parallel to the fault surface, the fault is a *strike-slip fault*. Relative motion is horizontal, offsetting the blocks laterally in one direction or the other. Note that it does not matter which side you start on to determine which way the opposite block moved.

D   How Are Faults Expressed in Landscapes?

07.02.d1 Borah Peak, ID

07.02.d2 Echo Cliffs, AZ

07.02.d3 Buckskin Mtns., AZ

Hanging Wall

Footwall

1. When fault movement offsets Earth's surface, as during an earthquake, it can cause a step in the landscape, called a *fault scarp* (▲). The fault scarp shown here uplifed the mountain range relative to the valley.

2. We commonly recognize faults because of *offsets* or abrupt *terminations* of layers. Also, rocks along faults are highly fractured and easily eroded, so they erode into linear topographic notches (▲). This fault truncates layers and forms a linear notch.

3. Up close, faults have other distinctive characteristics (▲). They can fold the truncated layers, juxtapose two very different rocks against one another, or display shattered and scratched rocks along the fault zone. This fault has all these diagnostic features.

4. Faults commonly break the crust into large, fault-bounded blocks, called *fault blocks*. If normal faults downdrop one fault block relative to blocks on either side, the resulting feature is called a *graben*. Graben can refer to the downdropped block or to the resulting valley.

Horst

Graben

Horst

07.02.d4

5. A block that is uplifted relative to blocks on either side is called a *horst*. Parts of the American Southwest, including the aptly named *Basin and Range Province*, have fault-block horsts (mountains) and grabens (valleys).

Before You Leave This Page

✓ Sketch and describe the formation and landscape expression of joints and faults.

✓ Sketch and describe three types of faults, showing the relative displacement and the type of stress.

✓ Sketch a horst and a graben.

7.2

How Are Folds Expressed in Landscapes?

DEFORMATION CAN FOLD ROCK LAYERS on such a grand scale that a single fold occupies an entire mountain range. Other folds are so small and elegant that we carry them home in our pockets. Folds can form at depth and at the surface. The folded layers may be bent in gentle arcs or squeezed tightly into sharp angles. We classify and name folds based on their shape, in order to have a convenient way to describe what we observe in landscapes and in smaller exposures of rocks. For many people, folds are the most interesting features in landscapes.

A What Is a Fold and What Are the Main Types of Folds?

07.03.a1

1. Before folding (◄), most rock layers are horizontal because most sedimentary and volcanic layers form with a more or less horizontal orientation.

2. Compression causes shortening, often accommodated by folding of the layers. When you scrunch up a rug, the folds (creases) are perpendicular to the direction of shortening, as shown here for rocks (▶). Compression can form folds, reverse faults, or both. It can occur in any tectonic setting but is most common along convergent plate-tectonic boundaries.

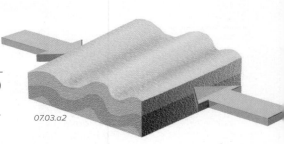

07.03.a2

Anticlines and Synclines

07.03.a3 Tibet

3. If rock layers warp up in the shape of an *A*, the fold is generally called an *anticline* (◄). In an anticline, the *oldest* rocks are in the center of the fold. Within a fold, the place where the layers are mostly tightly bent or reverse their direction of tilt is the fold *hinge*. The hinge in this fold is along the top of the hill.

07.03.a4 Tibet

4. In the opposite case, where rock layers fold down (◄) in the shape of a *V* or *U*, the fold is generally called a *syncline*. In a syncline, the *youngest* rocks are in the center of the fold. The hinge of this fold is in the center, where the layers are most tightly folded and change direction. The tilted parts of the fold on either side of the hinge are called the *limbs* of the fold.

0.25 km

07.03.a5

5. Synclines and anticlines occur together, usually as part of a series of folds. The upward fold of the layers on the left of diagram is an anticline. For now, note how it has been eroded. More on this later in this chapter.

6. The downward fold of layers on the right side of the diagram is a syncline. The beds that dip to the right in the center of the diagram are part of both folds. More specifically, they are a limb of both folds. Note also the landscape formed by the syncline.

Dome

7. Layers that are uplifted in a circular or elliptical area and dip away in all directions form a structural *dome*. Erosion exposes deeper and older rocks in the center of this dome.

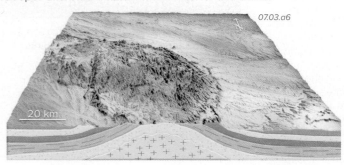

07.03.a6

20 km

Basin

8. A *basin*, formed by folding, is the opposite of a dome. Layers dip toward the center of the basin from all directions. The center of a basin usually contains younger layers than surrounding areas.

07.03.a7

10 km

Monocline

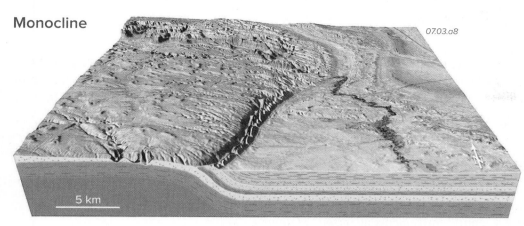

07.03.a8

07.03.a9 Grand Junction, CO

9. In some folds, nearly flat layers bend down (dip) in one direction and then flatten out again. This type of fold is a *monocline,* a name that indicates that the fold has only one dipping segment (limb). Monoclines can be tens of kilometers long or exposed in a small outcrop. Some monoclines have great names, such as the Coxcomb fold, Waterpocket fold, and Comb Ridge in Utah and Arizona. The monocline shown here is along the San Rafael Swell in central Utah.

10. This photograph shows a medium-sized monocline in sandstone layers. Horizontal layers on the left bend down in the center of the image (dip to the right) and then fold back to horizontal on the right side of the image.

B How Do Tilted and Folded Layers Erode?

07.03.b1 Raplee Anticline, San Juan River, UT

07.03.b2 Sideling Hill, MD

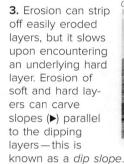

07.03.b3 Moab, UT

1. Some mountains, including this one (▲), are large folds, with the shape of the mountain mimicking the shape of the fold. In some cases, the folding and formation of the mountain are occurring today, but generally folding occurred sometime in the past, followed by erosion that later uncovered an existing fold, as in this case.

2. Commonly, erosion does just the opposite—it erodes the uplifted layers first, leaving behind a ridge of down-folded layers. In other words, the ridges contain synclines whereas the valleys were anticlines. This is very common in the Appalachian Mountains of Maryland (▲), Pennsylvania, and elsewhere.

3. Erosion can strip off easily eroded layers, but it slows upon encountering an underlying hard layer. Erosion of soft and hard layers can carve slopes (▶) parallel to the dipping layers—this is known as a *dip slope*.

4. A dip slope can follow planar dipping layers or gently curving ones. If a ridge has a dip slope on one side, it is called a *cuesta* or a *hogback* (▶), like this one north of Denver.

07.03.b4 Lyons, CO

5. The Appalachian Mountains of the eastern U.S. are famous for their large folds and the way they were eroded. This map (▼) has a satellite image superimposed on topography for the area near the Susquehanna River in southeastern Pennsylvania. The image includes curving mountains and ridges (green) alternating with lowlands (pinkish brown). The river cuts downhill across the ridges and lowlands.

6. Some ridges and valleys are straight, but others curve across this distinctive region, known as the *Valley and Ridge Province.* The ridges and valleys represent the eroded hinges and limbs of folds. Can you identify examples of each? The hinges are curves and the limbs are more straight. Some anticlines are ridges and others are valleys; some synclines are valleys but others are ridges.

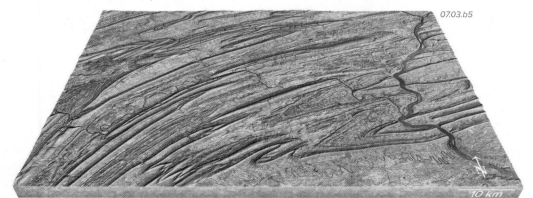

07.03.b5

Before You Leave This Page

☑ Sketch a cross section of an anticline, syncline, and monocline, labeling the hinges and limbs.

☑ Sketch or describe a dome and a basin.

☑ Summarize or sketch the ways that folds are expressed in the landscape.

7.3

7.4 What Is an Earthquake?

ONE DRAMATIC EXPRESSION OF DEFORMATION is an earthquake, which occurs when energy stored in rocks is suddenly released. Most earthquakes are produced when stress builds up along a fault over a long time, eventually causing the fault to slip. Similar kinds of energy are released by volcanic eruptions, explosions, and even filling of human-constructed reservoirs or pumping pressurized fluids down drill holes.

A How Do We Describe an Earthquake?

When an earthquake occurs, it releases mechanical energy, some of which is transmitted through rocks as vibrations called *seismic waves*. These waves spread out from the site of the disturbance and travel through the interior or along the surface of Earth. Scientists record the waves using scientific instruments at *seismic stations*.

The place where the earthquake is generated is called the *hypocenter* or *focus*. Most earthquakes occur at depths of less than 100 km (60 mi), some occur as shallow as several kilometers, and some subduction-zone earthquakes occur as deep as 700 km (430 mi). The *epicenter* is the point on Earth's surface directly *above* where the earthquake occurs (directly above the hypocenter). If the seismic event happens on the surface, such as during a human-caused surface explosion, then the epicenter and hypocenter are the same.

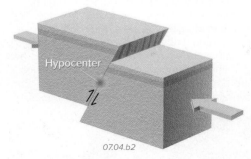

07.04.a1

Seismic waves, once generated, spread in all directions. The curved bands show the peaks of waves radiating from the hypocenter. The intensity and duration of waves are measured by seismic stations (locations 1 and 2). A seismic station closer to the hypocenter (station 1) will detect the waves sooner than those farther away (station 2).

B What Causes Most Earthquakes?

Most earthquakes are generated by movement along faults. When rocks on opposite sides of a fault slip past one another abruptly, the movement generates seismic waves, while materials near the fault are pushed, pulled, and sheared. Slip along any type of fault can generate an earthquake.

Normal Faults

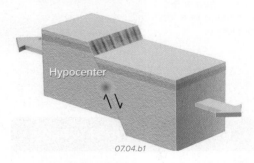

07.04.b1

In a *normal fault*, the rocks above the fault (the hanging wall) move down with respect to rocks below the fault (the footwall). The crust is stretched horizontally, so earthquakes related to normal faults are most common along divergent plate boundaries, such as oceanic spreading centers, and in continental rifts.

Reverse and Thrust Faults

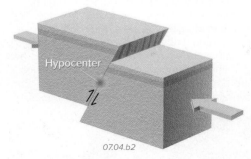

07.04.b2

Many large earthquakes are generated along *reverse faults*, especially the gently dipping variety called *thrust faults*. In thrust and reverse faults, the hanging wall moves up with respect to the footwall. Such faults are formed by compressional forces, like those associated with subduction zones and continental collisions.

Strike-Slip Faults

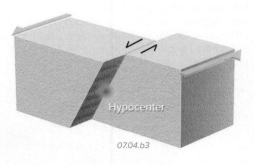

07.04.b3

In *strike-slip faults,* the two sides of the fault slip horizontally past each other. This can generate large earthquakes. Most strike-slip faults are near vertical, but some have moderate dips. The largest strike-slip faults are transform plate boundaries, like the San Andreas fault in California.

C How Do Volcanoes and Magma Cause Earthquakes?

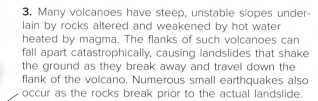

1. Volcanoes generate seismic waves and cause the ground to shake through several processes. An explosive volcanic eruption causes compression, transmitting energy as seismic waves (shown here with yellow lines).

2. Volcanism can be accompanied by faulting and associated earthquakes. Volcanoes add tremendous weight to the crust, and this loading can lead to faulting and earthquakes. The fault shown here caused an earthquake at depth, downdropping the volcano relative to its surroundings.

3. Many volcanoes have steep, unstable slopes underlain by rocks altered and weakened by hot water heated by magma. The flanks of such volcanoes can fall apart catastrophically, causing landslides that shake the ground as they break away and travel down the flank of the volcano. Numerous small earthquakes also occur as the rocks break prior to the actual landslide.

4. As magma moves beneath a volcano, it can push rocks out of the way, causing earthquakes. Magma can push rocks sideways or open space by fracturing adjacent rocks and uplifting the earth's surface. The emplacement of magma can cause a series of small and distinctive earthquakes, called *volcanic tremors*. All types of magma-related earthquakes are closely monitored by geologists and seismologists (scientists who study earthquakes) because they can signal an impending eruption.

07.04.c1

D What Are Some Other Causes of Seismic Waves?

Landslides

Catastrophic landslides, whether on land or beneath water, cause ground shaking. On the Big Island of Hawaii, lava flows form new crust that can become unstable and suddenly collapse into the ocean. Seismometers at the nearby Hawaii Volcanoes National Park often record seismic waves caused by such landslides and by fractures opening up on land in response to the sliding of the land toward the sea.

07.04.d1

Explosions

Mine blasts and nuclear explosions compress Earth's surface, producing seismic waves measurable by distant seismic instruments. Monitoring compliance with nuclear test-ban treaties is done in part using a worldwide array of seismic instruments. These instruments recorded detonation of a nuclear bomb. Seismic waves generated by a blast, such as the one shown on the top graph, are more abrupt than those caused by a natural earthquake.

07.04.d2

Earthquakes Caused by Humans

Humans can cause earthquakes in several ways. Reservoirs built to store water fill rapidly and load the crust, which responds by flexing and faulting. After Lake Mead behind Hoover Dam in Nevada and Arizona was filled, hundreds of moderate earthquakes occurred under the reservoir between 1934 and 1944. Similarly, very shallow (less than 3 km deep) earthquakes occur near Monticello Reservoir in South Carolina. Worldwide, scientists have identified dozens of cases of earthquakes associated with dams. Most of the seismic activity occurs during the initial filling of a reservoir by water, which adds additional stress to underlying rocks.

Humans have also caused earthquakes by injecting wastewater underground into a deep well (drill hole) at the Rocky Mountain Arsenal northwest of Denver. This caused more than a thousand small earthquakes and two magnitude 5 earthquakes, which caused minor damage nearby. When the waste injection stopped and some waste was pumped back out of the ground, the number of earthquakes decreased. Similar small earthquakes are interpreted to have been caused by the disposal of wastewater associated with the process of *hydraulic fracturing* ("fracking"), where drilling fluids are injected in drill holes to increase fluid pressure in order to open fractures and other openings enough to allow the extraction of oil and gas from shale and other rocks.

Before You Leave This Page

☑ Explain what a hypocenter and epicenter each represent.

☑ Sketch and describe the types of faults that cause earthquakes.

☑ Describe how earthquakes and seismic waves are caused by volcanoes, landslides, and humans.

7.4

7.5 How Does Faulting Cause Earthquakes?

MOST EARTHQUAKES OCCUR because of movement along faults. Faults slip because the stress applied to them exceeds the ability of the rock to withstand the stress. Rocks respond to the stress in one of two ways—they either flex and bend, or they break and slip. Breaking and slipping causes earthquakes.

A What Processes Precede and Follow Faulting?

Before faulting, rocks change shape (i.e., they *strain*) slightly as they are squeezed, pulled, and sheared. Once stress builds up to a certain level, slippage along a fault generally happens in a sudden, discrete jump. Faulting reduces the stress on the rocks, allowing some of the strained rocks to return to their original shapes. This type of response, where rocks return to their original shape after being strained, is called *elastic behavior*.

Pre-Slip and Elastic Strain

1. An active strike-slip fault (▶) has modified the appearance of a landscape for hundreds of thousands of years, causing a linear trough along the fault. Some segments of streams follow the fault. At the time shown here, the strength of the fault is greater than the tectonic forces working to slide the blocks past each other. The rocks strain and flex, but the stresses are not great enough to make the rocks break. Friction along the fault helps keep it from moving. The sizes of the yellow arrows represent the current magnitude of the stress that is building along the fault.

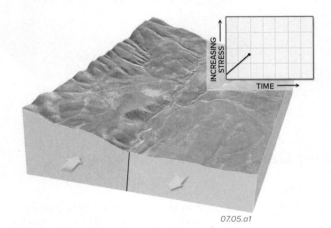

07.05.a1

2. With time, stress increases along the fault as depicted by the upward-sloping line on this graph (◀), which plots stress as a function of time. In response, the rocks may deform *elastically*, changing shape slightly without breaking. The fault might not be obvious at the surface because it is beneath the stream or covered with loose rocks, sand, and soil. One clue that the fault exists is its expression on the landscape, in this case a break in slope along the hillside.

Slip and Earthquake

3. Over time, stress along the fault (represented by the yellow arrows) becomes so great that it exceeds the fault's ability to resist it. As a result, the fault slips (▶) and the rocks on opposite sides of the fault rapidly move past each other. A large earthquake occurs (at the orange dot on the front of the block), generating seismic waves (not shown) that radiate outward from the fault.

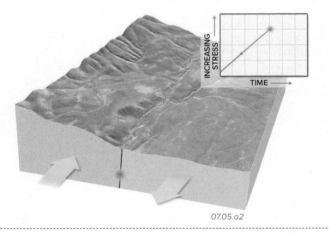

07.05.a2

4. In the stress-versus-time graph (◀), the point at which the earthquake occurred is shown as an orange dot. At this point, the rocks were no longer strong enough and there was not sufficient friction along the fault surface to prevent movement. Much of the built-up stress will be relieved almost instantly as the fault slips.

Post-Slip

5. With the stress partially relieved (▶), the rocks next to the fault relax by elastic processes and largely return to their original, unstrained shape. The movement that has occurred along the fault, however, is permanent. It is not elastic and is recorded by a new break in the topography (a fault scarp). After the earthquake, stress again begins to slowly build up along the fault (as represented by the smaller yellow arrows). The new, subtle break along the straight part of the stream is a clue that something happened here.

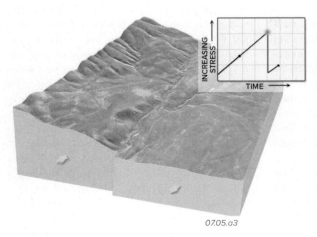

07.05.a3

6. In the stress-versus-time graph (◀), the release of stress after the earthquake is only temporary. The black dot at the end of the line is the current state of stress, and the cycle of stress buildup and release will continue. In this way, the rock strains elastically before the earthquake, ruptures during the earthquake, and mostly returns to its original shape afterwards. However, the blocks on opposite sides of the fault have moved permanently. This sequence is called *stick-slip behavior* because the fault sticks (does not move) and then slips.

B How Do Earthquake Ruptures Grow?

Most earthquakes occur by slip on a preexisting fault, but the entire fault does not begin to slip at once. Instead, the earthquake rupture starts in a small area (the hypocenter) and expands over time.

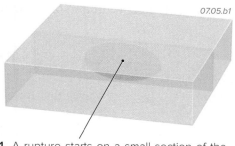
07.05.b1

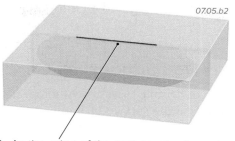
07.05.b2

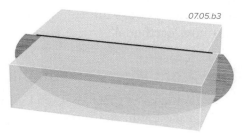
07.05.b3

1. A rupture starts on a small section of the fault below Earth's surface and begins to expand along the preexisting fault plane. Some rocks break adjacent to the fault, but most slip occurs on the actual fault surface, which is weaker than intact rock.

2. As the edge of the rupture migrates outward, it may eventually reach Earth's surface, causing a break called a *fault scarp*. Seen from above, the rupture migrates in both directions, but it may expand farther in one direction than in the other.

3. The rupture continues to grow along the fault plane and the fault scarp lengthens. The faulting relieves some of the stress, and rupturing will stop when the remaining stress can no longer overcome friction along the fault surface. At that point, the earthquake stops.

Earthquake Ruptures in the Field

07.05.b6 Hebgen Lake, MT

07.05.b4 Landers, CA

4. The 1992 Landers earthquake ruptured across the Mojave Desert of California, forming a fault scarp (◄). In this photo, the scarp is cutting through granite. The fault had mostly strike-slip movement, with some vertical movement. The fault is part of a zone of strike-slip faults that are related to the San Andreas transform boundary, but farther east into the continent. The zone is called the *East California Shear Zone*, which is discussed and shown on a map later in this chapter.

5. The 1959 Hebgen Lake earthquake in southern Montana just outside Yellowstone National Park formed a several-meter-high fault scarp (◄), expressed as a step in the land surface. The steep, relatively recently formed scarp has less soil and vegetation than adjacent, less steep parts of the slope. The earthquake and fault scarp were generated by slip along a normal fault.

6. Scarps are most obvious soon after they form, as in the exmaple above, but become more obscure over time. Erosion rounds off the top edge of the scarp, and sediment accumulates at the base of the scarp, producing a rounded step in the topography (◄). With enough time, the scarp develops soil and a covering of vegetation, so it becomes more subtle.

07.05.b7 La Jencia scarp, Socorro, NM

Buildup and Release of Stress

When a fault slips, it relieves some of the stress on the fault, causing the stress levels to suddenly drop. Gradually, the stress rebuilds until it exceeds the strength of the rock or the ability of friction to keep the fault from slipping. The figure below shows a conceptual model of how the amount of stress changes over time.

On this plot, the magnitude of the stress imposed on the fault builds up gradually. When the *amount of stress* equals the *strength of the fault,* the fault slips, and the stress immediately decreases to the original level. In this manner, the amount of stress on a fault forms a zigzag pattern on the graph. It increases gradually (sloping line), and then decreases abruptly (vertical line) when an earthquake occurs. This process is called the *earthquake cycle* and it is one explanation for why a fault commonly produces earthquakes of a similar size. The average time between repeating earthquakes is called the *recurrence interval.*

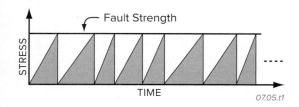

Fault Strength

STRESS

TIME

07.05.t1

Before You Leave This Page

✓ Describe how the buildup of stress can strain and flex rocks, leading to an earthquake.

✓ Describe or sketch how a rupture begins in a small area and grows over time and ruptures Earth's surface.

✓ Describe some characteristics of fault scarps and ruptures.

✓ Sketch and describe how stress changes through time along a fault according to the earthquake-cycle model.

7.5

How Do Earthquake Waves Travel?

EARTHQUAKES GENERATE VIBRATIONS that travel through rocks as *seismic waves*. The word *seismic* comes from the Greek word for earthquake. Geoscientists who study earthquakes are *seismologists*. Geophysical instruments record and process information on seismic waves, and these data allow seismologists and geologists to understand where and how earthquakes occur.

What Kinds of Seismic Waves Do Earthquakes Generate?

Earthquakes generate several different types of seismic waves. Seismologists study *body waves,* which are waves that travel inside Earth, and *surface waves*, which travel on the surface of Earth.

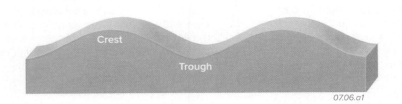

07.06.a1

1. *Shapes of waves* — To describe seismic waves, we begin by defining waves in general. Most waves are a series of repeating crests and troughs (◄). Whether moving through the ocean or through rocks, waves can travel, or *propagate*, for long distances. However, the material within the wave barely moves. Sound waves travel through the air and thin walls, but the wall does not move much. Think of a seismic wave as a pulse of energy moving through a nearly stationary material.

Body Waves

2. Most earthquakes occur at depth, so they first produce seismic waves that travel through the Earth as *body waves*. The waves propagate (move outward) in all directions. There are two main types of body waves, *P-waves* and *S-waves*, which propagate in different ways.

3. One type of body wave is called a *primary wave*, or simply a *P-wave* (►). A P-wave compresses the rock in the same direction as the wave propagates. It is like a sound wave that com-

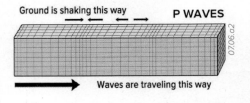

presses the air through which it travels. P-waves can travel through solids and liquids because these materials can be compressed and then released. The P-wave is the fastest seismic wave, traveling through rocks at 6 to 14 km/s depending on the properties of the rock. For comparison, sound waves in air travel at an average of 0.3 km/s; P-waves are more than 20 times faster.

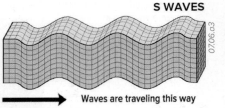

4. *Secondary waves*, also called *S-waves*, shear the rock side to side or up and down (◄). This shearing movement is perpendicular to the direction of travel. The wave shown here propagates to the right, but the material shifts up and down. It could also shift side to side, but the motion would still be perpendicular to the propagation direction of the wave. S-waves are slower (3.5 km/s) than P-waves and cannot travel through liquids, such as magma. When S-waves fail to pass through some part of the Earth, like the outer core, we infer that this region is mostly molten, rather than a solid.

Surface Waves

5. When body waves reach Earth's surface, some energy is transformed into new waves that only travel on the surface (*surface waves*). There are two main kinds of surface waves: *Rayleigh waves* and *Love waves*. Surface waves cause the damage during an earthquake.

6. One type of surface wave is a *Rayleigh wave*, also called a *vertical surface wave* (►) because it displaces the surface in a vertical (up and down) direction. A Rayleigh wave is similar to an ocean wave,

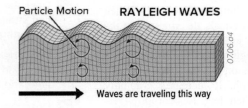

in that material moves up and down in an elliptical path. These earthquake waves propagate in the direction of the large arrow, or perpendicular to the crests of the waves. Rayleigh waves cause most of the damage during an earthquake, in part because they accelerate buildings up and then down, literally shaking and flexing the building apart. During large earthquakes, observers sometimes can see these waves rippling across the surface.

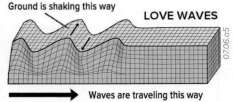

7. The second type of surface wave is a Love wave (◄), which is also described as a *horizontal surface wave* for the way material vibrates horizontally and shuffles side to side. Like an

S-wave, the motion of material in a Love wave is perpendicular to the direction in which the wave travels. This motion, if strong enough, can cause buildings to slide sideways off their foundations. When strengthening buildings for earthquakes, one goal is to more firmly secure the building to its foundation, countering the motion of both types of surface waves. To remember the difference between the two types of surface waves, consider that two *lovely* puppies often walk side by side.

B How Are Seismic Waves Recorded?

Sensitive digital instruments called *seismometers* are able to precisely detect a wide range of earthquakes. The recorded seismic data are uploaded to computers that process signals from hundreds of instruments registering the same earthquake. These computers calculate the location of the hypocenter and the magnitude or strength of the earthquake. From these data, we gain insight about how and where earthquakes occur.

1. A *seismometer* detects and records the ground motion during earthquakes.

2. A large mass is suspended from a wire. It resists motion during earthquakes.

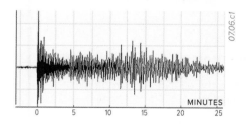

07.06.b1

3. The mass hangs from a frame that in turn is attached to the ground. When the ground shakes, the frame shakes too, but the suspended mass resists moving because of inertia. As the ground and frame move under the mass, a pen attached to the mass marks a roll of slowly rotating recording paper. As a result, the pen draws a line that records the ground movement over time.

4. This device only records ground movement parallel to the red arrows, so it only records a single direction or *single component* of motion.

5. A modern seismic detector, called a *seismograph*, contains three seismometers oriented 90° from each other to record *three components* of motion (north-south, east-west, and up-down). From these three components, seismologists can determine the source and strength of the seismic signal.

6. Seismologists place seismographs away from human noise and vibration, and bury them to reduce wind noise. Seismic waves (in yellow) can come from any direction.

07.06.b2

C How Are Seismic Records Viewed?

1. Prior to the 1990s, seismic waveforms were mostly represented as curves on a paper *seismogram,* which is a graphic plot of the waves recorded by a seismometer. Seismologists developed this plot to better visualize the ground shaking caused by earthquakes. Today, most seismic data are recorded by digital instruments and displayed on computer screens. The seismogram for a single earthquake is shown below.

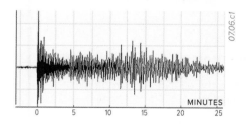

07.06.c1

2. This diagram (seismogram) shows the record of an earthquake as recorded by a seismometer. It plots vibrations versus time. On seismometers, time is marked at regular intervals so that we can determine the time of the arrival of the first P- and S-waves.

3. Background vibrations unrelated to the earthquake commonly look like small, somewhat random squiggles on seismograms.

4. After an earthquake, P-waves arrive first, marked by the larger squiggles. In this example, the earthquake occurred at 8:00 a.m., and the time of the P-wave's arrival was 2.5 minutes later.

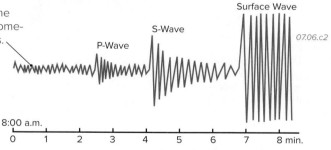

07.06.c2

5. The S-wave arrives later. The delay between the P-wave and the S-wave depends primarily on how far away the earthquake occurred. The longer the distance from the earthquake, the greater the delay.

6. Surface waves arrive last and cause intense ground shaking, as recorded by the higher amplitude squiggles on the seismogram.

Amplitude and Period

Seismic waves are characterized by how much the ground moves (*wave amplitude*) and the time it takes for a single wave to pass (*period*). Period is related to the wavelength and velocity of the wave. Both amplitude and wavelength can be measured from a seismogram. Amplitude is critical when estimating the strength and damage potential of an earthquake. The period can also be a critical component in assessing potential damage, because buildings vibrate when shaken by earthquakes. Every building has a natural period that can match, or *resonate* with, the earthquake wave. Resonance can cause intensified shaking and increased damage.

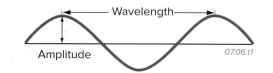

07.06.t1

7.6

7.7 How Do We Determine the Location and Size of an Earthquake?

EARTHQUAKES OCCUR DAILY AROUND THE WORLD, and a network of seismic instruments records these events. Using the combined seismic data from several instruments, seismologists calculate where an earthquake occurred and its magnitude (how large it was). This can be done automatically using computers.

A How Do We Locate Earthquakes?

Seismologists maintain thousands of seismic stations that sense and record ground motions. When an earthquake occurs, parts of this network can record it. Large earthquakes generate seismic waves that can be detected around the world. Smaller earthquakes are detected only locally.

1. Seismometer Network Senses a Quake

Seismometers in the U.S. National Seismic Network (some of which are shown below) represent a fraction of all seismometers.

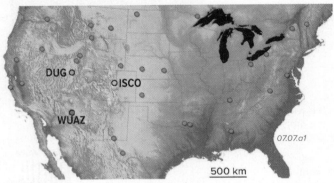

07.07.a1

500 km

On October 1, 2005, a moderate earthquake is felt in Colorado. Three seismic stations (labeled DUG, WUAZ, and ISCO) record wave arrivals and are chosen to locate the epicenter. Seismic stations are given abbreviated names that reflect their locations—for instance, ISCO is the code for Idaho Springs, Colorado.

2. Select Earthquake Records

Records from at least three stations are compared when calculating an earthquake location. Ordinarily, records from many stations are used in an automated, computer-based process.

P-waves travel faster than S-waves, and so reach a seismic station some time before the S-wave arrives. The time interval between arrival of the P-wave and S-wave is called the *P-S interval*. The farther a station is from the earthquake, the longer the P-S interval will be. Identifying the arrival of the P-wave and S-wave on these graphs is not always easy, but it can be done by seismologists or by computer.

The three seismograms show differences in the P-S interval. Based on the P-S intervals, ISCO, which has the shortest P-S interval, is the closest station to the earthquake, followed by DUG and WUAZ.

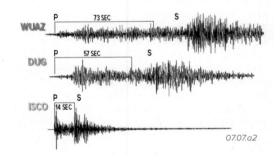

07.07.a2

3. Estimate Station Distance from Epicenter

The P-S interval is proportional to the distance from the epicenter to the seismic station, although slightly affected by the types of materials through which the waves pass. This relationship is shown on a graph as a *time-travel plot*.

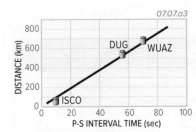

07.07.a3

P-S intervals are measured from the seismograms shown in part 2 and then plotted on the graph. This gives the distance from each station to the earthquake's epicenter.

Station	Distance (km)
WUAZ	670
DUG	540
ISCO	65

From the graph, the distance from each station to the epicenter is now known, but not the direction.

4. Triangulate the Epicenter

The distance from each station to the earthquake can be plotted as a circle on a map to find the epicenter.

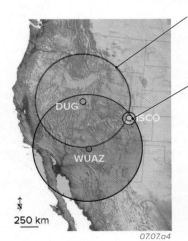

A circle is drawn around each station, with a radius equal to the distance calculated from plotting the P-S interval on the graph.

The intersection of three circles is the epicenter of the earthquake. If more circles were plotted, they should intersect at the same point.

We calculate the depth of the earthquake's hypocenter in a similar way, using the interval between the P-wave and another compressional wave that forms when the P-wave reflects off Earth's surface near the epicenter.

250 km

07.07.a4

B   How Do We Measure the Size of an Earthquake?

The *magnitude* of an earthquake is a measure of the released energy and is used to compare the sizes of earthquakes. There are several ways to calculate magnitude, depending on the earthquake's depth. The most commonly mentioned scale, called the "Richter" or "Local" magnitude (Ml) scale, is illustrated here.

Measuring Amplitude

1. Seismometers are calibrated so that the measurements made by two different instruments are comparable.

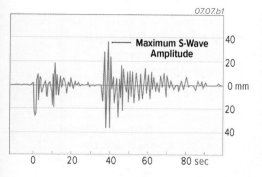
07.07.b1

2. The maximum height (amplitude) of the S-wave is measured on the seismogram. It is proportional to the earthquake energy. This measure is used for shallow earthquakes.

Magnitude

3. The amplitude of S-waves decreases as a wave propagates. We plot the relationship between distance and S-wave amplitude on a graph (▶) called a *nomograph*.

4. For each seismic station, we draw a line connecting the distance and amplitude of the S-wave.

5. The earthquake's magnitude is where each line crosses the center column. These three lines for the 2005 Colorado earthquake all agree, and yield a 4.1 local magnitude (Ml). Magnitude is a logarithmic scale, so a one-unit increase in magnitude represents a tenfold increase in ground motion.

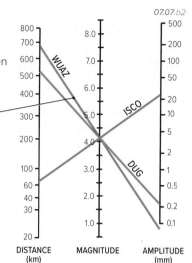
07.07.b2

C   What Can the Intensity of Ground Shaking Tell Us About an Earthquake?

Some of the most damaging earthquakes occurred before seismometers were invented. For such events, we rely on reports of damage and shaking *intensity* as a way to classify the relative sizes of earthquakes.

The *Modified Mercalli Intensity Scale*, abbreviated as *MMI*, describes the effects of shaking in everyday terms. A value of "I" reflects a barely felt earthquake. A value of "XII" indicates complete destruction of buildings, with visible surface waves throwing objects into the air.

A series of very large earthquakes in 1811 and 1812 shook Missouri, Arkansas, Tennessee, and the surrounding areas. Shaking was felt over a wide region. The intensities on this map, numbered from III to XI, indicate what people in different areas felt and saw when the earthquake happened. Some of the intensities are listed to the right of the figure.

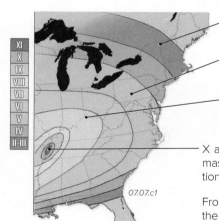
07.07.c1

500 km

III. Felt strongly by persons indoors, especially on upper floors of buildings.

V. Felt by nearly everyone; many awakened. Some dishes and windows broken. Unstable objects overturned.

VI. Felt by all, many frightened. Some heavy furniture moved. Some plaster on walls and ceilings cracks and falls. Damage slight.

X and XI. Some well-built wooden structures destroyed. Most masonry and frame structures destroyed, along with foundations. Bridges destroyed. Rails bent. Damage extensive.

From such maps of intensity, the earthquake is generally near the bulls-eye in the center of the worst damage, but other factors, such as the type of soil, locally influence the intensity.

Energy of Earthquakes

The Richter magnitude describes the amount of ground motion, but the scale is logarithmic. The ground motion increases by a factor of 10 from a magnitude 4 to a 5, from a 5 to a 6, and so on. The amount of energy released increases more than 30 times for each increase in magnitude, so a magnitude 8 releases approximately 30 times more energy than a magnitude 7.

Another common measure of earthquake energy is *moment magnitude,* or *Mw*, which is calculated from the amount of slip (displacement) on the fault and the size of fault area that slipped. Moment magnitude is applicable for both large and small earthquakes and is widely used by seismologists and other geoscientists. How do earthquakes compare to other energy releases with which we are familiar? An average lightning strike (*Mw* ~2) is miniscule compared to a small earthquake. However, an average hurricane is larger than the energy released by the largest historic earthquakes, such as the ones that struck Chile in 1960, Alaska in 1964, Sumatra in 2004, and Japan in 2011.

Before You Leave This Page

☑ Observe different seismic records of an earthquake and tell which one was closer to the epicenter.

☑ Describe how to use arrival times of P- and S-waves to locate an epicenter.

☑ Explain or sketch how we calculate local magnitude.

☑ Explain what a Modified Mercalli intensity rating indicates.

7.8 Where Do Most Earthquakes Occur?

MOST EARTHQUAKES OCCUR ALONG PLATE BOUNDARIES, and maps of earthquake locations outline Earth's main tectonic plates. There are some regions, however, where *seismicity* (earthquake activity) is more widespread, reaching far away from plate boundaries and into the middle of continents. Where do earthquakes occur, and how can we explain the distribution of earthquakes across the planet?

A Where Do Earthquakes Occur in the Eastern Hemisphere?

This map and the one on the next page show the worldwide distribution of earthquake epicenters, colored according to depth. Yellow dots represent shallow earthquakes (0 to 70 km), green dots mark earthquakes with intermediate depths (70 to 300 km), and red dots indicate earthquakes deeper than 300 km. Examine these two maps and observe how earthquakes are distributed. Note how this distribution compares to other features, such as edges of continents, mid-ocean ridges, sites of subduction, and continental collisions. Go ahead, it's an interesting exercise!

07.08.a1

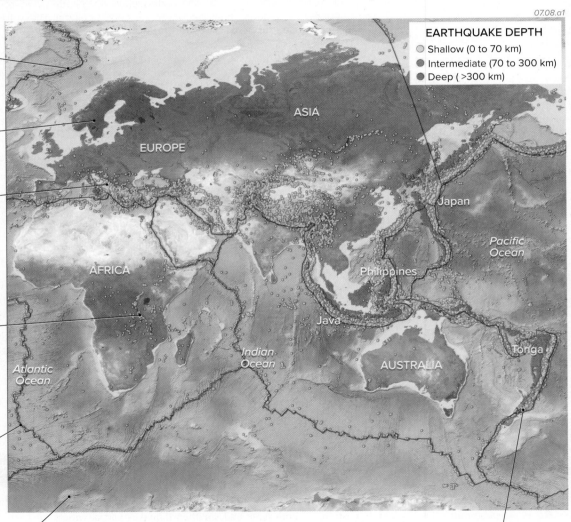

EARTHQUAKE DEPTH
- Shallow (0 to 70 km)
- Intermediate (70 to 300 km)
- Deep (>300 km)

1. Most earthquakes occur in narrow belts that coincide with plate boundaries. The belt of earthquakes north of Iceland marks a divergent plate boundary along a mid-ocean ridge.

2. Between the belts of earthquakes are some large regions with relatively few earthquakes, like the northern part of Europe.

3. A seismically active zone stretches along the southern part of Europe and continues eastward into Asia. This activity follows a series of mostly convergent boundaries, including continental collisions, that are occurring from the Mediterranean Sea to Tibet.

4. A diffuse zone of seismic activity cuts across eastern Africa, following the East African Rift, a region of elevated topography, active volcanism, and faulted blocks. This region is a continental rift within Africa.

5. Mid-ocean ridges, such as the one south of Africa, only have shallow earthquakes (only yellow dots on this map). In these locations, rifting and spreading of two oceanic plates produces faulting and magmatic activity, both of which cause earthquakes.

6. Large regions of the ocean lack significant seismicity because they are not near a plate boundary. Some seismicity beneath the oceans occurs away from plate boundaries and is mostly related to volcanic activity or to minor faulting that accompanies cooling and subsidence of the oceanic lithosphere.

7. Seismicity is concentrated in the western Pacific, with the main zones of seismicity being associated with oceanic trenches and volcanic islands near Tonga, Java, the Philippines, and Japan. These zones run parallel to oceanic trenches and mark subduction zones. Worldwide, approximately 90% of significant earthquakes occur along subduction zones. Subduction zones have shallow, intermediate-depth, and deep earthquakes, with deep and intermediate-depth earthquakes being common only along subduction zones. Note that there is a consistent pattern of shallow earthquakes close to the trench and progressively deeper earthquakes farther away. What do you think causes this pattern? We address this topic on the next page.

B Where Do Earthquakes Occur in the Western Hemisphere and Atlantic Ocean?

The map below shows the Western Hemisphere, including North and South America and adjacent parts of the Pacific and Atlantic Oceans. Observe the distribution of earthquakes, especially how earthquakes compare to the edges of continents, mid-ocean ridges, and sites of subduction.

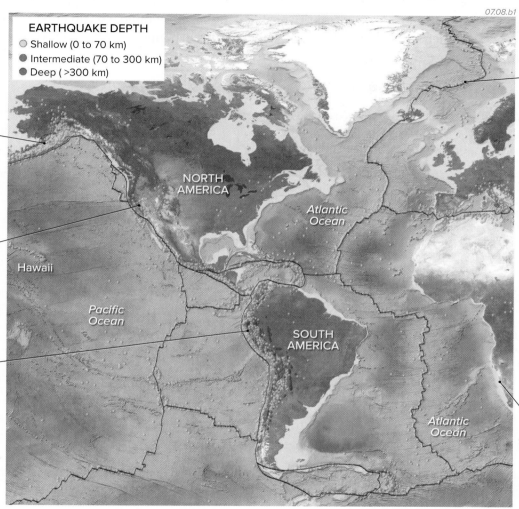

1. A belt of strong seismic activity occurs along the southern part of mainland Alaska and the Aleutian Islands to the west. This belt parallels an oceanic trench and contains shallow and intermediate-depth earthquakes. It marks a subduction zone where the oceanic Pacific plate subducts beneath Alaska. This belt is a continuation of the activity in Japan and the western Pacific (i.e., the Pacific Ring of Fire).

2. Earthquakes follow the west coast of North America and extend into the mountains of the West. These earthquakes reflect diverse types of faulting (strike-slip, normal, and thrust faulting), as well as volcanism.

3. Intense seismic activity follows the western coasts of Central America, including Mexico, and South America. Included in this activity are deep and intermediate-depth earthquakes along subduction zones, especially the one beneath western South America. Shallow earthquakes are closer to the trench, and deep ones are farther away.

EARTHQUAKE DEPTH
○ Shallow (0 to 70 km)
● Intermediate (70 to 300 km)
● Deep (>300 km)

NORTH AMERICA

Atlantic Ocean

Hawaii

Pacific Ocean

SOUTH AMERICA

Atlantic Ocean

07.08.b1

6. A belt of shallow earthquakes follows the Mid-Atlantic Ridge, a mid-ocean ridge formed where the North and South American plates spread westward from the Eurasian and African plates. The pattern of earthquakes mimics the shape of the flanking continents, and the shape of the mid-ocean ridge is largely inherited from the time when these continents rifted apart in the Mesozoic.

7. Note the relative lack of seismicity along the west coast of Africa and east coasts of North and South America.

07.08b2

4. A deep oceanic trench flanks the western coast of South America, marking a subduction zone where oceanic plates subduct beneath the western side of the continent. Observe the pattern of earthquakes for this area on the large map above before examining the figure to the right.

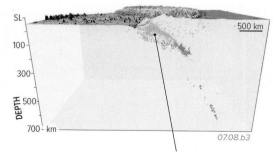

SL-
100-
300-
500-
700- km
DEPTH
500 km
07.08.b3

5. In a side view, subduction-related earthquakes, shown as dots, are shallower to the west (near the trench) and deeper to the east, recording the descent of the oceanic plate. This pattern follows, and helps define, the position of the subducted slab, which is inclined from the shallow to the deep earthquakes.

Before You Leave This Page

☑ Summarize some generalizations about the distribution of earthquakes, especially the relationship to plate boundaries.

☑ Sketch and explain how you could recognize a subduction zone from a map showing earthquakes colored according to depth, and how you could infer which way the subduction zone is inclined.

7.8

What Causes Earthquakes Along Plate Boundaries and Within Plates?

DIFFERENT TECTONIC SETTINGS have different types of earthquakes. Earthquakes formed along a plate boundary generally record the relative movement of the two plates along this boundary (divergent, convergent, or transform) or reflect other processes, such as magmatism, associated with the boundary. Other earthquakes occur in the middle of plates, for example during continental rifting.

A How Are Earthquakes Related to Mid-Ocean Ridges?

Earthquakes are common along mid-ocean ridges, where two oceanic plates spread apart. Most of these earthquakes form at relatively shallow depths and are small or moderate in size. Some earthquakes reflect spreading of the plates, whereas others record motion as the two plates slide by one another on transform faults.

1. Seafloor spreading forms new oceanic lithosphere that is very hot and thin. Stress levels increase downward in Earth, but in mid-ocean ridges the rocks in the lithosphere get very hot at a shallow depth, too hot to fracture (they flow instead). As a result, earthquakes along mid-ocean ridges are relatively small and shallow, with hypocenters less than about 20 km (12 mi) deep.

2. Many earthquakes occur along the axis of the mid-ocean ridge, where spreading and slip along normal faults downdrop blocks along the narrow rift. Numerous small earthquakes also occur due to intrusion of magma into fissures.

3. As the newly created plate moves away from the ridge, it cools, subsides, and bends. The stress caused by the bending forms steep faults, which are associated with relatively small earthquakes.

4. Strike-slip earthquakes occur along transform faults that link adjacent segments of the spreading center. Largely because of the typically thin lithosphere, earthquakes along these oceanic transform faults are small and shallow.

07.09.a1

B How Are Earthquakes Related to Subduction Zones?

A subduction zone, where an oceanic plate underthrusts beneath another oceanic plate or a continental plate, undergoes compression and shearing along the plate boundary. It can produce very large earthquakes.

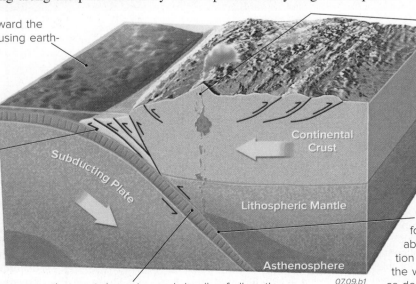

1. As the oceanic plate moves toward the trench, it is bent and stressed, causing earthquakes in front of the trench.

2. Larger earthquakes occur in the accretionary prism as material is scraped off the downgoing plate. Shearing within the prism causes slip and earthquakes along numerous thrust faults.

3. Large earthquakes occur along the entire contact between the subducting oceanic plate and the overriding plate. The plate boundary is a huge thrust fault called a *megathrust.* Earthquakes along megathrusts are among the most damaging and deadly of all earthquakes. During large earthquakes, the megathrust can rupture upward all the way to the seafloor, displacing the seafloor and unleashing destructive waves in a tsunami.

5. Earthquakes can also occur within the overriding plate due to movement of magma and from volcanic eruptions. Compressive stresses associated with plate convergence can cause thrust faulting inland from the magmatic arc.

4. The downgoing oceanic plate is relatively cold and so continues to produce earthquakes from shearing along the boundary, from downward-pulling forces on the sinking slab, and from abrupt changes in mineralogy. Subduction zones are typically the only place in the world producing deep earthquakes, as deep as 700 km (430 mi). Below 700 km, the plate is too hot to behave brittlely or to cause earthquakes.

07.09.b1

C How Are Earthquakes Related to Continental Collisions?

During continental collisions, one continental plate underthrusts beneath another. Collisions can be extremely complex, as different parts collide at different times and rates. Collisions cause large tectonic stresses that shear and fault a broad zone within the overriding and underthrusting plates. As a result, earthquakes are widely distributed along the collision zone.

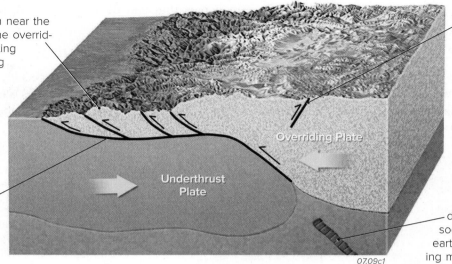

07.09c1

1. Large thrust faults form near the plate boundary in both the overriding plate and underthrusting plate (not shown), causing large but shallow earthquakes. These earthquakes can be deadly in heavily populated areas, such as India, Nepal, Pakistan, and Iran.

2. Large, deadly earthquakes are produced along the plate boundary, or megathrust, between the two continental plates.

3. Thrust faults also form within both continental plates, causing moderately large earthquakes. The immense stresses associated with a collision can reactivate older faults within the interior of either continent, as is presently occurring in Tibet and China. Strike-slip faults and normal faults may be generated as entire regions are stressed by the collision zone or are shoved or sheared out of the way.

4. Any oceanic plate material that was subducted prior to the collision is detached, so actual subduction and associated earthquakes stop. A few deep earthquakes have resulted from the sinking motion of such detached slabs.

D How Are Earthquakes Generated Within Continents?

In addition to continental collisions, earthquakes occur in other tectonic settings within continents. These settings include continental rifts, continental strike-slip faults, magmatic areas, and reactivated preexisting faults.

1. Continental rifts generally produce normal faults, whether the rift is a plate boundary or is within a continental plate. The normal faults downdrop fault blocks into the rift, causing *normal-fault earthquakes*. Such earthquakes are typically moderate in size.

2. A transform fault can cut through a continent, moving one piece of crust past another. The strike-slip motion causes earthquakes that are mostly shallower than 20 to 30 km (10 to 20 mi), but some of these strike-slip earthquakes can be quite large. The San Andreas fault of California is the best-known example of a continental transform fault, but large, destructive earthquakes also occur along continental transform faults in Turkey, Pakistan, Nicaragua (Central America), and New Zealand.

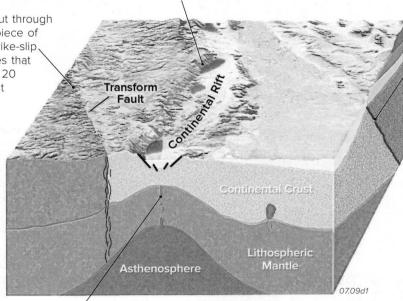

07.09d1

3. Intrusion of magma (shown here in red) within a plate can cause small earthquakes as the magma moves and creates openings in the rock. Moving magma can produce distinctive earthquakes, which are unlike those produced by movement along faults. Heat from the magma can substantially weaken the crust, causing even more rifting and seismic activity.

4. Preexisting faults in the crust can readjust and move as the continental plate becomes older and is subjected to new stresses, such as from distant plate boundaries. Reactivation of these structures can occur in the interior of a plate and produce large earthquakes, like those in Missouri in 1811.

Before You Leave This Page

☑ Sketch and explain earthquakes along mid-ocean ridges, including oceanic transforms.

☑ Sketch and explain earthquakes associated with subduction zones, including earthquakes in the overriding plate.

☑ Summarize how continental collisions cause earthquakes.

☑ Describe some settings in which earthquakes can occur within a continental plate.

7.9

How Do Earthquakes Cause Damage?

MANY GEOSCIENTISTS SAY that "earthquakes don't kill people, buildings or tsunamis do." This is because most deaths from earthquakes are caused by the collapse of buildings or other structures. Destruction and collapse may result from ground shaking during an earthquake, or it can occur later due to landslides, fires, and floods caused by the earthquake. Earthquakes on the seafloor can spawn large ocean waves that can devastate coasts near and far.

A What Destruction Can Arise from Shaking Due to Seismic Waves?

Direct damage from an earthquake results when the ground shakes because of seismic waves, especially surface waves near the epicenter of the earthquake. Damage can also occur after the earthquake, such as from fires and flooding, that are triggered by the earthquake. In the example below, most damage occurred during or shortly after the earthquake.

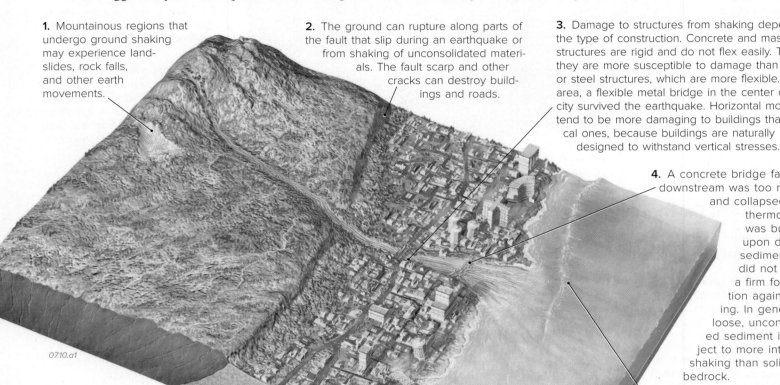

1. Mountainous regions that undergo ground shaking may experience landslides, rock falls, and other earth movements.

2. The ground can rupture along parts of the fault that slip during an earthquake or from shaking of unconsolidated materials. The fault scarp and other cracks can destroy buildings and roads.

3. Damage to structures from shaking depends on the type of construction. Concrete and masonry structures are rigid and do not flex easily. Thus, they are more susceptible to damage than wood or steel structures, which are more flexible. In this area, a flexible metal bridge in the center of the city survived the earthquake. Horizontal motions tend to be more damaging to buildings than vertical ones, because buildings are naturally designed to withstand vertical stresses.

4. A concrete bridge farther downstream was too rigid and collapsed. Furthermore, it was built upon delta sediments that did not provide a firm foundation against shaking. In general, loose, unconsolidated sediment is subject to more intense shaking than solid bedrock.

5. A *tsunami* is a giant wave that can rapidly travel across the ocean. An earthquake that occurs undersea or along coastal areas can generate a tsunami, which can cause damage along shorelines thousands of kilometers away.

07.10.a1

8. Historically, most deaths from earthquakes are due to collapse of poorly constructed houses and buildings, such as ones composed of mud, loosely connected blocks, and earthen walls. Even modern reinforced concrete, like that in freeways and bridges, can fail (▼).

7. Ground shaking of unconsolidated, water-saturated sediment causes grains to lose grain-to-grain contact. When this happens, the material loses most of its strength and begins to flow, a process called *liquefaction*. This can destroy anything built on top (▼).

6. *Aftershocks* are smaller earthquakes that occur after the main earthquake, but in the same area. Aftershocks occur because the main earthquake changes the stress around the hypocenter, and the crust adjusts to this change with more faulting. Aftershocks are very dangerous because they can collapse structures already damaged by the main shock. Aftershocks after a tsunami can cause widespread panic.

07.10.a2 Oakland, CA

07.10.a3 San Francisco, CA

B What Destruction Can Happen Following an Earthquake?

Some damage results from *secondary effects* triggered by an earthquake, including earthquake-caused fires.

07.10.b1

07.10.b2 San Fernando, CA

07.10.b3 Sumatra

Fire is one of the main causes of destruction after an earthquake. Natural gas lines may rupture, causing explosions and fires. The problem is compounded if water lines also break during the earthquake, limiting the amount of water available to extinguish fires.

Flooding may occur due to failure of dams as a result of ground rupturing, subsidence, or liquefaction. Near Los Angeles, 80,000 people were evacuated because of damage to nearby dams during the 1971 San Fernando earthquake (Mw 6.7).

Earthquakes may cause both uplift and subsidence of the land surface by more than 10 m (30 ft). Subsidence accompanied the 2004 Sumatra earthquake, causing areas that had been dry land before the earthquake to become inundated by seawater, flooding buildings and trees. Subsidence and uplift can occur during or after an earthquake.

C How Can We Limit Risks from Earthquakes?

The probability that you will be affected by an earthquake depends on where you live and whether that area experiences tectonic activity. The risk of earthquake catastrophe depends on the number of people living in the region, how well the buildings are constructed, and individual and civic preparedness.

1. Earthquake hazard maps (▼) show zones of potential earthquake damage. Near Salt Lake City, Utah, the risk is greatest (areas colored red) near active normal faults along the Wasatch Front, the mountain front east of the city. Living away from the fault is less risky.

07.10.c2 Salt Lake City, UT

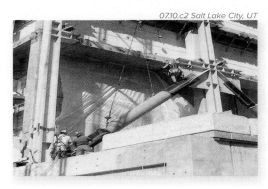

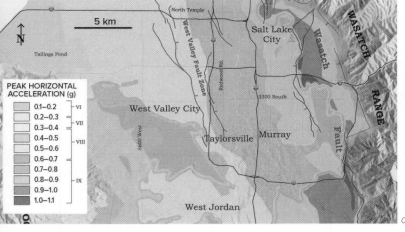

PEAK HORIZONTAL
ACCELERATION (g)

0.1–0.2	VI
0.2–0.3	VII
0.3–0.4	
0.4–0.5	VIII
0.5–0.6	
0.6–0.7	
0.7–0.8	
0.8–0.9	IX
0.9–1.0	
1.0–1.1	

07.10..c1

2. Some utility companies and hospitals have computerized warning systems that are notified of impending earthquakes by seismic equipment. The system will automatically shut down gas systems (to avoid fire) and turn on back-up generators to prevent loss of electrical power.

3. Earthquakes have different periods, durations, and vertical and horizontal ground motion. This makes it difficult to design earthquake-proof buildings. Some rest on sturdy wheels or have shock absorbers (▲) that allow the building to shake less than the underlying ground.

What to Do and Not Do During an Earthquake

There are actions you can take during an earthquake to reduce your chances of being hurt. If an earthquake strikes, you can seek cover under a heavy desk or table, and protect your head. You can also stand under door frames or next to inner walls, as these are the least likely to collapse. If possible, stand clear of buildings, especially those made of bricks and masonry.

During the shaking, stay away from glass windows and doors, bricks, and heavy objects that could fall. Always keep a battery-operated flashlight handy. Avoid using candles, matches, or lighters, since there may be gas leaks. Earthquakes may interrupt electrical and water service. Keeping 72 hours' worth of food, water, and other supplies in a backpack is a prudent plan for any type of natural disaster.

Before You Leave This Page

✓ Describe how earthquakes can cause destruction, both during and after the main earthquake.

✓ Describe some ways to limit our risk from earthquakes.

✓ Discuss ways to reduce personal injury during an earthquake.

7.10

How Does a Tsunami Form and Cause Destruction?

AN EARTHQUAKE BENEATH THE OCEAN can cause a large wave called a *tsunami,* which can wreak havoc on coastal communities, such as those in Japan. Most of Earth is covered by oceans, so many earthquakes, landslides, and volcanic eruptions occur beneath the sea. Each of these events can generate a deadly tsunami.

A How Are Tsunamis Generated?

Tsunamis are waves generated by a disturbance in the sea or much less commonly in a lake. They are generated by abrupt changes in water level in one area relative to another, such as occurs when seafloor is unevenly uplifted or downdropped during an earthquake. A large tsunami requires a large disturbance, like the rupture of a major fault that has a large amount of slip and significant displacement of the seafloor. Unlike typical ocean waves, which affect only the upper part of the sea, a tsunami can affect an entire body of water from top to bottom.

07.11.a1

1. Subduction zone megathrusts (◄) can lock for long periods of time, causing the seafloor above the overriding plate to bulge, strain, and flex up or down as it accommodates the forces of convergence. This upward and downward flexing is typically most prominent near the trench.

07.11.a2

2. When the megathrust finally ruptures in an earthquake (along the red asterisks), the bulging plate changes shape catastrophically. The water above the plate responds by lifting up from the ocean bottom toward the surface, forming a ridge of higher water (◄). Formation of this large wave and accompanying troughs in the water can cause the ocean to retreat from the shoreline, which was observed in the 2004 Sumatra earthquake-caused tsunami and in other deadly tsunamis. A sudden retreat of the ocean along a coastline is a warning sign that a tsunami may be coming.

07.11.a3

3. A tsunami, or a series of tsunamis, radiates away from the disturbance (◄), traveling at speeds between 600 to 800 km/hr (370 to 500 mi/hr). In deep water, the wave energy is distributed over the entire water depth, forming a wave only a meter or so high but more than 700 km (435 mi) across (in wavelength). If you were in the open ocean, you probably would not notice its passing. As the wave approaches the shore, its energy concentrates in shallower and shallower water. The velocity of the front of the wave decreases to 30 to 40 km/hr (20 to 25 mi/hr), causing the following water to pile up in a higher wave. Near shore, the tsunami becomes a massive, thick wave, like the front wall of a plateau of water. It may be a series of such waves.

Tsunamis Triggered by Landslides

4. A large mass of rock entering the water, or shifting from one part of the water to another, can catastrophically displace the water, generating a tsunami that radiates outward. This has occurred repeatedly off the west side of the Big Island of Hawaii, where huge landslide-debris deposits (shown in light green below) sit on the ocean floor.

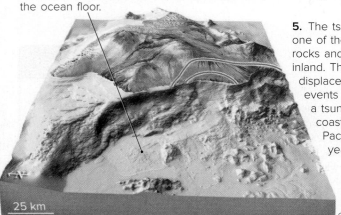

5. The tsunami generated by one of these slides carried rocks and coral 6 km (3.7 mi) inland. The volume of water displaced during these events probably produced a tsunami that struck coastlines around the Pacific about 120,000 years ago.

25 km

07.11.a4

Tsunamis Caused by Eruptions

6. The 1883 eruption of Krakatau in Indonesia, and the collapse of its immense caldera, generated a series of huge tsunamis that killed 36,000 people. A single catastrophic volcanic explosion produced the loudest sound ever heard, and most of Krakatau Island was demolished. The tsunami was as high as 40 m (more than 130 ft), and some effects of the tsunami were recorded 7,000 km away! The painting below is an overly dramatic representation. Most tsunamis are not simply large, curled waves as shown here.

07.11.a5

B What Kind of Destruction Can a Tsunami Cause?

Like the one associated with the Tohoku earthquake off northeastern Japan in 2011, tsunamis cause death and destruction along coastlines where human populations are concentrated. On May 22, 1960, the largest earthquake ever recorded on a seismometer (Mw 9.5) occurred in the subduction zone (megathrust) offshore of southern Chile. The tsunamis that followed flattened coastal settlements in Chile and traveled across the Pacific to devastate coastlines in Hawaii and Japan.

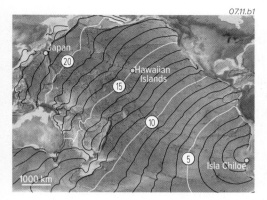

07.11.b1

Chile, May 22, 1960

◄ 1. During this earthquake, tsunamis were generated parallel to the coast. One headed in toward the shoreline, quickly striking Chile and Peru. Another set of tsunamis swept out across the Pacific Ocean at 670 km (420 mi) per hour. Each stripe equals one hour of travel time.

► 2. In Chile, the first tsunami struck 15 minutes after the earthquake. On Isla Chiloe, a 10-meter-tall surge of water swept over towns. The waves killed at least 2,000 people along the coasts of Chile and Peru.

07.11.b2

Hawaii, May 23, 1960

3. About 15 hours after the earthquake in Chile, the tsunami related to the earthquake hit Hilo and other parts of Hawaii (▼). A tsunami 11 m (36 ft) high killed 61 people, damaged buildings, and caused $23 million in damage. Seven hours later, the tsunami killed 140 people in Japan.

07.11.b3

Hokkaido, Japan, 1993

4. In 1993, a magnitude 7.8 earthquake occurred off the west coast of Hokkaido. Within five minutes, a tsunami struck the coastline. The tsunami killed at least 100 people and caused $600 million in property loss. It swept these boats inland across a concrete barrier built along the shoreline (▼).

07.11.b4

Papua New Guinea, 1998

5. In 1998, a magnitude 7.1 earthquake and associated underwater landslides generated three tsunami waves that destroyed villages along the country's north coast, killing 2,200 people. A 10-meter-high surge of water destroyed a row of populated houses along the coast shown here (▼).

07.11.b5

Tsunami Warning System

In an international effort to save lives, the United States National Oceanic and Atmospheric Administration (NOAA) maintains two *tsunami warning centers* for the Pacific Ocean. Twenty-six nations participate in this effort. Informed by worldwide seismic networks, these centers broadcast warnings based on an earthquake's potential for generating a tsunami. After the huge loss of life from the 2004 Sumatran tsunami, the United Nations implemented a warning system in the Indian Ocean. Scientists deployed warning buoys, like the one shown to the right, which can relay tsunami data by satellite. These buoys relay small changes in sea level detected by ocean-bottom sensors as a tsunami passes through the water column above the sensor.

07.11.t1

Before You Leave This Page

✓ Describe the different mechanisms by which tsunamis are generated.

✓ Summarize the kinds of damage tsunamis have caused.

✓ Briefly describe how tsunamis are monitored to provide an early-warning system.

7.11

7.12 What Were Some Recent Large Earthquakes?

THE WORLD HAS ENDURED a number of large and tragic earthquakes. These earthquakes have struck a collection of geographically and culturally diverse places, causing many deaths and extensive damage. Most large earthquakes have occurred along or near plate boundaries, especially along subduction zones. Some happened on faults that are close to, but not actually on, a plate boundary. Major earthquakes occurred recently in Japan, Haiti, and New Zealand. Each was destructive, but in different ways.

A What Happened During the Catastrophic Tohoku Earthquake of 2011?

A huge and catastrophic earthquake struck off northeastern Japan on March 11, 2011. It had an extremely large magnitude of 9.0 (Mw), making it one of the five largest earthquakes ever measured. Ground shaking during the earthquake caused extensive damage, especially on the large island of Hokkaido, nearest to the epicenter. The earthquake and resulting tsunami destroyed 125,000 buildings, left 24,000 dead or missing, and caused more than $300 billion in economic damages, making it the most expensive natural disaster in history.

1. The epicenter of the earthquake, shown here as a red dot, was on the seafloor east of Japan, along an oceanic trench that marks where the Pacific plate subducts beneath Hokkaido. The earthquake occurred at depth, along the subduction zone (plate boundary) that dips beneath Japan. The main hypocenter was 32 km deep, so this is classified as a shallow earthquake. In many regards, it was a typical, but very large, *megathrust* earthquake. It was followed by many aftershocks (shown as small yellow dots), which show how large an area of the plate boundary slipped during and after the main earthquake. Some of the aftershocks were large (magnitude 6), causing damage to already weakened buildings and raising fears of a new tsunami.

07.12.a1

2. The earthquake represented a sudden upward movement of rocks at depth. Based on the distribution of aftershocks and other seismic data, the fault slipped over an area of 300 km parallel to the trench and by 150 km perpendicular to the trench, nearly spanning the entire distance between the trench and the coastline. The fault rupture grew upward from the hypocenter toward the seafloor, uplifting a large swath of seafloor by up to 3 m. The displaced water formed into an extremely damaging tsunami that locally was higher than 10 m (33 feet), perhaps several times that in small stretches of the coast. A tsunami warning system warned, and saved, many residents, but thousands of people were trapped by the fast-moving, rising waters that spread far inland.

07.12.a2

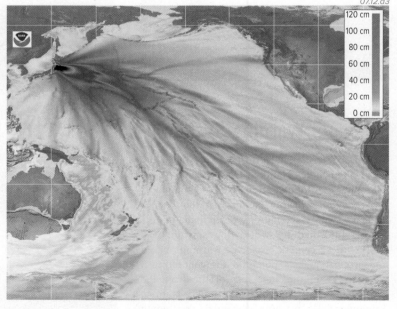
07.12.a3

3. News reports showed dramatic footage of seawater spilling onto the land and then gradually rising higher and higher, flooding low areas, destroying most buildings in its path, and pushing wrecked ships far inland (▲). As the tsunami moved farther inland, it became a slower-moving wall of sludge containing automobiles, boats, parts of destroyed buildings, and other debris. When it was done, the tsunami had inundated nearly 500 km² of Japan. It destroyed entire towns and heavily damaged the Fukushima Nuclear Power Plant, which had a reactor-core meltdown because the tsunami destroyed the cooling system, backup power generation, and other parts of the facility.

4. This map (▲) shows the forecasted maximum wave amplitudes modeled for the 2011 earthquake by tsunami forecasters at NOAA. As predicted, a tsunami of measurable height reached most Pacific shorelines, including Hawaii and the west coasts of North and South America.

B How Were Recent Earthquakes in Haiti and New Zealand Similar and Different?

Large earthquakes struck Haiti in 2010 and New Zealand in 2010 and 2011. The Haiti earthquake was a magnitude 7.0 (Mw), the 2010 Canterbury (New Zealand) earthquake was magnitude 7.1 (Mw), and the 2011 Christchurch (New Zealand) earthquake was magnitude 6.3 (Mw). All of these earthquakes were near, but not on, a plate boundary that mostly consists of strike-slip (transform) movement. However, the three earthquakes varied greatly in the amount of damage and death they inflicted. Why is this so? Let's examine the "tale of three earthquakes."

Haiti, 2010

1. The Haiti earthquake occurred on land 25 km west of the capital, Port-au-Prince. The epicenter (shown in red) is near, but not on, an active strike-slip fault that cuts east-west across the country. On this map, aftershocks are yellow dots, green lines are strike-slip faults, and red lines have thrust movement. The area is within a zone of complex faulting near the boundary between the Caribbean plate to the south and the North American plate to the north. The earthquake flattened more than 300,000 buildings in and around Port-au-Prince and killed perhaps 200,000 people. The poverty of the country, combined with a devastated infrastructure and an inefficient response by relief agencies, led to hunger, suffering, and disease after the quake.

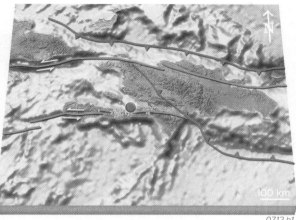

07.12.b1

07.12.b2

2. Most deaths during the Haiti earthquake were caused by collapsing buildings. Haiti is the poorest country in the Western Hemisphere, and most houses and buildings were poorly constructed. Most fared worse than the National Palace (▲), which was only partly collapsed.

New Zealand 2010, 2011

3. The two main earthquakes in New Zealand had epicenters (shown as red dots) on a broad coastal plain that lies east of the rugged Southern Alps and the Alpine fault, a mostly transform plate boundary that runs down the length of the island. Both earthquakes were very shallow (less than 15 km), were not on the plate boundary, and occurred on two different, but probably related, faults. Both were followed by abundant aftershocks (shown in yellow and orange), but only the 2010 quake ruptured the surface and was on a known fault (green line on map).

Christchurch
2010   2011
07.12.b3

07.12.b4

4. The Canterbury quake was a magnitude 7.1 (Mw), caused moderate damage, and only injured two people. The 2011 Christchurch quake was smaller, but killed nearly 200 people. It destroyed or damaged 100,000 buildings, some by liquefaction of the soil and associated expulsion of water from the ground (▲). The main difference in the amount of destruction was that the 2011 epicenter was very near Christchurch, New Zealand's second-largest city, and the quake had more vertical motion, destroying already weakened buildings.

Deadly Earthquakes

Earthquakes kill about 10,000 people per year on average. Most earthquake-related deaths are due to collapse of poorly built structures in cities and villages. Earthquake-generated tsunamis account for a large part of the destruction. The table to the right shows some deadly earthquake events. The highest death tolls are due to a deadly combination of high population densities, substandard construction practices, and being situated along subduction zones or other high-risk areas. Earthquakes discussed in this chapter are not included on this table.

Fatalities	Mw	Year	Location
830,000	8	1556	Shaanxi, China
11,000	6.9	1857	Naples, Italy
70,000	7.2	1908	Messina, Italy
200,000	7.8	1920	Ningxia, China
143,000	7.9	1923	Kanto, Japan
200,000	7.6	1927	Tsinghai, China
32,700	7.8	1939	Erzincan, Turkey
66,000	7.9	1970	Colombia
23,000	7.5	1976	Guatemala
242,000	7.5	1976	Tangshan, China
31,000	6.6	2003	Bam, Iran
88,000	7.9	2008	Sichuan, China

Before You Leave This Page

☑ Briefly summarize the four earthquakes presented here, including their tectonic settings and how each caused destruction.

☑ Discuss why the amount of damage and death varied among the quakes.

7.12

7.13 What Were Some Major North American Earthquakes?

SOME VERY LARGE AND DAMAGING EARTHQUAKES have struck North America in the last several centuries. Here, we discuss seven important earthquakes chosen not because they are all the largest, but because they illustrate a range of processes, damage, and locations.

1. This map of the conterminous United States has yellow dots showing the locations of earthquakes with a magnitude greater than 4 that occurred during the last several decades. The red lines on the map are faults that are interpreted to have slipped during the last 2 million years. Compare the distribution of earthquakes and these relatively young and active faults. Most active faults are in the western states, and most large earthquakes are in these same areas. Earthquakes have occurred elsewhere in the country, but most of these were too small to break the surface and form a fault scarp.

07.13.a1

07.13.a2

Alaska, 1964

2. A magnitude 9.2 (Mw) earthquake, one of the three or four largest earthquakes ever recorded, struck southern Alaska in 1964. It killed 128 people, triggered landslides, and collapsed parts of downtown Anchorage and nearby neighborhoods. This event was caused by thrust faults associated with the Aleutian Islands subduction zone. Most deaths and much damage were from a tsunami generated when a huge area of the seafloor was uplifted. The photograph above shows damage from the tsunami. This earthquake, like Alaska, is not shown on the map.

07.13.a3

San Francisco, 1906

3. A huge earthquake occurred when the San Andreas fault ruptured near San Francisco. The earthquake was likely a magnitude 7.7 to 7.8 (Mw) although not directly measured on seismometers. The earthquake ruptured the surface, leaving behind a series of cracks and open fissures. Within San Francisco, ground shaking destroyed most of the brick and mortar buildings. More than 3,000 people were killed and much of the city was devastated by fires that broke out after the earthquake. Geologists determined that 470 km (290 mi) of the fault ruptured during the event.

Mexico City, 1985

5. A magnitude 8.0 (Mw) earthquake occurred at a subduction zone along the southwestern coast of Mexico, well west of Mexico City (not shown on this map). It damaged or destroyed many buildings in Mexico City and killed at least 9,500 people. Destruction was so extensive partly because Mexico City is built on lake sediments deposited in a bowl-shaped basin. This geologic setting amplified the seismic waves and caused intensified and highly destructive ground shaking. Surface waves, which caused the most damage, traveled 200 km (120 mi) from their source.

Northridge, Los Angeles Area, 1994

4. This magnitude 6.7 (Mw) earthquake was generated by a thrust fault northwest of Los Angeles. The earthquake killed 60 people and caused $20 billion in damage. A section of freeway buckled, crushing the steel-reinforced concrete slabs. The thrust is not exposed on the surface, but when it ruptured it lifted up a large section of land. Geologists are concerned about a similar fault causing this type of earthquake right below downtown Los Angeles.

07.13.a4

07.13.a5

07.13.a6

Hebgen Lake, Yellowstone Area, 1959

6. This magnitude 7.3 (Mw) event was generated by slip along a normal fault northwest of Yellowstone National Park. Ground shaking set loose the massive Madison Canyon slide (◄), which buried 28 campers and formed a new lake, aptly named *Earthquake Lake*.

New Madrid, 1811–1812

7. New Madrid, Missouri, experienced a series of large (Mw 7.8–8.1) earthquakes generated over an ancient fault zone in the crust. The 1811–1812 earthquake death toll was relatively low because of the sparse population at the time. The New Madrid zone has a high earthquake risk and, as shown on the earthquake-hazard map below, is one of two areas in the eastern United States that are predicted to experience strong earthquakes in the future. Memphis lies in this zone, yet most of its buildings are not constructed to survive large earthquakes.

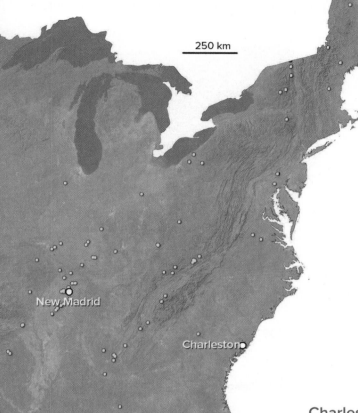

250 km

New Madrid

Charleston

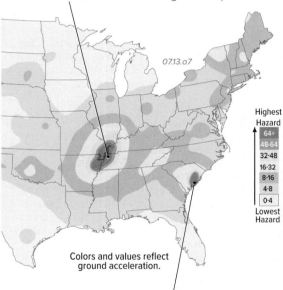

07.13.a7

| Highest Hazard |
| 64+ |
| 48-64 |
| 32-48 |
| 16-32 |
| 8-16 |
| 4-8 |
| 0-4 |
| Lowest Hazard |

Colors and values reflect
ground acceleration.

Charleston, 1886

8. This earthquake occurred at the highest risk area along the East Coast, near Charleston, South Carolina. It had an estimated magnitude of 7.3 (Mw), the largest ever recorded in the southeastern United States. Buildings incurred some damage (►), and 60 people died. The tectonic cause for this earthquake is still debated by geologists. The East Coast, including Washington, D.C., occasionally experiences earthquakes strong enough to be felt.

07.13.a8

Earthquakes in the Interiors of Continents

Why do large earthquakes like the ones at New Madrid, Missouri, occur in the middle of continents? Although the interior of North America is not near a plate boundary, the region is subjected to stress generated at far-off plate boundaries. In this case, the stresses are probably generated by a plate-driving force, known as *ridge push,* that originates along the Mid-Atlantic Ridge. These stresses can reactivate ancient faults that lie buried beneath the

cover of sediment. In the case of New Madrid, seismic and other geophysical evidence suggest that the area is underlain by an ancient rift basin that formed about 750 million years ago during the breakup of the supercontinent of Rodinia. Modern-day stress related to the current plate configuration is interacting with the ancient faults, occasionally causing them to slip and trigger earthquakes.

Before You Leave This Page

✓ Describe some large North American earthquakes and how they were generated.

✓ Summarize the various ways these earthquakes caused damage.

✓ Describe evidence that the eastern United States has earthquake risks.

7.13

What Is the Potential for Earthquakes Along the San Andreas Fault?

THE SAN ANDREAS FAULT is the world's best-known and most extensively studied fault. It runs across California from the Mexican border to north of San Francisco and is responsible for many destructive earthquakes. What has happened along the fault in the recent past, and what does this history say about its likelihood of causing large earthquakes? The USGS and others have forecasted a 99% probability that California will have a magnitude 6.7 or larger earthquake in the next 30 years.

Recent Earthquake History of Different Segments of the San Andreas Fault and Related Faults

1. The San Andreas fault has distinct *segments* that behave differently. These segments vary in the size and frequency of earthquakes. As a result, the earthquake hazard varies along the fault. This map shows some of the major segments of the San Andreas fault that have caused earthquakes in California. Circles show epicenters of some of the more important earthquakes. The San Andreas fault accounts for the largest quakes, but there are many other recently active faults (shown in green). Some of these have caused damaging, moderate-sized earthquakes.

07.14.a1

07.14.a2

2. The *northern segment* of the San Andreas fault was responsible for the famous 1906 earthquake that destroyed much of San Francisco. The earthquake had a magnitude of 7.7 (Mw) and ruptured 430 km (270 mi) of the fault, from south of the city all the way to the north end of the fault (the part that ruptured is shown in red). Damage (◄) was caused by ground shaking, fires, and liquefaction of water-saturated soils in areas that had originally been part of San Francisco Bay.

07.14.a3

3. The southern part of this segment ruptured in 1989 in the magnitude 7.1 (Mw) Loma Prieta earthquake, which was centered south of San Francisco. This earthquake is famous for disrupting a World Series baseball game. Ground shaking and liquefaction collapsed buildings (◄) and parts of bridges and freeways.

4. The next segment to the south, shown in blue, is the *central creeping segment*. The two sides of the fault creep past one another somewhat continuously and slowly, rather than storing up energy for a large earthquake. Creep continues to the north along the Hayward fault, also colored blue, through Oakland. The Hayward fault was the site of a ruinous earthquake in 1868, with an estimated magnitude of 7 (Mw).

5. South of the creeping segment is the *Parkfield segment*, a short segment included here as part of a larger orange-colored segment discussed below. It produces moderate-sized earthquakes that occur, on the average, every couple of decades. The Parkfield segment receives special scrutiny from geologists and seismologists because the frequent earthquakes provide an opportunity to study the behavior of a fault before, during, and after an earthquake.

6. The San Andreas continues to the southeast through a segment (shown in orange) that last ruptured during the great Fort Tejon earthquake of 1857. This earthquake ruptured 300 km (190 mi) of the fault, from Parkfield all the way to east of Los Angeles. The earthquake was approximately magnitude 8 (Mw), but damage was limited because the area was much less populated than it is now. This part of the San Andreas commonly is called the *locked segment* because it has not ruptured since 1857. It has the potential to cause a great earthquake, commonly called "the big one."

Features Along the San Andreas Fault

The San Andreas fault generally has a clear expression in the landscape. It is marked by a number of features that are common along active faults. Some of these features can also form in ways unrelated to active faulting.

Pond — Offset Drainage Channel — Linear Ridge — Scarp — Linear Valley — Drainage Parallel to Fault — Spring — Faults

07.14.a4

07.14.a5 Parkfield, CA

10. Geoscientists explore the fault to find localities that preserve a record of past faulting. Detailed studies of trenches dug across the fault (▶) help geoscientists unravel hundreds or thousands of years of the fault's movement history.

11. The aerial photograph to the right (▶) shows the same part of the San Andreas fault as depicted in the figure above. Can you match some of these features between the two images?

07.14.a6 Carrizo Plain, CA

1872

9. North and east of the San Andreas is a series of faults, called the *East California Shear Zone*. This zone caused several magnitude 7 or higher earthquakes in the 1900s and the large 1872 Owens Valley earthquake on the eastern side of the Sierra Nevada. The zone continues from the eastern side of the Sierra Nevada southward through the Mojave Desert, where it unleashed the 1992 Landers earthquake (Mw 7.3) and the 1999 Hector Mine earthquake (Mw 7.1).

8. On the map, note that the San Andreas fault has a distinct curve or bend in the middle of the southern locked (orange) segment. The bend causes regional compression and thrust faults, some of which are not exposed at the surface. These thrust faults caused the 1994 magnitude 6.7 (Mw) Northridge earthquake in metropolitan Los Angeles, and they have uplifted the large mountains, like the San Gabriel Mountains, north and northeast of the city.

East California Shear Zone

MOJAVE DESERT

Los Angeles

7. East of Los Angeles, the San Andreas branches southward into several faults. Some of these experienced several moderate-sized earthquakes in the 1900s, including some close to important agricultural areas. The fault scarps for these events are colored pink and lavender on this map.

Before You Leave This Page

☑ Briefly summarize the main segments of the San Andreas fault and whether they have had major earthquakes.

☑ Summarize features that might help you recognize the fault from the air.

7.14

Where Did This Earthquake Occur, and What Damage Might Be Expected?

THIS COASTAL REGION CONTAINS TWO FAULTS, an active volcano, and several steep-sided mountains prone to landslides. Any of these features could cause ground shaking. You will use seismic records from a recent earthquake to determine which feature caused the observed shaking. From this information, you will decide what hazards this earthquake poses to each of the small towns in the area.

Goals of This Exercise:

- Examine the large illustration and read the text boxes describing the types of features that are present.
- Use three seismograms to determine which feature is likely to have caused the earthquake.
- Consider potential earthquake hazards to determine what dangers each small town would face from the earthquake.
- Decide which town you think is the safest from earthquake-related hazards and justify your decision with supporting evidence.

Procedures

The area has several small towns and three seis-mometers, each named after the nearest town. Seismograms recorded at each seismic station during a recent earthquake are shown at the top of the next page. Use the available information to complete the following steps and enter your answers in the appropriate places on the worksheet or online.

1. There is a deep ocean trench along the edge of the continent. Ocean drilling encountered fault-bounded slices of oceanic sediment.

2. Along one part of the coastline, there is a thin, steep beach, called *Roundstone Beach*, that rises upward to some nearby small hills. The seafloor offshore is also fairly steep as it drops off toward the trench.

3. The town of *Sandpoint* is built upon land that was reclaimed from the sea by piling up loose rocks and beach sand until the area was above sea level.

A. Observe the features shown on the three-dimensional perspective. Read the text associated with each location, and think about what each statement implies about earthquake hazards.

B. Inspect the seismograms for the three seismic stations to determine where the earthquake probably occurred. You can get an idea from simply comparing the time intervals between the arrivals of P-waves and S-waves for each station.

C. Use the graph next to the seismograms to determine the distance from each station to the epicenter. This will allow you to more precisely locate the epicenter. Detailed instructions for this procedure are listed in topic 7.7 earlier in this chapter. For plotting your results, a map view of the area is included on the next page and a larger version is on the worksheet.

D. From the general location of the earthquake, infer which geologic feature is likely to have caused the earthquake.

E. Use the information about the topographic and geologic features of the landscape to interpret what types of hazards the recent earthquake posed for each town. From these considerations, decide which three towns are the least safe and which two are the safest for this type of earthquake. There is not necessarily one right answer, so explain and justify your logic on the worksheet, if asked to do so by your instructor.

07.15.a1

11. Offshore is a coral reef that blocks larger waves, creating a quiet lagoon between the reef and the shore.

Seismograms

12. These seismograms (►) represent the time period from just before the earthquake to 1.5 seconds after it occurred. The first arrivals of P-waves and S-waves are labeled for each graph, along with the P-S time intervals.

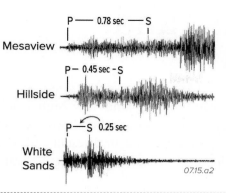

Mesaview — P — 0.78 sec — S

Hillside — P — 0.45 sec - S

White Sands — P — S 0.25 sec

07.15.a2

13. Use this graph (►) to determine the distance from each seismic station to the earthquake's epicenter. Find the appropriate time on the horizontal axis, follow it upward to the line, and read off the corresponding distance on the vertical axis. Then plot your results as circles on the map of the area (lower right), following the steps presented earlier in the chapter.

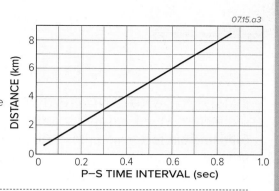

07.15.a3

4. A picturesque town, called *Hillside*, lies inland of some small mountains. The town is built on a flat, open area flanked by hills with fairly gentle slopes. It is a little higher in elevation than the nearby towns of Cascade Village and Riverton. The Hillside Seismic Station, shown by a triangle symbol on the map on the lower right, lies just to the east of the town.

5. In the northern part of the area, there is a flat-topped mountain, known as *Red Mesa*, surrounded by steep cliffs. A new landslide lies along the southern flank of the mountain. A small town and a seismic station, both called *Mesaview*, lie between the mesa and a high volcano.

6. A volcano called *Lava Mountain* rises above the region. It has steep slopes and is surrounded by layers of volcanic ash that appear to have erupted quite recently. Every so often, the volcano releases steam and makes rumbling noises. The shaking triggers landslides down the hillsides. The small town of Ashton is on the flanks of the volcano and has a picturesque setting with huge, colorful blocks of volcanic rocks near the town.

7. The *Gray Cliffs* form a nearly vertical step in the landscape. Streams pour over the cliffs in pleasant waterfalls, each taking a jog to the left after crossing the cliffs. The small settlement of *Cascade Village* is located next to one of the waterfalls. Rocks along the cliffs are fractured and shattered.

8. The small village of *Cliffside* lies next to a gray cliff. It was built on a marshy area that was underlain by soft, unconsolidated sediments. Several streams drain into the area, but no streams are able to leave because the area is lower than the surrounding landscape. As a result, the soil is commonly very soft and people sink in as they walk.

9. *Riverton*, a picturesque town, is built near a river at the head of a sandy bay. The seafloor slopes out to the bay at a gentle angle. Muddy waters from the river prevent reefs from growing offshore in front of the bay.

10. *White Sands* is a resort town along a white, sandy beach. The sand comes from the offshore coral reef. There is a seismic station, shown by a triangle symbol, with the same name as the town.

07.15.a4

7.15

Mountains, Basins, and Continental Margins

EARTH'S LAND AND SEAFLOOR VARY GREATLY in elevation. The land has high, rocky mountains, deep sediment-covered basins as valleys, and regional plains with very little topographic relief. Continental margins are fundamental features, representing the transition from continent to deep ocean basin. Present and past continental margins provide much of the world's oil, natural gas, and salt. How do mountains, basins, and continents form, and why are continental margins so rich in natural resources?

Beneath Monterey Bay, off the coast of central California, the seafloor displays a puzzling feature—a large submarine canyon. In this image, satellite data are shown for land, and computer-shaded and colored data show seafloor depths.

What features are present along continental margins, and how do these features form?

Adjacent to the shoreline, the land varies in topography from mountains and hills to relatively flat areas in valleys. Some of the inland hills are small, local features, but the mountain ranges continue along the coast for long distances.

What causes mountain ranges to be higher in elevation than adjacent valleys, and what processes form mountains and valleys?

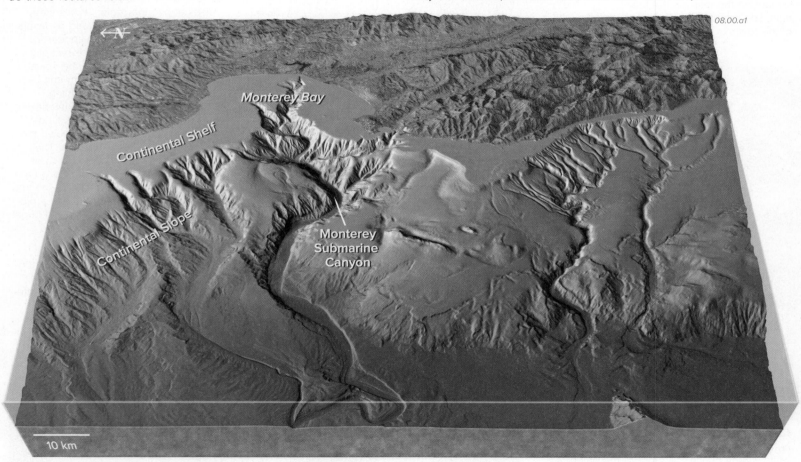

08.00.a1

Monterey Bay

Continental Shelf

Continental Slope

Monterey
Submarine
Canyon

10 km

A broad continental shelf flanks the coast, with relatively shallow water (less than about 100 m) extending out kilometers to tens of kilometers from shore. The area is a prized marine ecosystem and is the site of the Monterey Bay National Marine Sanctuary. Farther out, the relatively shallow bench of the continental shelf drops off toward deeper seafloor along the *continental slope* and then onto the even-deeper *continental rise*.

What are continental shelves, slopes, and rises, and what is happening across this transition to cause this observed increase in depth to the seafloor?

Monterey Submarine Canyon is enormous. It is similar in scale to the Grand Canyon. The canyon bottom is as much as 1,800 m (nearly 6,000 ft) below the rim and, in this deep segment, the canyon is 20 km (12 mi) wide. It resembles many valleys on land; it curves, goes from higher to lower areas, and has smaller side valleys (tributaries) that merge with the main channel.

What processes carve submarine canyons?

TOPICS IN THIS CHAPTER

Submarine Canyons and Fans

1. Continental shelves and slopes are blanketed by sediments, most of which are unconsolidated and weak. The combination of weak materials and a relatively steep angle on the slope and on the walls of submarine canyons causes some slopes to fail due to the force of gravity. Failure may be triggered by earthquakes, large storms, or overloading by newly deposited sediments. As sediments collapse during a slope

failure, they can break up and incorporate seawater between the grains. This forms a dense mixture of water and sediment (mostly clay, silt, and sand), such as this mass (◄) produced in a laboratory. These mixtures are more dense than normal seawater and flow downslope as fast-moving slurries, or *turbidity currents*. The dense, cloud-like slurry of a turbidity current travels through the water until the current slows and the grains progressively settle on the seafloor, larger grains first. Turbidity currents have destructive potential and are capable of eroding rock, even underwater. They help carve many submarine canyons, including the parts of Monterey Canyon.

2. This diagram (►) illustrates a turbidity current, shown in gray, beginning on the continental slope and flowing down a submarine canyon. As a turbidity current exits the steep canyon, it spreads out and slows down. Sand grains and any pebbles can no longer be suspended by the turbulence and settle out. As the current slows further, it deposits silt followed by clay particles. This sequence of depositing finer particles with time forms *graded beds*.

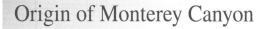
08.00.a3

3. As the turbidity current slows and spreads out across the continental slope and rise, it deposits its load of sediment in a fan-shaped deposit, or a *submarine fan*. A submarine fan can be hundreds to more than a thousand kilometers wide, and typically consists of mud and other deep-marine sediment that alternate with sandy turbidite deposits with graded bedding.

4. Underwater slopes can also fail as *submarine landslides*. A landslide mass can contain large, fairly coherent blocks or can come apart as it detaches from the slope and moves downhill. A landslide commonly forms distinctive lumps on the seafloor and may leave behind a ragged scar on the slope above. Landslides were key to forming and widening some submarine canyons.

Origin of Monterey Canyon

We do not expect to find huge canyons beneath the sea. When and how did Monterey Canyon form, and what processes are going on today in and around the canyon? Scientists explore the submarine canyon by bouncing sound waves off the seafloor, dredging and drilling rock samples from the bottom, and diving to the bottom in small submarines.

The formation and evolution of the canyon reflect the complicated plate-tectonic events that have affected California during the last 20 million years. Geoscientists have concluded that the upper part of the canyon was originally carved by streams when its granitic base was

above sea level, prior to 10 million years ago. Horizontal motion between the North American and Pacific plates shaved off this granitic slice and transported it northward up the coast of North America. During this movement, the canyon was submerged below sea level and filled by sediments, which were later eroded by landslides and underwater currents.

For the past several million years, dense slurries of sediment-rich water, called *turbidity currents*, have flowed down the canyon, scouring the channel and undercutting the canyon walls. The canyon widens as the steep, unstable walls collapse downward in underwater landslides and

debris flows. The turbidity currents carry sediment more than 200 km (120 mi) down the canyon and into deeper water, where the sediment is deposited in a broad feature called a *submarine fan*. The lower part of the canyon, like many submarine canyons, was never above sea level and has been carved entirely by turbidity currents and landslides. The position of the lower channel has shifted over time, as segments of the canyon have been offset by faulting or buried by submarine landslides.

8.0

8.1 Why Are Some Regions High in Elevation?

SOME REGIONS ARE MUCH HIGHER THAN OTHERS. Many mountains are not only steep but are also high in elevation. Elsewhere, huge regions of land are barely above sea level. What accounts for these differences? Regional variations in elevation primarily reflect the tectonic processes that occurred in the region and the nature of the crust and mantle at depth. A change in the subsurface can cause a region to be uplifted (rise in elevation) or to subside (drop in elevation). Is the region in which you live high, near sea level, or somewhere in between? What is the actual measured elevation where you live? You can easily look this up.

A What Controls Regional Elevation?

Regional elevations are controlled primarily by the thickness of the crust, but they can also be influenced by the density and temperature of materials in the crust and upper mantle.

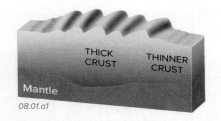

08.01.a1

Regions with thick crust are higher than those with thinner crust. In other words, mountain ranges have deep crustal roots.

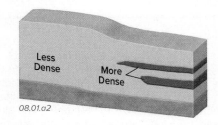

08.01.a2

Regions underlain by less dense crust will be higher in elevation than areas with a similar thickness of more dense crust.

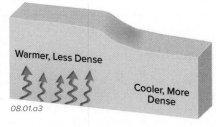

08.01.a3

Temperature of the crust and mantle also affects elevation. Warm rocks are less dense than cooler rocks, so areas with warm rocks are higher than areas with cool rocks.

B What Causes Variations in Crustal Thickness?

Differences in crustal thickness between regions reflect differences in their geologic histories. Such differences include whether the crust is continental or oceanic, and whether it has been deformed, eroded, or buried.

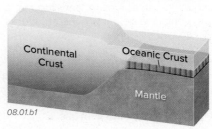

08.01.b1

1. Continents have relatively thick crust (▲), generally 30 to 50 km thick, and so are higher than ocean basins, which are underlain by oceanic crust that is much thinner, typically about 7 km thick.

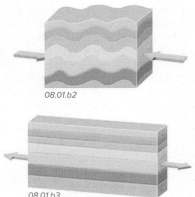

08.01.b2

08.01.b3

2. Crust thickens if compressed strongly enough from the sides (◄). It can respond to compression by folding or faulting.

3. Crust thins if it is stretched in a horizontal direction (◄), either by ductile stretching at depth or by normal faulting in the upper crust.

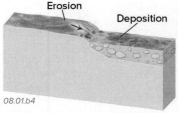

08.01.b4

4. Crust that loses material, such as by erosion (▲), will become thinner, whereas crust that gains material by deposition of sediment or volcanic rocks will become thicker.

C How Is Regional Elevation Decreased?

Normal faulting can thin the crust by displacing higher rocks from lower ones. This decreases crustal thickness and causes a region to subside.

Crustal thickness can be reduced if material is eroded from the top, as is common in many mountain belts.

Rocks contract when they cool, so subsidence results from cooling of large regions of the crust or upper mantle.

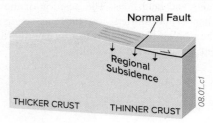

08.01.c1

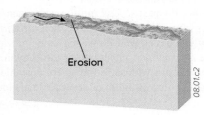

08.01.c2

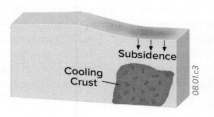

08.01.c3

D How Is Regional Elevation Increased?

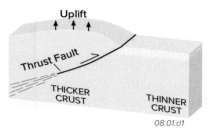

1. Crust that is compressed and shortened by thrust faults also thickens. This thickening causes the region to be uplifted. The thrust fault can also uplift rocks, forming a mountain.

08.01.d1

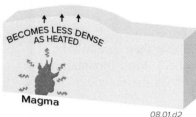

2. If the crust or mantle beneath a region is heated, the rocks expand and become less dense. As a result, the region can increase in elevation.

08.01.d2

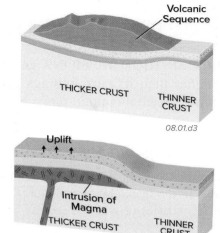

3. Crust can thicken by the addition of material to the surface, whether it is sediment or volcanic units, perhaps lava in huge volcanic fields. The added material builds up the surface and pushes down the crust in response to the weight.

08.01.d3

4. Magma can add to the crust at depth, and this addition of material thickens the crust. Several processes may operate together: magma can add material and also heat the crust at the same time.

08.01.d4

E What Is the Influence of the Thickness of the Lithosphere?

The lithosphere is, on average, about 100 km thick, but it varies in thickness from nearly zero at mid-ocean ridges to more than 150 km beneath some ancient continental interiors. These variations greatly influence elevation because the mantle part of the lithosphere is more dense than the asthenosphere.

1. A region with thin lithosphere, such as a mid-ocean ridge, will be higher than an adjacent region with thicker lithosphere, even if they have the same type and thickness of crust.

2. As a new oceanic plate moves away from a ridge, the asthenosphere cools enough to become lithosphere. The plate, therefore, thickens, becomes more dense, and subsides as it cools.

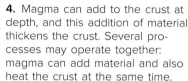

08.01.e1

3. Lithosphere is generally thicker in the central, ancient parts of continents, far away from modern plate boundaries, but these regions can have moderate elevation because of relatively thick continental crust.

4. Continental lithosphere can be thinned near plate boundaries by heating and other plate activity. The affected region can rise in elevation because dense lithosphere is replaced by less dense asthenosphere.

The Discovery of Isostasy

Isostasy is the principle that regional elevations adjust to the types and thicknesses of rocks at depth. It was discovered through observations made by George Everest while surveying India around 1850. Surveyors at the time understood that a weight suspended on a line (to level the surveying equipment) was deflected from vertical a very small amount by the gravitational attraction of nearby mountains. When taking this into account, Everest noted an unexplained discrepancy in positions on his survey. He found that the deflection of the weight from vertical was less than predicted.

To explain the discrepancy, a mathematician calculated the expected gravitational attraction of the Himalaya. Astronomer George Airy then used an analogy with floating icebergs and other common objects to suggest that higher mountains had thicker crustal roots.

By this model, lower density crustal material in the roots attracts the suspended weight less than would the denser mantle material that the crustal root has displaced. This case illustrates how observations related to one topic (surveying) can lead to a scientific discovery (isostasy) in another discipline.

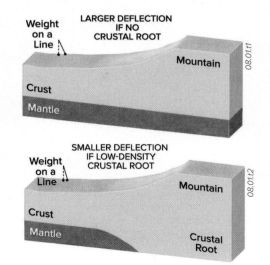

Before You Leave This Page

☑ Summarize or sketch the factors that control regional elevation.

☑ Summarize or sketch what causes variations in crustal thickness.

☑ Summarize several ways to increase elevation and to decrease elevation.

☑ Explain the observation that led to the discovery of isostasy.

8.1

8.2 Where Do Mountain Belts and High Regions Form?

MOUNTAIN BELTS AND OTHER HIGH REGIONS generally owe their high elevation to thick continental crust. Less commonly, a region is higher than its surroundings due to processes originating in the mantle. Where are the world's main mountain belts, why is each mountain range higher in elevation than its surroundings, and why did mountains form in these places?

A In Which Tectonic Settings Do Regional Mountain Belts Form?

Regional mountain ranges are hundreds or thousands of kilometers long. They are large enough that they can only be explained by major variations in the thickness, temperature, and density of the crust and lithosphere. Most mountain ranges occur near convergent plate boundaries or where there has been large-scale movement of material in the mantle.

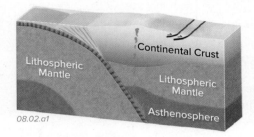

08.02.a1

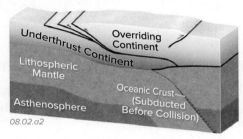

08.02.a2

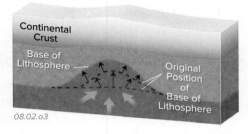

08.02.a3

Subduction Zones—Convergent margins are high in elevation largely because the crust is thickened by magmatic additions from the subduction zone and by crustal shortening. Also, in these regions, lithosphere is heated and replaced by less dense asthenosphere.

Continental Collisions—Collision zones have high elevations due to an increase in crustal thickness as one continent is shoved beneath another. In these settings, crustal thickening occurs by thrusting, folding, and other forms of deformation.

Mantle Upwellings—Less dense asthenosphere can move upward into the lithosphere, causing regional uplift. This occurs near hot spots, plate boundaries, and in some other settings, such as continental rifts. It is partly responsible for uplift of parts of the western United States.

B What Causes These Regions to Have High Elevation?

1. *Western Canada* has been a convergent margin for most of the last 100 million years. Its mountain ranges overlie crust thickened from major thrust faulting, from magmatic additions, and from collisions with island arcs and pieces of continental material.

2. The *Alps* mountain range of southern Europe is high because it has thick crust due to crustal shortening and collisions between Europe and smaller continental blocks that came from the south.

3. The *Tibetan Plateau* and the *Himalaya* are extremely high because of very thick crust that resulted when the Indian continent collided with, and was partly shoved beneath, the continental crust of Asia.

08.02.b1

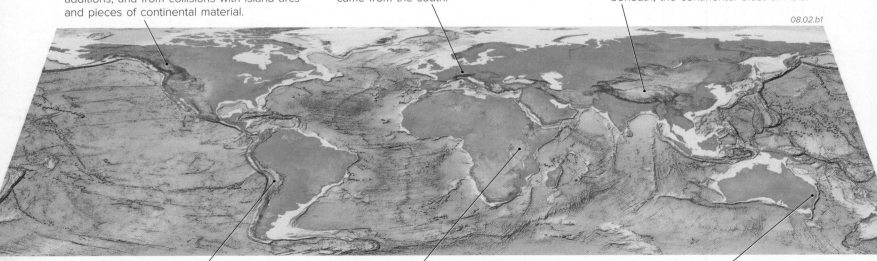

4. The *Andes* of South America are above a subduction zone. The underlying crust is hot and thick because of magmatic additions and crustal shortening.

5. The *East African Rift* is higher than most of Africa because of magmatic heating of the crust, thinning of the lithosphere, and the presence of a hot spot leading to mantle upwelling.

6. The *Great Divide Range* forms the eastern flank of Australia. There is currently no plate boundary here, and geoscientists continue to investigate the age and cause of uplift.

C What Happens During the Erosion of Mountain Belts?

Mountains, once formed, are subjected to weathering and erosion. These processes wear mountains down but are countered by uplift related to *isostasy*. Uplift is driven by buoyancy due to the root of underlying thick crust.

Early Mountain Building

1. As a mountain belt forms, uplift is commonly faster than erosion, and the mountain becomes higher and more rugged over time. A high mountain belt results from uplift that is faster than erosion.

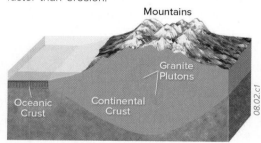

2. As soon as a mountain starts forming, weathering and erosion begin to wear it down, contributing sediment to streams and rivers. Sediment will be transported to adjacent low areas, perhaps in nearby oceans or other types of basins.

Erosion and Isostatic Rebound

3. As material erodes from a mountain belt, there is less weight holding down the thick crustal root. The buoyant crust can uplift, a process called *isostatic rebound*.

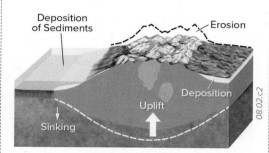

4. Sediment derived from the mountain is deposited in nearby basins, typically on both the sea and continental sides. The added weight of the sediment depresses the crust (isostasy) in these regional basins, making room for more sediment.

Late Stages of Evolution

5. Erosion and isostasy cause rocks deep in the crust to be uplifted and exposed at the surface. As a result, many mountain belts expose metamorphic and plutonic rocks.

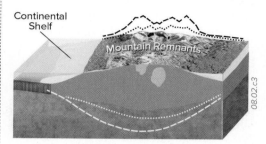

6. Through simultaneous erosion and isostasy, the mountain is eroded down and the thick crustal root is gradually reduced in size. Material eroded from the mountain ends up in adjacent basins, increasing the crustal thickness beneath the basins.

D What Controls Regional Elevations in North America?

The vertically exaggerated topographic profile below illustrates how elevations vary from east to west across the United States. It does not show the full thickness of crust, only the elevation of the land and depth of seafloor.

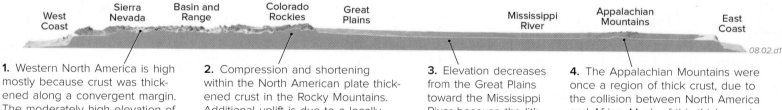

1. Western North America is high mostly because crust was thickened along a convergent margin. The moderately high elevation of the Basin and Range is largely due to very thin lithosphere.

2. Compression and shortening within the North American plate thickened crust in the Rocky Mountains. Additional uplift is due to a locally thin lithosphere and upwelling asthenosphere associated with rifting.

3. Elevation decreases from the Great Plains toward the Mississippi River because the lithosphere is cooler and thicker to the east.

4. The Appalachian Mountains were once a region of thick crust, due to the collision between North America and Africa. Much of this thickness has been lost due to erosion, so the range has lost elevation over time.

Rule of Thumb for Elevations

Regional elevations are relatively low for regions with thinner crust and relatively high for regions with thicker crust, but by how much? A rule of thumb is that increasing the thickness of the crust by 6 km will result in an increase in elevation of 1 km (~3,300 ft). Here is an example from Arizona.

Phoenix sits at an elevation of 300 m (1,000 ft), whereas Flagstaff is at more than 2,100 m (7,000 ft). This difference is about 2 km, so the crust beneath Flagstaff should be 12 km thicker than the crust beneath Phoenix (2 × 6 = 12). Geophysical measurements show that the crust beneath Phoenix is about 28 km thick, whereas crust beneath Flagstaff is about 40 km thick. The difference is 12 km, the value we would predict.

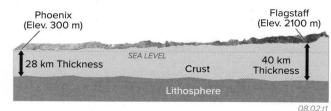

Before You Leave This Page

☑ Sketch and explain the main tectonic settings of high regions, providing an example for each setting.

☑ Summarize the settings of the world's high mountains and plateaus.

☑ Explain how erosion and isostasy help expose deeply formed rocks in eroded mountain belts.

☑ Summarize differences in regional elevation across North America.

8.2

8.3 How Do Local Mountains Form?

THE DISTINCTION BETWEEN LOCAL MOUNTAINS and regional mountain ranges is important. Regional mountain ranges are hundreds to thousands of kilometers long, contain many peaks, and typically involve uplifted, thickened crust or some other regional cause of the uplift. Other mountains are *local* features, too small to be accompanied by regional increases in crustal thickness. Instead, such mountains simply rest upon—and are supported by—the crust.

A How Does Volcanism Form Local Mountains?

A local mountain may be formed by a volcanic eruption that piles lava, ash, and scoria onto the crust. Such mountains vary in size from small scoria (cinder) cones to large shield and composite volcanoes.

08.03.a1 Flagstaff, AZ

08.03.a2 Mt. Hood, OR

08.03.a3 Castle Dome Mtns., AZ

Volcanism creates mountains by piling volcanic materials on a preexisting surface. Some of the smallest volcanic mountains and hills are scoria cones. They are clearly local features, not requiring regional changes in the thickness of the underlying crust.

Composite volcanoes consist of lava flows, variably compacted volcanic ash, and debris in mudflows and landslides. They commonly make lofty and steep mountains that have a typical volcano shape, like the one shown here.

Some mountains are not the actual volcano, but represent the eroded remnants of a volcano. The mountain above consists of volcanic rocks, but it does not have its original volcano shape. Instead, it is a mountain because the volcanic rocks are hard and resist erosion.

B How Do Faults Build Mountains?

Local mountains can also arise through faulting. Reverse faults create mountains by thrusting one fault block up and over another. Normal faults also form local mountains, even though they stretch and thin the crust in a region.

Mountains Formed by Reverse Faulting

1. Reverse faulting will make a mountain if the overthrust block is uplifted faster than it is eroded, or if it is composed of erosion-resistant rocks like granite and other crystalline rocks.

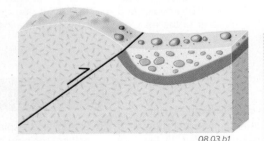
08.03.b1

08.03.b2 Garden of the Gods, CO

2. The Front Range, the high mountains west of Colorado Springs and Denver, Colorado, was uplifted along a series of reverse faults. As the range was uplifted, rock layers adjacent to the fault were tilted and folded, producing the artistically eroded red rocks in the Garden of the Gods and at Red Rocks.

Mountains Formed by Normal Faulting

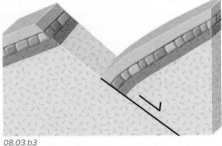
08.03.b3

3. During normal faulting, one block slips down, forming a basin. The other block remains high or is moved upward, and it can form a local mountain if it is not eroded away.

4. In the region near Death Valley, California, normal faulting down-dropped the valley floors down relative to the mountains, forming basins with the mountains on either side. The basins trap sediment eroded from the ranges. The floor of Death Valley is locally below sea level, but it is not connected to the sea.

08.03.b4 Panamint Range, CA

C How Does Folding Build Mountains?

Another way to make local mountains is by folding. Folding can warp and uplift Earth's surface as well as the underlying rock layers. Uplift and erosion of a folded, hard layer can create a topographical high that remains long after the folding has stopped.

1. Folding can form mountains and hills by deforming the land surface and near-surface rocks, as is happening near Los Angeles, California.

08.03.c2 Dinosaur NP, CO/UT

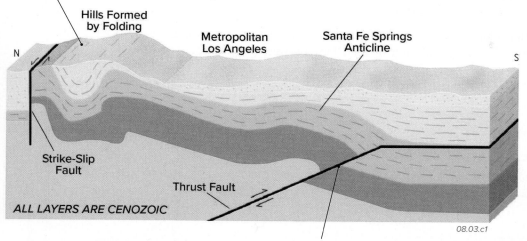

Hills Formed by Folding

Metropolitan Los Angeles

Santa Fe Springs Anticline

N   S

Strike-Slip Fault

ALL LAYERS ARE CENOZOIC

Thrust Fault

08.03.c1

2. Some oil well pipes near Los Angeles were being crushed and bent beneath a fold, called the Santa Fe Springs anticline. No one knew why until the large 1994 Northridge earthquake revealed thrust faults in the area, including ones hidden in the subsurface. This fault was breaking and bending the pipes as it folded and uplifted sedimentary rocks. Recent studies have identified additional faults in the area, including some directly below downtown Los Angeles.

3. Some mountains, including this one (▲) in Dinosaur National Park in Utah and Colorado, owe their existence to folding followed by erosion. In this area, folding ended more than 45 million years ago. Erosion downcut through the folded rocks until it encountered these folded layers of hard, light-colored sandstone. Soft rocks underlie the valley and were eroded away more easily, leaving the folded sandstone as a mountain.

D How Can Differential Erosion Form a Local Mountain?

08.03.d3 Caineville, UT

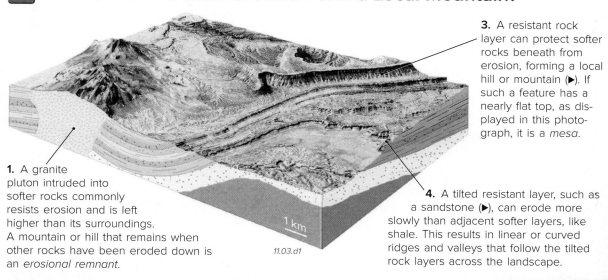

3. A resistant rock layer can protect softer rocks beneath from erosion, forming a local hill or mountain (▶). If such a feature has a nearly flat top, as displayed in this photograph, it is a *mesa*.

08.03.d4 San Rafael Reef, UT

1. A granite pluton intruded into softer rocks commonly resists erosion and is left higher than its surroundings. A mountain or hill that remains when other rocks have been eroded down is an *erosional remnant*.

1 km

11.03.d1

4. A tilted resistant layer, such as a sandstone (▶), can erode more slowly than adjacent softer layers, like shale. This results in linear or curved ridges and valleys that follow the tilted rock layers across the landscape.

08.03.d2 Stone Mtn., GA

2. *Stone Mountain* (◀) in Georgia consists of granite that solidified at a depth of 10 km. It was then uncovered by erosion, which removed the overlying and flanking softer rocks. The mountain, therefore, is an erosional remnant resulting from differential erosion.

Before You Leave This Page

✔ Describe how volcanism forms mountains.

✔ Sketch and describe how reverse faulting, normal faulting, and folding can each build mountains.

✔ Describe some ways that erosion can result in a mountain, ridge, or mesa.

8.3

8.4 Where Do Basins Form?

BASINS ARE LOW RELATIVE TO THEIR SURROUNDINGS and commonly trap sediment and water. They form in many tectonic settings, both on land and beneath the oceans, and can accumulate different kinds of sediment, depending on their geologic environments.

A In What Tectonic Settings Do Basins Form?

Basins form on both oceanic and continental plates and along plate margins. Some basins are as large as an ocean, while others are smaller, depressed areas, commonly near a local fault or fold. Do you live in a basin?

Passive Margin

The largest type of basin is a *passive margin*, a continental margin that is not a plate boundary. A passive-margin basin includes the continental shelf (lightest blue in the image above), continental rise, and continental slope, and generally is underlain by thin, previously rifted crust. It receives sediment from the continent and provides shallow-water environments for diverse life, such as offshore of North Carolina.

Continental Rift

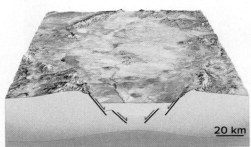

Continental rifts form when forces try to pull apart a continent, forming normal faults that downdrop some fault-bounded blocks. The downdropped blocks can accumulate coarse, continental sediment, fine-grained lake beds, and salt deposits. If the rifting progresses to seafloor spreading, a continental rift evolves into a passive margin. The rift shown here is similar to the Rio Grande Rift that runs north from Texas and New Mexico into Colorado.

Normal Fault Blocks

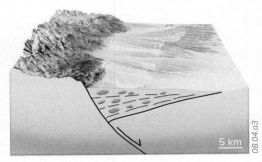

Normal faulting can downdrop a block, forming a basin that fills with sediment. Steep topography along the faulted mountain front produces coarse sediment that is delivered to the basin by debris flows, landslides, and steep, rocky streams. Finer grained sediment and evaporites can accumulate in lakes. *Normal-fault basins* can occur on land, for example Death Valley, or along rifted margins, like those that flank the Atlantic Ocean.

Reverse and Thrust Faults

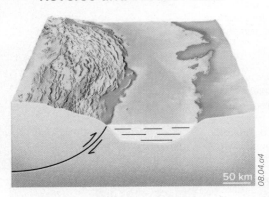

A *foreland basin* occurs when crust (either continental or oceanic) is depressed by the weight of thrust sheets. The basin develops as a depression in front of the thrusts because the extra weight causes the crust to warp downward. The Persian Gulf is a foreland basin. Thrust faults are also common in the accretionary prism (not shown) between a magmatic arc and trench, and a basin in this setting is a *forearc basin* (in front of the arc).

Strike-Slip Faults

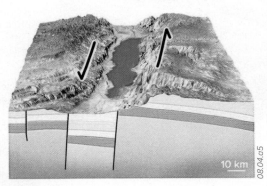

Basins can develop along a strike-slip fault if motion along the fault downdrops one block relative to another. Such downdropping is most common where the fault takes a bend across the surface, as shown here for a basin similar to the Dead Sea of the Middle East. Downdropping can also occur where strike-slip motion along several nearby faults causes the crust to pull apart, dropping a block in between as a *pull-apart basin*.

Regional Subsidence

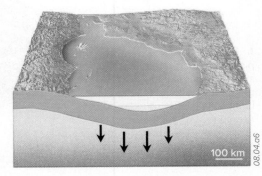

Huge basins hundreds or thousands of kilometers wide form due to *regional subsidence*, where a broad region drops in elevation. Causes of subsidence include regional cooling of the crust and mantle, lateral or vertical movement of the underlying crust and mantle, or conversion of less dense minerals in the lower crust and upper mantle to more dense ones. The Michigan Basin, a large basin in Michigan, resulted from regional subsidence.

B What Formed These Basins?

North America contains many basins, some regional in size and others kilometers across. Some basins are tectonically active and accumulating sediment, whereas others are ancient and obvious only when geologists compare the thicknesses and types of sedimentary rocks from one area to another. The map below shows old and new basins that contain more than 5 km of sedimentary and volcanic units. It color-codes basins as a function of age.

1. A basin sits between the Cascade volcanic arc and an offshore trench that marks subduction beneath the continent. This *Cascade forearc basin* receives abundant sediment from major rivers, like the Columbia River, that drain into the sea. The *San Joaquin Basin* of central California is an older version of a forearc basin, and it formed in front of the Sierra Nevada magmatic arc.

2. In southern California, small but locally deep basins formed along the San Andreas fault, a complex zone of mostly strike-slip movement. Some basins are pull-apart basins, and others are related to local thrusting or normal faulting where the fault takes a bend.

3. The interior of the western United States contains a passive margin formed during the Paleozoic by rifting of the western edge of North America. It locally accumulated more than 10 km (6 mi) of sediment. Much of this same region has more recent basins related to normal faulting (not shown).

4. The Michigan and Illinois basins formed within the continent, probably due to Paleozoic collisional tectonics in the Appalachians and from other deep processes.

5. The Appalachian Mountains and nearby areas contain thick sedimentary sequences deposited along the Paleozoic continental margin and in other basins before the Appalachian Mountains were formed.

6. The Gulf Coast contains thick sequences of sedimentary rocks, mostly related to Mesozoic rifting, as South America, the Yucatan, and other continental pieces rifted away from this region. During and after rifting, the continental margin subsided and became a passive margin. The Gulf Coast and the Permian Basin in west Texas are sites of important oil and gas resources.

08.04.b1

Columbia River — San Joaquin — Paleozoic Passive Margin — Los Angeles — Williston — Powder River — Green River — Mid-Continent Rift — Rio Grande Rift — Permian — Gulf Coast — Michigan — Illinois — Appalachian

Cenozoic Basin
Mesozoic and Cenozoic Basin
Paleozoic Basin

500 km

The Michigan Basin

A deep basin beneath Michigan contains a fairly complete column of sedimentary rocks deposited during the early and middle parts of the Paleozoic Era. The geologic map presented here shows the bedrock geology that would be present if we removed the glacial deposits that currently cover most of the area. Note that the rock layers form a bull's-eye pattern around the roughly circular basin, with the youngest layers (yellow and green) occurring in the center of the basin. A geologic cross section across the basin (below) shows that the layers are thicker in the center of the basin. This indicates that the basin was subsiding during deposition of the sediments. The origin of the basin is somewhat enigmatic and possibly involves several causes. The basin probably formed during an episode of continental rifting, but it may also have subsided partly because of flow, thinning, and cooling of the hot lower crust. It is an unresolved question being actively investigated.

08.04.t1

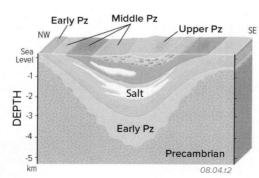

NW — Early Pz — Middle Pz — Upper Pz — SE
Sea Level
Salt
Early Pz
Precambrian
DEPTH
-1 -2 -3 -4 -5 km

08.04.t2

Before You Leave This Page

☑ Describe the different ways in which a basin can form.

☑ List some basins in the United States and describe what caused each to form.

☑ Describe the Michigan Basin and possible causes of subsidence.

8.4

8.5 How Do Mountains and Basins Form at Convergent Continental Margins?

AT SUBDUCTION ZONES BENEATH CONTINENTS, various processes create mountains and basins. Magmatic additions to the crust, along with crustal compression, cause thickening of the crust and the formation of a central mountain belt. Basins can form in front of, within, and behind the mountain belt.

A What Processes Accompany Ocean-Continent Convergence?

Along convergent boundaries, an oceanic plate subducts beneath a continental plate. Subduction causes melting in the mantle beneath the continent and generally also leads to compression and thickening of the continental crust. Such margins are generally dominated by a regional mountain belt.

1. As the oceanic plate approaches the convergent margin, it flexes and bends downward into the inclined subduction zone. An *oceanic trench* forms as a result and acts as a deep oceanic basin that traps sediment eroded from the adjacent mountain belt. The area between the trench and the mountain front is close to or below sea level because it is underlain by thin continental crust and oceanic material sliced off the downgoing plate; it is a *forearc basin*.

2. An *accretionary prism* forms along the upper parts of the subduction zone as sediment is contributed by the adjacent continent and scraped off the downgoing slab. It is a structurally complex zone of faults, folds, and rocks under various metamorphic conditions. As more material is stuffed under the prism, the prism thickens and is uplifted, but generally remains below sea level.

3. Convergence of the two plates generally causes horizontal compression within the continent. This results in thrust faults and other structures that thicken the crust and cause further uplift of the mountain belt. A belt of thrust faults and related folds, called a *fold and thrust belt*, can form behind the main mountain belt, pushing rocks over the interior of the continent.

4. The weight of the thrust sheets causes the continent to flex downward, forming a basin in front of the thrust belt. This basin is called a *foreland basin* because it occurs in front of the mountain belt. It receives sediment from the mountain belt and other parts of the continent. In some convergent margins, the subduction zone only weakly stresses the overriding plate, so there is less compression, no fold and thrust belt, and no foreland basin.

5. Magma generated along the subduction zone rises into the crust. It thickens the crust by erupting as volcanic rock on the surface and by solidifying at depth. The highest parts of most subduction-related mountain belts are near the areas with the greatest volcanic activity.

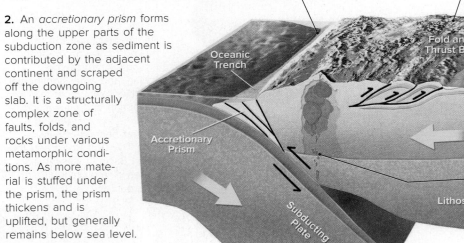

08.05.a1

B What Determines If the Overriding Plate Is Shortened or Extended?

Subduction is not always accompanied by compression and thrust faulting. Several factors influence whether the plate above a subduction zone experiences compression or extension, including the factors presented below.

Compression and horizontal shortening are common in subduction zones where the continental plate moves toward the subduction zone relative to the asthenosphere. This movement pushes against the subducted slab, which is difficult to move sideways through the solid mantle. As a result, the continent experiences compression, as is occurring in parts of the Andes of South America.

Extension is common when the overriding plate is not moving toward the slab relative to the asthenosphere, or is even moving away. The slab tends to pull back by itself, and the continent extends as its edge is pulled toward the ocean by the sinking slab. This is occurring along subduction zones in the western Pacific near Japan and the Philippines.

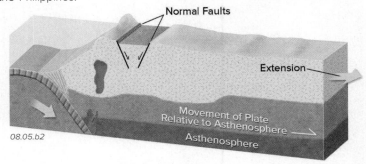

08.05.b1

08.05.b2

C What Features Accompany Continental Collisions?

Continental collisions involve the convergence of two tectonic plates that each carry continental crust. A continent generally is too buoyant to be subducted deeply, so one continent is shoved beneath the edge of the other continent, and the whole region is uplifted. The collision transmits large stresses to the plates on either side, forming thrust faults and thickened crust.

1. During a collision, one continental plate is shoved, or *underthrust,* beneath another plate. A foreland basin forms in front of the collision zone, and the basin sediments can be overridden by or incorporated into the thrust faults.

2. Collisions form high mountain belts composed of faulted, folded, and cleaved rocks. Uplift and erosion bring metamorphic and intrusive igneous rocks up to the surface. In some cases, the collision forms a high continental plateau, such as in Tibet.

3. Behind the collision zone, rocks can be folded and thrust away from the mountain belt. The weight of the thrust sheets pushes down adjacent crust, forming sedimentary basins in front of the thrust sheets.

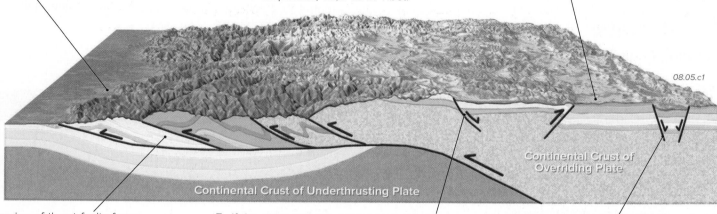

08.05.c1

Continental Crust of Overriding Plate

Continental Crust of Underthrusting Plate

4. A series of thrust faults forms along the collision zone and thickens the crust by shoving one slice of crust over another.

5. If the crust gets too thick or too hot, it may begin to spread under its own weight, flowing sideways. At the surface, such spreading can form normal faults and associated basins.

6. Stresses associated with the collision can cause other types of tectonic features to form hundreds to thousands of kilometers away from the actual plate boundary.

Profile Across the Tibetan Plateau and Adjacent Regions

7. The topographic profile below shows the high Tibetan Plateau viewed to the west. The high mountains on the left edge of the plateau are the Himalaya. To depict the topographic features at this regional scale, the topography is vertically exaggerated by 10 times.

8. The Himalaya is a spectacular mountain range (▶) that rises along the southern edge of the Tibetan Plateau, part of which is in the foreground of this photograph. The Himalaya are the world's highest mountain range, with many peaks more than 8 km (>26,000 ft) above sea level.

08.05.c3 Himalaya Mtns., Tibet

India   Himalaya   Tibetan Plateau   Tarim Basin

100 km   08.05.c2

9. Most of India, to the south of the Himalaya, has much lower elevation and relief, and it is tectonically stable away from the mountain front. Close to the Himalaya are foreland basins associated with thrust faults.

10. Mount Everest is the world's highest mountain, rising 8,850 m (29,035 ft) above sea level. It straddles the border between Nepal (to the south) and Tibet (to the north), and climbers can approach the mountain from either side.

11. The Tibetan Plateau is the largest, highest, and flattest plateau on Earth. Its average elevation is 5 km (over 15,000 ft), which is higher than any peak in the United States, except for some mountains in Alaska.

12. The Tarim Basin is a large desert north of the plateau. It is 3,000 m lower than the plateau and is partially filled by sediment derived from the adjacent highlands.

Before You Leave This Page

☑ Summarize how mountains and basins form in an ocean-continent convergent margin.

☑ Summarize one factor that favors shortening versus extension in a plate above a subduction zone.

☑ Summarize how mountains and basins form in a continental collision, using the Himalaya-Tibet region as an example.

8.5

8.6 # How Does Continental Extension Occur?

DURING CONTINENTAL EXTENSION, continental crust is thinned and stretched horizontally, typically causing the region to subside. Continental extension also breaks the crust into faulted blocks, forming mountain ranges and sedimentary basins. By studying sedimentary sequences, geologists can determine when a basin was active and how fast the sediments within it accumulated. The manifestations of continental extension are expressed throughout much of the western United States.

A How Do Continents Accommodate Crustal Extension?

When continental crust is extended, the upper part responds by breaking into discrete blocks bounded by normal faults. If the fault blocks do not rotate during extension, only a small amount of extension can occur. If the blocks and faults rotate, greater amounts of extension can take place.

Non-Rotating Fault Blocks

1. In some extended areas, adjacent normal faults dip in opposite directions and cut the crust into wedge-shaped fault blocks (▶).

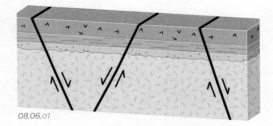

08.06.a1

2. Movement along the faults downdrops some blocks, forming sedimentary basins. These can be thousands of meters deep and tens of kilometers wide (▼).

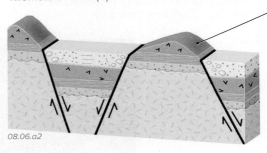

3. The upthrown block, called a *horst*, becomes a mountain bounded on both sides by faults. Erosion of the mountain contributes sediment to the basins. Such a downdropped block is a *graben*.

08.06.a2

4. Over time, the basins fill with sediment unless streams carry most of it away.

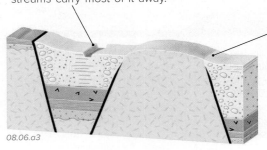

5. The mountains are gradually eroded down, and basin sediments may overlap the edges of the range, burying the fault along the edge of the range.

08.06.a3

Rotating Fault Block

6. In other extended areas, adjacent normal faults dip in the same direction and cut the crust into book-shaped fault blocks (▶).

08.06.a4

7. During fault movement, the blocks and faults both rotate, like books sliding on a shelf (▼).

8. The corner of a block that is rotated down becomes a basin.

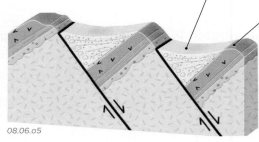

9. The corner that is rotated up becomes a mountain or ridge. The mountains and ridges commonly are linear, following the strike of the layers.

08.06.a5

10. As faulting and extension continue, units are tilted to steep dips. The oldest layers dip more steeply than more recent layers.

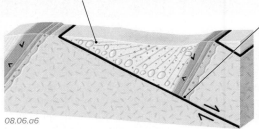

11. Faults are rotated to gentle dips and can have kilometers of displacements, allowing large amounts of crustal extension.

08.06.a6

08.06.a7 Death Valley, CA

12. Death Valley in eastern California is a classic example of extension and normal faulting within a continent. This photograph (◀), taken looking north from an overlook called Dante's View, shows the basin with mountain ranges on both sides. The white and gray units in the center of the basin are salt deposits, formed by evaporation of water within this hot, closed basin. A still-active normal fault runs along the steep mountain front in the foreground, downdropping a fault block to form the basin. The mountain in the distance is a corner of the same fault block as the basin and has been rotated upward as a rotating fault block.

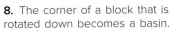

B What Happens When Extension Accompanies Subduction?

Some regions experience crustal extension and rifting in spite of being near a convergent boundary. In these cases, the region may be fairly low in elevation, except for the large volcanoes. Rifting, if it continues, can form a small ocean basin behind the arc.

1. Extension can accompany subduction of one oceanic plate beneath another oceanic plate or beneath a continental plate. In some subduction zones, extension occurs in front of the arc, causing the crust to thin by normal faulting. Thinning of the crust helps the region stay below sea level, forming a *forearc basin* between the arc and the trench.

2. Extension can occur behind or near the arc, where the crust is hot and weak. This causes normal faulting and thins the crust. The region subsides to lower elevations (near or below sea level) than is typical for a continental arc.

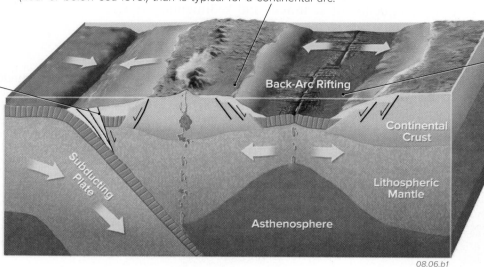

08.06.b1

3. Extension behind the arc may result in normal faults and downdropped blocks, and large amounts of extension will form a new ocean basin behind the arc. This *back-arc basin* will contain land-derived sediment along its margins and normal deep-ocean sediment in its center. Upward flow of underlying mantle continues to bring heat and material to the region, allowing the extension to continue. A well-developed back-arc basin has a somewhat small-scale version of a mid-ocean ridge.

C How Do We Determine the Age of a Basin?

Geoscientists use a variety of techniques to determine when a basin formed. They describe and measure layers in the basin, perform isotopic dating of volcanic rocks, or find key fossils. The age, thickness, and character of sediments record when and how fast a basin, like the one below, formed. Begin reading from the bottom left.

3. A unit *younger* than a basin may lie flat and overlap the edge of the basin and its faults. It shows that the basin had stopped forming by the time the unit was deposited.

2. Units deposited *during* formation of a basin may be very thick and contain coarse sediments that record steep slopes along the flanks of the basin.

1. Units *older* than a basin typically have the same thickness across the area because the basin did not yet exist. These older units were then tilted and faulted when the basin formed.

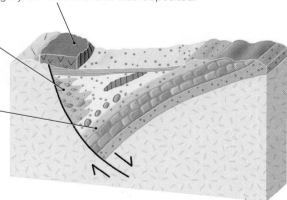

08.06.c1

4. We can calculate the rate of deposition for each unit by dividing the thickness of the unit by the time during which the unit was deposited. This plot (▼) for units in the deepest part of the basin shows that sediment accumulated most rapidly after 15 million years ago. This indicates that the basin began forming about 15 million years ago.

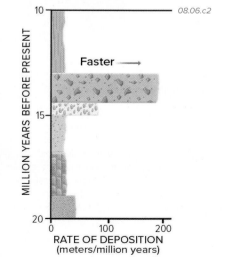

08.06.c2

Before You Leave This Page

☑ Describe and sketch the formation of non-rotating and rotating fault blocks.

☑ Summarize where extension can occur in a plate above a subduction zone.

☑ Describe or sketch how we can determine the age of a basin.

8.6

8.7 What Features Characterize the Interiors and Margins of Continents?

THE INTERIORS OF CONTINENTS tend to be tectonically stable, largely because they are far from plate boundaries. They typically contain a foundation of igneous and metamorphic rocks overlain by nearly flat-lying sedimentary rocks. The edges of continents mark the transition between continental and oceanic crust, but they are hidden beneath the sea and generally covered by a thick sequence of sediment and sedimentary rocks.

A What Features Are Common in Continental Interiors?

Continents commonly display a similar pattern, with a central region of older crystalline rocks surrounded by a relatively thin veneer of younger, nearly flat-lying sedimentary layers.

1. Some continents, including North America, have a central region called a *continental shield*. A shield consists of relatively old metamorphic and igneous rocks, commonly of Precambrian age. The crystalline (metamorphic and igneous) rocks exposed in the shield represent the kinds of rocks that underlie much of the continent, and are called the *crystalline basement*.

2. Surrounding the shield is a broad region called the *continental platform*. It is characterized by nearly horizontal sedimentary rocks that were deposited in various environments, including shallow seas, beaches, rivers, lakes, and sand dunes. The layers in the rocks are nearly flat lying because they are far from plate boundaries, but commonly are warped into broad basins and uplifts. Erosion across the gently dipping layers on the flanks of these structures exposes higher and lower rocks.

4. The boundary between the flat-lying platform sedimentary rocks and the underlying crystalline basement is a major *unconformity*. It separates rocks with very different ages, structural geometries, and geologic histories.

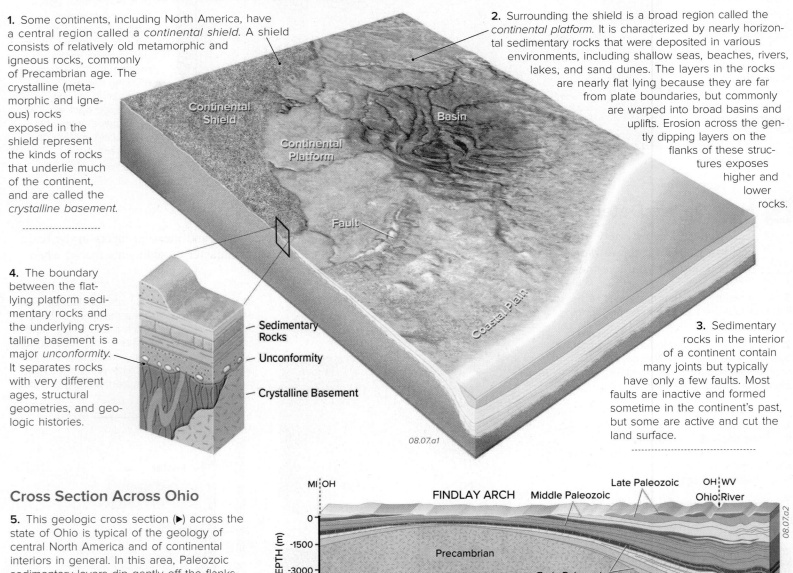

- Sedimentary Rocks
- Unconformity
- Crystalline Basement

08.07.a1

3. Sedimentary rocks in the interior of a continent contain many joints but typically have only a few faults. Most faults are inactive and formed sometime in the continent's past, but some are active and cut the land surface.

Cross Section Across Ohio

5. This geologic cross section (▶) across the state of Ohio is typical of the geology of central North America and of continental interiors in general. In this area, Paleozoic sedimentary layers dip gently off the flanks of a dome, called the Findlay Arch. The section is vertically exaggerated, so true dips are less than shown here, and the thicknesses of the layers are greatly exaggerated.

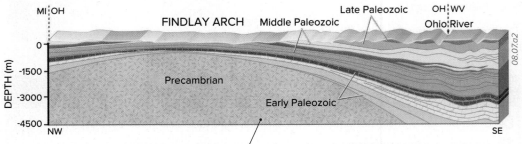

08.07.a2

6. Igneous and metamorphic rocks of the Precambrian crystalline basement rest beneath the sedimentary layers and come within less than 800 m of the surface, but are not exposed. They have been encountered in deep drill holes.

B What Features Are Typical of Continental Margins?

Some continental margins, such as the western coast of South America, are *active plate boundaries* where oceanic crust subducts beneath the edge of the continent. Many continental margins are not plate boundaries, and instead are *passive margins*. Both active and passive margins share some features.

100 km

08.07.b1

1. A *continental shelf* is a gently sloping surface that surrounds nearly all continents. On passive margins, it can extend from the shoreline as far as 1,500 km (930 mi) seaward, but it is typically narrow along active margins. The gentle slopes of most shelves, such as this one along the northeastern United States, are interpreted to have developed during the last Ice Age, when sea level was lower.

2. A *continental slope* connects the shelf with the truly deep ocean. Here, the ocean floor slopes down at angles typically between 5° and 25°. The slopes are greatly exaggerated in this figure.

3. The *continental rise* is farther out from the continental slope. Sediment transported off the continental slope accumulates here, forming a broad, gently sloping underwater plain.

4. The continental shelf and slope are locally cut by submarine canyons, including Baltimore Canyon, shown here, offshore of New York and New Jersey. The Monterey Submarine Canyon offshore of California, discussed in the opening pages of this chapter, is a classic example of a submarine canyon.

The Transition from Continent to Ocean on a Passive Margin

The transition from continent to deep ocean reflects progressively thinner continental crust and an abrupt change to oceanic crust. The thinned crust along most passive margins records rifting apart of the continent. Sediment on the continental margin varies greatly in thickness across the shelf, slope, rise, and abyssal plain.

1. Sediment is generally thinnest near the shoreline and on nearby parts of the continental shelf, which is underlain by continental crust with a close-to-normal thickness.

2. There are normal faults farther out, beneath the continental shelf and slope. These formed during the initial continental rifting that formed the margin. Normal faulting helped thin the crust, leading to deeper seafloor.

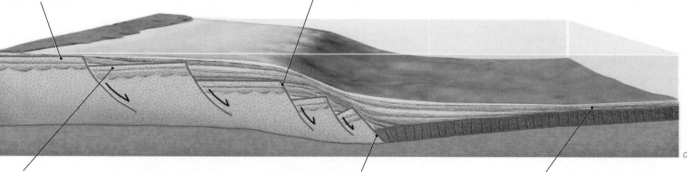

08.07.b2

3. Thick sediment accumulated over the downdropped fault blocks beneath the shelf and slope. The sedimentary layers can host important oil and gas resources.

4. The continental slope and rise mark the abrupt change from thinned continental crust of granitic composition to even thinner oceanic crust composed of basalt and gabbro.

5. The abyssal plain is farther from land and sources of land-derived sediment. It is underlain by oceanic crust and has a thin sediment cover composed of small particles of clay and other fine-grained material.

Before You Leave This Page

✓ Summarize or sketch features that are common in continental interiors.

✓ Sketch and describe the features, rocks, sediments, and geologic structures that occur along a typical continental margin.

8.7

8.8 How Do Marine Evaporite Deposits Form?

SALT AND OTHER EVAPORITE DEPOSITS OF MARINE ORIGIN occur along many continental margins, forming layers, irregularly shaped masses, and structural domes. In addition to salt, gypsum is a mineral common in evaporite deposits. Marine evaporite deposits form only in specific geologic settings, especially sites where seawater evaporates. They are important sources of salt and sulfur, and they can be instrumental in trapping petroleum. How and where do marine evaporite deposits form? Geologists commonly call evaporite deposits, whether marine or nonmarine in origin, by their shorthand term *evaporites*.

A How Do Evaporite Deposits Occur Along Continental Margins?

08.08.a1 Paradox, CO   08.08.a2 Carlsbad, NM

Natural salt (▶) is mostly composed of the sodium chloride mineral *halite* ($NaCl$), the mineral that makes up common table salt. Halite can be associated with *gypsum* (a calcium sulfate mineral), *sylvite* (a potassium salt), and other minerals that have high solubilities in water. Most evaporite deposits originally form layers, as is typical for any sedimentary rock. Such layers can be thinner than a centimeter or can comprise a layered sequence several kilometers thick. Outcrops of evaporite deposits are very soluble and so are relatively uncommon at Earth's surface. Salt recrystallizes at relatively low temperatures, so it can form large crystals, as shown here.

Salt and other evaporite deposits are very weak geologic materials, flowing easily when subjected to the stresses associated with deep burial and tectonics. They are much less dense than other kinds of rocks, and so they commonly flow as solid but soft masses, like the folded layers of gypsum shown here (▲). Upward-flowing masses of salt and gypsum can push up overlying layers, forming folds and domes, or in some cases even pierce through the overlying layers. Folds and other structures formed by moving salt and gypsum are commonly sites where petroleum (oil and gas) accumulates in significant quantities, in addition to deposits of sulfur minerals.

B How Do Evaporites Form Near Continental Margins?

Many evaporite deposits form when seawater evaporates, leaving behind a residue of salt, gypsum, and other minerals from chemical components that were dissolved in the water. Such evaporation is especially efficient in warm, dry climates, which have a high rate of evaporation. Formation of evaporite deposits requires that evaporation occur in water bodies with limited connection to the oceans, so that water made salty by evaporation cannot simply flow into the open ocean. Evaporite accumulations formed in such marine settings are *marine evaporite deposits*.

1. Most marine evaporite deposits form in narrow seas, especially inland seas formed when two pieces of continent begin to separate during continental rifting, and in narrow seas formed during the early stages of seafloor spreading. They also accumulate along shorelines, where seawater spills onto low areas next to the sea.

2. Evaporites can accumulate on tidal flats and other broad, flat areas adjacent to the sea. These areas are close to sea level and can be periodically flooded during high tides, storms, or when sea levels are high.

3. Most smaller coastal bodies of water can receive input from rainfall, runoff from the land, and inflow of seawater from an adjacent ocean. If enough of this water evaporates, it deposits evaporites along the shoreline and on the floor of the water body. Inflow of water from the land and sea can effectively replace the water lost to evaporation, permitting evaporation and salt deposition to continue over a long time.

4. In some cases, a low barrier of rock or sediment restricts the flow of water into and out of a body of water. The water becomes more salty due to evaporation, causing evaporites to precipitate. If sea level drops or if the barrier is uplifted by tectonics, the inflow of seawater can decrease or cease, causing widespread deposition of salt and other evaporite minerals as all or most of the trapped water evaporates.

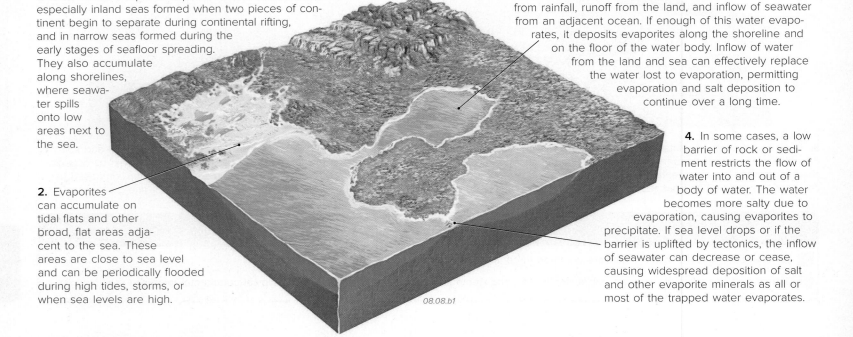

08.08.b1

C What Structures Do Salt Deposits Form?

Salt and other evaporites are such structurally weak rock that they can form their own unique kinds of geologic structures. They also can greatly influence how faults and folds develop in overlying rocks.

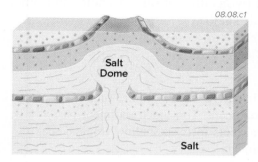

08.08.c1

Salt is less dense than most rocks; when buried, it can buoyantly flow toward the surface in steep, pipe-like conduits. The resulting structure is a *salt dome*.

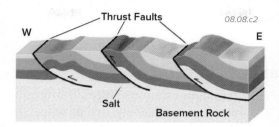

08.08.c2

When a region containing a thick salt layer is deformed, the salt can slip and flow, allowing overlying rocks to fold and fault. This cross section shows part of the Jura Mountains near the French-Swiss border. The folds and faults are underlain by a weak layer of salt.

08.08.c3 Iran

Where salt reaches the surface, such as in a salt dome or an anticline, it can flow downslope under the influence of gravity and form a *salt glacier,* such as this one.

D What Salt Structures Occur Along the Gulf Coast of the United States?

The Gulf Coast of the southern United States is world famous to geologists because it contains many salt structures, both on land and offshore. The salt structures have played a key role in the formation of the region's large oil fields and provide important sources of salt and sulfur minerals. For these reasons, they have been extensively studied by seismic surveys and by expensive drilling, sometimes in thousands of meters of water.

1. This diagram shows the land and seafloor in the Gulf of Mexico offshore of the Texas-Louisiana coast. An interpretation of the subsurface geology, drawn on the sides of the block, is based on studies by many geoscientists and billions of dollars of drilling for petroleum.

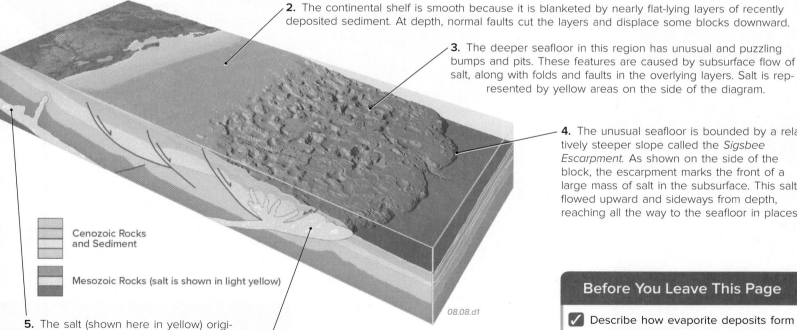

2. The continental shelf is smooth because it is blanketed by nearly flat-lying layers of recently deposited sediment. At depth, normal faults cut the layers and displace some blocks downward.

3. The deeper seafloor in this region has unusual and puzzling bumps and pits. These features are caused by subsurface flow of salt, along with folds and faults in the overlying layers. Salt is represented by yellow areas on the side of the diagram.

4. The unusual seafloor is bounded by a relatively steeper slope called the *Sigsbee Escarpment.* As shown on the side of the block, the escarpment marks the front of a large mass of salt in the subsurface. This salt flowed upward and sideways from depth, reaching all the way to the seafloor in places.

Cenozoic Rocks and Sediment

Mesozoic Rocks (salt is shown in light yellow)

08.08.d1

5. The salt (shown here in yellow) originally was deposited in a thick layer when continental rifting during the Mesozoic formed narrow basins. The basins at times had limited connection with the sea, causing evaporation of seawater and deposition of the salt layer. The salt was later buried by sediments, shown in light green, tan, and beige.

6. As the salt was buried and subjected to increased pressure, it flowed sideways and rose up through the overlying sedimentary layers. Movement of the salt folded and domed the layers. In places, it formed steep, pillar-shaped salt domes, shown here as finger-like yellow masses. The salt domes and associated folded rock layer trapped oil and gas, for which the Gulf Coast is well known.

Before You Leave This Page

✓ Describe how evaporite deposits form near continental margins.

✓ Describe how salt can occur in salt domes, some folded mountain belts, and salt glaciers.

✓ Describe how salt structures are expressed in the Gulf Coast region.

8.8

8.9 How Do Reefs and Coral Atolls Form?

REEFS ARE SHALLOW, MOSTLY SUBMARINE FEATURES, built primarily by colonies of living marine organisms, including coral, sponges, and shellfish. Reefs can also be constructed by accumulations of shells and other debris. Corals thrive in many settings, as long as the seawater is warm, clear, and shallow.

A In What Settings Do Coral Reefs Form?

Corals are a group of invertebrate animals that form calcium carbonate structures. To thrive, corals require nutrients, warmth, sunlight for photosynthesis, and water that is relatively free of suspended sediment. Too much sediment partially blocks the sun, can bury the coral, or can clog openings in the tiny organisms. Coral reefs form in shallow tropical seas with relatively clear water. Large waves batter many reefs, producing carbonate sediment.

1. Reefs (▼) are buildups on the seafloor, constructed by coral and other marine organisms. Some reefs occur along the edges of continents, forming *barrier reefs* offshore from the main coastline. Reefs and islands protect a continent from large waves. They enclose a lagoon on the landward side but have a side that faces the open ocean and is exposed to large waves and storms. Erosion of the reefs can form low, sandy islands with beaches covered by white sand produced by erosion and reworking of pieces of reef, shells, and other carbonate materials.

2. Reefs and other carbonate accumulations can form broad, shallow *platforms,* like the Bahama Islands east of Florida. In some cases, older reef deposits and dunes rise slightly above sea level. Between most islands, the water is shallow and the seabed is composed of white, carbonate-rich sand derived from wave erosion of reefs and the land.

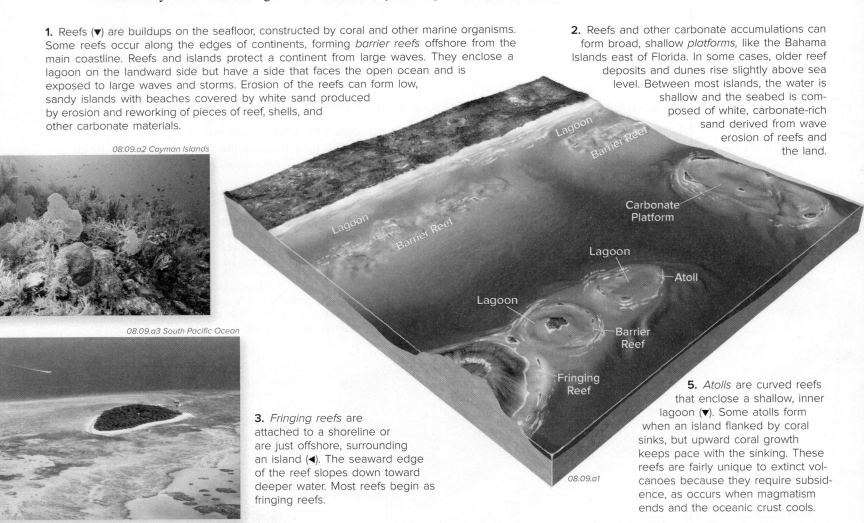

08.09.a2 Cayman Islands

08.09.a3 South Pacific Ocean

08.09.a1

3. *Fringing reefs* are attached to a shoreline or are just offshore, surrounding an island (◄). The seaward edge of the reef slopes down toward deeper water. Most reefs begin as fringing reefs.

5. *Atolls* are curved reefs that enclose a shallow, inner lagoon (▼). Some atolls form when an island flanked by coral sinks, but upward coral growth keeps pace with the sinking. These reefs are fairly unique to extinct volcanoes because they require subsidence, as occurs when magmatism ends and the oceanic crust cools.

08.09.a4 Great Barrier Reef, Australia

4. The *Great Barrier Reef* (◄) is along the eastern coast of Australia and has a unique history. Its base was formed along the edge of a shallow platform during the last Ice Age (17,000 years ago) when sea levels were lower. As sea levels returned to normal and began to drown the platform, the corals grew upward, keeping themselves in shallow water. Over time, the reef formed the largest organic buildup on Earth, one that is easily visible from space.

08.09.a5 Nukuoro Atoll, Federated States of Micronesia

B How Do Atolls Form?

Charles Darwin proposed a hypothesis for the origin of atolls after observing a link between certain islands and atolls during his research aboard the ship *Beagle* from 1831 to 1836. According to his model, shown below, atolls form around a sinking landmass, such as a cooling or extinct volcano. Another model (not shown) interprets some atolls as being the result of preferential erosion of the less dense center of a carbonate platform.

08.09.b1–3

Stage 1: A volcanic island forms through a series of eruptions in a tropical ocean, establishing a shoreline along which corals can later grow and construct a fringing reef.

Stage 2: After volcanic activity ceases, the new crust begins to cool and sink. Coral reefs continue building upward as the island subsides, forming a barrier reef some distance out from the shoreline.

Stage 3: The volcano eventually sinks below the ocean surface, but upward growth of the reef continues, forming a ring of coral and other carbonate material. This forms an *atoll*, with a central, shallow lagoon.

C Where Do Reefs Occur in the World?

Most of the world's reefs are in tropical waters, located near the equator, between latitudes of 30° north and 30° south. Reef corals are more diverse in the Pacific, probably because many species went extinct in the Atlantic Ocean during the last Ice Age. The map below shows coral reefs as red dots.

08.09.c1

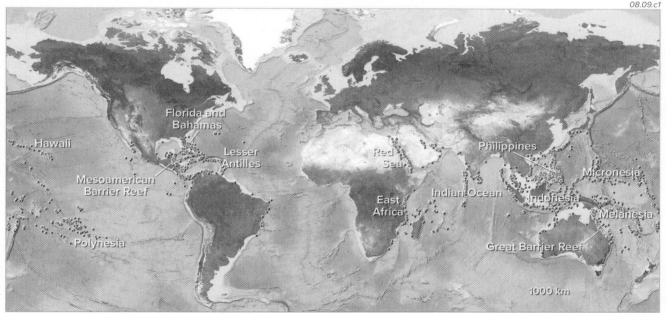

1000 km

4. Reefs in the *Philippines* cover an estimated 25,000 square kilometers and consist of fringing reefs with several large atolls. Reefs also flank *Indonesia* and nearby *Malaysia* (not labeled).

5. The *Great Barrier Reef*, along the northeastern flank of Australia, is the largest reef complex in the world. The world's second largest reef is in *New Caledonia,* a series of islands east of Australia and south of Micronesia.

1. The central and southwestern Pacific, including *Polynesia* and *Micronesia*, has many atolls and reefs, including a wide variety of barrier and fringing reefs. Farther north, Hawaii is also warm enough for reefs.

2. Well-known reefs are present throughout much of the *Caribbean* region, including *Florida,* the *Bahamas,* and the *Lesser Antilles.* The longest barrier reef in the Caribbean extends some 250 km (150 mi) along the Yucatan Peninsula, from the north of Belize, southward to Honduras (the Mesoamerican Barrier Reef).

3. Reefs occur along the continental shelf of *East Africa,* such as in Kenya and Tanzania. Other reefs encircle islands in the Indian Ocean and the shoreline of the Red Sea.

Before You Leave This Page

✓ Describe the different kinds of reefs and where they form.

✓ Describe the stages of atoll formation.

✓ Name some locations with large reefs.

8.10 What Are Tectonic Terranes?

EMBEDDED WITHIN CONTINENTS are pieces of crust that have a different geologic history than adjacent regions. These exotic pieces, called *tectonic terranes*, originate in a variety of tectonic settings. Many are structurally added to the edges of continents during tectonic collisions. There likely are terranes in the region where you live. They are common east of the Appalachians, along the Pacific Coast, and buried beneath sediments along the Gulf Coast.

A How Do We Recognize a Terrane and Where Do Terranes Originate?

A tectonic terrane is defined as being bounded by faults and having rocks, structures, fossils, and other geologic aspects that are unlike those in adjacent regions.

1. The boundaries between a terrane and the adjacent regions are major faults or other zones of shear. The fault-bounded nature of a terrane, such as this volcanic terrane, means that the terrane has no continuous link with the rocks around it.

2. A tectonic terrane has a different sequence of rocks than adjacent regions. The terrane on the left has pillow basalt overlain by shale, limestone, and conglomerate, but the continental rocks to the right have none of these units. Adjacent terranes usually also have different ages of rocks and different types of structures. These discrepancies imply that the two pieces of crust had different geologic histories.

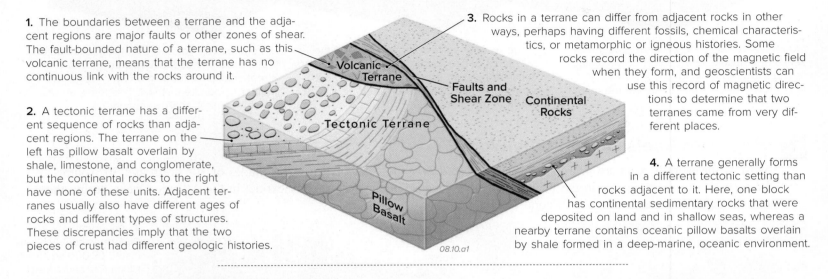

3. Rocks in a terrane can differ from adjacent rocks in other ways, perhaps having different fossils, chemical characteristics, or metamorphic or igneous histories. Some rocks record the direction of the magnetic field when they form, and geoscientists can use this record of magnetic directions to determine that two terranes came from very different places.

4. A terrane generally forms in a different tectonic setting than rocks adjacent to it. Here, one block has continental sedimentary rocks that were deposited on land and in shallow seas, whereas a nearby terrane contains oceanic pillow basalts overlain by shale formed in a deep-marine, oceanic environment.

08.10.a1

Some Original Settings of Terranes

5. Some terranes contain pillow basalt, deep-sea sediment, and other attributes that indicate they originated as oceanic crust. Such terranes must be later added to the continent, or we would not see them today.

6. Many terranes consist of andesitic volcanic rocks and volcanic-derived sedimentary rocks that formed as island arcs. Island arcs are very common terranes because they may move across the ocean until they collide with, and become part of, another landmass.

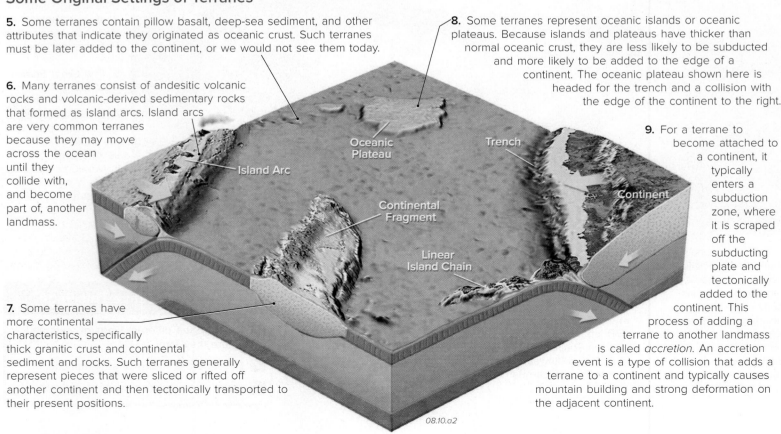

8. Some terranes represent oceanic islands or oceanic plateaus. Because islands and plateaus have thicker than normal oceanic crust, they are less likely to be subducted and more likely to be added to the edge of a continent. The oceanic plateau shown here is headed for the trench and a collision with the edge of the continent to the right.

9. For a terrane to become attached to a continent, it typically enters a subduction zone, where it is scraped off the subducting plate and tectonically added to the continent. This process of adding a terrane to another landmass is called *accretion*. An accretion event is a type of collision that adds a terrane to a continent and typically causes mountain building and strong deformation on the adjacent continent.

7. Some terranes have more continental characteristics, specifically thick granitic crust and continental sediment and rocks. Such terranes generally represent pieces that were sliced or rifted off another continent and then tectonically transported to their present positions.

08.10.a2

B How Old Is North America?

The crust of North America varies widely in age. The oldest parts of the Precambrian shield formed as early as 3.8 to 4.0 billion years ago, but most of the continent was added later as a series of terranes, mostly during Precambrian, Paleozoic, and Mesozoic time. Read clockwise, starting from the upper left.

1. This map of North America shows the ages of different rocks exposed at the surface. Darker browns and reds are Precambrian rocks, purple and blue colors show Paleozoic rocks, greens represent Mesozoic rocks, and yellows and tan show Cenozoic rocks.

2. The eastern half of Canada contains a vast area of Precambrian rocks, called the *Canadian Shield*. These rocks, colored brown and red on this map, are mostly 3.0 to 1.7 billion-year-old metamorphic and igneous rocks.

3. Around 1.1 billion years ago, a continent-sized terrane, called the *Grenville Province* (rusty yellow on the map), was added to the southeastern edge of the shield.

4. Easternmost Canada and nearby parts of the United States contain the *Avalon terrane*, which collided with North America in the Paleozoic.

11. Alaska and western Canada are a mosaic of terranes, including slices of the North American continent and oceanic terranes that were formed far south of the equator. The terranes were added throughout the Paleozoic, Mesozoic, and Cenozoic.

10. Paleozoic rocks (blue) and Mesozoic rocks (green) cover the center of the continent and represent the continental platform. They are underlain by Precambrian crystalline rocks that are locally exposed at the surface in the Rocky Mountains.

9. Western North America contains many terranes; some formed far away. These terranes were accreted onto the west coast of the continent during Paleozoic, Mesozoic, and Cenozoic times.

5. A number of late Precambrian and Paleozoic terranes are present in the *Appalachian Mountains* and in the *Piedmont Province* to the east. These terranes are thought to be pieces of continents and island arcs that collided with North America. Their accretion occurred primarily during the Paleozoic formation of the Appalachian Mountains.

08.10.b1

500 km

8. In the Southwest and the Southern Rockies, several large Precambrian provinces were added onto the southern edge of North America between 1.9 and 1.6 billion years ago.

7. Mexico largely consists of terranes added to North America from the Paleozoic onward. The largest terranes are Paleozoic and Mesozoic island arcs that collided with the west coast of the Mexican mainland during the Mesozoic.

6. The tan, yellow, and green areas along the southern and southeastern edge of North America represent the *Coastal Plain*. This low-lying region is covered by Late Mesozoic and Cenozoic sediments that were deposited after early Mesozoic rifting thinned the crust and blocked out this edge of North America. The southern half of Florida has a terrane (a piece of Africa) at depth.

Before You Leave This Page

☑ Summarize the characteristics used to recognize a terrane and a few of the main tectonic settings in which terranes originate.

☑ Describe how terranes are added to crust, and identify regions of North America that represent such added pieces of crust.

8.10

8.11 How Did the Continents Join and Split Apart?

CONTINENTS SHIFT THEIR POSITIONS over time in response to plate tectonics. They have rifted apart and collided, only to rift apart again. Where were the continents located in the past, and which mountains resulted from their motions? The story of the movement of the continents is the same story as the origin of the modern oceans. But here, we emphasize which continents were joined and how they separated. We start with 600 million years ago and work forward to the present.

600 Ma: The Supercontinent of Rodinia

1. The images on these pages show one interpretation of where the continents were located in the past. Geologist Ron Blakey created the artistic renderings of the continents, mountains, and oceans. For most time periods, he created two views, one focused on the western hemisphere (image on the left) and one on the eastern hemisphere (image on the right), generally with some overlap. We begin here with a single image, centered on the South Pole.

2. Before the Paleozoic, in the last part of the Precambrian, all of the major continents were joined. This supercontinent is called *Rodinia*. Nearly all of the other side of the globe is a huge ocean.

3. North America was in the initial stages of rifting from Rodinia. This rifting outlined the western margin of North America, but geologists are not certain which continent was adjacent to North America. Options include Australia, Antarctica, and Asia.

4. Large parts of Rodinia were near the South Pole. There is evidence of widespread glaciation in Rodinia, but geologists are debating the extent and timing of glaciation. Geologists are also actively investigating how the continents were arranged during this time by trying to more precisely match the ages and sequences of rocks between different continents.

08.11.a1

500 Ma: Dispersal of the Continents

5. At 500 Ma, in the early part of the Paleozoic, North America and Europe were separate, moderate-sized continents that had not yet joined.

6. Antarctica, Australia, South America, and Africa were joined in the Southern Hemisphere, together forming the southern supercontinent of *Gondwana*, which was mostly located in the Southern Hemisphere (mostly out of view on these figures). Gondwana was separated from the northern continents by some width of ocean.

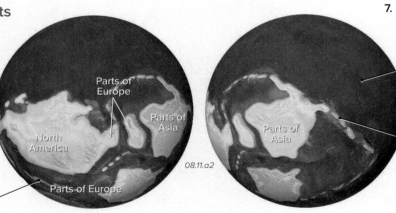

08.11.a2

7. With the continents still mostly clustered together, the other side of Earth was a single large ocean, much larger than the size of the present-day Pacific.

8. Island arcs surrounded Europe and parts of Asia. Some of the arcs would later collide with the continents, adding tectonic terranes.

370 Ma: Before Pangaea

9. In the middle of the Paleozoic, at 370 Ma, North America and parts of Europe were joined but were not connected with Asia, which lay to the north. North America had collided with a microcontinent called *Avalonia*. This created mountains in what is now the northern Appalachians.

10. North America was approaching Africa and South America along a convergent margin. The continents were on a collision course as the intervening ocean became narrower over time.

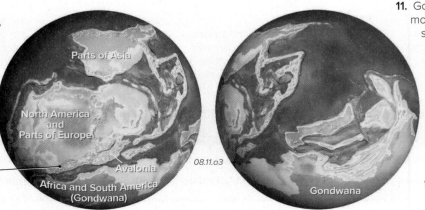

08.11.a3

11. Gondwana remained mostly intact, except for some slices of continental crust that probably were rifted away from the larger supercontinent. Avalonia was probably an example of one of these rifted pieces, but it had broken away sometime before 370 Ma, the time pictured here.

280 Ma: The Supercontinent of Pangaea

12. In the late Paleozoic, around 280 Ma, a continental collision between North America and the northern edge of Gondwana (South America and Africa) formed the Appalachian Mountains along the East Coast and the Ouachita Mountains (not labeled) in the southeastern United States. The collision also affected parts of Europe.

13. After this collision and a series of smaller collisions, all the continents were joined in a supercontinent called *Pangaea*.

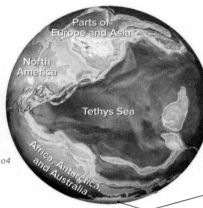

08.11.a4

14. A wedge-shaped ocean, the *Tethys Sea*, separated Asia from landmasses farther to the south.

15. Southern Africa, Australia, and Antarctica were close to the South Pole and so at this time were partly covered with ice.

150 Ma: Gondwana and Laurasia

16. At 150 Ma, in the late Jurassic, North America had separated from Africa and South America. The Atlantic Ocean now existed, and continents on either side of the Atlantic were moving away from each other due to seafloor spreading. The left globe is rotated so that the central Atlantic Ocean is in the center of the image.

17. During this time, North America was still joined with Europe and Asia, forming the northern supercontinent of *Laurasia*. South America had not yet rifted away from Africa.

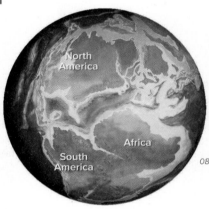
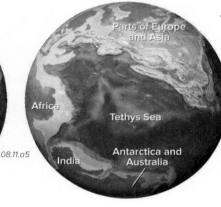

08.11.a5

18. Antarctica was still attached to the southern tips of Africa and South America. India was attached to the northern edge of Antarctica. These continents were soon to be rifted apart, which would mark the end of *Gondwana*.

Present

19. This view shows the present-day configuration of continents and oceans. Examine these globes and think about the present-day plate boundaries while envisioning which way the continents are moving relative to one another. Use the relative motions to predict where the continents were likely to have been in the past. Check your predictions by examining each set of previous globes as you step backward in time.

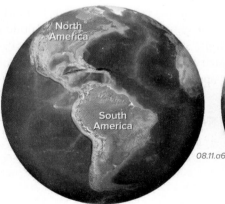

08.11.a6

20. When you are done working backward in time, start at 600 Ma and track the position of every continent you can and watch each ocean open. By viewing this sequence, you are observing how our present-day world came to be.

21. Finally, think about how the position of the continents will change if present-day plate motions continue into the future. What collisions are yet in store for North and South America?

Before You Leave This Page

☑ Briefly summarize the general positions of the continents in the past, especially since 280 Ma.

☑ Identify times when the continents were joined in the supercontinents of Gondwana, Laurasia, Pangaea, and Rodinia.

8.11

How Do Oil and Natural Gas Form?

OIL AND NATURAL GAS, together called *petroleum*, form naturally when sediment rich in organic material is deposited, buried, and heated to slightly elevated temperatures. Once formed, petroleum can escape to the surface or be trapped at depth, where it can be discovered and extracted through drilling, providing energy and fuels.

A What Is Petroleum and Where Does It Come From?

Naturally occurring petroleum is an organic substance, largely composed of carbon chemically bonded with hydrogen and smaller amounts of other elements. The dominance of hydrogen and carbon atoms is the reason we use the term *hydrocarbons* to refer to oil and natural gas, as well as to their refinery-produced derivative products, including gasoline and diesel fuel. The organic material that turns into hydrocarbons comes from several sources, including organism-rich reefs, aquatic plants, and especially microorganisms. When heated, these organic materials convert to a succession of other hydrocarbons, including oil, a process called *maturation*, which generally takes millions of years.

08.12.a1

08.12.a2

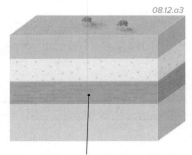
08.12.a3

The first stage in the formation of oil and gas is accumulation of organic material, perhaps in a layer of dark, organic-rich mud. A rock that contains enough organic material to produce petroleum is referred to as a *source rock*. At low temperatures on or near Earth's surface, the organic material is relatively unordered or still retains some of the structure of the animals or plants from which the material was derived.

To end up as oil, the organic material must be preserved before it can decompose. This usually involves being deposited in oxygen-poor conditions and buried under other layers of sediment. When buried to shallow depths and heated to less than about 60°C, the organic starting material is converted into *kerogen*, a thick substance composed of long chains of hydrocarbons.

Over time, source rocks can be buried by more sedimentary layers, becoming heated by the temperature increase with depth. When heated to 60°C to 120°C, long hydrocarbon chains in kerogen break down into heavy and light *oils*. At these and greater temperatures (up to about 200°C), the oily hydrocarbons convert into *natural gas*.

B Where Do Oil and Gas Reside in Conventional and Unconventional Resources?

Oil and gas can occur in rock in two main ways: conventional oil and gas, and unconventional oil and gas. Most oil produced in the past was from conventional resources, but the recent energy boom in North America mostly involves unconventional resources. The main distinction is that conventional oil and gas are mobile and migrate from the source rock toward the surface. Oil and gas in unconventional settings, such as shale gas and shale oil, are less mobile and have to be explored and extracted by newer drilling techniques, such as *hydraulic fracturing*. Important in understanding petroleum are the concepts of *porosity*, which is the amount of open space in a material, and *permeability*, the ability of a material to allow fluid to pass through.

Setting of Conventional Oil and Gas

1. In *conventional* oil and gas, rocks have enough porosity and permeability to allow oil or gas to accumulate and flow through the pore spaces, fractures, and other openings in the rocks. Oil and gas, being lighter than water, rise upward, either becoming trapped below the surface or reaching the surface in seeps.

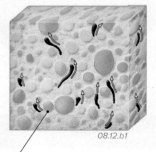
08.12.b1

2. In the example here, a conglomerate has well-rounded pebbles and cobbles in a matrix of mostly sand. Such a rock has abundant porosity (open space) and permeability (ability to transmit fluid), allowing oil and gas to accumulate and flow upward.

Setting of Shale Gas and Shale Oil

3. In *shale gas* and *shale oil*, a shale source rock contains sufficient organic material to generate oil and gas, but its very low permeability does not allow the natural gas and oil to flow out of the shale. As a result, any gas and oil generated within the shale remain trapped in a number of small-scale sites.

4. Some gas and oil remain attached with the original organic material or occupy other sites that allow the gas and oil to form small, isolated masses.

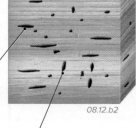
08.12.b2

5. Somewhat larger, but still small, accumulations of natural gas and oil form in openings along bedding surfaces and in small cross-cutting fractures. In either case, the openings are not interconnected, so the gas and oil remain trapped within the shale in which they formed.

C Where Do Oil and Gas Migrate?

Once oil and gas form, what happens? Both are mobile, fluid materials and can travel along fractures and through pore spaces between grains. In many situations and in some rock types, such as shale, oil and gas remain within, or fairly close to, the source rock where they originated. In other cases, they migrate far from where they formed.

1. As you can observe for yourself by placing several drops of any kind of oil in a bowl of water, oil is lighter (less dense) than water. It floats on the surface of water and so will buoyantly rise through groundwater toward the surface. Water under pressure can force oil and gas upward or laterally (sideways) through the rock. Gas is even lighter than oil.

2. Oil and gas, like groundwater, can flow through rocks that are permeable. Some rocks, like many sandstones, have open spaces between the grains and along fractures, and so are relatively permeable. Oil and gas may move up through a permeable layer, such as the inclined sandstone layer shown here.

3. Other rocks are less permeable and block the flow of oil, gas, and groundwater. A rock unit can be relatively impermeable if it lacks interconnected pore spaces and through-going fractures. Rocks that are typically impermeable include (1) shale, which has very small pore spaces, (2) unfractured granite, which has crystals that generally fit tightly together, and (3) salt, which flows easily to close up any open spaces.

6. If oil flows into sandy sediments, it can form *oil sands* or *tar sands*. Large deposits in Alberta, Canada, are mined in large pits to extract the hydrocarbons from the sandy host. Oil can migrate upward until it reaches the surface, where it flows out onto the surface as an *oil seep* (▶).

08.12.c2 Southwestern UT

5. To trap oil and gas at depth, a rock unit must have no through-going pores or fractures to provide an easy pathway to the surface. Severely deformed and fractured rocks, therefore, generally are less able to trap hydrocarbons than undeformed rocks. Some faults, however, effectively block the flow of fluids because the faulting has produced finely crushed rock fragments that filled open pore spaces.

08.12.c1

Migration of Oil

Petroleum

Migration of Oil

4. Oil and gas will be prevented from reaching the surface if they become trapped at depth by impermeable rocks, like a shale. Oil and gas rise as far as they can, floating on top of water within the rock (▶). Gas is lighter than oil, so it floats on top of the oil, which floats on top of the water.

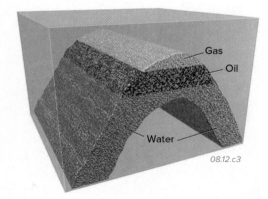
Gas
Oil
Water
08.12.c3

Directional Drilling and Hydraulic Fracturing

An energy boom involving shale gas and shale oil has occurred in the last several decades. The boom is the result of two innovative exploration techniques: *directional drilling*, which only became technologically feasible in the 1980s, and *hydraulic fracturing*, which has been in practice since the 1920s.

In exploring for conventional oil and gas resources, most drill holes are near-vertical, reaching the target depth with the shortest possible drill hole. In exploring for shale gas and shale oil, the drillers guide the drill hole along a curved path so that it becomes more and more horizontal (parallel to the layers). When the drilling reaches a layer of organic-rich shale, as much as several millions of gallons of water and sand, with some chemicals, are pumped down into the hole, pressurized, and

allowed to flow out into the shale from the drill hole. The high fluid pressures cause bedding planes and existing fractures to open up slightly and cause new fractures to form, the process of *hydraulic fracturing* (fracking). A goal is to form an interconnected network of opened fractures and bedding planes that will allow gas and oil trapped in the shale to escape. Well-rounded sand grains in the pressurized fluid are included to keep the fluid pathways propped open after the fluid pressures are reduced. As with any energy source, hydraulic fracturing presents trade-offs between economic benefits and societal concerns, including whether fluids and chemicals from hydraulic fracturing will contaminate fresh groundwater above the shale layer. Studies by the EPA and USGS indicate that such contamination has not occurred.

Before You Leave This Page

- ✓ Summarize the natural sources of oil and gas and how they form by burial and heating.

- ✓ Explain the difference between conventional and unconventional petroleum resources and in how they are drilled.

- ✓ Sketch or describe how oil and gas move through rocks and how they can be trapped at depth or end up on the surface.

8.12

Where Will Mountains and Basins Form in This Region?

The figure below shows part of a continent and adjacent ocean. There are no plate boundaries now, but a subduction zone will form along the western coast of the continent, and the eastern part of the continent will be rifted away. You will use the typical patterns that form along such boundaries to predict where mountains and basins will form once the new plate boundaries are fully developed.

Goals of This Exercise:

- Observe the continent and ocean below, and read the descriptions of the types of features that will form in the future.
- Use your understanding of plate boundaries and the settings in which mountains and basins develop to predict where mountains and basins will form. Sketch your predictions on a diagram of the region.
- Predict what the regional topography will be like in different parts of the region, identifying whether an area will rise or subside, and what changes on the surface, within the crust, or in the mantle would cause this change in elevation.

This view shows a continent and ocean at some time, which we will call *Time 1*. The western part of the region is a typical ocean basin and has no trenches, mid-ocean ridges, or hot-spot-related islands.

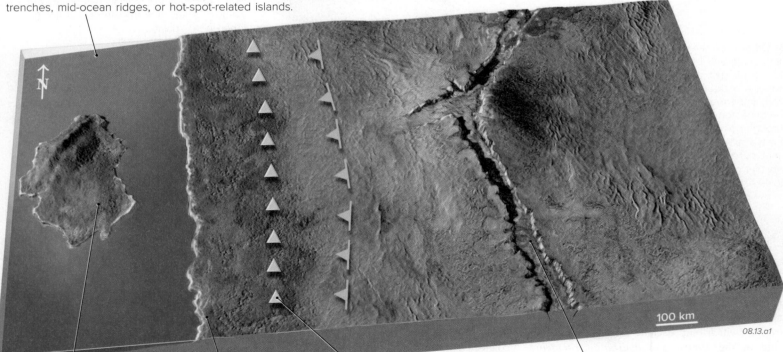

100 km

08.13.a1

A small piece of continent lies offshore in the middle of the ocean. When the oceanic plate begins to move, this piece of continent will be carried toward and will collide with the main continent.

The ocean-continent edge is currently a passive margin, not a plate boundary. It will become an ocean-continent convergent boundary, and the oceanic material will be subducted eastward below the continent.

Once plate convergence begins, a magmatic belt will form inland from the coast, near the position of the yellow triangles. Farther inland, a thrust belt will form as shown by the blue dashed line with teeth. In the thrust belt, the western part of the continent will be thrust eastward over the central part of the continent.

A continental rift has formed, with three arms radiating out from a high central region, which is a hot spot marked by voluminous volcanism. This rift will split the continent into two pieces. At some later time, the piece of continent to the right will break away completely, and seafloor spreading will form a new ocean basin. Even later, at a time we will call *Time 2*, the edge of the continent will have evolved into a passive margin (not a plate boundary), and the spreading center will be out of the region.

Procedures

Use the data to complete the following steps, entering your answers in the worksheet or online.

A. Observe the regional features shown on the figure on the left page, which represents the situation at *Time 1*. Read the descriptions associated with that figure and decide what each statement implies about the *future topography* (elevations) of the area.

B. For each feature (subduction zone, thrust belt, etc.) that will form by *Time 2*, think about how that feature is typically expressed in the topography. Does it form a mountain range, a basin, or a mountain with a nearby basin?

C. On the worksheet, sketch your predictions about the area's topography for *Time 2* on the simplified figure below, which shows the same area as the figure on the previous page. The figure shows the overall shape of the continent but not the topography. Use the following letters: *O* for an oceanic trench, *A* for an accretionary prism, *M* for mountains, *V* for volcanoes in the continental magmatic belt, *B* for a basin, and *P* for a passive margin. Feel free to sketch some simple lines to portray the locations of the features. Your instructor may have you predict other features that might develop, such as a tectonic terrane or features related to a collision.

D. On the map below are letters A–D. A is along the coast, B is at the future position of the magmatic belt, C is within the future fold and thrust belt, and D is along the coast from which the other piece of the continent was rifted. In the worksheet, predict what will happen to the crustal thickness in each of the four locations, and identify the processes that could cause thickening or thinning of the crust or the mantle part of the lithosphere beneath each site.

Perspective of the Region in the Future (Time 2)

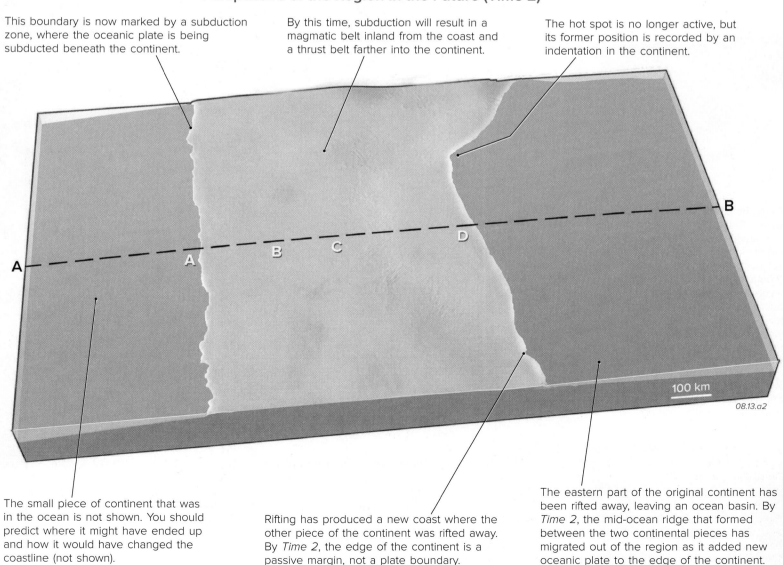

This boundary is now marked by a subduction zone, where the oceanic plate is being subducted beneath the continent.

By this time, subduction will result in a magmatic belt inland from the coast and a thrust belt farther into the continent.

The hot spot is no longer active, but its former position is recorded by an indentation in the continent.

100 km

08.13.a2

The small piece of continent that was in the ocean is not shown. You should predict where it might have ended up and how it would have changed the coastline (not shown).

Rifting has produced a new coast where the other piece of the continent was rifted away. By *Time 2*, the edge of the continent is a passive margin, not a plate boundary.

The eastern part of the original continent has been rifted away, leaving an ocean basin. By *Time 2*, the mid-ocean ridge that formed between the two continental pieces has migrated out of the region as it added new oceanic plate to the edge of the continent.

Sculpting Landscapes

ONCE NEAR EARTH'S SURFACE, the foundation of rocks, sediment, and soil is subjected to sunlight, moisture, rain, running water, wind, and other processes that affect the character and location of these materials. The process of weathering tends to disintegrate and loosen earth materials, physically, chemically, or usually both at the same time. Once the materials are weakened and loosened, the process of erosion removes them from their original locations and begins transporting them away. How do we recognize these processes and their effects in landscapes? In this chapter, we focus on the processes that sculpt natural landscapes and also affect materials in your neighborhood.

This large perspective view shows satellite data superimposed over topography for the area of central Utah around Moab, a city famous for its wealth of outdoor activities, including hiking, rock climbing, mountain biking, off-road driving, river running, and landscape-oriented tourism. This area contains Canyonlands National Park, Arches National Park, Dead Horse Point State Park, famous red-rock canyons, and countless other beautiful sites.

What combination of factors have resulted in such a naturally endowed place?

09.00.a3 Green River in Labyrinth Canyon, UT

Arches National Park, north and west of Moab, is world renowned for its graceful rock arches (▶), naturally eroded into cliffs of red sandstone.

How do arches and natural bridges form, and how do these two differ?

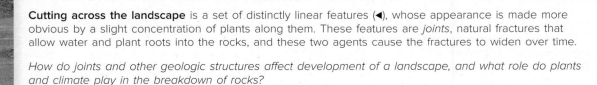

09.00.a2 Arches NP, UT

Labyrinth Canyon of the Green River

Dead Horse Point

Green River

Colorado River

Confluence of the Colorado and Green Rivers

Canyons of the Colorado River and Green River offer some spectacularly scenic views (▲). High vistas, such as Dead Horse Point, overlook canyons cut by the two rivers and their tributaries, exposing beautifully layered sequences of rocks. Both rivers flow toward the south and join at The Confluence, below which is Cataract Canyon, one of the most exciting stretches of white-water rafting anywhere.

What natural processes are affecting these layered rocks, and why does the landscape have a stair-stepped appearance?

Cutting across the landscape is a set of distinctly linear features (◀), whose appearance is made more obvious by a slight concentration of plants along them. These features are *joints*, natural fractures that allow water and plant roots into the rocks, and these two agents cause the fractures to widen over time.

How do joints and other geologic structures affect development of a landscape, and what role do plants and climate play in the breakdown of rocks?

09.00.a4 The White Rim, Canyonlands NP, UT

TOPICS IN THIS CHAPTER

09.00.a5 Arches NP, UT

09.00.a1

Arches National Park

Colorado River

Moab

09.00.a6 Labyrinth Canyon, UT

The red-rock landscapes have a wealth of unusual shapes that remind us of more familiar objects, such as the aptly named "Three Gossips" (▲). In addition, there are sheer rock walls and spires, such as the Fisher Towers and Castle Valley, that attract rock climbers from all over the world.

How does nature produce natural rock formations shaped like humans, animals, castles, and fanciful places?

The rivers have cut steep-walled canyons and deposited loose sediment along their channels. The steep cliffs and slopes are worn back, falling into the canyon, mostly a few rocks at a time, but sometimes as huge, sudden landslides. The debris shed off the cliffs partially cover underlying slopes.

What causes rock pieces to detach and fall off a steep rock face, what is their appearance in landscapes, and what happens to these pieces over time?

Moab: A Natural Wonder

Moab, located in central Utah, has some of the most fantastic scenery on the planet. Visitors come from all over the world to hike or gaze into the canyons and to photograph the arches, spires, and strange shapes. Access to the canyon is afforded by hiking trails, bike trails, four-wheel drive roads, and river running along the Colorado and Green rivers. The canyons have names that try to convey their essences, such as Labyrinth Canyon, Stillwater Canyon, and Cataract Canyon. It would be difficult to find any place on Earth with such a wealth of natural beauty. How did this area turn out this way?

This part of Utah was in part of North America that experienced many changes in environment during the last 300 million years, and these different environments deposited a striking sequence of rock layers—the foundation for the scenery. The rock layers were mostly deposited on land or in shallow water, where their iron-bearing minerals could be turned red by oxidation. The different layers were mostly deposited by ancient streams, along shorelines, and in huge sand dunes. After they had been buried and hardened into sedimentary rock, the entire region was uplifted.

Once the layers were near the surface, they could be affected by the atmosphere, moisture in the soil, and other components of the environment that tend to break down rocks. When loosened, these materials can be stripped off the land surface and transported away by running water and wind, eventually being deposited someplace else. The unusual shapes, such as arches, resulted from a combination of the thickness and hardness of rock layers, the presence of joints and other geologic structures, and the way in which each rock unit wears away.

9.0

9.1 What Can We Observe in Landscapes?

EARTH'S GEOLOGIC HISTORY IS RECORDED in its rocks and landscapes. To understand this history, we often begin by observing a landscape to determine what is there. Most geologic landscapes display a variety of features, such as different rock layers that we can distinguish by color, texture, and the way the rocks fracture. These pages provide a guide for observing a landscape and reading its story.

A What Features Do Landscapes Display?

Observe the top photograph, trying to identify distinct parts of the scene and then focusing on one part at a time. After examining the photograph, read the accompanying text.

09.01.a1 Canyonlands NP, UT

Color commonly catches our attention. These rocks are various shades of red, tan, and gray. Close examination of these rocks by geologists reveals that the rocks consist of consolidated sand and mud, and therefore are *sedimentary rocks*.

Another thing to notice is that this hill has different parts. A small knob of light-colored rocks sits on the very top.

Below the knob is a reddish and tan slope and a small reddish cliff.

There is a main, light-colored cliff, the upper part of which has a tan color and is fairly smooth and rounded.

Some parts of the cliff have horizontal lines that can be followed around corners of the cliff. These lines are the outward expression of *layers* within the rock. We call such layers in sedimentary rocks *beds* or *bedding*. These beds originally extended across the area prior to more recent erosion.

Lower parts of the cliff have a darker reddish-brown color and display many sharp angles and corners. Some of these corners coincide with vertical cracks, or *fractures*, that extend back into the rock. The reddish-brown color is a natural stain on the outside of the rocks.

Below the cliff is a slope that has pinkish-red areas locally covered by loose pieces of light-colored rock. A reasonable interpretation is that the loose pieces have fallen off the main cliff.

09.01.a2 Canyonlands NP, UT

In this figure, color overlays accentuate different features and parts of the hill. Compare these features with the photograph above.

The uppermost three rock units (numbered 1, 2, and 3) are shaded tan, yellow, and orange. Rocks of the main cliff (4) are shaded light purple. On the lower slope, the reddish rocks are shaded light orange (5), whereas the covering of loose, light-colored rocks is shaded gray (6).

Brown lines highlight beds in the rock units.

Black lines mark fractures cutting the rocks.

Simplifying this scene into a few types of features makes it easier to observe, describe, and understand the landscape. In this landscape, we observe only a few beds and rock layers but many fractures. Some layers are more resistant to weathering and form cliffs, whereas less resistant ones form slopes. Weathering has rounded off corners on the top of the cliff, removed the reddish-brown stain, and loosened pieces that fell off the cliff, covering a slope of underlying, reddish rocks.

Reexamine the top photograph. Do you look at the scene differently? Try this strategy when observing features close to where you live.

B. What Are Some Strategies for Observing Landscapes?

Observe the photograph of Monument Valley below and try to recognize the types of features, such as layers and fractures, described on the previous page. After you have made your observations, read the text, which describes aspects to observe and some helpful strategies for looking at any landscape. Start by observing the photograph.

1. Most landscapes have a fairly complex appearance when viewed in their entirety, so a useful approach is to focus on one part of the landscape at a time. In the scene below, examine the left side of the image and compare it to the center. What similarities and differences do you observe?

2. Another approach is to let the geology guide your observations from one part to another. In this scene, spend some time looking only at the cliff, and then focus on the reddish slope below the cliff. Next, pay attention to the piles of loose rocks that rest on the reddish slope.

3. Next, try focusing on one *type of geologic feature* at a time. In this photograph, start by concentrating on the fractures in the cliff. Are they steep, and are they evenly spaced? How do they affect the appearance of the cliff? Use the same approach to look at the ledges that cross the reddish slope.

09.01.b1 Monument Valley, AZ

4. Color is one of the first things we notice in any scene. Rocks, sediment, and soils have a range of colors depending on the composition of the materials and the environmental conditions imprinted on those materials. Some colors are integral to the rock, but others are a natural stain on the outside surfaces of the rock. The rocks of Monument Valley are reddish brown to tan inside, but are locally coated by a darker brown stain.

5. Some rock types are more *resistant to erosion* than others and have more dramatic expressions in the landscape. Cliffs and ledges generally represent rock types that are hard to erode, as shown by the cliffs of hard sandstone in this photograph. Slopes or soil-covered areas contain weaker materials, such as the loose windblown sand in the foreground of this image.

8. To visualize different components of a landscape, draw a sketch that captures the main features but leaves out less important details. Compare the sketch below with the photograph above. Note how the sketch changes the way you look at the photograph.

7. The *shapes* of eroded rocks depend on the hardness of the rock, thickness of layers, spacing of fractures, and many other factors. Landscapes change over time, so shapes seen today will evolve into different shapes on timescales of years to millions of years.

6. Obvious features in many landscapes are *layers* in the rocks. The cliff represents a thick layer of sandstone, whereas the underlying slopes and ledges are the expression of dozens of layers. In this location, the layers are nearly horizontal, but layers can be tilted or even folded. These differences in orientation have a great impact on the appearance of the resulting landscape. To understand how layers influence the landscape, you first observe the layers and recognize how they are oriented.

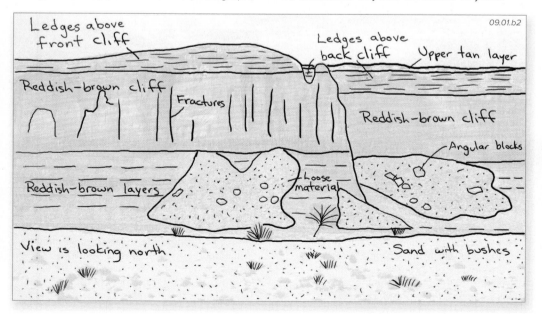

09.01.b2

Ledges above front cliff
Ledges above back cliff
Upper tan layer
Reddish-brown cliff
Fractures
Reddish-brown cliff
Angular blocks
Reddish-brown layers
Loose material
View is looking north.
Sand with bushes

Before You Leave This Page

- ✓ Draw a simple sketch of a landscape photograph, identifying the main components, like those shown on these pages.

- ✓ Sketch and summarize the different aspects or features you can observe in a landscape.

What Physical Processes Affect Earth Materials Near the Surface?

ROCKS AND OTHER EARTH MATERIALS at and near Earth's surface are subjected to processes that break them apart and alter their components. These processes, and the material's response to them, are called *weathering* and can change the color, texture, composition, or strength of the materials. Such processes result from physical and chemical weathering. *Physical weathering* breaks rocks into smaller fragments without causing any change in their chemical makeup. These smaller fragments can then be attacked by chemical reactions, the process of *chemical weathering,* or they can be moved from the original site by the process of *erosion*.

A What Is the Role of Joints in Weathering?

Joints are fractures, or very fine cracks, in rocks that show no significant offset. Joints help break rocks into smaller pieces and permit water and roots to penetrate into the rock, thereby promoting weathering.

09.02.a1

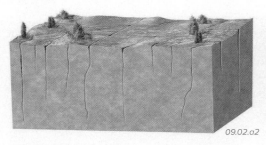
09.02.a2

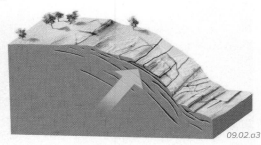
09.02.a3

Most joints form in rocks at depth and may later be uplifted to the surface. The orientation and spacing of preexisting joints and faults help determine the rates of physical and chemical weathering at the surface. More closely spaced joints promote more rapid weathering.

Some joints form as a result of expansion due to cooling or to a release of pressure as rocks are uplifted to the surface. These *expansion joints* can be difficult to distinguish from preexisting joints that formed by other processes.

As Earth is sculpted by erosion, the topography influences stresses that build up when the weight of overlying rocks is *unloaded*. During unloading, expansion joints can form that mimic topography, peeling off thin sheets of rock, a process called *exfoliation*.

B How Are Joints Expressed in the Landscape?

Joints greatly influence how a landscape develops. Joints affect the strength of a rock, help control its resistance to weathering and erosion, and influence whether pieces of rock are pried loose from the landscape.

09.02.b1 Connecticut Valley, MA

09.02.b2 Black Hills, SD

09.02.b3 Enchanted Rock SNA, TX
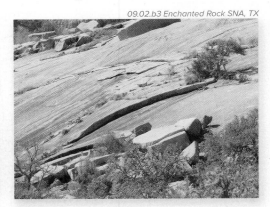

Joints are the dominant features of this roadcut, but the amount of jointing is not uniform. The less-jointed areas are more resistant to weathering than the highly jointed ones.

The spacing and orientation of joints, along with rock type, determine how fast a rock will weather and which parts of the landscape will be most easily eroded. Joints play a prominent role in the weathering of the granite shown in this photograph.

Exfoliation joints can be nearly horizontal or can mimic topography, in either case shaving off thin, curved slices of rock parallel to the surface. They can form natural stair steps, curved rock faces (as shown here), or large, dome-shaped landforms.

C What Other Physical Processes Loosen Rocks and Other Earth Materials?

Joints, which formed by processes at depth or by expansion of rock near the surface, play a major role in weathering. Other processes may also help break rock and loose materials into smaller pieces.

1. As rocks are heated and cooled, different minerals expand and contract by different amounts. This daily and seasonal *thermal expansion* imposes stresses on the boundaries between minerals and causes microfracturing in and along mineral grains, which physically loosens the mineral grains.

2. When water in a fracture freezes, it expands 8% and exerts a strong outward-directed force on the walls of the fracture. This process of *frost wedging* can widen and lengthen the fracture and pry off loose pieces of rock.

3. Water percolating through fractures and pore spaces may precipitate crystals of salt, calcite, and other minerals. As they grow, the crystals exert an outward force that fractures or weakens the rock. This process is called *mineral wedging.*

4. Burrowing organisms, including rodents, earthworms, and ants, bring material to the surface where it can be further weathered and eroded. As such, these creatures are agents of physical weathering.

09.02.c2 Baja California Sur, Mexico

5. Plant roots can extend into fractures and grow in length and diameter, expanding preexisting fractures (▶). This process is *root wedging*, which wedges apart rock exposures.

09.02.c1

D How Does Fracturing a Rock Affect Weathering?

Weathering affects rock surfaces that are exposed to air and water, so rocks weather from the outside in. Physical weathering breaks rocks into pieces, providing more surface area where chemical weathering processes can operate.

Surface Area of a Cube of Rock

1. If joints and other fractures in rock form a three-dimensional network, the rock may be broken into box-shaped pieces bounded by fractures, maybe like this cube. What is the total amount of exposed surface area on the sides of the cube?

09.02.d1

2 cm

2 cm

2. To calculate the surface area of one face (side), we multiply the height by the width of that face.

2 cm × 2 cm = 4 cm²

3. There are six faces on a cube, so we multiply the area of one face by 6 to get the total surface area.

4 cm² × 6 sides = 24 cm²

1 cm

1 cm

09.02.d2

Fracturing a Cube into Pieces

4. What happens to the total surface area if we fracture the same cube into eight pieces? First, we calculate the surface area for each smaller cube.

1 cm × 1 cm = 1 cm² for each side

1 cm² × 6 sides = 6 cm² for each cube

But there are eight such cubes.

6 cm² × 8 cubes = 48 cm²

5. Therefore, this fracturing has doubled the exposed surface area, providing more surfaces where weathering can operate. The rock will therefore weather faster.

09.02.d3 Bluff, UT

6. Physical weathering of steep outcrops can loosen pieces that fall, tumble, or slide downhill and accumulate on the slopes below. These piles of angular blocks are *talus,* and such slopes are *talus slopes.* The largest talus blocks here are 1 m across. The blocks have much more surface area than the same amount of rock in the smooth cliff from which they were derived.

Before You Leave This Page

☑ Describe several ways that joints form.

☑ Describe how joints are expressed in the landscape.

☑ Sketch or describe physical weathering processes.

☑ Sketch or explain why fracturing aids weathering.

9.2

9.3 How Do Chemical Processes Affect Earth Materials Near the Surface?

CHEMICAL WEATHERING alters and decomposes rocks, soils, and other earth materials, principally through chemical reactions involving water. When chemical and physical weathering processes combine to break down and alter earth materials, they produce minerals that are more stable in surface conditions than the original minerals. They transform rocks into clay, sand, and other materials.

A How Does Changing a Rock's Environment Promote Weathering?

Many rocks and minerals form deep within Earth. When they are brought near the surface by uplift and erosion, they encounter conditions very different from those in which they formed and may become unstable.

1. Minerals that crystallize in high-temperature magmas are generally unstable when subjected to the low-temperature conditions that characterize Earth's surface. Most magma temperatures are above 700°C, whereas surface temperatures range from minus 40°C to plus 45°C (minus 40°F to plus 122°F).

2. During metamorphism, some minerals crystallize beneath the surface in dry, high-pressure and high-temperature environments. Once such rocks reach Earth's wet, low-pressure and low-temperature surface, they can change to different minerals that are more stable at the new conditions.

09.03.a1

3. Oxygen (O_2) is abundant in the atmosphere and as a dissolved component in rain and most surface water. This oxygen chemically reacts with rocks, causing some minerals to oxidize (rust).

4. Liquid water is more abundant on and near Earth's surface than at depth. Water, especially when it is slightly acidic, is a chemically active substance that can break the bonds in many minerals. It increases the rate of chemical weathering.

B What Happens When Rocks Dissolve?

The main agents for chemical weathering are water and weak acids formed in water, such as carbonic acid (H_2CO_3). These agents dissolve some rocks, loosen mineral grains, form clay minerals, and widen fractures.

1. Limestone (below) and other rocks rich in calcium carbonate or magnesium carbonate are soluble in water and in acids. They dissolve and form pits and cavities.

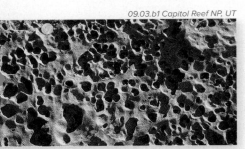
09.03.b1 Capitol Reef NP, UT

2. Over time, the pits deepen, widen, and may interconnect, forming furrows (small troughs).

09.03.b2

3. Fractures can widen as water flows through them and dissolves material from the walls of the fracture, as in the limestone outcrop below. Caves can form by dissolution of limestone and other soluble rocks at depth.

09.03.b3 Austrian Alps

One Way That Calcite Chemically Dissolves in Water

4. Limestone is a relatively soluble rock because it is composed of calcite, which is soluble in weak acids. The most common weak acid in surface water is carbonic acid, produced when rainwater reacts with carbon dioxide (CO_2) in the atmosphere, soil, and rocks. The chemical reaction for the dissolution of calcite in carbonic acid is:

$$CaCO_3 + H_2CO_3 \longrightarrow Ca^{2+} + 2(HCO_3)^-$$

Calcite ... Carbonic acid ... Calcium ion ... Bicarbonate ion in solution

5. Acids in water produce unbonded H^+ ions, each of which is a proton without a balancing electron and is available to make other chemical bonds. H^+ ions are small and can easily enter crystal structures, releasing other ions, like calcium, into the water.

C What Happens When Earth Materials Oxidize Near Earth's Surface?

Oxygen (O_2) is common near Earth's surface and reacts with some minerals to change the oxidation state of an ion. This is common in iron-bearing minerals because iron (Fe) has several oxidation states.

1. Many mafic igneous rocks contain dark, iron-bearing minerals, such as pyroxene. Iron in pyroxene can become oxidized, producing iron oxide minerals.

2. Hematite consists only of iron and oxygen and is more stable than pyroxene under oxygen-rich conditions. It commonly forms during oxidation and gives oxidized rocks a reddish color.

09.03.c1 Wilson Cliffs, NV

3. If iron-bearing rocks become oxidized, they generally take on a red color from the iron oxide mineral hematite. Reddish rocks can lose their reddish color if they interact with fluids that have less oxygen (▶).

$$4FeSiO_3 \ + \ O_2 \ \longrightarrow \ 2Fe_2O_3 \ + \ 4SiO_2$$

| Pyroxene | Oxygen | | Hematite | Silica (in water) |

D What Happens When Minerals Chemically React with Water?

When some minerals react with water they undergo a chemical reaction where the mineral combines with water to form a new mineral. This reaction, called *hydrolysis*, converts some minerals to clay.

1. One kind of feldspar, containing potassium (K), is called K-feldspar. When this mineral reacts with acids (waters that have free H^+ ions), it can be converted into clay minerals by hydrolysis, as expressed in the chemical equation below.

2. During the reaction, the H^+ ion moves into the crystalline structure, expelling the K^+ ion and some silica, which both get carried away in the water.

09.03.d1 Crete

3. If exposed to wet conditions, many rocks convert into clay minerals. The gray limestone shown here (▶) contained impurities that weathered into clay minerals and reddish hematite that accumulated between the blocks.

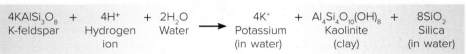

$$4KAlSi_3O_8 \ + \ 4H^+ \ + \ 2H_2O \ \longrightarrow \ 4K^+ \ + \ Al_4Si_4O_{10}(OH)_8 \ + \ 8SiO_2$$

| K-feldspar | Hydrogen ion | Water | | Potassium (in water) | Kaolinite (clay) | Silica (in water) |

E How Does Weathering Make the Ocean Salty?

Have you ever wondered why the ocean is salty or where the salt comes from? It turns out that most of the ocean's salts are derived from weathering and the dissolution of rocks and other earth materials on the land.

1. Rock, sediment, and soil on and near Earth's surface are exposed to the water and to oxygen in the atmosphere. Water that reacts with rocks and minerals can come from several sources, including rain and other forms of precipitation.

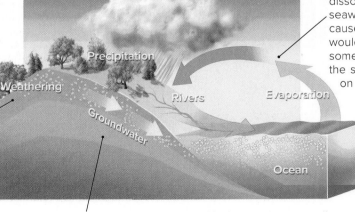

09.03.e1

Precipitation

Weathering

Rivers

Evaporation

Groundwater

Ocean

5. When seawater in the oceans evaporates, the dissolved salts remain in the water, increasing seawater's salt content (*salinity*). Such evaporation causes the seas to be saltier than streams. The seas would be even saltier if salt was not removed from some parts of the sea by deposition of salt beds. If the salt beds, once formed, are uplifted and exposed on land, they can also contribute dissolved salt to streams and ultimately back to the sea.

2. Some water infiltrates into the subsurface, where it may chemically react with the materials. During weathering, hydrolysis reactions commonly produce clay minerals and also drive out positive ions (cations) from the preexisting mineral structure. The dominant cations in feldspar, a very abundant mineral, are sodium and potassium, which can form common salts.

3. The dissolved cations, along with negative ions like chlorine, are carried by moving water, either in streams or in the subsurface by groundwater. Much of this water eventually finds its way to the oceans, where it contributes its salts and other ions to seawater.

4. Modern oceans typically contain about 3.5% dissolved salt and are much saltier than water in streams. Streams contain some dissolved salts but only a small amount, so they are considered to be fresh water, not salt water. If oceans get their salt from streams, why are the oceans saltier?

Before You Leave This Page

✓ Describe several reasons why minerals formed at depth may not be stable at the surface.

✓ Summarize how limestone dissolves and what features are formed by dissolution.

✓ Briefly summarize the processes of oxidation and hydrolysis.

✓ Explain why oceans are salty.

9.3

9.4 How Does the Type of Earth Material Influence Weathering?

ALL MINERALS AND ROCKS can be thought of as *parent material* acted upon by physical and chemical weathering. How parent materials break down depends on a number of factors, including jointing, surface area, and the kinds of minerals that compose the rock. These differences cause weathered rock outcrops to have a variety of distinctive appearances. We can often recognize how materials are reacting to weathering just by observing them.

A What Controls How Different Minerals Weather?

Many factors determine how minerals weather, including the climate, how much time the mineral has been exposed to weathering, and the chemical composition and atomic structure of the mineral or rock.

Reactivity

09.04.a1 Durango, CO

09.04.a2 Miami, FL

Chemical Bonding

09.04.a3 Miami, FL

09.04.a4 Big Maria Mtns., CA

1. Sandstone is composed mainly of sand-sized grains of quartz. Most quartz grains weather by physical processes, rather than chemical processes.

2. Limestone is very soluble and prone to chemical weathering, especially dissolution and especially in wet climates. It also weathers by physical processes.

3. The bonds in some minerals allow them to be readily dissolved in water and weak natural acids, as in this dissolved limestone. Salt and gypsum also are very soluble.

4. Other minerals have stronger bonds that make them less soluble in water. Quartz in this quartzite has very strong bonds and is not very soluble in cold water.

Relative Resistance of Minerals to Weathering

5. The stability of minerals is in a very general way related to the order in which the minerals commonly crystallize from a magma. According to *Bowen's reaction series* (▶), mafic minerals and Ca-rich feldspar crystallize first (i.e., while the magma was at its hottest), followed by Na-rich and K-rich feldspar, muscovite, and quartz. In this illustration, minerals are arranged in their general crystallization order, from top to bottom. Note that olivine and pyroxene, dark-colored minerals, are the first to crystallize, while K-feldspar and quartz crystallize last. As different minerals weather at different rates, a rock can simply disintegrate into a collection of discrete mineral grains (▼). Quartz is especially resistant to chemical weathering and so survives as sand grains in beaches, sand dunes, streams, and soils.

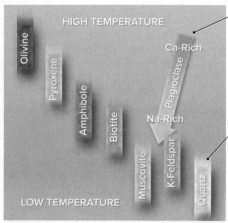
09.04.a5

6. As a magma cools, the first minerals to crystallize do so at the highest temperatures. These minerals are typically least stable when subjected to weathering at the low temperatures of Earth's surface. Crystallization order is also accompanied by changes in the percentage of SiO_4^{-4} bonds within the mineral.

7. Quartz crystallizes late according to Bowen's reaction series and is the silicate mineral most resistant to weathering. Although Bowen's reaction series is very idealized, it is one way to think about the stabilities of minerals during weathering. Minerals are most stable in the conditions closest to those under which they formed. Thus, high-temperature minerals are most unstable at Earth's surface and are the quickest to weather.

09.04.a6

Before You Leave These Pages

✓ Summarize the factors that control how different minerals weather.

✓ Explain the origin of the three main weathering products (sand grains, clay minerals, and dissolved ions).

✓ Describe how the character of a rock influences how it weathers.

B How Do Different Rocks Respond to Weathering?

Rocks are composed of minerals. Some minerals are hard and resist physical weathering, whereas others are weak and easily broken. Some minerals are chemically stable and resist chemical weathering, while others are less chemically stable. Some rocks are mostly a single mineral, but others contain many minerals. All these variations cause different granite, one of our most common classes of rock, to weather in different ways.

1. Granite and related igneous rocks contain feldspar, quartz, and smaller amounts of mica (biotite and muscovite), iron oxides, or amphibole. As these rocks weather, the different minerals respond in different ways.

2. Feldspar is the most abundant mineral in granites, forming the cream-colored crystals shown here. Feldspar chemically weathers by hydrolysis to form clay minerals. During this process sodium (Na), potassium (K), and other ions leach out of the feldspar. Some of the liberated ions are released and can be carried away by water.

3. Clay minerals weathered from granite and similar rocks can accumulate in soil or be eroded and transported away by water and wind. Some clay particles are washed out to sea, and others are deposited in lakes, floodplains, deltas, and other muddy environments (▶).

09.04.b1 Baja California Sur, Mexico

09.04.b2 Westwater Canyon, UT

09.04.b3 Great Sand Dunes NP, CO

4. Granite is at least 25% quartz, which is the medium-gray, partially transparent mineral here. Quartz is very resistant to chemical and physical weathering. As chemically reactive minerals break down around it, quartz weathers into intact grains.

5. Quartz grains eroded from granite typically become quartz sand (◀), which can be transported away by water and wind. Quartz sand accumulates along streams, in dunes, and on beaches. Feldspar can also form sand grains if the granite or other source of feldspar is not too chemically weathered, as can occur in dry climates or in areas of rapid erosion.

C How Does the Character of a Rock Influence Weathering?

Differences in mineral composition, particle size, and other rock properties play an important role in how a rock responds to weathering. Equally important are joints, bedding planes, and other discontinuities.

1. *Composition*—Weathering of a rock is influenced by the types of minerals it contains. Most sandstone, such as the one in this cliff, consists largely of quartz, a mineral that is very stable on Earth's surface; it mostly weathers by physical processes. In contrast, the recesses below the cliff contain fine-grained, clay-rich sedimentary rocks that are more easily chemically weathered and eroded.

2. *Variation in Composition*—Some outcrops have different parts with large contrasts in susceptibility to weathering. The more susceptible parts will weather faster than the more resistant parts. Such *differential weathering* can form alternating ledges and slopes, as shown here, or rocks with holes where less resistant material has been removed.

09.04.c1 Bluff, UT

4. *Discontinuities*—Joints, bedding planes, and other discontinuities provide pathways for the entry of water into a rock body. A rock with lots of these features will weather more rapidly than a massive rock containing few such discontinuities. For example, highly jointed parts of a cliff weather faster than less jointed parts. Rocks with thin layers generally break apart and weather more readily than rocks with thick layers. As usual, observing a landscape and asking why something looks the way it does leads to interesting throughts.

3. *Surface Area*—Rock that is already broken into pieces provides more surface area on which chemical weathering can act. Solid, unjointed bedrock provides less surface area and weathers more slowly.

9.4

9.5 How Do Climate, Slope, Vegetation, and Time Influence Weathering?

SPATIAL VARIATIONS in climate slope, vegetation, and time also impact rock weathering, and these factors are highly interdependent. Climate, for example, will influence the effectiveness of chemical and physical weathering, which impacts soil formation and the abundance of vegetation, which in turn influences wedging by roots and secretion of organic acids. Time is a key factor—more time increases the cumulative impact of weathering.

A How Does Climate Influence Weathering?

Abundant precipitation and higher temperatures cause chemical reactions to proceed faster. Thus, warm, humid areas generally have more highly weathered rock because chemical weathering operates faster than in cold or dry climates. Elevation influences temperature and precipitation patterns and is yet another influence on weathering.

1. This figure plots two climatic factors—precipitation (blue bar) and average annual temperature (orange bar)—as a function of latitude, from the tropics (on the left) to polar regions (on the far right). The values are all relative to the polar values, so no scale is needed.

2. The horizontal surface represents the weathering surface, and the brown boxes below the surface indicates the depth of weathering. Observe the various graphs and consider why there might be a relationship between the climatic factors and the depth of weathering.

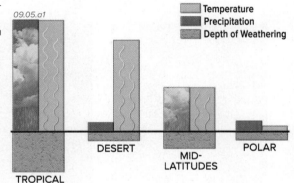

09.05.a1

Legend:
- Temperature
- Precipitation
- Depth of Weathering

TROPICAL · DESERT · MID-LATITUDES · POLAR

3. The depth of weathering (and thickness of soil) is greatest in the tropics because this climate has high temperatures, abundant precipitation, and vigorous plant growth, which contribute to weathering.

4. Weathering depths are shallowest in subtropics and polar areas, both of which have low amounts of precipitation. Although temperature is high in the subtropics (e.g., desert), there is little water to facilitate chemical weathering. Instead, physical weathering may dominate, depending on the specifics of the site.

B How Does Slope Influence Weathering?

How rocks weather is also controlled by geographic factors not related to the rock itself but to spatial variations in slope, specifically its aspect (the direction the slope faces) and its steepness.

1. *Windward Slopes*—Mountains are sites where air rises and cools, forming clouds on the mountaintop or on the windward side of the mountain. If there is enough condensation and deposition, precipitation will occur near the top or on the windward side.

2. *Slope Aspect*—The orientation of the slope, called the *slope aspect*, is an important factor in weathering. In addition to some slopes receiving more or less rain, slopes facing the Sun receive more light and heat than those facing away. Thus, sunny surfaces tend to be warmer and drier, to have more evaporation, and to have less chemical weathering, soil, and plants than slopes facing away from the Sun. In the Northern Hemisphere, south-facing slopes will receive more Sun, except in the tropics. Physical weathering is more important on Sun-facing slopes. Chemical weathering, however, will likely be dominant on the slope facing away from the Sun.

3. *Shaded Slopes*—Slopes of some orientations are more sheltered (shaded) from sunlight than slopes of other orientations. If a slope is sheltered from the direct rays of the Sun, it is cooler, can better retain its moisture, and may have more plants. Moisture, soil, and plants promote chemical weathering, so a slope facing away from the Sun generally has more soil than one that faces the Sun.

4. *Steepness of Slopes*—On steep slopes, rainfall runs off faster, and weathering products may quickly wash away by runoff. Soil and other loose materials can also slide down steep slopes. Weathering is slower in drier climates, on a steep slope, or for a more resistant rock, so soil may be less developed and hillslopes may be more rocky and barren.

5. On gentle slopes, weathering products can accumulate, and water may stay in contact with rock for longer periods of time, resulting in higher weathering rates. Once formed, soil and loose pieces remain in place longer.

09.05.b1

C How Does Vegetation Influence Weathering?

Weathering can also be caused by biological activity. Roots can pry open joints in rocks, a type of physical weathering. Animals of various sizes, from termites to mammals, burrow into the ground seeking food and shelter, and, in the process, loosen and mix sediment. Plants also contribute to chemical weathering through root secretion and through carbon dioxide (CO_2) given off during respiration.

09.05.c1 Selway, ID

1. *Lichens* consist of algae and fungi, living symbiotically. The alga provides food through photosynthesis while the fungus houses the algae, providing water and protection. Lichen can commonly obscure the underlying material (◄).

2. Lichens secrete oxalic acid, which effectively dissolves minerals, particularly carbonate rocks, like limestone and marble (►). This tombstone is almost entirely covered with lichens, which help make this monument's inscription illegible.

09.05.c2 Mount Pleasant, MI

09.05.c3

3. Plant root respiration produces carbon dioxide, which diffuses into the pore spaces between particles within soil (◄). The carbon dioxide, when combined with water, forms carbonic acid, causing a type of chemical weathering that is very destructive on chemically reactive limestone and marble.

4. The root zone for many trees (►) is far more extensive than most of us realize. This oak root system, for example, is double the extent of the crown, providing enormous opportunity for both physical and chemical weathering.

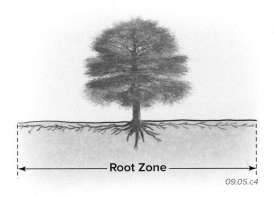

Root Zone

09.05.c4

D How Does Time Influence Weathering?

Time is a crucial factor in weathering. Physical and especially chemical processes take time, so the more time that is available, the more weathering will occur. The speed of weathering and the volume of material affected in a given time will depend on climate, slope aspect, vegetation, composition, and jointing of the rock or sediment. Due to great variability in these factors, and all the possible combinations, rates of weathering can range from rapid to extremely slow.

09.05.d1 Mount Pleasant, MI

09.05.d2 Mount Pleasant, MI

Both monuments featured in these photographs are composed of marble, yet the tombstone on the left, from the 1880s, is highly weathered while the tombstone on the right, from 1982, exhibits relatively little weathering. Since marble is prone to chemical weathering, it lost favor as a monument choice in the 1930s. These monuments provide a vivid example of the importance of time in weathering, for either natural or human-constructed objects.

Before You Leave This Page

✓ Sketch and explain how the rate and depth of weathering are affected by the type of climate.

✓ Summarize or sketch and explain how weathering is affected by slope.

✓ Describe ways that weathering is impacted by vegetation and time.

9.5

How Is Weathering Expressed?

PHYSICAL AND CHEMICAL WEATHERING affect materials below the surface, such as during the formation of soil, but they also sculpt rocks on the surface. Several guiding principles affect both domains—the subsurface and the surface. The results of weathering are obvious, once we know what to observe.

A Why Does Weathering Produce Rounded Features?

Weathering processes usually work inward from an exposed surface. This commonly results in rounded shapes in weathered outcrops, and weathering commonly generates loose, partially rounded blocks. This process of producing somewhat spherical shapes out of solid bedrock is called *spheroidal weathering*. The three figures below illustrate what can happen to a rock that has joints but lacks other types of discontinuities.

Rock that is newly jointed generally has sharp, angular edges. Weathering attacks edges from two sides and corners from three sides. These edges and corners wear away faster than a single smooth surface.

09.06.a1

Over time, faster weathering of the edges and corners of the rock will begin to smooth away these areas and any other parts of the rock that stick out.

09.06.a2

Weathered rocks can become moderately rounded, losing their sharp edges and angular features. Rocks dislodged from the bedrock will get smaller with time as they are weathered from all sides. Most rounding occurs while a rock is being transported, such as by streams.

09.06.a3

B How Is Spheroidal Weathering Expressed in the Landscapes?

Weathering exerts enormous control over the appearance of landscapes. Weathering helps define differences in landscapes from one region to another, from one side of a hill to the other, or between different rock types.

09.06.b1 Julian, CA

09.06.b2 Boulder, MT

09.06.b3 Baja California Sur, Mexico

1. Weathering mostly affects rocks from the outside in, so weathered rocks have an outer weathered zone and an inner unweathered zone. The outer zone is a *weathering rind*. As weathering continues, the weathering rind thickens and can in some cases be used to infer how long the rock has been on or near the surface and exposed to weathering.

2. As weathering attacks a jointed rock, differential weathering along the fractures can cause the intact but joint-bounded blocks to become rounded. The outer weathered rind of the blocks splits away from the stronger, less weathered rock in the center, forming rounded shapes in a process called spheroidal weathering.

3. Spheroidal weathering also affects rocks on the surface, such as the granite shown above. As rocks are weathered and uncovered by erosion, they commonly display rounded shapes. Corners of rocks exposed to the elements (rain, Sun, wind, etc.) are affected in the way described in the section above.

4. The rounding processes of weathering can also sculpt larger masses of rock, including those still attached to rocks underneath (▶). The feature in this photograph consists of several spherical shapes, separated by a recessed area along a nonresistant (easily weathered and eroded) sedimentary layer.

09.06.b4 Monticello, UT

09.06.b5 Bryce NP, UT

5. Not all weathered rocks end up as rounded shapes. The spectacular scenery of Bryce Canyon National Park in Utah contains spires, called *hoodoos*, produced by erosion of highly jointed rock layers. Two sets of intersecting joints formed columns, and weathering and erosion did the rest. Up close, the sides of the hoodoos are somewhat rounded.

C How Is Weathering Expressed in Your Town and Backyard?

You would be mistaken to believe that a trip to a distant national or state park is necessary to view features that result from weathering. Examples of physical and chemical weathering are numerous—you just have to know where to look. It is as simple as taking a short walk, starting with the shape of rocks around your home or campus. These photographs illustrate some of the everyday manifestations of weathering that can be observed in any town.

Oxidation

09.06.c1 Silverton, CO

1. Oxidation is expressed almost everywhere. It causes reddish colors in nearly all natural exposures of rock, although often less dramatically than the brightly colored hillside above. Steel, a common material in our tools and communities, is nearly all iron. When steel is exposed to oxygen in the atmosphere and in water, it can accumulate a reddish-orange coating of rust, which is mostly a collection of iron-oxide minerals, same as in the hillside.

Salt and Carbonate Crystallization

09.06.c2 Boulder, CO

2. Calcium carbonate is a constituent of mortar between bricks, concrete in side-walks, stucco on walls, and some building stones. Under wet conditions, the calcium carbonate, which is quite soluble (think of limestone, composed of calcium carbonate), can go into solution. As the solution is drawn to the surface, the carbonate precipitates into crystals that weaken the mortar holding the bricks together.

Biotic Weathering

09.06.c3 Mount Pleasant, MI

3. The roots of lichen secrete acids and wedge into monuments, especially marble tombstones. In the monument pictured above, most of the monument is highly weathered, in part due to the corrosive effects of lichen. The white streak is less weathered and is thought to be due to rainwater interacting with the metal, creating a solution poisonous to the lichen.

Differential Weathering

09.06.c4 Mount Pleasant, MI

4. Many materials in your hometown or on your campus are a composite of different materials, whether natural or made by humans. The different components usually weather at different rates, like rocks in a cliff do. In this boulder on a college campus, light-colored dikes (formed by magma in a joint) are more resistant to chemical weathering than are the dark rocks cut by the dikes. Consequently, the dikes weather out in relief, as if carved by a sculptor. In this case, though, the sculptor was natural weathering.

Root Wedging

09.06.c5 Baja California Sur, Mexico

5. Roots emanating from trees find their way into joints, whether they are in rocks or sidewalks. Most of us have seen a sidewalk tipped up and fractured by an underlying root. As the root gets bigger, the disturbance does, too.

Soil

09.06.c6 Arco, ID

6. Soil is a mixture of variably decomposed rock, sediment, and organic material, with water and air in the spaces between grains. Soil is most obvious in farms or gardens, but it also underlies most neighborhoods, although covered by streets, sidewalks, and houses. All soil is the product of weathering.

Before You Leave This Page

✓ Sketch and explain the process of spheroidal weathering and how it is expressed in landscapes.

✓ Describe some manifestations of weathering that can be observed almost anywhere.

9.7 How Are Landscapes Eroded?

WEATHERING AND EROSION begin with starting materials and then sculpt them into the landforms we observe. *Weathering* involves the *in situ* disintegration of rocks and other materials, and we subdivide it into two varieties—physical weathering and chemical weathering. *Erosion* is the scouring and stripping away of materials from a landscape, usually those that have been loosened by weathering. Together, weathering and erosion reshape a landscape and leave behind many signs of their actions.

A How Does Erosion Operate and What Are the Results?

During erosion, material is removed from a landscape and transported away, and such wearing away of the surface is called *denudation*. Erosion removes material loosened by weathering, but it can also scour and move unweathered rock.

Gravity—Rocks, soils, and other materials are under the constant pull of gravity. If they reside on a flat surface, they likely will remain in place or move very slowly, but material on slopes (▶) can fall, slide, or tumble down the slope and accumulate farther down the slope or near the base.

09.07.a1 Salmon, ID

09.07.a4 Carmel, CA

Waves—Waves along shorelines, either along the ocean or the edge of a lake, constantly crash against the shore (◀). Wave action can carry away sand and other sediment and can erode into solid bedrock and undercut cliffs, causing even more material to fall onto the surf zone.

Running Water—Streams use the power of running water and the scouring ability of sediment carried by the water to cause erosion. Running water can remove soil, sediment, and other loose materials from a landscape, or carve into solid bedrock (▶).

09.07.a2 Cascade Creek, Cascade, CO

09.07.a5

Wind—Blowing wind can carry dust and fine sand grains, which act as sandpaper, scraping against and sanding away any materials exposed to the wind, including solid rock. The strange shapes here (◀) were scoured by blowing sand, which acts as a scouring agent.

Ice—Glaciers are large, moving masses of ice on land that persist for many years. Along their sides and bases (▶) they scour away any loose material and even solid bedrock, causing ice in those parts to become darker because of all the eroded material. Some glaciers are relatively small features, perhaps a kilometer or two long, deepening the valley

09.07.a3 Glacier Bay, AK

09.07.a6 Badlands NP, SD

and scouring its sides. Other glaciers occur as huge ice sheets that cover large parts of a continent and grind away any topographic features in the way, leaving behind a somewhat flat and polished region after the ice sheets have melted away.

Wind Erosion—Strong winds can remove so much material that the ground elevation is lowered (▲). This wind erosion can scour away soils, causing the land to be less productive. In this photograph, wind has sculpted soft sedimentary rocks into two pillars, remnants of the sedimentary layers that form the ridge in the background. This process is most efficient with soft and weak materials, like the ones here.

B How Can We Envision the Slow Change of Landscapes Through Time?

Most landscapes evolve so slowly that we rarely notice significant changes in our lifetime. To get around this limitation, geoscientists use a strategy called *trading location for time*, which uses different parts of a landscape to represent different stages in the evolution of the landscape. In other words, we mentally arrange the different parts into a logical progression of how we interpret the landscape to have changed, or will change, through time. The three computer-generated models below illustrate this approach; each could represent a different place or a different stage in the evolution of a landscape. The earliest stage is on the left, and the most recent is on the right.

1. Erosion cuts into a sequence of rock layers, carving a mountain with steep sides and a broad top. Such a flat-topped mountain is called a *mesa.*

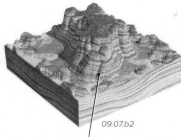
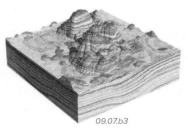
09.07.b1   09.07.b2   09.07.b3

3. With sufficient time, erosion continues to wear down the terrain, leaving a series of low, rounded hills and isolated knobs. Given enough time, a mesa can evolve first into a butte and then into one or more rounded hills and knobs.

2. With time, erosion wears away the edges of the mesa, forming a smaller, steep-sided mountain, which geologists commonly call a *butte.*

4. This photograph (▼) illustrates the approach of trading location for time. The photograph shows several buttes atop an isolated hill, surrounded by flat-topped mesas composed of the same rock layers. Although far from the cliffs, can you envision that these buttes were once part of the mesas?

5. Try out the strategy of trading location for time in the photograph below. Observe the three main features and recognize that each could represent a phase in the evolution of the landscape. Next envision the sequence of events where the mesa on the left could have been eroded into the center butte. Then envision removal of the remnant of the cliff-forming sandstone layer, resulting in the rounded hill on the right. When you have done this, reexamine the three figures above and observe how they are represented in this scene.

09.07.b4 Moab, UT

09.07.b5 Monument Valley, AZ-UT

The Long-Term Effects of Weathering and Erosion

09.07.b6 Sawtooth Mountains, ID   09.07.b7 Baja California Sur, Mexico   09.07.b8 Joshua Tree NP, CA

6. When a mountain forms and for some time afterwards, it is relatively steep as streams and glaciers carve down into the landscape faster than gravity and other processes can lower the peaks—topographic relief is high.

7. Over time, weathering and erosion lower the peaks and lessen the slopes, and produce an overall softer-appearing, less rugged land-scape. The steepest topographic features are typically removed first because they are the most unstable.

8. With more time, bedrock can be eroded down to gently sloping surface called a *pediment.* The low area in the foreground above is a pediment cut on granite bedrock, with knobs of granite remaining above the main erosion surface.

Before You Leave This Page

☑ Describe the agents that cause erosion and some of the results.

☑ Describe the strategy of trading location for time in visualizing the evolution of landscapes.

☑ Describe how topographic relief changes if weathering and erosion are dominant processes.

9.7

9.8 How Do Landscapes Record Transport and Deposition by Gravity, Streams, Ice, and Waves?

ONCE WEATHERED AND ERODED from a landscape, material can be transported farther away and later deposited. The main agents that transport earth materials are running water (streams), glaciers, waves in oceans and lakes, wind, and the ever-present force of gravity, which on its own moves materials down slopes and also drives or influences the other transport agents. Wind has some special characteristics and is discussed later.

A What Processes Transport and Deposit Sediment?

The same processes capable of erosion are also those that can simply continue transporting that material away. Gravity, streams, glaciers, waves, and wind are the main phenomena that transport material, and they are often called *agents of transport*. Each works in a different way, but each affects the material being transported.

Gravity on Slopes

1. All materials on Earth are subject to the downward pull of gravity. Material on slopes, especially on slopes that are relatively steep, can succumb to this force by falling, sliding, tumbling, or slowly creeping down the slope (▶). Gravity can be aided by moving water, mud, and ice.

09.08.a1 British Columbia, Canada

3. Material moved by gravity can come to rest on more gentle slopes and stop moving. In the example shown here (◀), a resistant ledge-forming sandstone continuously shed pieces downhill as the ledge was eroded back, mantling the slope with pieces, many of which are no longer moving.

09.08.a3 Acoma, NM

2. Many slopes are mantled (covered) by material that has moved downhill under the force of gravity. A covered slope, like this one (▶), is a *talus slope* and the material is called *talus*. Any material that moves downhill, even if it does not cover the slope, can also be referred to by the term *colluvium*.

09.08.a2 San Rafael Swell, UT

4. Material that has moved downhill under the force of gravity may not be obvious where mostly covered by soil or vegetation. Colluvium underlies nearly all of the forested hillsides in this photograph (◀) but is only exposed where the slope was undercut by an eroding stream.

09.08.a4 British Columbia, Canada

Water and Mud on Slopes

5. Water and mud can transport material, including huge boulders (▶), down slopes and into stream valleys and other areas below. Such slurries of mud and other debris are called *debris flows* and, as shown in this photo, can wreak havoc on anything in the way of the muddy torrent.

09.08.a5 Venezuela

6. Downhill, where the slope becomes less steep or adjoins a valley floor, the material being transported down the slope or in channels commonly accumulates at the bottom of a steep slope or at the base of the mountain, forming a cone-shaped feature known as an *alluvial fan* (◀).

09.08.a6 Vermilion Cliffs, AZ

Glaciers

7. Glaciers (▶) are typically loaded with rock debris collected along the way. Much of this material was eroded from bedrock and consists of angular clasts of various sizes. Glaciers also grind and pulverize material they transport, producing abundant particles of *silt* (between sand and clay in grain size).

09.08.a7 Mendenhall Glacier, AK

09.08.a8 Athabasca Glacier, Alberta, Canada

8. A glacier deposits material along its base, sides, or front, usually all at the same time. Glacially deposited materials can be blanket-like, covering the ground somewhat evenly, but generally are more irregular, forming curved ridges, hill-sized piles of sediment (◀), or irregular terrain with pits and mounds.

Running Water in Streams

09.08.a9 Rampart Creek, Alberta, Canada

09.08.a10 Icefields Parkway, Alberta, Canada

09.08.a11 Animas River, Durango, CO

9. Flowing streams are a powerful agent of transport, carrying huge volumes of sediment from higher elevations to lower ones. Some streams (▲) carry a wide range of clast sizes, including sand, pebbles, cobbles, and boulders. Clasts may be carried only a short distance or hundreds to thousands of kilometers.

10. In landscapes, sediment recently deposited by streams typically forms ribbons or blankets of debris near or within the stream channel (▲). These features are transient, being covered and rearranged by the next large flow of water down the stream. They generally include partially rounded cobbles, pebbles, and sand.

11. Even streams that seem to be moving in a tranquil manner (▲) are transporting sediment. A tan or brown color to such streams indicates that the stream is carrying abundant fine sediment, mostly mud within the visible water and sand along the streambed. This sediment forms muddy *floodplains* next to the stream.

Waves Along Shorelines

12. Waves along a shoreline can be very powerful as they crash onto bedrock or beach sediments along the shore. Through this process, they pick up sediment, transporting it back and forth within a relatively limited area, but over time they can transport material down the beach for long distances.

09.08.a12 Carmel, CA

09.08.a13 Naxos, Greece

13. Sediment moved by waves is mostly deposited along the beach. Depending on their setting and the types of materials that are available, beaches can be mostly covered by sand or by cobbles and other clasts, which became rounded when the waves pounded one clast against another.

Before You Leave This Page

☑ Summarize the main processes that transport and deposit sediment, explaining what drives the processes, what they do, and what features they produce in the landscape.

9.8

9.9 How Do Landscapes Record Transport and Deposition by Wind?

WIND TRANSPORTS AND DEPOSITS SEDIMENT, but it is different from the other agents of transport because it derives its transporting energy from the atmosphere, not directly from gravity, so it can carry material uphill. Wind is limited to carrying material that is sand sized or smaller, and it produces a number of distinctive landforms, including sand dunes of various sizes and shapes. It also transports and deposits glacially derived silt, producing additional distinctive landforms. Ancient wind deposits and directions are preserved in many layers of sandstone.

A How Does Wind Transport Sediment?

Wind is generated by differences in air pressure and at times is strong enough to transport material, but only relatively small and lightweight fragments, like sand and clay. Transport of these materials by the wind is most efficient in dry climates, where there is limited vegetation to bind materials together and hold them on the ground.

1. Wind is capable of transporting sand and finer sediment, as well as lightweight plant fragments and other materials lying on the surface. It generally moves material in one of three ways and can deposit sediment in various settings, some of which are shown in the photographs below.

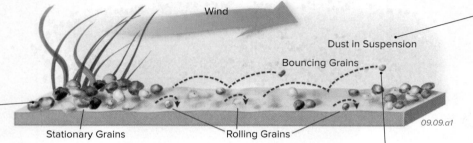

2. Most materials on Earth's surface are not moved by the wind because they are too firmly attached to the land (such as rock outcrops), are too large or heavy to be moved, or are both.

5. Wind can pick up and carry finer material, such as dust, silt, and salt. This mode of transport is called *suspension,* and wind can keep some particles in the air for weeks, transporting them long distances.

Wind

Dust in Suspension

Bouncing Grains

Stationary Grains

Rolling Grains

09.09.a1

3. If wind velocity is great enough, it can roll or slide grains of sand and silt and other loose materials across the ground.

4. Very strong winds can lift sand grains, carry them short distances, and drop them. This process is akin to bouncing a grain along the surface and is called *saltation*.

Shape of Sand Dunes and Cross Beds

09.09.a2 Namibia, Africa

09.09.a3

09.09.a5 Kanab Canyon, UT

6. The most obvious expression of material being transported and deposited by wind is in a sand dune (▲). Examine the features of this dune. The dune, like most others, is clearly asymmetrical, with a steep side facing to the right, down which sand is slipping—this is called the *slip face* and is on the downwind side (the *leeward side*). The left side of this dune has a more gentle slope, and is on the upwind side (the *windward side*). With time, the dune migrates from left to right, in the direction the wind blows.

7. When sand moves over a dune, it blows up the windward side and then is blown over or slides down the leeward side, depositing sand on the slip face. Over time, this produces a series of thin, curved layers on the downwind side of the dune, as shown above (▲). Such beds are formed at an angle and so are called *cross beds*. After the cross beds are formed, the top of the dune can be blown off, truncating the cross beds (as in the lower layer in the figure above). The image below (▼) shows truncated cross beds (lines) with a new layer of sand starting over the top.

09.09.a4 Namibia

8. In addition to being formed by wind, cross beds can also form from running water. Cross beds up to several meters high form within streams, deltas, and shorelines, but larger cross beds (▲) typically form in large sand dunes. The cross beds above are in a sandstone that was deposited 190 million years ago. There are several sets (thick layers) of cross beds, with sets being truncated at the base of overlying sets. Cross beds in each set preserve the curved profile of the front of the dune. From their shape, the dunes tell us that the winds were blowing from left to right at this ancient time (the Jurassic).

B What Landforms Does Wind Produce?

Windblown sand and silt accumulate in some distinctive features, including sand dunes of various types, each of which forms under different wind conditions. Wind generates other landforms and smaller features, which are most obvious in deserts and other arid lands. Processes and features related to the wind are described as *aeolian*.

1. Some dunes have a crescent shape (▶), with tails pointing in the prevailing downwind direction (from left to right in this case). This common type of dune is a *crescent dune* or *barchan dune*. Each dune migrates in the direction of the tails (the prevailing wind direction).

09.09.b1 Morocco

09.09.b4 Sunland Park, NM

4. Dunes that are actively moving and being reshaped by the wind are called *active dunes*. Some dunes are not so active, allowing plants to take hold (◀), provide a windbreak, and help anchor the sand in place. We commonly call such dunes *stabilized dunes*.

2. Other dunes are more linear or gently curved (▶), and can be many kilometers long. They are *longitudinal dunes* if they form parallel to prevailing winds or are *transverse dunes* if they form perpendicular to prevailing wind direction.

09.09.b2 Namibia

09.09b5 Namibia

5. Some winds are strong enough that sand is blown through, without accumulating in sizeable dunes. Winds can instead leave thin streaks of sand (◀) parallel to the prevailing wind. Such *wind streaks* can form on the downwind (leeward) side of small plants and stones.

3. Many dunes have more irregular, complex shapes, like these dunes (▶). If a dune has variably trending sand ridges radiating out from a central peak, it is a *star dune*, like these. Star dunes and other irregular dunes form where wind directions are highly variable over time.

09.09.b3 Namibia

09.09.b6 Tibet

6. *Loess* is wind-deposited silt and clay. Recently formed deposits can be a thin blanket over topography or thicker accumulations in valleys (◀). Loess is very common in parts of the midwestern United States, where it was the starting material for highly productive soils.

Landforms Influenced by the Wind

7. In many climates, especially deserts, rock surfaces develop a dark coating if left undisturbed for hundreds to thousands of years. This coating, called *rock varnish* or *desert varnish,* consists of iron-oxide and manganese-oxide materials, which are mostly derived from wind-blown dust. Rock varnish becomes darker the longer a rock is exposed at the surface, with very dark varnish requiring thousands of years. The darkness of varnish is therefore an indicator of how long that rock surface has been exposed. Rock varnish can be weathered or worn away, resetting the process. The dark varnished boulders shown here (▲) sat undisturbed on the surface for thousands of years, before parts of the varnish were scraped off by Native Americans to form artistic petroglyphs.

09.09.b7 Picture Rocks SP, AZ

09.09.b8 Granite Wash Mtns., AZ

8. In some settings, stones become concentrated on the surface through time, forming a feature called *desert pavement.* Over time, finer materials blow away, wash away, or move down into the soil, while pebbles and larger clasts remain on the surface or move up from just below the surface. If left undisturbed, the pavement becomes better developed over time, and exposed stones get coated with desert varnish, like the ones shown here. It takes more than 10,000 years to form a well-developed pavement with darkly varnished stones. Desert pavement can therefore be used as an indication of the age of that surface.

Before You Leave This Page

✓ Sketch and explain how wind transports material.

✓ Sketch and describe how wind blows sand in a dune and how this forms cross beds.

✓ Summarize some common landforms formed by the wind or features in which wind is involved.

9.9

9.10 How Do Arches and Natural Bridges Form?

ARCHES AND NATURAL BRIDGES ARE BEAUTIFUL but relatively rare, because they each require a special combination of rocks and circumstances to form. However, each type of feature nicely illustrates how weathering and erosion interact with rocks and geologic structures to form the landscapes we observe. On these pages, we describe and interpret the origin of these unusual but fascinating landscape features.

A How Do Natural Arches Form?

A naturally formed arch, called a *natural arch* or *rock arch* to distinguish it from human-constructed, architectural arches, is a large opening that goes all the way through a narrow panel of rock. Most arches have an opening that is gently curved along its top and can be curved or relatively flat along its base. Arches are large enough that we generally can walk through them. Similar, but smaller or more inaccessible, openings are often called *windows*.

09.10.a1 Wilson Arch, UT

09.10.a2 Arches NP, UT

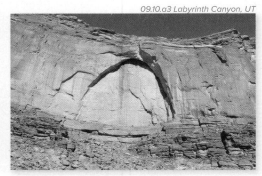
09.10.a3 Labyrinth Canyon, UT

The photograph above shows the typical characteristics of an arch. Note that it is cutting through a narrow ridge of rock, which allows the opening to go all the way through. The opening is curved along its top but has a less curved floor. Arches only form and persist in a relatively hard and resistant rock, in this case a sandstone.

The area shown above is a prime setting for formation of an arch. The upper cliff is a hard sandstone that has weathered into a narrow panel of rocks, commonly called a *fin*. A fin is the result of weathering controlled by joints on both sides. After a fin is formed, later fracturing can cut into or through the fin, leading to the formation of an arch.

In addition to joints that define the sides of a fin, cliffs are subject to failure when the downward pull of gravity exceeds the strength of the rock. Such failure occurs by fractures that start at the base of the strong layer and grow upward, generally with the shape of half an ellipse. The shape of such fractures is where most arches gain their initial arch-like shape.

Formation of a Natural Arch

09.10.a4

09.10.a5

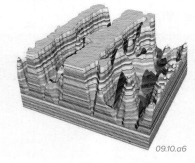
09.10.a6

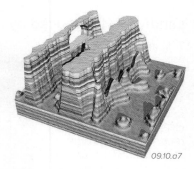
09.10.a7

Formation of an arch requires a strong starting rock. Most rock layers have one or more sets of joints that are vertical and in about the same orientation. As weathering attacks the rock, the joints are the weak link so they concentrate weathering, becoming linear notches.

With continued weathering and erosion, the joint-controlled notches become deeper and wider. The steep walls of the notches are subject to collapse, which widens the notches. The notches will also capture and channelize running water, further accentuating their erosion.

With enough time, typically millions of years, instead of a flat surface cut by notches, the landscape evolves into a set of narrow fins of rock, flanked by lower, more eroded areas and by isolated knobs, the remains of earlier fins. At this point, the stage is set for forming arches.

If part of a fin fails under the pull of gravity, an elliptical fracture can form, outlining the shape of an arch. Weathering and erosion remove the pieces broken by the fracturing, cleaning out the arch and extending it all the way through the fin. Later weathering and erosion refine the shape.

B How Do Natural Bridges Form?

A *natural bridge* is similar to an arch, and many natural bridges are arch shaped, so it is easy to confuse the two. The main distinction is that a natural bridge owes its existence to erosion by running water in a stream. The sequence of events that form a natural bridge are different than those for an arch and are intimately tied to the way in which the bridge-related stream carved its canyon.

09.10.b1 Natural Bridges NM, UT

09.10.b2 Rainbow Bridge, UT

09.10.b3 Tonto Natural Bridge, AZ

The most renowned collection of natural bridges is in Natural Bridges National Monument in southeastern Utah. Here, erosion of the sides of the channel (lateral erosion) by a downcutting stream breached the walls between two adjacent parts of the canyon, forming natural bridges of various sizes and shapes. The stream that initially formed the natural bridge still runs below the bridge.

Once a natural bridge is formed by a stream, it can be modified by later weathering and erosion, in some cases forming distinctly arch-like shapes, such as the world-famous Rainbow Bridge. This later weathering and erosion can smooth off edges, but parts of the bridge can still collapse due to gravity, forming sharp corners and faces, like those on the lower left side of Rainbow Bridge.

Natural bridges occur in various parts of the world, and each has its own story. The Natural Bridge of Virginia is regarded by many as America's first tourist attraction. Tonto Natural Bridge in Arizona (▲) is composed of travertine and has an unusual origin. It formed when springs deposited a dam of travertine across an already-formed canyon, but later the stream eroded beneath the travertine.

Formation of a Natural Bridge

09.10.b4

09.10.b5

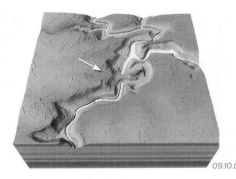
09.10.b6

Natural bridges form in canyons, and the formation of a bridge is most likely if the canyon follows a curved path, allowing it to erode through a wall from one segment of the canyon to another up or downstream. One way to form a markedly curved canyon is for a stream, before it cuts its canyon, to follow a curved path that meanders across the landscape, like the *meandering stream* shown above.

If the stream starts to downcut into the underlying rocks, this curved path becomes, for the most part, locked in (i.e., set in stone). Through erosion, the river cuts downward, deepening the canyon, and such curved bends are called *entrenched meanders*. In the entrenched meander in the center of the figure above, the river will erode laterally into the narrow ridge of rock between two segments, making it thinner with time.

At some point, the river can erode through the intervening ridge, but only at river level, leaving the overlying rock intact. It is most efficient for the river at this point to flow straight under the natural ridge, rather than following its original path, which was longer and more curved. The original curved path abandoned by the stream is called an *abandoned meander*. Most natural bridges have an associated abandoned meander.

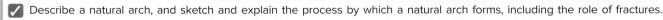

Before You Leave This Page

✓ Describe a natural arch, and sketch and explain the process by which a natural arch forms, including the role of fractures.

✓ Describe a natural bridge, sketch and explain the process by which a natural bridge forms, and list the other features likely to be associated with a natural bridge.

9.11 How Do Caves Form?

WATER IS AN ACTIVE CHEMICAL AGENT and can dissolve rock and other materials. Weathering near the surface and groundwater at depth can work together to completely dissolve limestone and other soluble rocks, leaving openings in places where the rocks have been removed. Such dissolution of limestone forms most caves, but caves form in many other ways. Once a cave is formed, dripping and flowing water can deposit a variety of beautiful and fascinating cave formations.

A How Do Limestone Caves Form?

Water near the surface or at depth as groundwater can dissolve limestone and other carbonate rocks to form large caves, especially if the water is acidic. Cave systems generally form in limestone and other carbonate rocks because most other rock types do not easily dissolve. A few other rocks, such as gypsum or rock salt, dissolve too easily—they completely disappear and cannot maintain caves. The figure below illustrates how limestone caves form.

1. Limestone is primarily made of calcite (calcium carbonate), a relatively soluble mineral that dissolves in acidic water. Rainwater is typically slightly acidic due to dissolved carbon dioxide (CO_2), sulfur dioxide (SO_2), and organic material. Water reacts with calcite in limestone, dissolving it. This *dissolution* can be aided by acidic water coming from deeper in the Earth, by microbes, and by acids that microbes produce.

3. Most caves form below the water table (the upper limit of groundwater), but some form from downward-flowing water above the water table. In either case, dissolution over millions of years can form a network of interconnected caves and tunnels in the limestone. If the water table falls, groundwater drains out of the tunnels and dries out part of the cave system.

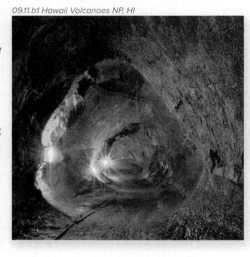
09.11.a1

2. Groundwater dissolves limestone and other carbonate rocks, often starting along fractures and boundaries between layers, then progressively widening them over time. Open spaces become larger and more continuous, allowing more water to flow through and accelerating the dissolution and widening. If the openings become continuous, they may accommodate underground pools or underground streams.

4. If the roof of the cave collapses, the cave can be exposed to the air. This can further dry out the cave. Such a roof collapse commonly forms a pit-like depression, called a *sinkhole*, on the surface.

5. Limestone caves range in size from miniscule to huge. The Mammoth Cave system of Kentucky is the longest cave in the world, with an explored length of over 640 km (400 mi) long and some part still unexplored. Carlsbad Caverns in New Mexico is also huge and spacious (▶).

09.11.a2 Carlsbad Caverns NP, NM

B What Are Some Other Types of Caves?

Most but not all caves developed in limestone. Caves in volcanic regions are commonly *lava tubes*, which were originally subsurface channels of flowing lava within a partially solidified lava flow. When the lava drained out of the tube, it left behind a long and locally branching cave. Such caves tend to have a curved, tube-like appearance with walls that have been smoothed and grooved (▶) by the flowing lava.

09.11.b1 Hawaii Volcanoes NP, HI

09.11.b2 Hueco Tanks SP, TX

Almost any rock type can host a cave, as long as it is strong enough to support a roof over the open space. Granite and similar igneous rocks (◀), which are not very soluble, can form caves, especially where physical and chemical weathering has enlarged areas along fractures, like in the example shown here. Many non-limestone caves are along a contact between a stronger rock above, which holds up the roof, and a weaker rock below, to form the opening.

C What Features Are Associated with Caves?

Caves are beautiful and interesting places to explore. Some contain twisty, narrow passages connecting open chambers. Others are immense tunnels full of cave *formations*. Caves can be decorated with intricate features formed by dissolution and precipitation of calcite and several other minerals.

1. Most caves form by dissolution of limestone. Certain features on the land surface can indicate that there is a cave at depth. These include the presence of limestone, sinkholes, and other features of karst topography. Collapse of part of the roof can open the cave to the surface, forming a skylight that lets light into the cave.

2. Caves contain many features formed by minerals precipitated from dripping or flowing water. Water flowing down the walls or along the floor can precipitate *travertine* (a banded form of calcium carbonate) in thin layers that build up to create formations called *flowstone* (▼).

09.11.c2 Carlsbad Caverns NP, NM

09.11.c3 Kartchner Caverns SP, AZ

6. Dissolution of limestone along fractures and bedding planes, along with formation of sinkholes and skylights, disrupts streams and other drainages. Streams may disappear into the ground, adding more water to the cave system.

5. In humid environments, weathering at the surface commonly produces reddish, clay-rich soil. The soil, along with pieces of limestone, can be washed into crevices and sinkholes, where it forms a reddish matrix around limestone fragments.

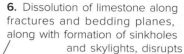
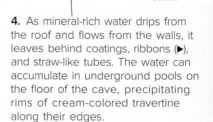

09.11.c1

3. Probably the most recognized features of caves are stalactites and stalagmites, which are formed when calcium-rich water dripping from the roof evaporates and leaves calcium carbonate behind. *Stalactites* hang tight from the roof. *Stalagmites* form when water drips to the floor, building mounds upward. The two can join, forming a column.

4. As mineral-rich water drips from the roof and flows from the walls, it leaves behind coatings, ribbons (▶), and straw-like tubes. The water can accumulate in underground pools on the floor of the cave, precipitating rims of cream-colored travertine along their edges.

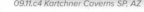

09.11.c4 Kartchner Caverns SP, AZ

Carlsbad Caverns

About 260 million years ago, Carlsbad, New Mexico, was an area covered by a shallow inland sea. A huge reef, lush with sea life, thrived in this warm-water tropical environment. Eventually, the sea retreated, leaving the reef buried under other rock layers.

While buried, the limestone was dissolved by water rich in sulfuric acid generated from hydrogen sulfide that leaked upward from deeper accumulations of petroleum. Later, erosion of overlying layers uplifted the once-buried and groundwater-filled limestone cave and eventually exposed it at the surface. Groundwater dripped and trickled into the partially dry cave, where it deposited calcium carbonate to make the cave's famous formations.

09.11.t1 Carlsbad Caverns NP, NM

Before You Leave This Page

☑ Summarize the character and formation of caves and sinkholes.

☑ Briefly summarize how stalactites, stalagmites, and flowstone form.

☑ Describe features on the surface that might indicate an area may contain caves at depth.

9.11

9.12 What Is Karst Topography?

LIMESTONE AND OTHER SOLUBLE ROCKS commonly respond to weathering by dissolving away, producing distinctive landscapes characterized by a somewhat disorganized appearance. Instead of a typical network of drainage systems, this type of landscape—*karst topography*—features sinkholes and other depressions, streams that disappear into the ground, gray rocks that look like they are dissolving away, and in some places exotically shaped pillars. Karst topography is common in many parts of the world, perhaps near where you live.

A What Are the Characteristics of Karst Topography?

Karst topography has diverse expressions depending on the topographic setting, climate, and other factors, but always indicates the presence of some type of soluble rock, especially limestone and dolomite, and, in some locations, rock salt. Karst leaves an imprint on the topography as a whole and in the appearance of individual exposures of rock.

09.12.a1

1. Examine the topography represented by this three-dimensional perspective (▶) of the area near Oolitic, Indiana. What are the features of this landscape?

2. One of the main characteristics of this terrain is the presence of numerous pits. These are sinkholes and shallow depressions formed by the partial collapse of the roofs of caves. The widespread distribution of the pits suggests that most of the area is underlain by caves. Since caves tend to form in limestone, we can infer that much of the area is underlain by limestone, as is indeed the case.

4. Some of the hills in this area do not have as many pits and instead have typical stream valleys. When these streams enter the karst, they cease to have any expression on the surface because their flow has been captured by the network of pits—such streams are called *disappearing streams* and are another characteristic of karst topography. They add water to the underground drainage system.

3. Another characteristic is the relative lack of a well-developed drainage system, such as smaller streams feeding into larger streams in a typical branching pattern. Instead, if we imagine rain falling in some area, the runoff is likely to flow into one of the pits instead of reaching a stream. This illustrates another attribute of karst regions—they commonly have a network of underground passages, including underground streams and lakes.

5. Sinkholes can be prominent pits (▶) or they can be small and subtle topographic depressions, perhaps only marked by the presence of a pond. Most sinkholes suggest collapse or settling of the roof of a cave below that spot. The photograph shown here was taken from inside a cave, looking up at a sinkhole.

09.12.a2 Longhorn Caverns, SP, TX

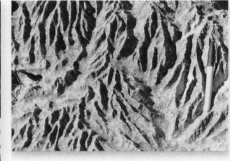
09.12.a4 Pedernales Falls SP, TX

7. Areas of karst display evidence of rocks that have partially dissolved (◀) due to their exposure to water from precipitation or from water that was in contact with the rocks, such as water in soil, groundwater, or in a stream. Such rocks typically have grooves, marking where rainwater or some other type of water dissolved material.

6. Sinkholes are a significant natural hazard. The one shown here (▶) destroyed cars and buildings in Winter Park, Florida, where the underlying bedrock is limestone. The collapse resulted from the lowering of the water table due to pumping from water wells. Underground water helps support the roof of a cave, so when the water is removed, the roof can no longer support itself and collapses. Features described as "sinkholes" in news reports can also form in other ways.

09.12.a3 Winter Park, FL

09.12.a5 Philippines

8. In some settings, especially in tropical climates where there is abundant rainfall, karst terrains can feature steep pillars, knobs, and oddly shaped rocks (◀), all sculpted by dissolution of limestone. These form spectacular tourist-destination landscapes in China, Cambodia, the Philippines, and other parts of Southeast Asia and adjacent islands. Some of these pillars are on land and others are surrounded by the ocean. They represent rocks that were somewhat more resistant to dissolution.

B What Is the Distribution of Karst?

The formation of karst relies on the presence of soluble materials at or near the surface, so a map showing the distribution of karst essentially shows the distribution of limestone, dolomite, and other soluble rocks. As the two maps below illustrate, karst is very widespread, as are its associated hazards.

1. This map shows the distribution of karst terrain in the world, excluding Antarctica, which is mostly covered with ice and too cold for liquid water. Karst is present on all the other continents and in all types of climates, from polar climates of the Arctic to tropical climates of central America, and all climates in between. Most of these areas indicate the presence of limestone.

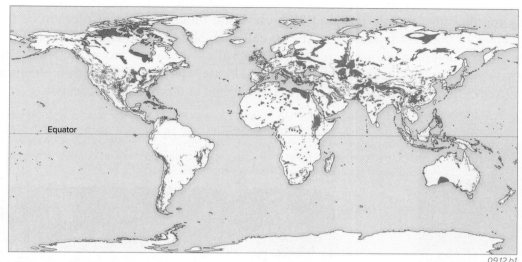

09.12.b1

2. Karst is especially widespread in Europe, such as in areas adjacent to the Alps mountain range, and in southern and central Asia. In this latter area, karst is present in the cold and relatively dry conditions in the high elevations of Tibet and in the hot and humid, low elevations of Southeast Asia. This illustrates that rock type, specifically limestone, is the primary control on karst.

3. This map depicts the distribution of karst terrain in the U.S. Purple areas are karst formed by limestone either at the surface (dark purple) or from limestone at depth (light purple). Orange areas indicate areas underlain by salt, gypsum, and other soluble rocks formed by the evaporation of water (and so are called *evaporite rocks*). Brown shows areas with other types of caves, such as lava tubes, in volcanic rocks.

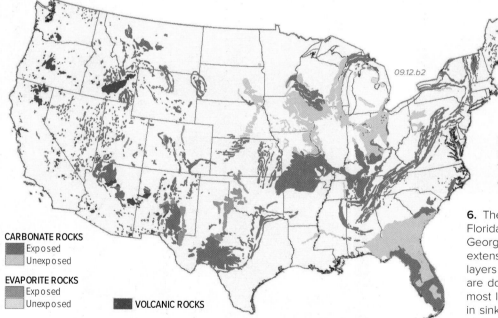

09.12.b2

CARBONATE ROCKS
▪ Exposed
▪ Unexposed

EVAPORITE ROCKS
▪ Exposed
▪ Unexposed

▪ VOLCANIC ROCKS

4. In the U.S., karst occurs in belts within and adjacent to the Appalachian Mountains, following the tilted and folded layers of limestone and dolomite.

5. Karst is widely distributed in Ohio, Indiana, Kentucky, Tennessee, Missouri, and other states.

6. The largest area of karst is in Florida and adjacent parts of Georgia, which are underlain by an extensive sequence of flat-lying layers of limestone. These areas are dotted with sinkholes and are most likely to be places described in sinkhole-related news stories.

09.12.b3 Miami, FL

7. In addition to sinkholes, evidence for the possible presence of karst includes exposures of limestone with smaller holes formed by dissolution (◄). Weathered limestone commonly contains fossils and also generally has a rough surface, with a feel of sandpaper. Such rocks receive colorful, very descriptive names, like "tear-pants weathering." If a prospective home buyer finds such weathered limestone on real-estate property, the possibility of karst should be investigated by consulting a geologic map of the region (which will show places where limestone is exposed, indicating potential karst issues).

Before You Leave This Page

☑ Describe the characteristics of karst topography and what features might indicate that karst is present.

☑ Briefly summarize the main locations of karst topography for both the world and the U.S.

9.12

9.13

What Formed Diverse Landscapes of the Rocky Mountains of Colorado?

THE COLORADO ROCKY MOUNTAINS are regarded as some of the most beautiful mountains of North America. When thinking of this region, most people envision high-elevation mountains, and Colorado certainly has its share of those, with 53 peaks above 14,000 feet in elevation, the so-called "Colorado fourteeners." In addition to the high peaks, Colorado has narrow, deep canyons and broad, relatively flat valleys between the mountains. The mountains are the headwaters for major rivers, including the Colorado, Arkansas, Platte, and Rio Grande. South-central Colorado presents a diverse suite of landscapes that illustrate weathering, erosion, and sculpting of landscapes.

09.13.a2 West-Central Colorado

09.13.a3 Rocky Mountain NP, CO

1. The main attraction of the Colorado Rocky Mountains is its high, rugged peaks and ridges, arranged one after another (◄). The high peaks have bowl-shaped depressions scoured into the bedrock, a signal that glaciers were involved in sculpting the landscape. The modern streams became established after the glaciers melted away and continued to carve into the valleys, eroding and transporting away the rocks, sediment, and soil.

2. An important control on weathering and erosion is the elevation above sea level. Many Colorado peaks are so high and cold year-round that trees are unable to grow, resulting in a sharp to gradational *tree line*—the upper limit of trees (▲). Above the tree line, there is commonly little soil and few large plants to bind loose material to the hillslopes, so rocks weather into loose, easily eroded pieces.

3. Many landscape scenes in Colorado consist of granite (◄), a relatively hard rock. Joints, like those seen here, play an important role in weathering and the resultant shapes. Other scenes consist of volcanic rocks (▶) or sedimentary rocks. Joints and other fractures are still a major influence, but so are layers.

09.13.a4 Florissant, CO

09.13.a5 San Juan Volcanic Field, CO

278

09.13.a6 Central Colorado

4. Beneath the forests and grassy flatlands is soil (▲), formed by in-situ decomposition of rocks and other materials loosened by physical weathering. Soil is best developed near the surface, and the effects of weathering decrease downward.

5. There are clear linkages between different parts of the landscape. The high, rocky peaks experience freezing conditions, so frost wedging and other processes contribute loose pieces of sediment that move down hillslopes under the influence of gravity. These materials pile up on the slope, forming talus slopes, which contribute sediment to stream channels farther downhill. Once the pieces are within the channel, they can become more rounded as colliding rocks chip off each other's corners. Examine the three photographs to the right, which illustrate different parts of the process. What do you observe, and what processes are responsible for the aspects you observe? After you have studied each photograph, read the short explanation in the text below the three photographs.

09.13.a7 Granite, CO

09.13.a8 Granite, CO

09.13.a9 Fairplay, CO

6. The upper photograph in this sequence shows the disintegration of a granite into quartz grains and feldspar grains by weathering (mostly physical weathering). Weathering of granite and related rocks produces much of the world's sand. In the middle photograph, weathered and fresh pieces of granite break away from granite bedrock, due to frost wedging and other physical weathering, and move downhill in a talus slope. The removal of this material from the bedrock is an expression of erosion. In the lower photograph, river stones, representing several varieties of granite and an assortment of metamorphic rocks, have become partially rounded as they were carried by a stream fed from above by talus slopes and smaller streams.

09.13.a1

Pikes Peak

Sangre de Cristo Range

San Luis Valley

Great Sand Dunes National Park

7. The broad valley in this region is the San Luis Valley, which is covered by sediment derived from the mountains. Sand is picked up by the prevailing winds, which blow west to east, and is blown against the adjacent mountains, which trap the sand, forming dunes (◄) in Great Sand Dunes National Park.

09.13.a10 Great Sand Dunes NP, CO

Before You Leave This Page

☑ Describe processes that sculpt landscapes in the Colorado Rocky Mountains and associated valleys, and how they are represented in what is observed today.

9.13

How Did These Landscapes Form?

THE MAIN GOAL OF EARTH SCIENCE is to gain an understanding of the processes that affect our planet, including those responsible for landscapes. In this exercise, you will observe some photographs of diverse landscapes, identify the main features, and interpret what processes are currently acting on that landscape. Then you will consider how that landscape formed and how it has changed in the past or will change in the future.

Goals of This Exercise:

- Using photographs, make observations of landscapes, identifying the main features that are present.
- Interpret what processes are occurring now and how these are affecting the present scene.
- Interpret and summarize the history of the landscape.

Procedures

For each photograph, use your observations to complete the following steps.

A. Observe the landscape to understand its various components. Identify and name the types of materials and main landscape features that are visible or can be inferred. The text next to each photograph describes some aspects about the landscape that might be helpful to know, but that you cannot determine with any certainty from afar.

B. From your observations, propose what processes are likely to have been occurring at the time the photograph was taken. Consider the physical processes, including the forces acting on the materials and the effect of slope. Try to infer what type of climate is represented by the photographs and how this climate might affect weathering, including chemical weathering, which is likely occurring but may not be visible from afar. Think broadly about the other processes that might be occurring. For steps 1 and 2, record your observations on the worksheet.

C. Think about how the landscape likely evolved over time. What did it look like in the past, and what is the likely sequence of events that produced the present scene? Finally, make predictions about how you think this landscape will appear several million years into the future, long enough to substantially change the scene. For step 3, draw sketches representing the history and future you propose.

09.14.a1

Landscape 1

HELPFUL INFORMATION—The rocks on top of the hill represent a layer of sandstone and conglomerate. The reddish rocks in the slope below are finer grained, mostly siltstone and fine-grained sandstone. The location is several thousand feet above sea level, but well below the tree line. The temperatures are very warm in the summer. There is a major river in the region, but it does not directly affect this location.

09.14.a2

Landscape 2

HELPFUL INFORMATION—The rocks that form the smooth cliff in the background are reddish sandstone. The same sandstone composes the rock that is precariously balanced in the foreground. The darker reddish rocks below the balanced rock are sandstone and siltstone. The location is several thousand feet above sea level, but well below the tree line. The temperatures are very warm in the summer. There is a major river in the region, but it does not directly affect this location. There are natural arches in the vicinity, carved into the same rock units as shown here.

09.14.a3

Landscape 3

HELPFUL INFORMATION—All of the rocks in this scene are the same rock, a granite. This scene is from the Colorado Rocky Mountains. The main mountain is below the tree line, with many trees and bushes between the outcrops, but the mountain in the distance is high enough to be above the tree line. Temperatures in this area are well below freezing at night during the winter. There is a stream at the base of the mountain, but it does not greatly affect the part of the hill shown here. It does transport away pieces of granite derived from the hill.

Soil and Unstable Slopes

WEATHERING PRODUCES SOIL, one of our most precious resources. Different types of soils form in different geographic settings, especially as a function of climate, starting material, and how long soil formation has been occurring. Soils and other materials can become unstable on slopes, and such slope instability is called *mass wasting*—the movement of material downslope in response to gravity. Mass wasting can be slow and barely perceptible, or it can be catastrophic, involving thick, fast-moving slurries of mud and debris. What factors determine if a slope is stable, and how do slopes fail? In this chapter, we explore the formation of soils, the process of mass wasting, and the importance of both phenomena to our lives.

The Cordillera de la Costa is a steep 2 km-high mountain range that runs along the coast of Venezuela, separating the capital city of Caracas from the sea. This image, looking south, has topography overlain with a satellite image taken in 2000. The white areas are clouds and the purple areas are cities. The Caribbean Sea is in the foreground. The map below shows the location of Venezuela on the northern coast of South America.

In December 1999, torrential rains in the mountains caused landslides and mobilized soil and other loose material as turbulent, flowing masses of muddy debris (flash floods) that buried parts of the coastal cities. Some light-colored landslide scars are visible on the hillsides in this image.

How does soil and other loose material form on hillslopes? What factors determine whether a slope is stable or is prone to landslides and other types of downhill movement?

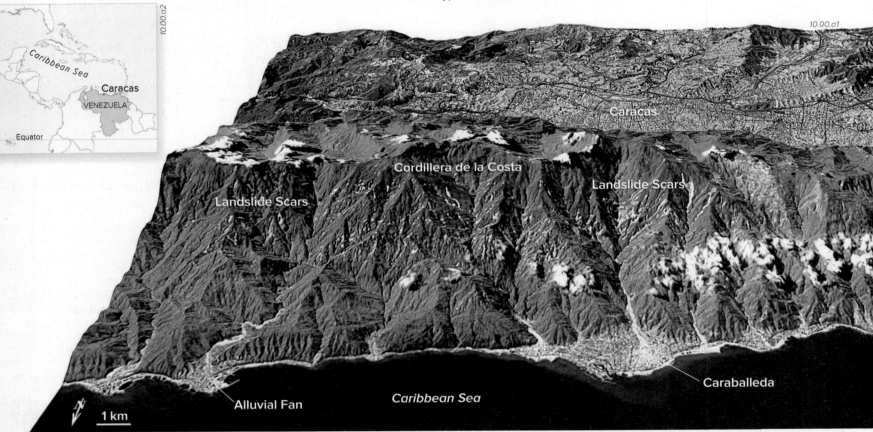

The mountain slopes are too steep for buildings, so people built the coastal cities on the less steep fan-shaped areas at the foot of each valley. These flatter areas are alluvial fans composed of mountain-derived sediment that has been transported down the canyons and deposited along the mountain front.

What are some potential hazards of living next to steep mountain slopes, especially in a city built on an active alluvial fan?

The city of Caraballeda, built on one such alluvial fan, was especially hard hit in 1999 by debris flows and flash floods that tore a swath of destruction through the town. Landslides, debris flows, and flooding killed more than 19,000 people and caused up to $30 billion in damage in the region. The damage is visible as the light-colored strip through the center of town.

How can loss of life and destruction of property by debris flows and landslides be avoided or at least minimized?

TOPICS IN THIS CHAPTER

Huge boulders smashed through the lower two floors of this building in Caraballeda and ripped away part of the right side (▼). The mud and water that transported these boulders are no longer present, but the boulders remain as a testament to the strength of the event.

10.00.a3 Caraballeda, Venezuela

1999 Venezuelan Disaster

A *debris flow* is a turbulent slurry of water and debris, including mud, sand, gravel, pebbles, boulders, vegetation, and even cars, houses, and other small structures. Debris flows can move at speeds up to 80 km/hr (50 mph), but most are slower. In December 1999, two storms dumped as much as 1.1 m (42 in.) of rain on the coastal mountains of Venezuela. The rain loosened soil on the steep hillsides, causing many landslides and debris flows that coalesced in the steep canyons and raced downhill toward the cities built on the alluvial fans.

In Caraballeda, the debris flows carried boulders up to 10 m (33 ft) in diameter and weighing 300 to 400 tons each. The debris flows and flash floods raced across the city, flattening cars and smashing houses, buildings, and bridges. They left behind a jumble of boulders and other debris along the path of destruction through the city.

After the event, USGS geoscientists went into the area to investigate what had happened and why. They documented the types of material that were carried by the debris flows, mapped the extent of the flows, and measured boulders (▼) to investigate processes that occurred during the event. When the scientists examined what lay beneath the foundations of destroyed houses, they discovered that much of the city had been built on older debris flows. These deposits should have provided a warning of what was to come.

10.00.a5 Caraballeda, Venezuela

◄ This aerial photograph of Caraballeda, looking south up the canyon, shows the damage in the center of the city caused by the debris flows and flash floods. Many houses were completely demolished by the fast-moving, boulder-rich mud.

10.00.a4 Caraballeda, Venezuela

10.0

10.1 What Is Soil?

EARTH'S LAND SURFACE is covered by a wide range of materials, some that are classified as soil and others that are not soil. How soil is defined varies depending on the perspectives of the definer. To some people, soil is any loose material on the surface, but to most geoscientists, such materials must also show evidence of having been affected by in-place weathering and other processes that produce layers and support the growth of plants.

A What Is a Soil and What Does It Contain?

Soil is a thin layer on Earth's surface that is capable of supporting life and consists of four components—minerals, organic material (living and dead), water, and air. Soil is truly at the interface between the atmosphere, lithosphere, hydrosphere, and biosphere, each of which is represented by one of the soil components. The four soil components interact with their environment to determine the overall properties of a soil.

Most soil consists of approximately half solid material and half *pore space*, the openings between and within grains. Most of the solid material consists of pieces of rocks and minerals, such as those pictured on the far upper right, with lesser but highly variable amounts of organic material. Pore space is filled with water and air.

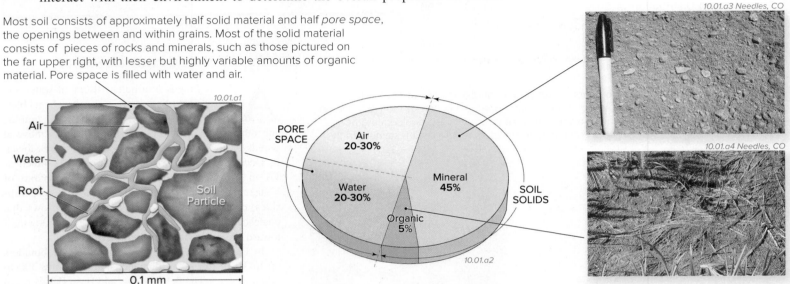

10.01.a3 Needles, CO

10.01.a4 Needles, CO

10.01.a1

10.01.a2

B What Is the Mineral Fraction of Soils?

The solid component of soil is comprised mostly of inorganic solids, especially several types of minerals. We subdivide minerals in soil into *primary minerals* and *secondary minerals*. Primary minerals, like quartz, are derived directly from the weathering and erosion of various types of rocks. Primary minerals undergo no chemical transformation during the soil-forming process, but they can be reduced in size or shape. Secondary minerals, including most silicate clays, are minerals that were not originally present but instead have been produced by fairly intense chemical weathering.

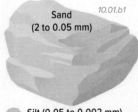

Sand (2 to 0.05 mm) 10.01.b1

Silt (0.05 to 0.002 mm)
Clay (Less Than 0.002 mm)

1. We subdivide the sizes of particles in soils into three main categories: sand, silt, and clay, from largest to smallest (◄). The sizes are greatly enlarged in this figure.

2. This graph (►) plots the abundance of different types of minerals that typically occur in each size of particle (smaller on the left, larger on the right).

3. If particles are very small, we refer to them as *clay* sized, or simply clay. Such small particles in a soil consist mostly of secondary minerals. Most of these will be silicate minerals, like clay minerals. Remember that the term "clay" refers to a family of sheetlike silicate minerals and to a very small grain size. Most clay minerals are also very small (clay sized). Clay particles consist of stacked flattened particles similar to a deck of cards, giving clays a slippery feel.

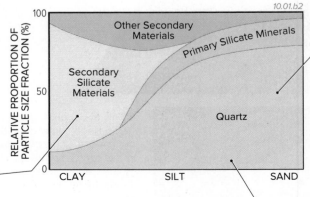

10.01.b2

4. The next larger particle is *silt*. Silt-sized particles mostly consist of quartz and other primary silicate minerals. They typically contain some secondary minerals, but only a minor amount of clay minerals.

5. If particles are sand sized, they are mostly quartz and other primary silicate minerals (e.g., feldspar and mica), with minor amounts of non-clay secondary minerals. Most sand also contains oxide minerals, such as magnetite, a magnetic iron oxide mineral that generally forms the dark grains observed in most sand.

C   What Is the Organic Fraction of Soils?

The organic fraction of soils is rich in carbon and contains plants and animals, some living, some dead, and some in various stages of decay. There can be a balance between organic matter that is accumulating and that which is lost due to erosion and microbial decay, or there can be a net gain or loss of organic material. The organic content of most soils is rather low, typically about 5%, and includes *humus*—a durable, partially decomposed form of organic matter. Humus and other organic matter are vital for soil health by improving infiltration, retaining water, decreasing evaporation, retaining and cycling nutrients, reducing erosion by binding soil together, and providing an energy source for soil organisms.

1. Organic matter in soils contains a wide variety of material, mostly variably decomposed parts of plants. It nearly always includes decomposed leaves, twigs, branches, and roots, and can also include, depending on the setting, bark from trees, pine needles and pine cones (in pine forests), mosses, mushrooms, and the woody interiors of cacti. If enough plant material accumulates and is not destroyed, like in marshy bogs, it can form *peat* (◄), a somewhat consolidated mass of partly decayed moss, wood, and other plant parts.

10.01.c1 Milwaukee, WI

10.01.c4 Black Hills, SD

3. Plants and animals form complex interrelationships (◄) based on nutrient and energy cycling. One gram of soil may contain millions of plants and animals, representing a food chain involving producers, consumers, and decomposers. The nutrients pass from the soil to grass and other plants and to small organisms, and then to higher level consumers, like deer.

2. Soil that is high in organic matter sticks together, as shown by the jar on the left (▶). The soil on the right, placed in the bottle just before this photograph was taken, is already separating rapidly and settling because it is low in organic matter.

10.01.c2

10.01.c3

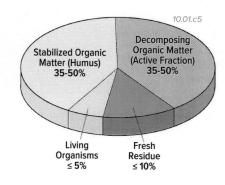

10.01.c5

Stabilized Organic Matter (Humus) 35-50%

Decomposing Organic Matter (Active Fraction) 35-50%

Living Organisms ≤ 5%

Fresh Residue ≤ 10%

4. Organic material occurs in soil in several forms. After land has been cultivated for a couple of decades, much of the decomposing organic matter—the *active fraction*—is consumed by animals, plants, and microbes; in the process, nutrients become available for plants. A more stable form of organic matter is humus, which can make up half the soil's organic matter. Humus acts like a sponge, absorbing up to six times its weight in water.

D   What About the Pore Space of Soil?

Soil is approximately half pore space filled with air and water. The amount of pore space in a material is its *porosity* and is expressed as a percentage (10% means one-tenth of the material is pore space). Soils with clay-sized particles have the greatest amount of pore space because clay particles, unless they are perfectly aligned, do not fit together very well. As a result, they prop up one another, producing countless tiny air pockets—abundant pore space. Sandy soils, with their rounded sand grains, also have a moderate amount of porosity, but not as much as unaligned clays.

1. In sandy soils, like the one shown here, there is commonly 35% to 40% porosity, whereas clay-rich soils can have well over 50% porosity. In either type of soil, pore space is reduced when soils are compacted, especially for clayey soils.

2. *Soil air* content is inversely proportional to water content—if there is more water in the pore space, there is less pore space remaining for air. Compared to atmospheric air, soil air typically has a higher CO_2 content, higher relative humidity, and lower oxygen content.

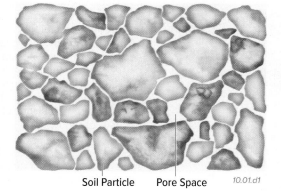

Soil Particle   Pore Space   10.01.d1

3. *Soil water* is not the free-flowing variety found in streams. Instead, it adheres to the surrounding grains. If the pore space is too small or the grains are too close, the soil water may move very slowly, if at all. Soil water contains nutrients for roots and determines rates of biological and chemical reactions, including how quickly a soil can form from the starting material.

Before You Leave This Page

☑ Explain what soil is.

☑ Describe the relationship between soil particle size and the predominance of primary and secondary minerals.

☑ List and describe the benefits of humus to soil health.

☑ Describe the relationship between air and water in soil pore spaces.

10.2 How Important Are Water and Organics in Soil?

WATER AND ORGANIC MATERIAL are critical ingredients in soils, as they are the main determinants of whether plants will grow or perish. Water enters and leaves soil through various pathways and remains in the soil in several ways as soil water. Too much water or too little can both be harmful for plants. Organic materials also enter and leave a soil, and the amount and form of organic material greatly influences a soil's productivity—how well a soil supports plant growth, especially crops.

A How Does Water Enter and Exit Soil?

1. This figure illustrates some aspects of the movement of water into and out of soil. Observe the various pathways shown here and then read the associated text in a counterclockwise direction.

2. *Precipitation* is the ultimate source of fresh water on land, and so it is also the source of soil water. If precipitation falls on the land, three things can happen to the water: it can flow downhill on the surface as *runoff*, sink into the soil through the process of *infiltration*, or *evaporate* back into the air.

3. Once in the soil, the water can remain there or continue flowing into the subsurface, driven by the force of gravity. Water percolating downward can leave the soil zone and become *groundwater*.

6. Once it reaches the above-ground parts of plants, the water is used in the process of photosynthesis to make food for the plant. Some of this water is lost as water vapor, mostly through the pores of leaves, through the processes of *transpiration*. The combination of evaporation and transpiration is *evapotranspiration*. The water vapor released to the atmosphere via evapotranspiration is then available for additional precipitation. The flow of water from one part of the environment to another is part of the *hydrologic cycle* (also called the water cycle).

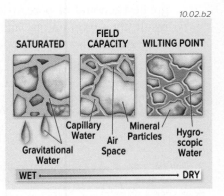

Precipitation
Transpiration
Evaporation
Runoff
Evaporation
Infiltration
Capillary Action
Runoff
Flow Into Groundwater
Water Up Roots
SOIL
BEDROCK
GROUNDWATER
10.02.a1

4. Soil water can instead be drawn back toward the surface by *capillary action*, which is made possible by the high surface tension of liquid water. Once near the surface, the water can evaporate and be lost from the soil.

5. Plant roots can extract water from the soil and carry it up through the roots and into stems and higher parts of the plant. This rise of water in plants is also due to capillary action.

B How Does Water Reside in Soil and How Much Is Available to Plants?

Water occurs in several settings in soils, but only some of this is available to plants. Some water exists in the crystalline structure of minerals and other earth materials, so this water is not mobile or accessible to plants. Water also resides in pore spaces in the soil, but not all of this water is available, either.

Settings of Soil Water

1. There are three main settings for soil water. Soil water can be a thin coating that adheres to soil particles. This is called *hygroscopic water,* which is typically so tightly held that it is unavailable to plants.

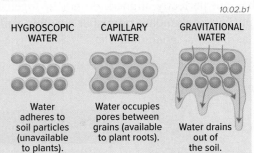

10.02.b1

HYGROSCOPIC WATER — Water adheres to soil particles (unavailable to plants).

CAPILLARY WATER — Water occupies pores between grains (available to plant roots).

GRAVITATIONAL WATER — Water drains out of the soil.

2. If extra soil water is available, the water occupies more of the pore spaces and is not attached to the soil particles. This water is available to capillary action, either through the soil or through the roots, and is therefore called *capillary water*. This is the setting of water that is most important to the long-term survival of plants with roots (some plants acquire moisture in other ways, such as directly from precipitation or from water vapor).

3. If incoming water is abundant, it can pass through the soil, pulled by gravity deeper into the subsurface. Water that follows this route is termed *gravitational water*. It can become groundwater.

Amount of Soil Water

4. The three different settings of soil water greatly influence whether the soil environment is hospitable or hostile to plant growth. If all the pore spaces are filled with water, the soil is considered to be *saturated*. Any additional water will accumulate on the surface (*ponding*), flow over the surface as overland flow, or percolate through the soil and into underlying rocks to become groundwater.

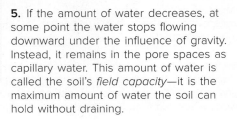

10.02.b2

SATURATED | FIELD CAPACITY | WILTING POINT
Gravitational Water | Capillary Water | Air Space | Mineral Particles | Hygroscopic Water
WET ——————————→ DRY

5. If the amount of water decreases, at some point the water stops flowing downward under the influence of gravity. Instead, it remains in the pore spaces as capillary water. This amount of water is called the soil's *field capacity*—it is the maximum amount of water the soil can hold without draining.

6. If the soil has even less water, it can have limited or no capillary water, only hygroscopic water. This condition is called the *wilting point* because at this low water content, plants begin to wilt and perish.

C What Nutrients Are Vital for Soil Productivity?

The vast majority of nutrients is in the top 10 cm of soil, right below the surface. Additional nutrients can be added to soils as *fertilizers*, such as those adding nitrogen (N), phosphorus (P), and potassium (K). These nutrients are often lacking in soils, which can limit productivity. Therefore, a label for fertilizer purchased at a store will list the contents of these chemical elements. Calcium (Ca), magnesium (Mg), iron (Fe), zinc (Zn), and sulfur (S) are other important soil nutrients.

1. Plants obtain their carbon from the atmosphere, but they need water and various nutrients from the soil (▶). Nitrogen is vital for protein and enzymes that are at the heart of all biological processes. A sign of N-deficient plants is a stunted growth and a yellowish green color (▼).

10.02.c1

10.02.c2

2. Potassium is necessary for water uptake and conservation within leaves. A sign of K-deficient plants is a burnt appearance of leaf edges.

3. Phosphorus is a difficult nutrient to maintain in the soil. Plants lacking in P are typically stunted, particularly in their root systems.

10.02.c3

4. Spreading N, P, and K on soils (▲) is a way to make up for nutrient deficiencies, but some soils will leach N, P, and K into local groundwater systems. These nutrients can then encourage unwanted algal blooms in local water supplies.

D What Role Does Organic Matter Play in Soil?

The organic fraction of soils contains living plants and animals and dead plants and animals in various stages of decay. Though the organic content is rather low in soils, typically 5%, it plays a vital role in soil health. Organic matter binds soil together, bonds nutrients, cycles nutrients vital for health, and serves as an energy source for organisms.

1. This figure (▶) illustrates the flow of material between the atmosphere, hydrosphere, lithosphere, and biosphere. Examine the figure and think of all the types of matter that might move from one sphere to another.

2. Plants lose leaves and branches, or they can die completely, providing material to the ground. This dead plant material on the ground, called *litter*, adds organic material to the soil. Litter can also change the soil chemistry, typically making it more acidic (lower pH).

3. Decomposition of dead organisms, carried out by microbes, returns nutrients from the dead plant materials to the soil. There can be a balance between organic matter accumulating and organic matter lost to microbial decay and erosion, but the addition and loss of organics can be out of balance, resulting in a change in the amount of organics over time.

DEAD LEAVES, OTHER PLANT AND ANIMAL MATTER

Plants grow

Water and air penetrate the soil

Rocks are broken down

Decomposers break down organic matter

ROCKY SUBSOIL

Minerals and nutrients are released into the soil

10.02.d1

4. Roots deliver water and nutrients from the soil to the plant. As they grow downward, roots move organic material deeper into the soil. They also break up the soil, creating pathways for downward moving air, water, nutrients, particles, and creatures. Over time, root growth causes a very slow churning effect in soils, generally improving soil productivity.

10.02.d2

5. Earthworms (▲) and other small organisms are an especially important part of this cycling of organic material. They improve soil by fertilizing it with their waste, mixing soil layers by their movement and aerating the soil with their burrows.

Before You Leave This Page

☑ Sketch and explain the hydrologic cycle as it relates to soil.

☑ Sketch and explain the settings in which water resides in soil and explain which water is available to plants.

☑ Explain the importance of the chemical elements N, P, and K in soils.

☑ Discuss the vital role organic matter plays in soil productivity.

10.3 How Does Soil Form?

SOIL BLANKETS MUCH OF THE LAND SURFACE, providing a place for plants, animals, microbes, and humans to live. Soil is affected by geologic, biologic, and hydrologic processes, and thus represents the interplay between the lithosphere, biosphere, hydrosphere, and atmosphere. The processes and factors that influence weathering also control how soil forms in different climates. What does soil contain, and how does it form?

A What Is Soil?

Soil is the unconsolidated material above bedrock and contains both mineral matter and organic matter (typically decaying vegetation) along with air and water. The incorporation of water and organic material into soil is what makes plant growth possible. Soil differs from sediment in that sediment is weathered rock that is transported or deposited by water, ice, wind, or gravity, whereas soil forms more or less in place.

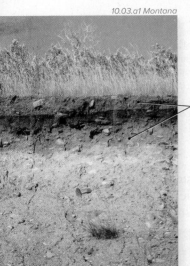

10.03.a1 Montana

1. What do you observe in this photograph (◀) of a vertical cut through soil layers? Pay special attention to the colors.

2. There are different zones or layers, called *horizons,* with rather gradational boundaries. These different layers are not the same as beds formed by sedimentation; instead, each horizon forms and grows in place by weathering of rock and sediment, and by the addition of material from plants, animals, and the atmosphere.

3. In this idealized soil profile (▶), each soil horizon is assigned a letter to denote its position or its character.

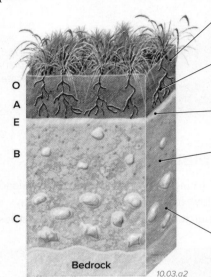

10.03.a2

4. An *O horizon* is a surface accumulation of organic debris, including dead leaves, other plant material, and animal remains.

5. An *A horizon* is topsoil, composed of dark gray, brown, or black organic material mixed with mineral grains.

6. An *E horizon* is a light-colored, leached zone, lacking clay and organic matter.

7. A *B horizon* contains little organic material, but it can have a red color due to the accumulation of iron oxide. In dry climates, the B horizon can be whitish or have whitish streaks due to calcium carbonate accumulations. It may also include gypsum and salt.

8. A *C horizon* is composed of either weathered bedrock or unconsolidated sediment, and it grades downward into unweathered bedrock or sediment.

B What Processes Occur During Soil Formation?

What happens to form soil from rock and other materials? Soil forms gradually over thousands of years and involves some of the same processes as weathering, including dissolution, oxidation, hydrolysis, and root wedging. Soil formation also involves the vertical transport of dissolved material up and down through the soil profile.

Where Material Comes From

1. Soil material is mostly derived from weathering of underlying rock and sediment, but some material is introduced by water and wind. Sediment washes onto the surface from adjacent hillslopes or arrives as windblown dust and salts.

2. Soil receives several types of material from the land surface. Leaves, pine needles, twigs, and other plant parts accumulate on the surface and are worked downward into the soil. Roots emit CO_2 gas, other gases come from the atmosphere, and moisture mostly arrives as rainfall and snowmelt.

3. Weathering weakens and loosens underlying bedrock, providing starting material to make soil. This material can be worked up into the soil, or the soil can gradually affect deeper and deeper levels of the bedrock. Some residual material remains in place at depth.

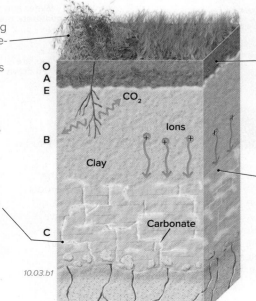

10.03.b1

How Material Moves

4. Soil material moves both down and up as it is carried by water, plants, animals, and gravity.

5. *Zone of Leaching*—The upper part of soil loses easily dissolved material downward. Water soaking into the soil *leaches* (dissolves and removes) soluble material liberated by chemical weathering, carrying it deeper. Clay minerals and other fine particles are carried downward by infiltrating water. Plant parts and other organic material are also worked downward into the soil.

6. *Zone of Accumulation*—Chemical ions leached from above may accumulate in an underlying zone, if the water does not carry them all the way to the water table, where they would enter the groundwater system. Clay minerals, iron and aluminum oxides, salt minerals, and calcium carbonate commonly accumulate in layers, depending on how much water and oxygen pass through the soil.

C   What Soil Profiles Are Typical for Different Climates?

Climate, especially temperature and moisture, strongly affects the type and rate of weathering, the abundance of plants, and the type of soil that results. Scientists, farmers, and other people classify soil in many different ways, but here we discuss three major soil types defined by climate. The top two types of soil (from moister climates) involve thicker sequences of soil than does the one for arid climates. They also contain more clay, an especially important component of soil. The accompanying photographs show the upper parts of the soil profiles. Soils can be assigned names, such as pedalfer and pedocal, suggestive of their main ingredients.

Tropical Climates (Laterite)

1. In humid, tropical climates, there is abundant rainfall and associated plant growth. Such areas include rain forests and swamps, both of which are characterized by dense plant growth.

10.03.c1 Natal, Brazil

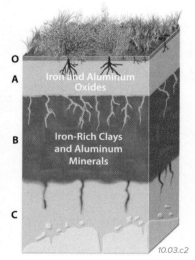

10.03.c2

2. In tropical climates, intense weathering and abundant soil moisture cause severe chemical leaching, leaving behind a soil rich in iron (Fe) and aluminum (Al) oxides, commonly giving the soil a deep red color. This extremely leached type of soil is a *laterite*, whose name comes from the Latin word for brick (clay-rich soils are used to make brick).

Temperate Climates (Pedalfer)

3. Temperate climates are cooler and mostly have less rainfall than tropical climates. Such areas contain savannas, grasslands, farms, or lush forests of leafy, deciduous trees or pine trees.

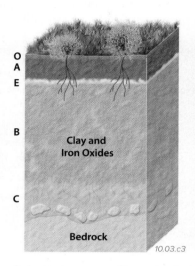

10.03.c4 Badlands NP, SD

10.03.c3

4. In cooler areas with moderate to high rainfall, the A and B horizons contain abundant insoluble minerals, including quartz, as well as iron oxide minerals. More soluble minerals like calcium carbonate are absent. Informal names for such soils are *grassland soil* or *forest soil,* depending on the type of vegetation. A soil with these characteristics is a *pedalfer,* named for the abundant aluminum (Al) in clay and oxide minerals and iron (Fe), mostly in oxide minerals.

Arid Climates (Pedocal)

5. Arid climates are dominated by overall dryness and sparse precipitation. They can be very hot, as in subtropical deserts, very cold, as in the Dry Valleys of Antarctica, or moderate in temperature, but still dry. Plants and animals are typically sparse.

10.03.c5 Sonora, Mexico

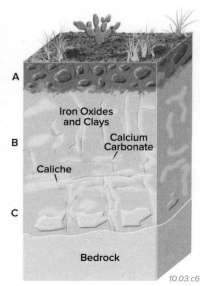

10.03.c6

6. In arid climates, there is limited vegetation, so there is little or no O horizon, and usually only a thin A horizon. Clay, iron oxide, and salts, all partly derived from windblown material, accumulate at various levels in the soil. Ca^{2+} and CO_3^{2-} ions are dissolved from upper soil horizons and chemically precipitated farther down as calcium carbonate ($CaCO_3$). The amount of water passing through the soil is not enough to completely remove these ions, so the amount of calcium carbonate increases with time, first coating clasts and eventually forming a discrete soil-carbonate layer called *caliche*. Soil formed in arid climates can be called a *desert soil* or a *pedocal* for the abundance of calcium carbonate.

Before You Leave This Page

☑ Describe what a soil is and the processes by which it forms.

☑ Sketch and describe the main soil horizons and the processes and materials that occur in each horizon.

☑ Discuss the different soils formed in different climates and the factors responsible for these differences.

10.3

10.4 How Do Terrain, Parent Material, Vegetation, and Time Affect Soil Formation?

SOIL CHARACTERISTICS are impacted by many factors, including climate, organic materials, topographic relief, parent material from which the soil was derived, and time. We refer to these factors with the acronym ClORPT (read the above sentence again, but with ClORPT in mind). We just addressed the role of climate, but here explore the "ORPT" part of ClORPT—organic material, relief, parent material, and time, beginning with slope and parent material.

A What Is the Relationship Between Slope and Soil Development?

Topography—variations in slope and elevation—affects soil properties by influencing erosion rates. Soil formation is favored when the rate of weathering exceeds the rate at which soil can be dislodged and carried away. Low-lying areas gather soil moisture and generally have more organic matter available to vegetative communities.

Slopes oriented to the north in the Northern Hemisphere are shadier, cooler, and wetter, and accumulate more organic matter than do slopes that face south. If there is abundant organic matter, as in humid climates, it will move downslope, rendering a darker color to soils in the lower parts and at the bottom of a slope.

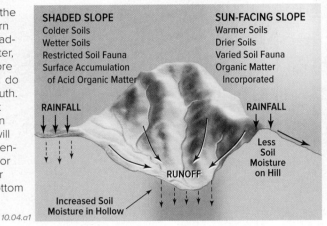

SHADED SLOPE
Colder Soils
Wetter Soils
Restricted Soil Fauna
Surface Accumulation of Acid Organic Matter

SUN-FACING SLOPE
Warmer Soils
Drier Soils
Varied Soil Fauna
Organic Matter Incorporated

RAINFALL

RAINFALL

Less Soil Moisture on Hill

RUNOFF

Increased Soil Moisture in Hollow

10.04.a1

Steeply sloping surfaces promote the rapid movement of water, soil particles, organic matter, and nutrients, so the in-situ formation of soils is limited. In contrast, these beneficial materials accumulate in bottomlands where they are less disturbed by movement, so they develop thicker soils. Such soils can be impacted by a high water table, but they can be productive if well drained.

10.04.a2

Note the difference in color and organic matter accumulation between a soil sample taken on a slope (left) and one taken at the bottom of the same slope (right).

B What Is the Relationship Between Parent Material and Soil Development?

Parent material is the rock from which a soil is derived. It can be derived in place or transported from another location. Parent material influences the rate at which weathering will occur, which in turn influences permeability, pH, and the types of nutrients available to plants. Soils that develop in place are called *residual soils*.

1. A well-jointed rock in a cold, moist climate is prone to freeze–thaw action. Here (▶), a rocky peak is being broken down by a series of fractures, some of which are produced by the expansion of freezing water. The resulting pieces serve as parent material for soil.

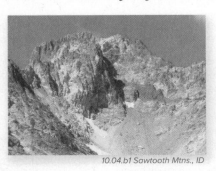

10.04.b1 Sawtooth Mtns., ID

2. Shale, a sedimentary rock composed mostly of clay-size particles (▶), is easily decomposed and weathers into a clay-rich soil that has abundant soluble cations, such as Ca⁺, Mg⁺, and K⁺. When combined with humus, these soils are quite fertile, but the fine-grained texture of clay soils hinders drainage.

10.04.b2 Durango, CO

10.04.b3 Gallup, NM

3. Soil derived from sand (▲) is coarse textured and dominated by the sand component. Such soils will exhibit good drainage and are easy to plow, but because water drains so easily, these soils tend to be drought prone. If weathered from predominantly quartz sandstone, soils will be relatively infertile as quartz (SiO_2) is inert, providing little in the way of soil nutrients.

10.04.b4

4. Soils formed from limestone, a chemically soluble rock, are often highly productive. Calcium carbonate weathered from the limestone (▶) can act as a buffer, preventing soil pH from becoming too low (too acidic). Limestone is often ground up and used as a soil additive (agricultural lime) to raise pH to the ideal 6.5–7.0 range. In dry climates, where there is less rainfall to dissolve limestone, a limestone parent rock can result in dense accumulations of soil carbonates and a very basic soil.

C | How Do Vegetation and Organisms Impact Soil Formation?

Climate helps determine the distribution of vegetation and organisms. Living organisms, in turn, contribute to the accumulation of organic matter. Organic matter helps in the formation of soil structure, is used to differentiate soil horizons, is vital in nutrient cycling, and reduces soil erosion by the protective effects of leaf litter, which minimizes rain-splash erosion and overland flow. In addition to the abundance of microorganisms involved in nutrient cycling, creatures such as gophers, moles, and earthworms act to transport and mix soil.

This illustration shows the interplay of climate and vegetation in soil formation. Weathering breaks down bedrock, providing an opportunity for vegetation and organisms to further wear down bedrock through root action and animal burrowing. Organic matter begins to accumulate, giving rise to the development of soil and soil horizons.

Observe the sequence of soil formation as represented by this illustration and think about the soils you have observed in the area where you live. The best place to observe the soils near you are in profiles in places where natural processes or humans have cut through the soils, like alongside a creek, in cuts next to a road, or where workers are digging an excavation, such as for a new building. How well developed are soils in your area?

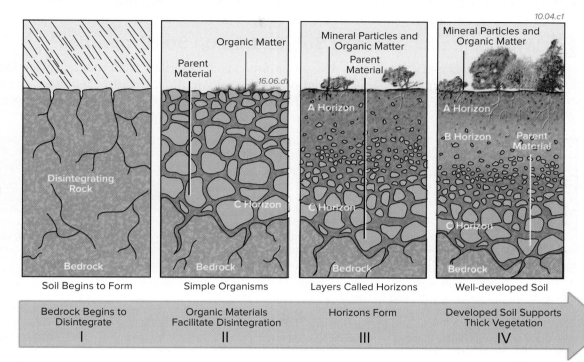

Soil Begins to Form | Simple Organisms | Layers Called Horizons | Well-developed Soil

Bedrock Begins to Disintegrate **I** | Organic Materials Facilitate Disintegration **II** | Horizons Form **III** | Developed Soil Supports Thick Vegetation **IV**

D | How Does Time Affect Soil Formation?

The longer a soil has been exposed to weathering processes and organic-matter accumulation, the better the soil development with distinguishable *horizons* (layers). Soils form within decades in tropical, moist environments, but over thousands of years in arid environments. In any climate or setting, soil progressively becomes better developed with more time, as long as it is not disturbed or eroded away. Time is the "T" in ClORPT.

This figure shows soil development as a function of time. Bedrock is weathered by physical and chemical processes to form weathered parent material. Transported parent material, such as stream gravels, will already be at this stage.

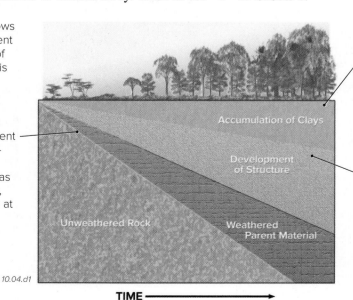

TIME ⟶

When vegetation takes hold, organic matter begins to accumulate in the surface layers of the soil. Binding action by organic matter helps in the creation of soil structure.

Continued weathering by inorganic and organic acids contributes to the formation of clays and further development of soil structure—better developed soil layers.

Before You Leave This Page

☑ Explain the fertility variations of soils formed on steep versus gentle slopes and of soils derived from sandstone versus shale.

☑ Describe how vegetation and organisms contribute to the formation of soil at different stages in its development.

☑ Outline how the passage of time contributes to the development of soil and soil layers.

10.4

10.5 What Are the Major Types of Soil?

MANY DIFFERENT CLASSIFICATION SYSTEMS for soils exist, but soil scientists and others generally use two systems. One system is based on the proportions of different grain sizes, and the other is called the *Soil Taxonomy*, which divides soils into 12 different types, or soil orders. We assign a soil to a soil order, mainly on the basis of the characteristics of the soil and on the soil-forming factors, which impart distinct properties to the different soil horizons, as viewed in the soil profile. Both systems of soil classification are very useful in quickly conveying information about soils.

A What Is the Influence of Soil Texture on Soil Properties?

One way we classify a soil is by its *soil texture*, which describes the average distribution of particle sizes. The size distribution in turn exerts a major control on the soil's fertility, stability, and other properties. Each particle size can behave in distinct chemical and physical fashions. Soil texture is an inherent property that changes little with land use and management.

Size Limits of Soil Particles	
Sand	2.00-0.05 mm
Silt	0.05-0.002 mm
Clay	< 0.002 mm

<- Mud ->

1. A key aspect of a soil is the size of particles it contains: the sand, silt, and clay. This table (◄) lists the sizes of these three particles in millimeters. The term *mud* is a general term that refers to both silt and clay.

10.05.a1

2. These three photographs show greatly enlarged views of three different sizes of particles. The left one is coarse sand, the middle one is fine sand and silt, and the right one is clay. Even at this scale, clay and silt look similar because both have grains too small to see without a magnifying glass. Different combinations can exist in soil. A soil could be mostly sand with some silt and clay. Another could be mostly silt.

3. The relative proportion of the three sizes (sand, silt, and clay) is the *soil texture*. We can describe soil texture by plotting the proportions of each particle size on a triangular diagram, like the one shown below. In this example, percent sand is along the bottom, percent clay is on the left side, and percent silt is on the right side. A soil that is 100% clay plots at the top of the triangle. One that is 100% sand plots in the lower left corner, and one that is 100% silt plots in the lower right corner. Most soils have a mix of the three particle sizes, so they plot somewhere in the triangle. Based on percentages, we can describe the texture with the words listed where the soil percentages plot. The names are not as important as understanding how texture is described and the implications to humans.

4. A soil with the percentages represented by this block of soil (►) contains 60% clay, 10% silt, and 30% sand. Note where it plots on the triangular diagram. We call such a soil texture a *clay* (yet a third usage for this term). A soil like this that contains too much clay often drains poorly, is hard to *aerate* (get air into it), and gets too compacted for most agriculture.

5. This soil has more sand than clay and silt (►). It has a mix of sizes but is quite sandy. Soils that are too sandy easily lose their water and nutrients, although sandy soils are easily plowed and are good for some construction and engineering purposes.

6. The soil block below depicts a soil that is nearly equal amounts sand, silt, and clay, so it plots near the center of the triangular diagram. This type of relatively even distribution of sand, silt, and clay is generally healthy and is called a *loam*, several varieties of which occupy the middle region of the diagram. A clay loam is near the very center. Loamy soils allow good drainage of water through the soil, nutrient retention, and aeration, all of which are optimal for plant growth.

7. This soil block is mostly silt and clay (◄). To a soil scientist, mud means a combination of silt and clay, like in this type of soil. What other people refer to as mud could be this or could be all clay.

8. This soil is mostly silt, with some clay and sand (◄), a silty loam. A silty textured soil, like this one, has properties that are intermediate between sand and clay. If the word *loam* is in a soil's name, it indicates that it is overall a good soil for agriculture.

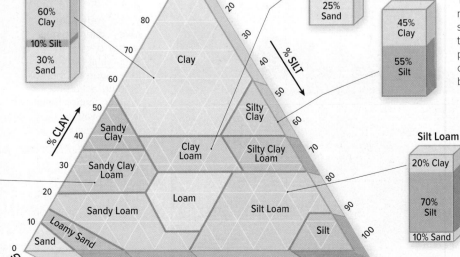

10.05.a2

B How Are Soils Classified Based on the Soil Taxonomy?

Using the Soil Taxonomy classification system, soils are subdivided on the basis of distinct physical, chemical, and biotic variations within soil horizons, usually to a depth of about 2 m. Examples of distinctive soil physical properties would be soil texture, color, and distinctiveness of soil horizons. Chemical properties would include pH and carbonate content. The amount, depth, and degree of decay of organic matter also helps differentiate soils. The resulting soils can be grouped into 12 soil orders, grouped into three categories: *site-dominated soils*, *developing soils*, and *soils strongly influenced by climate*. These three groups are differentiated in the table below. Following pages show a photograph of a representative soil profile for each soil order and a globe showing where each soil order is most common in the Americas.

Soil Taxonomy					
Type of Soil	Soil Order	Derivation of Name	Characteristic of Soil	Geographic Setting	% Land Coverage
Site-Dominated Soils	Andisol	"andesite," a common volcanic rock of Andes of South America.	Weathers into fertile soil with a high organic matter content.	Volcanic zones	0.8
	Histosol	Greek *histos* "tissue," from organic residue accumulates in wetland soil.	Thick accumulation of organic matter, poor drainage, little evidence of horizons.	Wetlands	1.2
Developing Soils	Entisol	"recENT," recently formed soil.	Recent or underdeveloped soils, little or no horizon development.	Highly variable	19.0
	Inceptisol	Latin *inceptum*, "beginning" for soil beginning to show layers.	Soils beginning to show evidence of horizonation.	Highly variable	16.1
Soils with Strong Climatic Influence	Alfisol	"Al" from aluminum, "F" from iron, important constituents of this order.	Clay accumulation in B horizon, moderate weathering, bases leached so liming needed.	Humid, mid-latitude forests	10.7
	Aridisol	Latin *aridus* "dry" for desert soil.	Desert soil lacking in organic matter accumulation, high pH, accumulation of carbonates at or beneath surface.	Hot deserts	12.5
	Gelisol	Latin *gelatio*, "freezing" for areas with permafrost (permanently frozen ground).	Permafrost close to soil surface, much frost action, high organic matter accumulation.	High latitudes or high elevations	9.6
	Mollisol	Latin *mollus*, "soft" for easy-to-till (plow) layer.	Rich accumulation of humus, proper balance of bases, nutrients, well-developed horizons.	Mid-latitude grasslands	7.4
	Oxisol	"oxide," for large amounts of iron and aluminum oxides.	Extensive weathering, red color, low pH, laterite formation possible.	Wet, tropical, hot locations	7.9
	Spodosol	Greek *spodos*, "wood ash" from the grayish appearance of bleached E horizon.	Bleached E horizon from extensive leaching, low pH, and accumulation of iron in B horizon.	Coniferous forests of Northern Hemisphere	3.7
	Ultisol	Latin *ultimus*, "ultimate" amount of leaching removes most bases.	More strongly weathered and redder than Alfisols, clay accumulation in B horizon.	Subtropical forests and savanna	8.5
	Vertisol	Latin *verto*, "to turn" from extensive clay-particle movement due to repeated clay expansion and contraction.	Rich in swelling clays, forms cracks upon drying, low in humus.	Subtropics and tropics	2.6

Note: Percentages of world coverage exclude ice-covered areas from calculation.

Before You Leave This Page

☑ Summarize how we describe the texture of a soil, including the main particle sizes and some examples of how texture affects plants.

☑ Describe the three general categories of soils, as listed in the Soil Taxonomy.

10.6 What Types of Soils Are Most Influenced by Their Climate?

CLIMATE IS AN EXTREMELY important factor in the formation of soils. Eight soil orders all bear the strong imprint of climatic influences. Of these eight soil orders, four form in relatively warm climates, whereas the other four form in temperate or polar climates. For each type of soil, the temperature, amount of rainfall, and other aspects of the climate play a strong role in what type of soil develops.

A What Types of Soils Form in Warm Climates?

Four soil orders form under relatively warm conditions—near the equator, in the tropics, or in the adjacent subtropics or warm parts of the temperate zone. Although these areas all have relatively warm temperatures in common, tropical areas are very wet, whereas those in the subtropics can be very dry. The combination of these conditions results in four soil orders: *Oxisols*, *Ultisols*, *Vertisols*, and *Aridisols*.

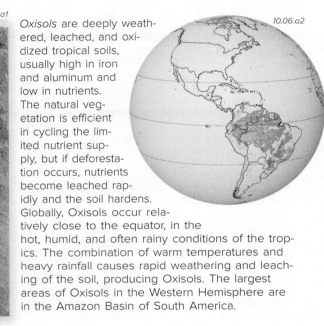

10.06.a1

10.06.a2

Oxisols are deeply weathered, leached, and oxidized tropical soils, usually high in iron and aluminum and low in nutrients. The natural vegetation is efficient in cycling the limited nutrient supply, but if deforestation occurs, nutrients become leached rapidly and the soil hardens. Globally, Oxisols occur relatively close to the equator, in the hot, humid, and often rainy conditions of the tropics. The combination of warm temperatures and heavy rainfall causes rapid weathering and leaching of the soil, producing Oxisols. The largest areas of Oxisols in the Western Hemisphere are in the Amazon Basin of South America.

10.06.a5

10.06.a6

Vertisols are clay-rich soils located in temperate and subtropical environments. Due to the high clay content, Vertisols swell when wet and shrink when dry, creating pronounced changes in volume and disrupting a normal layered structure. The high clay content and poor structure of Vertisols create problems with agricultural management. Their tendency to shrink and swell also causes engineering problems, such as cracked foundations. Vertisols are rare in the Western Hemisphere, occurring mostly along the western side of the Gulf of Mexico, but are more widespread in eastern Africa, in west-central India, and in the eastern half of Australia.

10.06.a3

10.06.a4

Ultisols form in the subtropics and warmer mid-latitudes, and are less highly weathered than Oxisols. Fairly high rainfall totals encourage downward movement of clay minerals into the B horizon. These soils tend to be acidic, which can limit their agricultural productivity. Some Ultisols occur in the tropics, but these soils extend farther away from the equator than do Oxisols. Ultisols are abundant in warm, humid parts of continents, like the southeastern U.S., tropical Central America, and parts of South America. If you live in the southeastern U.S., does this type of soil look familiar?

10.06.a7

10.06.a8

Relatively dry (arid) areas give rise to *Aridisols*. The lack of vegetation and water causes soils to form with little organic matter or leaching of nutrients. The lack of water restricts weathering, so soil horizons are poorly developed. Calcium carbonate, salt, and gypsum, which are easily leached in humid environments, tend to accumulate at the surface and in the subsurface as a whitish, caliche layer. Aridisols occur in dry conditions, so they are most abundant in the dry subtropics, especially in desert regions. In the Western Hemisphere, they are widespread in the Desert Southwest of North America and along the west side of South America.

B What Types of Soils Form in Temperate or Polar Climates?

Four soil orders form in cooler climatic regions, ranging from the relatively warm, semiarid lands of Africa to the coldest polar regions. Some of these regions are considered to be temperate, implying moderate average temperatures and significant variation from season to season. Other regions are polar, such as those near the Arctic Circle. These four soil orders reflect their specific climates. Several of these soils include much of the agricultural heartland of the central U.S. and south-central Canada, and much of Europe and Asia.

10.06.b1

Alfisols are usually fertile soils with ample accumulation of organic matter in the A horizon. A yearly moisture surplus assists in moving clay down into the B horizon, where the clays retain moisture and nutrients for plant growth. Alfisols have a natural tendency to become acidic, but this can be counteracted by regular lime applications to raise soil pH. Agriculturally, Alfisols are the second most productive soil. They show one of the widest distributions of any of the soil orders, which indicates that they form under a variety of climatic conditions. Patches of Alfisols are scattered across North and South America.

10.06.b2

Spodosols are formed under acidic forest litter in cooler portions of the mid-latitudes and in high latitudes. Acids deriving from the A horizon leach the soil of nutrients and organic matter, leaving a bleached, grayish E horizon. Because of the high acidity and short growing season, Spodosols tend to be fairly unproductive. Spodosols are largely restricted to colder climates, such as in the northern parts of North America, Europe, and Asia. These soils are particularly prominent in eastern Canada and along the western coast of Canada. They also cover much of Scandinavia, including most of Norway and Sweden.

10.06.b5

10.06.b6

10.06.b3

Mollisols are the most agriculturally productive soils. They are rich in calcium carbonate and organic material, with little leaching, and an organic-rich A horizon. These soils of the mid-latitudes experience some dry periods that prevent nutrients from being completely leached out of the root zone. Mollisols are generally found beneath tall and short prairie grass. They cover much of the agriculturally fertile Great Plains region of central North America and also cover relatively large parts of Mexico and southern South America (e.g., Argentina). They are also agriculturally important in other parts of the world, such as southern Russia.

10.06.b4

Gelisols are soils above permafrost, which is permanently frozen ground, and are themselves frozen for at least a portion of the year. They are often waterlogged and rich in organic matter. Gelisols are restricted to polar regions or to very high elevations. They cover nearly all the polar regions of northernmost Canada and Alaska and northernmost Asia (Siberia). A few small patches occur in the western U.S., marking the highest and coldest peaks of the Rocky Mountains. A thin belt of Gelisols in the mountains of South America marks the coldest part of the Andes. They also occur in high, cold Tibet and the Himalaya.

10.06.b7

10.06.b8

Before You Leave This Page

☑ Describe the main characteristics of Oxisols, Ultisols, Vertisols, Aridisols, Alfisols, Mollisols, Spodosols, and Gelisols, the settings in which they form, and where they occur in the Western Hemisphere.

☑ Explain which of these eight soil orders are most agriculturally productive, which ones are likely to be least productive, and why.

10.6

10.7 What Other Factors Control the Formation and Distribution of Soils?

SOME SOILS ARE DOMINATED by their local site, rather than by climatic factors, or are soils that are poorly developed, perhaps because they started forming relatively recently or are in sites that are continuously disturbed. The final four soil orders have these characteristics. Together with the climate-controlled soil types, the 12 soil orders occur in a somewhat patchy distribution on land, reflecting the various climatic and nonclimatic controls.

A What Types of Soils Are Site-Dominated or Poorly Developed?

Of the last four soil orders, two, Histosols and Andisols, have properties determined more by site location than climatic influences. The other two, Entisols and Inceptisols, are mostly young soils that have not been impacted significantly by soil-forming processes. With time, the influence of climate becomes increasingly important, influencing weathering rates and soil formation.

10.07.a1

Andisols are formed from volcanic ash and other volcanic material. Minerals derived from weathering of such materials can have high nutrient and water-holding capacity, making these soils extremely fertile. The productive nature of these soils, however, is tempered by the threat of hazards imposed by volcanic eruptions, landslides, and earthquakes. These soils naturally occur in volcanic areas, mostly around the Pacific Ring of Fire, a belt of active volcanoes that forms along the west coast of the Americas and down the Pacific coast of Asia. They are common in Japan, part of the Pacific Ring of Fire.

10.07.a2

Entisols are recently created soils with no B horizon and little or no profile development. Entisols can also form where parent material is resistant to weathering. Often, the parent material has been transported recently by streams, wind, glaciers, or mass wasting into "new" landscapes. Given the variety of environments where these processes take place, the geographic settings of Entisols are quite variable and widespread around the world. They are widespread in the western mountains and sandy areas of North and South America, including nearly all of Baja California. They are the dominant soil in sandy deserts.

10.07.a5

10.07.a6

10.07.a3

Histosols are wetland soils without permafrost. Poor drainage causes oxygen-poor conditions, and organic matter accumulates in a thick O horizon. When well drained, Histosols can be extremely productive agriculturally, but they are prone to compaction due to the low density of highly porous organic matter. Histosols mostly occur in high latitudes of the Northern Hemisphere, areas that were recently covered by continental ice sheets, and so have low relief and poorly developed drainage networks. Not enough time has passed since the last glacial advance for much soil development in such cold environments.

10.07.a4

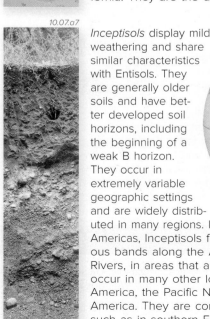

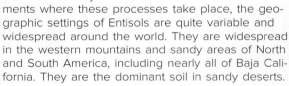

Inceptisols display mild weathering and share similar characteristics with Entisols. They are generally older soils and have better developed soil horizons, including the beginning of a weak B horizon. They occur in extremely variable geographic settings and are widely distributed in many regions. In the Americas, Inceptisols form obvious bands along the Amazon and lower Mississippi Rivers, in areas that are not often flooded. They occur in many other locations, including Central America, the Pacific Northwest, and western South America. They are common elsewhere in the world, such as in southern Europe and in eastern Asia.

10.07.a7

10.07.a8

B What Is the Global Distribution of Soil Orders?

Distribution of the 12 soil orders on the planet is best viewed with globes rather than a flat map, because globes more properly show the latitudinal setting of each soil order and do not have the distortions inherent in any projection of a sphere to a flat map. Their main disadvantage is that only half of the world is visible at once. Examine these three globes to compare the distributions of the 12 soil orders, but pay most attention to the broad patterns, not the local ones.

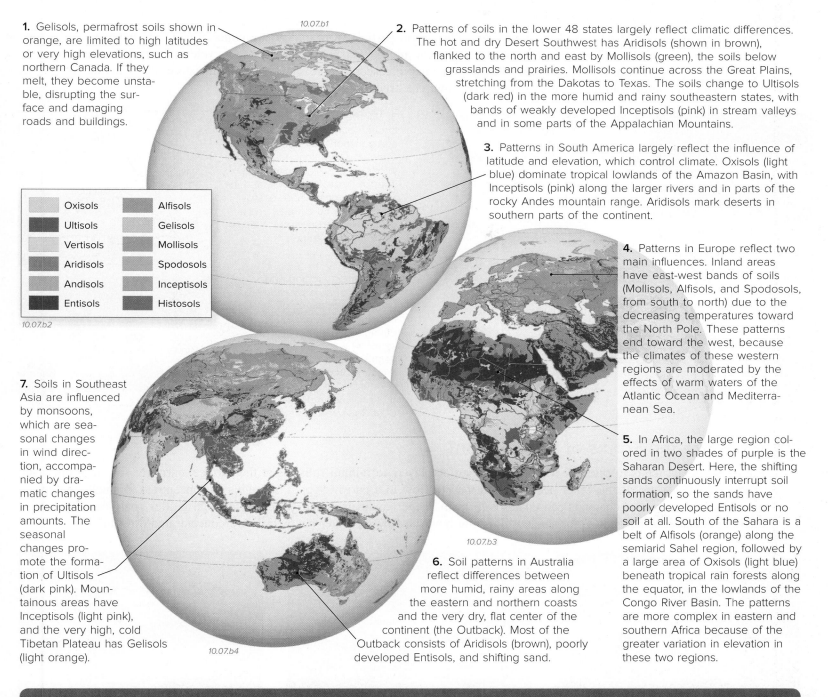

1. Gelisols, permafrost soils shown in orange, are limited to high latitudes or very high elevations, such as northern Canada. If they melt, they become unstable, disrupting the surface and damaging roads and buildings.

10.07.b1

2. Patterns of soils in the lower 48 states largely reflect climatic differences. The hot and dry Desert Southwest has Aridisols (shown in brown), flanked to the north and east by Mollisols (green), the soils below grasslands and prairies. Mollisols continue across the Great Plains, stretching from the Dakotas to Texas. The soils change to Ultisols (dark red) in the more humid and rainy southeastern states, with bands of weakly developed Inceptisols (pink) in stream valleys and in some parts of the Appalachian Mountains.

3. Patterns in South America largely reflect the influence of latitude and elevation, which control climate. Oxisols (light blue) dominate tropical lowlands of the Amazon Basin, with Inceptisols (pink) along the larger rivers and in parts of the rocky Andes mountain range. Aridisols mark deserts in southern parts of the continent.

Oxisols		Alfisols	
Ultisols		Gelisols	
Vertisols		Mollisols	
Aridisols		Spodosols	
Andisols		Inceptisols	
Entisols		Histosols	

10.07.b2

4. Patterns in Europe reflect two main influences. Inland areas have east-west bands of soils (Mollisols, Alfisols, and Spodosols, from south to north) due to the decreasing temperatures toward the North Pole. These patterns end toward the west, because the climates of these western regions are moderated by the effects of warm waters of the Atlantic Ocean and Mediterranean Sea.

7. Soils in Southeast Asia are influenced by monsoons, which are seasonal changes in wind direction, accompanied by dramatic changes in precipitation amounts. The seasonal changes promote the formation of Ultisols (dark pink). Mountainous areas have Inceptisols (light pink), and the very high, cold Tibetan Plateau has Gelisols (light orange).

10.07.b4

5. In Africa, the large region colored in two shades of purple is the Saharan Desert. Here, the shifting sands continuously interrupt soil formation, so the sands have poorly developed Entisols or no soil at all. South of the Sahara is a belt of Alfisols (orange) along the semiarid Sahel region, followed by a large area of Oxisols (light blue) beneath tropical rain forests along the equator, in the lowlands of the Congo River Basin. The patterns are more complex in eastern and southern Africa because of the greater variation in elevation in these two regions.

10.07.b3

6. Soil patterns in Australia reflect differences between more humid, rainy areas along the eastern and northern coasts and the very dry, flat center of the continent (the Outback). Most of the Outback consists of Aridisols (brown), poorly developed Entisols, and shifting sand.

Before You Leave This Page

☑ Describe the main characteristics of Andisols, Histosols, Entisols, and Inceptisols, and in what settings they commonly occur.

☑ Explain some of the most important factors that control the distribution of soil orders.

10.7

What Are the Causes and Impacts of Soil Erosion?

HUMANS OBTAIN A GREAT majority of their calories from the land, so good soil stewardship is vital. Yet much land throughout the world is suffering degradation from one of the biggest threats to soil—*soil erosion*. Human activities have caused increased soil erosion in many places, but we can limit much of this by employing conservation practices that are known to be successful. What causes soil erosion, and what can we do about it?

A What Are the Causes of Soil Erosion?

Two processes are involved in soil erosion—*detaching* a soil particle from underlying material and then *moving* (transporting) the particle. The main causes of detachment are rainsplash and freeze–thaw cycles. The main agents of transportation are surface runoff and wind. Erosion of most landscapes is natural, but humans cause or accelerate soil erosion through deforestation, farming, overgrazing, and construction projects.

10.08.a1

1. With each falling raindrop (◄), pore spaces in soils are increasingly sealed, which reduces percolation of surface waters down into groundwater. As a result, more water remains on the surface to detach particles as the water starts to move.

2. As water accumulates on the surface of this field (►) and begins to flow downhill as *runoff*, it has the potential to detach and transport soil particles that are not anchored by grass, roots, or surface litter. The top layer of soil can be eroded off one part of a field and deposited in another part, or it can be transported away. Erosion can also carve into the soil, removing some of the deeper layers as well.

10.08.a2 Iowa

3. Wind erosion (►) also occurs through particle disaggregation and transport. Heavier soil particles are rolled along the surface through *creep*, lighter particles bounce through *saltation*, and the lightest become airborne, *suspended* in wind currents and sometimes transported for thousands of kilometers.

10.08.a3

4. The most erosion and transport of soils by wind occurs during times of the strongest winds during storms. This dust storm (►) was from the central U.S. in the 1930s, a time of very warm temperatures and great drought known as the *Dust Bowl* years.

10.08.a4 Colorado

B What Are the Forms of Soil Erosion?

Raindrop impact tends to promote sealing of the surface layer, so rainfall runs off at the surface, carrying soil in a thin layer of unchanneled water. If the water becomes confined to flow in a narrower path, it begins to concentrate in small, shallow channels. These smaller channels can coalesce to form deeper channels called *gullies*. If enough gullies coalesce, they can form a *streambed*. The photographs below illustrate how the flow of water becomes more concentrated downhill.

10.08.b1 Austin, TX

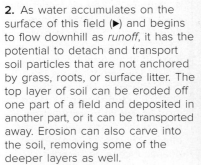

In the photograph above, water after heavy rains spreads out in a thin sheet across the land, rather than being confined to a stream channel. Such flow is classified as *sheetwash*.

10.08.b2 Austin, TX

This photograph shows sheetwash being concentrated into a shallow channel, flowing toward the viewer. As water enters the channel, it becomes more turbulent and able to erode.

10.08.b3 Austin, TX

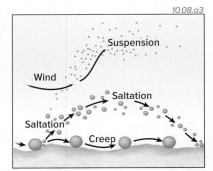

In this photograph, water from sheetwash and smaller channels merges and has more power to cut deeper into the land, forming a *gully* and eroding away the upper layers of soil.

C What Is the Impact of Soil Erosion in the United States?

Impacts of soil erosion are a loss of agricultural productivity, loss of wildlife habitat, filling in of reservoirs, impaired health due to the spread of bacteria and viruses, more fertilizer use, flooding, and mass wasting.

10.08.c1

1. Drought weakens plant roots so that the network of roots can no longer bind soil against wind erosion. Such erosion (◄) removes the most nutritionally rich A soil horizon. Overgrazing in dryland areas can be especially devastating, causing soil erosion and loss of productivity.

10.08.c2

2. Water-borne sediment derived from soil erosion carries fertilizer residue, (◄) which can cause overfertilization of water bodies, a condition dangerous to aquatic plants and animals. A muddy or cloudy water column also decreases photosynthesis, resulting in lower dissolved oxygen in the water column.

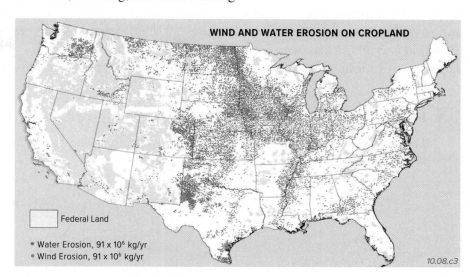

WIND AND WATER EROSION ON CROPLAND

Federal Land

• Water Erosion, 91 x 10⁶ kg/yr
• Wind Erosion, 91 x 10⁶ kg/yr

10.08.c3

3. In the U.S., wind erosion (marked in red) is more of a problem in High Plains states, whereas water erosion (marked in blue) is more common in the Midwest and the eastern and southern states (▲). While substantial reductions in soil erosion have occurred in the last several decades, it is still estimated that more than 80% of the nation's soils are being lost at a rate greater than the rate of formation.

D What Are Some Solutions to Soil Erosion?

Techniques for reducing erosion can be divided into two related categories: preventive and control measures. Preventive techniques include minimizing the impact of rainsplash erosion and runoff. Control techniques are designed to reduce the velocity of runoff after soil has already been dislodged, capturing sediment and nutrients before leaving the field.

Preventive Techniques

10.08.d1

10.08.d2 *Iowa-Minnesota Border*

2. *Strip cropping*, pictured here (◄), protects the vulnerable land between the row crop (corn), capturing water and slowing surface runoff. Note that the land is also contour plowed, preventing runoff from traveling quickly down the slope.

1. Managing the number of grazing animals (▲) prevents erosion due to overgrazing. Too many grazing animals per acre often results in loss of vegetative cover, which in turn leads to soil erosion.

10.08.d3

3. *No-till agriculture* retains last year's crop residue (◄) to reduce rainsplash erosion, control sheetwash, retain moisture, and protect against unseasonable early frosts.

Control Techniques

4. Trees and other plants are placed along waterways and designed to capture sediment before these sediments and the contaminants on sedimentary particles reach the water (▶). Also, if trees are planted in rows perpendicular to the predominant wind direction as *windbreaks*, they help slow wind and reduce wind's erosive force.

10.08.d4

Before You Leave This Page

☑ Compare erosion control techniques. Explain which would be most appropriate around your hometown. Why?

☑ Explain why some land uses result in faster soil erosion than others.

10.9 What Controls the Stability of Slopes?

GRAVITY PULLS MATERIAL DOWNHILL, and some rocks, soil, and other loose material are not strong enough to resist this persistent force. Downward movement of material on slopes under the force of gravity is called *mass wasting*, and it occurs to some degree on all slopes. Mass wasting can proceed very slowly or very quickly, sometimes with disastrous results. Mass wasting is an important part of the erosional process, moving material downslope from higher to lower elevations, and feeding sediment from hillslopes to streams, beaches, and glaciers.

A How Does Gravity Affect Slope Stability?

The main force responsible for mass wasting is *gravity*. The force of gravity acts everywhere within and on the surface of Earth, tending to pull everything toward Earth's center.

1. On a flat surface, the force of gravity acts on a block by pushing it vertically down against the base of the block. The block will not move under this force.

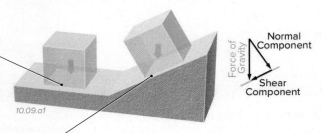
10.09.a1

2. On a slope, gravity acts at an angle to the base of the block. Part of the force pushes the block against the slope and another part pushes the block down the slope. These two parts of the force are referred to as *components*.

3. The part of the force pushing the block against the slope is the *normal component*.

4. The other component acts parallel to the slope, trying to shear the block down the slope. It is the *shear component*.

5. As the angle of slope becomes steeper, the shear component becomes larger while the normal component becomes smaller. If the slope angle is too steep, the shear component becomes enough to overcome friction, and it causes the block to slide.

B How Steep Can a Slope Be and Still Remain Stable?

The steepest angle at which a pile of unconsolidated grains remains stable is called the *angle of repose*. This angle is controlled by frictional contact between grains. In general, most loose, dry material has an angle of repose between 30° and 37°. This angle is somewhat higher for coarser material, for more angular grains, for material that is slightly wet, and for material that is partly consolidated. It is lower for material with flakes or rounded grains and for material that contains so much water that adjacent grains lose contact.

Dry Sand   Angle of Repose
10.09.b1

1. Dry, unconsolidated sand grains form a pile (◄), and the angle of the resulting slope is at the *angle of repose*. If more sand is added, the pile becomes wider and higher, but the angle of repose remains the same. If part of the pile is undercut and removed, the grains slide downhill until the pile returns to a stable slope at the angle of repose. If sand is slightly wet, surface tension between the grains and a thin coating of water enables the sand to be stable on steeper slopes.

2. Loose rocks and other loose material (talus) accumulate on some slopes (►) and at the bases of cliffs. Such talus material commonly forms a slope that is at the angle of repose for the particular sediment. The smooth *talus slope* shown here is at its stable angle. If part of the slope became steeper, material would slide down until the slope regained stability.

10.09.b2 Salmon, ID

15.10.b3 Morocco

3. Most slopes of sand dunes (◄) reflect the angle of repose for dry sand. Slopes can be more gentle than the angle of repose, but if they begin to exceed the angle of repose then the slope fails, slumping downhill. Walking up a sand dune causes the barely stable sand to slide beneath your feet.

4. The slopes of a scoria (cinder) cone (►) reflect the angle of repose because they are typically composed of loose, volcanic scoria. The angle of repose will be steeper for coarser scoria and for material that partially fused together during the eruption.

10.09.b4 Northern AZ

C   What Factors Control Slope Stability?

The main control on slope stability is the angle of repose for the material. Intact rock can form cliffs or steep slopes, but soil, sediment, and strongly fractured rock form slopes reflecting their angle of repose.

The addition of minor amounts of water increases the strength of soil, but too much water pushes grains apart and weakens the soil. Materials with high clay-mineral content can flow downhill when they become wet.

Fractures, cleavage, and bedding reduce the overall mechanical strength of a rock and may allow rocks to slip downhill. In this illustration, rock layers oriented parallel to the slope allow material to slide.

10.09.c1

10.09.c2

10.09.c3

D   What Triggers Slope Failure?

Slope failure occurs when a slope is too steep for its material to resist the pull of gravity. Some slopes slide or creep downhill continuously, but others fail because some event caused a previously stable slope to fail.

1. Precipitation can saturate sediment, weakening an unconsolidated material by reducing grain-to-grain contact. A slope that was stable under dry conditions may fail when wet. Slopes can also fail after wildfires, which destroy plants that help bind and stabilize the soil.

2. Hillslopes can fail when the load on the surface exceeds a slope's ability to resist movement. Humans sometimes build heavy structures on slopes, overloading the slope and causing it to fail. Areas with gentle slopes, such as near this town, are less prone to slope failures.

3. Modification of a slope by humans or natural causes can increase a slope's steepness so that it becomes unstable. Erosion along river banks, as shown here, or wave action along coasts can undercut a slope, making it unsafe.

10.09.d1

4. Volcanic eruptions can shake, fracture, and tilt the ground, unleashing landslides from oversteepened slopes. Eruptions can cover an area with hot ash and other loose material, causing melting of ice and snow. Melting can rapidly release large amounts of water and mobilize volcanic material in destructive debris flows.

5. A sudden shaking, such as tremors caused by an earthquake along this fault scarp, may *trigger* slope instability. Minor shocks from heavy trucks or human-caused explosions can also start a slope failure.

6. Oversteepening of cliffs or hillslopes during road construction can cause them to fail, especially if fractures or layers are inclined toward the road.

Slope Stability in Cold Climates

In cold climates, water is frozen much of the year, and ice, although solid, can flow. Freeze—thaw cycles, where ice freezes and then thaws repeatedly, cause ice to flow and can contribute to slope failure.

When water-saturated soil freezes, it expands, pushing rocks and boulders on the surface upward. When the soil thaws, the boulders move down again. This process, called *frost heaving,* is a large contributor to the downslope movement of material in cold climates. In addition, when the upper layers of soil thaw during the warmer months, the water-saturated soil may move downslope more easily. Frost heaving can form polygon-shaped outlines in the soil, called *patterned ground* (▼).

10.09.t1 Kongakut, AK

Before You Leave This Page

☑ Describe or sketch the role that gravity plays in slope stability.

☑ Describe the concept of the angle of repose and its landscape expressions.

☑ Describe some factors that control slope stability, and events that trigger slope failure.

10.9

10.10 How Do Slopes Fail?

THE RAPID DOWNSLOPE MOVEMENT of material, whether bedrock, soil, or a mixture of both, is commonly referred to with the general term *landslide*. Movement during slope failure can occur by falling, sliding, rolling, slumping, or flow. We classify slope failures by how the material moves and the type of material involved.

A What Are Some Ways That Slopes Fail?

Most people have seen evidence of slope failure when hiking, driving past a roadcut, or watching television news, nature shows, or movies. Slope failure can be as subtle as a small pile of rocks at the base of a hill or as dramatic as a mudflow that has destroyed a neighborhood or hillside in China or California. The photographs below show images of various types and sizes of slope failures.

10.10.a1 Yampa River, UT

10.10.a2 Denali region, AK

2. During an earthquake, brown masses of rock and soil (◄) slid down these steep slopes in Alaska, smashed apart, and flowed as avalanches of rock, soil, and ice across a white glacier in the valley below. Parts of the avalanche flowed across the valley and partway up hillsides on the other side of the valley.

3. On this hillside (►), millions of rock pieces fell from an upper cliff and slid downhill, accumulating below as a talus slope that partially covers an underlying dark shale. The pieces came to rest at the angle of repose, the specific angle at which those materials are stable. If the slope is steepened, such as by erosion of the base or by addition of new material on top, the talus will slide and adjust until it again attains its stable angle. Talus slopes generally accumulate a little at a time, from the gradual addition of one rock fragment after another, but some are produced or enlarged more rapidly, from rock falls and other types of slope failure.

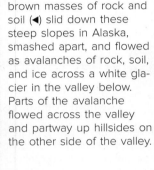
10.10.a3 Book Cliffs, UT

1. This rocky cliff (▲) failed after being undercut by a river. Large sandstone blocks, one the size of a building, collapsed downward. The falling block detached along a prominent joint surface, which has since accumulated a coating of brown rock varnish.

10.10.a4 El Salvador, Central America

4. This landslide in El Salvador flowed down a steep, unstable slope (◄) and cut a swath of destruction across a neighborhood. Adjacent slopes on this hill appear to be just as steep as the part that failed, and they pose a hazard to the remaining homes.

5. Undercutting of hillsides by coastal erosion and highway construction has made many slopes steep and unstable. In 2017, rocks and soil flowed and slumped downward, covering and blocking the Pacific Coast Highway near Big Sur (►), as shown in this computer-generated image. Such slope failures commonly occur during or after intense rainstorms.

10.10.a5 Pacific Coast Highway, Big Sur, CA

B How Are Slope Failures Classified?

Classification of slope failure is imprecise because the processes commonly grade into one another, and more than one mechanism of movement can occur during a single slope-failure event. All classifications consider *how* the material moves, what *types of material* move, and the *rate of movement* of the material.

10.10.b1 Colorado National Monument, CO

Mechanism of Movement

◄ Geoscientists classify slope failures primarily by how the material moved. Rocks and other material can *fall* off cliffs; *slip* along fractures, cleavage, or bedding planes; *topple* over; or do all three.

▶ Other slope failures involve the slow *creep* of the uppermost soil cover or the *flow* of material, as during turbulent flows of mud, rocks, and other debris. This brown mud flowed during an earthquake.

10.10.b2 Fort Funston, CA

10.10.b3 Monument Valley, UT

Type of Material

◄ Some slope failures involve slabs of *solid rock* or large pieces of broken rock derived from cliffs and rocky hillslopes. Such rocks can further break apart after they begin to move.

▶ Many slope failures mobilize *unconsolidated material* that is stripped from hillsides. Material can include soil, loose sediment, pieces of wood and other plant parts, boulders, and other types of loose debris.

10.10.b4 Grand Canyon, AZ

10.10.b5 Venezuela

Rate of Movement

◄ Fast rates—Another important factor is the *rate of movement* of the material. Some slope failures start in an instant and send material downhill at hundreds of kilometers per hour, or at least too fast for people to outrun.

▶ Slow rates—Other mass movements are more gradual and move downhill at rates that are imperceptible to an observer. This slow-moving mudflow carries trees, some tilted, along for the ride.

10.10.b6 Slumgullion, CO

Submarine Slope Failures

Slope failure is not restricted to the land; it can also occur on steep or even gentle slopes on the seafloor. Such *submarine slope failures* can be caused by overloading of sediment on a slope or in a submarine canyon. They can also be triggered by shaking during an earthquake, volcanic eruption, or storm. Various types of slope failure occurred off the southwestern coast of the Big Island of Hawaii, forming the large mass of debris shown here in green.

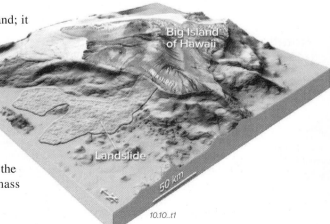
10.10..t1

Before You Leave This Page

✓ Describe slope failures and some ways they are expressed in the landscape.

✓ Summarize the classification of slope failures, and describe the different types of movement, types of material, and rates of movement.

10.10

10.11 How Does Material on Slopes Fall and Slide?

SOME SLOPE FAILURES involve materials *falling* off a cliff or *sliding* down a slope. These mechanisms commonly involve rock and pieces of rock, but they can also involve materials that are less consolidated. Rocky cliffs and slopes might appear to be immune to slope failure because they consist of hard bedrock, but they can fail spectacularly and catastrophically.

A What Happens When Rocks Fall or Slide?

Rocks and other material can fall from cliffs or can slide on fractures or other weak planes. Falling and sliding rock masses may begin as relatively intact blocks, but they commonly break apart as they begin moving or when they hit the bottom of a cliff. Some slides rotate as they move. Others simply slide down the hill. One type of rock failure can lead to another because a rock fall may remove some support, causing higher parts of a cliff to slide.

10.11.a2 Canyon Lake, AZ

Rock Falls and Debris Falls

In a *rock fall* (◀), large blocks or smaller pieces of bedrock detach from a cliff face and fall until they smash into the ground. Rock falls can be triggered by rain, frost wedging, thawing of ice that had held rocks to the cliff, an earthquake, river erosion, or human construction that undercuts a cliff. Less consolidated debris, including loose sand, can also fall off a cliff. Part of the cliff in this photograph (▶) has collapsed, producing a rock fall composed of large, angular blocks. As the rocks fell, and during impact with the ground, they fractured and broke apart, most remaining as a large mass of shattered rocks, while some loose blocks fell, rolled, and slid downhill.

10.11.a1

Rock Slides

10.11.a4 Naxos, Greece

In a *rock slide* (◀), a slab of relatively intact rock detaches from bedrock along a bedding surface, preexisting fault, joint, or other discontinuity that is inclined downslope. As it slides, parts of the slab typically shatter into angular fragments of all sizes, but large blocks can remain relatively intact. In the rock slide shown to the right (▶), road construction undercut these sedimentary rocks, which had layers and fractures that were inclined downslope, toward the road. At some point after construction, the rocks slid along the layers and into the road.

10.11.a3

Rotational Slides

10.11.a6 Capitol Reef NP, UT

In some slides, rock layers and other material rotate backward as they slide (◀). Such *rotational slides* move along one or more curved slip surfaces. This type of slide, also called a *slump,* can occur in bedrock or less consolidated material. Individual slices can remain relatively intact or can break and spread apart. The rocks in the middle of this photograph (▶) detached from the top cliff and slumped halfway down the slope. As the block moved downhill, it rotated, causing the layers to tilt to the left. Fractures within the block allowed parts of the block to shift up or down relative to one another.

10.11.a5

B What Is the Geometry of a Rock Slide?

A combination of geological circumstances is required to detach a slab of rock and create a rock slide. It requires sufficiently steep slopes along with bedding planes, fractures, or other flaws that are inclined downslope.

1. Many rock slides occur in bedrock with discrete layers that differ in rock type and therefore in strength. Such rock layers are most common in sedimentary rocks, but they are also present in many volcanic and metamorphic sequences. In this figure, a sequence of different sedimentary layers has beds inclined downhill, toward a small stream valley.

2. At their upper end, most rock slides detach along a series of preexisting joint or fault surfaces. In other cases, as in this example, the stresses that build up in the rock before it slides are enough to form new fractures, allowing the rock slide to detach from bedrock.

3. Detachment of the base of the rock slide from underlying bedrock commonly occurs along a layer of weak rock, such as shale, mudstone, or salt. This weak layer may allow the overlying slab to slide fairly easily and remain partially intact as it moves downhill.

4. To be able to slide down the dip of the layers, the upper layer must have space downhill in which to slide. That is, the layers will probably not slide if they are supported by more rocks in a down-dip direction. In this example, a stream has eroded a low area, giving the sliding rock slab somewhere to go.

5. Although not shown here, rock slides can slip along joints, fault surfaces, cleavage, or some other discontinuity, rather than bedding surfaces. Preexisting faults of the proper orientation are very susceptible to rock slides because they are planar, fairly continuous, and structurally weak.

6. Most slides leave a linear or curved scar, or *scarp*, on the hillslope, marking where the slide pulled away from the rest of the hill. This upper end is also called the *head* of the slide.

7. The leading edge of a slide, the *toe*, can overrun the land surface in front of the slide.

10.11.b1

The Vaiont Disaster, Italy

In 1960, a dam was built across the Vaiont Valley in northeastern Italy. This valley runs along the bottom of a syncline, where the rocks have been folded downward and dip into the valley from both sides. The rocks are mainly limestone but with interlayered thin beds of shale and sandstone. Some of the limestone beds contain caverns formed when groundwater dissolved the rock. Fractures in the rocks run both parallel and perpendicular to the bedding planes.

During August and September 1963, three years after the dam was completed, heavy rain fell in the area. One day in October, the south wall of the valley failed and slid into the reservoir behind the dam. The slide was 1.8 km high and 16 km wide with a volume of 240 million cubic meters. The slide moved along shale layers that parallel the bedding planes in the limestone.

As the slide moved into the reservoir, it displaced an equal volume of water, forcing a surge of water 240 m above lake level onto the village of Casso on the northern side of the valley. Waves within the reservoir killed 1,000 people. Waves 100 m high swept over the dam. Although the dam did not fail, the water that overtopped the dam killed 2,000 people living in villages below the dam.

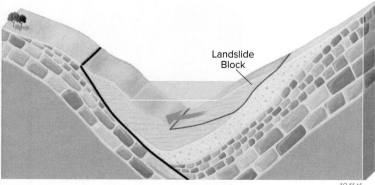

Landslide Block

10.11.t1

Before You Leave This Page

☑ Sketch and describe a rock or debris fall, a slide, and a rotational slide.

☑ Sketch how the geometry of layers, faults, and other features could allow a rock slide to begin.

☑ Describe the Vaiont landslide disaster and factors that caused it to happen.

10.11

10.12 How Does Material Flow Down Slopes?

SLOPE FAILURE CAN MOBILIZE WEAK MATERIAL, including soil, sediment, broken rock, and loose debris. Material flows downhill if it is poorly attached to a hillside, is internally too weak to resist the downward pull of gravity, and the slope is steep enough to allow flow. If the material incorporates some water, it becomes weaker and better able to flow. Such material can move rapidly, as in a debris flow, or can creep down the hill at a nearly imperceptible rate. All these types of slope failure involve the *flow* of material.

Creep and Solifluction

10.12.a1

Creep and *solifluction* are very slow, continuous movement of soil or weathered rock down a slope. Both processes occur on almost all slopes. At the surface, evidence for creep is expressed in bent or leaning trees (◀), warps in roads and fences, and leaning utility poles. At roadcuts, creep commonly causes bedding and other layers in the subsurface to bend downhill (▶). A type of creep at high latitudes or elevations is called *solifluction*. Here, poorly drained ground freezes for most of the year (permafrost) but near-surface ground thaws during the summer. This layer of saturated ground can move downslope in irregular, often overlapping lobes as pictured here (▶). This photograph of solifluction lobes nicely illustrates why the front of a slope failure commonly is called the *toe*, or in this case "toes."

10.12.a2 Colorado Springs, CO

10.12.a3 Norway

Debris Slides

10.12.a4

Soil, weathered sediment, or other unconsolidated material can move downslope as a *debris slide* (◀). Debris slides are usually less than 10 m thick and leave behind a low *scarp*. A debris slide moves downhill partly as a sliding, coherent mass and partly by internal shearing and flow. A debris slide can lose coherency as it moves, thereby evolving into an *earth flow* or a *debris flow*. The debris slide in this photograph (▶) has a clearly expressed scarp partway up the hill. At the base of the hill, the slide has wrinkled the ground as it moved downhill, forming a slightly steeper toe.

10.12.a5 Southern Montana

Earth Flows

Earth flows (◀) are flowing masses of weak, mostly fine-grained material, especially mud and soil. The material moves like thick, wet concrete, generally slowly enough to outrun, but it contains enough water to be slightly fluid. Earth flows contain more mud and other fine-grained material than they do rocks and also can be called *mudflows*. The earth flow in this photograph (▶) mobilized clay-rich altered volcanic and sedimentary rocks, as well as a surface veneer of angular rocks. The earth flow moved downhill until it reached and dammed a larger valley and completely flooded a small village.

10.12.a6

10.12.a7 Thunder Mtn., ID

Debris Flows

10.12.a8

Debris flows (◄) are wet, downhill-flowing slurries of loose mud, soil, volcanic ash, rocks, and other objects picked up along the way. Some contain only a little water, whereas others are water rich and flow like a thick soup. Debris flows, especially their mud-dominated varieties, are called *mudflows* by the media, and they can move rapidly. They often result from heavy rains that saturate the soil and other loose materials. Debris flows are thick and more dense than water, and can support and carry large boulders or even houses. In this photograph (▶), debris-flow deposits from the 1999 Venezuelan disaster, discussed in the opening pages of this chapter, consist of angular blocks, some several meters across. During transport, the rocks were enclosed in a matrix of wet mud and other debris.

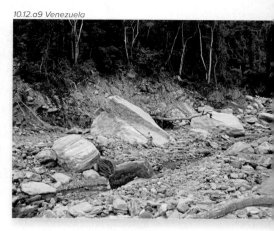

10.12.a9 Venezuela

Rock and Debris Avalanches

Debris avalanches (◄) are high-velocity flows of soil, sediment, and rock that result from the collapse of steep mountain slopes. A debris avalanche moves down the slope, in many cases traveling considerable distances down valleys and across relatively gentle slopes. A *rock avalanche* occurs when a rock mass falls off a cliff face and shatters on contact, sending a turbulent jumble of rock fragments, some bigger than cars, flowing downhill. They can flow at over a hundred kilometers per hour, fast enough to continue flowing uphill when they encounter a topographic obstacle. Both types of avalanches are often triggered by earthquakes and volcanic eruptions, and can kill thousands of people at a time. In 1970, an earthquake shook loose a large chunk of an Andean mountain, causing a rock and debris avalanche (▲) that buried a town (not visible in this photograph) and 18,000 of its inhabitants.

10.12.a11 Huascaran, Peru

10.12.a10

Classification of Slope Failures

10.12.a12

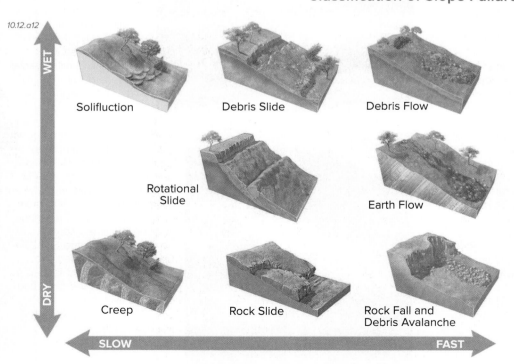

WET

DRY

Solifluction

Debris Slide

Debris Flow

Rotational Slide

Earth Flow

Creep

Rock Slide

Rock Fall and Debris Avalanche

SLOW

FAST

This diagram (◄) summarizes the classification of slope failures based on the type of material involved and the mechanism of movement, but viewed in the context of the *rates* of movement and whether movement occurred under *dry* or *wet* conditions. In this table, fast means you cannot outrun it, moderate means you have a chance to escape, and slow means you might have to stay for a while to see anything happen. For the slowest rates, you would not be able to see the movement in real time, only the results of months, years, or decades of accumulated movement.

Before You Leave This Page

☑ Sketch and describe what happens during the following: creep, debris slide, earth flow, debris flow, and debris avalanche. Compare how each of these features flows.

☑ Draw and explain a table summarizing different types of slope failures.

10.12

10.13 Where Do Slope Failures Occur in the U.S.?

LANDSLIDES AND OTHER SLOPE FAILURES have destroyed large parts of cities in some countries and have killed more than 30,000 people in a single event. Each year in the United States, landslides cause an average of 25 to 50 deaths and result in more than $3 billion in damage. What slope-failure events have affected the United States? What is the likelihood they will occur again? Which areas are most at risk in the future? Is your area safe?

A Where and How Have Large Slope Failures Affected the United States?

There have been countless landslides, debris flows, and other slope failures. The landslide that accompanied the 1980 eruption of Mount St. Helens was the largest landslide in U.S. history. It generated enough debris to fill 250 million dump trucks. Other slope failures are described below.

10.13.a1 central Alaska Range, AK

1. A large (7.9 magnitude) earthquake struck the Denali (Mount McKinley) region of south-central Alaska in 2002. Ground shaking associated with the earthquake caused huge slope failures off the region's steep mountains. Spectacular rock falls, rock slides, and debris avalanches slid down the steep slopes and flowed across and buried the Black Rapids Glacier. The region is prone to slope failures partly because tectonics has rapidly uplifted the mountains and formed an array of joints, faults, and tilted rock layers.

10.13.a2 Pacific Palisades, CA

2. Landslides and debris flows are common in southern California (◄), where some houses have been built on risky sites that have been undercut by waves, streams, and roads, causing unstable slopes to collapse downhill.

10.13.a4 Tully Valley, NY

4. The largest landslide to affect New York in 75 years moved through the Tully Valley, near Syracuse, in 1993. The earth flow mobilized over a million cubic meters of weak clays that had been deposited in a glacial lake. Similar glacial lakebeds occur in many parts of the region.

10.13.a3 Gros Ventre, WY

3. The Gros Ventre slide (◄) is one of North America's largest historic landslides. During this 1925 rock slide, a slab of limestone more than 1.5 km (1 mi) long broke loose along weak bedding planes that dipped toward, and were undercut by, the Gros Ventre River of Wyoming.

10.13.a5 Thistle, UT

5. The most costly landslide in U.S. history devastated the small Utah town of Thistle in Spanish Fork Canyon in 1983. The landslide moved slowly downhill, reaching a maximum rate of only 1 m (3 ft) per hour. It severed railroad service between Denver and Salt Lake City, buried or flooded two major highways, and formed a new lake where the town had been. A railroad tunnel now cuts through the light-colored landslide material.

B What Is the Potential for Landslides and Debris Flows in the United States?

All states receive some damage from landslides and debris flows, but not all areas have the same potential hazard. The potential for landslides is highest near steep mountains, such as in Colorado, the Appalachians, and other mountain areas that have weak, heavily weathered materials. It is also high in areas of recent tectonic activity.

Potential for Landslides and Debris Flows in the United States

1. This USGS map portrays the landslide hazards of the lower 48 states. Red areas represent the greatest hazards, followed by yellow and then green. Other areas have less potential for landslides or are unstudied. What landslide potential exists where you live?

2. Many parts in the Pacific Northwest experience landslides because of the many steep mountains, heavy rainfall, and rainfall that melts snow cover. The region also contains areas with high potential for debris flows, especially along slopes and valleys connected to the active Cascade volcanoes. An eruption on a snow-capped peak can unleash large debris flows.

3. The coastal parts of central and southern California have high landslide potential because of steep mountains, high potential for ground shaking during earthquakes, and coastal erosion that undercuts weak material along hillslopes overlooking the shoreline. Debris flows are also common, especially in the high mountains flanking Los Angeles, California. These recently uplifted mountains have very steep slopes and receive locally intense rainfall. Wildfires greatly worsen the situation.

4. A high potential for landslides and debris flows occurs in the Rocky Mountains and along the Wasatch Front, the steep mountain front that flanks Salt Lake City, Utah.

5. Landslide hazards in the central United States are mostly along steep bluffs that flank the rivers, or in areas, such as the northern Great Plains, underlain by weak materials.

6. In the east, landslides are common in the Appalachian Mountains, where landslides mobilize soil or occur along weaknesses in folded, faulted, and weathered rock layers. Shales, because of their inherently weak character, are especially prone to landslides.

7. Florida and the coastal plain of the southeast Atlantic seaboard have some of the lowest potential for landslides because the region lacks steep slopes, except near sinkholes.

10.13.b1

250 km

Landslides and La Conchita

The coastal community of La Conchita in southern California was partially overrun by a landslide in 2005, which was a repeat of one in 1995 (right). The landslide mobilized poorly consolidated sediment along the steep bluffs overlooking the town. The material flowed down into the community, burying and destroying a number of houses. Although the 1995 landslide had previously destroyed part of the town, houses were rebuilt in the area after 1995, only to be destroyed by the 2005 landslide. The 2005 landslide remobilized parts of the 1995 landslide and also incorporated new material farther back into the cliff. Would this be a good place to rebuild again? How much should hazard insurance cost people who live here?

10.13.t1

Before You Leave This Page

☑ Briefly describe factors involved in landslides in the United States.

☑ Summarize some factors that make some areas of the United States have high risks for landslides or debris flows.

☑ Identify whether you live in an area with a high potential for landslides, and list some possible factors why your area has this potential.

☑ Describe what happened at La Conchita. Was it a good idea to build or rebuild there?

10.13

10.14

How Do We Assess the Risk for Problem Soils and Future Slope Failures?

SOME SOILS AND SLOPES cause major problems. They can fail, causing buildings to crack or even collapse, or can otherwise pose a severe hazard to humans living in the area. We can limit damage caused by problematic soils and slope failures by first identifying the characteristics of the soil or the slope, and assessing whether those characteristics point to a potential hazard. We can then try to mitigate the problem or move out.

A How Do We Recognize Problem Soils?

Certain soils cause problems if they underlie buildings and roads. Some problem soils have limited strength when shaken, and others compact from the weight of objects constructed on top. Still others swell when they become wet and shrink when they dry out, causing roads and building foundations to crack from all the up and down movement.

10.14.a1 San Francisco, CA

10.14.a2

10.14.a3 St. Johns, AZ

Liquefaction occurs when loose sediment becomes saturated with water and individual grains lose grain-to-grain contact as water squeezes between them. *Quicksand* is an example of liquefaction. Liquefaction is especially common when loose, water-saturated sediment is shaken during an earthquake, as shown in the scene here.

In some soils, clay minerals start out arranged randomly, with much pore space between individual grains. If we build heavy things on top, the clay minerals and therefore the soil become compacted. *Soil compaction* causes problems for overlying structures, because some parts of the soil have more clay, and therefore compact more, than others.

Some soils contain a high proportion of certain clay minerals, called *swelling clays,* which increase in volume when wetted, expanding upward or sideways. As these clays dry out, they decrease in volume, causing the soil to shrink or compact. Repeated *expansion* and *compaction* during wet-dry cycles can crack foundations and require constant patching of roads.

B How Do We Recognize Prehistoric Slope Failures?

If you were going to buy some land out in the mountains, how would you tell if slope in the area has had some type of slope failure in the past? You might start by examining the topography, types of soils, the surface and subsurface distribution of rock types, and the condition of the rocks. If a slope has failed in the past, it can do so again.

10.14.b1 Mount Hood, OR

10.14.b2 Hat Creek, British Columbia, Canada

10.14.b3 Venezuela

This flank of Mt. Hood, a Cascade volcano, contains large deposits composed of broken, angular fragments of rock. These deposits formed when parts of the summit collapsed, slid, and tumbled downward at some time in the recent past. Note also the lack of soil.

Landslides and other slope failures can disrupt or even overrun existing topography, forming a random-looking assemblage of humps and pits. We call this type of landscape *hummocky topography,* and it is often indicative of some type of slope failure.

Each slope failure leaves behind characteristic deposits. Geoscientists observe modern deposits and then use these characteristics to recognize deposits of past events that were similar to the modern one. The deposits above were left by debris flows.

C How Do We Assess an Area's Potential for Slope Failure?

How would we try to assess the likelihood that a location will suffer the disastrous consequences of slope failure? We would first examine evidence of past slope failures in the area or in adjacent areas that have a similar setting. Then we would evaluate the steepness of slopes, including any recent changes, and any other factors leading to slope failure.

10.14.c1 Nepal

Evidence of Past Slope Failures

One of the best indications for potential slope failure is evidence of past failures. The more recent the failure, the more likely such an event will recur in the near future. This part of the Himalaya (◄), with steep hillslopes scarred with slope failures, looks very risky.

10.14.c2 Durango, CO

Situated in Area with Known Problems

The slope angle of hillsides and mountains (►) is clearly a key factor in the potential for slope failure but must be evaluated in the context of the types of materials and subsurface structures present.

Steepness of Slopes

10.14.c3 Fort Funston, CA

A site may be at risk if its geologic setting is similar to other slopes that have failed. Part of this cliff (◄) collapsed during an earthquake, and nearby parts have the same steep slopes, weak rocks, and position that can be undercut by the ocean waves.

10.14.c4 Zermatt, Switzerland

Recent Changes in Slope

Natural processes, over time, act to adjust slopes to the appropriate, stable angle, but this equilibrium can be upset if the slope is steepened or undercut by natural or human activities. This rock slide (►) failed in 1991, burying 31 houses.

10.14.c5 Capitol Reef NP, UT

Conditions of Material

Another factor is the nature of material on a slope (◄): whether material is loose sediment or solid bedrock, and is resistant to erosion or relatively weak. The presence, spacing, and orientation of geologic structures can weaken rocks and facilitate downslope slippage of materials.

10.14.c6 Augustine Volcano, AK

Potential Triggers

If other factors are equal, an area is at higher risk for slope failure if there are frequent events, such as volcanic eruptions or earthquakes, that could trigger slope failure. Steep slopes with loose material on an active, shaky volcano are big trouble (►).

The Slumgullion Landslide

The Slumgullion Landslide is a dramatic feature in the landscape of the San Juan Mountains of southwestern Colorado. The landslide begins at a steep scarp within some especially weak volcanic rocks and then flows downslope at meters per year. It slows down during the winter, when water within the flowing material freezes into ice, strengthening the otherwise unconsolidated debris. When the landslide reached the valley below, it spread out and blocked the Lake Fork of the Gunnison River, forming Lake San Cristobal, Colorado's second-largest natural lake. The landslide is among the most studied in the world, helping us understand the processes associated with such phenomena.

10.14.t1

Before You Leave This Page

☑ Describe some ways that geoscientists investigate slope failures.

☑ Summarize characteristics used to identify prehistoric slope failures.

☑ Summarize some aspects that might indicate that an area has a high potential for slope failure.

10.14

Which Areas Have the Highest Risk of Slope Failure or Problem Soils?

THIS GEOLOGICALLY DIVERSE PLACE has features that appear to be related to slope failure and problematic soils. Large, angular blocks occur in several different settings, and some of the hills may not be stable or safe. You will use descriptions and images of these features to determine what hillslope processes are occurring in different areas, and how they affect where people may live safely. The landscape is stylized and exaggerated to highlight potentially hazardous areas.

Goals of This Exercise:

- Observe the landscape to investigate the overall setting of different areas, and interpret the setting of each location from the figure and descriptions.
- Assess the hazards in different areas.
- Construct a map that shows areas that have a high risk for different types of slope failure.
- Identify locations that you think are most safe and moderately safe on which to build.

Procedures

Use the available information to complete the following steps, entering your answers in the appropriate places on the worksheet or online.

A. Observe the features shown on this landscape. Read the text boxes associated with each feature and decide what that statement implies about the setting of the area and how the landscape reflects the underlying geology.

B. Think about the description of each area and consider possible types of slope failure that could occur there. Provide a reasonable interpretation of what types of slope processes are occurring and what key observations led you to that conclusion.

C. On the figure in the worksheet, draw approximate boundaries around areas that you interpret as having the highest risk for each type of slope failure. Label each area with a few words to identify the main hazard you interpret to be present.

D. Draw the letters S and M on the map for sites where you think it probably would be safe to live. Write an (S) for one or more relatively safe places, an (M) for a moderately safe place to live. There is not a single best choice for any of these sites, so be prepared to describe your reasoning and to discuss your choice.

1. A series of small hills, referred to by local people as the *Bent Fence Hills,* contains trees that are tipped over at odd angles. Local farmers complain that they have to keep straightening their crooked fences on these hillslopes. For some reason, no one has ever built a house here.

2. A flat-topped hill, called *Flattop Hill,* is surrounded by a steep cliff formed by a resistant layer of basalt. The basalt is jointed and underlain by a weak layer of clay. Below the cliff are a series of large, angular blocks of basalt. A large, spoon-shaped scar scoops into part of the cliff.

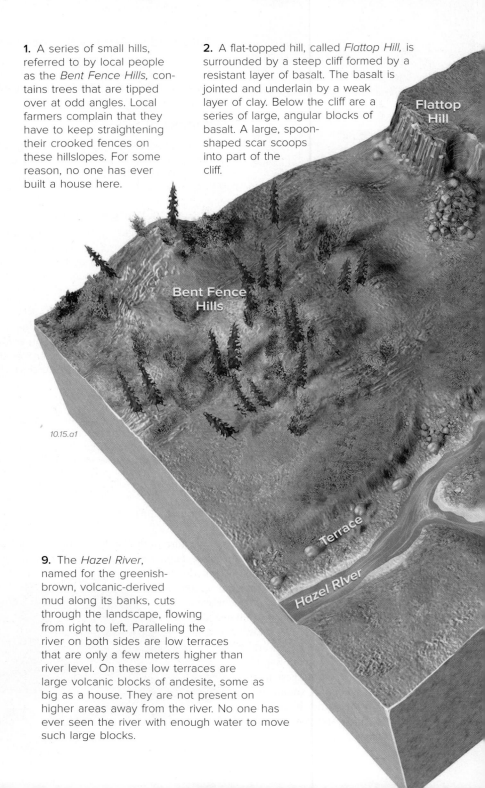

10.15.a1

9. The *Hazel River,* named for the greenish-brown, volcanic-derived mud along its banks, cuts through the landscape, flowing from right to left. Paralleling the river on both sides are low terraces that are only a few meters higher than river level. On these low terraces are large volcanic blocks of andesite, some as big as a house. They are not present on higher areas away from the river. No one has ever seen the river with enough water to move such large blocks.

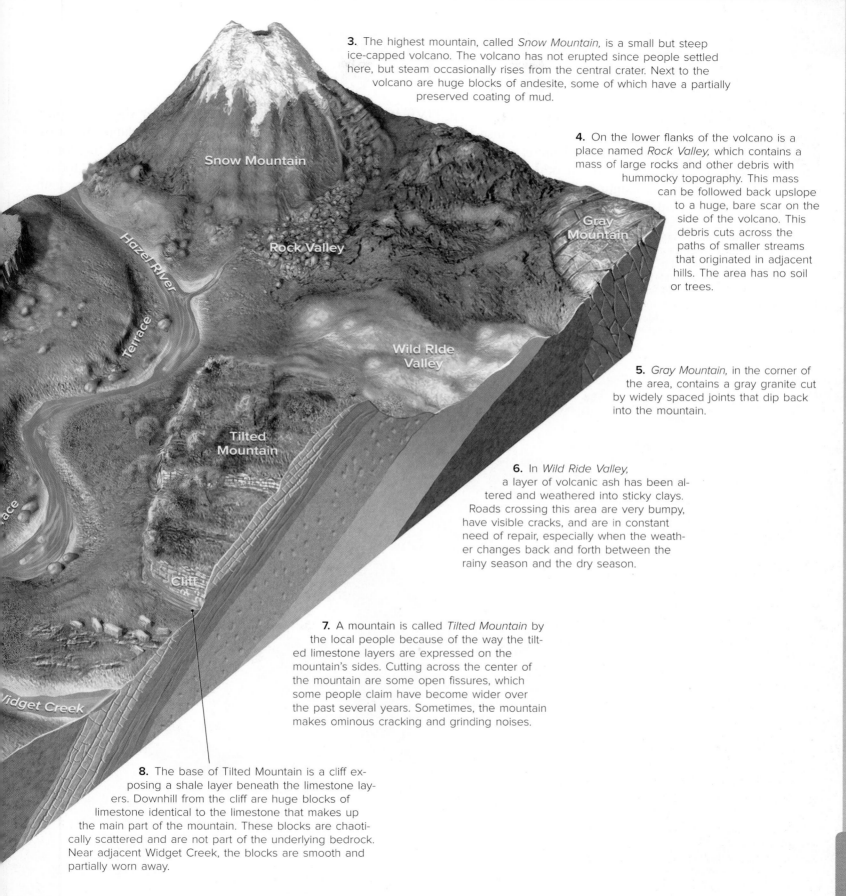

3. The highest mountain, called *Snow Mountain,* is a small but steep ice-capped volcano. The volcano has not erupted since people settled here, but steam occasionally rises from the central crater. Next to the volcano are huge blocks of andesite, some of which have a partially preserved coating of mud.

4. On the lower flanks of the volcano is a place named *Rock Valley,* which contains a mass of large rocks and other debris with hummocky topography. This mass can be followed back upslope to a huge, bare scar on the side of the volcano. This debris cuts across the paths of smaller streams that originated in adjacent hills. The area has no soil or trees.

5. *Gray Mountain,* in the corner of the area, contains a gray granite cut by widely spaced joints that dip back into the mountain.

6. In *Wild Ride Valley,* a layer of volcanic ash has been altered and weathered into sticky clays. Roads crossing this area are very bumpy, have visible cracks, and are in constant need of repair, especially when the weather changes back and forth between the rainy season and the dry season.

7. A mountain is called *Tilted Mountain* by the local people because of the way the tilted limestone layers are expressed on the mountain's sides. Cutting across the center of the mountain are some open fissures, which some people claim have become wider over the past several years. Sometimes, the mountain makes ominous cracking and grinding noises.

8. The base of Tilted Mountain is a cliff exposing a shale layer beneath the limestone layers. Downhill from the cliff are huge blocks of limestone identical to the limestone that makes up the main part of the mountain. These blocks are chaotically scattered and are not part of the underlying bedrock. Near adjacent Widget Creek, the blocks are smooth and partially worn away.

10.15

Glaciers, Coasts, and Changing Sea Levels

GLACIERS AND COASTS ARE GEOLOGICALLY ACTIVE PLACES that shape the landscape over short and long timescales. Glaciers sculpt the landscape and cause large changes in sea level as the volume of ice on land increases or decreases. Past sea levels have been more than 200 m (660 ft) higher than today, flooding large parts of the continent and any other low-lying land. At other times in the past, sea level was about 120 m (390 ft) lower than today, exposing large parts of the continental shelves. The most rapid sea level changes are related to changes in the extent of glaciers and continental ice sheets.

The northeastern United States and adjacent Canada, shown in this shaded relief map, have a striking collection of features on land and along the coasts. The map extends from North and South Dakota (in the northwestern corner of the map) across the Great Lakes as far east as Maine and Virginia. The northern part of the map includes southeastern Canada.

Huge, smooth troughs (each labeled with a **T** on the map) cut across the landscape. Examples are near the northwestern corner of the map in western Minnesota and the Dakotas, and in the area southwest of Lake Erie, in Ohio and Indiana.

What caused these smooth areas of the landscape, and is the process still occurring?

Curiously curved ridges (labeled with an **R** on the map) cross some of the smooth areas and are especially noticeable southwest of Lake Erie and Lake Michigan.

What are these ridges, how did they form, and what do they tell us about the geologic history of this region?

11.00.a1

TOPICS IN THIS CHAPTER

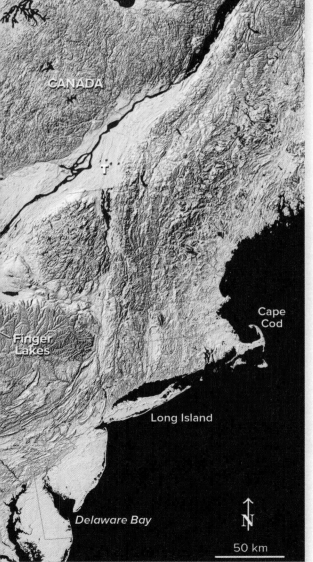

North of the Great Lakes, the landscape of Canada is rough on a small scale, containing many lakes.

Why does this region have so many lakes, and when and how did this landscape form?

The Maine coastline is very irregular, with many bays, where the ocean reaches inward like fingers into the land. Offshore are a variety of islands, for which the region is well known.

How did this coastline form, and has it always been this way?

Cape Cod, Massachusetts, protrudes into the ocean like a flexed arm or a boot with curled toes.

How did Cape Cod form, and what coastal processes caused it to have this shape?

Delaware Bay and Chesapeake Bay to the south connect to the ocean, but their shapes resemble river valleys.

How did these large bays form, and — if they started as river valleys — when and why did they become flooded by the ocean?

Ice Ages and Coasts

Landscapes in the Great Lakes area contain evidence that huge ice sheets once flowed across this part of the continent — in the recent geologic past. This conclusion arises from comparing the distinctive landscape features and their associated sedimentary deposits with those observed today near currently active glaciers. With glaciers, as with most geologic features, observing the present is the key to interpreting the past.

For the last two and a half million years, Earth has experienced an *ice age,* during which large regions of the Northern Hemisphere, as well as Antarctica and parts of South America, were covered year-round with ice and snow. Where the ice was thick enough, or rested on a steep enough slope, it moved downhill as a mass of flowing ice called a *glacier.* Some glaciers were small and restricted to mountain areas, whereas others covered large parts of the continents, forming *continental ice sheets.* Continental ice sheets flowed southward from Canada and smoothed off and carved grooves into the underlying landscape by grinding ice, rocks, and sand against the bedrock. The smooth troughs on this map were carved by continental glaciers that flowed southward from Canada. Some areas south of the ice had large lakes that formed in association with the glaciers.

As the climate warmed over the past 20,000 years, the ice sheets and glaciers melted and covered less area. Rocks and other sediment once carried in the ice were dropped along the front of the melting glaciers, forming a series of curved ridges south of the Great Lakes. The ice also left piles of glacially derived sediment on Long Island and Cape Cod.

Water released from melting ice carved new river valleys and flowed into the sea, causing global sea level to rise. The rising seas flooded coastlines and river valleys, forming the many inlets and bays along the Atlantic coast of the eastern United States.

11.0

11.1 What Are Glaciers?

GLACIERS ARE MOVING MASSES OF ICE, ranging in size from huge ice sheets that cover large regions to smaller glaciers that are restricted to a single mountain or valley. Most glaciers are primarily ice and snow, but they typically contain significant amounts of rocks and finer sediment that were incorporated into the glacier as it flowed from higher elevations to lower ones. Glaciers create spectacular scenery when they are still around and when they have melted away.

A What Are the Characteristics of Glaciers?

1. Glaciers form where snow and ice accumulate faster than they melt, so many glaciers begin in snowfields in higher elevations or at higher latitudes (closer to the poles). Glaciers only form if an area is cold and receives enough snowfall to allow ice and snow to accumulate faster than they melt.

2. As the snow gets buried, it compresses into ice, turning blue in the process. As a result, the icy parts of glaciers have a distinctly blue appearance. Most glaciers also have lines (grooves, ridges, and sediment-rich streaks) formed by flow within the glacier. These are fairly straight or gently curved if the glacier has a simple pattern of flow, but they are contorted and folded if the glacier experienced more complex patterns of flow.

3. Ice can cover broad areas, but glaciers can become confined within valleys as they flow from higher elevations to lower ones. As adjacent ice-filled valleys merge, so do the glaciers, producing a wider and commonly thicker mass of flowing ice.

4. Whether a glacier forms depends partly on the slope of an area. Areas with gentle slopes allow snow and ice to accumulate to sufficient thicknesses, whereas some mountain slopes are so steep that snow and ice slide downhill instead of piling up.

11.01.a1 Dry Valleys, Antarctica

11.01.a2 Patagonian Andes, Argentina

5. This glacier of blue ice flows down a steep valley and ends in a lake of meltwater from the glacier (◀). As a glacier moves, internal stresses cause its upper surface to break, forming fractures, each of which is called a *crevasse*. Crevasses are especially abundant and well developed where a glacier flows around a curve, like around a bend in a valley, or where the land beneath the glacier changes slope, either from a steep slope to a more gentle one or from a gentle slope to a steeper drop. The glacier shown here breaks apart, opening up crevasses as it flows around a bend and over a steep drop-off.

11.01.a3 Swiss Alps

6. As a glacier moves past bedrock, it plucks away pieces of rock, and its surface may be partially or totally covered by rock pieces derived from nearby steep slopes (◀). It grinds up some rock into a fine rock powder. The glacier carries away this material, coarse and fine, depositing it where the glacier melts. The glacier in the photograph above has dark fringes of rocky material on both sides and at its end.

Before You Leave These Pages

✓ Describe the characteristics of glaciers, including the three main types (ice sheets, valley glaciers, piedmont glaciers).

✓ Summarize which places have ice sheets and glaciers.

B   What Are the Types of Glaciers?

1. The largest accumulations of ice are in *continental ice sheets* (▶), regionally continuous masses of ice like those covering nearly all of Antarctica and Greenland. Such ice sheets tie up huge amounts of water when they form or when they increase in extent or thickness, resulting in a significant drop in global sea level, sometimes exceeding 100 m. Continental ice sheets release these same large volumes of water when they melt, and nearly all this water flows into the sea, resulting in a significant rise in global sea level. Clearly, understanding rises and falls of sea level generally requires understanding changes in continental ice sheets.

11.01.b1 Antarctica

2. Glaciers that flow down valleys (▶) are called *valley glaciers* or *Alpine glaciers* (named after the Alps). Valley glaciers tend to be fairly narrow (several kilometers wide) but can flow down valleys for tens of kilometers.

5 km
11.01.b2 Dry Valleys, Antarctica

3. As some valley glaciers or ice sheets flow out of the mountains into broader, less confined topography, they can spread out (▶), forming a *piedmont glacier* (piedmont means foot of the mountain).

5 km
11.01.b3 La Perouse Glacier, AK

C   Where Are Most Ice Sheets and Glaciers?

Glaciers form where snow and ice accumulate faster than they melt, so they form in cold climates. Most glaciers are therefore in high latitudes (closer to the North or South Pole), at high elevations, or some combination of high latitude and high elevation.

1. In the Southern Hemisphere, glaciers occupy high peaks of the Andes, especially in Patagonia, the most southerly (high latitude) part of the mountain range.

2. The largest ice mass on Earth is on the continent of Antarctica, which sits squarely over the South Pole. Ice and snow cover about 98% of the continent, mostly in the form of huge ice sheets. Valley glaciers form where an ice sheet flows into valleys.

3. Glaciers cover many of the highest parts of the Tibetan Plateau and the Himalaya, the highest mountain range on Earth, even though this region has a fairly low latitude (not close to the poles). Glaciers and ice sheets are present elsewhere in Asia, especially in islands and peninsulas along the Arctic Ocean.

4. Large ice sheets and smaller glaciers occupy 80% of Greenland and large areas of the neighboring islands, including Iceland.

5. In the main part of North America, glaciers are present in Alaska, northern Canada, the Rocky Mountains of the United States and Canada, the Coast Range of British Columbia, and on the larger volcanoes and other high peaks of the Cascade Range.

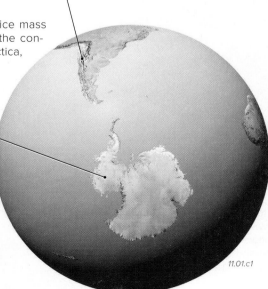
11.01.c1

11.01.c2

11.1

11.2 How Do Glaciers Form, Move, and Vanish?

GLACIERS FORM, MOVE DOWNHILL, AND EVENTUALLY MELT AWAY. How does a glacier form? Once formed, how does a glacier move across the landscape, and what happens to it as it flows downhill, toward warmer areas with more melting and generally less snowfall?

A How Do Snow and Ice Accumulate in Glaciers?

1. Glaciers, including the ones below, form by the accumulation of snow and ice. The snow is derived from snowfall, but can be moved around by the wind or by avalanches, which are masses of snow and ice that fall, slide, or flow downhill.

2. Snow falls as individual flakes. Loose snow can contain 90% air between the flakes. Once on the ground, flakes get pressed together by the weight of other snowflakes on top.

3. As more snow accumulates on top, snowflakes farther down are compressed, forcing out more than 50% of the air. The snowflakes become compressed into small, irregular spheres of more dense snow.

4. With increasing depth and pressure, the snow begins to recrystallize into small interlocking crystals, forming solid ice. Ice is a crystalline material and is considered to be a type of rock. Crystalline ice contains less air and commonly has a bluish color. The blue color is due to the way oxygen-hydrogen bonds in water molecules interact with light (the same reason why lakes, seas, and other water bodies are blue).

11.02.a1 La Perouse Glacier, AK

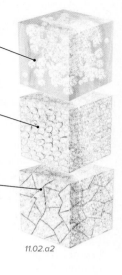
11.02.a2

B How Does a Glacier Form and Change as It Moves Downhill?

Glaciers form when the amount of snow and ice accumulating from snowfall exceeds the amount lost by melting and other processes. In this situation, the snow and ice pile up and may start to move as a glacier.

1. The upper part of the glacier or ice sheet, where snow and ice are added faster than they melt, is the *zone of accumulation*. Gravity, working on the weight of accumulating snow, causes the glacier to flow downhill.

2. As the glacier moves downhill, it loses more and more ice and snow by melting, by wind erosion, and by loss of ice molecules directly to the air, a process referred to as *sublimation*. At some point along the glacier, the losses of ice and snow exactly balance the amount of accumulation; this boundary is called the *equilibrium line*. The equilibrium line is sometimes, but not always, marked by a gradational boundary between snow-covered ice upslope on the glacier and exposed bluish ice downslope. The bluish ice formed at depth and became exposed at the surface as upper levels of ice and snow were removed. In some cases, the entire length of a glacier may be covered with snow, but blue ice can be observed at depth in fractures (crevasses) that cut the glacier's upper surface.

11.02.b1

Zone of Accumulation

Snow

Equilibrium Line

Blue Ice

Zone of Ablation

Land

1 km

Sea

4. At lower elevations, ice melts away faster than it can be replenished by downward movement of ice within the glacier and by snowfall. The lower part of the glacier where it is losing ice is the *zone of ablation*. This loss eventually causes the glacier to end or terminate, either on land or in the sea. The end (also called the front) of a glacier is called the *terminus*.

3. The valley glacier in the photograph to the right has an upper, snow-covered part (zone of accumulation) and a lower area of blue ice and dark rocks below the equilibrium line.

11.02.b2 Morteratsch, Swiss Alps

C How Do Glaciers Move?

Glaciers move downhill because the ice is not strong enough to support its own weight against the relentless downward pull of gravity. As glaciers move and spread downward, they move by internal shearing and flow of the solid ice, by simply sliding across the bedrock, or by some combination of these two mechanisms.

1. As gravity pulls the ice downhill, friction along the base of the glacier causes the bottom of the glacier to lag behind the upper, less constrained parts. The upper part of a glacier (▼) therefore flows faster than the lower part, causing internal shearing within the glacier.

2. If the interface between the glacier and the underlying bedrock is very irregular and is relatively dry, the base may become locked to the bedrock and not move at all. Only the coldest glaciers are completely frozen at their bases.

11.02.c2

3. If the bedrock—glacier interface is less irregular (i.e., smoother) or contains water from melting ice, the glacier may be able to slide over the bedrock. Such glaciers can move relatively rapidly.

11.02.c3

4. The rates at which glaciers move are extremely variable. Many glaciers move about a meter per day, but some move centimeters per day. The fastest ones move more than 30 meters per day.

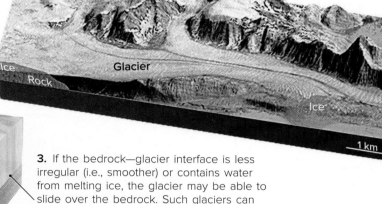
Glacier
Ice
Rock
Ice
1 km
11.02.c1

D What Happens When a Glacier Encounters the Sea or a Lake?

1. When a glacier reaches the ocean or a lake, it may float on the water if the sea or lake is deep enough. Ice, even the dense blue variety within glaciers, floats because it is less dense than either fresh water or salt water.

2. As ice along the leading edge of a glacier floats, it tends to spread or be pulled apart, forming large crevasses within the ice. These allow large blocks of ice (▶) to collapse off the front of the glacier, a natural process called *calving*.

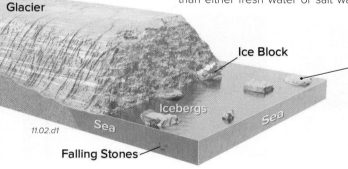
Glacier
Ice Block
Icebergs
Sea
Sea
Falling Stones
11.02.d1

3. As the blocks of ice fall into the water, they float, forming *icebergs*. As much as 90% of an iceberg is beneath the water. As icebergs melt, rocks and other sediment within them drop into the water. Some ice sheets and glaciers flow into the sea with such large quantities of ice that they form a large *ice shelf* that floats on seawater (▶). These can be hundreds of kilometers wide.

11.02.d2 Glacier Bay, AK

11.02.d3 Antarctica
Ice Shelf
Sea Ice

Glaciers, Snowfields, and Sea Ice

Not every large mass of ice on Earth's surface is a glacier. Some accumulations of snow and ice never move, and these are simply called *snowfields*. Large masses of ice also form when the upper surface of a lake or the sea freezes. In the ocean, such ice is called *sea ice*. In all but the coldest places, like parts of the Arctic Ocean, sea ice freezes in the winter and thaws in the spring or summer. Freezing excludes most salt from the crystalline structure of ice, so sea ice melts to form water that is largely fresh (not salty). In the photograph below, broken sheets of sea ice surround small, rocky islands in Antarctica.

11.02.t1 Antarctica

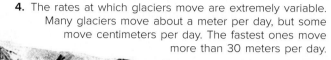

11.2

How Do Glaciers Erode, Transport, and Deposit?

GLACIERS ARE CAPABLE OF INCREDIBLE amounts of erosion, often gouging into landscapes hundreds of meters deep. Some of us picture glaciers as uniformly white and free of debris, but most glaciers incorporate abundant rocky debris and act as conveyor belts transporting the debris many kilometers. Once deposited, glacially carried and deposited debris forms distinct landforms that are used for agriculture, recreation, and urbanization. What processes are involved in glacial erosion, transport, and deposition? How can we identify these processes and glacial deposits in the landscape?

A How Do Glaciers Erode?

Ice is not a hard material, but the base and sides of a glacier contain rocks and other material that can gouge (pluck) and scrape (abrade) the underlying land surface, smoothing off rough edges and removing rocks and other sediment. Once plucked and abraded from the bedrock, the debris can become incorporated into the ice, transported some distance, and eventually deposited. Meltwater at the glacier's base and front (terminus) can also cause erosion.

1. Glaciers cause erosion in three main ways: plucking, abrasion, and from glacial meltwater. In the example here, a glacier is moving over a small hill (▶) that predated the advance of the glacier.

2. At a glacier's base, pressure is great enough that ice melts, forming a thin film of water through a phenomenon known as *pressure melting*. This water can refreeze inside joints that were either preexisting or where reduced pressure (*unloading*) at a bedrock step creates jointing by expansion. As the glacier moves, rock is torn away (plucked) from the joint and incorporated into the glacier.

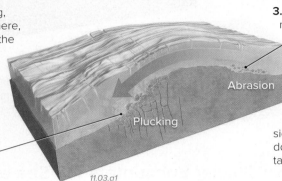

11.03.a1

3. On the upflow side of the hill, the motion of the glacier is pressing material against the bedrock, which is in the way. As a result, this side of the hill is heavily abraded and smoothed off, resulting in a more streamlined, almost aerodynamic, shape that is less resistant to the flowing ice. With abrasion concentrated on the upflow side and plucking concentrated on the downflow side, the glacially eroded hill takes on an asymmetric shape.

Abrasion

4. Rock and smaller sized sediment at the glacier's base scrapes at underlying bedrock through a process called *abrasion*.

Glacial Grooves — 11.03.a2 Kelley's Island, OH

Striations — 11.03.a3 Marquette, MI

5. When ice sheets flow across the surface, they smooth and polish rocks over broad areas. They typically carve the top of bedrock into a relatively smooth, polished surface, which is gouged by scratch marks, called *glacial striations*. Such polished surfaces and scratch marks are evidence that a glacier once moved across the area (note, however that fault movement can form similar features). If the gouge marks are large and deep, they are *glacial grooves*, which provide evidence of erosion by dragged sediment and boulders at the base of a glacier. Glacially derived water can help carve or accentuate some glacial grooves.

Plucking

6. *Plucking* is concentrated on the downflow side of irregularities in the underlying bedrock, but it can occur anywhere beneath a glacier.

Roche Moutonnée — 11.03.a4 Rockwood, CO

Direction of Ice Flow

7. Plucking can occur where rocks become loosened by and incorporated into ice at the base of a glacier. The combination of plucking and abrasion can sculpt an asymmetric feature—a *roche moutonnée* (▲). Abrasion dominated on the right side, grinding a smooth and polished surface. Plucking occurred on the left side, resulting in a steep, more angular face. Plucking liberates large rocks that are left behind, perhaps hundreds of kilometers from their origin, when the glacier melts. These rocks typically are compositionally different than the local bedrock and so seem out of place; such a boulder (▼) is a *glacial erratic*.

Erratic — 11.03.a5 Pinedale, WY

Glacial Meltwater

8. Part of the glacier can melt along its base from pressure melting or on the surface from sunlight, in either case forming meltwater.

Subglacial Channel — 11.03.a6 Mendenhall Glacier, AK

9. The meltwater can flow in a channel along the base of the glacier. The pressurized, flowing meltwater is heavily laden with rocks and other glacial sediment and so exerts strong erosive power on the bedrock underlying glacial sediment. The water typically leaves the glacier at its front.

Before You Leave These Pages

✓ Describe how glaciers erode, transport, and deposit material.

B How Do Glaciers Transport Material?

Once material is eroded by glacial abrasion and plucking, glaciers can incorporate and transport vast amounts of debris, ranging from house-sized boulders to microscopic clay particles. Debris is carried on the surface of the ice, within the ice, and at the base of the ice. Together these processes make glaciers effective conveyor belts of debris transport.

Transport Within Ice

11.03.b2 Glacier Bay, AK

1. Material can be encased within the ice, carried along with the moving ice. The glacier can retain this debris somewhere within the glacier, or internal shearing along the inclined shear planes shown here can bring debris within the ice upward to the surface of the glacier, where it is then carried on top. Such inclined shear planes develop when moving ice segments encounter immobile ice and "ride" over and past this obstruction.

Transport on Top of Ice

2. Some debris is transported on top of the ice (▶), especially debris added through erosion of steep slopes flanking a glacier. Such debris can fall into crevasses (cracks), becoming encased in the ice.

11.03.b3 Muldrow Glacier, AK

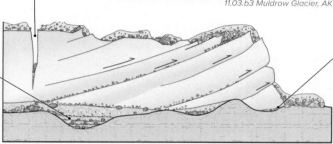

11.03.b1

Transport Near Base of Ice

11.03.b4 Mendenhall Glacier, AK

3. Sediment transport also occurs at the ice—bedrock interface (▲), as exposed in the meltwater tunnel in the photograph above. In some glaciers, ice moves along a "soft" bed of deformable sediment. Here, finely ground sediment, called *rock flour*, turns meltwater gray and is carried out to the glacier's margin.

C How Do Glaciers Deposit Material?

Any sediment carried by ice, icebergs, or meltwater is called *glacial drift*. In any of these three cases, the debris eventually comes to rest. If deposited directly by ice, the material is generally unsorted (particles are not segregated by size) and unstratified (lacking layers)—it is called *till*. If deposited by meltwater, the glacial deposits are sorted and stratified (layered) in appearance.

1. This glacier is directly depositing dark till at its terminus. There is debris on top and along the sides of the glacier (▶), and there is more debris inside and at the base of the glacier. If the terminus of a glacier remains at about the same place, the till piles up into an irregular mass (the lumpy hills at the downhill end of the glacier).

2. This material (▶) was directly deposited by ice and is a glacial till. It is characteristically unsorted and unstratified. Most fragments (clasts) are generally angular, but others become somewhat rounded as their corners get knocked off and ground (abraded) away as they are carried by the ice. Note the people for scale in the lower right.

11.03.c1 Marcus Baker Glacier, AK

Terminus

11.03.c2 Athabaska Glacier, Alberta, Canada

Till

11.03.c3

Glaciofluvial Outwash

11.03.c4 Milwaukee, WI

Glaciofluvial Deposits

3. In this photograph (◀), meltwater in the foreground is issuing from the glacier in the background and carries abundant sediment. Deposition of sediment by glacial streams is called *glaciofluvial deposition*, where the term refers to the involvement of glaciers and streams (fluvial).

4. Glaciofluvial deposition involves running water and so creates sorted and stratified deposits, as in the sediments shown here (◀). Running water can sort larger, heavier clasts, such as pebbles, from those that are smaller and lighter, such as sand, producing a layered deposit composed mostly of glacial debris.

11.3

11.4 What Are the Landforms of Alpine Glaciation?

AS GLACIERS MOVE, the ice scours underlying rock and unconsolidated materials, picking up the pieces and carrying debris toward lower elevations. In mountainous areas, glaciers pluck rocks from peaks and ridges, producing some distinctive landforms that we can use to recognize landscapes that are glacially carved. Glaciers and ice sheets grind into the underlying land surface, wearing down hills and other topographic high points, and locally polishing smooth surfaces onto bedrock.

A How Does Glacial Erosion Modify Landscapes?

Glaciers occupy and modify landscape features that existed before glaciation and imprint into the landscape clues that glaciers were once there. Glacial erosion reshapes a valley, typically changing it from a pre-glacial, stream-carved, V-shaped valley to a glacially carved *U-shaped valley*. Glaciers deposit the eroded material locally or farther away.

11.04.a1 · During Glaciation

11.04.a2 · U-Shaped Valleys · After Glaciation

11.04.a3 Denali NP, AK · U-Shaped Valley

In these two computer-generated perspectives of the San Juan Mountains of Colorado (▲), glaciation in the left image (an artist's interpretation) results in the present-day landscape of the second image (a satellite image combined with topography).

One result of glaciation (▲) is formation of a U-shaped valley (i.e., a "U" shape in profile), which contrasts with V-shaped stream valleys.

B What Landforms Do Valley Glaciers Form When They Deposit Sediment?

Valley glaciers erode and transport debris along their sides and bases, but flow within the glacier and the merging of adjacent glaciers distributes debris throughout much of the glacier. The sides contain especially abundant sediment because the ice receives loose materials from the mountainous slopes and streams flanking the glacier. The base of a glacier is also relatively rich in sediment because it plucks away pieces of bedrock and any loose materials over which the glacier moves. An accumulation of sediment that was carried and deposited by a glacier is a *moraine*. That is, moraine is an accumulation of till that can form ridges or be a somewhat flat sheet of till covering the land surface. We classify moraines into different types, according to where they form.

1. A *lateral moraine* forms along the sides of the glacier and is expressed as a dark fringe of rocks and other debris (▶). When the glacier melts, lateral moraines commonly form low ridges along what were the edges of the glacier. The computer-generated image to the far right displays lateral moraines from three now-gone glaciers.

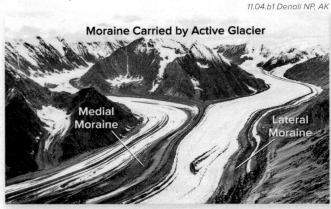
11.04.b1 Denali NP, AK · Moraine Carried by Active Glacier · Medial Moraine · Lateral Moraine

11.04.b2 Sawtooth Mountains, ID · Lateral Moraine · Moraine Left Behind After Glacial Melt · Terminal Moraine · N→ · 1 km

2. A *medial moraine* is a sediment-rich belt in the center of the glacier. A medial moraine (▲) forms where two glaciers join, trapping their lateral moraines within the combined glacier. Medial moraines may not be well preserved.

3. A *terminal moraine* forms at the termination of a glacier and marks the glacier's farthest downhill extent (▶).

C What Alpine Landforms Form in Bedrock?

In mountains, glacial erosion produces distinctive landforms chiseled out of bedrock, including bowl-shaped basins flanked by steep ridges, and U-shaped valleys carved by the moving ice. We can use these features to recognize landscapes carved by glaciers, even after the ice melts away. The three-dimensional terrain below shows many of these features from the aptly named Glacier National Park in Montana, and photographs of examples are below the main figure.

1. Near the uppermost end of an alpine glacier, the ice plucks pieces from the bedrock, excavating a bowl-shaped depression called a *cirque*. When the ice melts, it exposes the cirque.

2. A lake within a glacially scoured depression in a cirque is referred to as a *tarn*. Some lakes within cirques are connected, one after another, by a stream down the glacially carved valley.

3. Hard bedrock ridges that flank cirques are commonly narrow, sharp, and jagged, like the ridges shown in the figure below. Such a ridge is an *arête* and is jagged because it has been glacially eroded from both sides.

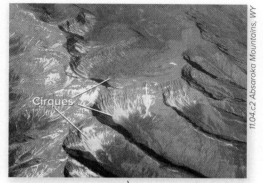
11.04.c2 Absaroka Mountains, WY
Cirques

11.04.c3
Tarn

14.04.c4 Sierra Nevada, Independence, CA
Arête

11.04.c1
Cirque
Horn
Tarn
Arête
Cirque
Hanging Valley
U-Shaped Valley
1 km
11.04.c6 Silverton, CO

11.04.c5 Zermatt, Switzerland
Horn

Hanging Valley

4. Glaciers from smaller valleys can merge with a larger, thicker glacier flowing down a main valley. The larger glacier scours deeper into the bedrock, so the main valley is deepened more than the side valleys, forming a U-shaped glacial trough. When the glaciers melt away, the side valleys are higher than the main valley, and we refer to one of these as a *hanging valley* (▶). A U-shaped valley eroded below sea level and subsequently flooded by a rise in sea level is a *fjord* (▼), famous examples of which are in Norway, Alaska, and the Arctic.

5. Where three or more cirques merge by headward erosion, they form a pyramid-like feature called a *horn*. The most famous of these is the Matterhorn (▲) near Zermatt, Switzerland.

11.04.c7 Baffin Island, Canada
Fjord

Before You Leave This Page

☑ Explain how glacial erosion can modify a landscape.

☑ Describe the characteristics and origins of landforms formed by alpine glaciation.

11.4

11.5 What Are the Landforms of Continental Glaciation?

CONTINENTAL ICE SHEETS can cover large areas of continents, completely reshaping the landscape by scouring down the surface and depositing sheets and lumps of sediment beneath, within, and in front of the ice sheet. When the ice melts away, the entire surface tends to have low relief because any protruding topography was planed off and many low points are filled with sediment. Accumulations of glacially affected sediment form distinctive features.

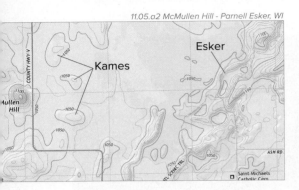
11.05.a2 McMullen Hill - Parnell Esker, WI

1. Meltwater carves tunnels through and along the bases of many glaciers, depositing sediment within the tunnels and out in front of the glacier. When the glacier melts back, sediment sorted and deposited along these meltwater channels forms long, sinuous ridges called *eskers*. The sinuous ridges on the map to the left and the photograph to the right are eskers, left behind by a retreating glacier. Eskers are an "endangered species" of glacial landforms in that many have been removed because they are highly prized as a source of gravel for roadbeds, construction fill, and other uses.

14.05.a3 West Dundee, WI

11.05.a4 McMullen Hill, WI

2. *Kames*, such as this one in Wisconsin (▲), are moundlike hills that are believed to have formed where meltwater in stagnant ice deposited sediment in ice crevasses or in the space between the glacier and the land surface. In the map above, a kame forms a fairly round hill. Like eskers, kames are often excavated for gravel.

11.05.a5 Great Plains, ND

3. These flat to gently rolling plains (◄) are composed of sediment deposited from the base of the ice as *ground moraine*, which occurs in many parts of the Great Lakes region and upper Midwest of the U.S. The ground moraine has enriched the soils for farming, and the smoothened topography makes the area just well-drained enough, while not too susceptible to erosion.

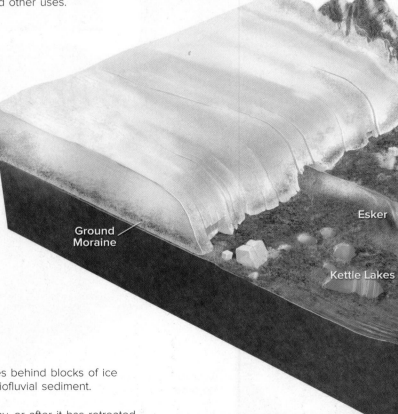

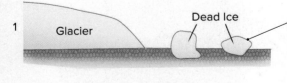

4. As a glacier retreats, it leaves behind blocks of ice encased in the glacial and glaciofluvial sediment.

5. As the glacier remains nearby, or after it has retreated farther, the ice block can become partially or totally buried by glacial outwash deposited by glacial streams.

6. When an ice block melts, as in the bottom diagram, it creates a small depression, called a *kettle*, within the sediment. This depression can fill with water, becoming a *kettle lake*, as in the photograph to the right (►). If the kettle does not fill with water, it may simply look like a pit in the ground. Kettles can also form in till, not just outwash. Kames and kettles commonly reside together, forming a rugged terrain of steep hills, ponds, and poor drainage.

11.05.a6

11.05.a7 Northeastern SD

7. On the 3D perspective and photograph below and on the map to the right (▶) are some curiously shaped hills that resemble teardrops, each with its blunt, steep end pointing to the direction from which the ice flowed (from the north in all three cases). Each of these streamlined hills is composed of till and glaciofluvial deposits and is called a *drumlin*. A drumlin forms as a moving glacier sculpts these soft materials into a shape designed to minimize drag, similar to the shape of a submarine. Drumlins form in groups or fields, called *drumlin fields*, as shown in the 3D perspective below. The greatest concentrations of drumlins are in eastern Wisconsin and in central New York. Drumlins are the only glacial feature on this page formed by advancing ice. All the rest are formed when the ice is stagnant or melting back.

Drumlin

11.05.a9 Central Wisconsin

11.05.a10 Marion, NY

8. *Recessional moraines* form as the front of the glacier melts back and stagnates for a while in one location, depositing a pile of sediment along the front of the glacier. The shape and distribution of a recessional moraine replicate the shape of the front of the glacier when it stagnated. Recessional moraines form curvilinear patterns around the Great Lakes (▶). Between moraines are flatter terrain consisting of ground moraine and outwash plains. Most of the curved ridges displayed in the large figure in the opening pages of this chapter are recessional moraines.

11.05.a11

11.05.a8

9. A *terminal moraine* represents the maximum forward extent of the front of the glacier, whether it is a continental ice sheet or an alpine glacier. It has the same shape and character as a recessional moraine, but it is farthest in front. Areas between a terminal moraine and the present-day glacier, if it still exists, were once covered with ice. Moraines, such as this one pictured to the right (▶), can have noticeable relief, up to 100 m above the surrounding landscape. The steep terrain is usually forested, providing opportunities for hiking, skiing, and wildlife habitat. Flatter outwash plains and ground moraine are used for agriculture and urban settlement.

Moraine (Ridge)

11.05.a12 Kettle Moraine SF, WI

10. Melting ice sheets produce large braided streams that carry glaciofluvial sediment away from recessional or terminal moraines and deposit it either nearby or some distance away. The landform on the map below (▼) is a glacial outwash plain, which can be pitted by kettles that form depressions or ponds.

Recessional Moraines

Terminal Moraine

Drumlins

Kame

Glacial Outwash Plain

Roche Moutonnée

Ground Moraine

11.05.a1

11.05.a13

N BLUE SPRING LAKE DR

LITTLE PRAIRIE RD

Blue Spring Lake

Kettle

S SHORE DR

DAHLIN LN

Moraine

TAMARACK RD

Outwash Plain

Before You Leave This Page

☑ Summarize where continental glaciers carry and deposit sediment, explaining the three main types of moraine.

☑ Sketch and describe the features associated with continental ice sheets, and explain how each type of feature formed.

11.5

11.6 What Features Are Peripheral to Glaciers?

REGIONS WITH COLD CLIMATES exhibit other features that are either related to glacial processes or are simply related to freezing of the ground. Some other features form in the cool, wet climates that accompany continental glaciation but in regions too warm for glaciers.

A What Types of Deposits Are Related to Glacial Episodes?

Glaciers produce an abundance of sediment and water, so glacially derived sediment can accumulate over wide regions and can be transported far from the actual glaciers by streams, wind, and waves.

11.06.a1 Tibet

11.06.a2 Shaanxi Province, China

Glacially produced sediment of all sizes mixes with abundant glacial meltwater to form large streams with waters loaded with sediment. The streams deposit sediment on broad *outwash plains* in front of the glaciers. The gravel deposits shown here were deposited during the recent Ice Age and are being eroded into by a modern stream.

Glaciers pulverize entrained rocks, producing abundant silt-sized material that can be blown away by the wind. Accumulations of windblown silt are called *loess*, and many loess deposits are glacially derived, such as these soft, tan deposits that drape over topography. This particular deposit formed during the recent glacial episode.

B What Is Permafrost and Where Does It Occur?

In cold regions, below some depth, water in and below the soil remains frozen year after year, a condition called *permafrost*. The uppermost parts, called the *active layer*, thaw during summer.

11.06.b1 Denali NP, AK

1. The ground below the surface in this photograph (◄) is permanently frozen and so it is *permafrost*. During warm months, the top few meters of the ground (the active layer) thaw, allowing some vegetation to grow. Permafrost commonly does not allow trees to grow. Trees on the edge of permafrost are typically short and stunted.

2. In North America (▶), large areas of continuous permafrost are restricted to northern Canada and Alaska. In other areas, permafrost is either discontinuous or occurs in high, cold mountains. When frozen, permafrost is a very hard material, but it weakens considerably if it thaws.

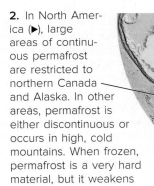

PERMAFROST ZONES
- Continuous
- Discontinuous
- Sporadic
- Isolated Patches

11.06.b2

11.06.b3 Kolyma River, Siberia, Russia

3. The smooth, lower part of this ledge (◄) is permafrost, and it has a hard, icy appearance. On top is a thin, brownish layer that is not always frozen (the active layer). Trees on permafrost commonly lean in various directions because permafrost keeps the roots shallow, and thawing of the active layer limits how well the roots support the trees.

11.06.b4 Aerial photograph, North Slope, AK

4. *Patterned ground* consists of geometric patterns, such as polygons and circles, above permafrost. It has several different expressions, but it forms when expansion and contraction from frost action concentrates gravels, stones, or boulders at the surface. These geometric patterns repeat over wide swaths of continuous and discontinuous permafrost.

C What Types of Lakes Were Associated with Glacial Times?

The cool, wet climates that favor glaciers also favor an increase in precipitation and the formation and maintenance of lakes. In the western United States, huge lakes existed at times during the recent ice age, but largely dried up when the climate changed about 15,000 years ago. Evidence of the extent and height of these lakes is still visible.

1. An ice-age lake filled low, interconnected basins in the Rocky Mountains of western Montana. This lake, named *Glacial Lake Missoula,* caused catastrophic flooding, as described in more detail below. Shorelines from this lake were etched as horizontal lines into the hills surrounding Missoula, Montana (▼).

11.06.c2 Missoula, MT

2. A large ancestral lake, named *Lake Lahontan,* filled the low basins of western Nevada. The lake was up to 240 m (790 ft) deep about 13,000 years ago, and some modern lakes in the area are remnants of this larger ice-age lake.

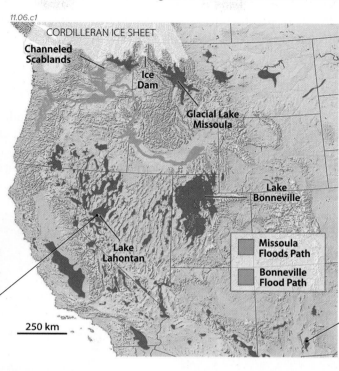

11.06.c1

CORDILLERAN ICE SHEET

Channeled Scablands

Ice Dam

Glacial Lake Missoula

Lake Bonneville

Lake Lahontan

Missoula Floods Path

Bonneville Flood Path

250 km

11.06.c3 Bonneville Salt Flats, UT

3. The Great Salt Lake of Utah is a remnant of a much larger ice-age lake named *Lake Bonneville.* As the large lake dried up, it left the Bonneville Salt Flats (▲), home to rocket testing, land-speed records, and many square miles of salt.

4. Smaller lakes formed in closed basins across much of the American Southwest. In places, they left salt flats, fine-grained lake deposits, and shorelines carved onto hillsides.

Lake Missoula and the Channeled Scablands

A famous story among geologists is the history of Lake Missoula and a peculiar topographic region in eastern Washington known as the *Channeled Scablands.* The scablands are so named because the area is crossed by many gorges, which curiously do not contain streams large enough to have carved the gorges.

During the 1920s, geologist J. Harlan Bretz proposed a hypothesis to explain this mystery, but it took decades to be accepted by the larger geologic community. According to this hypothesis, on more than one occasion, glacial Lake Missoula breached the glacial dam holding back its waters, and catastrophic torrents of water raced across the landscape to the west, carving the scablands. The huge floods carried gigantic boulders, carved smooth depressions (potholes) into the bedrock (▶), and formed enormous ripples

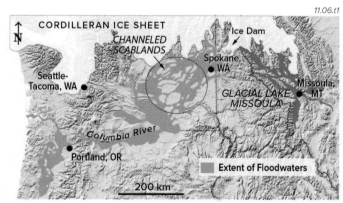

11.06.t1

CORDILLERAN ICE SHEET

CHANNELED SCABLANDS

Ice Dam

Spokane, WA

Seattle-Tacoma, WA

GLACIAL LAKE MISSOULA

Missoula, MT

Columbia River

Portland, OR

Extent of Floodwaters

200 km

11.06.t2 West Bar, Columbia River, WA

(above right). The shapes of the ripples record immense currents moving downstream, from right to left in this view. These ripples are so large that they were hard to recognize as such.

11.06.t3 Dry Falls, WA

Before You Leave This Page

☑ Describe the characteristics of different deposits related to glacial episodes and how each type forms.

☑ Describe permafrost and where it occurs.

☑ Describe several large ice-age lakes and some of the features they formed, either while full or while emptying.

11.6

11.7 What Happened During Past Ice Ages?

OVER THE PAST TWO AND A HALF MILLION YEARS, huge ice sheets and glaciers intermittently covered large areas of the Northern and Southern Hemispheres. The ice sheets and glaciers increased during some time periods we call *glacial periods*, and decreased during other times, called *interglacial periods*. Overall, the past two and a half million years had a marked increase in glacial periods compared to most other times in Earth's geologic history. As a result, geologists call this time period the *Ice Age*. The Ice Age coincides with the Pleistocene Epoch of the Quaternary Period of the geologic timescale.

A What Parts of the Earth Were Covered with Ice During the Ice Age?

By examining landscapes for glacially polished surfaces, moraine, and other indications of glaciation, geologists can infer which areas were covered by ice in the past and which ones were not. We also can use sedimentary records from lakes, determining whether a lake existed at some time or was still covered by ice. By considering other factors, including elevation and latitude, we can extrapolate the actual observations to produce maps that show the interpreted extents of glaciers and ice sheets at different times during the Ice Age.

This figure, centered on Greenland, shows the present-day distribution of ice in the Northern Hemisphere. It also shows sea ice that covers much of the Arctic Ocean between North America and Asia. Note that the only large land areas covered by ice are Greenland, adjacent islands in northern Canada, and parts of Alaska and westernmost Canada.

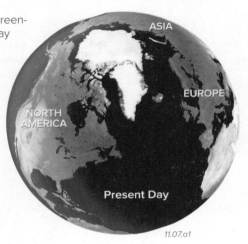

Approximately 28,000 years ago, ice sheets and glaciers were much more extensive on Earth, including in the Northern Hemisphere. Note that a large ice sheet covered nearly all of Canada and much of the northern parts of the United States. Ice sheets and glaciers also covered much of northern Asia, northern Europe, and the Alps of southern Europe.

11.07.a1

11.07.a2

B What Parts of North America Were Covered by Ice and When Did the Ice Retreat?

The maps below show interpreted ice cover for North America at two times. The left map shows a time when the ice cover was close to a maximum, approximately 20,000 years ago, whereas the map on the right shows the position of ice at about 10,000 years ago, after the ice sheets had retreated and as the Ice Age was ending. Was the area where you live or go to school covered by ice in the recent geologic past?

1. Most of Alaska was not covered by ice. This allowed people and animals from Asia to migrate into North America.

2. The center of the ice sheet was in northern Canada. Here, the ice is interpreted to have been several kilometers thick.

3. Ice sheets extended from Canada into the northern parts of the United States, covering New England, the Great Lakes region, and the upper Midwest. Glaciers also occupied parts of the Rocky Mountains but are not shown.

4. When we say that the ice sheets *retreated*, we mean that ice within the sheets was still flowing forward, but the front of the ice sheet moved back (retreated) because ice melted faster than it could be replenished.

5. By 10,000 years ago, the ice sheets had melted back, covering only part of north-central Canada.

6. As the ice retreated, the northern United States emerged from beneath the ice, the Great Lakes formed, and river systems like the upper Mississippi began to develop.

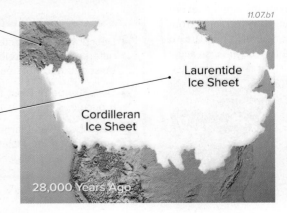

11.07.b1

Laurentide Ice Sheet

Cordilleran Ice Sheet

28,000 Years Ago

11.07.b2

11,000 Years Ago

What Record Did Past Glaciations Leave in the Northern United States?

Glaciers once covered northern parts of the conterminous United States, leaving behind evidence of which areas they covered and which ones they did not. The shaded-relief map below shows the area south of the Great Lakes. The small inset map highlights the locations of ridges of moraine left behind as ice sheets retreated from the region. Try to match the moraines on the small map to the same feature on the large map.

1. A continental ice sheet once covered much of the area south of the Great Lakes, in the upper Midwest of the United States. The largest features formed by glaciation are smooth troughs that in this area trend from northeast to southwest, or locally north to south. These smooth areas were once covered by ice sheets, which smoothed off the underlying landscape as they moved southwest and south out of Canada.

2. As the ice sheets melted away, they left a veneer of glacial sediment on the landscape. Ridges, representing piles of glacial sediment, mark the position of the front of the ice sheet as it melted back. Most of these ridges, highlighted in red on the inset map, are recessional moraines.

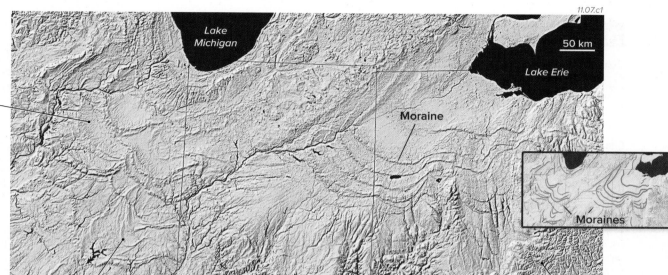

3. Areas with rougher topography south of the troughs, including those in the lower right corner of this map, were never glaciated. They are called *driftless areas* because they were not glaciated and so do not have a covering of glacial drift (glacial sediment).

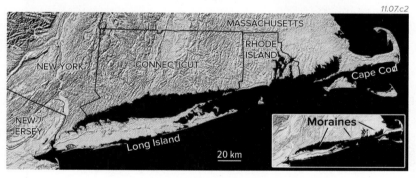

4. Glaciers also covered New England (▲). They piled up glacial sediment, forming ridges of moraine, here shaded red, on Long Island, Rhode Island, Cape Cod, and the islands of Martha's Vineyard and Nantucket.

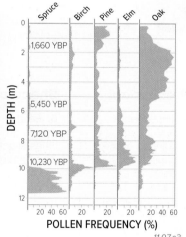

5. Glaciers left behind other evidence in the Northeast and Great Lakes area, including outcrops that display glacial scratches and polish. As the glaciers retreated, meltwater collected in low areas, forming numerous small and large lakes. Sediment accumulating in lakes and in other settings contain a record, in the form of pollen, of the types of plants that grew at different times. These pollen records (◄) document a dramatic shift from spruce trees to leafy trees as glaciers retreated and the climate warmed.

The Ice-Age Hypothesis

The idea that huge ice sheets covered the land was not intuitive, but arose to explain an ever-growing number of hard-to-understand observations made in Europe and North America. As discussed in A. Hallam's book *Great Geological Controversies,* these observations included bones of reindeer and Arctic birds in southern France, large out-of-place boulders (erratics) scattered across much of Europe, and the presence of scratched and polished bedrock far beyond the existing glaciers in the Alps. European naturalists of the time debated how to explain these curious features. In the 1820s, they hypothesized that widespread, prehistoric glaciations had occurred in the Alps and northern Europe. Today, we call this interval of time the *Ice Age,* and we recognize that glaciers grew and shrank many times during the last two and a half million years.

Before You Leave This Page

☑ Describe what parts of the Northern Hemisphere and North America were covered with ice during the Ice Age.

☑ Discuss evidence for past glaciations in the United States.

11.7

11.8 What Starts and Stops Glaciations?

TO UNDERSTAND THE REASONS for glacial and interglacial times, we need to further explore what causes global changes in climate. Earth's climate system is very complex, and human history is short compared to the timescales on which global climate change occurs, so we do not completely understand all the causes.

A What Variations in Earth's Tilt and Orbit Influence Global Climate?

Some variations in climate, those that occur over a timescale of thousands to tens of thousands of years, are likely controlled by the amount of solar radiation reaching Earth. Milutin Milankovitch, a Serbian astronomer and geophysicist, recognized that Earth's climate could be influenced by changes in the amount and direction of Earth's tilt and the shape of Earth's orbit. We call such changes *Milankovitch cycles*.

Changes in Earth's Tilt (Obliquity)

Over time, Earth's axis of rotation changes its tilt relative to its plane of orbit around the Sun. About every 40,000 years, the tilt changes from 22.5° to 24.5° and back again. The amount of tilt toward the Sun affects Earth's climate because tilt affects how much summer sunlight strikes higher latitudes. The diagrams below are for winter in the Northern Hemisphere.

1. The *maximum tilt angle* of Earth's rotation axis is 24.5°. This amount of tilt increases the effects of the seasons. When combined with other climatic factors, it can lead to a decrease in glacial activity because warmer summer temperatures melt more polar ice.

2. The present-day position of Earth's tilt is 23.5°. Earth's tilt is currently adjusting back from the maximum 24.5° tilt, which last occurred near the end of the Pleistocene glacial advance.

3. When Earth's tilt is at its *minimum tilt angle* (22.5°), high latitudes receive less direct sunlight during the summer, causing cooler summers, less melting, and an increase in glaciers.

Wobble of Rotation Axis (Precession)

4. *Precession* is similar to what happens when a spinning top slows down and wobbles. As Earth wobbles, its spin axis changes from pointing at the North Star (Polaris), where it currently points, to pointing at the star Vega and back again.

5. Precession causes December to be winter in the Northern Hemisphere during some times in Earth history, but June to be winter at other times in Earth history. If the effects of precession are added to other astronomical factors, these phenomena could affect global climate. The precession cycle lasts about 23,000 years.

Shape of Orbit (Eccentricity)

6. *Eccentricity* is the term for the noncircular shape of Earth's orbit around the Sun. Earth's orbit changes over long timescales, sometimes being more circular and sometimes slightly more elliptical. These changes cause variations in the amount of insolation (solar energy) reaching Earth.

7. This change from a more circular to a less circular orbit is thought to only slightly affect climate, but when added to other astronomical cycles, its effect might be significant. The eccentricity cycle lasts about 100,000 years.

8. Scientists have computed the effects of each of these factors (tilt, precession, and eccentricity) and then combined the effects to investigate how they interact, both in the past and in the future. When used to try to reconstruct past events, the calculated effects can be compared with records of past climate, like the isotopic composition of ice cores. Such comparisons support the hypothesis that Milankovitch cycles can explain many past climatic variations, including the waxing and waning of glaciations. Computer models can also be used to predict future Milankovitch-related climate changes, but these results are currently being debated by the scientific community.

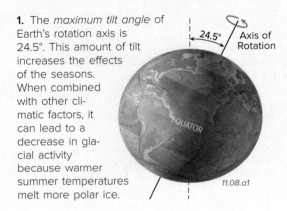

11.08.a1 11.08.a2 11.08.a3

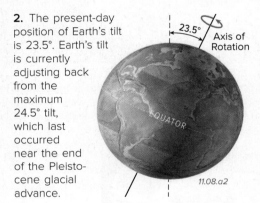

11.08.a5

11.08.a4

B Can Variations in Solar Heating and Atmospheric Composition Influence the Intensity of a Glacial Episode?

Several factors may not, by themselves, cause the onset or demise of a glacial episode, but they can influence how severe the resulting glaciation is. For example, the amount of energy given off by the Sun is not constant, and the composition of Earth's atmosphere is not constant. As a result, the amount of energy that can pass through the atmosphere and be retained by Earth is variable and can lead to a warmer or cooler climate.

1. Every 11 to 14 years, the level of *sunspot activity* on the Sun increases, producing very small changes in solar energy output. Fluctuation in solar energy can influence Earth's climate system, affecting temperature, precipitation, and other weather phenomena. When sunspot activity declines, evidence shows that Earth's climate cools by a small amount.

2. During glacial episodes, snow and ice cover more of Earth's surface, and cloud cover also increases. Both of these increase the reflectivity, also called the *albedo*, of Earth so that more insolation gets reflected off Earth's surface and is lost to space. This loss of heat to space makes the climate cooler. A cooling climate can result in more snow, ice, and clouds, leading to more cooling. In this way, the system reinforces itself—a positive feedback.

3. Volcanoes release millions of tons of carbon dioxide (CO_2) into the atmosphere every year, and plants and marine life extract some of this CO_2. An increase in the amount of CO_2 and methane (CH_4), both greenhouse gases, tends to warm the planet. The amounts of CO_2 and CH_4 were relatively lower during glacial episodes and higher during interglacial periods.

4. Large, explosive volcanic eruptions can add significant quantities of volcanic ash, dust, and sulfur dioxide (SO_2) (which is converted to sulfuric acid aerosols) into the atmosphere. These aerosols reflect insolation back into space, allowing less sunlight to reach Earth's surface. A major volcanic eruption will increase the amount of ash and dust in the atmosphere, perhaps resulting in global cooling. It is thought that volcanic eruptions in the tropics during the 13th to 15th centuries, along with decreased solar activity, may have triggered an unusually cold period in Europe.

11.08.b1

C What Is the Role of Ocean Currents and Continental Positions in Glaciations?

Ocean currents transport water of different temperatures from one part of the ocean to another. Such currents can warm or cool land areas, helping to increase or decrease glaciation on land.

1. Upwelling currents can bring deep cold water to the surface. Cold currents help cool the land, perhaps allowing glaciers to form, if there is sufficient precipitation.

2. Cold currents, however, can also inhibit the growth of glaciers because they put less moisture into the atmosphere, leading to less snowfall.

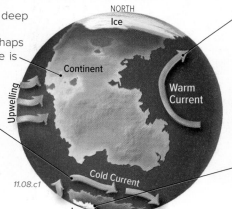

11.08.c1

3. Warm currents bring tropical waters northward, waters that have been heated by the Sun. These currents can warm adjacent parts of a continent, inhibiting glaciations. Warm currents also bring warm, moist air, increasing precipitation, which can allow more ice to accumulate if the temperature is cold enough.

4. The position of continents affects the geometry of ocean currents, deflecting currents in certain directions or blocking the connection between different oceans. In this way, continental positions influence ocean currents, which impact regional and local climates.

5. A continent located at or near the North or South Pole provides a large landmass on which continental ice sheets can form.

Causes of Ice Ages

The ice age lasting from approximately 2.6 million years ago to the present day included many glacial and interglacial periods. This time span lies mostly in what has traditionally been called the Pleistocene epoch, which ended with the end of the last major glaciation about 12,000 years ago. Major glacial periods commonly lasted about 40,000 to 100,000 years, with interglacial periods lasting about 10,000 years on average. Changes between the two conditions apparently could occur rapidly in geologic terms, in some cases over less than several thousand years. Some scientists regard the present time as an interglacial.

The cause of ice ages remains controversial, but the factors described on these two pages are among the culprits that can help instigate an ice age or influence how pronounced one is. Some factors, such as the role of CO_2, are complex, because they can affect climate but also respond to changes in climate.

Before You Leave This Page

✓ Describe how variations in Earth's tilt and orbit influence global climate.

✓ Describe how global climate can be affected by atmospheric gases, volcanic ash, and the amount of snow, ice, and cloud cover.

✓ Describe the role of ocean currents and continental positions on glaciations.

11.8

11.9 What Processes Occur Along Coasts?

COASTS ARE THE INTERFACE BETWEEN LAND AND WATER and so respond to processes that arise from both sides and from changes in sea level. Waves and tides affect a coast from the water side, while rivers, wind, and other transport agents contribute sediment from the land. Together these processes sculpt the coast, redistribute sediment, and present challenges for people who live along a coast.

A What Types of Processes Affect Coasts?

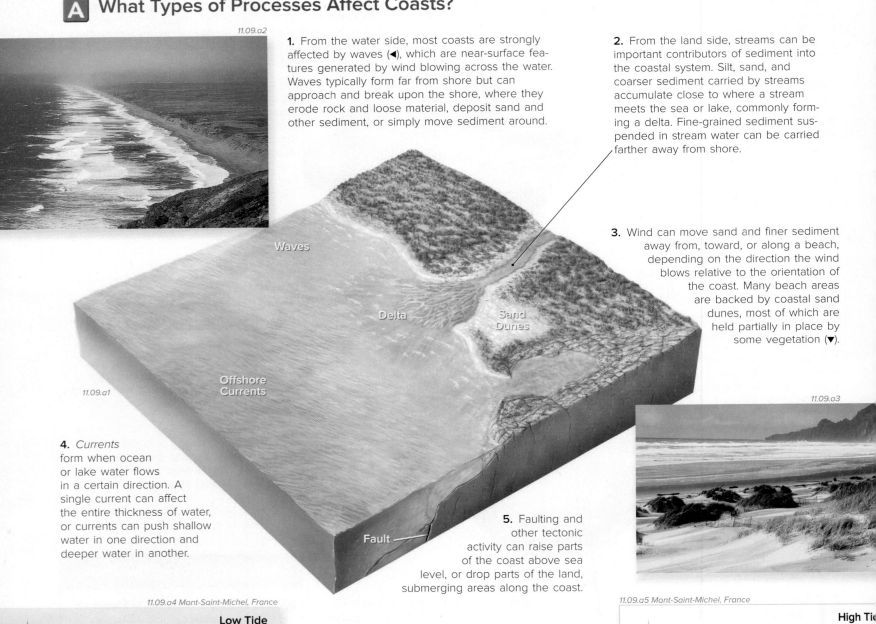

11.09.a2

1. From the water side, most coasts are strongly affected by waves (◄), which are near-surface features generated by wind blowing across the water. Waves typically form far from shore but can approach and break upon the shore, where they erode rock and loose material, deposit sand and other sediment, or simply move sediment around.

2. From the land side, streams can be important contributors of sediment into the coastal system. Silt, sand, and coarser sediment carried by streams accumulate close to where a stream meets the sea or lake, commonly forming a delta. Fine-grained sediment suspended in stream water can be carried farther away from shore.

3. Wind can move sand and finer sediment away from, toward, or along a beach, depending on the direction the wind blows relative to the orientation of the coast. Many beach areas are backed by coastal sand dunes, most of which are held partially in place by some vegetation (▼).

11.09.a3

4. *Currents* form when ocean or lake water flows in a certain direction. A single current can affect the entire thickness of water, or currents can push shallow water in one direction and deeper water in another.

Waves

Delta

Sand Dunes

Offshore Currents

11.09.a1

Fault

5. Faulting and other tectonic activity can raise parts of the coast above sea level, or drop parts of the land, submerging areas along the coast.

11.09.a4 Mont-Saint-Michel, France

Low Tide

11.09.a5 Mont-Saint-Michel, France

High Ti

6. Changes in sea level greatly affect coasts. In most places, tides raise and lower sea level relative to the land twice a day. Longer term changes in sea level are primarily due to changes in climate and tectonics. Tidal flats, such as the one surrounding Mont-Saint-Michel in France, shown in both of these photographs, are uncovered by low tides (◄) and flooded by high tides (►) and storms. Such low areas could be submerged by an overall rise in sea level accompanying climate change.

B | What Factors Affect the Appearance of a Coast?

Coasts around the world have diverse appearances, from sandy white beaches to dark, craggy cliffs that plunge vertically into the sea, with no beach at all. A number of factors control these differences, including orientation of the coast, slope of the seafloor, hardness of the rocks, and contributions of sediment from the land.

Factors on the Water Side

1. The appearance of a coast is greatly influenced by the strength of the waves and tides that impact the shore. Stronger waves will typically cause greater erosion and move larger clasts of sediment along the coast.

2. The size and intensity of storms influence the appearance of a coast because storms bring with them large waves, strong winds, and intense rainfall. Some coasts are ravaged by hurricanes, whereas others rarely experience the erosive effects of powerful storms.

3. The slope of the seafloor is also a factor. Steep slopes can allow large waves to break directly against rocks along the shore, whereas more gentle slopes cause waves to break a short distance offshore.

4. The orientation of a coastline is also important, because waves typically approach from specific directions in response to prevailing winds. The dominant wave direction may change with the season (summer versus winter or dry versus rainy seasons). Also, some parts of the coast will receive less wave action because they are sheltered in a bay or are protected by an island, barrier reef, or other offshore feature. The coast below is rocky and affected by strong waves, especially during powerful storms.

Factors on the Land Side

5. On the land side, the appearance of the coast reflects the hardness of the bedrock along the coast. Hard rock that resists erosion tends to form rocky cliffs (▶), whereas erosion sculpts softer sediment and rock into more gentle slopes and rounded hills.

11.09.b3 Scotland

6. Coastal landscapes also reflect the amount and size of available sediment. A coast cannot be rocky if the only materials present are soft and fine grained. Streams provide a fresh influx of sediment into the coastal environment.

7. Coasts undergoing uplift have a different appearance than those where the land has dropped relative to water level. A rise in sea level flooded stream valleys along the North Carolina coast, producing a coastal outline marked by long, narrow inlets and bays (▼).

8. Climate is a major factor influencing coastal landscapes. Wet climates provide abundant precipitation for erosion, formation of soil, and the growth of vegetation. Vegetation stabilizes soil and limits the amount of material that can be picked up by wind and water or moved downslope by gravity. Dry climates result in less vegetation, less soil, and less stable slopes.

Storm

Rocky Coast

Waves

Coast Composed of Soft Sediment

Stream

11.09.b1

Delta

Sheltered Bay

Waves

Flooded River Valley

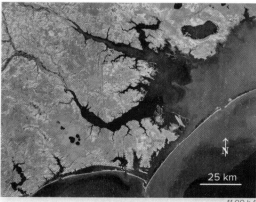
25 km
11.09.b4

11.09.b2 Acadia NP, ME

Before You Leave This Page

✓ Summarize or sketch the types of processes that affect coasts.

✓ Summarize or sketch how different factors, from the water side and from the land side, affect the appearance of a coast.

11.9

11.10 What Causes High Tides and Low Tides?

THE SEA SURFACE MOVES UP AND DOWN across the shoreline, generally twice each day. These changes, called *tides*, are observed in the oceans and in bodies of water, such as bays and estuaries, that are connected to the ocean. What causes tides to rise and fall, and why are some tides higher than others?

A What Are High and Low Tides?

Tides are cyclic changes in the height of the sea surface, generally measured at locations along the coast. The difference between high and low tide is typically 1 to 3 m, but it can be more than 12 m or almost zero.

High Tide   **Average Sea Level**   **Low Tide**

11.10.a1   *11.10.a2*   *11.10.a3*

During *high tide*, the height of water in the ocean has risen to its highest level relative to the land. At this point, the water floods onto the shoreline, covering more of the beach. In most places, high tide occurs every 12 hours and 25 minutes.

Following high tide, the water level begins to fall relative to the land—the tide is going out. At some time, water level will reach the average sea level for that location, but it keeps falling on its way to low tide.

When the water level reaches its lowest level, it is at *low tide*, and more of the shore is exposed. Low tide in most places also occurs every 12 hours and 25 minutes. Water level begins to rise again after low tide, and the tide is coming in. Rising tide spreads water across the land.

B What Causes High and Low Tides?

Tides rise and fall largely because water in the ocean is pulled by the gravity of the Moon and to a lesser extent the Sun. As Earth rotates on its axis, most coasts experience two high and two low tides in each 25-hour period.

1. This figure depicts Earth and the Moon as if looking directly down on Earth's North Pole. It shows the Moon much closer to Earth than it would be for the size of the two bodies. Earth rotates (spins) counterclockwise in this view and, relative to the Sun, completes a full rotation once every 24 hours.

2. The Moon makes one complete orbit around Earth each 29 days, also counterclockwise in this view. Due to this motion, it takes 24 hours and 50 minutes for a point on Earth facing the Moon to rotate all the way around to catch up with and again face the Moon. The extra 50 minutes is because the Moon will have orbited 1/29th of the way around Earth after 24 hours.

Moon's Gravity

Earth's Rotation

11.10.b1

3. The Moon exerts a gravitational pull on Earth and its water. This pulls the water in the ocean toward the Moon, causing it to mound up on the side of Earth nearest to (i.e., facing) the Moon. Coastal areas beneath the mound of water experience high tide. On this figure, the thickness and mounding of the (blue) water are greatly exaggerated.

4. On parts of Earth that are facing neither toward nor away from the Moon, sea level is lower as water is pulled away from these regions toward areas of high tide. Coastal areas here experience low tide. At this place, for the next six hours and 12 minutes, tides will be rising—a situation called *flood tide*.

5. On the side of Earth opposite the Moon, the water is relatively far from the Moon and feels less of the Moon's gravitational pull (recall that the force of gravity decreases with distance). The water bulges out, called the *tidal bulge*, and the side of Earth facing directly away from the Moon therefore experiences high tide. At this place, for the next six hours and 12 minutes, tides will be falling—a situation called *ebb tide*.

6. Earth rotates much faster than the Moon orbits Earth, so it is best to think of the mounds of water—but not the water itself—as remaining fixed in position relative to the Moon as Earth spins. During a complete rotation of Earth, a coastal area will pass through both mounds of water, causing most coasts to have two high tides (and two low tides) in each 24-hour and 50-minute period. Odd coastal configurations can cause local differences in the number of daily tides.

C   Why Are Some High Tides Higher Than Others?

From week to week, not all high tides at any location reach the same level—some are higher then average and others are lower than average. Similarly, some low tides are very low and others are not very low. Such variations are related to the added influence of the Sun's gravity and follow a predictable pattern that repeats about every 15 days.

Spring Tides

1. Like the Moon, the Sun exerts a gravitational pull on Earth and its water. The Sun is larger and more massive than the Moon, but it is so much farther away from Earth that the Sun's gravitational effect on Earth is about half that of the Moon.

11.10.c1

Time 1: New Moon

Sun's Gravity

Moon's Gravity

Time 2: Full Moon

3. The Moon orbits all the way around Earth in about 29 days. About once a month, here labeled Time 1, the Moon and the Sun are on the same side of Earth. The gravity of the Sun and Moon then pull the ocean in the same direction, causing the high and low tides to be more extreme than normal. These extreme tides are called *spring tides,* but they occur during all months of the year.

2. The Sun's gravity attracts a mound of water on the side of Earth facing the Sun and causes another mound on the side facing away from the Sun. These thin mounds, shown in dark blue, are always in the same position relative to the Sun. Locations on Earth rotate through each position once every 24 hours.

4. About two weeks later, here labeled Time 2, the Moon has moved around to the side of Earth opposite the Sun. However, the forces of the Moon and Sun are again aligned, so another spring tide occurs. Note that spring tides occur when the side of the Moon facing Earth is either not illuminated at all by the Sun (a *new moon*) or is fully illuminated (a *full moon*).

Neap Tides

5. Approximately seven days after each spring tide, the Moon journeys 1/4 of the way around Earth. In this position, the Moon's and Sun's gravity are pulling at right angles to one another, and one acts to cause a high tide while the other acts to cause a low tide.

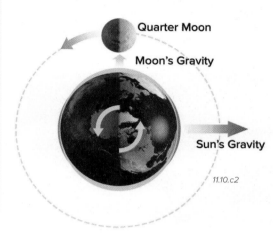

Quarter Moon

Moon's Gravity

Sun's Gravity

11.10.c2

6. At these times of the month, the effect of the Sun's gravity partially offsets the effects of the Moon's gravity, so the differences between high tide and low tide are less than average. These lower-than-average high tides and higher-than-average low tides occur every two weeks and are called *neap tides,* from an Old English word for "lacking." They happen when the Moon, as viewed from Earth, is only half illuminated by sunlight (called a *quarter moon*).

The Most Extreme Tides in the World

Some places have higher tides than others. The difference between high and low tide, or the *tidal range,* can be so small as to be nearly undetectable, or it can be so extreme as to be dangerous. The Mediterranean Sea and Gulf of Mexico have very little tide because they are largely surrounded by land, which limits the flow of water into and out of the basins. Much of the Caribbean has only one tide each day. Southampton in the United Kingdom has four high tides a day. The world's highest tides are in the Bay of Fundy, along a large embayment in the Atlantic coast, flanking Nova Scotia. In this place, the geometry of the coast and sea bottom funnel water in and out of the bay at just the right rate to cause a tidal range of as much as 16 m (52 ft).

High Tide

Low Tide

11.10.t1 Bay of Fundy, eastern Canada

Canadians use the large tides to generate electricity for Nova Scotia. These two photographs illustrate the extreme tidal range within the Bay of Fundy.

Before You Leave This Page

☑ Describe or sketch what tides are.

☑ Sketch and describe how tides relate to the position of the Moon and why.

☑ Sketch or summarize how the gravity of the Moon and Sun cause spring tides and neap tides.

11.10

11.11 How Do Waves Form and Propagate?

WAVES ARE THE MAIN cause of coastal erosion. Most ocean waves are generated by wind and only affect the uppermost levels of the water. Waves transport and deposit eroded material along the coast. How do waves form, how do they propagate across the water, and what causes waves to break?

A What Is an Ocean Wave and How Is the Size of a Wave Described?

1. *Waves* are irregularities on the surface of a body of water usually generated by wind. They vary from a series of curved ridges and troughs to more irregular bumps and depressions to breaking curls (▼).

11.11.a1 Two Points SP, ME

2. In deep, open water, many waves occur in sets, where adjacent waves are similar in size and shape to one another and follow one behind another.

3. The lowest part of a wave is the *trough*, the highest part is the *crest*, and the vertical distance between the two is the *wave height*. The distance between two adjacent crests is the *wavelength*. All of these features can vary greatly with location and time. A typical ocean wave is several meters high, with a wavelength of several tens of meters.

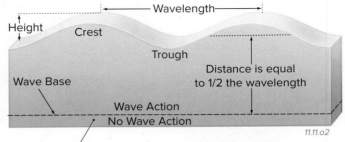
11.11.a2

4. Most waves are near-surface features. They affect only the surface of the water and depths typically down to several tens of meters. Below some level, called the *wave base*, the wave ceases to have any effect. The depth of the wave base is about half the wavelength, so if two wave crests are 20 m apart, then the wave base is about 10 m deep. The equation that expresses this relationship is:

$$depth\ of\ wave\ base = wavelength / 2$$

B How Do Waves Propagate Across the Water?

A set of waves can travel a long distance across an ocean or other body, but the water through which the wave passes moves only a short distance. To visualize this, imagine shaking a rope into a series of waves. The waves move along the rope, but any part of the rope stays about the same distance from your hand.

1. Water waves propagate in a manner similar to waves in a rope or to seismic surface waves. Water molecules move up and down and from side to side, but they mostly stay in about the same place. Compressive forces cause water within the wave to push against the water in front of it. The three figures shown here (▶) are snapshots of a set of waves propagating to the right. Examine the motion of water at points A, B, and C within the wave.

2. Here (▶), the waves have propagated through the water to the right, but the lettered reference points have moved only a short distance. Point A, which is close to the surface, moves more than point B, which is deeper. Point C is below the wave base and does not move at all.

3. Later, points A and B are beneath the wave trough (▶) and have nearly returned to their positions on the reference line. As the next wave crest approaches, they will move left and upward, and then right, returning to near their starting points.

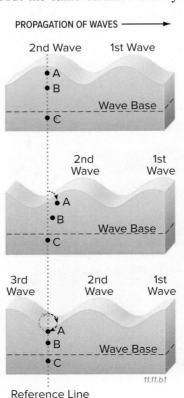
11.11.b1
Reference Line

4. During the passage of an entire wave (crest to crest), points A and B will each have followed a small, circular path. The figure below (▼) shows the circular paths of 20 different points at different depths and positions along the wave.

5. Water particles on the surface of the water travel the most, going up and forward and then down and back, in a circular motion, as the wave passes.

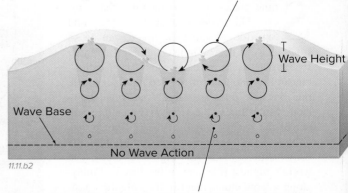
11.11.b2

6. Deeper within the wave, water particles travel smaller circular paths, and the paths become smaller with depth. Water that is right above the wave base barely moves, and water below the wave base does not move at all as the wave passes.

C How Do Waves Form and What Happens When They Reach Shallow Water?

As a wave moves from deep water into shallower water, it starts to interact with the bottom and changes in size and shape. A wave can also bend (refract) as one part of the wave encounters the bottom before other parts do.

How Waves Form

1. Most ocean waves are caused by wind blowing across the surface of the water.

2. When a gentle breeze is blowing, gas molecules in the air collide with the surface of the water, transferring some of their momentum to the water. This forms waves that are small in height and wavelength. Once a wave forms, it catches the wind even more and so can increase in height and wavelength.

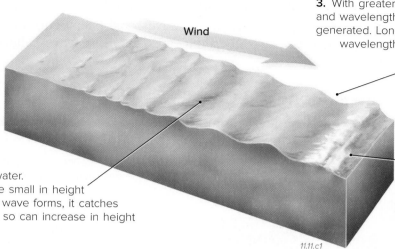

11.11.c1

3. With greater wind speed, waves get larger, both in height and wavelength. The stronger the wind, the larger the wave generated. Long-wavelength waves move faster than shorter wavelength ones and can travel farther before dying out.

4. Waves continue to move, even if the wind dies or the waves move away from the windy area. Such waves existing independently of winds or storms are called *swell*.

5. If the wind gets too strong, a wave becomes too steep to be stable and its top collapses (breaks), even if it is still out in open water. This collapse traps air and forms a white, foamy wave, a *whitecap*.

6. Most of the time, there are multiple sets of waves propagating in different directions, resulting in *wave interference*. Where wave crests from two sets join, the waves get higher. Interference results in a choppy sea surface with bumps and depressions.

How Waves Break Upon the Shore

7. As waves approach the shore, they encounter shallower water and change in a fairly systematic way. In this figure, waves are moving to the right, toward the shore.

8. In deep water, waves propagate unimpeded as long as the sea bottom is deeper than the wave base. The deep-water waves shown here have a longer wavelength and lower height than waves nearer shore.

Wave Direction → *Breaking Wave*

Increase in Height, Decrease in Wavelength

Wavelength

Wave Base
Wave Base = 1/2 Wavelength

11.11.c2

9. When a wave reaches water that is shallower than the wave base, its lower parts begin shearing against the bottom. The waves slow, crowd together, and get higher as their wavelengths get shorter.

10. Near shore, the wave becomes too high and steep to support itself. The top of the wave, which is moving faster than lower parts, topples over (breaks), forming a plunging *breaker* or a jumble of spilling water. In contrast to waves in the swell, these waves have horizontally moving water molecules. The area where waves break on the shore is the *swash zone*, or simply the *surf zone*.

How Waves Bend

11. Due to varying wind directions and underwater topography, waves typically approach the shore at an angle, rather than straight on. In this figure, waves propagate toward the shore obliquely, from left to right. In deep water, the wave crests are straight or only gently curved.

11.11.c3

Rip Current

12. As a wave begins to encounter the bottom, its side closest to shore is slowed more than the segment in deeper water. This difference in velocity causes the initially straight waves to refract (bend)—*wave refraction*.

13. Water flowing back out to sea (*backwash*) after the wave comes ashore moves perpendicular to the shore. The flow can take the form of fast and dangerous *rip currents*, which cause many drownings. Rip currents tend to disrupt the normal wave pattern. Always swim parallel to the shore if caught in a rip current.

Before You Leave This Page

☑ Sketch and label the parts of a wave, including the wave height, amplitude, wavelength, and wave base.

☑ Explain how the propagation of a wave differs from the motion of the water through which the wave travels.

☑ Sketch and explain why a wave rises and breaks as it reaches shallow water.

☑ Explain why waves bend if they approach the shore at an angle.

11.11

11.12 How Is Material Eroded, Transported, and Deposited Along Coasts?

SOME COASTS ARE ENERGETIC ENVIRONMENTS where solid rock is eroded into loose sand, pebbles, and other kinds of sediment. Sediment can also be brought in by streams, wind, waves, and ocean currents. Once on the beach, sediment moves in and out from the beach, and often laterally along the beach, as waves alternately transport and deposit sediment.

A How Do Waves Erode Materials Along the Coast?

Most erosion along coasts is done by waves. Waves crash against rocky shores and onto beaches, breaking off pieces of bedrock that can then be reworked by more wave action and ultimately transported away.

1. When waves break onto a rocky coast, they cause erosion by swirling away loose pieces of bedrock and by picking up and crashing these loose pieces back against the bedrock. They also grind away rocks by scraping sand back and forth against the bedrock. Salt crystals from the sea and ice (which expands as it forms) force pores open through *salt wedging* and *frost wedging*, respectively.

2. Crashing and grinding water and sediment wear away at the bedrock, especially at the level where wave action is strongest. Over time, this repeated erosion in the same place may carve a *wave-cut notch* into a rocky coast. The notch undercuts overlying rocks, leaving them unsupported and prone to collapsing into the sea.

3. As waves wash sand and stones back and forth across the sea bottom, they smooth off the underlying bedrock, carving a *wave-cut platform*. Knobs of resistant bedrock locally rise up above the platform, but they may eventually be worn away by the erosive action of waves.

4. Once pieces of rock are loose and within the surf zone, waves smash them together, rounding off angular corners and fracturing larger pieces into smaller ones. In this way, large, angular rocks derived from bedrock over time become the rounded and flattened stones that dominate many beaches. With further action, stones wear down into sand.

5. Through this mechanism, waves liberate and rework pieces of bedrock. In the process, they create stones and sand that help the waves erode and rework even more bedrock, and break other stones into sand.

11.12.a1

B How Does the Shape of a Coast Influence Wave-Related Erosion?

Most coasts are not straight but have curves, bays, and other irregularities. As a result, some parts of the coast are somewhat protected, while other parts bear the brunt of oncoming storms and waves.

1. Waves approach a coast from a specific direction, usually at an angle to the shore. Curves in the coastline form inward-curving bays, whose quiet waters are protected from the largest waves. Parts of the shore that are struck head-on by the waves will experience more wave action and erosion than those that are at an angle to the full force of the oncoming waves. From season to season, the prevailing wind may change direction, causing what was once protected to be subjected to strong waves.

2. A *promontory*, which is a ridge of land that juts out into the water, is in a vulnerable position. The steep sides of many promontories allow large, powerful waves to focus all their energy on the rock, instead of losing energy through interaction with a gently sloping bottom.

3. The seafloor flanking a promontory can cause waves to bend (refract) around the promontory and strike it from all sides. All other things being equal, more waves means more erosion, so promontories tend to be preferentially worn away, eventually resulting in a straighter coastline.

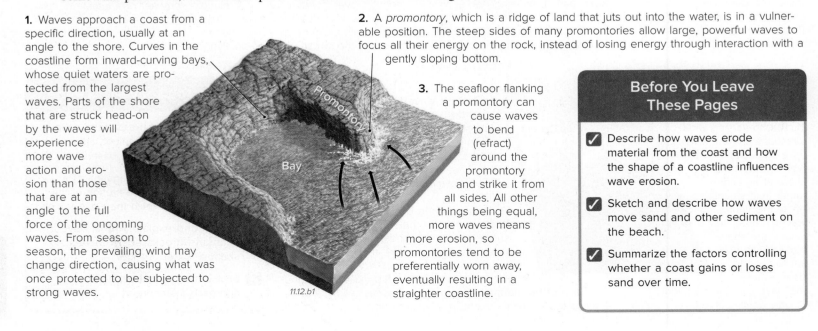

11.12.b1

C How Do Sand and Other Sediment Get Moved on a Beach?

1. During normal (non-stormy) conditions, waves wash sand and other sediment back and forth laterally near the beach. Sediment on the sea bottom is churned up and carried toward the beach by incoming waves.

2. After a wave breaks on the beach, most water flows directly downslope off the beach, carrying sediment back toward the sea. Sediment gets re-worked back and forth by the waves, but it may not be transported very far. The area on a beach where broken waves run up onto the beach is the *swash zone*. The return flow, back toward the sea, is the *backwash*.

3. During storms, large, vigorous waves can carry sediment farther up the beach than normal, depositing it out of the reach of smaller waves that characterize more typical, less stormy conditions. Storms can also erode material from the beach, carrying it farther out to sea. Some beaches lose sand and become rocky during the winter because of the increased energy of larger winter waves, but they regain sand and become sandy and less rocky in the summer.

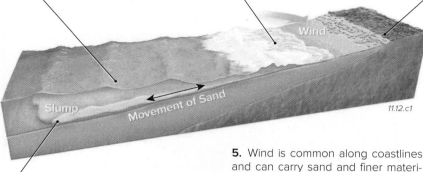

11.12.c1

11.12.c2

4. Sediment along beaches and farther offshore can slump downhill if the sea bottom is too steep to hold the sediment. Sediment will also slump if it is physically disturbed, perhaps by deep wave action during a storm, by shaking during an earth-quake, or if sediment piles up too fast.

5. Wind is common along coastlines and can carry sand and finer materials long distances along or away from the beach. Low- to moderate-strength wind cannot dislodge sand that is wet because surface tension from the water between sand grains tends to hold the grains together. Wind is more effective above the shoreline where the sand is dry and loose.

6. If waves approach the coastline at an angle, the sand and other sediment can be moved laterally along the coast (▲). Incoming waves move the sand at an angle relative to the coastline, then the sand washes directly downslope when the water washes back into the sea. By this process, the sand moves laterally along the coast (beach drift). Sand and sediment farther offshore can also move laterally due to ocean currents paralleling the shore, a process called *longshore drift*.

D What Determines Whether a Coast Gains or Loses Sand with Time?

A coastline can gain or lose sand, depending on the rate at which sand enters the system and the rate at which it leaves. Many coastlines retain approximately the same amount of sand over time. The amount of sediment available to the system is described as the *sediment budget*, and it controls many factors of the coast.

1. On most coasts, sand and other kinds of sediment largely derive from erosion taking place inland. Larger volumes of sediment are produced and carried to the ocean if the land receives sufficient precipitation to generate runoff and is not overly protected by vegetation that limits the effectiveness of erosion.

3. Streams provide an influx of sediment into the coastal system. Deltas formed at the interface between a stream and the sea can be reworked, contributing sediment that is transported offshore or along the coast.

4. Coastal sand dunes commonly consist of sand blown inland from the beach, representing a net loss of sand from the beach. Other dunes may derive their sand from the uplands and add sand to the coastal system. Many dunes, however, simply swap sand with the beach as wind alternately blows landward and seaward, or from season to season.

2. Sediment is generated by wave erosion and associated slumping of rocks along the coast, which adds to the sand budget. Waves can bring sand in from offshore, pick up and take sand out to deeper water, or simply swash it back and forth.

5. Most areas have an atmospheric circulation that steers a current flowing along the coast. This *longshore current* transports sediment parallel to the coast. Such currents can add sand to the beach system, remove sand, or add as much sand as they remove (so the amount of sand is more or less in equilibrium).

6. Waves erode reefs and offshore islands, especially during storms, and carry loose sediment toward the coast. Many white-sand beaches consist of calcium carbonate sand eroded from coral reefs and shells. These creatures build their shells from chemicals dissolved in the water, so they increase the amount of sand in the system when the shells are smashed and broken by the waves.

11.12.d1

11.12

11.13 What Landforms Occur Along Coasts?

COASTLINES CAN DISPLAY SPECTACULAR LANDFORMS. Erosion carves some of these landforms, while sediment deposition forms others. Erosion is dominant in high-energy coastal environments, whereas deposition is more common in low-energy environments. Another controlling factor is whether the coastal zone has been uplifted relative to sea level or has been submerged under encroaching seas.

A What Coastal Features Are Carved by Erosion?

In a high-energy coastal environment, the relentless pounding of waves wears away the coastline, eroding it back toward the land. Such erosional retreat is not uniform but is often concentrated at specific locations and certain elevations. This results in some distinctive landforms along the coast.

11.13.a1 Southern Australia

1. *Sea Cliffs*—Coasts composed of hard bedrock can be eroded into cliffs (◄) that plunge directly into the surf or that are fronted by a narrow beach. Sea cliffs are somewhat more common in regions with active tectonism, especially where the land has been uplifted.

11.13.a3 Southern Australia

3. *Sea Caves* and *Sea Arches*—Erosion concentrates in the tidal zone where waves can undercut cliffs, forming caves (◄). Erosion can cut through small promontories jutting out into the sea, forming arches or windows through weaker spots in the bedrock. Caves and sea arches are most common along uplifted coasts.

11.13.a2 Crete

2. *Wave-Cut Platforms*—Continued erosion at sea level can bevel off bedrock, forming a flat, wave-cut platform (◄). It may be covered by water at high tide but fully exposed at low tide. A platform can be uplifted above sea level, at which point we call it a *marine terrace*.

11.13. a4 Southern Australia

4. *Sea Stacks*—Erosion along a coastal zone is not uniform, and some areas of rock are left behind as erosion cuts back the coast (◄). Such remnants can form isolated, steep-sided knobs called *sea stacks*, such as these along Australia's famous Great Ocean Road.

Formation of a Sea Cave and Sea Stack

5. A sea stack forms where a promontory extends out into the sea. As waves approach the shore, they refract, focusing erosion on the front and sides of the promontory.

6. If parts of the rock are weaker than others, because of differences in rock type or the relative concentration of fractures, rock behind the tip of the promontory may erode faster than the tip, forming a sea cave.

7. Continued erosion can collapse the roof of a cave and carve a passage behind the former tip of the promontory. The more resistant knob of rock becomes surrounded by the sea, and is a sea stack.

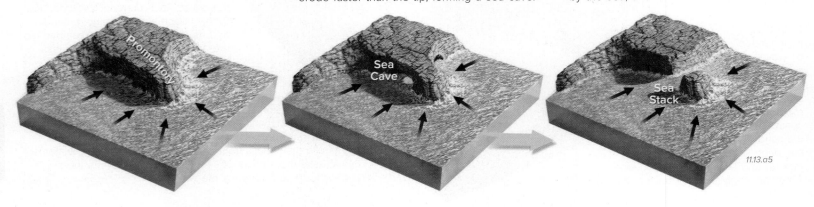

11.13.a5

B What Coastal Features Result from Deposition?

As sediment moves along a coastline, it can preferentially accumulate in places where water velocity is slower, forming a variety of low, mostly sandy features. These include sand bars, barrier islands, and sand spits.

11.13.b1 Goodland Bay, FL

11.13.b2 Dauphin Island, AL

11.13.b3 Tigertail Beach, FL

1. *Sandbar*—Offshore of many coasts is a low, sandy area, called a sandbar or gravel bar. Bars are typically submerged much of the time and can shift position as waves and longshore currents pick up, move, and deposit the sand.

2. *Barrier Island*—Offshore of many coasts are low islands that act as barriers, partially protecting the coast from large waves and rough seas. Many barrier islands are barely above sea level and consist of loose sand, including sand dunes, and saltwater marshes.

3. *Spit*—Along some coasts, a low ridge of sand and other sediment extends like a prong off a corner of the coast. Such a feature is a sand spit or a spit, and is easily eroded, especially by storm waves. A spit can change length over time, reflecting gains and losses in the sediment budget.

Formation of a Spit, Baymouth Bar, and Barrier Island

11.13.b4

4. A spit forms when waves and longshore currents transport sand and other beach sediment along the coast, building a long but low mound of sediment that lengthens in the direction of the prevailing longshore current.

5. If a spit grows long enough, it may cut off a bay, becoming a *baymouth bar.* This bar shelters the bay from waves, creating a *lagoon,* and may allow it to fill in with sediment, forming a new area of low-lying land, perhaps creating a *marsh.*

6. If sea level rises enough to submerge low-lying parts of the spit or baymouth bar, former spits and bars may become long, sandy barrier islands. Barrier islands may also form if mounds of sediment, deposited by streams when sea level is lower, become islands when sea level rises.

Cape Cod

Cape Cod sticks out into the Atlantic Ocean from the rest of Massachusetts like a huge, flexed arm. The "curled fist" is mostly a large spit. Other features are bars and barrier islands. Much of the sediment was originally deposited here by glaciers, which retreated from the area 18,000 years ago. As the glaciers melted, global sea level rose, flooding the piles of sediment and causing them to be reworked by waves and longshore currents.

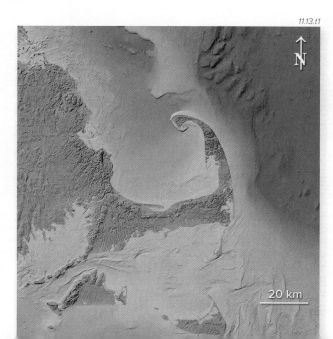
11.13.t1

Before You Leave This Page

✓ Describe the different types of coastal features.

✓ Sketch and summarize one way that a sea stack, spit, baymouth bar, and barrier island can each form.

✓ List the types of features that are present on Cape Cod, and discuss how these types of features typically form.

11.13

11.14 What Are Some Challenges of Living Along Coasts?

COASTS CAN BE RISKY PLACES TO LIVE because of their dynamic nature. Destruction of property and loss of life result from waves, storm surges, and other events that are integral to the coastal environment. Beaches and other coastal lands can be totally eroded away, along with poorly situated buildings. How can home buyers and investors identify and avoid such unsuitable and potentially risky sites?

A What Hazards Exist Along Coasts?

Most coastal hazards involve interactions between water and land, but strong winds also pose risks. Significant hazards accompany storms, which can produce large waves, strong winds, and surges of water onto the land.

11.14.a1 Pemaquid Point, ME

11.14.a2 Long Beach Island, NJ

11.14.a3

1. Waves are constantly present but are not always a threat to land, buildings, and people along coasts. The most damage from waves occurs during extreme events, such as hurricanes and other storms. Waves can erode land and undermine hillsides, causing slopes and buildings to collapse into the water.

2. A *storm surge* is a local rise in the level of the sea or large lake during a hurricane or other storm. A storm surge results from strong winds that pile up water in front of an approaching storm, inundating low-lying areas along the coast. Surges are accompanied by severe erosion, transport, deposition, and destruction.

3. Strong winds and rain accompany storms that strike the coast. Communities right on the coast are especially susceptible to these hazards because they often lack a windbreak between them and open water. Also, many coastal lands are flat, so structures built in low-lying areas are prone to rainfall-related flooding.

Before

11.14.a4 Topsail Island, NC

After

11.14.a5 Topsail Island, NC

4. These images document damage caused by Hurricane Fran in 1996 along the beach on Topsail Island, North Carolina. This photograph shows the area before the hurricane. White numbers mark two houses in both photographs. Compare these two photographs to observe what happened to the two houses, and to houses nearby, during the hurricane.

5. This photograph, taken after the hurricane, shows the loss of beach and destruction of houses caused by waves, storm surge, and erosion. The hurricane came ashore with sustained winds measured at 185 km/hr (115 mph) and a 4-meter-high (12-foot) storm surge. It caused more than $3 billion in damage.

B What Approaches Have Been Attempted to Address Coastal Problems?

Various tactics are used to minimize the impacts of natural coastal processes, including erecting barriers, trying to reconstitute the natural system, or simply avoiding the most hazardous sites.

11.14.b1 Dauphin Island, AL

1. One way to limit the amount of erosion is to construct a *seawall* along the shore. Such walls consist of concrete, steel, large rocks (◀), or some other strong material that can absorb the impacts of waves, especially during storms. As shown in this photograph, large rocks and other debris can be placed along the coast to armor it in an attempt to protect it from erosion. Material used in this way is called *rip rap*. Building a seawall commonly results in loss of beach in front of the wall because the supply of sand from inland is choked off.

11.14.b2 Clearwater, FL

2. Another type of wall, called a *jetty* (◀), juts out into the water, generally to protect a bay, harbor, or nearby beach. Jetties are commonly built in pairs to protect the sides of a shipping channel. In an attempt to protect one area of the coast, jetties, like many other engineering approaches to coastal problems, can have unintended and problematic consequences. Jetties and other walls can focus waves and currents on adjacent stretches of the coast. These directed waves and currents, deprived of their normal load of sediment by the wall, erode the adjacent areas as they try to regain an equilibrium amount of sediment.

3. Low walls, called *groins*, are built out into the water to influence the lateral transport of sand by longshore currents and by waves that strike at an angle to the coast. A groin is intended to trap sand on its up-current side, but it has the sometimes unintended consequence of causing the beach immediately down-current of the groin to receive less sand and to become eroded.

11.14.b3 Presque Isle SP, PA

4. A wall, called a *breakwater*, can be built out in the water to bear the brunt of the waves and currents. Breakwaters are built parallel to the coast to protect the beach from severe erosion and to cause sand to accumulate on the beach behind the structures. Some communities bring in sand to replenish what is lost to storms and currents. This procedure of beach nourishment is expensive and may last only until the next storm.

Avoiding Hazards and Restoring the Coastal System to Its Natural State

One approach preferred by some people, including many geologists, is simply not to build in those places that have the highest likelihood of erosion, coastal flooding, coastal landslides, and other coastal hazards. Geologists can map a coast and conduct studies to identify the most vulnerable stretches of coastline. With such information in hand, the most inexpensive approach—in the long run—is to forbid the building of houses or other structures in those areas identified as high risk. In the wake of the destruction of New Orleans and nearby communities by Hurricane Katrina, there is a debate about whether to rebuild those neighborhoods that are at highest risk, such as those that are well below sea level.

In many cases, such geologic concerns are either ignored or are overruled by financial and aesthetic interests of developers, communities, and people who own the land. Beachfront property is desirable from an aesthetic standpoint and so can be expensive real estate, which some people think is too precious to leave undeveloped.

Another approach is to try to return the system to its original situation, or at least a stable and natural one, rather than trying to "engineer" the coastline. Engineering solutions can be expensive, may not last long, or may have detrimental consequences to adjacent beaches. Returning the system to a natural state may involve restoring wetlands and barrier islands that buffer areas farther inland from waves and

wind. Examining the balance of sediment moving in and out of the system can help identify non-natural factors, such as dammed rivers, which if restored to original conditions would bring more sediment into the system and stabilize beaches, dunes, and marshes.

Before You Leave This Page

☑ Summarize some of the hazards that affect beaches and other coastlines.

☑ Describe approaches to address coastal erosion and loss of sand, including not building and trying to restore systems to a natural state.

11.14

11.15 What Happens When Sea Level Changes?

SEA LEVEL HAS RISEN AND FALLEN many times in Earth's history. A rise in sea level causes low-lying parts of continents to be inundated by shallow seas, whereas a fall in sea level can expose previously submerged parts of the continental shelf. Changes in sea level produce certain landscape features along the coast and farther inland, and they can deposit marine sediments on what is normally land. Changing sea level produces two kinds of coasts: *submergent coasts* and *emergent coasts*.

A What Features Form if Sea Level Rises Relative to the Land?

Coastlines adjust their appearance, sometimes substantially, if sea level rises or falls relative to land. A relative change in sea level can be caused by a global change in sea level or by tectonics that causes the local land to subside or be uplifted relative to the sea. Distinctive features form along a coast when sea level rises relative to the land.

1. *Submergent coasts* form where the land has been inundated by the sea because of a rise in sea level or subsidence of the land.

2. The shape of the land exerts a strong control on how the coastline will look after it is flooded by rising sea level. Examine this figure (▼) and predict what will happen to different features if sea level rises. Try it, it's an interesting exercise.

3. After the land is inundated, flooded stream valleys give the coast an irregular outline, featuring branching estuaries and other embayments.

4. Hills and ridges in the original landscape are surrounded by rising seas, forming islands along the shore.

5. Preexisting deltas and coastal dunes, when flooded, may become offshore bars or sandy barrier islands. Barrier islands may become totally submerged by rising seas.

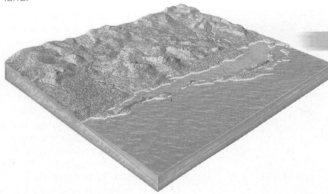

11.15.a1

6. An *estuary* is a coastal body of water that is influenced by the sea and by fresh water from the land. A common site for an estuary is a delta or a stream valley that has been flooded by the sea, either of which allows fresh water from the land to interact with salt water from the sea. Water levels in the estuary and the balance between fresh and salt water are affected by tides and by changes in the amount of water coming from the land. The satellite image below shows the Chesapeake Bay estuary. The bay was a valley originally carved by rivers, but it was flooded when sea level rose at the end of the last glacial episode (▼).

11.15.a3 Norway

7. The coasts of Norway (◄), Greenland, Alaska, and New Zealand all feature narrow, deep embayments called *fjords*. Fjords are steep-sided valleys that were carved by glaciers and later invaded by the sea as the ice melted and sea level rose.

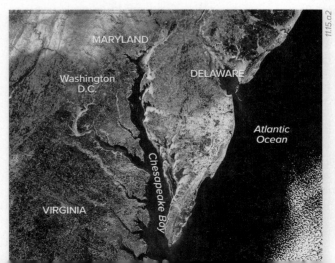

11.15.a4 North Carolina

8. Many barrier islands (◄) are interpreted to have been formed by rising sea level. Some barrier islands began as coastal dunes or piles of sediment deposited by streams. As sea level rose, the rising water surrounded the piles of sediment, resulting in new islands.

B   What Features Form if Sea Level Falls Relative to the Land?

Some features suggest a fall in sea level, or uplift of the land by tectonics or by isostatic processes. Tectonic processes can result in a gentle uplift of the shore, resulting in emergent features. Another mechanism for creating emergent features is through isostatic rebound where the shore rises after glacial retreat. A fall in sea level can expose features that were submerged and can greatly affect what happens on the adjacent land.

1. *Emergent coasts* form where the sea has retreated from the land due to falling sea level or due to uplift of the land relative to the sea. What would the area below look like if sea level dropped?

4. After sea level drops, erosion incises (cuts) valleys into the land. If sea level drops in a series of stages, emergent wave-cut notches form topographic steps on the land, and wave-cut platforms form a series of relatively flat benches, known as *marine terraces.*

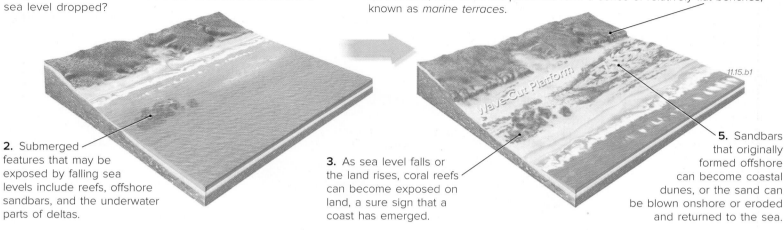

11.15.b1

2. Submerged features that may be exposed by falling sea levels include reefs, offshore sandbars, and the underwater parts of deltas.

3. As sea level falls or the land rises, coral reefs can become exposed on land, a sure sign that a coast has emerged.

5. Sandbars that originally formed offshore can become coastal dunes, or the sand can be blown onshore or eroded and returned to the sea.

11.15.b2 California

11.15.b3 Windley Key Fossil Reef State Geological Park, FL

6. Wave-cut platforms form within the surf zone along many rocky coasts and, when exposed above sea level, form relatively flat terraces on the land. The surface of such marine terraces may contain marine fossils and wave-rounded stones.

7. Coral reefs and other features that originally formed at or below sea level can be exposed when seas drop relative to the land. This fossil coral reef, now well above sea level, provides evidence of relative uplift of the land or lowering of sea level.

8. A wave-cut notch is an originally horizontal recess eroded into rock by persistent wave erosion at sea level along a coast. This photograph, taken at low tide, shows a notch cut by waves and by the soluble rocks being dissolved. Uplift of the land or a drop in sea level can leave a wave-cut notch high and dry, a hint of what occurred.

11.15.b4 Palau

Wave-Cut Notch

Before You Leave This Page

☑ Summarize what a submergent coast is and what types of features can indicate that sea level has risen relative to the land.

☑ Summarize what an emergent coast is and what types of features indicate that sea level has fallen relative to the land.

11.15

11.16 What Causes Changes in Sea Level?

SEA LEVEL HAS VARIED GREATLY IN THE PAST. Global sea level has been more than 200 m higher and more than 120 m lower than today. What processes caused these changes in sea level? Large variations in past sea level resulted from a number of competing factors, including the extent of glaciation, rates of seafloor spreading, and global warming and cooling. These processes operate at different timescales, from thousands to millions of years.

A How Does Continental Glaciation Affect Sea Level?

The height of sea level is greatly affected by the existence and extent of glaciers and regional ice sheets. At times in Earth's past, ice sheets were more extensive than today, and at other times they were less extensive or absent.

1. The ice in glaciers and continental ice sheets accumulates from snowfall on land. When glaciers and ice sheets are extensive, they tie up large volumes of Earth's water, causing sea level to drop (▶).

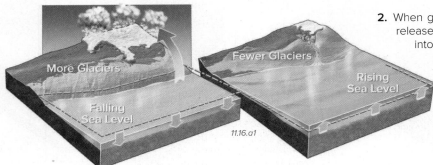

11.16.a1

2. When glaciers and ice sheets melt, they release large volumes of water that flow back into the ocean, causing sea level to rise (◀). This has caused sea level to rise in the last 15,000 years.

3. The growth and shrinkage of ice sheets and glaciers is the main cause of sea level change on relatively short timescales (thousands of years).

B How Do Changes in the Rate of Seafloor Spreading Affect Sea Level?

At times in Earth's history, the rate of seafloor spreading was faster than it is today, and at other times it was probably slower. Such changes cause the rise and fall of sea level.

1. The shape and elevation of a mid-ocean ridge and adjacent seafloor reflect the rate of spreading. As the plate moves away from the spreading center, it cools and contracts, causing the seafloor to subside, creating space for seawater.

2. If seafloor spreading along a ridge is slow, the ridge is narrower because the slow-moving plate has time to cool and contract before getting very far from the ridge. Slow seafloor spreading and narrow ridges leave more room in the ocean basin for seawater (▶). So over time, a decrease in the spreading rate causes sea level to fall.

3. If seafloor spreading along a ridge is relatively fast (10 cm/year or faster), the ridge is broad because still-warm parts of the plate move farther outward before cooling and subsiding. So an increase in the seafloor spreading rate is accompanied by broader ridges that displace water out of the ocean basins (◀), causing sea level to rise. In other words, faster spreading yields more young seafloor, and young seafloor is less deep than older seafloor, raising sea level.

11.16.b1

C How Do Changes in Ocean Temperatures Cause Sea Level to Rise and Fall?

Sea level is affected by changes in ocean temperatures, which cause water in the oceans to slightly expand or contract. Such effects accompany global warming or global cooling, and result in relatively moderate changes in sea level.

1. Water, like most materials, contracts slightly as it cools, taking up less volume. The amount of contraction is greatly exaggerated in this small block of water.

2. When ocean temperatures fall, water in the ocean generally contracts, causing sea level to fall (▶).

3. Water expands slightly when heated, taking up more volume. Again, the amount of expansion is exaggerated in this small block of water.

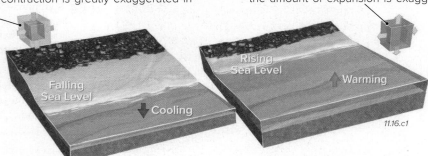

11.16.c1

4. When ocean temperatures increase, water in the ocean generally expands slightly, causing a small rise in sea level (◀). The percentage of expansion is small, but it can cause a moderate rise in sea level.

D How Does the Position of the Continents Influence Global Sea Level?

As a result of plate tectonics, continents move across the face of the planet, sometimes being near the North or South Poles and at other times being closer to the equator. These positions influence sea level in several ways.

1. Glaciers and continental ice sheets form on land, and so they require a landmass to be cold enough to allow ice to persist year round. This occurs most easily if a landmass is at high latitudes (near the poles) or is high in elevation.

2. At most times in Earth's past, ice ages occurred when one or more of the continents were near the poles (▶), like Antarctica is now. Continents at high latitudes usually include glaciers, so such a configuration of the continents can cause global sea level to drop because water is stored in glaciers.

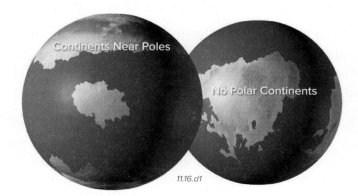

3. At other times in Earth's past, the larger continents were not so close to the poles (◀). This lower latitude position of continents minimized or eliminated widespread glaciation (the non-ice-age periods in Earth's history). This low-latitude configuration of the continents therefore tends to keep sea levels high.

E How Do Loading and Unloading Affect Land Elevations Relative to Sea Level?

Weight can be added to a landmass, a process called *loading*. A weight can also be removed, a process called *unloading*. Loading and unloading can change the elevation of a region relative to sea level.

1. Weight loaded on top of a region imposes a downward force that, if large enough, can downwarp the land surface beneath the load and in adjacent areas (▶). Loading, such as by continental ice sheets, lowers the loaded region relative to sea level. This can allow seawater to inundate regions near the ice sheets. The ice in this figure and the amount of subsidence are very stylized and vertically exaggerated.

2. If ice melts and the weight is unloaded from the land, the region flexes back upward, a process known as *isostatic rebound*. The uplifted, rebounding region rises (▶) relative to sea level.

3. Unloading and isostatic rebound can occur when continental ice sheets melt. Rebound begins as soon as significant amounts of ice are removed, but it can still be occurring thousands of years after all the ice is gone.

Ongoing Isostatic Rebound of Northeastern Canada

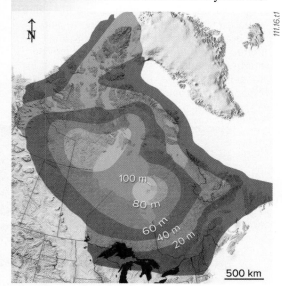

The northern part of North America has been covered by glaciers off and on for the last two and a half million years. The weight of these continental ice sheets loaded and depressed this part of the North American plate. When the ice sheets began melting from the area approximately 15,000 years ago, unloading caused the land, especially in Canada, to begin to isostatically rebound upward.

The amount of rebound has been measured both directly and indirectly. We can measure uplift directly by making repeated elevation surveys across the land and then calculating the amount of uplift (rebound) between surveys. Rates of rebound are typically millimeters per year, which is enough to detect with surveying methods. Satellite measurements (GPS) are also sensitive enough to measure such changes. The amount and rate of rebound can also be inferred more indirectly by documenting how shorelines and other features have been warped and uplifted. Contours on this map to the left indicate the amount of rebound interpreted to

have occurred in northeastern North America over the last 6,000 years. Some areas have experienced more than 100 m of uplift. Isostatic rebound can also occur around large lakes as they fill or empty over time, loading or unloading the land.

Before You Leave This Page

✔ Summarize how continental glaciation, rates of seafloor spreading, ocean temperatures, and position of the continents affect sea level.

✔ Explain how loading and unloading affect land elevations using the example of Canada.

11.16

11.17

What Coastal Damage Was Caused by These Recent Atlantic Hurricanes?

THE EAST COAST AND GULF COAST of North America routinely suffer damage from hurricanes and other storms. In 2011 and 2012, several large but not especially powerful hurricanes hit different coastlines of the U.S. and caused extensive damage. They included Hurricanes Irene (2011), Isaac (2012), and Sandy (2012). The damage caused by these three storms illustrates some issues of building along coasts.

A What Hazards Do Hurricanes and Other Storms Pose to Coastlines?

Hurricanes bring strong winds, heavy rains, and the risk of tornadoes, but the greatest damage is commonly due to storm surge, where the hurricane pushes a mound of water against the coastline. This temporarily raises sea level in affected areas and allows large waves to crash farther up the beach than is normal. As a result of the surge and waves, extreme amounts of erosion occur, damaging or destroying structures too close to the beach or too low in elevation. The U.S. Geological Survey (USGS) has an extensive program to assess the hazards of such storms, as presented here.

11.17.a1 Myrtle Beach, SC

1. Many people live along the East Coast (◄) and Gulf Coast of North America. Living near the coast has many advantages, but it also poses a risk to structures built in places where coastal erosion could occur, removing sand, roads, and buildings that are in the way. Areas farther back could be flooded if water rises high enough.

2. To assess how risky living along a particular stretch of coast is, the USGS coastal scientists developed a conceptual model summarized in this figure (►). It compares the maximum height of the swash zone (R_{high}) with the heights of the base (D_{low}) and top (D_{high}) of coastal dunes. If the swash is below the base of the dune, called the *swash regime*, little permanent erosion occurs.

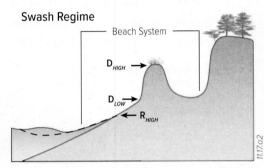
Swash Regime
11.17.a2

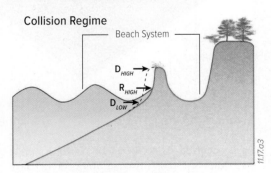
Collision Regime
11.17.a3

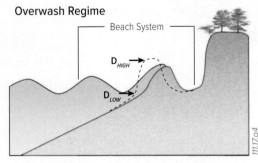
Overwash Regime
11.17.a4

Inundation Regime
11.17.a5

3. If the local sea level is higher, such as from a storm surge, the swash and waves can directly strike the lower parts of coastal dunes in what the USGS calls the *collision regime*. This causes erosion of the base of the dune, and the loss of sand can be permanent. Erosion is, however, somewhat limited to areas close to the shoreline.

4. In the next stage, the *overwash regime*, the maximum height of swash and waves reaches the top of the dunes. This causes erosion of the dunes, with the transport of sand toward the land. Through this process, the beach or entire barrier island can migrate toward the land. Houses that were on the beach may now be in the water.

5. The next and most damaging stage is when the water level is so high that the entire dune is submerged, the *inundation regime*. Large quantities of sand are typically transported toward the land, permanently moving the beach and burying areas farther inland in sand. If the site is a barrier island, the entire island may become submerged.

11.17.a6 Nags Head, NC

6. This photograph (◄) illustrates erosion of the front (seaward side) of a dune. Although the house was high enough to avoid being flooded, erosion is devouring the dune on which the house sits. The dune acted as a buffer between the house and the waves, but as the dune became narrower, the house became more threatened.

7. This photograph of a barrier island (►) documents the process of *overwash*, where material is moved from the side facing the sea (i.e., the waves in the foreground) to the side away from the sea. The seaward side lost beach material while the backside of the island grew.

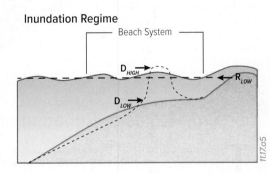
11.17.a7 Cone Banks, NC

Overwash

B What Damage Was Caused by Hurricanes Irene, Isaac, and Sandy?

Hurricane Irene (2011) struck the Outer Banks of North Carolina and adjacent areas, and Hurricane Isaac (2012) struck the Gulf Coast, including many of the same areas previously damaged by Hurricane Katrina (2005). Hurricane Sandy (2012) caused widespread damage along the East Coast, especially in New Jersey and New York.

11.17.b1
August 30, 2011

Hurricane Irene

11.17.b2
May 6, 2008

August 31, 2011

Hurricane Isaac

11.17.b3
August 8, 2012 Pre-Isaac

September 2, 2012 Post-Isaac

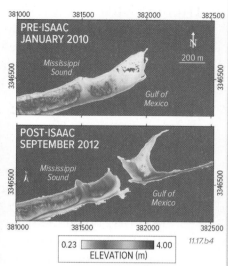
PRE-ISAAC JANUARY 2010
Mississippi Sound
Gulf of Mexico
200 m

POST-ISAAC SEPTEMBER 2012
Mississippi Sound
Gulf of Mexico

0.23 ▬▬▬ 4.00
ELEVATION (m)
11.17.b4

1. These three photographs (▲) are of the same part of the Outer Banks, North Carolina. The photograph on the top right shows the areas before Irene, whereas the other two show the conditions after Irene. Yellow arrows point to the same house. Observe changes that occurred from this single storm. Severe erosion of the beach obliterated any physical evidence that the house closest to the beach ever existed. Erosion cut a new channel through the landscape.

2. This photograph pair (▲) illustrates the damage done to barrier islands in Louisiana by Hurricane Isaac. The small island in the background was eroded to a much smaller size, but the low islands and sandbars in the foreground became totally submerged.

3. These figures (▲) show before and after topographic maps of another island, measured using a surveying technique (lidar) that uses lasers. Note how this storm breached the island and redistributed material from one area to another. Barrier islands clearly are transient features. The locations are indicated with Universal Transverse Mercator (UTM) coordinates.

Hurricane Sandy

4. When Hurricane Sandy struck the Northeastern U.S., one of the hardest-hit areas was in New Jersey, where the center of the storm came ashore. Compare the before and after scenes (▼).

11.17.b5
Before May 21, 2009

After November 5, 2012

5. These two lidar images (▼) document the changes in elevation in the same area. Note that houses and the bridge were destroyed by erosion during the storm surge. Elevation can be almost as important as proximity to the beach.

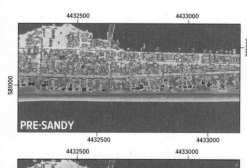
PRE-SANDY

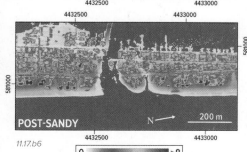
POST-SANDY   *N →*   *200 m*
11.17.b6

0 ▬▬▬ >8
ELEVATION (m)

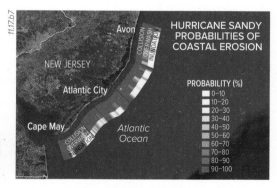
11.17.b7
HURRICANE SANDY PROBABILITIES OF COASTAL EROSION
Avon
COLLISION OVERWASH INUNDATION
NEW JERSEY
Atlantic City
Cape May
Atlantic Ocean
COLLISION OVERWASH INUNDATION

PROBABILITY (%)
- 0–10
- 10–20
- 20–30
- 30–40
- 40–50
- 50–60
- 60–70
- 70–80
- 80–90
- 90–100

6. As the storm neared, the USGS assigned extent-of-damage probabilities, which were reviewed when assigning evacuation routes.

Before You Leave This Page

☑ Sketch and describe the four regimes of beach erosion.

☑ Summarize the kinds of erosion and deposition that occurred during three Atlantic hurricanes, identifying areas you consider to be most vulnerable.

11.17

What Is Happening Along the Coast of This Island?

THIS PREVIOUSLY UNKNOWN ISLAND is being considered as a possible site for a small settlement and seaport to resupply ocean travelers with water and other supplies as they pass by. To help plan this process, you are asked to observe the coastline, identify the various coastal features, infer what processes are occurring, and identify possible sites for the port and nearby settlement. Your considerations should include any issues that could affect the operation of the port, such as avoiding natural hazards and keeping shipping channels from becoming blocked with sand. The area does experience hurricanes.

Goals of This Exercise:

- Observe the three-dimensional perspective of the newly discovered island in order to identify the coastal features and processes.

- Fully consider any factors that would impact the location of a port and small settlement for people operating the port.

- Propose an acceptable location for the port and settlement, and be prepared to defend your choices.

Geographic Setting

Listed below is some geographic information about the island. North is toward the top of the page.

- The island, surrounded by the ocean, is in the Northern Hemisphere, located at approximately 30° N latitude, at about the latitude of Florida.

- The island is somewhat elongated and nearly 10 km in its longest direction. It reaches an elevation of 1,000 m above sea level in two mountains.

- The climate of the island is warm during the summer, when it sometimes rains. It is wetter during the winter, when it receives enough rainfall to provide ships with water, as long as the water resources are managed intelligently. The ocean moderates the climate, keeping it a little cooler during the day and warmer at night relative to a place that is not next to the sea. The air is fairly humid, so the mountains sometimes cause and receive short summer rains.

- There is a moderate amount of vegetation on the island, except on the rocky mountains. The most dense vegetation is near a delta on the north side of the island and on a raised flat area on the western (left) side. Both sites are suitable for growing crops.

11.18.a2

11.18.a3

1. Along the western side of the island are cliffs and steep slopes that rise out of the sea, with only a narrow beach along the shoreline (◄). The beach is composed of sand, but it also contains many rounded pebbles and larger stones plus some large angular rocks that match the types of rocks in the cliff. There are also shells and fossils of other marine creatures.

2. Above the cliffs is a relatively flat area, like a shelf in the landscape (▲). This type of feature is only observed on this western side of the island. Scattered about on this flat surface are rounded pebbles and larger stones, along with abundant loose sand. On some parts of the surface there is a layer that contains fossil shells (▼) and fossils of other marine organisms identical to those that occur on the beach.

11.18.a4

11.18.a1

Procedures

Follow the steps below, entering your answers for each step in the appropriate place on the worksheet or online.

A. Observe the overall character of the island, noting what is along each part of the coast and the locations of streams (a source of fresh water). Read the information about the geographic setting of the island (on the previous page).

B. Observe the photographs and read the accompanying text describing different parts of the island. Use what you know about coasts to interpret what type of feature is present, what processes are typical for this type of feature, and what the significance of each feature is for a port and small settlement.

C. Synthesize all the information and decide on the best locations for the settlement and port, which should be fairly close to one another (no more than 1 km apart). Be able to logically support your choices. There is not a single correct answer or best site.

D. OPTIONAL: Your instructor may have you write a short report or prepare a presentation describing all the factors you considered and why you chose the sites you did. You may also be asked to interpret the natural history of the island.

3. Along the north shore of the island is a delta where the largest streams in the area reach the sea. This area has abundant trees and wetlands (▶) and is easily accessed from the east by walking along the beach.

11.18.a5

11.18.a6

4. The north shore of the island has gentle relief along the shore and has a well-developed beach that is mostly sand (▲). The beach slopes gently into the water, allowing people to wade quite a distance out into the sea. The sand is constantly in motion, moving up and back on the beach and also moving laterally along the coast. Lying on some parts of the beach are pieces of wood, similar in size to the trees observed near the delta to the west.

5. Cut into the eastern end of the island are a series of embayments, or bays, that branch farther inland. In the largest bay, the land slopes into the bay at a moderate angle, gentle enough to walk up. The bay projects toward two mountains that have some well-developed vegetation on their flanks. The mountains are not too steep to walk up. Small streams flow on every side of the mountains. The mountains are not volcanoes, but are rocky and without much soil.

11.18.a7

6. Extending off the land from the southern side of the island is a low ridge of relatively loose sand (▶). Some of the high parts of the ridge are clearly sand dunes. Some lack any vegetation, but others are anchored in place by grasses and other plants. To the east is a similar long island, again composed of sand, including dunes.

11.18

WATER IS THE MOST IMPORTANT RESOURCE provided by Earth—nearly all life on Earth needs water to live and thrive. We are most familiar with *surface water,* water that occurs in streams, lakes, and oceans. Yet, the amount of fresh water in these settings is much less than the amount of fresh water that is frozen in ice and snow or the amount of water that occurs in the subsurface as *groundwater.* This chapter is about surface water and groundwater, and the important ways in which they interact.

12.00.a2 Shoshone, ID

The Snake River Plain, shown in this large satellite-based image, is a curved swath of low, basalt-covered land that cuts through the mountains of southern Idaho. It contains a mixture of dry, sage-covered plains, water-filled reservoirs, green agricultural fields, and recent lava flows of dark-colored basalt, such as those at Craters of the Moon National Monument. Most of Idaho's population lives on the Snake River Plain near the streams and reservoirs.

The Big Lost River, Little Lost River, and adjacent streams that enter the plain from the north never reach the Snake River. Instead, the water from the streams seeps into the ground between the grains in the sediment and through narrow fractures in the basalt. For this reason, the rivers are called "lost."

Where does water that seeps into the subsurface go?

Within the Snake River Plain, the Snake River has eroded a canyon (▲) down into layers of ledge-forming basalt and slope-forming sediment. The farmlands of the canyon bottom are on fertile sediment deposited by the river, and they receive water from streams, springs, and wells drilled to extract groundwater.

Lost Rivers Area

Craters of the Moon National Monument

Snake River

Thousand Springs

Snake River

At Thousand Springs, huge springs gush from the steep volcanic walls of the Snake River Canyon (◄). The canyon includes 15 of the 65 largest springs in the United States, including those at Thousand Springs. The largest commercial trout farms in the United States use ponds fed by these springs.

What causes water from beneath the ground to flow to the surface as a spring, and where does the water in a spring come from?

12.00.a3 Thousand Springs, ID

TOPICS IN THIS CHAPTER

12.00.a4 Jackson Hole, WY

The Snake River begins its journey in Jackson Hole, Wyoming (▲), from streams that drain the Tetons and nearby Gros Ventre mountain range. The relatively higher rainfall and snowmelt in these highlands sustain the river as it flows westward across the dry plains. Downstream from Jackson Hole, streams entering the plain from the east and south flow directly into the Snake River, increasing its flow.

Where does the water in streams come from, and do most streams gain or lose water from groundwater?

25 km

12.00.a1

Many lakes and farms are situated next to the Snake River. Farmers irrigate millions of acres of agriculture with surface water derived from reservoirs, lakes, and rivers, and with groundwater pumped to the surface. Chemicals used by some farms cause contamination of groundwater and surface water.

What happens if groundwater is pumped from the subsurface faster than it is replaced by precipitation and other sources? What do we do if water supplies are contaminated?

Disappearing Waters of the Northern Snake River Plain

Groundwater beneath the Snake River Plain is an essential resource for the region, providing most of the drinking water for cities and irrigation water for farms and ranches away from the actual river. Geoscientists study where this water comes from, how it moves through the subsurface, and potential limits on using this resource.

Some water enters the subsurface from the Big and Little Lost Rivers, which flow into the basin from the north and then abruptly or gradually disappear as their water sinks into the porous ground. Other groundwater comes directly from the main Snake River and from tributaries that enter the basin from the south and east. Perhaps surprisingly, the largest influx of water to the subsurface is seepage from irrigated fields and associated canals.

The Snake River Plain slopes from northeast to southwest. The flow of groundwater follows this same pattern, flowing southwest and west through sediment and rocks in the subsurface. Groundwater derived from the disappearing rivers flows southwest, along the northern side and center of the basin. The groundwater does not flow like an underground river but as water between the sediment grains and within fractures in the rocks. Where the Snake River Canyon intersects the flow of groundwater, water reemerges on the surface, pouring out at Thousand Springs. This region illustrates a main theme of this chapter—surface water and groundwater are a related and interconnected resource.

12.0

12.1 Where Does Water Occur on Our Planet?

WATER IS ABUNDANT ON EARTH, occurring in many settings. Most water is in the oceans but is salty. Most fresh water is in ice and snow or in groundwater below the surface, with a smaller amount in lakes, wetlands, and streams. Water also exists in plants, animals, and soils and as water vapor in the atmosphere.

A Where Did Earth's Water Come from and Where Does It Occur Today?

Most water on Earth probably originated during the formation of the planet or from comets and other icy celestial objects that collided with the surface. Over time, much of this water moves to the surface, for example, when magma releases water vapor during eruptions.

1. *Oceans*—Of Earth's total inventory of surface and near-surface water, an estimated 96.5% occurs in the oceans and seas as *saline* (*salty*) *water*. The remaining 3.5% is *fresh water* held in ice sheets and glaciers, groundwater, and lakes, swamps, and other features on the surface.

2. *Streams*—Streams are extremely important to us and are the main source of drinking water for many areas. They contain, however, only a very small amount of Earth's fresh water.

3. *Lakes*—Water occurs on the surface in lakes of various sizes. Most are freshwater lakes, but those in dry climates are saline or *brackish* (between fresh and saline). Lakes contain a majority of the liquid fresh water at Earth's surface.

4. *Swamps and Other Wetlands*—These wet places contain water lying on the surface and water within the plants and shallow soil. They constitute about 11% of the liquid fresh water on the surface.

5. *Atmosphere*—A small, but very important, amount (0.001%) of Earth's water is contained in the atmosphere. It occurs as invisible water vapor, as water droplets in clouds, and as rain, falling snow, and other types of precipitation.

6. *Glaciers*—Nearly 69% of Earth's fresh water is tied up in ice and snow in ice caps, glaciers, and permanent snow. A small amount also exists in permafrost and ground ice.

7. *Soil Moisture*—Earth's soils contain about as much water as the atmosphere (not much), but like water in the atmosphere, soil moisture is crucial to our existence.

8. *Biological Water*—Water is tied up within the cells and structures of plants and animals. It is clearly important to us but represents an exceptionally small percentage of Earth's total water (0.0001%).

9. *Groundwater*—About 30% of Earth's total fresh water occurs as groundwater. Groundwater is mostly in the open pores between sediment grains or within fractures that cut rocks. Most groundwater is fresh, but some is brackish or saline.

10. *Deep-Interior Waters*—An unknown but perhaps very large amount of water is chemically bound in minerals of the crust and mantle. Some scientists think Earth's interior may contain more water than the oceans.

12.01.a1

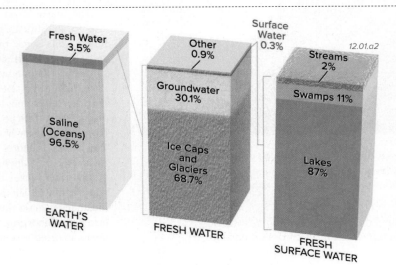

11. These bar graphs (▶) show USGS estimates of the distribution of water on Earth's surface and the uppermost levels of the crust. The left bar shows that the oceans contain 96.5% of Earth's total free water (water that is not bound up in minerals), but this water is saline. Only 3.5% of Earth's water is fresh. These graphs do not include water in Earth's deep interior.

12. The middle bar (◀) shows that most of Earth's fresh water occurs in ice caps and glaciers. Almost all the rest is groundwater. Less than 1% occurs as liquid surface water in lakes, swamps and wetlands, and streams.

13. The right bar (◀) shows where Earth's small percentage of fresh, liquid, surface water resides. Most is in lakes, followed by swamps and streams. Note that for each of two bars on the right, the percentages in that bar reflect the fraction of water in only the top part of the bar directly to its left (percentages of percentages of percentages for the rightmost bar, so not very much of the overall amount of water on Earth).

Fresh Water 3.5%

Saline (Oceans) 96.5%

EARTH'S WATER

Other 0.9%

Groundwater 30.1%

Ice Caps and Glaciers 68.7%

FRESH WATER

Surface Water 0.3%

12.01.a2

Streams 2%

Swamps 11%

Lakes 87%

FRESH SURFACE WATER

B | How Does Water Move from One Setting to Another?

Water is in constant circulation on Earth's surface, moving from ocean to atmosphere, from atmosphere back to the surface, and in and out of the subsurface. The circulation of water from one part of this water system to another is called the *hydrologic cycle*. From the perspective of living things, the hydrologic cycle is the critical system on Earth. It involves a number of important and mostly familiar processes, and is driven by energy from the Sun.

1. *Evaporation*—As liquid water is heated by the Sun, some of its molecules become energized enough to break free of the attractive forces binding them together. Once free, they rise into the atmosphere as water vapor. Most evaporation occurs from the oceans, which cover three-fourths of Earth's surface.

2. *Condensation*—As water vapor cools, like when it rises, water molecules join together. Through this process, water vapor becomes a liquid or turns directly into a solid (ice, hail, or snow). These water drops and ice crystals then collect and form clouds.

3. *Precipitation*—When clouds cool, perhaps when they rise over a mountain range, the water molecules become less energetic and bond together, commonly falling as rain, snow, or hail, depending on the temperature of the air. The resulting precipitation may reach the ground, evaporate as it falls, or be captured by leaves and other vegetation before reaching the ground.

4. *Sublimation*—Water molecules can go directly from a solid (ice) to vapor, a process called sublimation (not shown here). Sublimation is most common in cold, dry, and windy climates, like some polar regions.

5. *Infiltration*—Some precipitation seeps into the ground, infiltrating through fractures and pores in soil, sediment, and rocks. Some of this water becomes groundwater, some remains within the soil, and some rises back up to the surface. Water can also infiltrate into the ground from lakes, streams, canals, irrigated fields, or any body of water.

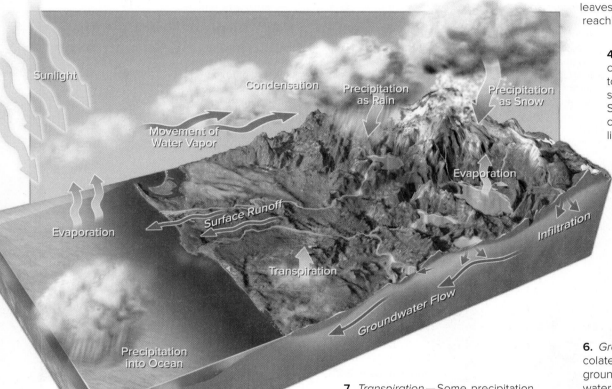

12.01.b1

6. *Groundwater Flow*—Water that percolates or infiltrates far enough into the ground becomes groundwater. Groundwater can flow from one place to another in the subsurface, or it can flow back to the surface, where it emerges in springs, lakes, and other features. Such flow of groundwater may sustain these water bodies during dry times.

7. *Transpiration*—Some precipitation and soil moisture is taken up by root systems and other water-collecting mechanisms of plants. Through their leaves, plants emit water vapor into the atmosphere by the process of transpiration. The combined loss of water via transpiration and evaporation from soil, leaves, and other parts of the land is called *evapotranspiration*.

8. *Surface Runoff*—Rainfall or snowmelt can produce water that flows across the surface as runoff. Runoff from direct precipitation can be joined by runoff from melting snow and ice, from lakes, and by the flow of groundwater onto the surface. The various types of runoff collect in streams and lakes. Most is eventually carried to the ocean by streams, where it can be evaporated, completing the hydrologic cycle.

9. *Ocean Gains and Losses*—Most precipitation falls directly into the ocean, but the ocean loses much more water to evaporation than it gains from precipitation. The difference is made up by runoff from land.

12.1

12.2 How Do We Use Fresh Water?

WE USE LARGE QUANTITIES OF WATER each day. We use water for a variety of purposes, especially power generation and irrigation of farms. How much water does each of our activities consume, and where does the water come from?

A What Are the Main Ways in Which We Use Fresh Water?

Every five years, the U.S. Geological Survey conducts a detailed study of water use in the United States. The most recent USGS compilations show that we use fresh water in six or seven main ways, depending on how we classify the useage. Water use in the United States is hundreds of billions of gallons per day, most of which is from surface waters. Examine the graph below and think about the ways in which you use water.

12.02.a2 Huntington, UT

1. *Thermoelectric Power*—Electrical power generation plants (◄) are the largest users of fresh water in the United States, using slightly more than 40%, but much of this water is returned to the environment, not actually lost. Some of this water is from recycled sources. Power plants are also the largest users of saline (salty) water. Such plants drive their turbines by converting water into steam, and they also use large amounts of water to cool hot components. The visible emissions from such plants are mostly steam.

2. *Irrigation*—Farms and ranches are the other large users of fresh water, using 37% of fresh water. Farms (►) use water from groundwater, streams, lakes, and reservoirs to irrigate grain, fruit, vegetables, cotton, animal feed, and other crops. Much of this water is lost through evaporation to the atmosphere before it can be used by plants.

12.02.a3 Pasco, WA

3. *Public and Domestic Water Uses*—The third-largest use of fresh water is by public water suppliers and other domestic uses (▼). We consume water by drinking, bathing, watering lawns, filling artificial lakes, and washing clothes, dishes, and cars. Much water from public water suppliers also goes to businesses. Most water for public and domestic use comes from streams and groundwater.

12.02.a5 West Driefontane, South Africa

FRESHWATER USAGE

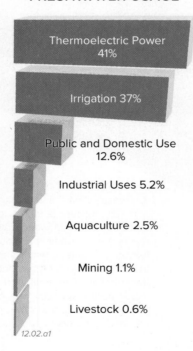

Thermoelectric Power 41%

Irrigation 37%

Public and Domestic Use 12.6%

Industrial Uses 5.2%

Aquaculture 2.5%

Mining 1.1%

Livestock 0.6%

12.02.a1

12.02.a4 Phoenix, AZ

4. *Industrial and Mining Uses*—Fresh water and saline water are extensively used by industries, including factories, mills, and refineries. Water is integral to many manufacturing operations, including making paper, steel, plastics, and concrete. Mining (▲) and related activities also use water in the extraction of metals and minerals from crushed rock.

12.02.a6 Tonto Creek, AZ

5. *Aquaculture*—According to the USGS study, we use approximately 2.5% of fresh water to raise fish (◄) and aquatic plants. Most of this use is in Idaho, near the Thousand Springs area. Such water is not totally consumed—much is released back into the Snake River.

6. *Livestock*—Watering of cows, sheep, horses (►), and other livestock accounts for only 0.6% of fresh water use, but much water is used for irrigation to raise hay, alfalfa, and other animal feed. Many ranches use small constructed reservoirs and ponds as the main water source for animals.

12.02.a7 Delta, UT

B How Do We Use and Store Surface Water?

12.02.b1 Grand Coulee Dam, WA

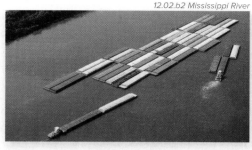
12.02.b2 Mississippi River

12.02.b3 Black Hills, SD

1. *Electrical Generation*—The movement of surface water can generate electricity. To do this, we build dams that channel water through turbines in a hydroelectric power plant.

2. *Transportation*—We use many large waterways, such as the Mississippi River, as energy-efficient transportation systems to transport agricultural products, chemicals, and other industrial products.

3. *Recreation*—People use surface water in lakes and rivers for many types of recreation, including swimming, tubing, rafting, boating, and fishing. We also use fresh water to fill ponds, fountains, and swimming pools.

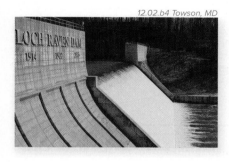
12.02.b4 Towson, MD

4. Surface waters are commonly stored in natural lakes and in constructed *reservoirs* behind concrete (◄) and earthen dams. Water for drinking and other municipal uses can be stored in underground or above-ground storage tanks.

5. We construct canals (►), raised aqueducts, and large and small pipelines to move fresh water from one place to another, especially to irrigate farms or to bring water to large cities. Canals also carry groundwater that has been pumped to the surface.

12.02.b5 Tempe, AZ

C How Do We Refer to Volumes of Water?

Water-resource studies typically report volumes of water in one of three units: *gallons*, *liters*, or *acre-feet*. Gallons and liters are familiar terms, but the concept of an *acre-foot* of water requires some explanation.

How big is an acre? An acre covers an area of 4,047 m² (43,560 ft²). If a perfect square, an acre would be 64 m (210 ft) on a side. An acre is equivalent to 91 yards of an American football field. There are 640 acres in a square mile.

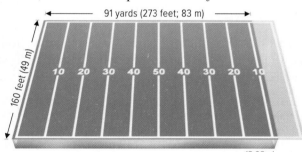
91 yards (273 feet; 83 m)
160 feet (49 m)
10 20 30 40 50 40 30 20 10
12.02.c1

An acre-foot of water is the volume of water required to cover an acre of land to a height of one foot. Imagine covering 91% of a football field (one acre) with a foot of water. An acre-foot is equivalent to about 326,000 gallons, or more than 1.2 million liters of water.

Drinking-Water Standards in the United States

The U.S. Environmental Protection Agency (EPA) sets standards for safe drinking water. Nearly all *public* water supplies in the United States meet these standards, which can be found at *www.epa.gov*. These standards set a limit on the concentrations of selected contaminants in water. Small municipalities commonly have more trouble meeting these standards than large cities because of limited budgets for water analysis, and for building and running facilities to remove contaminants. The EPA standards do not apply to private wells.

Many people prefer the taste and convenience of bottled water to public tap water, but there are generally no health reasons to buy bottled water so long as the public water provider meets all the federal, state, and local drinking water regulations. Commercially bottled water is monitored by the Food and Drug Administration (FDA) but is not as closely monitored as public water systems. The FDA requires a bottler to test its water source only once a year. Also, bottled water can cost as much as 1,000 times more than municipal drinking water.

Before You Leave This Page

☑ Describe ways we use fresh water, and which four uses consume the most.

☑ Describe how we use and store fresh water.

☑ Describe in familiar terms how much water is in an acre-foot.

☑ Describe what a drinking water standard is, who sets the limits, and to whom they do and do not apply.

12.2

12.3 What Is a Drainage Network?

STREAMS ARE CONDUITS of moving water driven by gravity, flowing from higher to lower elevations. The water in streams comes from precipitation, snowmelt, and springs. A stream drains a specific area and joins other streams draining other areas, forming a drainage network.

A What Is a Stream?

A stream carries flowing water through a single channel or through a number of interconnected channels. Such channels vary in size from small streams several meters wide to major rivers that are kilometers across.

1. The Potomac River (▼) winds its way along the border between Maryland and West Virginia on its way to Washington, D.C. A number of smaller streams join the river from both sides, forming a drainage network. Water flowing in streams can move rock fragments and dissolved minerals from high elevations to low elevations.

2. The amount of water that flows through a stream channel varies with time, mostly reflecting the influence of changes in the seasons (e.g., from winter to spring) and changes in the weather. At some times of the year and during rainy periods, the flow increases. It decreases at especially dry times of the year or during times of few storms. The amount of water flowing in a given amount of time is the *discharge*, which represents the volume of water that flows by in some amount of time. It has units of cubic meters per second or m³/s.

12.03.a1

3. We calculate discharge (represented by the letter Q) by multiplying the cross-sectional area of the channel by the velocity of the flow:

$$Q = stream\ depth \times stream\ width \times stream\ velocity$$

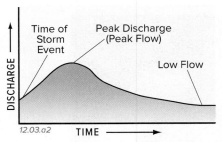

12.03.a2

4. A graph showing the change in the amount of discharge over time is a *hydrograph* (▲). This hydrograph shows that discharge increased and then decreased over time in response to a storm. The shape of the graph reflects how a stream responds to precipitation, telling us useful information about a stream and the area it drains.

B Where Does a Stream Get Its Water?

Each stream drains a naturally defined area, called a *drainage basin*. A basin slopes from higher areas, where the stream begins, to lower areas, toward which the stream flows. Runoff from rainfall, snowmelt, springs, or water leaving a lake will flow downstream and out of the drainage basin at its low point.

1. *Drainage Basin*—In this figure, each of two adjacent streams has a drainage basin (▼), shaded in different colors. Runoff from the red area drains into the stream on the left; runoff from the blue area drains into the stream on the right. The ridge between the two drainage basins is the boundary between water flowing into different drainage basins and is a *drainage divide*.

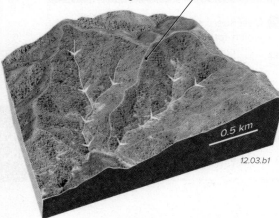

12.03.b1

2. *Basin Slope*—Overall slope of a drainage basin helps us determine how fast water in the basin empties after a heavy rain or after snowmelt, as shown by the graph below.

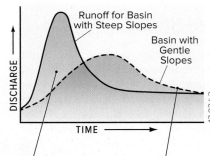

12.03.b2

3. Runoff from a steep drainage basin is fast, and much water arrives downstream at about the same time, yielding higher discharge values.

4. Runoff from a more gently sloped basin is spread out over time, leading to lower peak discharge values.

5. *Basin Size and Shape*—A drainage basin's size and shape influence its flow response to rainfall. The plot below (▼) shows a hydrograph for a single storm event, along with a simplified map of the basin's shape.

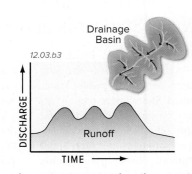

12.03.b3

6. Following a storm event, a hydrograph for a simple basin shows a single-peak increase in discharge with a gradual decrease, like the graph in part A (above). In contrast, a complex, three-part drainage basin, depicted here, may show a three-peak response to a single event. Total discharge in a larger or more complex basin will be higher but more spread out because some water travels short distances, while other water travels longer distances.

C What Are Tributaries and Drainage Networks?

Streams have a main channel fed by smaller subsidiary channels called *tributaries*. Each tributary drains part of the larger drainage basin, and the combination of tributaries and the main stream forms a *drainage network*. The response of a stream to precipitation is influenced by the number and size of its tributaries, as well as slope, soil conditions, amount of vegetation, and other factors.

12.03.c1

1. In this drainage network (◄), smaller tributaries join to form larger drainages, which join to form even larger drainage basins. The drainage network has a branched appearance similar to a tree, with numbers indicating the *stream order*. A *first-order stream* (1 on the figure) has no tributaries, but two such streams can join to produce a *second-order stream* (2), and so on.

2. Runoff in this type of drainage network, with many branches (stream orders), is carried by many channels. As a result, the system responds more slowly and with a smaller discharge peak to a precipitation event than a drainage system with fewer stream orders.

12.03.c2

3. A drainage network with fewer tributaries (stream orders) responds more quickly to an identical storm event (◄). The resulting increased discharge causes more erosion of sediment along the stream, more transport of sediment out of the area, and potentially more flooding.

D How Does Geology Influence Drainage Patterns?

The patterns that streams carve across the land surface are strongly influenced by the geology. Channels form preferentially in weaker material and so reflect differences in rock type and the geometry of folds, faults, joints, and other structural features. There are a number of drainage types, including the three shown below.

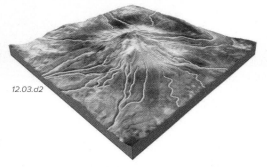

12.03.d1

12.03.d2

12.03.d3

1. *Dendritic Drainage Pattern*—Where rocks have about the same resistance to erosion, or if a drainage network has operated for a long time, streams can form a treelike, or dendritic, drainage pattern.

2. *Radial Drainage Pattern*—On a fairly symmetrical mountain, such as a volcano or resistant pluton, drainages flow downhill and outward in all directions (i.e., radially) away from the highest area.

3. *Structurally Controlled Pattern*—Erosion along faults, joints, or tilted and folded layers can produce a drainage that follows a layer or a structure and then cuts across a ridge to follow a different feature. The map pattern of the drainages can look like a trellis.

North American Drainages

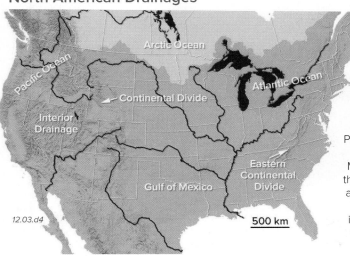

12.03.d4

4. Colors on this map show areas of the land that drain into different parts of the sea. Boundaries between colors are drainage divides, the best known of which is the *continental divide*, separating drainages that flow westward into the Pacific Ocean from those that flow east and south into the Gulf of Mexico. Other drainages flow into the Arctic Ocean, and some drainages in the western United States have interior drainage (they flow into low continental areas and do not reach the sea).

Before You Leave This Page

☑ Sketch and describe the variables plotted on a hydrograph and what this type of graph indicates.

☑ Describe how the shape and slope of a drainage basin affect discharge.

☑ Sketch or describe how the distribution of tributaries influences a stream's response to precipitation.

☑ Sketch three kinds of drainage patterns, and discuss what controls each type.

12.3

12.4 How Do Streams Transport Sediment and Erode Their Channels?

STREAMS ERODE BEDROCK and loose material, transporting the liberated material as sediment and as chemical components dissolved in the water. The sediment is deposited when the stream can no longer carry the load, such as when the current slows or the sediment supply exceeds the stream's capacity to carry the sediment.

A How Is Material Transported and Deposited in Streambeds?

Moving water applies force to a channel's bottom and sides, and can pick up and transport particles of various sizes: clay, silt, sand, cobbles, and boulders. The amount of sediment carried by the stream, including material chemically dissolved in solution, is the *sediment load*.

12.04.a1

1. Fine particles (silt and clay, collectively referred to as mud) can be carried *suspended* in the moving water, even in a relatively slow current. This material is the *suspended load*.

2. Sand grains can roll along the bottom or be picked up and carried down-current by bouncing along the streambed—the process of *saltation*.

3. Larger cobbles and boulders generally move by rolling and sliding, a process called *traction*, but they only move during times of high flow. The largest of these clasts (pieces of rock broken off from larger rocks) can be briefly picked up, but only by extremely high flows. The size of the largest particles that can be transported is called the stream's *competence*.

4. Material that is pushed, bounced, rolled, and slid along the streambed is the *bed load*. If the amount of sediment exceeds the stream's capacity to carry it, as when velocity drops, the sediment is deposited. The balance between transport and deposition shifts as conditions change, and grains are repeatedly picked up and deposited.

5. Some chemically soluble ions, such as calcium, sodium, and chlorine, are *dissolved* in and transported by the moving water. They constitute the *dissolved load*.

B What Processes Erode Material in Streams?

Moving water and the sediment it carries can erode bedrock, sediment, or other material that it flows past. Erosion occurs along the base and sides of the channel and can fragment and remove sediment within the channel. The silt, sand, and larger clasts carried by the water enhance its ability to erode.

1. Sand and larger clasts are lifted by low pressure created by water flowing over the clast tops. They can also be pushed up by turbulence. Once picked up, the grains move downstream and collide with obstacles, where they chip, scrape, and sandblast pieces off the streambed—the process of *abrasion*. Abrasion is concentrated on the upstream side of obstructions, such as larger clasts or protruding bedrock.

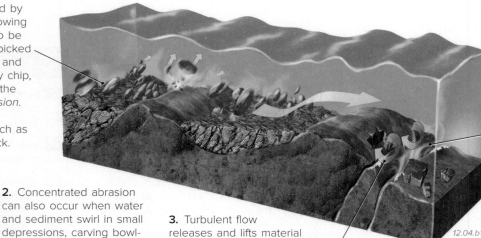

12.04.b2 McKinney Falls SP, TX

12.04.b1

2. Concentrated abrasion can also occur when water and sediment swirl in small depressions, carving bowl-shaped pits called *potholes* (◄). These are tens of centimeters across.

3. Turbulent flow releases and lifts material from the streambed, particularly pieces loosened by fracturing, boundaries between rock layers, and other discontinuities.

4. Soluble material in the streambed, such as salt, can be removed by *dissolution*. Most dissolved material in streams, however, comes from groundwater that has flowed into the stream. Other dissolved material mostly comes from soluble rock layers along the stream, but some also comes from farms and other sources.

C How Does Turbulence in Flowing Water Affect Erosion and Deposition?

Water, like all fluids, has *viscosity*—resistance to flow. Viscosity and surface tension are responsible for the smooth-looking surface of slow-moving streams. As the water's velocity increases, the flow becomes more chaotic or *turbulent,* and the water can pick up and move material within the channel.

1. All streams have parts that are more turbulent and parts that are less turbulent. Water moving smoothly in parallel layers is called *laminar flow*. In such smooth-appearing water, viscosity limits chaotic flow (turbulence). Fast flow and rough, obstruction-filled stream bottoms promote *turbulent flow*.

2. Moving water has momentum, so it tends to keep moving in the same direction unless its motion is perturbed, such as where the water encounters a bend in the stream. In many cases, water can flow smoothly over somewhat uneven surfaces.

Slower, Less Turbulent

Faster, More Turbulent

12.04.c1

3. As water velocity increases, viscosity is less able to dampen chaotic flow, and the water flow becomes more complex, or turbulent. As turbulence increases, swirls in the current, called *eddies,* form in both horizontal and vertical directions.

4. Fast-moving water has numerous eddies where flow strays from the downstream direction.

5. Near the bottom of the stream, up-current eddies can overwhelm gravitational force and lift grains from the channel. Turbulence, in general, increases the chance for grains to be picked up and carried in the flow.

D How Do Erosion and Deposition Occur in Streams Confined Within Bedrock?

Many streams, especially those in mountainous areas, are carved into bedrock and are referred to as *bedrock streams*. If the bedrock is relatively hard, the shape of the stream channel is controlled by the geology. If bedrock consists of softer material, such as easily eroded shale, then the material will have less control over the shape and character of the stream channel.

Erosion

1. The steep gradients and strong, turbulent flow typical of mountain streams (▶) erode down into the channel faster than the stream can erode outward, against its sides. The bed load of sand, cobbles, and boulders helps break up and erode the bedrock channel. Rapid changes in gradient, as occur along waterfalls and rapids, increase water velocity, turbulence, and the ability of the moving water to erode the channel.

12.04.d1 Nepal

2. As a result, steep bedrock streams commonly incise deep channels (▶). They can have relatively straight sections, initially controlled by the location of softer rock types, faults, or other zones that are more easily eroded than surrounding rocks. Once formed, such hard-walled canyons, like the aptly named Black Canyon of the Gunnison, make it difficult for the stream to shift its position.

12.04.d2 Black Canyon of the Gunnison, CO

Deposition

12.04.d3 Marble Canyon, AZ

3. Deposition in bedrock channels occurs where the water velocity decreases, for example along the stream banks during flooding or in pools behind rocks or other obstacles. Rocks and sediment constrict this river (◄), forming a pool of less turbulent water upstream. During floods, sediment is deposited in slow-moving eddies on the flanks of this pool, but such sediment is vulnerable to later erosion and is therefore very transient.

Before You Leave This Page

☑ Sketch and describe how a stream transports solid and dissolved material.

☑ Sketch and explain the processes by which a stream erodes into its channel and which sites are most susceptible to erosion.

☑ Sketch and describe turbulent flow.

☑ Describe some aspects of erosion and deposition in bedrock channels.

12.4

12.5 How Do Stream Systems Change Downstream or Over Short Time Frames?

STREAM SYSTEMS BECOME LARGER as more tributaries join the drainage network. As a stream flows downstream, it generally increases in size, discharge, and the amount of sediment it carries. A stream changes over short time spans, for example, after a storm, from winter to summer, from year to year, and from century to century.

A How Does the Gradient of a Stream Change Downstream?

The character of a stream changes as the stream flows downhill from its *headwaters*, where it starts, to its *mouth*, where it ends. The flow direction of a stream from high elevations to lower ones is referred to as being *downstream*.

Gradient

1. The change in a stream's elevation downstream, when viewed graphically, is called the stream's *profile*. The profile of most streams is steep in the headwaters, gradually becoming less steep downstream toward the mouth. The steepness is also called the *gradient*, which is defined as the change in elevation for a given horizontal distance.

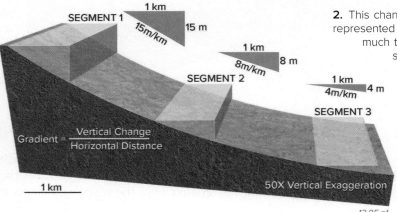

Gradient = Vertical Change / Horizontal Distance

SEGMENT 1 — 1 km — 15m/km — 15 m
SEGMENT 2 — 1 km — 8m/km — 8 m
SEGMENT 3 — 1 km — 4m/km — 4 m

1 km

50X Vertical Exaggeration

12.05.a1

2. This change in gradient downstream is represented by the blue triangles, which show how much the stream drops for a given length of stream (in this case, 1 km). A steeper gradient means the stream drops more over the same horizontal distance. We express gradient as meters per kilometer, feet per mile, degrees, or as a percentage (e.g., 4%). Here, gradient is calculated for three segments. It varies from 15 m/km to 4 m/km and decreases downstream. The vertical scale of the triangles is exaggerated compared to the horizontal scale.

B What Controls the Profiles of Streams?

Streams and other agents of transportation erode mountains and carry the sediment downhill, eventually depositing it in a basin or along the sea. The lowest level to which a stream can erode is its *base level*. The base level controls the topography along a stream, how a stream develops over time, and how it responds to change.

1. This terrain shows a typical drainage network consisting of mountainous headwaters, mid-elevation foothills, and a broad, low-elevation plain, ending at a shallow inland sea.

2. High above base level, steep gradients in the mountains cause streams to erode sharply into the bedrock. The terrain appears rough and may include deep canyons cut into bedrock.

3. Foothills in front of the mountains also experience erosion but have intermediate gradients and generally appear less rough.

4. Closer to base level, streams on the broad plain have a much lower gradient, and the surrounding landscape has less relief and appears relatively smooth. This plain has low relief because either it has been eroded down or its low parts have filled with sediment. In this case, it is some of both.

6. As shown by the side of the block, variations in roughness of the landscape reflect the decrease in gradient from the mountains to the broad plains. A profile down the channel of any stream in this area is less irregular than the rough topography defined by the ridges and canyons. Most rivers and large streams have a fairly smooth, concave-up profile.

5. A stream cannot erode below sea level. In this terrain, sea level represents the base level. In general, base level for a stream is the ocean, a lake, or the bottom of a closed land basin (where the area has an internal drainage). For most stream systems, the *ultimate base level* is sea level (the ocean).

12.05.b1

Headwaters

Foothills

Stream

Broad Plain

Sea

C How Does a Stream Change Downstream or Over Time?

As tributaries join a stream, the amount of water in the stream generally increases, and other factors, such as the size of the channel and velocity of the water, also change. In addition, the amount of precipitation, snowmelt, and influx from springs and groundwater varies, both during a single year and over longer timescales of decades to centuries. As a result, streams exhibit changes along their length and from one time to another.

Channel Size, Water Velocity, Discharge, and Sediment Load

1. Streams erode bedrock and other materials and then transport the sediment downstream. Sediment can be deposited anywhere along the way or can be carried all the way to the mouth of the stream. The system shown here has a main stream fed by three main tributaries, labeled T1, T2, and T3. Small graphs around the map plot how parameters change down the stream, from the headwaters (H), past each tributary (T1, T2, and T3), to the start of a delta (D), and the stream's mouth (M).

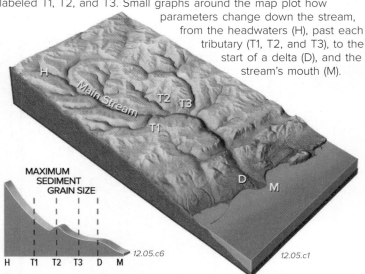

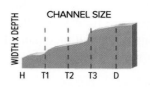
CHANNEL SIZE

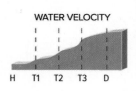
WATER VELOCITY

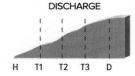
DISCHARGE

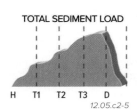
TOTAL SEDIMENT LOAD

12.05.c2-5

MAXIMUM SEDIMENT GRAIN SIZE

12.05.c6

12.05.c1

2. Downstream, the channel increases in overall *size* (◄), where size indicates a combination of the width and depth across the channel. Specifically, size means the *cross-sectional area* of the channel, obtained by multiplying the channel's width times its average depth.

3. The *velocity of water flow* increases downstream (◄), as a higher volume of water allows the water to flow more easily and faster through the channel. Velocity looks higher in steep streams, but *turbulence* and *friction* in such streams reduce the average velocity.

4. Since the cross-sectional area of the channel and the velocity of the water both increase, so does the total *volume of water* flowing through the stream (◄). The volume of water flowing through any part of the stream per unit of time is the *discharge* and is calculated by multiplying the water velocity times the cross-sectional area.

5. The total amount of sediment that the stream is carrying, the stream load, increases downstream (◄), until large amounts of sediment begin to be deposited within the delta and near the mouth.

6. As the gradient of the stream decreases from the headwaters to the mouth (▲), the *maximum size of sediment* that the stream carries decreases. Also, abrasion during transport reduces the size of clasts. For both reasons, coarse material is more common in the headwaters than it is near the mouth.

Changes in Flow Over Time

7. The amount of water flowing in a stream can vary throughout the year. Examine the hydrograph to the right (►), which plots discharge for one stream during different seasons. Other streams have other patterns than shown here, but this one is typical for streams in some parts of the United States.

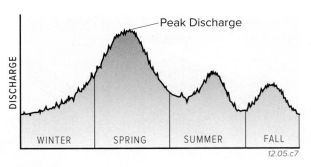
Peak Discharge
DISCHARGE
WINTER   SPRING   SUMMER   FALL
12.05.c7

8. In the graph above, the lowest discharge is in the winter, when some streams and lakes are frozen and precipitation occurs as snow and as gentle winter rains. Discharge is highest during spring, when the snow and ice melt, adding meltwater to the stream. The highest value on the plot (during the spring) is the *peak discharge*, an important aspect to consider when planning for floods or for the filling of reservoirs.

9. After the most intense period of snowmelt, discharge in the stream decreases throughout the rest of the spring and into the early summer. An increase in rainfall due to a yearly summer-thunderstorm season causes a short-lived increase in discharge in the middle of the summer and fall. This scenario is one possible explanation for a hydrograph with this shape.

10. A stream that flows all year, like the one plotted here, is a *perennial stream*. No place has rain all of the time to keep a stream flowing, so some water in a perennial stream must be supplied by groundwater flow into springs, by a melting snowpack, by a lake, or by some combination. Some streams do not flow during the entire year, but only during rainstorms and spring snowmelt. Such a stream that only flows intermittently is an *ephemeral stream*.

Before You Leave This Page

- ✓ Describe and sketch how to calculate a gradient for a stream and how gradient changes downstream.
- ✓ Describe how channel size and other factors change downstream, and explain why discharge might change from season to season.

12.5

12.6 Why Do Streams Have Curves?

ALL STREAMS HAVE CURVES OR BENDS, ranging from gentle deflections to tightly curved but graceful meanders. Why are streams curved? What is inherent in the operation of a stream that causes it to curve? Curves and bends are a natural consequence of processes that accompany the movement of water in a stream.

A What Is the Shape of Stream Channels in Map View?

All streams have bends, but not all bends are the same. Some are gentle, open arcs, where the stream veers slightly to one side and then the other, whereas others are tight loops. The shape of a stream in map view can be thought of as having two main variables: whether there are single or multiple channels and how curved the channels are. The amount that a channel curves for a given length is its *sinuosity*.

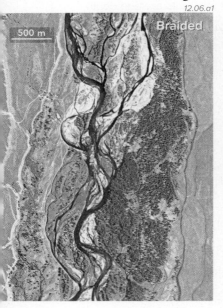
12.06.a1

 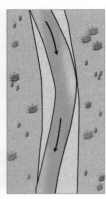 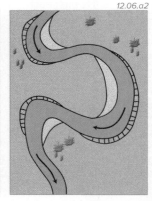
12.06.a2

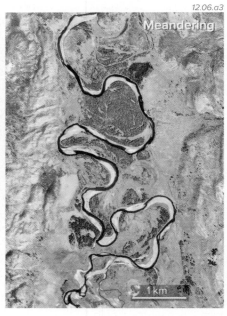
12.06.a3

▲ *Braided streams* are characterized by a network of interweaving, sinuous channels, but the overall channel can be fairly straight.

▲ Many streams consist of a single channel that is gently curved. This type of stream is referred to as having *low sinuosity*.

▲ *Meandering streams* have channels that are very curved, commonly forming tight loops. Such streams have *high sinuosity*, and this type of bend is a *meander*.

B What Processes Operate When a Stream Meanders?

Channels in soft materials, especially for streams with low gradients, generally do not have long, straight segments. Instead, they flow along sinuous paths. Curves or meanders cause differences in water velocity in the channel and reflect a balance between deposition and erosion, as illustrated below for a meandering stream.

1. Small graphs show profiles across the channel in different locations. Arrows in the channel show relative velocity (longer arrows are faster). In fairly straight segments, the channel is nearly symmetric (not deeper on one side than the other). The current is fastest in the center of the channel and slowest along the banks. In such straight segments, sediment can be deposited along the channel margins where velocity is lowest, and erosion can occur in the middle of the channel where velocity is highest.

2. Where the stream is curved, the channel becomes asymmetric (is shallower on one side than the other). The channel is shallower and the water velocity is lower on the inside of a bend. This causes sediment to be deposited on the inside of the bend in a crescent-shaped deposit of loose sediment called a *point bar* (◄).

3. The channel is deeper, and water flows faster on the outside of the bend. Also, inertia causes the force of the water to be directed toward the outside of a bend. These factors cause the outside bend to be eroded into a steep streambank called a *cutbank* (▼). Erosion of the cutbank can balance deposition on the point bar, keeping the channel width fairly constant.

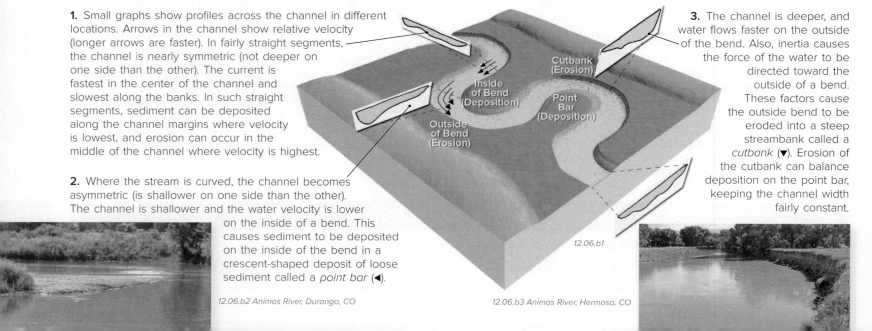

12.06.b1

12.06.b2 *Animas River, Durango, CO*

12.06.b3 *Animas River, Hermosa, CO*

C   How Do Meanders Form and Migrate?

Meanders are landforms produced by migrating rivers and smaller streams, and are extremely common in streams that have low gradients. Meanders have been extensively studied in the field and simulated in large, sand-filled tanks. In the laboratory, water is initially directed down a straight channel in fine sand. Almost immediately, the water begins to transform the straight channel into a sinuous one, similar to the sequence shown below.

1. A curve starts to form when a slight difference in roughness on the channel bottom causes water to flow faster on one side of the channel than on the other.

2. The side of the channel that receives faster flow erodes faster, creating a slight curve. The faster moving current slightly excavates the channel bottom, deepening the outside of the bend, forming deeper areas called *pools*.

7. Once formed, a curve continues to affect the flow by causing faster flow and increased erosion on the outside of the bend. Some secondary currents develop in the bend area and further excavate the pools, speeding flow and enhancing the cutbank. This type of system, where a feature or process, once started, affects the system in such a way that it results in even more of the same (in this case erosion), is called a *positive feedback*.

10. Meanders sometimes join as they migrate toward each other, in the direction of the yellow arrows. This cuts off the meander.

11. The narrow neck of a looping meander can also get cut off during a flood event, when the stream rises above the channel and across the floodplain, connecting two segments of the stream. In either case, the part of the meander that is abandoned is a *cutoff meander*.

3. The overall discharge in the stream is constant, so the deeper channel on the outside of a bend takes more water, leaving less water for the other side. The water on the inside of the bend becomes shallower and slower.

4. The sediment carried by the slower water on the inside of the bend is dropped and deposited on a *point bar*, depicted here as sand-colored material.

5. Erosion scours the opposite (outside) bank of the channel, forming a *cutbank*.

6. Through this process, each meander begins to preferentially erode its banks toward the outside. This causes the stream to migrate toward the sides and downstream, as shown by the small orange arrows.

8. As meanders migrate back and forth across the lowlands, they continuously erode and deposit the loosely bound sediment in the floodplain and channel. This is the main way in which a *floodplain* forms, and the old meanders remain as curved *scars* on the floodplain or on the point bar.

9. Meanders migrate until they encounter a resistant riverbank, until the volume and velocity of flow drop too low for erosion to continue, or until two parts of a meander intersect.

12. Cutoff meanders formed in either way (10 or 11) are initially filled with water, forming isolated, curved lakes, called *oxbow lakes*.

12.06.c1

Interfering with Sinuosity

Streams attain, through natural processes, their characteristic sinuosity, which represents the interplay between variations in channel depth, water velocity, erosion, deposition, and transport of sediment. Humans can upset this balance by straightening rivers and eliminating their natural variability. These engineering solutions (▶) often cause trouble downstream because they upset the dynamics and equilibrium of the system. Streams that have been channelized may exit the channelized segment with a higher velocity, lower sinuosity, and less sediment than is natural. Areas downstream of the channelized segment, therefore, can experience extreme erosion and destruction

of riverbank property. In the photograph below, find a point bar and a cutbank, and note which part of the bend (outside or inside) is being "protected." The view is downstream, where more erosion will occur on the cutbank.

12.06.t1 Animas River, Hermosa, CO

Before You Leave This Page

✓ Sketch and describe the difference between braided, low-sinuosity, and high-sinuosity (meandering) streams.

✓ Sketch or describe how velocity and channel profile vary in a meandering stream, and what features form along different parts of bends.

✓ Sketch or describe the evolution of a meander, including how a cutoff meander forms and how it can lead to an oxbow lake.

12.6

12.7 What Features Characterize Steep Streams?

STREAMS CAN BEGIN in almost any setting, from a small hill in a pasture to a large, snowy mountain. In either case, the place where a stream starts is its *headwaters*. In its headwaters, a stream is fed by rain, snowmelt, groundwater, or some combination of these. The headwaters of many large rivers are in mountains, where streams are steep and actively erode the land with turbulent, fast-moving water, producing distinctive landforms.

A How Do Streams Start?

A stream does not start with a fully formed channel full of water, but instead it grows incrementally as surface runoff becomes concentrated into channels. Smaller channels join others until a stream forms.

Rainwater causes *splash erosion* as it hits the ground, and water flowing over the surface as *runoff* causes erosion. Water tends to accumulate in natural cracks and low spots, such as these small channels, rather than spreading uniformly across the land.

Concentrated flow erodes or dissolves materials, especially those that are weak or loose, eventually carving a small channel or *gully*. Once formed, a channel accommodates runoff within its small drainage basin. The increased flow causes further *gully erosion* and channel deepening. *Headward erosion* cuts into the area between the channels and can lead to *stream capture*, the natural diversion of water from one stream into another.

Channels occur at all scales. Microscopic channels feed into small channels that feed into larger ones, ultimately forming a stream. The spacing and geometry of the channels are influenced by the steepness of the slope, type of material in the slope, type and density of vegetation, and other factors.

12.07.a1

B What Landforms Characterize the Headwaters of Mountain Streams?

Mountain streams begin in bedrock-dominated areas with relatively high relief and, in many cases, high elevation. The energetically moving water wears rock down and sculpts the bedrock into steep landforms with moderate to high relief.

12.07.b2 Beartooth Plateau, WY

Headwaters

12.07.b1

Lake

1 km

12.07.b7 Ice Lake, Silverton, CO

1. The headwaters of many streams are in high mountainous areas (▲), where the streams derive their water from rainfall, melting ice and snow, or mountain springs. Other streams originate in lower, flatter areas and are supplied by precipitation, lakes, springs, or the joining of small, local channels.

5. Lakes are common in mountains where water is impounded by some obstruction, such as a landslide, or where water fills a natural low spot (▲). If a lake is created by a human-constructed dam, it is a *reservoir*.

4. In mountains, streams generally have an abundant supply of sediment, much of which moves down steep slopes flanking the stream (▼).

2. Streams in mountains typically have steep gradients that result in fast-moving, turbulent flow, as shown in the lower left photograph. Many have eroded down, or *incised*, into bedrock, and they commonly are energetic, able to carry large clasts, like boulders (▼).

3. With time, mountain streams begin to widen their valleys and deposit a thin, winding ribbon of stream gravels along which they flow (▼).

12.07.b3 Boulder Canyon, MT

12.07.b4 Central CO

12.07.b5 ANWR, AK

12.07.b6 Tibet-Nepal Border Region

C How Do Rapids and Waterfalls Form?

Streams that flow in an area of mountains and hills commonly have *rapids*, a segment of rough, turbulent water along a stream. Some segments can be so steep that water cascades through the air, forming a *waterfall*.

12.07.c1 Grand Canyon, AZ

12.07.c2 Marble Canyon, AZ

12.07.c3 Gullfoss, Iceland

Most rapids (▲) develop when the gradient of a stream steepens or the channel is constricted by narrow bedrock walls, large rocks, or other debris that partially blocks the channel. Many rapids form where tributaries have deposited fans of debris that crowd or clog the main channel. These obstructions cause water to flow chaotically over and around obstacles, creating extreme turbulence.

The obstruction forming a rapid generally causes water upstream to slow down and "pile up," producing a relatively smooth and slow-moving stream segment called a *pool*. The photograph above, looking downstream, shows a pool upstream of a rapid marked by whitewater. Many streams have this characteristic alternation of rapids and pools, and so are called *pool-and-riffle streams*.

Most waterfalls are awe-inspiring, as water cascades over a ledge or cliff (▲). Where a stream has an abrupt change in gradient, such as at a waterfall, it is called a *nickpoint*. A nickpoint can be distinct or more subtle, and generally indicates some difference in erosion above and below the nickpoint. With continued erosion, the nickpoint, and in this case a waterfall, migrates upstream over time.

D What Features Are Characteristic of Steep, Sediment-Rich Streams?

As mountain streams flow toward lower elevations, they interact with tributaries and commonly decrease in gradient as they pass through foothills or mountain fronts. Steep, sediment-rich streams commonly have a braided appearance if they have enough room to spread out, such as when they leave the mountains. They are called *braided streams*.

12.07.d1 New Zealand

When a mountain stream reaches less confined spaces, it commonly spreads out in a network of sediment-filled channels that split apart and rejoin, producing a *braided pattern* (◀). Such a stream is a *braided stream*. These channels are not strongly incised, so the stream spreads out. As it does, the stream deposits sediment along its channel and over a broad floodplain.

Braided streams form when the stream has a relatively high sediment load dominated by sand and larger sediment. Sand and gravels are the dominant clasts in this braided stream (▶), but some braided streams also carry finer materials, such as mud and silt derived from glaciers and other sources. Overall, braided streams are relatively mud-poor when compared with streams that have meanders, unless a braided stream drains glaciers.

12.07.d3 Savage River, AK

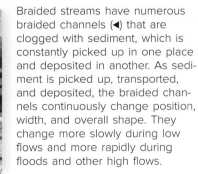
12.07.d2 Waiapu River, New Zealand

Braided streams have numerous braided channels (◀) that are clogged with sediment, which is constantly picked up in one place and deposited in another. As sediment is picked up, transported, and deposited, the braided channels continuously change position, width, and overall shape. They change more slowly during low flows and more rapidly during floods and other high flows.

These braided-stream deposits contain partially rounded cobbles and pebbles in a sand-dominated matrix. The stream can transport these large clast sizes because it has a steep gradient and carries large amounts of turbulent water during the spring snowmelt. These deposits also contain silt-size particles, but most silt is carried farther downstream.

12.07.d4 Athabasca Parkway, Alberta, Canada

Before You Leave This Page

☑ Describe some of the landforms associated with the headwaters of mountain streams, including rapids and waterfalls.

☑ Describe the characteristics of braided streams and the types of sediment they carry and deposit.

12.7

12.8 What Features Characterize Low-Gradient Streams and Deltas?

IF A STREAM CROSSES areas of low relief, the gradient of its channel decreases and the stream may spread out once it is no longer confined by a narrow canyon or valley. As a gradient of a stream decreases, the stream deposits its coarser material but continues to carry mostly clay to sand-sized sediment. The stream deposits and reworks (picks up and transports) this sediment, producing distinctive landforms. A similar process occurs in a delta, where a slowing stream spreads out into many channels and deposits sediment on the gentle landscape.

A What Landforms Typify Streams with Low Gradients?

Many streams flow across plains that have gentle overall slopes. Such streams reflect their environs, being dominated by the erosion, transport, and deposition of relatively fine-grained sediment. The features characteristic of these streams, which mostly have a single, main channel, occur at all scales, from those along small creeks to those along the Mississippi River. Features include meanders, floodplains, and low stream terraces. On these two pages, we explore two *meandering rivers*: the Animas River of southwestern Colorado and the Mississippi River of the central U.S.

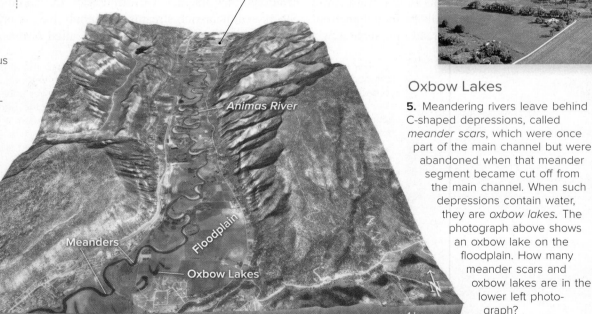

12.08.a3 Pecatonica, IL

One Main Channel

1. Streams on gentle plains usually occupy a single channel rather than being braided. This single-channel characteristic is linked to the gentle downstream gradient of the stream and its floodplain. This figure shows the low-gradient Animas River, near Durango, Colorado. This river occurs on a gentle plain within a mountainous region, so it is important to focus on the characteristics of a stream rather than its surrounding environment. Farther upstream, this river is steep and confined to a narrow and deep bedrock canyon.

Meanders

2. Rivers on gentle plains typically flow in noticeably curved paths, or meanders (▼). Meanders can be gentle curves or can sharply loop back on themselves, with the two meander segments separated only by a narrow strip of land.

Floodplain

4. All streams on gentle plains have floodplains (▶) beside the channel. Floodplains are areas adjacent to a stream channel regularly inundated by floodwaters, which deposit sand, silt, and clay particles. The channel is not necessarily in the middle of the floodplain.

Animas River

Meanders

Floodplain

Oxbow Lakes

12.08.a1 Animas River, Hermosa, CO

1 km

Oxbow Lakes

5. Meandering rivers leave behind C-shaped depressions, called *meander scars*, which were once part of the main channel but were abandoned when that meander segment became cut off from the main channel. When such depressions contain water, they are *oxbow lakes*. The photograph above shows an oxbow lake on the floodplain. How many meander scars and oxbow lakes are in the lower left photograph?

Stream Terraces

6. Many streams have older, stranded floodplains that form gently sloping *terraces* perched above and outside the current floodplain (▼).

12.08.a4 Thompson River, British Columbia, Canada

12.08.a2 Ikpikpuk River, AK

Point Bars

3. Most meandering rivers have crescent-shaped deposits of sand and gravel that parallel the inside bend of a meander (◀). Such point bars are typically visible as patches of bare, recently deposited sediment. How many point bars can you identify in the photograph to the left?

Meander Scars

12.08.a5 Pecatonica, IL

7. Meander scars are exposed as low, curved ridges, lines of vegetation, or curved dry or water-filled depressions (▲).

12.08.a6

8. On this computer-generated perspective (▲), the very broad floodplain of the Mississippi River has countless crescent-shaped scars of ancient meanders, abandoned by the shifting of the river. Note also the sand-colored point bars next to the active channel.

9. The dark, C-shaped areas are cutoff meanders (meander scars) that are filled with water—oxbow lakes. Many of these lakes are kilometers long.

B | What Happens as a Stream Approaches Base Level?

Several landscape-building processes occur when a stream enters the ocean, lake, or a temporary base level. Large rivers, like the Mississippi River, pump fresh water far into the ocean and carry fine sediment out to sea. They deposit coarser sediment as soon as the current slows, forming a *delta* along the shoreline.

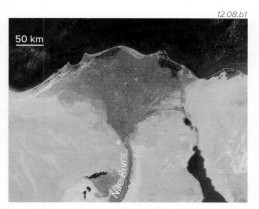

12.08.b1

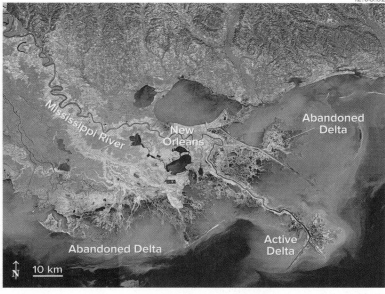

12.08.b2

4. Over the last 7,000 years, the Mississippi has created and then abandoned at least six huge mounds of sediment, each of which marks a former location of the river mouth and its associated delta; some of these abandoned deltas are labeled on the figure. A new delta, the active delta, is forming where the Mississippi River currently enters the Gulf of Mexico. Eventually, the river will shift and abandon this bird-foot-shaped delta too.

1. What is a delta? This satellite view (▲) shows the green, triangular-shaped delta formed where sediment from the Nile River is deposited out into the Mediterranean Sea.

2. A delta also forms where the Mississippi River meets the Gulf of Mexico near New Orleans, as shown in the large satellite image above (▲). As it approaches the coast, the river slows down and changes from a meandering river within a broad floodplain to a series of smaller channels that branch and spread out in various directions. This branching drainage pattern is a *distributary system*.

3. Dark blue colors on this image indicate clear, deeper waters of the Gulf, whereas lighter blue areas contain suspended sediment and mostly have shallower water. Sediment from the river accumulates and builds up the delta, which is eroded by waves and by underwater slumps of the steep, unstable delta front.

Before You Leave This Page

☑ Sketch or describe the features that accompany low-gradient rivers, explaining how each forms.

☑ Describe the characteristics of a delta and how one forms when a stream approaches its base level, such as an ocean or lake.

12.8

12.9 What Features Are Associated with Streams?

IN ADDITION TO FLOODPLAINS, streams are commonly flanked by other features, including *levees* and *stream terraces*. Some of these features, such as a natural levee, are formed by deposition, whereas others form when a stream erodes downward, or incises, into the land. Such downward erosion can form intricate, winding canyons, called *entrenched meanders*. Dams constructed across stream valleys cause the stream to change behavior above and below the dam.

A How Are Levees and Stream Terraces Formed?

Many streams are flanked by natural embankments, called *levees*, and by stream terraces, which are relatively flat benches that are perched above a stream and that stair-step up and outward from the active channel. Most terraces are composed of river-derived sediment and are essentially abandoned floodplains and alluvial plains, but others are cut directly into bedrock and form by erosion.

Levees

1. Along the edges of many channels is a raised embankment, or *levee*. Natural levees are created by the river, and humans construct artificial levees to try to keep floodwaters from spilling onto the floodplain.

2. During flooding, sediment-carrying floodwater rises above the channel and begins to spread out. As it does, the current slows and so deposits sediment in long mounds next to and paralleling the channel.

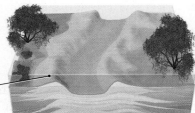

12.09.a1

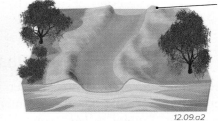

12.09.a2

3. When the flood recedes, sediment that was piled up next to the channel remains as an elevated rise or levee. Levees are barriers to water flow from the channel to the floodplain, and from the floodplain back into the channel after a flood. Human-constructed levees contain rocks, dirt, and cement, but they can fail.

Highest Terrace

Middle Terrace

Floodplain

12.09.a3 Moose, WY

Stream Terraces

4. Terraces form a series of flat to gently sloping benches or steps, flanked by steeper slopes (risers). Terraces successively step up and away from the channel. The highest terrace may be tens of meters or more above the active channel. A low terrace may be only a meter or so above the channel and therefore often flooded, perhaps every couple of years.

5. This series of terraces (▶) are numbered from highest (1) to lowest (3). The modern floodplain is labeled (F). After examining the sequence of figures at the bottom of the page, revisit this image and determine which of these terraces formed first and which one formed last.

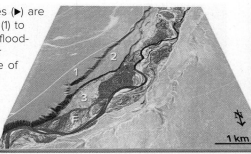

1 km

12.09.a4

| First Stage (oldest) | Second Stage | Last Stage (youngest) |

12.09.a5

6. The first stage in terrace formation is deposition of sediment, such as on the floodplain (1) shown above. At this stage in its history, the river is nearly at the same level as the floodplain (i.e., is not incised). The flat surface of the floodplain will later become the flat part of a terrace.

7. A change in conditions, such as a drop in base level, causes the river to downcut through its floodplain deposits, forming a second, lower floodplain (2). Remnants of the first floodplain are stranded on both sides of the river (1) and are now a terrace. If high enough, the terrance is unlikely to be flooded again.

8. With further downcutting, the river abandons the second floodplain (2), creating a third, even lower one (3). The oldest floodplain (1) is now high and dry. The series of downcutting events creates a stair-step appearance to the land.

B | How Are Entrenched Meanders Formed?

The landforms we know as meanders form only in loose sediments, like those on floodplains. However, in the Four Corners region of the American Southwest and in areas west of the Appalachian Mountains, meanders with typical sweeping bends are deeply incised in hard bedrock, forming some puzzling canyons. What do these deeply incised, winding canyons, called *entrenched meanders*, tell us about the history of rivers and streams in these areas?

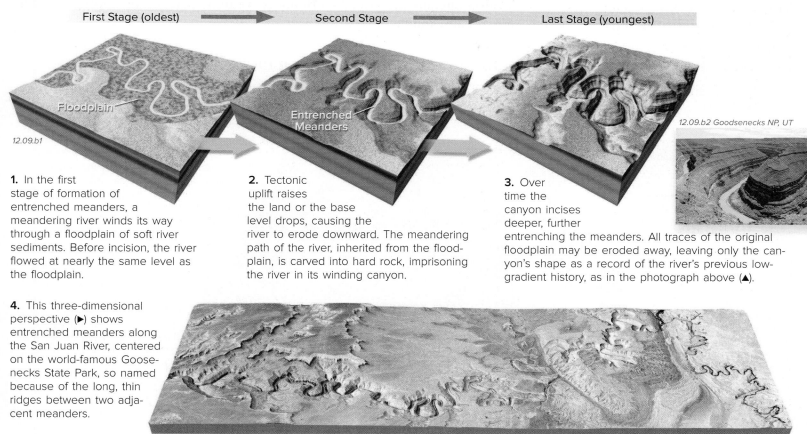

First Stage (oldest) ⟶ Second Stage ⟶ Last Stage (youngest)

Floodplain

12.09.b1

Entrenched Meanders

12.09.b2 Goosenecks NP, UT

1. In the first stage of formation of entrenched meanders, a meandering river winds its way through a floodplain of soft river sediments. Before incision, the river flowed at nearly the same level as the floodplain.

2. Tectonic uplift raises the land or the base level drops, causing the river to erode downward. The meandering path of the river, inherited from the floodplain, is carved into hard rock, imprisoning the river in its winding canyon.

3. Over time the canyon incises deeper, further entrenching the meanders. All traces of the original floodplain may be eroded away, leaving only the canyon's shape as a record of the river's previous low-gradient history, as in the photograph above (▲).

4. This three-dimensional perspective (▶) shows entrenched meanders along the San Juan River, centered on the world-famous Goosenecks State Park, so named because of the long, thin ridges between two adjacent meanders.

12.09.b3 San Juan River, UT

C | What Are the Depositional and Erosional Consequences of Dams?

Human-constructed dams provide hydroelectric power generation, water storage, or flood control, but they stop a stream's normal flow and transport of sediment. The reservoir behind the dam represents a *temporary base level* and so causes the stream to deposit sediment behind the dam, limiting the dam's longevity.

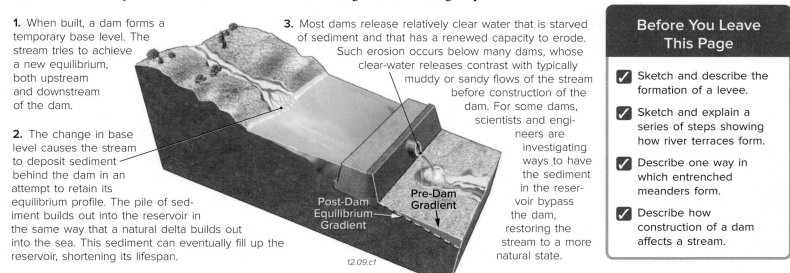

1. When built, a dam forms a temporary base level. The stream tries to achieve a new equilibrium, both upstream and downstream of the dam.

2. The change in base level causes the stream to deposit sediment behind the dam in an attempt to retain its equilibrium profile. The pile of sediment builds out into the reservoir in the same way that a natural delta builds out into the sea. This sediment can eventually fill up the reservoir, shortening its lifespan.

3. Most dams release relatively clear water that is starved of sediment and that has a renewed capacity to erode. Such erosion occurs below many dams, whose clear-water releases contrast with typically muddy or sandy flows of the stream before construction of the dam. For some dams, scientists and engineers are investigating ways to have the sediment in the reservoir bypass the dam, restoring the stream to a more natural state.

Post-Dam Equilibrium Gradient

Pre-Dam Gradient

12.09.c1

Before You Leave This Page

- ☑ Sketch and describe the formation of a levee.
- ☑ Sketch and explain a series of steps showing how river terraces form.
- ☑ Describe one way in which entrenched meanders form.
- ☑ Describe how construction of a dam affects a stream.

12.9

12.10 What Is and What Is Not a Flood?

THROUGHOUT HISTORY, PEOPLE HAVE LIVED along streams, especially along large rivers. Streams are sources of water for consumption, agriculture, and industry, and provide transportation routes and energy. Stream valleys offer a relatively flat area for construction and farming, but people who live along streams are subject to an ever-changing flow of water. High amounts of water flowing in streams often lead to flooding. In most parts of the world, flooding is a common and costly type of natural, and perhaps, human-caused disaster.

A What Is the Difference Between a Flood and a Normal Flow Event?

Streams are dynamic systems, and they respond to changes in the amount of water entering the system. When more water enters the system than can be held within the natural confines of the channel, the result is a *flood*.

12.10.a1 Hermann, MO

1. A flood occurs when there is too much water for the channel to hold. As a result, water spills out onto the adjacent land, usually inundating parts or all of the floodplain.

2. Flow in a channel, even when there is not a flood, may cause riverbank erosion. Such erosion can destroy structures built close to the stream and make the stream change position over time, turning what was floodplain into channel, and what was channel into floodplain. Many people will call such a destructive event a flood.

3. Human-constructed levees can sometimes protect property from flooding during large flood events. The levees, whether human constructed or natural, can also trap water on the floodplain after the peak flooding ends. The solution to this is to sometimes breach the levee after the flood, allowing the trapped water on the floodplain to flow back into the channel.

4. Large floods can expand the width of the floodplain by burying preexisting rocks and material with sediment deposited by the stream. Sediment beneath the floodplain includes old channel deposits in addition to floodplain silt and clay.

Normal, Bank-Full Flows

5. Normal (i.e., non-flooding) flows in streams can range from nearly dry to *bank-full*. Although there may be abundant water flowing down the channel, it is generally not considered a flood unless the water overflows the banks. A stream's natural floodplain is an excellent place to contain excess floodwaters—as long as it remains undeveloped by humans. Low-intensity development, such as a park, is often an appropriate use of a floodplain.

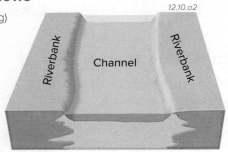
12.10.a2
Riverbank   Channel   Riverbank

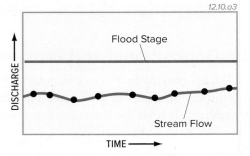
12.10.a3

6. This hydrograph shows a typical non-flood flow. The line labeled *Flood Stage* shows the amount of discharge required for the stream to overtop its banks and spill out onto the floodplain (i.e., a flood). During extended times of dry conditions, or at least weather that is normal for the region, hydrographs may show little change in stream flow over time, as shown here.

Flows During a Flood

7. When the amount of water in a stream exceeds the channel capacity, a flood occurs, inundating the floodplain. This hydrograph shows prolonged precipitation or snowmelt upstream that causes a flood event downstream, as represented by discharge greater than flood stage.

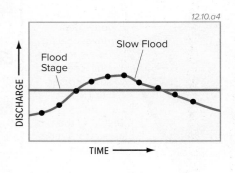
12.10.a4
DISCHARGE
Flood Stage
Slow Flood
TIME

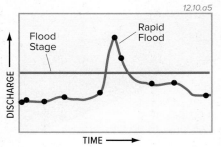
12.10.a5
DISCHARGE
Flood Stage
Rapid Flood
TIME

8. Intense rainfall can unleash a brief *flash flood*, with a rapid rise in water levels and an increase in discharge that lasts only a short time. Similarly, rapid onsets of flooding result from failure of a natural or constructed dam, but the resulting flows last longer.

B What Are the Causes of Flooding?

What causes discharge to exceed the channel's capacity? A simple answer is that there is more water in the channel than can be accommodated. This can be the result of natural processes or human-caused events.

Snowmelt

1. Flooding occurs when warming temperatures or rainfall melt snow and ice somewhere in the drainage basin.

112.10..b1 Norway

2. In the Northern Hemisphere, flooding from melting ice and snow occurs in the spring, from March to May. Heavy rain that coincides with melting can cause even worse flooding.

Local Heavy Precipitation

3. Some floods are caused by heavy rainfall over a short period of time, causing a brief, but dangerous, flash flood.

12.10.b2 White Canyon, UT

4. A thunderstorm upstream of this site sent a fast-rising, muddy flash flood down this desert drainage. Vehicles attempting to cross such floods can be washed downstream. Such a flash flood may only last minutes.

Regional Precipitation

5. Regional floods occur when abnormally high precipitation falls over a large area over days, weeks, or months.

12.10.b3 Tucson, AZ

6. Heavy regional rains caused by moisture from a former hurricane caused this normally dry river to destroy offices built in a risky place—on loose sediment of the floodplain and on an outside bend.

Effects of an Ice Dam

7. During the winter, streams in cold climates can freeze, forming large volumes of ice. When warmer temperatures cause melting, the ice can block stream flow, causing a flood.

12.10.b4 Grand Forks, ND

8. Spring snowmelt and effects of ice dams often cause flooding along the Red River of the North, here still frozen before a flood.

Volcanic Eruption

9. If volcanic peaks are covered with snow when the volcano erupts, the snow will melt and cause flooding or catastrophic mudflows.

12.10.b5 Muddy River, WA

10. A volcanic eruption on snowy Mount St. Helens caused flooding and mudflows downstream, destroying this bridge.

Dam Failure

11. Dams occur as both natural and human-constructed features. Poorly engineered dams have failed, releasing floodwaters into downstream channels.

12.10.b6 Eastern ID

12. Catastrophic release of water during failure of the earthen Teton Dam, Idaho, in 1976 flooded towns downstream.

Urbanization

12.10.b7 Dalian, China

13. When urban growth replaces natural lands, the area responds differently to precipitation and snowmelt. Urbanization increases runoff by increasing the amount of impermeable surfaces.

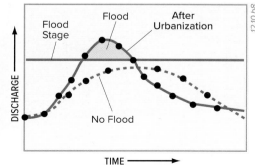

12.10.b8

14. This hydrograph shows that stream flow, for the same amount of water, became more abrupt and extreme after urbanization, causing a stream to rise above flood stage.

Before You Leave This Page

☑ Sketch and describe a flood that overflows the channel versus a flow that stays within the channel. Include hydrographs in your sketches.

☑ Sketch the difference between a hydrograph showing a regional flood and a local one of brief duration.

☑ Summarize some causes of flooding.

12.10

12.11 Where Is Groundwater Found?

A HUGE SUPPLY OF GROUNDWATER lies beneath Earth's surface. Groundwater occurs beneath all areas of the world but is far below the surface in some areas and very near the surface in others. Where does this water come from, and how does it find room to accumulate in the solid Earth beneath us?

A What Is Groundwater?

Groundwater is free water (exists as a liquid rather than being chemically bonded in minerals) that is beneath Earth's surface. Surface and near-surface rocks and soil can be relatively dry, but deeper parts are generally saturated with water. Groundwater is present in three different settings that reflect the types of rocks, sediment, and geologic features that host the water.

1. In sediment or sedimentary rocks, adjacent grains do not fit together, so there is some space between the grains. These spaces, called *pore spaces,* hold groundwater. Here, the tan objects are grains, the brown represents pore spaces, and the blue indicates pore spaces that are saturated (filled) with groundwater (▶).

12.11.a2

12.11.a1

3. Most groundwater occurs in pore spaces and fractures, but some resides in subterranean openings and caves. Caves can be filled or partially filled with water, or they can be completely dry. The rock shown here is a soluble limestone with wide fractures, bedding planes, and small cavities. Water passing through the rock dissolved soluble materials, widening fractures and bedding planes, ultimately forming open cavities (▶).

2. All types of rocks have fractures that provide openings (▲) in which groundwater can accumulate. If fractures are interconnected, the groundwater can flow. Here, a gray rock is cut by fractures that are unsaturated (brown) or filled with water (blue).

12.11.a3

B How Does Groundwater Accumulate?

Groundwater originates from precipitation and snowmelt that seeps from the surface down into the subsurface. The water accumulates in pores, fractures, and cavities within soil, loose sediment, and rock.

1. When rain falls on the surface or snow melts, the water can either evaporate, be absorbed by plant roots, flow downhill as runoff, or seep into the subsurface.

5. The water table can be deep below the surface, or it can intersect the surface in lakes, streams, or swamps.

12.11.b1

2. Water that soaks into the soil first passes through a part of the subsurface where most of the pore spaces are filled with air rather than water. This upper part, called the *unsaturated zone,* can be centimeters thick or can continue to depths of hundreds of meters. It can become completely dry during long periods without rain.

3. As water penetrates deeper into the subsurface, it eventually enters a zone where water fills nearly all the pore spaces and fractures. This zone is the *saturated zone* and is where most water occurs in the subsurface.

4. The top of the saturated zone is the *water table,* shown as a dashed red line. Below the water table, water fills and can flow through the interconnected pore spaces. Above the water table, some air remains in the pore spaces, and water within the pores can seep downward but is not connected enough to flow laterally.

C What Controls How Water Flows Through Rocks?

Water flows downhill, so the rate of groundwater flow is controlled by the steepness of the water table and two important properties of the material—*porosity* and *permeability*. These two properties are related to one another, but they are not the same thing. Porosity is a measure of how much water a rock can hold, but permeability indicates whether or how easily groundwater can flow through the rock.

Porosity

1. *Porosity* is the proportion of the volume of rock that is open space (pore space). It ranges from less than 1% to more than 50% and determines how much water a material can contain.

2. Well-rounded and well-sorted sediment usually has higher porosity than angular or poorly sorted sediment because round grains do not fit together as tightly. This jar of marbles (◄), analogous to well-rounded cobbles or sand grains, shows that a lot of pore space exists in such materials, provided the space is not filled with a natural cement.

12.11.c1

3. Sediment that is poorly sorted (►) tends to have less porosity because smaller grains fill the spaces between larger grains. Lower porosity also typifies sediment that has angular grains, whose corners help fill open spaces, or sedimentary grains that are held together by a natural cement, which fills in pore spaces.

12.11.c2

4. Clay-rich sediments and sedimentary rocks, like shale, consist of small particles shaped like plates or sheets (◄) that do not fit tightly together. There is abundant open space (porosity) between them, but such pores, like the clay particles, are very small, making movement of water difficult. Clay particles can become compacted or can swell when wet, reducing porosity.

12.11.c3

5. In igneous and metamorphic rocks, porosity is usually low because the minerals are tightly intergrown (►), leaving little free space. Some igneous rocks have less than 1% porosity. Fractures cutting any rock, however, open up narrow spaces and increase the porosity by some amount.

12.11.c4

Permeability

6. *Permeability* is a measure of the ability of a material to transmit a fluid. It is related to the size and interconnectedness of the pore spaces. Materials with low porosity usually have low permeability.

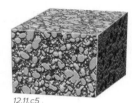

7. Loosely cemented gravel and sand commonly contain interconnected pore spaces (◄) that allow relatively easy groundwater flow. Such materials have *high permeability* and host groundwater in many areas, as represented by the blue in this figure.

12.11.c5

8. Fractures cut most rocks (►), opening spaces that typically represent a small volume of the rock and only slightly increase porosity. Well-connected fractures, however, allow water to flow and provide *higher permeability*. Fractures are the only significant permeability in granite and most other igneous rocks.

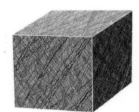

12.11.c6

12.11.c7

9. When clay particles compact, they tend to become aligned parallel to one another (◄). This decreases the porosity and causes the pore spaces to be very small. Shales and similar rocks will have very *low permeability*, or perhaps no permeability.

10. It is possible to have a highly porous rock with little or no permeability. A good example is a vesicular volcanic rock (►). The bubbles that once contained gas give the rock a *high porosity*, but most vesicles are not connected, so the rock has *low permeability*.

12.11.c8

11. Below are examples of high-permeability rocks. The conglomerate on the left has abundant sandy matrix between the pebbles and cobbles, and the rock on the right is permeable sandstone cut by fractures that are interconnected. Both examples allow water to accumulate in large quantities and move easily through the material. Permeability can be measured in the laboratory or tested in drill holes, and is expressed mathematically using an equation called *Darcy's law*. Geologists and hydrologists use this equation to model the flow of groundwater.

12.11.c9 Goldfield Mtns., AZ

12.11.c10 Sedona, AZ

Before You Leave This Page

☑ Sketch how groundwater accumulates and occurs in rock and sediment.

☑ Sketch and describe what the water table represents.

☑ Distinguish porosity and permeability, providing examples of materials with high and low values of each.

12.11

12.12 How and Where Does Groundwater Flow?

GROUNDWATER FLOWS BENEATH THE SURFACE in ways that are controlled by several key principles. The direction and rate of groundwater flow are largely controlled by the slope of the water table, the permeability of the materials, and the geometry and nature of the subsurface rock. Some rock types allow easy groundwater flow, whereas others essentially preclude any significant movement. In the figure below, examine how the geometry of the water table compares to the overlying surface topography.

A What Is the Geometry of the Water Table?

The water table is usually not a horizontal surface but instead has a three-dimensional shape that mimics the shape of the overlying land surface. The shape of the water table commonly has the equivalents of slopes, ridges, hills, and valleys. The shape of the water table controls which way the upper levels of groundwater flow.

1. In most environments, the water table typically has the same shape as the overlying land surface but is more subdued. Where the land surface is high, the water table is also high, but not quite as high. The similarity in shape between topography and the water table is less straightforward in some arid environments and in places where humans have pumped out groundwater faster than it can be replenished by precipitation.

2. The water table generally slopes from higher to lower areas. It generally is deeper below the surface under mountains than under lowlands, so its slope is less steep than that of the land surface. The shape of the water table is largely independent of the geometry of rock units through which the water table passes.

3. Groundwater just below the water table flows down the slope of the water table. Beneath the main hill in this example, it flows from left to right, from areas with a higher water table to areas with a lower water table. The blue arrows show flow directions of water right below the water table. Groundwater can also flow up or down in response to gravity and water pressures imposed from nearby.

12.12.a1

4. Where the water table is horizontal, for example near this lake, groundwater may flow very slowly or not at all. Deeper water may flow in directions different from near-surface water but this is not shown.

5. The terminology used to describe features of a water table is derived from terms we use for topography. A high part of the water table separating parts sloping in opposite directions is called a *groundwater divide*. Groundwater flows in opposite directions on either side of a groundwater divide.

6. Where the water table intersects the land surface, there may be lakes, wetlands, or a flowing stream. The stream in this figure occurs where the water table is at the surface. However, streams do not all necessarily coincide with the water table, because some flowing streams are underlain by unsaturated materials.

B What Controls the Rate of Groundwater Flow?

1. The rate of groundwater flow is typically measured in meters per day, but it can be much slower or faster. Rate is primarily controlled by permeability, which can vary by 12 orders of magnitude. The rate is controlled to a lesser extent by the steepness of the water table because groundwater flow is driven by the force of gravity. Other factors being equal, groundwater flows faster down a steep water-table slope and slower down a more gentle one. The slope of the water table is also called the *hydraulic gradient*.

12.12.b1

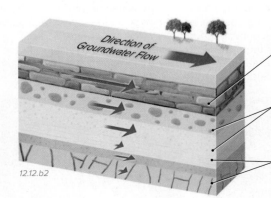

12.12.b2

2. The rate of groundwater flow is strongly controlled by the *permeability* of the rock type. In this diagram, flow is fastest in highly permeable cavernous limestone.

3. Flow is moderately fast in a porous conglomerate or well-sorted sandstone.

4. Flow is slower in shale, which has small pores, and in a granite with poorly connected fractures.

C   What Is an Aquifer?

An aquifer is a large body of permeable, saturated material through which groundwater can flow well enough to yield significant volumes of water to wells and springs. To be a good aquifer, a material must have high permeability, as occurs in poorly cemented sand and gravel, most sandstone, cavernous limestone, or highly fractured rocks of nearly any type. A material with low permeability does not make a good aquifer.

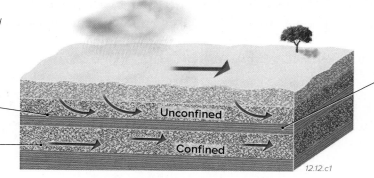

1. The most common type of aquifer is an *unconfined aquifer* where the water-bearing unit is open (not restricted by impermeable rocks) to Earth's surface and atmosphere. Rainwater or surface water can seep unimpeded through the unsaturated zone and into an unconfined aquifer.

2. A *confined aquifer* is separated from Earth's surface by rocks with low permeability. Here, a permeable sandstone aquifer is bounded above and below by layers of low-permeability gray shale.

3. A low-permeability unit, such as this layer of shale, can restrict flow and is referred to as an *aquitard*. An *impermeable* unit blocks flow completely. Such units are the opposite of an aquifer.

12.12.c1

D   How Are Wells Related to the Water Table?

A well is a hole dug or drilled typically deep enough to intersect the water table. If the well is within an aquifer, water will fill the open space to the level of the water table. This freestanding water can be drawn out by buckets or pumps.

12.12.d1

This well has been drilled from the land surface downward past the water table. The aquifer is unconfined and has filled the well with water to the height of the water table.

12.12.d2

In dry seasons, or during periods of high groundwater use, some wells run dry. This occurs when the water table drops and the well is no longer deep enough into the aquifer.

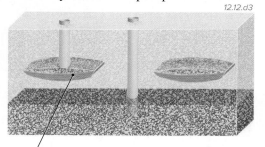

12.12.d3

Perched water sits above the main water table and generally forms where a discontinuous layer or lens of impermeable rock, in this case gray shale, blocks water infiltrating into the ground and causes groundwater to collect above the lens.

Artesian Wells and Water

We often hear the word *artesian*, commonly in the context of bottled water or certain beverages. What does this term imply? Does it mean that the water is better tasting, more natural, or more healthy? The short answer to these three questions is no, or at least not necessarily.

The term *artesian* means that groundwater is in a confined aquifer and is under enough water pressure that the water rises some amount within a well. The water does not have to reach the surface for the well to be called artesian, but many artesian systems have enough pressure to force the water all the way to the surface, creating a well or spring. Although it is a catchy advertising term, the term *artesian* is not indicative of how the water tastes, whether it is more natural than other types of groundwater, or whether it is healthy. It only means that the groundwater is confined and under pressure, and as a result rises some amount in the well.

In studying artesian wells, springs, and other aspects of groundwater, geoscientists use the concept of a *potentiometric surface*, an imaginary surface to which water would rise if allowed. If the potentiometric surface is above the ground, water in an artesian well will flow out onto the surface on its own, without pumping.

12.12.t1

Before You Leave This Page

- ☑ Sketch and describe the typical geometry of the water table beneath a hill and a valley, showing the direction of groundwater flow.

- ☑ Summarize two factors that control the rate of groundwater flow.

- ☑ Sketch and describe an unconfined, confined, and artesian aquifer.

- ☑ Sketch and describe the origins of perched water.

12.12

12.13 What Is the Relationship Between Surface Water and Groundwater?

SURFACE WATER AND GROUNDWATER ARE NOT ISOLATED SYSTEMS. Rather, they are highly interconnected with water flowing from the surface to the subsurface and back again. Most groundwater forms from surface water that seeps into the ground, and some streams and lakes are fed by groundwater.

A How Does Water Move Between the Surface and Subsurface?

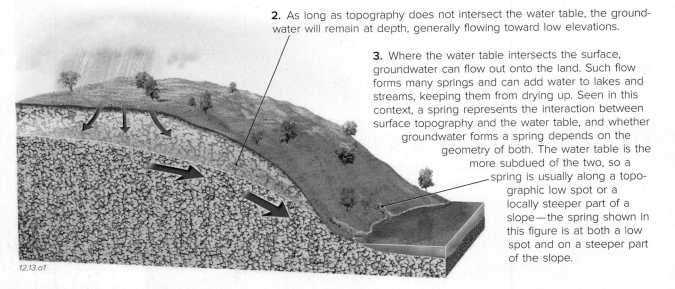

1. Surface water can soak into the subsurface and become groundwater if the surface material is permeable and the water table is deep enough so there is an unsaturated zone into which water can seep. Percolation of water into the groundwater system helps replenish or recharge water lost by wells, springs, or other parts of the system. Such replenishment is referred to as *groundwater recharge*.

2. As long as topography does not intersect the water table, the groundwater will remain at depth, generally flowing toward low elevations.

3. Where the water table intersects the surface, groundwater can flow out onto the land. Such flow forms many springs and can add water to lakes and streams, keeping them from drying up. Seen in this context, a spring represents the interaction between surface topography and the water table, and whether groundwater forms a spring depends on the geometry of both. The water table is the more subdued of the two, so a spring is usually along a topographic low spot or a locally steeper part of a slope—the spring shown in this figure is at both a low spot and on a steeper part of the slope.

12.13.a1

B What Causes Groundwater to Emerge as a Spring?

A spring represents a place where groundwater flows out of the ground onto the surface. At most springs, the water table intersects the surface. This can occur in a variety of geologic settings, some of which are summarized below. Some groundwater is heated by hot rocks before coming to the surface in warm springs, in hot springs, and in a *geyser*, a kind of hot spring that intermittently erupts fountains or sprays of hot water and steam.

Springs Related to Rock Units

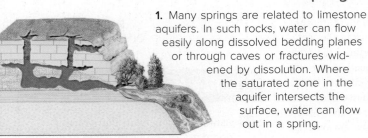

1. Many springs are related to limestone aquifers. In such rocks, water can flow easily along dissolved bedding planes or through caves or fractures widened by dissolution. Where the saturated zone in the aquifer intersects the surface, water can flow out in a spring.

12.13.b1

2. Many springs are related to contacts between units, like along unconformities. In this example, a sedimentary unit that is a good aquifer lies above an unconformity, separating it from a less permeable crystalline rock (granite). Groundwater flows to the surface in springs along the boundary. Such springs can form along any type of geologic contact.

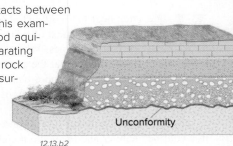

Unconformity

12.13.b2

Springs Related to Geologic Structures

3. Faults can serve as conduits for groundwater that feeds springs. Most faults are zones of intense fracturing and are therefore permeable. Groundwater can flow through a rock unit until it encounters the permeable fault, emerging as springs along the fault.

12.13.b3

4. Some springs form where groundwater in permeable rock encounters a less permeable obstacle. Here, groundwater flowing down the hydraulic gradient rises to the surface under pressure upon encountering a less permeable rock along a fault. Aquifers with such springs can be unconfined, as shown here, or confined.

12.13.b4

C How Are Lakes Related to Groundwater?

Lakes can have various relationships to groundwater. Most lakes occur where the water table intersects the ground surface, but some have a different setting. Most wetlands represent the interaction between rainfall, surface water, and groundwater and may be nourished by groundwater flow.

1. Some lakes are perched above the water table. These lakes can be transient, lasting only a short time after precipitation. A perched lake can be permanent if the inflow of water into the lake is at least equal to the amount lost by outflow to the ground, by evaporation to the air, or by other means.

2. Most lakes mark where the water table intersects and rises above the land surface. A lake can be fed entirely or partially by inflow of groundwater.

3. Many lakes are along the bottoms of valleys where groundwater is commonly close to or at the surface. Such lakes may be nearly in equilibrium with the adjacent groundwater, neither gaining nor losing water.

4. Wetlands can form peripheral to lakes, commonly at the same level as the water table. Other lakes and wetlands are perched on uplands that contain clay or other less permeable material close to the surface. The low permeability can trap precipitation and runoff, slowing the infiltration of water into the ground, forming a wetland from the ponded water.

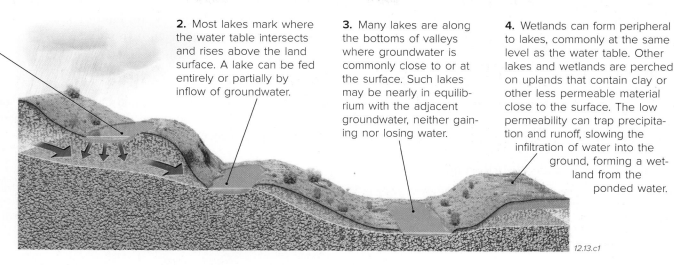

12.13.c1

D How Do Streams Interact with the Water Table?

Water in many streams decreases to a trickle and disappears entirely farther down the drainage. In other cases, a stream flows even when there has not been rain or snowmelt in a long time—what is the source of this water? These situations are a result of interactions of the stream with groundwater.

12.13.d1

1. Some streams are lower in elevation than the water table next to the stream, so groundwater flows into the stream as shown here by blue arrows, which show the direction of groundwater flow below the water table. A part of a stream that receives water from the inflow of groundwater into the channel is said to be *gaining* or to be a *gaining stream.*

12.13.d3

3. Some losing streams disappear when they cross from hard, less permeable rocks onto softer, more permeable materials. The water seeps into the ground, where it may continue to flow at a shallow depth in the loose sand and gravel in the basin.

12.13.d2

2. Other stream channels flow across an area where the water table is at some depth below the surface. The part of the stream that loses water from outflow to groundwater is said to be *losing* or to be a *losing stream*. The blue arrows show that groundwater below the water table flows down and away from the channel.

Before You Leave This Page

☑ Sketch and describe how the interaction of the water table with topography causes water to flow between the surface and subsurface.

☑ Sketch or describe what is required to form a spring and possible settings where this occurs.

☑ Sketch and describe ways that lakes and wetlands relate to groundwater.

☑ Describe gaining and losing streams and how a stream can lose its water entirely.

12.13

12.14 What Problems Are Associated with Groundwater Pumping?

THE SUPPLY OF GROUNDWATER IS FINITE, so pumping too much groundwater, a practice called *overpumping*, can result in serious problems. Overpumping can cause neighboring wells to dry up, land to subside, and gaping fissures to open across the land surface. It can draw salt water into wells that started with fresh water.

A What Happens to the Water Table if Groundwater Is Overpumped?

Demands on water resources increase if an area's population grows, the amount of land being cultivated increases, or open space is replaced by industry. Groundwater is viewed as a way to acquire additional supplies of fresh water, so new wells are drilled or larger wells replace smaller ones when more water is needed.

Minor Groundwater Withdrawal

1. A simple case illustrates the problems with overpumping. The two figures below show what happens when an unconfined aquifer is pumped, first by a small-volume pump and later by a larger pump. The topography of this area is fairly flat, there are no bodies of surface water, and a single type of porous and permeable sediment composes the subsurface.

2. As people move into a nearby town, they drill a small well down to the water table to provide fresh water. The small well pulls out so little groundwater that the water table remains as it has for probably thousands of years—nearly flat and featureless. The well remains a dependable source of water because its bottom is well below the water table.

3. Across the entire area, groundwater flows from right to left, down the gentle slope of the water table. The blue arrow shows the direction of flow for groundwater in the saturated part of the aquifer, right below the water table. This arrow is drawn above the water table so the arrow is visible, but water above the water table is not connected enough to flow in this way.

12.14.a1

Increased Groundwater Withdrawal

12.14.a2

4. As more people move into the surrounding area, the town drills a larger well to extract larger volumes of water to satisfy the growing demand. The new, larger well pumps water so rapidly that groundwater around the well cannot flow in fast enough to replenish what is lost. This causes the local water table to drop, forming a funnel-shaped *cone of depression* around the well.

5. The direction of groundwater flow changes dramatically across the entire area. Instead of flowing in one direction, groundwater now flows toward the larger well and into the cone of depression from all directions. The change in flow direction has unintended consequences. It may cause serious safety issues, since waste-disposal sites, like landfills, are generally planned with the original groundwater-flow direction in mind. The change in flow direction can bring contaminated water into previously fresh wells.

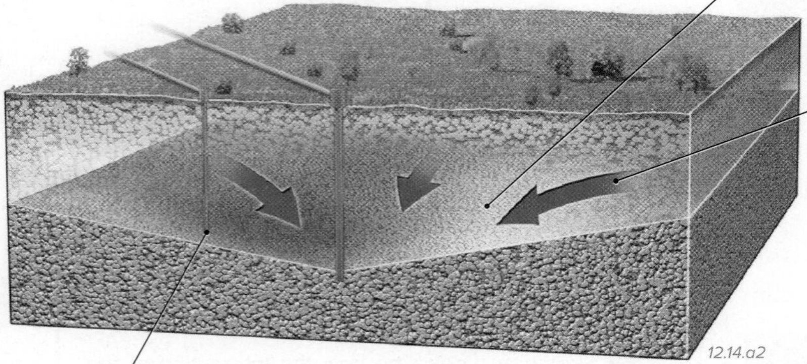

6. The original small well dries up because it no longer reaches the water table, which has been lowered by the larger well's cone of depression. A cone of depression is common around nearly all wells, but overpumping of the aquifer can accentuate the situation, causing a large cone of depression that can have drastic consequences. It can dry up existing wells, change the direction of groundwater flow, and contaminate wells. In addition, overpumping can dry up streams and lakes if they are fed by groundwater, or can cause the roofs of caves to collapse. Overpumping can affect any type of aquifer, including a shallow unconfined aquifer (shown here), a deep, confined aquifer, or a perched aquifer.

Before You Leave These Pages

☑ Sketch a cone of depression in cross section, describing how it forms and which way groundwater flows.

☑ Describe how a cone of depression can cause a well to become polluted.

☑ Sketch or describe some other problems associated with overpumping, including subsidence, fissures, and saltwater incursion.

B What Problems Are Caused by Excessive Groundwater Withdrawal?

In addition to changing the configuration of the water table, overpumping can cause the ground surface to subside if sediment within the underlying aquifer is dewatered and compacted. In certain settings, subsidence causes wide fissures to open on the surface.

Before Groundwater Pumping

1. Many areas have settings similar to this one: mountains composed of bedrock flank a valley or basin underlain by a thick sequence of sediment. Most water is pumped from the sediment-filled basin and used by people in the valleys.

12.14.b1

2. Bedrock has interconnected fractures that give it some permeability, but it has a much lower overall porosity and water than sediment in the basin.

3. The water table slopes from the mountains toward the basin, across the boundary between bedrock in the mountains and sediment beneath the valley.

After Groundwater Pumping

4. If wells overpump groundwater, the water table will drop over much of the area. In some cases it has dropped more than 100 m (~330 ft).

5. As the water table drops, the thickness of the unsaturated zone increases as the upper part of the aquifer is dewatered. Sediment within and below the dewatered zone compacts because water pressure no longer holds open the pore spaces.

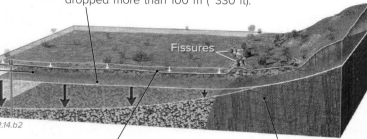
12.14.b2

Fissures

12.14.b3 Chandler Heights, AZ

6. Compaction of the sediment causes the overlying land surface to subside by several meters. Once the sediment compacts, the subsidence and loss of porosity are permanent and will not be undone if pumping stops and water levels rise again. In many places, like the San Joaquin Valley of California and the valleys of southern and central Arizona, the land surface has dropped as much as ten meters due to subsidence related to groundwater withdrawal. Along the coast, such subsidence could lower an area below sea level, as has occurred in New Orleans.

7. The granite cannot compact, so open fissures develop across the land surface along the boundary between land that subsided and land that did not (in the mountains). The earth fissure pictured here (▶) formed by this type of subsidence.

C How Can Groundwater Pumping Cause Saltwater Incursion into Coastal Wells?

Some wells are by necessity near the coasts of oceans and seas. These wells have a special threat—overpumping can draw salt water into the well, a process referred to as *saltwater incursion* or *saltwater intrusion*.

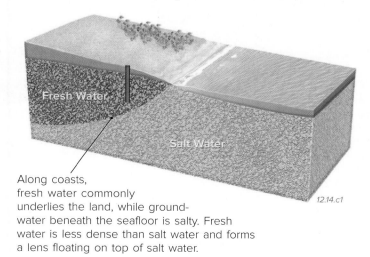

Fresh Water

Salt Water

12.14.c1

Along coasts, fresh water commonly underlies the land, while groundwater beneath the seafloor is salty. Fresh water is less dense than salt water and forms a lens floating on top of salt water.

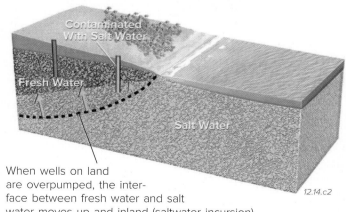

Contaminated With Salt Water

Fresh Water

Salt Water

12.14.c2

When wells on land are overpumped, the interface between fresh water and salt water moves up and inland (saltwater incursion). Wells closest to the coast will begin to pump salt water and will have to be shut down due to overpumping.

12.15 How Can Water Become Contaminated?

CONTAMINATION OF SURFACE AND SUBSURFACE WATER SUPPLIES is a major problem facing many communities. Some contaminants are natural products of the environment, whereas others have human sources, the direct result of our modern lifestyle. What are some main sources of water contamination?

1. Examine this figure, trying to recognize every potential source that could contaminate surface water and groundwater. Then read the accompanying text blocks. Go ahead, it's a great figure.

2. Water contamination can have natural causes. Weathering of rocks releases chemical elements into surface water and groundwater—some of these elements are beneficial and others are not. Rocks, especially those that have been mineralized by hot fluids, may contain lead, sulfur, arsenic, or other potentially hazardous elements. Mining activities and natural erosion move mineralized rocks away from where they formed, further spreading these contaminants.

3. We use large amounts of petroleum and coal, which have to be discovered, extracted, transported, and processed. Any of these activities potentially cause pollution. Some of the worst disasters are leaks from pipelines and supertankers, fires at refineries and storage tanks, and leaks from wells.

4. Old landfills are the repositories for countless discarded items, many of which contain hazardous substances. Such items include diapers, toxic liquids from household or commercial use, compact fluorescent bulbs (which contain toxic mercury), old tires, some types of batteries, and other garbage. If not properly sited and sealed from the environment, landfills can be major sources of pollution. Landfills along streams, such as this one, can be breached by lateral erosion of channels. Supposedly impermeable linings beneath the landfill can crack during settling and from daily landfill operations, allowing a toxic stew to seep into the underlying groundwater.

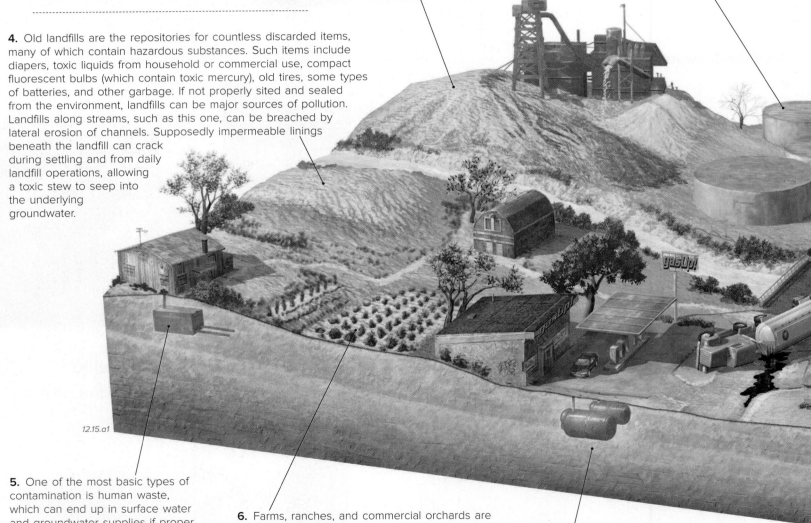

12.15.a1

5. One of the most basic types of contamination is human waste, which can end up in surface water and groundwater supplies if proper sanitary procedures are not followed. Contamination of this sort comes from septic tanks, accidental spills from wastewater treatment facilities, or, in less affluent parts of the world, from waste disposal in open sewers and trenches. It also includes medicines, detergents, and other household chemicals we use.

6. Farms, ranches, and commercial orchards are contributors of chemical and organic contamination. Chemical contaminants include fertilizers that contain nitrates, insecticides to control pests, herbicides to combat invasive weeds, and defoliants to remove leaves before harvesting crops like cotton. Irrigated fields build up salts as water evaporates, and much of this gets carried into ditches by excess irrigation water. Animal waste, which contains harmful bacteria, hormones, and feed additives, is also a potential problem.

7. Gas stations contaminate water because of leaks from underground storage tanks and spills that occur while filling vehicles. Gas stations frequently go out of business if they have to dig up leaking underground storage tanks. Spills from tanker trucks, railroad cars, and trucks delivering fuel from distribution hubs may cause water contamination if there is an accident.

8. To manufacture the items we use in our daily lives, factories use many different raw materials and chemicals. Plastic products, for example, are everywhere around us: containers for soda and bottled water, plastic bags for groceries and other purchases, and many parts of our cars. These plastics are mostly produced from petroleum, which must be refined and processed in refineries and plastic factories. Petroleum and various chemicals, along with the waste produced during the manufacturing process, can accidentally escape, causing an industrial site to become heavily contaminated, as shown in the photograph here. In the past, liquid contamination was pumped down "disposal wells," often ending up in the groundwater. Ponds intended for temporary storage can leak, contaminating surface water and groundwater. Fumes and particles emitted from smokestacks settle back to the ground or are washed down by rain and snowfall, possibly contaminating air, plants, buildings, soils, surface water, or groundwater.

12.15.a2 Russia

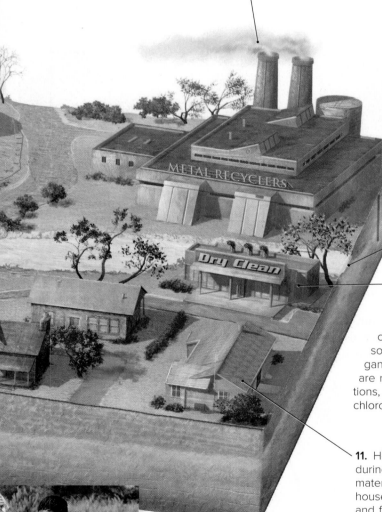
12.15.a3 Newcastle, England

9. Even if a community is careful with wastes, contamination can be carried into the area by rivers that drain polluted areas upstream. Polluted surface water can seep into groundwater, and groundwater can flow into and pollute streams. Soils can contaminate water, which pollutes the stream, which then pollutes the next town downstream.

10. In the past, dry cleaners were sources of groundwater pollution because of the chemical solvents used to clean clothes without water. Such solvents have names from organic chemistry and commonly are referred to by their abbreviations, such as PCE for perchloroethylene ("perc" for short).

11. Houses cause water pollution during the production of the materials used to build the house, from actual construction, and from day-to-day activities that include the use of fertilizer, termite treatment, and household pesticides. Oil and gas spilled from cars and other machines, along with oil improperly disposed of during do-it-yourself oil changes, can contaminate large volumes of fresh water.

12. To determine whether water is contaminated, we collect and analyze samples of surface water (◄) or groundwater. We may discover the contamination only if we drill into it or if an unusual health issue appears in a local population.

Water, Arsenic, and Bangladesh

Bangladesh, in south Asia, is a geologically challenged country. Much of it consists of lowlands that are flooded by storm surges in the sea and by the Ganges, one of the world's largest rivers.

One of Bangladesh's worst problems, however, is water contamination. For centuries, poor sanitation in this impoverished nation polluted the rivers and other surface-water sources with cholera, dysentery, and other diseases. To provide a new source of water, people sank more than 10 million tube wells (created by pounding tubes into the soft sediment). Unfortunately, the sediment and groundwater have a high content of naturally derived arsenic, many times the recommended limit, causing arsenic poisoning on a scale never before seen. To help solve the problem, geologists from the U.S. Geological Survey and the Geological Survey of Bangladesh conducted a large project, sampling the well waters (►), studying the surface and subsurface geology, drilling wells into a deeper aquifer (▼), and evaluating whether bacteria can be used to reduce the arsenic concentrations.

12.15.t1

12.15.t2

Before You Leave This Page

☑ Describe the many ways that surface water and groundwater can become contaminated.

12.15

12.16 How Does Groundwater Contamination Move and How Do We Clean It Up?

WATER CONTAMINATION CAN BE OBVIOUS OR SUBTLE. Some streams and lakes have oily films and give off noxious fumes, but some contamination occurs in water that looks normal and tastes normal but contains hazardous amounts of a natural or human-related chemical component. How does contamination in groundwater move, how do we investigate its causes and consequences, and what are possible remedies?

A How Does Contamination Move in Groundwater?

As contamination enters groundwater, it typically moves along with the flowing groundwater. Contamination can remain concentrated, can spread out, or can be filtered by passage through sediment and rocks.

Groundwater contamination typically moves with the groundwater down the slope of the water table.

Contamination is drawn out parallel to the direction of groundwater flow.

Some contamination can be naturally filtered by materials through which the contaminated groundwater flows.

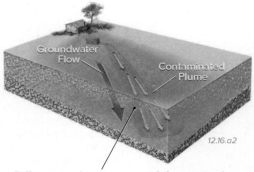

Contamination from this septic tank will move to the right away from the water well. The direction of groundwater flow is clearly important in deciding where to put the septic tank relative to the well.

Diffusion and mixing spread the contaminated zone as it migrates away from the source. Consequently, the shape of most contamination spreads out like smoke from a chimney and is called a *plume*.

Contamination from the septic tank on the left will be filtered by slow movement through sandstone, whereas contamination from the septic tank on the right will flow rapidly away, unfiltered, through permeable, cavernous limestone.

B How Do We Depict the Water Table?

To understand the geometry of the water table, which controls the direction of groundwater flow, we collect data from existing wells, from new drilling, and from geophysical surveys, such as measuring how gravity varies across an area. From such data, especially the elevations of the water table, we can produce very informative maps and diagrams.

The most important piece of information about groundwater is a map showing variations in the elevation of the water table. The first step in constructing such a map is to collect and plot elevations of the water table in all available wells. Each number on this map is the elevation (in meters above sea level) of the water table at a well in that location. High numbers mean the water table is higher than in sites with lower numbers.

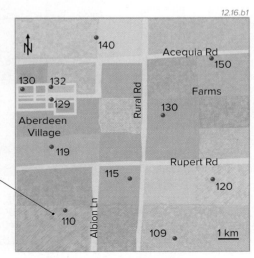

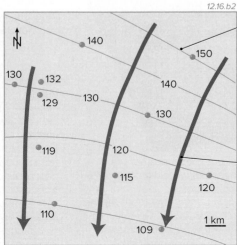

We then draw contours, like those used on a topographic map, to show the elevation of the top of the water table (in the subsurface). The contours shown here indicate the elevation of the water table in meters. Each contour follows a specific elevation on the water table. Arrows drawn perpendicular to the contours show the direction of groundwater flow, which is down the slope of the water table, from higher contours to lower ones.

How Is Groundwater Contamination Tracked and Remediated?

Once groundwater contamination is identified, what do we do next? We compile available information to compare the distribution of contamination with all relevant geologic factors. One commonly used option to clean up, or *remediate*, a site of contamination is called "pump-and-treat." Some contamination can be mostly remediated, but remediation is much more expensive than not causing the problem to begin with.

1. The first step to remediation is to properly understand the situation—what is the nature of the contamination, where is the contamination now, where did it come from, where is it going, and what are the geologic controls?

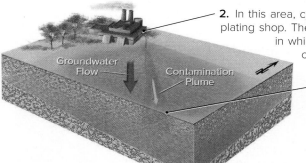

12.16.c1

2. In this area, contamination consists of chromium released by a chrome-plating shop. The water table slopes to the southeast, so this is the direction in which the upper levels of groundwater will flow. We predict that contamination will move in this same direction.

3. Chromium ions are carried away by groundwater flow and also chemically diffuse through the water, albeit at a slower rate than the groundwater flows. The combination of flow and diffusion causes the contamination to spread out like smoke from a chimney, forming a plume of contamination. There is no contamination upflow (northwest) of the shop, but the plume of contamination will spread to the southeast.

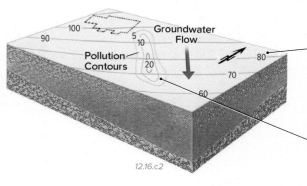

12.16.c2

4. To investigate the situation, we contour elevations of the water table to more precisely determine which way groundwater is flowing. In this case, the contours decrease in elevation to the southeast. Groundwater flows to the southeast, perpendicular to the contours (and toward lower elevation contours).

5. We draw a second set of contours based on chemical analyses of the concentration of contamination, in this case chromium. For example, areas within the 5 mg/L contour have at least 5 mg/L chromium, and those within the 10 mg/L contour have at least 10 mg/L. The EPA limit for chromium is 0.1 mg/L, so these values are well above EPA standards.

6. From these maps, we can now determine where the contamination is, which way it is moving, and where it will go in the future (down the slope of the water table). If from interviews or historical records we can determine how long ago the contamination occurred, we can use simple calculations (distance/time) to get the rate of flow. We also can use computer simulations to model past and future movement.

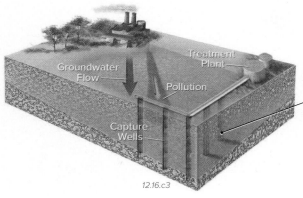

12.16.c3

7. Finally, we try to clean up the contamination. One strategy is to drill wells in front of the projected path of the contamination to contain, capture, and extract the contaminated water. Pumping brings contaminated water to the surface, where it is processed with carbon filters or other appropriate technology to separate the contaminant from the water. The cleaned water is typically reinjected into the ground, allowed to evaporate in evaporation ponds, or channeled to flow down streams.

A Civil Action

Woburn, Massachusetts, a small town 10 miles north of Boston, was the site of a classic legal case involving groundwater contamination. The case was made famous in the book *A Civil Action* by Jonathan Harr and in a movie of the same name starring John Travolta.

The trouble began in the 1960s when the city drilled two new groundwater wells for municipal water supplies. The wells were drilled into glacial and river sediments that had filled an old valley. After the wells were installed, some residents complained that the water tasted odd and had a chemical odor. Over the next 20 years, residents began to show a high incidence of leukemia and other serious health problems. Chemical analyses showed that the groundwater was contaminated with trichloroethylene (TCE) and other volatile organic compounds. Local families filed a lawsuit against several chemical companies that were potentially responsible. The verdict remains complex, but the site is a classic example of the interaction of geology, water, health, and environmental law.

Before You Leave This Page

☑ Sketch a plume of contamination, showing how it relates to the source of contamination and the direction of groundwater flow.

☑ Describe some ways in which geoscientists investigate groundwater contamination.

☑ Sketch how chemical analyses define a plume of contamination and one way a plume could be remediated.

12.16

12.17 What Is Going On with the Ogallala Aquifer?

THE MOST IMPORTANT AQUIFER IN THE UNITED STATES lies beneath the High Plains, stretching from South Dakota to Texas. It provides groundwater for about 20% of all cropland in the country, but it is severely threatened by overpumping. The setting, characteristics, groundwater flow, and water-use patterns of this aquifer connect many different aspects of water resources and illustrate their relationship to geology.

A What Is the Setting of the Ogallala Aquifer?

1. The *Ogallala aquifer,* also called the *High Plains aquifer,* covers much of the High Plains area in the center of the United States. The blue-outlined area on this map shows the extent of the main part of the aquifer. The aquifer forms an irregularly shaped north-south belt from South Dakota and Wyoming through Nebraska, Colorado, Kansas, the panhandles of Oklahoma and Texas, and eastern New Mexico.

2. The Ogallala aquifer covers about 450,000 km² (174,000 mi²) and is currently the largest source of groundwater in the country. It provides 30% of all groundwater used for irrigation in the U.S. In 1980, near the height of the aquifer's use, 17.6 million acre-feet of water were withdrawn to irrigate 13 million acres of land. The water is used mostly for agriculture and rangeland. The main agricultural products include corn, wheat, soybeans, and feed for livestock.

12.17.a1

3. The aquifer is named for the Ogallala Group, the main geologic formation in the aquifer. The formation was named by a geologist in the early 1900s after the small Nebraskan town of Ogallala.

4. Much of the Ogallala Group consists of sediment deposited by streams and wind during the last half of the Cenozoic, mostly between 19 and 5 million years ago. Braided streams carried abundant sediment eastward from the Rocky Mountains, spreading over the landscape and depositing a relatively continuous layer of sediment. Deposition stopped when regional uplift and tilting caused the streams to downcut and erode rather than continuing to deposit sediment. Present-day streams continue to erode into the aquifer and drain eastward and southward, eventually flowing into the Gulf of Mexico.

The Aquifer in Cross Section

5. This vertically exaggerated cross section shows the thickness of the aquifer from west to east. It shows the aquifer in various colors; rocks below the aquifer are shaded bluish gray. Note that the aquifer is at the surface and is an *unconfined aquifer.*

7. The upper part of the aquifer (shaded yellow) is above the water table and in the *unsaturated zone.*

12.17.a2

8. Blue colors show levels of the water table for 1950 and 2000, and purple shows the predicted levels for 2050. Note that water levels in the aquifer have fallen due to overpumping. The western part is predicted to be totally depleted by 2050 (no purple).

6. The irregular base of the aquifer is an unconformity that reflects erosion of the land before deposition of the aquifer.

B   Where Does Groundwater in the Aquifer Come from and How Is It Used?

1. Most of the water going into the aquifer is from local precipitation. This map shows the amount of precipitation received across the area, with darker shades indicating more precipitation. The western part of the aquifer receives much less precipitation (rain, snow, and hail) than the eastern part.

2. Areas of the aquifer that receive the least amounts of precipitation—the south-western parts—are also those predicted to go dry by 2050.

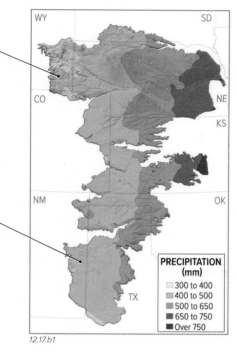

PRECIPITATION (mm)
- 300 to 400
- 400 to 500
- 500 to 650
- 650 to 750
- Over 750

12.17.b1

3. This graph shows the water balance for the Ogallala aquifer. Water going into the aquifer (▼) is shown above the axis, whereas water being lost by the aquifer is below the axis. Some groundwater recharge occurs where water from precipitation seeps into the aquifer, especially in areas that receive higher amounts of precipitation, as either rain or snow.

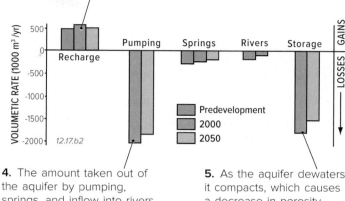

12.17.b2

- Predevelopment
- 2000
- 2050

4. The amount taken out of the aquifer by pumping, springs, and inflow into rivers greatly exceeds the recharge, so most parts of the aquifer are being dewatered.

5. As the aquifer dewaters it compacts, which causes a decrease in porosity and a loss of pore space (in which to store water). This cannot be undone.

C   How Has Overpumping Affected Water Levels in the Ogallala Aquifer?

The USGS estimates that the aquifer contains 2.9 billion acre-feet of water. That is enough to cover the entire lower 48 states with 1.5 feet of water. How much has overpumping affected the aquifer's water levels, and what will happen to the region and to the country if large parts of the aquifer dry up?

1. This map shows the thickness (in meters) of the *saturated zone* within the aquifer. In some of its northern parts, more than 300 m (1,000 ft) of the aquifer is saturated with water, whereas less than 60 m (180 ft) remain saturated in the southern parts.

2. This map shows how many feet the water table dropped in elevation between 1980 and 1995 as a consequence of overpumping. The largest drops, exceeding 10 m, occurred in south-western Kansas and the northern part of Texas. Compare this map to the one for precipitation.

3. *Future Predictions* — It is uncertain what will happen, but hydrogeologists are conducting detailed studies of key areas to try to predict what will happen in the next decades. Projections of current water use, combined with numerical models of the water balance, predict that some parts of the aquifer will go dry by 2050. This will have catastrophic consequences for the local farmers, ranchers, and businesses, and for people across the country who depend on the aquifer for much of their food. Subsidence related to groundwater withdrawal and compaction of the aquifer will be an increasing concern. What do you think would happen to the region if this aquifer were partly pumped dry?

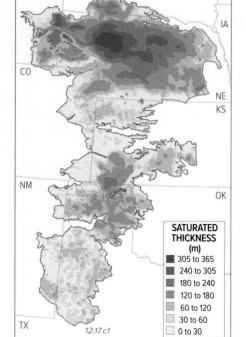

SATURATED THICKNESS (m)
- 305 to 365
- 240 to 305
- 180 to 240
- 120 to 180
- 60 to 120
- 30 to 60
- 0 to 30

12.17.c1

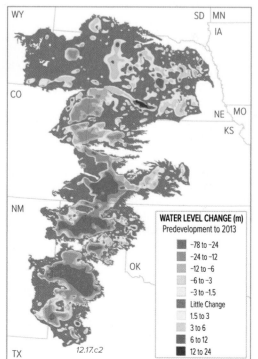

WATER LEVEL CHANGE (m)
Predevelopment to 2013
- −78 to −24
- −24 to −12
- −12 to −6
- −6 to −3
- −3 to −1.5
- Little Change
- 1.5 to 3
- 3 to 6
- 6 to 12
- 12 to 24

12.17.c2

Before You Leave This Page

✓ Summarize the location, characteristics, and importance of the Ogallala aquifer.

✓ Summarize the water balance for the aquifer and how water levels have changed in the last several decades.

12.17

12.18

Who Polluted Surface Water and Groundwater in This Place?

SURFACE WATER AND GROUNDWATER IN THIS AREA are contaminated. You will use the geology of the area, along with elevations of the water table and chemical analyses of the contaminated water, to determine where the contamination is, where it came from, and where it is going. From your conclusions, you will decide where to drill new wells for uncontaminated groundwater.

Goals of This Exercise:

- Observe the landscape to interpret the area's geologic setting.
- Read descriptions of various natural and constructed features.
- Use well data and water chemistry to draw a map showing where contamination is and which way groundwater is flowing.
- Use the map and other information to interpret where contamination originated, which facilities might be responsible, and where the contamination is headed.
- Determine a well location unlikely to be contaminated in the future.

Procedures

Use the available information to complete the following steps, entering your answers on the worksheet or online.

A. This figure shows geologic features, rivers, springs, and human-constructed features, including a series of wells (lettered A through P). Observe the distribution of rock units, sediment, rivers, springs, and other features on the landscape. Compare these observations with the cross sections on the sides of the terrain to interpret how the geology is expressed in different areas.

B. Read the descriptions of key features and consider how this information relates to the geologic setting, to the flow of surface water and groundwater, and to the contamination.

C. The data table on the next page shows elevation of the water table in each lettered well. Use these data and the base map on the worksheet to construct a groundwater map with contours of the water table at the following elevations: 100, 110, 120, 130, 140, and 150 meters. On the contoured map, draw arrows pointing down the slope of the water table to show the direction of groundwater flow.

D. Use the data table showing concentrations of a contaminant, purposely unnamed here, in groundwater to shade in areas where there is contamination. Use darker shades for higher levels of contamination.

E. Use the groundwater map to interpret where the contamination most likely originated and which facilities were probably responsible. Mark a large X over these facilities on the map, and explain your reasons in the worksheet or online.

F. Determine which of the lettered well sites will most likely remain free of contamination, and draw a circle around one such well.

G. Devise a plan to remediate the groundwater contamination by drilling wells in front of the plume of contamination; mark these on the map with the letter R.

1. The region contains a series of ridges to the east and a broad, gentle valley to the west. Small towns are scattered across the ridges and valleys. There are also several farms, a dairy, and a number of industrial sites, each of which is labeled with a unique name. Hydrogeologists studied one of these towns, Springtown, and concluded that it is not the source of any contamination.

2. A main river, called the Black River for its unusual dark, cloudy color, flows westward (right to left) through the center of the valley. The river contains water all year, even when it has not rained in quite a while. Both sides of the valley slope inward, north and south, toward the river.

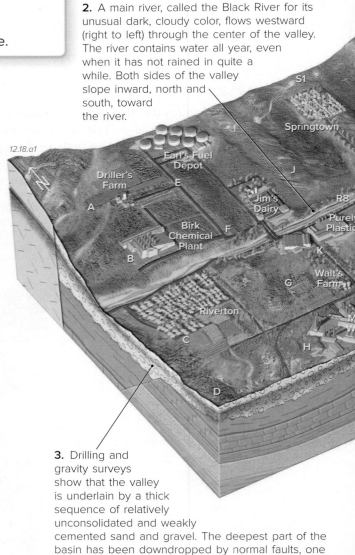

12.18.a1

3. Drilling and gravity surveys show that the valley is underlain by a thick sequence of relatively unconsolidated and weakly cemented sand and gravel. The deepest part of the basin has been downdropped by normal faults, one of which is buried beneath the gravel.

6. Bedrock units cross the landscape in a series of north-south stripes, parallel to the strike of the rock layers. One of the north-south valleys contains several large coal mines and a coal-burning, electrical-generating plant. An unsubstantiated rumor says that one of the mines had some sort of chemical spill that was never reported. Activity at the mines and power plant has caused fine coal dust to be blown around by the wind and washed into the smaller rivers that flow along the valley.

7. A north-south ridge is composed of sandstone, called the lower sandstone. Slidetown, a new town on this ridge, is not a possible source of the contamination because it was built too recently. A few nice-tasting, freshwater springs issue from the sandstone where it is cut by small stream valleys.

Stratigraphic Section

Gravel—Unconsolidated sand and gravel in the lower parts of the valley

Upper Sandstone—Well-sorted, permeable sandstone

Upper Shale—Impermeable, with coal

Sinkerton Limestone—Porous, cavernous limestone

Middle Shale—Impermeable shale

Lower Sandstone—Permeable sandstone

Lower Shale—Impermeable shale

Basal Conglomerate—Poorly sorted with salty water

Granite—Sparsely fractured; oldest rock in area

12.18.a2

8. The highest part of the region is a ridge of granite and sedimentary rocks along the east edge of the area. This ridge receives quite a bit of rain during the summer and snow in the winter. Several clear streams begin in the ridge and flow westward toward the lowlands.

9. A company built a coal-burning power plant over tilted beds of a unit named the Sinkerton Limestone, so called because it is associated with many sinkholes, caves, and karst topography. The limestone is so permeable that the power plant has had difficulty keeping water in ponds built to dispose of waste waters, which are rich in the chemical substances (including the contaminant) that are naturally present in coal.

10. The tables below list water-table elevations in meters and concentrations of contamination in milligrams per liter (mg/L) for each of the lettered wells (A–P). This table also lists the concentration of contamination in samples from four springs (S1–S4) and eight river segments (R1–R8). The location of each sample site is marked on the figure. Wells M, N, and P are deep wells, drilled into the Sinkerton Limestone aquifer at depth, although they first encountered water at a shallow depth. The chemical samples from these wells were collected from deep waters.

5. From mapping and other studies on the surface, hydrogeologists have determined the sequence of rock units, as summarized in the stratigraphic section in the upper right corner of this page. These studies also document a broad anticline and a syncline beneath the eastern part of the region. Note that contamination can flow through the subsurface, following limestone and other permeable units, instead of passing horizontally through impermeable ones, like shale.

4. Based on shallow drilling, the water table (the top of the blue shading) mimics the topography, being higher beneath the ridges than beneath the valleys. Overall, the water table slopes from east to west (right to left), parallel to the regional slope of the land. All rocks below the water table are saturated with groundwater.

Slidetown

R1

Cornhead Coal Mine

S4

M

R2

R3

N

R4

Coal Power Plant

R7

Kellogg

R5

CQ Coal Mine

Midnight Coal Mine

R6

P

200 m

O

S3

L

Well	Elev. WT	mg/L
A	110	0
B	100	0
C	105	0
D	110	20
E	120	10
F	115	0
G	120	0
H	120	50

Well	Elev. WT	mg/L
I	130	30
J	130	0
K	120	0
L	130	0
M	150	50
N	150	0
O	140	0
P	150	0

Spring	mg/L
S1	50
S2	0
S3	0
S4	0

River	mg/L
R1	0
R2	20
R3	0
R4	0

River	mg/L
R5	0
R6	0
R7	5
R8	5

12.18

Energy and Matter in the Atmosphere

ALMOST ALL NATURAL SYSTEMS on Earth derive their energy from the Sun, but not all areas receive the same amount of sunlight. Instead, the amount of energy reaching Earth's surface varies from region to region and from time to time. The interaction of energy with the Earth's atmosphere and surface determine the climate, weather, and habitability of an area.

13.00.a2 Namibia, Africa

Sunlight warms the land and oceans, which in turn warms our atmosphere (◄), making some regions, such as this spectacular desert in Namibia, warmer than others.

What type of energy is in sunlight, and does all of the Sun's energy make it to Earth's surface?

The Sun rises and sets each day (►), except in some polar places where the Sun shines 24 hours a day during the summer. In other places, on the opposite pole of the planet, there is total darkness during the same 24 hours.

What causes variations in the number of daylight hours, both from place to place and from season to season?

13.00.a3 Philippines

13.00.a4 Sapporo, Japan

Many regions have seasons, changing from the warm days of summer to the cold, snowy times of winter (►).

What causes the change from season to season, and do all areas experience summer at the same time?

13.00.a5 Banda Island, Indonesia

Tropical areas like Indonesia (►) do not have a distinct summer and winter but may have a rainy season and a dry season.

Why do some regions experience summer and winter but others do not?

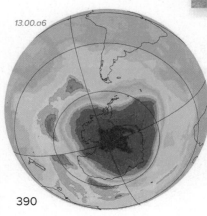

13.00.a6

Antarctica, during its winter, has a dramatic thinning of the overlying ozone layer in the atmosphere, shown here in purple (◄).

What is ozone, what causes this thinning, and why is there so much global concern about this phenomenon when ozone makes up less than 0.001% of all the gas in the atmosphere?

TOPICS IN THIS CHAPTER

Ancient people, such as the builders of Stonehenge (▶) 4,000 years ago, used changes in the position of the Sun over time to schedule important activities, such as the planting of crops.

What causes the seasons, and why do temperatures change from season to season?

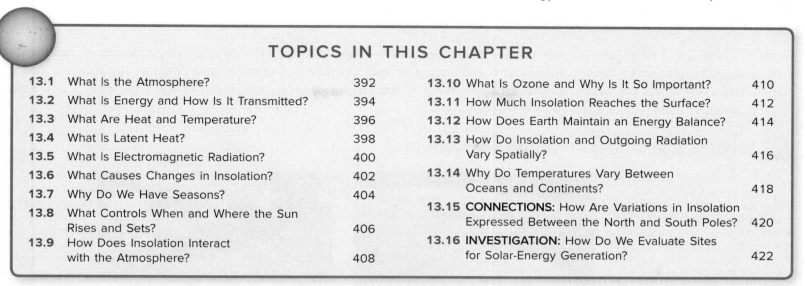

13.00.a7 Stonehenge, Wiltshire, England

13.00.a8 West-Central IL

Greenland

Arctic Ocean

ASIA

NORTH AMERICA

Atlantic Ocean

AFRICA

SOUTH AMERICA

Atlantic Ocean

Indian Ocean

Southern Ocean

ANTARCTICA

13.00.a1

Clouds, as in the thunderstorm above (▲), consist of small drops of water and ice crystals. The water to make the drops and crystals evaporated from the surface using energy from the Sun.

How much energy is needed to cause evaporation, and where does that energy ultimately go?

13.00.a9 Earth and Moon

This amazing view of Earth (▲) combines two types of images taken from a spacecraft orbiting the Moon, one image of Earth and other images taken of the lunar surface.

Where does the light coming from Earth originate, and does this light indicate that Earth is emitting energy?

13.0

13.1 What Is the Atmosphere?

A RELATIVELY THIN LAYER OF GAS—the *atmosphere*—surrounds Earth's surface. The atmosphere shields us from harmful high-energy rays from space, is the source of our weather and climate, and contains the oxygen, water vapor, and other gases on which all life depends. What is the character and composition of the atmosphere, and how does it interact with light coming from the Sun?

A What Is the Character of the Atmosphere?

Examine this view of Earth, taken from a spacecraft orbiting high above the Earth. As viewed from space, Earth is dominated by three things: the blue oceans and seas, the multicolored land, and clouds. If you look closely at the very edge of the planet, you can observe a thin blue fringe that is the atmosphere. From this perspective, the atmosphere appears to be an incredibly thin

13.01.a1 Arabian Peninsula

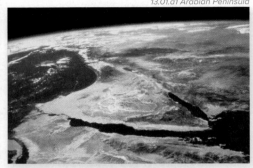

13.01.a2 San Luis Valley, CO

As viewed from the ground, the atmosphere mostly appears as a blue sky with variable amounts of clouds, which commonly are nearly

layer that envelops our planet, separating us from the dark vastness of space. Clouds, which are so conspicuous in any image of Earth taken from space, mostly circulate within the lower atmosphere, bringing rain and snow. The winds that move the clouds are also within the lower atmosphere.

white or some shade of gray. During sunset and sunrise, the sky can glow reddish or orange. The colors of the sky and clouds are due to the way sunlight interacts with matter in the atmosphere.

B What Is the Structure of the Atmosphere?

The atmosphere extends from the surface of Earth upward for more than 100 km, with some characteristics of the atmosphere going out to thousands of kilometers. The atmosphere is not homogeneous in any of its attributes, but instead has different layers that vary in temperature, air pressure, and the amount and composition of gases. Each layer has the term "sphere" as part of its name, referring to the way each layer successively wraps around the Earth with a roughly spherical shape. Examine the figure below and then read the text from the bottom left, starting with the lowest and most familiar part of the atmosphere.

4. The top layer is the *thermosphere*, derived from the Greek word for heat because this layer, surprisingly, can become very hot (more than 1,500°C) as gas particles intercept the Sun's energy. It is the altitude where the spectacular *auroras* (i.e., "Northern Lights") originate from interactions of solar energy and energetic gas molecules.

3. Above the stratosphere is the *mesosphere*, where "meso" is Greek for "middle," as this layer is in the middle of the atmosphere. The mesosphere starts at 50 km, the top of the stratosphere, and goes up to more than 80 km (~50 miles) in altitude. The upper part of the mesosphere is very cold (–85°C, –120°F), and is considered by many scientists to be the coldest place within the Earth system. It is within the mesosphere that most small meteors burn up, producing the effect called "shooting stars." Radio waves from Earth bounce off this layer and the overlying layer, allowing us to hear radio stations from far away.

2. The next layer up is the *stratosphere*, beginning at an altitude of about 10 km above sea level, at about the elevation of Earth's highest peaks. The name is derived from a Latin term for spreading out, referring to its layered (not mixed) character. Temperatures are also stratified, varying from cooler lower altitudes to warmer upper ones. The lowest part of the stratosphere is an altitude at which many commercial jets fly because the air offers less resistance to motion, allowing appreciable fuel savings.

1. The lowest layer is the one with which we surface-dwellers interact. It contains the air we breathe, clouds, wind, rain, and other aspects of weather. This layer is the *troposphere*, with the name "tropo" being derived from a Greek word for turning or mixing, in reference to the swirling motion of clouds, wind, and other manifestations of weather.

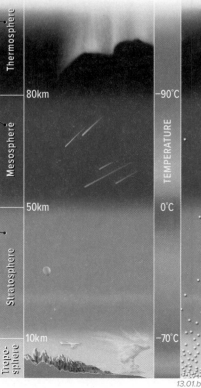

13.01.b1

5. On the right side of this figure are scattered bright dots that represent gas molecules in the atmosphere. The molecules are infinitely smaller and more abundant than shown here. Note that the molecules are concentrated lower in the atmosphere and become much more sparse upward. Over 70% of the mass of the atmosphere is in the lowest 10 km, that is, within the troposphere. The mesosphere and thermosphere contain only a few tenths of a percent of the atmosphere's mass.

C   What Is the Composition of the Atmosphere?

The atmosphere is not completely homogeneous in its vertical, horizontal, or temporal composition, but chemists and atmospheric scientists have estimated its average composition, as represented in the graph and table below.

1. The atmosphere is held in place by the balance between gravity (which is directed downward and keeps the gases close to the surface) and a buoyancy force (which is directed upward and exists because material tends to flow toward the vacuum of outer space). Near the surface, the greater weight of the overlying atmosphere results in more molecules being tightly packed close to the surface of the Earth and a rapid thinning of the number of molecules with distance upward in the atmosphere, away from the surface. This effect of gravity also influences the composition of the atmosphere, with a higher proportion of heavier gases, like oxygen, low in the atmosphere (in the troposphere) and a higher proportion of lighter gases, such as hydrogen, higher up in the atmosphere.

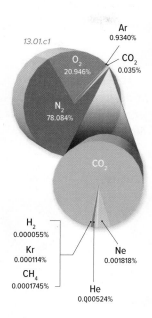

13.01.c1

Composition of Atmosphere (by volume)	
Gas	**%**
Nitrogen (N$_2$)	78.084
Oxygen (O$_2$)	20.946
Argon (Ar)	0.934
Water vapor (H$_2$O)	<0.01 to 0.400
Carbon dioxide (CO$_2$)	0.039
Neon (Ne)	0.002
Helium (He)	0.001
Less than 0.001% each	
Methane (CH$_4$)	
Krypton (Kr)	
Hydrogen (H$_2$)	
Nitrous oxide (N$_2$O)	
Xenon (Xe)	
Ozone (O$_3$)	
Nitrogen dioxide (NO$_2$)	

2. As shown by this diagram and table, the two dominant gases are nitrogen (78%) and oxygen (21%), followed by argon. Several other gases, such as a variable amount of water vapor, along with carbon dioxide (CO$_2$), methane (CH$_4$), nitrous oxide (N$_2$O), and ozone (O$_3$) play significant roles in global climate through their interaction with energy emitted by the Sun and re-emitted by the Earth.

3. The atmosphere also contains various types of solids and liquids called *aerosols*, such as dust, industrial pollutants, and tiny drops of liquid from volcanic eruptions. Aerosols play an important role in the energy balance of the Earth and aid in the formation of clouds and precipitation.

D   How Does the Atmosphere Interact with Energy from the Sun?

Gas in the atmosphere interacts with visible light and other energy radiated by the Sun, as well as energy reflected from and radiated from Earth's surface. An understanding of the possible types of interactions helps explain what we see every day, such as colors, as well as the underlying causes of weather and climate. Here we discuss four types of interactions: *transmission*, *reflection*, *absorption*, and *scattering*. The discussion will emphasize light, but the principles are applicable to other forms of radiant energy, such as ultraviolet energy.

1. *Transmission*—An object can be entirely or mostly transparent to light and other forms of radiant energy, in the way that a clear glass sphere permits most light to pass through. Allowing such energy to pass through is called *transmission*. The atmosphere is largely transparent to visible light, allowing it to pass through and illuminate Earth's surface. For some types of radiant energy the atmosphere is only partially transparent, because certain molecules interact with some of the incoming radiation. Energy from the Sun is efficiently transmitted through space, which is nearly a vacuum, lacking many molecules that could interact with the energy.

4. *Absorption*—Objects retain some of the energy that strikes them, and this process of retention is called *absorption*. Objects have varying degrees of absorption. This dark, dull-textured sphere is highly absorbent, retaining most of the light that strikes it, accounting for the sphere's dark color (not much light coming back off the surface). Energy that is absorbed by an object can be released back in another form (like heat from the dark ball). The giving out of radiant energy is called *emission*.

13.01.d1

2. *Reflection*—Instead of passing through an object, some light can bounce off the object, the process of *reflection*, as illustrated by light reflecting off a polished metal sphere. Not all light and energy bounce off real objects, so reflectivity can be thought of as a continuum from objects being perfectly reflective to having no reflectivity. For visible light, objects that are white or made of shiny, polished metal are highly reflective, whereas dark, rough objects have low reflectivities. In the summer Sun, a white car, which reflects much of the Sun's heat, is cooler than a dark car that is not so reflective. If an object, like a leaf, preferentially reflects green light, we see the object as green.

3. *Scattering*—Most objects reflect some light, but in a way that disperses the energy in various directions, *scattering* the light. The shiny but rough sphere shown here has a highly reflective surface, but the roughness causes light to be scattered in various directions. Such scattering occurs when light and other energy strike the land, and it is also caused by certain types of aerosols and gases, like water vapor, in the atmosphere. Gases in the atmosphere preferentially scatter blue light, which spreads out in all directions through the atmosphere, causing the dominantly blue color of the sky.

Before You Leave This Page

- ☑ Describe what the atmosphere is, including its average composition.

- ☑ Sketch and describe the layers in the atmosphere.

- ☑ Describe four ways that matter interacts with light and other radiant energy.

13.1

13.2 What Is Energy and How Is It Transmitted?

THE TRANSMISSION OF ENERGY, and the interactions between energy and matter, define the character of our planet and control weather, climate, and the distribution of life, including humans. Here, we examine the fundamentals of energy, including what it is, where it comes from, and how it is moved from one place to another.

A What Is Energy?

All matter contains *energy*, which is the capability of an object to do work, such as pushing or pulling adjacent objects, changing an object's temperature, or changing the state of an object, as from a liquid to a gas. How is such energy expressed at the scale of atoms and molecules (combinations of atoms)?

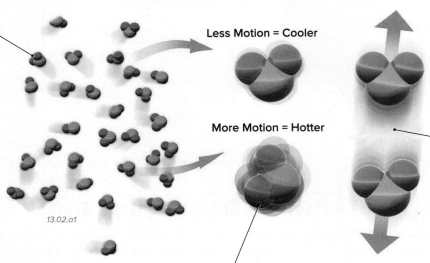

1. Energy is expressed at an atomic level by *motions* of atoms, molecules, and their constituent parts. These motions include changes in position and in-place vibrations and rotations. The more motion the atoms and molecules display, the more energy the system contains. Atoms and molecules in a gas can move at over 200 m/s, or about 500 mi/hr. These fast-moving objects collide with adjacent ones, causing atmospheric pressure in a similar fashion to the way in which air molecules hold out the walls of an inflated balloon.

Less Motion = Cooler

More Motion = Hotter

13.02.a1

3. Energy can also be tied up *within* the atomic or molecular structure of matter. When we heat a liquid, we impart energy to the constituent molecules, causing some of the molecules to move apart, escaping as vapor. But these molecules also carry stored energy, which can be released when the molecules recombine into a liquid. This type of energy is called *potential energy*, because it is not being expressed directly, but could potentially be released. Where potential energy is related to a change in the *state of matter,* such as from a liquid to a gas, as in the example presented here, it is called *latent energy* (latent means hidden).

2. The temperature of an object is a measure of the average energy level (motions) of its molecules and atoms. The molecules of all objects on Earth, regardless of temperature, are moving, some more than others. This type of energy, due to motions of objects, is called *kinetic energy*.

B How Does Energy Relate to the State of Matter?

In our everyday world, matter exists in three different forms or *states*—solid, liquid, and gas. The energy content of the material determines which of these states of matter dominates at any time and place.

1. When matter is in a *solid state*, the constituent atoms and molecules are bound together, such as in the structured internal architecture of a crystal. The energy levels (motions) of the atoms and molecules are low enough that the solid can withstand the vibrations and other motions without coming apart. As a solid is heated up by an external energy source, like the flame, the motions become more intense.

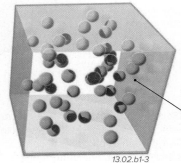

13.02.b1-3

2. If enough energy is added to a solid, the motions begin to break the bonds that hold the solid together. The material begins to melt, turning into a *liquid*, which is a collection of mobile atoms and molecules that more or less stay together but are not held into the rigid form of a solid. For example, adding energy increases the vibrations of water molecules in an ice crystal until the bonds holding the ice crystal together begin to disintegrate and the ice melts.

3. If even more energy is added, such as by placing the liquid water on a hot burner, the added energy causes even more energetic motions of water molecules in the liquid, allowing more and more molecules to break free of the liquid and enter the air as a *gas*—water vapor. As a result, atoms and molecules in their gaseous state are more energetic than in their solid or liquid equivalents. Not only are the molecules in a gas vibrating and rotating intensely, molecules can now move freely through the air at high speeds, having a large amount of kinetic energy.

C What Occurs During the Processes of Warming and Cooling?

When something changes temperature, we say that it is warming or cooling, but what is happening during a change in temperature? For warming or cooling to occur, energy must be transferred from one object to another, such as from a flame to the air around it.

When an object increases in temperature, it is *warming*. Warming occurs where an object *gains energy* from the surroundings, resulting in the increase in temperature. You gain energy when you feel the warming benefit of a campfire (▶). There is no need to touch the flames to feel the warming effect. Energy is transmitted from the fire to the surroundings. When you heat up water in a pan on the stove, energy is transferred from the burner to the pan and then from the pan to the water. If you add enough energy, the water molecules become so warm and energetic that they start to escape the pan during the process of boiling.

13.02.c1

13.02.c2

Cooling is the *loss* of energy to the surroundings. When you are out in cold weather, you lose energy to the cold air, causing you to feel cold. A similar process occurs when you add warm water to ice cubes in a glass dish (◀). Initially, the water is warmer than the ice, but once the water and ice are in contact, molecular motion in the water is transmitted to the cooler ice. This transfer of energy adds more energy to the ice, eventually melting it. As the surrounding water loses energy to the ice, it cools, resulting in cold water. Cooling indicates a loss of energy, whereas warming indicates a gain of energy.

D What Are the Four Types of Energy Transfer?

Heat is thermal energy transferred from higher temperature to lower temperature objects. Heat transfer, also called *heat flux* or *heat flow*, results when two adjacent masses have different temperatures. The four mechanisms of heat transfer are *conduction, radiation, convection,* and *advection*. An understanding of the various mechanisms of energy transfer is crucial in understanding weather, climate, evolution of landscapes, and many of our daily activities.

1. *Conduction*—A water-filled pan on a burner gets hot as thermal energy is transferred by direct contact between the burner and pan, and the pan and water. Heat transfer by direct contact is *conduction*, which involves transferring thermal energy from the warmer object (with more energy) to the cooler one (with less energy). Energy can only be transferred from the more energetic one to the less energetic one. Molecules are most densely packed in solids and least densely packed in gases, so conduction is a more important mode of transfer in solids, whereas gases conduct energy much less efficiently.

13.02.d1

4. *Advection*—Moving a pan full of hot water away from the stove also transfers heat from one place to another. Energy transfer by the *horizontal movement* of a material, such as moving the pan sideways off the burner, is called *advection*. Fog along many coastlines results from warm air flowing sideways and mixing with cooler air over the coast. This horizontal transfer of energy via the moving air is advection.

2. *Radiation*—A hot burner on a stove can warm your hands a short distance away. Such warming occurs because heat from the burner radiates through the air, a process called *radiation* or *radiant heat transfer*. Radiation is energy transmission by means of electrical and magnetic fields. All objects constantly emit various kinds and quantities of this particular type of electrical and magnetic radiation, which is known as *electromagnetic radiation*. This form of energy is even capable of passing through a vacuum.

3. *Convection*—Energy is conducted through the base of the pot and into the lowest layer of water contacting the pot. These molecules move faster, requiring more room to do so, which causes the volume occupied by the warmer water to increase and its *density* (mass per unit volume) to decrease. Being less dense than the overlying cooler water, the warm water rises. When the rising water reaches the surface, it cools and flows back down the sides of the pot. This type of vertical heat transfer by flow of a gas, a liquid, or a weak solid is *convection*. If the material flows around a circular path, as in the pan, we use the term *convection cell*.

Before You Leave This Page

☑ Describe what constitutes the energy in an object.

☑ Describe how energy relates to states of matter.

☑ Describe the four major mechanisms of energy transfer and provide an example of each.

13.2

13.3 What Are Heat and Temperature?

THE TERMS HEAT AND TEMPERATURE are used every day, but what do they actually mean? *Temperature* is a measure of the object's internal kinetic energy—the energy contained within molecules that are moving—and *heat* is thermal energy transferred from one object to another. Moving molecules drive many processes in the Earth-ocean-atmosphere system, such as evaporation, precipitation, and erosion.

A What Is Sensible Heat?

1. The term heat is used in two ways. Scientists use *heat* to refer to the transfer of thermal energy from a warmer object to a cooler one or to the energy that is transferred in this way. The amount of heat is specified in a unit called a *Joule*, a measure of work or energy. Two common examples illustrate heat nicely.

2. What happens when you hold a cup of hot tea? Your hand feels heat coming from the cup. You are feeling the transfer of thermal energy from the cup to your hand. For this to happen, water molecules were heated and made to move. Once in the cup, the moving molecules collided with the inside of the cup, warming it. That heat is transferred through the cup and against your hand via conduction—from the burner to the bottom of the kettle, to the water, to the cup, and finally to your hand. Conduction and convection both help distribute heat within the kettle and the cup.

13.03.a1

3. What do you think happens when you hold a cold glass of ice water? In this case, the molecules in your hand are more energetic than the ones in the cold drink, so heat is transferred from your hand to the glass. Your hand feels cold because you are losing thermal energy to the cold glass.

4. This type of heat, which changes the temperature of two objects through exchange, is called *sensible heat* because we can sense it. But do we sense the actual temperature of an object or just the heat gain or loss? Try this experiment: Find a metal object and a wooden or plastic object in the same place. Place your hand on each and observe what you feel. Go do it, and then come back and continue reading. There, you no doubt sensed that the metal felt colder than the wood or plastic, but both have been in the room for a while and so are exactly the same temperature. This experiment shows that we sense heat gain and loss more than the actual temperature. In your experiment, metal conducted heat away from your hand faster, and so felt colder, but it wasn't.

B What Is Temperature and How Do We Measure It?

Temperature is a quantitative measure of the average kinetic energy of molecules in an object—in other words, the hotness or coldness. Measurement of temperature of an object, whether a solid, liquid, or gas, involves the transfer of sensible heat from the object to some type of measuring device, usually a thermometer. Official temperatures are measured in a variety of ways, depending on the accuracy that is required and the location where temperature is to be measured.

13.03.b1   13.03.b2 Redoubt volcano, AK   13.03.b3   13.03.b4

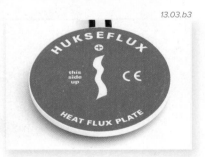

The mercury-in-glass thermometer is the most familiar tool to measure temperature. Mercury is a convenient element for this task because many of its physical properties remain consistent over the range of temperatures experienced on Earth. As the mercury's temperature increases, it expands and fills more of the tube. When it cools, the mercury contracts and withdraws down the tube.

Infrared thermometers calculate the temperature of a solid or liquid surface by pointing the sensor at the surface and measuring a range of wavelengths of energy emitted by that surface. Equations then relate wavelength to energy and energy to temperature. Infrared thermometry is a form of remote sensing, and is convenient when the surface is too far away or too dangerous to be measured directly.

When we need to measure sudden and slight temperature changes very precisely, we use special thermometers in which differences in energy content cause a thermoelectric response that can be wired to a computerized data recorder. The temperature can be calculated using specific equations that derive the amount of heat as a function of the amount of electrical current and the resistance of the electrical circuit.

Temperatures in the atmosphere are usually measured by weather balloons. These include instrument packages called *radiosondes* that measure a range of variables at various heights as the balloon ascends. Temperature is measured using thermoelectric principles. Wind, humidity, and other variables are also measured and relayed to the ground via radio signal.

C How Does the Fahrenheit Scale Relate to the Celsius and Kelvin Scales?

Most Americans are familiar with the *Fahrenheit temperature scale*, in which 32° represents the freezing point of water and 70° is a comfortable temperature. Nearly all nations except the U.S. use the *Celsius scale*, and scientists use the Celsius scale or a related scale called the *Kelvin scale*. We typically compare the scales with reference to the temperatures at which water freezes or boils, called the *freezing point* and *boiling point*, respectively.

1. In the Fahrenheit scale (°F), the boiling point of water is 212° (at sea level, but lower at higher elevations). In the Celsius scale (°C), the boiling point is 100° (at sea level).

2. Typical room temperature is 70°F, which is equivalent to 21°C.

3. The freezing point of pure water is 32°F, which is equivalent to 0°C. The Celsius scale was calibrated to the freezing and boiling points of water, with 100°C separating the two.

4. The two scales correspond to one another at −40° (that is, −40°F equals −40°C).

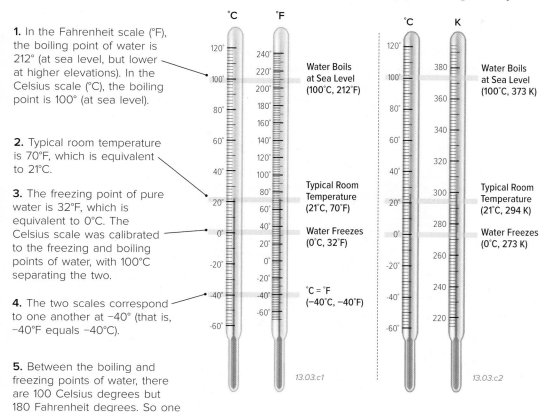

Water Boils at Sea Level (100°C, 212°F)

Typical Room Temperature (21°C, 70°F)

Water Freezes (0°C, 32°F)

°C = °F (−40°C, −40°F)

13.03.c1

Water Boils at Sea Level (100°C, 373 K)

Typical Room Temperature (21°C, 294 K)

Water Freezes (0°C, 273 K)

13.03.c2

6. Both the Celsius and the Fahrenheit scales are "arbitrary" in the sense that zero degrees doesn't mean that there is a lack of internal energy. Likewise, a doubling of the Fahrenheit temperature does not mean that there is twice as much internal energy. In scientific calculations, we need a temperature scale that allows us to relate changes in internal energy to the absolute amount of heat gained or lost by a system.

7. The Kelvin temperature scale (K) was devised as an "absolute" temperature scale to remedy these problems. In the Kelvin system, 0 K corresponds to the temperature at which no internal energy exists and all molecular motion theoretically ceases. This temperature is known as *absolute zero*, and is −273°C or −460°F. Doubling the internal energy of molecules would double their motion and be associated with a doubling of the Kelvin temperature.

8. In the Kelvin system, water freezes at 273 K and boils at 373 K. Converting from Celsius to Kelvin temperature is easy:

$$K = C + 273$$
or
$$C = K - 273$$

Conversions between Fahrenheit and Kelvin can be made by converting first to Celsius and then to Kelvin. Note that we do not use a degree symbol with the Kelvin scale.

5. Between the boiling and freezing points of water, there are 100 Celsius degrees but 180 Fahrenheit degrees. So one Fahrenheit degree is only 100/180 (or 5/9) of a Celsius degree. This fact, along with the different "starting point" (i.e., the freezing point of water) forms the basis of converting between Fahrenheit and Celsius. To convert from Fahrenheit to Celsius, we must first subtract 32 degrees from the Fahrenheit temperature to allow for the fact that the starting point is offset by 32 degrees in the two systems. Then we must multiply by 5/9 to allow for the differences in the value of a degree on each scale. The equations for converting back and forth are as follows:

$$C = 5/9 \times (F - 32)$$
$$F = (C \times 9/5) + 32$$

D How Many Stations Report Temperature?

This globe (▶) shows part of the worldwide distribution of weather stations that have temperature data sets for at least many decades. The distribution is uneven, with most stations being on continents. Most stations are concentrated in densely populated areas, especially in the lowlands of more developed regions, such as the eastern U.S. Other regions, such as the center of South America (the Amazon rain forest), have very few stations to represent rather large areas. In recent decades, remote-sensing techniques have allowed truly global temperature coverage, but these data sets are not available as far back in time.

13.03.d1

Before You Leave This Page

- ☑ Explain sensible heat.
- ☑ Explain what temperature is.
- ☑ Explain how various instruments to measure temperature work.
- ☑ Describe the strengths and weaknesses of the various temperature scales.
- ☑ Convert back and forth between the three temperature scales.
- ☑ Characterize the distribution of temperature-monitoring stations.

13.3

13.4 What Is Latent Heat?

WATER OCCURS IN ALL THREE PHYSICAL STATES—solid, liquid, and gas—at temperatures common on Earth. Although the chemical structure of water remains unchanged from state to state, the three states, also called *phases*, are differentiated by the physical spacing and connections of the water molecules. Considerable quantities of energy, contained as *latent heat*, are involved in these changes of state and act as moderators of global climate.

A What Are the Forms of Latent Heat?

The chemical substance *water* consists of two hydrogen atoms bonded to one oxygen atom, with a chemical formula of H_2O. The change in state between any two of these phases requires an addition of energy or involves release of energy, depending on which direction the change is occurring (e.g., liquid to solid versus solid to liquid). In the figure below, a blue-to-red arrow indicates that a change (e.g., melting) requires energy to proceed, whereas a red-to-blue arrow indicates that energy is released.

1. When ice is placed in warmer surroundings, like an ice cube on a kitchen counter, energy from its environment flows into the ice, increasing the internal motions of water molecules in the solid, crystalline structure. At first the ice heats up (increases in temperature), but once it reaches a certain temperature (its melting point), it begins to melt. *Melting* requires energy to be added.

2. During melting, energy input into the system is stored (absorbed) in water molecules of the liquid—as *latent heat*. The latent heat associated with melting is called the *latent heat of fusion*. If enough energy is removed for the liquid water to be converted back into ice during *freezing*, the latent heat stored in the water molecules is released back into the surroundings as heat (also called *latent heat of fusion*). For freezing to continue, this released energy must be dispersed into the surroundings. The warm flow of air coming from the back or bottom of a household freezer is the heat being dispersed from the cooling system.

3. Conversion from a liquid phase to a vapor phase, *evaporation*, also requires an input of energy from the surroundings, as when you boil water in a pan. During evaporation, energy added to the liquid breaks the bonds holding the water molecules together, allowing molecules to escape as a gas. The liberated gas molecules are moving fast and so have increased kinetic energy and carry energy stored as *latent heat*. When the gas molecules are cooled and recombine into a liquid through the process of *condensation*, the latent heat is released back into the surroundings. The latent heat associated with evaporation and condensation is the *latent heat of vaporization*.

4. Water can also go directly from the solid state to a vapor, the process of *sublimation*. Sublimation requires energy from the surroundings and stores latent heat in the gas molecules—the *latent heat of sublimation*. The reverse process, converting water vapor directly into ice, is called *deposition* and is the main way snowflakes form. Deposition releases the latent heat back into the environment, but when it is snowing it is cold enough that we would not easily notice any addition of latent heat to the cold air.

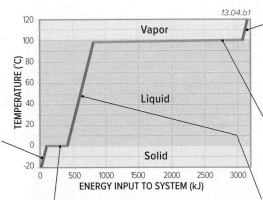

13.04.a1

B What Happens to Temperature During Melting and Boiling?

1. An interesting thing happens to temperature when we melt ice and evaporate water, as shown by the graph to the right. This graph plots the energy input into the system (measured in a unit of energy called a kilojoule) versus the resulting temperature if we start with a kilogram of ice. The process begins in the lower left corner, with ice at −20°C, well below its melting temperature (0°C).

2. The initial input of energy into the system causes ice to increase in temperature, represented by the first inclined, red part of the line. The increase in energy is expressed as sensible heat.

6. Once all the water has become vapor, further heating causes the steam to increase in temperature (inclined, red line). In this entire process, more energy was used to change states (brown horizontal lines) than was used to increase temperature (the inclined, red lines). Of the total amount of energy used, less than 20% went to change temperature and more than 80% was used to change state!

5. As the water starts to boil, it does not increase in temperature (long, horizontal brown line). All increase in energy is used to convert liquid into the vapor phase and is stored as latent heat.

3. When the temperature reaches 0°C, the melting point, the ice starts to melt. The temperature does not change during melting (as shown in the short, horizontal brown line). Instead, all the increase in energy is going into breaking the bonds and is stored as latent heat.

4. Once all the ice is melted, the increase in energy again causes an increase in temperature (sensible heat), as shown by the second inclined red line. The temperature of the water increases until it reaches the boiling point (100°C).

C   What Does Latent Heat Do to the Surroundings?

Latent energy added to the water molecules or released by the water molecules allows the phase change to proceed, but it also impacts the temperature of the surrounding environment. More than five times the energy is involved in evaporating or condensing water (the latent heat of vaporization) than in raising the temperature of the same mass of water from the freezing point to the boiling point. The large quantities of latent heat have a huge role in many aspects of our world, including changes in atmospheric temperature.

Water: Vapor, Liquid, and Ice

1. This graph shows conditions under which each phase occurs for water. For reference, a pressure of 1.0 bar is the average atmospheric pressure at sea level, and 21°C is a typical room temperature.

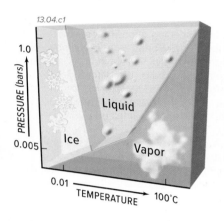

13.04.c1

2. Ice occurs at low temperatures, whereas liquid and water vapor are favored by higher temperatures. Higher pressure acts to hold the water molecules within the liquid rather than allowing them to escape into the air. If we cool vapor, we get liquid at higher pressures and ice at lower ones, causing the formation of clouds or precipitation, both of which can consist of liquid (drops of water) or ice.

3. Phases, such as vapor, that require more energy to form are called *high-energy states*, whereas less energetic ones are *low-energy states*.

Cooling and Heating the Air

4. When water in the atmosphere changes state, it releases or takes in thermal energy, heating or cooling the surrounding air. This diagram illustrates the change in air temperature during phase changes. Red arrows indicate that phase change in this direction causes the surrounding air to heat up. Blue arrows indicate that the surrounding air must provide heat to the phase change, so the air cools.

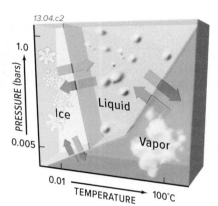

13.04.c2

5. Energy must be released into the surroundings for water to go from a higher energy state to a lower energy state. Heat is released (red arrows) when water vapor forms droplets or ice crystals, or when liquid water freezes—this warms the surrounding air.

6. Heat is taken in from the surroundings (blue arrows) when ice melts or water evaporates, or when ice sublimates. Any of these processes cool the air.

D   What Are Examples of Latent Heat in the Environment?

13.04.d1 Hurricane Dolores

1. In 2015, Hurricane Dolores formed in the Eastern Pacific and came onshore in Southern California, where it caused heavy rainfall, flooding, and landslides. Hurricanes release a huge amount of latent heat as water vapor condensed into water drops in the clouds and as rainfall. In a single hurricane, the amount of latent heat released during condensation is equivalent to much more than all the electrical energy consumed by the world since electricity was first harnessed for human use.

2. A cold beverage in a can, bottle, or glass warms as a result of conduction from its warmer surroundings, but it warms even more quickly because condensation (as expressed by growing water drops) releases latent heat on the outside of the container.

13.04.d2

3. Severe freezes are a serious threat to the citrus industry in the southern and southwestern U.S. Sustained subfreezing temperatures may ruin the crop and even destroy the tree. To prevent extensive losses, grove owners spray the trees with water, which freezes on the crop. At first sight the weight of the accumulated ice only seems to add to the damage by breaking off fruit and limbs, but each kilogram of water that freezes on the trees releases enough energy to the fruit and plant to prevent cold temperatures from destructively freezing water within the cells of the plant.

13.04.d3 Florida

Before You Leave This Page

☑ Explain the relationship between latent heat and changes in the state of water.

☑ Compare the quantities of energy involved in the latent heats of fusion and vaporization, compared to changes in temperatures.

☑ Sketch and describe how latent heat affects the environment, providing some examples.

13.4

13.5 What Is Electromagnetic Radiation?

ELECTROMAGNETIC RADIATION is one of the fundamental forces of nature. It dominates our daily interactions with the world, determining the color of objects and the physical characteristics of air, water, and land. Electromagnetic radiation from the Sun controls our weather and climate and is essential to all life on Earth's surface.

A What Is the Character of Electromagnetic Radiation?

Electromagnetic radiation (EMR) is all around us, with its most obvious expression being visible light. EMR is also what causes sunburns, is the heat felt from a heat lamp, and is the basis of wireless communications, including TV, radio, WiFi, and microwaves. It consists of interacting electrical and magnetic fields radiated from charged particles.

The Nature of EMR

1. Although EMR consists of electrical and magnetic fields, we can visualize EMR as a series of waves of electrical and magnetic energy that are moving from one place to another, in this case, from left to right (▶). Like any waves, some parts are higher and some are lower. The direction in which a wave is moving is the *direction of propagation*. The wave shown here is propagating from left to right. To envision how such a wave moves, think about what happens when you shake the end of a rope or string. The rope curves into the series of waves, which move from your hand outward to the end of the rope.

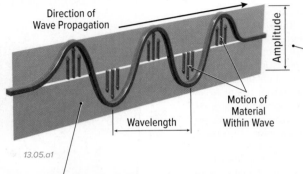
13.05.a1

2. In this type of wave, the motion of any part of the wave is mostly up and down, parallel to the small arrows on this figure—the rope stays in your hand, so it is not moving away from you, although the waves are. Motion within the wave (up and down) is perpendicular to the direction of propagation of the wave (left to right).

3. In describing such waves, whether in a rope or EMR, the term *amplitude* refers to the height of the wave, from trough to crest. The greater the difference in height between the top and bottom of the wave, the greater the amplitude. Surfers like large-amplitude waves.

4. We use *wavelength* to describe the distance between two adjacent crests (tops) or two adjacent troughs (bottoms). The longer the distance between two adjacent crests or troughs, the greater the wavelength. If we know the speed of the wave, we can easily calculate the *frequency*—the number of waves per second passing a point.

Generation of EMR

5. Electromagnetic radiation is emitted by atoms and molecules. Atoms have a tiny central core—the *nucleus*—that is so much smaller than the entire atom that it cannot be shown here. The red and yellow spheres are electrons. Groups of electrons travel around the nucleus at different distances, called *electron shells*, each of which has a different level of energy, increasing away from the nucleus. If an electron in an outer, higher energy shell drops into an inner, lower energy shell, it emits the extra energy as EMR that radiates out in all directions (shown here as one direction for simplification).

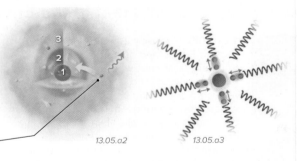
13.05.a2 13.05.a3

6. Electrons within atoms and the bonds (shown here as springs) within molecules vibrate back and forth, changing positions slightly. This motion emits EMR, and the faster such motions occur, the higher the frequency of the EMR that is emitted.

Relationship Between Temperature and EMR

7. Not all molecules in an object vibrate or move at the same speed, so an object emits energy with some variation in the wavelengths and amounts of energy it emits. Some molecules vibrate or move faster, emitting EMR with shorter wavelengths and higher energies. Other molecules vibrate or move more slowly, resulting in EMR with longer wavelengths and lower energies. Temperature is a measure of the average energy content of all molecules in an object.

8. Examine this figure, which shows three blocks of the same material but at different temperatures, with the blue block being cool, the tan block being warmer, and the red block being the hottest. Coming off each block are arrows depicting the amount and wavelength of EMR being emitted. Observe how the amounts and wavelengths relate to temperature.

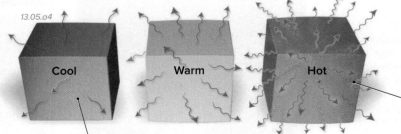
13.05.a4

9. If an object is relatively cool, the motion of its constituent atoms and molecules will be relatively slow. As a result of this slower motion, the block emits relatively low amounts of EMR, and the emitted EMR has relatively long wavelengths.

10. As the temperature of an object increases, the atoms and molecules begin moving faster, and they change energy states more often, producing EMR with a shorter wavelength. As a result, a hotter block emits more EMR, and the wavelength of the EMR becomes shorter at higher temperatures. If you shake a rope rapidly, you make shorter waves than if you shake it slowly. Try it if you are curious.

B   What Is the Electromagnetic Spectrum?

Objects in the universe emit a diversity of wavelengths of electromagnetic radiation. EMR varies from relatively long-wavelength, low-energy radio waves to short-wavelength, high-energy X-rays and gamma rays. The different wavelengths of electromagnetic radiation are arranged in what is called the *electromagnetic spectrum*, shown below. Inspect this figure and read the associated text. In this figure, the longer wavelengths (measured in meters) are at the top. Accordingly, shorter frequencies, which indicate the number of waves per second (measured in Hertz), are also at the top.

13.05.b2 Manti, UT

1. The longest wavelengths of EMR are *radio waves*. These include the waves that carry signals from typical FM and AM radio stations, and also include even longer radio waves called VLF (for very low frequency), with wavelengths of up to more than 100 km. VLF waves can penetrate significant depths of water and so are used for communication with submarines.

2. *Microwaves* are another form of EMR, with shorter wavelengths, higher frequencies, and higher energies than radio waves. Microwave ovens use a specific frequency of microwave that energetically excites (heats) water molecules.

3. Next on the scale is *infrared energy* (IR). Although we cannot see IR, our skin is sensitive to it, so we often think of infrared radiation as heat. Infrared energy is incredibly important on Earth, playing a key role in keeping our planet a hospitable temperature. There are several types of IR, including *thermal-IR*, which is close to microwaves in wavelength, and *near-IR*, which is near to visible light, the next entry on the spectrum.

4. *Visible light* occupies a relatively narrow part of the spectrum. It varies from red colors at long wavelengths to violet colors at short ones. Orange, yellow, green, and blue are in between red and violet.

5. Next to, and with shorter wavelengths (higher frequencies) and more energy than visible light, is *ultraviolet light* (UV). Ultraviolet is more energetic than visible light and is known to cause skin cancers and possible genetic mutations.

6. The shortest wavelength (highest frequency and therefore most energetic) waves are *X-rays* and *gamma rays*. These are potentially harmful. The energy in X-rays is used in medical technology as it will pass through soft tissue, but not bone.

7. The wavelengths of visible light are less than a millionth of a meter (10^{-6} m), which is a measure called a *micrometer* or *micron* for short and depicted as μm. The different colors of visible light are different wavelengths of energy, ranging between 0.4 μm (violet) and 0.7 μm (red). The component wavelengths of visible light can be observed (▲) when the light is split by an optical prism or by waterdrops in a rainbow.

8. The human eye is sensitive to radiation of the wavelengths between violet and red, and this is why this portion of the EMR spectrum is known as "visible light." The Sun's wavelengths are concentrated at 0.5 μm. This wavelength is in the middle of visible light, coinciding with blue light. It is not a coincidence that the human eye developed to detect EMR of the wavelengths that carry most energy from the Sun.

Spectrum diagram:

INCREASING FREQUENCY ↓

(Hz)
1
10^2
10^4
10^6
10^8
10^{10}
10^{12}
10^{14}
10^{16}
10^{18}
10^{20}
10^{22}
10^{24}

(m) — INCREASING WAVELENGTH ↑
10^8
10^6
10^4
10^2
10^0
10^{-2}
10^{-4}
10^{-6}
10^{-8}
10^{-10}
10^{-12}
10^{-14}
10^{-16}

Long Radio Waves · AM Radio · FM Radio · Microwaves · Infrared · Ultraviolet · X-Rays · Gamma Rays

13.05.b1

Visible Spectrum — INCREASING WAVELENGTH IN MICROMETERS (μm) ↑
0.7
0.6
0.5
0.4

9. The Sun produces huge quantities of energy per square meter of its surface, as indicated by the yellow area on the graph to the right. It also produces most of its energy near 0.5 μm, centered on the wavelength of visible light; these wavelengths are called *shortwave radiation*. The Sun emits a very low proportion of its energy below 0.1 μm or above 1.0 μm in wavelength.

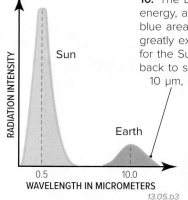

RADIATION INTENSITY

Sun

Earth

0.5   10.0
WAVELENGTH IN MICROMETERS
13.05.b3

10. The Earth emits low quantities of energy, as indicated by the relatively small blue area on the graph, which is here greatly exaggerated relative to the curve for the Sun. Most of that energy is emitted back to space at wavelengths of around 10 μm, called *longwave radiation*. As we will discover, the fact that Earth receives its energy at short wavelengths, and returns it at wavelengths 20 times longer, is of fundamental importance to maintaining temperatures on Earth favorable for life.

Before You Leave This Page

☑ Describe why an object emits a range of EMR wavelengths.

☑ Describe the relationship among molecular motions, wavelengths of emitted EMR, and temperature.

☑ Summarize the electromagnetic spectrum, indicating the relative order of different types of EMR on the spectrum.

13.5

13.6 What Causes Changes in Insolation?

THE ENERGY TRANSMITTED from the Sun to Earth, called incoming solar radiation, or *insolation*, has varied only slightly during the short time for which we have accurate measurements from satellites. It does vary somewhat from year to year and between different times of the year, and it varies greatly as a function of latitude.

A How Does the Amount of Insolation Change During a Year?

Earth receives a relatively consistent amount of insolation, equivalent to 1,366 watts per square meter (W/m^2), an amount called the *solar constant*. The amount of insolation varies during the year because Earth is closer to the Sun at certain times of the year, but not the times you would expect.

1. Although the average Earth-Sun distance is 150 million km, Earth's orbit is slightly elliptical rather than circular. The orbit is so nearly circular that it is virtually impossible to notice the elliptical shape in an accurately drawn view looking perpendicular to the orbital plane (▶). However, Earth is slightly closer to the Sun at some times of year than others, with the Earth-Sun distance varying about 3% during the course of the year.

2. The date of closest approach is called the *perihelion*, where "peri" is a Greek word for "around" and "helio" refers to the Sun. Perihelion currently occurs in early January, but the exact date varies a bit each year and moves forward by one day each 65 years. Over several hundred thousand years, it may change season completely, influencing our climate.

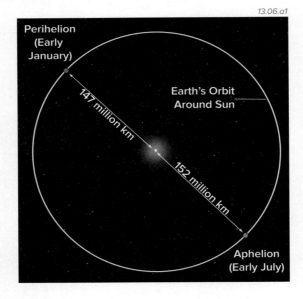

13.06.a1

Perihelion (Early January)

147 million km

Earth's Orbit Around Sun

152 million km

Aphelion (Early July)

3. The date of farthest approach is the *aphelion*, where "apo" comes from a Greek word meaning "from." The aphelion is early in July, varying slightly from year to year. In other words, we are farthest from the Sun during our summer—the seasons cannot be explained by our changing distance from the Sun.

4. This variation in Earth-Sun distance, from perihelion to aphelion and back again, constantly changes the average amount of insolation Earth's atmosphere receives. When Earth is closest to the Sun (during the Northern Hemisphere winter) it intercepts more of the insolation. The total amount of variation in energy related to Sun-Earth distance is about 7%.

B Does the Sun's Output of Energy Change from Year to Year?

Like most natural systems, the Sun's activity varies, sometimes increasing and other times decreasing. The Sun exhibits cycles that repeat over the course of a decade or so, and it has longer fluctuations that occur over multiple decades. These cycles influence the Sun's total output of energy, called the *total solar irradiance* (or simply TSI).

1. This top pair of images (▶) shows the Sun at two different times, approximately 9 years apart. In the left image, the Sun has a number of dark spots, termed *sunspots*. These are places that are slightly cooler than the rest of the Sun, and so show up darker. In the image to the right, the Sun lacks any sunspots. The number of sunspots varies from year to year.

2. This second pair of images shows the Sun at the same two times as in the top images, but displays energy at ultraviolet wavelengths (▶) rather than visible light. They indicate that the Sun is much more active and emitting more overall energy when there are more sunspots. So more sunspots means more energy output from the Sun. Satellite data show small (<1%) monthly changes in the output of solar energy and a correlation between sunspot activity and TSI.

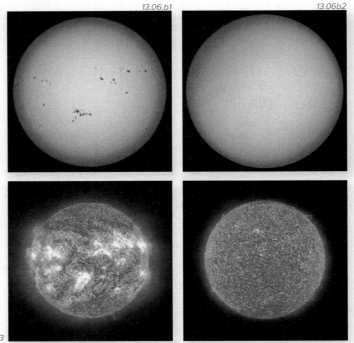

13.06.b1

13.06b2

13.06.b3

13.06.b4

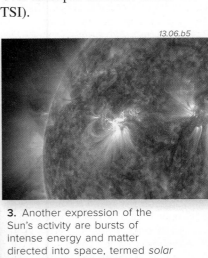

13.06.b5

3. Another expression of the Sun's activity are bursts of intense energy and matter directed into space, termed *solar flares* (▲). The bright regions producing solar flares in the image above are much larger than Earth. Solar flares pose a hazard to people and equipment in space and affect wireless communications on Earth.

C How Does the Angle of Incidence Affect the Amount of Insolation Received?

On Earth, the amount of sunlight varies from time to time and from place to place. In the summer, sunlight feels more intense than it does in winter. Also, there are huge ranges in temperature between the warm, tropical regions of the world and the cold, polar regions. A main factor controlling the maximum possible amount of insolation is position on the earth, specifically the latitude.

1. When thinking about insolation, an important aspect to consider is the orientation of energy with respect to the surface being irradiated. In the figure to the right, a flat surface is illuminated by two identical lights, each with an energy equal to the amount of solar energy striking the top of the atmosphere. The left light is shining directly down on the surface, so the surface receives the same amount of energy per square meter as the light emits.

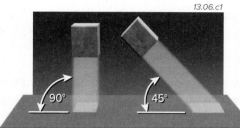
13.06.c1

2. The second light is angled 45° relative to the surface, so when its light strikes the surface, the light is spread out over a larger area. As a result, the amount of energy per area is less, and the area would be less brightly lit. For an angle of 45°, the energy is spread out over an area 1.41 times longer than for the left light, so the resulting energy is about 70% less. The angle between the direction of energy emission and the surface is called the *angle of incidence,* and is 90° for the left light and 45° for the right one.

3. This figure (▶) shows the Sun at four different positions, corresponding to four different angles of incidence, each measured at noon, for four different latitudes. The 10 thin orange rays of sunlight represent equal amounts of energy (the solar constant). The brackets on the ground indicate the width over which that same amount of sunlight strikes the surface.

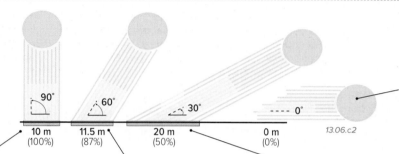

13.06.c2

7. For a location directly at Earth's North or South Pole (the Sun position on the far right), the angle of incidence can be 0°, and none of the 10 rays of sunlight hit the ground. Brrrrrrrrrrr!

4. In this example, if the angle of incidence at noon is 90°, as happens near the equator and in the tropics, the 10 rays of sunlight intersect Earth's surface over a width of 10 meters. The light is perpendicular to the surface, so the area receives the maximum amount (100%) of energy possible.

5. In a subtropical latitude, with a 60° angle of incidence, the 10 rays of sunlight are spread out over a wider area, so the surface receives less insolation per square meter compared to a site at the equator, but not much less (87%).

6. If the angle of incidence is smaller, like 30° in a high-latitude location, like in Canada, the 10 rays of sunlight are spread out over 20 m (twice the area as at the equator), so any area receives only 50% of the sunlight per square meter that the equator does.

8. The figure below illustrates how equal rays of solar energy strike different parts of the earth, varying in angle of incidence, from a position along the equator to ones nearer the pole. It only shows the southern half of the planet, but the same relationship applies to the northern half, too.

13.06.c3

9. Some insolation strikes the outside of Earth's atmosphere perpendicularly, like a light pointed directly toward a surface. This part of the atmosphere is receiving the maximum insolation possible, or an amount of energy equivalent to the solar constant, at the "top" of the atmosphere.

10. In the mid-latitude regions, insolation strikes the outside of Earth's atmosphere at a slightly oblique angle and is distributed over a wider area than at the equator. Thus, the amount of energy is distributed over a larger area compared to the equator, and the energy per area is less. Insolation arriving above polar regions strikes Earth's atmosphere at a very low angle of incidence, so the energy is spread over a relatively large area and is less intense.

11. The atmosphere reflects and absorbs some energy, so the amount of insolation that reaches Earth's surface is influenced by the angle of incidence (▶). Insolation that comes directly down at the surface, like a vertical light, passes through the atmosphere in the shortest distance possible, so less energy is lost. Energy that comes at a lower angle of incidence has to pass through more atmosphere, so more energy never reaches the surface.

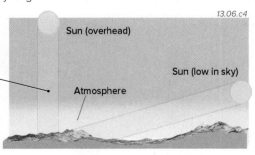
13.06.c4

12. The angle of incidence varies as a function of how high the Sun is in the sky. The angle of incidence varies with the slope of the land, but to simplify things, we will often consider a flat land surface. When the shadow of an object is directly underneath the object (▶), the angle of incidence is 90°.

13.06.c5

Before You Leave This Page

✓ Explain why the amount of insolation coming from the Sun varies during a year and from year to year, including the variation in sunspot activity.

✓ Sketch and explain how the angle of incidence affects how much insolation hits different parts of the atmosphere and reaches the surface.

13.6

13.7 Why Do We Have Seasons?

MOST LOCATIONS progress through different seasons, from warmer summers to cooler winters and back again. The progression from season to season accompanies changes in the position of the Sun, such as from higher in the sky during the summer to lower in the sky during the winter. Except at the equator, the durations of daylight and darkness also vary, with longer days during the summer and longer nights during the winter. What causes most parts of Earth to have different seasons? Why do tropical areas not have summer or winter?

A What Causes the Seasons?

The annual "march of the seasons" indicates variations in the amount of sunlight received at different latitudes during the course of a year. The main cause of these variations is that Earth's axis of rotation, about which our planet spins once during a 24-hour day, is tilted relative to the plane in which we orbit the Sun, the *orbital plane*.

1. Earth's axis of rotation is tilted relative to our planet's orbital plane, as shown by the figure below, which is a sideways perspective of Earth's nearly circular orbit. The axis remains fixed in orientation with respect to the orbital plane and the stars throughout the year—the tilt does not change during a year. As Earth orbits the Sun, the Northern Hemisphere faces in the direction of the Sun during some times of the year, while the Southern Hemisphere faces the Sun at other times. Observe the relationship between the orientation of the rotation axis and the direction to the Sun.

7. The seasons, therefore, are controlled by the relationship between the tilt of Earth's rotation axis and the orbit of Earth around the Sun. Each season is bounded by a solstice and an equinox, which in turn represent days of either a maximum or minimum amount of apparent tilt relative to the Sun.

2. During December, the North Pole and Northern Hemisphere face away from the Sun. The Northern Hemisphere therefore receives less direct insolation and so experiences the cooler temperatures of winter. Rotation about the axis brings most parts of the hemisphere into the sunlight for part of the day.

6. In late June, the North Pole and Northern Hemisphere more directly face the Sun and so start to experience summer. The *June Solstice* (Summer Solstice in the Northern Hemisphere) marks the date when the North Pole's rotation axis is pointed most directly toward the Sun. At the same time, the South Pole and Southern Hemisphere face away from the Sun and start to experience winter.

23.5°

September 21

(COMPONENTS NOT TO SCALE)

December 21

June 21

23.5°

23.5°

23.5°

13.07.a1

March 21

3. In contrast, the South Pole and Southern Hemisphere more directly face the Sun and receive more direct insolation. The Southern Hemisphere therefore has its summer (at the same time the Northern Hemisphere has its winter). The *December Solstice* (Winter Solstice in the Northern Hemisphere, Summer Solstice in the Southern Hemisphere) is the date when the North Pole (marking Earth's rotation axis) is pointed farthest away from the Sun.

4. As Earth orbits around the Sun, the rotation axis no longer points directly away from the Sun. In March (and later again in September), Earth's axis is pointing sideways relative to the Sun—neither the North Pole nor the South Pole is inclined toward the Sun. In this position, neither hemisphere receives more insolation than the other, and so both experience the more moderate temperatures of spring and fall. The *March Equinox* and *September Equinox* are times when the axis is exactly sideways and the durations of daylight and darkness are equal (equinox is Latin for "equal night").

5. Between March and June, Earth's continued orbit causes the North Pole and Northern Hemisphere to increasingly face the Sun and warm up as summer approaches. During this time, the South Pole and Southern Hemisphere begin to face away from the Sun, cooling down on the way to winter.

8. In thinking about the seasons, it is important to remember that Earth's orbit is nearly circular, as shown in this view looking straight down on Earth's orbital plane (▶). In this figure, as in the one above, the sizes of the Earth and Sun are greatly exaggerated relative to the size of the orbit, and the Earth is shown much larger than it actually is relative to the Sun.

9. The nearly circular geometry of Earth's orbit reinforces the fact that seasons are not caused in any way by differences in distance from the Sun. In fact, Earth is slightly closer to the Sun during the northern winter (perihelion) and slightly farther away during the northern summer (aphelion), opposite to what would be required to explain the seasons.

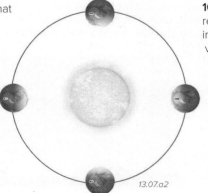

13.07.a2

10. In each globe (◀), Earth's rotation axis is represented by the small purple rod protruding from the North Pole (pointing toward the viewer and a little toward the left). Note that the axis points in the same direction throughout the year—it points toward Polaris, the North Star (not shown). Over thousands of years, the axis varies in orientation, not always pointing at Polaris. The amount of tilt varies by a degree either way (22.5° to 24.5°) over 40,000 years, but we are most interested here in the current tilt of 23.5°, an angle that will reappear throughout this book.

B What Factors Determine the Temporal and Spatial Variations in Insolations?

Earth's rotational axis is tilted relative to the plane in which we orbit the Sun (the orbital plane). The tilt of the axis is currently 23.5° from vertical to the orbital plane, an angle that is reflected in important geographic features on our planet, including how we define the Tropic of Cancer, Tropic of Capricorn, Arctic Circle, and Antarctic Circle. For this discussion, we use the concept of sun angle, which is the angle of incidence at noon for a location and specific day.

The Tropics

1. These two globes show the Earth on the solstices—days when one of the poles is most exposed to the Sun and the other pole most faces away. During the December Solstice, the noon Sun is no longer directly overhead at the equator but instead is 23.5° to the south, directly over the *Tropic of Capricorn*. On this day, the equator has a Sun angle of 66.5° (90°–23.5°).

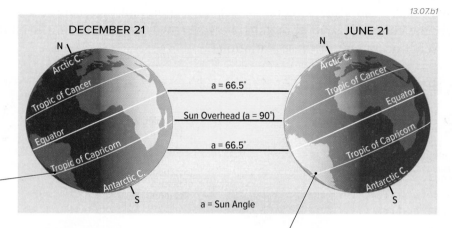

13.07.b1

4. The rest of the year, the overhead Sun migrates between the Tropics of Cancer and Capricorn, delivering relatively intense and constant insolation between these latitudes, causing a warm region called the *tropics*. In contrast, mid-latitude and polar regions receive a highly variable supply of energy and exhibit more variability during the year—that is, they exhibit more *seasonality*.

2. During the December Solstice, the Southern Hemisphere receives the maximum amount of insolation it will receive all year, marking the start of the southern summer. Also on this day, the Northern Hemisphere receives the least amount of insolation for the year, marking the beginning of the northern winter.

3. During the June Solstice, Earth is on the opposite side of the Sun, and the North Pole faces toward the Sun by its maximum amount. The overhead Sun, and the position of maximum insolation, have migrated to a latitude 23.5° north of the equator, the *Tropic of Cancer*. On the day of the June Solstice, the Northern Hemisphere receives the most insolation of the year, shifting into summer, while the Southern Hemisphere receives the least amount and moves into winter.

The Polar Circles

5. These two globes show the same two days as those above, the solstices. During the December Solstice, the northward limit of sunlight is the *Arctic Circle*, located at a latitude of 66.5° N, or 23.5° south of the North Pole. Along the Arctic Circle, the Sun will barely appear on the horizon at noontime and then disappear for a very long night. All places north of that latitude receive no insolation on this day—they have 24 hours of darkness.

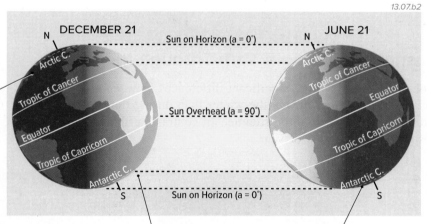

13.07.b2

8. At this time, the area within the Arctic Circle (66.5° N) is fully illuminated. Note that the same angles keep reoccurring—23.5° and 66.5° (which is 90° minus 23.5°). These values, for the two tropics (Cancer and Capricorn) and for the two polar circles (Arctic and Antarctic), are the same as the tilt angle of the rotation axis. The tilt of Earth's axis controls the locations of the tropics and polar circles.

6. Note that the entire *Antarctic Circle*, in the Southern Hemisphere, is illuminated, but with very oblique sunlight (low Sun angles). As the Earth completes its daily rotation, the South Pole and nearby areas remain in the sunlight all 24 hours—these places have 24 hours without darkness.

7. During the June Solstice, locations in the Southern Hemisphere now face away from the Sun. The southern limit of sunlight is the Antarctic Circle, at latitude 66.5° S. Here, the Sun will appear on the horizon for a brief moment at noon.

City	Latitude	Equinox	Dec. Sol.	June Sol.
Saskatoon, Canada	52.0° N	38.0°	14.5°	61.5°
Jacksonville, FL	30.5° N	59.5°	36.0°	83.0°
San José, Costa Rica	10.0° N	80.0°	56.5°	76.5°
Quito, Ecuador	0.0°	90.0°	66.5°	66.5°

9. Using the 23.5° angle and an area's latitude, we can calculate the Sun angle for any place on Earth. Examine this table (◄), which lists Sun angles for the equinoxes and both solstices for several cities. Note that the Sun angle varies by 47° from solstice to solstice—47° is two times 23.5°.

Before You Leave This Page

☑ Sketch and explain how the position of the Earth with respect to the Sun at various times of the year explains the seasons.

☑ Explain the solstices and equinoxes in terms of Earth-Sun position, seasons, and locations of tropics and polar circles.

13.7

13.8 What Controls When and Where the Sun Rises and Sets?

THE SUN RISES EACH MORNING and sets each evening, but at slightly different times from day to day. Also, the Sun does not rise or set in exactly the same direction every day, although the changes from day to day are so gradual as to be unnoticeable. Over the course of several months, however, we can notice significant changes in where and when sunrise and sunset occur. What accounts for these variations?

A Why Does the Sun Rise and Set?

1. At any moment in time, half of the Earth's surface area is sunlit, experiencing day, and half is in the darkness of night. The dashed line encircling the world separates the lighted and dark halves and is called the *circle of illumination*. When viewed straight on, as in these figures, the circle of illumination appears as a line, but it has a curved shape from any other perspective. The colored areas on these globes are time zones.

2. In the left globe, North America and western South America are on the side of the Earth hidden from the Sun, and so it is night. At the same time, eastern South America and Africa are on the side facing the Sun and so are in daylight. With a rotating Earth, this view is only an instantaneous snapshot of which areas are in sunlight and which are in darkness.

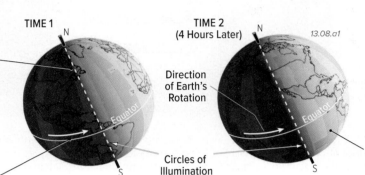

4. The *circle of illumination*, the boundary between day and night, moves westward across the surface as the planet rotates eastward. As this occurs, the Sun has not changed position—the Earth has simply rotated. It finishes a complete rotation in 24 hours.

3. In the right globe, the Earth has rotated an additional four hours—the globe rotates to the right when viewed in this perspective, or counterclockwise when viewed from above the North Pole. At this later time, South America and eastern North America have rotated into the sunlight (it is morning), but the west coast of the U.S. is still in the last hour of night.

B Why Does the Length of Daylight Vary Through the Year?

In high-latitude parts of the world, there are significant differences between the length of daylight from season to season. In such regions, days are noticeably shorter during the winter than during the summer. In accordance with this, nights are longer during the winter and shorter during the summer. In contrast, at the equator, there are always 12 hours of day and 12 hours of night, irrespective of the time of year.

1. These globes show the circle of illumination at three different times of year: the December Solstice, an equinox, and the June Solstice. To help us visualize the circle of illumination, the three larger globes are depicted as if being observed by the small figure next to the corresponding small globes. The axial tilt remains fixed in orientation as Earth orbits the Sun, but here it is portrayed from different perspectives.

2. During an equinox (the large, center globe), the tilt axis is oriented neither toward nor away from the Sun. So, the pattern of light and dark is symmetrical with respect to the equator and other lines of reference. It takes the same amount of time for every location to rotate in and out of sunlight. At equinox, every location on Earth has 12 hours of sunlight and darkness.

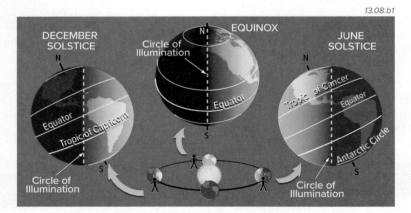

3. At other times of the year, Earth's axis appears tilted toward or away from the Sun, and so the circle of illumination is not symmetrical relative to lines of reference, such as the tropics. In the left globe, representing the December Solstice, any line of latitude in the Southern Hemisphere, such as the Tropic of Capricorn, is more in sunlight than in darkness. Therefore, days are longer and nights are shorter. The opposite is true for any latitude in the Northern Hemisphere, which is more in the dark than in the light, causing nights to be longer than days.

5. The variation in the lengths of day versus night from season to season increases with increasing latitude, either north or south of the equator. Such seasonal changes in day length are absent at the equator and greatest at the poles.

4. During the June Solstice and adjacent months, the opposite is true—more of the Northern Hemisphere is in sunlight than is in darkness. As a result, days are longer than nights in the Northern Hemisphere during this time (the northern summer). In contrast, more of the Southern Hemisphere is in darkness, so days are shorter and nights are longer during this time (the southern winter).

C Why Do Arctic Areas Sometimes Have 24 Hours of Sunlight or Darkness?

The most extreme variations in the lengths of daytime and nighttime occur in the highest latitudes, including the Arctic region around the North Pole and the Antarctic region around the South Pole. North of the Arctic Circle and south of the Antarctic Circle, summer days can have more than 24 hours of straight daylight. During the winter, it can remain dark for all 24 hours, night after night. Either condition can last for months. How is this so?

13.08.c1 Arctic Circle

13.08.c2

This image combines different photographs to show the path of the Sun during several hours at a location north of the Arctic Circle. The Sun remains low in the sky and dips toward the horizon at midnight, but never actually sets—this location has 24 hours of sunlight during the middle of summer. The low Sun angle means that insolation striking the land is spread out and so is relatively weak, and it has a long and attenuated path through the air.

The figure above shows sunlight on the north polar region during the December Solstice. All the area within the Arctic Circle is in darkness (it is on the side opposite to the Sun). As the Earth rotates about its axis, the entire area within the circle remains out of the sunlight—24 hours of darkness. In days following the solstice, sunlight begins to creep into the Arctic Circle, so less of the Arctic has 24-hour nights.

The opposite situation occurs during the June Solstice, shown here. Note that the entire Arctic Circle faces the Sun and will remain in sunlight as the Earth rotates about its axis. On days before and after the solstice (when the North Pole less directly faces the Sun), areas just inside the Arctic Circle would not be in constant sunlight, and so would have slightly less than 24 hours of sunlight and would have minutes to hours of nighttime.

D What Controls the Time and Direction of Sunrise and Sunset?

Except at the equator, the times of sunrise and sunset shift slightly from day to day, but typically by less than a few minutes every day. From month to month, however, we notice significant differences in the times of sunrise and sunset, and therefore in the duration of day and night. The changes in time are accompanied by gradual changes in the direction from which the Sun rises and the direction in which it sets.

1. This figure depicts where and when the Sun rises and sets at a location at 45° N latitude, which is halfway between the equator and the North Pole, like Minneapolis, Minnesota. Observe this figure and note how the locations of sunrise and sunset change by date.

2. The December and June dates, marking the two solstices, show the two extremes. The locations and times of sunrise and sunset fall between these two extremes for all other dates. Either equinox is halfway between the solstices. Each day, the Sun's path defines a circle, which lies on a plane that is inclined at an angle that is equal to the site's latitude (45° in this case).

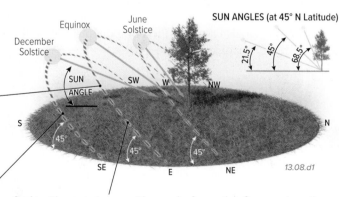
13.08.d1

3. At the December Solstice, the Northern Hemisphere faces away from the Sun the maximum amount, and so the Sun is as low in the sky as it ever gets. The Sun rises and sets as far south as any day of the year, rising in the southeast and setting in the southwest. This is the shortest day and longest night of the year.

4. At either equinox, neither pole faces the Sun, and so there are 12 hours of daylight and 12 hours of darkness at every location on Earth. Everywhere on Earth, the Sun rises due east of the site and sets due west on an equinox. If you wanted to determine directions without a compass and it happened to be the date of an equinox, you could precisely determine an east-west direction by drawing a line from the direction of sunrise to the direction of sunset, as was done by ancient cultures. Also, you could determine your latitude by measuring the Sun angle at noon on this day—your latitude is equal to 90° minus the Sun angle.

5. At the June Solstice, the Northern Hemisphere faces the Sun, and so the Sun is the highest in the sky it will be all year. The Sun rises and sets as far north as any day of the year, rising in the northeast and setting in the northwest. This is the longest day and the shortest night of the year.

Before You Leave This Page

☑ Sketch, label, and explain why the length of day and night vary during the year, and which dates have the longest and shortest daylight hours.

☑ Sketch, label, and explain why a polar region can have 24 hours of daylight or 24 hours of darkness.

☑ Sketch, label, and explain where the Sun rises and sets, and has the highest and lowest Sun angles, for different times of the year.

13.8

How Does Insolation Interact with the Atmosphere?

INSOLATION REACHES THE EARTH but has to pass through the atmosphere before it reaches us. The atmosphere does not transmit all of the Sun's energy; some wavelengths of energy are partially or completely blocked by atmospheric components, such as gas molecules. The interactions between insolation and the atmosphere explain many aspects of our world, like blue skies, red sunsets, and even the existence of life.

A What Are the Principal Components of the Atmosphere?

The atmosphere is composed of gas molecules, especially nitrogen and oxygen; small solid particles, including dust; and drops of water and other liquids. These particles and drops are together called *aerosols* to convey that they are suspended in the atmosphere, floating with the moving air. Most gases and aerosols are produced by natural processes, but some are introduced into the atmosphere by activities of humans.

13.09.a1 Iceland

13.09.a2 Mount Etna, Italy

13.09.a3 Cedar Mesa, UT

Gases—The atmosphere is nearly all nitrogen (N_2) and oxygen (O_2), with lesser amounts of argon (Ar), and water vapor (H_2O), which occurs within the steam shown above. It contains trace amounts of other molecules, such as carbon dioxide (CO_2), methane (CH_4), nitrous oxide (N_2O), and sulfur dioxide (SO_2).

Solid Particles—If tiny enough, solid particles can be suspended in the atmosphere. Such particles include volcanic ash (▲), and wind-blown dust, salt, and pollen. They also include soot and smoke from natural and human-caused fires. Some small particles are produced by chemical reactions in the air.

Drops of Liquid—The atmosphere contains drops of water with much smaller amounts of other liquids. Most water drops are tiny enough to remain suspended in the air as in clouds (▲). If small drops combine or otherwise grow, they may become too heavy to remain suspended, falling as rain.

B How Do Atmospheric Components Affect Insolation?

Insolation, like all types of electromagnetic radiation, can be affected by material through which it passes. As solar energy attempts to pass through the atmosphere, it interacts in various ways with the different atmospheric components. The types of interaction that occur depend on the size and physical nature of the component (e.g., solid versus a gas), the wavelength of the electromagnetic energy (blue versus red light, for example), and other factors.

13.09.b2

Reflection and Absorption

1. *Reflection*—Some atmospheric components can *reflect* incoming insolation, such as by this snowflake. Reflected energy can be returned directly into space or can interact with other atmospheric components and remain in the atmosphere.

Snowflake

2. *Absorption*—An atmospheric component, such as this gas molecule, can instead *absorb* the energy, converting the incoming electromagnetic energy into kinetic energy expressed as motions of the molecule.

Gas Molecule
13.09.b1

Scattering

3. *Scattering*—Insolation can be *scattered* by atmospheric components (▶), which send the energy off in various directions. Some processes of scattering affect shorter wavelengths of EMR more than longer ones; for example, blue light is scattered more than red light. Other processes affect all wavelengths of solar energy equally.

4. *Sky Color*—As insolation enters the atmosphere, blue and violet light are preferentially scattered by gases, and this scattered light causes us to see the sky as blue. The remaining light that passes through gives the Sun a yellowish white color. When sunlight passes through the atmosphere at a low angle, as during sunrise and sunset, most colors, except orange and red, have been scattered out. Scattering of the remaining orange and red light produces the orange and red glow at these times.

C   How Do Different Layers of the Atmosphere Interact with Insolation?

If the atmosphere consisted of gases that did not interact with insolation, we would expect that its temperature would decline with distance from Earth. However, atmospheric temperatures exhibit some surprising changes with altitude, as a direct result of the interactions between insolation and atmospheric components, specifically gases.

1. Observations from high-flying planes, balloons, and rockets reveal that the atmosphere is divided vertically into four layers distinguished by their thermal properties. These layers are, from bottom to top, the *troposphere, stratosphere, mesosphere,* and *thermosphere*. In the troposphere and mesosphere, temperatures decline with increasing altitude, as expected, displaying a *normal temperature gradient*. Temperatures in the stratosphere and thermosphere, however, actually increase with altitude—a *temperature inversion* (reverse gradient). Why?

2. The unexpected change in temperature gradient in some layers is due to the absorption of insolation of some wavelengths of the electromagnetic (EM) spectrum, but not others that arrive at the outside of the atmosphere. Most energy is in visible-light wavelengths, with significant amounts of adjacent wavelengths of ultraviolet (UV) and infrared (IR) energy. Some wavelengths of energy interact with certain atmospheric molecules, transferring energy as they do so. Different interactions occur in different layers, accounting for the differences in temperature gradient, and therefore the distinctions among the different layers. The four atmospheric layers are separated by distinct breaks called *pauses*.

3. The very shortest and most energetic of EM radiation from the Sun are X-rays and gamma rays, with wavelengths approaching the size of gas molecules. These incoming wavelengths of EMR are effectively intercepted by the few molecules of nitrogen (N_2) and oxygen (O_2) in the uppermost parts of the atmosphere—the *thermosphere*. The greatest number of interceptions (and transfer of energy) occurs at the first opportunity for the gamma and X-rays to encounter these gases, at the outermost parts of the thermosphere. This causes the outer thermosphere to warm up, but so few molecules exist at such levels that you would freeze to death instantly, even at temperatures approaching 1,200°C. A progressively smaller proportion of these energetic rays penetrates lower in the thermosphere, so fewer energy exchanges occur, and temperatures decline downward across the thermosphere.

4. The next layer down, the *mesosphere*, possesses no particular properties to intercept wavelengths of insolation, so it displays a normal temperature gradient (temperature decreases upward). The boundary at the top of the mesosphere is the *mesopause*.

5. The *stratosphere* has relatively high concentrations of the trace gas *ozone* (O_3), which effectively absorbs UV wavelengths. The same principle prevails as in the thermosphere, with the greatest amount of absorption occurring near the top of the stratosphere, which is therefore relatively warm. Progressively less absorption of incoming UV occurs downward, resulting in a decrease in temperatures downward—a temperature inversion. The top of the stratosphere is the *stratopause*.

6. Conditions in the troposphere can be very complex with the presence of various sublayers of air and the influence of clouds, but temperatures usually decrease upward (a normal temperature gradient). The top of the troposphere is the *tropopause*.

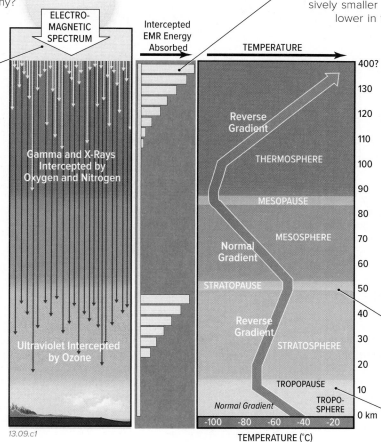

13.09.c1

7. The greatest amount of the Sun's emitted energy is at wavelengths of visible light, along with UV and IR. Interactions between solar energy and various components of the atmosphere (such as N_2) cause some wavelengths of energy to be intercepted through absorption, scattering, or reflection in the atmosphere. Therefore, the spectrum of EM energy that reaches the surface is different in detail from that which enters the top of the atmosphere, as depicted by the curve showing the amount of insolation at sea level on this graph (◄). Certain wavelengths of EMR are absorbed by certain components, such as water (H_2O), oxygen (O_2), and carbon dioxide (CO_2), as expressed by dips in the curve for insolation received at sea level.

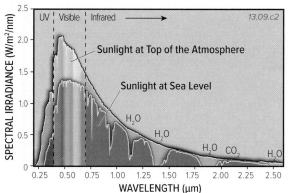

Before You Leave This Page

- ☑ Summarize the principal components of the atmosphere.

- ☑ Sketch and explain reflection, absorption, and scattering, and how they explain the color of the sky.

- ☑ Sketch, label, and explain the four layers in the atmosphere, how they interact with insolation, and how these interactions explain the different temperature gradients.

13.9

13.10 What Is Ozone and Why Is It So Important?

OZONE IS AN ESSENTIAL GAS in the atmosphere, shielding life on the surface from deadly doses of ultraviolet radiation from the Sun. In the past several decades, there has been major concern about the loss of ozone in our atmosphere, particularly a seasonal decrease in ozone above Antarctica. How can a gas that generally constitutes less than one molecule in every 10 million in the air be so important, and why does the Antarctic region experience severe ozone loss?

A What Is Ozone and Where Does It Occur?

Ozone is a molecule composed of three oxygen atoms bonded together (O_3), instead of the much more common arrangement of two oxygen atoms in a molecule of oxygen gas (O_2). More than 90% of ozone occurs in the stratosphere, but ozone also occurs in lesser amounts in the troposphere and mesosphere.

1. This graph (▶) plots the concentration of ozone in the atmosphere as a function of altitude above Earth's surface. Note that concentrations are very low overall, measured in parts per million (ppm—the number of ozone molecules for every one million molecules of atmosphere). The maximum concentration is in the middle of the stratosphere, a zone called the *ozone layer*, but even here the ozone concentration is only 6 to 8 ppm. Ozone concentrations decrease rapidly into the overlying mesosphere.

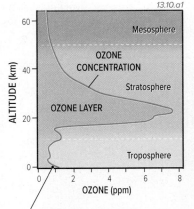

13.10.a1

2. There is a slight increase in concentrations low in the troposphere, just above the surface. Whereas ozone in the ozone layer (stratospheric ozone) is beneficial to life, ozone near the surface (*ground-level ozone*) is a harmful air pollutant produced by sunlight striking hydrocarbons, such as car exhaust.

13.10.a2 Los Angeles, CA

13.10.a3

3. Ozone is a component of *smog* (◀), a type of air pollution common in larger cities, especially under certain atmospheric conditions. The term smog was derived from air pollution that looks like smoke and fog.

4. Ozone is also produced naturally due to ionization of oxygen gas molecules in the air near lightning. In such settings, we can sense ozone's distinctive smell, and this is the origin of the term *ozone*, after the Greek verb for "to smell."

B How Is Ozone Produced and Destroyed in the Stratosphere?

1. Electromagnetic radiation with a wavelength of visible light or shorter (e.g., ultraviolet) has the ability to affect the chemical bonds in a compound, a process called *photodissociation*. Such processes occur throughout the atmosphere, but particularly in the stratosphere, where (1) much of the EMR arriving from the Sun has not yet been absorbed and (2) the concentration of molecules becomes high enough to allow photodissociation and subsequent formation of ozone. This figure shows how photodissociation forms and destroys ozone.

2. Oxygen in the atmosphere is mostly oxygen gas (O_2), composed of two oxygen atoms bonded together. These molecules absorb EMR from the Sun, including several types of ultraviolet (UV) energy. They are strongly affected by an energetic type of UV called UV-C, which has relatively short wavelengths. UV-C can break apart the two oxygen atoms, liberating a free (unbonded) oxygen atom (O).

3. The freed oxygen atom can quickly bond with another free oxygen atom, forming a new molecule of oxygen gas (O_2).

4. Alternatively, a freed oxygen atom can combine with an existing oxygen gas molecule to form a molecule of ozone (O_3). This process is the way ozone is produced in the atmosphere.

UV-C

UV-C & UV-B

Ozone

13.10.b1

5. Ozone molecules are capable of absorbing a longer wavelength form of ultraviolet called UV-B, in addition to UV-C. UV-B has lower energy than UV-C, but more of it reaches Earth's surface. UV-B is the wavelength of ultraviolet radiation that causes sunburn and contributes to skin cancer. It also produces vitamin D in our bodies when it interacts with our skin.

6. UV-C and UV-B can photodissociate an ozone molecule, leaving a molecule of oxygen gas and a free oxygen atom. These can recombine with other atoms and molecules to form another molecule of ozone, so the process of formation and destruction of ozone is a cycle, with ozone and oxygen gas continuously being broken apart, combined, and broken apart again. If the rates of formation and destruction of ozone molecules are equal, the ozone concentration will remain constant. Variations in atmospheric conditions, however, change the relative rates, causing concentrations to change over time and from place to place.

C What Is the Distribution of Ozone in the Atmosphere?

1. These two globes show the amount of ozone in the atmosphere at two different months, as measured from a satellite. In general, ozone concentrations are relatively low (green) over the equator, and are higher (yellows and oranges) in middle and high latitudes. The difference between the two globes shows that ozone concentrations change with the season, responding to changes in the patterns of insolation. Ozone amounts are expressed as Dobson units (DU).

OZONE (DU)

100   500

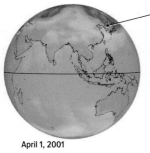

April 1, 2001

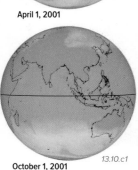

October 1, 2001   13.10.c1

2. In April, the highest concentrations are in higher latitude regions of the Northern Hemisphere.

3. In October, the highest values are in high latitudes of the Southern Hemisphere, but there is a huge area with very low concentrations centered over the South Pole (at the bottom of the globe). This low exists because total darkness during the Antarctic winter lasts for several months, during which time there is no insolation to form ozone, but ozone continues to be chemically destroyed (see below).

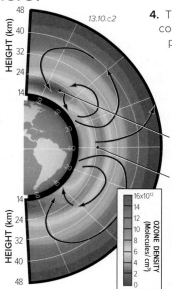

13.10.c2

HEIGHT (km)

OZONE DENSITY (Molecules/cm³)
16x10¹²
14
12
10
8
6
4
2
0

4. This figure shows ozone concentrations in the main part of the stratosphere. Ozone is most concentrated (red zones on the figure) in the lower middle stratosphere—the *ozone layer*. The highest concentrations are in high latitudes, and below an altitude of about 25 km. In equatorial regions, the zone of maximum ozone is at higher altitudes, typically more than 25 km. Production of ozone is greatest in the tropical equatorial regions, which receive the most insolation, including UV, but a very slow-moving circulation pattern in the stratosphere moves the ozone upward and laterally toward the poles, accounting for the higher concentrations there.

D What Is Causing Depletion of Ozone and Formation of the "Ozone Hole"?

Ozone is critical to life because it shields us from dangerous UV-C and UV-B radiation. In the past half century, we became aware that our protective ozone shield was being depleted by human activities, especially the production of chlorofluorocarbons (CFC), chemicals released from aerosol cans, air conditioning units, refrigerators, and polystyrene. As a result, production of CFCs was limited by international agreement, the Montreal Protocol, adopted in the late 1980s.

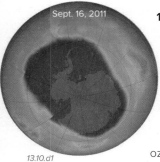

Sept. 16, 2011

13.10.d1

13.10.d3

1. This image shows ozone concentrations in the Southern Hemisphere as measured by a NASA satellite, with the smallest amounts in purple. The image, taken in September, shows 10.6 million square miles of the Antarctic region with severe ozone depletion. This depletion is described as "thinning of the ozone layer."

2. Most ozone is destroyed by natural processes, but humans have introduced chemicals that accelerate such losses. Chlorofluorocarbons (CFCs) contain *halogens*, elements like chlorine and bromine that easily bond with another element or molecule. Halogens can break apart ozone by attracting one of the oxygen atoms away, forming a new molecule, such as one with chlorine and oxygen (ClO).

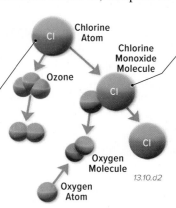

Chlorine Atom

Chlorine Monoxide Molecule

Ozone

Oxygen Molecule

Oxygen Atom

13.10.d2

3. The ClO molecule is short lived, and the chlorine atom breaks away to combine with other O atoms, typically outcompeting O₂ for bonding with O, thereby breaking apart ozone or keeping it from forming. On average, a single Cl may be responsible for the destruction of up to 10,000 ozone molecules. More than 70% of halogens in the atmosphere were introduced by humans.

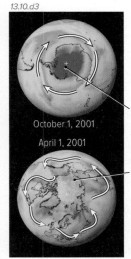

October 1, 2001

April 1, 2001

4. Halogens are regarded as a major factor in the depletion of the ozone layer in the Southern Hemisphere, shown in the top globe. The bottom globe shows the Northern Hemisphere at an equivalent time of year (spring in both places). Why is there thinning of the ozone layer over Antarctica but less so over the North Pole?

5. The contrasting geography of the two polar regions plays a major role. Antarctica, in the center of the top globe, is a continent surrounded by ocean. This arrangement produces a zone of rapid circumpolar winds, which effectively exclude the import of ozone from nonpolar areas.

6. In contrast, the Arctic is an ocean surrounded by irregularly shaped continents (Eurasia and North America). The alternating pattern of oceans and continents induces far greater north-south movement and complex wind patterns, which encourage the exchange and mixing of gases, including ozone, into and out of the northern polar atmosphere.

Before You Leave This Page

☑ Explain what ozone is, where it is, and how it protects us.

☑ Describe the natural and human-related processes that contribute to ozone formation and destruction.

13.10

13.11 How Much Insolation Reaches the Surface?

NOT ALL INSOLATION reaches Earth's surface. Much of it is intercepted (absorbed, scattered, or reflected) by the atmosphere. Measurements and models allow us to account for the destination of insolation globally. Approximately 69% of the energy arriving at the top of the atmosphere remains in the Earth's system, of which 20% is stored (absorbed) in the atmosphere and 49% is absorbed by, and heats, the Earth's surface. The rest (31%) is lost back into space. This accounting of insolation is termed the *global shortwave-radiation budget*.

A How Is Insolation Intercepted in the Atmosphere?

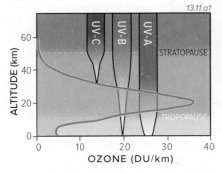
13.11.a1

13.11.a2

13.11.a3

The atmosphere absorbs radiation of various wavelengths, including ultraviolet (UV) light, an energetic shortwave EMR emitted by the Sun. This graph depicts the absorption of different wavelengths of UV (in Dobson units), along with the concentration of ozone (the red curve). UV is absorbed chiefly by oxygen gas, ozone, and nitrogen gas. As shown in this graph, the shortest-wavelength, highest-energy UV (UV-C) is absorbed first, followed by longer-wavelength, lower-energy UV-B. Nearly all UV-A, the longest-wavelength, least-energetic UV, is transmitted to the surface. When a wavelength of EMR passes through a substance, such as the atmosphere, without being intercepted, we say that the substance is "transparent to the energy." The atmosphere is nearly transparent to UV-A.

Clouds can absorb, scatter, and reflect insolation, bouncing it back higher into the atmosphere and even into space. Clouds in the lower part of the troposphere, *low-level clouds*, can be highly reflective, like the ones shown here. The dark underside of the clouds indicates that little light is transmitted.

Clouds higher in the atmosphere, such as these wispy *cirrus clouds*, are typically less reflective, but still reflect about 50% of insolation striking them. These *high-level clouds* occur in the middle to upper troposphere and are typically composed of ice crystals, unlike low-level clouds, which contain mostly water drops.

B What Happens to Insolation That Reaches Earth's Surface?

Earth's surface is diverse in its topography, rocks, soil, bodies of water, types of vegetation, and amount of human development. Each type of surface material has certain colors and other characteristics, so some surfaces reflect relatively little insolation and absorb much heat, while others reflect most insolation and absorb less heat.

1. The Apollo astronauts saw the Earth rise above the Moon's horizon (▼). The human eye is only sensitive to EMR in the visible (0.4–0.7 μm) wavelengths, and the Earth emits most of its energy at wavelengths 20 times longer than this (infrared). What the astronauts saw was visible light originating from the Sun and reflecting from the Earth. All objects reflect some insolation, and the percentage of insolation that is reflected by an object is termed its *albedo*. Albedo varies from 0% for a theoretical black, rough object that reflects no light to 100% for a perfectly white, smooth one; no common objects have these end-member values. The average albedo for the entire Earth-ocean-atmosphere system is 31%.

2. Different land surfaces have different albedos. When humans use the land surface, they generally change the albedo compared to its natural state, modifying the energy balance, with local and perhaps regional implications. Examine the albedo values for different types of land cover and think how changes in the land cover would affect the energy balance.

13.11.b1

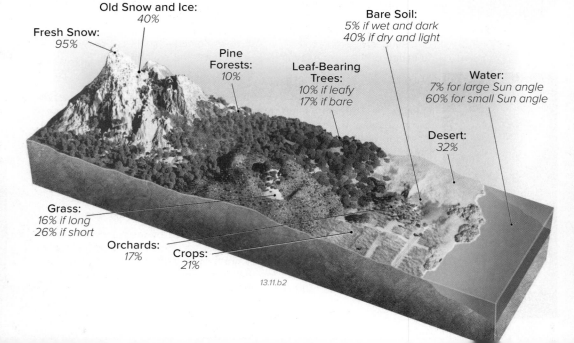

Old Snow and Ice: 40%

Fresh Snow: 95%

Pine Forests: 10%

Bare Soil: 5% if wet and dark 40% if dry and light

Leaf-Bearing Trees: 10% if leafy 17% if bare

Water: 7% for large Sun angle 60% for small Sun angle

Desert: 32%

Grass: 16% if long 26% if short

Orchards: 17%

Crops: 21%

13.11.b2

C What Percentage of Insolation Goes Where?

We can measure and model how much insolation is intercepted in the air versus how much reaches the surface. Some insolation that reaches the surface is reflected upward into the atmosphere, and some of this reflected energy goes all the way back into space. Most energy is absorbed by the water, land, and vegetation. The figure below shows the global shortwave-radiation budget—how much insolation goes where, a key component of the global energy balance.

1. We begin by considering the total amount of insolation arriving at the top of the atmosphere as being represented as 100%. Then, we can examine what percentages of this total amount end up where. The left side of this figure depicts reflection and scattering of insolation, whereas the right side represents absorption.

2. *Scattering* by various air components (gas molecules, dust, etc.) returns 7% of total insolation back to space. Such scattering causes blue skies, red sunsets, and red sunrises, and downward-scattered radiation is what illuminates the surface in shady areas and during overcast days. Scattering decreases the amount of insolation that reaches the surface.

3. The amount of energy *reflected by clouds* is variable and controlled by many complex factors. Different types of clouds have different albedos, and some lower-level clouds are obscured below other, higher clouds. The location of a cloud is also important—tropical clouds receive more direct overhead sunlight and so have more available insolation to reflect, whereas low clouds over the poles receive little direct light, or no light at all during winter. Also, the amount of the world covered by clouds varies, depending on the season and other aspects of weather. Considering all the factors, clouds on average reflect 20% of the planet's total insolation back to space.

4. Chemical constituents of the atmosphere, such as O_2 and N_2 gas, absorb EMR of various wavelengths, intercepting 17% of the total insolation. Events on Earth's surface may cause considerable changes in the number of aerosols in the troposphere. Volcanic eruptions emit gases, volcanic ash, and other aerosols into the atmosphere, and these effects persist for several years. These additions increase the amount of absorption in the atmosphere.

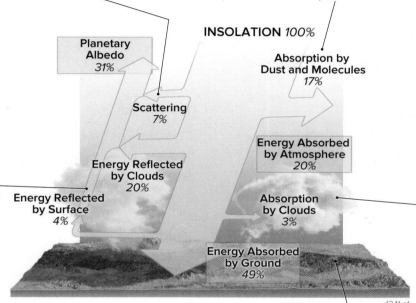

INSOLATION *100%*

Planetary Albedo *31%*

Absorption by Dust and Molecules *17%*

Scattering *7%*

Energy Absorbed by Atmosphere *20%*

Energy Reflected by Clouds *20%*

Energy Reflected by Surface *4%*

Absorption by Clouds *3%*

Energy Absorbed by Ground *49%*

13.11.c1

5. Clouds absorb only a small percentage of insolation, about 3% of total insolation. The complex interaction between reflection, scattering, and absorption by various layers of clouds makes their impact on climate difficult to quantify—the net effect of clouds is one of the most difficult aspects to represent in global climate models. Some types of clouds tend to hold in heat, warming the surface, whereas others have an overall cooling effect.

7. So of the total insolation, 20% is converted to sensible heat or latent heat after absorption in the atmosphere, causing an increase in temperature or a change in state (mostly evaporation of water), respectively. A total of 49% of the insolation is retained by the land as what is called *ground heat*. The remaining 31% is reflected back into space at various levels, accounting for the remainder of the 100% of insolation (20% + 49% + 31% = 100%).

6. Earth's surface albedo is fairly low but varies widely. Snow is highly reflective (high albedo), but many rocks are dark and rough (low albedo). Water generally has a lower albedo than land, and more than 70% of the Earth is covered with water. Due to reflection, scattering, and absorption in the atmosphere, only 53% of insolation reaches Earth's surface. But 4% of this energy is reflected back to space once it hits the surface, leaving 49% to be absorbed.

Changing the Global Radiation Budget

13.11.c2 *Luzon Island, Philippines*

In June 1991, Mount Pinatubo in the Philippines erupted (◄), throwing volcanic ash and gases up to 34 km into the troposphere and stratosphere. The ash circled the Earth, reducing insolation by about 1 watt per square meter (W/m^2), and cooling global temperatures 0.5 C°. The image to the right shows, in red, the dust circulating around the tropical regions of the world the following December (more than a half year later). The eruption was also accompanied by a decrease in global vegetation, as measured by satellite, suggesting extra absorption of insolation by volcanic ash. The effects lasted for about two years until the ash either fell out of the atmosphere or was washed out by precipitation.

13.11.c3

Before You Leave This Page

☑ Summarize how the atmosphere intercepts insolation.

☑ Describe albedo, give five examples, and summarize the importance for Earth's energy balance.

☑ Sketch, label, and explain the interactions of insolation with the atmosphere and surface, noting the percentages of insolation lost and stored in various ways.

13.11

13.12 How Does Earth Maintain an Energy Balance?

SIXTY-NINE PERCENT OF INSOLATION received at the outside of the Earth's atmosphere is available for sensible, ground, and latent heating. Ultimately, all of this energy must be returned to space as longwave radiation in order to attain a balance between incoming and outgoing radiation. A greater loss to space would cool the global system, and a smaller loss would increase global temperatures. Just as there is a shortwave radiation budget, there is a budget of *global outgoing longwave radiation*. By interacting with outgoing longwave radiation (OLR), greenhouse gases help maintain Earth's hospitable temperature. Earth would be a very different planet if the greenhouse effect operated in a different manner or if the amount of greenhouse gases were different.

Sensible and Latent Heat Flux from Earth's Surface

1. There are various ways that Earth's surface and atmosphere transfer longwave energy. This page examines losses via the transfer of sensible and latent heat, and the facing page deals with losses through emission of longwave radiation. Read both pages in a counterclockwise order.

2. For this discussion, we set the amount of insolation hitting the top of the atmosphere at 100 units. Of the total insolation (shortwave) that enters the top of Earth's atmosphere, 31 units (31%) is reflected or scattered by the atmosphere directly to space. That leaves 69 units within the surface and atmosphere.

3. Gas molecules, clouds, and various particles absorb 20% of insolation (20 units). This energy will later be released back into space.

4. The other 49 units (49%) of insolation is transmitted through the atmosphere and then absorbed by Earth's surface, including land, oceans, and other bodies of water. This energy must somehow escape the surface, or else the surface would keep heating up indefinitely as the Sun continued to transmit shortwave energy.

5. If 49% of insolation reaches Earth's surface, this is more than twice the amount (20%) that is absorbed in the atmosphere. An implication of this is that the Sun heats Earth's surface more than it does the atmosphere, and in turn, the surface heats the atmosphere. Warming of the atmosphere from below in this way is one reason why air temperatures generally decrease upward with increasing altitude.

6. As the atmosphere is heated from below, the warmer air near the surface starts to rise upward, inducing convection in the troposphere. Energy equivalent to 7% of total insolation is transmitted, mostly by convection, to the adjacent air as sensible heat. This flow of energy is called the *sensible heat flux*.

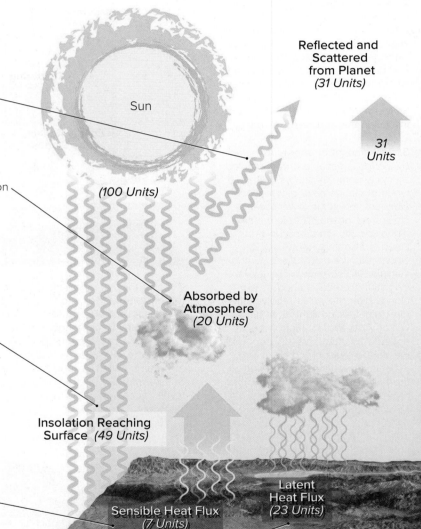

Reflected and Scattered from Planet *(31 Units)*

31 Units

Sun

(100 Units)

Absorbed by Atmosphere *(20 Units)*

Insolation Reaching Surface *(49 Units)*

Latent Heat Flux *(23 Units)*

Sensible Heat Flux *(7 Units)*

7. Most of the earth is covered by ocean, and many land areas include lakes, wetlands, and heavily vegetated regions, so much of the energy reaching the surface goes to *latent heat flux* (melting ice, evaporating water, and transpiration from plants). Melting of ice only transfers energy between different parts of the surface (ice sheet to sea, for example), so it does not directly impact the atmospheric energy budget—but evaporation does. As the warm air rises, it carries aloft the recently evaporated water vapor into the ever cooler air at higher altitudes. Eventually the moist air cools sufficiently to condense into water drops and form clouds, which then release the latent heat into the atmosphere. Almost half of all the energy reaching the surface of Earth (49 units) is returned to the atmosphere in this way (23 units). The combined contributions of sensible and latent heat carry about 30 of the 49 units of the shortwave radiation stored at the surface into the atmosphere.

Longwave Energy Flux From Surface and Atmosphere

12. We started with 100% insolation, so the 69% plus the 31% shortwave insolation reflected (the planetary albedo) provide a perfect balance of input and output of energy to and from the earth's land-ocean-atmosphere system. Keep in mind that these values are average annual values for the globe. Any individual place is unlikely to experience such a balance. Circulation of the atmosphere and oceans transfers energy from places that have an excess relative to the global average to those areas that have a relative deficit. Note that the numbers on these two pages add up to slightly more than 100 units (100%) because values have been rounded to whole numbers. When carried out with more precise numbers, it all adds up to 100 units.

11. A total of 69 units of longwave energy go back into space: 57 units emitted by the atmosphere and the 12 units emitted directly from the surface.

Emitted Directly to Space (12 Units)

10. The 57 units in the atmosphere are eventually emitted to space in the form of longwave radiation.

69 Units

Emitted Directly Atmospheric Molecules (57 Units)

9. At this point, the atmosphere has 50 units of energy—20 units from insolation initially absorbed by the atmosphere and 30 units it received from the surface via sensible and latent heat flux. This heat energy cannot be transmitted to space directly via conduction or convection because space is essentially a vacuum. Instead, the atmosphere releases this energy by emitting longwave radiation in all directions: upward to space, sideways to other parts of the atmosphere, and downward to the surface. There is a back-and-forth exchange of radiant energy between the atmosphere and surface, largely controlled by the greenhouse effect. As a result of these interactions, there is a net flux of 7 units upward. These 7 units join the 50 units already in the atmosphere for a total of 57 units.

Longwave Energy Flux (Net Flux 7 Units)

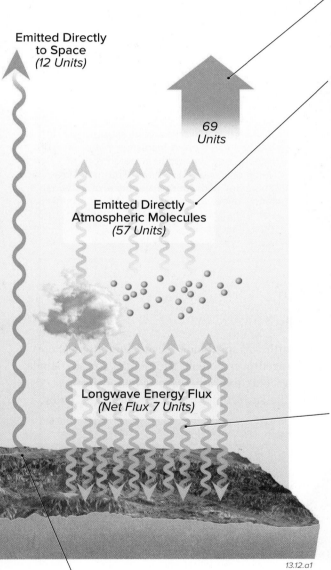

13.12.a1

8. Of the 49 units of shortwave radiation that reach Earth's surface, 12 units are emitted directly from the surface (water and land) to space as longwave radiation, without significant interactions with the atmosphere. This is possible because of the *atmospheric window* that allows certain IR wavelengths to radiate upward through the atmosphere with only minimal losses to absorption, reflection, and scattering.

A World without the Greenhouse Effect

Imagine two Earths, identical in all other ways except the presence of a greenhouse effect. In the figure shown here, the lower globe has greenhouse gases in its atmosphere, whereas the upper globe does not.

From surface and satellite observations, average global surface temperatures on Earth currently are about +15°C (+59°F). Well-understood equations relating the emission of energy to temperature allow us to estimate the average surface temperature of our imaginary world at −18°C (−9°F), significantly colder than the currently observed temperature of Earth.

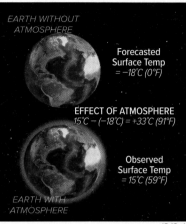

EARTH WITHOUT ATMOSPHERE

Forecasted Surface Temp = −18°C (0°F)

EFFECT OF ATMOSPHERE 15°C − (−18°C) = +33°C (91°F)

Observed Surface Temp = 15°C (59°F)

EARTH WITH ATMOSPHERE

13.12.t1

Some of the difference between these two estimates must be caused by greenhouse gases, which are keeping Earth's current temperatures warmer, thereby allowing water to exist in a liquid state, a key factor in supporting life as we know it. Calculating the actual contribution (in degrees of warming) of greenhouse gases is too complex to pursue here, because such calculations involve many other factors.

From these rough calculations, we can see why there is concern over the impacts of increased concentrations of greenhouse gases due to human-related emissions of carbon dioxide (CO_2), methane (CH_4), and nitrous oxide (N_2O). We address the changing concentrations of greenhouse gases and the broader topic of *climate change* in a later chapter on climate.

Before You Leave This Page

☑ Sketch, label, and explain the nature of the Earth's overall energy balance, including the longwave radiation balance.

☑ Explain the linkage between the longwave radiation balance and the "greenhouse effect" on Earth.

13.13 How Do Insolation and Outgoing Radiation Vary Spatially?

FLOWS OF ENERGY into and out of Earth's system vary spatially, depending on latitude, whether it is inland or over the ocean, cloud cover, and many other factors. The pattern of insolation also changes over several timescales, from daily rotation of the planet to the longer changes in season, causing spatial and temporal imbalances — zones of surplus energy and zones with an energy deficit, relative to the planetary average. These energy imbalances provide the driving force for global weather and climate.

A How Does Insolation Vary Spatially?

Insolation striking the top of the atmosphere, when averaged over a year, shows a smooth gradient, from higher amounts over the equator to much lower amounts at the poles. A more complex pattern emerges if we examine how much insolation actually reaches the surface during the course of a year, as shown in the figure below.

1. At the broadest scale, latitude controls insolation. The highest amounts are in low latitudes, and the lowest amounts are near the poles. Purples on this globe are as low as 120 W/m², whereas orange and red are more than 250 W/m².

2. Although insolation is strong in the tropics it is reduced somewhat due to absorption, reflection, and scattering by the abundant clouds that characterize these same regions.

3. The highest values are in subtropical deserts.

4. Universally low values mark the Antarctic and Arctic, which have days to months of total darkness. With such low Sun angles, the lengthy summer days do not make up for the dark winters.

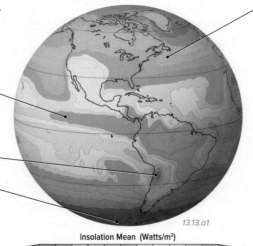

13.13.a1

Insolation Mean (Watts/m²)
120   150   190   230   260   300

5. The mid-latitudes are between 30° and 60° latitude, between the lines shown on this globe. At these latitudes, there is a relatively steep gradient, with the amount of annual insolation decreasing toward the pole.

6. There is a marked contrast between the gradual variation in insolation patterns over oceans and the irregular patchwork of variations over continents. The more irregular character over land is because of its variable land covers and elevations, which in turn influence the albedo, cloud cover, and other factors.

B How Does Outgoing Longwave Radiation Vary Spatially?

Using satellites, we measure the amount of longwave radiation that leaves the Earth system (land, oceans, and atmosphere). The figure below shows the amount of outgoing longwave radiation (OLR), averaged over a year. The picture for any single day is more complex, with more irregular weather-related patterns. The amount of outgoing longwave radiation from a region greatly reflects the amount of incoming shortwave insolation that strikes the surface, so it is lowest near the poles.

1. The quantity of energy emitted from an object depends on its surface temperature. Latitude strongly controls annual insolation, so it also controls larger patterns of both temperature and OLR.

2. The highest amounts of OLR are in the subtropics, which receive more insolation than regions farther from the equator. Some of this "extra" energy is directly returned as OLR from the surface and from the atmosphere.

3. The pattern of high OLR in the subtropics is complicated over land, such as a zone of lower OLR along the western side of South America. This is due to a combination of a high mountain range (the Andes) and a cold ocean current (the Humboldt Current) along the west coast, both of which reduce the temperature. Reduced temperatures affect OLR.

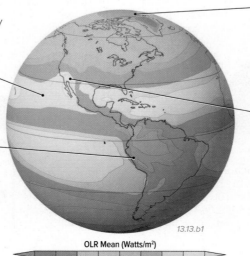

13.13.b1

OLR Mean (Watts/m²)
150   180   220   250   280

4. The Arctic and Antarctic differ from one another, although it is difficult to see from this perspective. The oceanic Arctic emits more OLR, appearing warmer, than the continental Antarctic. A similarly low OLR is associated with land in Greenland, compared to the adjacent ocean, which emits more OLR because it is warmer.

5. Conversely, some of the highest continental values of OLR are between 20° and 30° latitude, on either side of the equator. These are the global deserts, such as the Mojave Desert of the Southwest. In addition to having high temperatures, the absence of water vapor allows OLR to escape through the atmosphere.

6. The patterns in this globe and the one above change with the season, as the latitude of maximum insolation shifts north and south.

C How Does the Balance Between Incoming and Outgoing Radiation Vary Spatially?

Tropical regions receive much more insolation than do mid-latitude or polar regions, and this results in an unequal distribution of energy as a function of latitude—tropical regions have more Sun-induced energy than the poles. Energy flows from regions of relative surplus toward regions with relative deficits.

Radiative Balances

1. This graph shows the average amount of insolation at the top of the atmosphere versus the amount of outgoing energy (OLR) emitted, as a function of latitude, from the equator on the left to the North Pole on the right. A graph constructed from the equator to the South Pole would be similar.

2. Like a bank account, the difference between incoming and outgoing radiation is a location's *radiation balance*. On average, regions near the equator, tropics, and subtropics receive more energy from insolation than they emit in OLR—they have a positive balance or a *radiation surplus*. The region of surplus extends as far north (and south) as about 35° latitude, well past the Tropics of Cancer and Capricorn. The amount of insolation and OLR barely change across the tropics (leftmost part of the graph).

3. Poleward of the tropics, insolation declines rapidly, while the amount of outgoing longwave radiation decreases more gradually.

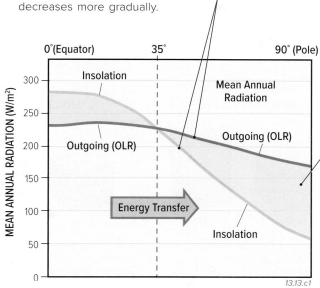

4. The two curves cross around 35° latitude, represented by the vertical dashed line. At this location, the amount of outgoing radiation (OLR) equals the amount of incoming radiation, on average—at this latitude, there is a radiation balance of zero.

5. Poleward of 35°, regions emit more OLR than the shortwave radiation that they receive from insolation—they have a *radiation deficit*. Insolation continues declining rapidly at mid-latitudes, reaching very low values in the polar regions. The energy balance becomes increasingly negative toward the poles.

6. Energy is transferred from areas of surplus to those of deficit, toward the poles.

Global-Scale Patterns of Radiative Surplus and Deficit

7. This globe shows the mean annual radiation balance based on satellite observations. Orange, red, and dark pink represent zones of surplus, whereas yellow, green, and blue are regions of deficit. The white dotted lines show 35° N and 35° S. What patterns do you notice?

8. As expected, the regions with the highest energy surplus, depicted in red and orange, are concentrated along the equator, extending out across the tropics on both sides of the equator.

9. The patterns on land are locally more complex than the patterns in the ocean, suggesting the complicating influence of variations in elevation and the resulting complex patterns of clouds, snow, wind directions, etc.

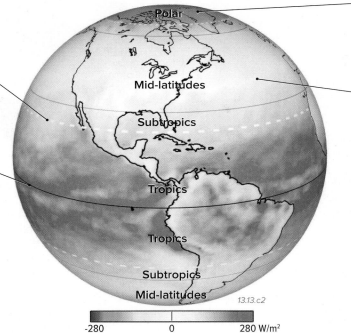

10. The regions with the most severe energy deficits are near the poles. These regions lose much more energy to space than they gain from low-angle sunlight during the summer, and they receive no insolation during the winter.

11. The mid-latitudes, between the tropics and the polar circles, exhibit a slight radiative deficit, represented here by a pale pink color. They gain energy transferred poleward from the tropics.

12. More than half of the planet's surface area has an energy surplus, and less than half has an energy deficit. The total amount of surplus energy equals the amount of energy deficit, and the planet is in an overall energy balance. As we will explore later, the transfer of energy from areas of radiative surplus to those of deficit drive much of our wind patterns and weather.

Before You Leave This Page

✓ Sketch and explain the global patterns of insolation and outgoing longwave radiation.

✓ Sketch and explain global patterns of surplus and deficit energy using a graph or map, and explain how these are expressed in global-scale patterns in the regional distribution of areas of energy excess and deficits.

13.14 Why Do Temperatures Vary Between Oceans and Continents?

WATER EXHIBITS VERY DIFFERENT thermal properties from those displayed by the rocks and soil. These differences in thermal properties cause oceans and land to warm and cool at different rates, leading to significant temperature variations between oceans and land. Such differences help explain major patterns of global temperature and climate.

A How Do Water and Earth Materials Respond to the Same Changes in Energy?

Heat Capacity

1. In evaluating how water, rocks, and other materials heat up or cool down, an important consideration is how much energy is needed to heat up an object, and how much heat that object can retain. A physical attribute called *heat capacity* expresses how much heat is required to change a volume's temperature by one Kelvin.

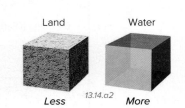

HEAT CAPACITY

Less   13.14.a1   *More*

SPECIFIC HEAT

Land   Water

Less   13.14.a2   *More*

Specific Heat

3. To compare the inherent thermal responses of different materials, irrespective of how much of the material is present, we use a property called *specific heat capacity*, or simply *specific heat*. Specific heat is the amount of energy needed to increase a kilogram mass of a substance by 1 K (or 1 C°).

2. The heat capacity of an object is determined by the kind of material in the object, such as rock versus water, and by the size of the object. The larger block above has a greater heat capacity than the smaller block, as long as both are composed of similar materials.

4. The specific heat of water is four times that of most rocks and materials. This means that it takes four times more energy to heat water than it takes to heat an equivalent mass of rock.

Thermal Responses of Water Versus Other Earth Materials

5. Due to their differences in specific heat, oceans (water) heat up during the day differently than land. If the same amount of insolation strikes water and land, the land will increase in temperature 4 degrees (K or C°) for every degree the water increases. As a result, land warms up much faster than the ocean under the same environmental conditions, but some of this difference is offset by increased losses of heat from the land to the air.

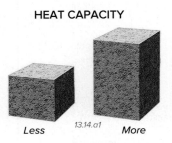

13.14.a3

1°C   4°C

Water   Land

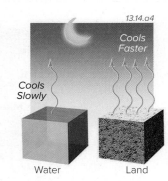

13.14.a4

Cools Faster

Cools Slowly

Water   Land

6. At night, the water and land both lose energy to the cool night air. For the same energy loss, land cools more than does water. Also, the land became hotter during the day and hot objects radiate more energy than cool ones, so the land loses energy faster than the water. As a result, land cools off much faster than water at night. Air heats more quickly over land and more slowly over water.

Depth of Heating, Cooling, and Mixing

7. Another factor in how water and land respond to changes in insolation and air temperatures is how deeply heat is able to enter each material. Nearly all water allows at least some transmission, so shortwave radiation can penetrate to depths of tens of meters or more. In contrast, rocks and soil are largely opaque, so insolation is confined to the surface and heat must move downward into the land by conduction, which it does by only a meter or two during the day.

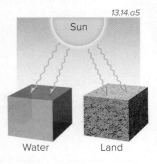

13.14.a5

Sun

Water   Land

8. Some materials, such as water, are relatively mobile, which allows them to flow and mix. Under calm ocean conditions (left column), limited mixing causes surface waters warmed by the Sun to remain near the surface, so there is a strong temperature contrast with depth. Surface winds induce waves (center column), resulting in turbulence, which carries warm waters downward, mixing them with cooler waters at a depth. Salt water (right column) is more dense than fresh water, so any waters that are saltier than normal, such as from partial evaporation, can sink, causing mixing of the water column. Mixing allows heat to be carried deeper into the water column (much faster than heat is conducted) and brings up cooler water that gives off energy to the atmosphere more slowly (because it is cool). As a result, mixed water heats up more slowly than does land, which experiences almost no vertical mixing.

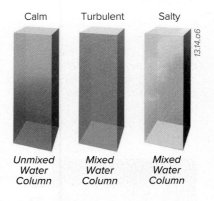

Calm   Turbulent   Salty

13.14.a6

Unmixed Water Column   *Mixed Water Column*   *Mixed Water Column*

Latent Heat

9. Water has another unique capacity relative to land—it can store abundant energy as latent heat. Insolation that strikes water is transformed into one of three different fluxes of energy. Some energy goes into heating the water (ground heat), some heats the air (sensible heat in the atmosphere), but a large amount goes into latent heat produced by evaporation. In contrast, insolation striking land goes mostly into ground heat and into sensible heat in the atmosphere. Land contains some water, but lesser amounts of its insolation go into latent heat.

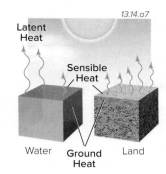

13.14.a7

Distribution of Continents and Oceans

10. Another factor that influences the global energy budget, and the balance of energy that falls on land versus the oceans, is the difference between the Northern and Southern hemispheres. As can be observed on any map or globe, the Northern Hemisphere has the majority of the planet's landmasses, whereas the Southern Hemisphere is dominated by oceans. As a result, an equal amount of insolation striking both hemispheres will result in more latent heat being generated in the Southern Hemisphere than in the Northern Hemisphere. In December, when the Southern Hemisphere more directly faces the Sun, more insolation will fall on water than during June.

13.14.a8

B How Do Temperatures Reveal Thermal Differences Between Ocean and Land?

1. These various factors, from specific heat to latent heat, cause land and water to respond very differently to insolation and to the change from day to night. Land, with its relatively small specific heat, limited mixing, and limited amount of latent heating, warms up more quickly than water and reaches higher temperatures. At night, land's higher daytime temperatures cause it to lose heat more rapidly than does water. This keeps the night warm for a while, but eventually the cool night air dominates.

2. Water, with its large specific heat, partial transparency, and ability to mix vertically, heats up more slowly and does not reach as high a temperature. Also, much insolation is converted into latent heat that is transferred to the atmosphere via evaporation and condensation, so this energy is not available to heat the body of water. Large bodies of water therefore experience smaller temperature variations and more moderate temperatures overall, compared to land. Land areas adjacent to the water can partly experience the moderate temperatures caused by the unique thermal properties of water.

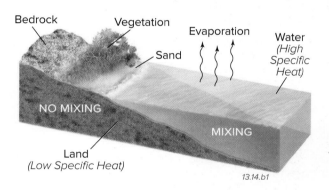

13.14.b1

Mean Annual Temperature

13.14.b2

3. These three globes show the yearly average temperature, typical January temperature, and typical July temperature for a typical year. Observe the temperatures shown on each globe and use concepts presented in this chapter to try to explain the main patterns. Aspects to consider include variations in insolation due to latitude and clouds, land-sea contrasts, and the distribution of continents. These globes show data for the different times of year, so they express seasonal variations. In each globe, red and orange are hotter, blue and purple are colder, and yellow and green are intermediate in temperature. The overall average temperatures (◄) range from less than −50°C in Antarctica to locally greater than +20°C in some tropical and subtropical regions.

Typical January Temperatures

13.14.b3

4. Typical January temperatures (◄) go from less than −50°C near the North Pole to more than +25°C in the tropics.

5. Typical July temperatures (►) are higher in the Northern Hemisphere than in the Southern Hemisphere.

Typical July Temperatures

13.14.b4

Before You Leave This Page

☑ Sketch, label, and explain all the factors that cause land and ocean at the same latitudes to heat and cool differently, identifying how these factors affect temperatures.

☑ Summarize some of the main patterns in the global distribution of average, maximum, and minimum temperatures.

13.14

13.15

How Are Variations in Insolation Expressed Between the North and South Poles?

VARIATIONS IN INSOLATION, both as a function of latitude and from season to season, help explain many aspects of our world—average temperatures, hours of daylight, type of climate and weather, type of landscape, and overall livability of a place. For a transect down the west coasts of the Americas, from the Arctic to the Antarctic, we examine the average monthly amounts of insolation, length of day, and temperature as a way to connect concepts in this chapter with actual places. Examine the photographs, graphs, and text for each place, and think about what explains the patterns for that place and the variations from one place to the next. For each place, the graph on the left shows variation in insolation from month to month at the top of the atmosphere, whereas the graph on the right shows average number of daylight hours (red bars) and average monthly temperature (blue curve).

13.15.a1 ANWR, AK

Northern Alaska and Canada

The North Pole is located in the Arctic Ocean, but parts of Alaska and Canada are north of the Arctic Circle (66.5° N). Compare the graphs to the right, which show the monthly variation in insolation (first graph), number of daylight hours (bar graph) and average temperatures (line graph). Note that during parts of winter, there is no daylight.

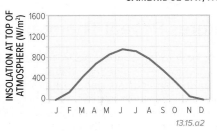

13.15.a2

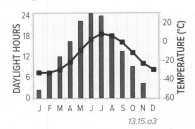

CAMBRIDGE BAY, NUNAVUT, CANADA

13.15.a3

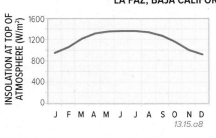

13.15.a4 British Columbia, Canada

Pacific Northwest

The northwestern part of the mainland U.S. and adjacent parts of British Columbia, Canada, straddle the famous 49th parallel (49° N latitude). They are squarely in the mid-latitudes. Like most of the places on this page, the region is near the ocean, so its temperature variations are moderated somewhat by the waters of the adjacent ocean.

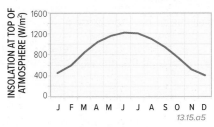

13.15.a5

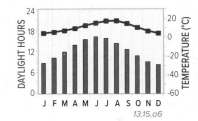

VANCOUVER, BRITISH COLUMBIA, CANADA

13.15.a6

13.15.a7 Baja, Mexico

Baja California Sur, Mexico

Baja California, part of Mexico, is a desert peninsula bordered by the Pacific Ocean to the west and the Gulf of California to the east. La Paz, the capital of Baja California Sur, is at a latitude of 24° N, just north of the Tropic of Cancer (23.5° N). Note that the graphs for Baja and the two previous places display a maximum centered on June to August (summer).

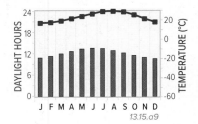

13.15.a8

LA PAZ, BAJA CALIFORNIA SUR, MEXICO

13.15.a9

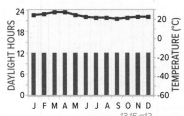

13.15.a10 Ecuador

Ecuador

The South American country of Ecuador is named for its position on the equator. It is on the west coast of the continent and contains parts of the Andes, Amazon rain forest, and the Galápagos Islands, famous for their active volcanoes and unusual animals. Note the pattern on the length-of-day bar graph and the minor variation in monthly insolation for near the equator.

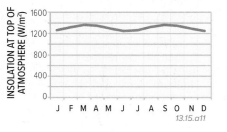

13.15.a11

GALÁPAGOS ISLANDS

13.15.a12

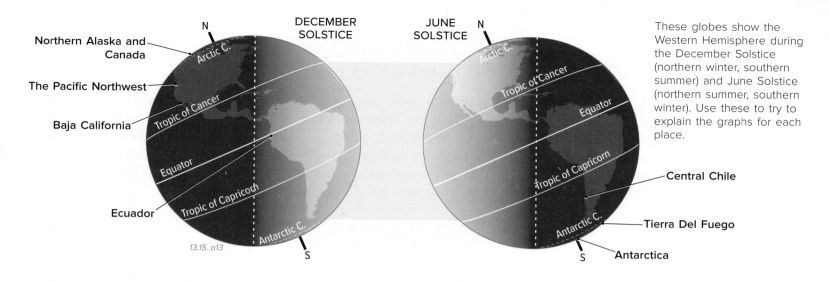

DECEMBER SOLSTICE

JUNE SOLSTICE

Northern Alaska and Canada

The Pacific Northwest

Baja California

Equator

Ecuador

Arctic C.

Tropic of Cancer

Tropic of Capricorn

Antarctic C.

13.15..a13

Arctic C.

Tropic of Cancer

Equator

Tropic of Capricorn

Antarctic C.

Central Chile

Tierra Del Fuego

Antarctica

These globes show the Western Hemisphere during the December Solstice (northern winter, southern summer) and June Solstice (northern summer, southern winter). Use these to try to explain the graphs for each place.

13.15.a14 Santiago, Chile

Central Chile

About halfway down the western coast of South America is the central part of Chile. The country's capital of Santiago is at a latitude of 33.5° S, south of the Tropic of Capricorn. It is inland and higher than the coast. Note that in the Southern Hemisphere, the graphs of insolation, hours of daylight, and temperatures now have June through August troughs rather than peaks. You know why, don't you?

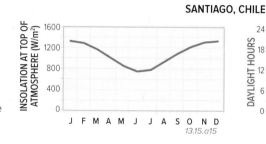

SANTIAGO, CHILE

13.15.a15

13.15.a16

13.15.a17 Ushaia, Argentina

Tierra del Fuego

Tierra del Fuego is the southernmost tip of South America, in southern Chile and Argentina. Ushuaia, Argentina, at a latitude of 54.8° S, is called the southernmost city in the world. It is not within the Antarctic Circle, so every day has some daylight and darkness. The southern tip of South America, surrounded by cold seas, is a frigid, stormy place.

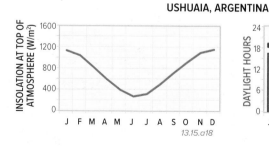

USHUAIA, ARGENTINA

13.15.a18

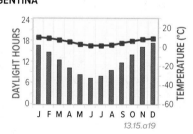

13.15.a19

13.15.a20 Antarctica

Antarctica

Antarctica, centered over the South Pole, consists of vast plains of ice with some impressive mountain ranges. The main part of the continent is within the Antarctic Circle, so it has months of darkness. The South Pole has continuous darkness from April to September (the southern winter), but 24 hours of sunlight from October to February. It is always a very cold place.

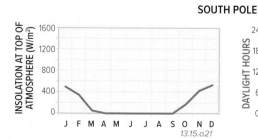

SOUTH POLE

13.15.a21

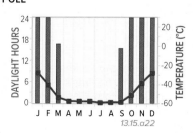

13.15.a22

Before You Leave This Page

✓ Sketch the general patterns of average insolation, number of daylight hours, and temperatures for three sites: one in the Northern Hemisphere, one in the Southern Hemisphere, and one near the equator. Explain the patterns in terms of latitude.

13.15

13.16 How Do We Evaluate Sites for Solar-Energy Generation?

SOLAR ENERGY IS A RENEWABLE SOURCE of light, heat, and electricity. Solar energy is typically collected with a solar panel, which can generate electricity or can heat air, water, or some other fluid. You have an opportunity to evaluate the solar-energy potential of five sites in South America, using concepts you have learned about insolation in this chapter. To do this, you will use Sun angles at each site to determine how directly sunlight will strike solar panels at different times of the year. You will then consider some regional factors that influence how much insolation reaches the ground. From these data, assess how efficient each site will be at generating solar energy.

Goals of This Exercise:

- Understand the importance of latitude in constraining the maximum possible amount of insolation striking the surface, to help evaluate the solar-energy potential of a site.
- Use factors controlling regional amounts of insolation, along with characteristics of the geographic setting of a site, to evaluate how much insolation is likely to actually arrive at Earth's surface, the key factor in the solar-energy potential of each site.

1. The globe below is a view centered on South America, and it shows the South Pole and adjacent parts of the Southern Hemisphere. It locates five possible sites you will consider for their solar-energy potential: (1) the Galápagos Islands of Ecuador; (2) Macapá, a small settlement in Brazil; (3) La Serena, a coastal town of Chile; (4) near Mar Chiquita, in central Argentina, and (5) Ushuaia, Argentina. Some characteristics of each site are described below.

2. The Galápagos are a series of islands in the Pacific Ocean, far west of South America. They are known for their exotic animal species, made famous by the visits of Charles Darwin. Although surrounded by the ocean, many parts of the islands receive little rain.

3. La Serena is a popular tourist site along the Pacific coast of Chile. It has a cool, desert climate, caused in part by a cold ocean current (the Humboldt Current) that limits the amount of moisture in the air. It is on the fringe of the Atacama Desert, the driest place on Earth (the tan strip along the coast to the north of 30° S on the globe).

4. Ushuaia, the southern-most city in the world, is near Cape Horn. Cape Horn is surrounded on three sides by ocean, some of which is coming from near Antarctica and is very cold. As a result, Ushuaia has a relatively cold and humid climate. It is close to 55° S latitude, but for this exercise we will consider it to be at 60° S to simplify our calculations.

Site	Lati-tude	Sun Angle		
		Equi-nox	Dec. Sol.	June Sol.
Galápagos	0°	90°	66.5° S	66.5° N
Macapá	0°	90°	66.5° S	66.5° N
La Serena	30° S	60°	82.5°	36.5°
Mar Chiquita	30° S	60°	82.5°	36.5°
Ushuaia	~60° S	30°	53.5	6.5°

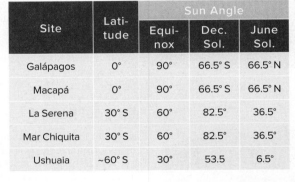

7. From the latitude of each site, we can calculate the Sun angle at each site for an equinox and both solstices. These results are presented in the table above.

6. Macapá is a small community in Brazil, near the mouth of the Amazon River, just inland from the sea. It is next to the Amazon River and is part of the huge Amazon rain forest, which stretches from the Atlantic Ocean westward all the way to the foothills of the Andes. As part of the rain forest, Macapá is a humid, rainy place.

5. Mar Chiquita is a small inland lake on the plains of central Argentina, far from the oceans but at a relatively low elevation. The proposed site is north of the lake, far enough that the lake does not significantly affect the local climate. This part of Argentina is east of the Andes, the huge mountain range along the western coast of South America. The Andes block moisture from the west, causing the plains of Argentina to be relatively dry.

13.16..a1

Procedures

A. Read descriptions of each site and consider how the geographic setting might impact the favorability of the site for solar-energy production.

B. For each site, use the Sun angles to look up the maximum percentage of solar energy that is theoretically available.

C. Consider how insolation varies with latitude and from season to season as an important consideration for the suitability of each site. Graphs from the previous two-page spread (the Connections spread) will be helpful here.

D. List the pros and cons of building a solar-energy facility at each site.

E. OPTIONAL EXERCISE: Your instructor may provide you with simple formulas for calculating Sun angle from latitude and have you determine Sun angles and solar favorability for other sites, some near where you live.

8. In considering the various sites, an important factor is that the solar panel will be horizontal. So we can use Sun angle as a measure of whether the sunlight is coming in perpendicular to the solar panel, the optimum orientation. Although this orientation works fine near the equator, at other latitudes solar panels are inclined at some other angle so as to maximize the amount of sunlight (keeping the panel perpendicular to the direction of sunlight, wherever possible).

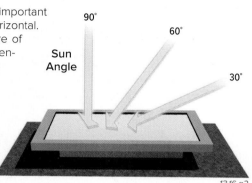

Sun Angle

90° 60° 30°

13.16.a2

9. Use the Sun angle for each site to determine the maximum percentage of energy that can be produced by the solar panel for that Sun angle, relative to the maximum amount it could produce if it were perfectly perpendicular to the Sun. Write down your results to guide your considerations in evaluating each site.

| Percentage of Solar Energy Available ||
Sun Angle	Percent
90°	100%
80°	98%
70°	94%
60°	87%
50°	76%
40°	64%
30°	50%
20°	34%
10°	17%
0°	0%

10. For each location, use your angle calculations to predict how the amount of insolation arriving at the top of the atmosphere will vary from season to season. Key times in the seasonal changes will be the solstices (December and June) and the equinoxes (March and September).

11. Since your calculations represent the amount of insolation reaching the top of the atmosphere, now consider how the climate of each site might decrease this amount, such as from excessive cloudiness. Use the globes below to complete this step. Finally, combine your angle calculations with results from the globes below to make a list showing the pros and cons of each site. Choose the best site and be able to defend your conclusions.

12. Downward Shortwave Radiation Flux This globe shows the amount of shortwave radiation (insolation) that actually reaches the surface. Red, orange, and yellow show larger amounts of insolation reaching the surface, whereas blue and purple show smaller amounts. The generally greater amounts of insolation in the tropics versus the low amounts near the South Pole show the effects of latitude. The deviations from the broad pattern, such as the green and bluish green low over the Amazon, are due to processes in the atmosphere that reflect, absorb, or scatter insolation. Use what you know about the geographic setting of each site to propose what processes explain the pattern on this globe. The large globe on the previous page shows the topography (mountains versus low, flat regions) and land cover (vegetation versus rocks). These variations affect factors like cloudiness and the amount of rain.

13. Outgoing Longwave Radiation Flux (OLR) This globe shows the amount of OLR emitted by Earth's surface. Such radiation is not useful for generating solar energy because it is longwave and, therefore, less energetic than shortwave radiation, and because it is going upward, not down. It is a useful indication, however, of processes going on in the atmosphere because it suggests, for example, the abundance of water vapor (a greenhouse gas) and clouds, which can absorb longwave radiation before it exits the atmosphere, thereby reducing OLR. On this globe, orange represents large amounts of OLR, whereas purple and blue represent much smaller amounts. What do the patterns indicate, and what are the implications of both globes for solar energy? Again, examine the large globe on the previous page for clues.

13.16.a3   13.16.a4

Atmospheric Motion

MOTION OF THE EARTH'S ATMOSPHERE controls climate, rainfall, and weather patterns, all of which greatly affect our lives. It is driven largely by differences in insolation with influences from other factors, including topography, land-sea interfaces, and especially rotation of the planet. These factors control motion at local scales, like between a mountain and valley, at larger scales encompassing major storm systems, and at global scales, determining the prevailing wind directions for the entire planet. All of these local-to-global circulations are governed by similar physical principles.

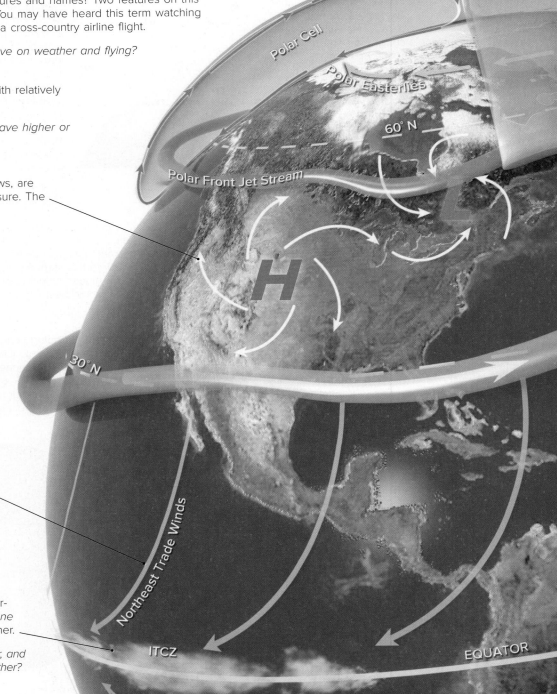

Large-scale patterns of atmospheric circulation are shown here for the Northern Hemisphere. Examine all the components on this figure and think about what you know about each. Do you recognize some of the features and names? Two features on this figure are identified with the term "jet stream." You may have heard this term watching the nightly weather report or from a captain on a cross-country airline flight.

What is a jet stream and what effect does it have on weather and flying?

Prominent labels of H and L represent areas with relatively higher and lower air pressure, respectively.

What is air pressure and why do some areas have higher or lower pressure than other areas?

Distinctive wind patterns, shown by white arrows, are associated with the areas of high and low pressure. The winds are flowing outward and in a clockwise direction from the high, but inward and in a counterclockwise direction from the low. These directions would be reversed for highs and lows in the Southern Hemisphere.

Why do wind patterns develop around areas of high and low pressures, why do these patterns spiral, and why are some spirals clockwise and others counterclockwise?

North of the equator, prevailing winds (shown with large gray arrows) have gently curved shapes. For most of human history, transportation routes depended on local and regional atmospheric circulation. These winds were named "trade winds" because of their importance in dictating the patterns of world commerce. The trade winds circulate from Spain southwestward, causing Christopher Columbus to land in the Bahamas rather than the present U.S.

What causes winds blowing toward the equator to be deflected to the west?

Prevailing winds from the north and south converge near the equator. This zone of convergence, called the *Intertropical Convergence Zone (ITCZ)*, is a locus of humid air and stormy weather.

What causes winds to converge near the equator, and why does this convergence cause unsettled weather?

TOPICS IN THIS CHAPTER

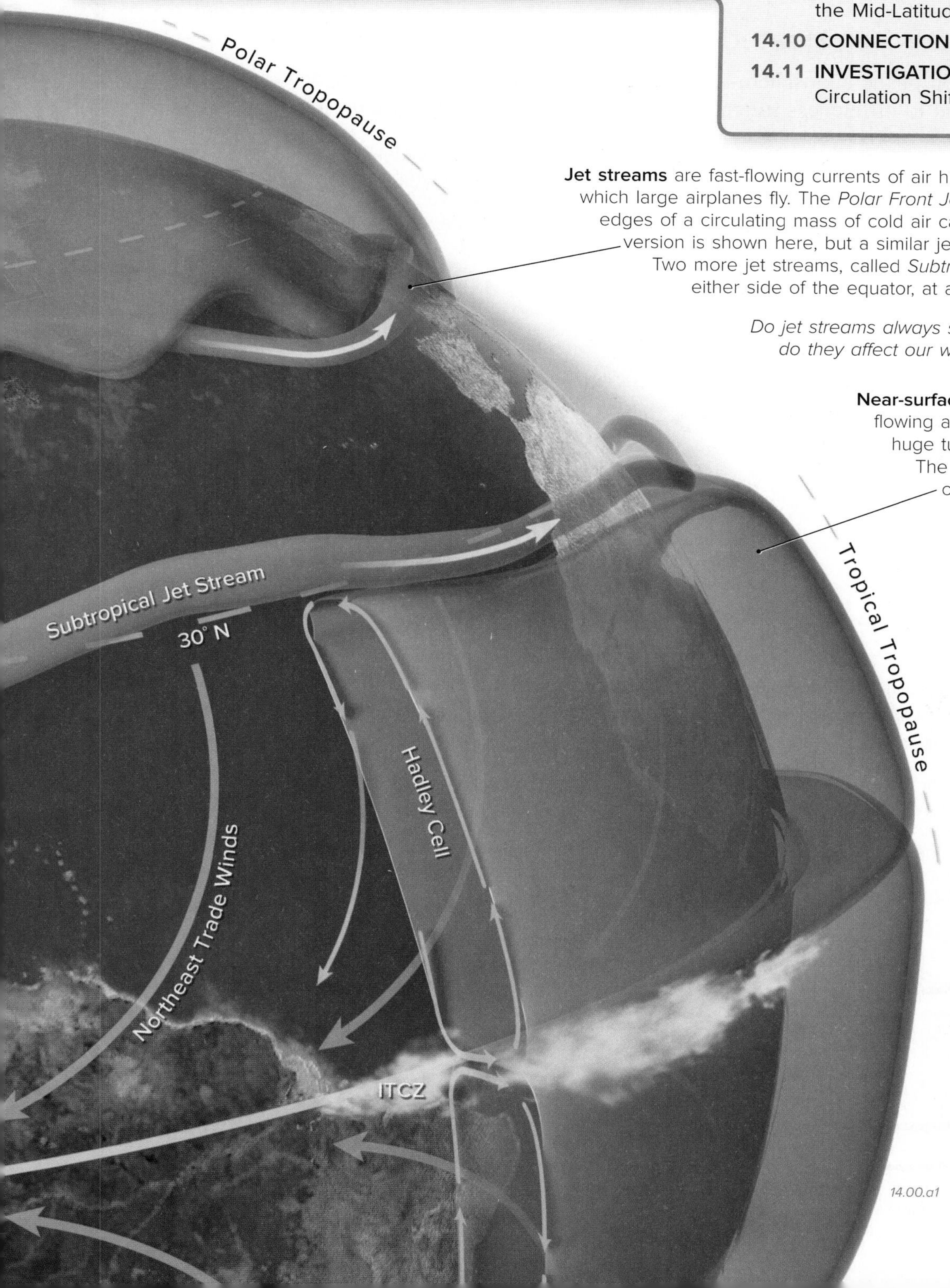

Jet streams are fast-flowing currents of air high in the troposphere, near the altitude at which large airplanes fly. The *Polar Front Jet Stream* occurs at high latitudes, along the edges of a circulating mass of cold air called the *Polar Cell*. The Northern Hemisphere version is shown here, but a similar jet stream occurs in the Southern Hemisphere. Two more jet streams, called *Subtropical Jet Streams*, occur in the subtropics on either side of the equator, at about 30° N and 30° S.

Do jet streams always stay in the same position, and how do they affect our weather?

Near-surface winds interact with upward- and downward-flowing air higher in the atmosphere, together forming huge tube-shaped air circuits called *circulation cells*. The most prominent of these are *Hadley Cells*, one of which occurs on either side of the equator.

What controls the existence and location of circulation cells, and how do the Hadley Cells influence global weather and climate?

Motion in the atmosphere affects us in many ways. It controls short-term weather and long-term climate, including typical average, maximum, and minimum temperatures. The large-scale patterns of air circulation, along with effects of local winds, cause some areas to be deserts and others to be rain forests, and cause winds to change direction with the seasons and from night to day. Regional air circulation affects the amount and timing of rainfall for a region, which in turn controls the types of soils, vegetation, agriculture, and animals situated in an area. Winds determine which areas of the U.S. are more conducive to wind-power generation than others. The result of these global, regional, and local atmospheric motions is a world in which the tropics are not too hot, the polar areas are not too cold, and no areas have too little moisture for life.

Polar Tropopause

Subtropical Jet Stream

30° N

Hadley Cell

Northeast Trade Winds

ITCZ

Tropical Tropopause

14.00.a1

14.0

14.1 How Do Gases Respond to Changes in Temperature and Pressure?

THE ATMOSPHERE CONSISTS LARGELY OF GASES, with lesser amounts of liquids, such as drops of water, and solids, such as dust and ice. By their nature, gases expand easily or contract in volume in response to changes in temperature and pressure. Variations in temperature and resulting changes in pressure are the main drivers of motion in the atmosphere.

A How Does a Gas Behave When Heated or Cooled?

The amount of insolation at the top of the atmosphere varies considerably from place to place and through time. These variations in insolation in turn lead to differences in temperatures, to which gases in the atmosphere respond.

14.01.a1 Namibia, Africa

1. Consider what happens when we want to make a hot air balloon rise (◄). Typically, a propane-powered burner heats ambient air, causing the air to expand in volume. This increase in volume inflates the balloon. Since the same amount of gas now occupies a much larger volume, the *density* of the heated air is less than the density of the surrounding air, so the balloon rises. So, as air increases in temperature, it tends to increase in volume and become less dense.

2. The figure below shows how a quantity of gas responds to either an increase in temperature (heating) or a decrease in temperature (cooling). The starting condition is represented by the cube of gas on the left.

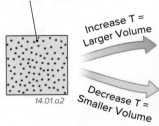

Increase T = Larger Volume

Decrease T = Smaller Volume

14.01.a2

3. An increase in the temperature of a gas means more energetic molecules, so a larger volume is needed to accommodate the same amount of gas.

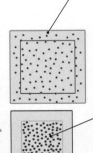

4. If a gas cools, the molecules within it have less kinetic energy (motions) and can therefore be packed into a smaller volume. The gas has a higher density and will tend to sink.

5. This example shows that temperature and volume of a gas are directly related—in fact, they are *proportional* if pressure is held constant. Such a proportional relationship means that if temperature is doubled, volume doubles too. If temperature decreases by half, volume does too. This specific relationship is called *Charles's Law*, which is one of the fundamental laws governing the behavior of gases, and it explains why a hot air balloon rises.

B What Happens When a Gas Is Compressed?

If a gas is held at a constant temperature but forced to occupy a smaller *volume*, the *pressure* of the gas increases. Pressure is proportional to the number of collisions of the molecules. If the same gas fills a larger volume, the collisions and amount of pressure both decrease. In both cases, if we instead change the pressure, the volume of the gas will adjust accordingly. A material, like a gas, that can be compressed is said to be *compressible*.

1. Molecules of gas in the sealed container in the left canister below are under pressure, represented by the two weights resting on top. At some temperature, the molecules have a corresponding amount of energy, and some of the moving molecules are hitting the movable lid, resisting the downward force of the attached weight.

2. Removing a weight reduces the downward pressure on the gas. However, the gas retains its same average energy level (temperature) and therefore exerts the same upward force on the movable lid as before. The upward force from the gas molecules exceeds the downward force of the weight and so raises the lid, increasing the volume occupied by the gas. In this way, a decrease in pressure results in an increase in volume, if the gas does not change temperature.

3. Increasing the downward pressure by adding a third weight on the original canister causes the lid to slide down. This increase in pressure causes a decrease in volume. As the gas is compressed into a smaller volume, the number of the molecules impacting the lid increases. When this upward force from the gas molecules equals the downward force from the weight, the lid stops moving, and the volume and pressure of the gas stop changing.

4. The relationship between pressure and volume of a gas, under conditions of constant temperature, is *inversely proportional*—if pressure increases, volume decreases. If pressure decreases, volume increases. Either pressure or volume can change, and the other factor responds accordingly, changing in the opposite direction by a proportional amount. That is, if the volume is cut in half, the pressure doubles. If the volume doubles, the pressure is cut in half. This inversely proportional relationship between pressure and volume under constant temperature is called *Boyle's Law*.

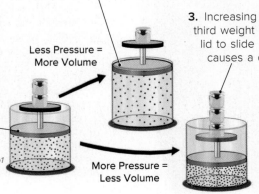

Less Pressure = More Volume

More Pressure = Less Volume

14.01.b1

C How Are Temperatures and Pressures Related?

Since Charles's Law relates volume to temperature, and Boyle's Law relates volume to pressure, we might suspect that we can relate temperature and pressure. Combining Charles's Law and Boyle's Law leads to the *Ideal Gas Law*, which relates temperature, pressure, and density (mass divided by volume). Basic aspects of the Ideal Gas Law help explain the processes that drive the motion of matter and associated energy in the atmosphere.

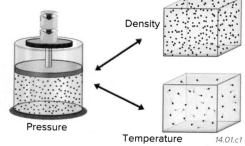

Density

Pressure

Temperature   14.01.c1

1. We can represent the Ideal Gas Law with a figure, with words, or with an equation. We begin with this figure (▶), which expresses the two sides of the equation. On one side of the equation (the left in this figure) is pressure. On the right side of the equation are density and temperature. The Ideal Gas Law states that if we increase a variable on one side of the equation (like increasing pressure), then one or both of the variables on the other side of the equation have to change in the same direction—density or temperature have to also change, or perhaps both do.

2. Examine this figure and envision changing any one of the three variables (pressure, density, or temperature), and consider how the other two variables would respond to satisfy the visual equation.

3. What happens if pressure increases? If temperature does not change, then density must increase. If pressure increases but density does not change, then temperature has to increase. Alternatively, temperature and density can both change. This three-way relationship partly explains why temperatures are generally warmer and the air is more dense at low elevations, where the air is compressed by the entire weight of the atmosphere, than at higher elevations, where there is less air. Higher pressure often results in higher temperatures.

4. What does the relationship predict will happen if a gas is heated to a higher temperature? If the density does not change, the pressure exerted by the gas on the plunger must increase. If the pressure does not change, the density must decrease. This is because density and temperature are on the same side of the equation, so an increase in one must be matched by a decrease in the other—if the other side of the equation (pressure) does not change. The relationship indicates that heated air can become less dense, which allows it to rise, like in a hot air balloon.

5. The Ideal Gas Law can also be expressed by the equation to the right:

$$P = R\rho T$$

where P is pressure, R is a constant, ρ is density (shown by the Greek letter rho), and T is temperature. Note how this equation roughly corresponds to the figure above.

D How Can Differences in Insolation Change Temperatures, Pressures, and Density in the Atmosphere?

The way gas responds to changes in temperature and pressure is the fundamental driver of motion in the atmosphere. Since temperature changes are largely due to insolation, we can examine how insolation affects the physical properties of gas and how this drives atmospheric motion.

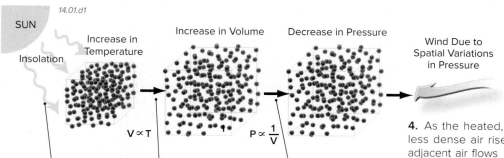

14.01.d1

SUN

Insolation

Increase in Temperature

Increase in Volume

Decrease in Pressure

Wind Due to Spatial Variations in Pressure

$V \propto T$

$P \propto \dfrac{1}{V}$

1. The Sun is the major energy source for Earth's weather, climate, and movements of energy and matter in the atmosphere and oceans. In the figure above, insolation strikes Earth's surface (land or water), which in turn heats a volume of gas in the overlying atmosphere.

2. The increase in temperature results in expansion of the gas because of the increased kinetic energy of the molecules in the gas; expansion is an increase in volume. If the same number of gas molecules occupy more volume, the density of the air decreases (the air becomes less dense).

3. The increase in volume can result in a decrease in pressure (less frequent molecular collisions). As a result, the air mass is now less dense than adjacent air that was heated less. The more strongly heated and expanded air rises because it is less dense relative to surrounding air (which was not heated as much and so is more dense).

4. As the heated, less dense air rises, adjacent air flows into the area to replace the rising air. The end result is vertical and lateral movement of air—vertical motion within the rising air, and lateral motion of surrounding air toward the area vacated by the rising air.

5. In this way, the response of gas to changes in temperature, pressure, and density (or volume), as expressed by the gas laws, is the primary cause of motion in the atmosphere. Variations in insolation cause changes in temperature, pressure, and density, which in turn cause air to move within the atmosphere.

Before You Leave This Page

- ☑ Sketch and explain why a gas under constant pressure expands and contracts with changes in temperature.

- ☑ Sketch and explain why a gas at a constant temperature expands or contracts with changes in pressure.

- ☑ Sketch and summarize the Ideal Gas Law and how the three gas laws explain motion in the atmosphere due to variations in insolation.

14.2 What Causes Winds?

WIND IS MOVEMENT OF AIR relative to Earth's surface. It forms in response to differences in air pressure from place to place and between different heights above the Earth's surface. Such differences in air pressure are generated primarily by uneven solar heating, especially as a function of distance from the equator, by temperature differences between different heights in the atmosphere, and by differences in how land and water respond to changes in temperature. Air flows from areas of higher pressure, where air sinks, to areas of lower pressure, where air rises.

A What Is Air Pressure?

Pressure is an expression of the force exerted on an area, usually from all directions. In the case of a gas, pressure is related to the frequency of molecular collisions that occur when a freely moving object or gas molecules collide with other objects, such as the walls of a container holding the gas. It is such collisions that keep a balloon, soccer ball, or bicycle tire inflated.

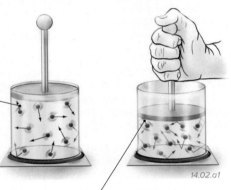

14.02.a1

14.02.a2

1. Molecules of gas in a sealed glass container move rapidly in random directions, and some molecules strike the walls of the container. The force imparted by these collisions is *pressure*. The more collisions there are, the more pressure is exerted on the walls of the container.

2. If we push down on the lid of the container, the same number of molecules are confined into a smaller space. Lower parts of the container walls are now struck by a greater number of the more closely packed gas molecules, so the pressure is greater. In other words, decreasing the volume of a gas increases its pressure.

3. What happens if we put a weight (▶) on top of the lid (center container) and then cool the container? What will happen if we heat the gas in the container?

4. If we cool the container by placing it in ice, the molecules become less energetic and strike the walls and lid of the container less often—the gas pressure decreases and the lid moves down.

5. If we instead heat the container, the gas molecules become more energetic and strike the walls and lid of the container more often—the gas pressure increases and lifts the lid.

B What Causes Atmospheric Pressure and How Is It Measured?

Earth's atmosphere is composed of gas molecules, including nitrogen (N_2) and oxygen (O_2), that have mass and are kept from escaping to space by Earth's gravitational field. The weight of these molecules exerts a force on Earth's surface and on other objects, including us. Pressure (force/area) is applied equally in all directions.

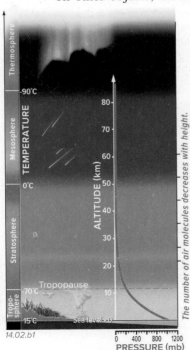

14.02.b1

The number of air molecules decreases with height.

PRESSURE (mb)

Pressure Decreases with Altitude

1. The density of molecules in the atmosphere decreases upward, and the force of Earth's gravity also decreases away from the surface. This results in a decrease in pressure (◀) as fewer gas molecules exert force on the surfaces below them. Atmospheric pressure is generally highest at low elevations, like near sea level, and is measured in *millibars* (1/1000 bar). It is much less at high elevations, where there is less air and gravity.

Measuring Pressure with a Barometer

2. We can measure air pressure with an instrument called a *barometer*. The barometer shown to the right (▶) is a sealed glass tube fixed in liquid mercury. Changes in air pressure cause the liquid level in the tube to rise or fall, allowing the measurement of relative pressure. Such barometers have units of *inches (or centimeters) of mercury*. Pressure is also reported in units of a *bar*, with one bar being approximately equal to the average air pressure at sea level. Modern digital instruments record pressure in millibars.

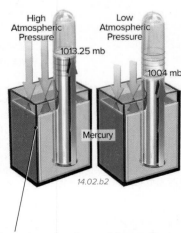

14.02.b2

3. When air pressure is higher, it pushes down harder and forces the mercury higher in the tube. The opposite happens if air pressure is lower.

C How Does Air Pressure Vary Laterally?

1. Air pressure also varies laterally from area to area, and from hour to hour, and these variations are typically represented on maps, like the one shown here (▶). Such maps either show the pressure conditions at a specific date and time or show pressure values averaged over some time period, like a month or a year. To allow us to compare different regions and see the larger patterns, the map uses pressure values that are corrected to sea level or their sea-level equivalent. In this way, we eliminate the effects of differences in elevation from place to place.

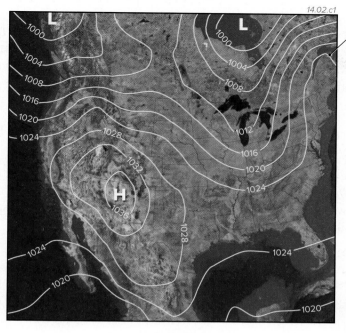

14.02.c1

2. Such maps of air pressure contain numbered lines, called *isobars*, that connect locations with equal pressure. If you could follow an isobar across the countryside, you would follow a path along which the pressure values, once corrected to their sea-level equivalents, would be equal. Successive isobars are numbered to represent different values of air pressure, usually in millibars (e.g., 1,020). Note that isobars do not cross, but can completely encircle an area. Such maps are commonly used to describe the daily regional weather conditions.

3. Most maps of air pressure feature the large capital letters H and L. An H represents an area of relatively higher pressure called a high-pressure area or simply a *high*. An L represents a low-pressure area, commonly called a *low*. An elongated area of high pressure can be called a *ridge* of high pressure and an elongated area of low pressure is a *trough*.

4. The map patterns change with time, corresponding to changes in air pressure that accompany changes in weather. Patterns typical for a region also change from season to season.

D How Do Variations in Pressure Cause Air to Move?

Air moves because there are variations in air pressures, in density of the air, or in both. Such pressure and density variations are mostly caused by differential temperatures of the air (due to differences in solar heating) or by air currents that converge or diverge. The atmosphere is not a closed container, so air can compress or expand. The resulting volume changes can make air pile up or spread out, causing variations in air pressure.

Movement of air occurs to equalize a difference in air pressure between two adjacent areas (▶), that is, a *pressure gradient*. Air molecules in high-pressure zones are packed more closely together than in low-pressure zones, so gas molecules in high-pressure zones tend to spread out toward low-pressure zones. As a result, air moves from higher to lower pressure, in the simplest case (as shown here) perpendicular to isobars.

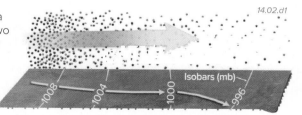

14.02.d1

Isobars (mb)

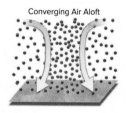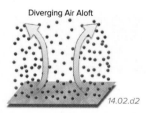

Converging Air Aloft   Diverging Air Aloft

14.02.d2

High-pressure zones and low-pressure zones can be formed by atmospheric currents that converge or diverge (▲). *Converging* air currents compress more air into a smaller space, increasing the air pressure. *Diverging* air currents move air away from an area, causing lower pressure.

Describing Wind Directions

Wind direction is conveyed as the direction *from which the wind is blowing*. Wind direction is commonly expressed with words, such as a northerly wind (blowing from the north). It can also be described as an azimuth in degrees clockwise from north. In this scheme, north is 0°, east is 090°, south is 180°, and west is 270°.

The atmosphere also has vertical motion, such as convection due to heating of the surface by sunlight. A local, upward flow is an *updraft*, and a downward one is a *downdraft*.

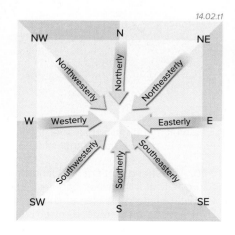

14.02.t1

NW   N   NE
Northwesterly Northerly Northeasterly
W Westerly   Easterly E
Southwesterly Southerly Southeasterly
SW   S   SE

Before You Leave This Page

☑ Explain what air pressure is and how we measure it.

☑ Explain how atmospheric pressure changes with altitude.

☑ Describe how air pressure is portrayed on maps, what low- and high-pressure zones are, and how variations in air pressure cause air to move.

14.2

14.3 What Causes Some Local and Regional Winds?

VARIATIONS IN AIR PRESSURE cause wind at all scales, from local winds affecting a relatively small area to regional ones affecting large parts of the globe. The origin and patterns of such winds reflect influences from unequal heating by sunlight, by heat related to the evaporation of liquid water and the condensation of water vapor, by differing thermal responses of land versus sea, and other factors. Such circulations contribute to the climate of a place, which in turn influence many aspects of geology.

A What Causes Breezes along Coasts to Reverse Direction Between Day and Night?

An example of how local conditions can cause local winds is experienced by people who live along coasts, where gentle winds sometimes blow in from the sea and at other times blow toward the sea. The gentle wind is called a *sea breeze* if it blows from sea in toward the land and a *land breeze* if it blows from the land out to sea. Such breezes are mostly due to differences in the way that land and sea warm up during the day and cool down at night. A key aspect of these winds is that heated air is less dense and rises, while air that is cooled is more dense and sinks.

The Sea Breeze (Daytime)

1. During daytime hours, land heats up significantly, particularly in summer. Land heats up more overall and more rapidly than water does. The hot air over the land surface rises, inducing a local low-pressure area over the land.

14.03.a2 Frankfort, MI

2. At the same time, air over the water body is cooled by the relatively cool water temperatures and by cooling associated with evaporation. The relatively cool air over the water sinks, inducing a local high-pressure area over the water.

14.03.a1

3. The difference between the low pressure over land and the high pressure over the water represents a pressure gradient, which pushes air near the surface from higher pressure to lower pressure. This flow is an *onshore breeze* or *sea breeze* that feels cool to people on the beach.

5. In this photograph (◄), heating of the land causes air over the land to rise, drawing in moist air from the adjacent water. Rising of the moist air along the coast forms thin, scattered clouds, but can draw in much thicker masses of clouds, forming coastal fog and overcast skies.

4. Air aloft moves in the opposite direction, from land to sea. This is a response to an upper-level pressure gradient caused by the "extra" air rising over land and less upper-level air over the sea, due to air sinking.

The Land Breeze (Nighttime)

6. At night, land cools significantly, particularly if there isn't much cloud cover. The relatively cool air over the land surface sinks, inducing a surface high-pressure area. Sometime in the evening hours, the surface low formed over the land during the day weakens and becomes a high.

14.03.a4 Philippines

7. The water body doesn't cool as much at night, so air over the water stays relatively warm compared to the air over the land. Therefore, this air rises, generating relatively low pressure over the water body.

14.03.a3

8. The difference between the high pressure over land and the low pressure over the water is a pressure gradient. The pressure-gradient force pushes near-surface air from higher to lower pressure, from the land to the water as an *offshore breeze* or *land breeze*.

11. In this photograph (◄), an early morning offshore flow, from the land to the sea, pushes moist air offshore, moving clouds out to sea and causing clear conditions along the shoreline.

9. Aloft, the air that rose over the low pressure over the water flows toward land in order to replace the air that sank to create the surface high.

10. The strength of sea and land breezes is proportional to the difference in temperature between the water and land. The land breeze circulation strengthens through the night, begins to weaken at sunrise, and stops sometime in the morning. Then, the wind reverses to a sea breeze, which strengthens until peaking in the afternoon.

B | Why Do Regional Wind Patterns Develop?

The Sun warms equatorial regions of Earth more than the poles, setting up a flow of warm air toward the poles. This flow of warm air is balanced by a flow of cold air from the poles toward the equator. Earth's rotation complicates the wind directions, producing curving patterns of circulating wind.

Rotation and Deflection

1. Earth is a spinning globe, with the equatorial region having a higher spin velocity than polar regions. As a result, air moving north or south has an apparent deflection sideways caused by the rotation, a response called the *Coriolis effect*.

2. Sunlight strikes equatorial regions more directly than it does areas closer to the poles and so preferentially heats the equatorial regions. As Earth rotates, the Sun's heat forms a band of warm air that encircles the globe and is re-energized by sunlight each day.

3. Warmed equatorial air rises and flows north and south, away from the equator. Air at the surface flows toward the equator to replace the air that rises. The Coriolis effect deflects this surface wind toward the west.

4. These flows of air combine into huge, tube-shaped cells of circulating winds, called *flow cells*. Some flow cells have surface winds flowing toward the poles. Others have winds flowing toward the equator.

Prevailing Winds

5. Recall that wind direction is referenced by the direction from which it is coming. A wind coming from the west is said to be a "west wind." A wind that generally blows from the west is a *westerly*.

6. Polar regions receive the least solar heating and are very cold. Surface winds move away from the poles, carrying cold air with them. *Polar easterlies* blow away from the North Pole and are deflected toward the west by Earth's rotation.

7. *Westerlies* dominate a central belt across the United States and Europe, so weather in these areas generally moves from west to east.

8. *Northeast trade winds* blow from the northeast and were named by sailors, who took advantage of the winds to sail from the Old World to the New World.

9. *Southeast trade winds* blow from the southeast toward the equator. Near the equator, they converge with the northeast trade winds in a stormy boundary called the *Intertropical Convergence Zone*.

10. *Westerlies* also occur in the Southern Hemisphere and are locally very strong because this belt is mostly over the oceans and has few continents to disrupt winds.

11. *Polar easterlies* flow away from the South Pole and deflect toward the west but are mostly on the back side of the globe in this view.

14.03.b1

Why the Coriolis Effect Occurs

12. How does the *Coriolis effect* deflect air (and water) movement on Earth's surface? Air, like the surface, is being carried around Earth by rotation. The surface has a faster velocity near the equator than at the poles because it has to travel a greater distance in 24 hours. Thick, blue arrows show the rate of rotation.

13. Therefore, as air moves toward the poles, it is rotating faster toward the east than the land over which it moves. It appears from the surface to be deflected to the east.

14. The opposite occurs as air moves toward the equator and encounters areas with a faster surface velocity. The air appears to lag behind, deflecting to the west as if it were being left behind by Earth's rotation.

14.03.b2

Before You Leave This Page

☑ Sketch and explain how surface temperature differences result in variations in air pressure and changing wind directions along coasts.

☑ Sketch and explain the directions and causes of global wind patterns.

☑ Sketch and explain the Coriolis effect and how it influences global wind patterns.

14.3

14.4 What Are Some Significant Regional Winds?

DIFFERENCES IN AIR PRESSURE cause a variety of regional to local wind conditions, such as those associated with storms, which are discussed in the chapter on weather. Some local winds are not so much related to weather systems as they are to differences in pressure that tend to occur at certain times of the year or after the establishment of an area of high pressure. These local to regional winds have interesting names, like Chinook winds or Santa Ana winds, and can have profound impacts on people.

A What Is a Chinook Wind and Where Do They Form?

The term *Chinook* originated in the Pacific Northwest and can refer to several types of winds. The most common usage is for a warm, dry wind that blows down the flanks of a mountain range. Chinooks are so warm and dry, they are called "snow eaters" for the way in which they can cause a sudden melting of snow and ice on the ground. The onset of a Chinook wind can cause a sudden rise in temperatures, especially during the winter. A Chinook in Loma, Montana, caused temperatures to rise from −48°C (−59°F) to 9°C (49°F) within a 24-hour period, the most change recorded for a single day in the U.S. In Spearfish, South Dakota, a Chinook off the adjacent Black Hills caused temperatures to rise 27 C° (49 F°) in two minutes, the world's fastest rise in temperature ever recorded!

1. This figure depicts the formation of a *Chinook wind*. The process begins when winds push moist air against the *windward* side of a mountain, where windward refers to the side from which wind is blowing. As the moist air rises up the mountain, it cools, causing the formation of clouds, a process that also releases latent heat. The heat warms the air, which continues rising toward the mountain peaks.

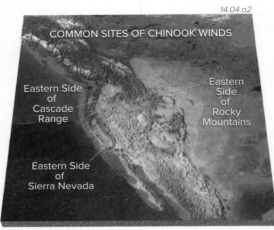
14.04.a2

COMMON SITES OF CHINOOK WINDS

Eastern Side of Cascade Range

Eastern Side of Sierra Nevada

Eastern Side of Rocky Mountains

14.04.a1

Windward Side

Leeward Side

Chinook Wind

2. Once the air reaches the peak, it begins flowing down the other side—the *leeward* side—of the mountain (the side opposite the windward side). As the air descends, it continues drying out and is compressed and heated. It was also warmed from the release of latent heat on the windward side. The warm, dry air descends from the mountain and spreads across the adjacent lowland, forming a Chinook wind.

3. This map shows the locations where Chinook winds are relatively common. As expected, Chinooks occur on the leeward side of mountain ranges (prevailing winds are from the west to east in this region), in Alaska, and elsewhere.

B How Do Katabatic Winds Affect Polar Regions?

1. A wind that blows downslope forming a wind of cold air is called a *katabatic wind*. Strong and regional katabatic winds affect Antarctica and Greenland, both of which have a high central landmass surrounded by ocean. Air over the middle of the landmasses is very cold and so is very dense, flowing off the central topographic highs and down the icy slopes.

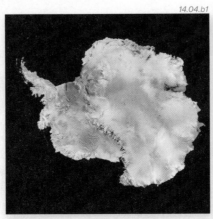
14.04.b1

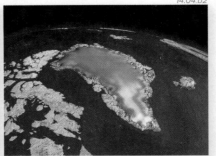
14.04.b2

14.04.b3 Antarctica

2. These two perspective views show that Antarctica (on the left) and Greenland (on the right) both have a broad, high area centered in the middle of the ice. Katabatic winds blow down off these high areas in all directions. These winds are especially strong where they are channeled down valleys, such as the famous Dry Valleys of Antarctica, so named because strong katabatic winds have stripped most ice and snow off the land surface.

3. Katabatic winds in Antarctica generally involve cold but dry air, but they can interact with clouds along the coast, sometimes creating stunning effects as the cold air and clouds spill off the highlands, similar to the scene in the photograph above.

C How Do Santa Ana Winds Affect Southern California?

Winds in Southern California typically blow from west to east (that is, they are westerlies), bringing relatively cool and moist air from the Pacific Ocean eastward onto land, especially in areas right along the coast, like Los Angeles and San Diego. These coastal cities also often have onshore sea breezes during the day and offshore land breezes at night. At other times, however, regional winds, called *Santa Ana winds*, blow from the northeast and bring dry, hot air toward the coast, causing hot, uncomfortable weather and setting the stage for horrendous wildfires.

Setting of Santa Ana Winds

1. *Santa Ana winds* are regional winds that blow from the northeast (▶), typically developing during spring and fall, when high pressure forms over the deserts of eastern California and Nevada. Circulation of air associated with the area of high pressure pushes winds south and westward toward the coast, in marked contrast to the normal onshore flow. This air is coming from the Mojave Desert and other desert areas to the north and east, so it is very dry.

14.04.c1

2. Santa Ana winds from the Mojave Desert are partially blocked by the mountains on the northern and eastern sides of Los Angeles. The winds spill through mountain passes, such as Cajon Pass northeast of Los Angeles, and are funneled down the canyons and into the Los Angeles basin. The funneling effect causes winds to be especially strong within the canyons. As the air moves from higher deserts (to the northeast) down toward Los Angeles, the air compresses and heats up. As a result, during an episode of Santa Ana winds, the coastal areas of Southern California experience much hotter and drier weather conditions than are normal. Due to this behavior of air flowing from high to low areas, Santa Ana winds are considered to be a type of katabatic wind.

Effect of Santa Ana Winds on Wildfires

14.04.c2 San Diego area, CA

3. The hills and mountains of coastal Southern California receive enough precipitation to be covered with thick brush (▲), such as oak, or by forests at higher elevations. During a Santa Ana wind, the fast winds dry out the brush, trees, and other vegetation, making it prone to wildfires.

4. The hills and mountains of Southern California experience some of the most spectacular but devastating wildfires of any place on the planet (▶). Santa Ana winds push these fires southwestward toward the cities and through neighborhoods in the foothills. Wildfires associated with Santa Ana winds can burn thousands of homes, causing hundreds of millions of dollars in damage. Pushed by the strong winds down the canyons, the fast-moving fires can cause the deaths of firefighters and people who did not evacuate in time.

5. This amazing image from NASA (▶) combines a satellite image of Southern California and adjacent states with the locations of fires (shown in red), as determined by processing a different kind of satellite data. The smoke produced by a number of fires trails off across the Pacific Ocean, clearly showing Santa Ana winds blowing from the northeast. These fires killed 9 people and injured dozens of others, destroyed more than 1,500 homes, and burned more than 2,000 km^2 of forest, brush, and neighborhoods. The region was declared a federal and state emergency, as more than a million people were evacuated, the largest such evacuation in California history. The especially large amount of destruction from these fires was due to the combination of strong Santa Ana winds and a prolonged drought that had dried out the natural vegetation.

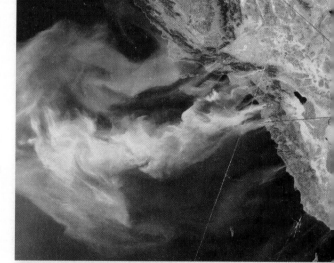

14.04.c4

14.04.c3

Before You Leave This Page

☑ Sketch and explain the origins of a Chinook and of a katabatic wind.

☑ Sketch and summarize the origin and setting of Santa Ana winds and why they are associated with destructive wildfires.

14.4

14.5 How Do Variations in Insolation Cause Global Patterns of Air Pressure and Circulation?

VARIATIONS IN INSOLATION from place to place and season to season cause regional differences in air pressure, which in turn set up regional and global systems of air circulation. These circulation patterns account for many of the characteristics of a region's climate (hot, cold, wet, dry), prevailing wind directions, and typical weather during different times of the year. Here, we focus on vertical motions in the atmosphere resulting from global variations in insolation and air pressure.

A What Pressure Variations and Air Motions Result from Differences in Insolation?

1. On this figure, the top graph plots the average amount of insolation striking the top of the atmosphere as a function of latitude. On the surface below, the large letters represent high- and low-pressure zones. Arrows show vertical and horizontal airflow and are color coded to convey the overall temperature of air.

2. The maximum amount of insolation striking the earth is along the equator and the rest of the tropics. This heats up the air, causing the warm air to expand and rise. The expansion and rising results in a zone of surface low pressure in equatorial regions. The upward flow of air helps increase the height of the tropopause over the equator, as shown by the dashed line.

3. Surface winds flow toward the low pressure to replace the rising tropical air. The rising tropical air cannot continue past the tropopause (the top of the troposphere), so as it reaches these heights, it flows away from the equator (EQ) to make room for more air rising from below. This upper-level air descends in the subtropics, near 30° latitude, where it forms a zone of high pressure.

4. The amount of insolation decreases away from the equator, with a relatively sharp drop-off across the mid-latitudes (30° to 60° latitude). Descending air and high pressure in the subtropics causes surface air to flow toward higher latitudes (to the right in this diagram).

5. The amount of insolation reaches a minimum near the poles. Air near the poles is very cold and dense, sinking to form a zone of high pressure near the surface. The descending air flows away from the pole, toward 60° latitude. In the upper atmosphere, air flows toward the poles to replenish the air that sank.

6. This global pattern of rising and sinking air and resulting low and high air pressures dominates the motion of Earth's atmosphere. The pattern results from variations in insolation and is compensated by horizontal flows, both near the surface and in the upper atmosphere.

14.05.a1

B How Does Sea-Level Air Pressure Vary Globally?

1. This map shows sea-level equivalent air pressure averaged for 1981 to 2010. Observe the main pattern and compare these patterns with the figure above. Can you explain the larger patterns on this map?

2. Two belts of high pressure (shown in light gray) encircle the globe at about 30° N and 30° S (the subtropics). Between these two is a belt of lower pressure (shown in medium gray) straddling the equator, in the tropics. The equatorial low pressure and flanking subtropical high-pressure zones are due to the large air current that rises in the tropics and descends in the subtropics. The large high over Asia is called the *Siberian high*.

3. A set of low-pressure areas occurs near 60° N. The lows are best developed in the oceans, and are poorly developed on land, reflecting differences in how water and land heat up and cool down. The strong low near Iceland is the *Icelandic low* and one southwest of Alaska is the *Aleutian low*.

4. A prominent air-pressure feature on this map is a belt of extremely low pressure (shown in dark gray) in the ocean just off Antarctica. This belt is so well developed in the Southern Hemisphere because of the abundant ocean surface, uninterrupted by continental landmasses at this latitude (60° S). An intense high-pressure belt (very light gray) occurs over continental Antarctica, in contrast to the Arctic, which is mostly ocean.

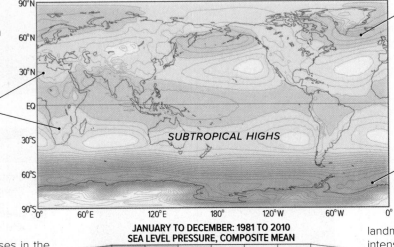

SUBTROPICAL HIGHS

JANUARY TO DECEMBER: 1981 TO 2010
SEA LEVEL PRESSURE, COMPOSITE MEAN

995   1012.5   1027.5   1045 mb

14.05.b1

What Influences the Patterns of Airflow Around High- and Low-Pressure Areas?

Once a gradient in air pressure starts air moving, the air is affected by the Coriolis effect and, if close to Earth's surface, is acted on by friction with the surface. The results of these interactions produce distinctive and familiar patterns of wind and circulation of air around areas of high and low pressure.

Low- and High-Pressure Areas

1. In an area of *low pressure*, the pressure-gradient force pushes air laterally into the low from all directions. If the Earth were not rotating and therefore had no Coriolis effect, this simple inward flow pattern would remain intact. Friction would likewise not perturb this pattern because friction does not change the direction of airflow, only the speed. Near the center of the low, the converging winds force some air upward, higher into the atmosphere.

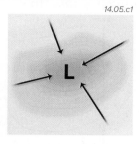
14.05.c1

2. In an area of *high pressure*, the pressure-gradient force pushes air laterally out in all directions. As with a low pressure, this outward flow of air is influenced by the Coriolis effect and by friction near Earth's surface. As air flows outward from the high, it is replaced by air flowing down from higher in the atmosphere.

14.05.c2

Northern Hemisphere

3. Earth does rotate, so the Coriolis effect, represented in this figure by green arrows, deflects the inward-flowing air associated with a low-pressure area. This deflection causes an inward-spiraling rotation pattern called a *cyclone*. In the Northern Hemisphere, the Coriolis deflection is to the right, causing a *counterclockwise* rotation of air around the low-pressure zone. This counterclockwise rotation is evident in Northern Hemisphere storms.

14.05.c3

4. In a Northern Hemisphere high-pressure area, the Coriolis effect deflects winds to the right, but friction again slows the winds. This results in an outward-spiraling rotation pattern called an *anticyclone*. In the Northern Hemisphere, circulation around an anticyclone is *clockwise* around the high-pressure zone.

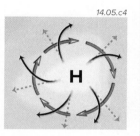
14.05.c4

Southern Hemisphere

5. For a low-pressure area in the Southern Hemisphere, the air flowing inward toward a low is deflected to the left by the Coriolis effect. This causes air within a cyclone in the Southern Hemisphere to rotate *clockwise*, opposite to what is observed in the Northern Hemisphere. In either hemisphere, friction causes wind patterns within cyclones to be intermediate between straight-inward winds and circular gradient winds.

14.05.c5

6. For a high in the Southern Hemisphere, the air flowing outward from the high is deflected to the left by the Coriolis effect. This causes the air within an anticyclone in the Southern Hemisphere to rotate *counterclockwise*, opposite to the pattern in the Northern Hemisphere. Therefore, cyclones and anticyclones can rotate either clockwise or counterclockwise, depending on hemisphere.

14.05.c6

14.05.c8　14.05.c7

7. This satellite image (◄) shows a cyclone near Iceland (Northern Hemisphere). It is characterized by a circular rotation of strong winds around the innermost part of the storm. It displays a distinctive inward-spiraling flow of the clouds and air around an area of low air pressure. Within a cyclone, air moves rapidly toward the very low pressure at the center of the storm. The clouds display an overall counterclockwise rotation of the storm.

| ≥250 km/hr |
| 210-249 km/hr |
| 178-209 km/hr |
| 154-177 km/hr |
| 118-153 km/hr |
| 63-117 km/hr |
| 0-62 km/hr |

8. The origins and paths of hurricanes in the North Atlantic and Caribbean basins in 2011 are shown here (▲). Note that few storms originate in the far southern part of this map, partly because the Coriolis effect is so small at these latitudes that it fails to impart the necessary rotation to the storms. Once formed, the paths of hurricanes are steered by the global wind patterns, which generally move Atlantic hurricanes west (guided by the trade winds) and then north and east once they enter the latitudes of the westerlies.

Before You Leave This Page

✓ Sketch and explain the global air-circulation patterns.

✓ Summarize the global patterns of air pressure, identifying key features.

✓ Sketch and explain circulation patterns around areas of low and high pressure, in both the Northern Hemisphere and Southern Hemisphere.

14.6 How Does Air Circulate in the Tropics?

TROPICAL CIRCULATION is driven by the intense solar heating of land and seas near the equator. The heated air rises and spreads out from the equator, setting up huge, recirculating cells of flowing air. The rising air results in a belt of tropical low pressure, and where the air descends back toward the surface is a belt of subtropical high pressure. What determines where the rising and sinking occur, and how does the Coriolis effect influence this flow?

General Circulation in the Tropics

1. Examine the large figure below and note the main features. What do you observe, and can you explain most of these features using concepts you learned from previous parts of the chapter? Tropical areas are known for their lush vegetation (▶), which in turn is due largely to relatively abundant and consistent insolation, warm temperatures, and abundant rainfall. After thinking about these aspects, read the rest of the text.

14.06.a2 Kakadu World Heritage Site, Australia

2. At the surface, winds generally converge on the equator from the north and south. The south-flowing winds in the Northern Hemisphere are apparently deflected to the right relative to their original path, blowing from the northeast. These winds are called the *northeast trade winds* because they guided sailing ships from the so-called Old World (Europe and Africa) to the New World (the Americas).

3. A belt of high pressure occurs near 30° N and 30° S, where air descends to the surface of the Earth. This air originally rose in the low pressure located near the equator as a result of excess heating.

4. The rising and descending air, and the related high- and low-pressure areas, are linked together in a huge cell of convecting air—the *Hadley cell*. One Hadley cell occurs north of the equator and another just south of the equator.

5. Note that the Hadley cell extends to approximately 30° north and south of the equator, so it generally encompasses all the tropics and some distance beyond.

6. In the Southern Hemisphere, winds blowing toward the equator are deflected to the left (west), resulting in winds blowing from the southeast, forming the *southeast trade winds*.

Counterclockwise Earth Rotation

Northeast Trade Winds

Southeast Trade Winds

Hadley Cell

14.06.a1

60°N
30°N
EQ
30°S
60°S

Formation of Hadley Cells

7. Insolation, on average, is most intense near the equator, in the tropics. The position of the overhead Sun migrates between the Tropic of Cancer and Tropic of Capricorn from season to season. The Sun-heated air rises from the tropics, forming a belt of low pressure at the surface. As the warm, moist air rises, the air cools somewhat, forming clouds; this accounts for the typical cloudiness and haziness of many tropical areas. Condensation of drops further heats the air, aiding its rise.

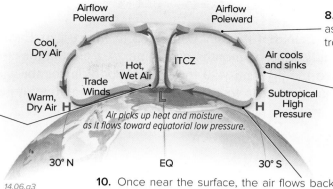

14.06.a3

8. After rising, this air spreads out poleward as it approaches the upper boundary of the troposphere (the tropopause).

9. Once the upper-level flow reaches about 30° N and 30° S latitude, it sinks, both because it begins to cool aloft and due to forces arising from the Earth's rotation. This sinking air dynamically compresses itself and the surrounding air, producing the subtropical high-pressure systems.

10. Once near the surface, the air flows back toward the equator to replace the air that rose. This return flow causes the trade winds aloft. The flow from the two hemispheres converges at the ITCZ.

Influence of the Coriolis Effect

11. As the air flows toward the equator in each hemisphere from the subtropical high to the ITCZ, the Coriolis effect pulls it to the right (in the Northern Hemisphere) or left (in the Southern Hemisphere) of its intended path, as shown by the arrows on the left side of this diagram. The Coriolis effect is weak near the equator, however, so the deflection is only slight. The result is surface air flowing from northeast to southwest in the Northern-Hemisphere tropics (the northeast trade winds) and from southeast to northwest in the Southern-Hemisphere tropics (the southeast trade winds).

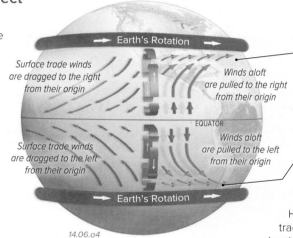

14.06.a4

12. In the Northern Hemisphere, as the air flows poleward after rising at the ITCZ, the weak Coriolis effect also pulls the air slightly to the right of its intended path. The result is that some of the upper-level air moves from southwest to northeast at the top of the Northern Hemisphere Hadley cell.

13. In the Southern Hemisphere, the Coriolis effect deflects the upper-level winds to the left of their intended path, causing a northwest-to-southeast flow at the top of the Southern Hemisphere Hadley cell.

14. As the seasons progress, the set of Hadley cells and the ITCZ migrate—to the Northern Hemisphere in Northern-Hemisphere summer and to the Southern Hemisphere in Southern-Hemisphere summer. If the trade-wind flow crosses the equator, the Coriolis deflection begins to occur in the opposite direction, and the winds can reverse direction (not shown).

Seasonal Variations in the Position of the Intertropical Convergence Zone

15. As the overhead Sun shifts north and south within the tropics from season to season, the ITCZ shifts, too. In the northern summer, it shifts to the north. The typical June position of the ITCZ is the reddish line on the figure below, and the December position is the blue line.

16. The ITCZ generally extends poleward over large landmasses in the hemisphere that is experiencing summer. This larger shift over the land than over the oceans is because of the more intense heating of land surfaces.

17. Unlike the ITCZ, the subtropical high pressure doesn't exist in a continuous belt around the Earth. The ocean-covered surfaces support high pressure better than land surfaces because land heats up too much at these latitudes, especially in summer. The heated air over the land rises, counteracting the tendency for sinking air in the Hadley cell. So the subtropical high pressure tends to be more vigorous over the oceans.

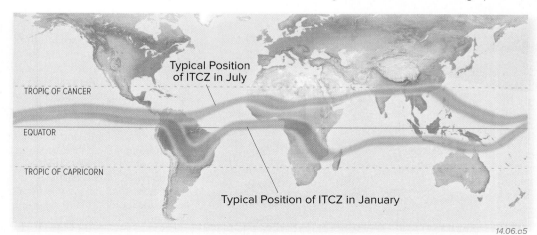

14.06.a5

Before You Leave This Page

☑ Sketch, label, and explain the main patterns of air circulation and air pressure over the tropics and subtropics.

☑ Sketch and explain air circulation in the Hadley cells.

☑ Locate and describe the Intertropical Convergence Zone and its seasonal shifts.

14.6

14.7 How Does Air Circulate in High Latitudes?

POLAR REGIONS RECEIVE LITTLE INSOLATION compared to the rest of Earth. As a result, the poles are very cold places that experience winter darkness for months at a time. Air circulation around the poles results from this relative lack of solar heating and also the proximity to the axis of rotation for the planet. The encroachment of polar air away from the poles can cause nearby areas to experience very cold temperatures. Airflow away from the poles results in a belt of relatively stormy weather near 45° to 60° N and 45° to 60° S.

General Circulation at High Latitudes

1. Examine the large figure below and observe the main features near the poles. Note the circulation directions near the surface versus those aloft. After you have made your observations, read the rest of the text.

2. Cold, dense air sinks near the North Pole. As it nears the surface, it then flows outward, away from the poles (to the south).

3. As the air flows south, it is deflected to the right by the Coriolis effect, which is very strong at these latitudes. As a result, surface winds generally encircle the North Pole, blowing in a clockwise direction when viewed from above the pole (▶). In the small globe to the right, the golden arrows show Earth's rotation and light-yellow arrows show surface winds.

4. The south-flowing air eventually begins to heat up and rise, usually somewhere between 60° and 45° latitude. This rising air causes a series of low-pressure areas at the surface, called the *subpolar lows* (L on this figure). Once the air rises to its maximum height, the flow turns back to the north, completing a circulating cell of cold air—the *polar cell*. The polar cell is represented here by the large blue arrows, with air rising near 60° N and descending at the pole.

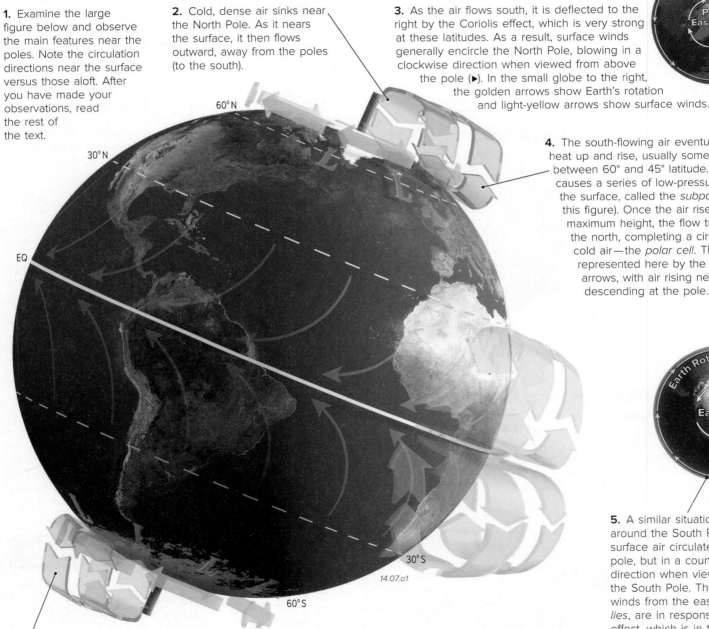

14.07.a1

5. A similar situation occurs around the South Pole, where surface air circulates around the pole, but in a counterclockwise direction when viewed from below the South Pole. These circular winds from the east, *polar easterlies*, are in response to the Coriolis effect, which is in turn caused by rotation of the Earth and enhanced by the comparative lack of surface friction with the ocean surfaces that dominate these latitudes. Remember that this view is from below the South Pole, a different perspective than you are used to.

6. As near the North Pole, cold air flowing away from the South Pole eventually heats up enough to rise, producing a belt of low pressure. The rising air aloft turns south and descends back near the pole, completing the polar cell. The polar cell involves very cold air at such high latitudes, causing the land to largely be covered year-round in ice and snow (▶).

14.07.a4 Antarctica

Circulation Around the Poles

7. The very cold air over the poles is so dense that it has a tendency to sink vigorously to the surface, creating high surface pressure—*polar highs*. The air then moves equatorward, because that is the only direction it can go from the pole.

8. The Coriolis effect is very strong at high latitudes, so the air deflects strongly and circulates around the pole, as shown here for the North Pole. Around the North Pole, the surface winds moving south deflect to the right of their intended path and so blow from the east—they are *polar easterlies*.

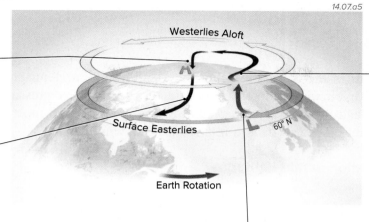
14.07.a5

10. At upper levels, the return flow of air northward toward the pole is also deflected to the right of its intended path (in the Northern Hemisphere). As it is turned to the right, it blows from the west (a westerly flow aloft). So not only is air flowing away from the pole near the surface and toward the pole aloft, the surface and upper-level airflows are rotating in opposite directions (clockwise near the surface, counterclockwise aloft). This is difficult to capture in a single perspective, which is why the polar flow is represented on this page with several figures.

9. As the air flows away from the pole, it warms and rises, producing low surface pressure—subpolar lows. In the winter, the subpolar lows are particularly intense over water bodies because the water is relatively warm at that time of year, relative to air elsewhere at these latitudes. The water warms the air, allowing it to rise.

Northern Hemisphere

11. The figure below shows the Northern Hemisphere polar cell, as viewed directly down on the North Pole. The gray paths depict surface flows (easterlies), whereas the brighter paths show upper-level flow (westerlies). Color gradations on arrows indicate whether air is warming (blue to red from tail to head) or cooling (red to blue).

12. High surface pressure is present at the pole, but shifts slightly in position from season to season. Low-pressure zones (the subpolar lows) occur over the adjacent oceans. The subpolar low in the Atlantic is the *Icelandic Low.* Another subpolar low, on the opposite side of the North Pole, occurs over the northern Pacific Ocean, and is the *Aleutian Low,* named for the Aleutian Islands west of mainland Alaska.

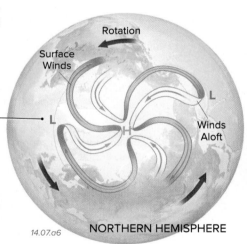
14.07.a6   NORTHERN HEMISPHERE

Southern Hemisphere

13. The polar cell in the Southern Hemisphere, shown below, is over the South Pole. Unlike the polar cell in the Northern Hemisphere, this one is centered over land—Antarctica. Antarctica is surrounded by uninterrupted oceans. The entire region is a very cold place, so even the air that is shown as warming is still very cold.

14. Surface winds flowing away from the pole are deflected to the left of their intended path, and so circulate counterclockwise around the pole when viewed from below (polar easterlies). Winds aloft move toward the pole and are deflected left of their intended path, flowing clockwise, in the opposite direction from the surface winds. The outward flowing surface air is balanced by the inward flowing air aloft. A continuous belt of low pressure occurs over the ocean.

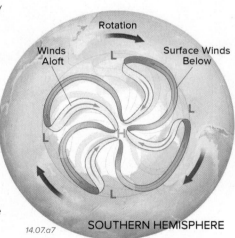
14.07.a7   SOUTHERN HEMISPHERE

Differences Between the North and South Polar Areas

The general patterns of atmospheric flow are similar for the North and South polar regions, except for the reversed directions of circulation (clockwise versus counterclockwise) related to different deflection directions for the Coriolis effect in the Northern and Southern hemispheres. One striking difference, however, is the difference between the land-sea distributions around the two poles. The North Pole is over ocean, although this ocean (the Arctic Ocean) is mostly frozen most of the year. The Arctic Ocean is nearly enclosed by the surrounding continents (North America, Asia, and Europe).

In contrast, the South Pole is over land, the frozen continent of Antarctica, and this continent is completely surrounded by open ocean. Differences in the responses of land and water to warming, cooling, and latent heat account for some difference in the specifics, such as having two main low-pressure areas in the oceans adjacent to the north polar region, but a continuous belt of low pressure in the ocean around Antarctica.

Before You Leave This Page

✓ Sketch, label, and explain the main patterns of air circulation and air pressure in the polar regions.

✓ Sketch and explain the polar cells, summarizing why circulation occurs.

✓ Explain the differences between polar circulation and pressures in Northern and Southern hemispheres.

14.7

14.8 How Does Surface Air Circulate in Mid-Latitudes?

THE MID-LATITUDES are regions in the Northern Hemisphere and Southern Hemisphere that lie between the tropics (23.5°) and polar circles (66.5°). Air circulation in the mid-latitudes is driven by pressures set up by circulation in the adjacent tropics and polar regions, and by the Coriolis effect. Surface winds within most of the mid-latitudes blow from west to east (westerlies), but the subtropics can have a relative lack of wind.

General Circulation at Mid-Latitudes

14.08.a2 Hells Canyon, ID

1. Examine the large figure below and observe the main features in the mid-latitudes, especially the region between 30° and 60° latitude (the dashed lines) in both the Northern and Southern hemispheres. The mid-latitudes are characterized by distinct seasons and changing weather patterns, and from forested areas (◄) to deserts. After observing the features on the large globe and reflecting on implications for weather, read the rest of the text.

2. Cold surface air flowing away from the poles warms and ascends along the poleward edge of the mid-latitudes. This rising air causes storminess and low pressure: the *subpolar lows* (L on this figure). This rising air is part of the cold *polar cell*.

3. In the Northern Hemisphere, air from the south flows toward the subpolar lows and away from high pressure in the subtropics (near 30–35° latitude). The Coriolis effect is relatively strong in the mid-latitudes and deflects this flow to the right, forming a belt of wind blowing from west to east—the *westerlies*.

4. Along the edge of the mid-latitudes, in the subtropics, air in the descending limb of the Hadley cell causes high pressure. These regions near 30° N and 30° S—the *horse latitudes*—can have weeks without wind, which posed a hazard to early sailing ships and their cargo.

5. The top part of the graph below shows how the combination of a northward-directed pressure-gradient force and eastward-directed Coriolis effect results in the westerlies in the Northern Hemisphere. The Coriolis effect strengthens with increasing latitude, turning winds even more to the east as they move north.

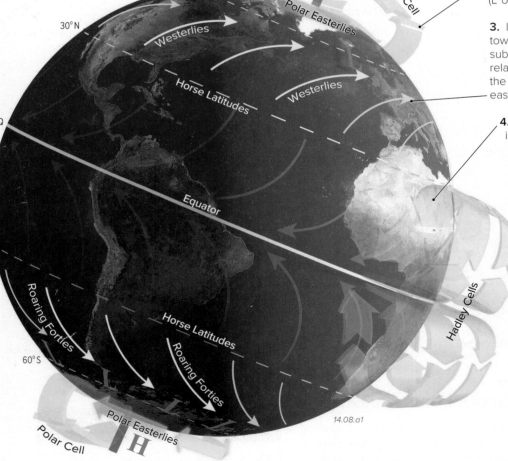

14.08.a1

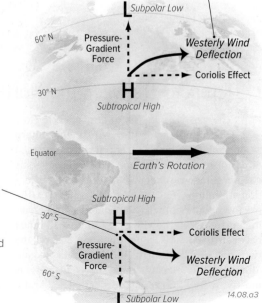

14.08.a3

7. The Southern Hemisphere has a belt of low pressure (the subpolar lows) caused by rising air in the polar cell. Air flows toward these lows and away from the subtropical highs, deflecting to the east and becoming westerlies. These westerlies are especially strong (there are no wind-blocking continents) and so are called the *roaring forties* (between 40° S and 50° S latitude).

6. In the Southern Hemisphere, air is driven south by high pressure in the subtropics. As this air flows south, it is deflected to the east by the Coriolis effect. The result is again west-to-east flow in the middle latitudes, forming a Southern Hemisphere belt of westerlies. Note that in both the Northern and Southern hemispheres, the Coriolis effect is deflecting flows in the mid-latitudes toward the east, in the direction the Earth is rotating.

Circulation Around Highs and Lows in the Northern Hemisphere

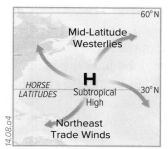

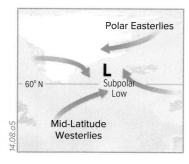

8. In the Northern Hemisphere, high pressure commonly occurs in the subtropics, along the horse latitudes. Winds around the highs are deflected in a clockwise direction by the Coriolis effect. Winds north of the high reinforce the westerlies, whereas winds to the south reinforce the easterly trade winds. As winds spin around the high, they push cold air equatorward and warm air poleward.

9. Circulation around lows, including subpolar lows, is counterclockwise in the Northern Hemisphere. The subpolar lows reside between polar easterlies to the north and westerlies to the south (in the mid-latitudes). The rising air and shearing between opposite-directed winds causes windy, stormy conditions.

Circulation Around Highs and Lows in the Southern Hemisphere

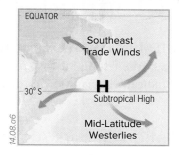

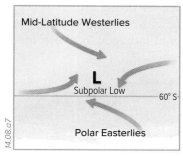

10. In the Southern Hemisphere, high pressure around the subtropics is accompanied by counterclockwise circulation. The highs are situated between southeast trade winds to the north and westerlies in the mid-latitudes to the south. As in the Northern Hemisphere, circulation around pressure systems redistributes warm and cold air and moisture.

11. Lows in the Southern Hemisphere, such as the subpolar lows, have clockwise circulation. The lows lie between the westerlies and polar easterlies, and this latitudinal belt is extremely windy and stormy. At this latitude, there are few large landmasses to disrupt the winds.

Migration of Pressure Areas in the Mid-Latitudes

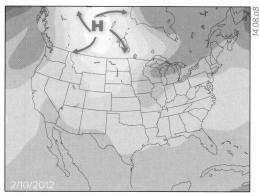

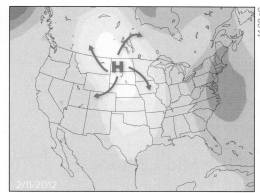

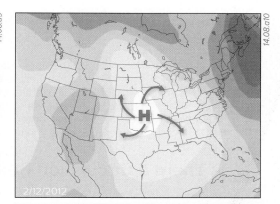

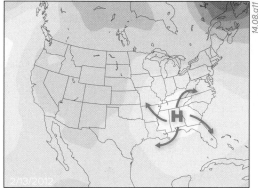

12. Embedded within the mid-latitude westerlies are traveling, spinning circulations of high and low pressure, much like a swirling flow that is embedded within the overall current in a stream. The four figures here show the movement of a high-pressure area (an *anticyclone*) guided by the mid-latitude westerlies. Anticyclones generally move equatorward as they migrate eastward, as shown in this series of daily images. Traveling low-pressure systems called *cyclones*, not shown here, tend to move poleward as they travel from west to east across the mid-latitudes; they are associated with stormy weather. As either type of pressure migrates through an area, it causes the wind direction to change relatively suddenly. As a result, wind and other weather patterns change more quickly in the mid-latitudes than in any other part of the Earth.

13. Note how the winds change in any area across which the pressure area migrates. In the case shown here, an area has northwesterly winds as the high pressure approaches from the northwest, followed by southeast winds once the high has passed to the east.

Before You Leave This Page

☑ Sketch, label, and describe the general circulation in the mid-latitudes, in both the Northern and Southern hemispheres, and explain the processes that cause air in the mid-latitudes to circulate generally from west to east.

☑ Sketch, label, and explain the circulation around areas of high and low pressure, and how they relate to the prevailing winds.

☑ Summarize the overall movement of mid-latitude cyclones and anticyclones.

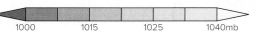

1000   1015   1025   1040mb

14.8

14.9 How Does Air Circulate Aloft over the Mid-Latitudes?

SURFACE WINDS IN THE MID-LATITUDES are generally from west to east in both hemispheres. Higher in the troposphere, the main features are two currents of fast-moving air—jet streams—that encircle the globe near the boundaries of the mid-latitudes. What factors determine the direction and speed of airflow aloft in the mid-latitudes, such as in the jet streams?

A What General Circulation Occurs Aloft over the Westerlies?

The main direction of surface air in the mid-latitudes is as curving arcs that move poleward away from the subtropics and then bend increasingly to the east, becoming westerlies. What happens aloft? The figure below shows the setting of upper-level airflows in the mid-latitudes, between the polar cell and the Hadley cell. The pole is to the left.

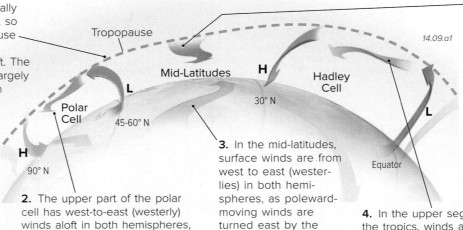

1. Circulation of air is essentially restricted to the troposphere, so the geometry of the tropopause (the top of the troposphere) influences the flow of air aloft. The height of the tropopause is largely controlled by temperatures in the underlying troposphere and whether the underlying air rises or sinks. As a result, the tropopause is shallower (8 km or less) over the poles, where cold, dense air sinks, than over the equator (16 km or more), where air is warmer and rises; it increases in height from the poles across the mid-latitudes.

2. The upper part of the polar cell has west-to-east (westerly) winds aloft in both hemispheres, in response to the strong Coriolis effect at high latitudes. Surface winds are easterlies.

3. In the mid-latitudes, surface winds are from west to east (westerlies) in both hemispheres, as poleward-moving winds are turned east by the Coriolis effect.

4. In the upper segment of the Hadley cell, over the tropics, winds aloft move southwest to northeast in the Northern Hemisphere, but northwest to southeast in the Southern Hemisphere; upper-level flows in both areas are therefore westerlies.

5. Surface winds in the mid-latitudes, and the winds on both sides, are moving overall from west to east (in both hemispheres). Thus, it is logical that the winds aloft over mid-latitudes will also be westerly, as shear from the adjacent circulations pull the air from west to east. Friction is less aloft than at the surface, so the upper-level westerlies over the mid-latitudes are stronger than westerlies at the surface.

B Is There a Circulation Cell in the Mid-Latitudes?

1. If there were a cell of circulating air in the mid-latitudes, as there is in the tropical and polar latitudes, then the sinking from the poleward edge of the Hadley cell would cause shear that would also induce sinking at the equatorward edge of the mid-latitudes.

2. Likewise, the rising motion at the equatorward edge of the polar cell (the subpolar lows) would induce rising next to the poleward edge of the mid-latitudes. This rising motion, coupled with the sinking next to the Hadley cell, would logically set up upper-level circulation over the mid-latitudes, with air moving from the pole toward the tropics. This hypothetical mid-latitude circulation system is known as the *Ferrel cell*.

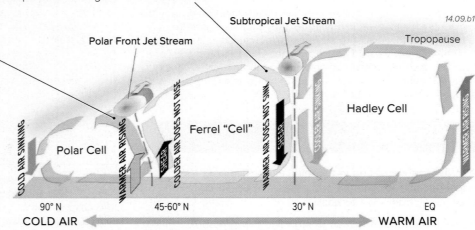

3. Descending air near the subtropical highs flows poleward along the surface and is turned toward the east by the Coriolis effect, contributing to the mid-latitude westerlies. Aloft, some rising air at the subpolar lows moves equatorward.

4. However, other factors cause the Ferrel cell to be developed very weakly or not at all. Thermal properties, which result in differences in air density, also dictate whether air will rise. Air on the subtropical side of the Ferrel cell (near 30° latitude) is relatively warm, so it should be rising, not sinking. Air on the polar side of the cell is colder, so it should be sinking, not rising. In other words, the thermal effects of the Ferrel cell's warmer air rising and cooler air sinking somewhat offset the dynamic effect of the shear from the adjacent Hadley and polar cells. The result is the near absence of a circulation cell in the mid-latitudes.

C What Are Jet Streams and How Do They Form?

Fast-moving, relatively narrow currents of wind, called *jet streams*, flow aloft along the boundaries of the mid-latitude air currents. One jet stream is located along the edge of the polar cell and another is along the edge of the tropical Hadley cell. How do such jet streams form and what controls their locations?

1. Jet streams are strong air currents that are long and relatively narrow, resembling a ribbon or tube wrapping around the planet. This figure shows two jet streams: the *polar front jet stream* and the *subtropical jet stream*. They are shown for the Northern Hemisphere, but equivalents are also present in corresponding positions in the Southern Hemisphere. The jet streams are typically a few hundred kilometers wide and about 5 km thick, and are located high in the troposphere, near the tropopause. They generally follow paths that curve or meander, and can locally split into separate strands that eventually rejoin. The four jet streams all blow from west to east as a result of Earth's rotation.

2. The subtropical jet stream circles around the globe at about 30° latitude (in both hemispheres), near the boundary between the Hadley cell and westerlies aloft in the mid-latitudes.

3. In both the Northern and Southern hemispheres, the polar front jet stream encircles the globe near the edge of the polar cell. It shifts position north and south from time to time, but typically resides between 45° and 60° latitude. As it shifts farther away from the pole, it brings cold air into the mid-latitudes, like the center of North America.

4. In addition to influencing weather, jet streams affect air travel in good and bad ways. A plane flying in the same direction as the jet stream (to the east) moves faster than normal, but a plane flying against the jet stream (west) is slowed. Jet streams also cause clear-sky turbulence on flights.

14.09.c1

D What Are Rossby Waves and How Do They Relate to Jet Streams?

1. Jet streams do not track around the planet along perfectly circular routes, but instead typically follow more irregular curved paths, like along the thick blue line in the polar projection below. These meandering paths, when viewed in three dimensions, resemble curving waves, and are called *Rossby waves* after the scientist who discovered them. The lines on this map, which is viewed directly down on the North Pole are highlighted by dark colors showing local areas of low pressure. Notice that these are concentrated in the subpolar areas.

2. Where the polar jet stream curves toward the pole, this part of the jet stream is called a *ridge*. Ridges bend poleward.

3. Where the polar jet stream curves away from the pole, we call that type of curve a *trough*. Troughs bend equatorward.

4. The size and position of ridges and troughs is not constant, but changes over days, weeks, and months. Troughs can get "deeper" or "shallower," while ridges get "stronger" or "weaker."

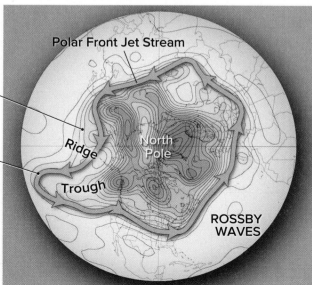

14.09.d1

5. Bends in the polar jet stream (Rossby waves) have a great influence on weather in nearby parts of the mid-latitudes. When a trough becomes more accentuated and shifts farther away from the pole (i.e., it deepens), it allows cold Arctic air to extend farther away from the pole, such as occurred during recent incursions of cold air southward in what was called the "polar vortex." Rossby waves influence surface high and low pressures near the polar jet and help strengthen, weaken, and guide mid-latitude cyclones as they migrate across the surface.

Before You Leave This Page

✓ Describe the flow of air aloft over the mid-latitudes.

✓ Sketch, label, and explain the two types of jet streams.

✓ Sketch and explain Rossby waves, including a trough and ridge.

14.9

What Causes Monsoons?

A COMMON MISCONCEPTION is that the word "monsoon" refers to a type of rainfall, but the word actually refers to winds that reverse directions depending on the season. One of these seasonal wind directions typically brings dry conditions and the other brings wet conditions. Monsoons impact a majority of the world's population.

A What Are the Features of the Asian Monsoon?

One way to characterize a monsoon is to compare maps showing wind directions for different times of the year. Such maps can then be compared to rainfall records to determine which seasonal wind directions bring dry conditions and which ones bring wet conditions. The maps below show climatological wind conditions, averaged over three decades, for two different months—January and July. Arrows show wind directions, and shading represents pressure at sea level, with light gray being high and dark gray being low. Examine the patterns of circulation for each month and then compare the patterns between the months.

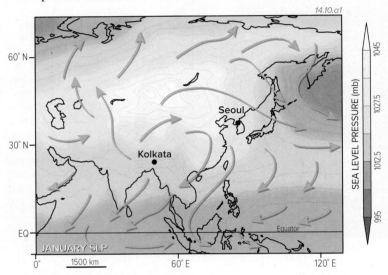

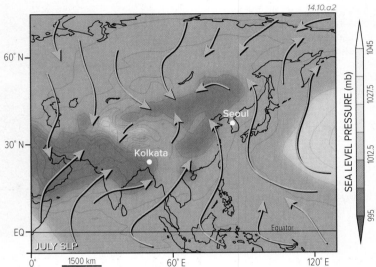

1. *January*—This map shows typical wind conditions for Asia during January. In the center of the map, winds define a region where flow is clockwise and outward, centered on the light-colored area of high pressure. This high-pressure area, the *Siberian High*, forms from cold, sinking air over Siberia. This circulation brings very dry air (from the cold interior of the continent) from the north over southern Asia and from the northwest across eastern Asia. We would predict from these wind patterns that little precipitation would occur in much of Asia at this time.

2. *July*—This map shows that wind conditions for the same region during July are totally different than they are for January. Circulation that marked the high pressure is gone, replaced by an area of inward and counterclockwise flow over Tibet (north of Kolkata). In the Northern Hemisphere, this pattern of circulation is diagnostic of a low-pressure area, which in this case is caused by warm, rising air that accompanies warming of the Asian landmass. This low is called the *Tibetan Low*. This circulation brings very humid air from the southwest over southern and southeastern Asia. How do you think this circulation affects rainfall?

Seasonal Variation in Precipitation

3. Observe these graphs showing average monthly precipitation amounts for two very different parts of Asia: Kolkata, one of the largest metropolitan areas of India, and Seoul, the capital of South Korea. For both cities, notice prominent precipitation peaks that occur during the summer—the wet season. The increase in precipitation during the wet season results from the flow of moist air from oceans onto land, toward the Tibetan Low. The dry season, during the winter months, reflects the flow of dry air from the land, flowing outward from the Siberian High.

Effect of the Monsoon on Vegetation

4. These satellite images show increased vegetation due to monsoon-related rains along the western coast of India. The left image is during the dry season, when wind patterns bring in dry air. The right image is from the end of the monsoon. Note the increase in plant cover (green areas) during the monsoon-caused rainy season.

14.10.a5 Western India   14.10.a6 Western India

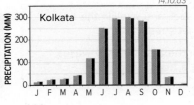

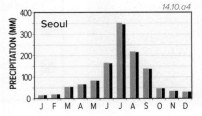

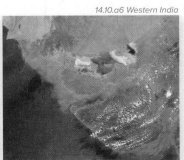

B | What Other Regions Experience Monsoon Circulations?

West Africa

January—In January, near-surface winds in West Africa largely flow from the northeast, bringing in dry air from inland areas, including the Sahara Desert, and carrying it southwest to coastal areas and farther offshore. Such *offshore flows* generally result in dry weather.

| ← Winter | ← Summer |

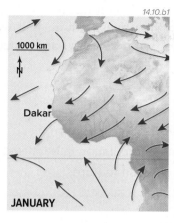
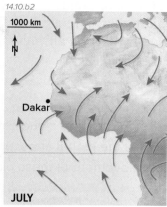

JANUARY   JULY

July—A shift in wind direction in July brings moist ocean air from several directions onto the very hot land where air has risen. This change in wind direction causes enormous differences in precipitation, as shown by the graph below for Dakar, Senegal. Along with the increase in precipitation comes an increase in the amount of vegetation. In Dakar and much of the region, precipitation is nearly nonexistent in January and adjacent months.

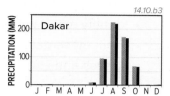

14.10.b3

Northern Australia

January—In January (the southern summer), winds over northern Australia bring moist air from the ocean onto the heated land surface.

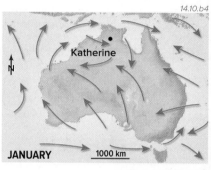
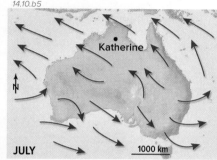

JANUARY   1000 km   JULY   1000 km

July—The wind shifts by July (winter) as the land surface cools, creating higher pressure over the land. This causes a large drop in precipitation, as shown by the graph below for Katherine, Australia. The monsoon flow in July results in little rain.

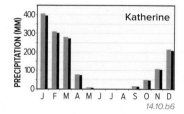

14.10.b6

Southwestern U.S.

January—Southwestern North America, most of which is desert, has a less dramatic, but still important, monsoon effect. In the winter months, winds blow from various directions, and winter precipitation in this region is from brief incursions of cold, wet air (i.e., cold fronts) from the northwest.

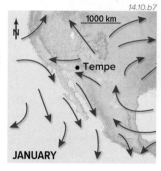

JANUARY   JULY   1000 km

July—During the late summer months, heating of the land surface and the resulting low pressure causes a shift in winds. Winds from the south bring moist air northward from the Gulf of Mexico and Gulf of California, and summer thunderstorms form when this air interacts with the heated land. These summer thunderstorms cause precipitation to peak in August, as shown by the graph below for Tempe, Arizona. Note the different scale needed to show the relatively small amounts of precipitation in this desert area versus the previous ones. Nearly as much precipitation falls in the winter from the cold fronts.

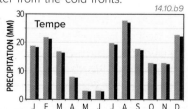

14.10.b9

The Effect of Monsoons on Cultures

Monsoons greatly influence the lives of people living in regions with seasonal shifts in wind. The main effects of a monsoon are seasonal variations in precipitation, which in turn affect water supplies, amount of vegetation, and overall livability for some normally dry landscapes. Many cultures plan their activities around these seasonal changes, conserving water during the dry season and taking advantage of the plentiful water during the wet season.

The monsoon pervades the psyches of people in southern and southeastern Asia, especially the region from India to Vietnam, in ways not fathomable to most North Americans or Europeans. The influence on agriculture, including the cultivation of rice, and on flooding and other natural hazards is obvious, but the monsoon also appears in literature, art, music, architecture, and nearly every other aspect of culture. Ceremonies commonly mark the anticipated start of the monsoon. In years when the monsoon rains arrive later than usual, people become very concerned that harvests will suffer. The date of the onset of the monsoon rains varies by location, but generally proceeds from south to north with the onset in April and May.

Before You Leave This Page

✓ Explain what causes a monsoon, using examples from Asia, West Africa, northern Australia, and southwestern North America.

✓ Describe some of the effects of shifting monsoonal winds.

14.11 What Occurs During Seasonal Circulation Shifts?

GLOBAL ATMOSPHERIC CIRCULATION responds directly to insolation. As the Sun's direct rays migrate seasonally, belts of winds, such as the westerlies, migrate too. In this investigation, you will examine the general circulation of the atmosphere, as expressed by data on air pressure, wind velocity, and cloud cover for two months with very different seasons—January and July.

Goals of This Exercise:

- Identify major patterns in air pressure, wind velocity, and cloud cover for each season.
- From these data, identify the major features of the global atmospheric circulation in each season.
- Assess and explain the degree of seasonal movement of these circulation features.

When examining broad-scale patterns of the Earth, such as global circulation patterns, a useful strategy is to focus on one part of the system at a time. Another often-recommended strategy is to begin with relatively simple parts of a system before moving to more complex ones. For this investigation, you will infer global patterns of air circulation by focusing on the Atlantic Ocean and adjacent lands (▶).

This globe is centered on the central Atlantic, and its top is slightly tilted toward you to better show the Northern Hemisphere. As a result of this tilt, Antarctica is barely visible at the bottom of the globe. The colors on land, derived from satellite data, depict rocks and sand in tan and brown. Vegetation is in various shades of green, with the darkest green indicating the thickest vegetation (usually forests). Shallow waters in the Caribbean region (on the left side of the globe) are light blue.

Observe the entire scene, noting which areas on land have the most vegetation and which ones have the least. Compare these vegetation patterns with large-scale patterns of atmospheric circulation and air pressure, like subtropical highs and the Intertropical Convergence Zone (ITCZ).

Consider what directions of prevailing winds would occur in different belts of latitude. For example, where in this globe are the two belts of trade winds (one north and one south of the equator)? How about the mid-latitude belts of westerlies in each hemisphere? Consider how these winds might blow moisture-rich air from the ocean onto land. After you have thought about these aspects, read the procedures below and examine the globes and text on the next page, which highlight average air pressure, wind velocity, and cloud cover for two months—January and July.

14.11.a1

Procedures

Complete the following steps on a worksheet provided by your instructor or as an online activity.

A. Study the two globes showing air pressure (on the next page), and note areas with high and low pressure. First, locate a belt of low pressure near the equator and adjacent belts of subtropical highs on either side. Next, locate the Icelandic Low and a high-area pressure to the south (called the Bermuda-Azores High), located in the Atlantic Ocean between Africa and North America. Determine for which season each is strongest or if there is not much difference between the seasons. Mark and label the approximate locations of these features on the globe on the worksheet.

B. Next, examine the two globes that show wind velocity. In the appropriate place on the worksheet, draw a few arrows to represent the main wind patterns for different regions in each month. Label the two belts of westerlies and the two belts of trade winds. If the horse latitudes are visible for any hemisphere and season, label them as well. Mark any somewhat circular patterns of regional winds and indicate what pressure feature is associated with each.

C. Examine the two globes that show the average cloud cover for each month. From these patterns, label areas that you interpret to have high rainfall in the tropics due to proximity to the ITCZ or low rainfall due to position in a subtropical high. Examine how the cloud patterns correspond to the amount of vegetation, pressure, and winds.

D. Sketch and explain how the different features of circulation and air pressure change between the two months. Answer all the questions on the worksheet or online.

Air Pressure

These two globes show average air pressure over the Atlantic and adjacent land areas during the months of January and July. Lighter gray indicates relatively high pressures, whereas darker gray indicates low pressures. The lines encircling the globe are the equator and 30° and 60° (N and S) latitudes.

Observe the main patterns on these two globes, noting the positions of high pressure and low pressure and how the positions, shapes, and strengths change between the two seasons. Then, complete the steps described in the procedures section.

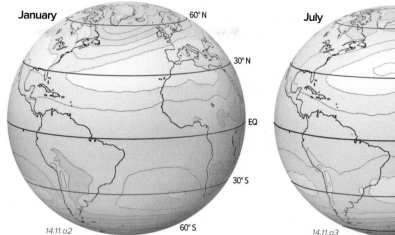

14.11.a2

14.11.a3

Wind Velocity

These globes show average wind velocities for January and July. The arrows show the directions, while the shading represents the speed, with darker being faster. In this exercise, the directions are more important than the speeds, but both tell part of the story.

Observe the large-scale patterns, identifying those patterns that are related to global circulation (i.e., westerlies) versus those that are related to more regional features, such as the Bermuda-Azores High and the Icelandic Low. Note also the position of where winds converge (ITCZ) along the equator and how this position changes between the two months.

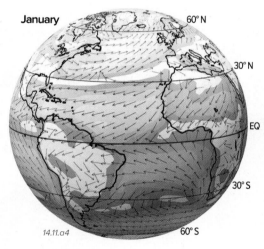
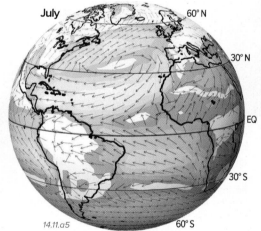

14.11.a4

14.11.a5

Cloud Cover

The clouds that form, move above Earth's surface, and disappear can be detected and tracked with satellites, shown here for January and July of a recent year. On these globes, light colors that obscure the land and ocean indicate more abundant clouds (and often precipitation), whereas the land shows through in areas that average fewer clouds.

Observe the large-scale patterns, noting which areas are cloudiest and which ones generally have clear skies. Relate these patterns of clouds to the following: amount of vegetation on the land, air pressure for that month, and average wind directions. Answer all questions on the worksheet or online.

14.11.a6

14.11.a7

14.11

MOISTURE IN THE ATMOSPHERE in the form of water vapor, liquid water, and ice, controls most aspects of our weather and climate. Moisture moves back and forth from Earth's surface to the atmosphere and, once in the atmosphere, is transferred vertically and laterally by moving air. Atmospheric moisture is expressed as clouds, precipitation, storms, weather fronts, and other phenomena. In this chapter, we explore how and where moisture occurs in the atmosphere, what happens when moisture moves with air currents, and how such motions are expressed as clouds, precipitation, and other aspects of weather.

Examine this central figure and observe the various features. Try to identify which features involve water in one form or another. Next, consider how moisture can move from one place to another, and from one state of matter (gas, liquid, solid) to another. After you have done this for each main feature shown, read the text surrounding the figure.

15.00.a2 Grand Cayman Islands

Our world has various types of clouds, some of which are shown here in the large figure and the photograph above. Some clouds are thin and wispy, whereas others are tall and puffy. Some are high in the troposphere, while others are close to Earth's surface. Some clouds are associated with precipitation (rain, snow, hail, or sleet), but most are not.

What is in a cloud, how do we classify and name different types of clouds, and what do different types of clouds tell us about what is going on in the atmosphere?

Condensation

Precipitation

Evapo-transpiration

Evaporation

Runoff

15.00.a1

15.00.a3 Plush, OR

Most clouds and precipitation are caused by cooling of moist air. Such cooling generally occurs when air rises into cooler parts of the atmosphere. Cooling of moist air can also occur when warmer air interacts with cooler land or water. As moist air cools, water vapor can change into a liquid (water drops) through the process of *condensation*. If cold enough, water vapor can instead form ice crystals through the process of *deposition*. Also, water drops and ice crystals can form from one another. In any case, the water drops and ice crystals are what form clouds and, under the right conditions, cause various types of precipitation (◄).

For each place where a cloud is shown in the large figure, what are some possible reasons why the air might be cooling at that location?

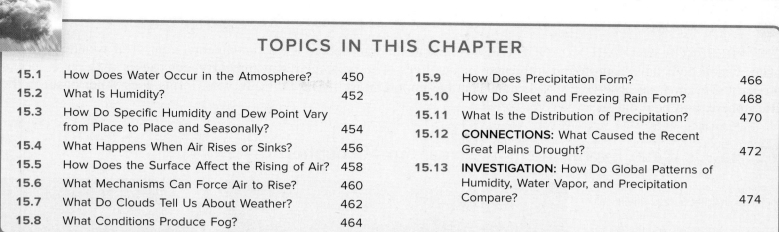

TOPICS IN THIS CHAPTER

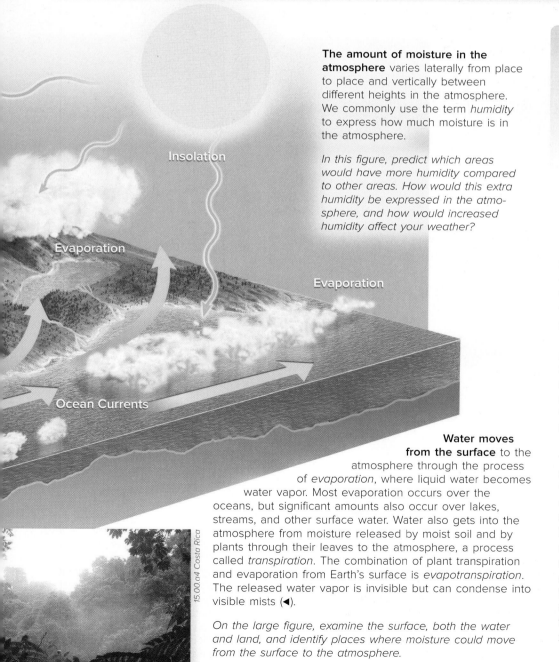

The amount of moisture in the atmosphere varies laterally from place to place and vertically between different heights in the atmosphere. We commonly use the term *humidity* to express how much moisture is in the atmosphere.

In this figure, predict which areas would have more humidity compared to other areas. How would this extra humidity be expressed in the atmosphere, and how would increased humidity affect your weather?

Insolation

Evaporation

Evaporation

Ocean Currents

Water moves from the surface to the atmosphere through the process of *evaporation*, where liquid water becomes water vapor. Most evaporation occurs over the oceans, but significant amounts also occur over lakes, streams, and other surface water. Water also gets into the atmosphere from moisture released by moist soil and by plants through their leaves to the atmosphere, a process called *transpiration*. The combination of plant transpiration and evaporation from Earth's surface is *evapotranspiration*. The released water vapor is invisible but can condense into visible mists (◀).

15.00.a4 Costa Rica

On the large figure, examine the surface, both the water and land, and identify places where moisture could move from the surface to the atmosphere.

Expressions of Moisture

The large figure on these two pages shows a wide variety of ways in which moisture is expressed on the surface and in the atmosphere. Water is easily identifiable in the large body of water (ocean) on the right, in the two lakes on land, and in the streams. This water is shown in liquid form, but under cold-enough circumstances there could also be ice, such as if the lakes were at least partially frozen or if the ocean had some sea ice (frozen seawater) or icebergs (large floating chunks of ice). We can easily recognize water in its solid form (ice) in the snow on the high mountains. There would also be a zone of especially moist air (more water vapor) over and adjacent to the ocean, lakes, streams, and wet parts of the land, but we cannot see it because water vapor is not visible.

Even less obvious is the water contained in the green plants and the underlying soil. This moisture moves from these sites to and from the atmosphere, such as rain that falls from the clouds, soaks into the soil, and is taken up by plants, which can then release this moisture back into the atmosphere via transpiration.

Moisture in the atmosphere is expressed in the various types of clouds. Most clouds are visible to us because they contain tiny drops of liquid water—even if there is no rain associated with the cloud. The wispy, high-level clouds consist of ice crystals, which are easily blown and streaked out by the strong upper-level winds. The higher parts of some of the taller lower-level clouds would also contain ice, which can remain aloft or eventually accumulate into such large particles that it falls to the surface as snowflakes or hail. In all, many aspects of our planet owe their existence to moisture in the atmosphere and to the exchange of moisture between the atmosphere and the surface.

15.0

15.1 How Does Water Occur in the Atmosphere?

THE PRESENCE AND ABUNDANCE OF WATER in the atmosphere are a fundamental control of weather and climate, which both have a profound influence on our lives. The molecular structure of water causes it to have special properties that we can observe every day and that are important to life on Earth. In what forms does water occur in the atmosphere, and how did the water get there?

A How Is Water Expressed In and Near the Atmosphere?

Examine this photograph of the Himalaya in Tibet (▶) and identify all the places where there are visible expressions of water. Are there places where water is likely to be present but not visible? Ponder this for a moment before reading on.

Water in the atmosphere, as on the Earth's surface and in its subsurface, occurs in three forms: as a gas (water vapor), a liquid (liquid water), and a solid (ice). In this scene, liquid water in the atmosphere is expressed as tiny drops in the clouds and as any raindrops falling from the clouds. Ice crystals, expressed as snow, are on the surface, but are also present as small crystals in some of the clouds. Water vapor is also present, but is not obvious—it always occurs as an invisible gas. Clouds, like the rest of the atmosphere, contain water vapor but only the airborne drops and ice crystals are visible.

15.01.a1 Himalaya, Tibet

15.01.a2 Galápagos Islands

Tiny water drops also form fog and mist, as in this photograph from the Galápagos Islands (▶). Such moisture sustains unusual plants that extract moisture directly from the atmosphere or from the liquid water (dew) that forms on leaves and other hard surfaces, including on the back of the giant tortoise (~1 m across) grazing on vegetation sustained by the mist.

B What Are Some Important Properties of Water?

Recall that water (H_2O) is a molecule composed of one oxygen atom strongly bonded to two hydrogen atoms. The asymmetrical arrangement of the hydrogen atoms causes the molecule to have a positive charge on the side with the hydrogen atoms and a negative charge on the opposite side, near the oxygen atom. A molecule with this charge distribution is said to be *polar*, and the polar nature of water is why it is such a good solvent (can dissolve other substances). This polar character has many other implications.

Hydrogen Bond

15.01.b1

The positive side of one water molecule is attracted to the negative side of an adjacent water molecule, forming a weak bond (i.e., a *hydrogen bond*) that tends to keep adjacent water molecules together. This attraction is what causes water to tend to stay together as a discrete drop, rather than flowing away (▶). This tendency for water to stay together with a discrete outer surface is called *surface tension*. Surface tension has to be overcome during evaporation, because it tends to hold water molecules within the liquid rather than letting them escape into the air. Surface tension also must be overcome to allow small drops of water to form in the atmosphere, as in clouds, and to allow these small drops to combine into larger drops, forming a raindrop.

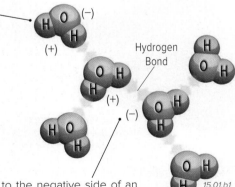
15.01.b2

15.01.b3

Surface tension allows water to attach itself to other objects, such as this damp cloth (▲). Note that the moisture has climbed up the cloth, higher than the level of the water. This ability of water to travel upward within small spaces is called *capillary action*. Capillary action is important in the upward motion of liquid water in soil, drawing soil water up toward the surface where it can evaporate. It is also important in plants, allowing water to rise from the roots through the branches and into the leaves.

How Do Water Molecules Move Between Liquid Water and Water Vapor?

Water molecules move between the three states—gas, liquid, and solid—in response to changes in the energy of the system, mostly to changes in thermal energy from insolation (or lack thereof). What actually occurs during such changes in state at the level of individual molecules? Here we take a closer look at the movement of water molecules between liquid and vapor, focusing on how the energy levels of the molecules instigate change.

1. This figure shows a totally closed container of water and air. Water molecules in each state are color coded for their energy levels, with yellow, orange, and red representing highest energy levels, and green, blue, and purple representing lower energy levels. Based on the colors in the container, molecules of water vapor (in the air) tend, on average, to have higher energy levels than water molecules in the liquid.

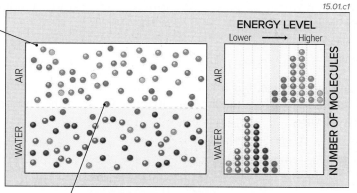

3. The graphs (histograms) on the right side of the figure display the frequency of molecules at various energy levels in the air (upper histogram) and water (lower histogram). In general, energy levels are higher in the air, as expressed by the peak of the air histogram being farther to the right (toward higher energy levels) than the peak for the water (liquid) histogram. This higher energy of the vapor is largely because of the *latent heat of vaporization*, which is energy the molecules gained primarily from insolation as they moved from liquid to vapor states. The critical

2. Note, however, that some molecules in the air have similar energy levels (purplish red on this figure) to those in the water. If the energy levels of vapor molecules are sufficiently low, these low-energy molecules will condense and join the liquid. Similarly, some molecules of the liquid will have sufficient energy to evaporate and pass into the vapor phase.

energy level where changes in state (evaporation or condensation) occur is indicated by the vertical orange band on the histograms. This system is in *equilibrium* when as many molecules are changing from liquid to gas as are moving in the opposite direction.

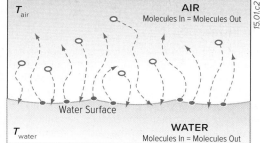

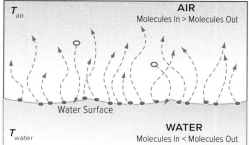

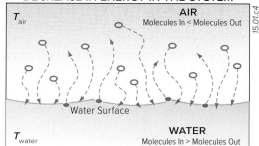

4. Due to the overlap in energies of molecules in the liquid and vapor, molecules are constantly moving from one state to the other. This figure illustrates that some molecules in the liquid attain high enough energy states to escape the water surface and become a molecule of water vapor (evaporation). In contrast, some vapor molecules will drop low enough in energy levels that they will join the liquid (condensation). If the system is in equilibrium, there is an equal exchange of water molecules between the liquid and vapor.

5. If the entire system is at higher overall energy levels, such as when the water and air are heated by the Sun or on a stove, many molecules in the liquid become more energetic, reaching energy levels high enough to allow them to escape into the air (i.e., evaporate). Fewer gas molecules in the air have low enough energy levels to condense into liquid. As a result of an increase in the energy of the system, increased evaporation causes the number of gas molecules (water vapor) to increase, while the liquid water loses mass.

6. If the system has lower overall energy levels, as when it is cooled, more gas molecules condense into liquid, while fewer molecules in the liquid evaporate. Since more water molecules condense onto the water surface than evaporate from it, the liquid gains mass. In contrast, the gas loses water molecules, causing a decrease in the amount of water vapor in the air. This transfer from vapor to liquid occurs when water drops condense on the outside of a cold glass or beverage can.

Movement of Water Molecules Into and Out of Ice

Similar processes occur when water molecules move between ice (the solid state) and liquid and gaseous states. When ice and liquid water are in contact, the energy levels of some molecules in the ice will overlap with some of those in the liquid. Some molecules will move from ice to water (*melting*) and others will move in the opposite direction (*freezing*). Likewise, when ice and vapor are in contact, some molecules in the vapor become solid ice (*deposition*), whereas some molecules in the ice move into the vapor (*sublimation*). How many move in each direction, and the resulting gains and losses of mass, depend on the overall energy level of the system. If there is equilibrium, the same number of molecules will move in opposite directions, but generally the ice, liquid, or vapor is losing mass to one of the other states.

Before You Leave This Page

☑ List some ways water is expressed in the environment.

☑ Sketch and describe the molecular structure of water and how it imparts important properties to water.

☑ Sketch and describe how overlap of energies causes molecules to move between liquid and vapor phases.

15.1

15.2 What Is Humidity?

THE AMOUNT OF WATER VAPOR in the air is referred to as *humidity*. Humidity is something we can sense, affecting whether the air feels humid or dry. We are most familiar with one measure of humidity—*relative humidity*, a term commonly used on daily weather reports. There are other measures of humidity, some of which are more useful for comparing the amount of moisture between different elevations, times, and regions. Understanding humidity leads to a much better understanding of weather and climate.

A What Are Humidity and Vapor Pressure?

The atmosphere is nearly all nitrogen and oxygen, but it contains a small but variable (about 1% to 4%) amount of water vapor (and other gases). The term *humidity* conveys the amount of water vapor in the atmosphere, and the amount of humidity can be represented in several ways.

Humidity and Vapor Pressure

1. Imagine two cubes, both the same size and partially filled with water but mostly filled with air (▶). In both cubes, some water molecules have evaporated from the liquid, becoming water vapor in the air. The amount of water vapor in the air is the humidity of the air. In this figure, molecules of water vapor are represented by small blue dots in the air. In the example shown here, the cube on the right has more water vapor than does the left cube, so the air in the right cube has a higher humidity than does air in the left cube.

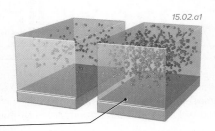
15.02.a1

2. Recall that the weight of the atmosphere pushing down causes atmospheric pressure. Some amount of this pressure is due to the weight of the water vapor that is part of the atmosphere, and this provides us with another way to represent humidity. The amount of pressure contributed by the water vapor is, not surprisingly, called the *vapor pressure*. Vapor pressure is expressed using the same units as atmospheric pressure, such as millibars (mb).

For a given volume of air, if there are more water vapor molecules present then the vapor pressure is higher. Vapor pressure is represented by the small letter *e*.

Water-Vapor Capacity

3. Air can include only a limited amount of water vapor, and the maximum amount it can include is called its *water-vapor capacity* (commonly depicted by e_s, where the *s* stands for saturation). The water-vapor capacity of air varies with the temperature of that air, as shown in the graph below. The blue curve represents the maximum amount of water vapor at different temperatures, with the water-vapor capacity increasing exponentially with increasing temperature—warmer air has a greater capacity for water vapor than does cooler air. Water-vapor capacity can be expressed in millibars (mb).

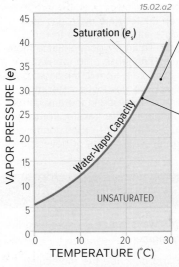

15.02.a2

VAPOR PRESSURE (e) — vertical axis (0 to 45)
TEMPERATURE (°C) — horizontal axis (0 to 30)
Saturation (e_s)
Water-Vapor Capacity
UNSATURATED

4. If the conditions of temperature and vapor pressure for some volume of air plot in the shaded area on this graph, even more water vapor can exist in the air and the air is *unsaturated*. These are the conditions we normally experience when the weather is fair (not raining or snowing).

5. If the conditions plot directly on the blue line, the air has as much water vapor as possible—it is *saturated* with respect to water vapor. Under these conditions, the water vapor begins to form droplets via the process of condensation or ice crystals. In other words, when the atmosphere reaches saturation, drops of water or crystals of ice form, such as in clouds and fog, perhaps followed by precipitation. In certain settings, conditions can be above the blue line, and the air is said to be *supersaturated*.

Relative Humidity

6. A common way to convey humidity is to express it as a ratio of how much water vapor is in the air relative to the maximum amount of water vapor that is possible (i.e., the water-vapor capacity). In the graph below, the green dot represents a vapor pressure of 20 mb and a temperature of 29°C. At that temperature, saturation would occur at about 40 mb, so the vapor pressure (humidity) represented by the green dot is half, or 50%, of the maximum possible. The atmosphere is unsaturated in this case. This percentage, representing the observed vapor pressure divided by the maximum possible vapor pressure (i.e., water-vapor capacity), is the *relative humidity*, the term we hear on daily weather reports. Relative humidity varies from single digits (e.g., 5%) in very dry places, such as in a desert, to 90% to 100% in very humid conditions.

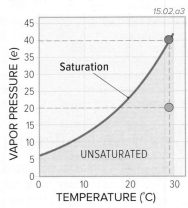

15.02.a3

VAPOR PRESSURE (e) — vertical axis (0 to 45)
TEMPERATURE (°C) — horizontal axis (0 to 30)
Saturation
UNSATURATED

Before You Leave These Pages

☑ Describe humidity, vapor pressure, relative humidity, and specific humidity, with graphs where needed.

☑ Describe what the dew point represents, using a graph to show how cooling can cause air to reach its dew point (saturation).

B What Is Specific Humidity?

Relative humidity helps describe how humid the air feels to us and indicates how close the air is to saturation, but relative humidity varies with temperature, even if the amount of water vapor in the air has not changed. Therefore, we also use another measurement of humidity, called *specific humidity*, that is not affected by such changes.

1. Specific humidity is expressed as the ratio of the mass of water vapor in some body of air to the total mass of that air. This cube (▶) contains molecules of water vapor (shown as blue dots) dispersed through a mass of atmosphere (filling the rest of the cube). In this example, the mass of all the water vapor is 12 grams (g), and the total mass of the air (including the water vapor) is 1 kilogram (kg). The specific humidity is therefore 12 g/kg, a typical value for air. We use the units of g/kg because the amount of water vapor is small compared to the total amount of air.

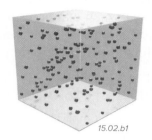

15.02.b1

2. Note that since specific humidity is calculated with the *masses* of water vapor and air, it is not directly dependent on the temperature or atmospheric pressure. The specific humidity can be changed by adding or subtracting water vapor, such as adding water vapor through evaporation of surface waters or by losing water vapor through the formation of precipitation.

C What is the Dew Point and Dew-Point Temperature?

Dew is expressed as drops of liquid water that condense out of the atmosphere and onto plants, rocks, walls, or any solid surface. It typically forms at night, in response to the air cooling. The temperature at which dew forms is another useful measure of humidity. The dew point is the temperature to which a volume of air must be cooled to become saturated with water vapor. If the air temperature is at the dew point, the air is so saturated with water vapor that vapor begins to condense as drops of liquid water, such as drops in clouds or rain, or as dew on solid surfaces.

1. To explore the formation of dew (▶), examine the sealed glass container on the left. It contains liquid water and water vapor (represented as blue dots). The container is warm enough that the water vapor can exist in the air without condensing—that is, the air is unsaturated.

15.02.c1

2. If we cool that same container with ice cubes, the temperature of the water and air will decrease. Recall that cold air has lower water-vapor capacity than warm air, so as the air cools it moves closer to saturation. At some temperature, the cooled air reaches saturation and can no longer allow so much vapor to exist in it. As a result, the water begins to condense as drops on the inside of the container, forming dew. As we know from our experience with cans or glasses of an icy beverage, drops of dew can also form on the outside of the container, especially if the surrounding air has high humidity.

3. To better visualize why cooling can cause the formation of dew, we return to a familiar graph, one that plots vapor pressure (e) versus temperature (▼). The curved blue line marks conditions where air is saturated in water vapor; it is labeled as e_s, where the s indicates saturation.

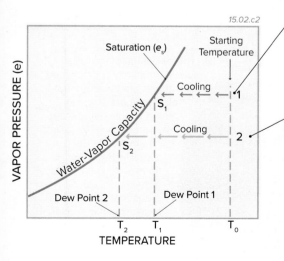

15.02.c2

4. We begin with an air mass at a certain temperature (T_0) and vapor pressure, marked as position 1, well within the unsaturated field of the graph. As this air cools, its position on the graph moves horizontally to the left (temperature is changing but vapor pressure is not), as indicated by the reddish arrows. It cools until it reaches saturation at position S_1, at which point the air has a relative humidity of 100%. The temperature at which saturation occurs is the *dew-point temperature* and can be read from the graph by drawing a vertical line down from S_1 to the temperature axis (T_1).

5. A second air mass, at position 2, starts at the same temperature (T_0) but has less water vapor in it (lower vapor pressure) than does air mass 1. If the air cools, as indicated by the green arrows pointing to the left, it also approaches the saturation curve. In order to reach saturation (at point S_2), however, it has to cool more than air mass 1. Air mass 2 has less water vapor than air mass 1, and as a result has a lower dew-point temperature (T_2).

6. To generalize, humid air at a given temperature has a higher dew-point temperature than does less humid air at the same temperature. If air is very humid, for example with a relative humidity of 80% to 90%, it may need to cool only a few degrees to reach its dew point. In contrast, air that is very dry has to cool significantly to reach its dew point, and on most nights it does not.

7. Using dew-point temperature as an index of moisture overcomes most of the difficulties of using relative humidity, because dew-point temperature does not change as much through the course of a "normal" day as relative humidity does. In the Arizona desert, for example, weather forecasters use dew-point readings to determine the official start of the summer monsoon. If the dew-point temperature is 55°F or higher for three straight days, summer rains and thunderstorms associated with the summer monsoon (change in wind directions), are likely.

How Do Specific Humidity and Dew Point Vary from Place to Place and Seasonally?

HUMIDITY OF THE AIR VARIES greatly from region to region and between different altitudes in the atmosphere. It also varies from one time to another, such as between different seasons. To compare different regions, altitudes, and seasons, we generally use specific humidity, which expresses the amount of water vapor in the air, independent of variations in temperature. Dew point is also useful since it expresses how humid the air feels, and can be used to predict the overnight formation of dew and the likelihood of precipitation.

A How Does Specific Humidity Vary Globally?

Specific humidity offers some advantages for comparing moisture of various places with different temperatures. The globes below depict specific humidity averaged over an entire year. On all the globes and maps on these next two pages, blues and greens represent higher values of specific humidity, whereas tan represents low values. Observe the globes below, noting areas that have unusually high or low specific humidity, and consider possible explanations for these humid and dry places, respectively.

1. The main pattern that emerges from inspection of these globes is that specific humidity varies primarily as a function of latitude—areas near the equator have much higher specific humidity than areas farther north and south (toward the poles).

2. A belt of high specific humidity straddles the equator, coinciding with the tropics. This region receives, on average, the most insolation, which in turn causes a relatively high amount of evaporation and thus a higher content of water vapor in the air. Also, warmer air has a relatively high capacity for water vapor (recall the blue curves from the previous pages).

3. The lowest values are near the poles, where cold temperatures and a cover of ice over the surface limit the amount of evaporation. Also, cold air has a lower capacity for water vapor. For all these reasons, there is less water vapor in polar regions than at lower latitudes.

4. The patterns of specific humidity are less complicated over the oceans than over land. The patterns over the oceans do not exactly follow latitude, mostly because of the influence of ocean currents that move warm and cold water north and south along the edges of continents. In the Atlantic Ocean, the warm Gulf Stream Current brings warm water northward along the East Coast of North America. The northward flow of warm water is accompanied by a northward expansion of moderate humidities toward Europe.

5. Patterns of specific humidity are more complicated on land, mostly reflecting the influence of topography, especially large mountain belts that interfere with prevailing winds, such as the Andes of South America.

15.03.a1

SPECIFIC HUMIDITY (g/kg)
3 6 9 12 15 18 21 24 27 30 33

6. Very low specific humidity also characterizes land areas along the subtropics, including the northern part of Africa. This region is the site of the Sahara Desert, the world's largest desert. We commonly associate deserts with dry air, and this is indeed reflected in the low values of specific humidity. The low humidity continues eastward across deserts of the Arabian Peninsula and onto southern Asia.

7. In the ocean west of southern Africa is another bend in the patterns of humidity. Lower specific humidities extend to the north offshore of the west coast of southern Africa and then bend westward into the South Atlantic Ocean. Think about what might cause this before reading on.

8. As you probably surmised, this pattern off southern Africa reflects another ocean current, but this time a cold current that brings cold water north, accounting for the lower humidity.

9. The annual average of specific humidity represented on these globes does not tell the whole story. In some regions, there are huge seasonal variations in humidity as the prevailing wind directions shift, such as in association with a monsoon. Before we explore some seasonal variations on the next page, think about how the large patterns on both these globes might change from January (the northern winter and southern summer) to July (northern summer and southern winter).

15.03.a2

B Which Parts of the U.S. Have the Highest and Lowest Dew Points?

Dew-point temperature is an important measure of the humidity of air, as well as a predictor of dew and precipitation. The maps below show average monthly dew-point temperatures and air temperatures across the conterminous U.S. for January and July. Blue colors represent low (cold) dew-point temperatures, whereas orange and red indicate high dew-point temperatures. While observing and comparing the two maps, identify areas that are unusually high or low and consider possible explanations for the patterns you observe. Find the values for the area where you live or would like to visit, and think about how this relates to the climate there and how humid it feels at different times of the year.

January

1. Extremely low dew-point temperatures, such as those that characterize the northern states in January, generally mean that the air is so dry that clouds do not form readily. Without cloud cover, night-time radiational cooling is intense, making January temperatures even more frigid.

2. Topography, such as mountain ranges in western North America, block moisture from penetrating to some regions, leading to low dew-point temperatures in the desert Southwest.

3. This map (▶) shows normal January air temperatures, averaged over the entire month. The warmest areas are in the Southwest and South Florida.

AVG. DEW-POINT TEMPERATURE (JANUARY)

AVG. TEMPERATURE (JANUARY)

15.03.c1

Legend:
- <-6
- -6 to -1
- 0 to 4
- 5 to 10
- 10 to 13
- 13 to 16
- 17 to 19
- 20 to 22
- >22 °C

4. January dew-point temperatures are extremely low in northern areas. Little insolation is available to drive the evaporation process under such conditions, so there is limited moisture in the air. When dew-point temperatures are below 0°C, they are called "frost-point" temperatures, and ice forms instead of dew.

5. Dew-point temperatures in eastern North America generally are not far below the air temperature, so air there does not have to cool very much to become saturated. Thus, clouds form even if there are no weather fronts. This is why many eastern states have overcast days in the winter.

6. January temperatures in the Southeast, especially southern Florida, are more moderate, so there is enough energy available to evaporate some water. This results in relatively high dew-point temperatures compared to the rest of the country.

July

7. Summer dew-point temperatures are much higher than in the winter. Increased insolation allows more energy to drive up temperatures and cause more evaporation, which increases humidity.

8. Areas in rain shadows of the moisture sources, such as the Intermontane West, have dew points that remain far below their air temperatures shown on the lower map. A slightly larger amount of moisture flows into southern Arizona from the Gulf of California in the summer, as represented by the small orange spots on the southwestern edge of this map.

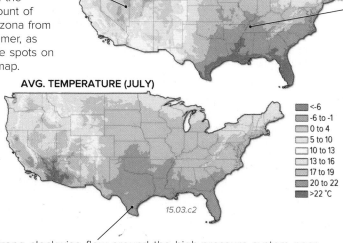

AVG. DEW-POINT TEMPERATURE (JULY)

AVG. TEMPERATURE (JULY)

15.03.c2

Legend:
- <-6
- -6 to -1
- 0 to 4
- 5 to 10
- 10 to 13
- 13 to 16
- 17 to 19
- 20 to 22
- >22 °C

10. Notice the steep gradient in dew-point temperature across the Great Plains, from moderately low dew points in the plains of eastern parts of New Mexico, Colorado, Wyoming, and Montana, to much higher dew points toward the Mississippi River. This gradient mostly reflects the northward flow of moisture from the Gulf of Mexico.

11. We can use the combination of these two maps to identify which parts of the country are hot and humid during the summer and those that are cool and less humid—useful in planning a summer vacation!

9. Summer dew-point temperatures are high in the Southeast, as depicted by the orange colors that fringe the Gulf Coast in the dew-point map for July. This high humidity, combined with high summer air temperatures, as shown on the map to the right, creates an oppressive summer climate. Strong clockwise flow around the high-pressure system near Bermuda (the Bermuda-Azores High) pumps moisture from the Gulf of Mexico, Caribbean Sea, and Atlantic Ocean deep into the heart of North America. There are no topographic barriers that would interfere with the movement of this moist air inland, so moisture from the ocean flows onto land, raising dew-point temperatures.

Before You Leave This Page

☑ Summarize the main global patterns in specific humidity and explain what causes the patterns.

☑ Describe the general patterns of dew point in the U.S., identifying some of the factors that contribute to high and low dew-point variations from place to place and January to July.

15.3

What Happens When Air Rises or Sinks?

THE ATMOSPHERE IS A DYNAMIC ENVIRONMENT, with upward, downward, and sideways motion. What happens to air that rises and encounters different temperatures and pressures? What happens when air sinks? The vertical motions of air cause clouds, precipitation, and many other weather phenomena.

A At What Rate Does an Air Parcel Cool As It Rises?

To explore the vertical motions of air, it is useful to track what happens to a discrete mass of air, called a *parcel* (used in the same way we refer to a lot of subdivided land as a "land parcel"). As we might expect, air changes temperature as it rises or sinks, largely in response to changes in air pressure and volume that accompany vertical motions through the air. Let's examine what happens as we follow a rising parcel of air, beginning with text at the bottom left (number 1).

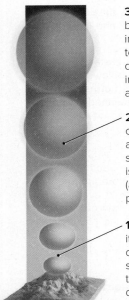

15.04.a1

3. When there is no energy exchanged between the parcel and its surroundings, as in the case shown here, the process is said to be *adiabatic*. When such an exchange does occur, the process is *diabatic*. Increasing insolation of the surface as noon approaches is an example of diabatic heating.

2. As the air rises higher, the air pressure continues to decrease. If the parcel does not acquire or lose additional energy with its surroundings, then the decrease in pressure is accompanied by (1) an increase in volume (as shown by the increasing size of the air parcel), and (2) a decrease in temperature.

1. If the parcel of air starts near the ground, its volume is confined by the weight of the overlying atmosphere, which exerts significant pressure. If the air begins rising, there is less air on top, which results in a decrease in air pressure.

4. If the rising motion can be assumed to be adiabatic, we can calculate how much a rising air parcel cools for a given rise in elevation, a rate called the *lapse rate*. Lapse rate is in units of degrees per vertical distance, usually in C°/km.

5. This graph (▶) plots the lapse rate for air that is rising adiabatically and remains unsaturated. The lapse rate is calculated simply as the change in temperature for the change in height, and reported as a ratio (C°/km). In the case shown here, the air cools 10 C° in one kilometer, or 10 C°/km.

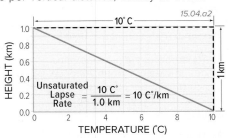

15.04.a2

Unsaturated Lapse Rate $= \dfrac{10\ C°}{1.0\ km} = 10\ C°/km$

6. If any parcel of air *rises* adiabatically and does not become saturated during its ascent, it always cools at 10 C°/km. This constant lapse rate, called the *unsaturated adiabatic lapse rate*, applies everywhere adiabatically rising air remains unsaturated. By convention, positive lapse rates are for temperatures that cool with height, as shown by the orange line on the graph above.

7. If a parcel *descends* adiabatically, it warms at the same rate that it cooled on ascent—10 C°/km. An adiabatically descending parcel of air will remain unsaturated, because the slightest bit of warming will increase the water-vapor capacity and therefore decrease the relative humidity. If the relative humidity is below 100%, the air is unsaturated.

B How Much Does a Saturated Air Parcel Cool as It Rises?

1. A different situation occurs if the atmosphere is saturated. As a saturated parcel of air rises, the cooling decreases the water-vapor capacity of the air and, since the air is already saturated, the moisture must come out of the vapor via condensation, freezing, or deposition, as illustrated in the diagram below.

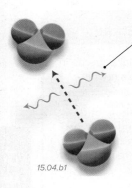

15.04.b1

2. As water vapor molecules condense, they release latent heat that warms the local environment. This warming counteracts some of the cooling caused by the expansion of the parcel during adiabatic ascent. As a result, a rising parcel of saturated air cannot cool as much with increasing height as does unsaturated air. The lapse rate for saturated air will be less than the unsaturated adiabatic lapse rate.

3. The rate at which saturated air cools with adiabatic ascent is called the *saturated adiabatic lapse rate*. This rate depends on temperature because at higher temperatures more water vapor is available for condensation, freezing, or deposition. The figure below plots unsaturated adiabatic lapse rate in red and the saturated adiabatic lapse rate in blue, for different starting temperatures on the surface (on the bottom of the graph).

4. The slope of the lapse rate for unsaturated air (red lines) is the same, regardless of starting temperature. The lapse rates for saturated air (blue lines), however, vary for different temperatures, as reflected by different slopes of the blue lines.

5. For any starting temperature, the blue line (saturated lapse rates) consistently plots to the right of the red line (unsaturated rate). This means that rising unsaturated air cools at a faster rate than saturated air (which is warmed by latent heat).

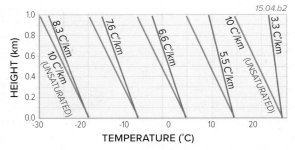

15.04.b2

6. Note that the slopes for the unsaturated and saturated air are only slightly different for low temperatures (not much water vapor), but become more different at higher temperatures (more water vapor). This means that with increasing starting temperatures (farther right on the graph), warmer saturated air parcels cool less with height.

C What Is the Lapse Rate of Air That Is Not Rising or Sinking?

Air adjacent to a rising parcel of air can have a totally different lapse rate than the rising air, and these rates can vary significantly from day to day and place to place, depending on the local weather, climatic setting, and other factors.

1. The temperature change with height for air surrounding the adiabatically moving air parcel is know as the *environmental lapse rate*. The environmental lapse rate can be quite variable. On this graph, each green line represents the environmental lapse rate under certain atmospheric conditions. The different lapse rates are expressed as different slopes (inclination of the lines). When dealing with lapse rates, we call a lapse rate a "steep" if the parcel cools a lot with height (as with the left line), even though that line does not look steep on this kind of graph. An environmental lapse rate is for air that is not rising or sinking, but instead represents the ambient conditions.

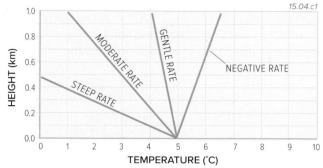

2. For the left line, the air temperature starts at 5°C at the surface but decreases 5 degrees (Celsius) by a height of 0.5 km. This yields an environmental lapse rate of 5 C°/0.5 km, or 10 C°/km. For this graph, the left line has the steepest lapse rate—its temperature cools the most rapidly with height.

5. The rightmost line has a *negative lapse rate*. It depicts a situation where temperature *increases* with increasing height. In such situations the environmental lapse rate is said to be *negative* because the temperature cools negatively (i.e., warms) with increasing height. A situation with a negative environmental lapse rate is called a *temperature inversion* (warmer air on top).

4. The third line (gentle rate) represents an even gentler lapse rate. This air starts at 5°C at the surface, but the temperature only decreases to 4°C one kilometer higher in the air. This equates to an environmental lapse rate of 1 C°/km. The gradual nature of this change is why this type of rate is called a gentle lapse rate (even though the line looks steep on this graph).

3. The second line from the left has a moderate (less steep) lapse rate. For this lapse rate, temperature decreases more slowly upward in the atmosphere. The air is 5°C at the surface, but temperatures cool to 1°C by one kilometer higher. The change in temperature is therefore 4 C° between the surface and one kilometer in altitude, yielding an environmental lapse rate of 4 C°/km.

D What Do the Different Types of Lapse Rates Indicate About the Stability of Air?

The three types of lapse rates—unsaturated adiabatic lapse rate, saturated adiabatic lapse rate, and environmental lapse rate—each indicate something about conditions in the atmosphere. By comparing the three types of lapse rates, we can predict whether air will rise, sink, or not move vertically. That is, we can tell how stable the air mass is. The rising of air due to its lower density than air around it is called *free convection*, or simply *convection*.

1. This graph compares different types of lapse rates for an adiabatically moving air parcel that starts from the surface at 10°C. The unsaturated adiabatic lapse rate is shown by the red line, and the saturated adiabatic lapse rate is depicted by the blue line. Green lines show different environmental lapse rates with varying steepness.

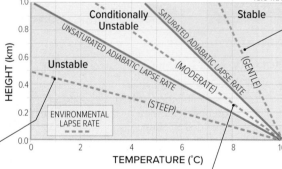

2. The green line in the lower left part of the graph has a steep environmental lapse rate, steeper than either the unsaturated or saturated lapse rates. If an air parcel starts to rise, it changes temperature according to one of the adiabatic lapse rates (red line or blue line, depending on whether it is unsaturated or saturated). Once either type of air has risen even a small amount, it will be warmer and less dense than the surrounding environment at a given height, so it will tend to continue rising. This will occur for either unsaturated or saturated adiabatically rising air if the environmental lapse rate falls within the orange region on this graph. In such situations, any rising air parcel will remain warmer than its surroundings and so will tend to continue rising, producing conditions that are called *unstable* (where air is rising).

3. A different situation occurs if the environmental lapse rate is moderate (the middle green line), plotting between the unsaturated and saturated lapse rates. In this case, if the adiabatically rising air is saturated (blue line), it will remain warmer than its surroundings (green line) and will tend to continue rising. If the rising air is unsaturated, however, it will become cooler than its surroundings (green line is to the right of the red line in this case) and tend to stop rising. For this reason, the situation represented by the yellow area is termed *conditionally unstable*. In this situation, the addition of moisture to the air increases the likelihood for rising motions and other unstable conditions.

4. The rightmost green line represents a gentle environmental lapse rate, where the ambient air cools relatively slowly with increasing height. As a result, an adiabatically rising air parcel (shown by the red or blue lines) would be cooler than the surrounding environment (green line). In this case, the air parcel would cease rising because it is cooler and has a higher density than the surrounding air. At any given height, it will more likely sink, a condition known as a *stable atmosphere*. This will happen any time the environmental lapse rate is more gentle than the saturated rate (the green-shaded area). If the environmental lapse rate is negative (sloping up to the right), it forms a temperature inversion—a very stable atmospheric setting.

Before You Leave This Page

☑ Sketch, describe, and contrast the unsaturated and saturated adiabatic lapse rates, explaining why the saturated adiabatic lapse rate is less than the unsaturated adiabatic lapse rate.

☑ Describe the relationship between the three lapse rates and atmospheric stability.

15.4

15.5 How Does the Surface Affect the Rising of Air?

EARTH'S SURFACE CONSISTS OF A VARIETY OF MATERIALS, including bare rock, soils, forests, cities, water, and ice. Each of these materials responds differently to insolation and to changes in temperature and humidity of the adjacent air. In turn, these surface materials can affect the temperature and humidity of that air. On land, these materials are referred to as *land cover*. Some changes in land cover are natural and others are caused by humans.

A How Does Human Development Impact Local Atmospheric Stability?

The amount of energy reflected and absorbed by different land surfaces varies and depends on several factors, including the type of land cover. Human modification of the land surface, especially changes in land cover, can affect the temperature of the land and overlying air. This can cause the environmental lapse rate to change, influencing whether air rises and affecting the stability of the local atmosphere.

Deforestation

1. Various human activities greatly affect the surface of the Earth and the type of land cover. One ongoing problem is *deforestation*, where native forests and other vegetation are cut down and removed as part of development. In deforestation, dark forest cover and other natural vegetation are often replaced with less heavily vegetated farms with abundant bare soil. The loss of plant cover commonly increases surface temperature by reducing shading and because insolation goes into heating the ground, rather than into evaporating water that was transpired by the forest.

In developed land, vegetated areas are removed and drainage systems remove standing water. Thus, after development, more of the incoming energy is used in sensible heating and less is stored as latent heat.

15.05.a1 West Virginia

Urbanization

2. Another major change in land cover occurs during *urbanization*, where natural lands, farms, and parks are replaced by asphalt, concrete, and buildings during the growth of cities and towns. Many aspects of development, including urbanization, cause a change in albedo (what percentage of insolation is reflected back into the atmosphere) and in the heat-retention characteristics of the land cover. If darker, rougher-textured materials are replaced with lighter colored, smoother-textured ones, the albedo increases, and more insolation is reflected off the surface. Often, however, dark asphalt and other materials more effectively absorb insolation, heating up more during the day and giving back this heat in the evening, raising city temperatures and forming a local warmer overall environment, called an *urban heat island* (UHI).

15.05.a2 Houston, TX

Effects of Development on Atmospheric Stability

3. Changes in land cover due to development can result in changes to surface temperatures and therefore to environmental lapse rates. In the figure below, an area of land originally consisted of natural vegetation, with trees and grass, and a small pond. After development, the same area was covered with asphalt, concrete, and bare dirt.

4. Deforestation and urbanization both occurred in this area, so we predict that more insolation will become sensible heat and less will be latent heat. This should cause more heating of the land surface, resulting in higher surface temperatures, both during the day and especially at night, as part of UHI. If the surface temperature increases but the temperature higher in the atmosphere does not change as much, there will be a steeper environmental lapse rate—a larger decrease in temperature from the surface upward to some height.

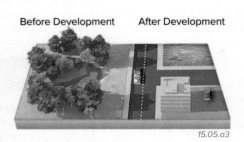
Before Development   After Development
15.05.a3

5. We can represent these changes with a graph (▶) showing changes to surface temperatures and environmental lapse rates. The graph shows the environmental lapse rate in green and the unsaturated adiabatic lapse rate in red, for times before development (Time 1, on the left) and after development (Time 2, on the right). Examine this graph, note the changes between the two times, and consider what the implications might be for this somewhat extreme example (temperature changes this large due to development are rare).

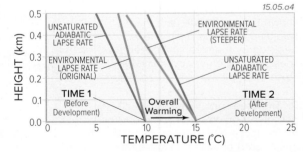
15.05.a4

6. The most obvious change is a shift to the right of the starting points of the two lapse-rate lines for Time 2 (after development). This shift reflects an increase in the local average temperature. Another change is a steepening of the environmental lapse rate (green line), because the surface warmed up a lot, but air farther up did not warm as much; thus, there is a larger change in temperature from the surface to that height (a steep rate). Finally, comparing the slopes of the lapse rates indicates that the atmosphere was stable (rising air was cooler than ambient air) before development but is unstable (rising air is warmer than ambient air) after development. This can change the amount of lifting (vertical motion), which may enhance the likelihood of precipitation if sufficient moisture is present.

How Do Water and Ice on the Surface Affect Stability?

Water has different thermal properties than land. Water has a higher *specific heat capacity* than air, meaning it can hold more heat. Water has more *thermal inertia*, which means that it reacts more slowly to changes in temperature imposed on it by the Sun, atmosphere, or land. In general, water surfaces stay cooler than inland areas in summer and remain warmer than inland areas in winter. Warm ocean currents and cold currents can influence the temperatures and stability profile of the overlying atmosphere. Ice and snow also affect surface temperatures, lapse rates, and atmospheric stability.

Warm Ocean Current

1. Warm ocean currents generally flow from lower latitudes into higher latitudes, bringing in water that is warmer than that outside the current (▶). Consider what a warming of the ocean surface due to a warm current might do to conditions in the overlying atmosphere.

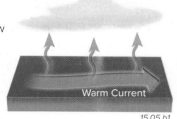

15.05.b1

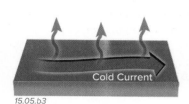

15.05.b2

2. This figure (▶) shows the environmental lapse rate and the unsaturated adiabatic lapse rate for the atmosphere over a warm ocean current. Near-surface air temperatures are relatively warm because they are heated by the underlying warm water.

3. The warm water has less of an effect higher in the atmosphere, so the temperature difference between the near-surface air and higher air is relatively large. This results in a steep environmental lapse rate. The slope of the unsaturated adiabatic lapse rate is not influenced by the water temperature (it is always the same, everywhere). If the environmental lapse rate plots to the left of the unsaturated adiabatic lapse rate, adiabatically rising air will remain warmer than its surroundings and will continue to rise, causing an unstable atmosphere. The rising, moist air can form clouds or rain.

Cold Ocean Current

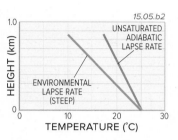

15.05.b3

4. Cold ocean currents generally bring colder water from higher latitudes into warm environments in lower latitudes (◀). We would predict that a cold current would cool the air near the surface, but what other effects will this have?

5. This figure (◀) shows the same types of lapse rates for the atmosphere over a cold ocean current. Near-surface air temperatures are relatively colder because of the underlying cold water, as reflected in a shift of both curves to the left on this graph compared to the previous graph for the warm current.

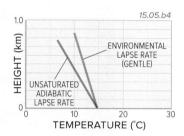

15.05.b4

6. The cooling of near-surface air results in a relatively small difference between near-surface and higher air. The environmental lapse rate is therefore gentle (which plots as a steeper line on this type of graph). The relatively gentle environmental lapse rate is now to the right of the unsaturated adiabatic lapse rate, indicating that air rising adiabatically will become cooler (and more dense) than the surrounding air, so it will cease rising. As a consequence, the atmosphere over a cold current tends to be stable. This stability and the limited water-vapor capacity of cold air will tend to inhibit vertical motion, which will restrict vertical cloud development and precipitation.

Snow and Ice on the Surface

Insolation is reflected by surface.

Cooled surface air does not rise.

15.05.b5

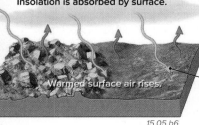

Insolation is absorbed by surface.

Warmed surface air rises.

15.05.b6

7. Snow and ice have a major impact on atmospheric temperatures, and not just because they are cold. Snow and ice both have very high albedo, as high as 90%, which is much higher than the albedo of rocks, plants, and other land cover. Also, snow and ice can appear rapidly, such as during a snowstorm, and can melt away nearly as fast. This appearing and disappearing can result in major changes in the thermal character of the surface over quite short timescales.

8. These two figures illustrate some differences between surfaces that are covered by snow and ice, versus those that are not. If snow and ice cover the ground and any frozen lakes, most of the insolation is reflected, keeping the surface cold. Cold air is dense, so the air over an ice or snow surface tends to sink. This increases the tendency for high pressure to exist over the ice-snow surface. The sinking air can trap cold air near the surface. The air near the surface is cooled by the ice and snow, but the air above it is not, so the atmosphere is stable. Such conditions can result in a temperature inversion, a very stable condition.

9. Once the snow and ice melt, the albedo decreases, allowing insolation to warm the surface. The warmer surface warms the air, ending the inversion, changing the lapse rate, and making the air less stable.

Before You Leave This Page

☑ Explain how human development can affect local surface temperatures and atmospheric conditions.

☑ Sketch and explain how warm and cold ocean currents can affect near-surface atmospheric stability and the amount of clouds and precipitation.

☑ Summarize how snow and ice on land and water can influence atmospheric stability and what happens when the snow and ice melt.

15.5

15.6 What Mechanisms Can Force Air to Rise?

AIR RISES FOR VARIOUS REASONS, some caused by differences in density between the air and its surroundings, and others a result of externally imposed factors, like a mountain. If an air parcel rises because atmospheric conditions are unstable, *free convection* results. If air is forced to rise due to external factors, it is called *forced convection*. Forced convection can occur in a number of settings, even if the atmospheric conditions are stable.

A How Do Convergence and Divergence of Air Cause Rising?

As air moves laterally across the surface of the earth and at higher levels, it can encounter obstacles that impede continued motion. Such obstacles can be mountains or other features on the land surface, or they can be other air masses moving in an opposing direction.

Convergence of Low-Level Air Masses

15.06.a1

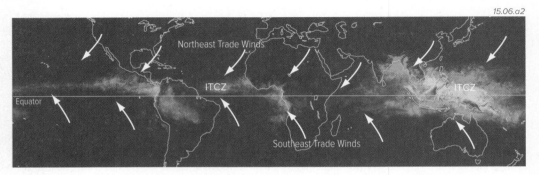

15.06.a2

1. Low-level convergence occurs when winds from two systems are moving on a collision course. The easiest way for the colliding air to move is upward. In such cases, the air parcel must rise whether it is stable or unstable. If the air reaches its dew point as it ascends and cools, water vapor condenses into tiny drops that form clouds and perhaps precipitation.

2. The best example of a location where low-level convergence is important at continental scales is the *Intertropical Convergence Zone* (ITCZ), which encircles the Earth, more or less centered on the equator. In the satellite image above (▲), the ITCZ appears as the belt of moist air and clouds, shown in red for its position during the northern summer and blue for its position during the southern summer. Along the ITCZ, the northeast trade winds, coming from north of the equator, converge with the southeast trade winds coming from south of the equator. The two sets of trade winds converge at the ITCZ, causing cloud cover as moist air is uplifted to a height at which it cools to its dew-point temperature.

Interaction of Moving Air with the Land Surface

15.06.a3

15.06.a4

15.06.a5

3. Low-level convergence also occurs when a wind is slowed down, such as by friction with the surface. The slowed air causes faster moving air behind it to begin to pile up, forcing some air to rise. A common scenario for this involves a sea breeze, where winds blowing across a large water body slow down as they encounter the more irregular land surface.

4. Mountainous areas represent obstructions to low-level winds. As air encounters topography, it slows down, piles up, and rises, called an *orographic effect*. As the air rises, it cools by adiabatic expansion, forming a cloud if it cools to the dew-point temperature. This orographic effect is why thunderstorms are usually more common over mountain peaks than over the adjacent valleys.

5. During daytime, the Sun heats the surface more efficiently than the air higher up. As the surface of a mountain warms, the adjacent air warms, which then flows upslope. When winds flowing up opposite sides of the mountain converge at the summit, they rise more. This is yet another reason why cloud formation seems to occur preferentially over mountains.

Convergence and Divergence Caused by Rossby Waves

6. Convergence can also occur from the way that upper-level air circulates around bends in the polar front jet stream, as illustrated in the figure to the right. These bends in the jet stream are called *Rossby waves*.

7. In upper levels of the troposphere, air that turns around ridges of high pressure will tend to move slightly faster than air around low-pressure troughs. This is because air turning around ridges moves to the right in the Northern Hemisphere (as shown in this example), in the same rotation direction as the Coriolis effect. The two forces work together on this type of bend.

8. Around troughs, air turns to the left (in the Northern Hemisphere), in opposition to the Coriolis effect. Along these bends, the Coriolis effect works against the flow of air, and the air moves slightly more slowly.

9. The slowing down of air as it moves from the upper-level ridge of high pressure to the upper-level trough of low pressure causes air to *converge* aloft on the ridge-to-trough side of the Rossby wave. This piled-up air then sinks and warms adiabatically, causing high pressure near the surface.

10. On the opposite side of the trough, the air speeds up as it turns into the next ridge. This speeding up causes the air to *diverge* and spread apart. Air from below flows into this zone of divergence, causing low pressure near the surface. This promotes rising of the near-surface air. All of the directions of rotation shown here are opposite in the Southern Hemisphere.

RIDGE (Airflow Speeds Up)
RIDGE (Airflow Speeds Up)
TROUGH (Airflow Slows Down)
Rossby Wave (Jet Stream) at 300 mb
(Vertical Scale Exaggerated)
Convergence
Divergence
Anticyclone
Cyclone
15.06.a6

B   How Do Mountains Affect Adiabatic Temperature Changes?

Mountains interact with moving air masses in interesting ways. In the example below, moist air encounters the side of a mountain and is forced to move upslope. Once it reaches the top, the air flows down the other side. Examine this setting and think about what happens to the air on its way up and on the way down.

1. As the air encounters the base of the mountain and begins moving up the slope, it is initially unsaturated and cools according to the unsaturated adiabatic lapse rate, which is 10 C°/km everywhere.

2. As the air flows upslope, at some elevation it has cooled enough that it becomes saturated (reaches the dew point), and at this level clouds form. This height is called the *lifting condensation level*—it is the base of the clouds. As this air continues rising, it now cools at the saturated adiabatic lapse rate, which is somewhat less than the unsaturated adiabatic rate.

15.06.b1

3. By the time the air reaches the top, it has been warmed by latent heat. As it moves down the other side, it will compress and warm, and the air will become unsaturated (warm air has higher water-vapor capacity than cold air). It will remain unsaturated, thereby causing a drying effect on this side of the mountain. It warms according to the unsaturated adiabatic rate of 10 C°/km.

4. As a result of this process, the air at the leeward side will be warmer than the air at the same elevation on the windward side.

C   How Do Fronts Induce Rising Motion?

Whenever two air masses with significantly different temperatures meet (▶), the warmer air will be forced up and over the colder air because the warm air is less dense. This process is called *frontal lifting*, and the interface is called a *front*. As this warm air rises, it will cool adiabatically. If it has abundant moisture, it will not need to rise very high before it cools to the dew-point temperature. At that level, active condensation or deposition begins and a cloud forms. This can lead to precipitation, as shown by colors on this weather map (▶) near a weather front marked by the blue line. The teeth on the line indicate which way the front is moving. The weather associated with fronts is presented in the next chapter.

Cool Air
Warm Air
15.06.c1

L
L
15.06.c2

Before You Leave This Page

☑ Sketch and explain four settings in which convergence of air masses induces rising atmospheric motion.

☑ Describe how mountains influence the humidity, lapse rate, and temperature as air moves up one side and then down the other side of a mountain.

☑ Explain how fronts can force air to rise.

15.6

15.7 What Do Clouds Tell Us About Weather?

CLOUDS ARE ACCUMULATIONS of liquid water and ice suspended in the air. The types and amounts of clouds vary from place to place, from time to time, and from season to season. What are the different types of clouds, and how does each type form? Clouds provide clues not only about the amount and distribution of moisture at a given place and time, but also about conditions in the atmosphere, specifically the stability of the atmosphere and the type of convection that is occurring. This, in turn, informs us about important aspects of our weather.

We classify clouds based on three main factors: their form, their altitude, and whether they are associated with precipitation. Each of these aspects is summarized below, and examples of the main types of clouds are pictured in the large central figure.

Cloud Form

1. The three main forms of clouds are cumuliform, stratiform, and cirriform. *Cumuliform* clouds generally are taller than they are wide, or at least have a lumpy appearance. "Cumulus" in Latin means "a heap," and forms the root of words like accumulation. *Stratiform* clouds have the form of one or more layers (stratus means "to stretch" in Latin) that commonly form a continuous cover across the sky, *Cirriform* are feathery, wispy clouds, named after cirrus, the Latin word for a "lock of hair."

Cloud Altitude

2. We also subdivide clouds as to whether they are relatively low in the troposphere (low clouds), are near the top of the troposphere (high clouds), or are at an intermediate elevation (called mid-level clouds or designated with the prefix "alto"). For example, a mid-level cumuliform cloud is referred to as an altocumulus cloud. Some clouds extend from low levels up to high ones.

Precipitation

3. Clouds are composed of drops of liquid water, crystals of ice, or commonly some of each. Most drops and crystals are held aloft within the cloud because the buoyancy forces from unstable rising air exceed the gravitational force that pulls them toward Earth's surface. When the drops or crystals become too large to stay aloft, or the rising (buoyancy) force weakens, the drops and crystals fall to the surface as rain, snow, hail, or some other type of precipitation. Clouds experiencing precipitation are designated with the prefix "nimbo" or the suffix "nimbus." A stratiform cloud that is precipitating is a nimbostratus cloud, and a cumulus cloud that is precipitating is a cumulonimbus. Even with all the possible combinations of words for form, height, and precipitation, there are only about ten main types of clouds, which are illustrated here.

Cumuliform Clouds

4. *Cumulus clouds* all have a somewhat puffy aspect but have different appearances, depending on their altitude and how unstable the air is (i.e., how strongly the air is rising). Cumulus clouds commonly are produced by air rising generally under its own impetus (free convection) and indicate that air is unstable to some degree.

15.07.a2

6. Mid-level clouds (▲) contain both liquid water and ice particles, with the level at which ice can be present being lower in winter. Cumuliform clouds at these levels are termed *altocumulus*. They are puffy and only somewhat wispy.

15.07.a3 Durango, CO

7. Typical clouds (▲) on many days are low-level clouds with a puffy (cumuliform) appearance, or *cumulus clouds*, which can be widespread across the sky or localized over mountain peaks. When the atmosphere is only slightly unstable and water vapor is not too abundant, cumulus clouds are not very tall. Such clouds are typical during morning hours before surface heating causes air to rise more rapidly, and at times when conditions are only slightly unstable.

5. Clouds high in the atmosphere (▶), are composed almost entirely of ice. When such high-level clouds are partly cumuliform (lumpy) and partly cirriform (wispy), they are *cirrocumulus clouds*.

15.07.a4

8. Cumulus clouds can occur in discrete layers (▶), called *stratocumulus clouds*. Like normal cumulus clouds, stratocumulus clouds are typically low clouds that are composed entirely of tiny drops of liquid water.

15.07.a5

15.07.a6 La Sal Mtns., UT

9. If moisture is abundant and the atmosphere is very unstable, strong lifting in the atmosphere can form cumulus clouds with great vertical extent (▲). Tall cumulus clouds associated with precipitation are *cumulonimbus clouds*, and commonly are accompanied by lightning. A cumulonimbus cloud contains liquid water near its base but can contain ice particles near its top, permitting hail to form. If well developed, such clouds develop an anvil shape at the top that points downwind, as the stronger upper-level winds push the moisture and storm cloud sideways.

Cirriform Clouds

10. *Cirrus clouds* are high in the troposphere and have a distinctive wispy appearance (▶), like the tail of a horse. At these levels, air temperatures are well below freezing, so these clouds are composed of ice crystals, which are easily blown and "smeared" about by the strong, upper-level winds. The ice refracts (bends) the light, enhancing the wispy, shimmery appearance of most cirrus clouds. Cirrus clouds often indicate that stormier weather is approaching, usually within several days.

11. A high-flying jet can leave a linear streak of clouds behind it, called a *contrail* (short for condensation trail). These form when water vapor produced by the engines encounters the much colder outside air, causing condensation of water drops or the formation of ice crystals. Once formed, contrails can be dispersed and sheared by the wind. They also contribute to the formation of cirrus clouds.

15.07.a7

Stratiform Clouds

15.07.a8

12. Stratiform clouds have a layered or spread-out aspect (▲). They are the clouds we observe when skies are overcast, but they may obscure just a part of the sky. Such layers of clouds form when air is forced upward (forced convection), such as by surface convergence or movement of a frontal boundary. Stratiform clouds are oriented laterally rather than vertically, implying that vertical motion is resisted by the atmospheric conditions.

15.07.a1

13. Stratiform clouds in the upper troposphere are high enough to be composed solely of ice and so have a somewhat wispy (cirriform), sheetlike appearance (▶). Accordingly, they are called *cirrostratus*.

15.07.a9 Grand Junction, CO

15.07.a10 Yampa River, CO

14. Mid-level stratiform clouds contain both ice and liquid water and can be expressed as high overcast. These mid-level clouds are termed *altostratus*. Stratus clouds can form at several levels (▶).

15.07.a11 Maine

16. A stratiform cloud that is precipitating is a *nimbostratus* cloud, as shown here (◀). This is a typical cloud associated with regional storms. A low-level stratiform cloud that is not precipitating is simply called a *stratus* cloud. Stratus can form a low blanket of clouds that causes overcast conditions.

15. *Fog* is a cloud that hugs the ground surface. It is described in detail in the next two spreads.

Before You Leave This Page

✓ Sketch each of the main types of clouds, explaining for each type the reasons for its appearance and how it forms.

15.7

15.8 What Conditions Produce Fog?

FOG IS SIMPLY A CLOUD at ground level, so the same conditions that create a cloud also produce fog. Specifically, fog is produced by cooling of the air, by increasing the humidity of the air, or some combination, with the end result that water vapor content in the air reaches saturation. Fog forms in several distinct settings, each of which causes air at or near ground level to become saturated.

A What Is Fog?

15.08.a1 Marshall Point, ME

1. Fog is a ground-level cloud that obscures visibility, such as of this lighthouse (◄). It is most common along coastlines, over large water bodies, and in certain low areas that have the right setting to produce fog, but fog can form nearly anywhere, under the right conditions.

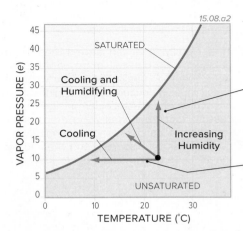

15.08.a2

2. Fog indicates that the relative humidity has reached 100% and the air temperature has reached the dew-point (or frost-point) temperature, causing saturation. One way in which air can reach saturation is by increasing the moisture content until vapor pressure reaches the water-vapor capacity, moving up on this graph (◄).

3. Another way to reach saturation is by cooling the air until it reaches the dew-point (or frost-point) temperature, represented on this graph as moving to the left.

4. A combination of cooling and increasing humidity can also cause saturation, as indicated by the upward sloping arrow on this graph.

B What Conditions Can Add Enough Humidity to the Air to Form Fog?

Fog can form where enough moisture is added to the air to cause saturation (100% relative humidity), but more commonly fog forms where the addition of moisture occurs simultaneously with the chilling of the air. Addition of moisture to the ambient air may happen when a nearby source of water for evaporation exists, or when precipitation evaporates as it falls through drier air.

Evaporation Fog

1. When a warm water body underlies colder air (◄), the environmental lapse rate is steep and the atmosphere is unstable. As the unstable air rises, it can incorporate water molecules that evaporate from the water body, increasing the humidity of the air. The increase in humidity and cooling causes the rising air to reach saturation, forming a type of fog called an *evaporation fog*. The instability causes an evaporation fog to rise in vertical columns, as shown in the photograph to the left. This phenomenon is common in fall, when water bodies are likely to remain warm from the long summer, but air above them begins to cool.

15.08.b1 Central Idaho

Valley Fog Produced by Increased Humidity

2. In rugged terrain, cold nocturnal (nighttime) downslope winds can trap atmospheric moisture in valleys. If moisture is added to the valley air, such as from wetlands, irrigated fields, canals, and lakes, the humidified and cooled air can reach the dew point, forming a *valley fog*. Perhaps the most famous valley fog in the U.S. is the tule fog of the Central Valley of California, the white area in the center of this regional satellite image (▶).

15.08.b2

Precipitation Fog

3. Fog can also form in association with precipitation, such as that which occurs along a warm front. A warm front is a mass of warm air that moves laterally, following a retreating mass of cold air. Along a warm front, warmer, less dense air is forced to rise at a gentle angle over more dense, colder air.

4. As the moist, warm air rises over the cold air, it is cooled and can reach saturation, forming clouds and rain. The released rain then begins to fall through the underlying cold air and toward the cold surface.

5. If the underlying colder air is dry, it can cause some or all of the falling raindrops to evaporate, increasing the humidity of the cold air. Since cold air has a lower water-vapor capacity than warm air, the increase in humidity can cause the cold air to reach its dew point, forming fog and other low-level clouds. Rain that reaches the ground can also evaporate, increasing the humidity of the near-surface air, potentially forming more fog. In both cases, evaporation of precipitation is causing a *precipitation fog*.

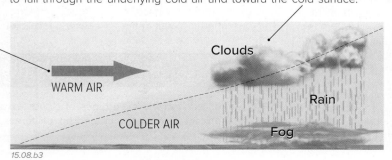

15.08.b3

C In What Settings Does Enough Near-Surface Cooling Occur to Form Fog?

Fog can also form where moist air is cooled enough to reach its dew-point temperature. This can occur where moist air moves over a colder part of Earth's surface, such as over snow or over a cold ocean current. It also occurs when moist air cools if it moves vertically. Several types of fog form primarily from cooling of moist air.

Radiation Fog

1. Although the term "radiation fog" sounds like something in the latest zombie movie, it is actually related to cooling of the ground surface by radiant heat loss. If skies at night are clear (i.e., cloud cover is sparse), and winds are light or calm, then the longwave radiation loss from the surface to the atmosphere will be relatively efficient, causing the surface to cool significantly.

2. Under these conditions, especially if near-surface air is humid, the surface temperature can drop to the dew point (or frost point) and cool the immediately overlying air, causing it to become saturated in water vapor. The condensation of water forms fog, in this case called a *radiation fog*. If the surface cools substantially, a temperature inversion will occur (colder air below warmer air), stabilizing the atmosphere and keeping the fog in place. Sometimes such fog can become concentrated in valleys and other topographically low places, therefore representing another way to form a valley fog.

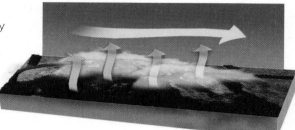
15.08.c1

Advection Fog

3. When warm air flows over a colder surface, the air loses some of its energy to the surface and therefore cools. If it cools to the dew point or frost point, the moisture in the air will condense and produce fog, as occurred when moist air moved over this glacier (▶). The movement of air laterally is called advection, so a fog produced in this manner is an *advection fog*.

15.08.c2 Athabasca Glacier, Alberta, Canada

4. A similar process produces fog along many coastlines. If moist air over an ocean or large lake blows onshore, it can encounter ground that is colder than the body of water. The cold ground lowers the temperature of the moist air, often enough to cause saturation and an advection fog. Advection fogs also form where warm, moist air blows over a cold ocean current (not shown). Advection fog is common along coastal California, which has a cold ocean current, called the California Current, offshore, bringing cold waters south along the coast.

15.08.c3

Upslope Fog

5. As air is forced uphill by the prevailing local wind, it will cool as it rises into lower atmospheric pressures. At some level above the surface, the air may cool to the dew-point (or frost-point) temperature, and a cloud will form. On the mountain slope, the cloud forms on the ground, so it is a fog (▼). This type of fog is commonly called an *upslope fog* to indicate that the fog is caused by the upslope motion of air.

15.08.c4

15.08.c5 Cumberland Gap, KY

Valley Fog Caused by Descending, Cold Air

6. *Valley fog* can form in various ways, some already described. In mountainous terrain, air near the surface cools at night because of longwave radiative loss. This air will flow downhill because it is more dense than warmer air. It may remain colder than the surrounding air, even as it warms adiabatically on descent. If so, the descending air may chill the ambient air in the valley. If the ambient air is cooled to the dew-point or frost-point temperature, it will become saturated and form a valley fog.

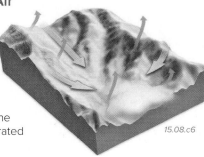
15.08.c6

Before You Leave This Page

✓ Sketch and describe three ways that increasing humidity, cooling, or some of both, can form fog.

✓ Sketch and describe four settings in which near-surface air can be cooled enough to form fog.

15.9 How Does Precipitation Form?

THE PROCESS OF PRECIPITATION is vital to life on Earth, helping to redistribute water from the oceans to the atmosphere to the land. Precipitation is the ultimate source of all the fresh water on the planet, which we depend on in our daily lives. How does precipitation occur? What is going on inside clouds that forms raindrops, snowflakes, and hail? It turns out there are two important mechanisms by which precipitation droplets form, one of which is somewhat surprising.

A What Is Precipitation?

1. Precipitation is the process whereby liquid droplets of water (raindrops), solid bits of ice (snowflakes and hail), or some combination of these fall from the sky. Examine this figure and consider all the processes that have to occur to cause rain, snow, or hail.

2. The cycle begins with evaporation of water in the oceans and other parts of Earth's surface, a process that puts water vapor into the atmosphere. Next, the water vapor forms the various types of clouds, which can contain tiny drops of water, ice crystals, or some of each.

3. For precipitation, some processes are occurring within the cloud that make the water droplets or bits of ice heavy enough that the pull of gravity can overwhelm the buoyancy forces (atmospheric instability) that uplift air within the cloud (the rising air is how most clouds form).

15.09.a1

4. Whether a cloud contains drops of liquid, ice crystals, or some combination depends primarily on the temperature of the cloud. If ambient temperatures near the cloud are warm, the cloud will contain mostly drops of liquid. Temperatures decrease upward within a cloud, however, so the upper levels can contain ice, while the lower levels contain drops.

5. Under cold ambient conditions, a cloud contains mostly ice throughout its vertical extent. Clouds at intermediate ambient temperature, or intermediate altitudes, will contain a mix of drops and ice.

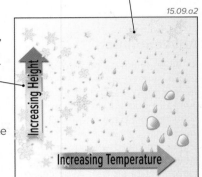

15.09.a2

B How Do Water Droplets Form and Grow?

1. An important factor in how water droplets in a cloud grow to become raindrops is the immense differences in size between the various players. Raindrops are huge compared to the size of water droplets that form a cloud, as shown by their relative sizes in the properly scaled diagram below.

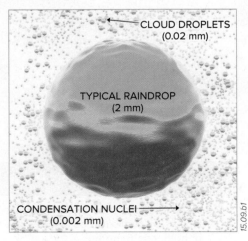

15.09.b1

2. It is energetically difficult for the tiny water droplets in clouds to just form by themselves, but it is easier for them to form if they condense around even tinier particles, such as dust, salt, and smoke. Due to this role, such particles are called *condensation nuclei*.

3. The figure to the right illustrates what can happen to a moving water drop that interacts with smaller cloud droplets around it.

4. Some larger drops form when liquid water droplets in clouds merge and grow to a size that can be pulled down by gravity. A water droplet begins to fall as soon as the downward-directed gravitational force exceeds the upward-directed buoyancy force caused by instability. This occurs sooner for larger droplets than for smaller droplets, and the larger drop overtakes the smaller droplets on their descent, making the falling drop even larger. Eventually, it falls as a raindrop.

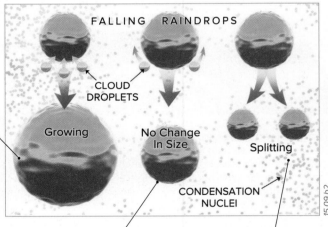

15.09.b2

5. In other cases, the smaller droplets simply slide past the falling drop, because the collision is not enough to break the surface tension that tends to keep each drop intact. In this case, the falling drop will not increase in size as it falls. This situation is actually very likely under many atmospheric conditions.

6. Alternatively, wind resistance can reshape the falling drop until it becomes easier to break the drop into separate drops than to retain its contorted shape. In this way, a falling raindrop becomes smaller, which can also occur if water molecules on the outside of the drop simply evaporate into the surrounding air.

How Do Ice, Liquid, and Vapor Interact Within a Cloud?

Many clouds are in environments in which the temperature is below freezing, even in summer. Exceptions are clouds in the tropics and those low in the atmosphere, or a cloud being warmed by latent heat during precipitation. Even when overall temperatures are below freezing, some liquid water droplets exist. This sets up a situation in which water exists in all three phases—vapor, liquid, and ice (solid). The interaction between these three coexisting phases allows precipitation to form much more easily than would otherwise be possible if only vapor and liquid were present.

1. The diagram to the right shows the familiar saturation curve for water vapor, but it focuses on conditions slightly above and slightly below freezing (0°C). At temperatures below freezing, the saturation curve for water vapor near ice (the dashed line) diverges from that for water vapor and liquid (the solid line). The divergence of the two curves begins at the freezing point (0°C) because ice can't easily exist above freezing.

2. The area between the two curves, shaded orange, depicts temperatures at which air is unsaturated with water vapor relative to a liquid but above saturation for air next to ice in the cloud. Another way to state this is that saturation is reached with slightly less water vapor when air is in contact with ice in the cloud than when it is in contact with liquid water, even at the same temperature.

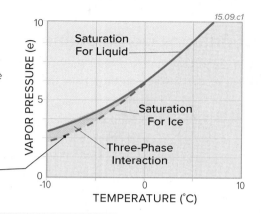

15.09.c1

3. If a parcel of air has the water-vapor capacity and temperature to plot in this orange area of the plot, vapor will become deposited on ice crystals, while liquid water drops are evaporating at the same time. The ice crystals can grow, while the drops become smaller.

4. This difference in saturation levels aids in the formation of precipitation. If ice, water vapor, and drops of liquid water coexist at the conditions specified by the orange field in the previous figure, then the air will be unsaturated next to the liquid drop. As a result, water molecules will evaporate from the drops, increasing the relative humidity of the surrounding air.

5. As the liberated water vapor molecules diffuse into the air, they cause an increase in water vapor adjacent to a nearby snowflake. The air near the snowflake therefore becomes saturated or supersaturated with respect to the ice. This in turn causes water vapor molecules to be deposited on the snowflake, enlarging it.

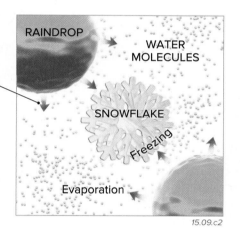

15.09.c2

6. Depending on the air temperature, the ice can remain as a solid or can melt, going directly from ice to liquid. In this way, liquid water becomes water vapor, which then becomes ice, which then becomes drops, an easier path (under many conditions) than trying to overcome surface tension to grow drops or the difficult energetics of nucleation required to make new drops. This process of precipitation is often called the *Bergeron process* after one of the scientists who discovered it.

How Cloud Seeding Works

Technology exists to "help" precipitation occur more efficiently in drought-stricken areas. Strategies generally involve injecting particles into clouds to enhance droplet or ice crystal growth, either via a ground-based delivery system or more commonly airplanes that fly through the clouds, as shown here. One strategy uses dry ice to cool the cloud to temperatures so low that vapor deposits onto ice more readily. A second strategy is to inject into the cloud microscopic solid particles that act as condensation nuclei. The idea is that water vapor can condense or deposit much more easily when it has something to "hold onto" during the phase change. Silver iodide has often been used for this purpose because its crystalline structure maximizes opportunities for vapor to attach to it. These cloud seeding efforts have achieved mixed results. In some cases, precipitation has been enhanced or shifted to other areas. Interventions into natural processes, however, often have unintended consequences, such as, in this case, silver-iodide pollution from the chemicals used to seed the cloud. Unintended consequences are always a concern when trying to affect a natural system.

15.09.t1

Before You Leave This Page

☑ Explain what precipitation is and how the size of a drop can change.

☑ Sketch the water-vapor capacity curve as a function of temperature, with respect to liquid water and ice, and describe how this leads to a Bergeron process for precipitation.

☑ Sketch and explain cloud seeding.

15.10 How Do Sleet and Freezing Rain Form?

RAIN AND SNOW ARE THE MOST COMMON forms of precipitation, but other types of precipitation are also important. The term *sleet* is used for a mixture of snow and rain or for precipitation as small partially frozen pellets that are too small to be called hail. *Freezing rain* is precipitation that reaches the ground as raindrops, but that immediately freezes upon contact with cold objects on the surface, such as the ground, roads, plants, and power lines. Can you envision the situations that could cause sleet or freezing rain?

A Under What Conditions Do Sleet and Freezing Rain Form?

Sleet and freezing rain are often confused, and they do have several similarities. Both result from snow and ice pellets that melt on their way down to the surface as they pass through a layer that is warmer than freezing.

Sleet

1. This graph shows a hypothetical temperature profile (the green line) that would allow sleet to form. Note that this temperature profile shows a subfreezing layer close to the surface, overlain by a zone of warmer temperatures (just above freezing)—a temperature inversion.

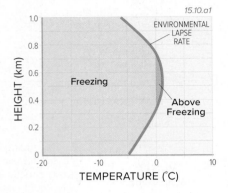
15.10.a1

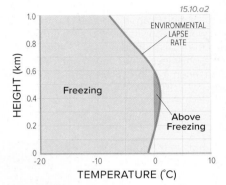
15.10.a2

Freezing Rain

3. This similar graph shows the conditions needed for freezing rain. As for sleet, the temperature profile displays a temperature inversion, but this time with a thinner zone of subfreezing temperatures near the surface. As with sleet, there is a temperature inversion with warmer temperatures (above freezing) above the near-surface air.

2. Freezing temperatures are present higher in the atmosphere, in the upper part of the graph. As snow falls, it crosses from freezing conditions in the top part of the graph through the zone of above-freezing temperatures (shaded in orange) and then back into freezing ones some distance above the surface. As a result, the snow will at least partially melt on the way down, forming a drop of water that then partially refreezes before it reaches the surface. This creates a tiny pellet that bounces on impact with the surface. This type of precipitation is called *sleet*.

4. As snow falls through these conditions, it will at least partially melt when it encounters the mid-level interval of warmer-than-freezing air. As the drop continues falling, it again encounters freezing temperatures, but not until it is at or right above the surface. As a result, the drop does not have enough time to refreeze, so it hits the surface as rain that sticks to the ground instead of bouncing like sleet. Upon encountering the cold ground and objects on the surface, the drop immediately freezes, forming coatings of ice on the ground and objects.

B Where Do Sleet and Freezing Rain Form?

The scenario that provides the best opportunity for the temperature inversion required for the formation of sleet and freezing rain is ahead of the warm front, as shown in the cross section below.

1. A warm front is a mass of relatively warm air that moves across the surface, following the retreat of cold air in its path. In the cross section shown here, the warm air is on the left side (on top) of the frontal boundary and cold air is on the right side of (below) the boundary.

6. Snow can be associated with a warm front if the warm air rises so high off the ground that it is below freezing. In this case, the entire profile remains below freezing, allowing snow to form aloft and reach the ground.

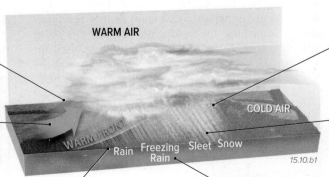
15.10.b1

2. The warm air is less dense than the cold air, so it is wedged up and over the colder air mass. If warm air resides above colder air, it creates a temperature inversion. As the warm air is forced to rise, it cools, forming clouds and rain above and along the boundary.

5. Even farther back from the surface warm front, rain produced aloft must pass through a thicker column of cold air. Thus, raindrops have enough time to freeze on the way down, forming sleet.

3. Close to the surface position of the front, rain produced along the front only falls through a short vertical interval of cold air. If raindrops pass through this zone without refreezing, they will reach the ground as normal rain. This is the type of precipitation that is closest to the warm front.

4. Farther back from the surface location of the front, the ground may still be at freezing temperatures. In this case, the rain freezes upon encountering the ground and other cold objects on the surface, producing freezing rain. As the warm front passes over an area hovering near freezing temperatures, therefore, normal rain follows freezing rain.

C What Regions Have a Higher Frequency of Freezing Rain?

1. This map of central North America (▶) shows the average annual number of days with freezing rain. Examine this map for regional patterns and try to identify whether the place where you live or visit has frequent freezing rain—you probably already know this answer, because freezing rain is not easily forgotten. A map of the frequency of sleet would show a similar distribution. Since latitude and elevation largely control temperature, they also influence where freezing rain occurs.

15.10.c1

3. Freezing rain is also frequent in the highland areas of the Northeast, including New England and some parts of the Appalachian Mountains. The freezing rain and sleet are the result of the same type of cyclonic storm system as in the Midwest. Sleet and freezing rain are also common in other humid uplands in mid-latitudes around the world, such as the Alps, Japanese highlands, and parts of the Andes of South America.

2. On this map, freezing rain is frequent in the Midwest, exhibiting a pattern that follows the typical paths of storms as they move eastward across the country. Such storms have a northward (counterclockwise) flow along their eastern, leading edge, bringing relatively warm air from the south that in many cases forms a warm front.

D How Does Terrain Contribute to the Distribution of Sleet and Freezing Rain?

Movement of air against mountains can also cause freezing rain and sleet. When a cold air mass is next to a mountain range, the circulation associated with it will cause air to move upslope as it encounters higher terrain, as shown here (▶). The movement of air can be in response to larger circulation systems, such as a cyclone.

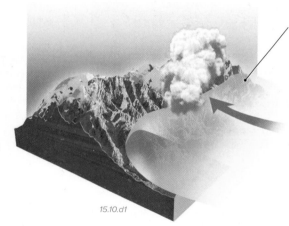

15.10.d1

A cold air mass beside the mountain is dense, while the air wedged above it may be warmer, leading to a temperature inversion. While the inversion would also stabilize the local atmosphere above the cold surface air mass, the movement upslope by the circulation around the pressure system may result in forced convection that produces rain at the lowest elevations, freezing rain at higher elevations, sleet farther upslope, and snow at the highest elevations. Mountains that block flow in this way are sometimes referred to as "cold air dams," as shown in the photograph below. Mountain-caused freezing rain, sleet, and snow are frequent next to the Rocky and Appalachian Mountains in winter.

15.10.d2

The Impacts of Freezing Rain

Freezing rain can cause extensive damage and poses severe hazards to animals, plants, and any object on the surface. Freezing rain can coat sidewalks and roadways with a layer of dangerously slippery ice, causing pedestrians to fall and automobiles to skid out of control. Freezing rain resembles rain as it falls, so people may believe that they are only driving in rain when in fact they are driving on an icy road. Even the rain that falls on windshields gives no clue that it will freeze on impact with the surface, because automobile defrosters

and friction created by windshield wipers prevent freezing that might alert the driver. The fact that sleet bounces informs motorists to its presence as ice, making it less dangerous.

Freezing rain adds considerably to the weight of trees, especially in fall or early winter, when leaves are still attached to tree limbs and branches. This may cause limbs to fall, perhaps onto power lines, which are themselves weighed down considerably by freezing rain. On average, freezing rain causes more deaths and property damage per hour than any other type of storm.

15.10.t1

Before You Leave This Page

☑ Sketch the temperature profiles that would lead to sleet and freezing rain.

☑ Sketch a warm front, showing where sleet and freezing rain are most likely to form.

☑ Characterize the spatial distribution of sleet and freezing rain in the U.S.

☑ Explain the impacts of freezing rain.

15.11 What Is the Distribution of Precipitation?

THE AMOUNT OF PRECIPITATION varies from region to region, season to season, and day to day. The daily variations are related to short-term changes in weather, but variations between regions and seasons reflect differences in the overall climatic setting, such as differences in latitude, prevailing wind directions, proximity to large water bodies, any nearby ocean currents, elevation of the land, and countless other factors. We explore some of these by examining regional variations in precipitation over North America and then the world.

A How Does Precipitation Vary Across North America?

January

1. Examine this map, which shows average amounts of precipitation for North America for January, averaged over several decades. In northern and high-elevation regions, much of this precipitation is as snow. Note any regional patterns and the precipitation that is typical for where you live.

2. Heavy rainfall along the coast of Canada and the Pacific Northwest is due to proximity to the ocean combined with orographic effects that cause moisture-laden Pacific air to move upslope under the influence of the mid-latitude westerlies.

3. Inland from the western mountains, precipitation totals decline dramatically. These areas are blocked from oceanic moisture sources by mountains both to the west and to the east.

4. Precipitation decreases inland away from moisture sources and northward with lower temperatures. Cold air has lower water-vapor capacity and therefore will not produce clouds with a sufficient amount of moisture to allow for much precipitation.

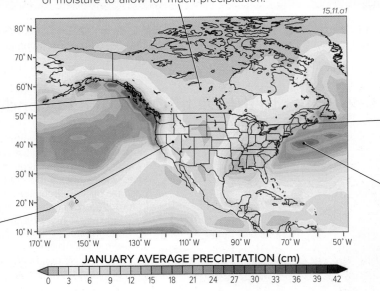

JANUARY AVERAGE PRECIPITATION (cm)

5. The Great Lakes are much warmer than the overlying air in winter. This generates a steep environmental lapse rate, which makes an adiabatically moving air parcel likely to be warmer than its surrounding environment. Thus, the lakes destabilize the atmosphere, increasing local "lake effect snow" totals downwind.

6. Precipitation in both the Atlantic and Pacific oceans at this time of year is focused near the boundary between cold air and warm air, in part because these regions experience frequent passage of fronts.

July

7. This map shows precipitation totals for July, the middle of summer. Note the regional patterns and then compare this map with the one for January.

8. July totals are higher than in January in most of North America, because local surface heating destabilizes the atmosphere. Along the West Coast, the cold California Current negates this effect in that region.

9. Summer precipitation totals are low in western North America and over the adjacent Pacific Ocean, even in places that are relatively wet in winter. This is because of the drying effects of sinking air on the eastern side of a high-pressure feature in the Pacific that makes its closest approach to North America in summer.

10. July totals are much higher than those in January in the far north. The warmer air has more energy to evaporate water, producing water vapor that can then precipitate. Also, the region is much closer to the cold-warm boundary in summer than in winter, facilitating frontal lifting.

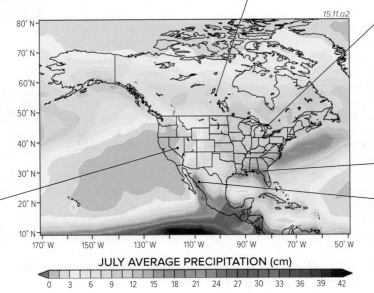

JULY AVERAGE PRECIPITATION (cm)

11. The Great Lakes are cool relative to the overlying air in July. This weak environmental lapse rate (or even an inversion) stabilizes the atmosphere, the opposite of winter conditions, limiting the amount of precipitation.

12. In summer, high pressure in the Atlantic reaches its northernmost extent. The clockwise circulation around this feature supplies warm, moist air into interior North America.

13. The prong of higher precipitation that occurs in western Mexico is related to a northward flow of moisture as part of the "Arizona monsoon." Near the mountains, intense surface heating and orographic lifting cause summer thunderstorms.

B How Do Annual Precipitation Totals Vary Globally?

The globes below show average annual precipitation at a global scale. Examine the regional patterns and consider what factors might be causing them. You may want to refer back to the globes showing specific humidity earlier in this chapter. An area that most people consider to be "wet" generally has high precipitation and high humidity. How dry an area seems is a function of low precipitation, low humidity, and evaporation rates that exceed precipitation rates. Read the text below in a counterclockwise direction.

1. Globally, precipitation is concentrated in equatorial oceanic areas, expressed as a belt of blue and green colors near the equator. Precipitation generally decreases poleward, inland, and downwind from major mountain ranges.

2. The tropical rain forests of Indonesia, along with adjacent parts of the Indian and Pacific oceans, are among the wettest places on Earth. They are part of an east-west belt of high precipitation that roughly follows the Intertropical Convergence Zone (ITCZ). Along the ITCZ, convergence of moisture-laden air carried by the northeast and southeast trade winds causes lifting and precipitation. The Hadley cell migrates with the seasons, so the position and influence of the ITCZ do, too.

8. The east coasts of mid-latitude continents are generally moderately wet, as cold and warm air meet frequently, producing frontal precipitation. In summer, convective precipitation supplements the totals.

7. The idea that Siberia receives extremely large amounts of snowfall is a myth. The intense Siberian High is driven by cold, dense air, and it produces a sinking motion. The cold surface also acts to decrease the environmental lapse rate, making it more likely that an adiabatically moving air parcel will remain colder than its surrounding environment, enhancing atmospheric stability.

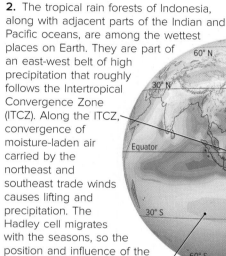

15.11.b1

15.11.b2

15.11.b3

3. Locations in the subtropics (30° N and S) tend to be very dry. This belt of latitudes contains many of the world's deserts, such as the Sahara and those of the Arabian Peninsula, southwest Africa, central Australia, and the American Southwest. Even though warm air would ordinarily tend to rise in these warm areas, this effect is counteracted by descending air in the downward-directed part of the Hadley cells. Some of these areas are especially dry because they are in the rain shadow of major mountain ranges. Exceptions to the dry subtropics occur along the east coast of continents (for example China), where there is an abundance of warm, moist air in association with warm surface ocean currents.

4. Polar areas are generally too cold to have much water in the atmosphere. Limited moisture, combined with the sinking of the intense cold air, opposes vertical cloud development. Furthermore, these areas are typically far from boundaries between warm and cold air, which can produce warm and cold fronts and their attendant storms. As a result, places near the poles receive very little precipitation.

5. The Amazon basin, with extensive tropical rain forests, is another region of high precipitation. This is due to the ITCZ and to orographic lifting of the moisture-laden trade winds over the eastern flank of the Andes Mountains. The southwestern coast of South America has the driest deserts in the world.

6. Relatively little was known about precipitation over the oceans until recent years when satellite data became available. Oceanic patterns are similar to continental ones, except for the absence of complications caused by topography.

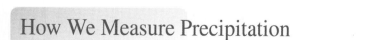

ANNUAL AVERAGE PRECIPITATION (cm)

0   36   73   109   146   183   219   255   292   328   365   402   438   474   511

How We Measure Precipitation

Rainfall is measured using a standard cylinder with a circumference that is 1/10 of the opening. The amount of rainfall can be read from graduated markings on the cylinder or from a measuring stick with hashmarks exaggerated 10 times. This allows for the observation of rainfall totals to the hundredths of an inch. Some rain gauges, like the one shown here, contain two cylinders, one for measuring typical rainfall amounts and a second larger cylinder used when extreme amounts of

15.11.t1

precipitation cause the first cylinder to overflow. When it snows, the snow is allowed to melt and the amount of liquid can be measured. Snow typically has an average density of 8% that of water, so the equivalent of one centimeter of rain will produce, on average, more than 12 cm of snow (an inch of rainfall will produce a foot of snow), but this ratio can vary widely, depending on local conditions.

Before You Leave This Page

✓ Summarize factors influencing the distribution of precipitation in North America during January and July.

✓ Summarize the main patterns of global precipitation, and their main causal processes.

15.12 What Caused the Recent Great Plains Drought?

PARTS OF THE GREAT PLAINS, stretching from Texas to Montana, experienced severe drought in 2012, resulting in huge crop losses and other problems. These losses occurred in spite of recent technological advances, such as improved soil and water management practices and the use of drought-resistant varieties of crops. Some effects, such as crop loss, resulted directly from the event, but others, such as less food for livestock, were indirect results of the drought. What caused this drought, and what were some primary and secondary impacts? This drought nicely illustrates the onset of such conditions over a relatively short time frame, whereas other, more recent droughts, such as one in California, are more persistent.

A What Was the Meteorological Setting of the Drought?

Examine each pair of maps. The top maps show averages of the daily pressure (as represented by the height of the 500 mb pressure) for May and July 2012. The bottom pair of maps shows specific humidity for the same two months. Compare the air pressure maps to the specific humidity maps and consider the relationship between the patterns of air pressure and humidity, if any exist.

Air Pressure

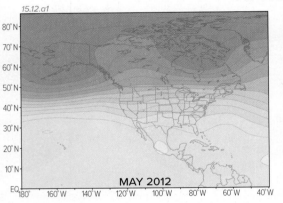

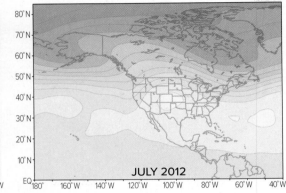

15.12.a1

15.12.a2

MAY 2012

JULY 2012

5440   5600   5720   5880
500mb GEOPOTENTIAL HEIGHT (m)

1. In May (▶), the air pressure patterns and upper atmosphere wind directions over the lower 48 states were essentially west-to-east, a condition meteorologists call a *zonal flow*. Pressures over the Great Plains were neither much higher nor lower than areas to the east and west. For most places, a zonal upper-level pattern results in typical weather, because it neither supports nor suppresses storms. For Texas, a west-to-east zonal flow brings in dry air from northwestern Mexico and the Desert Southwest (e.g., Arizona).

2. By July (◀), a strong ridge of high pressure had developed over the Great Plains states, with the entire southern U.S. under high pressure (high 500 mb heights). The sinking air associated with this high pressure caused abundant adiabatic warming and compression, and little cloudiness and precipitation. High pressure also blocked or deflected storm tracks away from the region. As a result of these factors, drought extended across most of the central U.S., including much of the Great Plains.

Specific Humidity

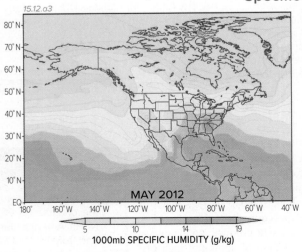

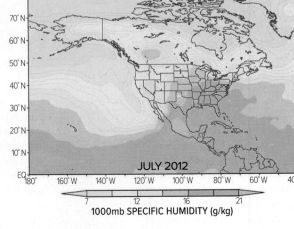

15.12.a3

15.12.a4

MAY 2012

JULY 2012

5   10   14   19
1000mb SPECIFIC HUMIDITY (g/kg)

7   12   16   21
1000mb SPECIFIC HUMIDITY (g/kg)

3. These two maps (▶) show specific humidity (the mass of moisture in a mass of air) for the same two months as shown above. Darker shades of green represent more moisture, whereas lighter shades indicate drier air. During May (the map on the left), dry air in the Desert Southwest and in the deserts of adjacent northern Mexico spread into west Texas, but moderately humid conditions existed in much of the central Great Plains, such as in Kansas and Nebraska.

4. The specific-humidity map for July (◀) is not too different from the one for May. The clockwise flow around the high pressure in the Atlantic has advected moisture into the interior of North America. A conclusion you can reach from comparing the four maps on this page is that the air-pressure conditions were very different, but the amount of moisture in the air was similar, although the patterns differ in detail.

B What Were the Patterns of Drought in the Middle of 2012?

The two maps below show the severity of the drought, as expressed by the Palmer Drought Index for the same two months as depicted on the previous page. Examine the patterns of drought on these maps and compare them to the maps of air pressure and specific humidity. From comparing each set of maps, think about whether the drought in the Great Plains was mostly caused by a lack of humidity or by upper-level air pressure and patterns of flow.

On these maps, areas experiencing severe and extreme drought are shown in orange and brown, respectively. Areas in light yellow are experiencing mid-range conditions, meaning not overly dry or overly moist. In May (▶), areas of the Great Plains with extreme drought are mostly restricted to isolated areas in northern Texas, eastern New Mexico, and especially southeastern Colorado. In the central and northern parts of the Great Plains, from Oklahoma to North Dakota (a north-south strip through the center of the map), conditions were in the mid-range. Note that California was also in a drought, which persisted for a number of years.

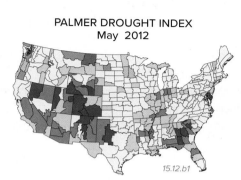

PALMER DROUGHT INDEX
May 2012

15.12.b1

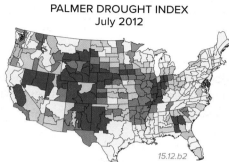

PALMER DROUGHT INDEX
July 2012

15.12.b2

By July (◀), extreme and severe drought had spread across more of northern Texas, eastern New Mexico, and into parts of Kansas, Nebraska, Wyoming, and the Dakotas. Nearly all of the Great Plains was experiencing at least moderate drought (dark yellow, orange, or brown). From your comparisons of these maps with those of air pressure and specific humidity (on the previous page), what do you conclude about the main cause of the drought inflicted on the Great Plains in the summer of 2012?

EXTREME DROUGHT	SEVERE DROUGHT	MODERATE DROUGHT	MID-RANGE	MODERATELY MOIST	VERY MOIST	EXTREMELY MOIST
-4.00 and Below	-3.00 to -3.99	-2.00 to -2.99	-1.99 to +1.99	+2.00 to +2.99	+3.00 to +3.99	+4.00 and Above

Primary and Secondary Impacts of Drought

In considering the effects of drought, some effects, such as drying out of agricultural fields and dying of crops, are due directly to the drought; these are *primary impacts*. Other effects are related to drought, but in a more indirect way, like an increase in the price of seeds that resulted from the initial dying of crops; such effects are *secondary impacts*. Here, we examine the primary and secondary impacts of the 2012 drought, focusing on Texas, but similar consequences of the drought were felt in other states.

An obvious primary effect of the drought was damage to crops due to the extreme dryness of the soil and lack of rainfall. In Texas, nearly half of the cotton crop was lost, and the corn harvest decreased by 40%. In a similar manner, it was estimated that up to 10% of the trees in Texas were damaged. Wildlife populations dwindled as their habitats were degraded. Ranching likewise suffered the effects of the drought, with a sharp decrease in the number of cows. Such losses were largely caused by a diminished supply of drinking water or an increase in water contamination as ponds and other sources of surface water dried up. Throughout the drought-stricken region, the supplies of surface water and groundwater decreased, with reservoirs estimated as 40% below normal.

There were also secondary impacts. The dry conditions caused grass and brush to dry out (a direct impact), causing a series of wildfires that burned over 3 million acres in Texas, resulting in more than $750 million in damage to homes (a secondary impact). Trees and plants stressed by the drought became more vulnerable to later disease and pest infestations. High temperatures and low water levels (primary impacts) led to low amounts of dissolved oxygen in some lakes, ponds, and streams, resulting in fish dying (a secondary impact). Reduced seed production altered the food chain, even for species that weren't affected by the drought directly. Shrinking supplies of agricultural products resulted in higher prices.

Other secondary impacts are even less obvious. Transportation was delayed or halted along highways because of wildfires. Tourism decreased as fewer people visited lakes, because the low water levels left an unpleasant and potentially dangerous situation for boaters. The amount of recreational fishing declined for the same reason and because of the previously mentioned fish kills. Reduced water supplies in cooling reservoirs used by power-generation plants caused a reduction in power output at the same time that energy demand peaked due to the higher temperatures (more air conditioning

needed). This resulted in rolling blackouts, providing an excellent example of a secondary impact. Geographers also use the concepts of primary and secondary impacts for other types of meteorological events, such as hurricanes and flooding. These terms are also considered in the context of global and regional changes that result from climate change and other events. Most geographers view problems from global, regional, and local perspectives.

The onset and end of a drought can be gradual or relatively abrupt—so can the impacts. A drought affected California from 2012 through 2016, but ended rather suddenly with heavy snows and rain in the winter of 2016-2017.

Before You Leave This Page

✓ Summarize the patterns of upper-level air pressure and specific humidity that accompanied the 2012 drought in the Great Plains.

✓ Distinguish between primary and secondary impacts of an event, like drought, providing examples of each type of effect from the 2012 drought in Texas.

15.13 How Do Global Patterns of Humidity, Water Vapor, and Precipitation Compare?

THE ABUNDANCE OF ATMOSPHERIC MOISTURE varies considerably from place to place on Earth. What are these variations, and what causes them? In this exercise, you will observe and compare four types of global data: specific humidity, amount of water vapor, average precipitation, and what material is on the surface (land cover). You will describe and explain the patterns of atmospheric moisture and how these are reflected in the land cover.

Goals of This Exercise:

1. Describe regional patterns of atmospheric moisture as measured using specific humidity, water vapor content, and precipitation, then propose explanations for these patterns.

2. Explain how the patterns of atmospheric moisture are reflected in variations in land cover on satellite images.

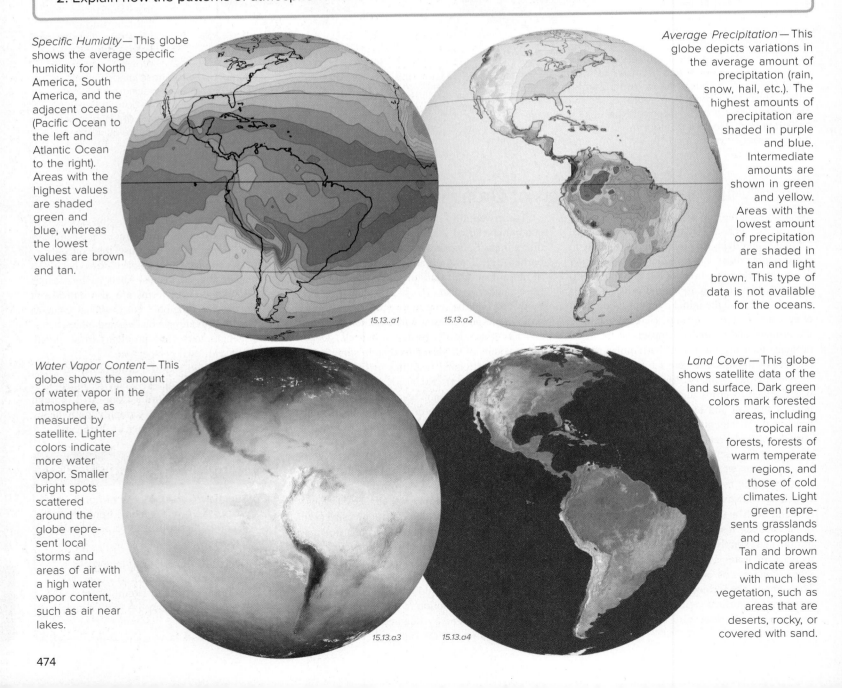

Specific Humidity—This globe shows the average specific humidity for North America, South America, and the adjacent oceans (Pacific Ocean to the left and Atlantic Ocean to the right). Areas with the highest values are shaded green and blue, whereas the lowest values are brown and tan.

15.13..a1

Average Precipitation—This globe depicts variations in the average amount of precipitation (rain, snow, hail, etc.). The highest amounts of precipitation are shaded in purple and blue. Intermediate amounts are shown in green and yellow. Areas with the lowest amount of precipitation are shaded in tan and light brown. This type of data is not available for the oceans.

15.13.a2

Water Vapor Content—This globe shows the amount of water vapor in the atmosphere, as measured by satellite. Lighter colors indicate more water vapor. Smaller bright spots scattered around the globe represent local storms and areas of air with a high water vapor content, such as air near lakes.

15.13.a3

Land Cover—This globe shows satellite data of the land surface. Dark green colors mark forested areas, including tropical rain forests, forests of warm temperate regions, and those of cold climates. Light green represents grasslands and croplands. Tan and brown indicate areas with much less vegetation, such as areas that are deserts, rocky, or covered with sand.

15.13.a4

Procedures

For each region (the Americas on the left page and Africa on the right page), complete the following steps. Write your observations and interpretations on the worksheet or answer questions online.

A. Observe the larger patterns on each globe. Where are the amounts of any factor highest, where are they lowest, and is there a general pattern for this part of the globe? Note any regions that do not appear to fit the regional pattern. Record your observations on the worksheet.

B. Propose interpretations for the large-scale features, such as variations north and south or east to west. For example, why do certain areas have more moisture? Propose explanations for local patterns, such as areas that differ somewhat from the regional patterns. What features are associated with these local patterns? Record your observations on the worksheet.

C. Compare the four globes, proposing explanations for why the patterns on one globe are consistent with those on the other globes. Pay special attention to how the amounts of precipitation and the kind of land cover relate to the two measures of atmospheric moisture (specific humidity and water vapor). Can you then explain some of the large-scale features of our planet? Write your observations and interpretations on the worksheet or answer questions online.

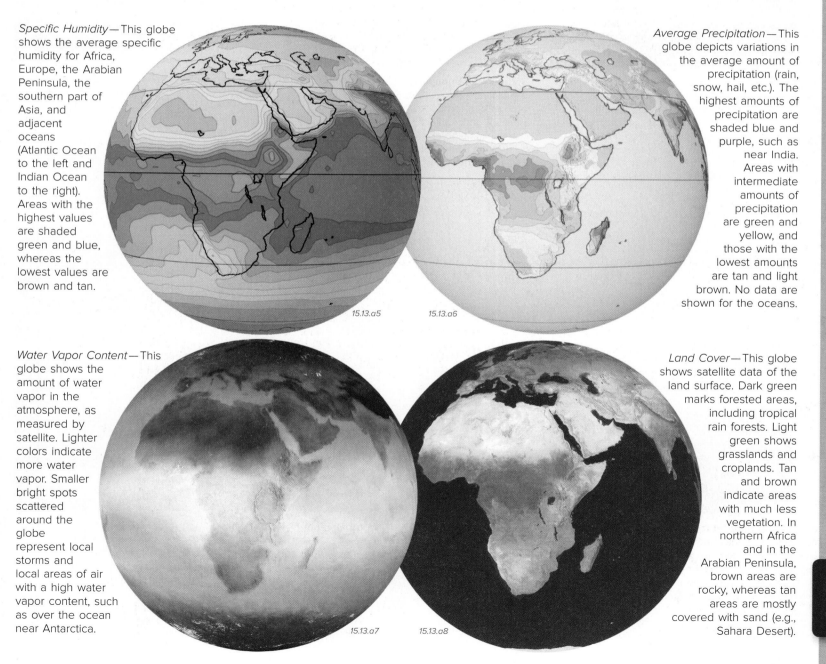

Specific Humidity—This globe shows the average specific humidity for Africa, Europe, the Arabian Peninsula, the southern part of Asia, and adjacent oceans (Atlantic Ocean to the left and Indian Ocean to the right). Areas with the highest values are shaded green and blue, whereas the lowest values are brown and tan.

Average Precipitation—This globe depicts variations in the average amount of precipitation (rain, snow, hail, etc.). The highest amounts of precipitation are shaded blue and purple, such as near India. Areas with intermediate amounts of precipitation are green and yellow, and those with the lowest amounts are tan and light brown. No data are shown for the oceans.

Water Vapor Content—This globe shows the amount of water vapor in the atmosphere, as measured by satellite. Lighter colors indicate more water vapor. Smaller bright spots scattered around the globe represent local storms and local areas of air with a high water vapor content, such as over the ocean near Antarctica.

Land Cover—This globe shows satellite data of the land surface. Dark green marks forested areas, including tropical rain forests. Light green shows grasslands and croplands. Tan and brown indicate areas with much less vegetation. In northern Africa and in the Arabian Peninsula, brown areas are rocky, whereas tan areas are mostly covered with sand (e.g., Sahara Desert).

15.13.a5   15.13.a6

15.13.a7   15.13.a8

15.13

16 Weather and Storms

OUR ATMOSPHERE IS DYNAMIC, featuring various types of weather systems, some relatively benign and beneficial, while others, like hurricanes and tornadoes, are dangerous and destructive. What types of weather do we experience, what causes weather, and what determines if a weather system is relatively gentle or severe? In this chapter, we explore the main types of weather systems on Earth.

16.00.a1 Eastern Colorado

Powerful tornadoes are among nature's most spectacular but frightening phenomena. A tornado forms beneath a thunderstorm as a swirling vortex of winds that can reach speeds of several hundred kilometers per hour. Tornadoes can have a tapered, almost delicate shape, or can be thick, dark, and ominous columns filled with debris.

How do tornadoes form, where do they occur, and how are they related to thunderstorms?

16.00.a2 Pickens County, AL

Tornadoes rampaged through the southeastern U.S. in April 2011, as part of the largest outbreak of tornadoes in history. Between April 25 and April 28, 358 tornadoes tore across Alabama and nearby states, completely destroying some houses, like the ones above, and killing more than 300 people.

How strong do tornadoes get, where do most occur, and what times of the day and year are they most likely to develop?

16.00.a3 Tuscaloosa, AL

The tornadoes carved swaths of destruction, totally leveling parts of neighborhoods (▲), while leaving nearby houses essentially undamaged except for uprooted and toppled trees, downed power lines, and broken windows. In addition to the destruction caused by the tornadoes, deaths and damage resulted from hail, lightning, and flooding.

How do hail and lightning form, and are they related to one another?

16.00.a4

The tornado outbreak of late April 2011 was part of a large weather system, shown in the satellite image above. A swirling storm centered over the upper Midwest (in the upper part of the image) has a distinctive comma shape, with a long tail curving to the south and southwest. The tornadoes formed in the cluster of bright clouds to the southeast—out in front—of the main system.

What is a weather system, how does one form, and why do some acquire a curved shape, like a comma or a coil?

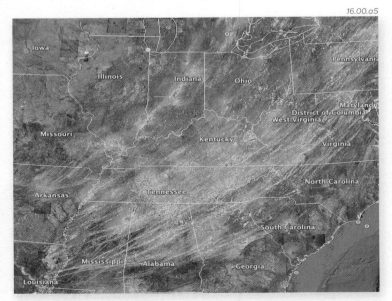

16.00.a5

Weather radar systems tracked each tornado as it moved across the region, as shown by this map depicting colorful lines (red signifies a very strong tornado) of tracks for each tornado. Some tornadoes stayed on the ground for hours and crossed multiple states. Radar images are a mainstay of daily weather reports.

How do radar systems work, what do they measure, and how do such data help us anticipate a storm's path and severity? Once a tornado or other severe storm is identified, how is the public notified?

TOPICS IN THIS CHAPTER

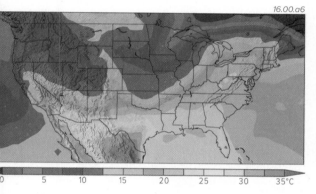

16.00.a6

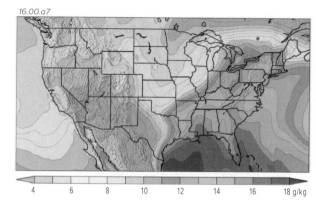

16.00.a7

A map of temperatures for the day of peak tornado activity (◄) shows a boundary between warm air (shown in green and yellow) in the Southeast against cold air (shown in purple and blue) in the upper Midwest. A map of specific humidity (►) for that day indicates that the warm air was also humid, but the cold air was relatively dry.

Why are some masses of air warm, while others are cold? How are boundaries between warm and cold air related to thunderstorms and tornadoes?

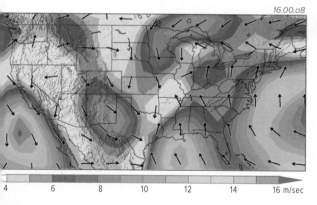

16.00.a8

Surface wind directions during this time (◄) show distinct patterns, with winds blowing from the south to the north in the area affected by the tornadoes (such as Alabama and Georgia). This wind direction brought warm, moist air northward. At higher levels, the winds turned to the right, moving storms from southwest to northeast. The colors on this map indicate wind speeds, with yellow and green representing the fastest winds.

Do regional wind patterns help us predict which way a storm will move?

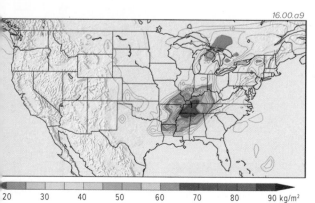

16.00.a9

Heavy precipitation was concentrated in the same region as the tornadoes (◄). In this map, purple and blue indicate the largest amounts of precipitation, which for these storms was as much as 40–50 cm (15 to 20 in.). The heavy rains, falling in a relatively short time period, produced severe flooding that destroyed roads and buildings and caused additional deaths.

What types of storms can produce very heavy precipitation?

What Is Weather?

When we use the term *weather*, we are referring to conditions in the atmosphere at some specific time and place, whether it is right now, an hour ago, or sometime next week. Weather refers to the temperature, humidity, and windiness, and whether it is fair or there is precipitation. It also refers to whether an area is experiencing severe storms, such as a thunderstorm, tornado, hurricane, or snowstorm, or is in a time of exceptional dryness. Weather refers to the various aspects featured on these two pages, and much more! Examine the four maps on this page and figure out what the weather was like on April 27, 2011, where you live.

Weather affects all of our lives, from deciding what clothes to wear, what week to plant crops, or how soon to take shelter from some type of severe weather. In this chapter, we explore the fundamental processes that control weather and then examine the main types of weather phenomena, including rain and snowstorms, freezing rain, thunderstorms, hail, lightning, hurricanes, and, of course, tornadoes. Along the way, we explain features you can observe, like the types of clouds, to understand and predict the weather around you.

16.0

16.1 Why Does Weather Change?

THE ATMOSPHERE IS SO THIN that a single regional mass of cold or warm air can dominate the entire vertical extent of the troposphere. With time, such cold or warm air masses can expand or shrink, become colder or warmer, and can remain stationary or move great distances. In the narrow zone where the two air masses meet, weather conditions can change drastically across short distances and times, sometimes dangerously.

A What Are Some Fundamental Controls of Weather?

Weather is controlled by the atmospheric processes described in previous parts of this book, and in fact those processes were described to get to this point—an understanding of weather and climate. Weather is a response to variations of insolation and resulting energy imbalances, to large-scale motion in the atmosphere, and to the important role of atmospheric moisture, including latent heat. Here, we briefly review the global-scale circulation patterns. Only the Northern Hemisphere is shown below, but similar patterns are also present in the Southern Hemisphere.

1. The most important features that control weather are the huge circulation cells that essentially divide each hemisphere into thirds (by latitude, not by area). Areas at high latitudes, poleward of approximately 60°, are under the effect of the *polar cell* and *polar easterlies*. High surface pressure dominates the pole.

2. The edge of the polar cell is marked by the *polar front jet stream*, a belt of high-speed winds that encircle the globe from west to east, high above the surface. The surface in this area has changeable and stormy weather due to rising air and shifting of the jet stream. Bends in the polar front jet stream are called *Rossby waves*, which exert a strong control on many weather systems.

3. A second jet stream, called the *subtropical jet stream*, encircles the globe (also west to east) high above the surface near 30° latitude. Between this jet stream and the equator, near-surface winds blow from the northeast, forming the *northeast trade winds*. Along the *equator*, high amounts of insolation-related heating cause rising air and associated low pressure. Winds converge from both sides of the equator, forming the *Intertropical Convergence Zone* (ITCZ).

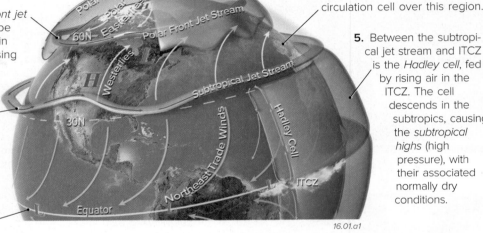

16.01.a1

4. Between the two jet streams, in the mid-latitudes and part of the subtropics, the general circulation is from the west—the *westerlies*. There is not a well-established, upper-level circulation cell over this region.

5. Between the subtropical jet stream and ITCZ is the *Hadley cell*, fed by rising air in the ITCZ. The cell descends in the subtropics, causing the *subtropical highs* (high pressure), with their associated normally dry conditions.

B What Is an Air Mass?

The pattern of global circulation helps form large bodies of air, called *air masses*, that have relatively uniform temperatures and moisture over large horizontal distances. We identify different types of air masses according to their temperature, moisture, and geographic characteristics.

The air-temperature map to the right depicts a cold air mass, represented by purple and blue, over the northern part of North America, including the northern Great Plains. A warmer air mass exists farther south, covering most of the southern and eastern U.S. The boundary between the two air masses is along the steep color gradient (a large change in color over a short distance) between the purple and green areas.

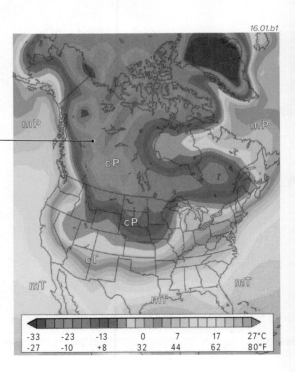

16.01.b1

	Dry	Humid
Cold	**A, cP:** Stable, often with surface inversions; fair weather	**mP:** Unstable, often with cold air overlying warmer air adjacent to oceans
Warm	**cT:** Unstable, with hot surface and cooler air aloft, but lack of moisture	**mT, E:** Often unstable, with abundant moisture, especially E

The classification system for air masses uses a two-letter abbreviation. The first letter is lowercase and signifies the moisture characteristics, where "c" is for continental (dry) and "m" is for "maritime" (humid). The second letter is uppercase and refers to the temperature characteristics: "A" for Arctic (or Antarctic), "P" for polar (somewhat warmer than Arctic), "T" for tropical, and "E" for equatorial. An "A" air mass is inherently continental (dry), because water in such areas is often frozen. An "E" air mass is inherently maritime, because so much of the air at equatorial latitudes is over or near a source of water.

Continental Air Masses

1. Hot, dry air masses, typically occurring over land, are continental tropical (cT) air masses. Note that cT air masses occur in the Desert Southwest of North America, the Sahara and Arabian Deserts, Tibet, and central Australia. In areas under the influence of cT air masses, conditions can be oppressively hot and dry for many months during the year.

2. *Arctic or Antarctic (A) air masses* are formed near the North and South Poles. Not only are they extremely cold, but they are also very dry because the water-vapor capacity is low in such cold air. Cold, dense air has a tendency to sink, and therefore, these air masses reinforce the region of polar high pressure (polar high) that resides over each pole. Surface winds near the pole move east-to-west (easterlies), so the surface A air masses tend to be steered westward as they drift away from the poles. The paths of A air masses can also be affected by curves in the upper-level polar front jet stream (Rossby waves), which can steer an A air mass toward the northern U.S. or into central Eurasia. These are the "Arctic blasts" that bring very cold air into the northern U.S.

3. *Continental polar (cP) air masses* are located farther away from the poles than A air masses. They are cold and dry, but less so than A air masses. cP air masses form over cold, inland areas, like northern Canada, Siberia, Mongolia, and north-central Europe. The southern edge of cP air masses is the polar front. In the winter, cP air masses can move significant distances equatorward, bringing cold air into the U.S. and southern Eurasia. In the summer, cP air masses are not as cold and do not reach as far south.

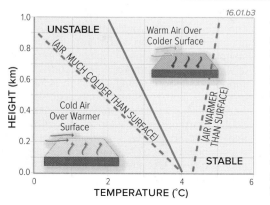

16.01.b2

Mollweide Projection

Maritime Air Masses

4. Due to their proximity to the equator, *equatorial (E) air masses* are very warm and moist. The high temperature and moisture support instability to great heights in the atmosphere, resulting in tropical clouds and storms. Storm development is aided by the general circulation of the atmosphere, which in equatorial regions promotes rising of the air, especially along the ITCZ.

5. *Maritime tropical (mT) air masses* form over warm, oceanic regions. mT air masses are warm and moist, and they have influence over much of the Earth because so much of the world (particularly in the low latitudes) is covered by ocean. For example, mT air tends to make most of the Atlantic Ocean and adjacent eastern U.S. warm and humid in summer. mT air masses are commonly near the subtropics.

6. *Maritime polar (mP) air masses* are cool and humid. They are not as cold as cP air masses because oceanic regions are not as cold as continental areas at the same latitude, particularly in winter. Such mP air masses are a cold, damp influence in places like Seattle and most of western Europe, as the westerlies push the cool, damp air eastward from their oceanic source region.

16.01.b3

[Graph: HEIGHT (km) vs TEMPERATURE (°C)]
UNSTABLE — Warm Air Over Colder Surface — (AIR MUCH COLDER THAN SURFACE) — Cold Air Over Warmer Surface — (AIR WARMER THAN SURFACE) — STABLE

7. Any air mass acquires characteristics of the region where it forms—its *source region*. Air masses that form inland or over frozen ocean areas (e.g., cP) are dry, whereas air masses that form over oceans (e.g., mT) are humid. Air masses that form in the low latitudes are generally warmer than those that form over high latitudes. When an air mass migrates from its source region across a different region (◄), its temperature and moisture will be modified, which can in turn affect the lapse rate, stability of the air, and how much mixing occurs within the air mass.

Before You Leave This Page

✓ Summarize the characteristics of each air mass type, where it is typically formed, and how it influences a region's climate.

✓ Explain factors that influence the rate at which an air mass acquires its characteristics and is modified as it migrates.

16.1

16.2 What Are Fronts?

THE NARROW ZONE separating two different air masses is called a *front* and is often the site of rising atmospheric motion. Whenever different air masses meet along a front, the less dense, warmer air mass is pushed up over the more dense, colder one. If the rising air cools to its dew-point temperature, cloud formation will begin, perhaps followed by precipitation. There are several types of fronts, depending on the manner in which one air mass is displacing the other and whether the front is moving. Fronts are associated with many of the storms most regions experience.

A What Processes Occur Along Cold and Warm Fronts?

1. The map below shows several fronts as decorated lines, among the areas of high pressure (H) and low pressure (L). The thin gray lines are isobars (lines of equal pressure), and the background of the map is a satellite image showing cloud cover. Examine each of the features on this map and note any correspondence between the fronts, high and low pressure, and cloud cover. Thick, white lines and letters show the locations of cross sections on the bottom of the page.

2. The blue lines represent *cold fronts*, the leading edges of cold air masses. Blue triangles along a cold front indicate the direction that the cold air mass is pushing. The east-facing triangles on the long cold front in the center of the U.S. indicate that the cold front is pushing eastward. Cold fronts are usually associated with *cP* but can also occur along the leading edge of an *A* or *mP* air mass.

3. The red lines represent *warm fronts*, the leading edges of warm air masses. The red semicircles along a warm front indicate the direction in which the warm air mass is moving. The warm front over northwestern North America is moving east. Warm fronts are usually associated with *mT* or *cT* air masses.

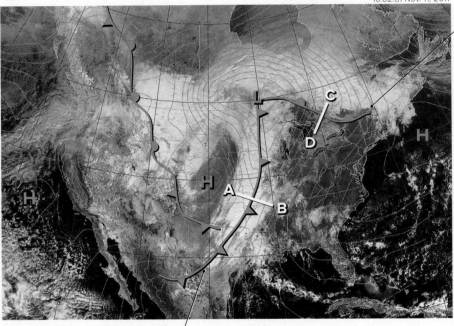

16.02.a1 Nov. 11, 2011

4. Let's examine the weather in the eastern half of the U.S., beginning with the long cold front. As the front and associated mass of cold, dense air move east, the less dense warmer air that is east of the front is displaced and forced upward.

6. To the north is a north-moving warm front. Air south of the warm front (like at "D") is warm and less dense, so it gets wedged over the more dense, cooler air north of it (at "C"). The gentle slope of this rising motion causes a shield of stratiform clouds over eastern Canada.

5. Warmer air ahead of the cold front (such as at B) moves up and over the colder air, creating cumuliform cloud cover. The air west of the cold front (at "A" for example) is cold and dense, so it tends to sink, associated with increasing surface pressure (H). The pressure gradient is very steep, as reflected by closely spaced isobars near the cold front.

Cold Fronts

7. The figure below shows a cross section through a cold front along line A-B on the map above.

The cold air mass displaces the warm air mass. Cold air is more dense and forces the warm air upward. If the rising warm air cools to its dew-point temperature, cloud cover and perhaps precipitation occur along the front.

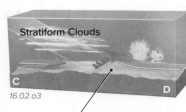

16.02.a2

Warm Fronts

8. The figure below shows a cross section through a warm front along line C-D on the map above.

As the warm air mass follows and displaces the cool air mass, it slides up over the colder air. If the rising air cools to the dew-point temperature, stratiform clouds and perhaps precipitation occur along the warm front, but spread out over a significant width.

16.02.a3

Comparing Warm and Cold Fronts

9. Compare the cold-front and warm-front figures, observing similarities and differences, then read the text below.

The slope at which a warmer air mass is pushed upward along a cold front is much steeper than the slope at which it is forced upward along a warm front. This tends to support cumuliform cloud formation along cold fronts, but stratiform cloud cover along warm fronts. As a result, precipitation along cold fronts tends to be intense, localized, and short-lived, whereas precipitation along warm fronts tends to be light, widespread, and long-lived. Use these tips the next time a weather front crosses your hometown.

B What Is a Stationary Front?

Sometimes, for a period of several hours or days, neither the cold air mass nor the warm air mass is displacing the other. There is a temporary stall in motion of the front, so such a stalled front is a *stationary front*. On a weather map, a stationary front, such as occurs here along the Atlantic Coast of North America, is represented by alternating blue and red line segments with alternating blue triangles and red semicircles. The blue triangles point away from the cold air, and the red semicircles point away from the warm air, as shown in the cross section along E–F on the map. Stationary fronts are associated with stratiform and cumuliform clouds.

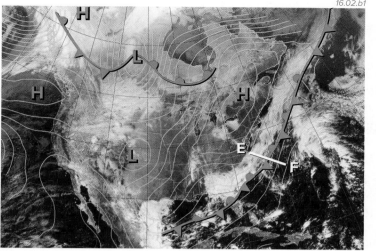
16.02.b1

16.02.b2

On this stationary front, the blue triangles point east, so the cold air is trying to push east and must therefore be on the opposite side of the front (on the west side). The warm air is therefore on the east side. A band of clouds and precipitation is along the front, mostly on the cold air side (west of the front). Since a stationary front is not moving, prolonged precipitation can occur as the front sits over the same place for an extended time.

C What Types of Weather Are Characteristic of Cold, Warm, and Stationary Fronts?

1. On this map, examine the fronts, locations of high and low pressure, and the underlying satellite image showing the distribution of clouds. Find a cold front, warm front, and stationary front, and for each one, sketch a simple cross section across the front, showing which way it is moving. After you have done this, read the rest of the text around the figure.

2. For the storm system in the Southwest, the blue line marks a cold front that is moving to the south. The red line is a warm front moving to the northeast, but a warm front slopes opposite to the way it is moving (see the cross section on the previous page), to the southwest in this case.

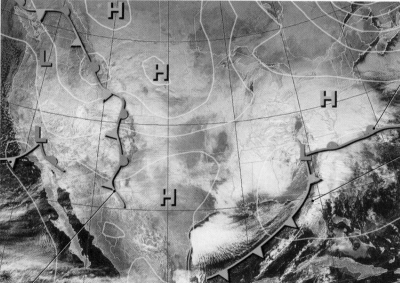
16.02.c1

4. In the East is an east-west oriented red line that marks a warm front. The semicircles point north, so this is the way the front is moving.

5. In the Southeast, the blue line is a cold front. The cold air is on the northwest side, and the front is moving to the southeast. The warm front and cold front join at an area of low pressure, marked by the red L. Symbols for four types of fronts are shown below. We discuss occluded fronts later.

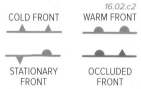
16.02.c2

COLD FRONT   WARM FRONT

STATIONARY   OCCLUDED
FRONT   FRONT

3. The red and blue line along the mountains in the west-central part of the country is a stationary front. The cold air is trying to move to the west, so it must be on the opposite side (east). Based on the cross section above, the front slopes away from the cold air.

6. This figure (▶) focuses on the low-pressure area in the East where the cold front (blue) meets the warm front (red). In this situation, the two fronts divide the region into three fairly distinct areas, called *sectors*. We reference each sector by the temperature of the air in that region relative to the two other sectors. The *cold sector* is behind the cold front and naturally has the coldest air.

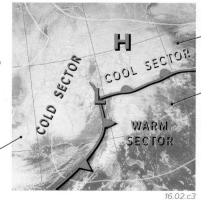
COLD SECTOR   COOL SECTOR   H

WARM SECTOR
16.02.c3

7. The region north of the warm front is intermediate in temperature and is referred to as the *cool sector*.

8. The air in front of the cold front is the warmest and so is called the *warm sector*. The counterclockwise rotation around the low is bringing air north from lower latitudes and from oceans in the warm sector. Clouds along the cold front are clumpy and probably cumuliform, but clouds in other areas are less clumpy and so are probably mostly stratiform.

Before You Leave This Page

✓ Identify cold, warm, and stationary fronts on a weather map and suggest what types of air masses could be on each side of the front.

✓ Sketch a cold, warm, and stationary front in cross section, including typical cloud types and the weather likely to be associated with each front.

16.2

16.3 Where Do Mid-Latitude Cyclones Form and Cross North America?

WEATHER FRONTS generally do not exist in isolation, but typically form, migrate, and fade away as part of a larger system called a *mid-latitude cyclone*. Cold and warm fronts usually trail from a central core of low pressure—a *cyclone*. While popular culture uses the term "cyclone" to refer to a tornado or other form of windstorm, a cyclone is really any enclosed area of low pressure. Mid-latitude cyclones migrate across Earth's surface.

A | What Surface and Upper-Level Conditions Form Mid-Latitude Cyclones?

Cyclones originate in various places as long as there are favorable conditions in the lower and upper atmosphere. Commonly, however, they develop downwind of mountain ranges, just offshore of cold land, or along bends in the polar front jet stream—Rossby waves.

Leeward Side of Mountains

As air moves downslope on the lee side of a mountain, it becomes stretched vertically as the depth of the atmosphere increases. In doing so, it acquires an increasing tendency for counterclockwise spin, or *vorticity*, wherever cold and warm air masses meet. In the U.S., westerlies cause cyclones to form east of large mountain ranges, such as the Rocky Mountains.

Offshore of Cold Land

Cyclones also form over water, offshore of cold land, especially during colder times of the year, when the coastal land has cooled more rapidly than the adjacent water. Rising air over the ocean forms an area of low pressure just offshore. This, plus winds blowing toward the pole, can lead to the formation of a cyclone along the cold-warm interface along the coast.

Polar Front Jet Stream

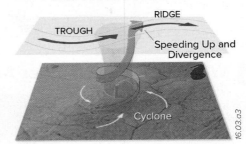

Along the polar front jet stream, a bend to the south in the Northern Hemisphere is called a *trough* and a bend to the north is called a *ridge*. As the fast-moving air passes from a trough to a ridge, it usually accelerates and spreads out, drawing underlying air upward. The resulting low pressure near the surface can develop into a cyclone.

B | Where Are Mid-Latitude Cyclones Common?

There are several common origins and paths of cyclones, influenced by the position of mountains and coastlines. Once formed, cyclones are guided by large-scale patterns of atmospheric circulation, so they follow similar paths—or *storm tracks*—across the surface. The combinations of specific sites of formation and common tracks produce storms with similar characters, paths, and associated weather, like those shown for North America below.

1. Low pressure off the west coast of North America, in the Gulf of Alaska, can spawn cyclones that bring cold, moist air from the ocean inland across Canada and the Pacific Northwest. These storms can deliver heavy rain along the coast and snow at higher elevations. Due to orographic effects, precipitation is especially heavy along west-facing slopes of coastal mountains.

2. Cyclones also form on the leeward sides of major mountain ranges, like the Canadian Rockies in British Columbia and Alberta. A storm formed here commonly incorporates cold Arctic air and is called an *Alberta Clipper*. Similar storms form east of the southern Rocky Mountains in or near Colorado, but are not as cold.

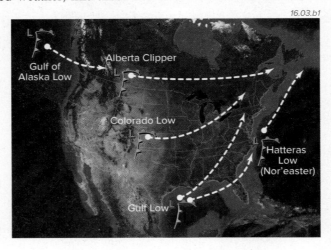

3. Wintertime lows, called *Gulf Lows*, form in the Gulf of Mexico because of the contrast between the cold land and the waters of the Gulf, which remain fairly warm.

5. Note that although the storm tracks converge near New England, when the storm systems arrive, they are often older, having spent most of their latent energy before they track to the New England area. So, New England has many cloudy days but less severe weather than places experiencing less mature storms.

4. During winter, low pressure commonly occurs over the Atlantic Ocean, often near Cape Hatteras in the Outer Banks of North Carolina. Storms formed at this general area, called *Hatteras Lows*, move northward, where they are also called *Nor'easters*. They are challenging to forecast because a slight variation in track can mean the difference between a foot or more of snow and none at all in the heavily populated areas of the Northeast.

C What Are the Three Stages in the Life Cycle of a Mid-Latitude Cyclone?

Formation

1. This map shows several fronts as decorated lines, areas of high and low pressure, and isobars as thin lines. A mid-latitude cyclone begins when a relatively cold air mass and a warmer one meet along a frontal boundary. Here, cold and warmer air masses meet along a stationary front near "Cyclone A."

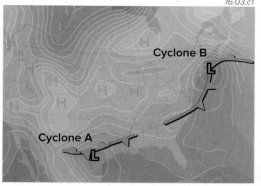

16.03.c1

2. At some point along this boundary, a small amount of surface convergence occurs due to local circulation features that push one air mass into the other because of topographic influences, or by some other local mechanism. This convergence causes the low pressure to intensify.

Maturity

3. At a later time, the low pressure has moved east and intensified, causing more surface air to converge on the low from all directions. The Coriolis effect causes the air to have an apparent deflection to the right, resulting in counterclockwise flow that pushes cold air southward, turning part of the front into an eastward-advancing cold front.

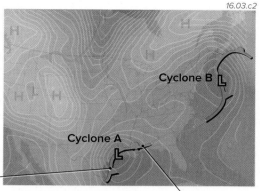

16.03.c2

4. On the east side of the low, warm air pushes northward, forming a north-moving warm front. Movement of the two air masses causes a clear bend in the front, an indication that cyclone A has matured into a mid-latitude cyclone. Cyclone B has an even more curved, more mature shape.

Occlusion

5. With time, the westerlies push both weather systems toward the east or northeast. Eventually, the cold fronts will catch up with, and even overtake, the warm fronts as both flow counterclockwise around the low-pressure areas. When this happens, the warmer air mass is lifted above the surface, because it is less dense than both the cold air and the "cool" air. This process is called *occlusion*, and a front that has experienced occlusion is an *occluded front*. In occlusion, the moisture is moved counterclockwise around the low (in the Northern Hemisphere), so the heaviest precipitation may be northwest of the low.

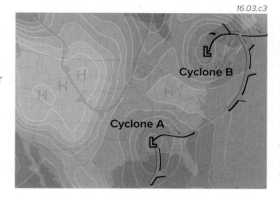

16.03.c3

6. Occluded fronts, like that near cyclone A on the map at left, are purple on weather maps with the alternating triangles and semicircles pointing in the direction that the system is moving. Though many mid-latitude cyclones never occlude, occlusion often, but not always, signals the end of the mid-latitude cyclone's life. This is because by this time, the cold air has warmed and the warm air has cooled. Some occluded fronts can continue to strengthen for some time.

7. This figure (▶) contains three cross sections that capture different stages in the process of occlusion. Occlusion begins nearest to the low-pressure core because there the cold front has the least distance to catch up to the warm front, and it proceeds away from the low over time. To visualize what occurs, move through the three cross sections, which represent different places, but also successive stages in the occlusion process.

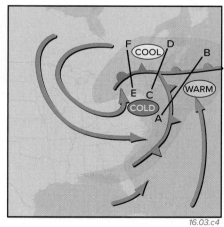

16.03.c4

OCCLUSION PROCESS

Phase 1: Cold Front Catching up to Warm Front

Phase 2: Meeting of Cold and Warm Fronts

Phase 3: Occlusion

8. In this cross section, the cold front is moving to the east (to the right) and the warm front is moving to the north, but the cold front is moving faster, so it is getting closer to the warm front.

9. Here, the cold front has caught up with the warm front, trapping the warm, moist air above the mass of cold air and the mass of cool air, which are in the process of joining together.

10. At segment EF, the two air masses are fully beneath the warm air, lifting it higher. As the warm air is forced up, it cools, forming clouds and precipitation.

Before You Leave This Page

☑ Explain how surface features and upper-level flow can form mid-latitude cyclones, including where these factors form cyclones in North America and what tracks those cyclones commonly follow.

☑ Sketch and describe the three stages in the life cycle of a mid-latitude cyclone (initial formation, maturity, and occlusion).

16.3

16.4 What Conditions Produce Thunderstorms?

THUNDERSTORMS ARE COLUMNS of moist, turbulent air with variable amounts of rain, strong wind, lightning, and hail. They are perhaps the most fundamental of all organized weather systems. They can provide needed rainfall for crops, but they can be accompanied by severe weather. We begin by examining the formation, growth, and decay of an individual thunderstorm that is independent of larger storm systems—the *single-cell thunderstorm*—and then explore the formation of thunderstorms that occur in clusters or in large, consolidated masses.

A What Stages of Development Characterize Single-Cell Thunderstorms?

Cumulus Stage

1. The initial development of a single-cell thunderstorm—the *cumulus stage*—occurs when the surface heats more rapidly than the atmosphere above it. This causes the heated near-surface
16.04.a1
air to rise relative to adjacent air, forming an *updraft*. Usually, such conditions begin to appear in mid-to-late morning, particularly in summertime. If the air is moist, one or more isolated *cumuliform* clouds will begin to form, starting out as a cumulus cloud. No rain is falling during this stage—the cloud is still growing, as indicated by the puffy, cauliflower-shaped top. Most developing cumuliform clouds never get past the first (cumulus) stage, either because the atmosphere is not sufficiently unstable, or is not moist enough, or both.

Mature Stage

2. The beginning of rainfall signals the onset of the next stage—the *mature stage*. At this stage, the updrafts that allowed the cloud to grow in the cumulus stage are accompanied by downdrafts induced by falling
16.04.a2
precipitation. Most *lightning* occurs during this stage. If the thunderstorm is too tall, its top can flatten against the tropopause, where the very stable conditions in the temperature inversion of the stratosphere forbid higher vertical development. The strong upper-level winds produce an anvil-shaped cloud top. Evaporation of some cloud water drops cools the downdrafts, particularly on hot summer afternoons. Upon hitting the surface, the downdraft will move laterally in the direction of winds that are pushing the storm, producing cool wind gusts—the *gust front* we feel before a storm.

Dissipating Stage

3. Eventually, enough rain falls to cool the surface, removing the unstable conditions that caused rising motions. This *dissipating stage* is characterized by continued rainfall until much of the
16.04.a3
moisture is rained out of the cloud. At this point, the cloud begins to disappear, taking on a ragged, feathery appearance, as it mixes with dry air from outside the cloud and evaporates into the surrounding air. Such mixing in of other air is called *entrainment*. Water in the cloud that does not precipitate may eventually evaporate into the surrounding drier air. Windstorms can form during collapse of the storm.

4. The entire life cycle of an individual single-cell thunderstorm (▶) is on the order of 1 to 2 hours, but some last much longer, and several single-cell thunderstorms can occur in sequence. Unless they occur in association with larger systems like a mid-latitude cyclone, single-cell thunderstorms are far more likely to occur in summer than in winter. They are also far more likely to occur sometime between late morning and late afternoon than during any other time of day. Longer-lived and more severe thunderstorms can occur if conditions are especially favorable, such as abundant moisture or some other atmospheric phenomenon that enhances and strengthens the thunderstorm. Such thunderstorms can cause extensive damage, as described on subsequent pages.

16.04.a4 Wind River Mtns., WY

Cloud Mergers

5. A single-cell thunderstorm can form or grow when two cumuliform clouds are forced together by winds. In this figure, adjacent fair-weather cumuliform clouds are supported by uplift associated with an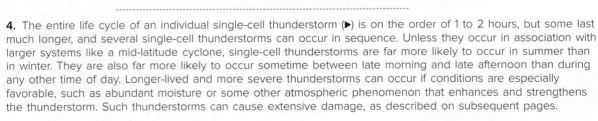
16.04.a5
16.04.a6
unstable atmosphere, but no cloud has enough moisture and uplift to be dangerous by itself. When the clouds collide, the instability and associated uplift beneath the collision point dictate that the moisture from both clouds must move upward. This causes very rapid vertical development and a sudden storm that may be difficult to anticipate.

Upper-Wind Effects

6. A thunderstorm that forms beneath strong winds aloft, such as the polar front jet stream, has an increased likelihood for strong development or intensification. Any increase in horizontal wind speed with height is called *vertical wind shear*. The fast winds in the top part of a tall thunderstorm, driven by the jet stream, may allow some of the rain from the storm to fall in areas separated from the updrafts. This reduces the possibility that the downdrafts associated with the falling precipitation will get in the way of, and cool, the updrafts that are sustaining the system. If vertical wind shear allows the precipitation to remain separated from the zone of uplift, the surface remains warm, allowing updrafts and overall atmospheric instability to continue for an extended time. This in turn allows the thunderstorm to persist and possibly grow stronger, increasing the possibility of more severe weather.

B   What Conditions Form Large Thunderstorms or Clusters of Thunderstorms?

Thunderstorms can also occur in clusters, called *multicell thunderstorms*, or can coalesce into a much larger single storm, called a *supercell thunderstorm*. Clusters of thunderstorms can form along a front or when the individual storms are sufficiently close to reinforce and strengthen one another. In either case, the thunderstorms can cause severe weather.

1. This figure (▶) shows two thunderstorms that are close enough to affect one another. In well-organized thunderstorms, downdrafts from one thunderstorm can contribute to updrafts in an adjacent thunderstorm. This happens because the downdraft of cold, dense air spreads out when it hits the surface, pushing warmer air upward, thereby strengthening a nearby thunderstorm or even triggering a new one. The downdrafts and resulting outflow can cause warm air ahead of the outflow (A) to be displaced upward to such an extent that it triggers another thunderstorm, or even a row of thunderstorms, as

16.04.b1

Enhanced Uplift at "A" Causes Cloud Growth

A

shown in the radar image to the left. These types of systems, where downdrafts coming from a line of storms produce another line of storms, proceed in a "conveyor belt" fashion and can affect huge regions. These types of storms are called *multicell thunderstorms*.

16.04.b2 Binghamton, NY

16.04.b3

cP

CF

mT

2. Lines of thunderstorms commonly form along cold fronts, especially where the two air masses have very different moisture characteristics. This often happens as hot, dry (*cT*) air masses from the West interact with warm, moist (*mT*) air masses from the Southeast (◀). This interaction can be driven by movement of a cold front and mid-latitude cyclone, but both of these air masses will be in the warm sector of the cyclone, as shown here. Such air masses meet at a typically north-south boundary called the *dry line*. Along the dry line, the more humid air mass will be lifted up and over the drier air mass, assuming that both are at the same temperature, because humid air is less dense than dry air. The resulting storms can have an ominous front (▶), a *squall line*, and so are called *squall-line thunderstorms*. They can be very severe due to the abundant moisture in the humid air mass, often even more severe than the storms generated by the subsequent passage of the cold front.

16.04.b4 Belleville, KS

3. Some single-cell thunderstorms develop into tall and powerful *supercell thunderstorms* (▶). Like other large thunderstorms, supercell thunderstorms commonly have a flat-topped, anvil shape. The long point of the anvil generally indicates the direction the storm is moving (left to right in the figure below), because the strong upper-level winds push the entire surface storm in the same direction that they push its top. The anvil indicates the presence of vertical wind shear (increasing wind speed with height), which assists storm development by enhancing uplift and allowing precipitation to fall away from the updrafts.

16.04.b5

16.04.b6 Humboldt, NE

4. Intense rain and hail form within supercells and fall toward the ground, often in brief but heavy downpours. Note that hail can form in other types of storms, too. *Lightning* and *thunder* are also present in most supercells. Lightning may discharge in the air or reach the ground.

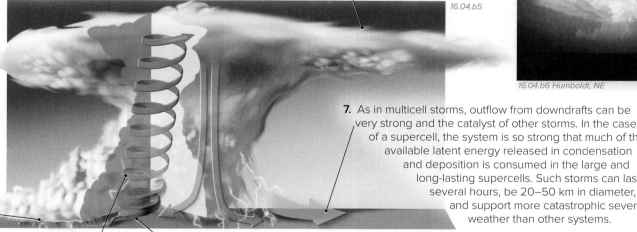

7. As in multicell storms, outflow from downdrafts can be very strong and the catalyst of other storms. In the case of a supercell, the system is so strong that much of the available latent energy released in condensation and deposition is consumed in the large and long-lasting supercells. Such storms can last several hours, be 20–50 km in diameter, and support more catastrophic severe weather than other systems.

5. Supercells feature a rotating vortex called a *mesocyclone*. All supercells have at least one mesocyclone, and some supercells may support several. The rotating vortex is interpreted to have started out horizontal, where it was formed by horizontal shearing of the wind against the surface. Later, updrafts within the thunderstorm distort the vortex into a steep orientation.

6. Subsequent *updrafts* can spiral into the mesocyclone, strengthening it by tightening its rotation and ultimately leading to formation of a tornado. About one-third of mesocyclones form tornadoes.

Before You Leave This Page

☑ Sketch and explain the processes that occur during the life cycle of a single-cell thunderstorm, and what can enhance them.

☑ Describe the formation of either multicell or supercell thunderstorms.

16.4

Where Are Thunderstorms Most Common?

THUNDERSTORMS CAN OCCUR as isolated single-cell storms, forming over mountains and other local features, or can be embedded in larger weather systems, like mid-latitude cyclones or multi-cell thunderstorms. By knowing how thunderstorms form, can you predict where they should be most common?

A How Does Thunderstorm Frequency Vary Globally?

Space-based detections of lightning can be used to determine the global distribution of thunderstorms. Optical sensors on satellites use high-speed cameras to look into the tops of clouds and detect flashes of lightning. Observe the map below, which shows the average yearly counts of lightning flashes per square kilometer based on NASA satellite data.

1. These data represent lightning observed over 8 years. Places with the highest frequency of lightning strikes are red or black. Places where fewer than one flash occurred (on average) each year are white, gray, or purple, and places with moderate numbers of flashes are green and yellow. Observe the main patterns before you read the associated text.

2. The globes show that lightning strikes are much more common on land than over the oceans. Large landmasses can heat up during the day and cause free convection, resulting in thunderstorms. Above the oceans, thunderstorms are more frequent near coastlines than in areas far from a coast; this reflects the sharp contrasts in air masses that occur along coastlines.

4. North America exhibits frequent thunderstorms, reflecting a supply of warm, moist air from the Gulf of Mexico, abundant solar heating of the land, and the frequent passages of fronts. Europe commonly lacks much contrast between air masses, as the westerlies push a continual parade of *mP* air masses across the continent. Also, the mountains mostly trend east-west, minimizing the interaction between very hot and very cold air masses. As a result, Europe has fewer thunderstorms than might be expected, considering the frequency of frontal passages and rainy days.

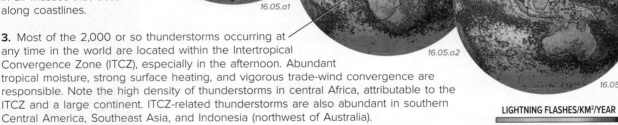

16.05.a1

16.05.a2

16.05.a3

3. Most of the 2,000 or so thunderstorms occurring at any time in the world are located within the Intertropical Convergence Zone (ITCZ), especially in the afternoon. Abundant tropical moisture, strong surface heating, and vigorous trade-wind convergence are responsible. Note the high density of thunderstorms in central Africa, attributable to the ITCZ and a large continent. ITCZ-related thunderstorms are also abundant in southern Central America, Southeast Asia, and Indonesia (northwest of Australia).

5. Polar regions, including Antarctica, are too cold to have moisture sufficient for the condensation and deposition required to support thunderstorms, even in the summer. Both poles are among the places least likely to experience thunderstorms and lightning.

LIGHTNING FLASHES/KM²/YEAR

| 0.1 | 0.4 | 1.4 | 5 | 20 | 70 |

Seasonal Variations

6. These maps show the variations in thunderstorm-associated lightning between summer and winter. The top map is for winter in the Northern Hemisphere and summer in the Southern Hemisphere. The bottom map shows the opposite seasons (northern summer; southern winter). Examine these maps and compare the number of thunderstorms in summer versus winter for different regions.

16.05.a4

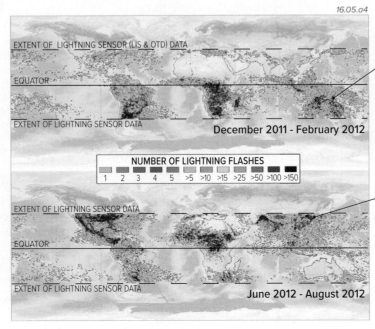

EXTENT OF LIGHTNING SENSOR (LIS & OTD) DATA

EQUATOR

EXTENT OF LIGHTNING SENSOR DATA

December 2011 - February 2012

NUMBER OF LIGHTNING FLASHES

| 1 | 2 | 3 | 4 | 5 | >5 | >10 | >15 | >25 | >50 | >100 | >150 |

EXTENT OF LIGHTNING SENSOR DATA

EQUATOR

EXTENT OF LIGHTNING SENSOR DATA

June 2012 - August 2012

7. From December through February, there are many more thunderstorms south of the equator than north of it. Note, for example, the difference in Australia and Southeast Asia between the two maps. During this time, thunderstorms are also abundant in southern Africa and in South America.

8. During June through August, the locus of thunderstorm activity shifts to the Northern Hemisphere, where it is summer. The difference in thunderstorm activity between the two maps is especially noticeable in North America and Central America. Thunderstorms in mainland Asia are tied closely to the summer monsoon, when surface heating combines with moist onshore flow to create strong convection. Central Africa, straddling the equator, reports thunderstorms during both seasons. Sinking air in the descending limb of the Hadley cell, combined with limited moisture, suppresses the development of thunderstorms in the Sahara Desert and Arabian Desert.

B How Does Thunderstorm Frequency Vary Throughout the Continental U.S.?

1. This map shows the frequency of thunderstorms in different parts of the U.S., as expressed by the number of days per year that an area has thunderstorms.

2. Few thunderstorms occur on the Pacific Coast, even though there are many rainy days. Surface heating is seldom strong enough to create free convection. Also, the difference between air masses across frontal boundaries is too slight to generate vigorous uplift. Cold air masses are not bitterly cold, and warm air masses are not too hot due to the moderating influence of the Pacific Ocean. Orographic uplift can generate an occasional thunderstorm, but even these are not too frequent as most of the coastal ranges are not very high and the highest ones are farther north, next to cold ocean waters.

3. Thunderstorms are somewhat more common elsewhere in the West, even though moisture is relatively scarce. More thunderstorms occur on the peaks of the highest mountains, where surface daytime heating moves air upslope from all sides, converging at the peaks and forcing additional uplift.

4. In the Midwest and Northeast, thunderstorms increase in frequency from north to south. Water-vapor capacities in the northern states are so low most of the year that they limit the supply of atmospheric moisture available for release as latent heat during condensation, which is essential to generate severe storms.

ANNUAL THUNDER DAYS

	<5
	5-10
	11-20
	21-30
	31-40
	41-50
	51-60
	61-70
	>71

16.05.b1

5. The Southeast has the highest frequency of thunderstorms in the U.S., with the Florida peninsula experiencing the most thunderstorms in the country. Winter and early spring thunderstorms mostly originate from cold fronts trailing from Gulf Lows and Colorado Lows. Spring, summer, and, to some extent, fall thunderstorms occur because of convective afternoon thunderstorms in an unstable atmosphere. When the surface heats up and air rises, a *double sea breeze* causes onshore flow from both the Gulf of Mexico and the Atlantic to fill in the low pressure over Florida. Late summer and fall also bring tropical cyclone–induced thunderstorms. The region has a combination of warm seawater, warm air, a sea–land contrast, and a position near storm tracks—a perfect recipe for thunderstorms.

Seasonal Variations

6. These maps show the frequency of thunderstorms for different months. In March and April, thunderstorms occur inland along the Lower Mississippi Valley because cold and warm air masses come into contact with each other at strong cold fronts, accompanied by moisture from the Gulf of Mexico. Read counterclockwise.

7. During May, the peak in Texas results from the combination of early summer convection on some days with late-season cold-front activity on others.

8. In June, the location of thunderstorm activity shifts farther north and inland, onto the Great Plains. Land surfaces are starting to warm up, but cold air is nearby in Canada and sometimes flows into the plains. Cyclonic activity enhances the likelihood of thunderstorms in the central U.S. Thunderstorm activity begins to pick up in Florida, with the arrival of summer.

MAR

SEP

APR

AUG

MAY

JUL

16.05.b2

JUN

THUNDER DAYS

<0.5		6.5-8.4	12.5-5.4
0.5-2.4		8.5-2.4	>15.5
2.5-6.4			

10. The September peak in Florida is the result of tropical cyclones (of strengths varying from tropical waves to hurricanes), enhancing the convectional thunderstorm activity. Hurricane season peaks in August and September. The rest of the southeastern U.S. has the same ingredients for thunderstorm formation as Florida, except for the double sea breeze.

9. Most of the U.S. sees a summer peak in thunderstorm days. By July, surface heating allows for destabilization of the atmosphere during afternoon hours. Summer cold fronts supplement these convectional thunderstorms in the Northeast and north-central states, while tropical cyclones can add to the frequency of thunderstorms in the Southeast. Increased thunderstorm activity in the Southwest and intermountain states of the West largely reflects the onset of the Southwest monsoon.

Before You Leave This Page

✓ Summarize variations in thunderstorm frequency around the world, explaining changes with seasons and from land to oceans.

✓ Describe which parts of the U.S. experience frequent thunderstorms, explaining the factors influencing the frequency and seasonality.

16.5

16.6 What Causes Hail?

HAIL IS A BALL OF ICE that forms under freezing temperatures within a cumulonimbus cloud and that subsequently falls toward the surface. Large hail that reaches the surface can do so at high speeds and with enough force to smash windows, dent cars, and destroy entire fields of crops. How does hail form, what controls the size of individual chunks of hail, and where and when is hail most common?

A How Does Hail Form?

Most of the largest hail forms in association with a supercell thunderstorm, but hail can also occur with other types of thunderstorms. The photograph to the right shows pea-sized pieces of hail, or *hailstones*, and the figure below illustrates the interesting way in which hail forms. Examine the figure and then begin reading clockwise from the lower left as we follow a single hailstone up and down through the storm. Hail, lightning, and thunder are commonly associated with one another, so the thunderstorms that cause these phenomena are commonly called *thunderheads*.

16.06.a2 South Mtns., AZ

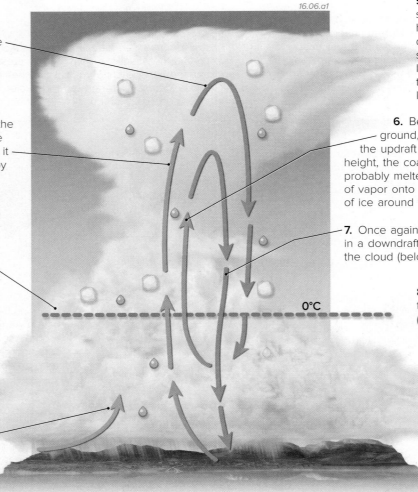
16.06.a1

4. Eventually the hailstone becomes influenced by a nearby downdraft, shown here as the beginning of a downward-arcing flow path.

3. As the central part of the storm continues to rise, more vapor deposition occurs and the hailstone grows. The hailstone will continue rising as long as it is not too heavy to be lifted by the updraft.

2. Once air rises high enough that it enters the subfreezing part of the cumulonimbus cloud (above the red line on this figure), the Bergeron process causes water vapor (not shown) to deposit onto an ice particle, causing it to grow. The resulting hail begins as a small, more-or-less-spherical object.

1. Supercell thunderstorms and other strong cumulonimbus clouds have strong updrafts capable of lifting raindrops, chunks of ice, and even airplanes unfortunate enough to fly into the storm. Such updrafts are associated with the mature stage of a thunderstorm, when air rises against gravity because of a very unstable atmosphere.

0°C

5. If the downdraft causes the hailstone to be carried below the freezing height (the red line), the outer surface of the ice can partially melt. At this stage, a core of ice is coated with a layer of liquid water. While temperatures remain above freezing, more liquid water can coat the hailstone.

6. Before the hailstone can fall to the ground, it may get into another updraft. After the updraft lifts the hailstone above the freezing height, the coating of liquid water, most of which was probably melted ice, can refreeze. More deposition of vapor onto ice occurs, accumulating another ring of ice around the hailstone.

7. Once again, the developing hailstone gets caught in a downdraft. After falling to the warmer parts of the cloud (below the red line), more melting occurs.

8. This cycle can repeat many times—freezing and deposition of ice (via the Bergeron process) onto the hailstone above the freezing level, alternating with melting of an outside layer of ice when the hailstone is below the freezing level. Each trip up and down results in another layer of ice added to the hailstone. Eventually the developing hailstone becomes too heavy to be lifted by the updraft, and gravity pulls it to the surface. The more vigorous the updraft and the greater the instability, the larger the hailstone will be before it is finally pulled to the ground by gravity. The larger the hailstone and the faster it hits the ground, the more force it has and the more damage it can cause.

9. A number of factors influence the process of hail formation. For example, the freezing level is higher in summer than in winter. Hail formation is most likely at the time of year when the cloud extends substantially both above and below freezing. If a cloud is entirely warmer than freezing (which is unlikely), it will only produce rain. If the entire cloud is below freezing temperatures, it will produce snow. Other important factors include the amount of moisture, vertical extent of the cloud, the strength of the updrafts and downdrafts, and the temperature contrast between air masses across a front.

B Where and When Does Hail Tend to Occur?

Hail is associated with thunderstorms, so its distribution across the globe and in the U.S. is similar to the distribution of thunderstorms. As in the case of most other forms of severe mid-latitude weather, the U.S. experiences more than its share of hail compared to other locations on Earth. These maps show the geography and seasonality of hail in the contiguous U.S. Observe the general patterns shown on the large map below, then examine the patterns for a certain month in the series of smaller maps below. Try to explain the main patterns before reading the associated text.

1. This map depicts the average number of days per year when a county experiences hail. From the legend, we can see that some areas average less than one day of hail per year, whereas others average more than 10.

2. Frequencies are low on the Pacific Coast, where contrasts between air masses are typically weak. Frequencies are also low in most of the interior of the West, where the air is usually dry.

3. Frequencies are slightly higher in a north-south band through Arizona, which experiences thunderstorms during a summer monsoon.

4. Hail occurs most frequently over the Great Plains, in the center of the map, in a wide, north-south band stretching from central Texas to the Canadian border (and beyond). The average size of hailstones is larger over the Great Plains than anywhere else in the U.S., and perhaps anywhere else in the world. Most damage from hail is from larger hailstones (e.g., >5 cm), and the Great Plains often experiences such hail sizes.

5. Frequencies in the eastern U.S. are intermediate and spotty in distribution. The size of hailstones generally becomes smaller toward the coast.

6. Florida and areas along the Gulf Coast have frequent thunderstorms, but low to intermediate frequencies of hail. Here, thunderstorms are seldom tall enough and well-organized enough to form hail. Furthermore, temperature contrasts between air masses are usually not as great as over the Great Plains.

MEAN ANNUAL NUMBER
OF HAIL DAYS
(60 Year Average)

1–2
2–4
4–7
7–15

16.06.b1

Seasonal Variations

7. The four maps below show hail frequencies for four different months, each representative of a season (winter, spring, summer, and fall, from left to right). How does hail vary from season to season?

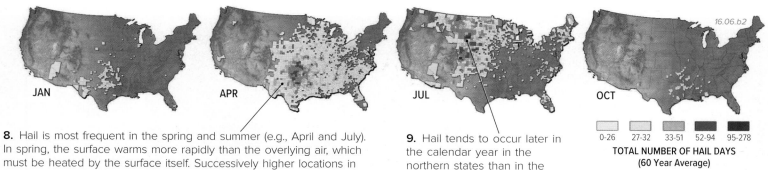

JAN   APR   JUL   OCT

16.06.b2

0-26   27-32   33-51   52-94   95-278

TOTAL NUMBER OF HAIL DAYS
(60 Year Average)

8. Hail is most frequent in the spring and summer (e.g., April and July). In spring, the surface warms more rapidly than the overlying air, which must be heated by the surface itself. Successively higher locations in the atmosphere require longer to heat. This sets up a steep environmental lapse rate, in which the surface can be quite warm while the atmosphere is still cold from winter. This destabilizes the atmosphere, setting the stage for the cumulonimbus clouds that are required for hail to form. In fall, the pattern is reversed as the surface cools more rapidly than the air above it, and the atmosphere is seldom so unstable.

9. Hail tends to occur later in the calendar year in the northern states than in the southern states. This is no surprise, because the boundary between the cold and warm air masses migrates northward as summer approaches. In fall (October) and winter (January), the surface is simply too cold in most of the U.S. to support the atmospheric instability needed for frequent cumulonimbus clouds and intense uplift.

Time-of-Day Variations

10. This graph plots average frequency of hail versus time of day for five cities. What do you observe? All of these cities show a strong tendency for hail in the afternoon, when the ground has warmed enough to destabilize the atmosphere. These patterns are very similar to those for severe thunderstorms.

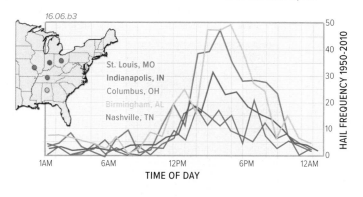

16.06.b3

St. Louis, MO
Indianapolis, IN
Columbus, OH
Birmingham, AL
Nashville, TN

HAIL FREQUENCY 1950-2010

1AM   6AM   12PM   6PM   12AM

TIME OF DAY

Before You Leave This Page

☑ Sketch and explain the formation of hail within a cumulonimbus cloud, identifying all essential components.

☑ Explain the factors contributing to the geography, seasonality, and time-of-day variations for hailstorms in the continental U.S.

16.6

16.7 What Causes Lightning and Thunder?

LIGHTNING IS A DANGEROUS but intriguing weather phenomenon. It results from electrical currents within a storm cloud, releasing tremendous amounts of energy that can strike the ground. In the U.S., more than 80 deaths per year are due to lightning. Although this number is much less than those killed simply by extreme heat and extreme cold, lightning is responsible for more deaths than any other type of severe storm in the U.S.

A How Is Lightning Generated?

Lightning is a sudden and large discharge of electrical energy generated naturally within the atmosphere, typically expressed as a bolt of electricity that flashes briefly and then is gone. Most lightning occurs as *cloud-to-cloud* lightning or occurs within a single cloud, but about 25% of lightning strikes the ground (called *cloud-to-ground lightning*). Some lightning is less discrete, appearing to form diffuse sheets, called *sheet lightning*.

16.07.a1

16.07.a2 west-central IL

1. Most lightning occurs as a single discrete bolt that can be relatively straight or extremely irregular with many obvious branches (◄). Lightning is associated with cumulonimbus clouds, the kind that form thunderstorms (►). Ice particles in the upper part of such clouds have a positive electrical charge, but the charge is mostly negative in the lower cloud where there is no ice. This charge separation is believed to happen because of the Bergeron process and different physical properties of ice particles and liquid water, but the exact process is unknown. The charge imbalance between the top and bottom of the cloud sets in motion a specific sequence of events, depicted in the series of figures below.

2. As a result of this charge imbalance, there is an excess of positive charges in the top of the cloud relative to the bottom of the cloud. Usually the Earth has a negative charge too, but as the cloud passes over a place, the negatively charged cloud base induces a positive charge directly below, and for several kilometers around, the cloud.

3. Recall that conduction requires energy transfer from one molecule to another, and therefore is most efficient in metals and other conductive materials. Air is a poor electrical conductor, so the flow of electricity through air cannot occur until the electrical potential becomes great enough to overcome the insulating effect of air.

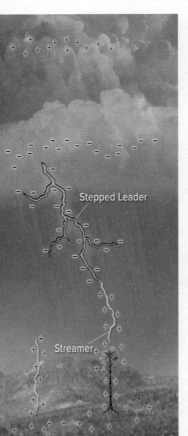

4. Lightning starts with negative charges near the base of the cloud. The negative charges begin to form an invisible channel of ionized molecules through the air, from the base of the cloud toward the positively charged ground. As the channel develops, it separates into a series of branches, all still invisible, and repeatedly starts and stops, advancing in a zigzag path resembling a staircase. For this reason, it is called a *stepped leader*.

5. As the negatively charged stepped leader approaches the ground, positively charged streams of energy, called *streamers*, move upward to join the leader, commonly reaching up from the highest object on the ground, such as a tree, pole, building, or other tall object.

6. When the streamer makes contact with the stepped leader, it establishes a pathway of ionized molecules through which electricity can flow between the cloud and the ground.

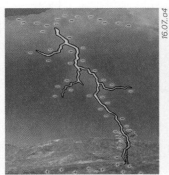
16.07.a4

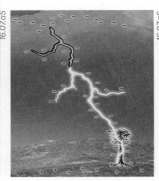
16.07.a5

16.07.a6

7. Once the streamer and stepped leader meet, an electrical connection is established and electrons flow from the base of the cloud to the ground, but this flow is still invisible.

8. As the negative charge approaches the ground, the positive charge begins to flow upward in what is called a *return stroke*. The return stroke begins near the surface.

9. The return stroke is nearly instantaneous, lighting up the sky with the discharge of energy. This is the main, bright flash we observe and recognize as lightning.

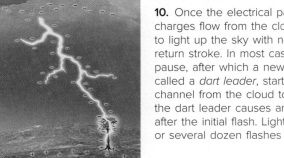

10. Once the electrical pathway is established, negative charges flow from the cloud to the ground, continuing to light up the sky with no obvious interruption after the return stroke. In most cases, there is then a short pause, after which a new negatively charged leader, called a *dart leader*, starts down the established channel from the cloud to the ground. The discharge of the dart leader causes another flash of light, distinctly after the initial flash. Lightning strikes can have several or several dozen flashes from dart leaders.

16.07.a3

16.07.a7

Thunder

11. *Thunder* results from rapid heating and expansion of the air along the path of the lightning bolt (◄). Around a lightning discharge, the air temperature approaches 15,000–30,000°C, which is much hotter than the surface of the Sun. The extreme heat causes this air to expand quickly away from the lightning stroke, as shown by the red arrows on this figure.

12. The expansion creates a momentary vacuum around the stroke. Almost instantly after the expansion, air rushes in from all sides toward the location where the lightning stroke passed. This is depicted by the green arrows. The sound of the matter in the various air streams colliding is what makes the sound of thunder.

16.07.a8

Using Thunder to Estimate Distances to Storms

13. Lightning travels at the speed of light. At about 300,000 km/s (186,000 mi/s), it can be considered instantaneous. But thunder only travels at the speed of sound (1,225 km/hr, or 760 mi/hr). Since there are 3,600 seconds in an hour, this means that the sound of thunder travels about 1.7 km or 1 mile in 5 seconds. So if you start counting when you see lightning, and can count to five before you hear the thunder, you are one mile away from the lightning. But beware: lightning does not necessarily flow straight down the middle of a storm. The strike could be farther ahead or behind the cloud from which it originated.

Elves, Sprites, and Jets

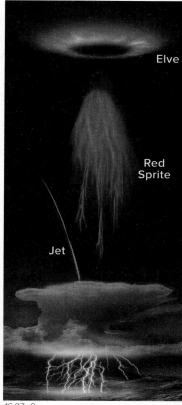

14. Intense lightning can cause a flat disk of dim reddish light, called an *elve*, to form about 96 km above the Earth. The light of an elve radiates outward in every direction, spreading over a huge area of sky.

15. Immediately after a very energetic bolt of lightning strikes the ground, ghostly red lights, called *red sprites*, may shoot straight up from the top of a thunderhead. Some red sprites soar up to 100 km into the atmosphere.

16. *Blue jets* are dim, blue streaks of light. They look like quick puffs of smoke that burst out of a thunderhead, arc upward, and then fade away. Blue jets can climb as high as 30 km into the atmosphere.

16.07.a9

Safety Tips Around Lightning

Lightning is extremely dangerous. The map below shows the number of lightning-related fatalities for each state from 1959 to 2014. How dangerous are states with which you are familiar? What factors contribute to this spatial pattern? How at risk are you for lightning?

Conduction of electricity is easiest in solids and most difficult in gases. Therefore, direct contact with any solid object, particularly a tall one that conducts electricity well (such as a tall metal post), is extremely dangerous in a thunderstorm. Tall objects minimize the distance that the electrical energy needs to travel through a gas (air) during lightning.

A house or other enclosed building offers the best protection from lightning. Open shelters on athletic fields, golf courses, and picnic areas provide little or no protection from lightning. When inside a building, stay away from windows and doors, and avoid contact with plumbing, corded phones, and electrical equipment. Basements are generally safe havens, but avoid contact with concrete walls that may contain metal reinforcing bars. A car is a relatively safe place, because the metal body and frame carry the electrical current if struck.

If no safe shelter is available, stay away from trees and other tall objects, and crouch down with your weight on your toes and your feet close together. Lower your head and get as low as possible without touching your hands or knees to the ground. Do not lie down!

Consider the 30-30 rule: if the time between seeing a flash and hearing the thunder is 30 seconds or less, the lightning is close enough to hit you, so seek shelter immediately. After the last flash of lightning, wait 30 minutes before leaving a shelter.

16.07..t2

WHEN THUNDER ROARS GO INDOORS!

NUMBER OF LIGHTNING FATALITIES BY STATE, 1959–2014

Alaska - 0
D.C. - 5
Hawaii - 0
Puerto Rico - 33

16.07.t1

STATES RANKED BY FATALITY RATE PER MILLION PEOPLE

| | 1–10 | | 11–20 | | 21–30 | | 31–52 |

Before You Leave This Page

✓ Sketch and explain how and why lightning and thunder form, and describe the factors that set the stage for their formation.

✓ Summarize some strategies for remaining safe from lightning and the reason for using each strategy.

16.7

16.8 What Is a Tornado?

A TORNADO IS THE MOST INTENSE type of storm known, expressed as a column of whirling winds racing around a vortex at speeds up to hundreds of kilometers per hour. The May 1999 tornadoes in Oklahoma City, Oklahoma, were estimated to have wind speeds over 500 km/hr. Fortunately, most tornadoes are less than several hundred meters wide, so they usually only damage narrow areas, but larger tornadoes can remain on the ground for hours and create a path of near-total destruction across one or more states.

A What Is a Tornado and How Does One Form?

A *tornado* is a naturally occurring, rapidly spinning vortex of air and debris that extends from the base of a cumulonimbus cloud to the ground. Tornadoes typically have diameters of tens to hundreds of meters, but the largest ones can be 2–3 km wide. Most individual tornadoes remain on the ground for minutes to tens of minutes and have paths on the ground of less than a few kilometers. Some very rare, long-lived tornadoes track over 100 km, last a few hours, and destroy everything in their direct path.

16.08.a1

16.08.a2

Formation of a Tornado

2. The formation of a tornado is complicated and not fully understood. Different sets of circumstances are present in different tornadoes, but tornado formation generally begins with horizontal winds moving across the surface. Wind speeds are usually higher aloft and decrease downward due to friction with the surface. This difference in wind speeds causes a *vertical wind shear* within the wind column. Wind shear can result in a rotating, subhorizontal vortex (▲). Vertical wind shear commonly generates this type of rotation, but by itself does not lead to severe weather.

1. This photograph (▲) shows the characteristics of a well-developed tornado, with the classic shape of a downward-tapering funnel. The tornado is visible because of the condensation of moisture as water drops within the vortex. A brown cloud of debris picked up from the ground typically surrounds the base of the tornado, extending out farther than the vortex of strongest winds. As in the movies, if you are within the debris field, you are in trouble. Some tornadoes have wider and less elegant shapes, like a stubby, thick pillar.

16.08.a3

3. In an unstable atmosphere, rising air (an updraft) can perturb the rotating tube of turbulent air (◄). With enough lifting (►), the tube will become more vertically oriented. As the tube is lifted, it generally splits into several short segments, some of which may even be rotating in opposite directions.

16.08.a4

4. Eventually, a segment of mechanical turbulence may become vertical, at which point it merges with the original and unstable updraft. Once it spans the distance from the cloud to the ground, it is a *tornado*. If a funnel-shaped cloud does not reach the ground, it is a *funnel cloud*. In the Northern Hemisphere, the parent storm system often rotates counterclockwise, so the tornado also spins counterclockwise, except in a few rare cases. As a tornado tightens its rotation, its wind speeds increase, making it a stronger, more powerful, and potentially more dangerous storm.

5. This figure (►) shows the essential aspects needed to form a tornado. The main ingredient is a well-developed cumulonimbus storm cloud, especially one associated with a supercell thunderstorm. A favorable regional setting is another prerequisite, such as a mid-latitude cyclone that puts cold air adjacent to much warmer, moister air. The position and motion of the cyclone are in turn controlled by such factors as the position of the polar front jet stream.

6. Another necessary ingredient is vertical wind shear near the surface, caused by higher speed winds aloft and slower winds near the surface. Mountains and other topographic irregularities tend to disrupt a regular wind-shear pattern, which is one reason why tornadoes are more common in areas that have relatively low relief (little topography) than in areas with mountains.

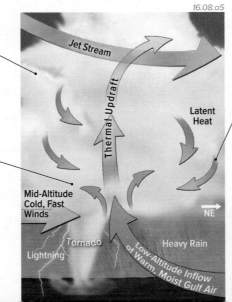
16.08.a5

7. Formation of a tornado—and its host thunderstorm—requires an unstable atmosphere. Without it, there will not be enough lifting to form the cloud and to realign a vortex into a vertical orientation.

8. A storm must have strong updrafts in order to lift the horizontally spinning tube of air. The host storm must also possess strong internal rotation. Such rotation, which can be observed on weather radar, is one aspect weather forecasters use to gauge whether or not a severe thunderstorm is likely to spawn tornadoes. Storms that spawn tornadoes commonly have a *hook echo* pattern on radar (▼).

16.08.a6 Lacombe, LA

B What Weather Systems Produce Tornadoes?

Tornadoes are related to strong thunderstorms, which form in a number of settings. Most tornadoes occur in association with multicell or large supercell thunderstorms, which are generally associated with cold fronts. Others occur embedded in tropical cyclones or with more isolated single-cell thunderstorms.

Associated with Cold Fronts

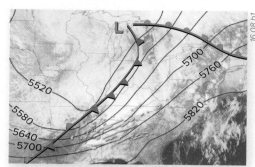

16.08.b1

Most catastrophic tornadoes are associated with multicell or supercell thunderstorms of the Great Plains and Southeast, usually in the warm sector of cold fronts. On the map above, a cold front (in blue) stretches south from a cyclone (marked by an "L") over the Great Lakes. In front of the cold front are yellow lines showing the tracks of tornadoes. These tornadoes formed in association with squall-line thunderstorms along the dry line. Such tornadoes would more likely occur in the afternoon and early evening when thunderstorms are most active.

Tropical Cyclones

16.08.b2 Hurricane Isaac, 2012

Tropical cyclones, like hurricanes, spawn tornadoes too, but a large percentage of these are not terribly strong because the upper-air pattern favored for tornado development is obliterated by tropical cyclone circulation. In the Northern Hemisphere, tornadoes are most common in the northeastern quadrant of a storm, such as embedded in the thicker swirls of clouds in the northeastern (upper right) part of Hurricane Isaac (▲). Tornadoes closer to the center of the tropical cyclone may occur at any time of day and are generally weaker than those farther from the center, which mostly occur in the afternoon.

Single-Cell Thunderstorms

16.08.b3

Single-cell thunderstorms can spawn tornadoes, but the tornadoes are generally small and weak, in part because the storm system is itself typically not very large or powerful. Tornadoes formed in this setting are generally restricted to the afternoon, when surface heating destabilizes the atmosphere and causes a peak in thunderstorm activity. Tornadoes formed in any setting can have a distinctive, sharp, low-forming cloud, called a *wall cloud* (▲), indicating that a tornado may be near. Storm chasers (people who purposely try to get close to a storm to observe and photograph it), are on the lookout for wall clouds.

C How Are Tornadoes Classified and What Type of Damage Do They Cause?

Tornadoes are classified using the *Enhanced Fujita scale* (EF-Scale), which is based on estimated wind speed as evidenced by damage, as shown in the table below. The amount of damage a tornado does is controlled by its strength (i.e., its EF rating), how long it stays on the ground, and what type of features are in the way (e.g., a city versus farmland).

16.08.c1 Colorado

This tornado is an awe-inspiring sight, but such thinly tapered tornadoes are typically not the strongest. The blue sky in the background suggests this may be related to a single-cell thunderstorm. It is passing over agricultural fields, so it may cause relatively limited damage, except to crops.

16.08.c2

This photograph illustrates the type of near-total destruction that tornadoes can cause. Wood-frame houses can be completely disassembled by strong winds within a tornado, and vehicles, trailers, and other objects that are not firmly attached to the earth can be picked up and destroyed.

Enhanced Fujita Scale			
Strength	3-Second Wind Gust		Examples of Damage
	(km/hr)	(mi/hr)	
EF0	105–137	65–85	Branches/small trees fall
EF1	138–178	86–110	Mobile homes overturned
EF2	179–218	111–135	Roofs blown off houses
EF3	219–266	136–165	Walls blown off houses
EF4	267–322	166–200	Strong frame homes leveled
EF5	>322	200+	Everything obliterated

Before You Leave This Page

✓ Sketch and explain what a tornado is and how one forms.

✓ Describe the weather systems that produce tornadoes.

✓ Summarize the Enhanced Fujita classification of tornado damage.

16.8

16.9 Where and When Do Tornadoes Strike?

THE INTERIOR OF NORTH AMERICA has by far the most frequent tornado activity in the world. Tornadoes also occur, but far less frequently, in Europe, in the southeastern parts of South America and South Africa, and in a few other regions. Large parts of some continents, like all of North Africa and northern Asia, do not report any tornadoes. Where and when should we expect tornadoes in the U.S., and what causes the huge disparity between the central U.S. and most of the rest of the world?

A How Does Tornado Frequency Vary Across the Continental U.S.?

The continental U.S. is the site of numerous tornadoes each year, especially in a north-south belt through the middle of the country, mostly corresponding to the Great Plains. Most of these tornadoes occur in the warm sector of a mid-latitude cyclone, ahead of an associated cold front that has great thermal contrasts between the warm and cold sectors. Such situations are most common in the mid-latitudes.

1. Examine this map, which shows the frequency of tornadoes by county across the continental U.S. From this map, three areas experience the highest incidence of tornadoes: one in the center of the country, one along the Gulf Coast, and one in Florida. The patterns reflect the typical types of fronts and other weather activity that occur in the different regions. The geography of North America also contributes to the high frequencies. The major mountain ranges are oriented from north to south rather than east-west, which allows warm air and cold air to meet in the center of the continent.

16.09.a1

TOTAL NUMBER OF
REPORTED TORNADOES
FROM 1950-2012

Legend:
0
1–30
31–60
61–120
121–180
181–220
221–252

3. Florida experiences frequent tornadoes, many forming in association with single-cell thunderstorms or with tropical cyclones. These are mostly weak tornadoes and are overall not as deadly. For tornadoes related to tropical cyclones, people have already evacuated from less sturdy structures because of earlier warnings about the approaching larger cyclone.

2. The region of most tornadoes is along a north-south belt, running from north Texas northward to the Dakotas and into Canada—this is justly called *Tornado Alley*. Many of these tornadoes are strong, but many storms lack enough moisture to produce violent tornadoes.

4. Another concentration of tornadoes appears along the Gulf Coast and adjacent inland areas—a region called *Dixie Alley*. Dixie Alley tornadoes are often more deadly than their Tornado Alley cousins, in part because they are often hidden in thunderstorms amid abundant Gulf of Mexico moisture.

Seasonal Variations

5. The maps below show frequencies of tornadoes for four different months, each representing a season, from 1950–2010. Compare the frequencies in different seasons and try to explain regional patterns.

6. In winter, tornadoes are rare except in Dixie Alley and Florida—areas near relatively warm waters of the Gulf of Mexico. Boundaries between warm and cold air masses are commonly located in this part of the Southeast, more so than anywhere else at this time of year.

7. In spring, the land starts to warm, but warming of the atmosphere lags behind, resulting in an unstable atmosphere. The polar front jet stream tends to be strong and displaced southward at this time, which gives upper-level support for tornadoes in the form of wind shear. Tornadoes become more common.

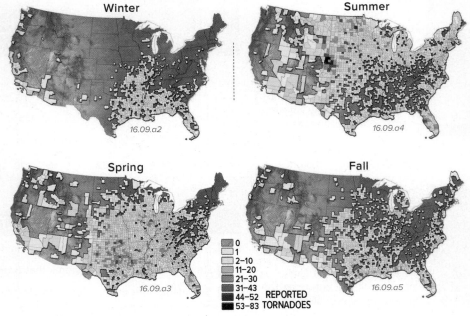

Winter
16.09.a2

Summer
16.09.a4

Spring
16.09.a3

Fall
16.09.a5

Legend:
0
1
2–10
11–20
21–30
31–43
44–52
53–83
REPORTED TORNADOES

8. Warming continues in the summer, and the boundary between warm and cold air migrates northward. Tornado activity follows into the northern Great Plains and areas to the east. In addition to increased surface heating and convection, moisture flows in from the south.

9. The frequency of tornadoes decreases in fall because lingering summer warmth in the atmosphere remains while the land begins to cool quickly. This stabilizes the atmosphere. Tropical cyclones as a source of tornadoes remain a threat along the Gulf Coast and Florida.

B How Does Tornado Frequency Vary Globally?

Tornadoes occur in other parts of the world, but are less frequent than in the tornado-prone parts of the U.S. Examine the globes below, where each dot represents an area where tornadoes occur, 'and think how it compares to the previous ones displaying the global distribution of thunderstorms, as expressed by lightning measurements.

1. The U.S. clearly experiences the highest frequency of tornadoes. Tornadoes mostly occur in the Great Plains and the Southeast.

2. Tornadoes also occur in Hawaii (not shown but close to the western edge of the globe) and also in the prairies of southern Canada and near the Great Lakes.

3. Unlike North America, Europe has large, east-west-oriented mountain ranges that inhibit the meeting of very warm and very cold air. Therefore, tornado frequencies are lower in Europe than in North America. Nevertheless, a mid-latitude location, a high population density, and sophisticated detection abilities ensure that most tornadoes that form are sighted.

4. Central and northern Asia report few tornadoes, but part of this may be due to under-reporting in the vast, less populated areas. Tornadoes occur in the Indian subcontinent, where strong storms accompany the inland flow of marine moisture during the monsoon.

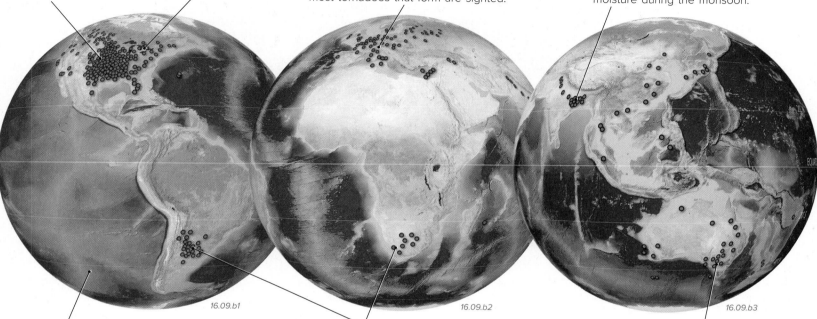

16.09.b1   16.09.b2   16.09.b3

5. Few tornadoes are known to occur over open seas because surface heating is far less intense over oceans than over land. The apparent lack of tornadoes is partly due to a lack of radar coverage over the oceans to detect them. Regions covered by sophisticated detection systems, such as the U.S. and Canada, tend to report more tornadoes.

6. The lowlands of mid-latitude South America and Africa also have some tornadoes. Recall the maps showing lightning-indicated thunderstorms here.

7. The high frequency of tornadoes in south Asia and parts of Australia is largely attributable to their proximity to a supply of warm, moist air. These areas also showed a high density of activity on the lightning-strike map.

Time-of-Day Variations

16.09.b4

168 Tornadoes in the Area of St. Louis, Missouri

NUMBER OF TORNADOES (y-axis: 5, 10, 15, 20, 25)
TIME OF DAY (x-axis: 1AM, 6AM, 12PM, 6PM, 12AM)

16.09.b5 Salina, KS 2012 (EF4)

8. This graph shows the average frequency of tornadoes versus time of day, using 60 years of data for St. Louis, Missouri. Here, tornadoes typically occur in the afternoon but continue into the evening and night. These nocturnal tornadoes are often the most deadly, for two reasons. First, people are asleep or otherwise less likely to be alert for tornado watches and warnings. Second, nocturnal tornadoes are proportionately more likely to occur in association with storm systems that are well organized, long-lived, and strong, such as multicell storms. Other locations in Tornado Alley, Dixie Alley, and Florida display similar temporal patterns.

9. The tendency for tornadoes to strike in the afternoon and early evening partly accounts for the relatively dark and foreboding aspect (◀) of most photographs of tornadoes—the Sun is low in the sky or has started to set. The thick, dark clouds help too, especially when the tornado occurs in multicell or supercell thunderstorms.

Before You Leave This Page

☑ Sketch or summarize the frequency and seasonality of tornadoes across the U.S., identifying the main factors that explain the observed patterns.

☑ Briefly describe the factors contributing to variations in tornado frequency around the world.

☑ Draw a simple graph of tornado frequency by time of day, and explain the reasons for the pattern.

16.9

16.10 What Are Some Other Types of Wind Storms?

SEVERE WEATHER INCLUDES VARIOUS WIND STORMS in addition to tornadoes and cyclones. Some of these other storms—with interesting names like waterspouts, microbursts, haboobs, and derechos—can be extremely dangerous. What are their characteristics, how do they form, and where do they occur?

Waterspouts

1. A *waterspout* is a rotating, columnar vortex of air and small water droplets that is over a body of water, usually the ocean or a large lake. There are two kinds: tornadic waterspouts and fair-weather waterspouts. Waterspouts are usually less intense than tornadoes, with smaller funnel diameters, weaker wind speeds, shorter lives, and shorter path lengths.

2. Some waterspouts resemble tornadoes (▶), and some are actual tornadoes that either formed over the water or formed over land and later moved over the water. These large, storm-related tornadic waterspouts form in the same way tornadoes on land form—strong updrafts associated with a powerful thunderstorm, such as a supercell thunderstorm. Such waterspouts, like any tornado, are dangerous, but are generally not as powerful because the cooler surface temperatures over water limit the atmospheric instability and the amount of lifting. This type of tornadic waterspout can be recognized by its association with a thunderstorm, and is much larger and stronger than a fair-weather waterspout.

16.10.a1   16.10.a2

3. The other type of waterspout is associated with normal cumulus clouds, and not typically with thunderstorms. Since many of these form when it is not raining, they are called fair-weather waterspouts. A fair-weather waterspout is a rotating column of air that starts from wind shear near the water that gets lifted upward by mild updrafts that are also forming the overlying cumulus cloud. The water in the spout was not picked up from the underlying water, but is condensation produced as the air is rising and cooling. This type of waterspout is most common in the tropics, such as in the Florida Keys, but it can also form in mid-latitudes, like on the Great Lakes.

Derechos

4. A *derecho*, whose name is derived from the Spanish word for "straight," is a regional wind storm characterized by strong, nonrotating winds. The winds can advance across huge areas and cross several states in a matter of hours. Wind velocities can be hurricane-strength and damage trees, power lines, and other structures.

5. This map (▶) shows a series of radar images (strongest echoes in red) capturing the progression of a single derecho that started near the Great Lakes (in the upper left) and moved east with time. As the derecho moved, it spread, forming a radar echo in the shape of a bow (like an archer's bow). Such a *bow echo* is characteristic of a derecho. Note how rapidly this derecho traveled, moving from Lake Michigan to the Atlantic Ocean in about 10 hours, at an average speed of about 150 km/hr.

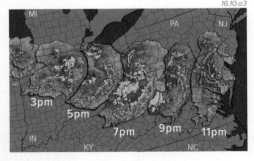

16.10.a3

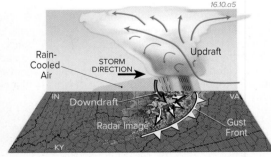

16.10.a5

7. Derechos are caused by downdrafts (red arrow in this figure) from a band of thunderstorms in a multicell storm, where a downdraft from one thunderstorm forms or strengthens an adjacent thunderstorm. Such downdrafts can be sustained for a long time and over long distances, forming a derecho. Embedded within a derecho may be other types of wind storms, including tornadoes, squall lines, and microbursts.

6. On this map (▶), contours depict the number of derechos experienced in different areas. The Ohio Valley, Upper Midwest, and Great Plains of the U.S. may experience more derechos than any place in the world. Summer is the most common season for derechos. They are long lived and can occur at any time of day or night.

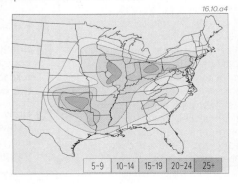

16.10.a4

| 5–9 | 10–14 | 15–19 | 20–24 | 25+ |

16.10.a6 Oklahoma

8. Derechos can be accompanied by a very distinctive and ominous type of cloud called a *shelf cloud*, a spectacular example of which is shown in this photograph (◀). Similar-appearing clouds can form in other ways, such as near tornadoes and dry lines.

Microbursts and Other Powerful Downdrafts

9. *Microbursts* are brief but strong, downward-moving winds (downbursts) that can flatten trees or structures. Unlike tornadoes, microbursts produce winds that move in a straight line, rather than rotating, so sometimes people refer to microbursts as *straight-line winds*. A microburst can be classified as being wet or dry, depending on whether it is embedded within a thunderstorm. Microbursts can be more severe than some tornadoes, and although only lasting a few seconds, are a major aviation hazard.

16.10.a7 Austin, TX

10. This photograph (◄) shows a wet microburst within a thunderstorm. It is expressed as a downward burst of moist air that, in this case, is carrying an increased amount of rain. Note how the rain is being spread out in all directions as it nears the ground. Microbursts generally affect a relatively small area, but they can extend over areas as large as several counties. In either case, their duration is short (measured in seconds or minutes).

11. Dry microbursts are downdrafts, generally not associated with rain. Under these conditions, they may pick up dust and loose materials, moving them as a small dust storm, or *haboob,* described below. Both wet and dry microbursts can be strong enough to cause significant damage, with or without dust.

12. Under some circumstances, a diffuse column of raindrops extends down from a cloud but evaporates before the drops reach the ground. This phenomenon, called *virga,* can occur for a variety of reasons, including downdrafts. Where associated with downdrafts, the descending air can compress and heat up as it is pushed downward, helping evaporate the water drops before they reach the ground.

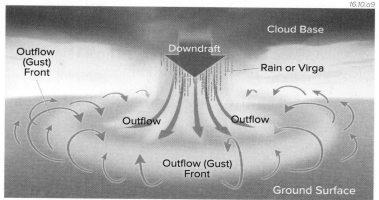
16.10.a9

13. The figure above illustrates the setting in which most microbursts occur. Late in the life cycle of a thunderstorm, the lifting power of the convecting air begins to be overwhelmed by the downward force of precipitation and downdrafts. These cool the lower parts of the cloud, the air below, and the surface, limiting the heat available for additional convection. At this point, the thunderstorm can start to collapse (the stage shown in the figure above). This can cause a very strong downdraft, which upon reaching the surface causes a microburst. Microbursts can move out from a dying thunderstorm in several directions or mostly in one direction. The tell-tale sign of a microburst is a "starburst" damage pattern (►), where features are blown radially, rather than a pattern of damage that indicates rotation around a central core (as in a tornado).

16.10.a10 Burnett County, MN

16.10.a8 Hermosa Cliffs, CO

Haboobs and Dust Devils

14. A name applied to severe downdraft-related, nonrotating dust storms is *haboob* (▼), derived from an Arabic word for "blasting." These are formed in a similar way to microbursts, at the collapse of a dying thunderstorm. They can pick up dust from desert areas and plowed fields, producing a thick cloud of dust that advances rapidly across the landscape. They can damage windows, cause respiratory issues, and create very hazardous driving conditions as visibility drops to a few meters in a matter of seconds.

16.10.a12 Phoenix, AZ

15. Often overlooked as a hazardous weather phenomenon is a *dust devil* (▲), also called a whirlwind or simply a dust storm. Like other rotating storms, these develop when the atmosphere is extremely unstable and upper-level patterns encourage rising motion. The upward motion at the surface creates strong pressure gradients, which strengthen the system seemingly uncontrollably. Eventually, the energy is expended, friction slows the wind, and the dust devil dissipates.

16.10.a11 Richland, WA

16. Dust devils and haboobs form in arid regions, so their strength is limited by the generally small latent energy released in condensation and deposition. But the thick dust, in addition to reducing visibility, can remove topsoil and erode the landscape.

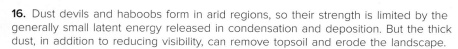

Before You Leave This Page

☑ Describe two types of waterspouts and settings in which each forms.

☑ Sketch and describe a derecho, explaining how one forms and the type of associated thunderstorm.

☑ Sketch and explain how most microbursts form, and summarize their varied manifestations.

☑ Summarize how haboobs and dust devils form.

16.10

16.11 What Is a Tropical Cyclone?

HURRICANES AND OTHER TROPICAL CYCLONES are some of nature's most impressive spectacles. These immense seasonal storms form over warm, tropical waters and can cause extremely heavy precipitation, all the while deriving energy from latent heat released by condensation and deposition. In the Atlantic and eastern Pacific, the most intense type of tropical cyclone is called a *hurricane,* while in the western Pacific and Indian oceans, the terms used are *typhoon* and *cyclone,* respectively. Tropical cyclones take days to traverse an ocean and spread energy as they move across the globe. How do such storms form, and how are they sustained?

A What Are the Characteristics of a Tropical Cyclone?

Tropical cyclones, hurricanes, and *typhoons* are characterized by low atmospheric pressure, large areas of strongly rotating winds, locally elevated sea levels near the storm, high wind-driven waves, coastal flooding and erosion, and flooding along streams as the storm passes inland. Follow a tour through a tropical cyclone in a counterclockwise manner.

1. Tropical cyclones are huge, circulating masses of clouds and warm, moist air. They are zones of low atmospheric pressure that cause air to rise and condense, creating locally intense rainfall.

2. Warm tropical ocean waters fuel tropical cyclone formation. Water vapor evaporated from the sea surface mixes with the air, rises, cools, and produces clouds, releasing energy during condensation, which warms the air around it. This energy further enhances the upward motion, drawing more surface moisture to replace the rising air. In this way, tropical cyclones, once formed, provide their own fuel (energy).

3. If the tropical cyclone encounters additional warm water in its path, extra energy is input to the storm, winds increase, and the storm grows in strength and size. A tropical cyclone dissipates when it passes over land or cold water, or when it mixes with much drier air.

8. In the Northern Hemisphere, the right front quadrant is the most destructive part of the storm. This is the zone where the counterclockwise winds push the waves, called a *storm surge,* onshore with additional momentum.

7. The strongest winds and most severe thunderstorms are usually found within the *eyewall* — the area immediately surrounding the eye.

6. Dry air flows down the center of the storm where it compresses and evaporates clouds, forming a cylinder of relatively clear, calm air, called the *eye.* Sinking formed this eye, a hole down into the clouds, in Hurricane Ike in 2008 (▼).

4. Differences in atmospheric pressure between the surrounding environment and the center of the tropical cyclone cause sea level to bow up by several centimeters. This effect is greatly exaggerated in the figure above to make it more visible. This mound of water can add to flooding problems when the cyclone reaches land, an event called *landfall.*

5. Like all enclosed low-pressure zones (cyclones), the Coriolis effect causes tropical cyclones to spiral counterclockwise in the Northern Hemisphere and clockwise in the Southern Hemisphere. The storm's overall track is steered by air pressures and currents high in the atmosphere. The track is greatly influenced by ridges and other areas of high pressure, which can block or deflect a cyclone, either toward or away from land.

16.11.a1

16.11.c

B Why Do Tropical Cyclones Rotate?

Tropical cyclones, like mid-latitude cyclones, are a result of multiple forces that influence where and how a cyclone moves. The pressure-gradient force acts to move air into the lower pressure from all sides, but this flow of air is deflected by Earth's eastward rotation—the Coriolis effect. The Coriolis effect is fairly weak at low latitudes but increases with the linear velocity of an object, like fast-moving winds associated with a tropical cyclone. This figure is a labeled satellite image of Hurricane Irene in 2011, a very large hurricane that caused extensive damage in the Caribbean, Gulf Coast, and the East Coast of the U.S.

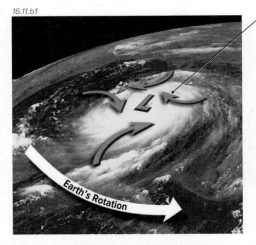

16.11.b1

The Coriolis effect deflects moving air to the right of its trajectory in the Northern Hemisphere (as shown by the red arrows on this figure) and to the left of its trajectory in the Southern Hemisphere. This produces a vortex shape and rotation of the entire cyclone system, with the now-familiar counterclockwise spin in the Northern Hemisphere and a clockwise spin in the Southern Hemisphere.

Earth's Rotation

C Where and How Are Tropical Cyclones Formed?

Tropical cyclones begin in the tropics as relatively weak disturbances in normal patterns of air pressures and wind directions. Few disturbances develop further, but some are dramatically strengthened by the environments through which they pass, attaining low-enough pressures and high-enough wind speeds to become a tropical cyclone. Here, we examine a common scenario that generates tropical cyclones over the Atlantic Ocean.

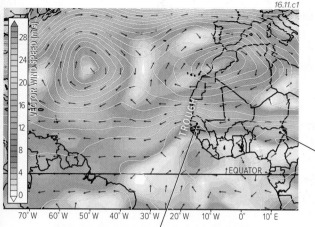
16.11.c1

1. Tropical cyclones in the Atlantic Ocean often begin over or near Northwest Africa in association with a west-flowing stream of air called the *African Easterly Jet* (AEJ), shown on the map to the left. The lines on the map reflect air pressures, and colors indicate upper-level wind speeds (red, yellow, and green are fastest). The relatively fast winds (green belt) of the African Easterly Jet flow across the arid regions of Africa. Note that the jet has wavelike bends north and south, similar to those observed for the polar front jet.

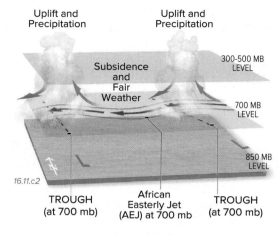
16.11.c2

4. This figure (▲) shows that the east side of a northward bend in the AEJ (a trough) is associated with lifting and unsettled weather; the west side experiences fair weather. The explanation for this is similar to that for Rossby waves.

2. The bends, or waves, in the African Easterly Jet are termed *easterly waves* because the bends propagate from east to west (i.e., coming out of the east). A bend to the north has relatively low pressure and is called a *trough*. Troughs are associated with low pressure, lifting air, and cloudy or stormy weather. If a westward-traveling wave encounters conditions favorable for further strengthening, such as warm sea temperatures, it can incrementally grow into a tropical depression, tropical storm, and hurricane.

3. This color-enhanced satellite image (▶) shows a belt of thunderstorms (in red) associated with westward-moving easterly waves and tropical depressions. The easternmost wave (on the right side of the image) is still over Africa, where it formed, but ones to the left have been carried westward over the Atlantic Ocean by the west-blowing trade winds. As the thunderstorms move over the warm waters of the Atlantic, they can strengthen.

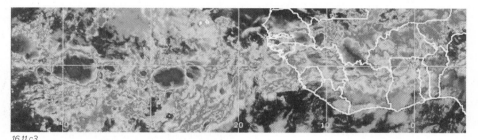
16.11.c3

Site of Origin and Tracks

5. A tropical cyclone forms under specific conditions limited to certain regions and then travels along a path called a *storm track*. These globes (▼) display global storm tracks of tropical cyclones in various parts of the world, colored according to strength. Tropical storms originate over warm water and mostly travel between 10° and 15° latitude and the subtropics—they cannot cross the equator. Note a general lack of tropical cyclones right along the equator, due to the absence of the Coriolis effect there. The locus of activity shifts seasonally, following the migration of the overhead Sun and the regions of the warmest water.

8. Tropical cyclones in the western Pacific and the northern Indian Ocean mostly form from June to November, but they can occur at any time of year. Most in the Southern Hemisphere form from January to March, the southern summer. Cyclones rotate clockwise and turn south and southwest in the Southern Hemisphere, but rotate counterclockwise and turn north and northeast in the Northern Hemisphere, reflecting the different deflections of the Coriolis effect on either side of the equator.

6. Tropical cyclones in the Atlantic Ocean, the ones that affect the Gulf Coast and East Coast of the U.S., mostly start west of Africa (related to the AEJ) from June 1 to November 30, the *Atlantic Hurricane season.*

7. Tropical cyclones, including hurricanes, also form in the tropical Pacific Ocean, but typically begin earlier, from May 15, and go through November.

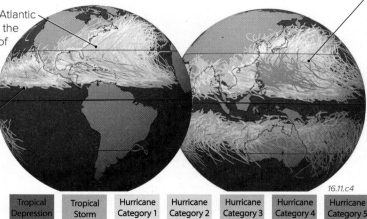
16.11.c4

| Tropical Depression | Tropical Storm | Hurricane Category 1 | Hurricane Category 2 | Hurricane Category 3 | Hurricane Category 4 | Hurricane Category 5 |

Before You Leave This Page

✓ Sketch, label, and explain the features and rotation of a tropical cyclone.

✓ Sketch and explain how an easterly wave can initiate a tropical cyclone, and describe where tropical cyclones originate, including seasonal variations.

16.11

16.12 What Affects the Strength of a Tropical Cyclone?

AS A TROPICAL CYCLONE MOVES, it responds quickly to changes in the environment it encounters. In some cases, environmental changes will cause a tropical cyclone to strengthen, perhaps becoming a hurricane or becoming a stronger hurricane. In other cases, the new environmental conditions cause a cyclone to weaken, dissipate, and eventually disappear. Consider the conditions under which a tropical cyclone forms and what factors might strengthen or weaken it, such as sea-surface temperatures and whether they are over water or land.

A What Conditions Strengthen or Weaken a Tropical Cyclone?

Changes in Water Temperature

16.12.a1

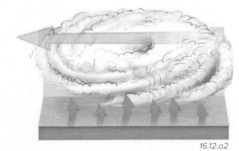

16.12.a2

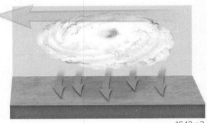

16.12.a3

1. These figures track the changes that occur when a tropical cyclone migrating across the surface encounters different conditions. The initial cyclone (shown above) is still growing, as expressed by its cumuliform top. The water below is just warm enough to sustain growth.

2. If the cyclone encounters even warmer sea temperatures, it may strengthen. The warmer water adds heat and additional moisture to the cyclone, moisture that will later generate even more heat during condensation (latent heat), a key factor in cyclone growth.

3. If the cyclone instead encounters cooler water, as when an Atlantic hurricane turns north in mid-ocean, it can lose thermal energy to the water. Also, the cooler water will not contribute much moisture, and it can even remove moisture by cooling the air.

Encounters with Other Air Masses

4. If a cyclone advances into, or incorporates, a wetter air mass, the additional moisture can strengthen the cyclone. The incorporation of more moisture is unlikely, as an air mass supporting a tropical cyclone is often at or near saturation already.

16.12.a4

16.12.a5

5. More commonly, a cyclone advances into, or incorporates, drier air. As the dry air is mixed in (entrained), cloud development ceases and cloud cover thins, eventually disappearing due to evaporation into the drier air. The loss of moisture robs latent heat from the storm, and this is how many cyclones dissipate.

Encounters with Land

6. A migrating cyclone will often cross onto land. Once it is over land, the cyclone no longer has an unlimited supply of empowering moisture. Also, the land can be dry and cold, so it will remove moisture and energy from the storm.

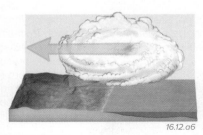

16.12.a6

16.12.a7

7. A cyclone can interact with surface topography. This increases friction, slowing the winds. Mountains also remove large amounts of moisture in the form of precipitation due to orographic effects. Such settings are among the main causes of cyclone-related flooding and fatalities.

Upper-Level Wind and Pressure

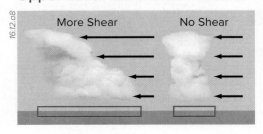

16.12.a8

More Shear   No Shear

8. Vertical wind shear is the difference in wind speed or direction between higher and lower levels in the atmosphere. While strong vertical wind shear assists in extra-tropical (not in the tropics) thunderstorm development, it generally tears apart or dissipates tropical cyclones. Fast winds in the upper troposphere will shear off the top of the developing tropical cyclone, as the near-surface trade winds push the base of the storm in the opposite direction. Note that vertical wind shear can spread out a storm's latent heat release over a larger area (as represented by the red boxes), limiting buildup of the storm.

9. Although not illustrated here, *upper-level pressures* can act to strengthen or weaken a tropical cyclone, like they do for a mid-latitude cyclone. They also influence the path of the storm.

B What Types of Damage Are Associated with Tropical Cyclones?

16.12.b1 Mississippi

16.12.b2 Long Beach, NY

16.12.b3 Gulfport, MS

1. Much damage from hurricanes and other tropical cyclones is from the high winds (▲), which can reach hundreds of kilometers per hour. Also, tornadoes and other strong local winds embedded within the larger storm can cause additional localized damage.

2. In addition to being subjected to strong winds, shorelines are afflicted by rough surf associated with the very large waves. The waves can erode parts of the beach, making houses even less protected, dumping sand far into neighborhoods (▲).

3. The coastal flooding associated with tropical cyclones also inundates low-lying areas along the coast. Much damage and death, however, also result from the flooding that occurs farther inland from streams having waters swollen by very high amounts of rainfall.

4. Recall that tropical cyclones form a mound of water beneath them, in response to the very low air pressure. When the storm approaches shallow coastal waters, this mound forms a wave, called a storm surge (▼), that rises as it approaches the shore. A storm surge can be devastating, flooding coastal areas with over 9 m (30 ft) of water. If the storm surge arrives at the same time as high tide, it is superimposed on the already-high water level. Also, the strong winds around cyclones push water in front of them, so the amount of storm surge can increase if a shoreline is in the direct path of the winds (northeast side of storms hitting the U.S.). Storm surges can bring marine objects onshore, including ships (▶).

16.12.b5 Staten Island, NY

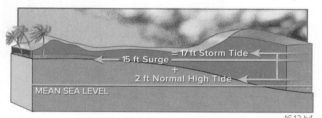

= 17 ft Storm Tide
15 ft Surge
2 ft Normal High Tide
MEAN SEA LEVEL

16.12.b4

Strengths of Tropical Cyclones

Meteorologists and newscasters commonly refer to a "category 3 hurricane," but what does this term actually indicate? This table shows how we categorize the strength of tropical cyclones.

The least energetic type of tropical cyclone considered dangerous is a *tropical depression*, so named because it is an area of low pressure. A tropical depression has at least one closed isobar on an air-pressure map, and successive tropical depressions are numbered sequentially in a given ocean basin each year (1, 2, 3, etc.). Once a tropical cyclone gains wind speeds of 64 km/hr (39 mi/hr), it is called a *tropical storm* and given a name. The first storm of the season in a given ocean basin receives a name that begins with an A, the second begins with B, and so on. If a named storm strengthens to wind speeds of 119 km/hr (74 mi/hr), it is termed a hurricane, typhoon, or cyclone, depending on the part of the world where it occurs. Atlantic and eastern Pacific hurricanes are then further classified according to their intensity on the *Saffir-Simpson scale* (named after the two people who devised it). The weakest hurricane is a category 1, and the strongest is a category 5. The amount of expected damage increases nonlinearly with storm strength (e.g., a category 4 storm generally produces more than four times the damage of a category 1 storm). Storms of Category 3–5 are often called "major" hurricanes, typhoons, or cyclones.

Tropical Storm Category	Maximum Sustained Winds	Anticipated Storm Surge	Examples
Depression	37–63 km/hr (23–38 mi/hr)		
Storm	64–118 km/hr (39–73 mi/hr)	> 1.0 m (> 3 ft)	Allison (2001)
Category 1	119–152 km/hr (74–95 mi/hr)	1.0–1.7 m (3–5 ft)	Irene (2011); Sandy (2012)
Category 2	153–176 km/hr (96–110 mi/hr)	1.8–2.6 m (6–8 ft)	Frances (2004); Ike (2008)
Category 3	177–209 km/hr (111–130 mi/hr)	2.7–3.8 m (9–12 ft)	Katrina (2005); Wilma (2005)
Category 4	210–250 km/hr (131–155 mi/hr)	3.9–5.6 m (13–18 ft)	Galveston (1900); Charley (2004)
Category 5	> 251 km/hr (> 155 mi/hr)	> 5.6 m (>18 ft)	Andrew (1992); Camille (1969)

Before You Leave This Page

✓ Sketch and explain the conditions that strengthen or weaken a tropical cyclone.

✓ Summarize the types of damage associated with tropical cyclones.

✓ Explain how we categorize tropical cyclones.

16.12

16.13 What Happened During Hurricane Sandy?

SEVERE STORMS are extremely dangerous, capable of killing thousands of people and inflicting incomprehensible damage. Severe weather—and the more benign varieties—develop, operate, move, and dissipate according to fundamental principles presented in this chapter. Here, we examine Hurricane Sandy, which in 2012 tracked up the East Coast of the U.S., before coming onshore in New Jersey.

16.13.a1

1. Hurricane Sandy began as a storm in the tropics (tropical cyclone) and became a hurricane that skirted the East Coast of the U.S. before coming onshore in New Jersey on October 29, 2012, as what many in the media and public called "Superstorm Sandy." On this satellite image (◄), Sandy appears as a swirl of clouds off the East Coast. Clearly visible is a well-defined center of the storm—the eye of the hurricane. To the west of Sandy is a long band of clouds, stretching from Florida to New England. What type of weather system do you think these clouds represent? Sandy later merged with this second weather system, which consisted of a cold front associated with a mid-latitude cyclone centered over New England. Sandy was weakening from hurricane status to tropical storm status as it came on shore, but the interactions with the second weather system had catastrophic results: over $75 billion in damages and nearly 150 deaths, about half in the U.S.

16.13.a2

2. The superstorm was large compared to typical hurricanes, with winds of 65 km (40 miles) per hour extending across an area more than 1,500 km (900 miles) in diameter. In this satellite image (▲), clouds associated with the storm cover the entire northeastern U.S. and large areas of adjacent Canada and the Atlantic Ocean.

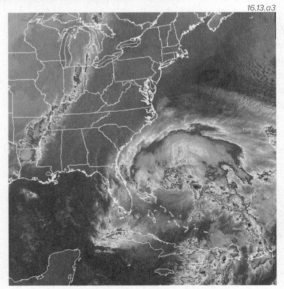

16.13.a3

3. Enhanced infrared satellite images, like this colorful map (◄), are a mainstay of daily weather reports. In this map, clouds are shown in light gray where they are relatively thin but blue, green, yellow, and red where thicker, with red being the thickest. We can extract estimates of cloud thicknesses from an infrared image because the temperature of the top of a cloud is mostly a function of how high it is—colder typically means a higher top, greater overall thickness, and more precipitation. Examine this scene and consider what each feature is, why it has the shape it does, and which way it is likely to be moving. Also think about what processes are occurring in each place.

16.13.a4 Oceansat-2

5. The image above shows wind speeds near the ocean surface as measured by radar on a satellite orbiting Earth. Wind speeds exceeding 65 km (40 miles) per hour are yellow, those above 80 kph (50 mph) are orange, and those above 95 kph (60 mph) are dark red. At the time shown in this image, strong winds along the western side were striking the Outer Banks of North Carolina, which is an especially vulnerable area for flooding due to storm surges associated with hurricanes.

4. Sandy, still a hurricane at this time, is the large area of intense clouds and rainfall on the right side of the image (▲). The storm has a coiled shape, reflecting rotation around a cyclone due to the Coriolis effect. It is moving north, as many storms do in this part of the Atlantic. Note a second line of clouds and precipitation to the west on land. These clouds formed along a cold front, a boundary between cold air to the west and warm air to the east. Sandy is within the *warm sector* of the storm. On this image, the different shades of gray on either side of the front reflect the temperature contrast between the two air masses, colder (lighter gray) to the west and warmer (medium gray) to the east.

Movement and Evolution of Hurricane Sandy

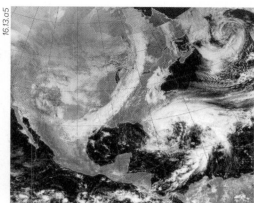

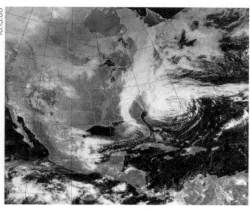

6. Hurricane Sandy originated in the tropics and then migrated northwest toward the North American continent before turning north and tracking along the East Coast. This sequence of three satellite images shows the evolution of Sandy with time, with the earliest image on the left and the latest image on the right. In the earliest image (▲), Sandy is leaving the tropics as a somewhat dispersed, poorly organized storm. Note the curved cold front crossing the center of the U.S.

7. In this second image (▲), Sandy has moved farther north, to a position off the Outer Banks. It has become more tightly coiled and better organized, which causes the hurricane to intensify and the destructive winds to speed up. As the hurricane and the mid-latitude cyclone to the west get closer, they begin to interact. Note that the southern "tail" of the cold front is curving toward the east, as it gets drawn into the counterclockwise circulation around the hurricane.

8. In this final image (▲), Sandy and the cold front have collided, causing exceptional storminess across the Northeast. Many hurricanes turn northeast and head out to sea as they come up the coast, but Sandy turned northwest, directly into New Jersey. This unusual track was caused by an area of high pressure located over easternmost Canada, which blocked Sandy from moving northeast. The interaction between Sandy and the second storm made the situation much worse.

Hurricane Sandy

Hurricane Sandy began as a tropical disturbance in the Caribbean. The system strengthened considerably because it remained over very warm waters and was in a region with limited vertical wind shear. Sandy evolved into a named tropical storm and then a hurricane. The storm moved over Jamaica and Cuba, where it was weakened due to inter-actions with the land. It was downgraded to a tropical storm as it moved into the Atlantic. Once over open, warm waters of the Gulf Stream, it was reinvigorated, partly because of the approaching second storm to the west.

Several aspects of Sandy were very unusual, accounting for the excessive amount of damage. First, Sandy was a huge hurricane by any standards, affecting essentially the entire eastern half of the U.S. in one way or another, even though its wind speeds were not very strong by comparison to other hurricanes. A second factor was the way Sandy interacted with the mid-latitude cyclone to the west. Sandy formed fairly late in the hurricane season, and the front, with its accompanying very cold air, was fairly early for that type of winter storm. The interac-tion between the tropical-derived moisture and very cold air caused extremely heavy precipita-tion, including snow, ice, and blizzards, aspects we do not typically associate with hurricanes. Third, Sandy struck a part of the coast that does not experience frequent hurricanes, so some structures, especially those along the beach, were not built to withstand such storms.

The three photographs included here repre-sent a small sample of the destruction caused by Sandy. The aerial image to the left shows damage to the Mantoloking area of New Jer-sey, including destruction of a highway and critical bridge, as well as severe damage to houses, especially those facing the ocean (on the right side of the aerial photograph). The two photographs on the right are from Coney Island, New York, and the coast of New Jersey. Consider what factors resulted in such severe damage to all three sites.

16.13.t1 Mantoloking, NJ

16.13.t2 Coney Island, NY

16.13.t3 New Jersey coastline

Before You Leave This Page

✓ Summarize how Hurricane Sandy formed, how it changed over time, and how it differed from a typical hurricane.

16.13

Where Would You Expect Severe Weather?

IN PLANNING A CROSS-COUNTRY TRIP, you have been asked to provide an overview of weather conditions across the U.S. To do this, you will study maps showing weather conditions at the surface and then determine where and how different air masses are interacting. In addition, you will identify areas where some type of severe weather is possible and identify what type of severe weather is likely.

Goal of This Exercise:

- Apply principles from this chapter to evaluate the weather conditions and make a forecast for where severe weather is most likely to occur.

Initial Procedures

Follow the steps below, entering your answers for each step in the appropriate place on the worksheet or online.

A. Observe and characterize the main weather features, including cloud cover, fronts, types of air masses, air pressure, moisture, and location of the polar front jet stream.

B. Consider all the surface and upper-level data to assess the likelihood of severe weather, identifying areas where different types of severe weather were likely to occur on two successive days.

These two pages display different types of maps, each showing certain weather features. All of the maps cover the main part of North America, but they do not cover precisely the same area.

Surface Conditions

These satellite images show the conditions for two successive days (Day 1 on the left, Day 2 on the right). Observe these images and propose what weather features (such as storms and fronts) each shows. Then, compare the two images to determine how each feature is moving. Complete this phase before continuing on to examine other maps. Copies of these maps are on the worksheet.

DAY 1

16.14.a1

DAY 2

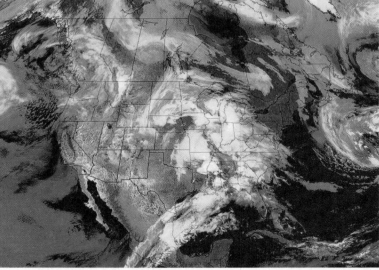

16.14.a2

Temperature

These maps show the average air temperature at the surface for each day. Red and orange show the warmest temperatures, purple and blue show the coldest, and green and yellow depict intermediate temperatures. Identify the boundaries between air masses of different temperature and propose what types of fronts those boundaries represent.

DAY 1   16.14.a3

DAY 2   16.14.a4

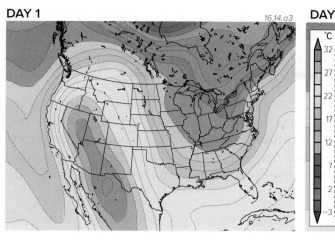

Atmospheric Moisture

These maps represent the amount of water in a column of atmosphere (in kg/m²) that would fall as precipitation if all of the water vapor condensed. Darker areas have abundant water and lighter ones have less water available to precipitation. Using the temperature and moisture maps, identify different types of air masses (e.g., cold and dry, versus warm and humid).

DAY 1   16.14.a5

DAY 2   16.14.a6

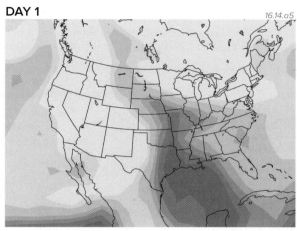
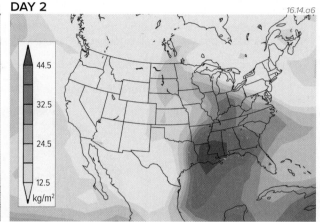

Conditions Aloft

These maps display white contour lines representing upper-level air-pressure conditions. Arrows show the general directions of airflow, which in these upper levels are parallel to the contours. Colors show estimated wind speeds, with red and orange representing the fastest speeds, purple and white showing the slowest, and green and yellow depicting intermediate wind speeds. Based on the high wind speeds and close spacing of the contours, where

DAY 1   16.14.a7

DAY 2   16.14.a8

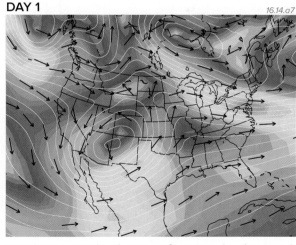
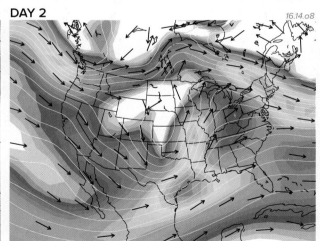

is the polar front jet stream? Which areas have upper-level support for storm development based on likely positions of upper-level divergence in the Rossby waves?

Final Procedure

c. Using the various data presented here, divide the map area into different regions, each characterized by a particular air mass. Describe the types of boundaries that are likely between these air masses, based on their location and direction of movement. Shade in regions where severe weather was possible for Day 1 and for Day 2. Your instructor may have you answer questions on a worksheet or online.

16.14

Oceans and Their Interactions with Other Earth Systems

OVER 70% OF THE PLANET'S surface is covered by oceans, which exchange energy and moisture with the overlying atmosphere. Oceans move in response to three main factors—winds moving over the top of them, spatial variations in the density of water, and the Coriolis effect. Oceans in turn affect the temperatures and pressures of the adjacent atmosphere. These two major systems of the earth—oceans and atmosphere—are closely connected and constantly interacting, and these interactions control many aspects of our climate.

These two globes represent temperatures during two time periods, one from 1964–1976 and one from 1978–1990. Specifically, the globes show how average land temperatures and sea-surface temperatures (SST) during these two time periods departed (differed) from long-term averages. For each globe, the average temperature for that time period is compared to average long-term temperatures. If the temperatures during that time were warmer than the long-term average, the values are positive (shown in red, orange, and yellow). If the temperatures during that time period were colder than the long-term average, the values are negative (shown in purple, blue, and green). Referencing a data set to long-term averages produces a comparison we call a "departure" or "anomaly." Observe the two globes for patterns and for changes between the two time periods.

What patterns do you observe and what happened to cause such different patterns over time?

Temperature Departure from Average (1964–1976)

Northwestern North America, especially Alaska, was colder than normal from 1964 to 1976, as shown by the purple and dark blue colors over land, The adjacent waters of the Pacific were also relatively cold, but mostly along the western coast of North America. The colder than normal temperatures continued south into southwestern North America and adjacent waters. In contrast, waters farther out in the Pacific were warmer than usual, as depicted by the large region colored red and orange.

17.00.a1

-0.92 -0.68 -0.44 -0.18 0.06 0.30

What causes warming or cooling of sea-surface temperatures, and how can one part of the ocean be colder than normal while another part is warmer than normal?

Temperature Departure from Average (1978–1990)

A shift in regional temperatures occurred around 1977, as shown by the different patterns between the globes. Alaska and adjacent waters of the eastern Pacific warmed significantly, as reflected by red and orange areas over Alaska. In contrast, cooling occurred in the middle of the northern Pacific, as indicated by the blue and green colors. The patterns in these two globes are nearly opposites, representing two common modes in temperatures in this region: a *cool phase,* in which SST are cooler than usual in the eastern north Pacific near Alaska and warmer than usual in the middle of the north Pacific, and a *warm phase*, with the opposite pattern. The switch from one mode to another is named the *Pacific Decadal Oscillation* (PDO).

17.00.a2

-0.32 -0.20 -0.08 0.06 0.18 0.30

Why are the patterns on these two globes the opposite of each other, and what causes these shifts as part of the PDO?

Regional changes in SST can be quantified by statistically comparing SST in the eastern and western Pacific and calculating an index (called the *PDO index*) that represents the difference. When the waters off Alaska are warmer than those in the central Pacific, the PDO index is positive. The top graph to the right plots the PDO index for several years and shows a major shift in climate at about 1977—the same temperature shift represented by the difference in the two globes. Red bars represent the warm phase.

What is an ocean oscillation, and do similar oscillations occur in other parts of the Pacific or even in other oceans?

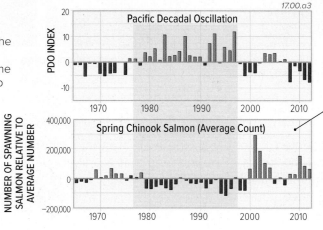

17.00.a3

The Pacific Decadal Oscillation, including its warm and cool phases, was first recognized by a fisheries scientist trying to explain year-to-year variations in how many salmon return to their spawning grounds in streams along the Pacific, as shown in the bottom graph. Comparison of the graphs reveals a striking correlation—the number of spawning salmon is higher than average during cool phases of the PDO and less during warm phases.

What causes these warm and cool phases in an ocean?

TOPICS IN THIS CHAPTER

17.00.a4 Mount Baker, WA

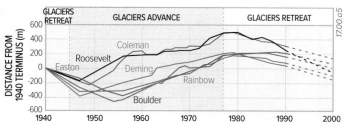

17.00.a5

This shift in climate in the northern Pacific correlates with a change in glaciers on Mount Baker (◄), one of the large volcanic mountains in the Cascade Range. As shown in the graph above, glaciers were advancing before this shift (during the cool phase) and retreating after it (during the warm phase). As shown on the graph of PDO index on the previous page, sometime around 1998 the PDO went back into a cool phase, heralding the return of cooler temperatures.

Do ocean oscillations affect the amount of precipitation and global amounts of ice?

The PDO and Climate

In 1977, a major shift occurred in the patterns of sea-surface temperatures (SST) in the North Pacific. Before that time, waters in the Gulf of Alaska had been relatively cool, while waters in the central north Pacific had been warmer than normal. After 1977, the patterns switched, with warmer than normal waters along the coasts of Alaska and British Columbia, but colder than normal waters in the central north Pacific. This change, sometimes called the *Great Pacific Climate Shift,* had profound implications for conditions on land in western North America and in the Pacific. This shift correlated with changes in glaciers, temperatures on land, and the number of spawning salmon.

In the past several decades, this climate shift has been recognized as part of a series of cyclical changes in broad-scale patterns in air pressure, wind directions, and SST, where the Pacific Ocean has two dominant modes: a *warm mode* and a *cool mode.* From data collected over the last century, scientists have been able to establish that the Pacific Ocean stays in the warm mode for several decades before switching to the cool mode for the next several decades: thus the "decadal" part of the Pacific Decadal Oscillation name. The term "oscillation" refers to a climatic condition, such as air pressure or SST, that shifts back and forth between two distinct modes or phases. The Pacific Decadal Oscillation is but one such oscillation. Similar oscillation occur in the equatorial Pacific, Atlantic Ocean, and Indian Ocean, each greatly impacting regional climate.

Wind and Pressure (1964–1976)

Wind and Pressure (1978–1990)

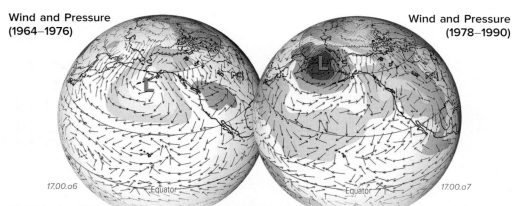

17.00.a6

17.00.a7

Shifts in the PDO are accompanied by changes in winds and air pressure. The two globes above show anomalies (departure from average) in wind directions (arrows) and surface air pressure (shading, where dark gray is lower pressure). During a PDO cool phase (left globe), the Aleutian Low is very weak and the counterclockwise winds weaken (clockwise anomalies). This gives the Alaska coast the colder temperatures during the cool phase. During a PDO warm phase (right globe), the Aleutian Low is much lower, and stronger counterclockwise winds around the Aleutian Low bring warm SST to Alaska's coast. Since wind directions are controlled largely by differences in air pressure, the patterns of air pressure also differ between these two phases.

How do changes in SST relate to changes in air pressure and wind direction? Do the oscillations cause the wind patterns or vice versa, or are both related to changes in air pressure?

17.0

17.1 What Causes Ocean Currents?

SURFACE WATER OF THE OCEANS circulates in huge currents that generally carry warmer water toward the poles and colder water toward the equator. Extremely large quantities of energy are stored in the uppermost 100 meters of the global oceans, and surface ocean currents redistribute this energy from one part of the ocean to another, carrying it from zones of excess energy to those with an energy deficit. What causes ocean currents, and can we make some general predictions about which way ocean currents would likely flow?

A What Causes Surface Waters of the Ocean to Move and Which Way Do They Go?

Various processes cause surface water to move, but the primary driving force is wind. Wind blowing across a body of water transfers some of its momentum onto the surface of the water, causing it to move. On a smaller scale, this causes waves, and on a global scale, this sets in motion large currents that can traverse an entire ocean.

1. As wind blows across water, it pushes against the surface, causing the uppermost water to begin flowing.

2. This water in turn transfers some of its momentum to waters just below the surface, setting up an overall shearing of the water column, with the strongest shear force (depicted by arrows) at the surface.

3. The effects of wind stress, and therefore current velocity caused by the wind, decrease with depth. They vanish at a depth of about 100 m, a depth known as the *null point*.

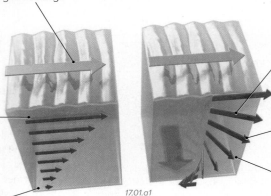

17.01.a1

4. If the planet were not rotating, water above the null point would flow parallel to the direction in which the wind is blowing. On Earth, however, as soon as the water starts moving it is subject to the Coriolis effect, so it has an apparent deflection away from its initial direction, turning to the right in the Northern Hemisphere.

5. As the effects of the shearing force decrease downward toward the null point, the direction of flow turns farther to the right with depth. In this way, the direction and velocity of flow, when plotted as a series of arrows, define a downward spiraling shape—an *Ekman spiral*.

6. As a result of these competing influences, the overall direction of flow of the uppermost ocean, when averaged between the surface and the null point, is at an angle to surface winds.

B Which Way Do Surface Winds Blow Against the Ocean Surface?

If wind patterns drive the flow of the upper part of the ocean, we should be able to anticipate the directions of ocean currents by considering the directions of wind.

1. This globe reviews the main belts of winds encircling Earth. Examine the globe and consider what pattern of ocean currents might result from these wind directions. Try this before reading on.

2. As you thought about likely directions of currents, you probably realized that eventually any current that flows east or west will at some point run into a continent—the continents are in the way, so an ocean current must turn to avoid the continent.

3. Recall that winds are generally not very strong along the equator, the so-called *doldrums*. They are also very weak in areas of descending air in the subtropics, at 30° N and 30° S latitude—the *horse latitudes*.

4. Another factor to consider is that the apparent deflection of moving objects by the Coriolis effect is opposite across the equator. The apparent deflection is to the right north of the equator and to the left south of it.

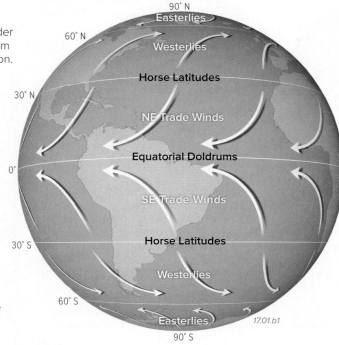

17.01.b1

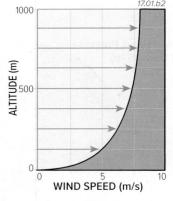

17.01.b2

5. This graph (▲) shows the general relationship between wind speed and altitude. Winds are slower near the surface, and they increase upward. Winds slow down near the surface because some of their energy of motion (kinetic energy) is transferred from the wind to the surface, producing waves and currents. As there is over land, the atmosphere has a friction layer over water.

C What Is the Anticipated Pattern of Global Surface Ocean Currents?

Given what we know about the global distribution of surface pressure cells and the direction of surface winds, it is reasonable to make a first approximation of the direction of surface ocean currents as a series of interacting circular motions, or *gyres*, like the ones in the stylized ocean shown below.

1. A key characteristic of this hypothetical planet is that the central ocean extends from pole to pole, so it crosses all of the regional wind belts, from the equatorial doldrums to the polar easterlies. The adjacent continents also nearly extend from pole to pole, effectively confining the ocean on both sides.

2. Surface waters flowing from a lower latitude (closer to the equator) toward a higher latitude (closer to the poles) transport water with greater stored energy than expected at that latitude and are considered "warm" currents. Those flowing in the reverse direction, from higher to lower latitudes, have less stored energy than expected and are termed "cold" currents. The terms *warm* and *cold* should only be considered relative to the environment through which they are passing and are indicative of whether they are bringing energy to or removing it from an environment. Accordingly, warm currents are colored red and cold currents are colored blue. Note that a single gyre brings warm water to the coast of one continent but cold water to the coast of the other continent, as described below.

East Coast of Continent

3. 45° N–60° N: Cold surface waters originating from north polar regions.

4. 25° N–35° N: Warm surface waters originating from equatorial regions and moving northward.

5. 25° S–35° S: Warm surface waters originating from equatorial regions. Note that this happens with winds from the opposite direction compared to the Northern Hemisphere.

6. 45° S–60° S: Cold surface waters originating from south polar regions. Note that this happens with winds from the opposite direction compared to the Northern Hemisphere.

West Coast of Continent

7. 45° N–60° N: Warm surface waters originating from subtropical regions and moving northward.

8. 25° N–35° N: Cold surface waters originating from mid-latitude regions.

9. Equatorial Latitudes: Mostly warm waters flowing east to west.

10. 25° S–35° S: Cold surface waters originating from mid-latitude regions and flowing north.

11. 45° S–60° S: Warm surface waters originating from subtropical regions.

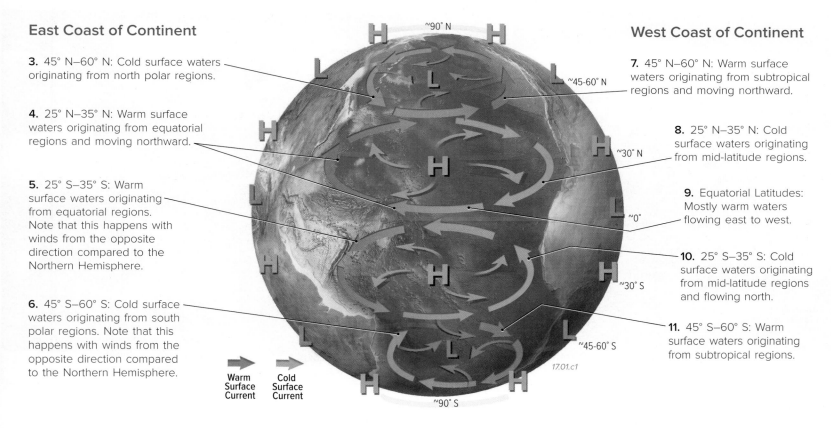

Warm Surface Current   Cold Surface Current

17.01.c1

12. To remind us that generalized models are just that, we include a numeric model of actual ocean currents (▼), whose squiggly lines bring to mind the style of painter Vincent van Gogh.

17.01.c2

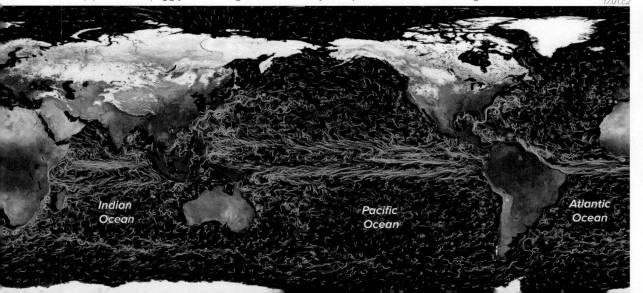

Indian Ocean   Pacific Ocean   Atlantic Ocean

Before You Leave This Page

☑ Sketch and explain how motion is transferred from the atmosphere to the ocean surface.

☑ Sketch and explain the anticipated directions of surface ocean currents in a simple one-ocean, two-continent global model, identifying warm and cold currents.

17.1

17.2 What Is the Global Pattern of Surface Currents?

OCEAN CURRENTS on Earth follow the main patterns we would predict from our understanding of global wind directions, but irregular shapes of the continents cause important complexities, such as isolating some protected seas from the larger circulation patterns. The global figure below depicts a summary of major surface currents and gyres in the oceans. The smaller globes and accompanying text each highlight major currents in different regions. The major currents have profound implications for the climate, weather, ecology, and the lives of the inhabitants, so we recommend that you take some time to learn their names.

1. This large map shows the locations, flow directions, and characteristics of ocean currents on a global scale. Warm currents are shown in red, and cold currents are shown in blue.

17.02.a1

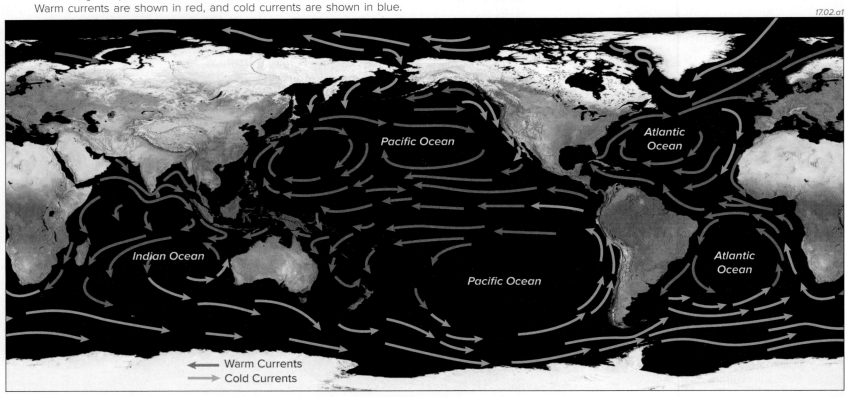

Pacific Ocean

Atlantic Ocean

Indian Ocean

Atlantic Ocean

Pacific Ocean

← Warm Currents
→ Cold Currents

Indian Ocean

2. Most of the Indian Ocean lies south of the equator. Surface circulation in the Indian Ocean is dominated by the flow around the southern subtropical gyre, consisting of (1) the warm, south-flowing *Mozambique Current* east of Africa, (2) the cool, north-flowing *Western Australian Current*, and (3) a west-flowing segment of the gyre along the equator to complete the loop. South of 40° S, waters become entwined with the *Antarctic Circumpolar Current*. Surface flow patterns vary seasonally according to the winds of the strong Asian monsoon and other factors.

S. Indian

Mozambique

W. Australian

Antarctic Circumpolar

17.02.a2

Western Pacific Ocean

3. The largest ocean basin on Earth, the Pacific basin, displays many of the characteristics expected from the model. It contains two subtropical gyres, one in the Northern Hemisphere and one in the Southern Hemisphere.

4. Northwestern Pacific: Water enters the western Pacific driven by westward flow on either side of the equator. As waters north of the equator approach the Philippines and mainland Asia, they turn north, carrying warm water past southern China and Japan in the *Kuroshio Current*. At about 40° N, this flow then turns east as the *North Pacific Current*.

5. Southwestern Pacific: South of the equator, the west-flowing current turns south down the east coast of Australia as the warm *East Australian Current*. Farther south, it becomes enmeshed with the strong *West Wind Drift*, driving eastward toward South America.

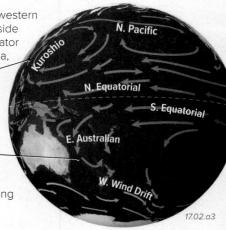

N. Pacific

Kuroshio

N. Equatorial

S. Equatorial

E. Australian

W. Wind Drift

17.02.a3

Northern Pacific Ocean

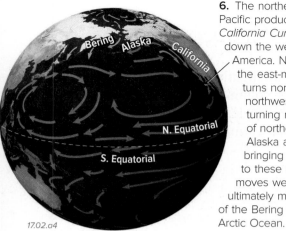

17.02.a4

6. The northern subtropical gyre in the Pacific produces a cold current, the *California Current*, that flows southward, down the west coast of North America. Near Alaska, a portion of the east-moving, mid-latitude water turns north as it approaches northwestern North America, turning northward along the coast of northern British Columbia and Alaska as the *Alaska Current*, bringing somewhat warmer waters to these northerly latitudes. As it moves west, the Alaska Current ultimately mixes with the cold waters of the Bering Current coming from the Arctic Ocean.

Southeastern Pacific Ocean

7. The cold waters of a current (the West Wind Drift) flowing east from Australia turn north along the west coast of South America. During this long track across the southern Pacific, the waters turn cold and form a very pronounced cold current, called the *Humboldt Current* (also locally known as the *Peru* or *Chile Current*). As the Humboldt Current approaches the equator, it turns back to the west toward Australia to complete the circuit of this huge gyre. The west-flowing segment is the *South Equatorial Current*.

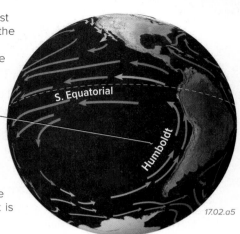

17.02.a5

Northern Atlantic Ocean

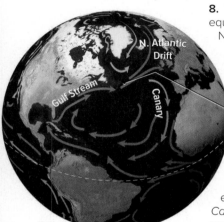

17.02.a6

8. The Atlantic Ocean north of the equator consists of a wide part between North America and Africa (the central Atlantic) and a narrower part between North America and Europe (the north Atlantic). Within the central Atlantic is a subtropical gyre that includes the northerly flowing *Gulf Stream*, which transports warm water up the East Coast of the U.S. To the north, the Gulf Stream gives way to the *North Atlantic Drift*, which brings warmth and moisture to northwestern Europe. From Europe, the cooler *Canary Current* flows south along the coast of northwest Africa.

Southern Atlantic Ocean

9. The part of the Atlantic Ocean south of the equator, between South America and Africa, is the south Atlantic. Within this region is the Southern Hemisphere gyre, which includes the cold *Benguela Current*, which flows north along the western coast of southern Africa. After flowing west, the gyre turns south along the eastern coast of South America, becoming the warm *Brazil Current*. To the south, this current is deflected to the east by its interactions with the *Antarctic Circumpolar Current*, flowing eastward.

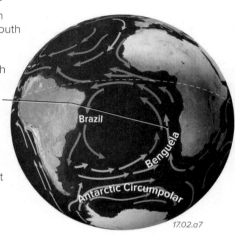

17.02.a7

Arctic Ocean

17.02.a8

10. Unlike its southern equivalent, the northern polar region, north of 80° N, is occupied entirely by ocean—the Arctic Ocean. The major outlets for the Arctic Ocean include the Pacific basin by means of the cold *Bering Current*, and the Atlantic via the *Labrador* and *Greenland currents*, which bring cool water down the east coast of Canada and Greenland. A major flow of warm ocean waters into the Arctic Ocean is accomplished by the North Atlantic Drift flowing northward along the coast of northwestern Europe.

Southern Ocean

11. Ocean waters surrounding Antarctica are regarded as the southern parts of the Pacific, Indian, and Atlantic oceans. No land interruptions, other than the southern tip of South America and New Zealand, exist in this region. The entire polar zone is occupied by continental Antarctica. Ocean-continent contrasts here are parallel to latitude. Strong pressure gradients and relative lack of land interruptions induce extremely strong westerly winds at these latitudes—known by sailors as the "roaring forties" and the "screaming sixties." Circling around Antarctica is the strong *Antarctic Circumpolar Current*.

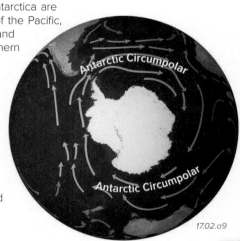

17.02.a9

Before You Leave This Page	✓ Sketch and explain the main patterns of ocean currents in the Northern Hemisphere versus those in the Southern Hemisphere, noting similarities and differences with a simple one-ocean model.

17.2

17.3 How Do Sea-Surface Temperatures Vary from Place to Place and Season to Season?

SEA-SURFACE TEMPERATURES (SST) vary greatly, from bath water warm to slightly below freezing. Early data on SST were collected from ships, but since the 1970s, satellites have collected voluminous SST data, documenting variations in temperature from region to region, season to season, and decade to decade. SST data have become even more important as climatologists investigate the causes and consequences of global warming and other types of climate change. How do SST vary on Earth, and what do such data tell us?

Average Sea-Surface Temperatures

1. These two globes both show SST averaged over the entire year and over several decades. Red and orange represent the warmest temperatures, purple and blue indicate the coldest temperatures, and green and yellow show intermediate temperatures.

2. Observe the global patterns and propose possible explanations based on how insolation varies from place to place. Then, examine the smaller patterns in the context of ocean currents (their position, direction of motion, and warm-versus-cold character). After you have done this for both globes, read the associated text.

3. The most obvious pattern in the SST data, not surprisingly, is the large temperature change between tropical latitudes and polar regions. Polar regions are green, blue, and purple (from cold to coldest) on this globe—they are the coldest seas on Earth.

4. Note, however, that the warmest seas (in red) are not necessarily along the equator. If you expected the equator to be warmest, how could you explain this apparent discrepancy? Part of the explanation is that seas at the equator are overlain by rising air and so commonly have cloud cover that reflects some insolation back into space. Also, some of the warmest seas are in the subtropics, where descending air dries out the surface, limiting cloud cover and permitting more sunlight to reach the sea.

5. An aspect to consider is whether an area of sea has a wide-open connection with open ocean. If so, that water can mix with or move to parts of the ocean where the waters are deeper and colder. On this globe, the warmest waters are in the relatively shallow parts of the Caribbean Sea and Gulf of Mexico, areas whose connection with the Atlantic Ocean is somewhat hindered by the many islands and peninsulas, so the Sun-warmed water mostly remains in those regions.

17.03.a1

-29 -25 -21 -17 -14 -10 -6 -2 2 6 10 14 17 21 25 29

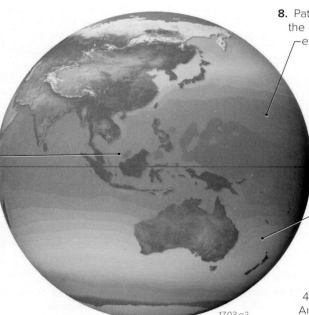

6. This globe shows most of the Indian Ocean (on the left) and about half of the Pacific Ocean (on the right). What patterns do you observe on this side of the globe, and what are some differences between this side and the other side of the planet, as shown in the top globe?

7. The semi-enclosed, shallow seas between the mainland of southeastern Asia and the islands of Indonesia and the Philippines are relatively warm, as is the Indian Ocean. As discussed later in this chapter, SST in this region, while always warm, can vary somewhat in response to an important ocean oscillation called the El Niño-Southern Oscillation, or ENSO for short. More on this important oscillation later.

8. Patterns are more regular in the middle of the Pacific Ocean, away from the complicating effects of continents, which influence weather patterns and steer ocean currents along their coasts. The SST data form bands of colors that mostly follow along lines of latitude. The patterns are more complicated closer to the continents, such as near Japan.

9. The southern parts of the Indian Ocean and Pacific Ocean have similar, latitude-parallel gradations. Again, this pattern largely reflects the lack of complicating landmasses. Only the southernmost tip of South America (not shown), Tasmania (the island south of mainland Australia), and the South Island of New Zealand extend south of 40° S latitude. Between these places and Antarctica, the ocean continues unimpeded.

17.03.a2

January Sea-Surface Temperatures

10. These two globes focus on the Atlantic Ocean, but at two different seasons. The left globe shows average SST for January, which is winter in the Northern Hemisphere but summer in the Southern Hemisphere. The right globe shows the same data for July. Observe the main patterns and compare the two globes.

11. In January, the colors of SST shift slightly but noticeably to the south, following the shift of maximum insolation as summer temperatures increase in the Southern Hemisphere. Also, whichever hemisphere is having summer experiences an overall expansion of warm SST.

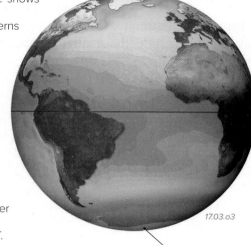

17.03.a3

12. In January (the southern summer), the waters around Antarctica are not as cold as they are in July (the southern winter; the right globe). The opposite pattern occurs near the North Pole—waters there are colder in January, the northern winter, than during July, the northern summer.

July Sea-Surface Temperatures

13. Compare the patterns of SST on the two sides of the ocean. On the western side of the ocean, next to the Americas, warm waters spread relatively far to the north and south, away from the equator. This larger expanse of warm waters reveals the effects of ocean currents, like the Gulf Stream, moving warm water away from the equator, along the eastern coasts of North and South America. Warm waters of the Gulf Stream continue toward Europe.

17.03.a4

14. Compare this with the pattern on the eastern side of the ocean, adjacent to Europe and Africa. Here, relatively cooler waters reach farther toward the equator. This pattern is especially obvious along the west coast of southern Africa, where the north-flowing Benguela Current brings cold water northward along the coast. A similar current, the Canary Current, brings cold water south along the coast of northwestern Africa. As a result of the opposite flows on the east versus west sides of the Atlantic Ocean, the width of warm water is much less near Africa than it is near South America and the Caribbean Sea.

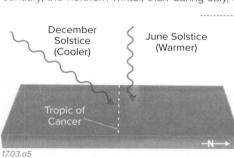

17.03.a5

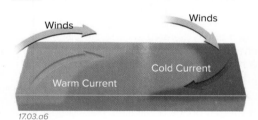

17.03.a6

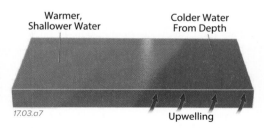

17.03.a7

18. SST can greatly affect adjacent land. This photograph (▶) shows desert along the western coast of southern Africa with the Atlantic Ocean in the distance. The dryness is partly due to the cold, stabilizing Benguela Current offshore and to prevailing winds that blow offshore.

Factors Influencing Sea-Surface Temperatures

15. From the four SST globes on these pages, the main pattern that emerges is the strong correlation of warm waters to low latitude. In the tropics, the overhead rays of the Sun (large Sun angle) strongly heat the oceans, especially during summer for that hemisphere (◀). Much less solar heating occurs near the pole (low Sun angle); none for much of the winter.

16. Another influence is ocean currents, such as the Gulf Stream (◀), which mostly move warm water away from the equator on the western side of oceans from about 20° to 50° north and south. In contrast, cold currents flow toward the equator on the eastern sides of oceans, as occurs with the Benguela Current off Africa, the Humboldt Current along the western side of South America, and the California Current down the western side of North America. Currents largely reflect regional wind patterns.

17. A third important factor is the flow of cold, deep waters toward the surface, or *upwelling* (◀). Cold-water upwelling occurs along the California coast, off the western coast of South America, and in other places. Offshore winds push water away from the coast, causing deeper water to upwell to replace this water. Upwelling leaves only a subtle imprint on global temperature maps, but it strongly influences the biological productivity of the ocean and the activities in many coastal communities.

17.03.a8 Namib Desert, Namibia

Before You Leave This Page

☑ Summarize the main patterns in sea-surface temperatures as observed using average temperatures.

☑ Describe and explain the main differences between SST in January and those in July, giving specific examples.

☑ Sketch and explain some factors that affect SST, and how ocean currents affect SST of western versus eastern sides of an ocean, providing examples from the Atlantic.

17.3

17.4 What Causes Water to Rise or Sink?

MOVEMENT WITHIN THE OCEANS arises from other factors in addition to shearing induced by the wind. Such factors include variations in the density of water in response to changes in its temperature and salinity. Changes in temperature and salinity occur from flows of fresh water into and out of the oceans as a result of precipitation, evaporation, ice formation and melting, and inputs from streams. The temperatures of ocean waters and the relative amounts of fresh water control the density of water, and these differences in density can alter the circulation of ocean waters.

A How Does the Temperature of Water Control Its Density?

Like most substances, water changes its density when subjected to increases or decreases in temperature. Unlike other substances, however, water exhibits some peculiar density changes as it approaches freezing and begins to solidify into ice.

Polarity of the Water Molecule

1. A key attribute of water that controls its behavior is its polarity. Recall that in a water molecule, the two hydrogen atoms are on one side of the oxygen, giving the molecule a polarity, which causes water molecules to be attracted to charged atoms (*ions*). This allows water to interact with other chemical substances, such as salt.

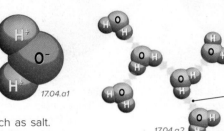

Hydrogen Bonding

2. Water molecules are attracted to each other as well as to ions. In water, a bond called a *hydrogen bond* forms between one molecule's hydrogen atom and another molecule's oxygen atom.

3. Hydrogen bonds form when the polarities of two adjacent water molecules attract one another. The hydrogen bond is responsible for some of water's unique properties (e.g., viscosity and surface tension).

Density of Water as a Function of Temperature

4. This graph (▶) shows how the density of fresh water changes with temperature. The curve on the right side shows how liquid water changes density in temperatures slightly above freezing. The vertical break marks 0°C, the temperature at which fresh water freezes. The angled line on the left shows how ice changes density with temperature. Examine the main features of this graph before we explore it further. Read clockwise around the graph, beginning on the other side.

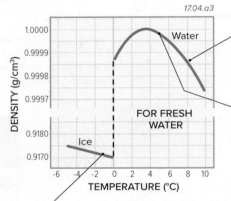

5. At temperatures above 4°C, water, like most substances, becomes more dense as it cools, as shown by moving from right to left up the sloping line on the right side of the graph. With cooling temperature, individual molecules become less energetic and can occupy a smaller volume, thereby increasing the water's density. Fresh water reaches its maximum density (1 g/cm³) at a temperature of approximately 4°C (39°F).

6. At a temperature of 4°C, however, fresh water begins to decrease in density, as is shown by the downward bend in the graph between 4°C and 0°C. This decrease in density occurs because, at temperatures below 4°C, the strength of the hydrogen bonds starts to exceed the disruptive random motions of the less energetic molecules, and the molecules start to organize themselves into a more regular arrangement that occupies more volume and decreases density.

7. As ice (the solid state of water) forms, the crystalline structure of the hydrogen bonds dominates, placing adjacent molecules about 10% more distant, on average, from their nearest neighbors than when in the liquid state at 4°C. This causes the density of ice to drop to about 0.917 g/cm³ (at 4°C, liquid has a density of 1.00 g/cm³). This decrease in density when water freezes causes ice to float on water—less dense materials, like wood and ice, float on a denser liquid, like water. Our world would be a very different place if ice were denser than water. Ice would begin to grow upward from the ocean floor over time until it occupied a very large percentage of the oceans.

Sinking and Rising of Water as a Function of Temperature

8. These density changes of water as a function of temperature, as illustrated on the graph to the right, have huge implications for our natural world, as discussed later in this chapter.

9. At temperatures between freezing and 4°C, slightly warmer waters are denser than cooler waters (if they have the same salinity), and will therefore sink, helping to redistribute warmth downward in the ocean. Once ice forms, it is even less dense than warmer waters, so it floats. In most substances, the solid state is denser than the liquid one, but not for water. Ice is approximately 9% less dense than liquid water.

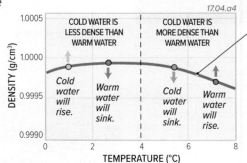

10. At temperatures above 4°C, water behaves like most other substances, becoming less dense as it warms. At temperatures above 4°C, cooler water is denser than warmer water and will therefore sink.

B How Does Change in Salinity Affect the Density of Water?

How Salt Occurs in Water and Changes Its Density

1. When we add a crystal of table salt (NaCl) to water, the crystal dissolves, separating into its constituent ions (Na$^+$ and Cl$^-$). The ions diffuse away from the dissolving crystal, becoming dispersed throughout the water. A molecule of water (H$_2$O) consists of an oxygen atom bonded with two hydrogen atoms, arranged off to one side. This geometry causes a water molecule to have a polarity in its electrical charges, with a positive charge on the side with the hydrogen atoms and a negative charge on the side with the oxygen atom.

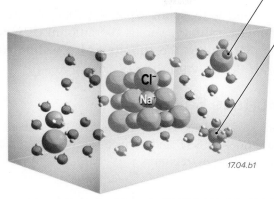

17.04.b1

2. A negatively charged chlorine ion (Cl$^-$) is attracted to the positive (H) end of the water molecule.

3. A positively charged sodium ion (Na$^+$) is attracted to the negative side of any adjacent water molecule.

4. The negatively charged chlorine anion (Cl$^-$) is surrounded by the positive sides of the encircling water molecules, while the positively charged sodium ion (Na$^+$) will be surrounded by the negative sides of water molecules. This arrangement helps keep the sodium and chlorine ions separated and limits their ability to rejoin to form a salt crystal. Adding more dissolved salt to water increases its *salinity*.

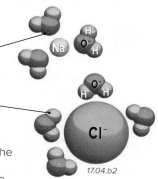

17.04.b2

5. As the salinity of water increases so does its density, in a reasonably linear fashion. Elevated saline content in water increases its density, so the saline water will sink. Less saline water will rise. The exact nature of the relationship between salinity and density also depends upon temperature.

Decreasing the Salinity of Water

Sunlight
17.04.b3
Less Saline

17.04.b4
River
Less Saline

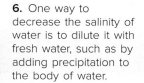
17.04.b5
Melting Ice
Less Saline   More Saline

6. One way to decrease the salinity of water is to dilute it with fresh water, such as by adding precipitation to the body of water.

7. Dilution of salt content also occurs from the introduction of fresh water, as from a stream. Most rivers contain some dissolved salt, but much less than the sea.

8. Ice consists of fresh water because it does not easily incorporate salt into its crystalline structure when it freezes. When ice melts, it adds fresh water to the body of water, diluting the salt content of the water and decreasing the salinity. Ice melting on land likewise provides abundant fresh water to streams, most of which drain into some body of water.

17.04.b6 *Lemaire Channel, Antarctica*

9. Several types of ice occur in water. The most familiar type is an *iceberg* (◄), which is a piece of ice broken off a glacier that flowed into a sea or lake. The second type is *sea ice* (not shown), which forms from freezing of seawater into a smoother layer of ice coating part of the sea.

Increasing the Salinity of Water

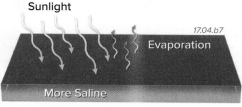
Sunlight
17.04.b7
Evaporation
More Saline

10. The most obvious way to increase the salinity of water is through evaporation, which removes water molecules but leaves the salt behind. The water vapor contains no salt and can be carried far away by motion in the atmosphere, eventually producing freshwater precipitation as rain, snow, sleet, or hail. This fresh water forms the streams and glaciers on land.

ICE-FORMING
17.04.b8
Ice Shelf
More Saline   Less Saline

11. Salinity also increases when ice forms on the sea. Fresh water is removed during freezing, leaving behind the salt in the water. As a result, the salinity of seawater increases below and adjacent to sea ice. If the sea ice melts, as can occur during the summer, it introduces the fresh water back into the ocean. We explore further the relationship between salinity, density, and temperature in the next two pages.

Before You Leave This Page

- ✓ Sketch a water molecule and explain why it has polarity.
- ✓ Sketch and explain the relationships between the temperature of water and its density, including changes at 4°C and 0°C.
- ✓ Describe how water dissolves minerals, and how salinity affects the density of water.
- ✓ Sketch and explain three ways to decrease the salinity of water and two ways to increase the salinity of water.

17.4

17.5 What Are the Global Patterns of Temperature and Salinity?

THE OCEANS VARY IN SALINITY from place to place and with depth. Spatial variations are due to differences in the amount of evaporation and formation of ice, which increase salinity, versus the amount of precipitation, input from streams, and melting of ice, all of which add fresh water that dilutes the salinity. Also, saline waters are denser than fresh water, so there are changes in salinity as a function of depth within the oceans.

A What Is the Observed Pattern of Sea-Surface Salinities?

The oceans vary in temperature and salinity, and therefore also in density, laterally across the surface and from shallower to greater depths. The maps below summarize the global variations in SST and surface salinity. Examine the patterns on each map and then compare the patterns between the two maps, region by region.

1. This map (▶), now familiar to you, shows typical SST, with red and orange being warmest and blue being coolest. Note where the warmest seawater is located, and think about what implication this might have for salinity. Do you think this warm water will be more or less saline than waters that are somewhat cooler, like those shown in yellow on this map?

2. The dominant feature of global variations in SST is the equator-to-pole contrast arising from the supply of insolation and the amount of energy available to heat the water.

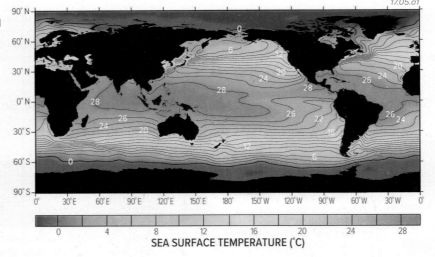

SEA SURFACE TEMPERATURE (°C)

3. Where are the coldest waters located? Naturally, they are near the poles. The cold end of the temperature scale below the map indicates that some of the coldest areas are at or even below freezing. We would expect ice to be present in these areas. Some ice could be forming and increasing the salinity of the seawater left behind, but salt causes water to resist freezing until below 0°C.

4. This map (▶) shows salinities of the ocean surface on the same day, with red areas having the greatest salinities and blue and purple having the smallest. First, observe the main patterns. Then, think about how you might explain these patterns by considering the processes described on the previous two pages, such as the amount of evaporation and cloud cover. After you have done these, read on.

5. The lowest salinities, represented by the regions in blue and purple, are mostly at high latitudes, such as areas closer to the poles. We can largely explain these low salinities by high-latitude regions being cold, with low amounts of insolation, and therefore low amounts of evaporation. Note, however, that some areas of low salinities are well away from the poles. One low-salinity region is in the very warm water that is southeast of Asia, adjacent to huge rivers. Therefore, the warmest areas of the ocean are not necessarily those with the highest salinity.

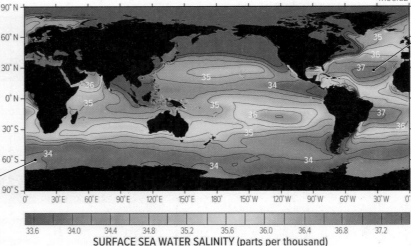

SURFACE SEA WATER SALINITY (parts per thousand)

7. High salinities occur in the subtropics, coinciding with locations of the subtropical high pressures. These areas have relatively clear skies and intense sunlight, which together cause high rates of evaporation. Also, the subtropics have low amounts of precipitation and dilution. The Mediterranean is very saline because it experiences high evaporation rates and has a limited connection to the Atlantic, with few high-discharge rivers flowing into it. Also note the presence of some moderate-salinity water in the northern Atlantic Ocean, near Greenland.

6. Low salinities away from high latitudes occur (1) where large rivers, like the Mississippi, Amazon, and Ganges-Brahmaputra rivers, deliver fresh water to the sea, (2) where cold currents, like the California Current and Humboldt Current, bring less saline waters away from the pole, and (3) in areas of high precipitation, like west of Central America and in the previously mentioned warm-water area southeast of Asia. Find all of these areas on the map.

B   How Do Temperature and Salinity Vary with Depth?

The oceans also vary in temperature and salinity as a function of depth. Seawater is warmest near the surface, where it is warmed by the Sun. The distribution of salinity is similar, with the most saline waters near the surface, where there is evaporation. Salinity has more complex patterns too, because saline waters are more dense than less saline water and therefore can sink. We present the temperature and salinity data in two pole-to-pole cross sections (North Pole on the right). The black, jagged shape at the bottom is the vertically exaggerated seafloor.

1. This pole-to-pole cross section (taken at 20° W longitude) shows Atlantic Ocean temperatures from the surface to the seafloor. Warmer surface waters overlie colder, deep ones. When this occurs, less dense water overlies more dense water, a highly stable arrangement.

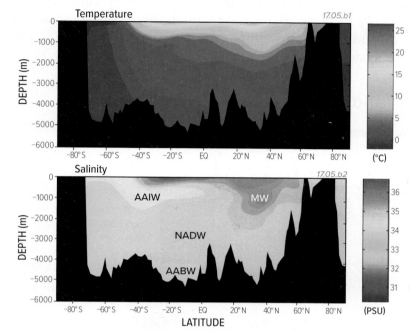

2. Only at polar latitudes, both in the far north (right) and far south (left), are surface and deep ocean water temperatures similar. When shallow and deep waters are similar in temperature and density, they can mix more easily, bringing deep water up toward the surface and causing shallow water to sink. Such mixing of waters from different depths is called *overturning* or *turnover*.

3. This cross section shows salinity (measured in a unit of salinity called PSU) along the same line, with the North Pole on the right. Examine the patterns on this cross section and compare it with the temperature cross section above. Predict which way some of the water is moving at depth.

6. Last, but not least, a key feature is an extensive tongue of saline polar waters that descends near Iceland near 60° N to 70° N and then extends well south of the equator as the *North Atlantic Deep Waters* (NADW). As we shall explore shortly, this flow of saline waters down into the depths of the ocean and to the south has a major role in moderating our climate.

4. In this cross section, a tongue of less saline waters from Antarctica, called the *Antarctic Intermediate Waters* (AAIW), forms a wedge below saline waters near the surface with somewhat more saline waters at depth. Along the bottom are some moderately saline, very cold and dense waters known as *Antarctic Bottom Waters* (AABW).

5. In the upper part of the middle of the cross section is a relatively thin veneer of extremely saline waters that formed from evaporation and limited rainfall in the subtropics. A thicker batch of salinity at 40° N marks saline waters (MW) exiting the Mediterranean Sea, which also explains the increased temperatures at 30° N to 40° N.

C   What Are the Regional Relationships Among Temperature, Salinity, and Density?

1. To explore how temperature, salinity, and density vary from region to region, this graph plots temperature on the vertical axis and salinity on the horizontal axis. Contours crossing the graph represent differences in density in parts per thousand, as calculated from the temperature and salinity. Greatest densities are in the lower-right corner. The yellow star represents the mean temperature and salinity of the world's oceans. Examine this graph and consider why a region plots where it does.

2. Equatorial waters are warm due to high insolation and are less saline than average seawater due to heavy tropical rainfall. Warm temperatures and low salinity produce low densities (i.e., numbers on the contours).

3. Mid-latitude waters are moderate in temperature and salinity. Compared to the average ocean, they are cooler because they receive less insolation than average, but less saline because precipitation generally exceeds evaporation.

4. Ice-free polar regions, predictably, have relatively cold seawater. They are also slightly less saline due to lower amounts of insolation and associated evaporation at high latitudes.

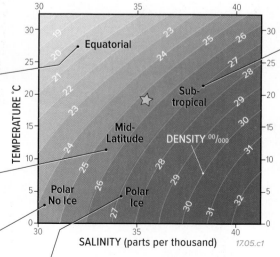

5. In polar zones where sea ice is actively forming, ice removes some fresh water from seawater, leading to greater salinities.

6. Subtropical waters are cooler than equatorial ones, but more saline, because surface evaporation exceeds precipitation. The salinity in particular leads to higher densities of these waters. The most saline waters in the world are in parts of the subtropics, especially where an area of the sea is mostly surrounded by continents and has somewhat restricted interchange with the open ocean. Such very saline waters are in the Red Sea and Mediterranean Sea.

Before You Leave This Page

☑ Summarize the global patterns of ocean salinity and temperature, how different regions compare, and what factors cause the larger patterns.

☑ Summarize how salinity and temperature vary as a function of depth, identifying the major distinct types of water at depth.

17.5

17.6 What Processes Affect Ocean Temperature and Salinity in Tropical and Polar Regions?

THE EXTREMES IN SEAWATER TEMPERATURE AND SALINITY occur at high and low latitudes. The warmest and most saline waters occur in the tropics, especially the subtropics. Very cold, less saline waters occur near the poles, away from where ice is actively forming. Here, we examine the processes occurring in some of the warmest and coldest parts of the ocean to illustrate how these important end members affect the planet's oceans and the planetary ocean-atmosphere-cryosphere system.

A What Are Warm Pools and How Are They Formed?

1. Within Earth's oceans are regions where the water temperatures are higher than in adjacent regions, and these warm regions persist, or re-form, year after year. These regions of warm seawater are called *warm pools*. The figure here illustrates how they are formed and maintained.

2. Excess insolation (compared to outgoing longwave radiation) generates warm air and warm sea-surface temperatures (29–30°C). Abundant precipitation falls on the surface of the tropical and equatorial oceans where, despite the high temperatures, precipitation exceeds losses to evaporation.

3. Warm, rising air carries large quantities of water vapor into the troposphere. It cools adiabatically but more slowly than the air around it and therefore continues to rise due to instability, forming very deep convection.

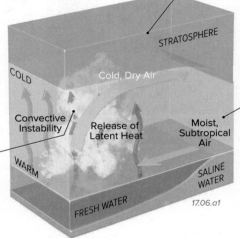
17.06.a1

4. Deep convection over tropical and equatorial regions leads to the release of latent heat and an expansion of the troposphere, causing the tropopause to be higher (17–18 km) over the tropics than over the rest of the planet.

5. Air at the tropopause and in the lower stratosphere is very cold (as low as −85°C) with extremely low moisture content (4 parts per million). The air rising from the tropics is no longer less dense than air around it so it ceases rising. Thus, air moves laterally toward the poles (the yellow arrow) as if along an invisible ceiling.

6. Air moves horizontally across the ocean surface to replace the unstable rising equatorial air, bringing water vapor evaporated from the subtropical regions, where evaporation exceeds precipitation. Once in the tropics, the air rises, condenses, and releases latent heat, which was derived from the subtropics. The tropics therefore accumulate excess energy to be released in this restricted area of rising air and instability. The ocean warms, forming a warm pool.

Location and Influence of Warm Pools

7. These two globes show SST, but they focus on areas of warm pools, which occur in the warmest waters, those shaded red and orange.

8. Earth's warmest and largest warm pool is near Southeast Asia, broadly between China and Australia. The warm pool is centered on the shallow seas that are partially enclosed by the islands of Indonesia. This warm pool is commonly called the *Western Pacific Warm Pool*. Warm waters in this region also extend into the adjacent Indian Ocean.

17.06.a2   17.06.a3

9. Another warm pool, often called the *Western Atlantic Warm Pool*, is centered on the Caribbean Sea, between North and South America. This warm pool becomes warmest and best developed in the summer and fall, when the full impact of summer warming has occurred. The warm water occupies the Caribbean Sea, Gulf of Mexico, and adjacent parts of the western Atlantic Ocean. The shapes of North America and South America enclose the Caribbean part of this warm pool on three sides.

17.06.a4 Raja Ampat, Indonesia

10. The warm waters of the Western Pacific Warm Pool sustain rich and diverse life. These seas are famous for their coral reefs and clear, warm waters, stretching from Indonesia (◄) to east of the Philippines, in the island republic of Palau (►).

17.06.a5 Palau

11. The Western Atlantic Warm Pool forms the warm waters of Caribbean vacation spots, like the Florida Keys, the Cayman Islands, and beaches near Cancún, Mexico (►).

17.06.a6 Akumal, Mexico

B What Processes Occur Around the Poles?

1. In terms of temperatures and humidity, the poles are the exact opposites of a warm pool. Here, we explore the processes near the poles, using a simple example where the entire polar region is ocean (no land).

2. Cold air aloft moving poleward is extremely dry. Near the pole, it subsides and warms adiabatically. The combination of extremely low moisture contents and warming of the sinking air results in very few clouds.

3. The lack of water vapor and clouds encourages the loss of outgoing longwave radiation. When combined with the sinking and warming air, this creates a very stable atmosphere due to a near-surface temperature inversion (warmer air over cooler air), which inhibits cloud formation and precipitation.

4. Ice floats on the surface of the ocean, due to the unusual physical properties of water. The high albedo (reflectance) of snow and ice reflects much of the already limited insolation back toward space, while the absence of clouds and water vapor ensures considerable heat loss through longwave radiation back toward space. As a result, the surface remains cold.

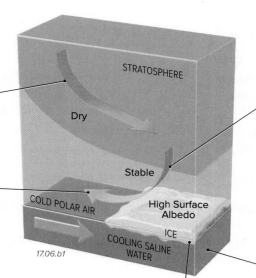

17.06.b1

7. At the same time, cold polar air blows outward across the surface away from the pole, ultimately replacing the warm rising air at the equator. In this way, the oceans and atmosphere are both moving energy and material toward and away from the poles.

6. As the dense, polar waters descend, surface waters from lower latitudes flow toward the poles to replace them. This flow of dense saline waters downward, and their replacement by poleward surface flows of warmer water, sets up circulation involving both the shallow and deep waters.

5. The formation of ice on the surface of the polar oceans increases the salinity of the underlying waters, turning them into *brine*—very saline water. This and the very cold water temperatures increase the density of the surface waters to the point that they attain a higher density than that of the deep ocean waters below them, causing surface water to descend.

17.06.b2 Antarctica

8. Ice shelves of Antarctica (◀) gain ice in three main ways, all of which result in ice that is fresh, not saline: (1) freezing of seawater below but excluding salt (called *brine exclusion*), (2) flow of glaciers from land to sea, and (3) snowfall and windblown snow on top. Brine exclusion makes adjacent water more saline, but melting of ice will release fresh water, decreasing the salinity. The resulting salinities indicate the relative contributions of freezing versus melting. The photograph to the right is of an ice shelf in Antarctica, but taken underwater, looking up at its underside, with a cooperative jellyfish for scale.

17.06.b3 McMurdo Sound, Antarctica

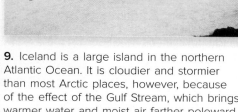

17.06.b4 Hekla Volcano, Iceland

9. Iceland is a large island in the northern Atlantic Ocean. It is cloudier and stormier than most Arctic places, however, because of the effect of the Gulf Stream, which brings warmer water and moist air farther poleward than is typical.

---

10. Many scientists believe that this global-scale connection between the poles—along with feedbacks between the oceans, atmosphere, and cryosphere—is responsible for the abrupt changes (vertical gray lines) in climate recorded by ice cores from

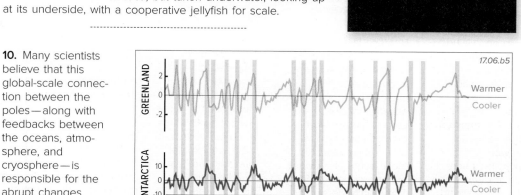

17.06.b5

Greenland and Antarctica (▲). Rapid drops in Greenland temperatures are strongly associated with simultaneous Antarctic warming. These changes occur over short geologic time periods of one to two thousand years, but not as fast as Hollywood would have us believe.

Before You Leave This Page

☑ Sketch and explain processes that form low-latitude warm pools, and identify where they are located.

☑ Sketch and explain processes that occur in polar areas and how they affect the salinity of water.

17.6

17.7 How Are the Atmosphere, Oceans, and Cryosphere Coupled?

THE OCEAN SURFACE marks the boundary between two of Earth's major systems—oceans and atmosphere. Both systems move mass and energy laterally and vertically, primarily in response to variations in density and to the equator-to-pole energy gradient. Deep-flowing abyssal waters are generally cooler and denser than surface waters, but they rise in some places and sink in others. The global, three-dimensional flow of matter and energy between the ocean, atmosphere, and cryosphere is a fundamental determinant of global climate.

A How and Why Do Temperature, Salinity, and Density of Seawater Vary?

The oceans vary in temperature and salinity, and therefore also in density, laterally across the surface and from shallower to greater depths. The figure below summarizes these variations from equator to pole, and between upper and lower levels of the ocean. Begin with the text above the figure, which describes the atmospheric processes imposed on the underlying sea, and then read how the sea responds to these changes.

1. Atmospheric motions result from the energy gained by insolation into the lower atmosphere, and by radiative cooling at higher levels. Warm, rising air is located over the seasonally migrating Intertropical Convergence Zone, centered at the equator. As the warm air rises in these regions, it cools and produces condensation, clouds, and precipitation.

2. In the subtropics, air is dominantly sinking in the descending limb of the Hadley cell. This forms semipermanent subtropical anticyclones, where high pressure causes warming and limits the cloudiness and precipitation.

3. The poleward edge of the mid-latitudes is another region where rising air is common. Upward atmospheric motions are caused by differences in density between cool polar and warm subtropical air masses, leading to low pressure and an increase in clouds and frontal precipitation. These latitudes receive less insolation than subtropical ones, which, together with the higher albedo of the clouds, reduces evaporation.

4. Near the poles, descending air combined with the limited insolation keeps polar regions cold and relatively dry. This in turn limits the amount of evaporation and causes a noticeable absence of precipitation.

5. Clouds in equatorial and tropical regions allow more precipitation than local evaporation, causing a decrease in salinities near the surface. The warmer, less saline waters have low density, forcing waters of intermediate density beneath them immediately around the equator.

11. As cold and saline waters near the pole sink deep into the oceans, they begin to flow back toward the equator as *abyssal waters*. In contrast, the flow of surface waters is generally toward the poles. Together, the poleward-flowing surface waters and equator-flowing deep waters form a circuit of flowing water that acts to "overturn" waters of the ocean.

Less Saline / THERMOCLINE / More Saline / Less Saline / SURFACE / ICE / More Saline / EQ / Warming / 30° N/S / 60° N/S / Cooling / POLE

17.07.a1

6. Unlike the atmospheric system, the oceanic system is heated from above. Insolation is absorbed in the upper layers of the oceans but does not penetrate far beyond a depth of at most 100 m. This upper zone is warmer and less dense, and is known as the *photic zone*. Deep ocean waters below these depths are consistently at between −1°C and 4°C throughout the world's oceans. These waters stay in liquid form at barely subzero temperatures because the salinity of the water decreases the freezing temperature (like antifreeze in your car) and because of the pressure exerted by the great depths of overlying ocean waters.

7. Warmer, less saline and less dense waters "sitting" on top of cooler, more dense waters is an extremely stable arrangement that discourages mixing between the two sets of water. The transition from the photic zone to deep ocean waters is usually marked by a very rapid change in density and in temperatures—a boundary called the *thermocline*. This figure represents the thermocline as the boundary between lighter blue (warmer) waters and darker blue (colder) waters.

8. Much insolation reaches the ocean surface in the subtropics, in part due to high Sun angles and a general absence of clouds. The combination of high evaporation and low precipitation leads to these subtropical waters becoming more saline. The salinity decreases with depth, especially near the thermocline.

9. In the mid-latitudes, lower insolation and the higher albedo of the clouds reduces evaporation, which, in combination with increased frontal precipitation, leads to a freshening of surface waters.

10. The formation of sea ice in polar latitudes preferentially extracts fresh water from seawater, leaving a concentration of dissolved salt in the very cold ocean waters. The combination of cool polar sea temperatures and high salinity produces surface waters of sufficient density to descend.

B What Is the Thermohaline Conveyor?

The flow of deep water and surface water combine to form a global oceanic circulation system, commonly called the *thermohaline conveyor* because it is driven by differences in water temperatures and salinities. This circulation system is among the most important on Earth. What is it, where is it, and how does it operate? The best way to understand it is to follow water through the system, beginning in the Caribbean Sea.

1. Ocean waters in the tropical Atlantic Ocean and Caribbean Sea, along with those in the Gulf of Mexico, heat up before entering the subtropics where their salinity increases through evaporation. Cold waters from the *Greenland Sea (1)* and the *Labrador Sea (2)* cool these salty waters in the North Atlantic around Iceland and the Maritime Provinces of Canada, making them sufficiently dense to descend deep into the Atlantic as the North Atlantic Deep Waters (NADW).

2. At depth, the NADW flow south through the Atlantic Ocean, threading their way between Africa and the Americas, crossing the equator, and turning east when they encounter Antarctica.

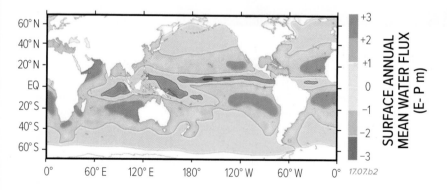

17.07.b1

3. Extensive ice shelves extending from Antarctica over the adjacent *Weddell Sea (3)* and *Ross Sea (4)* exclude salt water during their formation, causing the nearby waters to be very cold and saline. These Antarctic Bottom Waters (AABW) sink to join the NADW in moving eastward. Some deep waters branch northward into the Indian Ocean basin, while the remainder continues into the Pacific basin, mixing along the way and becoming slightly warmer and less dense.

5. Neither the Indian nor Pacific ocean basins extend sufficiently poleward to generate their own northern deep ocean waters, but deep waters upwell in the *Indian Ocean (5)* and join the surface flow.

4. Once in the Pacific, the deep waters turn north, again crossing the equator. The northward-moving deep waters are ultimately forced to rise, or upwell, at the Pacific basin's northern limit, south of *Alaska (6)*. Surface currents move these waters back south into the tropical and equatorial regions, ultimately returning to the Atlantic south of Africa and completing the loop to the Gulf Stream.

6. There is concern about how climate change could impact the operation of the thermohaline conveyor. To evaluate this issue, we can consider how climate change could impact global temperatures, and how these in turn would affect the amounts, distribution, and movement of water due to changes in evaporation, rainfall, and runoff. These factors affect temperatures, salinity, and density of water, which are critical in driving the thermohaline conveyor.

7. This map (◄) shows evaporation (E) minus precipitation over the oceans (Pm), expressing the flux of water from the surface to the atmosphere. Surface areas in red are losing the most water, and surface areas in purple are gaining water. Areas along the equator gain water from the large amounts of precipitation, whereas excess evaporation in the subtropics causes the surface to lose water. These fluxes are accompanied by changes in salinity of seawater. Added rainwater decreases the salinity, whereas evaporation increases salinity. Changes in salinity and temperature can influence the thermohaline conveyor, which in turn influences climates in North America, Europe, and other regions.

17.07.b2

8. This map (►) illustrates the pathways and changing water temperatures of the Gulf Stream as it works its way up the East Coast of North America. The stream is warm (orange) in the south where it starts, but cools (yellow to green) as it gets farther north and east, toward Europe.

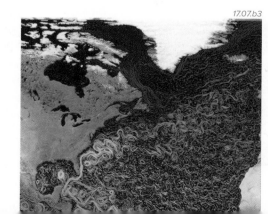

17.07.b3

Before You Leave This Page

✓ Sketch and explain what causes variations in temperature and salinity from the equator to the poles, and how this helps drive deep-ocean flow.

✓ Sketch and describe a simple map showing the thermohaline conveyor, identifying one location of descending water and one location of upwelling water and explaining why these occur.

17.7

17.8 What Connects Equatorial Atmospheric and Oceanic Circulation?

EQUATORIAL AND TROPICAL REGIONS are areas of greatest excess energy. The interaction of ocean and atmosphere systems there have great implications for climate and its variability in other parts of the world. There are large-scale, east-west connections within the equatorial atmosphere-ocean system. These connections are a major cause of global climate variability, including important phenomena called *El Niño* and *ENSO*.

A How Do Continents Impact Ocean Circulation?

The equator crosses two broad areas of land—South America and Africa—and a large number of islands of the maritime southeast Pacific, including those in Indonesia. These various landmasses block east-west flow in the oceans, turning ocean currents to the north or south. They also cause deep ocean currents to upwell along their flanks.

1. If there were no equatorial or tropical continents, the world would be circumscribed by a fairly homogeneous band of warm ocean water. The trade winds north and south of the equator would exert a small stress on the surface of the ocean, causing a net westward movement of the waters.

2. Continents and other landmasses are in the way of this westward flow of tropical ocean water, causing the warm water to accumulate, or pool, along the eastern side of the land, as expressed on this SST map.

3. Islands and peninsulas between Australia and Asia block the westward movement of warm surface waters, causing a ponding, or a warm pool, in the western equatorial Pacific.

4. In the Atlantic Ocean, westward movement of ocean waters along the equator is blocked by the band of land in Central America that joins the wider landmasses of North America and South America. Once blocked, these warm waters are directed northward into the Gulf Stream in the Atlantic and southward along the northeastern side of South America in the Pacific. The movement of these warm waters is clearly reflected in the map patterns of sea-surface temperatures.

−29 −25 −21 −17 −14 −10 −6 −2 2 6 10 14 17 21 25 29°C

17.08.a1

B How Do Warm Pools Influence Atmospheric Circulation Above Oceans?

1. Once formed along the eastern side of a continent, a warm pool can set up an east-west circuit of flowing air through the tropics.

2. Warm, moist air rises above the Western Pacific Warm Pool, where temperatures regularly exceed 28°C. As the air rises, it cools, producing condensation, clouds, and intense rainfall, as well as releasing large quantities of latent heat into the troposphere immediately above the warm pool.

3. Unable to penetrate vertically beyond the tropopause (top of the tropopause), the air moves horizontally away from the warm pool. As it moves, it mixes with the surrounding cool air and cools.

4. Having cooled sufficiently, the air descends over the eastern Pacific, warming as it descends.

5. This cooler, denser air from above turns west and passes back over the surface, joining the trade winds, warming and evaporating moisture from the ocean surface as it proceeds westward.

6. This pattern of east-west equatorial atmospheric circulation, initiated by warm air rising above warm pools, is known as the *Walker cell circulation*, named after the scientist who first recognized the pattern.

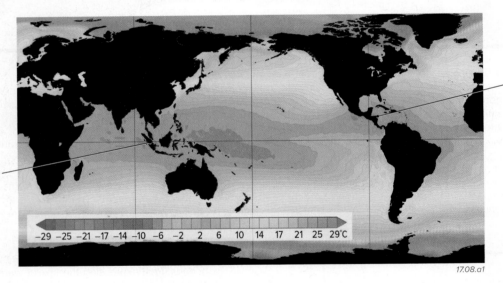

Equator

17.08.b1

C What Is El Niño?

The most well-known and publicized cause of climate variability, floods, droughts, hurricanes, heat waves, and landslides is a change in the strength of winds and ocean currents west of South America in what has become known as El Niño. El Niño is one expression of the ocean-atmosphere system operating over and within the equatorial Pacific Ocean. Phenomena associated with El Niño are more correctly termed ENSO because they involve oceanic (El Niño, EN) and atmospheric (Southern Oscillation, SO) components, which are linked.

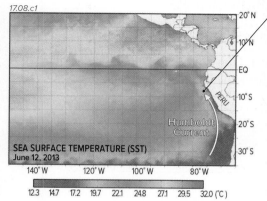

SEA SURFACE TEMPERATURE (SST)
June 12, 2013

12.3 14.7 17.2 19.7 22.1 24.8 27.1 29.5 32.0 (°C)

The waters off the coast of northern Peru are generally cold, as shown in this map of SST (◀), because the north-flowing Humboldt Current brings very cold water from farther south. The cold character of these waters is intensified by the upwelling of cold water near the shore, which occurs when warmer surface waters are pushed westward away from shore, partly by dry southeasterly trade winds. The cold upwelling limits evaporation, producing a dry desert coastline, but it brings to the surface nutrients that nourish plankton and cold-water fish.

Fishermen noted that the waters began warming (▶) toward the end of November and early in December (Southern Hemisphere summer), heralding the beginning of the Christmas season. They named this local phenomenon "El Niño," in reference to the boy child or "Christ child." Occasionally, the warming is exceptionally strong and persistent, referred to as a strong *El Niño* pattern. This brings abundant rains to the region, with severe flooding. The waters become less productive for sea life, and the fishing industry suffers.

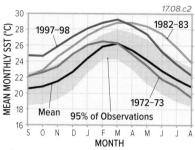

D What Is the Southern Oscillation?

1. Associated with changes in the strength of El Niño are fluctuations in air pressures in the Pacific. This map (▶) shows surface air pressures for part of the Pacific Ocean, centered on the equator. Low pressures are darker, whereas high pressures are lighter in color. When air pressure increases in the region near Darwin, Australia, air pressure typically decreases in Tahiti, a cluster of islands farther east in the central Pacific. When pressure decreases in Darwin, it generally increases in Tahiti. This change in relative strengths of atmospheric pressure is termed the *Southern Oscillation*.

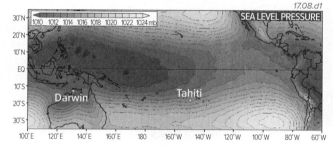

SEA LEVEL PRESSURE
1010 1012 1014 1016 1018 1020 1022 1024 mb
Darwin   Tahiti

2. Here is one way to envision these coordinated fluctuations in air pressures—the Southern Oscillation—as the ascending and descending limbs of a Walker cell over the Pacific. These fluctuations can be portrayed by a simple system (▶), where a change in the air pressures in one limb of the cell is matched by an opposite change in pressure at the other limb. For the Walker cell in the Pacific, conditions at Darwin, Australia, represent the ascending limb (near the warm pool, on the left) and those at the Pacific island of Tahiti represent the descending limb (on the right).

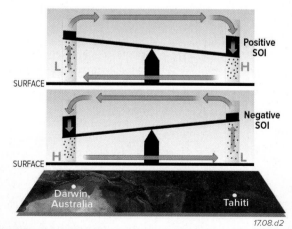

17.08.d2

3. To quantify these changes in pressure, we can calculate a *Southern Oscillation Index* (SOI), which represents the difference in air pressure between Tahiti and Darwin. We calculate it simply by subtracting the air pressure measured at Darwin from that measured at Tahiti (▲). If air pressure is higher at Tahiti than at Darwin, the SOI is positive (◀). If air pressure is higher at Darwin than Tahiti, the SOI is negative. Patterns of alternating fluctuations are apparent in the plot above, which compares pressures to the long-term mean. The graph shows that the SOI switches back and forth from positive to negative modes, generally every couple of years. We can anticipate that as air-pressure patterns change, wind directions should also change.

Before You Leave This Page

☑ Sketch the warm pools in each of the major ocean basins straddling the equator, and explain their distribution and relationship to the Walker cell over the equatorial Pacific Ocean.

☑ Sketch and explain what is meant by the terms El Niño and Southern Oscillation, and how they are expressed.

17.8

17.9 What Are the Phases of ENSO?

THE ATMOSPHERE-OCEAN SYSTEM in the equatorial Pacific is constantly changing. Although each year has its own unique characteristics, certain atmosphere-ocean patterns repeat, displaying a limited number of modes. We can use surface-water temperatures in the eastern equatorial Pacific to designate conditions as one of three phases of the *El Niño-Southern Oscillation* (ENSO) system—neutral (or "normal"), warm (*El Niño*), and cold (*La Niña*).

A What Are Atmosphere-Ocean Conditions During the Three Phases of ENSO?

El Niño and La Niña phases represent the end-members of ENSO, but sometimes the region does not display the character of either phase. Instead, conditions are deemed to be neither and are therefore assigned to the *neutral phase* of ENSO. To understand the extremes (El Niño and La Niña), we begin with the neutral situation.

Neutral Phase of ENSO

1. Warm, unstable, rising air over the western equatorial Pacific warm pool produces low atmospheric pressures near the surface.

2. Walker cell circulation in the equatorial troposphere brings cool, dry air eastward along the tropopause.

3. Cool, descending air over the eastern equatorial Pacific produces dominantly high atmospheric pressure at the surface and stable conditions in the atmosphere.

9. The warm, moist air above the warm pool rises under the influence of low pressures, producing intense tropical rainfalls that maintain the less saline, less dense fresh water on the surface of the warm pool.

4. Easterly trade winds flow over the Andes mountain range and then continue to the west across the ocean, pushing west against the surface waters along the coast of South America. The easterlies continue propelling the warm water westward toward Australia and southeast Asia, allowing the waters to warm even more as they are heated by insolation along the equator.

8. Warm waters blown to the west not only depress the thermocline to about 150 m below the surface, but also physically raise the height of the western equatorial Pacific compared to the eastern Pacific.

5. Offshore winds and westward displacement of surface waters induce upwelling of cold, deep ocean waters just off the coast of western South America. Abundant insolation under clear skies warms these rising waters somewhat, so there is no density-caused return of surface waters to depth.

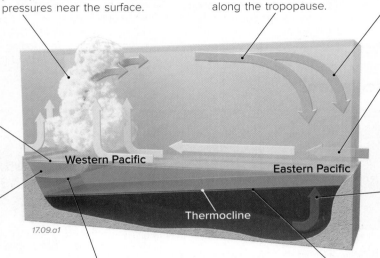

Western Pacific

Eastern Pacific

Thermocline

17.09.a1

7. In the western Pacific, surface waters are warm (over 28°C) and less saline because of abundant precipitation over the ocean and stream runoff from heavy precipitation that falls on land. The warm surface waters (the warm pool) overlie cooler, deeper ocean water—a stable situation.

6. The thermocline slopes to the west, being over three times deeper in the western Pacific than in the eastern Pacific. This condition can only be maintained by a series of feedbacks, including the strength of the trade winds.

Warm Phase of ENSO (El Niño)

10. During a warm phase (El Niño), the warm pool and associated convective rainfall move toward the central Pacific.

11. El Niño conditions are also characterized by weakened Walker cell circulation over the equatorial Pacific. This is expressed by decreased winds aloft and by a reduction in the strength and geographic range of the easterly trade winds near the surface.

15. For Australia, Indonesia, and the westernmost Pacific, El Niño brings higher atmospheric pressures, reduced rainfall, and westerly winds. The warm pool and associated convective rainfalls move toward the central Pacific, allowing cooler surface waters in the far west.

12. Upon reaching South America, the cool air descends over equatorial parts of the Andes, increasing atmospheric pressure, limiting convectional uplift, and reducing associated rainfall in Colombia and parts of the Amazon.

14. Changes in the strength of the winds, in temperatures, and in the movements of near-surface waters cause the thermocline to become somewhat shallower in the west and deeper in the east, but it still slopes to the west.

13. Weakening of the trade winds reduces coastal upwelling of cold water, which, combined with the eastern displacement of the descending air, promotes a more southerly location of the ITCZ in the southern summer and increased precipitation in the normally dry coastal regions of Peru and Ecuador.

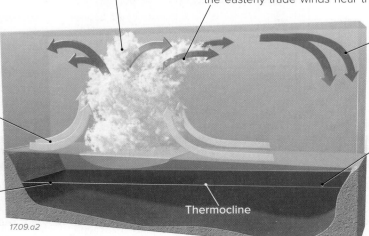

Thermocline

17.09.a2

Cold Phase of ENSO (La Niña)

1. In many ways, the cold phase of ENSO (La Niña) displays conditions opposite to an El Niño, hence the opposing name.

2. During a cold phase of ENSO (La Niña), Walker cell circulation strengthens over the equatorial Pacific. This increases winds aloft and causes near-surface easterly trade winds to strengthen, driving warmer surface waters westward toward Australasia and Indonesia.

8. The region of equatorial rainfall associated with the warm pool expands, and the amount of rainfall increases.

7. In the western Pacific, strong easterlies push warm waters to the west where they accumulate against the continent, forming a warmer and more expansive warm pool. In response, the thermocline of the western equatorial Pacific is pushed much deeper, further increasing the slope of the thermocline to the west.

3. Enhanced easterly trade winds bring more moisture to the equatorial parts of the Andes and to nearby areas of the Amazon basin. Orographic effects cause heavy precipitation on the Amazon (east) side of the mountain range (not shown).

4. Partially depleted of moisture and driven by stronger trade winds, dry air descends westward off the Andes and onto the coast. The flow of dry air, combined with the descending limb of the Walker cell, produces clear skies and dry conditions along the coast.

5. As surface waters push westward and the Humboldt Current turns west, deep waters rise (strong upwelling). The resulting cool SST and descending dry, stable air conspire to produce excessive drought in coastal regions of Peru.

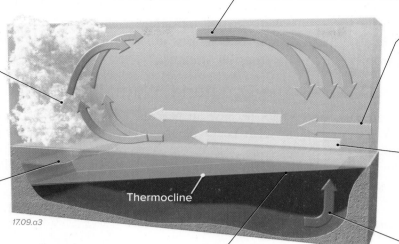

Thermocline

17.09.a3

6. The upwelling near South America raises the thermocline and causes it to slope steeper to the west. Cold water is now closer to the surface, producing favorable conditions for cold-water fish.

B | How Are ENSO Phases Expressed in Sea-Surface Temperatures?

As the Pacific region shifts between the warm (El Niño), cold (La Niña), and neutral phases, sea-surface temperatures (SST), atmospheric pressures, and winds interact all over the equatorial Pacific. These variations are recorded by numerous types of historical data, especially in SST. The globes below show SST for the western Pacific (near Asia) and eastern Pacific (near the Americas) for each phase of ENSO—neutral, warm, and cold. The colors represent whether SST are warmer than normal (red and orange), colder than normal (blue), or about average (light).

Neutral Phase of ENSO

17.09.b1-2

During the *neutral phase* of ENSO, SST along the equator in the Pacific are about average, with no obvious warmer or colder than normal waters near the Western Pacific Warm Pool (left globe) or South America (right globe). An area of warmer than normal SST occurs southwest of North America, but this is not obviously related to ENSO.

Warm Phase of ENSO (El Niño)

17.09.b3-4

During the *warm phase* of ENSO, a belt of much warmer than normal water appears along the equator in the eastern Pacific, west of South America. This warm water is the signature of an El Niño, causing the decrease in cold-water fishes. SST in the western Pacific are a little cooler than average, but an El Niño is most strongly expressed in the eastern Pacific (right globe).

Cold Phase of ENSO (La Niña)

17.09.b5-6

During the *cold phase* of ENSO (a La Niña), a belt of colder than normal water occurs along the equator west of South America, hence the name "cold phase." The western Pacific (left globe), however, now has waters that are warmer than normal. These warm waters are quite widespread in this region, extending from Japan to Australia.

Before You Leave This Page

☑ Sketch and explain atmosphere-ocean conditions for each of the three typical phases of ENSO, noting typical vertical and horizontal air circulation, sea-surface temperatures, relative position of the thermocline, and locations of areas of excess rain and drought.

☑ Summarize how each of the three phases of ENSO (neutral, warm, and cold) are expressed in SST of the equatorial Pacific Ocean.

17.9

17.10 What Are the Effects of ENSO?

WHILE THE IMMEDIATE IMPACTS OF ENSO are restricted to regions near the equatorial Pacific, shifts between different phases of ENSO cause climate variability well beyond the equator and the Pacific. When atmospheric conditions in one region affect a distant region, these distant associations are called *teleconnections*, which in this case are caused by interactions with the subtropical and polar front jet streams.

A What Is the Connection Between the Equatorial Warm Pool and the Extra-Tropics?

The effects of an ENSO can spread out of the tropics because air near the equator rises and flows northward as part of the Hadley cell. Motions in the Walker and Hadley cells occur simultaneously, so equatorial air can be incorporated into both circulation cells, spreading ENSO effects to distant parts of the globe (a teleconnection). To explore how this teleconnection operates, we follow a packet of the warm air that was produced in the Western Pacific Warm Pool during a cold-phase (La Niña) event. Begin reading on the lower left of the figure below.

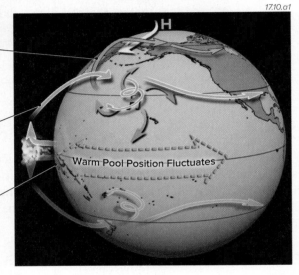

17.10.a1

3. Warmer air flowing poleward from the subtropics is forced to rise over the cold polar air along the polar front. This poleward-moving air also has an apparent deflection to its right (in the Northern Hemisphere), joining the rapidly east-flowing polar front jet stream.

2. When the rising air reaches the upper levels of the troposphere, it is transported poleward in the Hadley cell. It has an apparent deflection to the right (due to the Coriolis effect) and becomes part of the rapidly eastward-moving subtropical jet stream. Eventually, the air descends in the subtropics near 30° N latitude.

1. During a La Niña phase, excess warm water accumulates in the Western Pacific Warm Pool. As this excess energy transfers from the ocean to the atmosphere, convection begins to carry the warm, moist air upward.

Warm Pool Position Fluctuates

4. The different phases of ENSO are distinguished by the position of the western warm pool (west for La Niña, farther east for El Niño). As this position changes, moist tropical air is fed into the jet streams in different locations along their paths, affecting the subtropical and polar jets, particularly "downstream" (east) in the jet streams (in both hemispheres). Rising air from the warm pools also flows south into the Southern Hemisphere, affecting weather there as well.

ENSO and Patterns of Atmospheric Pressure

WARM PHASE
El Niño 90° E

180°   0°

90° W

COLD PHASE 90° E
La Niña

180°   0°

17.10.a2
90° W

5. These two views (◀) of the Northern Hemisphere summarize changes in atmospheric pressure that typically accompany the warm and cold phases of ENSO. For each phase, areas in blue typically have atmospheric pressures that are lower than normal and those in red typically have higher than normal atmospheric pressures. The colors do not indicate absolute pressure, just deviations from normal (anomalies).

6. During a warm phase (El Niño), areas of higher than normal pressure are typically centered over Hudson Bay, the north-central Pacific, southwestern Europe, and western Africa.

7. At the same time, lower than normal pressures form in the eastern Pacific, Mexico, and across the southern U.S.

---

8. During a cold phase (La Niña), the patterns of atmospheric pressure are nearly reversed from those of an El Niño. Lower than normal pressures extend across much of southern Europe, Greenland, and western North America.

9. During a La Niña, higher than normal atmospheric pressures are situated along the East Coast of the U.S. and cover most of the eastern Pacific. Recall that high pressure is associated with dry, sinking air, whereas low pressure is associated with rising air and unsettled weather. We would predict such conditions in the red (high-pressure) and blue (low-pressure) areas on these maps.

ENSO and Jet Streams

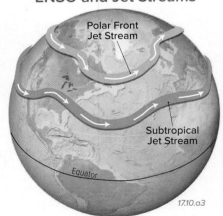

Polar Front
Jet Stream

Subtropical
Jet Stream

Equator

17.10.a3

10. The polar front and subtropical jet streams flow in a continuously changing wave-form around the globe, with the positions of the waves governed by locations of high and low pressure. Changes in the strength and location of the global pressure cells, as occur during ENSO events, therefore modify the usual pattern of the jet streams. The jet streams in turn control the typical paths of weather-generated storms and other weather patterns.

B What Are the Typical Responses to ENSO Around the Pacific Basin?

ENSO greatly affects weather conditions over the Pacific and in adjacent continents. The four maps below summarize the weather conditions that result from ENSO during summer and winter. Note that due to the increased energy gradient between the equator and the poles during winter, the "winter" hemisphere frequently shows greater changes by ENSO phase than the "summer" hemisphere. The details are less important than the main patterns, the general processes, and noting how ENSO affects where you live or visit.

Warm Phase (El Niño)

1. During the warm (El Niño) phase, a weakening of the Walker cell and easterlies (trade winds) shifts the Pacific warm pool eastward and limits upwelling next to South America.

2. As the Western Pacific warm pool shifts eastward, dry conditions spread across the far western Pacific (yellow). If prolonged, these conditions cause drought in this area.

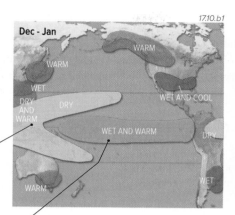

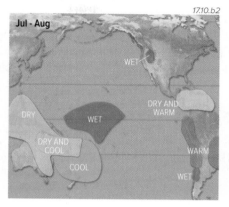

5. In July and August during El Niño, the central Pacific stays rainy and the western Pacific stays relatively dry.

6. During these months, effects of El Niño diminish in the Northern Hemisphere (where it is summer) but intensify in the Southern Hemisphere (where it is winter and jet streams are more robust).

3. The locus of precipitation follows the warm pool into the middle of the Pacific. The weakened easterlies also allow more December–January precipitation along the western coast of South America.

4. In North America, ENSO-related alteration of the path of the polar front in the winter affects the northern part of the continent, while changes in the subtropical jet stream affect conditions over the southern part of the continent.

7. The position of the southern subtropical jet stream controls the paths of precipitation-bearing frontal systems in the mid-latitudes of the Southern Hemisphere. Central Chile is most sensitive to shifts in the passage of these winter storms. The quantity of snow falling in the Andes controls water available for irrigation in Chilean and Argentinian rivers during the next dry summer season.

Cold Phase (La Niña)

8. During December–January of a cold (La Niña) phase, the Walker cell and easterlies intensify, pushing warm water westward into the Western Pacific Warm Pool, causing wet conditions across that region.

9. The westward ocean currents cause upwelling along the western coast of South America. The presence of cold waters in the eastern equatorial Pacific causes dry and cool conditions there. The position and intensity of the ITCZ changes, affecting the strength of trade winds, regional precipitation, and tropical storms in the Caribbean region.

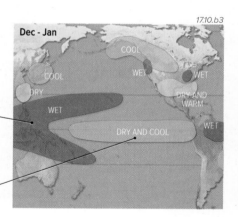

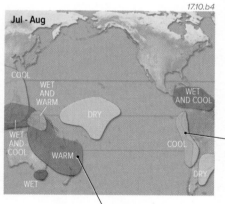

11. In July and August during La Niña, the western Pacific is warm and wet because of the presence of an extensive and strong warm pool. The cooler central Pacific stays relatively dry. Along the west coast of South America, the appearance of excessively cold water cools the coast during the southern winter, decreasing rainfall.

10. Effects of La Niña are abundant in North America. These include cooler and wetter than normal conditions next to the Gulf of Alaska, but dry and warm conditions in the southeast U.S.

12. In the Polynesian islands and northeastern Australia, warm, moist equatorial air clashes with cooler sub-Antarctic air along a northwest-southeast zone (shown in orange). Changes in the strength of the subtropical pressures shift this precipitation back and forth like a large door hinged over northeastern Australia.

Before You Leave This Page

☑ Sketch and explain the atmospheric connection of the Western Pacific Warm Pool and the subtropical and polar front jet streams.

☑ Explain how warm (El Niño) and cold (La Niña) phases of ENSO can affect the position of jet streams and other seasonal weather patterns, using examples from North America and one from elsewhere in the world.

17.10

17.11 What Types of Life Reside in the Oceans?

OCEANS COVER MORE THAN 70% OF EARTH'S SURFACE, so they are home to most of Earth's organisms. Oceans are among the most life-rich parts of the Earth, but the abundance and diversity of marine life varies greatly from region to region. Most parts have an overall lack of nutrients, resulting in relatively sparse life and organic productivity in those places. What factors cause these variations?

A How Does Marine Life Vary with Depth and Distance from Shore?

One way to classify creatures that live in the ocean is by whether an organism resides mainly suspended in water—is *pelagic*—or on the ocean floor—is *benthic*. In either case, marine organisms have adapted to different depths, temperatures, salinities, horizontal and vertical currents, substrate characteristics, distances to land, and nutrient sources.

17.11.a1 Point Lobos, CA

1. In the *coastal* or *intertidal zone* (◄), both terrestrial and aquatic organisms are challenged by water availability, wave action, temperature variations, and changes in substrate. Biodiversity is greater in more secure, rocky substrate than in less secure, shifting sand. Sea grass, crabs, starfish, and plankton are abundant in this zone.

2. The *pelagic zone* is located seaward of the coastal zone's low-tide mark and contains the vast open waters of the ocean. Life here can be classified according to location within the water column, mobility, and size. In the *epipelagic* (photic) zone, light penetrates and becomes differentiated by depth into colors. The *mesopelagic* and *hadal zones* are too deep for light to penetrate, a region called the aphotic zone.

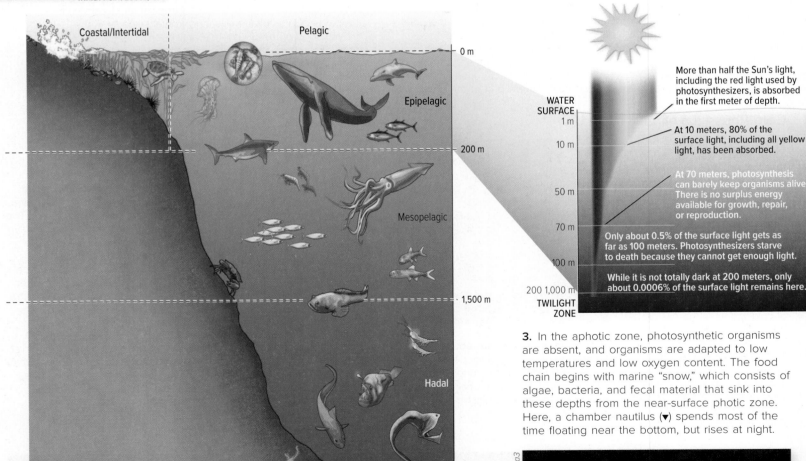

More than half the Sun's light, including the red light used by photosynthesizers, is absorbed in the first meter of depth.

At 10 meters, 80% of the surface light, including all yellow light, has been absorbed.

At 70 meters, photosynthesis can barely keep organisms alive. There is no surplus energy available for growth, repair, or reproduction.

Only about 0.5% of the surface light gets as far as 100 meters. Photosynthesizers starve to death because they cannot get enough light.

While it is not totally dark at 200 meters, only about 0.0006% of the surface light remains here.

WATER SURFACE
1 m
10 m
50 m
70 m
100 m
200 1,000 m
TWILIGHT ZONE

17.11.a2

3. In the aphotic zone, photosynthetic organisms are absent, and organisms are adapted to low temperatures and low oxygen content. The food chain begins with marine "snow," which consists of algae, bacteria, and fecal material that sink into these depths from the near-surface photic zone. Here, a chamber nautilus (▼) spends most of the time floating near the bottom, but rises at night.

4. The *benthic zone* contains all the habitats of the sea bottom, whether in coastal, continental shelf, or deep-sea environments. Organisms may live within the bottom material or on its surface. This zone is among the most poorly explored and understood on Earth.

B | Which Parts of the Ocean Have the Most Life and Why?

Coral reefs are formed by stony coral polyps living in symbiosis with green and red algae. Coral reefs occur in shallow water where sea surface temperatures range from 20°C to 36°C (68°F to 97°F). They are considered the "tropical rain forest" of the ocean, providing a habitat for some 25% of marine species. Unfortunately, reefs and their community of creatures are sensitive to overfishing, pollution, climate change, and other disturbances.

1. To estimate the spatial variation in the amount of life in the oceans, we can use satellites to measure the concentration of chlorophyll, which is a measure of organic productivity. Chlorophyll is produced mostly by *phytoplankton*, which are mostly microscopic photosynthesizing organisms, whose name is derived from the Greek words for "wandering plant." For the oceans, red and orange indicate the highest chlorophyll (most productive areas), purple and blue indicate the lowest productivity, and yellow and green represent intermediate values.

2. Observe the patterns on this map and identify which settings in the ocean generally have the highest and lowest productivity. Try to explain these patterns, using such factors as latitude, proximity to land, ocean currents, and areas of upwelling.

3. The highest chlorophyll concentrations, which indicate a high abundance of phytoplankton and high amounts of productivity, are along the edges of the continents and in shallow seas, such as the Baltic Sea in northern Europe.

4. In the open ocean, some of the highest chlorophyll concentrations are in the mid- and high-latitude oceans, such as between Europe and North America, adjacent to the Arctic Ocean, and east of Japan.

5. Much of the ocean is relatively unproductive (blue and purple), particularly in areas beneath the subtropical highs. Values are only slightly higher along the equator.

6. Locally high productivity occurs in areas where upwelling brings nutrients into shallow-enough levels for photosynthesis. One obvious area of upwelling is along the west coast of Africa.

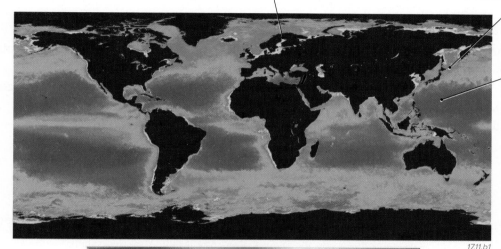

17.11.b1

CHLOROPHYLL A CONCENTRATION (mg/m^3)

17.11.b2

7. Coral reefs are areas of very high productivity. Corals are small, but not microscopic, marine animals that build colonies of calcium carbonate (of the mineral calcite and the rock limestone). They gain most of their food from photosynthetic algae that share their habitat, but they also have tentacles for obtaining food that floats or swims by. The close-up photograph above shows individual coral polyps with tentacles and a greenish coating of algae.

17.11.b3

9. High productivity also occurs in areas of upwelling. In the example here, water is being pushed parallel to the shore by wind. Cold, deep water *upwells*, replacing near-surface water that moved offshore and bringing nutrients up near the surface into the photic zone.

8. Corals have symbiotic algae that provide additional oxygen and nutrients for corals. In return, the corals provide CO_2 and a place to live. Reefs provide shelter and food for an entire community of creatures, including sponges, colorful fish (▶), crabs, and worm-like creatures. Over three-quarters of coral reefs are located in Asia and the Pacific Ocean. Coral thrive in these areas when water temperature is 20–25°C (68–77°F) and the water is clear. Nutrients are recycled efficiently.

17.11.b4

Before You Leave This Page

✓ Describe the zones of the ocean and their ability to support life.

✓ Explain the settings and creatures of coral reefs.

17.12

What Oceanic and Atmospheric Patterns Are Predicted for a Newly Discovered Planet?

A PLANET WAS DISCOVERED in a distant galaxy and is similar to Earth. It has oceans, an atmosphere very similar to Earth's, and ice at both poles. Since no astronaut has yet ventured to the planet, we currently only know things we can observe from a distance—the distribution of land, oceans, and sea ice, as well as the larger features on land. To guide future expeditions, you will use the known features to predict the global patterns of wind, ocean currents, and interactions between oceans, the atmosphere, and ice.

Goals of This Exercise:

- Observe the global distribution of continents and oceans on Planet 42.
- Create a map showing the predicted geometries of global winds, ocean currents, and major atmospheric features, like the ITCZ and polar front.
- Draw cross sections that portray the global patterns, predicting the locations of warm pools, zones of high and low precipitation, and possible sites for ENSO-type oscillations.
- Use your predictions to propose two habitable sites for the first two landing parties.

Main Procedures

Follow the steps below, entering your answers for each step in the appropriate place on the worksheet or online.

A. Observe the distribution of continents and oceans on the planet. Note any features that you think will have a major impact on the climate, weather patterns, and other aspects of atmosphere-ocean-cryosphere interactions. Descriptions of some features are on the next page.

B. Draw on the worksheet the predicted patterns of atmospheric wind circulation (trade winds, westerlies, etc.), assuming that the patterns are similar to those on Earth.

C. Draw on the worksheet the likely paths of ocean currents within the central ocean. Identify the segments of a current that will be a warm current, cold current, or neither.

D. Draw on the worksheet the likely locations of warm pools, and shade in red any area on land whose climate will be heavily influenced by the presence of this warm pool. Shade in parts of the ocean that are likely to be more saline than others.

E. From the pattern of winds and ocean currents, identify which side of a mountain range would have relatively high precipitation and which side would have lower precipitation because it is in a rain shadow. Color in these zones on the worksheet, using blue for mountain flanks with high precipitation and yellow for areas of low precipitation. Note, however, that precipitation patterns along a mountain range can change as it passes from one zone of prevailing winds into another, or if wind directions change from season to season.

Optional Procedures

Your instructor may also have you complete the following steps. Complete your answers on a sheet of graph paper.

F. Draw four cross sections across the central ocean, extending from one side of the globe to the other. Of these cross sections, draw one in an east-west direction across each of the following zones: (1) between the equator and 30° (N or S); (2) within a zone of westerlies (N or S); and (3) within the south polar zone, south of 60° S. The fourth cross section can be drawn at any location and in any direction, but it should depict some aspect of the oceanic and atmospheric circulation that is not fully captured by the other cross sections. On each cross section, draw the surface wind patterns, zones of rising or sinking air, and possible geometries of a Walker cell or other type of circulation, where appropriate. Be prepared to discuss your observations and interpretations.

G. Draw a north-south cross section through the oceans, showing shallow and deep flows that could link up into a thermohaline conveyor. Describe how this might influence the climate of the entire planet and what might happen if this system stops working.

Some Important Observations About the Planet

1. The planet rotates around its axis once every 24 hours in the same direction as Earth. Therefore, it has days and nights similar to Earth's. It orbits a sun similar to our own, at about the same distance. Its spin axis is slightly tilted relative to the orbital plane, so it has seasons.

2. The planet has well-developed atmospheric currents, with wind patterns similar to those on Earth.

3. The oceans are likewise similar to those on Earth. They are predicted to have ocean currents that circulate in patterns similar to those on Earth. The oceans are predicted to have variations in sea-surface temperature, salinity, and therefore density, but no data are available for these aspects.

4. The planet has ice and snow near both poles, some on land and some over the ocean.

1. A continent, simply called Polar Continent for now, is over the North Pole, but slightly off to one side. The western part of the continent extends south of 60°, but there are no data about whether this part has ice. There appear to be ice shelves and sea ice surrounding part of the continent.

2. There is a gap between Polar Continent and those continents to the south, providing a relatively land-free zone through which the ocean can circulate. Consider how this gap might influence speeds of winds and ocean currents here.

3. A nearly continuous mountain range runs along the coast of Eastern Continent.

5. Western Continent does not have significant mountains but, like Eastern Continent, has a curved coastline with large peninsulas that extend out into the ocean and embayments, where the ocean curves into the land.

4. There is extensive ice and snow over the South Pole, but the ice is floating on the ocean (there is no land at the pole). The sea ice extends to the Eastern and Western continents, both of which also have glaciers that flow downhill toward the sea, where they melt.

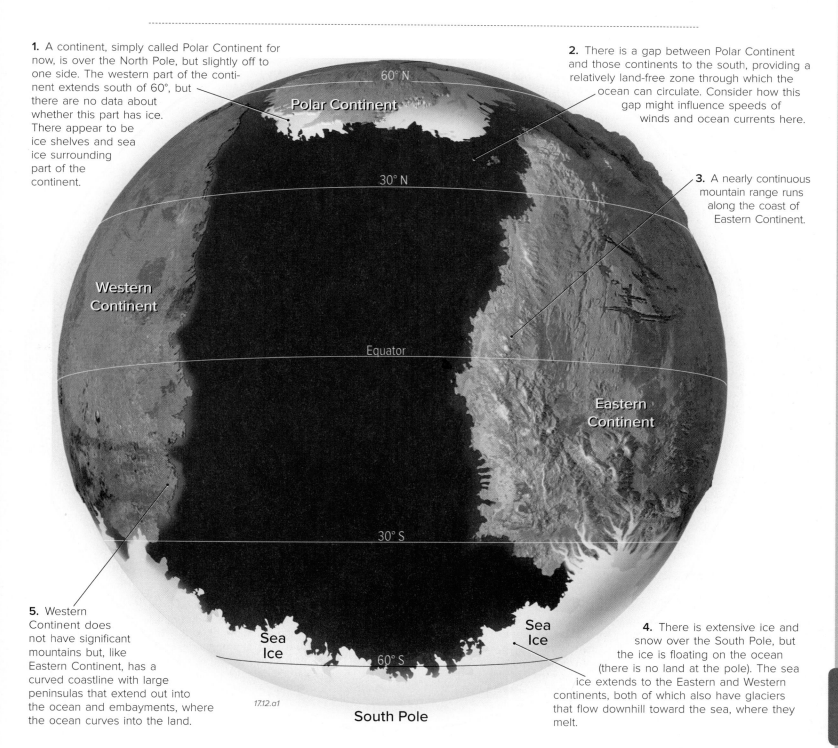

17.12.a1

18 Climates Around the World

CLIMATE IS THE LONG-TERM PATTERN of weather at a place, including not just the average condition but also the variability and extremes of weather. It is largely controlled by an area's geographic setting, especially latitude, elevation, and position relative to the ocean. Climate is also influenced by atmospheric pressure and moisture, prevailing and seasonal wind directions, and nearby ocean currents—in other words, everything we have explored so far in this book. Observed in this context, climate is the culmination of the various factors that move energy, air, and water (in all three states) in Earth's climate system.

The climate of a region can be described, measured, and represented in many ways, including maps of annual average temperature and precipitation, like the two globes below (temperature is on the left). To examine long-term climate, as opposed to short-term weather, we consider data averaged over years or decades. Climate is what we expect; weather is what we get. Remember that climate includes not only average values but also extremes and variability about the averages.

What is the difference between weather and climate, and how are the two related?

Maps of climatic variables, like the ones below, visually represent large amounts of data for the land, sea, or atmosphere. For numerous measures, like temperature and precipitation, maps can depict annual averages, values for a particular month (perhaps the warmest month) as averaged over decades, or some measures of seasonal extremes or variability from one decade to another.

What are some regional and global patterns in average temperature and precipitation on the two globes below?

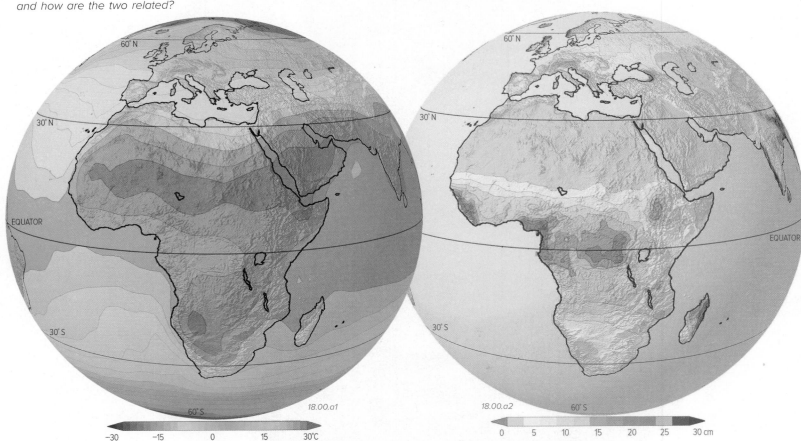

Average annual temperatures vary greatly across the planet (▲). In the map above, high average temperatures are in red and orange, and cooler ones are in green and blue. What patterns do you observe? Much of Africa, but especially the central and northern parts, are very warm, much warmer than Europe. As anyone who has traveled to these places knows, they have very different climates. Africa straddles the equator, so some parts of it are likely to be tropical, whereas Europe is situated well north of the equator. The hottest areas, however, are in the subtropics.

Do all areas with the same average temperature have the same climate? Are warm temperatures by themselves enough to define an area as tropical?

The average amount of precipitation (▲) is another factor most people consider when talking about the climate of an area. We might say that Scotland, the northern part of the British Isles, is a "cold, rainy place," but the Sahara of northern Africa is an "extremely hot and dry place." With such statements, we are expressing something about the average temperature or amount of precipitation but not what occurs on any given day (which is weather). Also, many regions have distinct wet and dry seasons, an important aspect of climate.

How can averages and variability in temperature and precipitation be combined to represent the climate of a region?

TOPICS IN THIS CHAPTER

By considering temperature and precipitation, both in terms of their averages and in terms of average maxima, average minima, and seasonal variations, geographers developed several climate classification systems, including the one portrayed on this globe (▶). The colors and letters indicate what type of climate characterizes an area. Examine the color patterns and letters on this globe and compare these to the globes showing average annual temperature and precipitation (on previous page). What correlations do you note?

How do we classify climates, and what do the letters assigned to a region indicate about its climate?

Oceans have characteristic climates, just like land areas. Note on this map that the climate of a land area generally continues out across an adjacent body of water, whether it is the Atlantic and Indian oceans or the Mediterranean Sea. Evaporation of water in the oceans provides energy for weather systems and other dynamic processes (like flooding), so assessing the climate of the oceans is important in understanding Earth.

On land, we consider types of plants in an area when talking about its climate, but how do we assign a climate zone to part of an ocean?

The climate of a region is influenced by many factors discussed in this book, including latitude and its relationship to Sun angle, elevation, topography, the locations of semipermanent pressure features, prevailing wind directions, common storm tracks, sea-surface temperatures, ocean currents, humidity, atmospheric oscillations, and teleconnections. These and other factors act in concert to produce the characteristic climates around the world, such as these in Africa and Europe.

How do Sun angle, topography, pressure features, humidity, storm tracks, ocean currents, and other factors interact to produce the observed variations in climate?

Climate is not static. Variations happen from year to year, and longer-term changes also occur. Witness the rise of global temperatures in the last 150 years, since the end of a very cold period in the 1800s. Human activities introduce materials, such as pollution and CO_2, into the atmosphere, oceans, and land. As natural lands or less densely populated areas are converted into cities and towns, long-term averages of temperature, precipitation, cloud cover, and local winds are affected—that is, urbanization changes the local climate.

In what ways do human activities affect local and global climate?

18.00.a3

18.0

18.1 How Do We Classify Climates?

CLASSIFICATION IS THE PROCESS of grouping similar items together and separating dissimilar items. Items that are similar in some ways can be different in others, and items that are in different groups can have certain similarities. Classifications allow us to examine general patterns, with the caveat that interpretations derived from classified groups are best done cautiously. Here, we examine a well-established method of classifying climates.

A What Is the Purpose of Classifying Climates?

Climates are classified to let us observe broad patterns and simplify communication about the characteristics of a region. For example, the terms tropical and polar efficiently convey some notions about such places with a single word. Historically, there have been thousands of meteorological (weather) stations collecting weather and climatic information, and in the past few decades this wealth of knowledge has expanded with the addition of large quantities of remotely sensed information. Classification helps us seek useful generalizations from this enormous amount of information. Climate can be classified in many ways, with the exact nature of the classification depending upon the research question that is of interest.

1. Colors on this map of an imaginary country represent mean annual temperatures (reds above 28°C, blue and purple below 0°C). The black contours represent mean annual precipitation in millimeters. Meteorological stations are shown as dots. Examine the patterns on this map and think about how you might classify different parts of the country (e.g., cold and wet).

2. The most southerly areas are hot, but we might classify stations in the southeast differently than those in the southwest because of the differences in precipitation (and therefore in humidity).

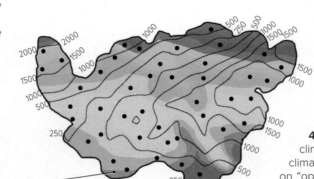

18.01.a1

3. An understanding of climate helps us understand other parts of the earth system. Knowing the typical amount and timing of precipitation helps us predict the magnitude of and timing of future floods. If winter temperatures are commonly below freezing, then we would also need to consider floods from snowmelt.

4. Most crops have fairly well-defined climatological requirements for their growth. A climate classification could be developed based on "optimal," "fair," or "poor" for growing that crop.

B What Climate Classification System Is Most Widely Used and Why Is It Useful?

Wladimir Köppen was a botanist interested in the global distribution of vegetation types. He surmised that the annual temperature and precipitation regimes determined the types of natural vegetation. Other influences on plant growth, such as which way a slope faces, the number of consecutive days without precipitation, and the types of soils, have more localized effects and so were not considered by Köppen.

1. Using the Köppen climate classification system or more recent modifications of this system, the lines separating different types of climate also separate different realms of natural vegetation. For each climate type, Köppen carefully selected thresholds of temperature and precipitation that preserved the boundaries for that characteristic assemblage of vegetation. The original scheme underwent modification as more information on global climates and vegetation became available. The resulting *Köppen climate classification* is the most widely used system; an example is portrayed on this climate map of South America (▶). Notice how the patterns of colors correlate with latitude, elevation, proximity to the ocean, and other aspects of position within the continent. Some of the boundaries would be recognizable on a satellite image.

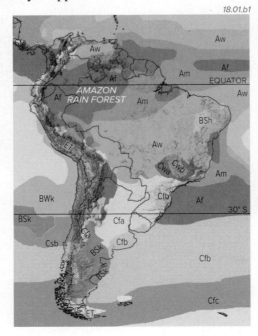

18.01.b1

2. In reality, boundaries between climatic types are not static from year to year and not as clearly delineated as the map would suggest. Instead, a boundary between vegetation and climate types usually occurs across a broad transition zone, known as an *ecotone*. In South America, an ecotone occurs between the Amazon rain forest and mountain climates of the Andes.

3. Köppen's system was designed to delineate vegetation realms, so it overlays well with worldwide natural vegetation zones—*biomes*. Both climate and vegetation are major controls on the types of soil that develop, leading to a notable similarity between the Köppen classification and maps of world soils. Köppen was interested in vegetation realms, so he did not classify the climate over the oceans. With satellite-based temperature and moisture data available for the oceans, the climate zones can be extrapolated over the ocean, as shown on the figures in this chapter.

C How Is the Köppen Classification System Organized?

The Köppen climate classification system has five major categories, represented by the first (capital) letter of the labels on the South America map. These are the following: A–*tropical climates*, B–*arid climates*, C–*temperate mid-latitude climates*, D–*harsh mid-latitude climates*, and E–*polar climates*, each represented by a photograph below. The five categories are further subdivided using a succession of criteria of temperature and the availability of water, as illustrated by the flowchart below. In the A, C, and D climate types, the second letter is lowercase and indicates whether the climate is wet year-round (f), has a dry summer (s), has a dry winter (w), or experiences a monsoon (m). The second letter is uppercase for arid and polar climates. Subsequent pages list key characteristics of each type of climate, but follow the flowchart below first.

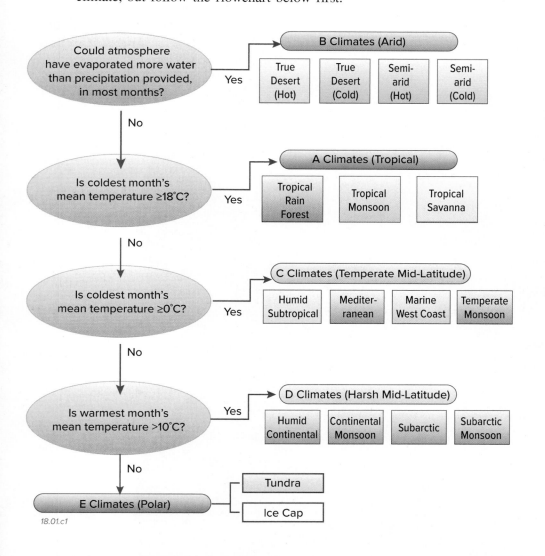

18.01.c1

B—Arid
18.01.c2 Namibia

A—Tropical
18.01.c3 Grand Cayman Islands

C—Temperate
18.01.c4 Germany Valley, WV

E—Polar
18.01.c6 Churchill, Canada

D—Harsh Mid-Latitude
18.01.c5 Northern British Columbia, Canada

Before You Leave This Page

✓ Explain the purpose of climatic classification.

✓ Describe Köppen's intent in developing his climatic classification system.

✓ Discuss the five main categories of climates in the Köppen climatic classification system.

18.1

18.2 What Are the Most Common Climate Types?

EACH TYPE OF CLIMATE, as defined by the Köppen climate classification, has specific characteristics that distinguish it from any other climate type. Once a climate is assigned to one of the five main groups (A–E), it is then categorized by factors such as whether the summer is hot, warm, or cool, and whether most precipitation falls in the summer, winter, throughout the year, or during a monsoon. The table on the right-hand page provides a reference for characteristics of each climate type. Carefully observe each globe, guided by the table, and then read the text blocks in clockwise order around the globes.

North and South America

1. The arrangement of climates is least complicated in the oceans, away from the land. Here, climate types are arranged in belts that roughly parallel latitude, progressing from A-type (tropical) near the equator to B, C, and E toward the poles. Some climate types, such as D-types (harsh mid-latitude), are more common over land than over the oceans, which moderate the temperatures.

2. Departures from this simple arrangement occur where ocean currents displace warm and cold water, and therefore climate types, north and south, like along the Gulf Stream.

Africa, Indian Ocean, and Antarctica

3. B-type climates are those where precipitation is not abundant enough to replenish water that could possibly be lost through evapotranspiration (*potential evapotranspiration*). B-type climates generally occur in the dry, sinking air that characterizes the subtropics. Note that B-type climates also occur over the ocean.

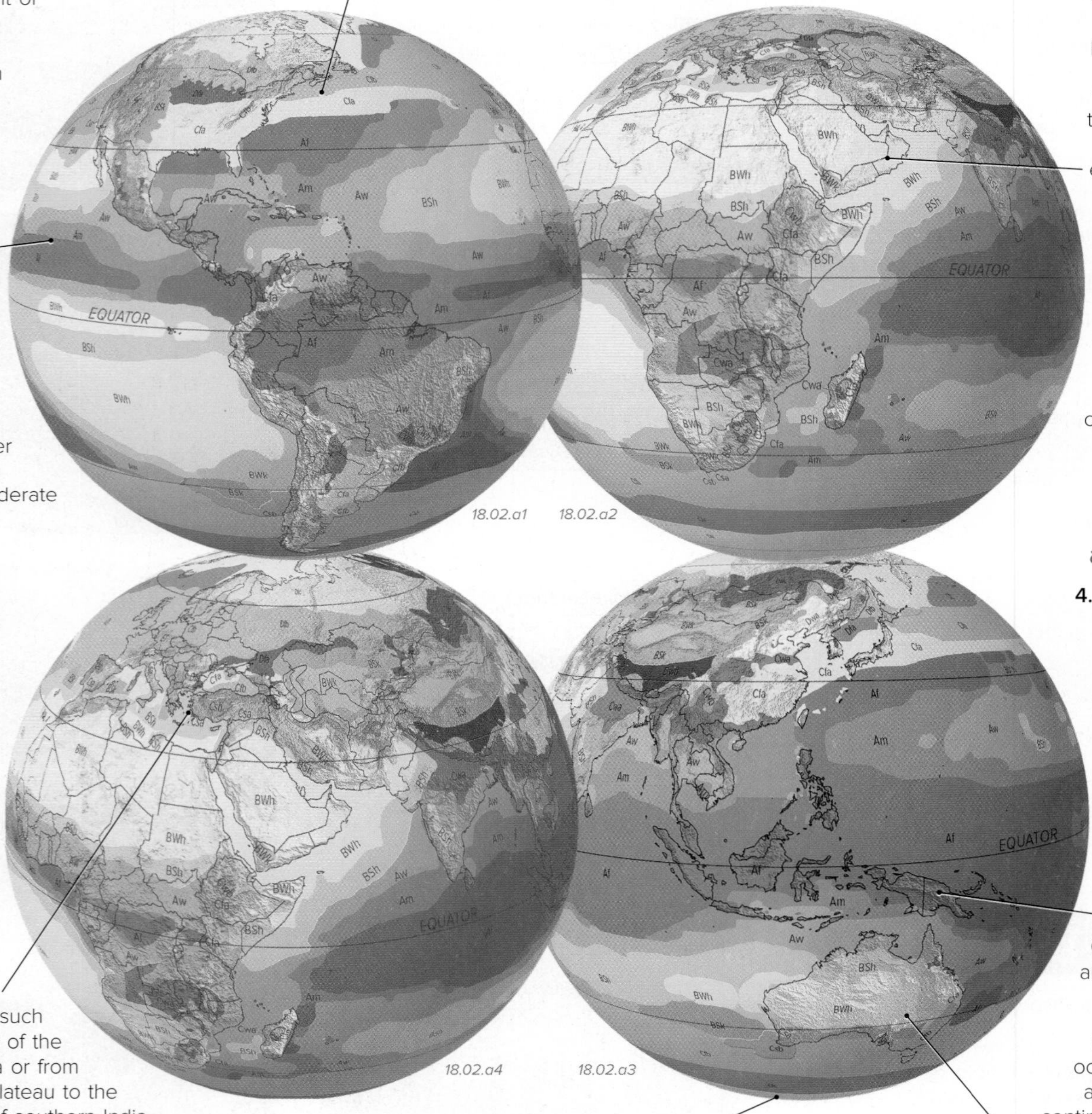

18.02.a1   18.02.a2

Eastern Asia and Australia

4. As shown by the Indian and Pacific oceans, most A-type climates occur over the ocean, and these conditions also encompass islands and adjacent landmasses, as in the region of Papua New Guinea, Indonesia, and the Philippines.

Europe and Western Asia

7. The patterns of climate types are most complex on land, especially in areas with large amounts of high relief and variations in topography. Steep gradients in climate types, as in western Asia, commonly reflect steep topographic gradients, such as those northeast of the Mediterranean Sea or from the high Tibetan Plateau to the tropical lowlands of southern India. Note that D-type climates over landmasses are largely restricted to the Northern Hemisphere, as in Siberia. An oceanic belt of D-type climates also encircles Antarctica (not shown in this view).

18.02.a4   18.02.a3

6. E-type (polar) climates are restricted to Antarctica and the surrounding ocean and to the Arctic Ocean and the surrounding land.

5. Climates in the ocean are generally asymmetric around continents, as on either side of Australia, reflecting the asymmetrical circulation of ocean currents, which bring cooler water and stable atmospheric conditions along one side and warmer water with unstable conditions along the opposite side.

Common Climate Types

First Letter	Second Letter	Third Letter	Climate Type	Descriptive Name	Description of Climate
A Tropical Climates	f		Af	Tropical Rain Forest	Generally hot with no cold season; wet year-round
	m		Am	Tropical Monsoon	Generally hot with no cold season; wet for most of the year
	w		Aw	Tropical Savanna or Tropical Wet-Dry	Generally hot with no cold season; wet for about half of the year
B Arid Climates	S	h	BSh	Hot Steppe	Semiarid; annual average temperature exceeds 12°C (54°F)
	S	k	BSk	Cold Steppe	Semiarid; annual average temperature is below 12°C (54°F)
	W	h	BWh	Hot Desert	Arid; annual average temperature exceeds 12°C (54°F)
	W	k	BWk	Cold Desert	Arid; annual average temperature is below 12°C (54°F)
C Temperate Mid-Latitude Climates	f	a	Cfa	Humid Subtropical	Wet year-round; long, hot summer and short, intermittent cold season
	f	b	Cfb	Marine West Coast	Wet year-round; warm summer and a cold but not severe winter
	f	c	Cfc	Marine Mild Summer	Wet year-round; mild summer and a cold but not severe winter
	s	a	Csa	Mediterranean	Wet in winter but dry in summer; long, hot summer and short, intermittent cold season
	s	b	Csb	Mediterranean	Wet in winter but dry in summer; warm summer and a cold but not severe winter
	w	a	Cwa	Temperate Monsoon	Wet in summer but dry in winter; long, hot summer and short, intermittent cold season
	w	b	Cwb	Temperate Monsoon	Wet in summer but dry in winter; warm summer and a cold but not severe winter
D Harsh Mid-Latitude Climates	f	a	Dfa	Humid Continental	Wet year-round; hot summer and severe winter
	f	b	Dfb	Humid Continental	Wet year-round; warm summer and long, severe winter
	w	a	Dwa	Continental Monsoon	Wet in summer but dry in winter; hot summer and severe winter
	w	b	Dwb	Continental Monsoon	Wet in summer but dry in winter; warm summer and long, severe winter
	f	c	Dfc	Subarctic	Wet year-round; short, mild summer and very long, severe winter
	f	d	Dfd	Subarctic	Wet year-round; very short, cool, intermittent summer, and very long, extremely severe winter
	w	c	Dwc	Subarctic Monsoon	Wet in summer but dry in winter; short, mild summer and very long, severe winter
	w	d	Dwd	Subarctic Monsoon	Wet in summer but dry in winter; very short, cool, intermittent summer, and very long, extremely severe winter
E Polar Climates	T		ET	Tundra	Cold, with no month having an average temperature above 10°C (50°F); freezes occur in every month
	F		EF	Ice Cap	Cold, with each month having an average temperature below freezing

First Letter: Climate Group; generally from A near the equator to E near the poles. Second letter generally describes precipitation amounts and seasonality.

Lowercase Second Letter (Groups A, C, and D): f = wet year-round; m = monsoon; s = dry summer; w = dry winter

Uppercase Second Letter for Group B: S = steppe; W = desert; Uppercase Second Letter for Group E: T = tundra; F = ice cap

Third Letter: h = hot; k = cold; a = hot summers; b = warm summers; c = mild summers; d = cool summers

Before You Leave This Page

☑ Summarize where on the planet the main climate groups (A–E) are most common, and explain how their locations compare to latitude, regional topography, and position in the interior of a continent versus next to the ocean.

☑ Summarize the factors considered (such as hot summers) when subdividing different types of climate within a main group.

18.2

18.3 What Is the Setting of Tropical Climates?

A-TYPE CLIMATES in the Köppen climatic classification system are "tropical," having consistently warm temperatures all year. Precipitation in these zones is primarily caused by the convergence of the trade winds along the Intertropical Convergence Zone (ITCZ), which shifts with the season to locations north and south of the equator. There are three types of A climates in the Köppen system—*Tropical Rain Forest*, *Tropical Savanna*, and *Tropical Monsoon*.

A Where Are the Various Types of Tropical Climates?

These globes show the distribution of each of the three types of Group A (tropical) climates. All three globes show all three types, but each type is discussed separately below the globes. Note that all three climate types are centered in the tropics, but they widen or narrow considerably from ocean to ocean and from continent to continent. Since the Köppen climate classification was defined on land and based on vegetation, the descriptions for each climate type below and on subsequent pages generally emphasize the characteristics of that climate on land.

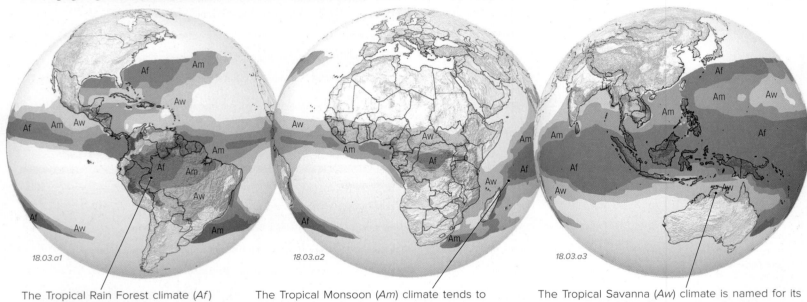

18.03.a1

18.03.a2

18.03.a3

The Tropical Rain Forest climate (*Af*) is characterized by abundant precipitation year-round, which on land supports lush vegetation. Notice that this type is surprisingly limited in extent over land, but it is more widespread over the oceans.

The Tropical Monsoon (*Am*) climate tends to occur between *Af* and *Aw* climates, as in parts of the western Pacific and Indian oceans that have wet and dry seasons, typically due to a seasonal switch in the dominant wind direction. The short dry seasons seldom allow soils to dry fully, so they support lush forests.

The Tropical Savanna (*Aw*) climate is named for its typical vegetation of grasslands and a few scattered trees. It occurs poleward of the other A-type climates and is therefore subject to the alternating influences of the subtropical highs and the ITCZ. Temperatures vary throughout the year somewhat more than in other A-type climates.

B Why Are Temperatures Consistent Throughout the Year in A-Type Climates?

A-type climates share a trait that we commonly associate with the word *tropical*—warm and relatively consistent temperatures throughout the year, a direct consequence of the way the Sun interacts with our tilted planet.

The noontime Sun is generally high over the tropics, as it migrates during the course of a year, from directly above 23.5° S latitude on the December Solstice to above 23.5° N latitude on the June Solstice, and back above 23.5° S again by the following December. As a result, it remains high over the region between these two latitudes (the tropics).

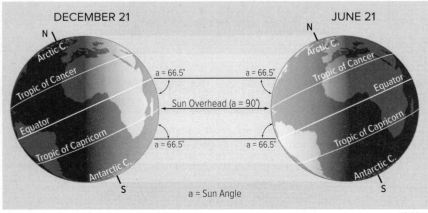

18.03.b1

This setting causes the areas between the tropics to be heated with similar intensity throughout the year, unlike the seasonal variations that occur beyond the tropics. Also, the nearly overhead position of the Sun means the lengths of day and night vary little seasonally. Locations near the equator experience approximately 12 hours of sunlight throughout the year, but not all areas in climate Group A do because they extend away from the equator.

C How Does Precipitation Differ Between the Three Types of Tropical Climates?

Tropical Rain Forest (Af)

18.03.c1 Palau

1. In a Tropical Rain Forest (*Af*) climate, precipitation exceeds losses through evaporation and transpiration in most months of the year. In addition to the lush vegetation, streams carry large volumes of water from the *Af* climates.

----

4. Trees become shorter and more widely spaced from climate *Af* to *Am* to *Aw*, from left to right in this figure (▶).

Tropical Monsoon (Am)

18.03.c2 Cambodia

2. For most months in a Tropical Monsoon (*Am*) climate, precipitation exceeds the needs of vegetation. The excess precipitation stored in the soil during these months allows rain forests to survive a few relatively dry months. Vegetation becomes slightly shorter and sparser in this zone compared to *Af*.

Tropical Savanna (Aw)

18.03.c3 Great Rift Valley, Kenya

3. The Tropical Savanna climate (*Aw*) is also known as a *Tropical Wet-Dry* climate because about half of the months receive abundant precipitation and the other half are very dry. The dry season is too long to support forests because the soil moisture becomes depleted during that time. Instead, the vegetation is grassland with scattered short, wide trees, such as acacia trees.

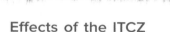
18.03.c4

Effects of the ITCZ

5. In an *Af* climate, the ITCZ brings rain most days throughout the year, although rainfall can be heavier in some months than others. *Af* climates remain under the influence of the ITCZ, even as it migrates north and south with the seasons.

6. For an *Am* climate, the ITCZ is a factor in most months, especially during the summer when it is typically overhead or very close. During low-Sun-season months, the ITCZ moves too far away to generate much precipitation.

7. In an *Aw* climate, the migrating ITCZ brings abundant summer rain. The low-Sun season is dry and warm as these areas come under the influence of the subtropical highs.

Examples

8. Group A climates are all tropical, so there is very little seasonality in temperature. Precipitation patterns do vary among the three types, displaying some seasonality. *Af* climates are closest to the equator and display the least variation in precipitation over the course of a year, whereas *Aw* climates are farthest away from the equator and experience the greatest seasonal changes. *Am* climates are typically in between in position and character. The plots below are *climographs*, each showing average temperatures as a line graph and monthly average as a bar graph. The star shows the location of that place. Observe the patterns and compare them with the information above about the ITCZ.

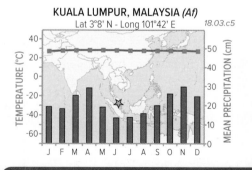
KUALA LUMPUR, MALAYSIA (Af)
Lat 3°8' N - Long 101°42' E
18.03.c5

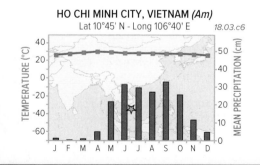
HO CHI MINH CITY, VIETNAM (Am)
Lat 10°45' N - Long 106°40' E
18.03.c6

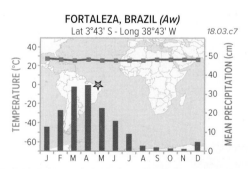
FORTALEZA, BRAZIL (Aw)
Lat 3°43' S - Long 38°43' W
18.03.c7

Before You Leave This Page

✓ Summarize where Group A climates generally occur, and the typical positions of *Af*, *Am*, and *Aw* climates relative to one another.

✓ Explain why the temperature remains nearly constant throughout the year in A climates, particularly in Tropical Rain Forest climates.

✓ Explain the relationship between the position of the ITCZ and the geographic location of each zone, and how this affects any seasonality, or lack thereof, in the amount of precipitation (i.e., wet and dry seasons).

18.3

18.4 What Conditions Cause Arid Climates?

ARID CLIMATES OCCUPY a greater portion of Earth's land surface than any other climate category. Arid climates are considered relatively dry places. They comprise Group B in the Köppen classification and are subdivided into desert climates and steppe climates, with the distinction being that deserts are more arid than steppes. Deserts and steppes are further sub-classified into hot or cold categories—not all deserts or steppes are hot! In the Köppen classification of Group B climates, *potential evapotranspiration* is taken into account in addition to precipitation; an area is classified as arid only if precipitation does not offset the loss of water through evaporation from the surface and transpiration through leaf surfaces.

A Where Are the Various Types of Arid Climates?

These globes show the distribution of the four types of arid climates—*Hot Desert* (*BWh*), *Cold Desert* (*BWk*), *Hot Steppe* (*BSh*), and *Cold Steppe* (*BSk*).

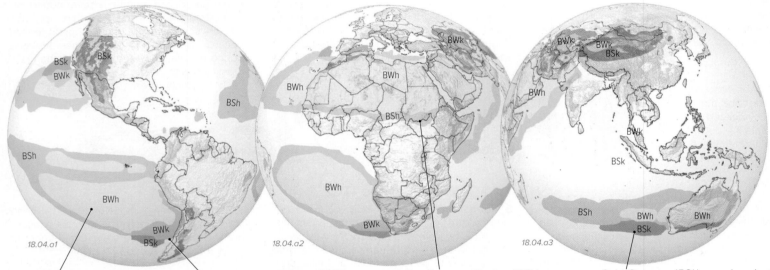

18.04.a1   18.04.a2   18.04.a3

A Hot Desert climate (*BWh*) covers huge areas of Africa (e.g., Sahara), the Arabian Peninsula, the interior of Australia, and parts of the American Southwest. It also extends over large areas of subtropical ocean.

A Cold Desert climate (*BWk*) is less common, occurring mostly in the interior of Asia (like southern Mongolia) and near cold ocean currents, like the Humboldt Current off the west coast of South America, as shown here.

Hot Steppe climates (*BSh*) generally surround the hot deserts and represent transitions to more humid climates, as across the Sahel region south of the Sahara and the area north and east of the hot deserts of Australia.

Cold Steppes (*BSk*) are abundant along the Great Plains and western interior of North America, in Tibet and other parts of central Asia, and in southern Australia. They cover relatively small parts of the ocean.

B What Controls Temperatures in Arid Climates?

18.04.b1 Big Maria Mtns., CA

18.04.b2 Sossusvlei, Namibia

1. Most deserts, like those in the American Southwest (◄), are in subtropical regions where the noon Sun is generally high, so insolation arrives nearly perpendicular to the surface. At the latitudes of most arid climates, the noontime Sun is high for several consecutive months.

2. Many, but not all, arid climates lack clouds and precipitation for much of the year. If there is little humidity, insolation is transmitted through the atmosphere (◄) with a minimum amount being absorbed in the atmosphere before it reaches the surface, promoting warm days.

3. At night, longwave radiation is lost to space efficiently because of the lack of clouds and water vapor to absorb outgoing longwave radiation, so temperatures fall rapidly.

18.04.b3

Incoming Insolation

Outgoing Longwave Radiation

Sun Angle

4. Cold Steppe (*BSk*) and Cold Desert (*BWk*) climates have the relative dryness of their hot counterparts, but have lower temperatures, like in Tibet and Mongolia. Tibet is cold because it has high elevations, averaging 4,500 m (about 14,800 ft). The limited amount of precipitation relative to potential evapotranspiration produces a characteristic sparseness of plants (▼).

18.04.b4 Tibet

C What Factors Contribute to the Lack of Precipitation in Arid Climates?

An area designated as an arid climate (Group B) in the Köppen classification system must have the *potential* to lose more moisture through evapotranspiration (called potential evapotranspiration) than it receives in precipitation. This condition can result from a limited amount of precipitation and a high demand for water by having intense insolation (i.e., high potential evapotranspiration); it is always some combination of the following factors.

Descending Air Along the Subtropics

The main cause of deserts and other arid areas is their position along the subtropics, where the descending limb of the Hadley cell brings dry air and high pressure. Note how the arid lands in the Sahara and the Arabian Peninsula (▶) form an east-west band along subtropical latitudes, a manifestation of the Hadley-caused high pressure. The descending air generally means that even if the air is humid, it cannot rise easily, which limits formation of clouds and precipitation. Arid lands of Australia and the American Southwest occupy similar subtropical settings.

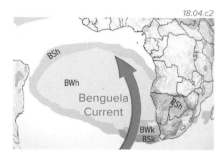
18.04.c1

Cold Ocean Currents

Other desert climate areas (*BWh* and *BWk*) and steppe climate areas (*BSh* and *BSk*) owe their existence partly or mostly to their proximity to cold ocean currents, which limit the amount of moisture in the air and promote atmospheric stability. One such area is the Namib Desert of southwest Africa (▶), adjacent to the cold Benguela Current (coming from nearer Antarctica). Another is the Atacama Desert along the southwest coast of South America, beside the cold Humboldt Current.

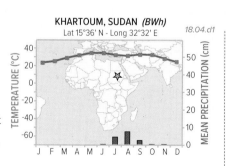
18.04.c2

Rain Shadow

An area can be dry because it is in the rain shadow of a mountain range. Southern and western sides of the Hawaiian Islands (▶) are on the rain-shadow side of the mountains. Moist air transported by the northeast trades moves up the northern and eastern slopes, causing heavy rain on those sides, but drying out the western, downwind sides.

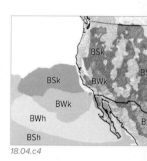
18.04.c3

Distance to Ocean

Many arid (Group B) areas are dry because they are in the interior of a continent, far from oceanic moisture sources, as in the case of the western interior of the U.S. (▶). Dry areas of the western U.S. are also in the rain shadow of mountains along the West Coast.

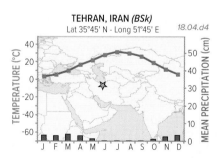
18.04.c4

D What Are Some Examples of Climate Conditions in Arid Climates?

The climographs below convey temperatures (the lines) and precipitation (bar graphs) in areas designated Group B (desert and steppe climates). As we might expect, precipitation totals are very small, and what precipitation falls evaporates quickly in the dry air, leaving a parched landscape.

Desert Climates

1. Khartoum (▶) is an example of a hot desert in the subtropics. The region receives essentially no rain for most of the year. The meager July–August rains result from a slight influence of the northern edge of the ITCZ. Khartoum is near the tropics, so its temperature seasonality is dampened and temperatures remain high, showing limited range.

KHARTOUM, SUDAN *(BWh)*
Lat 15°36' N - Long 32°32' E   18.04.d1

2. Few people live in the world's cold deserts. This town in China (▶) lies on the fringe of a desert climate. Temperatures vary markedly and are very cold during the winter, but precipitation is relatively low. Most cold deserts have even less precipitation than shown here.

URUMQI, CHINA *(BWk)*
Lat 43°43' N - Long 87°38' E   18.04.d2

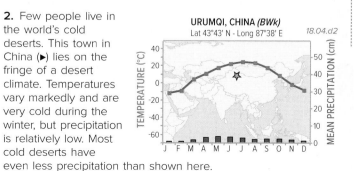

Steppe Climates

COBAR, AUSTRALIA *(BSh)*
Lat 31°29' S - Long 145°48' E   18.04.d3

TEHRAN, IRAN *(BSk)*
Lat 35°45' N - Long 51°45' E   18.04.d4

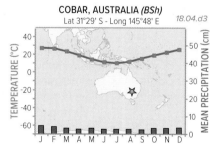

3. Steppe climates are more abundant and far more heavily populated than deserts. They are transition zones between desert climates and more humid climates. Precipitation in such environments is variable from year to year, with some years receiving desert-like precipitation totals and others experiencing precipitation more characteristic of humid climates. Like natural vegetation, human populations in these regions must be able to adapt to periodic dry spells and severe water shortages.

Before You Leave This Page

☑ Summarize the climate properties of Group B climates and where they generally occur.

☑ Sketch and explain why temperatures are high and precipitation is low in hot desert climates and other arid regions, providing examples.

18.4

18.5 What Causes Warm Temperate Climates?

TEMPERATE MID-LATITUDE CLIMATES, designated as Group C of the Köppen system, experience moderate temperatures and precipitation and so are climates where many people live. Some temperate climates are relatively warm, because they occur mostly in subtropical latitudes. These include Humid Subtropical, Mediterranean, and Temperate Monsoon climates.

A What Is the Spatial Distribution of the Various Temperate Mid-Latitude Climates?

The globes below depict distributions of six temperate mid-latitude climate types. These include *Humid Subtropical* climate (*Cfa*), two *Mediterranean* climates characterized by dry summers (*Csa* and *Csb*), and three other climates in which precipitation is predominantly in summer due to continental monsoon effects, for which they receive the "w" or "winter dry" designation (*Cwa*, *Cwb*, and *Cwc*).

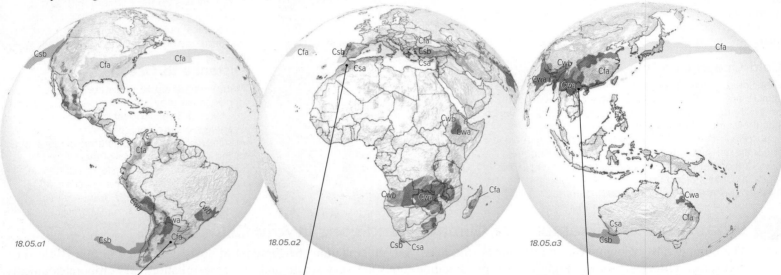

18.05.a1 18.05.a2 18.05.a3

Humid Subtropical climates (*Cfa*) are wet year-round and have hot summers with relatively short and intermittent cold seasons. They generally occur on the east coasts of continents and adjacent oceans—southeastern South America (shown here), southeastern U.S., eastern Asia (including much of Japan and part of China), and eastern Australia. Since *Cfa* climates are defined as having hot, rather than warm, summers, their extent over the oceans is somewhat limited.

Mediterranean climates experience wet winters but dry and hot (*Csa*), or dry and warm (*Csb*), summers. These are pleasant climates for vacationers or retirees. In addition to partly encircling the Mediterranean Sea, they generally occur near the west coasts of continents and in adjacent oceans, including along the Pacific coast of the U.S., southwestern South America, southern Eurasia, and the southwestern tips of Africa and Australia. Coastal California is famous for its moderate, Mediterranean climate.

Three other warm temperate climate types experience large variations in precipitation between seasons, generally associated with shifts in wind direction—a *monsoon*. Places with Temperate Monsoon climates receive most of their rainfall in the summer but are dry in the winter, and they can have hot (*Cwa*), warm (*Cwb*), or very rarely, mild (*Cwc*) summers. They are best known from India to Southeast Asia, but they also occur in southern South America, southern Africa, and northeastern Australia.

B What Affects Temperatures in Temperate Climates?

18.05.b1 Crete

1. When we think of a Mediterranean climate, we envision a place like Crete (◄), with warm summers, fair-weather cumulus clouds, and maybe a nearby ocean or sea. The pleasant temperatures are due largely to a location in or near the subtropics, where abundant insolation warms the land and water, which in turn warms the air.

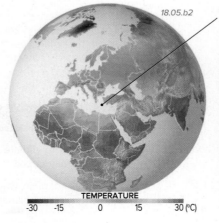

18.05.b2

TEMPERATURE

-30 -15 0 15 30 (°C)

2. On this globe showing average annual temperature, the Mediterranean (the nearly enclosed sea in the center) is a zone of transition, between very hot temperatures of the subtropics to the south and colder temperatures farther north. This characterizes a temperate climate—it is intermediate between tropical and harsh climates. Summers are warm to hot, because the Sun is high and day lengths are moderately long. Coastal areas have temperatures that are moderated by nearby bodies of water. Summers are influenced exclusively by tropical air masses, especially maritime tropical air masses and less often by continental tropical air masses. In winter, temperate climates are influenced by polar and tropical air masses, but the polar air masses are usually moderated by the time they reach these latitudes.

C What Affects Precipitation in Temperate Climates?

Temperate climates, especially those in or near the subtropics, are affected greatly by subtropical highs, like the *Bermuda-Azores High* in the north Atlantic and the *South Atlantic High*. When these high-pressure areas are strong and nearby, they suppress the formation of clouds and precipitation. Precipitation can occur more easily when the highs are farther away and when they are weaker.

1. Subtropical highs display a very important asymmetry—sinking motions descend to the surface on the eastern side (▶), which coincides with the *western side* of adjacent continents. However, the sinking does not extend to the surface on the western side, which corresponds to the *eastern side* of continents.

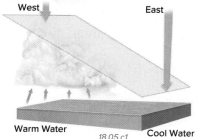

18.05.c1

2. The "tilting" of the subtropical highs is caused primarily by the difference in ocean currents on each side of the ocean. Cold currents on the eastern sides of ocean basins stabilize the atmosphere and allow sinking of air to the surface. Warm currents on the western margins destabilize the atmosphere by adding moisture and energy to the air, thereby encouraging rising atmospheric motion. Similar processes occur in both the Northern and Southern hemispheres, because in both settings cold currents are on the east side of the ocean, and warm currents are on the west side.

3. Hadley cells migrate and change width seasonally, following the position of maximum insolation northward during northern summer and southward in southern summer. Since the subtropical highs represent the descending limb of the Hadley cell, they shift and change in intensity and size too, as shown here.

4. In both hemispheres, the subtropical highs expand and make their closest approach to regions of temperate climates during the summer (June in the Northern Hemisphere and December in the Southern Hemisphere). The strong sinking action on the eastern side of a high, when it makes its closest approach in summer, stifles rain on the west coasts of the continents. Mediterranean (*Csa, Csb*) climates are characterized by this lack of rain in the summer.

5. Because sinking associated with a subtropical high does not extend to the surface on its western sides, the east sides of continents are places with Humid Subtropical climates. Here, clouds can grow vertically and summer rain falls in places like southeastern South America or the southeastern U.S. On this figure, note that Mediterranean climates (in light and dark pink) are more common on the west coasts of continents, but Humid Subtropical climates (shown in light green) are more common along the east coasts of continents—a direct response to the asymmetry of the subtropical highs.

18.05.c2

6. With the change in seasons, the Hadley cells migrate toward the opposite hemisphere (e.g., into the Southern Hemisphere during December, the northern winter). The subtropical high in the hemisphere experiencing winter also shifts toward the equator and becomes smaller and less intense, so it no longer blocks migrating weather systems or the southern advance of polar air masses. As a result, frontal precipitation from mid-latitude cyclones spreads across regions of temperate climate during the winter.

7. Thus, a Mediterranean climate owes its characteristic dry and warm or hot summer, but cooler, wet winter, to its subtropical latitude and to the interactions between migrating subtropical highs and ocean currents—a lot of things need to happen for a nice day on a California beach.

Examples

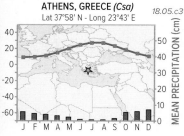

ATHENS, GREECE (Csa)
Lat 37°58' N - Long 23°43' E   18.05.c3

8. Athens, Greece, along the northern shore of the Mediterranean, experiences a typical Mediterranean climate, with hot and very dry summers.

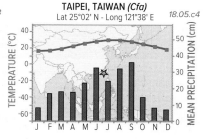

TAIPEI, TAIWAN (Cfa)
Lat 25°02' N - Long 121°38' E   18.05.c4

9. Taipei, Taiwan, has a Humid Subtropical climate, but winter precipitation is somewhat limited by wind patterns established after the end of the summertime Asian monsoon.

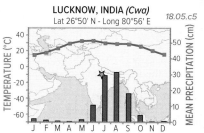

LUCKNOW, INDIA (Cwa)
Lat 26°50' N - Long 80°56' E   18.05.c5

10. Lucknow, India, is in the region affected by the South Asian monsoon, with dramatic variations in rainfall during the year, earning its monsoon designation (*Cwa*).

Before You Leave This Page

✓ Summarize the three varieties of Group C climates (Mediterranean, Humid Subtropical, and Temperate Monsoon) and where they generally occur.

✓ Describe factors that influence temperatures of a temperate climate.

✓ Sketch and explain how migration of subtropical highs produces different climates on west versus east coasts of a continent.

18.5

18.6 What Are the Settings of Mid-Latitude Climates?

NON-ARID, MID-LATITUDE CLIMATES tend to experience a relatively even distribution of precipitation year-round. Some are dominated by maritime air masses that moderate temperature swings and result in relatively mild winters and summers; they are within Group C in the Köppen designation. Other mid-latitude climates involve more severe winters because they are dominated by continental air masses, and these fall into the D classification. Mid-latitude regions are affected by westerlies, and precipitation patterns reflect the role of mid-latitude cyclones.

A Where Are Marine West Coast and Humid Continental Climates Found?

The globes below portray the distributions of three mid-latitude climate types—*Marine West Coast* climate (*Cfb* and *Cfc*), *Humid Continental* climate (*Dfa* and *Dfb*), and *Continental Monsoon* climate (*Dwa* and *Dwb*). As you can observe, most are in cold places.

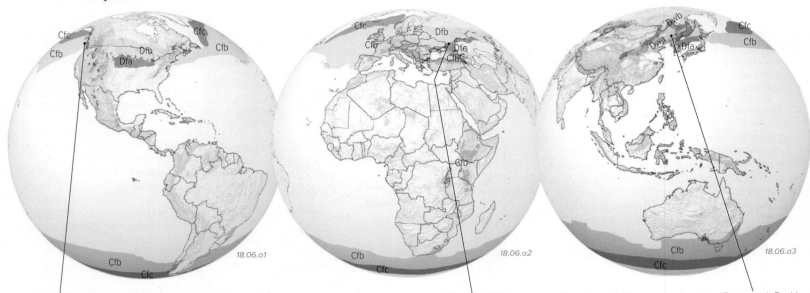

18.06.a1   18.06.a2   18.06.a3

The Marine West Coast climates (*Cfb* and *Cfc*) occur over or near oceans at moderately high latitudes, like the southwest coast of Canada. An abundant supply of moisture provides precipitation throughout the year. The influence of the ocean means that summers are warm (*Cfb*) or cool (*Cfc*), and winters are cold but not severe. Such climates occur in northern and southern parts of oceans and across most of western Europe.

Humid Continental climates (*Dfa* and *Dfb*) occupy the interior of continents, but only in the Northern Hemisphere—southern continents are not wide enough at these latitudes to allow these climates. Humid Continental climates are wet year-round and feature hot (*Dfa*) or warm (*Dfb*) summers but long and cold (*Dfa*) to severe (*Dfb*) winters. They are the climates of southern Russia and the upper Midwest and Great Lakes region of the U.S.

Continental Monsoon climates (*Dwa* and *Dwb*) are restricted in distribution, occurring mostly in Asia, in the interior of China and north of the Korean Peninsula. In both locations, precipitation varies greatly during the year, partly because of nearby monsoons. As the name implies, *Dwa* climates have dry winters and hot, wet summers. *Dwb* occurs farther north and has warm (not hot) summers.

Examples

18.06.a4 Vancouver, British Columbia

Mid-latitude climates experience long days in summer and short ones in winter. The influence of the ocean (◄) causes marine climates (*Cfb* and *Cfc*) to have less seasonality in temperatures than continental ones.

Humid Continental (*Dfa and Dfb*) climates are wet year-round, but they are farther from any oceanic influences and so are more seasonal (▼), experiencing warm or hot summers but much colder winters.

18.06.a5 Parnell Tower, WI

Climate	Distance from Ocean	Temperature Regime
Marine West Coast (warm or mild summer)	Immediately downwind of ocean, in mid-latitude westerlies	Warm to cool summers, mild winters
Humid Continental (hot or warm summer)	Far from the moderating influence of the oceans	Hot or warm summers, cold winters
Continental Monsoon (hot or warm summer)	May be near oceans, but circulation during part of year makes oceanic air a non-factor	Warm summers, dry winters

B What Affects Precipitation in Non-Arid Mid-Latitude Climates?

The mid-latitude westerlies bring a parade of mid-latitude cyclones to areas of Marine West Coast climates (*Cfb* and *Cfc*) and Humid Continental climates (*Dfa* and *Dfb*). This ensures abundant precipitation throughout the year in marine climates and adequate precipitation throughout the year in continental ones. Humid Continental climates of Asia experience little precipitation in winter, but the climates are not arid because potential evapotranspiration is low in winter there, so water is not removed readily from the surface. What other factors might affect the precipitation climatology of Marine West Coast and Humid Continental climates?

Marine West Coast

1. The two mid-latitude marine climates (*Cfb* and *Cfc*) are either over relatively cool oceans or over land areas where winds bring oceanic air across a continent, as along the Pacific coast of Canada and in western Europe. Even though the subtropical high-pressure zones are far away, they still make their nearest approach to these regions in the summer and suppress precipitation somewhat, particularly on the southern fringes of the Marine West Coast climate. As a result, precipitation tends to peak in October (when the oceans are warmest) and December (when the subtropical high is farthest away) in Northern Hemisphere areas under these two marine climates.

Humid Continental

2. Humid Continental (*Dfa*, *Dfb*) climates receive precipitation throughout the year, including snow in the colder months. Summers are somewhat wetter than the rest of the year in most Humid Continental climates because of convective thunderstorms on hot summer afternoons and the high water-vapor capacity in warm air. In North America, the clockwise flow of Gulf of Mexico moisture around the expanded Bermuda-Azores High adds to the summer precipitation potential. In winter, the combination of moisture from the Great Lakes and bitter cold air results in heavy snowfall (lake-effect snows) on lands downwind of the lakes.

Influence of the Westerlies

3. The mid-latitude westerlies, including the polar front jet stream (shown here in idealized form), push maritime polar air masses from the Atlantic deep into the interior of Europe. This is possible in part because south of Scotland, Europe lacks north-south-oriented mountain ranges to confine the moist air to the coastal zones. As a result, most of northwestern Europe is in a Marine West Coast climate (*Cfb*).

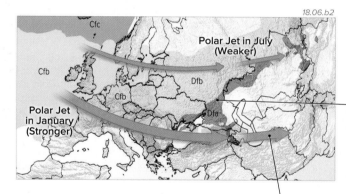

5. The east-west-oriented Alps often prevent Arctic and polar air masses from reaching southern Europe. Nothing prevents such air masses from penetrating farther southward in eastern Europe, so the Humid Continental climates extend farther south in this region. Mid-latitude cyclones that occur over these continental-climate areas produce less precipitation than might be expected because of the increased distance from the source of moisture (i.e., the Atlantic Ocean and Mediterranean Sea).

4. The effects of this moist air fade farther eastward across the continent, allowing continental air masses to exert more control. As a result, countries of eastern Europe, like Belarus and Ukraine, experience Humid Continental climates (*Dfb*).

Examples: West-to-East Changes Across Eurasia

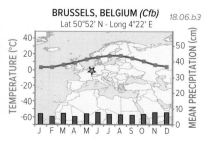

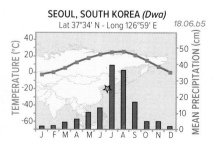

6. Brussels, Belgium, is very mild for its latitude. This is characteristic of Marine West Coast climates.

7. Kyiv (Kiev) Ukraine, farther east, has a Humid Continental climate (*Dfb*), with greater variations in temperature than Brussels.

8. Seoul, South Korea, has more precipitation seasonality than the *Dfb* climates because it is affected by the East Asian monsoon.

Before You Leave This Page

☑ Summarize the varieties of mid-latitude climates (*Cfb*, *Cfc*, *Dfa*, *Dfb*, *Dwa*, and *Dwb*) and where they occur.

☑ Explain the factors that contribute to temperatures and precipitation seasonality for mid-latitude climates.

☑ Sketch and explain how and why climates vary across Eurasia.

18.6

18.7 What Causes Subarctic and Polar Climates?

SUBPOLAR AND ADJACENT CLIMATES occur at very high latitudes in both the Northern and Southern hemispheres. Climates of polar regions, assigned to Group E in the Köppen system, are extremely cold and subdivided into *Ice Cap* (*EF*) and *Tundra* climates (*ET*). Farther from the pole, the Tundra climate gives way to other *Group D* climates—the *Subarctic* climate (*Dfc, Dfd*), and *Subarctic Monsoon* (*Dwc, Dwd*) climate. The Subarctic and Group E climates are characterized by frigid temperatures and low precipitation totals.

A Where Are Subarctic, Tundra, and Ice Cap Climates Found?

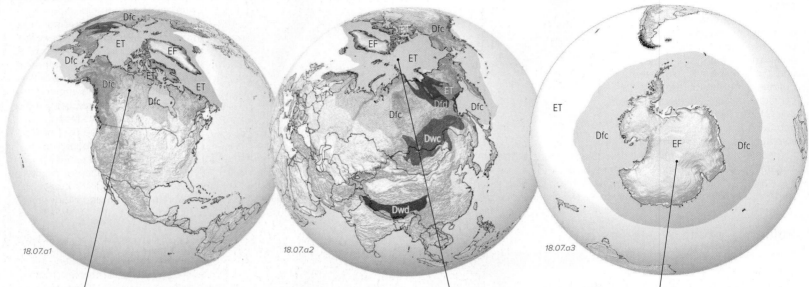

18.07.a1   18.07.a2   18.07.a3

The Subarctic climate (*Dfc* and *Dfd*) is located in a band away from the poles. The line separating it from more polar climates, the 10°C (50°F) isotherm in the warmest summer month, separates a climate that provides enough energy for trees to survive from one that does not. This boundary, called the *tree line*, is visible from satellites. The Subarctic climate occurs farther south at high elevations. A related subarctic climate, the Subarctic Monsoon climate (*Dwc* and *Dwd*) in Asia, receives even less winter precipitation than places with a Subarctic climate.

The polar (Group E) climates, Ice Cap (*EF*) and Tundra (*ET*), are centered around both poles (although only the North Pole is shown on this globe). In the Northern Hemisphere, the Ice Cap climate is restricted to Greenland, but the Tundra climate covers the Arctic Ocean, high Arctic islands, parts of the North Atlantic, and northernmost parts of North America, Asia, and Europe.

Antarctica is mostly covered by ice and snow and so possesses an Ice Cap climate. The surrounding ocean is classified as a Subarctic climate. True tundra (polar vegetation with no trees) occurs in some islands and other ice-free lands.

B What Causes Low Precipitation Amounts in Subarctic and Polar Climates?

1. Polar and adjacent Subarctic climates are characterized by low precipitation, except locally. This is largely because these are very cold places, as depicted on this globe of average annual temperatures. Very low temperatures have a number of implications for precipitation.

2. Very cold air has a low water-vapor capacity, so cold air can carry only limited amounts of water vapor. Without much water vapor, it is difficult to form clouds and even more difficult to generate precipitation. Also, much energy goes into latent heat during melting of ice and thawing of upper parts of the soil during the summer, so the surface doesn't heat up enough to generate convective precipitation.

3. For most of the year, Subarctic and polar regions lie far from the zone where warm and cold air masses meet, so frontal precipitation is minimal. Tropical cyclones can never penetrate far enough poleward and inland to influence Subarctic, Subarctic Monsoon, and polar climates.

4. Another limiting factor is that much of the region of Subarctic and polar climates is inland, or near frozen seas. Humidity flowing into the area must primarily originate over unfrozen seas away from the pole, but persistent high pressure over the pole causes wind to generally blow away from the poles, driving any moist air away. The high pressure and sinking air also limit atmospheric instability.

5. The most likely scenario to produce precipitation (especially over land) occurs in summertime, when the boundary separating cold and warm air masses—the polar front—makes its nearest approach to subarctic and polar latitudes. Marine areas often have precipitation peaks in fall, as the seasonal warming is delayed over oceans.

TEMPERATURE   18.07.b1

-30   -15   0   15   30 (°C)

C What Controls the Temperature Regime in Subarctic and Polar Climates?

Limitations in Solar Heating

18.07.c1 Greenland

1. *Low Sun Angles*—The latitude of these climates dictates that Sun angles are very low during summer (◄) and below the horizon for much of winter (24 hours of darkness). A low Sun angle results in less sunlight per unit area and a long path length through the atmosphere, which increases the amount of insolation absorbed, reflected, and scattered in the air before it reaches the surface.

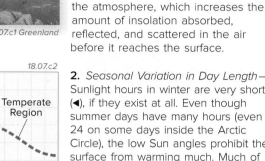

18.07.c2

2. *Seasonal Variation in Day Length*—Sunlight hours in winter are very short (◄), if they exist at all. Even though summer days have many hours (even 24 on some days inside the Arctic Circle), the low Sun angles prohibit the surface from warming much. Much of this energy is used to melt ice cover and thaw the frozen soil layer, so less energy is available for sensible heating.

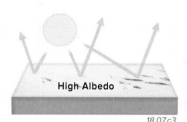

18.07.c3

3. *High Albedo*—Snow and ice surfaces (◄) have high albedo (i.e., reflect a high percentage of the incident solar energy), so a high percentage of the limited insolation is reflected directly back to space without any impact upon these areas. All these insolation-related factors maintain low polar and subpolar temperatures.

Atmospheric Circulation

4. *Continentality*—Northern polar oceanic areas are ice-covered for much, if not all, of the year (►). Extensive sea ice, which can surround the coastline and cover most of the Arctic Ocean, almost functions like a single large continent during some times of the year. The setting causes the North Pole to be characterized by intense continentality, with bitter low temperatures. The South Pole, centered over Antarctica, is a large, cold continent, with a fringe of sea ice that grows in the winter (e.g., July) and shrinks in the summer (December). The South Pole is higher in elevation than the North Pole and is by far the coldest place on Earth.

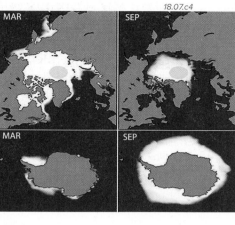

18.07.c4

5. *Atmospheric Circulation*—Strong upper-level westerlies circumnavigate the pole in each hemisphere (►). Air descending from aloft at the polar high is very cold, ensuring that Arctic and Subarctic climates are cold and dry. Arctic and Antarctic (A) air masses form near the poles and can migrate toward the equator as the westerlies dip equatorward in a trough, as is shown here over the eastern U.S.

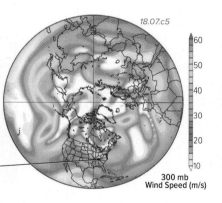

18.07.c5

300 mb
Wind Speed (m/s)

Examples

KRASNOYARSK, RUSSIA *(Dfc)*
Lat 56°1' N - Long 93°4' E

18.07.c6

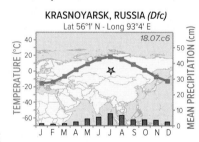

6. Notice the extreme continentality in this example of a Subarctic climate in Siberia. These climates have the greatest annual temperature range on Earth. Precipitation is low but peaks distinctly in summer at most sites.

18.07.c9 Churchill, Canada

NUUK, GREENLAND *(ET)*
Lat 64°11' N - Long 51°44' E

18.07.c7

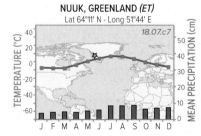

7. Greenland is mostly covered by an ice sheet, but a thin strip of coastal land is exposed. This site, Greenland's capital, is located on a peninsula and influenced by relatively warm local ocean currents for its latitude. It gets more precipitation than most other places with a Tundra climate.

9. Tundra climates also include ice and snow, but they can have unfrozen water during the summer (◄).

10. Antarctica is mostly covered by ice, but rocks protrude through the ice in some mountains (►) and along the coast.

VOSTOK STATION, ANTARCTICA *(EF)*
Lat 78°28' S - Long 106°50' E

18.07.c8

8. Found throughout Antarctica, interior Greenland, and the tops of the highest mountains, Ice Cap climates have unimaginably low temperatures and very low precipitation. This site near the South Pole has never received measurable precipitation.

18.07.c10 Antarctica

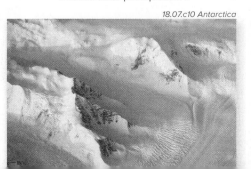

Before You Leave This Page

✓ Describe or sketch the distribution and characteristics of polar and Subarctic climates.

✓ Explain the major differences between Group D and Group E climates in the Köppen system.

✓ Summarize the factors that limit the amount of precipitation in Subarctic and polar climates.

✓ Explain what factors contribute to the very cold temperatures in Subarctic and Polar climates.

18.7

18.8 What Is the Role of Carbon in the Climate?

OF THE 92 STABLE ELEMENTS, only 17 are essential for all plants, and a few more are vital for animals. The nutrients formed by these elements must be recycled continuously through all of the "spheres" of the Earth to sustain life—forming biogeochemical cycles. Despite its very minute abundance on Earth, carbon (C) is the element that forms the building block for life. Organic molecules contain the carbon-hydrogen bond and include carbohydrates, lipids, proteins, and nucleic acids, all found in the many life-forms in ecosystems.

A What Processes Are Involved in Carbon Storage and Cycling?

1. This figure depicts the global carbon cycle as a series of arrows that show the direction in which carbon moves between components of the environment. The number with each arrow indicates how much carbon moves along that pathway (the flux), and numbers next to the main storage sites (stores) of carbon (e.g., oceans) indicate how much carbon is in that reservoir. In both cases, the amounts are in gigatons (10^{15} gm or a quadrillion grams). Some of these values are not known precisely, so they are educated guesses.

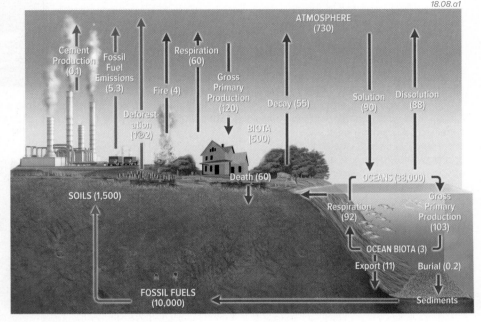

18.08.a1

5. The C emitted by human activities (red arrows) to the atmosphere, while small, causes imbalance in the system. Of this anthropogenic CO_2, some is cycled back to the surface by enhanced photosynthesis by plants, some remains in the atmosphere, and some goes into the ocean.

4. The C going into and out of the oceans would balance naturally, but two units of the extra units produced by people are absorbed by the ocean. Over long timescales, this extra carbon is cycled to the ocean floor as sediments.

2. Over time, Earth's atmospheric CO_2 was moved to the lithosphere as carbonate rocks formed over geologic time in a cycle so slow that it is considered to be in "storage" rather than in a cycle. So the amount stored in the lithosphere is much greater than that stored in the atmosphere. When considering the flux of carbon, we mostly focus on the amount in fossil fuels, not the entire lithosphere.

3. Notice that the exchange of carbon between two reservoirs is essentially balanced. For example, 120 gigatons goes from the atmosphere to land in the form of gross primary production (photosynthesis) and almost the same amount goes from the land to the atmosphere.

B How Do Humans Impact and Become Impacted by the Carbon Cycle?

The production of greenhouse gases through fossil fuel combustion is the most obvious way that humans impact the C cycle, but deforestation and other land-use changes also add carbon to the atmosphere. Aside from climate change and its associated impacts, elevated atmospheric CO_2 also increases weathering and decay rates by forming carbonic acid.

18.08.b1

18.08.b2 Central British Columbia, Canada

18.08.b3 Washington

18.08.b4

1. Fossil fuel combustion for industrial, transportation, and domestic uses amounts to about 95% of the C that people emit to the atmosphere.

2. Most of the other 5% comes from cement production, which grinds up and processes carbonate rocks (limestone and marble), releasing C as a by-product.

3. Deforestation reduces the amount of plants, which in turn reduces the amount of carbon that is extracted from the atmosphere. Changes in the rate of deforestation can affect the carbon balance.

4. Crops remove C from the atmosphere, but the problem is that they are usually only seasonal. Farmlands often replace forests that would have removed much more atmospheric carbon.

C How Does the Carbon Cycle Regulate Earth's Long-Term Temperature?

The massive quantities of stored C may change on timescales of millions of years through feedbacks tied to the global water and rock cycles. As atmospheric CO_2 concentration changes, so does the global climate.

1. Falling rain or snow moves CO_2 from the atmosphere to the lithosphere after it reacts with the precipitation to form a weak acid called carbonic acid (H_2CO_3). This loss of CO_2, a greenhouse gas, can cool the atmosphere.

7. By this process, some of the stored carbon gradually moves back to the atmosphere, which then slowly intensifies the greenhouse effect and warms the Earth again.

18.08.c1

2. If the acidic precipitation falls on land, the H_2CO_3 causes increased rates of chemical weathering. This type of chemical weathering (hydrolysis) releases ions of calcium and other elements into surface waters and groundwater.

6. CO_2 dissolved in magma rises with the magma toward the surface. As the magma nears the surface, the reduced pressures allow the CO_2 to escape, either as a slow outgassing or in a quick release during an explosive volcanic eruption. In either case, the CO_2 goes back into the atmosphere.

3. Overland flow and streams move the ions to the ocean. As they grow, plankton and shell-building organisms convert the calcium ions to calcium carbonate ($CaCO_3$). When they die, the $CaCO_3$ falls to the ocean floor to become limestone, chalk, or other types of seafloor sediments.

4. Plate tectonics moves limestone and its carbonates very slowly over millions of years to subduction zones. Most of the sediment is plastered against the overriding plate, effectively taking the enclosed carbon out of the cycle for a while.

5. Some of the oceanic sediments are taken to depth and heated, liberating CO_2 from the hot rock. Some CO_2 rises toward the surface, mostly in dissolved magma.

8. How effective the system is—and whether it leads to warming or cooling overall—is affected by several factors. If most limestone and other carbonate-bearing rocks end up being scraped off the subducting plate, this carbon is removed from the system by the combined action of the hydrosphere (rivers, groundwater, and oceans), biosphere (microorganism and shell builders), and lithosphere.

9. Periods in geologic history with abundant mountain building may offer more opportunity for chemical weathering, which stores extra carbon in the lithosphere. This would limit atmospheric CO_2 and keep Earth cool, but these natural changes are extremely slow. In the last 50 million years, the buildup of the Himalaya due to plate tectonics is linked to a global temperature drop that has occurred since very warm time periods between 50 and 120 million years ago. During that time, global temperature and sea level were much higher than today.

D How May the Carbon Cycle Be Impacted at Shorter Timescales?

Various schemes have been suggested to manipulate the carbon cycle, especially because of concerns about the climatic effects of CO_2. Three are most discussed.

Process 2: Ocean Fertilization—Scattering iron on the ocean surface may promote phytoplankton growth, which would increase the efficiency of Step 3 above. Would this be a logical and safe means of removing atmospheric CO_2?

Process 3: Invigorated Vegetation—Higher temperatures may increase the rate of carbon uptake by plants under longer growing seasons with more favorable conditions for photosynthesis. The increased rates of decay in a warmer world may liberate nutrients from dead organisms quickly to allow for increased carbon fixation by plants. Would this be a possible strategy for decreasing the amount of CO_2 in the atmosphere?

18.08.d1 Bushveld, South Africa

Process 1: Mineral Carbonation—Ultramafic igneous rocks (those with very high magnesium and iron, and low silica content), like the one shown here, absorb atmospheric CO_2 efficiently during mineral formation. Humans may be able to accelerate this slow natural process either by heating the rock or by injecting it with water rich in CO_2. In the U.S., ultramafic rocks are most common along the West Coast and near the Appalachian Mountains, near major population centers. Might this spatial distribution, which resulted from a complex sequence of geologic events unrelated to humans, be helpful?

18.08.d2 Angkor Wat, Cambodia

Before You Leave This Page

✓ Sketch, label, and explain the carbon cycle, identifying the main carbon reservoirs and fluxes, mentioning some possible human impacts.

✓ Sketch and explain how CO_2 regulates global temperatures.

✓ Describe some possible ways to decrease atmospheric CO_2.

18.9 What Is the Evidence for Climate Change?

EARTH'S CLIMATE HAS CHANGED, both over geologic timescales and in the past 150 to 200 years. The planet has experienced much colder temperatures than today, such as during the Ice Ages, and much warmer ones too, tens of millions of years ago. More recently, we have measured atmospheric temperatures for several centuries, and this record shows an overall increase in temperatures—*global warming*. What is the evidence that Earth's climate is changing?

A What Is the Evidence for Recent Climate Change?

Climate change can include global trends in warming, cooling, precipitation, wind directions, and other related measures. Global warming means *increasing* global atmospheric and oceanic temperatures from some point in the past to the present, and is usually expressed as the temperature compared to the mean global temperature for some time period (e.g., averaged from 1961 to 1990). Scientists examine various records of Earth's climate to investigate past changes, including recent ones.

Thermometer Record

1. Thermometers provide a direct measurement of air temperature. These measurements are compared with average temperatures for some reference time to calculate what is called an *anomaly* (a positive anomaly is warmer than the reference time and a negative anomaly is colder). The graph below plots recent temperatures relative to the reference time period. Most such calculations use a reference time period in the 1980s and 1990s.

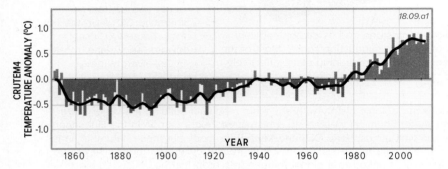

2. From these calculations, temperatures were cooler than average (a negative anomaly) in the late 1800s and from the 1940s to the 1970s. Overall, average air temperatures have increased over the last century, especially since the 1970s. Global and U.S. temperatures remain well above the long-term average.

Sea-Surface Temperatures

3. Another direct measurement of temperature is sea-surface temperature (SST), which is collected from buoys, ships, and more recently by satellites. Observe the SST graph below and then compare it to the air-temperature data to the left.

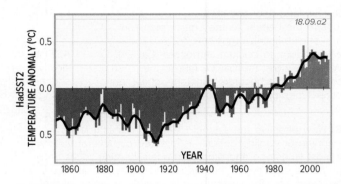

4. SST data show a relatively cold period (a negative anomaly) prior to 1940, then overall warming until 1998, when there was a strong El Niño. As with air temperatures, the rate of SST warming decreased but temperatures remained high, well above the long-term average (a positive anomaly).

Using Satellites to Investigate Temperature Changes

5. Increasingly, satellites are used to more reliably measure and monitor changes in temperature and other expressions of climate with instruments designed to measure different aspects of change. For example, some satellites measure temperature, moisture, or cloud cover in the lower atmosphere, whereas others measure temperature, abundance of different gases, and attributes of higher parts of the atmosphere, like the stratosphere. One advantage of using satellites is that they can provide a truly global data set as they systematically orbit the planet. Data sets of land temperatures must rely on weather stations that are not uniformly distributed (most are near centers of population), and some large regions, such as Antarctica and central Africa, have few such stations. As a result, land-based temperature data must be processed to estimate temperatures in areas with sparse data coverage. Satellite data involve much computer processing too, but they start with a global data set.

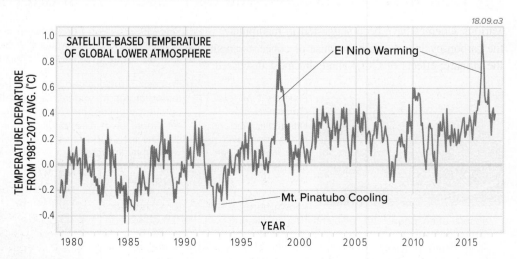

6. This plot (▲) shows monthly averages of satellite measurements of temperatures in the lower atmosphere, along with a red curve fit through the data. The data show an overall warming trend, with shorter periods of cooling and warming. The largest upward increases in temperature, including warm peaks around 1998 and 2016, occurred when there was a strong El Niño event, a phase of ocean and atmospheric circulation where warm water spreads from the western Pacific into the central and eastern Pacific, warming the region and the globe.

B   How Do We Use Proxy Data to Investigate Past Changes in Climate?

Earth's climate has been changing since the planet formed 4.55 billion years ago, varying over decades to billions of years. Geoscientists examine the record of climate in rocks, fossils, tree rings, ice cores, and many other observable or measurable characteristics of rocks and fossils. Such types of data are called *proxy evidence*, because they are not an actual measurement of the past climate but are some attribute that can be used to infer past climatic conditions. Some proxy data consist of direct observations or measurements of features, like the width of tree rings, whereas others employ chemical analyses, such as measuring the isotopes of oxygen and hydrogen, referred to as *stable isotopes*. The International Panel on Climate Change (IPCC) and U.S. Geological Survey (USGS) have summarized the kinds of proxy data that are available.

1. The lengths of most of the world's glaciers (▶) have decreased dramatically over the last 200 years, due to melting caused by higher temperatures. The reductions in length are directly observable, but scientists can model how much warming would be required to explain the melting of glaciers farther back in time, and this is proxy data.

18.09.b1 Icefields Parkway, Alberta, Canada

4. Stalactites (◀) and other cave formations have a sequence of thin layers that record successive stages of growth. We can measure isotopes and other chemical elements from these layers to infer conditions under which the cave formations grew. A similar approach is used with coral reefs, providing a proxy for temperatures of past shallow seas.

2. Modern climatic zones, such as the tropics, have certain characteristic plants and animals. If we find fossils of similar organisms in ancient rocks, we can infer that the region had a similar type of climate at that time. The structure of fossil leaves (▶), such as how straight or jagged the edges are, also varies with climate, so fossil leaves are climate proxy.

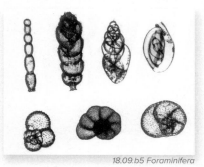

18.09.b2 Florissant Fossil Beds NM, CO

18.09.b5 Foraminifera

5. Marine organisms, including microscopic ones (◀), reflect the conditions under which they lived, so isotopic measurements of the shells or skeletons of such creatures provide a proxy for past conditions. This approach provides a long record of climate, since we have fossils for millions of years in outcrops and in drill holes into seafloor sediments.

3. In climates that have seasons, trees produce alternating dark and light bands (▶) of tree rings when they grow. We can use measurements of the width of tree rings and the density of wood to infer past climatic conditions, such as average temperature. We also measure variations in carbon and oxygen isotopes from ring to ring as a proxy for temperature.

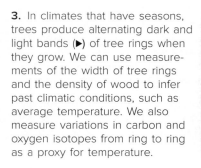

18.09.b3 Tree Ring

18.09.b6 Ice Core

6. The accumulation of glacial ice produces seasonal bands (◀) that can be counted and measured for stable isotopes in the ice and in trapped bubbles of atmospheric gases. Ice cores from drilling sites in Antarctica and Greenland provide an excellent record of past climate over the past 100,000 years or more. These are important proxies of past temperatures.

Weather, Climate, and a Scientific Approach

There is much discussion, and controversy, about one important aspect of climate change—*global warming*. As shown on these two pages, evidence for global warming since the mid-1800s is clear, but there is active debate about how much of the warming is attributable to human introduction of greenhouse gases (mostly CO_2) and how much is due to natural causes. Scientists are always testing, refining, or refuting new, old, or even widely accepted hypotheses—it is part of the scientific method. One misleading and nonscientific tactic used by members of the general public and media on both sides of the global warming debate is to point to a specific weather event, like a single hurricane or tornado, a Midwestern blizzard, or a lack of snow in some region, and say that this proves or disproves, or is a direct consequence of, global warming or the lack thereof. Such short-term events are *weather*, not *climate*, and have a complex and still incompletely understood relationship to climate change that is being investigated, largely through the use of complex computer models.

Before You Leave This Page

☑ Describe what climate change means, some ways it can be directly measured, and evidence indicating global warming in the last 100 years.

☑ Summarize some of the main types of proxy data, including physical or chemical characteristics that can be measured.

18.9

18.10 What Factors Influence Climate Change?

CLIMATE CHANGE IS ALWAYS OCCURRING, including global warming since the mid-1800s. There are many natural causes of climate change, including changes in Earth's orbit around the Sun and changes in solar activity. Many scientists conclude that human activities, including the burning of fossil fuels and the clearing of forests, contribute to climate change by releasing greenhouse gases to the atmosphere. Other factors may lead to *global cooling*, including ash from large volcanic eruptions and an increase in certain aerosols in the atmosphere. Here, we examine some of the factors that can influence our climate.

A What Processes Influence Atmospheric Temperature Change?

Earth's surface temperatures are dominated by energy from the Sun (insolation). Insolation heats the oceans, land, and atmosphere, but several factors influence how much of this energy reaches the surface and how much is retained.

Interaction of Insolation with Earth's Atmosphere, Oceans, and Land

1. Nearly all of Earth's heating at the surface comes from insolation, which heats the atmosphere, land, and oceans. Most of this energy eventually escapes back into space in the form of longwave infrared energy. The rest is delayed by interactions with Earth, keeping the planet warm by a process called the *greenhouse effect*. The amount of insolation hitting Earth varies regularly by a small amount, due to orbital fluctuations and changes in the Sun's energy output, as expressed by changes in *sunspot* activity. Sunspots are the darker areas that, on average, appear and disappear on the surface of the Sun in an 11-year solar cycle.

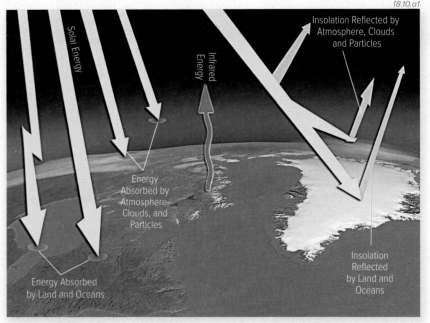

2. Some insolation is *absorbed* by the atmosphere (shown as orange disks in the figure), and some is *reflected* off the atmosphere. Much of the reflected insolation returns to space without heating Earth or its atmosphere, as depicted by the blue arrows.

3. Insolation is also absorbed by clouds, by soot from burning, and by fine particles (aerosols), which are produced by volcanoes, industry, and automobiles. Some of this absorbed energy radiates back into space as *infrared* (longwave) *energy* (shown by the wavy red arrow). Clouds and particles also reflect some insolation.

4. Some insolation is reflected back to space from the land surface and oceans. Ice in continental glaciers is an effective reflector. As glaciers melt, darker land or ocean is uncovered. This increases the amount of energy that is absorbed by Earth and subsequently re-radiated back to the atmosphere.

5. Some insolation is absorbed by the land and the oceans, both of which then radiate infrared energy back into the atmosphere. Some of this infrared energy is absorbed by atmospheric gases, such as water vapor (H_2O), carbon dioxide (CO_2), methane (CH_4), and nitrous oxide (N_2O), which are called *greenhouse gases*. Some portion of these gases is produced naturally and some is produced by human activities.

Greenhouse Gas Production

6. Several gases in Earth's atmosphere absorb infrared radiation emitted by Earth. This causes them to vibrate and heat up, and then to emit infrared radiation. This radiation can escape into space or be absorbed by other greenhouse gases, mostly lower in the atmosphere where the concentration of greenhouse gases is the highest. Since 1957, atmospheric scientists have collected air samples on the high peaks of Hawaii. The data for Mauna Loa are plotted on this graph (▶), which shows CO_2 content in the atmosphere as a function of time. The concentration of CO_2 in air, represented by the red line, has increased by 20% in the last 40 years. Most of this increase is attributed to humans, especially through the burning of fossil fuels.

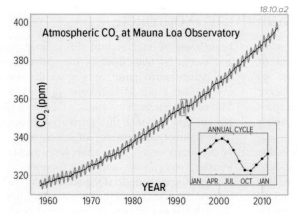

7. Our oceans are a huge reservoir for dissolved gases and other chemical components, including CO_2. As the CO_2 of our atmosphere increases, a large amount of this CO_2 goes into the oceans, where it affects seawater chemistry, especially the acidity. *Ocean acidification* is being studied to assess its impact on sea life. Seawater can contain less CO_2 when it is warm, so it releases CO_2 back to the atmosphere as it warms, as is occurring now. So some increase in the atmospheric CO_2 could be a result of recent warming.

Greenhouse Gases and Temperature Change Records from Ice Cores

8. Long-term records of CO_2 and other greenhouse gases are contained within ice cores and can be traced back hundreds of thousands of years. Ice cores from Antarctica have been analyzed for CO_2, isotopic temperature, and isotopic age. The data, shown on this graph (▶), allow comparison between prior natural variations and changes in the last several hundred years.

9. How well do CO_2 concentrations in the atmosphere track temperature change? On this graph, the two trends are very similar. Graphs of other greenhouse gases, such as CH_4 versus temperature and N_2O versus temperature, show a similar correspondence. These large fluctuations in the past have natural causes.

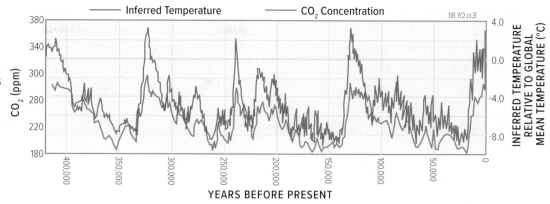

10. Scientists have used data from ice cores and other proxies to interpret how atmospheric gases changed in the past few hundred years. These data indicate that levels of CO_2, CH_4, and N_2O have increased since the start of the Industrial Revolution of the last part of the 18th century. Many scientists infer that these increases in greenhouse gases are partly responsible for the recent increase in temperature, but there remains debate about how much our human activities have affected, or will affect, our climate.

Orbital Variation

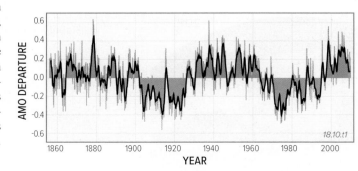

11. Earth's orbit is affected almost exclusively by the Sun and Moon, causing it to cycle from more elliptical to more circular paths, with one cycle completed over about 100,000 years. This influences Earth's climate, but the current global warming has occurred in less than 200 years. This is too short a time period for orbital changes to have caused the observed recent episode of warming. The tilt and direction of Earth's axis of rotation also change, which changes the contrast of the seasons, but these effects occur over tens of thousands of years.

Variations in Solar Radiation

12. The Sun is a candidate for causing climate change. We can directly measure the amount of solar radiation reaching Earth and also use proxies for solar activity, including the number of sunspots (▶) and isotopic data from ice cores and tree rings. There have been large variations in the number of sunspots, but most other data suggest that the Sun's energy emission has varied only slightly over the last 2,000 years, so direct warming by the Sun is interpreted to not be the main cause of recent warming.

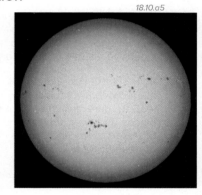

Ocean Oscillations, Climate Change, and Climate Models

Recent discoveries have changed our view of the relationship between ocean currents and climate. Studies of the oceans have documented that separate ocean basins (e.g., the Pacific Ocean) display variations in temperature that occur over years or decades—*ocean oscillations*.

One long-recognized oscillation is the El Niño–La Niña effect and is called the El Niño–Southern Oscillation (ENSO for short). More recently, we have discovered other ocean oscillations, including the Pacific Decadal Oscillation (PDO) and the Atlantic Multi-decadal Oscillation (AMO). As their names imply, these oscillations may last decades, whereas ENSO cycles last a year or two.

Variations in sea-surface temperatures (SST) and atmospheric pressure can be used to calculate a numerical *index* for each oscillation as a way to represent the changes. The graph here shows values for the Atlantic Ocean (AMO), relative to its long-term average. The AMO represents SST in the Atlantic basin—high numbers indicate warm SST. This graph shows that Atlantic SST undergoes cycles that last several decades. These cycles also relate to periods of abundant or rarer tropical cyclones. They also show some similar patterns to global temperatures, so because of such correlations, climate scientists are increasingly examining whether oscillations are primarily a cause or an effect of climate change.

Before You Leave This Page

☑ Summarize the major factors that influence atmospheric temperatures.

☑ Summarize how atmospheric greenhouse gas concentrations can correspond to temperature changes.

18.10

18.11 What Are the Consequences of Climate Change?

CLIMATE CHANGE HAS MANY POTENTIAL IMPACTS, ranging from obvious ones like an increase in global temperatures, to less obvious ones, such as a possible increase in certain diseases. Some consequences are highly probable, whereas others are very speculative. On these pages, we briefly introduce some of the most likely results of climate change, specifically those related to global warming. Abundant information is available elsewhere, including reports by governmental and nongovernmental organizations, if you are interested in following up on this topic.

A What Are Some Possible Consequences of Climate Change to the Environment?

1. The figure below illustrates some consequences of global warming, but many more are possible. Examine the figure and think about what you know about each feature being depicted and how it might respond to an increase in average temperatures. Then read the text.

2. The most obvious result of global warming would be, well, an *increase in global temperatures*. However, computer models of climate indicate that many areas, especially those near the poles, will indeed increase in temperature, but other regions might get colder.

3. Higher temperatures would be predicted to cause more *melting* of snow and ice, resulting in glaciers melting back and becoming less extensive, as has been observed.

4. Higher temperatures and altered ocean circulation could lead to more drought in some places and even the expansion of desert areas, the process of *desertification*.

5. Changes in global temperatures and accompanying changes in precipitation patterns can affect the *distribution of communities* of plants and animals. In response to warming, such communities may shift to higher elevations or higher latitudes to stay within an optimal temperature range. Climate change is predicted to decrease the geographic range of some communities, while increasing the range of others. Included in this consideration are croplands, some of which will benefit from global warming, while others will suffer.

6. An increase in global temperatures is predicted to increase *evaporation* from warmer surface waters. This would increase the amount of water vapor in the air. Warmer temperatures and more humidity may increase *precipitation*. More precipitation, along with more melting of glaciers, would lead to more runoff from streams, with an associated increase in the amount of flooding. Increased runoff also means a larger influx of fresh water into the ocean.

18.11.a1

7. Climate change has the potential to change the frequency and intensity of *severe weather*. For various reasons, global warming can either increase or decrease the frequency of certain types of severe storms. These are presented on the next page.

8. Global warming is predicted, and observed, to cause an increase in global sea-surface temperatures (SST) and a decrease in the amount of *sea ice*, at least in the Arctic.

9. The various consequences have complex interactions, as between SST, humidity, precipitation, runoff, and influx of fresh water into the ocean. Such interactions could result in other changes, like a change in ocean currents, including those involved in the *thermohaline conveyor*, an important moderator of global climate.

10. The computer-generated images below show the decrease in the amount of sea ice in the Arctic from 1979 (left globe) to 2015 (right globe), the period of the most rapid global and regional warming. In response to weather patterns, there are some year-to-year increases but an overall decrease.

11. Since the middle of the 1800s and before, sea level has been slowly increasing, rising 0.2 meters in 200 years. Currently, there is scientific debate about whether the rate of sea level rise is remaining the same or is accelerating with time.

1979

18.11.a2

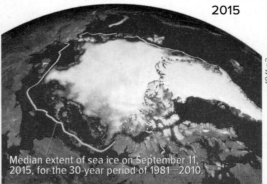

2015

18.11.a3

Median extent of sea ice on September 11, 2015, for the 30-year period of 1981–2010.

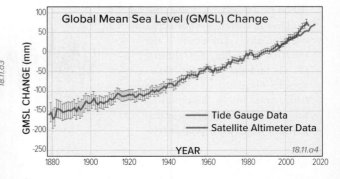

Global Mean Sea Level (GMSL) Change

GMSL CHANGE (mm)

—— Tide Gauge Data
—— Satellite Altimeter Data

YEAR

18.11.a4

B How Might Climate Change Promote More Severe Weather?

People have speculated that we are moving into an era of more numerous and destructive tropical cyclones and tornadoes, with the increase attributed to climate change. What do the actual science data say about these important questions? Observed increases in the frequency of all forms of severe weather are at least partially the result of recent construction and retirement of more land-based, weather-radar stations and other improved remote-sensing technology, which allow us to detect storms more easily. One approach is to compile data on severe storms over time to examine whether the frequency and severity of storms correlates with observations of climate change. But remember that simple correlation does not demonstrate causation.

1. Observational and modeling studies suggest that climate change will increase SST, which in turn might increase the intensity and duration of the strongest storms. Warming might also lengthen the severe tropical cyclone season. In theory, increased SST would allow an increase in evaporated water, which increases the amount of warming in the atmosphere. Increased evaporation can result in more energy for storms. The end result is larger, taller clouds fueling more powerful storms.

2. Warmer sea and land surfaces are also likely to affect atmospheric stability. Warming of the land and water surfaces causes increased rising of the adjacent air, which can enable clouds to continue growing upward. Taller clouds, fed by more rising air, can increase the severity of the associated storms.

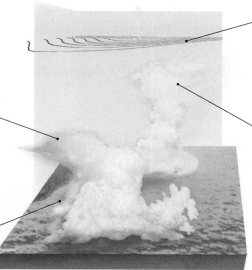

18.11.b1

3. Global warming could cause changes to upper-level circulation, which in turn could cause underlying areas to experience more frequent precipitation or drought. Changes in upper-level circulation patterns might make the upper levels of the atmosphere more or less favorable for storm development. Specifically, if climate change displaces or decreases the flow of upper-level westerlies, fewer developing tropical cyclones may have their tops sheared off as they move westward. Alternatively, global warming could increase vertical wind shear, which might cause tropical cyclones to be less severe. What do the actual data say?

Frequency of Tropical Storms

4. The graph below plots the number of Atlantic storms for each year. The time periods represented on the graph correspond to the interval of time when global temperatures have demonstrably increased (although in detail temperatures have gone up and down during several approximately 30-year cycles). Research using these and similar data suggests that short-lived Atlantic tropical storms may have become more frequent over the years, but moderate-duration storms have not.

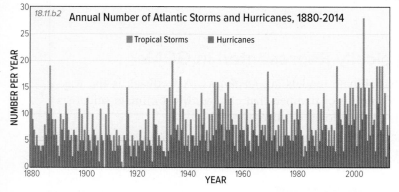

5. The research on this question is ongoing, with some studies predicting an increase in intensity, while others imply a decrease in intensity. Other studies acknowledge that changes in tropical cyclones under climate change scenarios will be small in relation to other factors that create variability in tropical cyclones, such as ENSO.

Frequency of Tornadoes

6. Another question to evaluate is whether global warming has caused an increase in the number of tornadoes. Some scientists have proposed that warming provides additional energy that might enhance the environment for such severe storms. Examine the graph below, which plots the number of tornadoes per year during the time when most warming occurred.

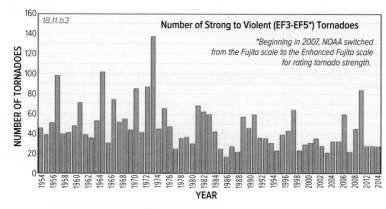

7. The data indicate that the number of strong-to-violent tornadoes in the U.S. has not increased over time. One possible explanation is that increasing surface temperatures would presumably be accompanied by increases in upper-level temperatures, leading to no net change in the number of storms.

Before You Leave This Page

✓ Sketch and explain some possible consequences of climate change, specifically global warming, to the environment.

✓ Discuss some ways that climate change could increase the frequency and strength of severe weather, and present some relevant data.

18.11

18.12 How Do We Use Computers to Study Climate Change?

COMPUTER MODELS are used to investigate issues related to climate, like global warming. The simplest climate models compute energy (temperature), water, or momentum for a small area over a limited time. The most complicated models use numerical methods and principles of physics to compute atmospheric conditions at many vertical levels across the entire Earth for a long period of time. These are called *general circulation models* (GCMs). A key issue in climate modeling and prediction is understanding the nature of positive and negative feedbacks in the climate system.

A What Are GCMs and How Do They Approach Climate Problems?

Modeling Approach

1. As in weather forecasting models, data are collected from a variety of sources, including aircraft, satellite imagery, balloon launches, and surface stations.

2. Data are quality controlled to eliminate obvious errors.

3. Equations are run to simulate atmospheric variables at each point for a set time into the future.

4. The "answer" at each grid point becomes the starting point for the next iteration of calculations for another time into the future.

5. After a prediction far enough into the future is generated, the results are scrutinized and plotted on maps.

Modeling Procedures

6. The same general data sources are used for climate prediction as for weather forecasting. However, because small-scale weather features can create quite a bit of havoc on short timescales, it is generally more important to have a finer mesh of data for weather forecasting models. Both types of models are generally so complicated that they are run on supercomputers or clusters of networked, high-powered computers.

7. Spatial analysis techniques are used to interpolate the data on a three-dimensional grid, with several vertical levels. The gridded data are run through the model, using the same seven basic equations that are used in weather forecasting models.

8. Some models use grid points initially, but then resolve atmospheric motion into a series of curves, which then become amplified or dampened based on the input of energy, matter, and momentum. The results of the analysis then become reinterpolated on a grid.

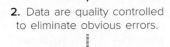

18.12.a1

B What Are Feedbacks and How Are They Typically Represented in GCMs?

Among the most important—and complicated—aspects of global climate models are *feedbacks*. A feedback is the way a system responds to a change in conditions, which in turn acts to amplify or dampen that change. Most climate scientists would agree that feedbacks are among the least understood parts of the global climate system.

1. These two figures were presented and discussed elsewhere, but they provide a useful review of feedbacks in the context of modeling. The left figure illustrates that we can think of natural phenomena as systems containing matter and energy, with individual parts that interact in some ways more than others. In studying and modeling a system, we want to understand the *inputs* to some part of the system, in this case the input of moisture into the atmosphere from the water and ice (shown by the upward arrows). Then we want to model, using physical principles, the *response* of this change to the system, such as to increased precipitation.

18.12.b1

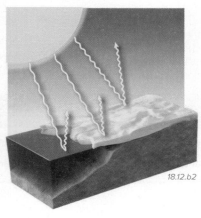

18.12.b2

2. One way the system can respond to the imposed changes is through positive and negative feedbacks. A *positive feedback* amplifies those changes (causes more change in the same direction), while a *negative feedback* drives the system in the opposite direction, undoing some or all of the changes. As an example, a decrease in the amount of sunlight and an accompanying decrease in temperature cause an increase in the amount of snow and ice, which reflect more insolation due to their high albedos, causing even more cooling, a positive feedback.

C What Are Some Positive and Negative Feedbacks in the Climate System?

GCMs consider feedbacks, but they are difficult to understand and model because a chain reaction of effects may result from a single change. Shown below are some important feedbacks modeled in modern GCMs. Some of these feedbacks are not as straightforward, nor as well understood, as these simple figures imply.

Changes to the Surface

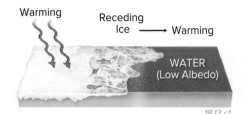

18.12.c1

1. Warming temperatures could melt more ice. Ice has a higher albedo than water, so replacing ice with water causes the surface to absorb more heat, causing even more warming in a positive feedback.

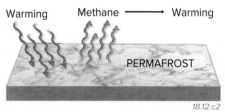

18.12.c2

2. Polar permafrost contains methane, a greenhouse gas. Warming the permafrost releases the methane, which causes more warming by absorbing more outgoing longwave radiation.

Water Vapor and Clouds

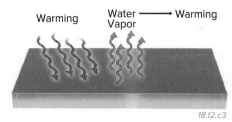

18.12.c3

3. Warming the oceans evaporates more water, the most abundant greenhouse gas. The extra water vapor could cause more warming by absorbing outgoing longwave radiation, which could evaporate more water vapor, which . . . you get the idea.

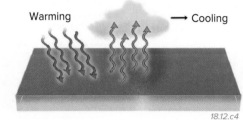

18.12.c4

4. But more water vapor should produce more low-level clouds, which reflect insolation. More water vapor leading to more clouds, therefore, could cause cooling, a negative feedback.

CO$_2$ In and Out of the Oceans

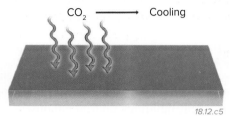

18.12.c5

5. Carbon dioxide (CO_2) goes in and out of the oceans, which hold over 50 times more CO_2 than the atmosphere. Any CO_2 that is emitted into the atmosphere but goes into the oceans reduces the rate of atmospheric warming.

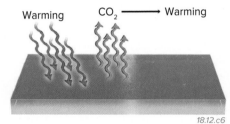

18.12.c6

6. If the oceans warm, however, CO_2 in the oceans may eventually be released back to the atmosphere through other processes, leading to more warming, a positive feedback.

The Importance of Feedbacks in Climate Models

To model past climate change and try to predict future climate change, GCMs and other climate models have to incorporate the effects of both positive and negative feedbacks, like those described above. The introduction of additional CO_2 into the atmosphere, such as by burning coal and other fossil fuels, has well-understood effects in the atmosphere, resulting in some amount of warming. Doubling the amount of CO_2 is predicted to cause approximately 1 C° of warming, if there are no feedbacks. Such a modest temperature increase by itself might not be a cause for concern, since temperature fluctuations of this and much larger magnitudes have occurred many times in Earth's past, especially when viewed over geologic timescales.

A critical question involves how the Earth system will respond to such an increase in temperatures. If the warming results in a large amount of melting of ice, the resulting change in albedo can cause even more warming, a positive feedback. If the original warming causes the formation of more low-level clouds, the reflective nature of such clouds could be a negative feedback, limiting or even reducing the amount of warming. It is fair to say that if we don't understand feedbacks, it is impossible to model the behavior of Earth's complex climate system accurately.

For this reason, much scientific research has been and currently is investigating the feedbacks that are likely to have the largest influence on our climate. Scientists are studying the amount of ice in ice sheets and glaciers on land, and in sea ice that forms from freezing of the surface of the ocean in polar climates. Some of this research can be done using satellites, but some is done by physically going onto ice-covered regions and observing and measuring what is happening. Atmospheric scientists are using satellites and other methods to measure the amount of water vapor at different levels in the atmosphere, since water vapor is also a greenhouse gas and is much more abundant than CO_2. If warming causes an increase in water vapor, the warming will be amplified, a positive feedback. As we better understand the feedbacks, we will better understand, and be able to model more accurately, the incredibly complex system that controls our climate.

Before You Leave This Page

✓ Explain what a GCM is, what kind of data it uses, and how the data are used in the model.

✓ Sketch and describe some of the major feedbacks incorporated in most GCMs.

✓ Explain the importance of feedbacks to GCMs and other climate models.

18.12

What Are Non-Fossil Fuel Sources of Energy?

USING FOSSIL FUELS FOR ENERGY has many conveniences, but it also has drawbacks, such as influencing our climate. As a result, considerable research and development have gone into ways to produce energy that do not involve putting more CO_2 into the atmosphere and oceans. The goal is to find energy sources that are least harmful to the environment and that are *renewable resources*, meaning that their supply is essentially limitless and using them doesn't remove something irreplaceable from Earth. There are a number of non-fossil fuel approaches.

A What Is Geothermal Energy and What Sites Are Most Favorable for Its Use?

Geothermal energy uses Earth's natural heat as an energy source. Geothermal power plants convert natural hot water to steam to power electrical generators. Naturally warm water can be piped from the ground to places where it can be used to heat buildings and greenhouses, or to keep streets and sidewalks free of ice and snow.

1. Temperature increases with depth, so water circulating through the crust can become heated at depth and then rise to the surface. This is how *hot springs* form.

2. The ideal combination of high temperatures and relatively shallow depths is most common in areas of recent volcanic activity, commonly within a collapsed caldera formed by eruption of volcanic ash.

3. Shallow magma chambers or recently solidified magma can heat water to high temperatures, exceeding 200°C to 300°C (~500°F). Although water of these temperatures would boil on the surface, it generally does not boil at depth because water pressures work against the great expansion in volume required to convert liquid water into gaseous steam.

4. To generate electricity, hot water is piped to the surface and into power plants. In the plant, the confining pressure on the overheated water is released and the hot water flashes into steam, driving the turbines in the electrical generators.

5. Regions with recent faulting can also be promising sites for geothermal energy. Faults disrupt the continuity of aquifers and provide a conduit for heated water to rise to the surface and issue from hot springs.

6. If rocks are hot at a shallow depth but dry, water derived from some other source can be pumped down drill holes to be heated by the rocks and then pumped back to the surface and used for heating or power generation.

18.13.a1

B How Is Electricity Generated by Hydroelectric Methods?

Hydroelectric power, almost all from dams, provides about 6% of electrical energy for the United States and about 16% of the world's electricity. It is the main source of electricity for some topographically rugged western states, like Idaho, Washington, and Oregon, as well as nations including Norway and Iceland. Electricity can also be produced by harnessing ocean currents and by temporarily damming seawater as it alternates between low and high tide.

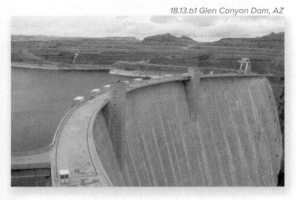
18.13.b1 Glen Canyon Dam, AZ

To generate hydroelectric power, dams capture water from rivers or streams, storing it in a *reservoir* behind the dam (◄). The dam is constructed out of concrete or compacted clay, rock, and other material. Large steel pipes or concrete tubes within the dam guide water to flow from higher elevations to lower ones, under the constant—and free—force of gravity.

When we need electrical power, operators of a dam allow water to flow through pipes and tubes within the dam and then into a powerhouse, where the moving water turns turbines or blades in electrical generators. The amount of energy produced is related to the velocity and volume of the water passing through the turbines. An ideal dam, from an energy perspective, is high and has a high-volume reservoir. An advantage of hydroelectric energy is that the water can be stored until we need the electricity (which cannot easily be stored). If the demand for electricity goes up, like during the day, operators of the dam let out more water and generate more electricity. At night, when there is less demand, operators let out less water.

Before You Leave These Pages

✓ Describe some favorable factors for producing geothermal energy.

✓ Summarize how we produce electricity with hydroelectric dams.

✓ Summarize how we produce electricity using nuclear power.

✓ Summarize electricity production from solar energy.

✓ Summarize electricity production using the wind.

C How Does Nuclear Fission Produce Energy?

Present-day nuclear power plants are based on the process of nuclear *fission*, during which an unstable isotope of a radioactive element splits into two parts. There is a significant amount of research being done to develop a reactor based on a sustained *fusion* reaction, involving atoms that collide to produce a larger and heavier element. Earth's radioactive heat arises from *fission*, whereas the Sun's energy comes from *fusion*. Here, we discuss *fission*, which is used in power plants today.

1. Uranium is a large and heavy element, with an average atomic mass of 238. The main isotopes of uranium, ^{238}U and ^{235}U, both decay by fission. ^{235}U decays more rapidly than ^{238}U, so it produces more energy, but it is much less abundant (more than 99% of uranium is ^{238}U).

2. When a uranium atom splits apart by fission, it releases relatively large amounts of energy, partly in the form of heat. In a reactor, ^{235}U atoms are bombarded by neutrons, and this induces fission. The heat produced by fission converts water to steam, which is then used to turn the turbines in electrical generators.

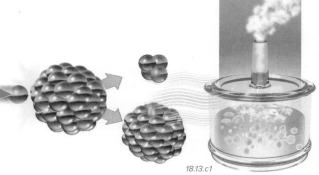

18.13.c1

3. Most reactors are designed to keep the uranium and its decay products isolated from the water that is converted to steam. Only heat is exchanged between the two parts of the system, so the steam being released does not contain radioactive materials.

D How Is Solar Energy Used?

Solar energy involves using the Sun's free electromagnetic energy to heat buildings and generate electricity. There are many strategies for using solar energy, including *passive solar*, *active solar*, and *photovoltaic panels*. All solar-energy approaches work best in sunny climates and in sites with unrestricted views of the Sun.

1. In *passive solar* (▶), light and infrared energy from the Sun enter a space through glass windows, naturally heating the inside air. Passive solar does not use any moving parts (hence the name *passive*) and is as easy as designing a house with large windows facing south (in the Northern Hemisphere) to collect the winter sun. Overhangs shield the windows in the summer.

2. *Active solar* implies that there are moving parts and some use of electrical energy, such as a fan for moving heated air or an electric pump for circulating heated fluids from the solar panel to the interior of the building.

18.13.d1

18.13.d2

3. *Photovoltaic panels* convert sunlight directly into electricity. Such panels, although expensive to produce and install, provide nonpolluting renewable energy, even to remote locations and small sites.

E How Is Electricity Produced from the Wind?

Wind is another clean and renewable energy source. It currently provides very little of the world's and North America's power requirements, but it is one of the fastest-growing sources of energy, both here and abroad.

1. Large-scale generation of electricity from wind requires a site that has strong winds much of the time. Important geologic factors to consider are how the surface topography interacts with or controls wind, and whether the materials beneath a site are suitable for building the necessary facilities. Each wind turbine has its own small electrical generator.

18.13.e1 Jacumba, CA

2. An advantage of wind power is that it is renewable, is nonpolluting, and can be used in remote locations and in areas that have little other infrastructure.

3. One disadvantage of wind power is that winds, and the resulting power, are variable. If there is no wind, there is no electricity generation. Wind turbines affect the *aesthetics* of the site, being large and conspicuous, even if they are painted to help blend in with the environment. They can be noisy, are relatively expensive to maintain, and kill birds, including endangered species. The downsides are weighed against the obvious benefits of clean, renewable power.

18.13

18.14 What Kinds of Climate and Weather Would Occur in This Place?

PLANET G is a hypothetical replica of Earth. It has plate tectonics, oceans, and an atmosphere very similar to Earth's. You will map the climatic zones on this planet, identifying cold and warm ocean currents, prevailing winds, potential locations of rain forests and deserts, places where the climate and topographic setting would be suitable for agriculture, and sites at risk for hurricanes.

Goals of This Exercise:

- Observe the general map patterns and photographs from different parts of the planet. Note any clues about climate.
- Create a map showing probable climatic conditions, based on ocean currents and prevailing winds.
- Locate likely sites for rain forests, deserts, agricultural areas, and locations where storms will strike land.

Procedures

Follow the steps below, entering your answers for each step in the appropriate place on the worksheet or online.

A. Observe the distribution of continents and oceans on the map below and on the larger version on the worksheet. Examine the photographs of the various areas and infer the environments and climates that are represented. Each photograph has a letter that corresponds to a letter on the map. Some photographs are on the next page.

B. Carefully examine the various types of data on the next page. For each data set, examine the climatic, weather, or geologic implications for different parts of the planet.

C. Draw on the worksheet your interpretations of whether each ocean current is warm, cold, or neither, and whether prevailing winds will bring warmth or coolness, and cause dryness or precipitation. Alternatively, answer questions about these factors online.

D. Map the main climatic zones. Include the likely locations of rain forests, deserts, and areas suitable for croplands. Think about which areas are at risk for hurricanes and other major tropical storms. Be prepared to discuss your observations and interpretations.

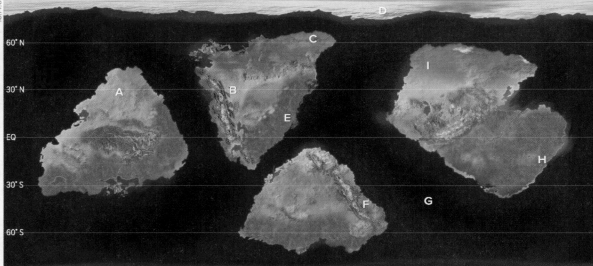

Some Important Observations About the Planet

1. The planet revolves around its axis once every 24 hours in the same direction Earth does. Therefore, it has days and nights similar to Earth's. It orbits a sun similar to our own, at about the same distance. Its spin axis is slightly tilted relative to the orbital plane.

2. The process of plate tectonics is operating on this planet at rates similar to those observed on Earth. Volcanoes, mid-ocean ridges, and mountain-building processes are also similar to those on Earth. There are continents, ocean basins, and polar ice caps.

3. The oceans are likewise similar to those on Earth. They show similar variations in sea-surface temperature and have ocean currents that circulate around the edges of the main ocean basins and through gaps between the continents.

4. The planet has well-developed atmospheric currents, with wind patterns similar to those on Earth.

Preliminary Interpretation of Ocean Currents

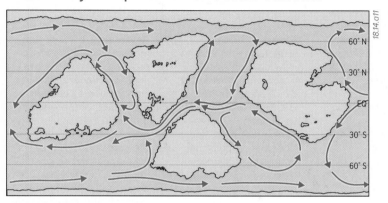

This map shows the directions of ocean surface currents. From these data, identify which currents are probably cold, warm, or neither. Water temperatures will be key in inferring probable climates.

Preliminary Interpretation of Wind Data

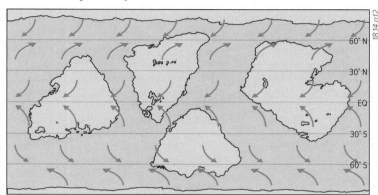

These data show the directions of prevailing winds, which are similar to the patterns on Earth. Different wind directions occupy bands encircling the planet, parallel to the equator.

Ocean Water Temperatures and Sea-Surface Pressures

This map depicts satellite data for sea-surface temperatures (SST) during late summer. Orange and red show warmer waters, whereas blue and purple show colder waters. The numbers give the temperature values in degrees Celsius (°C).

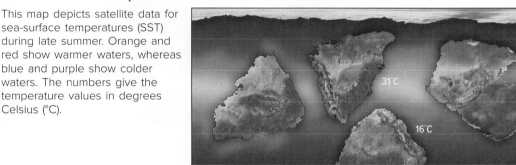

Climate Analysis

On the version of this map on the worksheet, show the probable locations of the following features by writing the associated letters on the map: rain forests (R), deserts (D), potential agricultural areas (A), and lands at highest risk for hurricanes (H). You can use colored pencils to shade in the extent of land associated with each letter. Your instructor may have you provide justification for your interpretations.

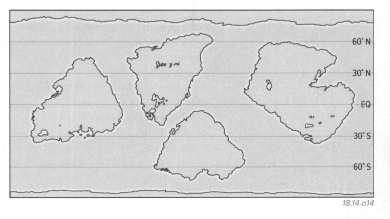

18.14

19 Our Solar System

OUR PLANET HAS NEIGHBORS, including our Moon and the Sun. Interesting features are exposed on the surfaces of our solar system's four innermost planets, on our own Moon, and on moons of the outer gas planets. The planets and moons highlighted in this chapter provide a brief portrait of the most important or interesting bodies in our solar system. Compared to illustrations in this chapter, the planets and moons are vastly farther apart and are more different in size than can be shown.

The four inner planets are called *terrestrial planets* because they have solid rocky surfaces (*terra* means earth). These planets also have similar overall compositions but not similar histories.

Mercury, closest to the Sun, is a small, heavily cratered planet with almost no atmosphere. It has a 650°C difference between night and day temperatures.

Why is the planet so heavily cratered?

Venus has a thick atmosphere of carbon dioxide that captures much of the solar radiation that reaches the planet. This extreme greenhouse effect causes surface temperatures to reach 450°C. The planet is shown here as if it did not have its thick atmospheric shroud.

What is the land surface like beneath the clouds?

Earth has plate tectonics and also a strong magnetic field caused by its rotating, molten iron outer core. Oceans cover 71% of the planet's surface. The abundant surface water sustains a diversity of life.

Why is Earth so different from the other inner planets?

Mars has been explored recently, and new data show that water once flowed on the Martian surface. But Mars lost most of its atmosphere sometime in the past, and now it is so cold that liquid water cannot exist in large quantities on its surface.

What is the evidence for past movement of water on Mars?

Asteroids are rocky fragments concentrated in an orbit between Mars and Jupiter. They are similar in composition to certain meteorites and are interpreted to be fragments left over from the formation of the solar system, probably with some pieces of small planetary objects that broke apart. The largest object in the asteroid belt is *Ceres*, an icy world classified as a *dwarf planet*, like Pluto. NASA's *Dawn* spacecraft began orbiting Ceres in 2015, capturing spectacular images. The asteroid belt is a prime region for future space missions.

Mercury

19.00.a1

Venus

Earth

Mars

How the Solar System Formed

The Sun formed about 5 billion years ago from the remnants of previous stars and cosmic dust, all of which had a beginning in what is called the "big bang." According to modern theories, the entire universe arose from the *big bang* 13.8 billion years ago, so the universe was 9 billion years old before our solar system began to form. Current theories for the formation of our solar system suggest that the Sun and planets condensed from a *nebula*, a cloud of gas and dust. Particles of dust clung together to form small chunks and then larger pieces, eventually ending up as planets. The Sun, meantime, continued to attract more material and became massive enough to begin atomic fusion and emit protons and electrons in a *solar wind*. The solar wind reached the inner planets, blowing away hydrogen, helium, and other light elements near the surface, leaving only heavy elements and other materials. Later, Earth gradually reacquired its supply of hydrogen and other light elements. The outer planets were less affected by solar wind and had enough gravity to retain hydrogen and helium, so they remain gaseous.

During the early stages of its formation, Earth is thought to have collided with another large object that was not quite yet a planet. This catastrophic collision ripped away part of Earth, forming our Moon. It also likely knocked Earth off its original axis of spin, giving the planet the present 23.5° tilt of its spin axis. In other words, we probably have a moon and seasons because of this immense collision nearly 4.5 billion years ago.

Jupiter, the largest planet in the solar system, consists of hydrogen and helium with a small rocky core, making it compositionally more similar to the Sun than to Earth. It has a banded, swirling atmosphere and many interesting moons.

Do Jupiter's moons look like ours, and are they all the same?

Jupiter

Saturn is similar to Jupiter, but it has a beautiful system of delicate, icy rings around the planet. Our spacecraft are actively exploring Saturn and its moons.

What have our spacecraft found so far?

Saturn

Uranus

Uranus is large, but smaller than Jupiter and Saturn. It is much farther out in the solar system, being as far from Saturn as Saturn is from the Sun. Uranus has arguably the oddest moon in the solar system.

Neptune

Neptune is very similar in size and composition to Uranus. Both planets are gaseous and have a blue color.

Why are Neptune and Uranus blue?

Pluto is a tiny body with an icy surface and an unusual orbit. Once considered to be the ninth planet, Pluto is no longer classified as a true planet, leaving our solar system with only eight true planets.

Pluto

The four large outer planets—Jupiter, Saturn, Uranus, and Neptune—are known as the *gas giants* because of their large size and gas-rich character, which is quite different from the inner, terrestrial planets. All four planets have their own moons and some type of rings. One way to think of the solar system is as an inner zone of rocky planets, an outer zone of giant, gaseous ones, and finally a zone of small, distinct objects, like Pluto, dominated by ice. This outward progression is related to how the solar system formed and evolved.

Outside Our Solar System

Our galaxy, the *Milky Way Galaxy*, is only one of countless immense galaxies in the universe. Each galaxy is composed of millions of stars, like the Sun, and many of these stars are orbited by planets. Astronomers have captured some amazing images of other galaxies and of nebulae, which are large accumulations of space dust and stars. Below are the Whirlpool Galaxy (top) and the Eagle Nebula (bottom), a birthplace for stars.

19.00.a2

19.00.a3

19.0

19.1 How Do We Explore Other Planets and Moons?

EXPLORING THE GEOLOGY OF OTHER PLANETS and moons is not as easy as studying geology on Earth, even when compared to Antarctica or other remote parts of our planet. Nearly all exploration of other planets and their moons has to be done remotely (from a distance) by examining them with telescopes and other instruments, either on Earth or in orbit around Earth. We gain additional information by sending spacecraft to visit these distant objects and orbit them or land and explore their surfaces. Our Moon is the only place other than Earth where geologists and other humans have walked on the surface, making observations and collecting rock samples.

A What Can We Observe with Telescopes on Earth and in Earth Orbit?

Historically, most investigations of other planets and moons in the solar system have used Earth-based telescopes. Astronomers still rely heavily on Earth-based telescopes, but they also use telescopes in orbit around Earth or launched into space to avoid the distorting effect of Earth's unpredictable atmosphere.

1. These images of Mars (▼) were taken several months apart from the *Hubble Space Telescope,* which orbits Earth.

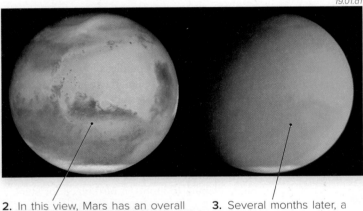

19.01.a1

2. In this view, Mars has an overall orange-red color but contains dark rocky areas, as well as ice caps at the north and south poles.

3. Several months later, a planet-wide dust storm had covered most of Mars, obscuring its surface.

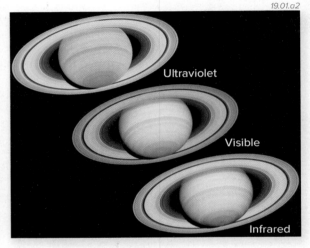

19.01.a2

4. Common telescopes view the night sky using visible light, but astronomers use telescopes that can also observe other parts of the *electromagnetic spectrum.* These three views of Saturn from the Hubble Space Telescope show how the planet appears when imaged in visible light (center), short-wavelength ultraviolet light (top), and long-wavelength infrared light (bottom). Each type of image provides data about different aspects of an object.

B What Do Radar Observations Tell Us About a Planetary Surface?

We can observe the surfaces of other planets and moons remotely by using radar or other techniques that can penetrate the atmosphere of a planet and reveal topographic features and roughness of the surface.

1. In this technique, a satellite transmits radar waves down and sideways toward a planetary surface. A sensor on the satellite then measures the amount of the radar signal reflected back from the surface.

2. If the planetary surface is rough or slopes toward the satellite, the surface will reflect more radar waves back toward the satellite. The area will appear bright on the radar image (lots of returned energy).

3. If the planetary surface is smooth and slopes away from the satellite, most of the radar energy will bounce away rather than return to the instrument. As a result, these types of areas will be relatively dark on the radar image.

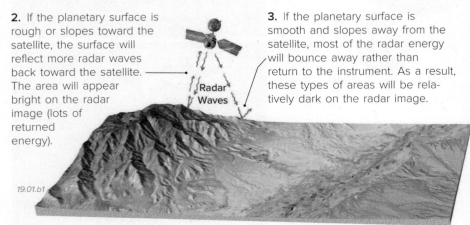

19.01.b1

4. This radar image from Venus shows a crater produced by a meteoroid impact. Much of the interior of the crater is dark because the surface there is relatively smooth. Broken rocks ejected from the crater form an apron of ejected rock around the crater and appear bright because they form rough topography.

19.01.b2

C How Can We Remotely Observe Temperature and Composition of Surfaces?

Geologists and astronomers working with engineers develop sophisticated instruments that allow us to measure the temperature of, and to infer the composition of, a planetary surface from afar. One technique measures the energy given off by rocks in the *infrared* part of the electromagnetic spectrum.

1. When rocks and other materials are heated by the Sun, they re-emit some of the energy as thermal energy with infrared wavelengths. By measuring how much infrared energy is given off, we can calculate the temperature of the surface, even at night. We can then infer what types of material are present by how well the material holds heat. Unconsolidated sediment and other low-density materials lose their heat faster than solid rock and materials that are relatively dense.

19.01.c1

2. This image of an impact crater on Mars combines visible and infrared measurements. The colors depict night-time temperatures: reds show warmer areas and blues show colder ones. The bright colors on this and similar images in this chapter are not the actual colors of the Martian surface.

3. The floor of the crater is bluish in this image, showing that the materials cooled relatively quickly after sunset. Geologists therefore infer that the crater contains some loose, unconsolidated material, probably sediment.

D How Have We Explored the Surfaces of the Moon and Some Planets?

In addition to observations from spacecraft, NASA and other space agencies have landed or intentionally crashed probes on several planets, moons, and asteroids. Astronauts have walked on the Moon with the expressed intent of collecting rock samples and observing other aspects of its geology.

19.01.d1

19.01.d2

19.01.d3 Curiosity Rover

Astronaut and geologist Harrison (Jack) Schmitt collects a sample near the Apollo 17 landing site on the Moon. What an amazing field site in which to do geologic field work!

Humans have landed spacecraft on the surfaces of Venus, Mars, Earth's moon, a moon of Saturn, and even an asteroid. Such spacecraft, called *landers*, collect images and various types of data during their descent and after they have landed.

In recent years, NASA has explored the geology of the Martian surface using *rovers*, which are small-wheeled vehicles that drive around on the surface, following commands issued from Earth. The rovers take images, collect infrared and other data, and even drill through the surface coating of rocks.

Encounters with Asteroids

Farther from the Sun than Mars, but this side of Jupiter, is a belt of hundreds of thousands of rocky chunks drifting in space. Some are more than 500 km (300 mi) across, others are meters across. Most asteroids are thought to be debris left over from the formation of the solar system 4.5 billion years ago, but some are probably parts of objects that broke up. Several space missions have passed close enough to asteroids to take detailed photographs or even land on the surface, as did the Near Earth Asteroid Rendezvous–Shoemaker mission. NASA's *Dawn* spacecraft captured this stunningly detailed image of an asteroid called *Vesta*. There are plans to land humans on an asteroid within the next decade or two. There are many asteroids within the main asteroid belt

between Mars and Jupiter, and others that reside or travel closer to Earth.

19.01.t1 Vesta

Before You Leave This Page

☑ Briefly explain why we put telescopes in orbit to better observe space.

☑ Sketch or describe what radar and infrared observations indicate about a planetary surface.

☑ Describe ways we collect information by landing spacecraft on an object, including those with scientists or rovers.

☑ Describe what asteroids are and where most are located.

19.2 Why Is Each Planet and Moon Different?

THE PLANETS AND MOONS OF OUR SOLAR SYSTEM exhibit a remarkable diversity of atmospheric and geologic characteristics. Some are gaseous giants with thick atmospheres, whereas others are like cratered snowballs. What processes affect the surfaces of planets and moons, and what causes these differences?

A What Determines the Kinds of Materials and Features on a Planet or Moon?

The surface environments of planets and moons are governed by their position in the solar system. The distance from the Sun to a planet or moon influences what material the planet or moon contains and how much solar heating it experiences. The planetary objects are equally influenced by their size, whether they have an atmosphere, and whether liquid water currently exists, or ever existed, on the planet or moon.

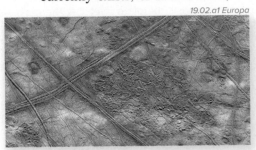
19.02.a1 Europa

19.02.a2 Jupiter

19.02.a3 Mars (Thermal Infrared Image)

Crater

Ejecta

1. *Composition*—The appearance and dominant processes of a planetary object are influenced by its composition, including its chemical composition, and the proportions of ice (▲), rock, liquid, and gas.

2. *Atmosphere*—The presence of a thick atmosphere can obscure the planet's surface. It can block incoming solar radiation, trap in heat, and lead to other phenomena, such as rain and erosion.

3. *Impacts*—All planetary surfaces have at least some craters formed by the impact of meteoroids and other objects. Some planets and moons contain many craters, whereas others have few preserved craters.

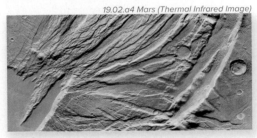
19.02.a4 Mars (Thermal Infrared Image)

19.02.a5 Olympus Mons, Mars

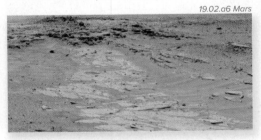
19.02.a6 Mars

4. *Tectonics*—Faulting and other tectonics modify planetary surfaces, causing variations in topographic relief. Only Earth exhibits our style of plate tectonics, but some other objects have tectonics.

5. *Volcanism*—Magma erupts onto the surface and may form a silicate rock, like on this basaltic shield volcano on Mars, may be sulfur rich, as on Jupiter's moon Io, or may be liquid water on an icy moon.

6. *Erosion and Deposition*— Wind, water, ice, and gravity can erode some areas and redeposit eroded material elsewhere. In this image, sedimentary layers are being eroded, revealing their upturned edges. Blowing sand partly covers these layers. Erosion, deposition, volcanism, and tectonics can remake a surface, a process called *resurfacing*.

B How Are Impact Craters Formed?

The dominant geologic process in our solar system is the formation of impact craters. Impact craters of all sizes are abundant on the surfaces of most planets and moons, except in areas that have been resurfaced. Impacts were more frequent early in the solar system's history because there was more debris in space.

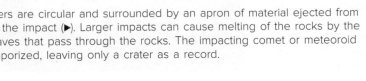

19.02.b2 Mars

An impact crater forms when a meteoroid, comet, or other object from space strikes a planetary surface at an extremely high velocity, blasting open a crater and fracturing adjacent rocks. Pieces of the Shoemaker-Levy 9 comet collided with Jupiter in 1994, as recorded by this series of images (◄). These collisions caused a disruption of Jupiter's gaseous surface, but did not form visible craters.

Most impact craters are circular and surrounded by an apron of material ejected from the crater during the impact (►). Larger impacts can cause melting of the rocks by the intense shock waves that pass through the rocks. The impacting comet or meteoroid may be totally vaporized, leaving only a crater as a record.

19.02.b1 Jupiter

C What Determines Whether an Object Has Active Tectonics and Volcanism?

Why do some planets and moons have active tectonics and volcanism, whereas others are inactive and heavily cratered? The main cause of this difference is the size of the object, which controls how fast it loses heat.

1. The size of an object affects heat loss. As the radius (r) of a sphere increases, surface area increases by r^2 (area of sphere $= 4\pi r^2$), whereas volume increases by r^3 (volume of sphere $= 4/3\pi r^3$). Therefore, larger objects have less surface area relative to their volume than do smaller objects. Here, we examine three objects, which are shown at their correct relative sizes.

2. Smaller objects, like the planet Mercury, have a large surface area relative to their volume. This causes them to lose heat more quickly, which in turn causes them to solidify, shutting off volcanism and tectonism.

19.02.c2

3. The solar system has a few smaller objects that remain volcanically and tectonically active. *Io,* an inner moon of Jupiter, is active because gravitational effects of Jupiter and its other moons provide additional heat energy to Io.

19.02.c1

4. Larger rocky planets, like Earth, have a relatively small surface area compared to their inner volume. This allows them to retain heat generated from their initial formation and from post-formation radioactive decay. As a result, larger planets remain volcanically and tectonically active longer than smaller planets.

19.02.c3

D How Can Water and Wind Modify a Planet's Surface?

Water is currently abundant on Earth, and scientists have demonstrated that water was present in the past on Mars and perhaps on other planets and moons. Water promotes chemical and physical weathering, erosion, transportation, and deposition of sediment. Flowing water leaves telltale landforms, including drainage networks, that point to its earlier existence. Wind forms dunes and other distinctive landforms, and can bury or modify parts of a planet's surface.

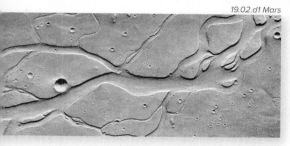

19.02.d1 Mars

19.02.d2 Mars

19.02.d3 Mars

Weathering and Erosion—These channels on Mars are interpreted to have been eroded by running water because they are similar to stream channels on Earth. If a planetary object has no water or wind, there will be little or no erosion of that object's surface.

Deposition—Material can be moved and deposited by various processes, including running water, wind, flowing ice, and slope failure. This channel on Mars has a curved meander floored with sedimentary deposits that look like a typical point bar.

Wind—If a planet has an atmosphere and strong winds, sand and dust can be blown across the surface, forming distinctive dunes and covering up what lies beneath. Much of the surface of Mars is obscured by windblown dust and sand dunes, like those shown above.

Using Crater Density to Estimate the Age of a Surface

Across the solar system, the degree to which the surface of a planet or moon is cratered varies widely. Planetary geologists use the density of craters to estimate the *age* of a planetary surface. The underlying principle is that the longer a surface is exposed, the more impact scars it receives. Surfaces that have remained undisturbed by tectonics, volcanism, and deposition for a long time will be more heavily cratered than those that have been more recently resurfaced by these processes. So a region that has a high density of craters is interpreted as being older than a region with fewer craters. Which part of this image of the Copernicus area of Earth's moon (◄) is youngest? The foreground, consisting of dark lava flows, is less cratered and is younger than the older, more heavily cratered part in the background.

19.02.t1 Moon

Before You Leave This Page

☑ Summarize the factors and processes that affect the appearance of a moon or planet.

☑ Summarize the characteristics of an impact crater and how one forms.

☑ Explain why smaller objects are more likely to be tectonically or volcanically inactive than larger objects.

☑ Explain how crater density can be used to estimate the age of a planetary surface.

19.2

19.3 What Can We Observe on the Inner Planets?

THE FOUR PLANETS CLOSEST TO THE SUN are the *inner planets*: Mercury, Venus, Earth, and Mars. These four *terrestrial planets* all have a rocky surface. The geology of these four planets differs primarily because of their different sizes and different atmospheres. For each planet, observe the images first, noting what features are most obvious, before you read the accompanying text.

A What Is on the Surface of Mercury, the Closest Planet to the Sun?

Mercury is a small and rocky planet that is relentlessly baked by its proximity to the Sun. It speeds through space at about 50 km/s and orbits the Sun in 88 days (that is, a Mercury-year is only 88 days long).

19.03.a1

1. Mercury (◄) has almost no atmosphere, and the temperature can reach 460°C (860°F) during the day and an extremely cold minus 80°C (–112°F) at night.

2. With so little atmosphere, there is no erosion by water or wind, so the surface is not modified by these processes. As a result, Mercury's surface is covered with numerous craters produced when meteoroids and other planetary debris collided with the planet's surface. Because of its extremely thin atmosphere, impacts are more common on Mercury than on the other terrestrial planets. If planets have a thick atmosphere, meteoroids and other objects often burn up due to frictional heating in the atmosphere before they reach the planet's surface.

19.03.a2

3. In this close-up view of part of Mercury, the surface is covered by impact craters of various sizes (◄). The surface appears so heavily cratered because tectonic and volcanic activity on this small planet ceased early in the solar system's history, when there still was an abundance of debris to collide with the planet. The subsequent lack of volcanoes, wind, or water means that the craters have not been eroded or covered by lava or sediment—that is, the surface of Mercury has not been recently resurfaced.

B What Is Beneath the Atmospheric Shroud of Venus?

Venus, the second planet out from the Sun, is similar to Earth in size, mass, and composition, but has a thick atmosphere that hides the planet's surface. It has experienced volcanism and tectonism in the last billion years, but there is no evidence of active plate tectonics.

19.03.b1

1. In this telescopic view of Venus (◄), the surface of the planet is obscured by a thick atmosphere of clouds and toxic gases that swirl around the planet. The atmosphere consists mostly of carbon dioxide and droplets of sulfuric acid, with almost no water vapor. The atmosphere exerts a stifling amount of air pressure. It allows the Sun's energy in, but keeps the heat from escaping back to space, causing the atmosphere to heat up, like a closed greenhouse. As a result of these extreme greenhouse effects and Venus' proximity to the Sun, surface temperatures exceed 450°C (840°F).

19.03.b2

2. Planetary scientists used radar to image the surface of Venus (◄) through the thick atmosphere. The radar image shows that the planet has significant topographic relief, including large, continent-sized high areas and vast, low plains. This map is shaded by elevation: blues are low regions and reds are high regions.

19.03.b3

3. This radar image (◄) shows bright areas that have a rough surface. The rough areas are interpreted as lava flows that flowed across a linear ridge formed by faulting or some other type of deformation.

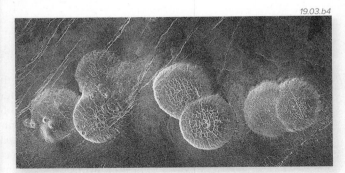

19.03.b4

4. These 25-km-wide, pancake-shaped mountains are interpreted to be thick lava domes (▲), demonstrating that volcanism has played an important role in the history of Venus.

C What Makes Earth Unique?

Earth is very different from its neighbors because it is just the right distance from the Sun to contain abundant water and allow a thick, but not too thick, atmosphere to develop, along with oceans, streams, lakes, and life. It has the right conditions to have water in all three phases, making it more suitable for life. A planetary object under these conditions is said to be in the "Goldilocks zone"—not too hot, not too cold, but just right for life.

19.03.c2 Sahara Desert, Africa

1. In this computer-rendered view of Earth, the planet is dominated by its blue oceans but also shows green and brown continents and white clouds and ice. Even when viewed over the short time frame of hours or days, Earth is a dynamic place, with clouds, oceans, wind, streams, and other features being in constant motion. Over the longer perspective of geologic time, mountains rise and are eroded down, continents break apart and travel great distances, and oceans widen or shrink with time due to seafloor spreading and subduction.

19.03.c1

2. Earth's surface conditions allow water to exist as water vapor, liquid, or ice (in the atmosphere, oceans, and glaciers, respectively). Moving water and ice erode the land and deposit sediment, modifying or completely resurfacing the land surface.

3. On Earth, weathering produces abundant sediment, which can be transported by water and atmospheric winds. A thin veneer of sediments and sedimentary rocks covers most of the land and seafloor. This image shows a large dust storm moving many tons of sediment across northern Africa.

4. Earth is big enough and generates enough heat to allow plate tectonics to operate and form the large-scale features observed today. The existence of tectonic activity on Earth, especially plate tectonics, is largely responsible for Earth's uniqueness. Tectonic-related volcanoes, like in the steam-emitting, divergent rift shown here (▶), release water vapor and other gases that are an essential part of our atmosphere. Subduction takes carbon-rich crustal materials, like limestone and organic shale, to great depths, removing the carbon from the near-surface environment. Without plate tectonics, Earth could have a thick, stifling, CO_2-rich atmosphere like that of Venus, and might be lifeless. It might not have a temperature or atmospheric chemistry that would support life as we know it.

19.03.c3 Krafla, Iceland

5. Without tectonics to dewater and degas Earth's interior, Earth might not have oceans, lakes, streams, or other parts of the hydrosphere. Without oceans and rain, it is likely that wind and dust storms would be the dominant agents of erosion, transport, and deposition of sediment, as is the current situation on Mars.

D What Is on the Surface of Mars, the Red Planet?

Compared to Earth, Mars is smaller and has a thinner atmosphere. The color of the planet is due to an abundance of reddish, windblown dust. Mars has been orbited by spacecraft with sophisticated cameras and instruments, and has been visited by landers and rovers. It is the focus of the *Connections* pages near the end of this chapter.

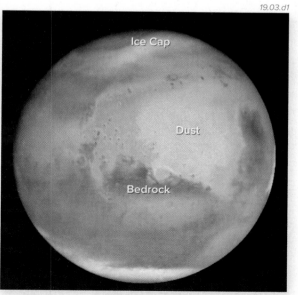

19.03.d1

1. Mars has enough atmosphere to maintain ice caps on the south and north poles, as well as patches of ice in some shaded areas. The ice caps shrink and grow with the Mars seasons.

2. Much of the surface consists of reddish dust that has been weathered from rocks and blown around the surface by strong Martian winds. The reddish color is from iron oxide minerals, like hematite.

3. Geologists interpret Martian bedrock to consist of basaltic lavas, including some with recently photographed columnar joints. Other areas have layered rocks that many geologists think are sedimentary in origin.

Before You Leave This Page

✓ Explain why the surface of Mercury is so heavily cratered.

✓ Describe why radar was required to investigate the surface of Venus, and what types of features we found.

✓ Discuss factors that make the surface of Earth so different from its neighbors.

✓ Summarize the materials and features present on the surface of Mars.

19.3

19.4 What Is On the Surface of Our Moon?

OUR NEAREST NEIGHBOR IN SPACE is the Moon. It is much closer than any other object in the solar system and can be observed in detail with the simplest of telescopes or binoculars. It was the first object in the solar system to be systematically studied by astronomers and geologists, who observed the Moon with telescopes. Subsequently, geologists mapped the topography, geology, and composition by sending spacecraft to orbit the Moon, and finally walked and drove on the Moon's surface, observing the features and collecting rock samples to bring back to Earth for detailed study.

A What Are the Main Geologic Features of the Moon?

Observe this large image of the Moon. The surface of the Moon is not all the same. It has lighter colored areas, with some dark, somewhat-circular patches. With binoculars, we can observe individual craters.

1. This view of the Moon shows the features on the side of the Moon that always faces Earth, called the *near side*. The other, or *far side*, cannot be seen from Earth but has been photographed by spacecraft.

19.04.a1

2. Much of the near side and nearly all of the far side consists of light-colored material that is heavily cratered (▶). This material mostly occurs at higher elevations and is called the *lunar highlands* or the *cratered highlands*.

19.04.a2

3. From samples collected by astronauts and other information, we know that the cratered highlands contain igneous rocks that are light colored because they consist almost entirely of feldspar. Rocks from the highlands (▶) have been dated to more than 4 billion years.

19.04.a3

4. The dark patches on the Moon (▼) are lower, flatter, and less cratered; they are *maria* (plural of *mare*, meaning sea). The maria have far fewer craters than the highland and so are much younger. The maria consist of dark basalt erupted as lava flows that buried and filled craters and other low spots that existed in the older, lunar highland material.

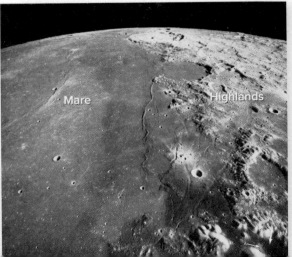

19.04.a4

6. The other obvious features on the Moon are *impact craters,* some of which have bright rays of material radiating outward. The rays overlie and cut across the top of the maria. Such *rayed craters* are some of the youngest features on the Moon, in some places probably being less than 100 million years old. Samples collected from lunar craters are mostly breccia containing angular rock fragments (▶) generated during impacts.

19.04.a6

5. Samples from maria consist of basalt lava (▼), mostly dated at 3.8 to 2.5 billion years old. At these past times, the Moon retained enough heat to allow volcanism. Vesicles in the basalt record gas in the magma.

19.04.a5

B What Other Features Are Observed on the Moon?

The Moon is tectonically and volcanically inactive, and has been so for more than two billion years. The vast majority of craters on the Moon are due to impacts of meteoroids and other objects onto the Moon's surface. Impact craters are associated with some familiar features. In addition, the Moon contains more water than we originally thought.

After a crater is excavated by an impact, the shattered walls of the crater are commonly too steep to withstand the Moon's gravity, even though it is only 1/6 that of Earth. Slope failures are common along the walls, like the rotated slump blocks, debris avalanches, and other loose debris shown here.

19.04.b1

19.04.b2

Several independent observations have confirmed that the Moon has abundant water as ice beneath shadowy craters and in minerals in the Moon's interior. The image here represents measurements of the amount of neutrons coming from the Moon's South Pole, which can be related to the amount of water. The purplish area in the center is interpreted to have the most water, and NASA deliberately crashed part of a rocket into this area and measured water in the ejected debris.

C What Causes the Phases of the Moon?

Every month the Moon appears to change its illumination on a regular schedule, going from completely dark to fully lit and back to dark again. Why is this cycle, called *phases of the Moon*, happening?

19.04.c1

19.04.c2

19.04.c3

1. At times, the side of the Moon facing Earth (near side) appears fully illuminated by sunlight. This is a *full moon*.

2. Seven days later, only half of the near side can be seen from Earth. This is a *quarter moon*.

3. Six days later, a thin sliver of the near side is illuminated. The next day, none is illuminated, which is a *new moon*.

4. Half of the Moon is always illuminated by the Sun, but from Earth we may not be able to see the entire sunlit half because of the Moon's position. The Moon orbits Earth in 28 days, and we see different amounts of the Moon's sunlit side on different nights.

5. During a new moon, the side being illuminated by sunlight is away from our view. The Moon appears dark to us, but the other side of the Moon is still completely sunlit. The yellow arrows depict the rays of sunlight, but the Sun is not shown.

6. At other times, we see half of the sunlit half of the Moon, so it is a quarter moon. When the moon is in this position, we can tell that the phases of the Moon are not in any way related to Earth's shadow.

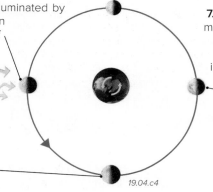

7. During a full moon, the side of the moon being illuminated by sunlight is facing Earth, so we see all of it. The other side is dark.

19.04.c4

A Model for the Formation of the Moon

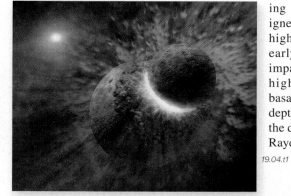
19.04.t1

Where did the Moon come from, and how did it form? Geologists and other scientists have investigated this question by examining several types of data. They calculate the age of the Moon by dating actual lunar samples. The chemical composition of these samples, including isotopic analyses, showed some unexpected similarities to rocks on Earth. This led to a hypothesis, currently favored by many scientists, that the Moon formed when a Mars-sized object collided with Earth early in the history of the solar system. The collision ejected a huge part of Earth's mantle into space, where it later aggregated under the force of gravity and formed the Moon.

As the Moon formed, and soon thereafter, it became hot enough for large parts to melt. As the magma began to solidify, heavier crystals sank downward (crystal settling), while less dense crystals, especially feldspar, floated upward. The floating crystals accumulated near the surface, forming the light-colored igneous rocks of the highlands. Frequent, early, and intense impacts cratered the highlands. Later, basaltic magmas from depth erupted, forming the dark-colored maria. Rayed craters formed even later, from more recent impacts. This proposed history of the Moon came from diverse sources of information, including telescope observations, field excursions to the lunar surface, and lots of laboratory measurements and computer simulations.

Before You Leave This Page

☑ Summarize the physical characteristics and rock compositions of the lunar highlands, maria, and craters, and explain how each feature formed.

☑ Sketch and describe what causes the phases of the Moon.

☑ Summarize one model for how the Moon and its different parts formed.

19.4

19.5 What Is Observed on Jupiter and Its Moons?

JUPITER, THE LARGEST PLANET IN THE SOLAR SYSTEM, is a gas giant more than three times farther from the Sun than Mars. Jupiter is orbited by, at present count, 67 officially named moons, including the largest moon in the solar system. To geologists, the icy and rocky moons are of greater interest than the gas-dominated planet itself because of their solid surfaces, spectacular geologic features, and wide diversity. Many astronomers and atmospheric scientists are more interested in Jupiter's interesting and dynamic atmosphere.

Jupiter is nearly 780,000,000 km from the Sun, but it is so large that we can see it on most clear nights. It is about 2.5 times more massive than all the other planets combined, and it contains more than 300 times the mass of Earth. Examine the large, page-spanning image, which was computer generated by wrapping actual images of Jupiter around a sphere. What do you observe on the surface of the planet?

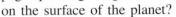
19.05.a2

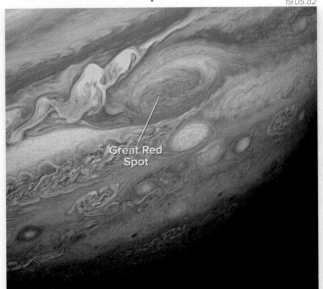

Great Red Spot

19.05.a3

Io

Jupiter

Europa

Ganymede

Callisto

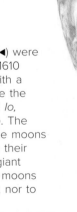

1. Jupiter is so far from the Sun that it takes nearly 12 Earth-years to complete an orbit—a Jupiter year is more than 4,300 Earth-days long. As viewed from Jupiter, the Sun appears much smaller and dimmer than it does from Earth.

2. The dominant features of Jupiter are the colorful bands and swirls of the planet's atmosphere. The atmosphere is mostly hydrogen with lesser amounts of helium, and trace amounts of methane, ammonia, and other gases. The interior of the planet is interpreted to consist of hydrogen in liquid and liquid-metallic forms, surrounding a solid core of mostly iron silicate minerals. Most of the planet is gas, so its overall density is less than that of Earth.

3. One of the most distinctive features in Jupiter's atmosphere (◀) is the Great Red Spot, which is a storm that has existed for hundreds of years. This spot is three times wider than the diameter of Earth but is currently shrinking in diameter.

4. Jupiter's four largest moons (◀) were discovered by Galileo Galilei in 1610 when he observed the planet with a telescope. These four moons are the *Galilean moons*, and are named *Io, Europa, Ganymede,* and *Callisto.* The dramatic differences between the moons are largely due to differences in their distance from the massive gas giant around which they revolve. The moons are not as close to one another, nor to Jupiter, as shown here.

19.05.a4

5. This image (◀) shows Jupiter's moon Io and the shadow of Io on Jupiter's surface. The image was taken by the Hubble Space Telescope, which orbits Earth. The area under the shadow is experiencing a solar eclipse.

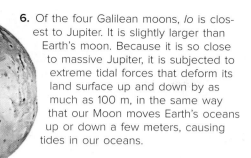

19.05.a5

6. Of the four Galilean moons, *Io* is closest to Jupiter. It is slightly larger than Earth's moon. Because it is so close to massive Jupiter, it is subjected to extreme tidal forces that deform its land surface up and down by as much as 100 m, in the same way that our Moon moves Earth's oceans up or down a few meters, causing tides in our oceans.

7. Pulling and squeezing of rocks by tidal forces generates heat, making Io the most volcanically active object in the solar system. Sulfur-rich lava flows cover its surface. NASA's *Galileo* spacecraft photographed one such eruption of lava (▶).

19.05.a6

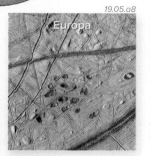

19.05.a7

8. *Europa* is farther away from Jupiter but is still heated by the tidal forces from Jupiter and the other Galilean moons. These forces allow Europa to remain tectonically active longer than would be merited by its size. Tectonic processes have extensively reworked the surface, accounting for the linear features (◀) and the nearly complete lack of craters. Beneath the icy crust is probably an ocean of liquid water.

19.05.a9

9. The surface of Europa is a crust of ice (mostly frozen water) marked by intersecting lines (◀). These lines appear to be fissures that allowed liquid water to erupt onto the surface.

19.05.a8

10. Parts of Europa's surface are covered by huge blocks of ice (▲) that broke apart and then froze in place, like icebergs in a frozen sea.

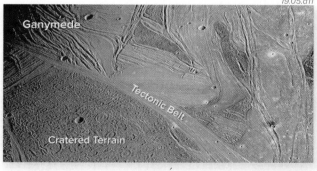

19.05.a11

11. *Ganymede* (▲), the largest moon in the solar system, is thought to consist of a rocky core with a water-ice mantle and a crust of water-ice and rocks.

19.05.a10

12. Ganymede's surface contains dark, heavily cratered patches (▲) that are relatively old. Younger patches and belts cross-cut the older surfaces and contain tectonic features similar to those seen on Europa, including some interpreted to be water-erupting fissures.

19.05.a1

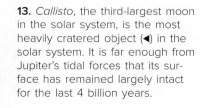

13. *Callisto*, the third-largest moon in the solar system, is the most heavily cratered object (◀) in the solar system. It is far enough from Jupiter's tidal forces that its surface has remained largely intact for the last 4 billion years.

19.05.a12

Before You Leave This Page

☑ Summarize the key characteristics of Jupiter, such as its size, internal composition, and atmospheric composition.

☑ Briefly summarize the main characteristics of each Galilean moon.

19.5

19.6 What Is Observed on Saturn and Its Moons?

SATURN IS A BEAUTIFUL PLANET, a gas giant encircled by a set of spectacular rings. Saturn is the second-largest planet in the solar system and is orbited by more than 50 named moons. Saturn's moons are quite diverse, reflecting differences in the materials and in the role of different geologic processes.

Saturn is farther from the Sun than Jupiter, in fact, nearly twice as far. Saturn and Jupiter are very far apart, with the distance between them being greater than the distance between the Sun and Mars. The large photograph was taken by the *Voyager* spacecraft, and colors have been enhanced in the smaller photograph to accentuate the bands.

1. Saturn, like Jupiter, consists mostly of hydrogen and helium, which make up the gaseous atmosphere. The gases become liquid as they are compressed closer to the center of the planet. The center of the planet is interpreted to have a solid core of rock and metal. Saturn is more than 1.4 billion km from the Sun, and it takes 29.5 Earth-years to orbit the Sun.

2. Like Jupiter, Saturn is a mini solar system, orbited by a collection of large and small moons. The image to the right shows Saturn, its rings, and some of its moons (small light-brown spots), four of which are discussed here: Titan, Iapetus, Enceladus, and Mimas. Titan is the largest of Saturn's moons, and the other three are included not because they are the three next largest, but because they are geologically interesting and nicely illustrate the geologic diversity of Saturn's moons. Our knowledge of the geology of these moons has increased dramatically due to the arrival of the *Cassini-Huygens* spacecraft in 2004.

3. Saturn is best known for its rings (▼), which extend outward from the planet a distance nearly equal to the distance between the Earth and Sun. The rings consist of widely separated icy chunks floating in space. Most of the icy chunks are the size of sand, pebbles, and boulders, with some larger pieces. Close-ups of the rings display intricate details of concentric thick and thin rings separated by dark-colored, more-empty space, as viewed in the image below, which is colored by particle size. Purple indicates regions where particles are larger than 5 cm (pebble size and larger), green and blue indicate particles smaller than 5 cm, and white bands mark rings where particles were large enough or dense enough to block the radio signals used to determine size.

19.06.a1

19.06.a2

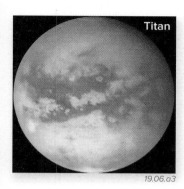
Titan

19.06.a3

4. *Titan* is the largest of Saturn's moons and the second-largest moon in the solar system, even larger than the planet Mercury. Its surface is obscured by a thick, cloudy atmosphere of mostly nitrogen and methane, but this image (◄) generated from various types of data shows some of Titan's surface features. The surface is inferred and observed to contain solid materials (ices) and liquids, including liquid methane and other hydrocarbons.

5. The *Cassini* spacecraft released the *Huygens* probe, which parachuted through Titan's atmosphere and softly landed on the surface. On the way down, it captured images of drainage networks and a lake or ocean. *Cassini* (►) has continued to capture radar images of the surface,

19.06.a4

Titan

confirming that liquids are widespread on Titan's surface. Once on the surface, the *Huygens* probe sent back an image of well-rounded icy boulders, presumably rounded by transport in flowing liquid.

Iapetus

19.06.a5

6. *Iapetus*, another moon of Saturn, is distinctive in that most of its icy, cratered surface is light colored (◄), but part is quite dark. The light-colored side is water ice; the darker side is interpreted to be a coating of dust that probably escaped from an adjacent moon and has been plastered on the leading edge of Iapetus as it orbits Saturn.

7. *Enceladus* (▼) is one of the lightest-colored objects in the solar system, possibly because an icy frost continuously forms on the surface.

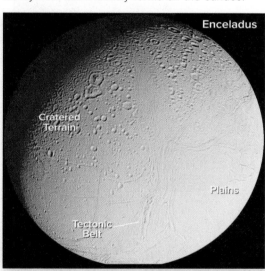
Enceladus

Cratered Terrain

Plains

Tectonic Belt

19.06.a6

8. The surface of Enceladus consists of at least three distinct types of terrain, the oldest of which is heavily cratered. Broad plains lie adjacent to the cratered terrain, and are much less cratered and therefore younger. They are interpreted to have been resurfaced by the eruption of water onto the surface.

9. The third type of terrain consists of tectonic belts that slice through the heavily cratered material and through the plains. These belts have linear ridges and troughs probably formed as fissures through which water erupted to the surface. The *Cassini* team discovered active ice geysers fountaining from the surface (►).

19.06.a7

10. *Mimas* (▼) is a relatively small moon whose pockmarked surface contains a large crater 130 km in diameter, with walls nearly 10 km (~33,000 ft) high. This crater formed from an impact that scientists have calculated nearly blasted the moon apart.

Mimas

19.06.a8

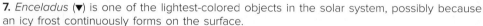

Before You Leave This Page

✓ Summarize the key characteristics of Saturn, such as its size and composition.

✓ Describe what materials compose the rings of Saturn.

✓ Summarize the main characteristics of the four moons of Saturn described here and the main geologic processes expressed on the surface of each.

19.6

19.7 What Do We Observe on the Outer Planets and Their Moons?

THE OUTER PLANETS OF THE SOLAR SYSTEM and their moons are less well known than those from Saturn inward to the Sun. Many of the observations are based on images taken by the *Voyager 2* spacecraft, which flew through the outer reaches of the solar system in the late 1980s. These images provided a wealth of new information about the planets Uranus and Neptune, which are called *ice giants*, and some of their moons.

A What Features Characterize Uranus and Its Unusual Moons?

The next large planet out from Saturn is another large, gaseous world called *Uranus*. This planet and some of its moons have some unusual characteristics and features.

1. The planet Uranus is nearly 2.9 billion km from the Sun. The distance from Uranus to Saturn, the next planet in, is comparable to the distance from Saturn to the Sun. It takes Uranus 84 Earth-years to orbit the Sun.

2. Uranus consists largely of liquid and icy materials, including water, methane, and ammonia. The atmosphere is a mixture of hydrogen, helium, and methane. The blue-green color is caused by methane, which absorbs red light and reflects blue light. Uranus does not appear to have a solid surface.

3. Uranus has rings and at least 27 moons, named after characters from the works of William Shakespeare and Alexander Pope, including Oberon, Titania, Juliette, Puck, Ariel, and Miranda.

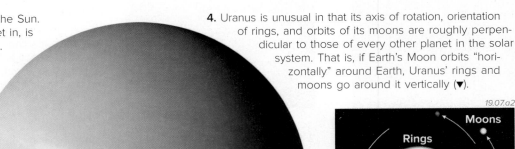
19.07.a1

4. Uranus is unusual in that its axis of rotation, orientation of rings, and orbits of its moons are roughly perpendicular to those of every other planet in the solar system. That is, if Earth's Moon orbits "horizontally" around Earth, Uranus' rings and moons go around it vertically (▼).

19.07.a2

Moons
Rings
Pole

Ariel

5. Ariel (▼) is a moderate-sized moon, approximately 580 km (360 mi) in diameter, and is covered by ice. The surface is cut by long fractures, some of which have been filled by upwelling liquid water. This moon is thought to be mostly inactive.

19.07.a3
Ariel

Miranda

6. To some geologists, Miranda, a small (236 km diameter) moon of Uranus, is the most bizarre world (▶) in our solar system. The surface of Miranda is covered with ice and displays several distinct types of terrain.

7. There is highly cratered terrain that is lighter colored and relatively old.

8. Disrupting the heavily cratered terrain are huge oval to angular features, each of which is a *corona*. The origin of these features is unresolved among planetary geologists, but they involve some normal faulting and probably upwelling of deeper materials.

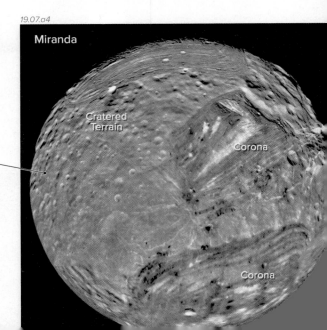
19.07.a4
Miranda
Cratered Terrain
Corona
Corona

B What Is the Geology of Neptune and Its Moon, Triton?

Neptune is the eighth and last planet from the Sun and is another gas giant similar in many ways to Uranus. The existence of Neptune was predicted mathematically before the planet was discovered by telescope.

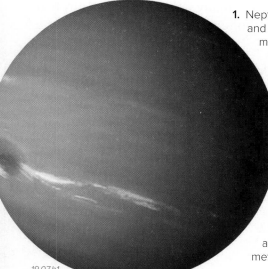

19.07.b1

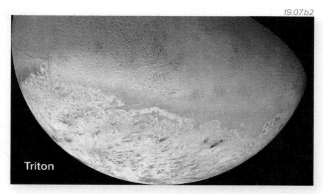

19.07.b2

1. Neptune is nearly 4.5 billion km from the Sun, and the distance from Uranus to Neptune is more than that between Uranus and Saturn. In other words, the distance between adjacent planets increases as one moves outward through the solar system. Neptune is so far from the Sun that it has not completed even one orbit since it was discovered. It takes 165 Earth-years to go around the Sun.

2. Neptune is about the same size as Uranus (~50,000 km in diameter) and has a similar composition, with ices and liquids inside and an outer atmosphere of hydrogen, helium, and methane. Its blue color is due to methane.

Triton

3. The surface of Triton (▲), Neptune's only large moon, consists of ices of nitrogen and carbon substances. It has two distinct halves; one part appears like the surface of a cantaloupe and is interpreted to represent activity from volcanic eruptions and active geysers.

C What Do We Know About Pluto and Its Companions?

Pluto was once considered to be the ninth and outermost planet, but astronomers recently reassigned Pluto to a type of solar system object called a *dwarf planet* or *plutoid*. So our solar system has eight planets, not nine, but this is still being debated.

Pluto was always an oddity compared to the eight planets. It is a relatively small, icy object, even smaller than Earth's moon. Pluto orbits the Sun in a very elliptical orbit that sometimes brings it closer to the Sun than Neptune. A circuit around the Sun takes 248 Earth-years. Pluto has a large companion named *Charon*, which is half Pluto's size, plus several very small moons.

19.07.c1

In 2015, NASA's *New Horizons* spacecraft visited Pluto (◄) and its moon Charon, providing much higher resolution images than previously existed. The spacecraft then headed out toward the *Kuiper Belt,* a disk-like zone of objects that lies beyond the orbit of Neptune. This belt has a number of objects that are similar to and far beyond Pluto. Some of the named Kuiper Belt objects are shown in this artist's conception (►).

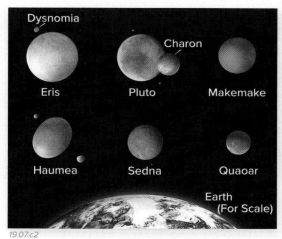

Dysnomia

Charon

Eris

Pluto

Makemake

Haumea

Sedna

Quaoar

Earth (For Scale)

19.07.c2

Comets

Comets are among the more interesting spectacles that sometimes grace the night sky. Comets are small, icy, and rocky objects with very elliptical orbits around the Sun. Some comets, such as *Halley's Comet*, visit the inner solar system regularly, whereas others visit at very long intervals. Comets are thought to come from a very outer part of the solar system, well beyond the orbit of Neptune. As a comet nears the Sun, gas and dust are stripped off by the solar wind and carried outward, forming a tail that always points away from the Sun. Spacecraft have visited and even crashed into comets to help us study them. The *Rosetta* space mission has given us the most comprehensive view of the geology of a comet, which is in this case is quite complex and odd (▼).

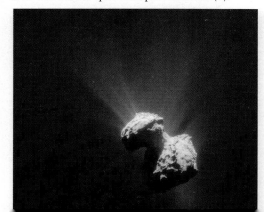

19.07.t1 Comet P67/C-G Rosetta Mission

Before You Leave This Page

☑ Describe some key features of Uranus and Neptune, and explain how they are similar.

☑ Describe unusual features on Ariel, Miranda, and Triton, and identify the materials that comprise the surfaces of these moons.

☑ Describe what is known about Pluto and its companions.

☑ Describe a comet and its tail.

19.7

19.8 What Have We Learned About Mars?

EXCITING DEVELOPMENTS IN PLANETARY GEOLOGY involve Mars, the *Red Planet*. Recently, Mars has been explored by orbiting spacecraft that carried sophisticated cameras and other instruments, many designed and controlled by planetary geologists. Spacecraft have landed on the planet and unleashed small, robotic rovers that travel across the surface, exploring and collecting data.

A What Have We Learned from Instruments Orbiting Mars?

Several spacecraft, including *Mars Odyssey*, *Mars Express*, and *Mars Reconnaissance Orbiter*, orbit the planet. As they pass over the planet's surface, they record images and take measurements designed to detect water and determine the composition of rocks, sediment, and ice. Using these spacecraft, we have made major discoveries.

19.08.a2 Valles Marineris, Mars

1. Mars contains a huge canyon system (◀), 4,000 km long—*Valles Marineris*. The canyon's length is equivalent to the width of the United States. The canyon, which began as a large rift, has been widened by inward collapse of the steep canyon walls and by other types of erosion.

4. Many parts of Mars have layered rocks (▶), which variably have a sedimentary or volcanic origin. Some are related to the polar ice caps. Layers in this image are colorized for emphasis. The surface of Mars is not this color.

19.08.a5

19.08.a3 Candor Chasm, Mars

19.08.a1

2. This image (▲) shows *Candor Chasma,* a part of the Valles Marineris system. The steep walls of the chasm have collapsed downslope, providing some of the most spectacular examples of slope failure in the solar system. Gullies carve into the cliffs and steep slopes.

5. Some parts of Mars have spectacular channels (▶), interpreted to have been formed by torrents of running water flowing on the surface some time in the past. Where the channels reached the gentle plains, they deposited piles of sediment, equivalent to deltas and alluvial fans on Earth.

19.08.a6

19.08.a7 Olympus Mons, Mars

19.08.a4 Victoria Crater, Mars

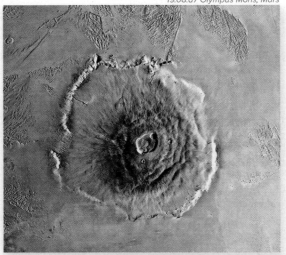

3. The Martian atmosphere is less dense than Earth's, but the winds are strong. Lace-like sand dunes occupy the center of beautiful Victoria Crater (◀). The crater was visited by the Mars Exploration Rover named *Opportunity.*

6. Mars has the solar system's largest volcano, *Olympus Mons* (◀). The volcano is 600 km across and 27 km high, nearly three times the height of Mount Everest. It is a large shield volcano like those on Hawaii, but is inactive. Large segments of the mountain collapsed in landslides and debris avalanches that moved downhill, spreading out to form areas of hummocky topography.

B What Have We Learned from Landers and Rovers on Mars?

NASA successfully landed three spacecraft that carried small rovers to navigate and photograph the Martian landscape. The three famous recent rovers, *Spirit*, *Opportunity*, and *Curiosity*, provided a wealth of new geologic information, including a few surprises.

19.08.b1

19.08.b2

19.08.b3

1. *Spirit* and *Opportunity* landed on the surface by cushioning themselves with large air bags that inflated just before the spacecraft bounced onto the surface. Then the rovers, shown here in an artist's conception, rolled off to explore nearby parts of the planet, taking photographs and collecting various data.

2. The rovers rolled across the surface on wheels, stopping to inspect outcrops or interesting rocks. They spun their wheels to dig up sediment on the surface or used a tool to scratch at the rocks. They have cameras and scientific instruments to measure composition, temperature, and other aspects.

3. The *Phoenix Mars Lander*, shown in an artist's conception, landed south of the polar ice cap to investigate the presence of water ice. The lander verified the presence of ice by observing patches below the spacecraft and in shallow trenches dug by the robotic arm, and by chemical analyses of samples.

19.08.b4

19.08.b5

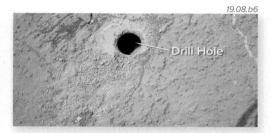
19.08.b6

Drill Hole

4. *Layers* — Many rock exposures in Mars have layers, such as these exposed on the floor of a crater. Geologists interpret some of these layers as indicating that liquid water existed on Mars' surface sometime in the past, and the water was flowing, like in a stream. Such streams are no longer flowing.

5. *Blueberries* — Within the layered rocks and weathering out onto the surface are millimeter-sized spherical objects, nicknamed *blueberries*. Measurements document that these contain the mineral *hematite*, which some planetary scientists interpret to have formed in the presence of water.

6. *Habitability for Life* — The *Curiosity* rover carries the Mars Science Laboratory, which has various scientific instruments and tools, including a small drill, to investigate the geology and climate of Mars. The aim is to evaluate the past habitability of the planet. The drilling has discovered minerals and chemical elements favorable for supporting past life, if it existed.

Choosing a Landing Site

How do researchers choose where to land? For the *Opportunity* rover, they chose the site on the basis of *infrared measurements* that, to geologists, indicated the presence of abundant hematite (an iron oxide mineral) in the area. On Earth, hematite most commonly forms under wet, oxidizing conditions. The geologists therefore concluded that if you were looking for water on Mars, this would be a good place to start. When *Opportunity* rolled off its platform to explore, it confirmed the presence of hematite in blueberries lying on the ground and weathering out of rocks — a great place to land! This image shows areas with several percent hematite in blue to 20% hematite in red.

19.08.t1

0 %   20 %

Before You Leave This Page

☑ Summarize two of the ways that geologists have explored Mars.

☑ Describe some features found by orbiting spacecraft and what they imply about processes that have occurred on the planet's surface.

☑ Describe some features discovered by the rovers *Spirit* and *Opportunity*.

☑ Explain how *Opportunity's* discoveries were made possible by prior spacecraft measurements.

How and When Did Geologic Features on This Alien World Form?

TRAVELING THROUGH SPACE, YOU ENCOUNTER AN UNKNOWN WORLD. Your spacecraft orbits the planet and takes images and measurements of the different geologic regions and of the most interesting geologic features. You will use these images and some initial observations about the features to interpret how each region or feature formed, and in what chronological order they formed.

Goals of This Exercise:

- Observe the planet to identify large regions that have a similar geologic appearance.
- Examine close-up images of features and read descriptions for each, to interpret how each feature formed.
- Use several strategies to reconstruct the sequence in which the different features formed.
- Summarize the geologic features and history of this planet.

Procedures

Use the available information to complete the following steps, entering your answers in the appropriate places on the worksheet or online.

A. Observe the image of the entire planet on the next page. Identify regions that have different geologic characteristics, and locate their approximate boundaries.

B. Observe each of the close-up images and read the description that accompanies each, looking for further clues about what types of geologic features are present and how each feature might have formed.

C. Determine the relative ages of the different geologic regions and features using cross-cutting relationships and the density of impact craters.

D. Your instructor may have you draw a simple geologic map of the planet, on which each map unit is a different type of geologic region or geologic feature. If you do this, draw a legend to accompany your map that has (1) a small box with the color or pattern you chose to depict that geologic terrain, (2) the name of the geologic terrain or feature, and (3) a brief description of less than 30 words that conveys the key characteristics of this terrain and your interpretation of the terrain's origin.

E. Write a short report or list summarizing the geology of the planet and its geologic history. Your instructor will guide you about the length and detail expected. This report should demonstrate the breadth of knowledge you have gained in this course, not just the concepts from this last chapter. In other words, use this final investigation to bring together concepts you have learned throughout the course.

1. The large image on the next page shows one side of the planet as illuminated by the local sun. The surface contains different types of geologic terrain as well as several obvious large geologic features. North is up in this view, south is down, west is to the left, and east is to the right. Some observations about this place are listed below, labeled with letters corresponding to the name or character of the place. Corresponding letters mark the place on the large view of the planet.

2. *Western Terrain* (W)—The western side of the planet (▶) consists of a heavily cratered terrain with many large and small craters. Samples of the rocks are very shattered and contain many angular fragments of highly weathered basalt.

3. *Dark Terrain* (D)—A dark, wide strip curves across the planet from south to northwest. Radar measurements indicate that it has a rough upper surface. A few normal faults cut across the dark material. As shown in the image below (▼), the dark material locally protrudes into the adjacent, heavily cratered terrain, covering it and filling some craters. The dark material is partly weathered basalt. The dark terrain has some small impact craters, but fewer than the western terrain.

19.09.a2

4. *Chasm* (C)—Cutting across the highly cratered terrain is a deep chasm that narrows progressively toward the south. On the image of the entire globe, the chasm has some important relationships with the dark terrain and with a reddish-brown sedimentary area (S) to the north. The close-up below (▼) shows one wall of the chasm. What features do you observe?

19.09.a3

5. *Polar Ice* (P)—The north and south poles are covered with water ice year round. The close-up image to the left (◄) shows the edge of the layered ice overlapping a crater. The ice has almost no craters.

19.09.a4

6. *Sedimentary Terrain* (S)—Adjacent to the north pole and the northern ice cap is a distinctive reddish-brown region. The unit has layers and appears to be sedimentary in origin. Along the southern edge of the terrain, the soft-looking, loose sediment is in contact with terrain that is more heavily cratered, as shown in the detailed image to the right (►). The sedimentary region has very few craters. Similar material may be present near the south pole but is not visible in this view of the planet.

19.09.a5

19.09.a1

7. *Valleys* (V)—A few valleys or channels extend south from the sedimentary region. They appear to be filled with sediment, and there is a feature that looks like a delta or fan where one channel empties into a crater. The large crater is part of moderately cratered terrain that makes up much of the eastern part of the globe. A close-up view of one channel is shown below (▼).

19.09.a6

8. *Mountains* (M)—In the southeastern part of this view, three large mountains rise above the plain. The close-up below shows one of the mountains (▼). The mountain is cone shaped with a central crater. The flank is indented by small craters, and the lower part of the mountain appears to be missing on one side (upper left side in this view).

19.09.a7

9. *Eastern Terrain* (E)—Much of the eastern hemisphere of the planet consists of a moderately dark, moderately cratered terrain. This terrain has fewer craters than the heavily cratered, western terrain, but has more craters than are present in the valleys, sedimentary terrain, chasm, or dark terrain.

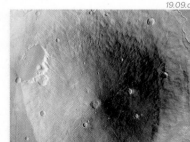

20 Our Universe

THE UNIVERSE IS AN AMAZING PLACE, with countless stars, galaxies, remnants of exploded stars, and many curious objects, some of which were only recently discovered. We can see stars and some galaxies simply by looking up into the evening sky or using binoculars and small telescopes. Astronomers use a variety of large telescopes to observe distant and faint objects, including telescopes that collect radio waves or X-rays from space. This chapter is about the universe, the fascinating objects it contains, how it formed and changes, and how we study it.

The Hubble Space Telescope, which is in orbit around the Earth, has taken many breathtaking images of distant stars, galaxies, and masses of gas and dust, but none more stunning than the image below. This image, called the *Hubble Ultra Deep Field*, was collected over many years and is of a tiny region of the sky that looks black and empty from Earth, but actually contains thousands of galaxies, each composed of millions to trillions of stars, like our Sun. The most distant of these formed early in the history of the universe, but the light took more than 10 billion years to reach us.

What are the implications of an image like this, which indicates that the universe contains a truly countless number of stars and galaxies, besides ours?

20.00.a1 Hubble Ultra Deep Field, Hubble Space Telescope

TOPICS IN THIS CHAPTER

20.00.a2 Andromeda Galaxy, Galaxy Evolution Explorer

A galaxy is a huge and massive system, typically containing millions to trillions of stars, many larger than our Sun. Some galaxies have a disk-like, spiral shape, like the one shown here (◄), but others have the shape of a stretched balloon or are very irregular with fuzzy boundaries. In addition to stars, galaxies contain large clouds of gas and space dust.

How does a galaxy form, and what accounts for the different shapes? What is the shape of our Milky Way Galaxy, and where within it are our Sun and solar system positioned?

20.00.a3 Small Magellenic Cloud, Hubble Space Telescope

The stars we observe in the night sky are each similar to our Sun, generating light and other energy by nuclear processes deep in the star's interior. As shown in this image (◄), some stars are yellow, but others are red, blue, or white. Some stars are smaller than our Sun, but others are thousands of times larger. Stars change during their lifetime, which can be billions of years, changing in size, color, and in other ways. Many stars, like our Sun, have planets orbiting them.

How does a star form, and what causes a star to change size and color over time?

20.00.a4 Planetary Nebula NGC 6302, Hubble Space Telescope

A nebula is a cloud of gas, dust, and other particles in space. Most are irregularly shaped masses, but some are shaped like rings, ragged spheres, or cones projecting in opposite directions (◄). One type of nebula, like the one shown here, forms when a star, during the later stages of its evolution, pushes away its outer layers of gases, forming huge cone-shaped jets of material moving in opposite directions. Other types of nebulae (plural of nebula) form when a massive star explodes, leaving behind a nebula as evidence of this event. Other types of nebulae were never anything else and are where stars are born.

What does the appearance of a nebula indicate about how it formed?

The Science of Astronomy

Astronomy is an ancient science, prompted by questions that earliest civilizations had about the Sun, Moon, other planets, stars, and their apparent movements across the sky. Early astronomers and other "sky watchers" made observations of the movements of these various objects across the sky, using some of these observations to guide the planting of crops and other activities. They named arrangements of stars — *constellations* — after familiar animals, such as *Taurus* the bull and *Pisces* the fish, or after personas and creatures from history or mythology, like *Pegasus* the winged horse or *Orion* the hunter.

Early scientists blended observations of the sky with newly developed concepts of math, such as geometry, to estimate the shape and size of Earth, the distance from Earth to the Sun, and the timing of eclipses that temporarily darkened either the Sun or the Moon. The new findings were used to try to answer key questions, including whether Earth went around the Sun or vice versa.

With time, our understanding of the solar system improved and the scientific field of astronomy became more sophisticated, as telescopes enabled early astronomers, like *Galileo*, to observe features and motions that were previously not visible. To calculate and explain the motions of astronomical objects, astronomers, physicists, and mathematicians developed new forms of mathematics, such as calculus by *Sir Isaac Newton*, and new scientific breakthroughs, including orbital motions by *Johannes Kepler*, and concepts of space and time by *Albert Einstein*. In these ways, astronomy changed our understanding of many aspects of the world, in addition to explaining objects and their motions in the sky.

20.0

20.1 How Do We Observe the Universe?

SOME ASPECTS OF THE UNIVERSE are readily observable by anyone, such as the progression of the Sun during the day or the Moon during a month. Other objects, including galaxies and faint clusters of stars, become visible when we peer through powerful binoculars or a small telescope. Observing even fainter objects requires larger and larger telescopes, including those that capture types of energy other than visible light.

A What Factors Affect How Much We Can Observe in the Sky Near Our Home?

We can observe many important astronomical features and events by simply going outside to some dark site and, weather permitting, gazing upward. A number of factors influence what and how much we see, including the amount of ambient light from our surroundings and the amount of clouds, dust, pollution, and water vapor in the atmosphere.

The best time for viewing stars and other objects is obviously at night, when there is less ambient light from the Sun. Even better are times of the night and month when the Moon is either not in view or is not fully illuminated by the Sun, as during a new moon. High-altitude sites are also preferred, since there we look through less of the atmosphere, which distorts our view of stars, causing them to brighten and dim (twinkle).

During the day, the stars and other astronomical objects are still overhead, but they are not visible to us because their light is overwhelmed by the much more intense sunlight. The Moon is often visible during the day, but we may not notice it because of the bright sunlight. Looking into the Sun without special protection can cause permanent damage to your sight. Sunspots may be visible, but only when viewed in a safe, protected way.

A major problem for most people's night-sky viewing is proximity to artificial light sources, such as street lights, that keep us from seeing dim stars and planets. Astronomers refer to the light from cities and other artificial sources as *light pollution* because such ambient light can affect a high-powered, optical telescope (i.e., one built for viewing visible light). To try to avoid this problem, astronomers site telescopes in remote places.

B How Do Optical Telescopes Work?

Telescopes are devices used to view or collect images and other types of data from distant objects. Simple versions are a tube with a lens and a viewing eyepiece. More complex versions use large mirrors, as much as 10 meters across, or multiple closely coordinated mirrors, spanning even greater diameters. Most telescopes operate under the same basic principles.

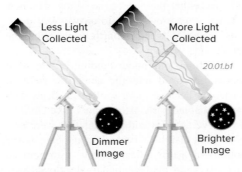

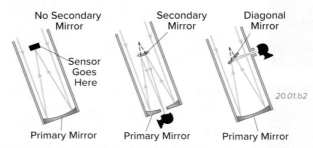

20.01.b3 Sofia Telescope mirror during tests

Light-Gathering Power—Our eyes and cameras are sensors, receiving light that is translated into images. The more light the sensor can receive, the dimmer the object we can detect. Telescopes operate like funnels, channeling more light toward the focal point. The bigger the funnel, the more light. A *refracting telescope* (▲) uses a lens to bend or *refract* the light toward the eyepiece or camera at the end of the tube.

Focusing the Light—One way to collect and focus the light is using a lens in front of the eyepiece, as shown in the previous figure. Another way is to use a mirror to focus the light to some point, called the *focus*, and then figuring out a way to guide that light toward our eyes or some other sensor. Several examples of how to do this are shown above. The sensor can be placed at the focus point, or the light can be redirected by another mirror to a location for the sensor that is out of the way. There are a number of advantages to using a mirror. A mirror does not have to support itself and can rest on an external support. In contrast, a lens must be supported by its edges, potentially causing it to deform under its own weight and thereby distort incoming light.

Resolving Power—Another important consideration is the size of details that can be resolved by a telescope. Larger telescopes can resolve finer details and dim objects. Light naturally generates distorting patterns as it passes through openings and reflects off surfaces, and such distortions limit the resolving power of a telescope.

C | How Can We Increase the Power, Resolution, and Clarity of Telescopes?

To observe faint objects, such as stars and galaxies on the far side of the universe, we need telescopes in very dark places and with high light-gathering powers and the ability to resolve details with clarity and lack of distortion. We use several approaches to satisfy these constraints.

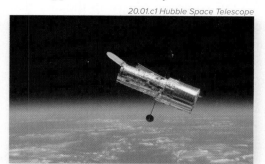
20.01.c1 Hubble Space Telescope

20.01.c2 Large Binocular Telescope, Mt. Graham, AZ

20.01.c3 Goldstone Radio Telescope, Goldstone, CA

1. One approach, used by many telescopes, is to avoid atmospheric effects by being above them—in space. The *Hubble Space Telescope* (▲) has captured dazzling photos of various astronomical objects, including distant galaxies, stars in the process of being formed, and dust storms on Mars. Many of the photographs in this book are from "Hubble" because they are so much clearer than images taken with ground-based telescopes.

2. We can use more than one telescope to collect images from slightly different positions, such as these twin mirrors. Using modern computing techniques, images from the different telescopes can yield an image with a quality comparable to one taken with a larger telescope. This can reduce atmospheric distortions using a technique called *interferometry*, which uses the separate images to smooth out any distortions.

3. Although our eyes see energy in wavelengths of visible light, objects in space emit other frequencies of electromagnetic radiation, including radio waves, ultraviolet energy (UV), X-rays, and gamma rays. Some types of energy, like radio waves, make it through Earth's atmosphere, so we can use ground-based radio telescopes (▲). Other wavelengths, like infrared, are mostly blocked by the atmosphere, so these telescopes must be in space.

4. Different approaches are used to collect different types of energy, in order to observe and study different types of features. In the figure below, the most energetic types of electromagnetic radiation (EMR) are on the left, with the energy levels decreasing from left to right. The energy level of different types of electromagnetic radiation is related to wavelength of the energy, with shorter wavelength energy being more energetic. The types of objects that are best viewed using a specific type of EMR are depicted along the top of the figure.

5. Gamma rays and X-rays from space have very short wavelengths and are blocked by gases in the upper parts of Earth's atmosphere, so they cannot be observed with ground-based telescopes. Instead, we use space-based telescopes to detect this radiation. These types of energy are produced by very hot and energetic objects.

6. Most types of ultraviolet (UV) energy, especially the most energetic types, are also blocked by the upper atmosphere. The atmosphere blocks most of this same dangerous UV from our own Sun. The UV that makes it through causes sunburns and helps heat the surface.

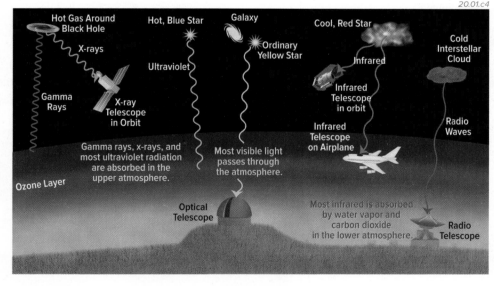
20.01.c4

9. Radio waves are the least energetic form of EMR shown here. This energy has very long wavelengths, which allows it to mostly pass through the atmosphere. Radio waves are used to study objects in space that are themselves not very energetic, like cold clouds of gas and dust between stars.

7. Electromagnetic radiation, in the form of visible light, passes through the upper atmosphere, but it can be blocked by clouds and pollution in the lower parts of the atmosphere. Our eyes are excellent at observing this type of energy, so many features are observed with visible light.

8. The lower atmosphere also partly blocks infrared (IR) energy, so it is best to observe this energy from space or high-flying planes.

Before You Leave This Page

☑ Sketch and explain factors, such as ambient light and the design of telescopes, that influence how we observe objects in space.

☑ Sketch and explain where we observe different types of electromagnetic energy and why this setting is preferred or required.

20.2 What Is Our Framework for Observing the Universe?

AS WE OBSERVE THE NIGHT SKY and find something interesting, how do we tell someone where in the sky that feature is located? Clearly, we need some kind of frame of reference, a way to describe a direction in which to observe the feature. Also, objects in the sky—including the Sun during the day, the stars and planets at night, and the Moon during the day and night—all rise from the east and have apparent movements across the sky, so the time of observation is also important. A frame of reference helps explain many features and events, including a solar or lunar eclipse.

A How Do We Describe a Direction in the Night Sky?

1. Stars and other astronomical objects are very far away from Earth, but it is sometimes helpful to envision their position relative to us by projecting them onto a nearby imaginary sphere around the earth (▶). Such an imaginary sphere is called the *celestial sphere*. From any place on Earth, whether we are on land or on a ship, the earth's surface obscures half of the celestial sphere, so we can think of the sphere as being a dome that is overhead. The boundary between the visible sky and the surface of the earth is called the *horizon*, a term also used in everyday discussions of landscapes, sunrises, and sunsets.

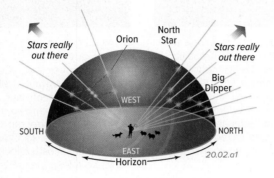

20.02.a1

2. To reference the position of an object on the celestial sphere, we begin by projecting the Earth's equator onto the sphere, referring to the resulting circle around the sphere as the *celestial equator*.

3. In a similar way, we can project Earth's North Pole and South Pole upward onto the celestial sphere, resulting in a *north celestial pole* and *south celestial pole*.

20.02.a2

4. Earth rotates around its spin axis through the North and South poles, completing one full rotation each 24 hours. This daily rotation causes all objects in space, including the Sun, Moon, and stars, to have an apparent movement across the sky. This movement follows circles that are parallel to the celestial equator, as represented by the orange dashed circles on the figure above. Since Earth's surface blocks half of our view, we observe this movement as an *arc* across the sky.

B How Does the Sun Move Through Constellations?

Another way to reference the location of a star or other astronomical object is by relating it to the plane in which the Earth orbits the Sun. Ancient sky gazers recognized this motion and noted the position of the Sun and Earth relative to *constellations*, each of which is a group or arrangement of stars that resembled some familiar shape, like a fish.

1. Earth *rotates* around its spin axis, as illustrated in part A above, but it also *revolves* (orbits) around the Sun. The plane in which the Earth orbits the Sun (the *orbital plane*) is not the same as the plane that defines Earth's equator (and thus the celestial equator). This is because the Earth's spin axis is tilted relative to the orbital plane.

2. The Earth does a complete orbit all the way around the Sun in a year (this is how a year is defined). During the course of a year, from a perspective on Earth, the Sun traces a line across the celestial sphere, represented by the dashed line in the figure to the right. Astronomers call this line the *ecliptic* because, as we shall soon observe, eclipses can only occur when the Moon crosses this line.

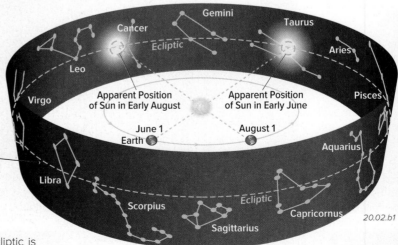

20.02.b1

3. The band of constellations that occur along the ecliptic is called the *zodiac* and has the names familiar to many people as the signs of the zodiac (Gemini, Virgo, etc.). *Astrology* is a nonscientific belief system that someone's personality is somehow influenced by the position of the Earth and Sun relative to the zodiac on the day that person was born. Astronomy is the scientific study of the universe.

4. As Earth orbits the Sun and our vantage point changes, different constellations will be observed behind the Sun (soon after sunset), at different times of the year (▲).

C   What Is a Solar Eclipse or Lunar Eclipse?

An eclipse is an astronomical phenomenon that occurs when the Sun, Moon, and Earth are all arranged in a straight line, which only occurs when the Moon is directly on the ecliptic. When this occurs, the Moon can block the Sun, casting a shadow on the Earth, causing a *solar eclipse*, or the Earth can temporarily cast its shadow on the Moon, causing a *lunar eclipse*. A lunar eclipse, which occurs at night, can be watched directly, but observing a solar eclipse, which occurs during the day, requires a special setup to prevent serious damage to your eyesight.

Solar Eclipse

1. During a *solar eclipse* (▶), the Moon is directly between the Earth and Sun, and so it casts its shadow on Earth. The Sun is much larger than the Moon, so the Moon's shadow only covers part of Earth. This is why some regions experience the eclipse (are within the Moon's shadow) but others do not. A location where the Moon totally blocks the Sun is said to experience a *total eclipse*, and a location where the Moon never completely blocks the Sun experiences a *partial eclipse*.

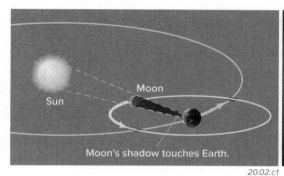

Sun · Moon · Moon's shadow touches Earth.

20.02.c1

What You See From Earth

20.02.c2

2. This is how a total solar eclipse appears from Earth. The direct light from the Sun is completely blocked, but a fringe of very hot gas streaming away from the Sun remains visible during the eclipse.

Lunar Eclipse

3. During a *lunar eclipse* (▶), Earth is directly between the Sun and Moon, and so the earth's shadow blocks sunlight from all or part of the Moon. Earth is larger than the Moon, so its shadow can easily cover the entire Moon. During the onset of a lunar eclipse, Earth's shadow gradually covers more and more of the Moon. After covering the entire Moon, the shadow then begins to gradually move off the Moon, allowing sunlight to again give the Moon its typical, silvery appearance.

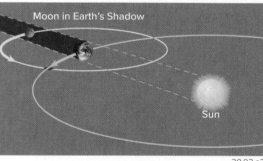

Moon in Earth's Shadow · Sun

20.02.c3

What You See From Earth

20.02.c4

4. During a total eclipse, the Moon takes on a reddish color, because the only sunlight reaching it has been bent and filtered by Earth's atmosphere. This filtering eliminates nearly all of the colors except red.

Why Solar and Lunar Eclipses Do Not Occur Every Month

5. Solar and lunar eclipses tend to occur within several weeks of one another, and they only occur two times a year. The explanation for this is that the plane along which the Moon orbits Earth is not exactly parallel to the plane in which Earth orbits the Sun. The planes are approximately 5° out of alignment, as shown in this figure (▼).

6. Solar and lunar eclipses require the Sun, Earth, and Moon to be precisely in a straight line, so that the Earth and Moon can cast their shadows on each other. During most of the year, the shadows of the Moon and Earth pass above each other, so no eclipse is possible.

7. At other times of the year, the shadows of the Moon and Earth are too low to shade the other object, so no eclipse occurs (▶).

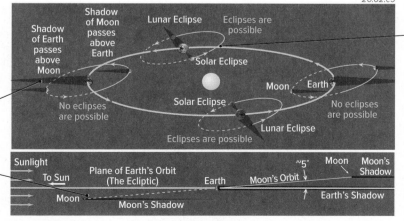

20.02.c5

Shadow of Earth passes above Moon · Shadow of Moon passes above Earth · Lunar Eclipse · Eclipses are possible · Solar Eclipse · Moon · Earth · No eclipses are possible · No eclipses are possible · Solar Eclipse · Lunar Eclipse · Eclipses are possible

Sunlight · To Sun · Plane of Earth's Orbit (The Ecliptic) · Earth · Moon · Moon's Orbit · ~5° · Moon · Moon's Shadow · Earth's Shadow · Moon's Shadow

8. An eclipse can only occur during those times of the year when the Moon crosses the orbital plane and at the same time is directly in line with the Sun and Earth. This occurs twice each year.

9. A lunar eclipse can only occur on nights when there is a *full moon*, when the side of the Moon facing the earth is fully illuminated by the Sun's light (when not being eclipsed). A solar eclipse can only occur during a *new moon*, when the side of the Moon visible from Earth is not illuminated by the Sun.

Before You Leave This Page

✔ Sketch and explain what is meant by the celestial sphere, celestial equator, celestial poles, ecliptic, and zodiac.

✔ Sketch and explain how both a solar eclipse and a lunar eclipse occur, and why they only occur two times each year.

20.3 How Does Temperature Influence the Type of Light an Object Emits?

STARS SHINE IN THE NIGHT SKY because they are emitting visible light, a type of electromagnetic radiation that our eyes can sense. Stars and other objects also emit other types of electromagnetic radiation (EMR), such as ultraviolet energy, which our eyes cannot sense, but our instruments can. The type of energy an object emits is related to its temperature, and we can use this relationship to estimate the temperature of stars and other objects, even though they are very, very far away. All EMR, including visible light, travels at the same speed, about 300 km/s in a vacuum.

A What Is Electromagnetic Energy?

Electromagnetic radiation (EMR) consists of energy radiated from charged particles and is manifested as interacting electrical and magnetic fields. Visible light is a type of EMR, as are radio waves. We can think of EMR as a series of waves radiating outward or we can consider it to be composed of particles. Both approaches are useful.

1. For now, consider EMR as a series of waves (▶) of electrical and magnetic energy that are moving from one place to another, in this case, from left to right. The direction in which a wave moves is the *direction of propagation*. The wave shown here propagates from left to right. Motion of any part of the wave is mostly up and down, parallel to the small arrows on this figure. The waves are part of an electrical field.

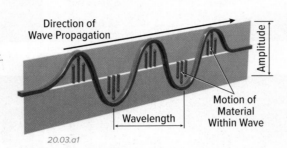

20.03.a1

2. In describing waves, the term *amplitude* refers to the height of the wave, from trough to crest. The greater the difference in height between the top and bottom of the wave, the greater the amplitude (strength of the electrical field). We use *wavelength* to describe the distance between two adjacent crests (tops) or two adjacent troughs (bottoms). The longer the distance between two adjacent crests or troughs, the greater the wavelength. If we know the speed of the wave, we can describe the wave's *frequency*—the number of waves per second passing a point.

3. Electromagnetic waves have electric and magnetic components, which can be envisioned as two mutually perpendicular planes. In this figure (▶), the electric field, containing the electrical component of the wave, is vertical (dark gray), whereas the magnetic field is horizontal (colored blue with the orange horizontal arrows).

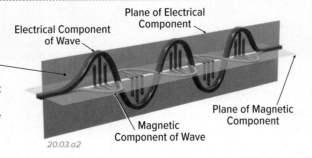

20.03.a2

4. The electrical and magnetic waves have the same wavelength, and the crests of the two types of waves are the same distance along the direction of wave propagation (i.e., the crests line up). Motion in both types of waves is perpendicular to the direction the wave propagates, as described in the previous figure. Electromagnetic waves get their name from these linked electrical and magnetic waves that move in unison.

5. Light is a type of EMR. We perceive light from the Sun as white light (▶), but by using a prism we can separate that light into a spectrum of colors. Each color represents a slightly different wavelength of visible light. Red light has longer wavelengths than does yellow light, and blue light has even shorter wavelengths. It is this variation in wavelengths that allows a prism to separate colors.

6. If we use a glass lens, we can refocus the light from a prism and recombine the colors back into white light (▶). When we see an object, like a red apple, as a single overall color, it is reflecting light in a narrow range of wavelengths of EMR, in this case red light. If something appears white, we are observing a wider range of EMR. If something appears nearly black, there is almost no light coming from it at any wavelength.

20.03.a3

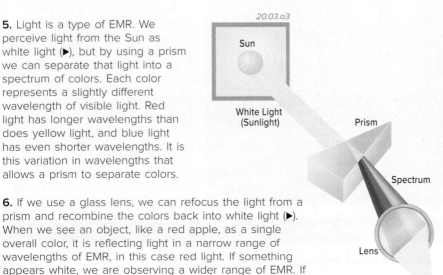

7. Visible light and other types of EMR can be generated by changes in energy within an atom, which consists of a central nucleus (▶) surrounded by electrons. Electrons can occur at various energy levels and locations around the nucleus, and these locations and energy levels are called *electron shells*. The atom shown here has three electron shells, numbered 1, 2, and 3.

20.03.a4

8. An electron in an outer electron shell carries more energy than one in an inner shell. If an electron moves from an outer shell to an inner shell (illustrated by the yellow arrow above), it must get rid of the excess energy by emitting it in the form of EMR. We can think of this EMR as outward-radiating waves or as a discrete particle of EMR, which we call a *photon*. We would have to add energy back to the atom to cause the electron to move back from a lower energy level to a higher energy level.

B What Influences the Type of Electromagnetic Radiation an Object Emits?

The temperature of an object is the main control over the type of electromagnetic radiation the object emits. Different objects have different temperatures, so they emit different wavelengths of EMR. Isolated atoms and molecules emit a limited number of wavelengths, but solids, liquids, and thick gases emit a broader range of wavelengths of EMR.

1. To explore the relationship between an object's temperature and the wavelengths of EMR it emits, consider what happens when you turn on a burner of an electric stove. As the burner starts to heat up, it begins to glow red. As the burner gets hotter, it begins to appear more orange than red. The shift in color is because the wavelengths of EMR being emitted change as the burner rises or falls in temperature. In other words, the wavelengths of EMR are controlled by an object's temperature, a relationship known as *Wien's law*.

The hotter burner glows more orange than the cooler burner, which glows red.

20.03.b1

20.03.b2

2. An object emits a range of wavelengths of EMR, but it emits more of some wavelengths than others. The curved lines on the figure to the left represent the wavelengths of EMR being emitted by stars of three different temperatures. The hottest star, on the left, emits its *peak wavelength* of EMR at short wavelengths (the blue peak left of the visible light spectrum). The maximum amount of EMR coming from the red star is at longer wavelengths (the red peak to the right of visible light). The yellow star, like our Sun, is in between, with its peak wavelength within visible light.

3. The universe contains various types of astronomical objects that vary greatly in temperature from one type to the next. These temperature differences in turn control the peak wavelength of EMR that type of object emits. This figure shows the main wavelengths of EMR emitted by some common astronomical objects. The hottest objects are on the left and the cooler ones are on the right.

4. At the bottom of the figure is the *electromagnetic spectrum*, which subdivides EMR into different types, based on wavelength. Shorter, more energetic wavelengths of EMR are to the left, and longer, less energetic wavelengths are to the right.

5. Visible light is in the middle of the electromagnetic spectrum, with wavelengths from about 400 to 700 nanometers (nm, or a billionth of a meter). Violet and blue colors have the shortest, most energetic wavelengths of visible light, whereas red colors have longer, less energetic wavelengths.

20.03.b3

| Gamma-Ray Burster | Pulsar | The Sun and Other Stars | Interstellar Cloud | Cosmic Microwave Background | Active Galaxy |

← INCREASING ENERGY →

← INCREASING WAVELENGTH →

| 0.001 nm | 0.1 nm | 10 nm | 1000 nm | 100 μm | 10 mm | 1 m | 100 m |

| Gamma Rays | X-Rays | Ultraviolet | Infrared | Microwaves | Radio Waves |

| | | CAT Scan | Tanning Lamp | TV Remote | Thermal Imager | Police Radar | Cell Phone | TV FM | Shortwave Radio | AM |

Visible Light

| 400 nm | 500 nm | 600 nm | 700 nm |

6. Radio waves have the longest wavelength and are the least energetic type of EMR. Gamma rays have the shortest wavelength and are most energetic.

Before You Leave This Page

☑ Summarize what electromagnetic radiation (EMR) is, how it travels through space, and what causes it.

☑ Describe factors that influence the type of EMR an object emits, and list the common types of EMR in order of increasing wavelength.

20.3

20.4 How Do We Use Spectra to Study the Universe?

ATOMS, MOLECULES, AND OTHER SUBSTANCES, whether they are in space or on our own planet, interact with the spectrum of light, in many cases selectively *emitting* light and other electromagnetic radiation (EMR) at certain wavelengths. These materials can also selectively block certain wavelengths of EMR, through the process of *absorption*, allowing other wavelengths to pass through. Astronomers use the patterns of emitted and absorbed EMR on spectra to identify chemical elements in stars, interstellar clouds, and other astronomical objects.

A What Causes a Substance to Emit Certain Wavelengths of Light?

Although atoms and other materials can emit or absorb various types of electromagnetic radiation, it is easier to explore how this occurs within the wavelengths of visible light, because we can see the results with our eyes. The wavelength of energy emitted or absorbed by an atom is related to its atomic structure and to changes in its electrons.

1. Electrons around the nucleus of an atom occur in discrete energy levels, or electron shells (▶). Higher energy levels for an electron are farther from the nucleus, in an outer shell. We designate the innermost, lowest energy electron level as number 1, and assign higher numbers to higher energy levels farther out from the nucleus. The example shown here is a hydrogen atom, the smallest and simplest type of atom. Hydrogen has a single proton in its nucleus and a single orbiting electron. These figures show four electron levels (shells).

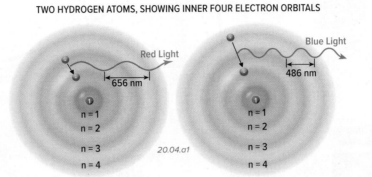

TWO HYDROGEN ATOMS, SHOWING INNER FOUR ELECTRON ORBITALS

Red Light
656 nm
n = 1
n = 2
n = 3
n = 4
20.04.a1

Blue Light
486 nm
n = 1
n = 2
n = 3
n = 4

3. An electron in an outer electron shell can shift downward two energy levels, such as from shell number 4 to number 2 (◀). This electron releases a different amount of light, and that light has a different wavelength than if an electron drops a single level (like from 3 to 2), as in the previous example. Where the electron drops from shell 4 to 2, the energy released is at wavelengths of blue light. If it drops from 3 to 1, the light is ultraviolet. Recall that blue light and ultraviolet have shorter wavelengths and are more energetic than red light.

2. When an electron drops from a higher energy level to an inner one, it must decrease its energy level, which it does by emitting that energy as electromagnetic radiation. If it drops from energy level 3 to the next one down (to the next shell inward), the energy released is at the wavelength of red light. So a hydrogen-rich star, with hydrogen atoms undergoing this process, would emit light at this wavelength.

4. Hydrogen and helium have only one or two electrons, respectively, and a limited number of electron shells, so they are relatively simple examples of this process. Even more complex elements, however, generate only a few characteristic wavelengths of energy as their electrons drop from a higher energy level to some lower one. We can, therefore, use emitted light to study the chemical composition of stars.

B How Do We Observe Emission Spectra?

1. We can observe these discrete wavelengths of emitted light by viewing a light source through a device called a *spectrometer*. A simple spectrometer (▶) allows incoming light to pass through a slit and then a prism, which separates the light into its different wavelengths. In the example shown here, electrons in hydrogen are emitting a limited number of wavelengths, including the red and blue ones we just explored.

A hydrogen atom emits at wavelengths set by its orbitals. If an electron drops from orbital 3 to 2, the atom emits red light. If it drops from 4 to 2, it emits blue light, etc. Therefore, only a few colors show up in the spectrum.

20.04.b1
5→2
3→2
4→2

Tube of Hot Hydrogen
Slit
Prism
Power Supply (Electricity heats hydrogen in tube.)
4→2
3→2
Hydrogen Emission Spectrum
6→2
5→2
Light emitted only some wavelengths.

Electron orbitals for helium differ from those of hydrogen, so they emit different wavelengths of light than hydrogen.

20.04.b2

Tube of Hot Helium
Slit
Prism
Power supply (Electricity heats helium in tube.)
Helium Emission Spectrum

3. Helium is the second element in the periodic table and the second lightest element, after hydrogen. It has two protons and two neutrons in its nucleus and two electrons. With this more complex structure, helium has more possible combinations of electrons dropping from one level to another. As a result, it emits more wavelengths of light (◀), including some in red, yellow, green, blue, and purple.

2. Astronomers use spectrometers with built-in sensors that record which wavelengths of light are present and how strong the light is at each wavelength. The recorded information is an *emission-line spectrum*.

4. You can tell, from these two examples, we should be able to identify whether a gas contains hydrogen or also has helium, simply by observing the emission spectra from the gas.

C Why Do Materials Absorb Certain Wavelengths of Light?

Just as a specific chemical substance, like hydrogen, emits light only at certain wavelengths, these same materials can absorb the same frequencies of light coming from someplace else but passing through the substance on the way to us. In this context, *absorb* means "taken into." From an outside viewer, this light is blocked. We observe and measure such patterns of light absorption using a spectrometer, which separates light into its various wavelengths.

1. When light from some external source, like a nearby star, passes through a hydrogen atom in an interstellar cloud, the light interacts with an electron orbiting a hydrogen nucleus (▶). Red light has a wavelength that can be absorbed by the electron, so such light does not make it through the hydrogen. The energy from the light cannot just go away, and so it must be somewhere in the system, a principle known as *conservation of energy*. In this case, the energy goes into the electron, which jumps to a higher electron level.

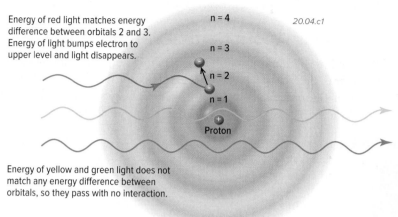

Energy of red light matches energy difference between orbitals 2 and 3. Energy of light bumps electron to upper level and light disappears.

20.04.c1

n = 4
n = 3
n = 2
n = 1
Proton

Energy of yellow and green light does not match any energy difference between orbitals, so they pass with no interaction.

2. In contrast, light in yellow and green wavelengths does not encounter an electron at appropriate energy levels for it to be absorbed, so these light frequencies pass through without being affected. Thus, if we observe that the light coming through an interstellar cloud is missing light in red wavelengths, but not in yellow and green wavelengths, we can conclude that this cloud contains hydrogen. We can observe and measure the wavelengths of light that are missing by careful examination of the resulting spectra.

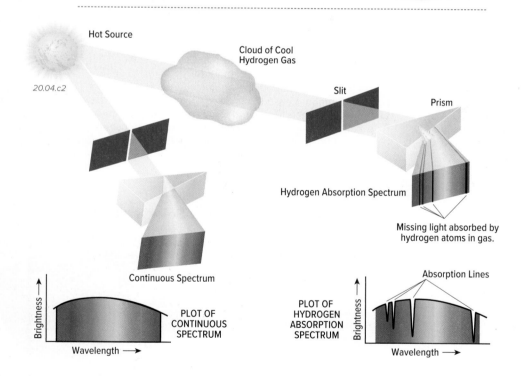

20.04.c2

Hot Source

Cloud of Cool Hydrogen Gas

Slit

Prism

Hydrogen Absorption Spectrum

Missing light absorbed by hydrogen atoms in gas.

Continuous Spectrum

Brightness →

PLOT OF CONTINUOUS SPECTRUM

Wavelength →

Absorption Lines

PLOT OF HYDROGEN ABSORPTION SPECTRUM

Brightness →

Wavelength →

3. This figure (◀) illustrates the concept of absorption and how it appears on a spectrometer. Some source of light, such as a star, is hot enough to emit light at the wavelengths of visible light. If the light travels through space without passing through any other objects, as is the situation for the light beam on the left, then we receive the original spectrum of light generated by the light source. If the light passes through some object on the way to us, then not all wavelengths of light will reach us.

4. If the light passes through a cloud of cool hydrogen gas, as does the beam of light on the upper part of the figure, the hydrogen will partially block certain wavelengths of light, especially the red ones. As we observe the light with a spectrometer, we can discern dark lines in the spectrum, marking the wavelengths of light that are being blocked. Which wavelengths are absorbed (or emitted) depends on temperature as well as composition.

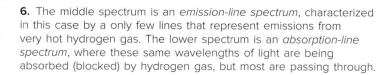

Continuous Spectrum

Emission-Line Spectrum (Hydrogen Gas)

Absorption-Line Spectrum (Hydrogen Gas)

20.04.c3

5. The three spectra to the left illustrate the three types of spectra we expect to find. The top spectrum, called a *continuous spectrum*, records light coming from a light source that is generating all frequencies of visible light, from violet on the left to deep red on the right. The source is probably also generating other frequencies of EMR, such as ultraviolet, but we cannot see these wavelengths.

6. The middle spectrum is an *emission-line spectrum*, characterized in this case by a only few lines that represent emissions from very hot hydrogen gas. The lower spectrum is an *absorption-line spectrum*, where these same wavelengths of light are being absorbed (blocked) by hydrogen gas, but most are passing through.

Before You Leave This Page

✓ Sketch and explain how an atom emits light at certain characteristic wavelengths and what the resulting light pattern is on emission spectra.

✓ Sketch and explain how an atom absorbs light at certain wavelengths and what the resulting light pattern is on absorption-line spectra.

20.4

20.5 What Controls the Motions of Objects?

OBJECTS IN THE UNIVERSE MOVE in various ways. Some move in a relatively straight line for long time periods, whereas others circulate in elliptical orbits, like the planets in our solar system. In addition, many objects spin about a rotation axis—Earth rotates once every 24 hours with respect to the Sun. These motions are mostly predictable once we consider the forces and principles that constrain those motions. The scientists who deciphered these principles, such as Johannes Kepler and Isaac Newton, forever changed our view of the universe.

A What Were Some Early Ideas About Earth's Setting in the Universe?

To appreciate our current understanding of motions in the universe, it helps to understand how these ideas evolved with time. For early peoples, the Earth was the central focus of their lives, providing shelter, water, and nutrition, so it was easy to think that Earth was the center of the universe. Such Earth-centric views were widely held in early civilizations, many of which thought the Sun was a god racing across the sky in a fiery boat or chariot.

Shape of the Earth

20.05.a1

1. In our modern world, it is easy to appreciate that the Earth is round (spherical), because we have actual photographs of our planet taken from orbiting satellites and elsewhere in space. Some early peoples held the view that the Earth was flat, because it looked flat to them, from their limited perspective of standing on the ground. Very early scientists, however, like Aristotle in the 4th Century BCE (Before Common Era), devised several ways to demonstrate that the Earth was a sphere. One clue was that the Earth's shadow during a lunar eclipse was round, as we can see today by superimposing multiple photographs (▲) taken at different times during a single lunar eclipse.

Distances to the Sun and Moon

3. Another aspect that intrigued early astronomers was estimating the distances between the Sun, Moon, and Earth. Most approaches to this problem involved basic principles of geometry. One approach, shown here (▶), reasoned that when the Moon was half lit from a perspective on Earth, a line from the Sun to the Moon and another line from Earth to the Moon would make an angle (β on this figure) of 90°. By then measuring the angle (α on this figure) between lines connecting Earth to the Moon and Earth to the Sun, we could calculate the relative lengths of the various lines. This demonstrated that the Sun was much farther away from us than the Moon, and therefore must also be much larger than the Moon.

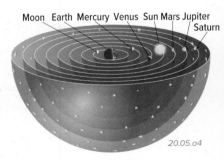

Half-Lit Moon (First Quarter)
β = 90°
α
Sun
Half-Lit Moon (Third Quarter)

20.05.a3

Size of the Earth

North Pole
Obelisk in Alexandria
~7°
Parallel Lines
Sunlight
~7°
Well in Syene

20.05.a2

2. After accepting that Earth was spherical, early scientists and mathematicians used early principles of geometry to estimate the size of the earth, Moon, and other objects. Eratosthenes (276-195 BCE), a Greek astronomer, mathematician, and geographer, used an ingenious approach. He observed and measured shadows of an obelisk (pillar) in Alexandria in northern Egypt and noted that light went straight down a water well in an Egyptian town (Syene) farther south, being careful to do his measurements at the same time of day and day of the year. He then used geometry to estimate how many degrees (▲) those two places were apart on a sphere. Knowing the ground distance between the two places, he could estimate the size of the entire sphere (Earth).

Center of the Solar System

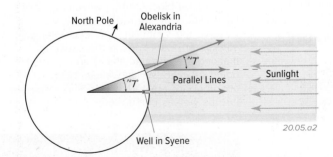

Moon Earth Mercury Venus Sun Mars Jupiter Saturn

20.05.a4

4. Another widely held early notion was that Earth was at the center of our solar system and indeed the center of the universe, as depicted here (▶). In this model, the Sun, Moon, planets, and all other objects in the universe revolved around Earth. Such a model is called *geocentric* (Earth-centered).
The stars were envisioned as all being about the same distance from Earth, and rotating on an outermost transparent sphere. This simple model had great difficulty explaining certain motions, especially of the other planets. This led Ptolemy, a second-century astronomer, to devise a scheme where planets did smaller loops, termed *epicycles*, as they went around the Earth. During the Renaissance, Polish physician Nicolaus Copernicus demonstrated that Earth and all the planets went around the Sun, a *heliocentric* (Sun-centered) model. Using his new model, Copernicus was able to calculate distances of each body from the Sun and explain planetary motions without the complicating epicycles. The insight of Copernicus was one of the great leaps in human knowledge.

B | What Are Kepler's Three Laws of Planetary Motion?

Johannes Kepler, a 17th-century German mathematician and astronomer, was an assistant to Danish astronomer Tycho Brahe, who had collected very detailed observations on motions of planets, especially those of Mars. From these data, Kepler calculated that Mars travelled around the Sun in an orbit that was elliptical (shaped like an ellipse), not circular as had been assumed. As a result of this new insight, Kepler was able to accurately calculate planetary motions, explaining patterns that circular orbits could not. In the process, he developed three new laws of planetary motion.

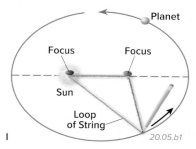

I   20.05.b1

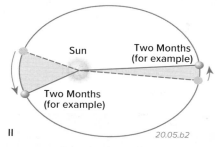

II   20.05.b2

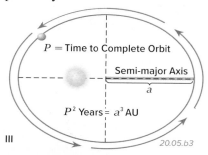

III   20.05.b3

An ellipse has the shape of a flattened circle, and can be drawn by looping a string around two thumbtacks and moving a pencil around in a somewhat circular motion. The two key points represented by the thumbtacks are each called a *focus*, and they, along with the length of the string, precisely define the ellipse. *Kepler's first law* is that a planet follows an elliptical orbit, with the Sun at one focus.

Kepler noted that a planet speeds up when it nears the Sun and slows down when it moves away from the Sun. *Kepler's second law* states that the speed of the planet varies in such a way that a line from the planet to the Sun sweeps across equal areas for equal amounts of time. In the figure above, the arrows represent how far the planet travelled in two equal intervals of time, and the two shaded patches are equal in area.

A planet with a larger orbit takes longer to complete a single revolution around the Sun, an amount of time called the planet's *period*. The size of an ellipse is related to half the length of its major axis, a length we call the *semimajor axis*. *Kepler's third law*, expressed in the equation above, proposes that the period squared is equal to the length of the semimajor axis cubed, where period is in years and length is in astronomical units (AU).

C | What Are Newton's Three Laws of Motion?

Isaac Newton (1642-1727), an English mathematician and physicist, is considered by many people to have been the most influential scientist who ever lived. He did pioneering studies of motion, gravity, and optics, and when mathematics of that time proved insufficient to solve certain problems, he invented calculus. He proposed three laws of motion, which along with Newton's *Law of Gravity*, explain motions of all objects, from falling objects on Earth to stars and galaxies.

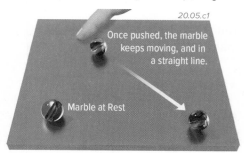

20.05.c1

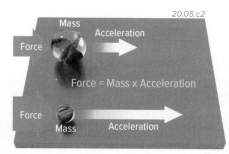

20.05.c2

20.05.c3

Newton, like Galileo, gave great importance to the concept of *inertia*, the tendency for an object at rest (like the left marble above) to remain at rest and for an object in motion (like the marble on the right) to stay in motion. *Newton's first law* says that a body at rest will remain at rest unless acted upon by a force, and a body in motion will remain in motion, moving in a straight line and at a constant speed, unless acted upon by a force, such as friction. This law applies to stars and marbles.

Newton's second law relates how much *acceleration* an object experiences when it is acted on by a force. By acceleration, we mean speeding up, slowing down, or changing direction. The second law, commonly abbreviated as $F = ma$ or $a = F/m$, states that the acceleration (a) an object receives is proportional to the net force (F) applied to it and inversely proportional to the mass (m). All other things being equal, a larger force or smaller mass means more acceleration.

Newton's third law states that when one body exerts a force on a second body, the second body also exerts that same amount of force, but in an opposite direction, on the first body. When one small skateboarder pushes another small skateboarder, both boarders move away (both are accelerated). If a small skateboarder pushes against a larger one, they also both move away, but the smaller boarder moves faster. Remember that $a = F/m$ (a larger mass means a smaller acceleration).

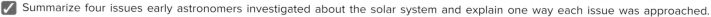

Before You Leave This Page

 Summarize four issues early astronomers investigated about the solar system and explain one way each issue was approached.

 Sketch and explain Kepler's three laws of planetary motion and Newton's three laws of motion.

20.6 How Do We Measure Distance, Motion, and Mass of Astronomical Objects?

GALAXIES, STARS, AND PLANETS do not stay in one place; they instead move long distances, commonly at very high speeds, much higher than those we experience on Earth. The distances from Earth to other astronomical objects are so great that we need completely new units for measuring distances. In addition to estimating speed and distance, we would like to know the size of such objects, so astronomers have devised clever ways to estimate diameters of distant objects and how much material an object contains, that is, its *mass*.

A What Units Do We Use to Measure Distances in Space?

In our everyday activities, we generally refer to distances using familiar units, like inches, feet, and miles, which are part of the *imperial system* of measurements, so named because they were defined

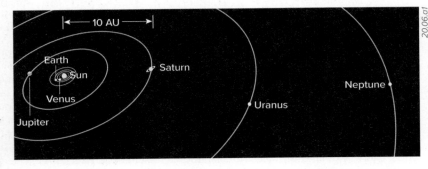

Object	Radius
Earth	6,371 km
Sun	696,00 km (~100 x radius of Earth)
Earth's Orbit	149.6 million km = 1 AU (~200 x radius of Sun)
Sun to Neptune	30 AU (~6,500 x radius of Sun)
Milky Way Galaxy	50,000 ly (3 x 10⁹ AU)

in England in the 1800s. Scientists use the *metric system*, in part because it is based on units of tens and hundreds (e.g., 100 centimeters to a meter). We use metric units to express the size of objects in the solar system, such as the radius of Earth. For the size of orbits and distances to other stars, we need different units. One unit astronomers use is an *astronomical unit* (AU), where one AU is the average distance from the Sun to Earth (▲). Saturn is 10 times farther from the Sun than is Earth, so the distance from the Sun to Saturn is 10 AU.

For even farther distances, we use a unit called a *light-year* (abbreviated "ly"), which is based on the distance that light travels in a year. The speed of light is nearly 300 million meters per second (299,792,458 m/s), which makes a light-year slightly less than 10 trillion kilometers (9.46×10^{15} m).

B How Do We Estimate Distances to Objects Far Away in Space?

We measure distances on Earth in various ways, such as using a tape measure to determine the size of a room. To measure distances in space, where we lack a long-enough tape measure, we use principles of geometry or consider how the brightness of a star will decrease if we are farther away.

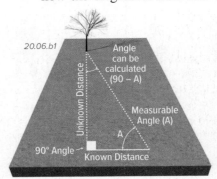

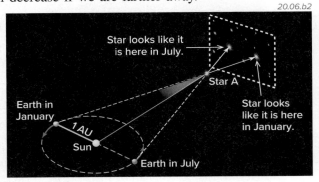

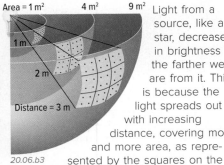

Light from a source, like a star, decreases in brightness the farther we are from it. This is because the light spreads out with increasing distance, covering more and more area, as repre-sented by the squares on the sides of several spheres (▲). In fact, the brightness decreases by a factor of the distance squared, a relationship called the *inverse-square law*. We can use several methods and assumptions to estimate how much light a star is emitting and then measure how bright the star appears to us from Earth. From the observed brightness, the predicted amount of light emitted, and the inverse-square law, we can estimate the distance to that star.

One method for determining distances uses basic geometric concepts, such as the interior angles of a triangle adding up to 180°. As the figure above illustrates, if we observe a distant object (a tree) from two locations that are a known distance apart and measure the various angles, we can calculate the distance to that object. This method, called *triangulation*, can be used in space, as on Earth.

One type of triangulation observes a star at two different times, such as when Earth is on opposite sides of its orbit around the Sun. From these two different perspectives, a star will appear to shift position relative to other objects in space. By using the Earth's orbit as the known distance (as in the tree example to the left), we can measure the angle of the apparent shift in position and then use geometry to calculate the distance to the star. This method is called *parallax*, which refers to the apparent shift in position of an object when viewed from different vantage points. This method only works for nearby stars.

How Do We Determine How Fast a Star Is Moving Toward or Away From Us?

In addition to measuring how far a star is from us, astronomers also want to estimate the direction in which stars and other objects are moving, and the velocity (speed) of such motions. There are several different approaches.

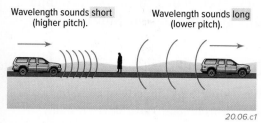

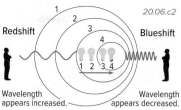

20.06.c1

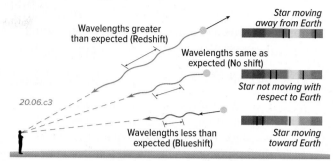

20.06.c3

When a car (or emergency siren or train) passes by us, the sound is different when the car is approaching than when it is moving away. This is because the velocity of the car is compressing or stretching sound waves. When a car approaches us, the sound waves are compressed into shorter wavelengths, so they produce higher frequencies, resulting in a higher pitched sound. When a car is moving away, waves are stretched, producing longer wavelengths and a lower frequency, lower pitched sound. This change in wavelength due to motion is a *Doppler shift*.

A Doppler shift is also observed with light waves. If a light source is moving toward us, the waves are compressed, causing light to shift toward shorter, blue wavelengths, called a *blueshift*. If the light source is moving away from us, the waves become longer, resulting in a shift toward longer, red wavelengths, or a *redshift*. We do not notice this shift in our daily activities, because the speed of normal objects is relatively slow.

Stars and other objects emit and absorb certain wavelengths of light, resulting in emission and absorption spectra. By comparing the actual spectra we observe to where those spectral lines should be (for example, hydrogen), we can calculate the amount of redshift or blueshift. The Doppler shift in turn allows us to determine the object's velocity along the line of sight, which is termed the *radial velocity*. The velocity along the line of sight is proportional to the shift in wavelength, but it may not be the actual velocity, since the object may not be moving directly away or toward us.

 ## How Do We Determine the Radius, Mass, and Other Properties of a Star?

The light coming from a star contains other information about that star, such as the star's average temperature. From the temperature and the amount of light the star is emitting, we can estimate its radius. For nearby stars, we can also measure the radius more directly by comparing its distance with how wide it appears in a telescope. We estimate how much mass a star contains by observing how it affects, and is affected by, its gravitational attraction to other objects.

20.06.d1

1. For some nearby stars, like Betelgeuse (◄), we can directly measure the star's radius from certain types of computer-processed images. Computer processing can help take a somewhat fuzzy appearance of a star and produce an image with sharper, more defined and measurable dimensions.

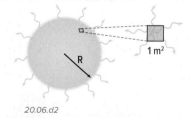

20.06.d2

2. An important characteristic of a star is how much energy it emits each second, which is called its *luminosity*. For example, a typical light bulb has a luminosity of 100 watts (a unit of energy per time). By measuring the wavelengths of light coming from a star, we can use *Wien's law* to estimate its temperature. If we know the temperature of an object, we can then calculate how much energy the object is emitting per square meter (its luminosity per meter) by using another well-established relationship called the *Stefan-Boltzmann law*. Thus, from the wavelength of the light coming from a star, we can determine the star's temperature and luminosity. If we know the luminosity and distance to the star, we can use the inverse-square law to determine how big the star needs to be to account for its measured brightness.

3. We would also like to know how much material, or mass, a star or other object contains. We measure mass in similar units to weight, such as kilograms. Your mass does not vary with your location, but your weight does; weight is related to your mass and to the strength of gravity at that place. Your weight would be less on the Moon than on Earth because the Moon has weaker gravity than Earth. The difference in gravity between the Moon and Earth is due to the Moon having less mass than Earth—more mass means stronger gravity.

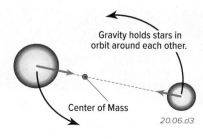

Gravity holds stars in orbit around each other.

Center of Mass

20.06.d3

4. Using Kepler's laws of motion and Newton's Law of Gravity, astronomers can determine the mass of a distant object by observing how its gravity affects other objects, such as its velocity as measured using the object's Doppler shift. In the figure above, two stars are close enough to have their motions controlled by one another's gravity. As a result, the two stars both follow each other around an elliptical orbit, moving around a point in space that represents the center of mass for the two stars. From the distance between the two stars and their velocity, we can calculate the masses of the stars.

Before You Leave This Page

✓ Explain the units we use to describe distances and sizes in space, and how we measure these.

✓ Sketch and explain the Doppler effect and how we use it to determine the velocity of motions of stars moving toward or away from us.

✓ Explain some methods for estimating the radius and mass of a star.

20.6

20.7 What Processes and Features Characterize Stars?

OUR SUN AND OTHER STARS are huge spheres of hot gas that generate vast quantities of light and other electromagnetic energy. In addition to bathing Earth with sunlight, the Sun also generates some spectacular displays, including huge streams of material ejected into space, occasionally toward Earth. What features characterize our Sun and other typical stars? What processes occur within stars to generate so much energy and action?

A What Features and Processes Occur On and Within Our Sun?

From Earth, the Sun appears to most people simply as a very bright light in the sky. We cannot look directly at the Sun without doing damage to our eyes, so it is difficult to get an idea of what the Sun is like. Astronomers have devised many ways to study the Sun to gain insight into how features on the Sun's surface change over time and what processes within the Sun's interior cause these features. The Sun is more than 100 times larger in diameter than Earth, but the density of Earth is nearly 4 times that of the Sun. The Sun is mostly composed of hydrogen and helium, the two lightest elements.

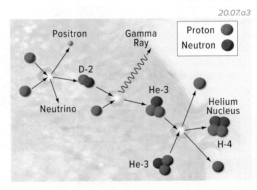

20.07.a1

1. This type of view is probably what most people envision when we think about the Sun—a round, glowing sphere with some dark spots, called *sunspots*. Astronomers view the Sun using visible light and other wavelengths of electromagnetic radiation, such as ultraviolet (UV). In addition, they study variations in the amount of energy emitted by the Sun, changes in the Sun's magnetic field, and many other aspects about the Sun. Some scientific satellites were designed specifically to examine the Sun.

2. The Sun has various features, both around it and within it. The visible surface of the Sun is called the *photosphere* and has an average temperature of ~6,000 K.

3. The Sun's atmosphere, directly above the photosphere, is a thin reddish zone called the *chromosphere*. It contains thin jets of energy and is reddish due to hydrogen.

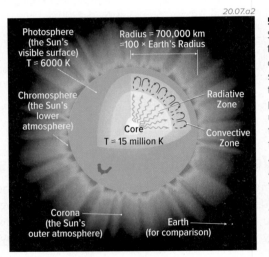

20.07.a2

Photosphere (the Sun's visible surface) T = 6000 K

Radius = 700,000 km =100 × Earth's Radius

Chromosphere (the Sun's lower atmosphere)

Core T = 15 million K

Radiative Zone

Convective Zone

Corona (the Sun's outer atmosphere)

Earth (for comparison)

5. The interior of the Sun is layered, with the center being the *core*. The core has such extremely high temperatures and pressures that it undergoes nuclear fusion, the source of the Sun's energy. Above the core is the *radiative zone*, through which energy from the core radiates toward the surface. The top layer is the *convective zone*, which churns via convection.

4. Reaching farther out from the Sun, and representing the outer part of the Sun's atmosphere, is an outward-directed stream of very hot gas, known as the *corona*.

20.07.a3

Positron

Gamma Ray

Proton
Neutron

D-2

He-3

Neutrino

He-3

Helium Nucleus

H-4

6. This figure (▲) shows the process of *nuclear fusion* within the Sun's core. In nuclear fusion, the nuclei (centers) of two atoms are forcibly combined (i.e., are fused), producing a new atom, but of a heavier element. In the Sun, through a complex process, two protons (the nuclei of hydrogen atoms) fuse to form an atom of helium, the next heavier element, which has two protons and two neutrons. This process generates large amounts of energy.

7. Like Earth, the Sun has a magnetic field generated by movement of electrically charged material within its interior. Earth's core is mostly iron, but the Sun's core consists of hot, compressed gases, predominantly hydrogen, which the Sun started with, and helium, most of which it started with and some of which it produced through nuclear fusion. The gases are *ionized*, meaning the nuclei have lost their electrons due to the extremely high temperatures.

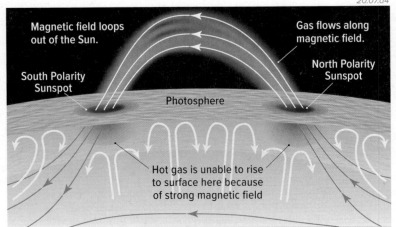

20.07.a4

Magnetic field loops out of the Sun.

Gas flows along magnetic field.

South Polarity Sunspot

North Polarity Sunspot

Photosphere

Hot gas is unable to rise to surface here because of strong magnetic field

8. Sunspots are features on the Sun's surface that appear darker than the rest of the surface because they are slightly cooler, but they are still incredibly hot (4,500 K). A sunspot forms where a strong flow of the Sun's magnetic field exits then enters the Sun's surface (◄). Ionized gas moves along this magnetic field, partially blocking the upward flow of even hotter gas from depth. As a result, the sunspots are cooler and appear darker than surrounding parts of the surface.

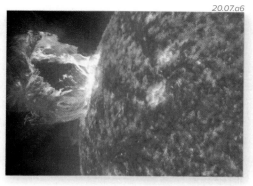

9. Most of us have heard of a solar phenomenon called a *solar flare*. A solar flare, photographed above, is a bright eruption of hot gas from the Sun's surface and into its lower atmosphere. Most solar flares last a few minutes or hours and are associated with an increase in emissions of radio waves and X-rays. These emissions can be strong enough to affect radio transmissions on Earth. They can also cause intense *auroras*, the green glow that many people call the "northern lights."

10. On many days, a huge mass of ionized solar material and energy is ejected from the Sun. Some people call such an event a *solar flare*, but it is more properly called a *coronal mass ejection* (abbreviated as CME). A CME can loop back toward the Sun's surface, as shown above, or can be ejected into space. CMEs and solar flares are both associated with sunspots and are interpreted to be caused by associated changes in the magnetic field.

11. The corona of the Sun also emits a more constant stream of energy and ionized particles, referred to as the *solar wind*. The particles are mostly electrons and protons that are behaving as separate entities rather than being held together, as in a typical atom. Such ionized material is known as a *plasma*, and this is what composes much of the Sun. The solar wind streams out from the Sun in all directions, compresses Earth's magnetic field, and helps shield us from cosmic rays.

B How Do Stars Vary in Size, Color, and Temperature?

Not all stars are exactly like our Sun. Instead, there is great variety in the sizes, colors, temperatures, and lifetimes of stars. As far as stars go, our Sun is considered to be a somewhat average star in terms of its size, temperature, and color. Such a relatively average to small star is known as a *low-mass star*, and a much larger one is a *high-mass star*.

1. This figure illustrates some of the main types of stars, based on their sizes and colors. To convey differences between stars, we usually describe how much mass a star contains, or we can describe its radius. To better grasp such comparisons, we reference the amount of mass to the amount in our Sun (one solar mass). Using this convention, a star with a mass equal to five Suns is said to have five solar masses.

2. Our Sun has an overall yellowish color, although our eyes have evolved to see such light as white. Our Sun is referred to as a *yellow star*. Recall that the color of light radiated by an object is related to its temperature, with red being cooler, blue being hotter, and yellow being in the middle. Our Sun has an intermediate temperature when compared to other stars.

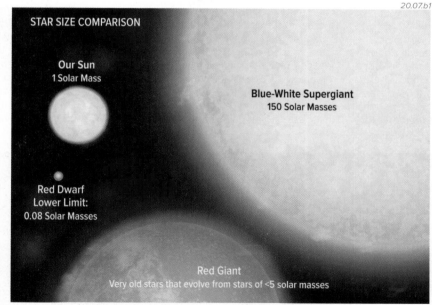

STAR SIZE COMPARISON

Our Sun
1 Solar Mass

Blue-White Supergiant
150 Solar Masses

Red Dwarf
Lower Limit:
0.08 Solar Masses

Red Giant
Very old stars that evolve from stars of <5 solar masses

5. Some stars are enormous, with a much larger radius and with much more mass than our Sun. Astronomers call these *supergiants*. There are very hot supergiants that are bluish or blue-white, and these can have a luminosity that is hundreds of thousands times the luminosity of the Sun. There are also red supergiants which are cooler than the bluish ones. Bluish supergiants can become red supergiants when they cool down. In fact, as we will soon learn, a single star can change from one type on this figure to another, as internal processes cause changes in temperature and size.

3. Some stars are much smaller than our Sun and are commonly called dwarfs. Most *dwarf stars* are cooler than the Sun, have a reddish color, and are called *red dwarfs*. Others are hotter and white, with a range of temperatures, and are called *white dwarfs*.

4. Some stars have a much larger radius than the Sun and commonly contain much more mass. Some of these stars are cooler than the Sun (hence the reddish color) and we call them *red giants*. There are also large yellowish stars, and these are *yellow giants*.

Before You Leave This Page

☑ Sketch and explain the features of the Sun, including its interior and exterior, and what they are interpreted to represent.

☑ Summarize some of the different types of stars, in terms of size, color, and temperature.

20.8 How Do Low-Mass Stars Change Over Time?

STARS EVOLVE OVER TIME, changing in size, temperature, color, and other properties. Astronomers observe various types of stars and interpret them to represent different stages in the life span of a typical star, a time that can be billions of years long or as short as tens of thousands of years. The changes experienced by a star are strongly controlled by the amount of material it contains—its mass. Stars that contain about the same mass as our Sun are considered *low-mass stars* and evolve in a certain series of steps, whereas stars that are much larger than our Sun, the *high-mass stars*, follow a very different progression. Here, we consider the evolution of low-mass stars.

A What Are the Main Chapters in the Life of a Low-Mass Star?

Stars do not start out as stars but form from some other type of starting material. Through the process of nuclear fusion, stars combine lighter elements into heavier ones, releasing light and other energy as a by-product of fusion. As they burn brightly, they are consuming ("burning") their supply of lighter elements—their fuel. The main changes stars undergo are related to their supply of fuel, what type of fuel they burn, and a competition between heat and gravity.

20.08.a1 Orion Nebula, Hubble Space Telescope

20.08.a2 Artist's conception

20.08.a3 Artist's conception

1. Scattered throughout the universe are *interstellar clouds*, which are irregularly shaped, somewhat diffuse masses of gas, plasma, and particles. Some interstellar clouds are relatively cold and so are hard to see, but others are hot and glowing. Such clouds are the birthplace of stars. To begin to form a star, some part of the cloud needs to have enough gravity to begin to draw material together.

2. As material in an interstellar cloud begins to draw together at hundreds of locations, forming a more dense, disk-shaped mass, it heats up, partly due to frictional heating and because materials heat up as they become compressed. When the mass is dense enough, it starts nuclear fusion and starts to glow. The resulting feature, called a *protostar*, emits jets of material from the rotating disk.

3. As the protostar becomes more massive, its gravity increases and it draws in even more material from the disk, forming a star. Such a disk is called an *accretion disk* because of the way the material accretes (gradually combines). Material that remains in the disk is the starting material for any planets that may form around the star. This is how our Sun and solar system were formed.

4. Once a star forms, it can evolve through various stages. This figure (▶) shows the inferred sequence of events for a typical low-mass star. The sequence begins on the upper left, with the evolution of an interstellar cloud into a protostar and then a star like our Sun. A star similar to our Sun is burning hydrogen, fusing it into helium, generating heat and other energy. The heated gas and plasma exert an outward-directed pressure that acts to make the star expand, while gravity pulls the material inward. If these two forces are in balance, called *hydrostatic equilibrium*, then the star remains stable.

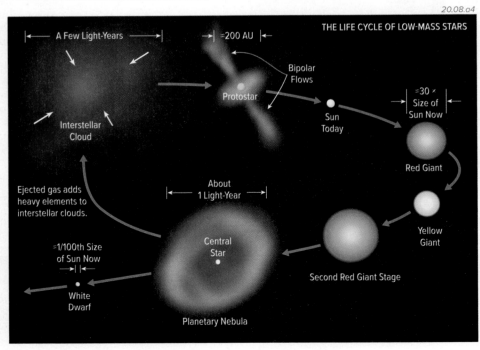

20.08.a4

THE LIFE CYCLE OF LOW-MASS STARS

A Few Light-Years

≈200 AU

Bipolar Flows

Protostar

Interstellar Cloud

Sun Today

≈30 × Size of Sun Now

Red Giant

Ejected gas adds heavy elements to interstellar clouds.

About 1 Light-Year

Yellow Giant

≈1/100th Size of Sun Now

Central Star

Second Red Giant Stage

White Dwarf

Planetary Nebula

5. As a low-mass star consumes its hydrogen fuel, the outward-directed pressure and inward-directed force of gravity are no longer in balance, and the star goes through a series of changes. It first increases in size to a *red giant* and then shrinks somewhat to become a *yellow giant*. Next, the star can again grow and go through a *second red giant stage*. In its final stages, a star begins ejecting its material and becoming a feature known as a *planetary nebula*. Finally, the nearly burned-out star becomes a white dwarf. Some of these steps are described in detail on the next page.

B What Causes a Low-Mass Star to Change Size and Color?

A star changes in size and temperature—and therefore in color and luminosity—in direct response to changes in the amount and type of fuel being burned (consumed by fusion) in its core. Three such changes are described below.

Red Giant Phase

1. When a star like our Sun has burned nearly all of its hydrogen fuel, its core shrinks and grows hotter as it becomes more compressed. Hydrogen just outside the core also begins to fuse. The higher temperatures within and just outside the core cause the gas and plasma to fuse more rapidly, causing the outward-directed pressures to increase and overwhelm the inward pull of gravity. As a result, the atmosphere of the star greatly increases in size, and the star becomes a giant. The increase in the volume of the atmosphere results in a decrease in density and temperature, so the color shifts from a hotter yellow color to a cooler (but still very hot) red one. Through these processes, a star like our Sun becomes a *red giant*.

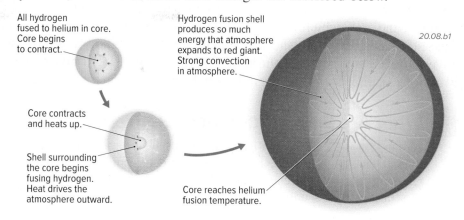

All hydrogen fused to helium in core. Core begins to contract.

Core contracts and heats up.

Shell surrounding the core begins fusing hydrogen. Heat drives the atmosphere outward.

Hydrogen fusion shell produces so much energy that atmosphere expands to red giant. Strong convection in atmosphere.

Core reaches helium fusion temperature.

20.08.b1

Yellow Giant Phase

2. As a star begins to run out of hydrogen, it begins fusing *helium* and then *carbon*. At this stage, the internal pressures decrease and the star contracts somewhat and heats up, becoming a *yellow giant*. Yellow giants can form from either low-mass stars or high-mass stars. Some yellow giants fluctuate in size, becoming larger and then smaller, and these are called *variable stars*. The figure to the right illustrates the changes that occur within a variable star (yellow giant) to cause it to increase and then decrease in size. This process can occur many times and on short enough timescales that astronomers can observe these changes. The fluctuation in size is related to whether radiation is trapped within the star or whether it can escape to space.

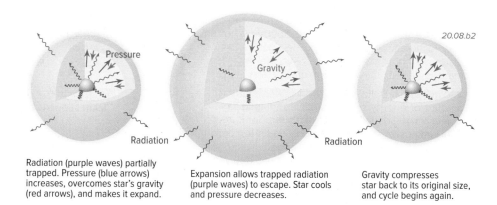

Pressure

Gravity

Radiation

Radiation

20.08.b2

Radiation (purple waves) partially trapped. Pressure (blue arrows) increases, overcomes star's gravity (red arrows), and makes it expand.

Expansion allows trapped radiation (purple waves) to escape. Star cools and pressure decreases.

Gravity compresses star back to its original size, and cycle begins again.

Planetary Nebula Phase

20.08.b3 Ant Nebula, Hubble Space Telescope

3. After a star has become a yellow giant, it is approaching its end. Most of the hydrogen has been burned, and fusion of helium (into carbon) releases less energy than does fusion of hydrogen (into helium), so the star begins to change again. The core compresses and heats up, but the atmosphere cools enough that small solid particles of carbon and silicon form. These particles are pushed outward, away from the star, by the outward flow of energy from the core, forming visible streams of material flowing away from the star. The resulting feature, shown in the image to the left, is called a *planetary nebula* (but has nothing to do with planets). Although beautiful, this type of nebula records the final stages in the life of a low-mass star and may only last thousands of years. After this material is ejected, the exposed core of the star cools and becomes a white dwarf. In the distant future, our Sun will change successively from its yellow phase into a red giant, yellow giant, planetary nebula, and white dwarf, but the process will take billions of years.

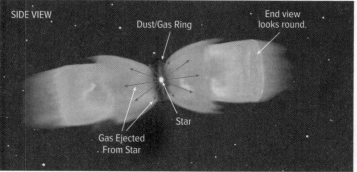

SIDE VIEW

Dust/Gas Ring

End view looks round.

Star

Gas Ejected From Star

20.08.b4

4. This figure (◄) illustrates that the opposite-directed streams of material ejected from a star during the planetary nebula phase are shaped like large funnels that open outward toward space. When viewed end on, the funnel-shaped features look like circular rings or have an oval shape.

Before You Leave This Page

✓ Sketch and explain the main changes a low-mass star experiences, from its initial formation to a white dwarf.

✓ Sketch and explain the red giant, yellow giant, and planetary nebula stages for a low-mass star.

20.8

20.9 How Do High-Mass Stars Change Over Time?

A STAR WITH MUCH MORE MASS than our Sun is a *high-mass star*, and its evolution through time is very different than that experienced by a low-mass star. The greater mass of a high-mass star produces higher densities in the center of the star, which in turn results in higher temperatures and an ability to fuse heavier chemical elements. A high-mass star has a relatively short life span, generally ending in a huge and extremely violent explosion, called a *supernova*. Several dozen supernovae (plural of supernova) have been observed by humans, some without a telescope, as very bright emissions of light. The debris left over from a supernova is a type of *nebula*.

A What Are Some Differences Between High-Mass and Low-Mass Stars?

A star is considered a *high-mass star* if it has at least eight times more mass than our Sun. A high-mass star grows from an interstellar cloud, just like a low-mass star does, but it accumulates much more material (mass). The greater mass within the resulting massive star causes differences in how nuclear fusion operates within the core of the star, and this in turn results in a history that is unlike that of a low-mass star like our Sun.

1. One way to portray differences among stars is by plotting the surface temperature of the star versus luminosity (▶), in what is called an *H-R diagram* (after the initials of the two astronomers who invented it). In such a diagram, stars that are both very hot and very luminous plot in the upper left part of the diagram, and these stars also have the highest masses (that is, they are high-mass stars).

2. The Stefan-Boltzmann law indicates that the amount of energy given out by a radiating object increases with the object's temperature. The luminosity of a star depends on its temperature and surface area (i.e., how large the star is). If two stars have the same temperature but one has a higher luminosity, then the more luminous star must be larger. On an H-R diagram, stars with the same temperature plot the same distance to the right on the diagram, along a vertical line.

3. On this diagram, most stars plot along a diagonal belt, called the *main sequence*, that connects very hot and luminous stars in the upper left with cooler, less luminous stars in the lower right. Note that temperature on this diagram increases to the left. A star that plots above the diagonal belt is more luminous than main-sequence stars of similar temperature, so it must be larger (i.e., have a higher surface area) to have this high a luminosity. Giant stars plot in the upper right part of the diagram, with cooler red giants plotting to the right of hotter blue giants.

4. A star that plots below the belt is less luminous than main-sequence stars of the same temperature, so it must be relatively small to have such a low luminosity. This is the area of the H-R diagram where white dwarfs plot.

5. This figure (▶) shows the inferred stages in the life cycle of a high-mass star. The sequence begins on the upper left, with an interstellar cloud that evolves into a protostar, as occurs for a low-mass star. The star that forms, however, is a high-mass star, much more massive than our Sun. The higher mass of this star causes its material to be more compressed and hotter, so the star is bluer and more luminous than our Sun. The star generates energy by burning hydrogen and fusing it into helium. As the star begins to run out of hydrogen, it expands and cools, forming a *yellow giant* then a *red giant*.

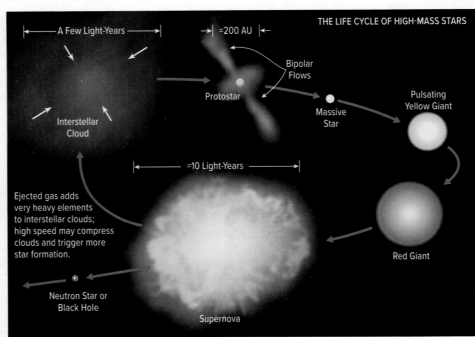

THE LIFE CYCLE OF HIGH-MASS STARS

A Few Light-Years

≈200 AU

Bipolar Flows

Protostar

Interstellar Cloud

Massive Star

Pulsating Yellow Giant

≈10 Light-Years

Ejected gas adds very heavy elements to interstellar clouds; high speed may compress clouds and trigger more star formation.

Red Giant

Neutron Star or Black Hole

Supernova

20.09.a2

6. The core of a high-mass star experiences much higher pressures and temperatures than does the core of a low-mass star, and these conditions allow a high-mass star to fuse heavier elements, such as helium into carbon, carbon into oxygen, and oxygen into silicon. Eventually, fusion produces an iron core, which, for several reasons, destabilizes the star. The immense gravity rapidly collapses the iron core, causing the outer material of the star to blow apart in a tremendous explosion called a *supernova*. The debris, including abundant heavy elements, ejected during the explosion forms a type of *nebula*.

B What Are the Expressions of a Supernova?

When a high-mass star in our galaxy explodes as a supernova, it can appear in the night sky as a bright flash of light that can last days or weeks. Some supernovae were reportedly visible during daylight. In addition to visible light, a supernova generates massive amounts of other types of energy and a huge number of high-speed particles.

A Supernova Event

1. This image (▶) shows the bright remnant and outward-flowing stream of material related to a supernova that occurred more than 300 years ago. This supernova occurred because of the collapse of a high-mass star, but supernovae can form for other reasons, including from a white dwarf. The typical supernova is, however, from the collapse of the core of a high-mass star and the resulting violent expulsion of material.

20.09.b1 Cassiopeia A, Spitzer Space Telescope

20.09.b2 Computer model

2. This image (◀) shows a computer model of a supernova. The bright flash in the center is from the collapsing high-mass star, and the material around the star is being ejected at very high speeds (10,000 km/s). During the bright flash, a supernova can generate more light than is generated by billions of stars in the same galaxy.

Supernova Remnants

3. The enormous quantity of material ejected during a supernova becomes an expanding interstellar cloud, or a *nebula*. A supernova-related nebula can begin with a somewhat spherical shape, like the one shown here (▶). Over time, however, it becomes more irregular as different areas of the outward-moving material encounter resistance from interstellar dust and other material that is in the way.

20.09.b3 Cassiopeia A, composite image

20.09.b4 Crab Nebula, HST

4. This image (◀) shows the Crab Nebula, a well-known nebula related to a supernova whose light reached Earth in the year 1054. Light takes a while to get from such a faraway source to the Earth. It is remarkable how much this nebula has expanded in a little more than a thousand years. Most of the material is outside the colored portion, but it is not visible in this type of image.

Generation and Dispersal of Heavy Elements

5. Due to their high mass and higher temperatures, massive stars are able to synthesize (produce), through nuclear fusion, chemical elements that are heavier than helium. One way to envision this process is as a series of concentric, spherical shells within the core of a massive star (▶). Each shell has the right combination of temperature, pressure, and density to make it capable of synthesizing a certain chemical element or range of elements. The production of heavy chemical elements by the nuclear fusion of two or more lighter elements is called *nucleosynthesis*. The table below shows some of the main fusion reactions.

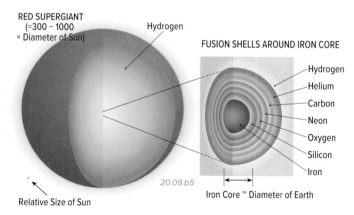

RED SUPERGIANT (≈300 – 1000 × Diameter of Sun)

Hydrogen

FUSION SHELLS AROUND IRON CORE

Hydrogen
Helium
Carbon
Neon
Oxygen
Silicon
Iron

Relative Size of Sun

20.09.b5

Iron Core ~ Diameter of Earth

7. This figure (◀) shows what element undergoes fusion in each shell around a high-mass star in its red giant phase. The lightest element, hydrogen, is fused into helium in the outermost, lowest temperature (but still very hot) shell. Successively heavier elements are fused inward, with the highest temperature shell fusing two silicon atoms to produce iron, which accumulates in the core. For various reasons, iron cannot be fused into heavier elements, so iron is the end of the process, and the formation of an iron core marks the beginning of the end for the star.

Major Fusion Reactions in Massive Stars		
Fusion Reaction	**Minimum Temp.**	**Duration of Reaction**
$4\,^{1}H \longrightarrow\ ^{4}He$	5,000,000 K	7,000,000 years
$3\,^{4}He \longrightarrow\ ^{12}C$	100,000,000 K	700,000 years
$^{12}C + ^{4}He \longrightarrow\ ^{16}O$	200,000,000 K	700,000 years
$^{12}C + ^{12}C \longrightarrow\ ^{20}Ne + ^{4}He$	600,000,000 K	300 years
$^{20}Ne + ^{20}Ne \longrightarrow\ ^{24}Mg + ^{16}O$	1,500,000,000 K	8 months
$^{16}O + ^{16}O \longrightarrow\ ^{28}Si + ^{4}He$	2,000,000,000 K	3 months
$^{28}Si + ^{28}Si \longrightarrow\ ^{56}Fe$	2,500,000,000 K	1 day

6. This table (◀) shows fusion reactions whereby two or more lighter atoms fuse into a heavier one. The left column shows the fusion reaction, the center column indicates the minimum temperature required for each reaction, and the right column lists how long that reaction occurs in a high-mass star with 25 times the mass of our Sun. Even heavier elements, along with these, can be produced during the explosion of a supernova.

Before You Leave This Page

☑ Sketch and explain the changes a high-mass star experiences, from its formation to its end as a supernova.

☑ Explain what the H-R diagram shows.

☑ Explain what happens during a supernova, what features it produces, and the process of nucleosynthesis.

20.9

20.10 What Objects Represent Remnants of Stars?

EVERY STAR HAS A LIFETIME, although the time that a star exists can be only several millions of years for a very massive star compared to many billions of years for a low-mass star. What happens to a star and its material when it is no longer a star? It turns out that the remnants of stars are some of the most interesting objects in the universe, with names that sound like they belong in a science fiction novel, like *neutron star*, *pulsar*, or *black hole*.

A What Remains When a Low-Mass Star Reaches the End of Its Life Cycle?

After a low-mass star goes through its various phases, including its time as a red giant, it sheds much of its outer material as a planetary nebula. What remains behind, the last remnant of the central star, is a white dwarf.

1. After a typical low-mass star reaches the normal yellow stage of our Sun, it can go through several phases, some of which are shown here. A star like our Sun will grow and cool into a red giant stage (and several other stages not shown here). Late in its history as a red giant, the internal pressures within the star overwhelm the inward pull of gravity, so the star starts driving material away from the central star.

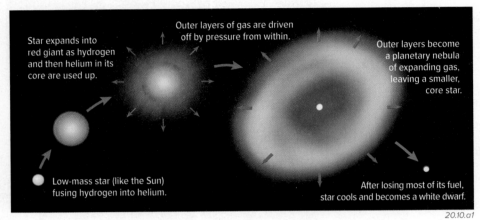

Star expands into red giant as hydrogen and then helium in its core are used up.

Outer layers of gas are driven off by pressure from within.

Outer layers become a planetary nebula of expanding gas, leaving a smaller, core star.

Low-mass star (like the Sun) fusing hydrogen into helium.

After losing most of its fuel, star cools and becomes a white dwarf.

20.10.a1

3. As a star loses material to the planetary nebula and through other processes, it shrinks in size due to the reduction in mass. The remnant is a white dwarf. A white dwarf consists mostly of carbon and oxygen that were produced by fusion in the core of the star in its earlier stages, with an outer layer of hydrogen and helium. The white dwarf can stay with the planetary nebula or it may be pulled away from the nebula by the gravity of another object.

2. The ejection of this material is guided by the magnetic field of the star, and the ejected material takes the shape of two outward-directed cones or funnels, forming a planetary nebula (for stars with 0.8–8.0 times the mass of our Sun). When viewed end-on, a cross section through the funnels looks like large circles or ovals, as shown in the figure above.

4. A white dwarf can remain solitary or it can interact with other objects in space. If it is gravitationally bound to another star (▶), called a *companion star*, we refer to the entire arrangement as a *binary star*. As the two stars orbit one another, locked in each other's gravity, one star can pull material from the other star. In the example illustrated here, the white dwarf is pulling hydrogen (and perhaps helium) from the atmosphere of a red giant. If the captured material makes the white dwarf abruptly increase in brightness, we call such a process a *nova*.

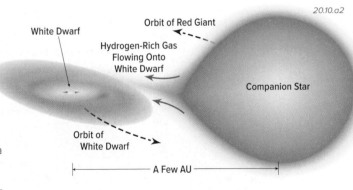

20.10.a2

White Dwarf

Orbit of Red Giant

Hydrogen-Rich Gas Flowing Onto White Dwarf

Companion Star

Orbit of White Dwarf

A Few AU

5. As a white dwarf pulls hydrogen away from the companion star, it is refueling. Due to some unusual behavior, called *degeneracy*, of highly compressed mass in the white dwarf, the addition of more mass actually causes the dwarf to shrink, but is accompanied by a less-than-normal increase in outward pressure. This degenerate behavior in the core has catastrophic consequences, causing a sequence of events depicted in the series of illustrations below.

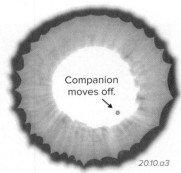

Companion star adds mass and new fuel to white dwarf.

When mass becomes ~ 1.4 solar masses, white dwarf collapses and heats up.

Heat and pressure cause white dwarf to explode.

Exploded remnants of white dwarf expand into space.

White dwarf begins with less than 1.4 solar masses.

Companion moves off.

20.10.a3

6. This figure shows events that can accompany the addition of hydrogen and other mass to a white dwarf, if it ends up with 1.4 times more mass than our Sun (a critical size limit for stability). The addition of mass causes the white dwarf to collapse, leading to a nuclear-driven explosion, a type of *supernova*. Material blasted outward by the explosion becomes an interstellar nebula. The fate of the companion star depends on many factors.

B What Other Features Represent the Remnants of Stars?

Due to their greater mass and the higher temperatures and pressures in their cores, high-mass stars produce distinctly different features when they reach the end of their life cycles. Such massive stars may end up producing a neutron star, X-ray star, or even an all-consuming black hole.

Formation of a Neutron Star

1. The figures to the right depict one route to a massive star's final moments as the star runs out of fuel to burn and its gravity overwhelms the diminishing outward-directed pressure. This sequence of events is predicted for a star that has between 8 and 20 times the mass of our Sun.

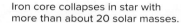

Iron core collapses in star with between about 8 and 20 solar masses.

Outer layers collapse toward core, where they heat up, are compressed, and then explosively driven off very high-density material in core.

Core generates huge burst of negatively charged nuclear particles (neutrinos).

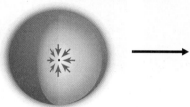

Outer layers explode, leaving a neutron star

20.10.b1

2. Late in its history, a massive star of this size has its iron core suddenly collapse due to its own very strong gravity. The collapse causes incredibly high pressures and temperatures that cause electrons and protons to merge into neutrons (a subatomic particle with a neutral charge). The collapse ejects the outer layers of the star, leaving the core as a mass of unattached neutrons rather than standard atoms. The collapse is considered a type of supernova. The resulting object—a neutron star—is incredibly dense and probably less than 20 kilometers across. If a neutron star is spinning and has a magnetic field, it can be a deep-space source of pulsing radio signals, termed a *pulsar*.

Formation of a Black Hole

3. A star that is even more massive, such as one more than 20 times the mass of our Sun, awaits a different fate. Such a star is so massive and has such intense gravity that it pulls material and energy into a smaller and smaller space, until not even light can escape, creating a *black hole*.

Iron core collapses in star with more than about 20 solar masses.

Inner material drawn into black hole at center, while outer material flattens into disk and shrinks.

Material collapses, heats up, and is driven off in jets, leaving behind a black hole.

Black Hole

20.10.b2

4. In this sequence of events, the iron core in a huge massive star begins to collapse inward. As more material is pulled inward, the density at the center of the core increases, resulting in an even stronger inward pull of gravity, which attracts even more material, which increases the density, which increases the gravity, and the cycle continues. Some material is driven off by the very high temperatures, forming jets on opposite side of the center. What remains behind is a small but very dense object that has such a strong gravitational force that nothing can escape, even light. The lack of any light radiating from or passing through the object is why we call such a feature a black hole.

The Role of Gravity in Forming a Black Hole

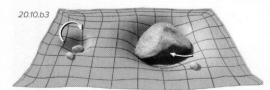

20.10.b3

5. To understand a black hole, we need to understand how gravity works. Imagine placing a small stone on a stretchy plastic sheet. The sheet represents a plot of the gravitational attraction, the stone represents the mass of a typical star, and the depression (called a *well*) in the sheet mimics how we can think of gravity as warping the fabric of space. If we roll a small ball, it is attracted into the gravity well. A larger mass, like a huge boulder, would create a larger gravity well, and the small ball rolls into the well faster. This analogy represents the way the gravity well of a massive object, like a black hole, attracts and captures nearby objects.

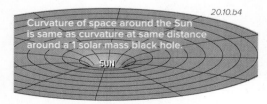

20.10.b4

Curvature of space around the Sun is same as curvature at same distance around a 1 solar mass black hole.

SUN

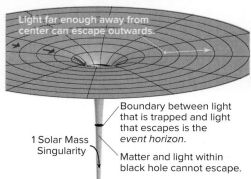

Light far enough away from center can escape outwards.

1 Solar Mass Singularity

Boundary between light that is trapped and light that escapes is the *event horizon*.

Matter and light within black hole cannot escape.

6. Our Sun creates a substantial gravity well that keeps the planets circling it, but a black hole has so much mass in such a small volume that it creates a very deep gravity well. It is so deep that matter cannot escape and no light can escape—a black hole.

Before You Leave This Page

☑ Sketch and explain what can happen to a low-mass star that turns into a white dwarf.

☑ Sketch and explain the sequence of events that cause a massive star to produce a neutron star or black hole.

20.10

20.11 What Are Galaxies, Including the Milky Way?

WE LIVE IN THE MILKY WAY GALAXY, an enormous region of stars, nebulae and other stellar remnants, interstellar clouds, planets, and countless other objects. Our Sun is only one of hundreds of billions of stars in our galaxy, and many of these stars have their own solar systems, with an unknown number of planets, comets, and asteroids. Likewise, the Milky Way Galaxy is only one of more than 100 billion galaxies. What is a galaxy, how does a galaxy form, how does our galaxy compare to others, and where are we located in our galaxy?

A What Is a Galaxy?

A galaxy is an enormous system of stars and other objects that tend to stay together because they are under the influence of each other's gravity. There are three main types of galaxies—*spiral*, *elliptical*, and *irregular*—with subdivisions of and variations within these three types.

20.11.a1 Galaxy cluster, Hubble Space Telescope

20.11.a2 Galaxy M51, Hubble Space Telescope

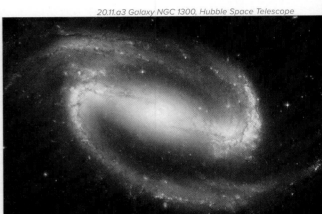
20.11.a3 Galaxy NGC 1300, Hubble Space Telescope

1. This image (▲) of deep space shows a number of galaxies, most of which contain many millions to hundreds of billions of stars. Some especially large galaxies in the universe are estimated to contain as much as 100 trillion stars. Observe this image and note the different shapes of the larger galaxies and whether they have well-defined shapes.

2. The picture most people have in their minds when they think of a galaxy is a *spiral galaxy*, like the one shown above (▲). A spiral galaxy has a coiled appearance, with arms curling around the center of the galaxy. The material in a galaxy rotates around the center of mass for the galaxy, and stars can move within the galaxy due to their interactions with the gravity of adjacent objects.

3. Some spiral galaxies have coiled arms, but many stars and other objects are arranged in a somewhat linear belt, or bar, though the center of the galaxy (▲). We call this type of spiral galaxy a *barred spiral galaxy*. There is much variation within spiral galaxies, from symmetrically coiled versions to distinctly barred ones. Likewise, some spiral galaxies have many arms and well-defined spiral shapes, while some, like the barred galaxy shown above, have only a few arms or barely have a spiral shape.

4. Some galaxies appear to be nearly circular or have the shape of an ellipse, like the bright one shown in the center of this image (▶). This type of galaxy is referred to as an *elliptical galaxy*. In three dimensions, an elliptical galaxy has the shape of an ellipsoid (the three-dimensional equivalent of a two-dimensional ellipse). Elliptical galaxies vary from nearly spherical in shape to very elongated ellipsoids. The most dense concentra-

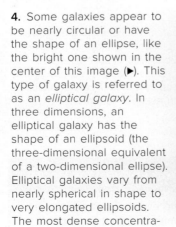

20.11.a4 Galaxies M60 and NGC 4647, Hubble Space Telescope

20.11.a5 Galaxy NGC 1569, Hubble Space Telescope

tion of stars is in the center of an elliptical galaxy, and the number of stars gradually decreases outward, giving the galaxy a fuzzy appearance, without a well-defined outer edge. The largest galaxies are elliptical, with some containing trillions of stars. They are mostly composed of old cool stars, hence the dominant reddish or yellowish color. Some are thought to have formed from the collision of two or more galaxies.

5. The third type of galaxy has an irregular (▲), poorly defined shape—it is an *irregular galaxy*. Such a galaxy may once have had a better defined spiral or elliptical shape, but it was disrupted by interaction with another galaxy. There are other peculiar shapes of galaxies, such as rings and somewhat linear galaxies, that are not described here.

B What Is the Milky Way Galaxy, and Where Within It Is Our Solar System?

Our Sun and its solar system are located within the Milky Way Galaxy, an average-sized spiral galaxy. To convey the sizes of galaxies and very long distances, astronomers use a unit of length called a *parsec*, which is 3.26 light-years. The Milky Way Galaxy is more than 30,000 parsecs across (greater than 100,000 light-years). Our solar system is not in the center of the galaxy, but is instead approximately 8,000 parsecs from the center and on one of the spiral arms.

20.11.b1

20.11.b2

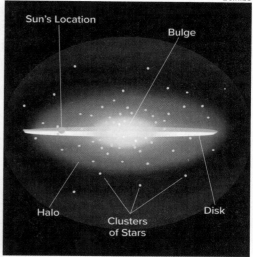

20.11.b3

The Milky Way was first recognized and described in ancient times as a somewhat fuzzy ("milky") belt of stars that extends across the night sky, as shown in the above illustration, which combines several types of images. The dark areas within the Milky Way are clouds of interstellar gas and dust that block the light of stars that are farther away. On a clear night, away from city lights, you can see the Milky Way without a telescope. The Milky Way is a disk-shaped, spiral galaxy that is slightly barred, but we only get an edge-on view from Earth, so it appears to us as a band of stars and dust across the sky.

The two illustrations above show the location of our Sun (and therefore Earth) within the Milky Way. The left view is looking perpendicular to the disk-shaped galaxy, so the spiral arms are visible, whereas the right figure shows a view from the side. Our Sun is located on one of the spiral arms, about two-thirds of the way out from the center. As can be observed in the side view (on the right), our galaxy, like most spiral galaxies, has a *bulge* of stars that protrude above and below the main disk. For the Milky Way, astronomers estimate that less than one-third of the galaxy's stars and gas are in the bulge.

C How Does a Galaxy Form, and How Can One Be Disrupted or Destroyed?

Spiral galaxies, like the Milky Way, are interpreted to form from the consolidation and organization of a huge rotating mass of dust, gas, and other material. Once formed, a galaxy can travel through space for billions of years, retaining its overall character, changing mostly as material is rearranged and as stars are born and die. A galaxy can experience catastrophic changes if, during its travels, it encounters another large object, specifically another galaxy.

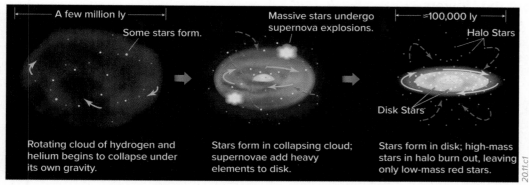

20.11.c1

20.11.c2 Antennae Galaxies

4. A galaxy can lose its shape and compactness if it interacts with other massive objects, such as when it collides with another galaxy. The two galaxies in this image (▶) are tearing each other apart.

1. A galaxy forms from a cloud of hydrogen and helium. Material in the cloud starts to rotate and rearrange itself in response to gravitational interactions between the various masses. The initial cloud is a few million light-years across, much larger than the resulting galaxy.

2. As the cloud contracts under its own gravity, it begins acquiring a disk-like shape. As the material gets more dense, stars form at a higher rate than before. Dying stars and supernovae add heavier elements to the material.

3. Over time, older stars change to yellow and red stars, while some high-mass stars burn out completely. New stars, shown in blue, continuously form in the parts of the galaxy with dense accumulations of mass. A galaxy is fully developed when it has a disk shape, spiral arms, and central bulge.

Before You Leave This Page

✓ Explain what a galaxy is and the main types of galaxies, including the Milky Way Galaxy.

✓ Describe how a galaxy forms and one way it can be disrupted.

20.11

20.12

How Did the Universe Form, and How Is It Changing Through Time?

GALAXIES, SOLAR SYSTEMS, AND PLANETS are but components of the larger universe. The branch of science that investigates the origin, evolution, and fate of the universe is *cosmology*, and a scientist who explores these topics is a *cosmologist*. Cosmology attempts to connect observations and interpretations from astronomy with those from various branches of chemistry and physics, including the physics of the interactions of matter and energy at a subatomic level. Here, we highlight some key questions that astronomers and cosmologists investigate.

Structure of the Universe

1. To understand how the universe formed and has changed over time, we start by observing the present-day structure of the universe. The computer-constructed image to the right shows a view of over a million galaxies, projected in such a way that this oval represents the universe in every direction. The dark band crossing the image represents the Milky Way Galaxy, which mostly blocks our view of objects deeper in space, such as distant galaxies.

2. From this and other data, we can conclude that the universe has a similar number of galaxies, and by inference other material, in every direction. In other words, the universe is relatively homogenous when viewed at a broad scale. There are some clumps or linear bands that have more material, but these are exceptions to the broader pattern. Models we propose and evaluate for the origin of the universe must take into account, and satisfactorily explain, these observations. Note that since the light in these galaxies can take millions to billions of years to reach us, our "present-day" view is really a look into different times of the past.

20.12.a1

Motions Between Galaxies Throughout the Universe

3. We also want to know the directions objects in the universe are moving with respect to us and to each other. Recall that we can use the Doppler shift of light (e.g., redshift) from stars and galaxies to determine an object's motion away from us. When we do this, we discover that almost every galaxy and other objects in the universe are moving away from us and away from each other. The easiest way to explain this is that the universe started at a single point in time and is expanding in all directions. To envision this, imagine

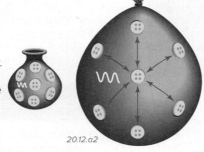

20.12.a2

dots (representing galaxies) painted on a balloon (▲) that is being continuously expanded. Every object would appear to be moving away from every other object, as we generally observe.

As space expands, it stretches the light waves, increasing their wavelengths.

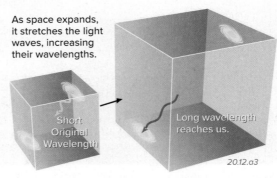

Short Original Wavelength

Long wavelength reaches us.

20.12.a3

4. One consequence of an expanding universe is that as the universe expands, so does the wavelength of light travelling across space. In the example here, blue light traveling from a distant galaxy to our own has its wavelength stretched as it passes through a part of space that is expanding. As a result, such light is redshifted by the time it reaches our telescopes. From any point in the universe, nearly all galaxies would appear to be moving away and would display redshifts. The stretching of light wavelengths would cause light formed during the early history of the universe to be stretched into longer wavelengths, such as microwaves. Today, we observe microwaves coming from every direction in space—the *cosmic microwave background*—and these microwaves are interpreted to be optical EMR from the early universe that has been stretched into longer microwave wavelengths.

Early History of the Universe

5. How old is the universe? Astronomers and cosmologists have several lines of evidence that help us constrain the age at which the universe began expanding. One very solid constraint is provided by the ages of actual samples from Earth, the Moon, and meteorites, each a representative of our solar system. The oldest of these materials is 4.6 billion years old. The universe must be much older than this age in order to have time to (1) produce Earth's abundant heavy elements in ancient stars and supernovae, (2) form the solar system, and (3) form the Earth, Moon, and meteorites. Astronomers can use the observed motions of galaxies and redshifts of light to estimate when the universe began expanding. There are other sophisticated approaches to estimating the age of the universe, but these are beyond the scope of an earth science textbook. The various age estimates, using different methods, data, and assumptions, all converge on an age of 13.8 billion years. Using physics research about the nature and behavior of high-energy particles, cosmologists have proposed a model, summarized in the figure below, of the steps interpreted for the history of the universe. Examine the figure and then read the associated text. Temperatures in Kelvin are shown above the figure, and times in seconds and years from the start of the universe are shown below the figure.

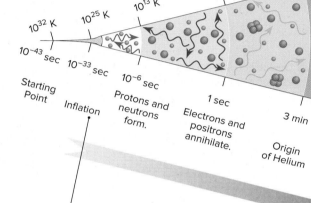

6. This model for the origin of the universe considers the universe to have originated at a single point in time, called the *Big Bang*. This model requires that the universe started very small and early on contained matter that was incredibly compact, dense, and hot, in a feature known as a *singularity*. According to this model, a number of events occurred very early in the history of the universe, and each occurred in a very short duration of time. Inflation started very early, within a tiniest fraction of a second, and occurred under scorching conditions of 10^{25} Kelvin, many orders of magnitude hotter than temperatures in any present-day star.

7. Matter, as most people think of it, did not exist early on, but was created from the vast amounts of energy. Protons and neutrons, the building blocks of the atomic nucleus, are interpreted to have formed before one second of time had passed. Nuclear fusion could occur in this hot, dense material, so early hydrogen atoms combined to form helium; the universe now contained its two lightest chemical elements. As the universe expanded, it cooled and began to emit successively longer wavelengths of energy, including visible light. Some of this energy reaches us as the cosmic microwave background.

8. The first stars and galaxies are interpreted to have formed some 100 million years after the Big Bang. The light from these early-formed galaxies and stars is just reaching Earth today, if the stars and galaxies are on the other side of the universe. The light from objects more distant than these has not yet reached us, so we do not know that they exist or what is out there.

9. Our solar system formed approximately 9 billion years after the formation of the universe, or 4.6 billion years before the present. Sometime in the future, many of the stars will run out of fuel, so they and the galaxies they form will begin to dim and then die out, gradually turning the entire night sky dark.

Before You Leave This Page

☑ Summarize models for the character, motions, and origin of the universe, explaining in general terms how the universe changed over time according to the Big Bang model of an inflating universe.

20.12

20.13

What Are These Astronomical Objects, and How Did They Form?

THE UNIVERSE CONTAINS DIVERSE OBJECTS, ranging from single gas molecules or grains of dust in an interstellar cloud to huge galaxies many times larger than our own Milky Way Galaxy. To better understand the universe, we need to be able to observe an object, identify what type of object it is, and interpret how it formed. In this investigation, you will observe seven different objects in the universe, identifying each, and explaining how each feature likely formed, based on the type of object it is.

Goals of This Exercise:

- Observe different astronomical objects and interpret what kind of feature each represents.
- Propose an explanation for how each feature formed, how it has changed over time, and how it might change in the future.

Procedures

A. For each of the seven objects on these two pages, first observe the image of that object and list as many observations as you can, describing the scene. After you have made your own observations, read the text accompanying the image.

B. From your observations and from information in this chapter, identify what type of feature an object likely represents. For example, is the object a certain type of galaxy, star, or stellar remnant?

C. Based on your identification of the object, explain how you think that object formed and what types of processes were occurring at the time the image was taken.

D. For each object, predict what will likely happen to it in the future.

Object 1

20.13.a1

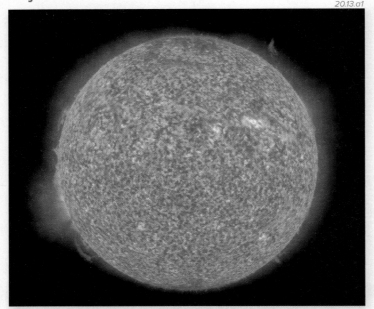

This feature, shown in an artist's version (not a photograph), is more than 10 times larger than the Sun. The red color is the true color of the light coming from the object. In addition to the main object, note the features around the object.

Object 2

20.13.a2

This photograph shows a very large object outside of our galaxy. In addition to the main object, observe the features on different parts of the object, including their color.

Object 3

20.13.a3

This photograph shows a very large object within our galaxy. In addition to the main object, there are many young stars nearby and within the object.

Object 4

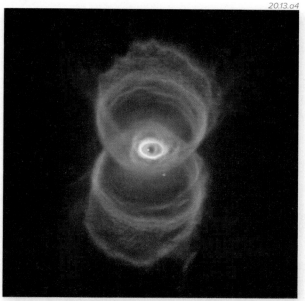

20.13.a4

This image shows material streaming from opposite sides of a star, which is the bright area in the center of the image. The lower material below the star is moving away, into the page, whereas material above the star is coming toward the viewer.

Object 5

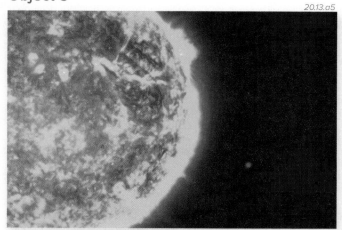

20.13.a5

This object, shown in an artistic interpretation (not an actual photograph), has a blue color. For comparison, the small, dim object to the right of the blue one is the size and brightness of our Sun relative to the blue object.

Object 6

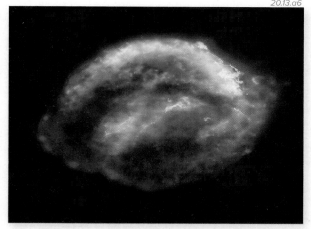

20.13.a6

This object is portrayed using an image that combines different types of energy being emitted. The material and energy sources are moving outward at a very high velocity away from the center of the object.

Object 7

The object in the center of the photograph to the right is a very large object in deep space. It is the largest example of this type of feature currently known. The view is oblique to the main shape of the object (that is, it is not looking directly down or from the side of the object).

20.13.a7

GLOSSARY

A

AA lava A blocky type of lava with a rough, jagged surface. (6.6)

abandoned meander An original curved stream path that has subsequently been abandoned by the stream, often in connection with the formation of a natural bridge. (9.10)

abrasion The mechanical wearing away of rock surfaces by contact with sand and other solid rock particles transported by running water, waves, wind, ice, or gravity. (11.3, 12.4)

absolute zero The temperature at which all molecular motion theoretically ceases (i.e., there is no internal kinetic energy); occurs at 0 Kelvin, or $-273°C$. (13.3)

absorption The concentration of electromagnetic energy within the molecular structure of matter. (13.1)

absorption-line spectrum A recording and display of the spectrum that results when EMR passes through a substance, such as a gas, that absorbs some wavelengths. Those wavelengths are displayed as gaps in the spectrum. (20.4)

abyssal plain A relatively flat, smooth, and old region of the deep ocean floor. (5.2)

abyssal waters Any water near the ocean floor, generally circulating toward the equator. (17.7)

acceleration A change in the speed or direction of an object. (20.5)

accretion The addition of tectonic terranes and other material to an existing landmass, usually along a convergent or transform plate boundary. (8.10)

accretionary prism A prism- or wedge-shaped structurally complex zone of faults, folds, and mostly metamorphosed rocks that form along the upper parts of a subduction zone; material derived from sediment contributed by adjacent volcanoes or continent and oceanic crust scraped off the downgoing slab. (5.7, 8.5)

accretion disk A disk-shaped mass of material that accretes onto a protostar. (20.8)

acre-foot The volume of water required to cover an acre of land to a height of one foot. (12.2)

active dune A dune of any type that is actively moving and being reshaped by the wind. (9.9)

active fraction The component of a soil consisting of decomposing organic matter. (10.1)

adiabatic A condition of a system in which energy does not enter or leave the system. (15.4)

advection The horizontal transfer of energy through a gas, liquid, or weak solid. (13.2)

advection fog A type of fog that occurs when warmer air flows laterally over a colder surface, causing it to chill from beneath to the dew-point temperature or frost-point temperature. (15.8)

aeolian Any process relating to the wind's ability to shape the landscape or a geomorphic feature, primarily caused by wind deposition. (9.9)

aerosol Any tiny solid or liquid particle suspended in the air. (13.1)

African Easterly Jet A west-flowing, fast stream of air in the northeast trade winds over or near northwestern Africa. (16.11)

aftershocks Smaller earthquakes that occur after the main earthquake and in the same area. (7.10)

agent of transport Any phenomenon, including gravity, streams, glaciers, waves, and wind, that moves material, which can later be deposited to create landforms. (9.8)

A horizon Topsoil composed of dark gray, brown, or black organic material mixed with mineral grains. (10.3)

air mass A body of air that is relatively uniform in its meteorological characteristics over distances of thousands of square kilometers. (16.1)

Alaska Current A south-to-north-flowing, warm surface ocean current in the eastern North Pacific Ocean. (17.2)

albedo The ratio of the amount of electromagnetic energy reflected by a surface to the amount of energy incident upon it. (11.8, 13.11)

Alberta Clipper A type of mid-latitude wave cyclone that forms on the leeward side of the Canadian Rockies and tracks generally east-northeastward. (16.3)

Aleutian Low A prominent area of semipermanent low atmospheric pressure (cyclone) in the subpolar North Pacific Ocean; is particularly strong in winter. (14.5)

Alfisol A fertile type of soil, typically with ample accumulation of organic matter in the A horizon and clay eluviation into the B horizon. (10.6)

alluvial fan A low, gently sloping mass of sediment, shaped like an open fan or a segment of a cone; typically deposited by a stream where it exits a mountain range or joins a main stream. (9.8)

alpine glacier A glacier that begins in mountainous terrain and flows down a valley as a narrow tongue of ice and rocks; syn. valley glacier. (11.1)

altocumulus cloud A type of mid-level, vertically oriented cloud that contains both liquid water and ice. (15.7)

altostratus cloud A type of mid-level cloud that is wide and layered, and composed of both liquid water and ice. (15.7)

amphibole A group of silicate minerals, including hornblende, characterized by double chains of silicate tetrahedra; most common in igneous and metamorphic rocks. (2.6)

amplitude The height between the trough and the crest of a wave, including electromagnetic, ocean, and seismic waves. (13.5)

andesite An intermediate-composition igneous rock that is the fine-grained equivalent of diorite; typically gray or greenish-gray, commonly with phenocrysts of cream-colored feldspar or dark amphibole. (3.10)

Andisol A site-dominated soil type formed from volcanic ash and other volcanic material. (10.7)

angle of incidence The angle between the direction of incoming energy and a surface. (13.6)

angle of repose The steepest angle at which a pile of unconsolidated grains remains stable. (10.9)

angular unconformity An unconformity (ancient erosion surface) in which the older, underlying strata dip more steeply or at a different angle than the younger, overlying strata. (4.2)

anion A negatively charged ion. (17.4)

Antarctic Bottom Waters The moderately saline, very cold, and very dense ocean waters in the middle to high latitudes of the Southern Hemisphere. (17.5)

Antarctic Circle The 66.5° parallel of latitude south of the equator; represents the most northerly latitude that experiences 24 continuous hours of daylight (on the December Solstice) and 24 continuous hours of darkness (on the June Solstice). (1.9)

Antarctic Circumpolar Current A west-to-east-flowing, surface ocean current in the high latitudes of the Southern Hemisphere; also known as the West Wind Drift. (17.2)

Antarctic Intermediate Waters A tongue of less saline ocean waters from near Antarctica that forms a wedge below saline waters near the surface in the Southern Hemisphere, centered around 40° S. (17.5)

anticline A fold, generally concave downward in the shape of an A, with the oldest rocks in the center. (7.3)

anticyclone An enclosed area of high atmospheric pressure. (14.5)

aphanitic rock An igneous rock that does not contain crystals visible to the unaided eye; can consist of microscopic crystals, fine-grained volcanic ash, volcanic glass, or a combination of these. (3.9)

aphelion The date of maximum distance between Earth and the Sun during the course of Earth's orbit, at present corresponding to a day on or near July 4. (13.6)

aphotic zone The part of a lentic system that does not have access to sunlight. (17.11)

aquifer A permeable body of saturated material through which groundwater can flow and which yields significant volumes of water to wells and springs. (12.12)

aquitard A low-permeability bed or other unit that retards but does not prevent the flow of water to or from an adjacent aquifer. (12.12)

Arctic (or Antarctic – A) air mass A body of air that is relatively uniform in its very cold and very dry characteristics for thousands of square kilometers, originating in high-latitude source regions that are either inland or over sea ice. (16.1)

Arctic Circle The 66.5° parallel of latitude north of the equator; represents the most southerly latitude that experiences 24 continuous hours of daylight (on the June Solstice) and 24 continuous hours of darkness (on the December Solstice). (1.9)

arête A hard, jagged bedrock ridge flanked by two or more cirques. (11.4)

arid (B) climate In the Köppen climate classification, a broad group of climatic types characterized by inadequate moisture availability, in which potential evapotranspiration typically exceeds precipitation. (18.4)

Aridisol A type of soil found in arid regions, characterized by a lack of organic matter, little leaching of nutrients, and limited horizonation; has a propensity for salinization and formation of a caliche layer. (10.6)

artesian Groundwater that is in a confined aquifer and is under enough water pressure to rise above the level of the aquifer. (12.12)

asbestos A general name for the fibrous varieties of a number of silicate minerals whose common characteristic is that they tend to form fibers. (2.6)

ash flow A fast-moving cloud of hot volcanic gases, ash, pumice, and rock fragments that generally travel down the flanks of a volcano; also known as a pyroclastic flow. (6.4)

asteroid One of countless rocky fragments in orbit around the Sun, mostly concentrated between the orbits of Mars and Jupiter. (19.0)

asthenosphere The area of mantle beneath the lithosphere that is solid but hotter than the rock above it and can flow under pressure; functions as a soft, weak zone over which the lithosphere may move. (5.1)

astronomical unit A unit of distance equivalent to the average distance between Earth and the Sun. (20.6)

Atlantic hurricane season The time of year when Atlantic tropical cyclones are most likely to occur—from June 1 through November 30. (16.11)

Atlantic Multidecadal Oscillation (AMO) An oscillation in sea-surface temperatures in the Atlantic Ocean, in which temperatures tend to be warmer than normal in much of the basin for many years and then colder than normal for many years. (18.10)

atmosphere The gaseous envelope that surrounds the solid Earth. (1.4)

atmospheric pressure The pressure, or force per unit area, exerted by the atmosphere; generally greater at sea level than in high elevations. (1.3)

atoll A curved to roughly circular reef that encloses a shallow, inner lagoon and low islands. (8.9)

atom Smallest possible particle of an element that retains the properties of that element. (2.10)

atomic mass The number of neutrons and protons in an atom. (2.10)

atomic number The number of protons in an atom. (2.10)

atomic symbol One or two letters representing the name of an element, as in the Periodic Table. (2.6)

atomic weight The sum of the weight of the subatomic particles in an average atom of an element, given in atomic mass units. (4.3)

aurora A light that originates from interactions of solar energy and energetic gas molecules, primarily in the thermosphere. (13.1)

B

back-arc basin The region adjacent to a subduction-related volcanic arc, on the side opposite the trench and subducting plate. (8.6)

badlands A distinctive type of topography with a soft, rounded appearance, mainly composed of shale and other easily eroded rocks. (3.13)

banded iron formation A sedimentary mineral deposit displaying layers or bands of iron-rich minerals and fine-grained, silica-rich material. (2.9)

barchan dune A crescent-shaped dune of windblown sand with its tails pointing in the direction of the prevailing wind; also known as a crescent dune. (9.9)

barometer An instrument used to measure atmospheric pressure. (14.2)

barred spiral galaxy A type of coiled galaxy in which the main mass of stars form a linear bar or belt in the center of the galaxy. (20.11)

barrier island A narrow island that forms along a coastline from the accumulation of sediment. (3.4)

barrier reef A long, commonly curved, coral reef that parallels the coast of an island or continent from which it is separated by a lagoon. (8.9)

basalt A fine-grained, dark-colored mafic igneous rock, with or without vesicles and phenocrysts of pyroxene, olivine, or feldspar. (3.10)

basaltic lava flow A lava flow composed of dark, basaltic material. (3.13)

base level The lowest level to which a stream can erode, commonly represented by the level of the sea, lake, or main river valley. (12.5)

base metal A relatively common and chemically active metal, including copper, lead, and zinc. (2.13)

basin A low-lying area or an area in which sediments or volcanic materials accumulate with no or a limited surface outlet; also can refer to a structural feature toward which rock strata are inclined in all directions. (5.6)

Basin and Range Province The physiographic region in the interior Southwest of the U.S., consisting of almost all of the state of Nevada and parts of adjacent states and characterized by alternating valleys and mountain ranges, commonly the expression of horsts and graben. (7.2)

batholith One or more contiguous plutons that cover more than 100 km². (6.11)

baymouth bar A bar of sand or gravel deposited entirely or partly across the mouth of a bay. (11.13)

beach A stretch of coastline along which sediment has accumulated. (3.4)

beach nourishment The procedure of bringing sand to a beach to replenish any sand lost to storms and currents. (11.14)

bed A distinct layer formed during deposition, generally in sediment or sedimentary rock; can be present in volcanic tuff and some metamorphic rocks. (9.1)

bed load Material, commonly sand and larger, that is transported along the bed of a river. (12.4)

bedding Layers or beds of varying thickness and character, generally in a sediment or sedimentary rock. (9.1)

bedrock The rocky materials that are solidly attached to the earth and its subsurface. (3.0)

bedrock stream A stream that is carved into bedrock, commonly in mountainous areas. (12.4)

Benguela Current A south-to-north-flowing, cold surface ocean current in the eastern South Atlantic Ocean. (17.2)

benthic Pertaining to an organism in the saltwater biome that resides mainly on the ocean floor. (17.11)

benthic zone The part of the saltwater biome that contains all of the habitats of the sea bottom, whether in coastal, continental shelf, or deep-sea environments. (17.11)

Bergeron process A process of precipitation formation that occurs by evaporation of liquid water, which allows the additional water vapor to deposit onto solid ice particles, enlarging the snowflake, which can then melt, forming a raindrop, or remaining solid on descent. (15.9)

Bering Current A small north-to-south-flowing, cold surface ocean current into the North Pacific Ocean through the Bering Strait. (17.2)

Bermuda-Azores High A prominent area of semipermanent, high atmospheric pressure (anticyclone) in the subtropical North Atlantic Ocean; is particularly strong in summer. (18.5)

B horizon A zone in the soil characterized by the accumulation of material, including iron oxide, clay, and calcium carbonate, depending on the climate and starting materials. (10.3)

big bang The modern theory that the universe had its origin in a single release of energy 10 to 15 billion years ago. (19.0)

binary star Two stars that are gravitationally bound to one another. (20.10)

biogeochemical cycle The movement of matter throughout the atmosphere, hydrosphere, lithosphere, and biosphere by interaction with living and dead organisms and involvement in chemical and physical processes. (1.5)

biome A number of adjacent ecosystems defined by the predominant types of vegetation and characterized by organisms that are adapted to that environment. (18.1)

biosphere The spherical zone that includes life and all of the places life can exist on, below, and above Earth's surface. (1.4)

biotic weathering Any physical or chemical weathering process caused by organisms. (9.6)

biotite A typically black or brown mica (sheet silicate mineral). (2.8)

black hole An incredibly dense object produced by the collapse of a giant star, whose gravitational force is so strong that nothing can escape, not even light. (20.10)

black smokers Hydrothermal vents on the seafloor in which hot water from within the rock jets out into cold seawater and forms a cloud of minerals, especially those rich in sulfur. (5.10)

block diagram A diagram that portrays in three dimensions the shape of the land surface and the subsurface distributions of rock units, folds, faults, and other geologic features. (1.8)

blue jet A dim, blue, smoke-like streak of light that bursts out above a thunderhead, arcs upward high into the atmosphere, and then fades away. (16.7)

blueshift A shift toward shorter blue wavelengths by a light source in motion. (20.6)

body waves Seismic waves that travel within Earth. (7.6)

boiling point The temperature at which a liquid boils; that is, the temperature at which the vapor pressure in a liquid increases to a value equal to the pressure of the gas adjacent to the liquid, thereby forcing the molecules in the liquid to leave the liquid spontaneously. (13.3)

bomb (volcanic) A large rock fragment representing either a large blob of magma or a solid angular block ejected during an explosive volcanic eruption. (6.3)

boulder A clast or rock fragment with a diameter exceeding 0.25 m. (3.6)

bow echo The arc-shaped pattern on Doppler radar that is characteristic of a derecho. (16.10)

Bowen's reaction series An idealized sequence in which minerals could crystallize from a magma as it cools. (9.4)

brachiopods A phylum of marine-invertebrate animals, commonly containing two valves and living on or close to the bottom of the seafloor. (4.9)

brackish Water that is intermediate in salinity between normal freshwater and seawater. (12.1)

braided stream A stream or river with an interconnecting network of branching and reuniting shallow channels. (12.6)

Brazil Current A north-to-south-flowing, warm surface ocean current in the western South Atlantic Ocean. (17.2)

breaker The top of a nearshore wave that tumbles over because its speed exceeds that of the rest of the wave. (11.11)

breakwater An offshore structure (such as a wall), typically parallel to the shore, that breaks the force of the waves to protect a shoreline or harbor. (11.14)

breccia A rock composed of large, angular fragments; typically formed in sedimentary environments, but can be formed by volcanic, hydrothermal, or tectonic processes. (3.7)

brine A highly saline water. (17.6)

brine exclusion The nonparticipation of salt in the freezing of seawater, which makes the surrounding water saltier. (17.6)

brittle deformation A type of structural deformation that is characterized by breaking of material, such as by jointing, faulting, and the formation of tectonic breccia; typical of shallow crustal conditions. (7.1)

burrow A commonly tubular opening formed when creatures wiggle or tunnel into mud; can be filled with a different type of sediment to form a trace fossil. (4.4)

butte An isolated, steep-sided feature that rises above the surrounding landscape. (9.7)

C

calcite A common rock-forming calcium carbonate mineral occurring in limestone and a variety of water-related deposits. (2.9)

caldera A large volcanic depression that is typically circular to elongated in shape and formed by collapse of a magma chamber. (6.3)

California Current A north-to-south-flowing, cold surface ocean current in the eastern North Pacific Ocean. (17.2)

calving The breaking off of a mass of ice from a glacier, iceberg, or ice shelf. (11.2)

Cambrian The earliest geologic period of the Paleozoic Era. (4.6)

Cambrian Explosion The widespread, relatively rapid appearance of diverse types of hard-shelled organisms near the beginning of the Cambrian Period. (4.9)

Canadian Shield A vast area of mostly crystalline rocks exposed in the eastern half of Canada and the Great Lakes region of the United States. (4.7)

Canary Current A north-to-south-flowing, cold surface ocean current in the eastern north Atlantic Ocean. (17.2)

capillary action The ability of water, aided by surface tension, to travel upward against the force of gravity. (10.2, 15.1)

capillary water Subsurface water that is within pore spaces between the grains but not adhering to the soil particles, and therefore is available to move upward through soil toward the surface. (10.2)

carbon cycle The movement of carbon between the biosphere, atmosphere, hydrosphere, and lithosphere. (1.5, 18.8)

carbonate mineral A mineral containing a significant amount of the carbon-oxygen combination called carbonate. (2.9)

carbonates Rocks or minerals that contain abundant carbonate. (2.6)

Carboniferous A Paleozoic geologic period, used mostly in Europe and Asia, for the combined duration of the Pennsylvanian and Mississippian Periods. (4.6)

Cascade forearc basin A basin that sits between the Cascade volcanic arc and an offshore trench, indicating subduction beneath the continent. (8.4)

cation A positively charged ion. (2.7)

celestial equator A projection of Earth's equator onto the celestial sphere. (20.2)

celestial sphere An earthbound viewer's perspective of the night sky as an imaginary sphere surrounding the earth. (20.2)

Celsius scale A system for indexing temperature calibrated such that water at sea level freezes at 0°C and water at sea level boils at 100°C. (13.3)

cement A natural material precipitated in the pore spaces between grains, helping to hold the grains together. Also, a processed, whitish powder, typically derived from calcium carbonate, that sets to a solid mass when mixed with water and usually other materials. (3.7)

cementation The precipitation of a binding material (the cement) around grains in rocks. (3.7)

Cenozoic Era A major subdivision of geologic time from 65 Ma to the present; characterized by an abundance of mammals. (4.5)

chalk A soft, very fine-grained limestone that forms from the accumulation of calcium carbonate from microscopic organisms that float in the sea. (3.7)

chemical bonding The process in which two atoms bond together by sharing, donating, or borrowing electrons from their outermost orbital shells. (2.11)

chemical weathering Chemical reactions that affect a rock or other material by breaking down minerals and removing soluble material from the rock. (3.5)

chert A sedimentary rock composed of fine-grained silica. (3.7)

Chile Current A local name for the Humboldt Current. (17.2)

chimney A hollow, circular column formed by sulfur-rich minerals precipitating around a submarine, hydrothermal vent. (5.10)

Chinook wind A warm, dry wind that descends the leeward side of mountain ranges in western North America. (14.4)

chlorofluorocarbon A type of chemical released from aerosol cans, air conditioning units, refrigerators, and polystyrene, which contain halogens that can destroy ozone molecules by attracting one of the oxygen atoms away from them. (13.10)

C horizon A zone in the soil composed of weathered bedrock or sediment, grading downward into unweathered bedrock or sediment. (10.3)

chromosphere The portion of the Sun's atmosphere that is directly above the photosphere. (20.7)

chrysotile A fibrous form of the silicate mineral serpentine; the most commonly used asbestos in the United States. (2.6)

cinder A loose black or red, pebble-sized material, often a type of basalt, that composes a cinder cone. (6.3)

cinder cone A relatively small type of volcano that is cone shaped and mostly composed of volcanic cinders (scoria); also known as a scoria cone. (6.3)

circle of illumination The circle that separates the half of the Earth that is experiencing daylight from the half that is experiencing darkness at a given time. (13.8)

circulation cell A large, somewhat rectangular, vertically oriented circulation system in which air rises, moves laterally, sinks, and then moves laterally to replace the air that rose. (14.0)

cirque An open-sided, bowl-shaped depression formed in a mountain, commonly at the head of a glacial valley; produced by erosion of a glacier. (11.4)

cirriform cloud A family of cloud types that is "feathery" or wispy and composed of ice crystals, usually high in the troposphere. (15.7)

cirrocumulus cloud A type of high-level cloud that is partly cumuliform (lumpy) and partly cirriform (wispy) and almost entirely composed of ice. (15.7)

cirrostratus cloud A type of high cloud that is somewhat wispy but with a continuous, sheetlike appearance. (15.7)

cirrus cloud A type of high-level cloud with a wispy appearance, composed of ice particles. (15.7)

clast An individual grain or fragment of rock, produced by the physical breakdown of a larger rock mass. (2.2)

clastic A material consisting of pieces (clasts) derived from preexisting rocks; usually formed on Earth's surface in low-temperature environments; syn. detrital. (2.2)

clay 1. Any fine-grained sedimentary particle that is finer than 1/256 mm. (3.6) 2. A family of finely crystalline, hydrous-silicate minerals with a two- or three-layered crystal structure. (2.8)

cleavage (mineral) The tendency of minerals to break along specific orientations of closely spaced planes. (2.3)

cleavage (rock) The tendency of a rock, especially a metamorphic rock, to split along mostly parallel planes. (3.11)

climate The long-term pattern of weather at a place, including not just the average condition but also the variability and extremes of weather. (18.0)

climograph A graph that shows average monthly temperature (with a line) and precipitation (with bars) for a location. (18.3)

closed system Any interconnected set of matter, energy, and processes that does not allow material and energy to be exchanged outside itself. (1.4)

cloud-to-cloud lightning The electrical charges that are exchanged suddenly but only between two clouds; comprises the majority of lightning strikes. (16.7)

cloud-to-ground lightning The electrical charges that are exchanged suddenly between a cloud and the surface; comprises about 25 percent of lightning strikes. (16.7)

coal The natural, brown to black rock derived from peat and other plant materials that have been buried, compacted, and heated; used for heating and to generate electricity. (3.7)

coastal dunes Sand dunes that form inland from beaches. (3.4)

cobble A rock fragment larger than a pebble and smaller than a boulder, having a diameter in the range of 64 to 256 mm. (3.6)

Cold Desert (BWk) climate In the Köppen climate classification, a climate type characterized by inadequate moisture availability, in which potential evapotranspiration typically exceeds precipitation by a large amount, and a mean annual temperature below 12°C (54°F). (18.4)

cold front The sloping boundary surface between an advancing mass of relatively cold air and a warmer air mass. (16.2)

cold phase (ENSO) A periodic cooling of the eastern equatorial Pacific surface ocean waters and reduction in precipitation adjacent to South America as the Southern Hemisphere's summer approaches; also known as La Niña. (17.9)

cold sector The part of a mid-latitude cyclone that is behind the cold front, characterized by cold air. (16.2)

Cold Steppe (BSk) climate In the Köppen climate classification, a climate type characterized by inadequate moisture availability, in which potential evapotranspiration typically exceeds precipitation by a small or moderate amount, and a mean annual temperature below 12°C (54°F). (18.4)

collision regime The scenario of coastal inundation during which the height of the swash zone exceeds the base of coastal dunes but does not exceed the top of coastal dunes. (11.17)

colluvium Sedimentary material that was transported to a location after being loosened and moved downhill via mass-wasting processes. (9.8)

columnar joint A distinctive fracture-bounded column of rock formed when hot but solid igneous rock contracts as it cools. (3.13, 6.12, 7.2)

compaction Process by which soil, sediment, and volcanic materials lose pore space in response to the weight of overlying material. (3.7)

companion star One star in a binary star pair. (20.10)

composite volcano A common type of volcano constructed of alternating layers of lava, pyroclastic deposits, and mass-wasting deposits, including mudflows; also known as a stratovolcano. (6.3)

compression The type of differential stress that occurs when forces push in on a rock. (7.1)

concept sketch A simplified sketch annotated with labels and complete sentences that describe Earth features and processes. (1.6)

conchoidal fracture A fracture in rock that has an irregular or smoothly curving surface. (2.8)

concretion A hard, compact accumulation of mineral matter in the pores of sedimentary or volcanic rocks; representing a concentration of constituents of the rock or cementing material. (4.4)

condensation A change in state from a gas to a liquid, which releases latent energy to the environment. (13.4)

condensation nuclei The solid aerosols that facilitate the conversion of water vapor to liquid water around them. (15.9)

conditional instability (atmosphere) A situation in which the atmosphere is stable if unsaturated but unstable if saturated; the environmental lapse rate is between the unsaturated adiabatic lapse rate and the saturated adiabatic lapse rate. (15.4)

conduction The transfer of thermal energy by direct contact, primarily between two solids. (13.2)

cone of depression A depression in the water table that has the shape of a downward-narrowing cone; develops around a well that is pumped, especially one that is over-pumped. (12.14)

confined aquifer An aquifer that is separated from Earth's surface by materials with low permeability. (12.12)

conglomerate A coarse-grained clastic sedimentary rock composed of rounded to sub-rounded clasts (pebbles, cobbles, and boulders) in a fine-grained matrix of sand and mud. (3.7)

conservation of energy The scientific principle that states that energy in a closed system cannot be created or destroyed. (20.4)

constellations Arrangements of visible stars in the night sky that were identified and given unique names by ancient observers. (20.0)

contact effects Evidence of baking, passage of hot fluids, or some other manifestation of the thermal and chemical effects of a magma chamber or a hot volcanic unit, as expressed in changes to adjacent wall rocks. (4.1)

contact metamorphism Metamorphism that principally involves heating of the rocks next to a magma or hot igneous material. (3.12)

continental collision A plate-tectonic boundary where two continental masses collide; also known as a continent-continent convergent boundary. (5.7)

continental crust The type of Earth's crust that underlies the continents and the continental shelves; average granitic composition, but includes diverse types of material. (5.1)

continental divide The boundary in North America between the area that drains into the Pacific Ocean and the area that drains into the Atlantic or Arctic oceans. (12.3)

continental drift The concept of the movement of continents and other landmasses across the surface of the Earth. (5.3)

continental ice sheet A large mass of ice, including glaciers, that covers a large part of a continent. (11.0)

Continental Monsoon (Dwa or Dwb) climate In the Köppen climate classification, a climate type characterized by adequate moisture in summer but not winter, a mean temperature of the coldest month below −3°C, and a hot or warm summer. (18.6)

continental platform A broad region that surrounds the continental shield and typically exposes horizontal to gently dipping sedimentary rocks. (8.7)

continental polar (cP) air mass A body of air that is relatively uniform in its cold and dry characteristics for thousands of square kilometers, originating in inland, subarctic source regions. (16.1)

continental rift A low trough or series of troughs bounded by normal faults, especially where two parts of a continent begin to rift apart. (5.6)

continental rise A gently sloping edge of a continental plate, connecting the continental slope and the abyssal plain; built up by shedding of sediments from the continental block. (8.7)

continental shelf A gently sloping, relatively shallow area of seafloor that flanks a continent, is underlain by thinned continental crust, and is covered by sedimentary layers. (5.2, 8.7)

continental shield A central region of many continents, consisting of relatively old metamorphic and igneous rocks, commonly of Precambrian age. (8.7)

continental slope A submarine slope that connects the continental shelf with deeper seafloor. (8.7)

continental tropical (cT) air mass A body of air that is relatively uniform in its hot and dry characteristics for thousands of square kilometers, originating in inland, subtropical source regions that are isolated from moisture sources. (16.1)

continent-continent convergent boundary A plate-tectonic boundary where two continental masses collide; also known as a continental collision. (5.7)

continuous spectrum A spectrum that displays all frequencies of visible light. (20.4)

contour line A line on a map or chart connecting points of equal value, generally elevation. (1.7)

contrail A condensation trail; a linear streak of clouds left behind by a jet. (15.7)

convection The vertical transfer of energy through a gas, liquid, or weak solid. (13.2)

convection cell A circulation system in a fluid that includes rising motion where the fluid is relatively hot, sinking motion where the fluid is relatively cold, and lateral movement between the areas of rising and sinking. (13.2)

convection current A flowing liquid or solid material that transfers heat from hotter regions to cooler ones, commonly involving movement of material in a loop. (5.14)

convective zone The top portion of the Sun's interior, which is marked by convective motions of gases. (20.7)

convergence (atmospheric) The horizontal piling up of air, either from two air streams colliding as they move laterally, or from an air current slowing down as it moves laterally. (14.2)

convergent boundary A plate-tectonic boundary where two plates have relative movement toward each other. (5.5)

cool sector The part of a mid-latitude cyclone that is ahead of the warm front, characterized by air that is intermediate in temperature, as compared to that ahead of and behind the cold front. (16.2)

Coriolis effect An apparent force resulting from Earth's rotation that causes moving objects to appear to be deflected to the right of their intended path in the Northern Hemisphere and to the left in the Southern Hemisphere. (14.3)

corona The outermost part of the Sun's atmosphere. (20.7)

coronal mass ejection A huge mass of ionized gases and energy that is ejected from the Sun. (20.7)

cosmic microwave background Energy found in every direction in space that is interpreted to be energy from the early universe that has been stretched into longer wavelengths. (20.12)

cosmic rays High-energy radiation that originates in space and plays a role in Earth's atmospheric processes. (1.3)

cosmologist A scientist who studies the origin, evolution, and fate of the universe. (20.12)

covalent bond A chemical bond created when two atoms share an electron. (2.11)

crater A typically bowl-shaped, steep-sided pit or depression, generally formed by a volcanic eruption or meteorite impact. (6.3)

cratered highlands *See* lunar highlands.

creep The slow, continuous movement of material, such as soil and other weak materials, down a slope. (10.12)

crescent dune A crescent-shaped dune of windblown sand with its tails pointing in the direction of the prevailing wind; also known as a barchan dune. (9.9)

crest The highest part of a wave, fold, hill, or other feature. (11.11)

Cretaceous The youngest geologic period of the Mesozoic Era. (4.6)

crevasse A fracture in a glacier, caused by internal stresses. (11.1)

cross bed A series of beds inclined at an angle to the main layers or beds. (3.8, 9.9)

cross section A diagram representing the geology as a two-dimensional slice through the land. (1.8)

cross-cutting relation The principle that a geologic unit or feature is older than a rock or feature that cross-cuts it. (4.1)

crust The outermost layer of Earth, consisting of continental and oceanic crust. (5.1)

crystalline A mineral that has an ordered internal structure due to its atoms being arranged in a regular, repeating way. (2.1)

crystalline basement The crystalline (metamorphic and igneous) rocks that underlie sedimentary and volcanic rocks in an area; widely exposed in the continental shield. (8.7)

crystalline rock A rock composed of interlocking minerals that grew together; usually formed in high-temperature environments by crystallization of magma, metamorphism, or by precipitation from hot water. (2.2)

cubic A common arrangement of atoms in a mineral or the tendency for a mineral to break along three perpendicular planes. (2.4)

cuesta Any ridge with a sharp summit and one slope inclined approximately parallel to the dip of layers, resembling in outline the back of a hog; also known as a hogback. (7.3)

cumuliform cloud A family of puffy cloud types that is taller than it is wide. (15.7)

cumulonimbus cloud A cumuliform cloud type that is actively precipitating and has an extensive vertical extent, composed of liquid water near the base and ice near the top; commonly associated with severe weather. (15.7)

cumulus cloud A type of cumuliform cloud that occurs in fair weather, without extensive vertical extent, but under free-convection conditions. (15.7)

cumulus stage The first stage in a single-cell thunderstorm's life cycle, characterized by updrafts that form and grow cumuliform clouds, but without precipitation. (16.4)

cutbank A steep cut or slope formed by lateral erosion of a stream, especially on the outside bend of a channel. (12.6)

cutoff meander A new channel formed when a stream cuts through the neck of a meander. (12.6)

cyclone An enclosed area of low atmospheric pressure; can also refer to the most intense type of storm of tropical origin in the Indian Ocean basin. (14.5)

D

Darcy's law An equation used to describe the flow of a fluid through a porous material. (12.11)

dart leader A second flow of electrical energy after the first lightning stroke that flows through approximately the same pathway as the original stroke between the cloud and the ground. (16.7)

data Valid observations that can be used to develop scientific explanations. (1.13)

daughter product The element produced by radioactive decay of a parent atom; syn. daughter atom. (4.3)

Daylight Savings Time A one-hour offset of standard time used in much of the United States for the purpose of skewing daylight hours toward evening hours on the clock. (1.10)

debris avalanche A high-velocity flow of soil, sediment, and rock, commonly from the collapse of a steep mountainside. (10.12)

debris flow Downhill-flowing slurries of loose rock, mud, and other materials, and the resulting landform and sedimentary deposit; also known as a mudflow. (9.8)

debris slide The downslope movement of soil, weathered sediment, or other unconsolidated material, partly as a sliding, coherent mass and partly by internal shearing and flow. (10.12)

December Solstice The day, on or near December 21, when the North Pole (marking where the Earth's axis of rotation intersects the Earth's surface in the Northern Hemisphere) is pointed most directly away from the Sun. (13.7)

decompression melting Melting of a rock or other material due to a decrease in pressure. (6.1)

deforestation The clearing of a forested landscape. (15.5)

deformation Processes that cause a rock body to change position, orientation, size, or shape, such as by folding, faulting, and shearing. (1.5)

delta A nearly flat tract of land formed by deposition of sediment at the mouth of a river or stream. (12.8)

dendrite One or more minerals that has crystallized in a branching pattern. (4.4)

dendritic drainage pattern A drainage network in which the pattern of streams is treelike; usually an indicator that underlying rocks have similar resistance to erosion or that the streams have been flowing for a long time. (12.3)

density A measure of how much mass is present per given volume of a substance. (1.12, 13.2)

denudation The wearing away of the surface after erosion and transportation remove material that may be loosened by weathering. (9.7)

deposition (atmospheric) A change in state from a gas to a solid, which releases latent energy to the environment. (13.4)

derecho A regional windstorm characterized by nonrotating winds that can advance over huge areas quickly. (16.10)

desertification The process by which a climate becomes increasingly arid, either through natural or anthropogenic causes. (18.11)

desert pavement A natural concentration of pebbles and other rock fragments that mantle a desert surface of low relief. (9.9)

desert soil A soil developed in a desert or semiarid region, generally characterized by an accumulation of abundant calcium carbonate and by a relative lack of organic material. (10.3)

desert varnish A thin, dark film or coating of iron and manganese oxides, silica, and other materials; formed by prolonged exposure at the surface; also known as rock varnish. (9.9)

detrital *See* clastic.

Devonian A geologic period near the middle of the Paleozoic Era. (4.6)

dew A condensed liquid water drop on a solid surface. (15.2)

dew-point temperature The temperature to which a volume of air must be cooled in order for it to become saturated with water vapor, thereby causing dew to form. (15.2)

diabatic A condition of a system in which energy is entering or leaving the system. (15.4)

diamond The hardest naturally occurring mineral, commonly used as a gemstone and industrial abrasive. (2.11)

differential weathering The breakdown of rock at differing rates because of differences in mineral composition, hardness, or other resistance to weathering. (9.4)

dike A sheetlike intrusion that cuts across any layers in a host rock, commonly formed with a steep orientation. (5.10)

diorite A medium- to coarse-grained, intermediate-composition rock; the phaneritic equivalent of andesite. (3.10)

dip The angle that a layer or structural surface makes with the horizontal, measured perpendicular to the strike. (7.2)

dip slope A slope caused by the erosion of hard and soft layers parallel to the dipping layers. (7.3)

directional drilling A horizontal drilling technique that, when used with hydraulic fracturing, permits the extraction of oil and natural gas from unconventional sources such as shale deposits. (8.12)

direction of propagation The direction in which a wave moves. (20.3)

disappearing stream A stream that vanishes suddenly from the surface as it begins to flow beneath the surface, often captured by the network of pits in karst terrain. (9.12)

discharge The volume of water flowing through some stretch of a river or stream per unit of time. (12.3)

disconformity An unconformity in which the bedding planes above and below the break are essentially parallel, but recording erosion or some other interruption in the deposition of layers. (4.2)

displacement The movement of a body in response to stress, either from one place to another, or manifested as a change in shape of the body. (7.1)

dissipating stage The final stage in a single-cell thunderstorm's life cycle, characterized by downdrafts, an end of precipitation, and decay of the system. (16.4)

dissolution The process by which a material is dissolved. (3.5)

dissolved load Chemically soluble ions, such as calcium and sodium, that are dissolved in and transported by moving water, as in a stream. (12.4)

distributary system The branching drainage pattern formed when a river branches and spreads out into a series of smaller channels. (12.8)

divergence (atmospheric) The horizontal spreading apart of air, either from two air streams moving laterally in opposite directions, or from an air current accelerating as it moves laterally. (14.2)

divergent boundary A plate-tectonic boundary in which two plates move apart (diverge) relative to one another. (5.5)

Dixie Alley The zone of high tornado frequency across the United States Gulf Coast region. (16.9)

doldrums The belt of weak winds along the Intertropical Convergence Zone, between the Northeast trade winds and the Southeast trade winds. (17.1)

dolomite A carbonate mineral containing calcium and magnesium. (2.9)

Doppler shift A change in an object's emitted wavelength due to its motion. (20.6)

drainage basin An area in which all drainages merge into a single stream or other body of water. (12.3)

drainage divide The boundary between adjacent drainage basins. (12.3)

drainage network The configuration or arrangement of streams within a drainage basin. (12.3)

driving force A force that causes a tectonic plate to move. (5.13)

drumlin A commonly teardrop-shaped hill formed when a glacier reshapes glacial deposits. (11.5)

drumlin field A group of drumlins in the same area. (11.5)

dry line The boundary between a warm, dry (cT) air mass and a warm, moist (mT) air mass, in the warm sector of a mid-latitude cyclone. (16.4)

ductile deformation A type of structural deformation that is characterized by the material flowing as a solid, without fracturing; typical of the crust below about 15 kilometers and of the asthenosphere. (7.1)

dust devil A small-scale, rotating dust storm. (16.10)

dwarf star A star that is much smaller than our Sun. (20.7)

dynamic equilibrium (systems) A condition of a system in approximate steady state maintained by equal quantities of matter and energy flowing into and out of a system. (1.4)

dynamic system Any interconnected set of matter, energy, and processes in which matter, energy, or both are constantly changing their position, amounts, or form. (1.4)

dynamo An electrical generator. (5.14)

E

earth flow A flowing mass of weak, mostly fine-grained material, especially mud and soil. (10.12)

earthquake cycle The gradual accumulation of stress on a fault followed by an abrupt decrease in stress during an earthquake. (7.5)

earth-system science An approach to earth science studies that focuses on Earth as a series of dynamic systems. (1.0)

East Australian Current A north-to-south-flowing, warm surface ocean current in the western South Pacific Ocean. (17.2)

easterly wave A bend in the trade-wind flow, where low atmospheric pressure from the Intertropical Convergence Zone protrudes farther poleward than elsewhere. (16.11)

East Pacific Rise A mid-ocean ridge (divergent boundary) in the South Pacific Ocean that crosses the eastern Pacific and approaches North America. (5.2)

eccentricity The barely noncircular shape of Earth's orbit around the Sun. (11.8)

ecliptic The line across the celestial sphere that the Sun traces in a year. (20.2)

ecotone A broad transition zone between two zones of natural vegetation (biomes), and therefore between two Köppen climate types as well. (18.1)

eddy A swirl in a fluid's current that can form in both horizontal and vertical directions, characteristic of turbulent flow, as in streams or the atmosphere. (12.4)

effervescence The potential of a mineral to have a vigorous bubbling reaction when a drop of dilute hydrochloric acid (HCl) is placed on it. (2.3)

effervescing A vigorous bubbling reaction that results when a drop of dilute hydrochloric acid (HCl) is placed on a mineral like calcite. (2.3)

E horizon A light-colored, leached zone of soil, lacking clay and organic matter. (10.3)

Ekman spiral The change in the direction of flow from the surface of the ocean downward to the null point, which turns from parallel to the wind direction at the surface to an increasingly rightward (in the Northern Hemisphere) deflection with depth. (17.1)

El Niño A periodic warming of eastern equatorial Pacific surface ocean waters accompanied by excessive precipitation and flooding adjacent to South America as the Southern Hemisphere's summer approaches; also known as a warm phase of ENSO. (17.8)

El Niño–Southern Oscillation (ENSO) The combined alterations of oceanic temperatures, atmospheric pressures, and resulting oceanic and atmospheric circulations in the equatorial Pacific Ocean. (17.9)

elastic behavior The ability of a material to strain a small amount and then return to its original shape when the stress is decreased. (7.5)

electromagnetic energy Various forms of energy, including light, infrared, and ultraviolet radiation. (1.3)

electromagnetic spectrum A range of electromagnetic radiation that includes visible light, infrared, ultraviolet, X-rays, and other wavelengths. (13.5, 19.1)

electron A stable, subatomic particle with a negative charge. (2.10)

electron cloud The area most likely to contain the electrons within an atom. (2.10)

electronegativity The measure of an element's ability to attract electrons. (2.11)

electron shell A discrete "level" away from a nucleus in an atom on which a maximum number of electrons can exist, with successively higher energy content existing on more outward levels. (2.10, 13.5)

element A type of atom that has a specific number of protons and chemical characteristics. (2.6)

elevation The vertical distance of an object above or below a reference datum (usually mean sea level); generally the height of a ground point above sea level. (1.8)

elliptical galaxy A circular or ellipsoid galaxy in which the concentration of stars diminishes with increasing distance from the galaxy's center. (20.11)

elve A flat disk of dim reddish light that forms high in the atmosphere after lightning strikes. (16.7)

emergent coast A coast that forms where the sea has retreated from the land due to falling sea level or due to uplift of the land relative to the sea. (11.15)

emission-line spectrum A recording and display of individual wavelengths in EMR emitted by a given source. (20.4)

energy The capability to do work. (1.5)

Entisol A type of recently developed soil, characterized by no B horizon and little or no profile development. (10.7)

entrainment The mixing of drier air with more humid air inside a cloud, characterized by the evaporation of cloud water and the decay of a cloud. (16.4)

entrenched meander A curved canyon that represents a meander carved into the land surface. (9.10, 12.9)

environmental lapse rate The rate of change of temperature with height in a static atmosphere. (15.4)

ephemeral stream A stream that has periods during the year when it does not flow. (12.5)

epicenter The point on Earth's surface directly above where an earthquake occurs (directly above the focus or hypocenter). (7.4)

epicycles The hypothetical secondary "loops" of planetary motion that Ptolemy added to his geocentric model to explain the apparent backward or retrograde motions of planets. (20.5)

epipelagic zone The part of the saltwater biome located seaward of the coastal zone's low-tide mark and containing the vast open waters of the ocean at shallow depth, above the mesopelagic zone, where light penetrates and separates into colors. (17.11)

equator The imaginary line of latitude, equidistant between the North and South poles, that divides the Earth into halves, and is defined as the 0° parallel. (1.9)

equatorial (E) air mass A body of air that is relatively uniform in its warm and very humid characteristics for thousands of square kilometers, originating in oceanic, equatorial source regions. (16.1)

equilibrium The state of a system characterized by an equal movement of atoms and molecules between two states of matter; also used in ecosystems and other contexts. (15.1)

equilibrium line The zone in a glacier where the losses of ice and snow balance the accumulation of ice and snow. (11.2)

era A main subdivision in the geologic timescale. (4.6)

erosion The wearing away of soil, sediment, and rock through the removal of material by running water, waves, currents, ice, wind, and gravity. (9.7)

erosional remnant A mountain or hill that remains when adjacent areas have eroded to lower levels. (8.3)

eruption column A rising column of hot gases, pyroclastic material, and rock fragments ejected high into the atmosphere during a volcanic eruption. (6.4)

esker A long, narrow, sinuous ridge composed of sediment deposited by a stream flowing within or beneath a glacier. (11.5)

estuary A channel where freshwater from the land interacts with salt water from the sea and commonly is affected by ocean tides. (11.15)

evaporation A change in state from a liquid to a gas, requiring an input of energy from the environment and the absorption of that energy at the molecular level as latent energy. (13.4)

evaporation fog A type of fog that occurs when cold air overlies a warm water surface, destabilizing the atmosphere and encouraging evaporation into the colder air. (15.8)

evaporite Any type of rock, such as rock salt, that forms from the evaporation of water. (9.12)

evapotranspiration The combined effect of evaporation and transpiration in moving liquid water from the surface up to the atmosphere in its gaseous form. (1.3, 10.2)

evolution The observed changes in the fossil record or in living organisms; also used to refer to theories that help explain the observed changes. (4.5)

evolutionary diagram A block diagram, cross section, or map that shows the history of an area as a series of steps, proceeding from the earliest stages to the most recent one. (1.8)

exfoliation The processes by which a rock sheds concentric plates, such as that which occurs due to the release of pressure during exposure. (9.2)

exfoliation joint A joint that forms during exfoliation and mimics topography. (9.2)

expansion joint A joint that forms as a result of expansion due to cooling or to a release of pressure as rocks are uplifted to the surface. (9.2)

external energy Energy that comes from outside Earth, especially from the Sun. (1.3)

extrusive rock An igneous rock that forms when magma is erupted onto Earth's surface. (6.2)

eye The center of a tropical cyclone, characterized by sinking air and clear and calm conditions. (16.11)

eyewall The area of a tropical cyclone immediately surrounding the eye. (16.11)

F

Fahrenheit scale A system for indexing temperature calibrated such that water at sea level freezes at 32°F and water at sea level boils at 212°F. (13.3)

failed rift A narrow trough (rift) at a triple junction, which becomes less active and fails to completely break up a continent into separate plates. (5.12)

far side (Moon) The side of the Moon that cannot be seen from Earth. (19.4)

fault A fracture along which the adjacent rock surfaces are displaced parallel to the fracture. (5.8, 7.2)

fault block A section of rock bounded on at least two sides by faults. (7.2)

fault scarp A step in the landscape caused when fault movement offsets Earth's surface. (7.2)

faunal succession The systematic change of fossils with age. (4.5)

feedback A reinforcement or inhibitor to a change in a system. (1.4)

feldspar A very common rock-forming silicate mineral that is abundant in most igneous and metamorphic rocks and some sedimentary rocks. (3.10)

felsic mineral A light-colored silicate mineral, such as feldspar and quartz, that predominates in the upper continental crust. (2.8)

felsic rock An igneous rock with a felsic composition, including granite, a light-colored igneous rock that contains abundant feldspar and quartz. (3.10)

Ferrel cell A hypothetical atmospheric circulation cell over the mid-latitudes in each hemisphere that appears only weakly if at all in the real world. (14.9)

field capacity The maximum amount of water (in centimeters or inches of precipitation equivalent) that a soil can hold. (10.2)

fin A very narrow panel of rock formed by fracturing and weathering. (9.10)

fissure A fracture or crack on the land surface, such as that which forms by differential subsidence. (12.14)

fissure (volcanic) A magma-filled fracture in the subsurface, which typically solidifies into a dike, or a linear volcanic vent erupting onto the land surface. (6.3)

fjord A long, narrow arm of the sea contained within a U-shaped valley, interpreted to be carved by a glacier and later invaded by the sea as the ice melted and sea level rose. (11.15)

flash flood A local and sudden flood of short duration, such as that which may follow a brief but heavy rainfall. (12.10)

flood The result of water overfilling a channel and spilling out onto the floodplain or other adjacent land. (12.10)

floodplain An area of relatively smooth land adjacent to a stream channel that is intermittently flooded when the stream overflows its banks. (9.8, 12.8)

flood stage The level at which the amount of discharge causes a river to overtop its banks and spill out onto the floodplain. (12.10)

flow cell A huge, tube-shaped cell of atmospheric circulation. (14.3)

flowstone Any deposit of calcium carbonate or other mineral formed by flowing water on the walls or floor of a cave. (9.11)

foci In an ellipse, the two central points that define the shape of the ellipse. (20.5)

focus The place where an earthquake is generated; also known as a hypocenter. (7.4)

fog A cloud at the surface of the Earth. (15.7)

fold and thrust belt A belt of thrust faults and related folds. (8.5)

foliation The planar arrangement of textural or structural features in metamorphic rocks and certain igneous rocks. (3.11)

force A push or pull that causes, or tends to cause, change in the motion of a body. (7.1)

forced convection The rising motion in the atmosphere caused by horizontal convergence, an orographic barrier, or a frontal boundary, rather than by unstable conditions. (15.6)

forearc basin A sedimentary basin that lies between the volcanic arc and the trench in a convergent plate boundary. (8.4)

foreland basin A basin that occurs when crust (either continental or oceanic) is warped by the weight of thrust sheets, especially when formed between a mountain belt and continental interior. (8.4)

forest soil A soil formed in temperate climates and in forests of deciduous trees or pine trees. (10.3)

formation A rock unit that is distinct, laterally traceable, and mappable; also used as an informal term for an eroded, perhaps unusually shaped, mass of rock. (3.8, 9.11)

fossil Any remains, trace, or imprint of a plant or animal that has been preserved from some past geologic or prehistoric time. (4.4)

fracking See hydraulic fracturing

fracture A break or crack in a rock, subdivided into joints and faults. (2.3)

framework silicates A group of silicate minerals in which tetrahedra share all four oxygen atoms, forming a structure bonded well in three dimensions. (2.7)

free convection The rising motion in the atmosphere caused by unstable conditions. (15.6)

freezing A change in state from a liquid to a solid, which releases latent energy to the environment. (13.4)

freezing point The temperature at which a substance changes its state from a liquid to a solid. (13.3)

freezing rain A form of precipitation in which a snowflake produced by the Bergeron process melts on its way down to the surface, but refreezes in a cold layer upon impact with the surface, causing it to stick to the surface. (15.10)

frequency The number of waves per second passing a point. (13.5)

fringing reef A reef that fringes the shoreline of an island or continent. (8.9)

front A narrow zone separating two different air masses. (16.2)

frontal lifting The rising motion in the atmosphere caused by a warmer, less dense air mass being pushed up over a colder, more dense one. (15.6)

frost heaving The uneven upward movement and distortion of soils and other materials due to subsurface freezing of soil water. (10.9)

frost-point temperature The subfreezing temperature to which a volume of air must be cooled in order to become saturated with water vapor, thereby causing frost to form. (15.3)

frost wedging Process by which jointed rock is pried and dislodged by the expansion of ice during freezing. (3.5, 9.2)

full moon The view of the Moon that we see from Earth when the side of the Moon facing us is fully illuminated by the Sun. (20.2)

funnel cloud A tornado-like circulation that does not reach down to the surface. (16.8)

G

gabbro A medium- to coarse-grained mafic igneous rock, the phaneritic equivalent of basalt. (3.10)

gaining stream The part of a stream or river that receives water from the inflow of groundwater. (12.13)

galaxy An enormous system of stars and other objects that stay together because they are under the influence of each others' gravity. (20.11)

galena A lead sulfide mineral with a high specific gravity and distinctive metallic-gray cubes. (2.9)

gamma ray The electromagnetic radiation of the shortest wavelengths (less than 10^{-11} micrometers). (13.5)

garnet A fairly common silicate mineral with a distinctive shape but nearly any color (red is most common). (2.8)

gas A state of matter in which the energy content is high enough that molecules break apart from each other and move freely at high speeds. (13.2)

gas giant A large, gas-dominated planet, including the four outer planets of our solar system (Jupiter, Saturn, Uranus, and Neptune). (1.11)

Gelisol A type of soil situated above permafrost, characterized by waterlogged conditions and organic matter that is slow to decompose. (10.6)

general circulation model (GCM) A type of complicated, computer-based model that predicts global atmospheric circulation using physically based and statistically based equations incorporating processes and feedbacks in the atmosphere, hydrosphere, cryosphere, lithosphere, and biosphere. (18.12)

geocentric model An early model of the solar system and universe in which Earth is at the center and all other objects revolve around it. (20.5)

geologic map A map that shows the distribution, nature, and age relationships of rock units, sediments, structures, and other geologic features. (1.7)

geology The study of planet Earth and other solid planetary objects, including their materials, processes, products, and history. (1.1)

geomagnetic polarity timescale A chronology based on the pattern and numeric ages of reversals of Earth's magnetic field. (5.14)

geoscientists Scientists such as geologists, geographers, climatologists, and oceanographers whose main focus is on the study of Earth systems. (1.1)

geothermal energy Energy that can be extracted from Earth's internal heat. (18.13)

geyser A type of hot spring that intermittently erupts fountains of hot water and steam. (12.13)

glacial drift Any sediment carried by ice, icebergs, or meltwater. (11.3)

glacial erratic A rock fragment carried by moving ice and deposited some distance from where it was derived. (11.3)

glacial groove A very large, deep scratch mark amid an otherwise smooth polished surface created by glacial abrasion. (11.3)

glacial period A time interval when glaciers were abundant. (11.7)

glacial striation A scratch mark amid an otherwise smooth polished surface created by glacial abrasion. (11.3)

glacier A moving mass of ice, snow, rock, and other sediment. (11.1)

glaciofluvial deposition Dropping of sediment by glacial streams. (11.3)

global cooling Decreasing global atmospheric and oceanic temperatures as measured or inferred from some point in the past to the present. (18.10)

global outgoing longwave-radiation budget The apportioning of upward-directed longwave radiation to various components of the global system on an annual basis. (13.12)

global shortwave-radiation budget The apportioning of shortwave radiation to various components of the global system on an annual basis. (13.11)

global warming Increasing global atmospheric and oceanic temperatures as measured or inferred from some point in the past to the present. (18.9)

Glossopteris A seed-bearing plant that lived before 250 million years ago, whose widespread fossils across many southern continents provided evidence supporting the theory of continental drift. (5.3)

gneiss A metamorphic rock that contains a gneissic foliation defined by a preferred orientation of crystals and commonly by alternating lighter and darker colored bands. (3.11)

gneissic structure A metamorphic foliation defined by a preferred orientation of crystals and generally by alternating lighter and darker colored bands representing varying percentages of different minerals. (3.11)

Gondwana Name given by geologists to the hypothetical combination of the southern continents into a single large supercontinent. (5.3, 8.11)

graben An elongate, downdropped crustal block that is bounded by faults on one or both sides. (7.2)

grade (ore) A value assigned to an ore deposit based on the concentration or percentage of the valuable commodity it contains. (2.13)

graded bed A sedimentary or volcanic layer that displays a gradational change in grain size from bottom to top. (3.8)

gradient The change in elevation for a given horizontal distance. (12.5)

granite A coarse-grained, felsic igneous rock containing mostly feldspar and quartz. (2.1)

graphite A soft, black, greasy-feeling carbon mineral. (2.11)

grassland soil A soil formed in a temperate climate beneath a surface of grass and other small plants. (10.3)

gravitational water Water within pore space in soil that is pulled by gravity deeper into the subsurface. (10.2)

gravity The force exerted between any two objects, such as that between the Sun, Earth, and Moon; syn. gravitational pull (1.3, 10.9)

Great Dying The mass extinction that marked the end of the Paleozoic Era; about 70% of all species went extinct. (4.9)

Great Oxygenation Event The first appearance of abundant free oxygen in Earth's atmosphere. (4.8)

Great Pacific Climate Shift The reversal of the Pacific Decadal Oscillation (PDO) from a cool phase to a warm phase around 1977. (17.0)

greenhouse effect The process that occurs when infrared energy radiating upward from a planetary surface is trapped by the atmosphere, warming the planetary body; greenhouse gases include water vapor, methane, and carbon dioxide. (1.3)

greenhouse gas An atmospheric gas that effectively absorbs infrared (longwave) radiation and emits it downward to the Earth, contributing to the warming of the lower atmosphere. (18.10)

Greenland Current A small, north-to-south-flowing, cold surface ocean current in the North Atlantic Ocean near the Greenland coast. (17.2)

greenstone A greenish, low-grade metamorphic rock derived from mafic or intermediate lava flows. (4.8)

Greenwich Mean Time (GMT) The reference point for world time; the time at the *Prime Meridian* in Greenwich, U.K. (1.10)

groin A low wall built out into a body of water to affect the lateral transport of sand by waves and longshore currents. (11.14)

ground heat The insolation absorbed at Earth's surface. (13.11)

ground-level ozone A harmful air pollutant produced by sunlight striking hydrocarbons, such as car exhaust, in the lower troposphere; also known as tropospheric ozone. (13.10)

ground moraine Sediment deposited from the base of the ice in a glacier and scattered around a large area. (11.5)

groundwater Water that occurs in the pores, fractures, and cavities in the subsurface. (12.0)

groundwater divide A relatively high area of the water table, separating groundwater that flows in opposite directions. (12.12)

Gulf Low A type of mid-latitude wave cyclone that forms near the Gulf of Mexico coast of the United States mainly in winter and tracks generally northeastward. (16.3)

Gulf Stream A south-to-north-flowing, warm surface ocean current in the western subtropical North Atlantic Ocean. (17.2)

gully A small channel eroded into the land surface, formed from rills that coalesce. (10.8, 12.7)

gully erosion Preferential erosion of a gully or channel, cutting it deeper. (12.7)

guyot A flat-topped seamount that has been leveled off by subsidence and wave erosion. (5.11)

gypsum A common calcium sulfate mineral, generally formed by the evaporation of water. (2.9)

gyre Any circular system of currents near the ocean's surface, caused by shearing effects of atmospheric flow around subtropical anticyclones or subpolar lows. (17.1)

H

haboob A severe, downdraft-related, non-rotating dust storm. (16.10)

hadal zone The part of the saltwater biome located seaward of the coastal zone's low-tide mark and containing the vast open waters of the ocean, at the greatest depth, below the mesopelagic zone (i.e., part of the aphotic zone). (17.11)

Hadley cell An atmospheric circulation cell with rising motion at the Intertropical Convergence Zone, poleward flow aloft to the subtropics in each hemisphere, sinking motion at the subtropical high-pressure belt, and surface motion back to the Intertropical Convergence Zone, with one cell in each hemisphere and the most vigorous motion occurring at longitudes experiencing noon. (14.6)

hailstone A ball of ice that forms in subfreezing temperatures within a cumulonimbus cloud and that subsequently falls toward the Earth's surface. (16.6)

half-life In radioactive decay, refers to the time it takes for half of the parent atoms to decay into a daughter product. (4.3)

halide mineral One of a family of minerals that consist of a metallic element, such as sodium or potassium, and a halide element, usually chlorine or fluorine. (2.9)

halite A salt mineral (NaCl) commonly formed by the evaporation of water; cleaves into cubes and has a distinctive salty taste. (2.4)

halogen Any of a group of five elements, including fluorine and chlorine, that easily bonds with another element or molecule. (13.10)

hand lens A small magnifying glass, commonly used in the field to examine a rock, mineral, or fossil. (2.1)

hand specimen A hand-sized piece of rock of a size for study, sampling, or for inclusion in a collection. (2.1)

hanging valley A glacial valley whose mouth is higher than the bottom of a larger glacially carved valley it joins. (11.4)

harsh mid-latitude (D) climate In the Köppen climate classification, a broad group of climatic types characterized by adequate moisture availability, the mean temperature of the coldest month less than −3°C, and the mean temperature of the warmest month above 10°C. (18.1)

Hatteras Low A type of mid-latitude wave cyclone that forms near the mid-Atlantic coast of the United States, mainly in winter, and tracks generally northeastward. (16.3)

hazard The existence of a potentially dangerous situation or event. (6.5)

headward erosion Erosion at the headwater of a gully as runoff water cuts the gully back toward the divide. (12.7)

headwaters The location or general area where a stream or river begins. (12.5)

heat The thermal energy transferred from one object to another. (13.3)

heat capacity The amount of energy required to change the temperature of a volume by one Kelvin, expressed in joules per cubic meter per Kelvin. (13.14)

heat flux The transfer of thermal energy from higher temperature to lower temperature objects. (13.2)

heft test An approach to determining the approximate density of a mineral by simply holding a mineral and noting how heavy it feels. (2.3)

heliocentric model A model of the solar system in which the Sun is at the center and all of the planets, including Earth, revolve around it. (20.5)

hematite An iron oxide metal that has a reddish streak and commonly forms under oxidizing conditions. (2.9)

high-mass star A star that is at least 10 times the size of our Sun. (20.7)

high-pressure area An area in the atmosphere characterized by relatively high atmospheric pressure, sinking air, and generally fair weather. (14.5)

high tide The maximum height to which water in the ocean rises relative to the land in response to the gravitational pull of the Moon; also refers to the time when such high levels occur. (11.10)

hinge The part of a fold that is most sharply curved. (7.3)

Histosol A site-dominated soil type formed in wetlands without permafrost, influenced by anaerobic conditions and the accumulation of organic matter in a thick O horizon. (10.7)

hogback Any ridge with a sharp summit and one slope inclined approximately parallel to the dip of layers, resembling in outline the back of a hog; also known as a cuesta. (7.3)

hoodoo A tall, thin rock formation produced by erosion of highly jointed rock layers, with somewhat rounded sides. (9.6)

hook echo The curved pattern on Doppler radar that is characteristic of a tornado. (16.8)

horizon (astronomy) The boundary between the visible sky and the earth. (20.2)

horizon (soil) A zone in soil that is distinct from adjacent zones, including differences in color, texture, content of minerals and organic matter, or other attributes. (10.3)

horizontal surface wave A type of surface wave in which material vibrates horizontally from side to side, perpendicular to the direction of wave propagation. (7.6)

horn A glacial erosional feature formed when three or more cirques merge by headward erosion. (11.4)

horse latitudes The regions near 30° N and 30° S. (14.8)

horst An elongate, relatively uplifted crustal block that is bounded by faults on two sides. (7.2)

Hot Desert (BWh) climate In the Köppen climate classification, a climate type characterized by inadequate moisture availability, in which potential evapotranspiration typically exceeds precipitation by a large amount, and a mean annual temperature above 12°C (54°F). (18.4)

hot spot A volcanically active site interpreted to be above an unusually high-temperature region in the deep crust and upper mantle. (5.11)

Hot Steppe (BSh) climate In the Köppen climate classification, a climate type characterized by inadequate moisture availability, in which potential evapotranspiration typically exceeds precipitation by a small or moderate amount, and a mean annual temperature above 12°C (54°F). (18.4)

H-R diagram A method of graphically displaying star attributes based on their temperature and luminosity. (20.9)

Hubble Space Telescope A remotely controlled telescope in Earth orbit that captures images without the interference of Earth's atmosphere. (20.1)

Humboldt Current A south-to-north-flowing, cold surface ocean current in the eastern South Pacific Ocean; also known locally as the Peru Current or the Chile Current. (17.2)

Humid Continental (Dfa or Dfb) climate In the Köppen climate classification, a climate type characterized by adequate moisture year-round, a mean temperature of the coldest month below −3°C, and a hot or warm summer. (18.6)

humidity The amount of water vapor in the atmosphere. (15.2)

Humid Subtropical (Cfa) climate In the Köppen climate classification, a climate type characterized by adequate moisture year-round, hot summers, and the mean temperature of the coldest month between −3°C and 18°C. (18.5)

hummocky topography A type of chaotic landscape characterized by randomly distributed humps and pits, commonly created by a landslide or less commonly by a pyroclastic eruption. (10.14)

humus A durable, partially decomposed form of organic matter. (10.1)

hurricane A regional term used in the Atlantic basin and eastern Pacific basin applied to the most intense type of storm of tropical oceanic origin. (16.11)

hydraulic fracturing A method of pumping pressurized fluids into shale deposits in order to extract natural gas and oil. (8.12)

hydraulic gradient The slope or gradient of the water table. (12.12)

hydrocarbon A gaseous, liquid, or solid organic compound composed of carbon and hydrogen. (8.12)

hydrogen bond A weak bond in water that forms between one molecule's hydrogen atom and another molecule's oxygen atom. (2.12, 15.1)

hydrograph A graph showing the change in the amount of flowing water (discharge) over time. (12.3)

hydrologic cycle The cycle representing the movement of water between the oceans, atmosphere, land, rivers and other surface water, groundwater, and organisms. (1.5)

hydrolysis A decomposition reaction involving water and commonly producing clays, as in soil. (3.5)

hydrosphere The part of Earth characterized by the presence of water in all its expressions, including oceans, lakes, streams, wetlands, glaciers, groundwater, moisture in soil, water vapor, and drops and ice crystals in clouds and precipitation. (1.4)

hydrostatic equilibrium A condition in which the outward force of energy generated within a star is balanced by the gravitational forces within the star. (20.8)

hydrothermal rock A rock that precipitated directly from hot water, either at depth or on the surface. (3.1)

hydrothermal vents *See* black smokers.

hygroscopic water Water within pore space in soil that adheres to soil particles; typically so tightly held that it is unavailable to plants. (10.2)

hypocenter The place where an earthquake is generated; also known as a focus. (7.4)

hypothesis A conception or proposition that is tentatively assumed, and then tested for validity by comparison with observed facts and by experimentation. (1.14)

I

Ice Age A period of time in which large regions of land were covered year-round with ice and snow, especially in the last 2 m.y. Coincides with the Pleistocene Epoch. (11.7)

Ice Cap (EF) climate In the Köppen climate classification, a climate type characterized by all 12 months having a climatological average temperature below freezing. (18.7)

iceberg A massive piece of ice floating or grounded in the sea or other body of water. (11.2, 17.4)

ice giants The planets Uranus and Neptune. (19.7)

Icelandic Low A prominent area of semipermanent low atmospheric pressure (cyclone) in the subpolar North Atlantic Ocean; is particularly strong in winter. (14.5)

ice sheet A mass of ice of considerable thickness and more than 50,000 km² in area, forming a nearly continuous cover of ice and snow over a land surface. (11.1)

Ideal Gas Law The relationship between pressure, density, and temperature in a gas. (14.1)

igneous rock A rock that formed by solidification of molten material (magma). (3.1)

imperial system The conventional English system of measurement that includes units of inches, feet, and miles. (20.6)

Inceptisol A type of recently developed soil characterized by mild weathering and the beginning of a weak B horizon, exhibiting similar characteristics to Entisols but generally older and having better horizonation. (10.7)

index contour A dark line on a topographic map that helps emphasize the broader elevation patterns of an area and allows easier following of lines across the map; on most topographic maps, every fifth line is an index contour. (1.7)

inertia The tendency for an object to remain in its current state of motion, whether it is in motion or at rest. (20.5)

infiltration Water and other fluids that seep into the ground through open pores, fractures, and cavities in soil and rocks. (12.1)

infrared energy A form of electromagnetic energy with longer wavelengths than visible light; much of the Sun's light that reaches Earth converts into this type of energy. (1.3, 13.5)

inner core The solid central part of Earth's core, extending from a depth of about 5,100 km to the center of the Earth (6,371 km); its radius is about one-third of the whole core. (5.1)

inner planets The four planets closest to the Sun (Mercury, Venus, Earth, and Mars); syn. terrestrial planets. (19.3)

inorganic Pertaining to a compound that is not produced by living organisms or that chemically contains no carbon. (2.1)

insolation The energy transmitted from the Sun to Earth, as incoming solar radiation. (13.6)

interferometry The use of separate telescopes to gather images and combine them into a single image to achieve the effect of a larger telescope. (20.1)

interglacial period A time during an ice age when glaciers are melting, retreating, or diminished in extent. (11.7)

intermolecular force A bond that occurs when several types of weak bonds attract a molecule (a combination of atoms) to another molecule. (2.11)

internal energy Energy that comes from within Earth and includes both the heat energy trapped from when the planet formed and the heat produced by radioactive decay. (1.3)

International Date Line The imaginary line that generally runs north-south but zigzags through the Pacific Ocean near the 180° meridian; locations west of this line are one day later in the calendar than locations east of this line. (1.10)

interstellar cloud A cloud of gases, plasma, and other matter from which stars may form. (20.8)

intertidal zone The area including the shoreline and a strip of adjacent land and water; specifically, the part of the saltwater biome nearest to the shore, characterized by wave action, temperature variations, and changes in substrate; also known as the coastal zone. (17.11)

Intertropical Convergence Zone (ITCZ) The belt of relatively low surface atmospheric pressure around the equatorial region, characterized by the collision of the northeast and southeast trade winds and rising motion. (14.3)

intrusive rock An igneous rock that solidified from magma below Earth's surface; syn. plutonic rock. (6.2)

inundation regime The scenario of coastal inundation during which the height of the swash zone exceeds the tops of coastal dunes. (11.17)

inverse-square law The law that states that the output we detect from a distant object, such as the brightness of its light, decreases by the square of its distance from us. (20.6)

ion A charged atom. (2.10, 17.4)

ionic bond Chemical bond formed because of the attraction of two oppositely charged ions, such as by the loaning of one or more electrons from one ion to another. (2.11)

ionized gases Gases whose nuclei have lost their electrons due to extremely high temperatures. (20.7)

iron formation A rock composed of millimeter- to centimeter-thick layers of iron-bearing minerals, especially iron oxide, commonly with quartz. (3.7)

irregular galaxy A galaxy with an irregular shape. (20.11)

island and seamount chain A series of islands and submarine peaks in a long, commonly linear belt. (5.2)

island arc A generally curved belt of volcanic islands above a subduction zone; also used as an adjective to refer to this setting. (5.2)

isobar A line connecting locations having the same atmospheric pressure; usually shown for sea-level-corrected pressures. (14.2)

isostasy The condition of equilibrium, comparable to floating, of the crust above the solid mantle. (5.1)

isostatic rebound Uplifting caused by the removal of weight on top of the crust, as when an ice sheet melts away or when erosion strips material off the top of a thick crustal root of a mountain. (8.2)

isotope One of two or more species of the same chemical element but differing from one another by having a different number of neutrons. (4.3)

isotopic dating The process of dating rocks using radioactive decay. (4.3)

J

jet stream A fast-flowing current of air, often near the tropopause. (14.9)

jetty An engineering structure built from the shore into a body of water to redirect current or tide, for example, to protect a harbor. (11.14)

joint A fracture in a rock where the rock has been pulled apart slightly, without significant displacement parallel to the fracture. (7.2)

Joule A unit of measurement of work or energy. (13.3)

June Solstice The day, on or near June 21, when the North Pole (marking where the Earth's axis of rotation intersects the Earth's surface in the Northern Hemisphere) is pointed most directly toward the Sun. (13.7)

Jupiter The largest planet in the solar system. (19.5)

Jurassic A geologic period in the middle of the Mesozoic Era. (4.6)

K

kame A hill formed where meltwater in stagnant ice deposited sediment in crevasses or in the space between the glacier and the land surface. (11.5)

karst terrain A topography characterized by sinkholes, caves, limestone pillars, poorly organized drainage patterns, and disappearing streams; generally formed from the dissolution of limestone or other soluble rocks. (3.13)

katabatic wind A local-to-regional-scale wind that blows from the peak of a hill or mountain down to the valley; also known as a mountain breeze. (14.4)

Kelvin scale An absolute system for indexing temperature in which all molecular motion ceases at 0 K, water freezes at 273 K, and water at sea level boils at 373 K. (13.3)

Kepler's first law of planetary motion The law that states that planetary orbits are elliptical, with the Sun at one focus. (20.5)

Kepler's second law of planetary motion The law that states that the speed of a planet in its orbit varies in such a way that a line from the planet to the Sun sweeps across equal areas in equal amounts of time. (20.5)

Kepler's third law of planetary motion The law that states that the square of the time needed for one revolution of a planet around the Sun (in years) is proportional to the cube of the semimajor axis of its orbit (in astronomical units). (20.5)

kerogen A thick substance composed of long chains of hydrocarbons. (8.12)

kettle A pitlike depression in glacial deposits, commonly a lake or swamp; formed by the melting of a large block of ice that had been at least partly buried in the glacial deposits. (11.5)

kettle lake A body of water occupying a kettle. (11.5)

kimberlite pipe A steep, cylindrical to funnel-shaped volcanic conduit that contains diamonds dispersed in an igneous rock. (2.14)

kinetic energy The energy possessed by a solid, liquid, or gas expressed by the motion of that object. (1.3, 13.2)

Köppen climate classification A scheme for categorizing climates based on the annual regime of temperature and precipitation, with the categories of climate types corresponding to the major realms of natural vegetation. (18.1)

K-P extinction Refers to the extinction of the dinosaurs and many other animals at the end of the Mesozoic Era, between the Cretaceous Period (K) and the Paleogene Period (P); traditionally referred to as the K-T extinction, separating the Cretaceous Period (K) and the Tertiary (T). (4.10)

K-T extinction *See* K-P extinction.

Kuiper Belt A zone of planetary objects beyond the orbit of Neptune. (19.7)

Kuroshio Current A south-to-north-flowing, warm surface ocean current in the western North Pacific Ocean. (17.2)

L

Labrador Current A small, north-to-south-flowing cold surface ocean current in the northern Atlantic Ocean between Greenland and eastern Canada. (17.2)

laccolith A bulge-shaped igneous body that has domed and tilted overlying layers and that is observed or interpreted to have a relatively flat floor. (6.12)

lagoon A relatively sheltered area of water near a shoreline. (3.4)

lahar A mudflow mostly composed of volcanic-derived materials and generally formed on the flank of a volcano. (6.7)

land breeze A local-to-regional-scale wind that blows from land to sea, typically at night. (14.3)

land bridges Hypothesized ridges, now barely submerged in the oceans, proposed to explain the worldwide migration of animals from their geographic origin; most proposed ridges have been disproved by more recent data. (1.14, 5.3)

land cover The natural or human-made features on the surface of the Earth, such as bare rock, water, soil, forest, or concrete. (15.5)

lander A spacecraft that lands on a planetary object and collects images and other data. (19.1)

landfall The arrival of a tropical cyclone on land. (16.11)

landslide A general term for the rapid downslope movement of soil, sediment, bedrock, or a mixture of these; also the feature or landform formed by this process. (10.10)

La Niña A periodic cooling of the eastern equatorial Pacific surface ocean waters and reduction in precipitation adjacent to South America as the Southern Hemisphere's summer approaches; also known as a cold phase of ENSO. (17.9)

lapse rate The change in temperature of an air parcel or the atmosphere in general, per unit change in elevation. (15.4)

latent energy The energy stored or released as a consequence of a change between the solid, liquid, or gaseous states of matter. (13.2)

latent heat flux The flow, or transfer, of latent heat in the atmosphere by convection and advection. (13.12)

latent heat of fusion The energy absorbed as a consequence of a change in state from a solid to liquid state of matter (melting), or released as a consequence of a change in state from a liquid to solid state of matter (freezing). (13.4)

latent heat of sublimation The energy absorbed as a consequence of a change in state from a solid to gaseous state of matter (sublimation), or released as a consequence of a change in state from a gaseous to solid state of matter (deposition). (13.4)

latent heat of vaporization The energy absorbed as a consequence of a change in state from a liquid to gaseous state of matter (evaporation), or released as a consequence of a change in state from a gaseous to liquid state of matter (condensation). (13.4)

lateral moraine Sediment carried in and deposited along the sides of a glacier. (11.4)

laterite A type of tropical soil rich in iron (Fe) and aluminum (Al) oxides, commonly giving the soil a deep red color. (10.3)

latitude A measure of position north or south of the equator; all locations sharing the same latitude in a hemisphere fall along an imaginary east-west line that encircles the Earth. (1.9)

Laurasia The inferred northern supercontinent that existed in the Mesozoic and included North America, Europe, and Asia. (8.11)

lava dome A dome-shaped mountain or hill of at least partly solidified lava, generally of felsic to intermediate composition. (6.4)

lava flow A volume of magma that erupts onto the surface and flows downhill from the vent of a volcano; also the solidified body of rock formed by this magma. (6.4)

lava fountain A shooting stream of molten lava propelled into the air by pressure and escaping gases during an explosive volcanic eruption. (6.4)

lava tube A long, tubular opening under the crust of solidified lava and representing a partially emptied subsurface channel of lava. (6.6)

leaching The separation or dissolution of soluble constituents from a rock, sediment, or soil by percolation of water. (10.3)

leeward The downwind side of a slope. (14.4)

levee A long, low ridge of sediment deposited by a stream next to the channel; some levees are built by humans to keep floodwaters from spilling onto a floodplain. (12.9)

lichens A symbiotic combination of algae and fungi. (9.5)

lifting condensation level The elevation at which an air mass next to a mountain reaches its dew point. (15.6)

lightning The bright flash observed at the instant that electrical energy is transferred through the atmosphere, after the return stroke is established. (16.7)

light pollution Artificial light sources that obscure an earthbound observer's view of features in the night sky. (20.1)

light-year A unit of distance equivalent to the distance light travels through a vacuum in one year. (20.6)

limb The planar or less curved parts of a fold on either side of the hinge. (7.3)

limestone A sedimentary rock composed predominantly of calcium carbonate, principally in the form of calcite, and which may include chert, dolomite, and fine-grained clastic sediment. (3.7)

lineation A linear structure in a metamorphic rock. (3.11)

liquefaction Loss of cohesion when grains in water-saturated soil or sediment lose grain-to-grain contact, as when shaken during an earthquake. (7.10)

liquid A state of matter in which a collection of mobile atoms and molecules more or less stay together but are not held in the rigid form of a solid. (13.2)

lithification The conversion of unconsolidated sediment or volcanic ash into a coherent, solid rock, involving processes such as compaction and cementation. (3.2)

lithosphere Earth's upper, rigid layer composed of the crust and uppermost mantle. (1.4)

lithospheric mantle The part of the uppermost mantle that is in the lithosphere. (5.1)

loading The process by which weight is added to the lithosphere. (11.16)

loam A soil texture consisting of a balance between particles of different sizes, enhancing the soil's ability to support agriculture. (10.5)

local mountain A mountain that is supported by the strength of the crust and is too small to be accompanied by a regional increase in crustal thickness. (8.3)

loess An essentially unconsolidated sediment consisting predominantly of silt, interpreted to be windblown dust, commonly of glacial origin. (9.9)

longitude An imaginary north-south-running line that passes through the North and South poles, encircles the Earth, and indicates how far east or west of the Prime Meridian a location is. (1.9)

longitudinal dune A type of sand dune that is linear or only very gently curving for many kilometers, oriented parallel to prevailing winds. (9.9)

longshore current A current, generally in an ocean, flowing more or less parallel to a coastline. (11.12)

longshore drift The process by which sand and other sediment move parallel to a coast with the prevailing longshore current. (11.12)

longwave radiation The electromagnetic energy at wavelengths greater than about 4 micrometers; corresponds to the wavelengths at which the Earth emits nearly all of its energy. (13.5)

losing stream Part of a stream or river that loses water from outflow to groundwater. (12.13)

Love wave *See* horizontal surface wave.

low-mass star A star, such as our Sun, that is in the small to average size range in the universe. (20.7)

low-pressure area An area in the atmosphere characterized by relatively low atmospheric pressure, rising air, and commonly stormy weather. (14.5)

low tide The lowest height to which water in the ocean drops relative to the land in response to the gravitational pull of the Moon; also refers to the time when such low levels occur. (11.10)

luminosity The amount of energy per second emitted by a star. (20.6)

lunar eclipse An astronomical phenomenon that occurs when Earth is directly between the Sun and the Moon, and Earth's shadow covers the Moon. (20.2)

lunar highlands High elevations on the Moon that contain a light-colored, heavily cratered material; syn. cratered highlands. (19.4)

luster The reflection of light from the surface of a mineral, especially its quality and intensity; the appearance of a mineral in reflected light. (2.3)

M

mafic mineral A generally dark-colored, silicate mineral with a high magnesium (Mg) and iron (Fe) content. (2.8)

mafic rock A generally dark-colored igneous rock with a mafic composition rich in magnesium and iron. (3.10)

magma Molten rock, which may or may not contain some crystals, solidified rock, and gas. (3.1)

magma chamber A large reservoir in the crust or mantle that is occupied by a body of magma. (6.11)

magnetic reversal A switch between normal polarity and reversed polarity of the Earth's magnetic field. (5.14)

magnetite An iron oxide mineral that is typically black and is strongly magnetic. (2.9)

magnetometer An instrument used to measure the direction and strength of magnetism in rocks and other materials. (5.14)

magnitude A measure of the amount of energy released by an earthquake; used to compare sizes of earthquakes. (7.7)

main sequence A diagonal central belt of star attributes on the H-R diagram that connects hot, luminous stars with cooler, less luminous stars. (20.9)

mantle The most voluminous layer of Earth; located below the crust and above the core. (5.1)

mantle convection Movement of mantle material in response to variations in density, especially those caused by differences in temperature. (5.13)

marble A metamorphic rock composed of recrystallized calcite or dolomite. (2.9, 3.12)

March Equinox The day, on or near March 21, when Earth's axis of rotation is pointed exactly sideways relative to the Sun, and the durations of daylight and darkness are equal throughout the world. (13.7)

mare A dark, low-lying, relatively smooth area on the Moon consisting of basalt. (19.4)

maria Plural of mare. (19.4)

marine evaporite deposit A salt accumulation formed when seawater evaporates, leaving behind a residue of material that was dissolved in the water. (8.8)

marine terrace A platform that was cut or constructed by waves but is now elevated above sea level; commonly covered by a thin veneer of marine sediment. (11.13)

Marine West Coast (Cfb or Cfc) climate In the Köppen climate classification, a climate type characterized by adequate moisture year-round and the mean temperature of the coldest month between −3°C and 18°C, with warm or mild summers. (18.6)

maritime polar (mP) air mass A body of air that is relatively uniform in its cool/cold and humid characteristics for thousands of square kilometers, originating in oceanic, subarctic source regions. (16.1)

maritime tropical (mT) air mass A body of air that is relatively uniform in its warm and humid characteristics for thousands of square kilometers, originating in oceanic, subtropical source regions. (16.1)

Mars Fourth planet out from the Sun. (19.3)

mass extinction The disappearance of many species and families of creatures in a geologically short period of time. (4.6)

mass spectrometer An instrument used to measure the abundance of different atoms and isotopes in a material, such as a rock or mineral to be numerically dated. (4.3)

mass wasting The downward movement of material on slopes under the influence of gravity. (10.9)

matrix The finer grained material enclosing or filling the areas between larger grains, crystals, or fragments of a rock. (3.7)

matter Any solid, liquid, or gas that occupies space and has mass. (1.5)

maturation The process by which organic material becomes petroleum upon burial and heating. (8.12)

mature stage The second stage in a single-cell thunderstorm's life cycle, characterized by continuing updrafts along with downdrafts, precipitation, lightning, and thunder. (16.4)

meander A sinuous curve or bend in the course of a stream or river. (12.6)

meandering stream A stream that has a strongly curved channel. (9.10, 12.6)

meander scar A crescent-shaped feature in the landscape that indicates the former position of a river meander. (12.8)

mechanical turbulence The vertical motion generated by friction or vertical wind shear. (16.8)

medial moraine Sediment carried in the center of a glacier, representing where two smaller glaciers joined; also refers to the deposited sediment and resulting landform. (11.4)

Mediterranean (Csa or Csb) climate In the Köppen climate classification, a climate type characterized by adequate moisture in winter but not in summer, hot or warm summers, and the mean temperature of the coldest month between −3°C and 18°C. (18.5)

megathrust A major thrust fault, representing the boundary between the subducted slab and overriding plate. (7.9)

melting A change in state from a solid to a liquid, requiring an input of energy from the environment and the absorption of that energy at the molecular level as latent energy. (1.5)

melting curve On a graph of temperature versus pressure, the line that separates conditions producing magma from those producing solid rock. (6.1)

Mercury The closest planet to the Sun. (19.3)

meridian Any line of longitude. (1.9)

mesa A broad, flat-topped and steep-sided, usually isolated hill or mountain. (6.3, 8.3)

mesocyclone A rotating vortex embedded within a supercell thunderstorm. (16.4)

mesopause The transition zone between the mesosphere and overlying thermosphere. (13.9)

mesopelagic zone The part of the saltwater biome located seaward of the coastal zone's low-tide mark and containing the vast open waters of the (epipelagic) zone, where light cannot penetrate ocean, at medium depth, below the photic zone (i.e., part of the aphotic zone). (17.11)

Mesosaurus A reptile that lived more than 150 million years ago, whose widespread fossils across many southern continents have provided evidence supporting the theory of continental drift. (5.3)

mesosphere The third layer of the atmosphere above the surface. (13.1)

Mesozoic Era A major subdivision of geologic time from 251 Ma to 65 Ma; characterized by dinosaurs. (4.5)

metallic bond A chemical bond formed when electrons are shared widely by many atoms. (2.11)

metamorphic rock A rock changed in the solid state by temperature, pressure, deformation, or chemical reactions that modified a preexisting rock. (3.1)

meteorite A fragment of a meteoroid that has fallen to a planetary surface. (1.14)

metric system The system of measurement used in science and in most countries that distinguishes units by multiples of 10. (20.6)

microburst The brief but strong, downward-moving winds in a "starburst" pattern out from a central point; sometimes referred to as straight-line winds. (16.10)

microwave The electromagnetic radiation of wavelengths between 1 millimeter and 10 meters. (13.5)

mid-latitude cyclone A large, migrating storm system composed of a central cyclone, with cold front and warm fronts which may become stationary at various times and possibly undergo occlusion, emanating from it. (16.3)

mid-ocean ridge A long mountain range on the floor of the ocean, associated with seafloor spreading. (5.2)

Milankovitch cycles Periodic variations in Earth's orbit and tilt, interpreted to influence Earth's climate. (11.8)

Milky Way Galaxy The spiral galaxy in which Earth is located; only one of countless galaxies in the universe. (19.0)

mineral A naturally occurring, inorganic, crystalline solid with a relatively consistent composition. (2.1)

mineral deposit A mass of naturally occurring rocks and other materials that are especially rich in some commodity that might be valuable. (2.13)

mineral wedging The growth of minerals that exert an outward force that can fracture rock or loosen grains. (3.5)

mineralize To convert to or impregnate with mineral material, as in the processes of ore deposition and of fossilization. (2.13)

mineralogists Geologists and other scientists who study minerals. (2.4)

Mississippian A geologic period near the middle of the Paleozoic Era. (4.6)

mnemonic A memorization device in which the first letter of each word in an easily remembered phrase represents the first letter of each term in an ordered list. (4.6)

Modified Mercalli Intensity Scale An earthquake intensity scale, recording the relative amount of damage and how the earthquake was perceived by people. (7.7)

Mohs Hardness Scale Consists of ten common minerals ranked in order of increasing relative hardness, from 1 to 10. (2.3)

Mollisol A fertile, mid-latitude type of soil, rich in calcium carbonates and organic material, with little leaching and a humus-rich A horizon. (10.6)

moment magnitude (Mw) A measure of the amount of energy released by an earthquake. (7.7)

monocline A fold defined by local steepening in gently dipping layers. (7.3)

moraine Sediment carried by and deposited by a glacier; also refers to the resulting landform. (11.4)

mouth The location where a stream, river, or canyon ends, such as where a river enters the sea. (12.5)

Mozambique Current A north-to-south-flowing, warm surface ocean current in the western Indian Ocean. (17.2)

mud A combination of silt and clay. (3.6)

mudflow Downhill-flowing slurries of loose rock, mud, and other materials, and the resulting landform and sedimentary deposit; also known as a debris flow or earthflow. (10.12)

multicell thunderstorm A cluster of individual thunderstorms that interact as part of an organized weather system to produce severe weather. (16.4)

muscovite A light-colored, sheet silicate mineral that is part of the mica family. (2.8)

N

native mineral Rock-forming mineral that contains only a single element. (2.6)

natural bridge An archlike opening in a panel of rock formed by the eroding actions of running water. (9.10)

natural cement *See* cement.

natural gas Hydrocarbons that exist as gas or vapor at ordinary temperatures and pressures. (8.12)

natural selection The process by which the organism best adapted to its environment tends to survive and transmit its genetic characteristics to the population; one theory for natural evolution. (4.5)

neap tide Lower-than-average high tides and higher-than-average low tides caused when the Sun's gravity partially offsets the effects of the Moon's gravity. (11.10)

near side (Moon) The side of the Moon that always faces Earth. (19.4)

nebula A shapeless cloud of gas and dust in space. (19.0)

negative feedback A response to a change in a system that diminishes and dampens more of the same type of change. (1.4)

negative magnetic anomaly A weak signal from magnetic instruments, indicative of rock formed during a time of reversed polarity. (5.14)

Neogene The youngest geologic period of the Cenozoic Era. (4.6)

Neptune The eighth and last planet from the Sun. (19.7)

neutral phase (ENSO) Any atmospheric and oceanic condition in the equatorial Pacific Ocean that is near-normal, without much evidence of anomalous warming (El Niño) or cooling (La Niña) conditions in the eastern equatorial Pacific. (17.9)

neutron A subatomic particle that contributes mass to a nucleus and is electrically neutral. (2.10)

neutron star The dense collection of neutrons that remains after the iron core of a high-mass star has collapsed and produced a supernova explosion. (20.10)

new moon Occurs when the side of the Moon facing Earth is not illuminated by the Sun. (20.2)

Newton's first law of motion The law that states that a body will remain in its current state of motion or rest unless acted on by a force. (20.5)

Newton's second law of motion The law that states that the acceleration an object receives will be proportional to the net force applied to it and inversely proportional to its mass. (20.5)

Newton's third law of motion The law that states that when one body exerts a force on a second body, the second body exerts an equal force on the first in the opposite direction. (20.5)

nickpoint A point at which a stream has an abrupt change in gradient. (12.7)

nimbostratus cloud A type of wide, flat cloud that is actively precipitating. (15.7)

nomograph A type of graph, used in seismology, to determine the local magnitude of an earthquake. (7.7)

nonclastic rock A sedimentary rock not composed of clasts. (3.7)

nonconformity An unconformity in which the older rocks below the unconformity are not layered. (4.2)

nonsilicate Mineral or other material that does not include silicon. (2.9)

Nor'easter A Hatteras Low that has tracked near or over the northeastern United States. (16.3)

normal component The part of the gravitational force that pulls a mass against the sloping surface on which it sits. (10.9)

normal fault A fault in which the block above the fault moves down relative to the block below the fault. (7.2)

normal-fault basin A low area that has been downdropped by one or more normal faults. (8.4)

normal polarity The orientation of the Earth's magnetic field such that the field flows from south to north, causing the magnetic ends of a compass needle to point toward the north. (5.14)

normal temperature gradient An atmospheric condition in which temperature decreases as height increases. (13.9)

North Atlantic Deep Waters The deep, very saline, polar waters that descend in the Atlantic Ocean near Iceland and flow at depth well south of the equator. (17.5)

North Atlantic Drift The northern extension of the Gulf Stream, which moderates the climate of northwestern Europe. (17.2)

north celestial pole A projection of Earth's North Pole onto the celestial sphere. (20.2)

North Pacific Current A west-to-east-flowing, surface ocean current in the high latitudes of the Northern Hemisphere. (17.2)

northeast trade winds The prevailing surface atmospheric circulation that exists in the tropical latitudes of the Northern Hemisphere, characterized by flow from northeast to southwest. (14.3)

no-till agriculture A technique to minimize soil erosion, whereby the previous year's crop residue is left intact and the field is not plowed into rows. (10.8)

nuclear fission The spontaneous breaking apart of atoms during radioactive decay, in the process releasing a large amount of energy. (1.11)

nuclear fusion The combination, or fusion, of two nuclei to form a heavier nucleus, in the process releasing a large amount of energy. (1.11)

nucleosynthesis The production of heavy chemical elements by the nuclear fusion of two or more lighter elements. (20.9)

nucleus A particle composed of protons and generally neutrons in the core of an atom. (2.10, 13.5)

null point The depth in the ocean at which the effects of wind stress vanish; usually at a depth of around 100 meters. (17.1)

O

obsidian A generally gray to black, shiny volcanic glass, usually of felsic composition. (3.10)

occluded front A narrow zone that evolves late in the life cycle of the mid-latitude cyclone, marking the location where the cold front "catches up with" the warm front and lifts the warmer air above the colder air. (16.3)

ocean acidification The process by which atmospheric carbon dioxide dissolves in seawater, reducing its pH. (18.10)

ocean-continent convergent boundary A plate-tectonic boundary where an oceanic plate converges with a continental plate, generally expressed by subduction of the oceanic plate beneath the continent. (5.7)

oceanic crust The type of thin, mafic crust that underlies the ocean basins. (5.1)

oceanic fracture zone A crack or step in elevation of the seafloor that formed as a transform fault along a mid-ocean ridge but is no longer a plate boundary. (5.2)

oceanic plateau A broad, elevated region in the ocean floor. (5.2)

ocean(ic) trench A narrow, steep-sided, elongate depression of the deep seafloor, formed by bending down of a subducting oceanic plate at a convergent plate boundary; includes the deepest parts of the ocean. (5.2)

ocean-ocean convergent boundary A plate boundary where two oceanic tectonic plates converge. (5.7)

ocean oscillation A temperature oscillation between two large areas of an ocean, such that warmer than normal sea-surface temperatures occur simultaneously in one part of the ocean while colder than normal sea-surface temperatures are occurring in the other; or a temperature anomaly over a single large part of the ocean, in which above-normal sea-surface temperatures persist for a period of many years followed by the dominance of below-normal sea-surface temperatures. (18.10)

octahedron A polyhedron with eight faces. (2.4)

O horizon An upper, organic-rich soil horizon composed of dead leaves and other plant and animal remains. (10.3)

oil sand A sand or other porous sediment impregnated by petroleum. (8.12)

oil seep The seepage of liquid petroleum at the surface. (8.12)

olivine A green iron-magnesium silicate mineral that composes much of the upper mantle and also occurs in mafic and ultramafic igneous rocks. (2.8)

open system Any interconnected set of matter, energy, and processes that allows material and energy to be exchanged outside itself. (1.4)

orbital plane The imaginary plane in which the Earth revolves around the Sun. (13.7, 20.2)

Ordovician A geologic period in the early part of the Paleozoic Era. (4.6)

ore A rock, sediment, or other material that can be mined for a profit. (2.13)

ore deposit A mineral deposit that contains enough of a commodity to be mined at a profit. (2.13)

original horizontality The principle that most sediments and many volcanic units are deposited in layers that originally are more or less horizontal. (4.1)

orographic effect The rising motion and enhancement of precipitation on the windward slopes of a mountain or other topographic feature. (15.6)

outer core The molten outer part of Earth's core, extending from a depth of 2,900 km to 5,100 km. (5.1)

outwash plain An area in front of a glacier where rivers and streams deposit glacially produced sediment. (11.6)

overpumping Pumping of groundwater at a rate that causes severe lowering of the water table in the aquifer. (12.14)

overturning The mixing of waters from different depths in a large body of water; also known as turnover. (17.5)

overwash The process by which a wave or storm surge moves material from the side of a landmass nearest to the sea to an area farther from the sea. (11.17)

overwash regime The scenario of coastal inundation during which the height of the swash zone reaches the tops of coastal dunes. (11.17)

oxbow lake An isolated, curved lake formed when a cutoff meander is filled with water. (12.6)

oxidation The chemical process during which a material combines with oxygen. (3.5)

oxide Any member of a group of minerals that consist of oxygen bonded with a metallic element, like iron. (2.6)

Oxisol A type of soil formed in wet tropical conditions and characterized by deep weathering, leaching, and oxidation, usually high in iron and aluminum and low in nutrients. (10.6)

oxygenation events Points in Earth's history during which levels of free oxygen in the atmosphere increased in a relatively short time. (4.8)

ozone A molecule consisting of three oxygen atoms, concentrated in the stratosphere where it absorbs harmful ultraviolet radiation, but also existing near the Earth's surface where it is a pollutant. (13.10)

ozone layer The lower part of the stratosphere where ozone is formed and photodissociated quickly. (13.10)

P

Pacific Decadal Oscillation (PDO) An oscillation in sea-surface temperatures in the North Pacific Ocean, in which temperatures are warmer than normal either near the Alaska coast or over the central North Pacific, while they are simultaneously below normal at the other location. (17.0)

Pacific Decadal Oscillation (PDO) index A numerical indicator of the degree to which the Pacific Decadal Oscillation is in its warm phase, with warmer than normal sea-surface temperatures in the eastern North Pacific near the Alaska coast (when the index is positive), or in its cool phase, which is characterized by cooler than normal sea-surface temperatures in the eastern North Pacific near the Alaska coast (when the index is negative). (17.0)

Pacific Ring of Fire The belt of frequent volcanic activity from the southwestern Pacific, through the Philippines, Japan, and Alaska, and then down the western coasts of North and South America, resulting from subduction of oceanic plates around the Pacific Ocean. (5.7)

pahoehoe A type of lava or lava flow that has a smooth upper surface or folds that form a "ropy" texture. (6.6)

Paleogene The oldest geologic period of the Cenozoic Era. (4.6)

Paleozoic Era A major subdivision of geologic time, beginning at the end of the Precambrian; from about 542 to about 251 million years ago. (4.5)

Palmer Drought Index A numerical indicator of the severity of drought conditions or of the excessiveness of water, customized to the place of interest at the time of year of interest. (15.12)

Pangaea An inferred supercontinent that existed from about 300 to about 200 million years ago and included most of Earth's continental crust. (8.11)

parallax A method of triangulation that takes advantage of the apparent shift in position of an object that is viewed from two different points. (20.6)

parallel Any line of latitude. (1.9)

parallel bedding A sequence of beds that are approximately parallel. (3.8)

parent atom An atom before it undergoes radioactive decay; syn. parent isotope. (4.3)

parent material Any minerals and rocks upon which physical and chemical weathering acts, and from which soils are derived. (9.4)

parsec A unit of astronomical distance with a length of 3.26 light-years. (20.11)

partial eclipse A location on Earth where the Moon's shadow only partially blocks the Sun during a solar eclipse. (20.2)

passive margin A continental edge that is not a plate boundary. (5.6)

passive solar The use of solar energy, involving light and infrared energy from the Sun entering a space through windows and naturally heating the inside air and mass. (18.13)

paternoster lakes A series of tarns connected by a stream. (11.4)

patterned ground Polygon-shaped outlines in the soil of Arctic areas, caused by frost heaving. (10.9, 11.6)

PDO index *See* Pacific Decadal Oscillation (PDO) index.

peak discharge The maximum volume of water per unit time, shown on a hydrograph. (12.5)

peat An unconsolidated deposit of partially decayed plant matter. (10.1)

pebble A small stone between 6 and 64 mm in diameter. (3.6)

pedalfer A temperate-climate soil whose A and B horizons contain abundant insoluble minerals, particularly iron and aluminum. (10.3)

pediment A gently sloping, low-relief plain or erosion surface carved onto bedrock, commonly with a thin, discontinuous veneer of sediment. (9.7)

pegmatite An igneous rock containing very large crystals, which may be centimeters to meters long. (3.9)

pelagic Pertaining to an organism in a saltwater biome that resides mainly suspended in the water. (17.11)

pelagic zone The part of the saltwater biome located seaward of the coastal zone's low-tide mark and containing the vast open waters of the ocean, consisting of the epipelagic, mesopelagic, and hadal zones at various depths. (17.11)

Pennsylvanian A geologic period in the latter part the Paleozoic Era. (4.6)

perched water Groundwater that sits above the main water table and generally is underlain by a layer or lens of impermeable rock that blocks the downward flow of groundwater. (12.12)

perennial stream A stream or river that flows all year. (12.5)

peridotite An ultramafic igneous rock generally containing abundant olivine, commonly with smaller amounts of pyroxene. (3.10)

perihelion The date of closest approach of Earth to the Sun during the course of its orbit, at present corresponding to a day on or near January 4. (13.6)

period 1. A time interval in the geologic timescale; a subdivision of an era. (4.6) 2. In astronomy, the time required for a planet to revolve once around the Sun. (20.5)

Periodic Table Table that organizes all the chemical elements according to the element's atomic number and electron orbitals. (2.6)

permafrost A condition in which water in the uppermost part of the ground remains frozen all or most of the time. (11.6)

permeability A measure of the ability of a material to transmit a fluid. (12.11)

Permian The last geologic period of the Paleozoic Era. (4.6)

petrified wood A piece of fossilized wood that has been replaced by silica, preserving some of the original structure of the wood. (4.4)

petroleum A general term for naturally occurring hydrocarbons, whether liquid, gaseous, or solid. (8.12)

phaneritic rock An igneous rock containing crystals that are visible to the unaided eye. (3.9)

phases of the moon The monthly cycle of the Moon in which its illumination changes on a regular schedule, going from completely dark to fully lit and back again. (19.4)

phenocrysts Crystals in an igneous rock that are larger than those around them, as in a porphyritic rock. (3.9)

photic zone The topmost layer of an aquatic system that has access to sunlight, generally to a depth of 100 meters or less. (17.7)

photodissociation The splitting apart of chemical bonds caused by exposure to radiation. (13.10)

photon A single particle of electromagnetic radiation. (20.3)

photosphere The visible surface of the Sun. (20.7)

photosynthesis The process by which plants produce carbohydrates, using water, light, and atmospheric carbon dioxide. (1.3)

photovoltaic panel A solar-energy device that converts sunlight directly into electricity. (18.13)

phyllite A shiny, foliated metamorphosed rock, intermediate in grade between slate and schist. (3.12)

physical weathering The physical breaking or disintegration of rocks when exposed to the environment. (3.5)

phytoplankton A broad group of mostly microscopic photosynthesizing organisms. (17.11)

piedmont glacier A broad glacier that forms when an ice sheet or valley glacier spreads out as it moves into less confined topography. (11.1)

pillow A rounded, pillow-shaped structure that forms when lava erupts into water. (5.10)

pillow basalt A basaltic lava flow that includes pillow structures. (5.10)

plagioclase A very common rock-forming feldspar mineral that contains sodium, calcium, or both elements. (2.8)

planet A celestial body that revolves around a sun in a solar system. (19.0)

planetary nebula A cloud formed when a dying star ejects its contents. (20.8)

plasma Ionized material such as free protons and electrons. (20.7)

plateau A broad, relatively flat region of land that has a high elevation. (5.2)

plucking The tearing away of rock by a glacier, particularly as a result of jointing caused by expansion during freeze-thaw action; common on the downflow side of a hill. (11.3)

plume An area of groundwater contamination that spreads out away from the source. (12.16)

plutoid A class of dwarf planet farther from the Sun than the orbit of Neptune. (19.7)

pluton A subsurface magma body or the mass of rock in which it solidifies; syn. intrusion. (6.11)

plutonic rock An igneous rock that solidified at depth rather than on the surface; syn. intrusive rock. (6.2)

point bar A series of low, arcuate ridges of sand and gravel deposited on the inside of a stream bend or meander. (12.6)

polar cell An atmospheric circulation cell with sinking motion near the pole, equatorward surface flow to subpolar latitudes, rising motion at the subpolar low-pressure belt, and motion aloft back to the pole, with one cell in each hemisphere and strong Coriolis deflection affecting these motions. (14.7)

polar (E) climate In the Köppen climate classification, a broad group of climatic types characterized by adequate moisture availability, the mean temperature of the coldest month less than −3°C, and the mean temperature of the warmest month below 10°C. (18.7)

polar easterlies The prevailing surface general circulation feature that exists between the pole and about 60° of latitude in each hemisphere, characterized by flow from east to west. (14.3)

polar front jet stream The fast-flowing current of air near the tropopause in each hemisphere at the boundary between the polar cell and the mid-latitudes. (14.9)

polar high The large permanent areas of high surface atmospheric pressure centered near the North Pole and South Pole. (14.7)

polar molecule A molecule, such as water, that has an uneven distribution of charges from one side to the other. (15.1)

ponding Water that accumulates in a concentrated area on the surface when the intensity of precipitation exceeds the rate of infiltration. (10.2)

pore space Any open space within rocks, sediment, or soil, including open space between grains in a sedimentary rock, within fractures, and in other cavities. (3.7)

porosity The percentage of the volume of a rock, sediment, or soil that is open space (pore space). (10.1, 12.11)

positive feedback A response to a change in a system that reinforces and amplifies more of the same type of change. (1.4)

positive magnetic anomaly A strong signal from magnetic instruments, indicative of more magnetic rock or of a rock formed during a time of normal polarity. (5.14)

potassium feldspar A very common silicate mineral that contains potassium; syn. K-feldspar. (2.8)

potential energy The energy stored within the molecular structure of a solid, liquid, or gas as a consequence of its position relative to where it could possibly move. (13.2)

potential evapotranspiration The maximum amount of water that could possibly be lost from the surface to the atmosphere over a given interval of time through the combined effect of evaporation and transpiration, if the water were abundantly available. (18.4)

potentiometric surface An imaginary surface to which groundwater would rise if allowed. (12.12)

pothole A bowl-shaped pit eroded into rock by swirling water and sediment. (12.4)

Precambrian A very long interval of geologic time, from the formation of the solid earth to the beginning of the Paleozoic; it is equivalent to 90% of geologic time. (4.5)

precession The changing orientation of the Earth's rotation axis over long periods of time. (11.8)

precious metal Gold, silver, or any minerals of the platinum group. (2.13)

precipitation (atmospheric) The water in solid or liquid form that falls from the atmosphere to the surface, including rain, freezing rain, sleet, snow, and hail. (1.5)

precipitation (chemical) The movement of a dissolved material out of the solution and into solid form. (3.7)

precipitation fog A type of fog that occurs when falling raindrops evaporate, increasing the humidity of the colder air beneath it, or when precipitation evaporates rapidly upon reaching the surface. (15.8)

pressure A force exerted on a unit area, expressed in a gas by the frequency of molecular collisions. (14.1)

pressure gradient The horizontal difference in pressure across some distance. (14.2)

pressure-gradient force The force that initiates movement of air from areas of higher atmospheric pressure to areas of lower pressure at right angles to the isobars. (14.3)

primary impact A result that is directly attributable to a specific phenomenon, such as a meteorological event. (15.12)

primary mineral Minerals in soil, such as quartz, that are derived directly from the weathering and erosion of various types of rocks but undergo no chemical transformation during the soil-forming process. (10.1)

primary wave (P-wave) seismic body wave that involves particle motion, consisting of alternating compression and expansion, in the direction of propagation. (7.6)

Prime Meridian The line of longitude running through Greenwich, England, which serves as the zero degree of longitude. (1.10)

principle of superposition The concept that a sedimentary or volcanic layer is younger than any rock unit on which it is deposited. (4.1)

profile (stream) The change in a stream's elevation from its headwaters to its mouth, viewed graphically. (12.5)

promontory A ridge of land that juts out into a body of water. (11.12)

proton Principal particle of an atomic nucleus with a positive charge. (2.10)

protostar A disc-shaped mass of gases and other matter in which nuclear fusion has begun. (20.8)

proxy evidence A clue about past climatic conditions during times when direct measurements are not available. (18.9)

P-S interval The time interval between the arrivals of the P-wave and the S-wave. (7.7)

pulsar A neutron star that produces pulsing radio signals. (20.10)

pumice Volcanic rock containing many vesicles (holes) formed by expanding gases in magma. (3.10)

pump-and-treat A commonly used option to clean up, or remediate, a site of groundwater contamination. (12.16)

punctuated equilibrium A hypothesis that new organisms or new characteristics of an existing organism appear rather suddenly in geologic terms instead of evolving gradually. (4.5)

pyrite A common, pale bronze to brass yellow, iron sulfide mineral, commonly called "fool's gold." (2.9)

pyroclastic eruption A volcanic eruption in which hot fragments and magma are thrown into the air. (6.4)

pyroclastic flow A fast-moving cloud of hot volcanic gases, ash, pumice, and rock fragments that generally travel down the flanks of a volcano; syn. ash flow. (3.9)

pyroxene One of a group of mostly dark, single-chain silicate minerals. (2.8)

Q

qualitative data Data that include descriptive words, labels, sketches, or other images. (1.12)

quantitative data Data that are numeric and typically visualized and analyzed using data tables, calculations, equations, and graphs. (1.12)

quartz A very common rock-forming silicate mineral, consisting of crystalline silica. (2.8)

quartzite A very hard rock consisting chiefly of quartz grains joined by secondary silica that causes the rock to break across rather than around the grains; formed by metamorphism or by silica cementation of a quartz sandstone. (3.12)

quicksand Loose sand saturated with water and undergoing liquefaction, rendering it unable to support weight. (10.14)

R

radial drainage pattern A drainage network in which the pattern of streams is generally outward away from a central high point. (12.3)

radial velocity The velocity of an object in motion along the line of sight. (20.6)

radiation (electromagnetic) The energy propagated from charged particles and manifested as interacting electrical and magnetic fields called electromagnetic waves; can be transferred either through matter or through a vacuum. (13.2)

radiation balance The difference between incoming and outgoing radiant energy. (13.13)

radiation deficit A property of matter at a given place and time characterized by a greater magnitude of outgoing than incoming radiant energy. (13.13)

radiation fog A type of fog that occurs when nocturnal longwave radiation loss cools the surface to the dew-point or frost-point temperature. (15.8)

radiation surplus A property of matter at a given place and time characterized by a greater magnitude of incoming than outgoing radiant energy. (13.13)

radiative zone The portion of the Sun's interior adjacent to the core, through which energy generated in the core radiates toward the surface. (20.7)

radioactive decay The spontaneous disintegration and emission of particles from an unstable atom. (4.3)

radiosonde An instrument package attached to a weather balloon that measures a range of variables at various heights. (13.3)

radio wave The electromagnetic radiation of the longest wavelengths (10 meters to 100 kilometers). (13.5)

rapid A segment of rough, turbulent water along a stream. (12.7)

rayed crater An impact crater on the Moon's surface that features visible ejected material radiating out from its center. (19.4)

Rayleigh wave *See* vertical surface wave.

recessional moraine A moraine that forms as the front of a glacier melts back and stagnates for some time in one location, depositing a pile of sediment. (11.5)

recharge The replenishment of water into a groundwater system, whether natural or done by humans. (12.13)

recurrence interval The time between repeating earthquakes. (7.5)

red dwarf A dwarf star that is cooler than our Sun. (20.7)

red giant A giant star that is cooler than our Sun. (20.7)

redshift A shift toward longer red wavelengths by a light source in motion. (20.6)

red sprite Faint red lights that sometimes project up from the top of a thunderhead after an energetic bolt of lightning strikes the ground. (16.7)

reefs Shallow, mostly submarine features, primarily built by colonies of living marine organisms, including coral, sponges, and shellfish, or by the accumulation of shells and other debris. (8.9)

reflecting telescope A type of telescope that magnifies an image by reflecting light from a curved mirror focused on an eyepiece. (20.1)

reflection The change in direction of electromagnetic energy when it interacts with matter, with the energy remaining concentrated before and after interaction. (13.1)

refracting telescope A type of telescope that magnifies an image by refracting light through lenses. (20.1)

regional metamorphism Metamorphism affecting an extensive region and related mostly to regional burial, heating, and deformation of rocks. (3.12)

regional mountain range A mountain range that is hundreds to thousands of kilometers long, contains many peaks, and typically involves uplifted, thickened crust. (8.2)

regional subsidence The process by which a region decreases in elevation, for example, subsidence due to crustal thinning. (8.4)

relative humidity The ratio of the atmospheric vapor pressure to the water-vapor capacity, usually expressed as a percentage. (15.2)

relief The difference in elevation of one feature relative to another; syn. topographic relief. (1.8)

remediate To remedy a fault or deficiency; for example, to clean up a site of contamination. (12.16)

renewable resource A resource that has a virtually unlimited supply and does not remove something irreplaceable when it is used. (18.13)

reservoir A lake that is created by a human-constructed dam. (12.7)

residual soils Soils that develop in place rather than being deposited into a place from elsewhere. (10.4)

resisting force A force that resists the motion of an object, such as resisting the movement of tectonic plates. (5.13)

resurfacing Remaking a surface through erosion, deposition, volcanism, or tectonics. (19.2)

return stroke The positively charged particles that flow upward from the ground as negatively charged particles approach the ground from the cloud, after an electrical connection is established in cloud-to-ground lightning. (16.7)

reverse fault A fault in which the block above the fault moves up relative to the block below the fault. (7.2)

reversed polarity The orientation of the Earth's magnetic field during many times in the geologic past, when the field flowed from north to south, causing the magnetic ends of a compass needle to point toward the south; the opposite of the present-day polarity. (5.14)

rhomb An oblique, equilateral parallelogram, with a shape like a sheared box. (2.5)

rhyolite A mostly fine-grained, felsic igneous rock, generally of volcanic origin; can contain glass, volcanic ash, pieces of pumice, and variable amounts of visible crystals (phenocrysts). (3.10)

ridge (atmospheric) An elongated area of high atmospheric pressure. (14.2)

ridge push A plate-driving force that results from the tendency of an oceanic plate to slide down the sloping lithosphere-asthenosphere boundary near a mid-ocean ridge. (5.13)

rift A narrow trough that runs along the axis of most mid-ocean ridges, formed when large blocks of crust slip down as spreading occurs. (5.6)

rip rap A layer of large, durable fragments of broken rock, concrete, or other material, placed to prevent erosion by waves or currents. (11.14)

risk An assessment of whether a hazard might have some societal impact. (6.5)

river A moving stream of water driven by gravity, flowing from higher to lower elevations. (12.3)

roaring forties The belt of particularly strong westerly winds aloft in the Southern Hemisphere between about 40° S and 50° S, where little land exists to provide friction to slow the winds. (14.8)

roche moutonnée An asymmetrical glacial erosional landform caused by abrasion on the upflow side and plucking on the downflow side. (11.3)

rock arch A naturally formed arch-shaped mass of rock above an opening through a narrow fin of rock. (9.10)

rock avalanche High-velocity, turbulent flow of soil, sediment, and rock that results from the collapse of a steep mountain front. (10.12)

rock cycle A conceptual framework presenting possible paths and processes to which an Earth material can be subjected as it moves from one place to another and between different depths within Earth. (1.5, 3.2)

rock fall A mass-wasting process whereby large rocks and smaller pieces of bedrock detach and fall onto the ground. (10.11)

rock salt A rock composed mostly or entirely of halite. (3.7)

rock slide A slab of relatively intact rock that detaches from bedrock and slides downhill, shattering as it moves. (10.11)

rock varnish A thin, dark film or coating of iron and manganese oxides, silica, and other materials; formed by prolonged exposure at the surface; also known as desert varnish. (9.9)

Rodinia An inferred supercontinent, consisting of all the continents joined, that existed near the Precambrian-Paleozoic boundary. (8.11)

root wedging The process of plant roots extending into fractures and growing in length and diameter, expanding preexisting fractures. (9.2)

Rossby wave Upper-tropospheric, mid-latitude waves of motion that are primarily west-to-east in each hemisphere, but with meanders toward the poles (ridges) and toward the equator (troughs), displaying wavelengths of hundreds to thousands of kilometers. (14.9)

rotation (tectonic) The tilting or horizontal rotation of a body in response to stress. (7.1)

rotational slide A slide in which shearing takes place on a well-defined, curved shear surface, concave upward, producing a backward rotation in the displaced mass; syn. slump. (10.11)

rover A small, remotely controlled wheeled vehicle used to explore the surface of a planet or moon. (19.1)

runoff Precipitation that collects and flows on the surface, such as in streams. (10.8)

S

Saffir-Simpson scale A classification system for the strength of Atlantic and eastern Pacific hurricanes, from Category 1 (weakest) to Category 5 (strongest). (16.12)

salinity The concentration of salt in water. (9.3)

saltation Transport of sediment in which particles are moved in a series of short, intermittent bounces on a bottom surface. (9.9, 12.4)

salt dome A structure formed when buried salt buoyantly flows to the surface in steep, pipelike conduits. (8.8)

salt glacier A gravitational flow of rock salt downhill on the surface. (8.8)

saltwater incursion Displacement of fresh groundwater by the advance of salt water, usually in coastal areas; syn. saltwater intrusion. (12.14)

sand A grain or rock fragment smaller than 2 mm and larger than 1/16 mm. (3.6)

sandbar A low, sandy feature, possibly submerged, offshore of a shoreline or within a sandy river. (11.13)

sand spit A low ridge of sand and other sediment that extends like a prong off a coast. (11.13)

sandstone A medium-grained, clastic sedimentary rock composed mostly of grains of sand, along with other material. (3.7)

Santa Ana wind A warm, dry regional wind of southern California that blows from the northeast, across desert areas, and over mountains, then warms as it descends the leeward slopes. (14.4)

satellite image Image taken by an artificial satellite and generally depicting the types of materials on the surface of Earth or another planetary object. (1.7)

saturated (atmosphere) A condition characterized by atmospheric vapor pressure that is equal to the water-vapor capacity; implies that no more water can evaporate at that temperature and that the relative humidity is 100%. (15.2)

saturated (soil) A condition of the soil in which all the pore spaces are filled with water. (10.2)

saturated adiabatic lapse rate The rate of change of temperature with height of an adiabatically rising air parcel with a relative humidity of 100 percent; the rate itself differs depending on the temperature but can be no more than 10 C°/km. (15.4)

saturated zone The area in the subsurface where water fills nearly all the pore spaces. (12.11)

Saturn The second-largest planet in the solar system, noted for the spectacular rings encircling it. (19.6)

scarp A linear or curved scar left behind by a landslide on a hillslope, marking where the landslide pulled away from the rest of the hill. (10.11)

scattering The change in direction of electromagnetic energy when it interacts with matter, characterized by the energy dispersing in different directions after interaction. (13.1)

schist A shiny, foliated, metamorphic rock generally containing abundant visible crystals of mica. (3.11)

schistosity A metamorphic foliation representing the parallel arrangement of mineral grains, especially mica in schist or other coarse-grained metamorphic rocks. (3.11)

scoria A dark gray, black, or reddish volcanic rock that contains abundant vesicles, usually having the composition of basalt or andesite; syn. volcanic cinders. (3.10)

scoria cone A relatively small type of volcano that is cone shaped and mostly composed of volcanic cinders (scoria); also known as a cinder cone. (6.3)

sea arch An opening through a thin promontory of land that extends out into the ocean. (11.13)

sea breeze A local- to-regional-scale wind that blows onshore in a coastal area, typically by day. (14.3)

sea cave A cave at the base of a sea cliff, usually flooded by seawater. (11.13)

sea cliff A cliff or steep slope situated along the coast. (11.13)

seafloor spreading The process by which two oceanic plates move apart and new magmatic material is added between the plates. (5.3)

sea ice Ice that forms from the freezing of seawater. (11.2, 17.4)

seamount A submarine mountain, in some cases flat-topped, that rises above the seafloor. (5.2)

sea stack An isolated, pillar-like, rocky island or pinnacle near a rocky coastline. (11.13)

seawall A human-constructed wall or embankment of concrete, stone, or other materials along a shoreline, intended to prevent erosion by waves. (11.14)

secondary impact A result that is indirectly attributable to a specific phenomenon, such as a meteorological event. (15.12)

secondary mineral A mineral produced by chemical weathering, such as in a soil. (10.1)

secondary wave (S-wave) A seismic body wave propagated by a shearing motion that involves movement of material perpendicular to the direction of propagation; an S-wave cannot travel through magma and other liquids. (7.6)

sediment Grains and other fragments that originate from the weathering and transport of rocks, and the unconsolidated deposits that result from the deposition of this material. (1.5, 3.1)

sedimentary rock Rock resulting from the consolidation of sediment. (3.1)

sediment budget The amount of sediment available to a system, such as along a shoreline. (11.12)

sediment load The amount of sediment, including material chemically dissolved in a solution, carried by a stream. (12.4)

seismic activity *See* seismicity.

seismicity Earth movements, either on the surface or at depth, caused by earthquakes. (7.8)

seismic station The location of a scientific instrument (seismometer) that measures seismic vibrations. (7.4)

seismic wave An elastic wave produced by earthquakes or generated artificially. (7.4)

seismogram The record made by a seismograph, an instrument that records seismic waves. (7.6)

seismologist A scientist who studies seismic waves by analyzing when and how these waves arrive and by using powerful computers and sophisticated programs to model subsurface parameters, such as density. (7.6)

seismometer An instrument that measures ground shaking or seismic activity. (7.6)

semimajor axis Half the length of the major axis in an ellipse. (20.5)

sensible heat A type of heat in a material detectable as its temperature. (13.3)

sensible heat flux The flow, or transfer, of sensible heat in the atmosphere by convection and advection. (13.12)

September Equinox The day, on or near September 21, when Earth's axis of rotation is pointed exactly sideways relative to the Sun, and the durations of daylight and darkness are equal throughout the world. (13.7)

shaded-relief map A map of an area whose relief is made to appear three-dimensional by simulating the shading on mountains, valleys, and other features. (1.7)

shale A fine-grained clastic sedimentary rock, formed by the consolidation of clay and other fine-grained material. (3.7)

shear The type of differential stress that occurs when stresses on the edge of a mass are applied in opposite directions. (7.1)

shear component The part of the gravitational force that pulls a mass parallel to the sloping surface on which it sits, and therefore may cause it to slide downhill. (10.9)

shear zone A generally tabular zone of rock that is more highly sheared and deformed than rocks outside the zone. (3.11)

sheeted dike. *See* dike.

sheet lightning The electrical charges that are exchanged in less discrete mechanisms than by a single stroke or series of strokes, causing a broad area of the sky to be illuminated. (16.7)

sheet silicates A group of silicate minerals, including micas, that have a distinctly sheetlike crystalline structure. (2.7)

sheetwash An unconfined flow of water in a thin layer downslope. (10.8)

shelf cloud A distinctive type of cloud that is characteristic of derechos. (16.10)

shield volcano A type of volcano that has broad, gently curved slopes constructed mostly of relatively fluid basaltic lava flows. (6.3)

shortwave radiation The electromagnetic energy at wavelengths less than about 4 micrometers; corresponds to the wavelengths at which the Sun emits nearly all of its energy. (13.5)

Siberian High An intense area of surface high atmospheric pressure (anticyclone) over north-central Asia in winter, caused by the sinking of very cold, dense air. (14.5)

silica Silicon dioxide (SiO_2), appearing either as a relatively pure form in a mineral (e.g., quartz) or as a component in more chemically complex minerals and rocks. (2.7)

silicates Minerals that contain silicon-oxygen tetrahedra; the most common mineral group on Earth. (2.6)

silicon The 14th element in the Periodic Table, having the atomic symbol Si. (2.7)

silicone A synthetic material in which carbon is bonded to silicon atoms to keep the material in long chains. (2.7)

sill A tabular igneous intrusion that parallels layers or other planar structures of the surrounding rock and usually has a subhorizontal orientation. (6.12)

silt A fine-grained rock fragment or clast, 1/256 to 1/16 mm in diameter. (3.6, 9.8)

Silurian A geologic period in the early part of the Paleozoic Era. (4.6)

single-cell thunderstorm The simplest category of thunderstorm, existing independently of other storm systems. (16.4)

singularity The hypothetical beginning of the universe as an extremely small, hot, and dense collection of matter. (20.12)

sinuosity The amount a river or stream channel curves for a given length. (12.6)

slab pull A plate-driving force generated by the sinking action of a relatively dense, subducted slab. (5.13)

slate A compact, fine-grained, low-grade metamorphic rock that possesses slaty cleavage. (3.12)

sleet A form of precipitation in which a Bergeron process–produced snowflake melts on its way down to the surface, but refreezes in a cold layer before hitting the surface, resulting in a small ball of ice that bounces from the surface. (15.10)

slip face The steep side of an asymmetrical mound, such as a sand dune, which is also the side that is downwind (leeward) of the prevailing atmospheric circulation. (9.9)

slope aspect The direction (orientation) that a slope faces. (9.5)

slope failure The sudden or gradual collapse of a slope that is too steep for its material to resist the pull of gravity. (10.9)

slump A slide in which shearing takes place on a well-defined, curved shear surface, concave upward, producing a backward rotation in the displaced mass; syn. rotational slide. (10.11)

smog A type of hazy air pollution caused by interactions between sunlight, hydrocarbons, and nitrogen oxides. (13.10)

snowfield A large area covered with snow and ice that, unlike a glacier, does not move. (11.2)

soil Unconsolidated material at and near the surface, produced by weathering; includes mineral matter, organic matter, air, and water, and is generally capable of supporting plant growth. (10.1)

soil compaction The reduction of pore space as clay particles begin to lie flat, often because of land uses such as overgrazing or passage of traffic. (10.14)

soil erosion The detachment of a soil particle from underlying material and subsequent transportation of the particle. (10.8)

Soil Taxonomy The most common soil classification system used in the United States. (10.5)

soil texture The average distribution of particle sizes in a soil, either sand, silt, or clay. (10.5)

soil water Water trapped within the pore spaces in a soil, either as hygroscopic water, capillary water, or gravitational water. (10.2)

solar constant The amount of energy emitted by the Sun each second to the "top" of the Earth's atmosphere, equivalent to 1,366 joules per square meter per second, or 1,366 watts per square meter. (13.6)

solar eclipse An astronomical phenomenon that occurs when the Moon is directly between the Sun and Earth, and the Moon's shadow covers part of Earth. (20.2)

solar energy 1. Electromagnetic energy from the Sun. (1.3) 2. An alternative energy technology that uses the Sun's electromagnetic energy to heat buildings or water, or to generate electricity. (18.13)

solar flare A burst of intense energy and matter directed from the Sun into space, which can affect wireless communication systems on Earth. (13.6)

solar nebula A rotating, disk-shaped collection of dust and gases from which our solar system is believed to have evolved. (4.8)

solar system The Sun, its eight planets, and other celestial bodies that orbit the Sun. (1.11)

solar wind The emission of energetic particles, mostly protons and electrons, by the Sun into space. (19.0)

solid A state of matter in which the constituent atoms and molecules are bound together tightly enough with low enough energy levels that the mass can withstand the vibrations and other motions without coming apart. (13.2)

solidification The process in which magma cools and hardens into solid rock, with or without the formation of crystals. (3.2)

solifluction The very slow, continuous movement of wet soil, regolith, or weathered rock down a slope. (10.12)

sorting A description of the size range of clasts in sediment. (3.6)

source area The place within the earth where magma is generated. (6.1)

source region An area of thousands of square kilometers with similar meteorological characteristics where an air mass forms and acquires its characteristics. (16.1)

source rock A rock or sediment that contains enough organic material to produce petroleum. (8.12)

South Atlantic High A semipermanent subtropical high atmospheric pressure system (anticyclone) in the Atlantic Ocean. (18.5)

south celestial pole A projection of Earth's South Pole onto the celestial sphere. (20.2)

southeast trade winds The prevailing surface atmospheric circulation feature that exists in the tropical latitudes of the Southern Hemisphere, characterized by flow from southeast to northwest. (14.3)

South Equatorial Current An east-to-west-flowing, surface ocean current south of the equator in the Pacific Ocean. (17.2)

Southern Oscillation (SO) The oscillation of surface atmospheric pressure between the area near Darwin, Australia, and the area near Tahiti; when pressure is higher at one of these places, it tends to be lower at the other. (17.8)

Southern Oscillation Index (SOI) A numerical indicator of the degree to which atmospheric pressure patterns differ between the western equatorial Pacific Ocean, near Darwin, Australia, and the eastern equatorial Pacific Ocean, near Tahiti; when the index is positive, pressures are higher near Tahiti and lower near Darwin; when it is negative, pressures are higher near Darwin and lower near Tahiti. (17.8)

specific gravity The ratio of the density of a substance to the density of freshwater. (2.3)

specific heat (capacity) The amount of energy required to change the temperature of a mass of one kilogram by one Kelvin, expressed in joules per kilogram per Kelvin. (13.14)

specific humidity The ratio of the mass of water vapor in a body of air to the total mass of that air, often expressed in grams per kilogram. (15.2)

spectrometer A device that separates a light source into its component wavelengths. *See also* mass spectrometer. (20.4)

spheroidal weathering A form of mostly chemical weathering in which concentric or spherical shells of decayed rock are successively separated from a block of rock. (3.13, 9.6)

spiral galaxy A galaxy that has a coiled appearance. (20.11)

spit A small point or low ridge of sand or gravel projecting from the shore into a body of water. (11.13)

splash erosion Erosion caused by raindrops as they strike the ground. (12.7)

Spodosol An agriculturally unproductive type of soil formed under acidic forest litter in cooler portions of the mid-latitudes and in high latitudes. (10.6)

spreading center A divergent boundary where two oceanic plates move apart (diverge). (5.6)

spring A place where groundwater flows out of the ground onto the surface. (12.13)

spring tides Higher-than-average high tides and lower-than-average low tides caused when the Sun's gravity adds to the effects of the Moon's gravity. (11.10)

squall-line thunderstorm A series of organized thunderstorms oriented along the dryline in the warm sector of a mid-latitude cyclone. (16.4)

stabilized dune A dune of any type that is stationary or barely moving, commonly with vegetation holding the sand in place. (9.9)

stable condition (atmosphere) A situation in which sinking atmospheric motion is encouraged because an adiabatically moving parcel of air is colder than its surrounding environmental air. (15.4)

stalactite A conical or cylindrical cave formation that hangs from the ceiling of a cave and is composed mostly of calcium carbonate. (9.11)

stalagmite A conical, cylindrical, or mound-like cave formation that is developed upward from the floor of a cave and is composed mostly of calcium carbonate. (9.11)

star dune A type of sand dune with an irregular, complex shape, and variably trending sand ridges radiating out from a central peak. (9.9)

states of matter The three major types of arrangements of atoms and molecules: solid, liquid, or gas. (13.2)

stationary front A narrow zone separating two air mass types in which neither air mass is displacing the other. (16.2)

Stefan-Boltzmann law A relationship linking the Kelvin temperature of an object to the amount of electromagnetic radiation it emits. (20.6)

stepped leader An invisible channel of charged particles that starts, stops, and zig-zags between a cloud and the ground immediately before a lightning strike. (16.7)

stick-slip behavior The sequence of a rock straining before an earthquake, rupturing during an earthquake, and then mostly returning to its original shape after the earthquake. (7.5)

storm surge The pileup of water from waves onshore as a severe storm, particularly a tropical cyclone, makes landfall. (11.14, 16.11)

storm track The path that a storm takes across the surface. (16.3)

straight-line winds *See* microburst.

stratiform cloud A family of cloud types that is layered, and wider than tall. (15.7)

stratigraphic section A columnar diagram that shows the sequence of rock units, generally in their approximate relative thicknesses. (1.8)

stratocumulus cloud A type of low, layered cloud that is composed solely of tiny drops of liquid water. (15.7)

stratopause The transition zone between the stratosphere and overlying mesosphere. (13.9)

stratosphere The second layer of the atmosphere above the surface, which contains most atmospheric ozone. (13.1)

stratus cloud A type of low-level stratiform cloud that is not precipitating. (15.7)

streak The color of powder a mineral leaves when rubbed against a porcelain plate. (2.3)

streak plate A piece of unglazed porcelain used to obtain a streak during mineral identification. (2.3)

streambed The zone of contact, or interface, between the water in a stream and the earth material over which it is passing; formed by the coalescence of gullies. (10.8)

stream capture The natural diversion of water from one stream into another. (12.7)

streamer The positively charged streams of energy that move upward from the surface to meet the stepped leader immediately prior to a lightning strike, in cloud-to-ground lightning. (16.7)

stream terrace A relatively flat bench that is perched above a river or stream and that was formed by past deposition or erosion of the river or stream. (12.9)

stress The amount of force divided by the area on which the force is applied. (7.1)

striation A series of straight, subparallel lines on the surface of a crystal. (2.8)

strike-slip fault A fault in which the relative movement between the material on either side is essentially horizontal, parallel to the strike (horizontal direction) of the fault surface. (7.2)

strip cropping The spatial alternating of different crops, each planted in a narrow strip often along the contours of the land, as a technique to minimize soil erosion. (10.8)

stromatolite A mound- or column-shaped feature of concentrically laminated carbonate materials, generally in ancient sedimentary rocks, interpreted to have been constructed by microscopic algae; also modern living examples. (4.4)

structurally controlled pattern A drainage network that tends to follow faults, joints, bedding, and similar geologic features. (12.3)

Subarctic (Dfc or Dfd) climate In the Köppen climate classification, a climate type characterized by adequate moisture year-round, a mean temperature of the coldest month below −3°C, and a mild or cool summer. (18.7)

Subarctic Monsoon (Dwc or Dwd) climate In the Köppen climate classification, a climate type characterized by adequate moisture in summer but not winter, a mean temperature of the coldest month below −3°C, and a mild or cool summer. (18.7)

subduction zone An area where subduction takes place, either referring to the actual downgoing slab and its surroundings, or to the region, including Earth's surface, above the subducting slab. (5.7)

sublimation A change in state from a solid to a gas, requiring an input of energy from the environment and the absorption of that energy at the molecular level as latent energy. (11.2, 13.4)

submarine fan A broad, fan-shaped accumulation of sediment on the seafloor, especially below the mouth of a large river or submarine canyon. (8.0)

submarine slope failure Slope failure that occurs on the seafloor. (10.10)

submergent coast A coast that forms where land has been inundated by the sea because of a rise in sea level or subsidence of the land. (11.15)

subpolar low A large, permanent area of low surface atmospheric pressure centered over the oceans near 60° N and 60° S, including the Icelandic Low and Aleutian Low in the Northern Hemisphere. (14.7)

subtropical jet stream The fast-flowing current of air near the tropopause in each hemisphere at the boundary between the Hadley cell and the mid-latitudes. (14.9)

sulfates A group of minerals that contain sulfur (S) bonded to oxygen. (2.6)

sulfides A group of minerals containing sulfur (S) bonded with a metal. (2.6)

sunspot A dark spot on the Sun that is slightly cooler than the rest of the Sun but associated with increased overall total solar irradiance. (13.6)

supercell thunderstorm A very large, single thunderstorm characterized by an organized set of features that allow the system to sustain itself for many hours and produce very severe weather. (16.4)

supergiant A giant star that is hundreds of times more massive than our Sun. (20.7)

supernova A violent explosion of outer material from a dying high-mass star. (20.9)

supersaturated A theoretical condition of the atmosphere characterized by vapor pressure that slightly exceeds the water-vapor capacity; implies that active condensation or deposition must be occurring. (15.2)

surface tension The tendency for molecules, such as water, to coalesce with a discrete outer surface. (15.1)

surface water Water that occurs in streams, rivers, lakes, oceans, and other settings on Earth's surface. (12.0)

surface waves Seismic waves that travel on Earth's surface. (7.6)

surf zone The area where waves break and spread water upon the shore. (11.11)

suspended load Fine particles, generally clay and silt, that are carried suspended in moving water. (12.4)

suspension A mode of sediment transport in which water or wind picks up and carries the sediment as floating particles. (9.9)

swash regime The scenario of coastal inundation during which the height of the swash zone remains below the bases of coastal dunes. (11.17)

swelling clays Soils, especially Vertisols, composed of a high percentage of certain clay minerals; they increase in volume when wet. (10.14)

syncline A fold, generally concave upward (in the shape of a U), with the youngest rocks in the center. (7.3)

synthetic Refers to material produced by humans. (2.7)

system An interconnected set of matter, energy, and processes. (1.4)

T

talus Loose rock fragments upon a steep slope, or an accumulation of such fragments. (9.2)

talus slope A steep slope composed of loose rock fragments, that is, talus. (9.2)

tarn A small lake, especially one within a cirque; a glacially scoured depression. (11.4)

tar sand *See* oil sand.

tectonic plate Any of the dozen or so fairly rigid blocks into which Earth's lithosphere is broken. (5.5)

tectonic terrane A fault-bounded body of rock that has a different geologic history than adjacent regions. (8.10)

teleconnection An association in atmospheric features, usually referring to opposing pressure conditions, between two distant places. (17.10)

temperate mid-latitude (C) climate In the Köppen climate classification, a broad group of climatic types characterized by adequate moisture availability, and the mean temperature of the coldest month between −3°C and 18°C. (18.5)

Temperate Monsoon (Cwa, Cwb, or Cwc) climate In the Köppen climate classification, a climate type characterized by adequate moisture in summer but dry winters; hot, warm, or mild summers; and the mean temperature of the coldest month between −3°C and 18°C. (18.5)

temperature A measure of the object's average internal kinetic energy—the energy contained within atoms or molecules that are moving. (13.3)

temperature inversion An atmospheric condition in which temperature increases as height increases. (13.9)

tension The type of differential stress where stress is directed outward, pulling the rock. (7.1)

tephra A pyroclastic material, regardless of size or origin, ejected during an explosive volcanic eruption; includes ash, pumice, and rock fragments. (6.4)

terminal moraine Glacially carried sediment that accumulates at the terminus (end) of a glacier and a landform composed of such material; generally marks the glacier's farthest downhill or most equatorward extent. (11.4)

termination The well-defined, commonly sharp end of a crystal. (2.4)

terminus The farthest extent, especially the downslope end, of a glacier. (11.2)

terrace A relatively level or gently inclined surface or bench bounded on one edge by a steeper descending slope. (4.11)

terrestrial planets Our solar system's four inner planets, which have solid rocky surfaces and include Earth, Mars, Venus, and Mercury. (1.11, 19.0)

tetrahedron A four-sided pyramid, as in a silica tetrahedron. (2.4)

texture The general physical appearance or character of a rock, especially the size, shape, and arrangement of minerals and other materials. (2.2)

theory An explanatory system of propositions and principles, supported to some extent by experimental or factual evidence and held to be true until contradicted or amended by new facts. (1.14)

thermal expansion The heating and cooling of material as it is heated, such as heating and cooling that cause different minerals, or different parts of a rock, to expand and contract by different amounts. (3.5)

thermal inertia The degree to which a particular type of matter remains at the same temperature when a change in energy is imposed on it. (15.5)

thermocline The level in the ocean at which temperature changes rapidly with depth. (17.7)

thermohaline conveyor The globally interconnected, three-dimensional oceanic circulation system. (17.7)

thermosphere The fourth layer of the atmosphere above the surface, characterized by a temperature inversion. (13.1)

thrust fault A reverse fault that has a gentle dip. (7.2)

thrust sheet The sheet of rock that has been displaced above a thrust fault. (7.2)

thunder The sound made when air rushes back into the channel where a lightning stroke passed. (16.7)

thunderhead A cumulonimbus cloud that is producing hail and lightning. (16.6)

Tibetan Low A prominent area of low atmospheric pressure in summer, centered over Tibet. (14.10)

tidal flat An area of shoreline that is flooded at high tide but exposed to air during low tide. (3.4)

tide A cyclic change in the height of the sea surface, generally measured at locations along the coast; caused by the pull of the Moon's gravity and to a lesser extent the Sun's gravity. (11.10)

till Unsorted, generally unlayered sediment, deposited directly by or underneath a glacier. (11.3)

time-travel plot A graph that plots the time difference between the arrivals of P-waves and S-waves as a function of the distance from the epicenter or other seismic disturbance to a seismic station. (7.7)

topographic map A map showing the topographic features of a land surface, commonly by means of contour lines. (1.7)

topographic profile A cross-sectional view across part of Earth's surface, showing variations in elevation or depth. (1.8)

topographic relief The difference in elevation of one feature relative to another; syn. relief. (1.8)

topography The general configuration of a surface, especially the land surface or seafloor, including its elevation, relief, and features; shape of the land. (1.8)

tornado A violent, rapidly rotating, funnel-shaped column of air that extends to the ground. (16.8)

Tornado Alley The swath from south-central Canada to central Texas where tornado frequency is among the highest on Earth. (16.9)

total eclipse A location on Earth where the Moon's shadow completely blocks the Sun during a solar eclipse. (20.2)

total solar irradiance (TSI) A measure of the Sun's total output of energy. (13.6)

trace fossils Features in rocks made by animals that moved across the surface or burrowed into soft sediment. (4.4)

traction The process by which particles roll, slide, or otherwise move on the surface by such transport agents as streams, wind, or waves. (12.4)

trading location for time A strategy that uses different parts of a landscape to represent different stages in the evolution of the landscape. (9.7)

transform boundary A plate boundary in which two tectonic plates move horizontally past one another. (5.5)

transform fault A strike-slip fault that accommodates the horizontal movement of one tectonic plate past another. (5.8)

transmission The passing of energy through matter, such as air. (13.1)

transpiration The transfer of water from the soil to the atmosphere via plant roots, stems, and leaves. (10.2, 12.1)

transverse dune A type of sand dune that is oriented linearly or only very gently curving for many kilometers, perpendicular to prevailing winds. (9.9)

travertine A variety of limestone that is commonly concentrically banded and porous. (9.11)

tree line On a mountain, the altitude beyond which trees cannot grow. (9.13)

triangulation A method of determining distances by forming a line between two reference points and measuring the angle between the line at each reference point and a distant object. The distance to the object is calculated using the known length of the first line and the two known angles. (20.6)

Triassic The earliest geologic period in the Mesozoic Era. (4.6)

tributary A secondary stream that joins or flows into a larger stream or lake. (4.11, 12.3)

trilobite A marine creature of the Paleozoic Era, characterized by a three-lobed external skeleton. (4.9)

triple junction The place where three tectonic plates and three plate-tectonic boundaries meet. (5.8)

tropical (A) climate In the Köppen climate classification, a broad group of climatic types characterized by adequate moisture availability and the mean temperature of the coldest month above 18°C. (18.3)

tropical cyclone A well-organized, rotating storm that originates in the tropical part of the Earth, ranging in strength from weak storms to hurricanes. (16.11)

tropical depression An enclosed area of low atmospheric pressure of tropical origin with at least one closed isobar. (16.12)

Tropical Monsoon (Am) climate In the Köppen climate classification, a climate type characterized by a mean temperature of the coldest month above 18°C, and by precipitation totals that exceed potential evapotranspiration in most months, but with enough water stored in the soils during the brief dry season to support the growth of forests. (18.3)

Tropical Rain Forest (Af) climate In the Köppen climate classification, a climate type characterized by a mean temperature of the coldest month above 18°C and precipitation totals that typically exceed potential evapotranspiration year-round. (18.3)

Tropical Savanna (Aw) climate In the Köppen climate classification, a climate type characterized by a mean temperature of the coldest month above 18°C, and by precipitation totals that exceed potential evapotranspiration for about half of the year, providing enough moisture for tropical grasslands and a few scattered trees, but not forests; also known as Tropical Wet-Dry climate. (18.3)

tropical storm A tropical cyclone with well-developed circulation features and wind speeds of 64 km/hr (39 mi/hr) to 118 km/hr (73 mi/hr). (16.12)

Tropic of Cancer The 23.5° parallel of latitude north of the equator; represents the most northerly latitude that experiences the directly overhead rays of the Sun (on or near June 21). (1.9)

Tropic of Capricorn The 23.5° parallel of latitude south of the equator; represents the most southerly latitude that experiences the directly overhead rays of the Sun (on or near December 21). (1.9)

tropics The region of the Earth between the Tropic of Cancer (23.5°N) and the Tropic of Capricorn (23.5°S). (13.7)

tropopause The transition zone between the troposphere and overlying stratosphere. (13.9)

troposphere The lowest layer of the atmosphere, which contains nearly all of the atmosphere's water. (13.1)

trough (atmosphere) An elongated area of low atmospheric pressure. (14.2)

trough (ocean) The lowest part of a wave. (11.11)

tsunami A large sea wave produced by uplift, subsidence, or some other disturbance of the seafloor, especially by a shallow submarine earthquake. (7.10)

tsunami warning centers An array of sensors in the ocean and a computerized infrastructure that broadcasts a warning upon the formation of a tsunami. (7.11)

tuff Volcanic rock composed of consolidated volcanic ash and other tephra, commonly including pumice, crystals, and rock fragments. (3.10)

Tundra (ET) climate In the Köppen climate classification, a climate type characterized by adequate moisture, the warmest month's temperature averaging below 10°C (50°F), and frosts possible in any month. (18.7)

turbidity current A current in water or air that moves downward because it is more dense than the adjacent water or air, especially applied to a swift, bottom-flowing current of water and suspended sediment on the seafloor or the bottom of a lake. (8.0)

turbulent Chaotic flow of water, air, or some other fluid. (12.4)

turnover The mixing of waters from different depths in a large body of water; also known as overturning. (17.5)

typhoon A regional term used in the western Pacific basin applied to the most intense type of storm of tropical oceanic origin. (16.11)

U

Ultisol A type of soil formed in the subtropics and warmer mid-latitudes that is less highly weathered than Oxisols, with the eluviation of clay minerals into the B horizon, and generally acidic. (10.6)

ultramafic A generally dark or greenish igneous rock composed chiefly of mafic minerals rich in magnesium and iron. (3.10)

ultraviolet light The electromagnetic radiation of wavelengths between about 10^{-8} and 0.4 micrometers. (13.5)

ultraviolet radiation The ultraviolet part of the electromagnetic spectrum. (1.3)

unconfined aquifer An aquifer where the water-bearing unit is open (not restricted by impermeable rocks) to Earth's surface and atmosphere. (12.12)

unconformity A boundary between underlying and overlying rock strata, representing a significant break or gap in the geologic record; an unconformity represents an interval of nondeposition or erosion, commonly accompanied by uplift. (4.2)

unconsolidated material Material sitting on top of solid rock, including soil, loose sediment, pieces of wood and other plant parts, boulders, and other types of loose debris. (10.10)

uniformitarianism The concept that the present is the key to the past; that is, geologic processes occurring today also occurred in the geologic past and can be used to explain ancient events and the geologic features they produced. (4.0)

unloading The process during which weight is removed from a landmass, such as by the melting of a glacier. (11.16)

unloading joint A joint formed from stresses that arise during uplift of buried rocks and that cause rocks to fracture due to reduced pressure. (7.2)

unsaturated (atmosphere) A condition characterized by atmospheric vapor pressure that is less than the water-vapor capacity; implies that more water could evaporate at that temperature under the right conditions and that the relative humidity is less than 100 percent. (15.2)

unsaturated adiabatic lapse rate The rate of change of temperature with height of an adiabatically rising air parcel with a relative humidity less than 100 percent; always amounts to 10 C°/km. (15.4)

unsaturated zone A part of the subsurface where most of the pore spaces are filled with air rather than water. (12.11)

unstable condition (atmosphere) A situation in which rising atmospheric motion is encouraged because an adiabatically moving parcel of air is warmer than its surrounding environmental air. (15.4)

upslope fog A type of fog that occurs when air cools to the dew-point temperature or frost-point temperature as it moves upslope. (15.8)

upwelling The flow of colder, deeper waters up toward the surface of the ocean, replacing surface ocean water that circulated elsewhere. (17.3)

Uranus The seventh planet out from the Sun. (19.7)

urban heat island (UHI) The phenomenon of increased air temperatures over cities as compared to their surrounding rural areas. (15.5)

urbanization The replacement of natural land cover with built-up land cover associated with the growth of cities and towns. (15.5)

U-shaped valley A steep-sided, rounded valley carved by a glacier. (11.4)

V

Valley and Ridge Province A physiographic region in the Appalachian Mountains of the eastern U.S., characterized by linear or curved ridges and valleys. (7.3)

valley fog A type of fog that occurs when cold nocturnal downslope winds trap atmospheric moisture in the valleys, causing the air to reach the dew-point temperature through the addition of humid air; or when cold winds move downslope into the valley, chilling the valley air to the dew-point or frost-point temperature. (15.8)

valley glacier A glacier that flows down a valley and tends to be narrow; syn. alpine glacier. (11.1)

vapor pressure The portion of the total atmospheric pressure exerted by water vapor molecules, usually measured in millibars. (15.2)

variable star A yellow giant star that fluctuates in size. (20.8)

varves Thin, alternating light-colored and dark layers of sediment that form in lakes because of seasonal variations of deposition and biologic activity. (4.7)

vein A generally tabular accumulation of minerals that filled a fracture or other discontinuity in a rock; formed by precipitation of material from fluids, especially hydrothermal fluids. (3.12, 7.1)

Venus The second planet from the Sun. (19.3)

vertical exaggeration Exaggerating the height of vertical features on a relief map so that viewers can see the topography more clearly. (1.16)

vertical surface wave A type of surface wave in which material moves up and down, perpendicular to the propagation direction of the wave. (7.6)

vertical wind shear A change in horizontal wind speed or direction with height. (16.4)

Vertisol A type of clay-rich soil located in temperate and subtropical environments, characterized by swelling when wet and shrinking when dry. (10.6)

vesicles Small holes found in a volcanic rock, representing gas bubbles in a magma that were trapped when the lava solidified. (3.9)

vesicular Adjective used to describe a rock containing vesicles. (3.9)

virga Any precipitation that evaporates before reaching the ground. (16.10)

viscosity A measure of a material's resistance to flow. (6.2)

viscous magma Magma that does not flow easily. (6.2)

volcanic ash Particles of volcanic tephra that are sand-sized or smaller, and accumulations of such material. (3.9)

volcanic bomb A rock fragment representing either a large blob of magma or a solid angular block ejected during an explosive volcanic eruption. (6.5)

volcanic breccia A volcanic rock containing angular fragments in a matrix of finer material. (3.9)

volcanic dome A dome-shaped volcanic feature, largely composed of solidified lava of felsic to intermediate composition. (6.3)

volcanic glass A natural glass produced by the cooling and solidification of molten lava at a rate too rapid to permit crystallization. (3.9)

volcanic neck A steep, typically butte-shaped topographic feature composed of volcanic materials that formed in the conduit within or beneath a volcanic vent and that were more resistant to erosion than surrounding materials. (6.12)

volcanic rock An igneous rock that formed from a volcanic eruption, including lava, pumice, volcanic ash, and other volcanic materials. (6.3)

volcanic tremor A small earthquake caused by the movement of magma. (7.4)

volcano A vent in the surface of Earth through which magma and associated gases and ash erupt; also the form or structure constructed from magma erupted from the vent. (6.3)

volume The amount of three-dimensional space, measured in cubic units of length, that a mass occupies. (14.1)

vorticity Any clockwise or counterclockwise horizontal rotation (spin) of air. (16.3)

W

Walker cell circulation Any east-west-oriented atmospheric circulation system in the equatorial part of the Earth characterized by relatively high pressure and subsiding air on one side, relatively low surface pressure and rising motion on the opposite side, east-to-west trade winds at the surface, and west-to-east flow of air aloft; often used to refer to the cell of this type over the equatorial Pacific Ocean. (17.8)

wall cloud A sharp, low-forming protrusion on a cumulonimbus cloud that is characteristic of tornadoes and other severe weather. (16.8)

warm front The sloping boundary between an advancing relatively warm air mass and a colder one. (16.2)

warm phase (ENSO) A periodic warming of eastern equatorial Pacific surface ocean waters, accompanied by excessive precipitation and flooding adjacent to South America as the Southern Hemisphere's summer approaches; also known as El Niño. (17.9)

warm pool A large region of ocean water warmer than the water surrounding it. (17.6)

warm sector The part of a mid-latitude cyclone that is between the warm front and the cold front, characterized by warm air. (16.2)

water table The surface between the unsaturated zone and the saturated zone, as in the top of groundwater in an unconfined aquifer. (12.11)

waterfall A steep descent of water within a stream, such as the place where it crosses a cliff or steep ledge. (12.7)

waterspout A rotating, columnar vortex of air and small water droplets that is over a body of water, usually an ocean or a large lake. (16.10)

water-vapor capacity The maximum atmospheric vapor pressure that can exist at a given temperature. (15.2)

wave An irregularity on the surface of a body of water. *See also* seismic wave. (11.11)

wave amplitude The height between the trough and the crest of a wave, including ocean waves and seismic waves. (7.6)

wave base The depth at which the action of a wave no longer has an effect. (11.11)

wave height The vertical distance between the trough and the crest of a wave. (11.11)

wave-cut notch A notch produced in rocks or sediment by continued wave action at a specific level along a coast. (11.12)

wave-cut platform A gently sloping surface or bench produced by wave erosion. (11.12)

wavelength The horizontal distance between two adjacent crests in a set of waves. (11.11)

weather The conditions in the atmosphere at some specific time and place. (16.0)

weathering Physical disintegration and chemical decomposition of rocks, sediment, and soil due to exposure to water and other atmospheric agents. (1.5, 3.2)

weathering rind A weathered, outer crust on a rock fragment or bedrock mass exposed to weathering. (9.6)

weight A measure of how much downward force the mass of an object exerts under the pull of gravity. (1.12)

westerlies The prevailing surface and upper-tropospheric general circulation feature that exists in the middle latitudes in each hemisphere, characterized by flow from west to east. (14.3)

Western Atlantic Warm Pool A large region of warm ocean temperatures centered on the Caribbean Sea. (17.6)

Western Australian Current A south-to-north-flowing, cold surface ocean current in the eastern Indian Ocean. (17.2)

Western Pacific Warm Pool The broad region of the warmest ocean temperatures on Earth, from China to Australia, spanning the western Pacific and eastern Indian Oceans. (17.6)

West Wind Drift A west-to-east-flowing, surface ocean current in the high latitudes of the Southern Hemisphere from southern Australia to South America; also known as the Antarctic Circumpolar Current. (17.2)

whitecap A white, foamy wave that forms when wind causes the wave to become too steep and to collapse. (11.11)

white dwarf A dwarf star of variable temperature. (20.7)

Wien's law A relationship between the temperature of a body and the wavelength at which it emits the most electromagnetic energy. (20.3)

wilting point The minimum amount of water (in centimeters or inches of precipitation equivalent) that is necessary in the rooting zone to permit extraction by plants; any less water in the soil will adhere to rock particles and resist evapotranspiration. (10.2)

windbreak A technique to minimize soil erosion, whereby a row of trees is planted perpendicular to the prevailing wind direction with the intent of slowing down the wind and therefore minimizing transportation of soil particles. (10.8)

wind streak A thin, small ridge of sand that occurs when winds in arid areas are strong enough that sand is blown through, without accumulating in sizeable dunes; often form on the downwind (leeward) side of small plants and stones. (9.9)

windward The upwind side of a slope. (14.4)

X

X-ray The electromagnetic radiation of wavelengths between about 10^{-11} and 10^{-8} micrometers. (13.5)

Y

yellow giant A giant star of intermediate temperature. (20.7)

yellow star A star that has an intermediate temperature. (20.7)

Z

zodiac The constellations that occur along the ecliptic. (20.2)

zonal flow A wind that is primarily blowing from west to east or east to west. (15.12)

zone of accumulation (glacier) The upper part of a glacier or ice sheet, where snow and ice are added faster than they are removed by melting and other processes. (11.2)

zone of accumulation (soil) A zone in the soil characterized by the accumulation of material, including iron oxide, clay, and calcium carbonate, depending on the climate and starting materials; also known as B horizon, zone of illuviation, or subsoil layer. (10.3)

zone of leaching A light-colored, leached zone of soil, lacking clay and organic matter; also known as E horizon. (10.3)

CREDITS

PHOTO CREDITS

Unless otherwise credited: ©Stephen J. Reynolds.

PREFACE

XVII: (10.00a3): Source: Photo by Lawson Smith, U.S. Army Corps of Engineers; (10.00.a4-a5): Source: Matthew C. Larsen/U.S. Geological Survey; (09.11.b1): ©Galyna Andrushko/Alamy; (14.10a5-a6): Source: Jacques Descloitres, MODIS Rapid Response Team, NASA/GSFC.

CHAPTER 1

Image Number: 01.00.a3: Source: Michael P. Doukas/U.S. Geological Survey; 01.00.a5: Photo by Cynthia Shaw; 01.02.d1: ©Karen Carr; 01.06.a1: Photo by Susanne Gillatt; 01.06.a2: Source: M.E. Yount/Alaska Volcano Observatory/U.S. Geological Survey; 01.06.a3: Photo by Susanne Gillatt; 01.06.a5: Photo by Cynthia Shaw; 01.06.b1: ©Getty Images RF; 01.08.b2: Source: Wendell Duffield/U.S. Geological Survey; 01.09.c1: Source: NOAA; 01.09.c2: ©National Air and Space Museum; 01.10.c2-3: Source: U.S. Geological Survey; 01.00.a10: Source: NASA/JHUAPL/SwRI; 01.12.a1: Source: Cyrus Read/Alaska Volcano Observatory/U.S. Geological Survey; 01.12.a2: Source: Kate Bull/Alaska Volcano Observatory/Alaska Division of Geological & Geophysical Surveys; 01.12.a3: Source: T.A. Plucinski/Alaska Volcano Observatory/U.S. Geological Survey; 01.12.b2: Photo ©J. Ramón Arrowsmith; 01.12.b3: ©Armstrong Flight Research Center/NASA; 01.12.b4: Photo by Hilairy Hartnett; 01.12.c1: Photo by Chris Marone; 01.12.c2: Courtesy of NASA/SDO and the AIA, EVE, and HMI science teams; 01.12.c3: Photo ©Ariel Anbar; 01.13.d1: Source: David E. Wieprecht/U.S. Geological Survey; 01.14.t1: Photo by Daniel Ball/Arizona State University; 01.15.a2: Source: Edwin L. Harp/U.S. Geological Survey.

CHAPTER 2

Image Number: 02.00.a2: ©Photo Take/Alamy; 02.01.a1: ©Dr. Parvinder Sethi; 02.01.a2: Photo by Michael Ort; 02.01.a3: Photo by Allen Glazner; 02.01.a4: Photo by Susanne Gillatt; 02.01.a6: ©Doug Sherman/Geofile RF; 02.01.b5: ©John A. Rizzo/Getty Images RF; 02.04.a1: ©McGraw-Hill Education, Bob Coyle, photographer; 02.08.b4: ©Doug Sherman/Geofile RF; 02.14.a3: Photo by Susanne Gillatt; 02.15.a6, 02.15.a7: Photos by Thomas Sharp/Arizona State University.

CHAPTER 3

Image Number: 03.00.a3: ©Digital Vision/Punchstock RF; 03.01.b4: ©Ariel Anbar; 03.03.a3: Photo by Jessica Barone; 03.04.a4: Photo by Cynthia Shaw; 03.04.a7: Source: Earth Science and Remote Sensing Unit, NASA Johnson Space Center; 03.04.a8: Source: U.S. Geological Survey; 03.08.a4: ©Sheila Terry/Science Source.

CHAPTER 4

Image Number: 04.04.a7: ©Icontec/Alamy RF; 04.05.b3, 04.05.b4: Photo by Susanne Gillatt; 04.07.a2: ©Dr. Parvinder Sethi; 04.07.b1: ©Daniel Griffin, University of Minnesota; 04.07.b2: Source: National Ice Core Laboratory/U.S. Geological Survey/National Science Foundation; 04.07.b3: Photo by Jack Ridge; 04.07.b4: Source: David R. Sherrod/U.S. Geological Survey; 04.07.t1: Photo by Daniel Ball/Arizona State University; 04.08.c1: Photo by Susanne Gillatt; 04.09.b2: ©Sinclair Stammers/Science Source; 04.09.b3: ©Charles Carter; 04.09.c1: "Ordovician Marine Environment," ©Karen Carr and courtesy Indiana State Museum Foundation; 04.09.c2: "Devonian Marine Environment," ©Karen Carr and courtesy Indiana State Museum Foundation; 04.09.c3: "Permian Riverside," ©Karen Carr and courtesy Indiana State Museum Foundation; 04.10.a1: "Triassic Landscape," ©Karen Carr; 04.10.a2: "Jurassic Landscape," ©Karen Carr; 04.10.a3: "Cretaceous Coastal Landscape," ©Karen Carr; 04.10.a4: "Cretaceous Marine Environment," ©Karen Carr; 04.10.b1: "Alaskan Dinosaurs," ©Karen Carr; 04.10.c1: "Micene River Landscape," ©Karen Carr; 04.10.c2: "North American Pleistocene Landscape," ©Karen Carr.

CHAPTER 5

Image Number: 05.01.a5: ©Dr. Parvinder Sethi; 05.08.b2: Source: Robert E. Wallace/U.S. Geological Survey; 05.10.c1: ©Dr. Ken MacDonald/Science Source; 05.10.t1: Source: NOAA Okeanos Explorer Program, Galapagos Rift Expedition 2011; 05.13.d3: Source: Peter J. Haeussler/U.S. Geological Survey; 05.14.b1: Gary Wilson-Paleomagnetic Research Facility, University of Otago.

CHAPTER 6

Image Number: 06.00.a1: Source: John Pallister/U.S. Geological Survey; 06.00.a2: Source: Donald Swanson/U.S. Geological Survey; 06.00.a4: Source: Austin Post/U.S. Geological Survey; 06.00.a5: Source: Lyn Topinka/U.S. Geological Survey; 06.03.a1: Source: J.D. Griggs/U.S. Geological Survey; 06.03.a2: Source: E. Klett/U.S. Fish and Wildlife Service; 06.04.a1: Source: J. Judd/U.S. Geological Survey; 06.04.a2: Source: Donald Swanson/U.S. Geological Survey; 06.04.a3: Source: J.D. Griggs/U.S. Geological Survey; 06.04.a4: Source: R. Clucus/Alaska Volcano Observatory/U.S. Geological Survey; 06.04.a5: Source: Alaska Volcano Observatory/U.S. Geological Survey; 06.04.a6: Source: M.E. Yount/U.S. Geological Survey; 06.04.c1: Source: John Pallister/U.S. Geological Survey; 06.04.c2: Source: Hawaiian Volcano Observatory/U.S. Geological Survey; 06.05.a2: Source: U.S. Geological Survey; 06.05.b1: Source: J.D. Griggs/U.S. Geological Survey; 06.05.b3: Courtesy of Henrik Thorburn; 06.05.c1: Source: Robert Krimmel/U.S. Geological Survey; 06.05.c2: Source: Peter Lipman/U.S. Geological Survey; 06.06.a1, 06.06.a4: Source: J.D. Griggs/U.S. Geological Survey; 06.06.a6: Source: U.S. Geological Survey; 06.06.a5, 06.06.b2, 06.06.b3: Source: J.D. Griggs/U.S. Geological Survey; 06.06.c2: Photo by Cynthia Shaw; 06.07.a2: Source: Jim Vallance/U.S. Geological Survey; 06.07.a3: Source: C.G. Newhall/U.S. Geological Survey; 06.07.a4: Source: Tom Casadevail/U.S. Geological Survey; 06.07.a5: Source: Game McGinsey/Alaska Volcano Observatory/U.S. Geological Survey; 06.07.b1: Source: T. Miller/Alaska Volcano Observatory/U.S. Geological Survey; 06.07.t1: Source: Steven Brantley/U.S. Geological Survey; 06.07.t2: Source: U.S. Geological Survey; 06.08.a2: Porojnicu Stelian/Shutterstock RF; 06.08.b2: William H. Rau/Boston Public Library; 06.08.c3: Source: Henry Glicken/U.S. Geological Survey; 06.08.t1: Source: Michael Doukas/U.S. Geological Survey; 06.09.c1: Source: EROS Center, U.S. Geological Survey; 06.12.a6: Photo by George H. Davis; 06.12.b6: Photo by Steve Semken; 06.13.a1: Source: Game McGimsey/Alaska Volcano Observatory/U.S. Geological Survey; 06.14.a2: Source: Lyn Topinka/U.S. Geological Survey.

CHAPTER 7

Image Number: 07.00.a2: Source: U.S. Geological Survey; 07.00.a3: U.S. Navy Photo by Photographer's Mate 3rd Class Jacob J. Kirk; 07.10.a2: Source: H.W. Wilshire/U.S. Geological Survey; 07.10.a3: Source: J.D. Nakata/U.S. Geological Survey; 07.10.b1: Source: U.S. Geological Survey; 07.10.b2: ©Karl V. Steinbrugge Collection, Earthquake Engineering Research Center, NISEE, University of California, Berkeley; 07.10.b3: Source: Guy Gelfenbaum/U.S. Geological Survey; 07.10.c2: Source: U.S. Geological Survey; 07.11.a5: ©Lynette Cook/SPL/Science Source; 07.11.b2: Source: NOAA/NGDC; 07.11.b3: Source: NOAA/NGDC, U.S. Navy; 07.11.b4: Source: NOAA/NGDC, International Tsunami Information Center; 07.11.b5: Hugh Davies/University of Papua New Guinea; 07.11.t1: Source: NOAA; 07.12.a2: MC3 Alexander Tidd/U.S. Navy; 07.12.a3: Source: NOAA Center for Tsunami Research; 07.12.b2: ©Logan Abassi/AFP/Getty Images;

07.12.b4: Commander Dennis J. Sigrist/International Tsunami Information Center/NZ Defence Force; 07.13.a2: Source: U.S. Army photo, U.S. Geological Survey Photographic Collection; 07.13.a3: Source: W.C. Mendenhall/U.S. Geological Survey; 07.13.a4, 07.13.a5: Source: Mechmet Celebi/U.S. Geological Survey; 07.13.a6: Source: J.R. Stacy/U.S. Geological Survey; 07.13.a8: J.K. Hillers/U.S. Geological Survey; 07.14.a2: Source: W.C. Mendenhall/U.S. Geological Survey; 07.14.a3: Source: C.E. Meyer/U.S. Geological Survey; 07.14.a5: Photo ©J. Ramón Arrowsmith; 07.14.a6: Source: Robert E. Wallace/U.S. Geological Survey.

CHAPTER 8

Image Number: 08.00.a2: National Center for Earth-Surface Dynamics; 08.03.a2: Photo by Scott U. Johnson; 08.09.a2: Photo by Cynthia Shaw; 08.09.a3: ©greenantphoto/Getty Images RF; 08.09.c4: ©Paul A. Souders/Corbis/VCG/Getty Images; 08.09.c5: Source: Earth Science and Remote Sensing Unit/NASA/Johnson Space Center; 08.11.a1, 08.11.a2, 08.11.a3, 08.11.a4, 08.11.a5, 08.11.a6: Map data ©Ron Blakey, Colorado Plateau Geosystems, Arizona USA.

CHAPTER 9

Image Number: 09.08.a5: Source: Matthew C. Larsen/U.S. Geological Survey; 09.09.b6: ©Julia Waterlow/Corbis Documentary/Getty Images; 09.11.b1: ©Galyna Andrushko/Alamy RF; 09.12.a3: Source: Tom Scott/Florida Geological Survey.

CHAPTER 10

Image Number: 10.00.a3: Source: Matthew C. Larsen/USGS; 10.00.a4: Photo by Lawson Smith, U.S. Army Corps of Engineers; 10.00.a5: Source: Matthew C. Larsen/USGS; 10.02.c3: ©FlairImages/iStock/Getty Images RF; 10.02.d2: www.boelke-art.de/Getty Images RF; 10.03.a1: ©Kenneth Fink/Science Source; 10.03.c1: ©W.K. Fletcher/Science Source; 10.04.a2: Photo courtesy of Patricia A. Trochlell/Wisconsin DNR; 10.06.a1, 10.06.a3, 10.06.a5, 10.06.a7, 10.06.b1, 10.06.b3, 10.06.b5, 10.06.b7, 10.07.a1, 10.07.a3, 10.07.a5, 10.07.a7: Source: USDA Natural Resources Conservation Service; 10.08.a2: Photo by Tim McCabe, USDA Natural Resources Conservation Service; 10.08.a4, 10.08.c1: Source: USDA Natural Resources Conservation Service; 10.08.c2: Photo by Lynn Betts, USDA National Resources Conservation Service; 10.08.d1: Photo by Ken Hammond, USDA Natural Resources Conservation Service; 10.08.d2: Photo by Tim McCabe, USDA Natural Resources Conservation

Service; 10.08.d3, 10.08.d4: Photo by Lynn Betts, USDA National Resources Conservation Service; 10.10.b3: Photo by Susanne Gillatt; 10.10.a2: Source: U.S. Geological Survey Earthquake Hazards Program; 10.10.a4: Source: Edwin L. Harp/U.S. Geological Survey; 10.10.a5: Source: J.T. McGill/U.S. Geological Survey; 10.10.b2: Source: D.M. Peterson/U.S. Geological Survey; 10.10.b5: Source: Matthew C. Larsen/U.S. Geological Survey; 10.12.a3: Old Humium, University of Oslo, Dept. of Geography and the University Centre in Svalbard; 10.12.a9: Source: Matthew C. Larsen/U.S. Geological Survey; 10.12.a11: ©Lloyd Cluff/Getty Images; 10.13.a1: Source: Peter J. Haeussler/U.S. Geological Survey; 10.13.a2: Source: J.T. McGill/U.S. Geological Survey; 10.13.a4: Source: Gerald Wieczorek/U.S. Geological Survey; 10.13.a5: ©John A. Karachewski RF; 10.13.t1: Source: R.L. Schuster/U.S. Geological Survey; 10.14.a1: Source: C.E. Meyer/U.S. Geological Survey; 10.14.a2: ©Victor de Schwanberg/Science Source; 10.14.b3: Source: Matthew C. Larsen/U.S. Geological Survey; 10.14.c6: Source: Jennifer Adleman/Alaska Volcano Observatory/U.S. Geological Survey.

CHAPTER 11

Image Number: 11.02.a1: Photo by Cynthia Shaw; 11.03.a2: Courtesy of Skye L. Rodgers, www.skyewriter.com; 11.03.a6: ©2012, Dan Trimble; 11.03.b3: Photo by Cynthia Shaw; 11.03.b4: Courtesy of Bixler McClure; 11.03.c1: Source: Bruce F. Molnia/U.S. Geological Survey; 11.03.c3: Source: U.S. Geological Survey; 11.04.a3, 11.04.b1: Photo by Cynthia Shaw; 11.04.c3: Source: Bruce F. Molnia/U.S. Geological Survey; 11.04.c6: Source: Lisa McKeon/U.S. Geological Survey; 11.04.c7: Source: Michael Studinger/NASA; 11.05.a3: ©Doug Sherman/Geofile RF; 11.05.a5: ©Thinkstock/Jupiter Images RF; 11.05.a7: Photo by Don Poggensee/USDA Natural Resources Conservation Services; 11.05.a12: Source: Wisconsin Department of Natural Resources; 11.06.a2: ©Pixtal/AGE Fotostock RF; 11.06.b3: Courtesy of Vladimir E. Romanovsky; 11.09.a2: ©John A. Karachewski RF; 11.09.a4, 11.09.a5: Photo by Cynthia Shaw; 11.10.t1: ©Bill Brooks/Alamy; 11.14.a2: ©Jeff Spielman/Photographer's Choice RF/Getyt Images RF; 11.14.a3: ©Stefan Witas/Getty Images RF; 11.14.a4, 11.14.a5: Source: U.S. Geological Survey; 11.14.b1: Source: Robert Morton/U.S. Geological Survey; 11.14.b2: Source: U.S. Army Corps of Engineers; 11.14.b3: Photograph by Ken Winters, U.S. Army Corps of Engineers; 11.15.a2: Source: NASA; 11.15.a4: Source: U.S. Geological Survey; 11.15.b2: ©John A. Karachewski RF; 11.15.b4: Photo

by Susanne Gillatt; 11.17.a1: ©Thinkstock/Alamy RF; 11.17.a6, 11.17.a7, 11.17.b1, 11.17.b2, 11.17.b3, 11.17.b8-9: Source: U.S. Geological Survey.

CHAPTER 12

Image Number: 12.00.a3: ©Peg Owens/Department of Tourism-Idaho; 12.02.a3: Photo by Cynthia Shaw; 12.02.b2: ©Glow Images; 12.07.b5: Photo by Susanne Gillatt; 12.07.d1, 12.07.d2: Photo by Michal Tal/National Center for Earth-Surface Dynamics; 12.07.d3: Photo by Cynthia Shaw; 12.08.a3: ©Doug Sherman/Geofile; 12.08.a2: ©Alan Morgan RF; 12.08.a5: ©Doug Sherman/Geofile RF; 12.08.b1: Source: Jacques Descloitres, MODIS Rapid Response Team, NASA/GSFC; 12.09.a3: Photo by Cynthia Shaw; 12.10.b4, Source: Kathleen Macek-Rowland/U.S. Geological Survey; 12.10.b5: Source: Lyn Topinka/U.S. Geological Survey; 12.10.b6: Source: U.S. Bureau of Reclamation; 12.10.b7: ©Dr. Parvinder Sethi; 12.15.a2: ©Antoine Gyori/Corbis/Getty Images; 12.15.t1, 12.15.t2: Source: U.S. Geological Survey; 12.15.a3: ©Colin Cuthbert/Newcastle University/Science Source.

CHAPTER 13

Image Number: 13.00.a3: Photo by Susanne Gillatt; 13.00.a4: ©Glow Images; 13.00.a5: ©Konstantin Kalishko/Alamy RF; 13.00.a7: ©Photographer's Choice/Getty Images RF; 13.00.a8: ©Bear Dancer Studios/Mark Dierker, photographer/McGraw-Hill Education; 13.00.a9: Source: NASA/JSC; 13.01.a1: ©InterNetwork Media/Getty Images RF; 13.01.a2: ©Bear Dancer Studios/Mark Dierker, photographer/McGraw-Hill Education; 13.01.c1: ©Don Farrell/Photodisc/Getty Images RF; 13.01.c2: ©Kevin Cavanagh, photographer/McGraw-Hill Education; 13.03.b1: ©Comstock/Getty Images RF; 13.03.b2: Source: Game McGimsey/Alaska Volcano Observatory/U.S. Geological Survey; 13.03.b3: Photo courtesy of Hukseflux USA, Inc.; 13.03.b4: Source: NOAA; 13.04.d1: Source: NASA-Goddard Space Flight Center, data from NOAA GOES; 13.04.d2: ©Photodisc/Getty Images RF; 13.04.d3: ©Marc Guitierrez/Flickr/Getty Images RF; 13.05.b2: ©Jim Reed/Getty Images RF; 13.06.b1, 13.06.b2, 13.06.b3, 13.06.b4: Images courtesy SOHO, the EIT Consortium, and the MDI Team; 13.06.b5: Courtesy of NASA/SDO and the AIA, EVE, and HMI Science Teams; 13.06.c5: ©Fuse/Getty Images RF; 13.08.c1: ©Photodisc/Getty Images RF; 13.09.a2: Image courtesy of the Earth Science and Remote Sensing Unit, NASA Johnson Space Center; 13.10.a2: ©Patrick Clark/Getty Images RF; 13.10.a3: ©Ingram Publishing; 13.11.b1: Source: NASA/JSC; 13.11.c2: Source: David

Walsh/U.S. Geological Survey; 13.15.a1: ©Sexto Sol/Getty Images RF; 13.15.a4, 13.15.a14: Photo by Susanne Gillatt.

CHAPTER 14

Image Number: 14.03.a2: ©Mark Francek; 14.03.a4: Photo by Susanne Gillatt; 14.04.b1: Source: U.S. Geological Survey; 14.04.b2: Source: NASA/Goddard Space Flight Center Scientific Visualization Studio; 14.04.c2: Source: U.S. National Park Service (NPS) photo by R.G. Johnsson; 14.04.c3: Source: U.S. National Park Service; 14.04.c4: Source: NASA/MODIS Rapid Response; 14.05.c7, 14.10.a5, 14.10.a6: Source: Jacques Descloitres, MODIS Rapid Response Team, NASA/GSFC.

CHAPTER 15

Image Number: 15.00.a2: Photo by Susanne Gillatt; 15.00.a3: Source: Laura J. Hartley/U.S. Geological Survey; 15.00.a4: Photo by Susanne Gillatt; 15.05.a2: Source: NASA; 15.07.a2: Source: NOAA National Weather Service; 15.07.a4: Source: Jay Madigan/NASA; 15.07.a5: ©Robert Rohli; 15.07.a8: Source: Ralph F. Kresge/NOAA; 15.08.b1: ©Ingram Publishing/SuperStock RF; 15.08.b2: Source: Jeff Schmaltz/MODIS Rapid Response Team, NASA/GSFC; 15.08.c5: ©Digital Vision/Getty Images RF; 15.10.t1: Source: NOAA/NWS Amarillo, Texas; 15.11.t1: Source: Henry Reges/CoCoRaHS HQ.

CHAPTER 16

Image Number: 16.00.a1: ©2010 Willoughby Owen/Getty Images; 16.00.a2: Source: NOAA/NWS/Atlanta National Weather Service Forecast Office; 16.00.a3: Source: NOAA/NOS/National Geodetic Survey/16.00.a4: Source: NASA image courtesty of the GOES Project Science Team; 16.00.a5, 16.04.b2: Source: NOAA; 16.04.b4: ©Ryan McGinnis/Getty Images; 16.04.b6: Source: NOAA; 16.07.a1: ©C. Wells Photography/Getty Images RF; 16.07.a2: ©Bear Dancer Studios/Mark Dierker; 16.08.a1: Minerva Studio/Shutterstock RF; 16.08.a6: Source: NOAA; 16.08.b2: Source: NASA image by Jeff Schmaltz, LANCE/EOSDIS Rapid Response; 16.08.b3: Source: NOAA Photo Library, NOAA Central Library OAR/ERL/Nation Severe Storms Laboratory (NSSL); 16.08.c1: ©Willoughby Owen/Getty Images RF; 16.08.c2: ©Alan Sealls/WeatherVideoHD.TV; 16.09.b5: ©Thomas Augustine; 16.10.a10: Source: Stacy Hopke, Burnett County Sheriff's Department/NOAA; 16.10.a11: Photo courtesy of Bill Cartwright; 16.10.a12: Source: NASA/University of Michigan; 16.11.a2: Image courtesy of the Earth Science and Remote Sensing Unit, NASA Johnson Space Center; 16.12.b1: Source: Barbara Ambrose NOAA/NODC/NCDDC/Department of Commerce; 16.12.b2: Source: Andrea Booher/FEMA; 16.12.b3, 16.12.b5: Source: FEMA/Tim Burkitt; 16.13.a1: Source: NASA Earth Observatory image by Robert Simmon with data courtesy of the NASA/NOAA GOES Project Science team; 16.13.a2, 16.13.a3: Source: NASA Goddard MODIS Rapid Response Team; 16.13.a4: Source: Isro/NASA/JPL-Caltech; 16.13.a5, 16.13.a6, 16.13.a7: Source: NASA-Goddard Space Flight Center, data from NOAA GOES; 16.13.t1: Source: Aerial photography courtesy of the NOAA Remote Sensing Division; 16.13.t2: Source: Andrea Booher/FEMA; 16.13.t3: Source: DoD/U.S. Air Force photo by Master Sgt. Mark C. Olsen; 16.14.a1, 16.14.a2: Source: NOAA/NWS.

CHAPTER 17

Image Number: 17.00.a4: Source: John Scurlock/U.S. Geological Survey; 17.06.a4, 17.06.a5, 17.06.a6: Photo by Susanne Gillatt; 17.06.b2: ©Brand X Pictures/Getty Images RF; 17.11.a1: Photo by Susanne Gillatt; 17.11.a3: ©Corbis/VCG/Getty Images RF; 17.11.b2, 17.11.b4: Photo by Cynthia Shaw.

CHAPTER 18

Image Number: 18.07.c1: Photo by Cynthia Shaw; 18.08.b1: ©Jupiter Images RF; 18.08.b3: ©Kent Knudson/PhotoLink/Getty Images RF; 18.08.b4: ©Roelof Bos/Getty Images RF; 18.08.d2: Photo by Susanne Gillatt; 18.09.b3: ©Imagemore/Glow Images RF; 18.09.b5: ©McGraw-Hill Education/Richard Gross, photographer; 18.09.b6: Source: Heidi Roop, NSF; 18.10.a5, 18.11.a2-a3: Source: NASA; 18.13.d1, 18.14.a2, 18.14.a6: Photo by Susanne Gillatt; 18.14.a7: ©BWAC Images/Alamy; 18.14.a8-a9: Photo by Susanne Gillatt; 18.14.a10: Source: Image produced by Hal Pierce, Lab for Atmospheres, NASA Goddard Space Flight Center.

CHAPTER 19

Image Number: 19.00.a2: Source: NASA and the Hubble Heritage Team (STSci/AURA), acknowledgement: N. Scoville (Caltech) and T. Rector (NOAO); 19.00.a3: Source: NASA, ESA, STSci, J. Hester and P. Scowen (Arizona State University); 19.01.a1: Source: NASA, James Bell (Cornell University), Michael Wolff (Space Science Institute), and the Hubble Heritage Team (STSci/AURA); 19.01.a2: Source: NASA and E. Karkoschka (University of Arizona); 19.01.b2: Source: NASA/JPL; 19.01.c1: Source: NASA/JPL-Caltech/Arizona State University; 19.01.d1-d2: Source: NASA; 19.01.d3: Source: NASA/JPL-Caltech; 19.01.t1: Source: NASA/JPL-Caltech/UCAL/MPS/DLR/IDA; 19.02.a1: Source: NASA/JPL/University of Arizona; 19.02.a2: Source: NASA/JPL; 19.02.a3-a4: Source: NASA/JPL-Caltech/Arizona State University; 19.02.a5: Source: NASA; 19.02.a6, 19.02.b2: Source: NASA/JPL-Caltech/MSSS; 19.02.d1: Source: NASA/JPL/Arizona State University; 19.02.d2: Source: NASA/JPL/MSSS; 19.02.d3: Source: NASA/JPL/University of Arizona; 19.02.t1: Source: NASA; 19.03.a1: Source: NASA/JPL/USGS; 19.03.a2: Source: NASA/JPL/Northwestern University; 19.03.b1: Source: NASA/JPL; 19.03.b2: Source: NASA/JPL/USGS; 19.03.b3-b4: Source: NASA/JPL; 19.03.c2: Source: Jacques Descloitres, MODIS Rapid Response Team, NASA/GSFC; 19.03.d1: Source: NASA, James Ben (Cornell University), Michael Wolff (Space Science Institute), and the Hubble Heritage Team (STSci/AURA); 19.04.a1: Source: NASA/JPL/USGS; 19.04.a2: Source: NASA; 19.04.a3: Source: NASA/JSC; 19.04.a4 Source: NASA; 19.04.a5: Source: NASA/JSC; 19.04.a6: Source: NASA; 19.04.b1-b2: Source: NASA; 19.04.c1, 19.04.c3: Photo by Donald Burt; 19.04.t1: Source: NASA/JPL/Caltech; 19.05.a2: Source: NASA/JPL; 19.05.a4: Source: John Spencer (Lowell Observatory) and NASA; 19.05.a5: Source: NASA/JPL/Ames Research Center; 19.05.a6: Source: NASA/JPL/University of Arizona; 19.05.a7: Source: NASA/JPL/DLR; 19.05.a8: Source: NASA/JPL/University of Arizona/University of Colorado; 19.05.a9: Source: NASA/JPL/University of Arizona; 19.05.a10: Source: NASA/JPL; 19.05.a12: Source: NASA/JPL/DLR; 19.06.a1-a2: Source: NASA/JPL; 19.06.a3: Source: NASA/JPL/University of Arizona; 19.06.a4: Source: NASA/JPL/ESA/University of Arizona; 19.06.a5: Source: CICLOPS/Space Science Institute; 10.06.a6-a8: Source: NASA/JPL/Space Science Institute; 19.07.a3: ©Calvin J. Hamilton; 19.07.a4, 19.07.b2: Source: NASA/JPL/USGS; 19.07.c1: Source: NASA; 19.07.t1: Source: ESA/Rosetta/NAVCAM; 19.08.a2: Source: NASA/JPL/Arizona State University; 19.08.a3: Source: NASA/JPL/USGS; 19.08.a4: Source: NASA/JPL-Caltech/University of Arizona/Cornell/Ohio State University; 19.08.a5-a6: Source: NASA/JPL/University of Arizona; 19.08.a7: Source: NASA/USGS; 19.08.b1: Source: NASA; 19.08.b2: Source: NASA/JPL/Cornell University; 19.08.b3: Source: NASA/JPL/UA/Lockheed Martin; 19.08.b4: Source: NASA/JPL-Caltech/MSSS; 19.09.b5: Source: NASA/JPL/Cornell/USGS; 19.08.b6: Source: NASA/JPL-Caltech/MSSS; 19.08.t1: Source: NASA/JPL/Arizona State University; 19.09.a2: Source: NASA/JPL-Caltech/Arizona State University; 19.09.a3: Source: NASA/JPL; 19.09.a4-a5:

Source: NASA/JPL-Caltech/Arizona State University; 19.09.a6-a7: Source: NASA/JPL/Arizona State University.

CHAPTER 20

Image Number: Page 582 (top): Source: NASA, ESA, and the Hubble Heritage Team (STSci/AURA); 20.00.a1: Source: NASA, ESA, H. Teplitz and M. Rafelski (IPAC/Caltech), A. Koekemoer (STSci), R. Windhorst (Arizona State University), and Z. Levay (STSci); 20.00.a2: Source: NASA/JPL-Caltech; 20.00.a3: Source: NASA, ESA, and the Hubble Heritage Team (STSci/AURA), 20.00.a4: Source: NASA, ESA, and the Hubble SM4 ERO Team; 20.01.b3: Source: NASA/Tom Tschida; 20.01.c1: Source: NASA/JSC; 20.01.c2: Source: ©2007 Large Binocular Corporation; 20.01.c3: Source: NASA; 20.03.b3-a: Source: NASA/CXC/University of Toronto/M. Durant, et. al.; 20.03.b3-b: Source: Images courtesty SOHO, the EIT Consortium, and the MDI Team; 20.03.b3-c: Source: NASA and the Hubble Heritage Team (STSci/AURA); 20.03.b3-d: Source: A. Caulet (ST-ECF, ESA) and NASA; 20.03.b3-e: Source: NASA; Source: 20.05.a1: ©Alan Dyer/Stocktrek Images/Getty Images RF; 20.06.d1: Source: Andrea Dupree (Harvard-Smithsonian CfA), Ronald Gilliland (STSci), NASA and ESA; 20.07.a1: Source: NASA's Scientific Visualization Studio, the SDO Science Team, and the Virtual Solar Observatory; 20.07.a5: Source: NASA/GSFC/SDO; 20.07.a6-a7: Source: Courtesy of NASA/SDO and the AIA, EVE, and HMI science teams; 20.07.b1: Source: NASA, ESA, and A. Field (STSc1); 20.08.a1: Source: NASA, ESA, M. Robberto (Space Telescope Science Institute/ESA) and the Hubble Space Telescope Orion Treasury Project Team; 20.08.a2: Source: NASA/JPL-Caltech/R. Hurt (SSC); 20.08.a3: Source: NASA/JPL-Caltech/T. Pyle (SSC); 20.08.b3: Source: NASA, ESA, and the Hubble Heritage Team (STSci/AURA); 20.09.b1: Source: NASA/JPL-Caltech/University of Arizona; 20.09.b2: Source: NASA/SkyWorks Digital; 20.09.b3: Source: NASA/JPL-Caltech/STSci/CXC/SOA; 20.09.b4: Source: NASA, ESA, J. Hester and A. Loll (Arizona State University); 20.11.a1: Source: NASA, ESA, and the Hubble Heritage Team (STSci/AURA); 20.11.a2: Source: X-ray: NASA/CXC/SAO Optical: Detlef Hartmann Infrared: NASA/JPL-Caltech; 20.11.a3: Source: NASA, ESA, and the Hubble Heritage Team (STSci/AURA); 20.11.a4: Source: NASA, ESA, and the Hubble Heritage Team (STSci/AURA)-ESA/Hubble Collaboration; 20.11.a5: Source: NASA, ESA, the Hubble Heritage Team (STSci/AURA), and A. Aloisi (STSci/ESA); 20.11.b1: Source: NASA, ESA, and Z. Levay (STSci); 20.11.c2: Source: NASA, ESA, and the Hubble Heritage Team

(STSci/AURA)-ESA/Hubble Collaboration; 20.12.a1: Source: Atlas image obtained as part of the Two Micron All Sky Survey (2MASS), a joint project of the University of Massachusetts and the Infrared Processing and Analysis Center/California Institute of Technology, funded by the National Aeronautics and Space Administration and the National Science Foundation; 20.13.a1: Source: NASA/KASK; 20.13.a2: Source: NASA, ESA, K. Kuntz (JHU), F. Bresolin (University of Hawaii), J. Trauger (Jet Propulsion Lab), J. Mould (NOAO), Y.H. Chu (University of Illinois, Urbana), and STSci; 20.13.a3-a7: NASA, ESA, STSci, J. Hester and P. Scowen (Arizona State University).

TEXT AND LINE ART

CHAPTER 1

Figure 01.02.c1, 01.02.c2: Martin Jakobsson/Stockholm Geo Visualisation Laboratory; 01.07.a4, 01.07.a5: Source: U.S. Geological Survey; 01.09.a4: After U.S. Geological Survey HA-743; 01.09.b2: U.S. Geological Survey; 01.13.d2: After U.S. Geological Survey Fact Sheet 100-03.

CHAPTER 5

Figure 05.04.a1: Source: U.S. Geological Survey; 05.04.b1: Source: Centers for Disease Control and Prevention/U.S. Geological Survey; 05.05.a1: Data from Michelle K. Hall-Wallace; 05.05.c1: Data from Michelle K. Hall-Wallace; 05.09.a1: After David Sandwell/Scripps Institution of Oceanography; 05.09.a2: After R. Dietmar Muller/School of Geosciences University of Sidney; 05.09.a3: After Gabi Laske, University of California, San Diego; 05.11.c5: Don Anderson/Seismological Laboratory, California Institute of Technology; 05.13.b1: Data from Michelle K. Hall-Wallace; 05.13.c1: Data from Michelle K. Hall-Wallace; 05.14.b2: After J. Kious/U.S. Geological Survey, 1996; 05.14.c1, 05.14.c2: Source: U.S. Geological Survey.

CHAPTER 6

Figure 6.08.a1: After L. Gurioli, *Geology*, 2005; 06.08.c1: After J. Verhoogen/California University, Department of Geological Sciences Bulletin; 06.08.c2: Source: U.S. Geological Survey; 06.09.d1: Source: U.S. Geological Survey Fact Sheet 100-03; 06.09.d2: Source: USGS Fact Sheet 2005-3024; 06.11.t1: Modified from King and H. Beikman/U.S. Geological Survey; 06.13.c1: Siebert L. and Simkin T. (http://www.volcano.si.edu/world/), 2002; 06.14.a3: Source: U.S.

Geological Survey Bulletin 1292; 06.14.c1: Source: U.S. Geological Survey Open-File Report 98-428; 06.14.t1: Source: U.S. Geological Survey Open-File Report 94-585.

CHAPTER 7

Figure 07.00.a1: Source: Travel time: Kenji Satake/Geological Survey of Japan, Rupture and Epicenters: Atul Nayak/Scripps Institution of Oceanography; 07.04.d1: After J. Zucca/Lawrence Livermore National Laboratory; 07.04.d2: Source: Jet Propulsion Laboratory/NASA; 07.07.a1: Source: U.S. Geological Survey; 07.07.a2, 07.07.a4: Data from National Earthquake Information Center/U.S. Geological Survey; 07.07.b1: Data from National Earthquake Information Center/U.S. Geological Survey; 07.07.b2: After N. Short, 2006/NASA; 07.07.c1: Modified from U.S. Geological Survey; 07.08.a1: Source: Paula Dunbar/NOAA; 07.10.c1: After C. Stover, 1993, U.S. Geological Survey Professional Paper 1527: I Wong/Utah Geological Survey Public Information Series 76; 07.12.a3: Source: NOAA; 07.13.a7, 07.14.a1: Source: Kathleen M. Haler/U.S. Geological Survey.

CHAPTER 8

Figure 08.00.a1: After G. Hatcher, Monterey Bay Aquarium Research Institute; 08.03.c1: Source: J. Shaw, *Science*, 1999; 08.04.b1: Source: U.S. Geological Survey Bulletin 2146-D; 08.04.t1: Source: King and H. Belkman/U.S. Geological Survey; 08.04.t2: Source: A. Bally, Geologic Society of America DNAG, 1989; 08.07.a2: Source: P. King, Princeton University Press, 1977; 08.07.b1: Source: D.L. Divins/National Geophysical Data Center; 08.08.c2: After T. Affoiter and J.P. Gratier, *Journal of Geophysical Research*, 2004; 08.08.d1: After F. Diegel, American Association of Petroleum Geologists, Memoir 65; 08.09.c1: Source: Chris Jenkins/Institute of Arctic & Alpine Research, University of Colorado at Boulder; 08.10.b1: Source: U.S. Geological Survey Tapestry of Time.

CHAPTER 9

Figure 09.12.b1: After http://web.env/auckland.ac.nz/our_research/karst; 09.12.b2: Source: U.S. Geological Survey.

CHAPTER 10

Figure 10.01.b2: Source: http://www.terragis.bees.unsw.edu.au/terraGIS_soil/sp_particle_size_fractions.html; 10.06.a2, 10.06.a4, 10.06.a6, 10.06.a8, 10.06.b2, 10.06.b4, 10.06.b6, 10.06.b8,

10.07.a2, 10.07.a4, 10.07.a6, 10.07.ab8: Source: UN/FAO; 10.08.c3: Source: USDA/NRCS; 10.11.t1: After G. Kiersch, *Civil Engineering*, 1964; 10.13.b1: Source: U.S. Geological Survey Open-File Report 97-0289.

CHAPTER 11

Figure 11.06.b2: After PACE21 Network, European Science Foundation; 11.06.t1: After L. Topinka, Cascade Volcano Observatory/U.S. Geological Survey; 11.07.c3: Source: Personal correspondence with Emi Ito, University of Minnesota, 2009; 11.16.t1: After S. Dutch/University of Wisconsin, Green Bay; 11.17.a2, 11.17.a3, 11.17.a4, 11.17.a5: Source: U.S. Geological Survey; 11.17.b4, 11.17.b6: Source: St. Petersburg Coastal and Marine Science Center/USGS; 11.17.b7: Source: U.S. Geological Survey.

CHAPTER 12

Figure 12.01.a2: After U.S. Geological Survey; 12.02.a1: Source: U.S. Geological Survey; 12.17.a2: Source: Texas Water Development Board; 12.17.b1: Source: U.S. Geological Survey; 12.17.b2: Source: Texas Water Development Board; 12.17.c1, 12.17.c2: U.S. Geological Survey.

CHAPTER 13

Figure 13.00.a6: Source: NASA; 13.03.d1: Source: NOAA; 13.04.c2, 13.09.c2: Source: Some data from Sellers, W.D., 1965: *Physical Climatology*, University of Chicago Press; 13.10.a1: Adapted from NASA/GSFC; 13.10.c1, 13.10.d1, 13.10.d3: Source: NASA/GSFC; 13.13.a1, 13.13.b1: Source: NCEP/NCAR Reanalysis dataset; 13.13.c1: Source: NOAA/NASA COMET Program; 13.13.c2, 13.14.b2, 13.14.b3, 13.14.bf4: Source: NCEP/NCAR Reanalysis dataset; 13.15.a3, 13.15.a6, 13.15.a9, 13.15.a12, 13.15.a16, 13.15.a19, 13.15.a22: Source: http://www.worldclimate.com/; 13.16.a3, 13.16.a4: Source: NCEP/NCAR Renalysis dataset.

CHAPTER 14

Figure 14.05.b1: Source: NCEP/NCAR Reanalysis dataset; 14.05.c8: Source: NOAA/NHC; 14.08.a8, 14.08.a9, 14.08.a10, 14.08.a11, 14.09.d1: Source: NCEP/NCAR Reanalysis dataset; 14.10.a3, 14.10.a4: Source: http://www.worldclimate.com; 14.10.b1, 14.10.b2: Source: NCEP/NCAR Reanalysis dataset; 14.10.b3: http://www.worldclimate.com; 14.10.b4, 14.10.b5: Source: NCEP/NCAR Reanalysis dataset; 14.10.b6: http://www.worldclimate.com; 14.10.b7, 14.10.b8: Source:

NCEP/NCAR Reanalysis dataset; 14.10.b9: Source: http://www.worldclimate.com; 14.11.a2, 14.11.a3, 14.11.a4, 14.11.a5: Source: NCEP/NCAR Reanalysis dataset; 14.11.a6, 14.11.a7: Source: NCEP/NCAR Renalysis dataset and NASA Earth Observatory.

CHAPTER 15

Figure 15.03.a1, 15.03.a2, 15.03.c1, 15.03.c2: Source: NCEP/NCAR Reanalysis dataset; 15.06.a2: Source: NASA/NOAA; 15.06.c2: Source: NOAA/NCEP; 15.10.c1: Source: Data from Changnon, S.A. and T.R. Rarl, 2003: *Journal of Applied Meteorology* 42:1302-1315; 15.10.a1, 15.10.a2, 15.11.b1, 15.11.b2, 15.11.b3, 15.12.a1, 15.12.a2, 15.12.a3, 15.12.a4: Source: NCEP/NCAR Reanalysis dataset; 15.12.b1, 15.12.b2: Source: NOAA/CPC.

CHAPTER 16

Figure 16.00.a6, 16.00.a7, 16.00.a8, 16.00.a9, 16.01.b1: Source: NCEP/NCAR Reanalysis dataset; 16.02.a1, 16.02.b1, 16.02.c1, 16.02.c3, 16.03.c1, 16.03.c2, 16.03.c3, 16.04.b3: Source: NOAA/NWS/HPC; 16.05.a1, 16.05.a2, 16.05.a3: Source: Data from NASA; 16.05.a4: Source: Modified from NASA; 16.05.b1, 16.05.b2, 16.06.b1, 16.06.b2, 16.06.b3: Source: NOAA/NCDC/SPC/National Severe Weather Database; 16.07.t1: Source: NOAA/NWS; 16.08.b1: Source: NOAA/NWS/HPC; 16.09.a1, 16.09.a2, 16.09.a3, 16.09.a4, 16.09.a5, 16.09.b1, 16.09.b2, 16.09.b3, 16.09.b4: Source: NOAA/NWS/SPC/National Severe Weather Database; 16.10.a3: Source: NOAA/NWS; 16.10.a4: Source: NOAA, modified from Coniglio, M.C. and D.J. Stensrud, 2004: *Weather*; 16.10.a5: Source: NOAA/NWS; 16.11.b1: Source: NASA; 16.11.c4: Source: NOAA; 16.13.a4: Source: NOAA; 16.14.a3, 16.14.a4, 16.14.a5, 16.14.a6, 16.14.a7, 16.14.a8: Source: NCEP/NCAR Reanalysis dataset.

CHAPTER 17

Figure 17.00.a1, 17.00.a2: Source: NCEP/NCAR Reanalysis dataset; 17.00.a3: Source: NOAA/Northwest Fisheries Science Center; 17.00.a5: Source: D'Aleo, J.D. and D. Easterbrook. Multidecadal tendencies in ENSO and global temperatures related to multidecadal oscillations. *Energy and Environment* 21(5), 437-460; 17.00.a6, 17.00.a7, 17.03.a1, 17.03.a2, 17.03.a3, 17.03.a4: Source: NCEP/NCAR Reanalysis Center; 17.05.a1, 17.05.a2: Source: NOAA/National Ocean Data Center; 17.05.b1, 17.05.b2: Source: http://www.soes.soton.ac.uk/teaching/courses/oa631/hydro.html#alt;

17.05.c1: Source: Webster, P.J.,1994: The role of hydrological processes in ocean-atmosphere interations. *Reviews of Geophysics*, 32(4), 427-476; 17.06.a2, 17.06.a3: Source: NCEP/NCAR Reanalysis Center; 17.06.b5: Source: Barker et al., 2011: 800,000 years of abrupt climate change. *Science* 334, 347-351; 17.07.b1: Source: Argonne National Laboratory; 17.07.b2: Source: Durack, P.J., S.E. Wijiffels, and R.J. Matear, 2012: Ocean salinities reveal strong global water cycle intensification during 1950 to 2000. *Science* 336, 455-458.

CHAPTER 18

Figure 18.00.a1-a3, 18.01.b1, 18.02.a1-a4, 18.03.a1-a3: Source: NCEP/NCAR Reanalysis dataset; 18.03.c5-c7: Source: http://www.worldclimate.com; 18.04.a1-a3, 18.04.c1-c4: Source: NCEP/NCAR Reanalysis dataset; 18.04.d1-d4: Source: http://www.worldclimate.com; 18.05.a1-a3, 18.05.b2, 18.05.c2: Source: NCEP/NCAR Reanalysis dataset; 18.05.c3-c5: Source: http://www.worldclimate.com; 18.06.a1-a3, 18.06.b1-b2: Source: NCEP/NCAR Reanalysis dataset; 18.06.b3-b5: Source: http://www.worldclimate.com; 18.07.a1-a3: Source: NCEP/NCAR Reanalysis dataset; 18.07.b1: Source: http://www.worldclimate.com; 18.07.c4: Source: National Snow and Ice Data Center, University of Colorado, Boulder; 18.07.c5: Source: NCEP/NCAR Reanalysis dataset; 18.07.c6-c8: Source: http://www.worldclimate.com; 18.08.a1: Source: Kaufmann, R.K. and C.J. Cleveland, 2008: *Environmental Science*, McGraw-Hill Higher Education; 18.09.a3: Source: National Research Council, 2006: Surface Temperature Reconstructions for the Last 2,000 Years, The National Academies Press; 18.11.b2-b3: Source: NOAA.

CHAPTER 19

Figure 19.02.b1: Source: R. Evans/NASA/HST Comet Science Team; 19.03.c1: Source: NASA/Johns Hopkins University; 19.05.a3: Source: NASA; 19.07.a2: Source: Erich Karkoschka/University of Arizona/NASA; 19.07.c2: Source: Modified from A. Field/NASA/ESA.

CHAPTER 20

Figure 20.01.b1-b2, 20.01.c4, 20.02.a1-a2, 20.02.b1, 20.02.c1-c5, 20.03.a3, 20.03.b1-b3, 20.04.a1, 20.04.b1-b2; 20.04.c1-c3, 20.05.a2-a4, 20.06.a1, 20.06.b2-b3, 20.06.c1-c3, 20.06.d2-d3, 20.07.a4, 20.08.a4, 20.08.b1-b2, 20.08.b4, 20.09.a1-a2, 20.09.b5, 20.10.a1-a3, 20.10.b1-b2; 20.10.b4, 20.11.b2-b3, 20.11.c1, 20.12.a1-a4: Source: Modified from Arny, Thomas. *Explorations: introduction to astronomy.* 7th ed. New York: McGraw-Hill Education, 2013.

INDEX

T

SPRING NIGHT SKY

This star chart is most accurate if used within an hour or so of the times listed and is plotted for observers located between 30° and 50° north latitude. All times are standard time; if daylight-saving time is in effect, add one hour.

To use this chart, hold it in front of you and rotate it so that the yellow label corresponding to the direction you are facing is positioned at the bottom, right-side up. The stars in the sky should match those depicted on the chart. The center of the chart is the zenith, the point in the sky directly overhead.

WHEN TO USE THIS STAR MAP

Early March:	2 a.m.
Late March:	1 a.m.
Early April:	midnight
Late April:	11 p.m.
Early May:	10 p.m.
Late May:	Dusk

SUMMER NIGHT SKY

This star chart is most accurate if used within an hour or so of the times listed and is plotted for observers located between 30° and 50° north latitude. All times are standard time; if daylight-saving time is in effect, add one hour.

To use this chart, hold it in front of you and rotate it so that the yellow label corresponding to the direction you are facing is positioned at the bottom, right-side up. The stars in the sky should match those depicted on the chart. Ignore all the parts of the map above horizons you are not facing.

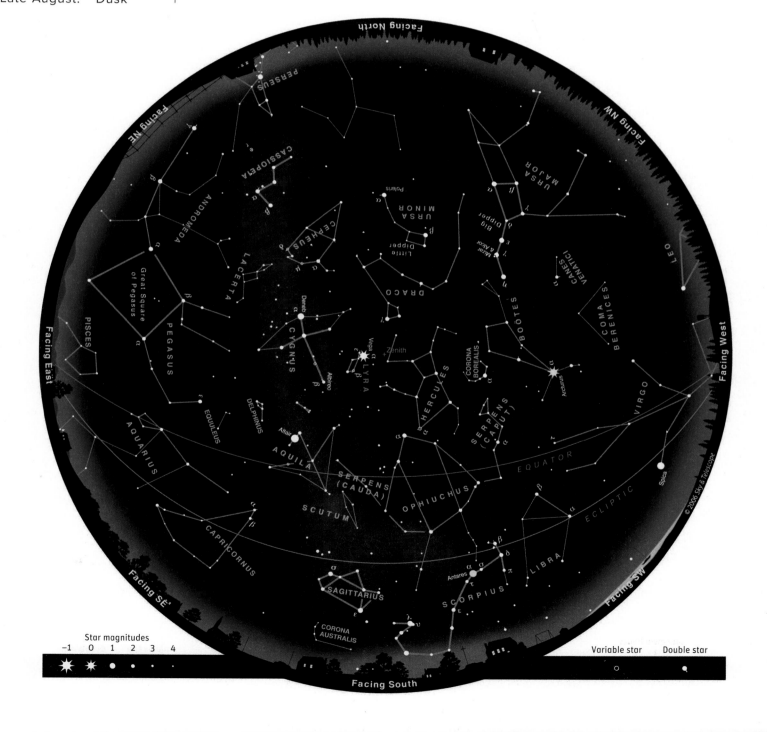

Star magnitudes
−1 0 1 2 3 4

Variable star   Double star

AUTUMN NIGHT SKY

This star chart is most accurate if used within an hour or so of the times listed and is plotted for observers located between 30° and 50° north latitude. All times are standard time: if daylight-saving time is in effect, add one hour.

To use this chart, hold it in front of you and rotate it so that the yellow label corresponding to the direction you are facing is positioned at the bottom, right-side up. The stars in the sky should match those depicted on the chart. The farther up from the map's edge they appear, the higher they'll be shining in your sky.

WHEN TO USE THIS STAR MAP

Early September:	midnight
Late September:	11 p.m.
Early October:	10 p.m.
Late October:	9 p.m.
Early November:	8 p.m.
Late November:	7 p.m.

Star magnitudes
−1 0 1 2 3 4

Variable star   Double star

WINTER NIGHT SKY

This star chart is most accurate if used within an hour or so of the times listed and is plotted for observers located between 30° and 50° north latitude. All times are standard time.

To use this chart, hold it in front of you and rotate it so that the yellow label corresponding to the direction you are facing is positioned at the bottom, right-side up. The stars in the sky should match those depicted on the chart. Remember that star patterns will look much larger in the sky than they do here on paper.

WHEN TO USE THIS STAR MAP

Early December:	midnight
Late December:	11 p.m.
Early January:	10 p.m.
Late January:	9 p.m.
Early February:	8 p.m.
Late February:	Dusk

Star magnitudes
−1 0 1 2 3 4

Variable star   Double star

© 2006 Sky & Telescope